Multivariate Statistics Beyond Normality

Multivariate Statistics Beyond Normality is a unique book that provides a modern and original introduction to multivariate statistics and then extends it beyond the multivariate normal distribution. Specifically, the extensions include spherical and elliptical distributions, the skew-normal distribution and related distributions, a detailed treatment of unified skew-elliptical distributions and their submodels, a study of weighted and selection multivariate distributions, and over 100 illustrative examples. Written by two leading specialists on multivariate statistics, this book includes the most recent and some novel results on skew-normal and related distributions, covering both singular and nonsingular cases in a unified way, and contains unpublished results on elliptical distributions from the first author's Ph.D. thesis. It presents illustrative data applications beyond normality that are relevant to both classical frequentist inference and Bayesian analysis. Designed for a broad readership by starting from basic fundamental concepts and leading to more advanced topics, the book includes 150 exercises, many original, to practice the concepts presented across the chapters, as well as 40 open problems that still need to be further researched.

Key Features

- Provides a modern and original introduction to multivariate statistics
- Extends classical results beyond normality
- Includes over 100 illustrative examples, 150 exercises, and 40 open research problems
- Uses color-coded highlights to facilitate learning
- Promotes both frequentist statistics and Bayesian analysis

Reinaldo B. Arellano-Valle is a Professor of Statistics at the Pontificia Universidad Católica de Chile (PUC) in Santiago, Chile. He received his Ph.D. in Statistics in 1994 from the University of São Paulo, Brazil. He is an international leader in the field of multivariate statistics, an elected member of the International Statistical Institute (ISI), and has received the 2019 Mahalanobis Award from the ISI for outstanding research in multivariate statistics.

Marc G. Genton is the Al-Khawarizmi Distinguished Professor of Statistics at the King Abdullah University of Science and Technology (KAUST) in Saudi Arabia. He received his Ph.D. degree in Statistics in 1996 from the Swiss Federal Institute of Technology (EPFL), Lausanne. He is a fellow of the American Statistical Association (ASA), of the Institute of Mathematical Statistics (IMS), and the American Association for the Advancement of Science (AAAS), an elected member of the International Statistical Institute (ISI), and he has received multiple awards.

Multivariate Statistics Beyond Normality

Reinaldo B. Arellano-Valle and Marc G. Genton

CRC Press
Taylor & Francis Group
Boca Raton London New York

CRC Press is an imprint of the
Taylor & Francis Group, an **informa** business
A CHAPMAN & HALL BOOK

Designed cover image: Reinaldo B. Arellano-Valle and Marc G. Genton

First edition published 2026
by CRC Press
2385 NW Executive Center Drive, Suite 320, Boca Raton FL 33431

and by CRC Press
4 Park Square, Milton Park, Abingdon, Oxon, OX14 4RN

CRC Press is an imprint of Taylor & Francis Group, LLC

ISBN: 978-1-032-96325-9 (hbk)
ISBN: 978-1-032-96916-9 (pbk)
ISBN: 978-1-003-59124-5 (ebk)

DOI: 10.1201/9781003591245

Typeset in Nimbus Roman
by KnowledgeWorks Global Ltd.

Publisher's note: This book has been prepared from camera-ready copy provided by the authors.

Access the Support Material:
https://www.routledge.com/Multivariate-Statistics-Beyond-Normality/Arellano-Valle-Genton/p/book/9781032963259

To Adelchi Azzalini and Heleno Bolfarine for their mentorship.

To my family and friends for their unconditional support and love.

Reinaldo

To Ying Sun with love from my heart.

Marc

Contents

Preface

Throughout our research careers, we have repeatedly encountered data whose structure, complexity, and inherent asymmetry defied the assumptions that classical multivariate models impose. The multivariate normal distribution, elegant and foundational as it is, offers only a partial view of the rich landscape of multivariate phenomena. Modern applications across science, engineering, finance, and the environmental and biological disciplines increasingly demand models that can accommodate skewness, heavy tails, nonlinear dependence, truncation, and selection mechanisms. These features are pervasive in real data but cannot be properly captured by Gaussian paradigms.

This book, *Multivariate Statistics Beyond Normality*, is the result of many years of collaboration, reflection, and engagement with these challenges. Our goal has been to provide a comprehensive, mathematically rigorous, and conceptually unified treatment of multivariate distributions that extend well beyond classical normality, while still retaining the structure and interpretability that practitioners value. We have sought to bring together, in one place, the main distributional families that have emerged in recent decades: spherical and elliptical models, skew-normal and skew-t distributions, unified formulations such as the SUN, SUT, and SUE families, and weighted and selection mechanisms, so that readers can understand both their theoretical foundations and their practical implications.

We begin in Chapter 1 with the essential tools for working in multivariate probability: cumulative distribution functions, densities, transformations, Jacobians, expectations, characteristic functions, and measures of information and geometry. Although these topics are classical, they form the basis for many of the advanced developments that follow, and our intention has been to present them with clarity, completeness, and an eye toward the constructions used later in the text. Chapter 2 enriches this foundation by examining symmetry in multiple forms (spherical, sign-invariant, and central) supported by geometric illustrations that we hope will provide insight to readers as they navigate increasingly complex models.

The multivariate normal distribution, revisited in Chapter 3, remains a cornerstone of our discussion. Its structure, geometry, and inferential properties motivate much of the general framework that follows. However, our purpose in revisiting the normal is not only historical: it is also to highlight its limits and to illustrate how later chapters generalize its strengths while overcoming its constraints.

Chapters 4 through 7 form the core theoretical contribution of this book. Chapter 4 develops spherical and elliptical distributions, which broaden the Gaussian structure through flexible radial densities while preserving useful geometric properties. Chapter 5 introduces the multivariate skew-normal distribution, tracing its origins to Azzalini's influential work and presenting its stochastic, hierarchical, and selection representations in a parameterization that is geometrically appealing. Chapters 6 and 7 extend these ideas substantially by introducing the unified skew-normal (SUN) and unified skew-t (SUT) families. These unified models synthesize many constructions scattered across the literature and offer a powerful framework that accommodates asymmetry, heavy tails, and complex dependence in a single system. We have placed particular emphasis on conditional structures, quadratic forms, and inferential tools, all of which play a crucial role in practical applications.

Chapter 8 broadens the scope to unified skew-elliptical distributions, integrating elliptical geometry with skewing mechanisms in an analytically tractable way. Chapter 9 discusses weighted and selection distributions, highlighting mechanisms that naturally generate skewness and heavy tails, and showing how they relate to many of the models treated earlier in the book.

Throughout our work, it has been important to us that theory and practice be closely connected. Chapter 10 therefore gathers a wide range of applications from environmental and spatial statistics to time-series modeling, classification, Bayesian inference, finance, and biomedical science, to name but a few. These examples demonstrate the practical benefits that non-Gaussian models can offer, and they illustrate how the distributions developed in this book arise naturally when modeling real phenomena. We have included numerous figures, diagnostic plots, and comparative analyses to help readers appreciate the differences between Gaussian and non-Gaussian fits and to showcase the flexibility of the unified families in particular.

A comprehensive appendix on linear algebra has been included in the book to ensure that the text remains self-contained. Each chapter concludes with exercises and open research problems intended to encourage further exploration; these reflect not only the current frontiers of the field but also the many opportunities for future development.

We have also applied color-coded highlights throughout the book to clarify concepts and provide visual intuition. Specifically, we have used the following colors for:

- Definitions
- Theorems
- Remarks
- Examples
- Exercises
- Open Research Problems

These colors also depict in the front cover the density contours of a bivariate $\mathcal{SUT}_{2,2}$ distribution with 4 degrees of freedom.

This book is intended for graduate students, researchers, and practitioners who wish to understand and apply modern multivariate distributions beyond the classical normal model. We hope that the material presented here will serve as both a reference and a source of inspiration for advancing statistical modeling in increasingly complex data environments. It is our wish that readers will find in these pages the same sense of curiosity, challenge, and intellectual excitement that has motivated our own work on multivariate statistics over the years.

The solutions to various selected exercises, as well as the code and data used for applications in Chapter 10, can be found on the book's associated website:
`https://marcgenton.github.io/2027AGbook/`

We take this opportunity to thank our many research collaborators over the years and all of the students who took our courses on multivariate analysis. They have all helped shape the content of this book.

Reinaldo B. Arellano-Valle
Pontificia Universidad Católica de Chile

Marc G. Genton
King Abdullah University of Science and Technology

January 2026

About the Authors

Reinaldo B. Arellano-Valle is a Professor of Statistics at the Pontificia Universidad Católica de Chile (PUC) in Santiago, Chile. He earned his Ph.D. in Statistics in 1994 from the University of São Paulo, Brazil. His Ph.D. thesis was awarded second place in the 1992-1994 Francisco Aranda-Ordaz Prize, a doctoral thesis competition sponsored by the Bernoulli Society. He is an international leader in the field of multivariate statistics. He is an elected member of the International Statistical Institute (ISI) and has received the 2019 Mahalanobis Award from the ISI for outstanding research in multivariate statistics. In 2024, his research achievements were also recognized by the Chilean Statistical Society.

Marc G. Genton is the Al-Khawarizmi Distinguished Professor of Statistics at the King Abdullah University of Science and Technology (KAUST) in Saudi Arabia. He obtained his Ph.D. degree in Statistics in 1996 from the Swiss Federal Institute of Technology (EPFL), Lausanne. He is a fellow of the American Statistical Association (ASA), of the Institute of Mathematical Statistics (IMS), and the American Association for the Advancement of Science (AAAS), and is an elected member of the International Statistical Institute (ISI). In 2010, he received the El-Shaarawi award for excellence from The International Environmetrics Society (TIES) and the Distinguished Achievement award from the Section on Statistics and the Environment (ENVR) of the American Statistical Association (ASA). He received an ISI Service award in 2019 and the Georges Matheron Lectureship award in 2020 from the International Association for Mathematical Geosciences (IAMG). He led a Gordon Bell Prize finalist team with the *ExaGeoStat* software for Supercomputing 2022. He received the Royal Statistical Society (RSS) 2023 Barnett Award for his outstanding research in environmental statistics and the prestigious 2024 Don Owen Award from the ASA's San Antonio Chapter. He led a team who was awarded the 2024 ACM Gordon Bell Climate Modelling prize for exascale climate emulators. During his 1996-2026 academic career, his research interests have included multivariate statistics, robustness, spatio-temporal data analysis, functional data analysis, and high-performance computing with applications in environmental and climate science, as well as renewable energies.

Chapter 1

Multivariate Distributions

A multivariate distribution is simply the joint probability distribution of the components of a random vector (or matrix), which are random variables defined on the same probability space. Every multivariate distribution can be described by its joint cumulative distribution. Two important cases are absolutely continuous and discrete multivariate distributions or random vectors, which can also be represented in terms of their joint probability density or mass function, respectively. The emphasis in this book is on the absolutely continuous case, for which a joint probability density function will exist almost everywhere.

1.1 Multivariate Cumulative Distribution Function

Let $\mathbf{X} = (X_1, \ldots, X_p)^\top \in \mathbb{R}^p$ be a random vector, i.e., a vector of p random variables. Its multivariate *cumulative distribution function (cdf)* is defined by

$$F(\mathbf{x}) = F(x_1, \ldots, x_p) = \mathrm{P}(X_1 \leq x_1, \ldots, X_p \leq x_p) = \mathrm{P}(\mathbf{X} \leq \mathbf{x}),$$

for any vector $\mathbf{x} = (x_1, \ldots, x_p)^\top$ of real numbers, where $\mathrm{P}(A)$ represents the probability that an event A occurs. Clearly, every multivariate cdf $F(\mathbf{x})$ satisfies the following well-known basic properties:

[1] $0 \leq F(\mathbf{x}) \leq 1, \forall \mathbf{x} \in \mathbb{R}^p$;

[2] $F(-\infty, x_2, \ldots, x_p) = F(x_1, -\infty, x_3, \ldots, x_p) = \cdots = F(x_1, \ldots, x_{p-1}, -\infty) = 0$;

[3] $F(+\infty, \ldots, +\infty) = 1$;

[4] $F(\mathbf{x})$ is monotone nondecreasing and right-continuous in each of its arguments;

[5] $0 \leq \mathrm{P}(\mathbf{a} < \mathbf{X} \leq \mathbf{b}) = \mathrm{P}(a_1 < X_1 \leq b_1, \ldots, a_p < X_p \leq b_p) = \sum_{\mathbf{t} \in \mathbf{a} \times \mathbf{b}} (-1)^{n_a(\mathbf{t})} F(\mathbf{t}), \forall \mathbf{a} < \mathbf{b}$, where $\mathbf{a} = (a_1, \ldots, a_p)^\top$ and $\mathbf{b} = (b_1, \ldots, b_p)^\top$ are vectors of real numbers, $\mathbf{a} \times \mathbf{b} = \times_{i=1}^{p} \{a_i, b_i\}$ is the Cartesian product of $\{a_1, b_1\}, \ldots, \{a_p, b_p\}$, and $n_a(\mathbf{t})$ is the number of a_i terms in $\mathbf{t} \in \mathbf{a} \times \mathbf{b}$.

Property [5] indicates that from the cdf we can compute $\mathrm{P}(\mathbf{X} \in (\mathbf{a}, \mathbf{b}])$ for any rectangle $(\mathbf{a}, \mathbf{b}] = \times_{i=1}^{p} (a_i, b_i]$ of $\mathbb{R}^p$ (which, of course, must be nonnegative), and therefore the *probability distribution* of $\mathbf{X}$, i.e., $\mathrm{P}(\mathbf{X} \in B)$, for every measurable (Borel) set B of $\mathbb{R}^p$. For example, for $p = 2$ in [5], we have for every pair of real numbers (a_i, b_i), with $a_i < b_i$, $i = 1, 2$, that

$$0 \leq \mathrm{P}(a_1 < X_1 \leq b_1, a_2 < X_2 \leq b_2) = F(b_1, b_2) - F(b_1, a_2) - F(a_1, b_2) + F(a_1, a_2).$$

Conversely, any scalar-valued function $F(\mathbf{x})$ that satisfies conditions [1]-[5] is a cdf. We are adopting here the partial order: $\mathbf{a} \leq \mathbf{b} \Longleftrightarrow a_i \leq b_i, \forall i = 1, \ldots, p$, and $\mathbf{a} < \mathbf{b} \Longleftrightarrow a_i < b_i, \forall i = 1, \ldots, p$.

Example 1 *In $\mathbb{R}^2$, consider the two bivariate functions:*

$$F(x_1, x_2) = \begin{cases} 0, & x_1 < 0 \quad or \quad x_2 < 0, \\ x_1, & 0 \leq x_1 < 1, \quad x_2 \geq 1, \\ x_2, & x_1 \geq 1, \quad 0 \leq x_2 < 1, \\ x_1 x_2, & 0 \leq x_1, x_2 \leq 1, \\ 1, & otherwise, \end{cases}$$

and

$$\tilde{F}(x_1,x_2) = \begin{cases} 0, & x_1 < 0 \quad or \quad x_2 < 0, \\ x_1, & 0 \le x_1 < 1, \quad x_2 \ge 1, \\ x_2, & x_1 \ge 1, \quad 0 \le x_2 < 1, \\ x_1x_2\{1+\theta(x_1-1)(x_2-1)\}, & 0 \le x_1,x_2 \le 1, \\ 1, & otherwise, \end{cases}$$

for $-1 \le \theta \le 1$. *It is straightforward to check that* F *and* $\tilde{F}$ *are two cdfs that satisfy the properties [1]-[5] above. Moreover,* $\tilde{F}(x_1,x_2) = F(x_1,x_2)$ *for* $\theta = 0$.

In general, if we have a rectangular matrix $\mathbf{W} = (W_{ij}) \in \mathbb{R}^{n\times p}$ for which all components $(W_{ij},\ i = 1,\dots,n,\ j = 1,\dots,p)$ are random variables jointly distributed on $\mathbb{R}^{np}$, we say that $\mathbf{W}$ is a *random matrix* distributed on $\mathbb{R}^{n\times p}$. The cdf of $\mathbf{W}$ is then defined as

$$F(\mathbf{W}^o) = F(w_{ij}, i = 1,\dots,n,\ j = 1,\dots,p) = \mathrm{P}(W_{ij} \le w_{ij}, i = 1,\dots,n,\ j = 1,\dots,p),$$

for any matrix $\mathbf{W}^o = (w_{ij})$ of real numbers. When the random matrix has repeated elements, such as a symmetric one, its cdf corresponds to the joint cumulative distribution of the non-repeated elements in the matrix. In what follows, however, we will adopt the classical notation that simply denotes the cdf of $\mathbf{W}$ by $F(\mathbf{W})$; see, e.g., Mardia et al. (1979), Muirhead (1982), Fang and Zhang (1990), Bilodeau and Brenner (1999), Anderson (2003), and Kollo and von Rosen (2005). This is of course an abuse of notation, since the random matrix and the variable argument of the cdf are denoted by the same letter, but this difference can be easily understood from the context. The same will be done with other probabilistic functions that are also used to characterize the distribution of $\mathbf{W}$, such as the multivariate probability density and characteristic functions.

1.2 Multivariate Probability Density Function

Let $F(\mathbf{x}) = F(x_1,\dots,x_p)$ be the cdf of a random vector $\mathbf{X} = (X_1,\dots,X_p)^\top \in \mathbb{R}^p$. If there is a non-negative function $f(\mathbf{x}) = f(x_1,\dots,x_p)$ such that

$$F(\mathbf{x}) = \int_{(-\infty,\mathbf{x}]} f(\mathbf{u})\mathrm{d}\mathbf{u} = \int_{-\infty}^{x_p} \cdots \int_{-\infty}^{x_1} f(u_1,\dots,u_p)\mathrm{d}u_1 \cdots \mathrm{d}u_p,$$

for every $\mathbf{x} = (x_1,\dots,x_p)^\top \in \mathbb{R}^p$, we say that $F(\mathbf{x})$ (or the corresponding random vector $\mathbf{X}$) has a multivariate *probability density function (pdf)* $f(\mathbf{x})$. In such a case, we say that $F(\mathbf{x})$ (or the respective random vector $\mathbf{X}$) is *absolutely continuous*, and we have

$$f(\mathbf{x}) = \frac{\partial^p F(x_1,\dots,x_p)}{\partial x_1 \cdots \partial x_p},$$

for all continuous points $\mathbf{x} = (x_1,\dots,x_p)^\top$ of $f(\mathbf{x})$. Also, every pdf $f(\mathbf{x})$ has the following basic properties:

[1] $f(\mathbf{x}) \ge 0$ for all $\mathbf{x} \in \mathbb{R}^p$;

[2] $\int_{\mathbb{R}^p} f(\mathbf{x})\mathrm{d}\mathbf{x} = 1$;

[3] $\mathrm{P}(\mathbf{X} \in B) = \int_B f(\mathbf{x})\mathrm{d}\mathbf{x}$ for any measurable set B in $\mathbb{R}^p$.

The property [3] indicates that $f(\mathbf{x})$ completely determines the probability distribution of $\mathbf{X}$. In particular,

$$\begin{aligned} \mathrm{P}(\mathbf{a} < \mathbf{X} \le \mathbf{b}) &= \mathrm{P}(a_1 < X_1 \le b_1,\dots,a_p < X_p \le b_p) \\ &= \int_{a_p}^{b_p} \cdots \int_{a_1}^{b_1} f(x_1,\dots,x_p)\mathrm{d}x_1 \cdots \mathrm{d}x_p \\ &= \int_{(\mathbf{a},\mathbf{b}]} f(\mathbf{x})\mathrm{d}\mathbf{x}. \end{aligned}$$

Recall also that the sample space or support of a random vector $\mathbf{X} = (X_1, \ldots, X_p)^\top$ is often a proper subset of $\mathbb{R}^p$, which may now be defined as $\{\mathbf{x} \in \mathbb{R}^p : f(\mathbf{x}) > 0\}$. Thus, since any cdf (or pdf) completely determines the probability distribution of a random vector, we may simply write $\mathbf{X} \sim F$ (or $\mathbf{X} \sim f$) to say that a random vector $\mathbf{X}$ has a distribution with cdf F (or pdf f). Similarly, if we have a random matrix $\mathbf{W} \in \mathbb{R}^{p \times q}$ with cdf $F(\mathbf{W})$ and pdf $f(\mathbf{W})$, we will write $\mathbf{W} \sim F$ or $\mathbf{W} \sim f$.

Example 2 *Computing the partial derivatives of the two bivariate cdfs of Example 1 yields the following bivariate pdfs:*

$$f(x_1,x_2) = \begin{cases} 1, & 0 \le x_1, x_2 \le 1, \\ 0, & \text{otherwise}, \end{cases}$$

and

$$\tilde{f}(x_1,x_2) = \begin{cases} 1+\theta(2x_1-1)(2x_2-1), & 0 \le x_1, x_2 \le 1, \\ 0, & \text{otherwise}, \end{cases}$$

for $-1 \le \theta \le 1$. *The pdf f describes the uniform distribution on the unit square* $0 \le x_1, x_2 \le 1$.

In Example 2, we note that the pdf $f(x_1,x_2)$ describes a uniform distribution on the unit square $S = [0,1] \times [0,1]$ in $\mathbb{R}^2$. In this case, we have for each event B in $\mathbb{R}^2$ that $\int_B f(x_1,x_2)\mathrm{d}x_1\mathrm{d}x_2 = \int_{B \cap S} \mathrm{d}x_1\mathrm{d}x_2 = \mathrm{vol}(B \cap S)$, the volume of $B \cap S$. In general, the construction of an absolutely continuous uniform distribution is based, of course, on the idea of volume measure. In fact, now consider any event S of $\mathbb{R}^p$ for which $0 < \mathrm{vol}(S) < \infty$, and let

$$I_S(\mathbf{x}) = \begin{cases} 1, & \text{if } \mathbf{x} \in S, \\ 0, & \text{if } \mathbf{x} \notin S, \end{cases}$$

be the indicator function of S. We say that a random vector $\mathbf{X} \in \mathbb{R}^p$ is uniformly distributed on S, denoted by $\mathbf{X} \sim \mathcal{U}(S)$, if its pdf is of the form

$$f(\mathbf{x}) = cI_S(\mathbf{x}), \quad \forall \mathbf{x} \in \mathbb{R}^p,$$

where $c = \mathrm{vol}(S)^{-1}$. Thus, if $\mathbf{X} \sim \mathcal{U}(S)$, then for any event B in $\mathbb{R}^p$, we have

$$\mathrm{P}(\mathbf{X} \in B) = \int_B cI_S(\mathbf{x})\mathrm{d}\mathbf{x} = \frac{\mathrm{vol}(B \cap S)}{\mathrm{vol}(S)}.$$

As usual, if $\mathrm{vol}(\partial S) = 0$, where ∂S denotes the boundary of S, which is the most frequent case, then the representations $\mathcal{U}(S)$, $\mathcal{U}(S^o)$ and $\mathcal{U}(\bar{S})$, where S^o and $\bar{S}$ denote the interior and the closure of S, respectively, are completely equivalent.

1.3 Marginal and Conditional Multivariate Distributions

1.3.1 Marginal multivariate distributions

Let $\mathbf{X} \in \mathbb{R}^p$ be a random vector with cdf $F(\mathbf{x})$. The distribution of any subvector of $\mathbf{X}$ is called marginal multivariate distribution. More specifically, suppose that $\mathbf{X}$ is partitioned as $\mathbf{X} = (\mathbf{X}_1^\top, \ldots, \mathbf{X}_n^\top)^\top$, where $\mathbf{X}_i \in \mathbb{R}^{p_i}$, $i = 1, \ldots, n$, with $\sum_{i=1}^n p_i = p$. Thus, the joint cdf of $\mathbf{X}_1, \ldots, \mathbf{X}_n$ is

$$F(\mathbf{x}) = \mathrm{P}(\mathbf{X} \le \mathbf{x}) = \mathrm{P}(\mathbf{X}_1 \le \mathbf{x}_1, \ldots, \mathbf{X}_n \le \mathbf{x}_n) = F(\mathbf{x}_1, \ldots, \mathbf{x}_n),$$

while their respective marginal cdfs, $F_1(\mathbf{x}_1), \ldots, F_n(\mathbf{x}_n)$ say, are given by

$$F_1(\mathbf{x}_1) = F(\mathbf{x}_1, +\infty, \ldots, +\infty), \ldots, F_n(\mathbf{x}_n) = F(+\infty, \ldots, +\infty, \mathbf{x}_n).$$

In the special case where $\mathbf{X}$ is absolutely continuous with *joint pdf* $f(\mathbf{x}) = f(\mathbf{x}_1,\dots,\mathbf{x}_n)$, it follows that $\mathbf{X}_1,\dots,\mathbf{X}_n$ are also absolutely continuous random vectors with *marginal pdfs* $f_1(\mathbf{x}_1),\dots,f_n(\mathbf{x}_n)$, respectively, given by

$$\begin{aligned} f_1(\mathbf{x}_1) &= \int_{\mathbb{R}^{p_n}}\cdots\int_{\mathbb{R}^{p_2}} f(\mathbf{x}_1,\mathbf{x}_2,\dots,\mathbf{x}_n)\mathrm{d}\mathbf{x}_2\cdots\mathrm{d}\mathbf{x}_n, \\ &\;\vdots \\ f_n(\mathbf{x}_n) &= \int_{\mathbb{R}^{p_{n-1}}}\cdots\int_{\mathbb{R}^{p_1}} f(\mathbf{x}_1,\dots,\mathbf{x}_{n-1},\mathbf{x}_n)\mathrm{d}\mathbf{x}_1\cdots\mathrm{d}\mathbf{x}_{n-1}. \end{aligned}$$

Example 3 *From the joint pdfs of Example 2, we obtain their marginal pdfs* $f_1(x_1) = \tilde{f}_1(x_1) = 1$, $0 \le x_1 \le 1$ *and* $f_2(x_2) = \tilde{f}_2(x_2) = 1$, $0 \le x_2 \le 1$. *Those marginal pdfs are all uniform on the interval* $[0,1]$, *although the two pdfs* f *and* $\tilde{f}$ *are different for* $\theta \neq 0$.

1.3.2 Conditional multivariate distributions

As is well known, the conditional probability that an event A_1 occurs given that we know that another event A_2 has occurred is defined by $\mathrm{P}(A_1 \mid A_2) = \mathrm{P}(A_1 \cap A_2)/\mathrm{P}(A_2)$, provided $\mathrm{P}(A_2) \neq 0$. Let now X_1 and X_2 be two random variables with joint pdf $f(x_1,x_2)$ and denote by $f_2(x_2)$ the marginal pdf of X_2. Hence, if we define the events $A_1 = \{X_1 \le x_1\}$ and $A_2 = \{x_2 \le X_2 \le x_2 + h_2\}$ $(h_2 > 0)$, then we have

$$\mathrm{P}(X_1 \le x_1 \mid x_2 \le X_2 \le x_2 + h_2) = \frac{\int_{-\infty}^{x_1}\int_{x_2}^{x_2+h_2} f(u_1,u_2)\mathrm{d}u_2\mathrm{d}u_1}{\int_{x_2}^{x_2+h_2} f_2(u_2)\mathrm{d}u_2},$$

provided that $\mathrm{P}(x_2 \le X_2 \le x_2 + h_2) \neq 0$. Thus, for x_2 such that $f_2(x_2) > 0$, we can define the conditional cdf of X_1 given that $X_2 = x_2$, $F(x_1|x_2)$ say, as the limit of $\mathrm{P}(X_1 \le x_1 \mid x_2 \le X_2 \le x_2 + h_2)$ as $h_2 \to 0$. That is,

$$F(x_1|x_2) = \int_{-\infty}^{x_1} f(u_1|x_2)\mathrm{d}u_1,$$

where $f(x_1|x_2) = f(x_1,x_2)/f_2(x_2)$ is called the *conditional pdf* of X_1 given $X_2 = x_2$.

In general, if $\mathbf{X}_1 \in \mathbb{R}^{p_1}$ and $\mathbf{X}_2 \in \mathbb{R}^{p_2}$ are two random vectors with joint pdf $f(\mathbf{x}) = f(\mathbf{x}_1,\mathbf{x}_2)$, and a marginal pdf $f_2(\mathbf{x}_2)$ for $\mathbf{X}_2$, the *conditional pdf* of $\mathbf{X}_1 \mid \mathbf{X}_2 = \mathbf{x}_2$ is defined as

$$f(\mathbf{x}_1|\mathbf{x}_2) = \frac{f(\mathbf{x}_1,\mathbf{x}_2)}{f_2(\mathbf{x}_2)},$$

provided that $f_2(\mathbf{x}_2) > 0$. When $f_2(\mathbf{x}_2) = 0$, $f(\mathbf{x}_1|\mathbf{x}_2)$ is defined arbitrarily. The conditional pdf of $\mathbf{X}_2|\mathbf{X}_1 = \mathbf{x}_1$ is defined in a similar way. Thus, every conditional pdf has the same basic properties as the unconditional one. For instance, $\mathrm{P}(\mathbf{X}_1 \in B_1 \mid \mathbf{X}_2 = \mathbf{x}_2) = \int_{B_1} f(\mathbf{x}_1|\mathbf{x}_2)\mathrm{d}\mathbf{x}_1$ for any measurable set B_1 in $\mathbb{R}^{p_1}$. In particular, for the conditional cdf of $\mathbf{X}_1|\mathbf{X}_2 = \mathbf{x}_2$, we have

$$F(\mathbf{x}_1|\mathbf{x}_2) = \mathrm{P}(\mathbf{X}_1 \le \mathbf{x}_1 \mid \mathbf{X}_2 = \mathbf{x}_2) = \int_{(-\infty,\mathbf{x}_1]} f(\mathbf{u}|\mathbf{x}_2)\mathrm{d}\mathbf{u}.$$

On the other hand, any joint pdf $f(\mathbf{x}_1,\mathbf{x}_2)$ may be represented in terms of its *marginal-conditional representations* as

$$f(\mathbf{x}_1,\mathbf{x}_2) = f(\mathbf{x}_1|\mathbf{x}_2)f_2(\mathbf{x}_2) \quad \text{or} \quad f(\mathbf{x}_1,\mathbf{x}_2) = f(\mathbf{x}_2|\mathbf{x}_1)f_1(\mathbf{x}_1).$$

A special situation occurs when the conditional pdf $f(\mathbf{x}_1|\mathbf{x}_2)$ (respectively, $f(\mathbf{x}_2|\mathbf{x}_1)$) does not depend on $\mathbf{x}_2$ (respectively, $\mathbf{x}_1$), which is described in the next section.

Example 4 *From the joint pdfs of Example 2 and the marginal pdfs of Example 3, we obtain for each $0 \leq x_2 \leq 1$ their conditional pdfs $f(x_1|x_2) = 1$, $0 \leq x_1 \leq 1$ and $\tilde{f}(x_1|x_2) = 1 + \theta(2x_1 - 1)(2x_2 - 1)$, $0 \leq x_1 \leq 1$. We see that $f(x_1|x_2)$ does not depend on x_2, whereas $\tilde{f}(x_1|x_2)$ does for $\theta \neq 0$.*

1.4 Independence

Two random vectors $\mathbf{X}_1 \in \mathbb{R}^{p_1}$ and $\mathbf{X}_2 \in \mathbb{R}^{p_2}$ with joint cdf $F(\mathbf{x}_1, \mathbf{x}_2)$ and marginal cdfs $F_1(\mathbf{x}_1) = F(\mathbf{x}_1, +\infty)$ and $F_2(\mathbf{x}_2) = F(+\infty, \mathbf{x}_2)$, respectively, are said to be *independent* if and only if (iff)

$$F(\mathbf{x}_1, \mathbf{x}_2) = F_1(\mathbf{x}_1)F_2(\mathbf{x}_2), \quad \forall (\mathbf{x}_1, \mathbf{x}_2).$$

If $\mathbf{X}_1$ and $\mathbf{X}_2$ have joint pdf $f(\mathbf{x}_1, \mathbf{x}_2)$, then they are independent iff for all $(\mathbf{x}_1, \mathbf{x}_2)$,

$$f(\mathbf{x}_1, \mathbf{x}_2) = f_1(\mathbf{x}_1)f_2(\mathbf{x}_2), \quad \text{or} \quad f(\mathbf{x}_1|\mathbf{x}_2) = f_1(\mathbf{x}_1), \quad \text{or} \quad f(\mathbf{x}_2|\mathbf{x}_1) = f_2(\mathbf{x}_2),$$

where $f_1(\mathbf{x}_1)$ and $f_2(\mathbf{x}_2)$ are the marginal pdfs (or multivariate probability mass functions (pmfs) for the discrete case) of $\mathbf{X}_1$ and $\mathbf{X}_2$, respectively.

A prerequisite for independence is that the support of the joint pdf $f(\mathbf{x}_1, \mathbf{x}_2)$ may be expressed as the Cartesian product of the supports of its marginal pdfs $f_1(\mathbf{x}_1)$ and $f_2(\mathbf{x}_2)$. To see this, consider the *joint support* $S = \{(\mathbf{x}_1, \mathbf{x}_2) \in \mathbb{R}^{p_1+p_2} : f(\mathbf{x}_1, \mathbf{x}_2) > 0\}$ and the *marginal supports* $S_1 = \{\mathbf{x}_1 \in \mathbb{R}^{p_1} : f_1(\mathbf{x}_1) > 0\}$ and $S_2 = \{\mathbf{x}_2 \in \mathbb{R}^{p_2} : f_2(\mathbf{x}_2) > 0\}$. Consider also the Cartesian product set $S_1 \times S_2 = \{(\mathbf{x}_1, \mathbf{x}_2) \in \mathbb{R}^{p_1+p_2} : \mathbf{x}_1 \in S_1 \text{ and } \mathbf{x}_2 \in S_2\}$. It is clear that if $\mathbf{X}_1$ and $\mathbf{X}_2$ are independent random vectors, then $S = S_1 \times S_2$. Hence, if $f(\mathbf{x}_1, \mathbf{x}_2)$ is a joint pdf (or pmf) and the set where $f(\mathbf{x}_1, \mathbf{x}_2) > 0$ is not a Cartesian product, then the random vectors $\mathbf{X}_1$ and $\mathbf{X}_2$ with joint pdf (or pmf) $f(\mathbf{x}_1, \mathbf{x}_2)$ are not independent.

We use the symbol $\perp\!\!\!\perp$ to denote independence. Also, we write $(\mathbf{X}_1 \perp\!\!\!\perp \mathbf{X}_2)|\mathbf{X}_3 = \mathbf{x}_3$ to say that $\mathbf{X}_1$ and $\mathbf{X}_2$ are *conditionally independent* given $\mathbf{X}_3 = \mathbf{x}_3$, which holds iff $F(\mathbf{x}_1, \mathbf{x}_2|\mathbf{x}_3) = F(\mathbf{x}_1|\mathbf{x}_3)F(\mathbf{x}_2|\mathbf{x}_3)$, $\forall (\mathbf{x}_1, \mathbf{x}_2)$.

Example 5 *From Examples 1-4, we see that the support of the bivariate random vector $(X_1, X_2)^\top$ is the same for f and $\tilde{f}$, namely, the unit square $[0,1] \times [0,1]$, and that it is the Cartesian product of the marginal supports. For f, the joint pdf is equal to the product of the two marginal pdfs; hence, X_1 and X_2 are independent random variables. For $\tilde{f}$, the joint pdf is not equal to the product of the two marginal pdfs for $\theta \neq 0$; hence, X_1 and X_2 are dependent.*

Example 6 *Let $(X_1, X_2)^\top$ be a random vector with uniform pdf on*

$$S = \left\{(x_1, x_2) \in \mathbb{R}^2 : 0 \leq x_1 \leq \frac{1}{2}, |x_2| \leq x_1\right\} \uplus \left\{(x_1, x_2) \in \mathbb{R}^2 : \frac{1}{2} < x_1 \leq 1, |x_2| \leq 1 - x_1\right\},$$

where $\uplus$ denotes the disjoint union; see Figure 1.1. Then, we have $f(x_1, x_2) = 2I_S(x_1, x_2)$, but S cannot be expressed as a Cartesian product, and so X_1 and X_2 are not independent.

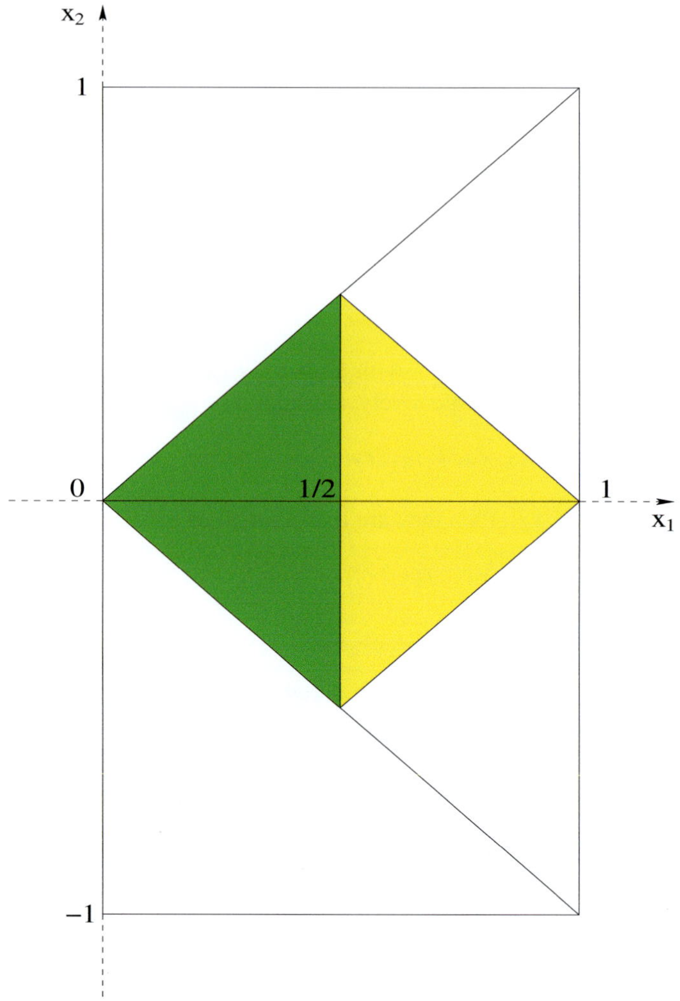

Figure 1.1 *Region S from Example 6 (disjoint union of green and yellow areas).*

In general, n arbitrary random vectors $\mathbf{X}_1, \ldots, \mathbf{X}_n$ with joint cdf $F(\mathbf{x}_1, \ldots, \mathbf{x}_n)$ and marginal cdfs $F_1(\mathbf{x}_1), \ldots, F_n(\mathbf{x}_n)$, respectively, are said to be ***mutually independent*** iff

$$F(\mathbf{x}_1, \ldots, \mathbf{x}_n) = \prod_{i=1}^{n} F_i(\mathbf{x}_i), \quad \forall (\mathbf{x}_1, \ldots, \mathbf{x}_n).$$

Moreover, if they have joint pdf (pmf) $f(\mathbf{x}_1, \ldots, \mathbf{x}_n)$ and marginal pdfs (pmfs) $f_1(\mathbf{x}_1), \ldots, f_n(\mathbf{x}_n)$, respectively, then $\mathbf{X}_1, \ldots, \mathbf{X}_n$ are independent iff

$$f(\mathbf{x}_1, \ldots, \mathbf{x}_n) = \prod_{i=1}^{n} f_i(\mathbf{x}_i), \quad \forall (\mathbf{x}_1, \ldots, \mathbf{x}_n).$$

In particular, a sequence of random vectors $\mathbf{X}_1, \ldots, \mathbf{X}_n$, all of the same dimension, is said to be a ***random sample*** of a pdf (pmf) f_1 (or cdf F_1), if they are ***independent and identically distributed (iid)*** with common pdf (pmf) f_1 (or cdf F_1). A more formal definition and some important consequences of the identically distributed property are given in the next section.

Example 7 *Let* $\mathbf{X} \in \mathbb{R}^n$ *be a random vector with uniform distribution on* $[\mathbf{0}_n, \mathbf{1}_n] = [0,1]^n$*, the unit hypercube of* $\mathbb{R}^n$*, whose pdf and cdf are*

$$f(\mathbf{x}) = I_{[\mathbf{0}_n, \mathbf{1}_n]}(\mathbf{x}) = \prod_{i=1}^{n} I_{[0,1]}(x_i), \quad \forall \mathbf{x} \in \mathbb{R}^n,$$

and

$$F(\mathbf{x}) = \prod_{i=1}^{n} \{x_i I_{[0,1]}(x_i) + I_{(1,+\infty)}(x_i)\}, \quad \forall \mathbf{x} \in \mathbb{R}^n.$$

Here we have that $X_1, \ldots, X_n$ *are iid* $\mathcal{U}([0,1])$*, the uniform distribution on* $[0,1]$*; or we write* $\mathbf{X} \sim \mathcal{U}([\mathbf{0}_n, \mathbf{1}_n])$*.*

1.5 Equality in Distribution

If two random vectors $\mathbf{X} \in \mathbb{R}^p$ and $\mathbf{Y} \in \mathbb{R}^p$ are *identically distributed*, we denote this fact by $\mathbf{X} \stackrel{d}{=} \mathbf{Y}$. More formally, if $\mathbf{X} \sim F$ and $\mathbf{Y} \sim \tilde{F}$, then

$$\mathbf{X} \stackrel{d}{=} \mathbf{Y} \Longleftrightarrow F(\mathbf{z}) = \tilde{F}(\mathbf{z}), \quad \forall \mathbf{z} \in \mathbb{R}^p.$$

An important byproduct of this fact is the following *invariance property*. Let $g_1, \ldots, g_q$ be measurable scalar-valued functions, where q is completely arbitrary. Then

$$\mathbf{X} \stackrel{d}{=} \mathbf{Y} \Longrightarrow (g_1(\mathbf{X}), \ldots, g_q(\mathbf{X})) \stackrel{d}{=} (g_1(\mathbf{Y}), \ldots, g_q(\mathbf{Y})).$$

To prove this, let $\mathbf{g}(\mathbf{x}) = (g_1(\mathbf{x}), \ldots, g_q(\mathbf{x}))^\top$, $\mathbf{x} \in \mathbb{R}^p$, and note that

$$\mathrm{P}(\mathbf{g}(\mathbf{X}) \in B) = \mathrm{P}(\mathbf{X} \in \mathbf{g}^{-1}(B)) = \mathrm{P}(\mathbf{Y} \in \mathbf{g}^{-1}(B)) = \mathrm{P}(\mathbf{g}(\mathbf{Y}) \in B),$$

where $\mathbf{g}^{-1}(B) = \{\mathbf{x} \in \mathbb{R}^p : \mathbf{g}(\mathbf{x}) \in B\}$ is the preimage (or inverse image) of B under $\mathbf{g}$. For instance,

$$\mathbf{X} \stackrel{d}{=} \mathbf{Y} \Longrightarrow (\mathbf{X}^\top \mathbf{A}_1 \mathbf{X}, \ldots, \mathbf{X}^\top \mathbf{A}_q \mathbf{X}) \stackrel{d}{=} (\mathbf{Y}^\top \mathbf{A}_1 \mathbf{Y}, \ldots, \mathbf{Y}^\top \mathbf{A}_q \mathbf{Y}),$$

where $\mathbf{A}_1, \ldots, \mathbf{A}_q$ are constant $p \times p$ matrices.

Example 8 *Let* X *be a random variable such that*

$$\mathrm{P}(X \le x) = \mathrm{P}(X \ge -x) = \mathrm{P}(-X \le x), \quad \forall x \in \mathbb{R}.$$

Then we have $X \stackrel{d}{=} -X$*, and* X *is called symmetric or said to have a symmetric distribution (around the origin). When* X *is absolutely continuous with cdf* F *and pdf* f*, we have* $X \stackrel{d}{=} -X$ *iff* $F(-x) = 1 - F(x)$*,* $\forall x \in \mathbb{R}$*, iff* $f(-x) = f(x)$*,* $\forall x \in \mathbb{R}$*.*

Example 9 *Let* $\mathbf{X} \sim \mathcal{U}([\mathbf{0}_n, \mathbf{1}_n])$ *and* $\mathbf{Y} = (1 - X_1, \ldots, 1 - X_n)^\top = \mathbf{1}_n - \mathbf{X}$*, where* $\mathbf{1}_n = (1, \ldots, 1)^\top \in \mathbb{R}^n$*. Then* $\mathbf{Y} \stackrel{d}{=} \mathbf{X}$*. In fact, denoting by* $\tilde{F}(\mathbf{y})$ *the cdf of* $\mathbf{Y}$ *and using straightforward properties of the indicator function* $I_A(x)$*, such as* $I_{[0,1]}(1-x) = I_{[0,1]}(x)$*,* $I_{A \uplus B}(x) = I_A(x) + I_B(x)$ *and*

$I_{A^c}(x) = 1 - I_A(x)$, where A^c is the complement of A, we have from Example 7 that

$$\begin{aligned}
\tilde{F}(\mathbf{y}) = \mathrm{P}(\mathbf{Y} \le \mathbf{y}) &= \mathrm{P}(\mathbf{X} \ge \mathbf{1}_n - \mathbf{y}) = \prod_{i=1}^{n} \{1 - F_i(1 - y_i)\} \\
&= \prod_{i=1}^{n} \{1 - (1 - y_i) I_{[0,1]}(1 - y_i) - I_{(1,+\infty)}(1 - y_i)\} \\
&= \prod_{i=1}^{n} \{1 - (1 - y_i) I_{[0,1]}(y_i) - I_{(-\infty,0)}(y_i)\} \\
&= \prod_{i=1}^{n} \{y_i I_{[0,1]}(y_i) + 1 - (I_{[0,1]}(y_i) + I_{(-\infty,0)}(y_i))\} \\
&= \prod_{i=1}^{n} \{y_i I_{[0,1]}(y_i) + I_{(1,+\infty)}(y_i))\} = \mathrm{P}(\mathbf{X} \le \mathbf{y}) = F(\mathbf{y}), \quad \forall \mathbf{y} \in \mathbb{R}^n.
\end{aligned}$$

1.6 Transformations and Expectations

1.6.1 Transformation of variables

Let the pdf of $\mathbf{X} \in \mathbb{R}^p$ be $f(\mathbf{x}) = f(x_1, \ldots, x_p)$ which is positive on a set $S \subset \mathbb{R}^p$. Suppose that the transformation

$$\mathbf{y} = (y_1, \ldots, y_p)^\top = (g_1(\mathbf{x}), \ldots, g_p(\mathbf{x}))^\top = \mathbf{g}(\mathbf{x})$$

is one-to-one from S onto $\tilde{S}$, where $\tilde{S}$ denotes the image of S under $\mathbf{g}$, a measurable function from $\mathbb{R}^p$ to $\mathbb{R}^p$. Hence, the inverse transformation

$$\mathbf{x} = (x_1, \ldots, x_p)^\top = (h_1(\mathbf{y}), \ldots, h_p(\mathbf{y}))^\top = \mathbf{h}(\mathbf{y})$$

exists for $\mathbf{y} \in \tilde{S}$. Assuming that the inverse functions $h_1, \ldots, h_p$ are continuously differentiable on $\tilde{S}$, it is well known that the pdf $\tilde{f}(\mathbf{y}) = \tilde{f}(y_1, \ldots, y_p)$ of the new random vector $\mathbf{Y} = \mathbf{g}(\mathbf{X})$ is

$$\tilde{f}(\mathbf{y}) = f\{\mathbf{h}(\mathbf{y})\} \mathrm{abs}\{J(\mathbf{y})\}, \tag{1.1}$$

where $\mathrm{abs}\{J(\mathbf{y})\}$ is the absolute value of $J(\mathbf{y})$, called *Jacobian (determinant) of the transformation of* $\mathbf{x}$ *to* $\mathbf{y}$, defined as

$$J(\mathbf{y}) = \begin{vmatrix} \frac{\partial x_1}{\partial y_1} & \frac{\partial x_1}{\partial y_2} & \cdots & \frac{\partial x_1}{\partial y_p} \\ \frac{\partial x_2}{\partial y_1} & \frac{\partial x_2}{\partial y_2} & \cdots & \frac{\partial x_2}{\partial y_p} \\ \vdots & \vdots & \ddots & \vdots \\ \frac{\partial x_p}{\partial y_1} & \frac{\partial x_p}{\partial y_2} & \cdots & \frac{\partial x_p}{\partial y_p} \end{vmatrix},$$

where $|\mathbf{A}|$ denotes the determinant of a matrix $\mathbf{A} \in \mathbb{R}^{p \times p}$.

Remark 1 *In some texts on probability and statistics, the Jacobian of the transformation of* $\mathbf{x}$ *to* $\mathbf{y}$ *is defined as the absolute value of the determinant* $J(\mathbf{y})$ *of the Jacobian matrix* $\partial \mathbf{x}/\partial \mathbf{y}^\top = (\partial x_i/\partial y_j)$.

Suppose now that the previous multivariate transformation is not one-to-one; that is, to each point of S corresponds exactly one point of $\tilde{S}$, but to each point of $\tilde{S}$ there may correspond more than one point of S. Assume that S can be represented as the union of k disjoint sets $A_1, A_2, \ldots, A_k$, such that the transformation $\mathbf{y} = \mathbf{g}(\mathbf{x})$ is a one-to-one mapping of each A_j onto $\tilde{S}$, $j = 1,2,\ldots,k$. That is, for each transformation of A_j onto $\tilde{S}$ there is a unique inverse transformation

$$\mathbf{x}_j = \mathbf{h}_j(\mathbf{y}), \quad j = 1, 2, \ldots, k,$$

each having a non-vanishing Jacobian, i.e., $J_j(\mathbf{y}) \neq 0$, $j = 1, 2, \ldots, k$. The pdf of $\mathbf{Y} = \mathbf{g}(\mathbf{X})$ is then given by

$$\tilde{f}(\mathbf{y}) = \sum_{j=1}^{k} f\{\mathbf{h}_j(\mathbf{y})\}\text{abs}\{J_j(\mathbf{y})\}. \tag{1.2}$$

In general, for every vector-valued measurable function $\mathbf{g} : \mathbb{R}^p \to \mathbb{R}^q$, the probability distribution of the transformed random vector $\mathbf{Y} = \mathbf{g}(\mathbf{X})$ can be calculated as

$$\mathrm{P}(\mathbf{Y} \in B) = \mathrm{P}(\mathbf{g}(\mathbf{X}) \in B) = \mathrm{P}(\mathbf{X} \in \mathbf{g}^{-1}(B)),$$

for any measurable set B in $\mathbb{R}^q$. In particular, if $\mathbf{X}$ and $\mathbf{Y}$ have pdfs $f(\mathbf{x})$ and $\tilde{f}(\mathbf{y})$, respectively, then

$$\mathrm{P}(\mathbf{Y} \in B) = \int_B \tilde{f}(\mathbf{y})d\mathbf{y} = \int_{\mathbf{g}^{-1}(B)} f(\mathbf{x})d\mathbf{x} = \mathrm{P}(\mathbf{X} \in \mathbf{g}^{-1}(B)).$$

Thus, when $p = q$ and the transformation $\mathbf{y} = \mathbf{g}(\mathbf{x})$ is one-to-one (except possibly on sets of Lebesgue measure 0 in the supports of $\mathbf{X}$ and $\mathbf{Y}$), the inverse transformation is $\mathbf{x} = \mathbf{h}(\mathbf{y})$ and, by the usual theory of transformation of multiple integrals, we have

$$\int_B \tilde{f}(\mathbf{y})d\mathbf{y} = \int_{\mathbf{g}^{-1}(B)} f(\mathbf{x})d\mathbf{x} = \int_B f\{\mathbf{h}(\mathbf{y})\}\text{abs}\{J(\mathbf{y})\}d\mathbf{y}.$$

Since this identity holds for any measurable set B in $\mathbb{R}^p$, the result (1.1) follows. The proof of (1.2) is similar by noting that $\mathrm{P}(\mathbf{Y} \in B) = \mathrm{P}(\mathbf{X} \in \biguplus_{j=1}^{k} A_j) = \sum_{j=1}^{k} \mathrm{P}(\mathbf{X} \in A_j)$, so that

$$\int_B \tilde{f}(\mathbf{y})d\mathbf{y} = \sum_{j=1}^{k} \int_{A_j} f(\mathbf{x})d\mathbf{x} = \sum_{j=1}^{k} \int_B f\{\mathbf{h}_j(\mathbf{y})\}\text{abs}\{J_j(\mathbf{y})\}d\mathbf{y} = \int_B \sum_{j=1}^{k} f\{\mathbf{h}_j(\mathbf{y})\}\text{abs}\{J_j(\mathbf{y})\}d\mathbf{y}.$$

Example 10 *In Examples 1-4, the absolute value of the Jacobian of the transformation* $(x_1,x_2) \to (y_1,y_2) = ((x_1+x_2)/2, (x_1-x_2)/2)$ *is* $\text{abs}\{J(y_1,y_2)\} = 2$. *Also, this transformation is clearly one-to-one from the unit square* $S = [0,1]^2$ *to the region* $\tilde{S} \subset \mathbb{R}^2$, *where*

$$\begin{aligned}\tilde{S} &= \left\{(y_1,y_2) : y_1 = \frac{x_1+x_2}{2}, y_2 = \frac{x_1-x_2}{2}, \textit{for } (x_1,x_2) \in [0,1]^2\right\} \\ &= \left\{(y_1,y_2) \in \mathbb{R}^2 : 0 \le y_1 \le \frac{1}{2}, |y_2| \le y_1\right\} \biguplus \left\{(y_1,y_2) \in \mathbb{R}^2 : \frac{1}{2} < y_1 \le 1, |y_2| \le 1 - y_1\right\}.\end{aligned}$$

Let $(Y_1,Y_2) = ((X_1+X_2)/2, (X_1-X_2)/2)$. *Thus, for* f *we see that* $(X_1,X_2)^\top$ *is uniform on* S *and that* $(Y_1,Y_2)^\top$ *is also uniform but on* $\tilde{S}$. *However, while* $X_1 \perp\!\!\!\perp X_2$, *the transformed random variables* Y_1 *and* Y_2 *are not independent (see Example 6).*

Example 11 *In Example 9, we see that the transformation* $\mathbf{Y} = (1-X_1,\ldots,1-X_n)^\top = \mathbf{1}_n - \mathbf{X}$, *where* $\mathbf{X} \sim \mathcal{U}([\mathbf{0}_n, \mathbf{1}_n])$, *is one-to-one from* $[\mathbf{0}_n, \mathbf{1}_n]$ *to* $[\mathbf{0}_n, \mathbf{1}_n]$. *The absolute Jacobian is trivially evaluated to be* $\text{abs}\{J(\mathbf{y})\} = 1$. *We then have* $\tilde{f}(\mathbf{y}) = f(\mathbf{y}) = \prod_{i=1}^{n} I_{[0,1]}(y_i)$, $\forall \mathbf{y} \in \mathbb{R}^n$.

Example 12 *Let* $(Y_1, Y_2) = ((X_1+X_2)/2, (X_1-X_2)^2/2)$*, where* X_1 *and* X_2 *have joint pdf*

$$f(x_1,x_2) = (2\pi)^{-1}e^{-(x_1^2+x_2^2)/2}, \quad -\infty < x_1 < \infty, \; -\infty < x_2 < \infty.$$

Note that X_1 *and* X_2 *are iid standard normal random variables, i.e.,* $\mathcal{N}(0,1)$*. We want to know the joint distribution of the random variables* Y_1 *and* Y_2*. In this case the transformation* $(x_1,x_2) \to (y_1,y_2) = ((x_1+x_2)/2, (x_1-x_2)^2/2)$ *is not one-to-one since to each point in* $\tilde{S} = \{(y_1,y_2) : -\infty < y_1 < \infty,\; 0 \le y_2 < \infty\}$ *there correspond two points in* $S = \{(x_1,x_2) : -\infty < x_1 < \infty, -\infty < x_2 < \infty\}$*. There are two sets of inverse functions:*
(1) $x_{11} = y_1 + \sqrt{y_2/2},\; x_{21} = y_1 - \sqrt{y_2/2}$;
(2) $x_{12} = y_1 - \sqrt{y_2/2},\; x_{22} = y_1 + \sqrt{y_2/2}$.
In both cases, the absolute Jacobian is $\text{abs}\{J_j(y_1,y_2)\} = \text{abs}(|\partial x_{ij}/\partial y_j|) = 1/\sqrt{2y_2}$, $j = 1,2$*. In addition, the definition of* y_2 *implies that there is one type of mapping when* $x_1 > x_2$ *and another when* $x_2 > x_1$*. Hence, we can define the disjoint sets*

$$A_1 = \{(x_1,x_2) \in \mathbb{R}^2 : x_2 > x_1\}$$

and

$$A_2 = \{(x_1,x_2) \in \mathbb{R}^2 : x_2 < x_1\}.$$

The line $x_1 = x_2$ *has been omitted since, when* $x_1 = x_2$ *we have* $y_2 = 0$*. However,* $\mathrm{P}(X_1 = X_2) = 0$*, thus excluding this line does not alter the distribution, and we can consider only* $S = A_1 \uplus A_2$*. Thus, (1) defines a one-to-one transformation of* A_2 *onto* $\tilde{S}$ *and (2) defines a one-to-one transformation of* A_1 *onto* $\tilde{S}$*. Therefore, the joint pdf of* (Y_1,Y_2) *is*

$$\begin{aligned}\tilde{f}(y_1,y_2) &= \frac{1}{2\pi}e^{-\frac{1}{2}\left[\{y_1+\sqrt{y_2/2}\}^2+\{y_1-\sqrt{y_2/2}\}^2\right]}\frac{1}{\sqrt{2y_2}} + \frac{1}{2\pi}e^{-\frac{1}{2}\left[\{y_1-\sqrt{y_2/2}\}^2+\{y_1+\sqrt{y_2/2}\}^2\right]}\frac{1}{\sqrt{2y_2}} \\ &= \frac{1}{\sqrt{\pi}}e^{-y_1^2}\frac{1}{\sqrt{2\pi}}y_2^{1/2-1}e^{-y_2/2}, \quad (y_1,y_2) \in \tilde{S}.\end{aligned}$$

This result shows that under the normality assumption, the random variables $Y_1 = (X_1+X_2)/2 = \bar{X}$ *and* $Y_2 = \sum_{i=1}^2 (X_i - \bar{X})^2$ *are independent, with* $Y_1 \sim \mathcal{N}(0,1/2)$ *and* $Y_2 \sim \chi_1^2$*.*

1.6.2 Jacobians and differentials: Properties and examples

The Jacobian's derivation is the difficult part in making transformations. We review some general results that can help with such derivations. Denote by $J_{(\mathbf{x}\to\mathbf{y})}$ the *Jacobian (determinant) of the transformation of* **x** *to* **y**. Some elementary properties are:

[1] (*Inversion rule*) $J_{(\mathbf{x}\to\mathbf{y})} = J^{-1}_{(\mathbf{y}\to\mathbf{x})}$;

[2] (*Chain rule*) If $\mathbf{y} = \mathbf{g}(\mathbf{x})$ and $\mathbf{z} = \mathbf{h}(\mathbf{y})$, then $J_{(\mathbf{x}\to\mathbf{z})} = J_{(\mathbf{x}\to\mathbf{y})}J_{(\mathbf{y}\to\mathbf{z})}$.

At this point, we should make some comments concerning the differential of a function of several variables. For a real-valued function $y = g(\mathbf{x})$, $\mathbf{x} \in \mathbb{R}^p$, the *differential* is defined by

$$\mathrm{d}y = \mathrm{d}g = \sum_{i=1}^{p} \frac{\partial g(\mathbf{x})}{\partial x_i}\mathrm{d}x_i$$

and so its *derivative* is

$$g'(\mathbf{x}) = \mathrm{D}g(\mathbf{x}) = \left(\frac{\partial g(\mathbf{x})}{\partial x_1}, \ldots, \frac{\partial g(\mathbf{x})}{\partial x_p}\right) = \frac{\partial g(\mathbf{x})}{\partial \mathbf{x}^\top}.$$

For a vector-valued function $\mathbf{y} = \mathbf{g}(\mathbf{x})$ of $\mathbf{x}, \mathbf{y} \in \mathbb{R}^p$, the *differential* is defined componentwise as

$$\mathrm{d}\mathbf{y} = \mathrm{d}\mathbf{g} = \begin{pmatrix} \mathrm{d}g_1 \\ \vdots \\ \mathrm{d}g_p \end{pmatrix} = \begin{pmatrix} \sum_{i=1}^p \frac{\partial g_1(\mathbf{x})}{\partial x_i} \mathrm{d}x_i \\ \vdots \\ \sum_{i=1}^p \frac{\partial g_p(\mathbf{x})}{\partial x_i} \mathrm{d}x_i \end{pmatrix} = \frac{\partial \mathbf{g}(\mathbf{x})}{\partial \mathbf{x}^\top} \mathrm{d}\mathbf{x},$$

where $\mathrm{D}\mathbf{g}(\mathbf{x}) = \partial \mathbf{g}(\mathbf{x})/\partial \mathbf{x}^\top = (\partial g_i(\mathbf{x})/\partial x_j) \in \mathbb{R}^{q \times p}$ is the usual *derivative (or Jacobian matrix)* of $\mathbf{y} = \mathbf{g}(\mathbf{x})$ with respect to $\mathbf{x}$, and $\mathrm{d}\mathbf{x} = (\mathrm{d}x_1, \ldots, \mathrm{d}x_p)^\top$. That is, $\mathrm{d}\mathbf{y}$ is a linear relation of $\mathrm{d}\mathbf{x}$ with Jacobian (determinant)

$$J_{(\mathrm{d}\mathbf{y} \to \mathrm{d}\mathbf{x})} = \left| \frac{\partial \mathbf{g}(\mathbf{x})}{\partial \mathbf{x}^\top} \right| = J_{(\mathbf{y} \to \mathbf{x})}.$$

For a matrix-valued function $\mathbf{W} = (w_{ij}) = (g_{ij}(\mathbf{V})) = \mathbf{G}(\mathbf{V}) \in \mathbb{R}^{r \times s}$, the *matrix differential* is defined componentwise as $\mathrm{d}\mathbf{W} = (\mathrm{d}w_{ij}) = (\mathrm{d}g_{ij}(\mathbf{V})) = \mathrm{d}\mathbf{G}(\mathbf{V})$; see also Section A.2.11 for matrix differentiation. For a scalar-valued function $w = g(\mathbf{V})$, $\mathbf{V} = (v_{ij}) \in \mathbb{R}^{p \times q}$, the differential is

$$\mathrm{d}w = \mathrm{d}g = \sum_{i=1}^{p} \sum_{j=1}^{q} \frac{\partial g(\mathbf{V})}{\partial v_{ij}} \mathrm{d}v_{ij} = \frac{\partial g(\mathbf{V})}{\partial \mathrm{vec}(\mathbf{V})^\top} \mathrm{dvec}(\mathbf{V}),$$

where the derivative $\mathrm{D}g(\mathbf{V}) = \partial g(\mathbf{V})/\partial \mathrm{vec}(\mathbf{V})^\top$ is a row vector in $\mathbb{R}^{pq}$ and vec is the vectorization operator (see Section A.2.10). Also, for $\mathbf{W} = \mathbf{G}(\mathbf{V}) \in \mathbb{R}^{r \times s}$, we have

$$\mathrm{dvec}(\mathbf{W}) = \mathrm{dvec}\{\mathbf{G}(\mathbf{V})\} = \frac{\partial \mathrm{vec}\{\mathbf{G}(\mathbf{V})\}}{\partial \mathrm{vec}(\mathbf{V})^\top} \mathrm{dvec}(\mathbf{V}),$$

where $\mathrm{D}\mathbf{g}(\mathbf{V}) = \partial \mathrm{vec}(\mathbf{W})/\partial \mathrm{vec}(\mathbf{V})^\top \in \mathbb{R}^{rs \times pq}$, which can be defined as *matrix derivative* of $\mathbf{W}$ with respect to $\mathbf{V}$. Hence, if $p = r$ and $q = s$ and the transformation $\mathbf{W} = \mathbf{G}(\mathbf{V})$ is one-to-one, then the Jacobian (determinant) from $\mathbf{W}$ to $\mathbf{V}$ is

$$J_{(\mathbf{W} \to \mathbf{V})} \equiv J_{(\mathrm{vec}(\mathbf{W}) \to \mathrm{vec}(\mathbf{V}))} = \left| \frac{\partial \mathrm{vec}(\mathbf{W})}{\partial \mathrm{vec}(\mathbf{V})^\top} \right|.$$

The extension to matrix transformations of the properties [1] and [2] of the Jacobian is direct. The immediate properties of the matrix differential are:

[d1] $\mathrm{d}(\mathbf{V} + \mathbf{W}) = \mathrm{d}\mathbf{V} + \mathrm{d}\mathbf{W}$;

[d2] $\mathrm{d}(c\mathbf{V}) = c\mathrm{d}\mathbf{V}$, where c is a constant;

[d3] $(\mathrm{d}\mathbf{V})^\top = \mathrm{d}\mathbf{V}^\top$;

[d4] $\mathrm{dtr}(\mathbf{V}) = \mathrm{tr}(\mathrm{d}\mathbf{V})$;

[d5] $\mathrm{d}(\mathbf{V}\mathbf{W}) = (\mathrm{d}\mathbf{V})\mathbf{W} + \mathbf{V}(\mathrm{d}\mathbf{W})$;

[d6] $\mathrm{d}(\mathbf{V} \otimes \mathbf{W}) = (\mathrm{d}\mathbf{V}) \otimes \mathbf{W} + \mathbf{V} \otimes (\mathrm{d}\mathbf{W})$, where $\otimes$ denotes the Kronecker product; see Section A.2.9.

Further details can be found in the following relevant references: p. 29 of Srivastava and Khatri (1979), Fang and Zhang (1990), p. 30 of Bilodeau and Brenner (1999), or Chapter 2 of Muirhead (1982). Nowadays, the monograph of Magnus and Neudecker (1999) has become the main reference in this area; for a more recent discussion, see also Section 1.4 of Kollo and von Rosen (2005). Examples of some important transformations are:

(1) *Vectorial-affine transformation*. Let $\mathbf{y} = \mathbf{A}\mathbf{x} + \mathbf{b}$, where $\mathbf{A} \in \mathbb{R}^{p \times p}$ is a nonsingular fixed matrix. Clearly, $\mathbf{x} = \mathbf{A}^{-1}(\mathbf{y} - \mathbf{b})$, so that $J_{(\mathbf{x} \to \mathbf{y})} = |\mathbf{A}|^{-1}$. Let $\mathbf{Y} = \mathbf{A}\mathbf{X} + \mathbf{b}$. If $\mathbf{X}$ has a pdf $f(\mathbf{x})$, then the pdf of $\mathbf{Y}$ is $f\{\mathbf{A}^{-1}(\mathbf{y} - \mathbf{b})\}\mathrm{abs}\{|\mathbf{A}|\}^{-1}$. When $\mathbf{b} = \mathbf{0}_p$, the transformation $\mathbf{y} = \mathbf{A}\mathbf{x}$ is said to be *linear*, although *linear* and *affine* are often used interchangeably.

(2) *Matricial-linear transformation.* Let $\mathbf{W} = \mathbf{AVB}$, where $\mathbf{W} \in \mathbb{R}^{p\times q}$ and $\mathbf{V} \in \mathbb{R}^{p\times q}$ are random matrices, and $\mathbf{A} \in \mathbb{R}^{p\times p}$ and $\mathbf{B} \in \mathbb{R}^{q\times q}$ are constant matrices. Since $\text{vec}(\mathbf{W}) = (\mathbf{B}^\top \otimes \mathbf{A})\text{vec}(\mathbf{V})$, it follows that $J_{(\mathbf{W}\to\mathbf{V})} = |\mathbf{B}^\top \otimes \mathbf{A}| = |\mathbf{A}|^q|\mathbf{B}|^p$. When $\mathbf{W} = \mathbf{AVA}^\top$, with $\mathbf{V} \in \mathbb{R}^{p\times p}$ symmetric, the Jacobian is $J_{(\mathbf{W}\to\mathbf{V})} = |\mathbf{A}|^{p+1}$.

(3) *Matricial inverse transformation.* Let $\mathbf{W} = \mathbf{V}^{-1}$, where $\mathbf{V} \in \mathbb{R}^{p\times p}$ is nonsingular and symmetric. Since $\mathbf{WV} = \mathbf{I}_p$, we have $(\mathrm{d}\mathbf{W})\mathbf{V} + \mathbf{W}(\mathrm{d}\mathbf{V}) = \mathbf{O}$, and therefore $\mathrm{d}\mathbf{W} = -\mathbf{V}^{-1}(\mathrm{d}\mathbf{V})\mathbf{V}^{-1}$. We find by the last result in (2) that $J_{(\mathbf{W}\to\mathbf{V})} = |\mathbf{V}|^{-(p+1)}$.

(4) *Conditional transformation.* If $\mathbf{y}_1 = \mathbf{g}_1(\mathbf{x}_1), \mathbf{y}_2 = \mathbf{g}_2(\mathbf{y}_1, \mathbf{x}_2), \dots, \mathbf{y}_n = \mathbf{g}_n(\mathbf{y}_1, \dots, \mathbf{y}_{n-1}, \mathbf{x}_n)$, where $\mathbf{x}_i, \mathbf{y}_i \in \mathbb{R}^{p_i}$ $(i = 1, \dots, n)$, then $J_{(\mathbf{y}_1,\dots,\mathbf{y}_n\to\mathbf{x}_1,\dots,\mathbf{x}_n)} = \prod_{i=1}^n \left|\frac{\partial \mathbf{y}_i}{\partial \mathbf{x}_i^\top}\right| = \prod_{i=1}^n J_{(\mathbf{y}_i\to\mathbf{x}_i)}$.

(5) *Polar transformation.* An example of (4) is the well-known transformation from rectangular coordinates $\mathbf{x} = (x_1, \dots, x_p)^\top$ to polar coordinates $(r, \boldsymbol{\theta}^\top)^\top = (r, \theta_1, \dots, \theta_{p-1})^\top$ given by:

$$\begin{aligned}
x_1 &= r\sin(\theta_1)\sin(\theta_2)\cdots\sin(\theta_{p-2})\sin(\theta_{p-1}),\\
x_2 &= r\sin(\theta_1)\sin(\theta_2)\cdots\sin(\theta_{p-2})\cos(\theta_{p-1}),\\
x_3 &= r\sin(\theta_1)\sin(\theta_2)\cdots\cos(\theta_{p-2}),\\
&\vdots\\
x_{p-1} &= r\sin(\theta_1)\cos(\theta_2),\\
x_p &= r\cos(\theta_1),
\end{aligned}$$

where $r > 0$, $0 < \theta_i \leq \pi$, for $i = 1, \dots, p-2$ and $0 < \theta_{p-1} \leq 2\pi$. The computation of the Jacobian $J_{(\mathbf{x}\to r,\boldsymbol{\theta})}$ is facilitated by considering the system of conditional transformations:

$$\begin{aligned}
y_1 &= x_1^2 + \cdots + x_{p-2}^2 + x_{p-1}^2 + x_p^2 = r^2,\\
y_2 &= x_1^2 + \cdots + x_{p-2}^2 + x_{p-1}^2 \qquad = y_1\sin^2(\theta_1),\\
y_3 &= x_1^2 + \cdots + x_{p-2}^2 \qquad = y_2\sin^2(\theta_2),\\
&\vdots\\
y_{p-1} &= x_1^2 + x_2^2 \qquad = y_{p-2}\sin^2(\theta_{p-2}),\\
y_p &= x_1^2 \qquad = y_{p-1}\sin^2(\theta_{p-1}).
\end{aligned}$$

Hence, we have $J_{(\mathbf{y}\to r,\boldsymbol{\theta})} = J_{(\mathbf{y}\to\mathbf{x})}J_{(\mathbf{x}\to r,\boldsymbol{\theta})}$, where it is straightforward to see that

$$J_{(\mathbf{y}\to\mathbf{x})} = 2^p\prod_{i=1}^p x_i = 2^p r^p \prod_{i=1}^{p-1}\{\sin(\theta_i)\}^{p-i}\prod_{j=1}^{p-1}\cos(\theta_j),$$

while by property (3)

$$\begin{aligned}
J_{(\mathbf{y}\to r,\boldsymbol{\theta})} &= J_{(y_1\to r)}J_{(y_2\to\theta_1)}\cdots J_{(y_p\to\theta_{p-1})}\\
&= 2^p r\prod_{i=1}^{p-1} y_i\sin(\theta_i)\cos(\theta_i)\\
&= 2^p r^{2p-1}\prod_{i=1}^{p-1}\{\sin(\theta_i)\}^{2(p-i)-1}\prod_{j=1}^{p-1}\cos(\theta_j);
\end{aligned}$$

hence

$$J_{(\mathbf{x}\to r,\boldsymbol{\theta})} = r^{p-1}\prod_{i=1}^{p-1}\{\sin(\theta_i)\}^{p-i-1}.$$

1.6.3 Expectation

Let $\mathbf{X} \in \mathbb{R}^p$ be a random vector with cdf $F(\mathbf{x})$. The *expectation (expected value or mean value)* of a scalar-valued function $g(\mathbf{X})$ is defined as

$$\mathrm{E}\{g(\mathbf{X})\} = \int_{\mathbb{R}^p} g(\mathbf{x}) \mathrm{d}F(\mathbf{x}),$$

where the integral is of the Riemann-Stieltjes kind. When $\mathbf{X}$ has a pdf f, the *expectation* of $g(\mathbf{X})$ is defined as

$$\mathrm{E}\{g(\mathbf{X})\} = \int_{\mathbb{R}^p} g(\mathbf{x}) f(\mathbf{x}) \mathrm{d}\mathbf{x}.$$

We assume that all necessary integrals converge so that the expectations are finite. In such a case, the random variable $g(\mathbf{X})$ is called *integrable*. It is well known that $g(\mathbf{X})$ is integrable iff $|g(\mathbf{X})|$ is integrable. In this case, we have by Jensen's inequality that $|\mathrm{E}\{g(\mathbf{X})\}| \leq \mathrm{E}\{|g(\mathbf{X})|\} < +\infty$. The expectation has the following additional properties:

[1] *Linearity property:*

$$\mathrm{E}\left\{\sum_{i=1}^{n} a_i g_i(\mathbf{X}) + b\right\} = \sum_{i=1}^{n} a_i \mathrm{E}\{g_i(\mathbf{X})\} + b,$$

where g_i, $i = 1, \ldots, n$, are integrable scalar-valued functions, and $a_i \in \mathbb{R}$, $i = 1, \ldots, n$, and $b \in \mathbb{R}$ are constants;

[2] If $g(\mathbf{x}) \geq 0$ for all $\mathbf{x} \in \mathbb{R}^p$, then $\mathrm{E}\{g(\mathbf{X})\} \geq 0$;

[3] If $g_1(\mathbf{x}) \leq g_2(\mathbf{x})$ for all $\mathbf{x} \in \mathbb{R}^p$, then $\mathrm{E}\{g_1(\mathbf{X})\} \leq \mathrm{E}\{g_2(\mathbf{X})\}$;

[4] If $a \leq g(\mathbf{x}) \leq b$ for all $\mathbf{x} \in \mathbb{R}^p$, then $a \leq \mathrm{E}\{g(\mathbf{X})\} \leq b$;

[5] Partition $\mathbf{X} = (\mathbf{X}_1^\top, \ldots, \mathbf{X}_n^\top)^\top \in \mathbb{R}^p$, where $\mathbf{X}_i \in \mathbb{R}^{p_i}$, $i = 1, \ldots, n$, with $\sum_{i=1}^n p_i = p$. Then, the expectation of a function $g_i(\mathbf{X}_i)$ of $\mathbf{X}_i$ may be computed in terms of only the marginal pdf $f_i(\mathbf{x}_i)$ of $\mathbf{X}_i$ as follows:

$$\mathrm{E}\{g_i(\mathbf{X}_i)\} = \int_{\mathbb{R}^p} g_i(\mathbf{x}_i) f(\mathbf{x}) \mathrm{d}\mathbf{x} = \int_{\mathbb{R}^{p_i}} g_i(\mathbf{x}_i) f_i(\mathbf{x}_i) \mathrm{d}\mathbf{x}_i, \quad i = 1, \ldots, n;$$

[6] If $\mathbf{X}_i$, $i = 1, \ldots, n$, are independent, then $g_i(\mathbf{X}_i)$, $i = 1, \ldots, n$, are also independent, and

$$\mathrm{E}\left\{\prod_{i=1}^{n} g_i(\mathbf{X}_i)\right\} = \prod_{i=1}^{n} \mathrm{E}\{g_i(\mathbf{X}_i)\}. \tag{1.3}$$

In particular, for any measurable sets B_i in $\mathbb{R}^{p_i}$, $i = 1, \ldots, n$, we have

$$\mathrm{P}(\mathbf{X}_1 \in B_1, \ldots, \mathbf{X}_n \in B_n) = \mathrm{P}\left(\bigcap_{i=1}^{n} \{\mathbf{X}_i \in B_i\}\right) = \prod_{i=1}^{n} \mathrm{P}(\mathbf{X}_i \in B_i). \tag{1.4}$$

Reciprocally, if the identity (1.3) (or (1.4)) holds for every integrable function g_i, $i = 1, \ldots, n$, (or for every measurable set B_i in $\mathbb{R}^{p_i}$, $i = 1, \ldots, n$), then $\mathbf{X}_i$, $i = 1, \ldots, n$, are independent.

Remark 2 *It should be noted that the expression (1.4) has usually been considered the most formal way to define the independence of random vectors.*

Example 13 *Let* $\mathbf{X} = (X_1,\ldots,X_n)^\top \sim \mathcal{U}([\mathbf{0}_n,\mathbf{1}_n])$ *and let* $Y = n^{-1}\sum_{i=1}^n Y_i$*, where* $Y_i = -\log X_i$*. We have* $\mathrm{E}(Y) = \mathrm{E}(Y_1)$ *by linearity and the fact that the* Y_i*s are equidistributed. Now* $\mathrm{E}(Y_1) = \int_{\mathbf{0}_n}^{\mathbf{1}_n}(-\log x_1)f(\mathbf{x})\mathrm{d}\mathbf{x} = \int_0^1(-\log x_1)\mathrm{d}x_1 = \int_0^{+\infty} e^{-y_1}\mathrm{d}y_1 = 1$*. Hence,* $\mathrm{E}(Y) = 1$*. Note also here that the cdf of* $\mathbf{Y} = (Y_1,\ldots,Y_n)^\top$ *is given by* $\tilde{F}(\mathbf{y}) = \mathrm{P}(Y_1 \le y_1,\ldots,Y_n \le y_n) = \mathrm{P}(-\log X_1 \le y_1,\ldots,-\log X_n \le y_n) = \prod_{i=1}^n \mathrm{P}(-\log X_i \le y_i) = \prod_{i=1}^n \mathrm{P}(Y_i \le y_i) = \prod_{i=1}^n \tilde{F}_i(y_i)$*, where* $\tilde{F}_i(y_i) = \mathrm{P}(-\log X_i \le y_i) = \mathrm{P}(X_i \ge e^{-y_i}) = e^{-y_i}I_{[0,+\infty)}$ *is the marginal distribution of* Y_i*,* $i = 1,\ldots,n$*. Hence, the* Y_i*s are iid exponential random variables.*

1.6.4 Conditional expectation

When two random vectors $\mathbf{X}_1 \in \mathbb{R}^{p_1}$ and $\mathbf{X}_2 \in \mathbb{R}^{p_2}$ are jointly distributed, expectations under the multivariate conditional distribution of $\mathbf{X}_1 \mid \mathbf{X}_2 = \mathbf{x}_2$ are important in many situations, for example, in regression analysis. When the multivariate conditional distribution of $\mathbf{X}_1 \mid \mathbf{X}_2 = \mathbf{x}_2$ has a conditional pdf $f(\mathbf{x}_1|\mathbf{x}_2)$, we can also define the *conditional expectation* of a scalar-valued function $g_1(\mathbf{X}_1)$ of $\mathbf{X}_1$ as

$$\mathrm{E}\{g_1(\mathbf{X}_1) \mid \mathbf{X}_2 = \mathbf{x}_2\} = \int_{\mathbb{R}^{p_1}} g_1(\mathbf{x}_1)f(\mathbf{x}_1|\mathbf{x}_2)\mathrm{d}\mathbf{x}_1,$$

provided that it exists. For simplicity, in what follows we denote $\mathrm{E}\{g_1(\mathbf{X}_1) \mid \mathbf{X}_2 = \mathbf{x}_2\}$ simply as $\mathrm{E}\{g_1(\mathbf{X}_1) \mid \mathbf{x}_2\}$. As this notation indicates, the conditional expectation $\mathrm{E}\{g_1(\mathbf{X}_1) \mid \mathbf{x}_2\}$ is a non-random function of $\mathbf{X}_2 = \mathbf{x}_2$, $\mathbf{x}_2 \in \mathbb{R}^p$. However, when we use the notation $\mathrm{E}\{g_1(\mathbf{X}_1) \mid \mathbf{X}_2\}$ we interpret the conditional expectation as a random function of $\mathbf{X}_2$. The conditional expectation has the following properties for scalar-valued functions g_1, g_2, and h_1:

[1] $\mathrm{E}[\{ag_1(\mathbf{X}_1)+bh_1(\mathbf{X}_1)+c\} \mid \mathbf{x}_2] = a\mathrm{E}\{g_1(\mathbf{X}_1) \mid \mathbf{x}_2\}+b\mathrm{E}\{h_1(\mathbf{X}_1) \mid \mathbf{x}_2\}+c$, where a,b,c are real constants;

[2] $\mathrm{E}\{g_1(\mathbf{X}_1)g_2(\mathbf{X}_2) \mid \mathbf{x}_2\} = \mathrm{E}\{g_1(\mathbf{X}_1) \mid \mathbf{x}_2\}g_2(\mathbf{x}_2)$;

[3] If $\mathbf{X}_1 \perp\!\!\!\perp \mathbf{X}_2$, then $\mathrm{E}\{g_1(\mathbf{X}_1) \mid \mathbf{x}_2\} = \mathrm{E}\{g_1(\mathbf{X}_1)\}$;

[4] $\mathrm{E}\{g_1(\mathbf{X}_1)\} = \mathrm{E}[\mathrm{E}\{g_1(\mathbf{X}_1) \mid \mathbf{X}_2\}]$;

[5] $\mathrm{var}\{g_1(\mathbf{X}_1)\} = \mathrm{E}[\mathrm{var}\{g_1(\mathbf{X}_1) \mid \mathbf{X}_2\}] + \mathrm{var}[\mathrm{E}\{g_1(\mathbf{X}_1) \mid \mathbf{X}_2\}]$;

[6] $\mathrm{E}\{g_1(\mathbf{X}_1)g_2(\mathbf{X}_2)\} = \mathrm{E}[\mathrm{E}\{g_1(\mathbf{X}_1) \mid \mathbf{X}_2\}g_2(\mathbf{X}_2)]$;

[7] Let $\mathbf{X}_1, \mathbf{X}_2$ and $\mathbf{X}_3$ be jointly distributed random vectors. We have the following:
(a) $\mathrm{E}[\mathrm{E}\{g_1(\mathbf{X}_1) \mid \mathbf{X}_2,\mathbf{X}_3\} \mid \mathbf{X}_2] = \mathrm{E}\{g_1(\mathbf{X}_1) \mid \mathbf{X}_2\} = \mathrm{E}[\mathrm{E}\{g_1(\mathbf{X}_1) \mid \mathbf{X}_2\} \mid \mathbf{X}_2,\mathbf{X}_3]$;
(b) If $\mathbf{X}_1 \perp\!\!\!\perp \mathbf{X}_2 \mid \mathbf{X}_3$, then $\mathrm{E}\{g_1(\mathbf{X}_1) \mid \mathbf{X}_2,\mathbf{X}_3\} = \mathrm{E}\{g_1(\mathbf{X}_1) \mid \mathbf{X}_3\}$.

1.7 Multivariate Characteristic Functions

The multivariate *characteristic function (cf)* of a random vector $\mathbf{X} \in \mathbb{R}^p$ is the function $\psi : \mathbb{R}^p \to \mathbb{C}$ defined by

$$\psi(\mathbf{t}) = \mathrm{E}\left(e^{\mathrm{i}\mathbf{t}^\top\mathbf{X}}\right) = \int_{\mathbb{R}^p} e^{\mathrm{i}\mathbf{t}^\top\mathbf{x}}\mathrm{d}F(\mathbf{x}), \quad \mathbf{t} \in \mathbb{R}^p,$$

where $\mathrm{i} = \sqrt{-1}$, the imaginary unit. If $\mathbf{X}$ admits a pdf f, then this becomes

$$\psi(\mathbf{t}) = \int_{\mathbb{R}^p} e^{\mathrm{i}\mathbf{t}^\top\mathbf{x}}f(\mathbf{x})\mathrm{d}\mathbf{x}, \quad \mathbf{t} \in \mathbb{R}^p.$$

As in the univariate case, the multivariate cf has the following properties:

[1] $\psi(\mathbf{0}_p) = 1$, $|\psi(\mathbf{t})| \le 1$ and $\psi(-\mathbf{t}) = \overline{\psi(\mathbf{t})}$, the complex conjugate;

[2] $\psi(\mathbf{t})$ is uniformly continuous on $\mathbb{R}^p$;

[3] The cf of $\mathbf{Y} = \mathbf{AX} + \mathbf{b}$ is $\tilde{\psi}(\mathbf{t}) = e^{\mathrm{i}\mathbf{t}^\top \mathbf{b}} \psi(\mathbf{A}^\top \mathbf{t})$, where $\mathbf{A} \in \mathbb{R}^{q \times p}$ and $\mathbf{b} \in \mathbb{R}^q$ are constant;

[4] (*Marginals*) Let $\psi_1(\mathbf{t}_1)$, $\psi_2(\mathbf{t}_2)$ and $\psi(\mathbf{t}_1, \mathbf{t}_2)$ be the marginals and joint cf of of $\mathbf{X}_1 \in \mathbb{R}^{p_1}$ and $\mathbf{X}_2 \in \mathbb{R}^{p_2}$, respectively. Then $\psi_1(\mathbf{t}_1) = \psi(\mathbf{t}_1, \mathbf{0}_{p_2})$ and $\psi_2(\mathbf{t}_2) = \psi(\mathbf{0}_{p_1}, \mathbf{t}_2)$;

[5] (*Independence*) $\mathbf{X}_1 \perp\!\!\!\perp \mathbf{X}_2$ iff $\psi(\mathbf{t}_1, \mathbf{t}_2) = \psi_1(\mathbf{t}_1)\psi_2(\mathbf{t}_2), \forall (\mathbf{t}_1, \mathbf{t}_2)$;

[6] Now let $p_1 = p_2 = p$ in [4]. We then have:
(a) (*Additivity*) If $\mathbf{X}_1 \perp\!\!\!\perp \mathbf{X}_2$, then the cf of $\mathbf{X}_1 + \mathbf{X}_2$ is $\psi_1(\mathbf{t})\psi_2(\mathbf{t})$, $\mathbf{t} \in \mathbb{R}^p$;
(b) (*Uniqueness*) $\mathbf{X}_1 \stackrel{d}{=} \mathbf{X}_2$ iff $\psi_1(\mathbf{t}) = \psi_2(\mathbf{t}), \forall \mathbf{t} \in \mathbb{R}^p$;
(c) (*Cramér-Wold theorem*) $\mathbf{X}_1 \stackrel{d}{=} \mathbf{X}_2$ iff $\mathbf{t}^\top \mathbf{X}_1 \stackrel{d}{=} \mathbf{t}^\top \mathbf{X}_2, \forall \mathbf{t} \in \mathbb{R}^p$.

We recall that the uniqueness property [6](b) means that the distribution of any random vector $\mathbf{X} \in \mathbb{R}^p$ is uniquely determined by its cf and vice versa. In turn, the Cramér-Wold theorem [6](c) establishes that the distribution (or cf) of $\mathbf{X} \in \mathbb{R}^p$ completely determines and is completely determined by the distributions (or cfs) of $\mathbf{t}^\top \mathbf{X}$ for every $\mathbf{t} \in \mathbb{R}^p$. In fact, if we let $\psi(\mathbf{t})$ and $\tilde{\psi}(s)$ be the cf of $\mathbf{X}$ and $\mathbf{t}^\top \mathbf{X}$, $\mathbf{t} \in \mathbb{R}^p$, respectively, we then have $\tilde{\psi}(s) = \mathrm{E}(e^{\mathrm{i}s\mathbf{t}^\top \mathbf{X}}) = \psi(s\mathbf{t})$ for all $s \in \mathbb{R}$ and $\mathbf{t} \in \mathbb{R}^p$; it follows from [6](b) that the distribution of $\mathbf{X}$ is uniquely determined by the distribution of $\mathbf{t}^\top \mathbf{X}$ for every $\mathbf{t} \in \mathbb{R}^p$ and vice versa. Moreover, since $\mathbf{t}^\top \mathbf{X}_i = \|\mathbf{t}\| \left(\mathbf{t}^\top / \|\mathbf{t}\|\right) \mathbf{X}_i$, $\forall \mathbf{t} \neq \mathbf{0}_p$, $i = 1, 2$, it follows that

$$\mathbf{t}^\top \mathbf{X}_1 \stackrel{d}{=} \mathbf{t}^\top \mathbf{X}_2, \forall \mathbf{t} \in \mathbb{R}^p \Longleftrightarrow \mathbf{t}^\top \mathbf{X}_1 \stackrel{d}{=} \mathbf{t}^\top \mathbf{X}_2, \forall \mathbf{t} \in \mathbb{S}^{(p-1)},$$

where $\mathbb{S}^{(p-1)} = \{\mathbf{x} \in \mathbb{R}^p : \|\mathbf{x}\| = 1\}$ is the subset of $\mathbb{R}^p$ that contains all vectors of length 1, i.e., the unit spherical shell of $\mathbb{R}^p$. The superscript $p-1$ in $\mathbb{S}^{(p-1)}$ refers to the dimension of this set. Hence, by [6](c), we have

$$\mathbf{X}_1 \stackrel{d}{=} \mathbf{X}_2 \Longleftrightarrow \mathbf{t}^\top \mathbf{X}_1 \stackrel{d}{=} \mathbf{t}^\top \mathbf{X}_2, \forall \mathbf{t} \in \mathbb{S}^{(p-1)}.$$

However, note that considering the unit vectors $\mathbf{e}_1 = (1, 0, \ldots, 0)^\top, \ldots, \mathbf{e}_p = (0, \ldots, 0, 1)^\top$ of $\mathbb{R}^p$ to show this last result is not enough, since they yield only the marginal components $\mathbf{e}_i^\top \mathbf{X} = X_i$, $i = 1, \ldots, p$, of $\mathbf{X}$.

The multivariate characteristic function is strongly related to the multivariate moment generating function when the latter exists. The multivariate *moment generating function (mgf)* of a random vector $\mathbf{X} \in \mathbb{R}^p$ is defined as $M(\mathbf{t}) = \mathrm{E}(e^{\mathbf{t}^\top \mathbf{X}})$, $\mathbf{t} \in \mathbb{R}^p$. The mgf of $\mathbf{X}$ is said to exist if it is finite in a neighborhood of the origin of $\mathbb{R}^p$, $\|\mathbf{t}\| < t_0$, $t_0 > 0$. In this case, we have $\psi(\mathbf{t}) = M(\mathrm{i}\mathbf{t})$ for all $\mathbf{t}$ or $M(\mathbf{t}) = \psi(-\mathrm{i}\mathbf{t})$ for $\|\mathbf{t}\| < t_0$.

The multivariate characteristic function, unlike the multivariate moment generating function, always exists, and $\psi(\mathbf{t})$ characterizes the multivariate distribution of any random vector $\mathbf{X}$. This fact is formalized in the so-called *multivariate Fourier's inversion theorem* or simply *inversion formula*, presented below for the absolutely continuous case; see Bilodeau and Brenner (1999) for a more general formulation. If the cf $\psi(\mathbf{t})$ is absolutely integrable, i.e., $\int_{\mathbb{R}^p} |\psi(\mathbf{t})| \mathrm{d}\mathbf{t} < +\infty$, then the corresponding multivariate distribution is absolutely continuous with pdf

$$f(\mathbf{x}) = \frac{1}{(2\pi)^p} \int_{\mathbb{R}^p} e^{-\mathrm{i}\mathbf{t}^\top \mathbf{x}} \psi(\mathbf{t}) \mathrm{d}\mathbf{t}.$$

Finally, the cf of a random matrix $\mathbf{W} \in \mathbb{R}^{p \times q}$ is defined by

$$\psi(\mathbf{T}) = \mathrm{E}\left(e^{\mathrm{i}\,\mathrm{tr}(\mathbf{T}^\top \mathbf{W})}\right), \quad \mathbf{T} \in \mathbb{R}^{p \times q},$$

where we observe that $\mathrm{tr}(\mathbf{T}^\top \mathbf{W}) = \mathrm{vec}(\mathbf{T})^\top \mathrm{vec}(\mathbf{W})$.

Example 14 *Let* $X = -\sum_{i=1}^{p} \log U_i$*, where* $U_1, \ldots, U_p$ *are iid* $\mathcal{U}([0,1])$ *random variables. Since* $X_i = -\log U_i$*,* $i = 1, \ldots, p$*, are also iid random variables, the characteristic function of* $X = \sum_{i=1}^{p} X_i$ *is then* $\psi(t) = \{\psi_1(t)\}^p$*, where*

$$\psi_1(t) = \mathrm{E}(e^{-\mathrm{i}t\log U_1}) = \int_0^1 e^{-\mathrm{i}t\log u_1}\mathrm{d}u_1 = \int_0^{+\infty} e^{-(1-\mathrm{i}t)x_1}\mathrm{d}x_1 = (1-\mathrm{i}t)^{-1}, \quad \forall t \in \mathbb{R}.$$

We find that X *has cf* $\psi(t) = (1-\mathrm{i}t)^{-p}$*,* $t \in \mathbb{R}$*, and hence that* X *has a* $\mathcal{G}amma(p,1)$ *distribution whose pdf is* $f(x) = x^{p-1}e^{-x}I_{[0,\infty)}(x)/\Gamma(p), \forall x \in \mathbb{R}$*, where* $\Gamma(a) = \int_0^\infty x^{a-1}e^{-x}\mathrm{d}x$*,* $a > 0$*, is the well-known gamma function. More generally, a random variable* X *with pdf* $f(x) = b^a x^{a-1}e^{-bx}I_{[0,\infty)}(x)/\Gamma(a), \forall x \in \mathbb{R}$ *is said to have a gamma distribution and denoted by* $X \sim \mathcal{G}amma(a,b)$*,* $a > 0$*,* $b > 0$*. The special cases* $X \sim \mathcal{G}amma(1,b) \equiv \mathcal{E}xp(b)$ *and* $X \sim \mathcal{G}amma(\nu/2,1/2) \equiv \chi^2_\nu$ *produce the exponential distribution and chi-squared distribution with* $\nu > 0$ *degrees of freedom, respectively. In this last case, the distribution of* $X^{1/2}$ *is denoted by* χ_ν*.*

Example 15 *Let* $\psi(\mathbf{t})$ *be the cf of a random vector* $\mathbf{X}$*. Clearly, the cf of* $-\mathbf{X}$ *is* $\mathrm{E}(e^{-\mathrm{i}\mathbf{t}^\top\mathbf{X}}) = \psi(-\mathbf{t})$*, where, by property [1],* $\psi(-\mathbf{t}) = \overline{\psi(\mathbf{t})}$*. Hence, we find* $-\mathbf{X} \stackrel{d}{=} \mathbf{X}$ *iff* $\psi(-\mathbf{t}) = \psi(\mathbf{t})$ *for all* $\mathbf{t}$*; i.e.,* $\psi(\mathbf{t})$ *is real-valued when* $\mathbf{X}$ *has a symmetric (around the origin) multivariate distribution.*

The following concept will be used later. A cf $\psi(\cdot)$ is said to belong to the *class* $\mathcal{E}$ if for any cf $\tilde{\psi}(\cdot)$ such that $\psi(t) = \tilde{\psi}(t)$ in a neighborhood of zero, it is true that $\psi(t) = \tilde{\psi}(t)$ for all $t \in \mathbb{R}$. A cf is said to belong to $\tilde{\mathcal{E}}$ if it does not belong to $\mathcal{E}$. As mentioned in Fang et al. (1990), the cf of a nonnegative (or nonpositive) random variable belongs to $\mathcal{E}$. For more details and key references on this concept, see the books by Fang et al. (1990) and Fang and Zhang (1990).

Let X, Y and Z be random variables such that (X,Z) and (Y,Z) have the same distribution. Then, it is obvious that $\mathbf{g}(X,Z)$ and $\mathbf{g}(Y,Z)$ will be equally distributed for any measurable (scalar or vector-valued) function $\mathbf{g}$; see Section 1.5. In particular, it follows that

$$(X,Z) \stackrel{d}{=} (Y,Z) \Longrightarrow X+Z \stackrel{d}{=} Y+Z \quad \text{and} \quad XZ \stackrel{d}{=} YZ.$$

Obviously, the marginal condition $X \stackrel{d}{=} Y$ by itself does not allow us to conclude that $(X,Z) \stackrel{d}{=} (Y,Z)$ for any other random variable Z. However, this is valid as long as Z is independent of (X,Y); therefore, we have

$$X \stackrel{d}{=} Y \text{ and } (X,Y) \perp\!\!\!\perp Z \Longrightarrow (X,Z) \stackrel{d}{=} (Y,Z) \Longrightarrow X+Z \stackrel{d}{=} Y+Z \quad \text{and} \quad XZ \stackrel{d}{=} YZ.$$

At this point, the natural question that arises is: If Z is independent of X and Y and $X+Z \stackrel{d}{=} Y+Z$ or $XZ \stackrel{d}{=} YZ$, can we conclude that $X \stackrel{d}{=} Y$? In response, we show that the following statements are true. For this, we denote by ψ_X, ψ_Y, and ψ_Z the respective cfs.

[$\mathcal{E}1$] Suppose that $Z \perp\!\!\!\perp (X,Y)$ and that one of the following conditions is satisfied:
(i) $\psi_Z(t) \neq 0$, almost everywhere;
(ii) $\psi_X \in \mathcal{E}$ (alternatively, $\psi_Y \in \mathcal{E}$).
Then

$$X+Z \stackrel{d}{=} Y+Z \Longrightarrow X \stackrel{d}{=} Y.$$

To prove this result, assume that $X+Z \stackrel{d}{=} Y+Z$. Then

$$\psi_X(t)\psi_Z(t) = \psi_{X+Y}(t) = \psi_{Y+Z}(t) = \psi_Y(t)\psi_Z(t). \tag{1.5}$$

If $\psi_Z(t) \neq 0$ for almost all real t, then $\psi_X(t) = \psi_Y(t)$ for almost all real t and therefore $X \stackrel{d}{=} Y$ by continuity of the cfs; but if $\psi_Z(t) \neq 0$ only along $|t| < \delta$ for some $\delta > 0$, then we can only claim that

$$\psi_X(t) = \psi_Y(t), \quad |t| < \delta. \tag{1.6}$$

That is, (1.6) cannot ensure that $\psi_X(t) = \psi_Y(t)$ for all $t \in \mathbb{R}$. However, if $\psi_X \in \mathcal{E}$ (or $\psi_Y \in \mathcal{E}$), then the fact that (1.6) holds implies that (1.5) also holds and therefore $X \stackrel{d}{=} Y$.

[$\mathcal{E}$2] Suppose that $Z \perp\!\!\!\perp (X,Y)$ and that one of the following conditions is satisfied:
(i) $\mathrm{P}(X > 0) = \mathrm{P}(Y > 0) = \mathrm{P}(Z > 0) = 1$ and $\psi_{\log Z}(t) \neq 0$ for almost all t or $\psi_{\log X} \in \mathcal{E}$;
(ii) $\mathrm{P}(Z > 0) = 1$ and $\psi_{\log Z}(t) \neq 0$ for almost all t or $\psi_{\log X} \in \mathcal{E}$;
(iii) $\mathrm{P}(Z > 0) = 1$ and $\psi_{(\log X|X>0)} \in \mathcal{E}$ and $\psi_{(\log(-X)|X<0)} \in \mathcal{E}$.
Then

$$XZ \stackrel{d}{=} YZ \Longrightarrow X \stackrel{d}{=} Y.$$

The proof of this result under condition (i) follows directly from [$\mathcal{E}$1]. In (i), the condition that $\psi_{\log X} \in \mathcal{E}$ is met, for example, if $\mathrm{P}(0 < X < 1) = 1$; see Lemma 5.1.3 in the book by Gupta and Varga (1993). For the proofs under conditions (ii) and (iii), see the book by Fang and Zhang (1990). An alternative formulation of the result in [$\mathcal{E}$2] is found in Lemma 1.2 of the book by Fang et al. (1990).

1.8 Multivariate Moments and Further Descriptive Measures

1.8.1 Mean and variance

Let $X \in \mathbb{R}$ be a real continuous random variable with pdf $f(x)$. The *expectation or mean* of X is defined as

$$\mu = \mathrm{E}(X) = \int_{-\infty}^{+\infty} x f(x) \mathrm{d}x,$$

and its *variance* as

$$\sigma^2 = \mathrm{var}(X) = \mathrm{E}\{(X-\mu)^2\} = \int_{-\infty}^{+\infty} (x-\mu)^2 f(x) \mathrm{d}x.$$

Obviously, the latter can also be computed as $\sigma^2 = \mathrm{E}(X^2) - \mu^2 = \int_{-\infty}^{+\infty} x^2 f(x)\mathrm{d}x - \mu^2$. Clearly, if this last integral is finite, then the variance σ^2, and hence the mean μ, are finite.

The mean $\mu = \mathrm{E}(X)$ and the variance $\sigma^2 = \mathrm{var}(X)$ are, if finite, location and scale measures, respectively, of the distribution of the random variable $X \in \mathbb{R}$. Two well-known properties are

$$\mathrm{E}(aX+b) = a\mathrm{E}(X) + b \quad \text{and} \quad \mathrm{var}(aX+b) = a^2 \mathrm{var}(X),$$

where $a \in \mathbb{R}$ and $b \in \mathbb{R}$ are constants.

1.8.2 Moments and cumulants

In general, and provided that they exist, we define the *k-th noncentral moment* of X by

$$\mu_{[k]} = \mathrm{E}(X^k) = \int_{-\infty}^{\infty} x^k f(x) \mathrm{d}x,$$

and its *k-th central moment* by

$$\bar{\mu}_{[k]} = \mathrm{E}\{(X-\mu)^k\} = \int_{-\infty}^{\infty} (x-\mu)^k f(x) \mathrm{d}x,$$

where $\mu = \mu_{[1]}$. When all the moments of X exist, its *cumulants* are defined to be the coefficients $\kappa_{[k]}$ in

$$\log \psi(t) = \sum_{k=0}^{+\infty} \kappa_{[k]} \frac{(\mathrm{i}t)^k}{k!},$$

where $\psi(t)$ is the cf of X. Well-known relations (see, e.g., Cramér (1946), Kendall et al. (1987), Bilodeau and Brenner (1999)) between the four first moments $\mu_{[k]}$, central moments $\bar{\mu}_{[k]}$, and the cumulants $\kappa_{[k]}$ are

$$\begin{aligned}
\kappa_{[1]} &= \mu_{[1]} = \mu, \\
\kappa_{[2]} &= \bar{\mu}_{[2]} = \mu_{[2]} - \mu_{[1]}^2 = \sigma^2, \\
\kappa_{[3]} &= \bar{\mu}_{[3]} = \mu_{[3]} - 3\mu_{[2]}\mu_{[1]} + 2\mu_{[1]}^2, \\
\kappa_{[4]} &= \bar{\mu}_{[4]} - 3\bar{\mu}_{[2]}^2 = \mu_{[4]} - 4\mu_{[3]}\mu_{[1]} - 3\mu_{[2]}^2 + 12\mu_{[2]}\mu_{[1]}^2 - 6\mu_{[1]}^4,
\end{aligned}$$

and

$$\begin{aligned}
\mu_{[1]} &= \kappa_{[1]} = \mu, \\
\mu_{[2]} &= \kappa_{[2]} + \kappa_{[1]}^2, \\
\mu_{[3]} &= \kappa_{[3]} + 3\kappa_{[2]}\kappa_{[1]} + \kappa_{[1]}^3, \\
\mu_{[4]} &= \kappa_{[4]} + 4\kappa_{[3]}\kappa_{[1]} + 3\kappa_{[2]}^2 + \kappa_{[2]}\kappa_{[1]}^2 + \mu_{[1]}^4.
\end{aligned}$$

1.8.3 *Skewness and kurtosis*

As measures of *skewness and kurtosis* of the distribution of the random variable $X \in \mathbb{R}$ we have the indexes γ_1 and γ_2, respectively, defined by

$$\gamma_1 = \bar{\mu}_{[2]}^{-3/2}\bar{\mu}_{[3]} = \kappa_{[2]}^{-3/2}\kappa_{[3]} \quad \text{and} \quad \gamma_2 = \bar{\mu}_{[2]}^{-2}\bar{\mu}_{[4]} - 3 = \kappa_{[2]}^{-2}\kappa_{[4]}.$$

The skewness index $\gamma_1 = 0$ for symmetric distributions and $\gamma_1 > 0$ (< 0) for right (left)-asymmetric distributions. The (excess) kurtosis coefficient $\gamma_2 = 0$ for the univariate normal model and $\gamma_2 > 0$ (< 0) for distributions that are heavier-tailed (lighter-tailed) than the normal one.

For more details on the performance of these indices to measure the skewness and kurtosis of a distribution, with some alternative proposals, see, for example, Groeneveld and Meeden (1984), Jones et al. (2011), and Siraj-Ud-Doulah (2021).

1.8.4 *Mean vector and variance matrix*

Let $\mathbf{X} \in \mathbb{R}^p$ be a random vector with pdf $f(\mathbf{x})$. If X_i and X_j are the i-th and j-th components of $\mathbf{X}$, respectively, then we may use the scalar-valued expectation definition to compute the *marginal mean of X_i* as

$$\mu_i = \mathrm{E}(X_i) = \int_{\mathbb{R}^p} x_i f(\mathbf{x}) \mathrm{d}\mathbf{x} = \int_{\mathbb{R}} x_i f_i(x_i) \mathrm{d}x_i, \quad i = 1, \ldots, p,$$

and the *marginal variance of X_i* as

$$\sigma_i^2 = \mathrm{var}(X_i) = \mathrm{E}\{(X_i - \mu_i)^2\} = \mathrm{E}(X_i^2) - \mu_i^2 = \int_{-\infty}^{\infty} x_i^2 f_i(x_i) \mathrm{d}x_i - \mu_i^2, \quad i = 1, \ldots, p,$$

as well as the *covariances between X_i and X_j* with bivariate pdfs f_{ij}, $i, j = 1, \ldots, p$, as

$$\sigma_{ij} = \mathrm{cov}(X_i, X_j) = \mathrm{E}\{(X_i - \mu_i)(X_j - \mu_j)\} = \mathrm{E}(X_i X_j) - \mu_i \mu_j = \int_{-\infty}^{\infty}\int_{-\infty}^{\infty} x_i x_j f_{ij}(x_i, x_j) \mathrm{d}x_i \mathrm{d}x_j - \mu_i \mu_j.$$

Note that $\sigma_i^2 = \sigma_{ii}$ since $\mathrm{var}(X_i) = \mathrm{cov}(X_i, X_i)$. Thus, the vector $\boldsymbol{\mu} = (\mu_1, \ldots, \mu_p)^\top \in \mathbb{R}^p$ and the matrix $\boldsymbol{\Sigma} = (\sigma_{ij}) \in \mathbb{R}^{p\times p}$ are called the *mean vector* and *variance matrix* of $\mathbf{X} = (X_1, \ldots, X_p)^\top$, respectively, and we write:

$$\boldsymbol{\mu} = \mathrm{E}(\mathbf{X}) = (\mathrm{E}(X_1), \ldots, \mathrm{E}(X_p))^\top \quad \text{and} \quad \boldsymbol{\Sigma} = \mathrm{var}(\mathbf{X}) = (\mathrm{cov}(X_i, X_j)).$$

The variance matrix $\boldsymbol{\Sigma} = \mathrm{var}(\mathbf{X})$ is sometimes called the *variance-covariance matrix* or simply the *covariance matrix* of $\mathbf{X}$. One-dimensional summaries of the matrix $\boldsymbol{\Sigma}$ are the *generalized variance* $|\boldsymbol{\Sigma}|$ and the *total variation* $\mathrm{tr}(\boldsymbol{\Sigma})$.

It is obvious that $\boldsymbol{\Sigma} = \mathrm{var}(\mathbf{X})$ is a *symmetric matrix* because $\mathrm{var}(\mathbf{X}) = (\mathrm{cov}(X_i, X_j)) = (\mathrm{cov}(X_j, X_i)) = \mathrm{var}(\mathbf{X})^\top$. Another important property of the mean vector $\boldsymbol{\mu} = \mathrm{E}(\mathbf{X})$ and the variance matrix $\boldsymbol{\Sigma} = \mathrm{var}(\mathbf{X})$ is:

$$\mathrm{E}(\mathbf{AX} + \mathbf{b}) = \mathbf{A}\boldsymbol{\mu} + \mathbf{b} \quad \text{and} \quad \mathrm{var}(\mathbf{AX} + \mathbf{b}) = \mathbf{A}\boldsymbol{\Sigma}\mathbf{A}^\top,$$

for every constant matrix $\mathbf{A} \in \mathbb{R}^{q\times p}$ and constant vector $\mathbf{b} \in \mathbb{R}^q$. In particular,

$$\mathrm{E}(\mathbf{a}^\top\mathbf{X} + b) = \mathbf{a}^\top\boldsymbol{\mu} + b \quad \text{and} \quad \mathrm{var}(\mathbf{a}^\top\mathbf{X} + b) = \mathrm{var}(\mathbf{a}^\top\mathbf{X}) = \mathbf{a}^\top\boldsymbol{\Sigma}\mathbf{a} \geq 0,$$

for every $\mathbf{a} \in \mathbb{R}^p$ and $b \in \mathbb{R}$. That is, while the mean vector $\boldsymbol{\mu}$ retains the linearity property, the variance matrix $\boldsymbol{\Sigma}$ is *positive semi-definite (psd)*, $\boldsymbol{\Sigma} \geq 0$. However, we could have that $\mathrm{var}(\mathbf{a}^\top\mathbf{X}) = \mathbf{a}^\top\boldsymbol{\Sigma}\mathbf{a} = 0$ for some $\mathbf{a} \neq \mathbf{0}_p$ iff $\mathbf{a}^\top\mathbf{X} = c$ with probability 1, where $c = \mathbf{a}^\top\mathrm{E}(\mathbf{X})$, i.e., iff the components $X_1, \ldots, X_p$ of $\mathbf{X}$ are linearly related random variables with probability 1. Consequently, the variance matrix $\boldsymbol{\Sigma} = \mathrm{var}(\mathbf{X})$ will be *positive definite (pd)*, $\boldsymbol{\Sigma} > 0$, iff the elements of $\mathbf{X}$ are linearly independent with probability 1.

On the other hand, since $\boldsymbol{\Sigma} \geq 0$, we can write $\boldsymbol{\Sigma} = \boldsymbol{\Gamma}\mathbf{D}_\lambda\boldsymbol{\Gamma}^\top$ with $\boldsymbol{\Gamma} \in \mathbb{R}^{p\times p}$ orthogonal and $\mathbf{D}_\lambda = \mathrm{diag}(\lambda_1, \ldots, \lambda_p)$, where $\lambda_1 \geq \cdots \geq \lambda_p \geq 0$ are the eigenvalues of $\boldsymbol{\Sigma}$. Hence, letting $\mathbf{D}_\lambda^{1/2} = \mathrm{diag}(\lambda_1^{1/2}, \ldots, \lambda_p^{1/2})$, we can define the *square root* of $\boldsymbol{\Sigma}$ as the matrix $\boldsymbol{\Sigma}^{1/2} = \boldsymbol{\Gamma}\mathbf{D}_\lambda^{1/2}\boldsymbol{\Gamma}^\top$. Consequently, we may always *normalize* any random vector $\mathbf{X}$ with mean vector $\boldsymbol{\mu}$ and variance matrix $\boldsymbol{\Sigma} > 0$ by letting $\mathbf{Z} = \boldsymbol{\Sigma}^{-1/2}(\mathbf{X} - \boldsymbol{\mu})$, the so-called *Mahalanobis transformation*. We find, obviously, that

$$\mathrm{E}(\mathbf{Z}) = \mathbf{0}_p \quad \text{and} \quad \mathrm{var}(\mathbf{Z}) = \mathbf{I}_p,$$

where $\mathbf{I}_p$ denotes the identity matrix in $\mathbb{R}^{p\times p}$. Another *normalized version* of $\mathbf{X}$ that satisfies this last fact is $\mathbf{Z} = \mathbf{D}_\lambda^{-1/2}\boldsymbol{\Gamma}^\top(\mathbf{X} - \boldsymbol{\mu})$. Note that this transformation of $\mathbf{X}$ considers three stages: The first relocates by $\boldsymbol{\mu}$, the second rotates by $\boldsymbol{\Gamma}^\top$ and the third rescales by $\mathbf{D}_\lambda^{-1/2}$.

Finally, any variance matrix $\boldsymbol{\Sigma} = (\sigma_{ij})$ induces a *correlation matrix* $\mathbf{R} = (\rho_{ij}) \in \mathbb{R}^{p\times p}$, where ρ_{ij} is the *correlation coefficient* between X_i and X_j and is defined as

$$\rho_{ij} = \frac{\sigma_{ij}}{\sigma_i\sigma_j} = \frac{\mathrm{cov}(X_i, X_j)}{\sqrt{\mathrm{var}(X_i)\mathrm{var}(X_j)}} \qquad i, j = 1, \ldots, p.$$

Clearly $\rho_{ii} = 1$ for all $i = 1, \ldots, p$. Also, the well-known *Cauchy-Schwarz inequality* $|\mathrm{cov}(X_i, X_j)| \leq \sqrt{\mathrm{var}(X_i)\mathrm{var}(X_j)}$ implies that $|\rho_{ij}| \leq 1$ for all $i, j = 1, \ldots, p$. Moreover, by defining the diagonal matrix $\mathbf{D}_\sigma = \mathrm{diag}(\sigma_1, \ldots, \sigma_p) \in \mathbb{R}^{p\times p}$, we can write

$$\boldsymbol{\Sigma} = \mathbf{D}_\sigma\mathbf{R}\mathbf{D}_\sigma.$$

The correlation matrix $\mathbf{R} \geq 0$ because $\boldsymbol{\Sigma} \geq 0$ and $\mathbf{D}_\sigma$ is symmetric; see Section 1.8.5 for further properties. Sometimes we will use the notation $\bar{\boldsymbol{\Sigma}} = \boldsymbol{\sigma}^{-1}\boldsymbol{\Sigma}\boldsymbol{\sigma}^{-1}$, where $\boldsymbol{\sigma} = \mathrm{diag}(\boldsymbol{\Sigma})^{1/2}$, to refer to the correlation matrix of a variance matrix $\boldsymbol{\Sigma} > 0$.

1.8.5 Higher-order multivariate moments

Further noncentral and central multivariate moments of order k for $\mathbf{X} = (X_1, \ldots, X_p)^\top \in \mathbb{R}^p$ may be defined, e.g., as the product-moments:

$$\mu_{[k_1,\ldots,k_p]} = \mathrm{E}\left(\prod_{i=1}^{p} X_i^{k_i}\right) \quad \text{and} \quad \bar{\mu}_{[k_1,\ldots,k_p]} = \mathrm{E}\left\{\prod_{i=1}^{p}(X_i - \mu_i)^{k_i}\right\},\ k_i \geq 0,\ i = 1, \ldots, p,\ \sum_{i=1}^{p} k_i = k.$$

For example, central product-moments for $p = 2$ are:

$$\begin{aligned}
\bar{\mu}_{[2,0]} &= \mathrm{E}\{(X_1 - \mu_1)^2\} = \sigma_1^2, \quad \bar{\mu}_{[0,2]} = \mathrm{E}\{(X_2 - \mu_2)^2\} = \sigma_2^2, \\
\bar{\mu}_{[1,1]} &= \mathrm{E}\{(X_1 - \mu_1)(X_2 - \mu_2)\} = \sigma_{12}, \\
\bar{\mu}_{[3,0]} &= \mathrm{E}\{(X_1 - \mu_1)^3\}, \quad \bar{\mu}_{[0,3]} = \mathrm{E}\{(X_2 - \mu_2)^3\}, \\
\bar{\mu}_{[2,1]} &= \mathrm{E}\{(X_1 - \mu_1)^2(X_2 - \mu_2)\}, \quad \bar{\mu}_{[1,2]} = \mathrm{E}\{(X_1 - \mu_1)(X_2 - \mu_2)^2\}, \\
\bar{\mu}_{[4,0]} &= \mathrm{E}\{(X_1 - \mu_1)^4\}, \quad \bar{\mu}_{[0,4]} = \mathrm{E}\{(X_2 - \mu_2)^4\}, \quad \bar{\mu}_{[2,2]} = \mathrm{E}\{(X_1 - \mu_1)^2(X_2 - \mu_2)^2\}, \\
\bar{\mu}_{[3,1]} &= \mathrm{E}\{(X_1 - \mu_1)^3(X_2 - \mu_2)\}, \quad \bar{\mu}_{[1,3]} = \mathrm{E}\{(X_1 - \mu_1)(X_2 - \mu_2)^3\}, \quad \text{etc.}
\end{aligned}$$

Moreover, when all the moments of $\mathbf{X}$ exist, the multivariate cumulants (or semi-invariants) are defined to be the coefficients κ in

$$\log \psi(\mathbf{t}) = \sum_{k_1=0}^{+\infty} \cdots \sum_{k_p=0}^{+\infty} \kappa_{[k_1,\ldots,k_p]} \frac{(\mathrm{i}t_1)^{k_1} \cdots (\mathrm{i}t_p)^{k_p}}{k_1! \cdots k_p!},$$

where $\psi(\mathbf{t})$ is the cf of $\mathbf{X}$. For instance, for $p = 2$, we have:

$$\begin{aligned}
\kappa_{[1,0]} &= \mu_1, \quad \kappa_{[0,1]} = \mu_2, \\
\kappa_{[2,0]} &= \sigma_1^2, \quad \kappa_{[0,2]} = \sigma_2^2, \quad \kappa_{[1,1]} = \sigma_{12}, \\
\kappa_{[3,0]} &= \bar{\mu}_{[3,0]}, \quad \kappa_{[0,3]} = \bar{\mu}_{[0,3]}, \quad \kappa_{[2,1]} = \bar{\mu}_{[2,1]}, \quad \kappa_{[1,2]} = \bar{\mu}_{[1,2]}, \\
\kappa_{[4,0]} &= \bar{\mu}_{[4,0]} - 3\bar{\mu}_{[2,0]}^2, \quad \kappa_{[0,4]} = \bar{\mu}_{[0,4]} - 3\bar{\mu}_{[0,2]}^2, \quad \kappa_{[2,2]} = \bar{\mu}_{[2,2]} - 2\bar{\mu}_{[2,0]}\bar{\mu}_{[0,2]} - \bar{\mu}_{[1,1]}, \\
\kappa_{[3,1]} &= \bar{\mu}_{[3,1]} - 3\bar{\mu}_{[0,2]}\bar{\mu}_{[1,1]}, \quad \kappa_{[1,3]} = \bar{\mu}_{[1,3]} - 3\bar{\mu}_{[0,2]}\bar{\mu}_{[1,1]}, \quad \text{etc.}
\end{aligned}$$

Additional results on multivariate moments of a random vector $\mathbf{X}$ are presented at the end of this chapter.

1.8.6 Conditional mean vector and variance matrix

Moments of $(\mathbf{X}_1 \mid \mathbf{X}_2 = \mathbf{x}_2)$ are called conditional multivariate moments. In particular, the conditional mean vector and the conditional variance matrix of $\mathbf{X}_1$ given $\mathbf{X}_2 = \mathbf{x}_2$ are, respectively,

$$\boldsymbol{\mu}_{1\cdot 2} = \mathrm{E}(\mathbf{X}_1 \mid \mathbf{x}_2) \quad \text{and} \quad \boldsymbol{\Sigma}_{11\cdot 2} = \mathrm{var}(\mathbf{X}_1 \mid \mathbf{x}_2).$$

The conditional expectation function $\boldsymbol{\mu}_{1\cdot 2}(\mathbf{x}_2) = \mathrm{E}(\mathbf{X}_1 \mid \mathbf{x}_2)$, defined on the support of $\mathbf{X}_2$, is called the regression curve of $\mathbf{X}_1$ on $\mathbf{X}_2 = \mathbf{x}_2$. If $\boldsymbol{\mu}_{1\cdot 2}(\mathbf{x}_2)$ is linear in $\mathbf{x}_2$, then the regression curve is called linear. The conditional variance matrix $\boldsymbol{\Sigma}_{11\cdot 2} = \mathrm{var}(\mathbf{X}_1 \mid \mathbf{x}_2)$ can also be a function of $\mathbf{x}_2$. When this matrix is constant, the regression curve of $\mathbf{X}_1$ on $\mathbf{x}_2$ is called homoscedastic. Two well-known relations involving marginal and conditional mean vectors and variance matrices are

$$\mathrm{E}(\mathbf{X}_1) = \mathrm{E}\{\mathrm{E}(\mathbf{X}_1 \mid \mathbf{X}_2)\} \quad \text{and} \quad \mathrm{var}(\mathbf{X}_1) = \mathrm{E}\{\mathrm{var}(\mathbf{X}_1 \mid \mathbf{X}_2)\} + \mathrm{var}\{\mathrm{E}(\mathbf{X}_1 \mid \mathbf{X}_2)\}.$$

1.8.7 *Expectation of a random matrix*

Let $\mathbf{W} = (W_{ij}) \in \mathbb{R}^{n\times p}$ be a rectangular random matrix, for which each component W_{ij} is a random variable with finite expectation. The *expected matrix* of $\mathbf{W}$ is defined as the rectangular matrix

$$\mathrm{E}(\mathbf{W}) = (\mathrm{E}(W_{ij})) \in \mathbb{R}^{n\times p}.$$

For $p = 1$, $\mathrm{E}(\mathbf{W})$ is a mean vector, while for $n = p = 1$, $\mathrm{E}(\mathbf{W}) = \mathrm{E}(W)$ is a scalar expected value. From this definition, we have the following properties:

[1] If $\mathbf{W} \in \mathbb{R}^{n\times p}$ is a random matrix with finite expectation, and $\mathbf{A} \in \mathbb{R}^{m\times n}$, $\mathbf{B} \in \mathbb{R}^{p\times q}$ and $\mathbf{C} \in \mathbb{R}^{m\times q}$ are constant matrices, then

$$\mathrm{E}(\mathbf{AWB}+\mathbf{C}) = \mathbf{A}\mathrm{E}(\mathbf{W})\mathbf{B}+\mathbf{C};$$

[2] If $\mathbf{W}_1 \in \mathbb{R}^{n\times p}$ and $\mathbf{W}_2 \in \mathbb{R}^{n\times p}$ are two random matrices with finite expectation, and a and b are scalar constants, then

$$\mathrm{E}(a\mathbf{W}_1 + b\mathbf{W}_2) = a\mathrm{E}(\mathbf{W}_1) + b\mathrm{E}(\mathbf{W}_2);$$

[3] If $\mathbf{W} \in \mathbb{R}^{n\times p}$ is a random matrix with finite expectation, and $\mathbf{A} \in \mathbb{R}^{p\times n}$, then

$$\mathrm{E}\{\mathrm{tr}(\mathbf{AW})\} = \mathrm{tr}\{\mathbf{A}\mathrm{E}(\mathbf{W})\}.$$

1.8.8 *Covariance matrix between two random vectors*

As a well-known expected matrix, we have the *covariance matrix between two random vectors* $\mathbf{X} = (X_1,\dots,X_p)^\top \in \mathbb{R}^p$ and $\mathbf{Y} = (Y_1,\dots,Y_q)^\top \in \mathbb{R}^q$, which is defined as the matrix

$$\mathrm{cov}(\mathbf{X},\mathbf{Y}) = (\mathrm{cov}(X_i,Y_j)) \in \mathbb{R}^{p\times q}.$$

It is clear that

$$\mathrm{cov}(\mathbf{X},\mathbf{Y}) = \mathrm{E}[\{\mathbf{X}-\mathrm{E}(\mathbf{X})\}\{\mathbf{Y}-\mathrm{E}(\mathbf{Y})\}^\top].$$

In particular, when $\mathbf{X} = \mathbf{Y}$, we have $\mathrm{cov}(\mathbf{X},\mathbf{X}) = \mathrm{var}(\mathbf{X}) = (\mathrm{cov}(X_i,X_j))$, so now we write

$$\mathrm{var}(\mathbf{X}) = \mathrm{E}[\{\mathbf{X}-\mathrm{E}(\mathbf{X})\}\{\mathbf{X}-\mathrm{E}(\mathbf{X})\}^\top].$$

From the expected matrix definition, the following are direct properties:

[1] $\mathrm{cov}(\mathbf{X},\mathbf{Y}) = \mathrm{E}(\mathbf{X}\mathbf{Y}^\top) - \mathrm{E}(\mathbf{X})\mathrm{E}(\mathbf{Y})^\top$ and, in particular, $\mathrm{var}(\mathbf{X}) = \mathrm{E}(\mathbf{X}\mathbf{X}^\top) - \mathrm{E}(\mathbf{X})\mathrm{E}(\mathbf{X})^\top$;

[2] $\mathrm{cov}(\mathbf{Y},\mathbf{X}) = \mathrm{cov}(\mathbf{X},\mathbf{Y})^\top$;

[3] $\mathrm{cov}(\mathbf{X}_1+\mathbf{X}_2,\mathbf{Y}) = \mathrm{cov}(\mathbf{X}_1,\mathbf{Y}) + \mathrm{cov}(\mathbf{X}_2,\mathbf{Y})$;

[4] If $p = q$, $\mathrm{cov}(\mathbf{X}\pm\mathbf{Y}) = \mathrm{cov}(\mathbf{X}) + \mathrm{cov}(\mathbf{Y}) \pm \mathrm{cov}(\mathbf{X},\mathbf{Y}) \pm \mathrm{cov}(\mathbf{Y},\mathbf{X})$;

[5] $\mathrm{cov}(\mathbf{AX},\mathbf{BY}) = \mathbf{A}\mathrm{cov}(\mathbf{X},\mathbf{Y})\mathbf{B}^\top$, where $\mathbf{A} \in \mathbb{R}^{r\times p}$ and $\mathbf{B} \in \mathbb{R}^{s\times q}$ are constant matrices;

[6] If $\mathbf{X} \perp\!\!\!\perp \mathbf{Y}$, then $\mathrm{cov}(\mathbf{X},\mathbf{Y}) = \mathbf{O}$, the zero matrix. However, the converse is generally not true;

[7] If $\mathbf{X} = (\mathbf{X}_1^\top,\mathbf{X}_2^\top)^\top \in \mathbb{R}^{p_1+p_2}$, then $\mathrm{E}(\mathbf{X}) = (\mathrm{E}(\mathbf{X}_1)^\top,\mathrm{E}(\mathbf{X}_2)^\top)^\top = (\boldsymbol{\mu}_1^\top,\boldsymbol{\mu}_2^\top)^\top$ and

$$\mathrm{var}(\mathbf{X}) = \begin{pmatrix} \mathrm{var}(\mathbf{X}_1) & \mathrm{cov}(\mathbf{X}_1,\mathbf{X}_2) \\ \mathrm{cov}(\mathbf{X}_2,\mathbf{X}_1) & \mathrm{var}(\mathbf{X}_2) \end{pmatrix} = \begin{pmatrix} \boldsymbol{\Sigma}_{11} & \boldsymbol{\Sigma}_{12} \\ \boldsymbol{\Sigma}_{21} & \boldsymbol{\Sigma}_{22} \end{pmatrix}.$$

1.8.9 *Variance and covariance for random matrices*

Let $\mathbf{W} \in \mathbb{R}^{n\times p}$ be a random matrix for which each of its components has finite second moments. From the above expected matrix definition, we find for the *mean matrix* of $\mathbf{W} = (W_{ij})$ that $\mathrm{E}(\mathbf{W}) = (\mu_{ij})$, where $\mu_{ij} = \mathrm{E}(W_{ij})$, $i = 1,\dots,n$, $j = 1,\dots,p$. The variance of $\mathbf{W} = (W_{ij})$ is, however, a quadruple indexed array consisting of all the covariances between the individual entries, say

$$\mathrm{var}(\mathbf{W}) = (\mathrm{cov}(W_{ij},W_{kl})) = (\sigma_{ijkl}).$$

To impose an order on this array, we may define the *variance matrix* of $\mathbf{W}$ as the matrix

$$\mathrm{var}(\mathbf{W}) \equiv \mathrm{var}\{\mathrm{vec}(\mathbf{W}^\top)\} \in \mathbb{R}^{np\times np}.$$

Under this definition, we should write the mean of $\mathbf{W}$ as $\mathrm{E}\{\mathrm{vec}(\mathbf{W}^\top)\} = \mathrm{vec}\{\mathrm{E}(\mathbf{W}^\top)\}$. Thus, since the row vectors of $\mathbf{W}$ are the column vectors, $\mathbf{X}_1,\ldots,\mathbf{X}_n$ say, of $\mathbf{W}^\top$, we have

$$\mathbf{W} = \begin{pmatrix} \mathbf{X}_1^\top \\ \vdots \\ \mathbf{X}_n^\top \end{pmatrix} \longrightarrow \mathbf{W}^\top = (\mathbf{X}_1,\ldots,\mathbf{X}_n) \longrightarrow \mathrm{vec}(\mathbf{W}^\top) = \begin{pmatrix} \mathbf{X}_1 \\ \vdots \\ \mathbf{X}_n \end{pmatrix},$$

and we find

$$\mathrm{var}(\mathbf{W}) = (\mathrm{cov}(\mathbf{X}_i,\mathbf{X}_j)) = (\boldsymbol{\Sigma}_{ij}), \quad \text{with} \quad \boldsymbol{\Sigma}_{ij} = (\mathrm{cov}(W_{ik},W_{jl})).$$

Similarly, we can define the *covariance matrix* between two random matrices $\mathbf{W}_1 \in \mathbb{R}^{n_1\times p_1}$ and $\mathbf{W}_2 \in \mathbb{R}^{n_2\times p_2}$ as the matrix

$$\mathrm{cov}(\mathbf{W}_1,\mathbf{W}_2) \equiv \mathrm{cov}\{\mathrm{vec}(\mathbf{W}_1^\top),\mathrm{vec}(\mathbf{W}_2^\top)\} \in \mathbb{R}^{n_1p_1\times n_2p_2}.$$

Two special cases are:

(a) $\mathbf{W}_1 = \mathbf{W}_2 = \mathbf{W} \in \mathbb{R}^{n\times p}$

$$\Longrightarrow \mathrm{cov}(\mathbf{W},\mathbf{W}) \equiv \mathrm{cov}\{\mathrm{vec}(\mathbf{W}^\top),\mathrm{vec}(\mathbf{W}^\top)\} = \mathrm{var}\{\mathrm{vec}(\mathbf{W}^\top)\} \equiv \mathrm{var}(\mathbf{W});$$

(b) $\mathbf{W}_1 = \mathbf{X} \in \mathbb{R}^p$, $\mathbf{W}_2 = \mathbf{W} \in \mathbb{R}^{n\times p}$

$$\Longrightarrow \mathrm{cov}(\mathbf{X},\mathbf{W}) \equiv \mathrm{cov}\{\mathbf{X},\mathrm{vec}(\mathbf{W}^\top)\}.$$

Finally, from the partition

$$\mathbf{W} = \begin{pmatrix} \mathbf{W}_1 \\ \mathbf{W}_2 \end{pmatrix} \Longrightarrow \mathbf{W}^\top = (\mathbf{W}_1^\top,\mathbf{W}_2^\top)^\top \Longrightarrow \mathrm{vec}(\mathbf{W}^\top) = \begin{pmatrix} \mathrm{vec}(\mathbf{W}_1^\top) \\ \mathrm{vec}(\mathbf{W}_2^\top) \end{pmatrix},$$

where $\mathbf{W}_i \in \mathbb{R}^{n_i\times p}$, $i = 1,2$, we can also find the following variance matrix:

$$\mathrm{var}\begin{pmatrix} \mathbf{W}_1 \\ \mathbf{W}_2 \end{pmatrix} \equiv \mathrm{var}\begin{pmatrix} \mathrm{vec}(\mathbf{W}_1^\top) \\ \mathrm{vec}(\mathbf{W}_2^\top) \end{pmatrix} = \begin{pmatrix} \mathrm{var}\{\mathrm{vec}(\mathbf{W}_1^\top)\} & \mathrm{cov}\{\mathrm{vec}(\mathbf{W}_1^\top),\mathrm{vec}(\mathbf{W}_2^\top)\} \\ \mathrm{cov}\{\mathrm{vec}(\mathbf{W}_2^\top),\mathrm{vec}(\mathbf{W}_1^\top)\} & \mathrm{var}\{\mathrm{vec}(\mathbf{W}_2^\top)\} \end{pmatrix}.$$

Example 16 *Let $\mathbf{X}_1,\ldots,\mathbf{X}_n$ be iid random vectors with mean vector $\boldsymbol{\mu} \in \mathbb{R}^p$ and variance matrix $\boldsymbol{\Sigma} \in \mathbb{R}^{p\times p}$; and let $\mathbf{W}$ be the data matrix:*

$$\mathbf{W} = \begin{pmatrix} \mathbf{X}_1^\top \\ \vdots \\ \mathbf{X}_n^\top \end{pmatrix} \Longrightarrow \mathrm{E}(\mathbf{W}) = \begin{pmatrix} \boldsymbol{\mu}^\top \\ \vdots \\ \boldsymbol{\mu}^\top \end{pmatrix} \quad \textit{and} \quad \mathrm{var}\{\mathrm{vec}(\mathbf{W}^\top)\} = \begin{pmatrix} \boldsymbol{\Sigma} & \mathbf{O} & \cdots & \mathbf{O} \\ \mathbf{O} & \boldsymbol{\Sigma} & \cdots & \mathbf{O} \\ \vdots & \vdots & \ddots & \vdots \\ \mathbf{O} & \mathbf{O} & \cdots & \boldsymbol{\Sigma} \end{pmatrix}.$$

We find:

$$\mathrm{E}(\mathbf{W}) = \mathbf{1}_n \otimes \boldsymbol{\mu}^\top = \mathbf{1}_n\boldsymbol{\mu}^\top \quad \textit{and} \quad \mathrm{var}(\mathbf{W}) = \mathbf{I}_n \otimes \boldsymbol{\Sigma},$$

where $\otimes$ denotes the Kronecker product (see Section A.2.9).

Example 17 *Let* $\mathbf{Z} = (Z_1,\ldots,Z_p)^\top$, *where* $Z_1,\ldots,Z_p$ *are iid* $\mathcal{N}(0,1)$ *random variables, where* $\mathcal{N}(0,1)$ *denotes the univariate standard normal distribution with pdf* $f(z) = (2\pi)^{-1/2}e^{-z^2/2}$, $z \in \mathbb{R}$. *It is well known that* $\mathrm{E}(Z_i) = \mathrm{E}(Z_i^3) = 0$, $\mathrm{E}(Z_i^2) = 1$, *and* $\mathrm{E}(Z_i^4) = 3$. *Hence,* $\mathrm{E}(\mathbf{Z}) = \mathbf{0}_p$, $\mathrm{var}(\mathbf{Z}) = \mathrm{E}(\mathbf{Z}\mathbf{Z}^\top) = \mathbf{I}_p$, $\mathrm{E}(Z_i\mathbf{Z}) = \mathbf{e}_i$, *and* $\mathrm{E}(Z_iZ_j\mathbf{Z}\mathbf{Z}^\top) = \delta_{ij}\mathbf{I}_p + \mathbf{e}_i\mathbf{e}_j^\top + \mathbf{e}_j\mathbf{e}_i^\top$, *where* $\delta_{ij} = 1$ *for* $i = j$ *and* 0 *otherwise, and* $\mathbf{e}_i = \mathbf{e}_{i:p}$ *is the i-th unit vector of* $\mathbb{R}^p$. *We then have* $\mathrm{cov}(Z_i\mathbf{Z}, Z_j\mathbf{Z}) = \delta_{ij}\mathbf{I}_p + \mathbf{e}_j\mathbf{e}_i^\top$, *hence (Magnus and Neudecker, 1999)*

$$\mathrm{var}(\mathbf{Z}\mathbf{Z}^\top) \equiv \mathrm{var}\{\mathrm{vec}(\mathbf{Z}\mathbf{Z}^\top)\} = \mathbf{I}_{p^2} + \mathbf{K}_p,$$

where $\mathbf{K}_p = \mathbf{K}_{pp} = (\mathbf{e}_j\mathbf{e}_i^\top)$ *is the commutation matrix of order* $p^2 \times p^2$. *Consider now the random matrix* $\mathbf{W} = \mathbf{A}\mathbf{Z}\mathbf{Z}^\top\mathbf{A}^\top$, *where* $\mathbf{A}\mathbf{A}^\top = \mathbf{\Sigma}$. *It is clear that* $\mathrm{E}(\mathbf{W}) = \mathbf{\Sigma}$, *while*

$$\begin{aligned}\mathrm{var}(\mathbf{W}) &\equiv \mathrm{var}\{\mathrm{vec}(\mathbf{A}\mathbf{Z}\mathbf{Z}^\top\mathbf{A}^\top)\} = \mathrm{var}\{(\mathbf{A}\otimes\mathbf{A})\mathrm{vec}(\mathbf{Z}\mathbf{Z}^\top)\} = (\mathbf{A}\otimes\mathbf{A})(\mathbf{I}_{p^2}+\mathbf{K}_p)(\mathbf{A}^\top\otimes\mathbf{A}^\top)\\ &= (\mathbf{A}\mathbf{A}^\top\otimes\mathbf{A}\mathbf{A}^\top) + (\mathbf{A}\otimes\mathbf{A})\mathbf{K}_p(\mathbf{A}^\top\otimes\mathbf{A}^\top) = (\mathbf{I}_{p^2}+\mathbf{K}_p)(\mathbf{\Sigma}\otimes\mathbf{\Sigma}),\end{aligned}$$

since $(\mathbf{A}\otimes\mathbf{A})\mathbf{K}_p = \mathbf{K}_p(\mathbf{A}\otimes\mathbf{A})$.

1.8.10 *Higher-order moments of a random vector*

In general, we can define the *k-th noncentral multivariate moment* of a random vector $\mathbf{X}$ (provided it exists) as follows:

$$\boldsymbol{\mu}_k(\mathbf{X}) = \begin{cases}\mathrm{E}(\mathbf{X}\otimes\mathbf{X}^\top\otimes\mathbf{X}\otimes\cdots\otimes\mathbf{X}\otimes\mathbf{X}^\top), & \text{if } k \text{ is even,}\\ \mathrm{E}(\mathbf{X}\otimes\mathbf{X}^\top\otimes\mathbf{X}\otimes\cdots\otimes\mathbf{X}^\top\otimes\mathbf{X}), & \text{if } k \text{ is odd,}\end{cases} \tag{1.7}$$

$k = 1, 2, \ldots$, where the Kronecker product required by the expectation operator involves k factors. For example, since $\mathbf{X}\otimes\mathbf{X}^\top = \mathbf{X}\mathbf{X}^\top$, the first four moments of $\mathbf{X}$ become:

$$\boldsymbol{\mu}_1(\mathbf{X}) = \mathrm{E}(\mathbf{X}),\quad \boldsymbol{\mu}_2(\mathbf{X}) = \mathrm{E}(\mathbf{X}\mathbf{X}^\top),\quad \boldsymbol{\mu}_3(\mathbf{X}) = \mathrm{E}(\mathbf{X}\mathbf{X}^\top\otimes\mathbf{X}),\quad \boldsymbol{\mu}_4(\mathbf{X}) = \mathrm{E}(\mathbf{X}\mathbf{X}^\top\otimes\mathbf{X}\mathbf{X}^\top).$$

An operational procedure to calculate these moments is to differentiate the characteristic function of $\mathbf{X}$:

$$\boldsymbol{\mu}_k(\mathbf{X}) = \begin{cases}\mathrm{i}^{-k}\dfrac{\partial^k\psi(\mathbf{t})}{\partial\mathbf{t}\partial\mathbf{t}^\top\partial\mathbf{t}\cdots\partial\mathbf{t}\partial\mathbf{t}^\top}, & \text{if } k \text{ is even,}\\ \mathrm{i}^{-k}\dfrac{\partial^k\psi(\mathbf{t})}{\partial\mathbf{t}\partial\mathbf{t}^\top\partial\mathbf{t}\cdots\partial\mathbf{t}^\top\partial\mathbf{t}}, & \text{if } k \text{ is odd,}\end{cases}$$

$k = 1, 2, \ldots$.

The moments of $\mathbf{X}$ can also be used to calculate *product-moments of quadratic forms* of $\mathbf{X}$. In fact, if $\mathbf{X}$ has $2k$-th moment and $\mathbf{A}_1, \ldots, \mathbf{A}_k$ are p-by-p fixed matrices, then

$$\begin{aligned}\mathrm{E}\{(\mathbf{X}^\top\mathbf{A}_1\mathbf{X})\times\cdots\times(\mathbf{X}^\top\mathbf{A}_k\mathbf{X})\} &= \mathrm{E}\{(\mathbf{X}^\top\mathbf{A}_1\mathbf{X})\otimes(\mathbf{X}^\top\mathbf{A}_2\mathbf{X})\otimes\cdots\otimes(\mathbf{X}^\top\mathbf{A}_k\mathbf{X})\}\\ &= \mathrm{E}\{(\mathbf{X}^\top\otimes\cdots\otimes\mathbf{X}^\top)(\mathbf{A}_1\otimes\cdots\otimes\mathbf{A}_k)(\mathbf{X}\otimes\cdots\otimes\mathbf{X}))\}\\ &= \mathrm{E}[\mathrm{tr}\{(\mathbf{A}_1\otimes\cdots\otimes\mathbf{A}_k)(\mathbf{X}\otimes\cdots\otimes\mathbf{X})(\mathbf{X}^\top\otimes\cdots\otimes\mathbf{X}^\top)\}]\\ &= \mathrm{tr}\{(\mathbf{A}_1\otimes\cdots\otimes\mathbf{A}_k)\mathrm{E}(\mathbf{X}\otimes\mathbf{X}^\top\otimes\mathbf{X}\otimes\cdots\otimes\mathbf{X}^\top)\}\\ &= \mathrm{tr}\{(\mathbf{A}_1\otimes\cdots\otimes\mathbf{A}_k)\boldsymbol{\mu}_{2k}(\mathbf{X})\}.\end{aligned} \tag{1.8}$$

In particular, since $\boldsymbol{\mu}_2(\mathbf{X}) = \mathrm{var}(\mathbf{X}) + \boldsymbol{\mu}_1(\mathbf{X})\boldsymbol{\mu}_1(\mathbf{X})^\top$, for $k = 1$, we have

$$\mathrm{E}(\mathbf{X}^\top\mathbf{A}_1\mathbf{X}) = \mathrm{tr}\{\mathbf{A}_1\mathrm{var}(\mathbf{X})\} + \boldsymbol{\mu}_1(\mathbf{X})^\top\mathbf{A}_1\boldsymbol{\mu}_1(\mathbf{X}). \tag{1.9}$$

Similarly, we can prove the following:

$$\mathrm{E}\{(\mathbf{X}^\top \mathbf{A}\mathbf{X})\mathbf{B}\mathbf{X}\} = \mathbf{B}\boldsymbol{\mu}_3(\mathbf{X})^\top \mathrm{vec}(\mathbf{A}), \tag{1.10}$$

where $\mathbf{A}$ and $\mathbf{B}$ are constant matrices of dimensions $p \times p$ and $q \times p$, respectively.

Using (1.7) with $\mathbf{X}$ replaced by the centered vector $\mathbf{X} - \boldsymbol{\mu}$, where $\boldsymbol{\mu} = \boldsymbol{\mu}_1(\mathbf{X})$ is the mean vector of $\mathbf{X}$, the k-th centered moment of $\mathbf{X}$ is defined, with $\boldsymbol{\mu}_1(\mathbf{X} - \boldsymbol{\mu}) = \mathbf{0}_p$ and $\boldsymbol{\mu}_2(\mathbf{X} - \boldsymbol{\mu}) = \mathrm{var}(\mathbf{X})$. Furthermore, if we set $\boldsymbol{\Sigma} = \mathrm{var}(\mathbf{X})$, additional special cases are obtained by assuming $\boldsymbol{\Sigma} > 0$ and replacing the matrix $\mathbf{A}$ with $\boldsymbol{\Sigma}^{-1}$. For instance, *Mardia's measures of multivariate skewness and kurtosis* (Mardia, 1970) for a multivariate distribution in $\mathbb{R}^p$ with mean vector $\boldsymbol{\mu}$ and variance matrix $\boldsymbol{\Sigma} > 0$ can be defined, respectively, as

$$\beta_{1,p} = \mathrm{E}[\{(\mathbf{X} - \boldsymbol{\mu})^\top \boldsymbol{\Sigma}^{-1}(\mathbf{X}' - \boldsymbol{\mu})\}^3] \quad \text{and} \quad \beta_{2,p} = \mathrm{E}[\{(\mathbf{X} - \boldsymbol{\mu})^\top \boldsymbol{\Sigma}^{-1}(\mathbf{X} - \boldsymbol{\mu})\}^2],$$

where $\mathbf{X}$ and $\mathbf{X}'$ are iid random vectors. These measures are invariant under affine transformations. We can also show that, in general, the measure of multivariate skewness $\beta_{1,p}$ is always nonnegative. When $p = 1$, we have $\beta_{1,1} = \gamma_1^2$ and $\beta_{2,1} = \gamma_2 + 3$. Finally, since

$$\beta_{1,p} = \mathrm{E}\{(\mathbf{Z}^\top \mathbf{Z}')^3\} = \sum_{i=1}^{p}\sum_{j=1}^{p}\sum_{k=1}^{p}\{\mathrm{E}(Z_i Z_j Z_k)\}^2 \quad \text{and} \quad \beta_{2,p} = \mathrm{E}\{(\mathbf{Z}^\top \mathbf{Z})\}^2 = \sum_{i=1}^{p}\sum_{j=1}^{p}\mathrm{E}(Z_i^2 Z_j^2),$$

where $\mathbf{Z} = \boldsymbol{\Sigma}^{-1/2}(\mathbf{X} - \boldsymbol{\mu})$ and $\mathbf{Z}' = \boldsymbol{\Sigma}^{-1/2}(\mathbf{X}' - \boldsymbol{\mu})$ are also iid random vectors with mean $\mathbf{0}_p$ and variance matrix $\mathbf{I}_p$, then proceeding in a similar way to the proof of (1.7), we arrive at the following alternative formulas to calculate these indices:

$$\beta_{1,p} = \mathrm{tr}\{\boldsymbol{\mu}_3(\mathbf{Z})^\top \boldsymbol{\mu}_3(\mathbf{Z})\} \quad \text{and} \quad \beta_{2,p} = \mathrm{tr}\{\boldsymbol{\mu}_4(\mathbf{Z})\}. \tag{1.11}$$

Investigating Mardia's measures of multivariate skewness and kurtosis for q-dimensional subvectors of $\mathbf{X}$, $1 \le q \le p$, has been proposed and studied by Chowdhury et al. (2022) in order to better describe the non-Gaussian features in lower subdimensions of the multivariate distribution; see also Balakrishnan et al. (2007) for a vectorial notion of skewness.

For a detailed study of key multivariate distributions, including their higher-order moments and relationships, see the book by Kollo and von Rosen (2005).

1.9 Measures of Information and Geometry

1.9.1 Entropy and mutual information

Entropy and mutual information (Shannon, 1948) are two fundamental concepts from information theory that quantify uncertainty and dependence for continuous random vectors.

The (differential) *entropy* of a continuous random vector $\mathbf{X} \in \mathbb{R}^p$ with pdf $f_{\mathbf{X}}(\mathbf{x})$ is defined as

$$H_{\mathbf{X}} = -\int_{\mathbb{R}^p} f_{\mathbf{X}}(\mathbf{x}) \log\{f_{\mathbf{X}}(\mathbf{x})\} \mathrm{d}\mathbf{x},$$

provided that the integral exists. It quantifies the uncertainty or information content of a continuous random vector. Unlike discrete entropy, it can take negative values and is not invariant under continuous transformations of variables.

Let $\mathbf{X} \in \mathbb{R}^p$ and $\mathbf{Y} \in \mathbb{R}^q$ be two continuous random vectors with joint pdf $f_{\mathbf{X},\mathbf{Y}}(\mathbf{x},\mathbf{y})$ and marginal pdfs $f_{\mathbf{X}}(\mathbf{x})$ and $f_{\mathbf{Y}}(\mathbf{y})$. The *mutual information* between $\mathbf{X}$ and $\mathbf{Y}$ is defined as

$$I_{\mathbf{X},\mathbf{Y}} = \int_{\mathbb{R}^p}\int_{\mathbb{R}^q} f_{\mathbf{X},\mathbf{Y}}(\mathbf{x},\mathbf{y}) \log\left\{\frac{f_{\mathbf{X},\mathbf{Y}}(\mathbf{x},\mathbf{y})}{f_{\mathbf{X}}(\mathbf{x}) f_{\mathbf{Y}}(\mathbf{y})}\right\} \mathrm{d}\mathbf{x}\mathrm{d}\mathbf{y}.$$

Mutual information measures the amount of information shared between $\mathbf{X}$ and $\mathbf{Y}$, i.e., how much knowing one of them reduces the uncertainty about the other. It can also be expressed in terms of differential entropies:

$$I_{\mathbf{X},\mathbf{Y}} = H_{\mathbf{X}} + H_{\mathbf{Y}} - H_{\mathbf{X},\mathbf{Y}},$$

where $H_{\mathbf{X},\mathbf{Y}}$ denotes the joint (differential) entropy. Mutual information is always nonnegative, $I_{\mathbf{X},\mathbf{Y}} \geq 0$, and equals zero if and only if $\mathbf{X}$ and $\mathbf{Y}$ are independent.

1.9.2 Divergence measures of information

Information-theoretic divergence measures are mathematical tools used to quantify how different one probability distribution is from another. They originate from information theory and provide a way to measure how much information is lost or gained when one distribution is used to approximate another.

The most common measure is the *Kullback-Leibler (KL) divergence* (Kullback and Leibler, 1951), which measures the "distance" (although it is not symmetric) between two distributions F and G:

$$D_{\mathrm{KL}}(F\|G) = \int f(\mathbf{x}) \log\left\{\frac{f(\mathbf{x})}{g(\mathbf{x})}\right\} \mathrm{d}\mathbf{x}.$$

It represents the expected amount of extra information required to represent data from F if we use G instead. A symmetric version is called the *Jeffreys divergence*: $J(F,G) = D_{\mathrm{KL}}(F\|G) + D_{\mathrm{KL}}(G\|F)$.

These measures are fundamental in statistics, machine learning, and signal processing for tasks such as model comparison, goodness-of-fit testing, parameter estimation, and feature selection. They provide a principled way to assess how close or dissimilar two probabilistic models are.

1.9.3 Distance measures of geometry

The *Wasserstein distance* (Vasershtein, 1969; Kantorovich, 1960) is a measure of the distance between two probability distributions that takes into account the geometry of the underlying space. Intuitively, it represents the minimal cost of transforming one probability distribution into another, where the cost is the amount of probability mass that must be moved times the distance it is moved. Formally, for two distributions F and G on a metric space $(\mathcal{X}, d)$, the q-th order Wasserstein distance is defined as

$$W_q(F,G) = \left(\inf_{\gamma\in\Gamma(F,G)} \int_{\mathcal{X}\times\mathcal{X}} d(\mathbf{x},\mathbf{y})^q \mathrm{d}\gamma(\mathbf{x},\mathbf{y})\right)^{1/q},$$

where $\Gamma(F,G)$ is the set of all couplings (or joint distributions) with marginals F and G, d is a metric on $\mathcal{X}$, and q is a positive integer. In this formulation, $\gamma(\mathbf{x},\mathbf{y})$ describes how much probability mass is transported from $\mathbf{x}$ to $\mathbf{y}$. The Wasserstein distance is a metric, i.e., a true distance (unlike the KL and J divergences).

The Wasserstein distance is widely used in statistics, machine learning, and optimal transport theory for comparing distributions because it reflects not just differences in mass but also the spatial structure of those differences.

1.10 Exercises

Exercise 1.1 *Let $X \sim F$ be an integrable random variable. Show that*

$$\mathrm{E}(X) = -\int_{-\infty}^{0} F(x)\mathrm{d}x + \int_{0}^{+\infty} \{1 - F(x)\}\mathrm{d}x.$$

Exercise 1.2 *Consider the linear functional* $\mathbf{a}^\top\mathbf{x} = \sum_{i=1}^p a_i x_i$ *for each (fixed)* $\mathbf{a} \in \mathbb{R}^p$ *and the Euclidean norm (length)* $\|\mathbf{x}\| = (\sum_{i=1}^p x_i^2)^{1/2}$. *Show for* $r > 0$ *that*

$$\mathrm{E}\left(|\mathbf{a}^\top\mathbf{X}|^r\right) < +\infty,\ \forall \mathbf{a} \in \mathbb{R}^p \Longleftrightarrow \mathrm{E}\left(|X_i|^r\right) < +\infty,\ \forall i = 1,\ldots,p \Longleftrightarrow \mathrm{E}\left(\|\mathbf{X}\|^r\right) < +\infty.$$

(Hint: $|x_i| \le \|\mathbf{x}\| \le \sum_{i=1}^p |x_i|$*). This result gives three equivalent ways of saying, for* $r > 0$*, that the r-th moment of a random vector* $\mathbf{X}$ *is finite. Show also that if for* $r > 0$ *the r-th moment of* $\mathbf{X}$ *is finite, then the s-th moment is also finite for any* $0 < s \le r$*. Hint:* $u^s \le 1 + u^r$ *for all* $u \ge 0$ *and* $0 < s \le r$.

Exercise 1.3 *(Hölder's inequality) For any univariate random variables X and Y, show that*

$$\mathrm{E}(|XY|) \le \{\mathrm{E}(|X|^r)\}^{1/r}\{\mathrm{E}(|Y|^s)\}^{1/s},\ r > 1,\ \frac{1}{r} + \frac{1}{s} = 1.$$

Exercise 1.4 *Let X and Y be random variables with joint pdf given by* $f(x,y) = 2\phi(x)\phi(y)I_S(x,y)$*, where* $\phi(z)$ *is the* $\mathcal{N}(0,1)$ *pdf and* $S = \{(x,y)^\top \in \mathbb{R}^2 : 0 \le y \le x < +\infty; -\infty \le x \le y \le 0; 0 \le -x \le y < +\infty; 0 \le x \le -y < +\infty\}$*. Show that X, Y,* $X+Y$*, and* $X-Y$ *are normally distributed, while the joint distribution of X and Y is not bivariate normal.*

Exercise 1.5 *Show that* $\mathrm{P}\{g(\mathbf{X}) < \varepsilon^2\} \ge 1 - \mathrm{E}\{g(\mathbf{X})^q\}/\varepsilon^{2q}$ *where* $g(\mathbf{x})$ *is a nonnegative function. Deduce that* $\mathrm{P}\{(\mathbf{X}-\boldsymbol{\mu})^\top\boldsymbol{\Sigma}^{-1}(\mathbf{X}-\boldsymbol{\mu}) < \varepsilon^2\} \ge 1 - \beta_{2,p}/\varepsilon^4$ *where* $\beta_{2,p}$ *is Mardia's measure of multivariate kurtosis.*

Exercise 1.6 *Let* $\mathbf{W} = (\mathbf{X}_1,\ldots,\mathbf{X}_n)^\top$*, where* $\mathbf{X}_1,\ldots,\mathbf{X}_n$ *are iid p-dimensional random vectors with mean vector* $\boldsymbol{\mu}$ *and variance matrix* $\boldsymbol{\Sigma}$*. Consider the transformation* $\mathbf{Z} = \mathbf{AWB}$*, where* $\mathbf{A}$ *and* $\mathbf{B}$ *are constant matrices of dimensions* $m \times n$ *and* $p \times q$*, respectively. Find* $\mathrm{E}(\mathbf{Z})$*,* $\mathrm{E}\{\mathrm{vec}(\mathbf{Z}^\top)\}$ *and* $\mathrm{var}\{\mathrm{vec}(\mathbf{Z}^\top)\}$.

Exercise 1.7 *Prove (1.10) and (1.11).*

Exercise 1.8 *The k-th moment of a random vector* $\mathbf{X}$ *is sometimes defined as the* $p^k \times 1$*-expected vector given by*

$$\mathrm{E}(\mathbf{X}^{\otimes k}) = \mathrm{E}(\mathbf{X} \otimes \mathbf{X} \otimes \cdots \otimes \mathbf{X}), \quad k = 1,2,\ldots.$$

For instance, $\mathrm{E}(\mathbf{X}^{\otimes 3}) = \mathrm{E}(\mathbf{X} \otimes \mathbf{X} \otimes \mathbf{X})$*. Provided it exists, prove that* $\beta_{1,p} = \|\mathrm{E}(\mathbf{Z}^{\otimes 3})\|^2$*, where* $\mathbf{Z}$ *is the standardized version of* $\mathbf{X}$*. Furthermore, if the fourth-order moment exists, prove that* $\beta_{2,p} = \mathrm{tr}\{\mathrm{E}(\mathbf{ZZ}^\top\mathbf{ZZ}^\top)\}$.

Exercise 1.9 *Consider Examples 16 and 17 with* $p = 1$ *and obtain classical univariate results.*

Exercise 1.10 *Let* $\mathbf{X}$ *be a random vector with an asymmetric Laplace distribution whose cf is*

$$\psi(\mathbf{t}) = \left(1 - \mathrm{i}\mathbf{t}^\top \boldsymbol{\mu} + \frac{1}{2}\mathbf{t}^\top \boldsymbol{\Sigma}\mathbf{t}\right)^{-1}, \quad \mathbf{t} \in \mathbb{R}^p.$$

Show that $\mathrm{E}(\mathbf{X}) = \boldsymbol{\mu}$ *and* $\mathrm{var}(\mathbf{X}) = \boldsymbol{\Sigma} + \boldsymbol{\mu}\boldsymbol{\mu}^\top$. *A reparameterized version of the distribution of* $\mathbf{X}$ *is defined by the following cf:*

$$\psi_r(\mathbf{t}) = \left\{1 - \mathrm{i}\mathbf{t}^\top \boldsymbol{\mu} - (\mathbf{t}^\top \boldsymbol{\mu})^2 + \frac{1}{2}\mathbf{t}^\top \boldsymbol{\Sigma}\mathbf{t}\right\}^{-1}, \quad \mathbf{t} \in \mathbb{R}^p.$$

Show for this case that $\mathrm{E}(\mathbf{X}) = \boldsymbol{\mu}$, $\mathrm{var}(\mathbf{X}) = \boldsymbol{\Sigma}$, $\beta_{1,p} = \theta\{\theta^2 - 6\theta + 3(p+2)\}$, *and* $\beta_{2,p} = 2(p+\theta)(p+2) - 3\theta^3$, *where* $\theta = \boldsymbol{\mu}^\top \boldsymbol{\Sigma}^{-1}\boldsymbol{\mu}$.

Exercise 1.11 *Show that in the univariate case of* $p = 1$, *the Wasserstein distance takes the following form for* $X \sim F$ *and* $Y \sim G$:

$$W_q(F,G) = \left(\int_0^1 |F^{-1}(t) - G^{-1}(t)|^q \mathrm{d}t\right)^{1/q}.$$

Exercise 1.12 *Let* $\mathbf{X} = (X_1, \ldots, X_p)^\top$ *be a random vector with mean vector* $\boldsymbol{\mu} = (\mu_1, \ldots, \mu_p)^\top$ *and such that* $X_1, \ldots, X_p$ *are independent random variables, with common variance* σ^2 *and common third and fourth moments* $\bar{\mu}_{[k]} = \mathrm{E}\{(X_i - \mu_i)^k\}$, $k = 3,4$, $i = 1, \ldots, p$. *If* $\mathbf{A} = \mathbf{A}^\top$ *is a* $p \times p$ *fixed matrix, prove that*

$$\mathrm{var}(\mathbf{X}^\top \mathbf{A}\mathbf{X}) = (\bar{\mu}_{[4]} - 3\sigma^4)\mathbf{a}^\top \mathbf{a} + 2\sigma^4 \mathrm{tr}(\mathbf{A}^2) + 4\sigma^2 \boldsymbol{\mu}^\top \mathbf{A}^2 \boldsymbol{\mu} + 4\bar{\mu}_{[3]}\boldsymbol{\mu}^\top \mathbf{A}\mathbf{a},$$

where $\mathbf{a} = (a_{11}, \ldots, a_{pp})^\top$ *is the* $p \times 1$ *vector of diagonal elements of* $\mathbf{A}$.

Exercise 1.13 *Prove that a* $p \times p$ *matrix* $\boldsymbol{\Sigma}$ *is the variance matrix of a random vector* $\mathbf{X} \in \mathbb{R}^p$ *if and only if it is positive semi-definite.*

Exercise 1.14 *Let* $\mathbf{W} = \mathbf{V}\mathbf{V}^\top$ *be a* $p \times p$ *positive definite matrix, where* $\mathbf{V}$ *is upper-triangular with positive diagonal elements. Prove that* $J_{(\mathbf{W} \to \mathbf{V})} = 2^p \prod_{i=1}^p v_{ii}^{p+1-i}$.

Exercise 1.15 *Let* $\mathbf{Y} = \mathbf{B}\mathbf{X}\mathbf{C}$, *where* $\mathbf{B} \in \mathbb{R}^{r \times p}$, $\mathbf{X} \in \mathbb{R}^{p \times n}$ *and* $\mathbf{C} \in \mathbb{R}^{n \times s}$. *Assuming that* $\mathbf{X} = (\mathbf{X}_1, \ldots, \mathbf{X}_n)$, *where* $\mathbf{X}_1, \ldots, \mathbf{X}_n$ *are iid random vectors with mean vector* $\boldsymbol{\mu}$ *and variance matrix* $\boldsymbol{\Sigma}$, *find* $\mathrm{E}(\mathbf{Y})$ *and* $\mathrm{var}(\mathbf{Y})$.

Chapter 2

Some Notions of Multivariate Symmetry

Symmetric multivariate distributions have occupied an important place in probability modeling, multivariate statistical analysis, and robustness studies. Unlike the one-dimensional case, where the notion of symmetry is rather clear, there are a number of "symmetry" notions in $\mathbb{R}^p$, $p \geq 2$, and one has to be explicit about what this means. For example, symmetry can be defined in terms of the structural properties of the cdf, the cf, or the pdf. Symmetry may also be interpreted as an invariance of the multivariate probability distribution of a random vector with respect to a specified group of transformations. A useful introduction to these and other approaches was provided by Fang et al. (1990), where at least eight ways to construct symmetric multivariate distributions were mentioned, several of which are discussed in this chapter. A characterization in terms of probabilities of half-spaces, and also covering certain broader versions of symmetry, was provided by Beran and Millar (1997). A more recent review of the topic can be found in Serfling (2006). In this chapter, we describe some types of symmetries that may be shared by a multivariate probability distribution, starting from the simple univariate case to obtain a better understanding of this concept. In order to simplify the writing, the symmetry (around 0) of a distribution is meant with respect to the origin, unless explicitly stated otherwise. Thus, on the real line, the symmetry of the distribution of a random variable X means simply reflection with respect to the origin, i.e., $\mathrm{P}(X \leq x) = \mathrm{P}(-X \leq x)$ for all x, which in turn means that $X \overset{d}{=} -X$.

2.1 Signs, Absolute Values, and Symmetric Distributions in $\mathbb{R}$

The key to understanding the symmetric distributions on the real line lies in the simple fact that any real number x can be decomposed into its absolute value, $r = |x|$, and its sign, $u = \mathrm{sgn}(x)$, where

$$\mathrm{sgn}(x) = \begin{cases} +1, & \text{for } x > 0, \\ -1, & \text{for } x < 0. \end{cases}$$

Because $x = ru$, the sign of $x = 0$ can be arbitrarily defined as $+1$ or -1. In view of the one-to-one correspondence between x and the pair $(r, u) = (|x|, \mathrm{sgn}(x))$, any distribution of a random variable X transfers uniquely into the joint distribution of $(R, U) = (|X|, \mathrm{sgn}(X))$ and vice versa. One may then use the marginal/conditional decompositions to represent the joint distribution of these random variables, for which there are two options:

[1] *Conditioning on the sign of a random variable.* One must specify the conditional distributions of $(R \mid U = +1)$ and $(R \mid U = -1)$, together with the marginal distribution of U which is completely determined by $\mathrm{P}(U = +1) = 1 - \mathrm{P}(U = -1)$. Thus, if F denotes the cdf of $X = RU$, it will be completely defined by

$$\begin{aligned} F(x) &= \mathrm{P}(X \leq x) = \mathrm{P}(RU \leq x) \\ &= \mathrm{P}(R \leq x \mid U = +1)\mathrm{P}(U = +1) + \mathrm{P}(R \geq -x \mid U = -1)\mathrm{P}(U = -1), \ \forall x \in \mathbb{R}. \end{aligned} \quad (2.1)$$

However, two natural conditions that one could also consider here are:

(i) $(R \mid U=+1) \stackrel{d}{=} (R \mid U=-1) \Longrightarrow R \perp\!\!\!\perp U$;

(ii) $\mathrm{P}(U=+1)=\mathrm{P}(U=-1)=\frac{1}{2} \Longrightarrow U \sim \mathcal{U}(\{-1,+1\})$.

Here $\mathcal{U}(\{a,b\})$ represents the discrete uniform distribution on $\{a,b\}$.

In this scenario, we find that $R=|X| \stackrel{d}{=} (X \mid X>0)$, so that its distribution completely determines the distribution of $X=RU$. In fact, if $\tilde{F}$ is the cdf of R, then by (2.1)

$$F(x) = \frac{1}{2}\{\mathrm{P}(R \le x)+\mathrm{P}(R \ge -x)\} = \begin{cases} \frac{1}{2}\{1-\tilde{F}(-x)\}, & x<0, \\ \frac{1}{2}\tilde{F}(x), & x \ge 0. \end{cases}$$

Obviously, the distribution of X always determines the distribution of $R=|X|$. When X is symmetric and absolutely continuous, this distribution becomes the half-F cdf $\tilde{F}(r)=2F(r)$, $r>0$. The corresponding half-pdf $\tilde{f}$ is related to the pdf f of F by $\tilde{f}(r)=2f(r)$, $r>0$. However, as described above, the reciprocal fact is true only if $R \perp\!\!\!\perp U$ and $U \sim \mathcal{U}(\{-1,+1\})$. This just means that the distribution of $X=RU$ is symmetric. In fact, since $U \sim \mathcal{U}(\{-1,+1\})$ is symmetric, then

$$\begin{aligned} U \stackrel{d}{=} -U, \quad U \perp\!\!\!\perp R \quad &\Longrightarrow \quad (R,U) \stackrel{d}{=} (R,-U) \\ &\Longrightarrow \quad X=RU \stackrel{d}{=} -RU=-X. \end{aligned}$$

[2] *Conditioning on the absolute value of a random variable.* Another way to characterize the distribution of X is by specifying the conditional distribution of its sign, $U=\operatorname{sgn}(X)$, on its absolute value, $R=|X|$, which is equivalent to specifying the conditional distribution of X given $|X|=r$, and then specifying a marginal distribution of R on the positive side of the real line. Since the random variable U is binary, the conditional distribution of $(U|R=r)$ is determined by

$$p(r)=\mathrm{P}(U=+1 \mid R=r)=\mathrm{P}(X>0 \mid |X|=r), \quad \forall r>0.$$

These probabilities may be specified arbitrarily considering that the function $p(\cdot)$ is such that $0 \le p(r) \le 1$ for all $r>0$. For a symmetric random variable $X \stackrel{d}{=} -X$, however, all the conditional distributions of X given $|X|=r$ are uniform on $\{x \in \mathbb{R} : |x|=r\}=\{-r,+r\}$, that is, $p(r)=1/2$, for all $r>0$, which implies that $U \perp\!\!\!\perp R$ and $U \sim \mathcal{U}(\{-1,+1\})$.

In other words, if we have two independent random variables U and R, where $U \sim \mathcal{U}(\{-1,+1\})$ and R is nonnegative, then we can always obtain a random variable X with the same distribution as RU, so $|X| \stackrel{d}{=} R$. Also, since its sign is assigned as $+1$ or -1 with the same probability and independently of the value of its modulus, such a random variable will be, of course, symmetrically distributed around the origin. That is,

$$X \stackrel{d}{=} -X \Longleftrightarrow X \stackrel{d}{=} RU, \quad \text{where} \quad R \stackrel{d}{=} |X| \perp\!\!\!\perp U \quad \text{and} \quad U \sim \mathcal{U}(\{-1,+1\}).$$

If, in addition, $\mathrm{P}(X=0)=0$, which holds automatically when the distribution of X is absolutely continuous, we may specifically let $U \stackrel{d}{=} X/|X|$ be the sign of X and show that if $X \stackrel{d}{=} -X$ then $\mathrm{P}(U=\pm 1)=1/2$. With this additional condition, we have

$$X \stackrel{d}{=} -X \quad \text{and} \quad \mathrm{P}(X=0)=0 \Longleftrightarrow X \stackrel{d}{=} RU, \quad \text{where} \quad R \stackrel{d}{=} |X| \perp\!\!\!\perp U \stackrel{d}{=} \frac{X}{|X|} \sim \mathcal{U}(\{-1,+1\}).$$

When a random variable X has a pdf f, then

$$X \stackrel{d}{=} -X \Longleftrightarrow f(x)=f(-x), \quad \forall x \in \mathbb{R}.$$

As an immediate consequence of the symmetry property, we then have for any integrable odd function g that

$$X \stackrel{d}{=} -X \Longrightarrow \mathrm{E}\{g(X)\}=0.$$

For instance, if $X \stackrel{d}{=} -X$ and $\mathrm{E}(|X|)<+\infty$, then $\mathrm{E}(X)=0$.

More generally, the symmetry of the distribution of a random variable X can be defined with respect to a constant $\xi \in \mathbb{R}$ simply by defining the centered random variable $X-\xi$ as symmetric with respect to the origin, that is, $X-\xi \stackrel{d}{=} -(X-\xi)$. If X is an integrable random variable, then we can define $\xi = \mathrm{E}(X)$. Considering also that the symmetry of a distribution is not affected by rescaling, we obtain $\omega^{-1}(X-\xi) \stackrel{d}{=} -\omega^{-1}(X-\xi)$, for all $\omega > 0$. If $\mathrm{E}(X^2) < +\infty$, then one can choose $\omega^2 = \mathrm{var}(X)$. That is, the random variable $X_0 = \omega^{-1}(X-\xi)$ can be interpreted as a standardized symmetric version of X. Hence, since $X = \xi + \omega X_0$, one can always construct a symmetric location-scale distribution for X considering a normalized and symmetric (at zero) distribution for X_0. In particular, if X_0 has a normalized pdf f_0, which is such that $f_0(-x) = f_0(x)$ for all x, then X will have a *symmetric location-scale pdf* given by

$$f(x) = \frac{1}{\omega} f_0\left(\frac{x-\xi}{\omega}\right).$$

Example 18 *Three well-known examples of univariate symmetric distributions are:*
(a) Continuous uniform: $X \stackrel{d}{=} -X \sim \mathcal{U}([-1,1]) \Longleftrightarrow |X| \sim \mathcal{U}([0,1])$.
(b) Double-exponential: $X \stackrel{d}{=} -X \sim f(x) = \frac{1}{2}e^{-|x|} \Longleftrightarrow |X| \sim \mathcal{E}xp(1)$.
(c) Standard normal: $X \sim \mathcal{N}(0,1) \Longleftrightarrow |X| \sim \sqrt{\chi_1^2} \equiv \mathcal{HN}(0,1)$ *(half-normal distribution).*

2.2 Symmetry as Invariance

A distributional property of a random vector associated with symmetry is that of invariance under a group of transformations. For instance, invariance under the simplest group of multiplicative transformations $\{-1,+1\}$ leads to symmetric distributions on the real line by defining the distribution of a random variable X to be symmetric about the origin if $X \stackrel{d}{=} -X$. It is natural to extend this notion of univariate symmetry to the multivariate case by defining the multivariate distribution of a random vector $\mathbf{X} \in \mathbb{R}^p$ to be symmetric about the origin if $\mathbf{X} \stackrel{d}{=} -\mathbf{X}$. However, as will be established in what follows, this idea of central symmetry, as well as other multidimensional forms of symmetries commonly considered in the literature, are induced by imposing invariance of the multivariate distribution of $\mathbf{X}$ under some specified groups of transformations acting on the sample space of $\mathbf{X}$. We adopt this approach to define multivariate symmetric distributions.

Let $\mathcal{G}$ be a set of transformations acting on a sample space $\mathcal{X}$ of random vectors into itself. The action of a transformation (operation) $\mathbf{G} \in \mathcal{G}$ on a sample $\mathbf{x} \in \mathcal{X}$ is $\mathbf{x} \longmapsto \mathbf{G}(\mathbf{x})$. A set of transformations $\mathcal{G}$ is called a *group* if the following two conditions are satisfied:

[1] $\mathcal{G}$ is closed under composition: If $\mathbf{G}_1, \mathbf{G}_2 \in \mathcal{G}$, then $\mathbf{G}_1\mathbf{G}_2 \in \mathcal{G}$, where $\mathbf{G}_1\mathbf{G}_2$ is defined by $\mathbf{G}_1\mathbf{G}_2(\mathbf{x}) = \mathbf{G}_1(\mathbf{G}_2(\mathbf{x}))$;

[2] $\mathcal{G}$ is closed under inversion: If $\mathbf{G} \in \mathcal{G}$, then there exists a transformation $\mathbf{G}^{-1} \in \mathcal{G}$ such that $\mathbf{G}\mathbf{G}^{-1} = \mathbf{G}^{-1}\mathbf{G} = \mathbf{I}$, where $\mathbf{I}$ is the identity transformation in $\mathcal{G}$.

Let now $\mathbf{G}(\mathbf{X})$ be the action of a transformation $\mathbf{G}$ from the group $\mathcal{G}$ on a random vector $\mathbf{X} \in \mathbb{R}^p$. We say that the multivariate distribution of $\mathbf{X}$ is *symmetric relative to $\mathcal{G}$* or simply that it is *$\mathcal{G}$-symmetric* if it is not affected by any transformation $\mathbf{G}$ from the group $\mathcal{G}$, which in turn means that the multivariate distribution of $\mathbf{X}$ is *$\mathcal{G}$-invariant*, i.e.:

$$\mathbf{X} \stackrel{d}{=} \mathbf{G}(\mathbf{X}), \quad \forall \mathbf{G} \in \mathcal{G}. \tag{2.2}$$

Evidently, this means that any function g on $\mathbb{R}^p$ characterizing a $\mathcal{G}$-invariant multivariate distribution is also invariant under $\mathcal{G}$, i.e., $g(\mathbf{x}) = g(\mathbf{G}(\mathbf{x}))$ for each $\mathbf{x} \in \mathbb{R}^p$ and each $\mathbf{G} \in \mathcal{G}$. In particular, if the random vector $\mathbf{X}$ has a pdf f, then its multivariate distribution is $\mathcal{G}$-symmetric iff

$$f(\mathbf{x}) = f(\mathbf{G}(\mathbf{x})), \quad \forall \mathbf{x} \in \mathbb{R}^p \text{ and } \mathbf{G} \in \mathcal{G}.$$

As in the univariate case, all the symmetry notions for the multivariate distribution of a random vector $\mathbf{X} \in \mathbb{R}^p$ considered in this chapter are defined with respect to $\mathbf{0}_p \in \mathbb{R}^p$. For $\boldsymbol{\xi} \in \mathbb{R}^p$, we say that the multivariate distribution of $\mathbf{X}$ is symmetric (in any of the senses described here) with respect to $\boldsymbol{\xi}$ if $\mathbf{X} - \boldsymbol{\xi}$ is symmetric (in the same sense) with respect to $\mathbf{0}_p$.

2.2.1 *Symmetry as orthogonal invariance*

A group of transformations from $\mathbb{R}^p$ into itself, commonly encountered in many situations of practical interest, is the *orthogonal group*. This group is identical to the set of all square orthogonal (also sometimes called orthonormal) matrices of dimension $p \times p$, which is denoted by

$$\mathcal{O}_p = \{\boldsymbol{\Gamma} \in \mathbb{R}^{p \times p} : \boldsymbol{\Gamma}^\top \boldsymbol{\Gamma} = \mathbf{I}_p\}.$$

Any orthogonal transformation $\boldsymbol{\Gamma}$ from the group $\mathcal{O}_p$ is hence a nonsingular square matrix whose inverse is equal to its transpose. In fact, an orthogonal square matrix $\boldsymbol{\Gamma}$ of dimension $p \times p$ is simply one that preserves the length of any vector $\mathbf{x} \in \mathbb{R}^p$:

$$\|\boldsymbol{\Gamma}\mathbf{x}\| = \|\mathbf{x}\|, \quad \forall \mathbf{x} \in \mathbb{R}^p.$$

This is equivalent to the fact that $\boldsymbol{\Gamma}$ preserves the inner product between two vectors $\mathbf{x}, \mathbf{y} \in \mathbb{R}^p$:

$$\begin{aligned} \|\boldsymbol{\Gamma}(\mathbf{x}+\mathbf{y})\|^2 = \|\mathbf{x}+\mathbf{y}\|^2, \forall \mathbf{x}, \mathbf{y} \in \mathbb{R}^p \quad &\Longleftrightarrow \quad (\boldsymbol{\Gamma}\mathbf{x})^\top (\boldsymbol{\Gamma}\mathbf{y}) = \mathbf{x}^\top \boldsymbol{\Gamma}^\top \boldsymbol{\Gamma}\mathbf{y} = \mathbf{x}^\top \mathbf{y}, \forall \mathbf{x}, \mathbf{y} \in \mathbb{R}^p, \\ &\Longleftrightarrow \quad \boldsymbol{\Gamma}^\top \boldsymbol{\Gamma} = \mathbf{I}_p = \boldsymbol{\Gamma}\boldsymbol{\Gamma}^\top. \end{aligned}$$

This last expression means, of course, that the columns (or rows) of the matrix $\boldsymbol{\Gamma}$ form an orthonormal basis of $\mathbb{R}^p$.

For $\mathcal{G} = \mathcal{O}_p$, the action of $\boldsymbol{\Gamma} \in \mathcal{O}_p$ on $\mathbf{x} \in \mathbb{R}^p$ is $\mathbf{G}(\mathbf{x}) = \boldsymbol{\Gamma}\mathbf{x}$, and then the group operation is the matrix multiplication. The subgroup $\mathcal{SO}_p = \{\boldsymbol{\Gamma} \in \mathcal{O}_p : |\boldsymbol{\Gamma}| = 1\}$ is known as the *special orthogonal group* or *rotational group* because it contains all rotations.

For a specified subgroup $\mathcal{G}_0$ of $\mathcal{O}_p$, one may obtain a different notion of symmetry in $\mathbb{R}^p$, by defining a multivariate distribution of a random vector $\mathbf{X} \in \mathbb{R}^p$ to be *$\mathcal{G}_0$-symmetric* (with respect to the origin) if equation (2.2) holds for all $\mathbf{G} \in \mathcal{G}_0$. In the following subsections, we describe well-known examples of symmetries indexed by different subgroups of the orthogonal group.

2.2.2 *Spherical symmetry*

A multivariate distribution of a random vector $\mathbf{X} \in \mathbb{R}^p$ is *spherically symmetric (SS)* if it remains invariant under any orthogonal transformation of the random vector $\mathbf{X}$:

$$\mathbf{X} \stackrel{d}{=} \boldsymbol{\Gamma}\mathbf{X}, \quad \forall \boldsymbol{\Gamma} \in \mathcal{O}_p.$$

This fact means that $\mathbf{X}$ (or its multivariate distribution) is *orthogonally invariant*, which in turn essentially means that rotations of $\mathbf{X}$ about the origin do not alter its multivariate distribution. Hence, spherical symmetry corresponds to $\mathcal{O}_p$-symmetry or, equivalently, to $\mathcal{SO}_p$-symmetry, since:

$$\mathbf{X} \stackrel{d}{=} \boldsymbol{\Gamma}\mathbf{X}, \quad \forall \boldsymbol{\Gamma} \in \mathcal{O}_p \Longleftrightarrow \mathbf{X} \stackrel{d}{=} \boldsymbol{\Gamma}\mathbf{X}, \quad \forall \boldsymbol{\Gamma} \in \mathcal{SO}_p.$$

Indeed, we should first note that any orthogonally invariant $\mathbf{X}$ is completely determined by its first coordinate (see Theorem 16 in Chapter 4 for details):

$$\mathbf{X} \stackrel{d}{=} \mathbf{\Gamma X}, \quad \forall \mathbf{\Gamma} \in \mathcal{O}_p \Longleftrightarrow \mathbf{s}^\top \mathbf{X} \stackrel{d}{=} X_1, \, \forall \mathbf{s} \in \mathbb{S}^{(p-1)}.$$

Then, for any $\mathbf{s} \in \mathbb{S}^{(p-1)}$ (the unit p-dimensional sphere), one can always find $\mathbf{\Gamma} \in \mathcal{SO}_p$ whose first row is $\mathbf{s}^\top$. Since $\mathbf{X} \stackrel{d}{=} \mathbf{\Gamma X}$, projecting on the first coordinate yields $X_1 \stackrel{d}{=} \mathbf{s}^\top \mathbf{X}$. Consequently, the terms *orthogonal invariance* and *rotational invariance* can be used interchangeably. Multivariate distributions satisfying this property are frequently called *multivariate spherical distributions* (see Chapter 4).

Equivalently, if ψ is the cf of $\mathbf{X}$, then

$$\begin{aligned} \mathbf{X} \stackrel{d}{=} \mathbf{\Gamma X}, \quad \forall \mathbf{\Gamma} \in \mathcal{O}_p \quad &\Longleftrightarrow \quad \psi(\mathbf{t}) = \psi(\mathbf{\Gamma t}), \quad \forall \mathbf{t} \in \mathbb{R}^p, \forall \mathbf{\Gamma} \in \mathcal{O}_p, \\ &\Longleftrightarrow \quad \psi(\mathbf{t}) = \varphi(\|\mathbf{t}\|^2), \quad \forall \mathbf{t} \in \mathbb{R}^p, \end{aligned}$$

for some scalar-valued function φ. If $\mathbf{X}$ has a pdf f, then this becomes

$$\begin{aligned} \mathbf{X} \stackrel{d}{=} \mathbf{\Gamma X}, \quad \forall \mathbf{\Gamma} \in \mathcal{O}_p \quad &\Longleftrightarrow \quad f(\mathbf{x}) = f(\mathbf{\Gamma x}), \quad \forall \mathbf{x} \in \mathbb{R}^p, \forall \mathbf{\Gamma} \in \mathcal{O}_p, \\ &\Longleftrightarrow \quad f(\mathbf{x}) = h(\|\mathbf{x}\|^2), \quad \forall \mathbf{x} \in \mathbb{R}^p, \end{aligned}$$

for some nonnegative scalar-valued function h. This fact means that the pdf of $\mathbf{X}$ is constant on every $(p-1)$-dimensional sphere of radius $r \geq 0$, given by

$$\mathbb{S}_r^{(p-1)} = \{\mathbf{x} \in \mathbb{R}^p : \|\mathbf{x}\| = r\},$$

hence, $\mathbf{X}$ is said to be spherically distributed.

It may be mentioned that multivariate spherical normal distributions (Chapter 3), multivariate spherical t distributions (Chapter 4), scale mixtures of multivariate spherical normal distributions (Chapter 4), and multivariate uniform distributions over the surface of a unit hyper-sphere are examples of spherical distributions. The multivariate uniform distributions on or inside unit hyper-spheres (see Example 19 below) are very important in the statistical analysis of directional data. Some real-world examples were considered in Mardia (1972).

Example 19 *An obviously rotationally invariant multivariate distribution is a spherical uniform. In fact, let $\mathbf{X} \sim \mathcal{U}(\mathbb{B}^{(p)})$ be the multivariate uniform distribution in the unit ball $\mathbb{B}^{(p)} = \{\mathbf{x} \in \mathbb{R}^p : \|\mathbf{x}\| < 1\}$ of $\mathbb{R}^p$. Since the random vector $\mathbf{X}$ has a pdf $f(\mathbf{x}) = cI_{\{\|\mathbf{x}\|<1\}}(\mathbf{x})$, where $c^{-1} = \mathrm{vol}(\mathbb{B}^{(p)}) = \frac{2\pi^{p/2}}{p\Gamma(p/2)}$, it is clear that $\mathbf{X}$ is rotationally invariant. Now, by "projecting" this multivariate uniform distribution onto its $(p-1)$-dimensional boundary, we have:*

$$\mathbf{U}^{(p)} \sim \mathcal{U}(\mathbb{S}_1^{(p-1)}) \Longleftrightarrow \mathbf{U}^{(p)} \stackrel{d}{=} \frac{\mathbf{X}}{\|\mathbf{X}\|}, \text{ where } \mathbf{X} \sim \mathcal{U}(\mathbb{B}^{(p)}).$$

It is clear that $\mathbf{U}^{(p)}$ is rotationally invariant, inheriting this property directly from $\mathbf{X}$. A rather surprising fact is that the multivariate uniform distribution of $\mathbf{U}^{(p)}$ is the only rotationally invariant distribution on the unit sphere $\mathbb{S}^{(p-1)}$.

Since the uniform random vector $\mathbf{U}^{(p)}$ is the only rotationally invariant distribution on the unit sphere $\mathbb{S}^{(p-1)} = \mathbb{S}_1^{(p-1)}$, another equivalent and well-known characterization of a multivariate spherical distribution (see Chapter 4) is:

$$\mathbf{X} \stackrel{d}{=} \mathbf{\Gamma X}, \, \forall \mathbf{\Gamma} \in \mathcal{O}_p \Longleftrightarrow \mathbf{X} \stackrel{d}{=} R\mathbf{U}^{(p)}, \text{ where } R \stackrel{d}{=} \|\mathbf{X}\| \perp\!\!\!\perp \mathbf{U}^{(p)} \text{ and } \mathbf{U}^{(p)} \sim \mathcal{U}(\mathbb{S}^{(p-1)}).$$

This fact means that a multivariate spherical distribution is completely determined by the distribution of the nonnegative *radial or generating random variable R*. In turn, this means that the multivariate conditional distribution of $\mathbf{X}$ given $R = r$ is uniform on $\mathbb{S}_r^{(p-1)}$, and therefore for any measurable subset A of $\mathbb{R}^p$, we have

$$\mathrm{P}(\mathbf{X} \in A) = \int_0^{+\infty} \mathrm{P}\{\mathbf{U}^{(p)} \in r^{-1}A \cap \mathbb{S}_r^{(p-1)}\}\mathrm{d}\tilde{F}(r),$$

where $\tilde{F}$ is the cdf of R.

As mentioned above, spherical symmetry can be easily described by the form of the characteristic function. More general notions of symmetry can be defined similarly: A random vector $\mathbf{X}$ is *α-symmetric* about the origin, $\alpha > 0$, if its characteristic function is of the form $\psi(|t_1|^\alpha, \ldots, |t_p|^\alpha)$, $\mathbf{t} = (t_1, \ldots, t_p)^\top \in \mathbb{R}^p$. This provides a natural way to extend the symmetry of univariate stable laws to the multivariate case. The 2-symmetric distributions are the spherically symmetric ones. For any $\alpha > 0$, this reduces in one dimension to the usual symmetry; see Cambanis et al. (1983) and Fang et al. (1990) for useful developments.

2.2.3 *Sign-invariant symmetry*

A multivariate distribution of a random vector $\mathbf{X} \in \mathbb{R}^p$ is *sign-invariant symmetric (SIS)* if

$$\mathbf{X} = (X_1, \ldots, X_p)^\top \stackrel{d}{=} (\pm X_1, \ldots, \pm X_p)^\top, \quad \text{for all } 2^p \text{ possible choices of } +1 \text{ and } -1. \tag{2.3}$$

Sign-invariant (or sign-permutation) symmetry corresponds to $\mathcal{G}_0$-symmetry, where $\mathcal{G}_0 = \mathcal{D}_p = \{\mathbf{D} \in \mathcal{O}_p : \mathbf{D} \text{ is diagonal and } \mathbf{D}^2 = \mathbf{I}_p\}$, i.e., the subgroup of the 2^p orthogonal square matrices of the form $\mathbf{D} = \mathrm{diag}(\pm 1, \ldots, \pm 1)$. In this case, the transformation $\mathbf{G} \in \mathcal{D}_p$ acts on the entries of each vector by leaving them unchanged or reversing their signs, i.e., $\mathbf{G}(\mathbf{x}) = \mathbf{D}\mathbf{x}$, $\mathbf{D} \in \mathcal{D}_p$. Hence, we can rewrite (2.3) by defining the random vector $\mathbf{X}$ (or its multivariate distribution) to be sign-invariant symmetric if

$$\mathbf{X} \stackrel{d}{=} \mathbf{D}\mathbf{X}, \quad \forall \mathbf{D} \in \mathcal{D}_p.$$

Recall that any vector $\mathbf{x} = (x_1, \ldots, x_p)^\top$ has a one-to-one representation in terms of its modulus, $|\mathbf{x}| = (|x_1|, \ldots, |x_p|)^\top = (r_1, \ldots, r_p)^\top = \mathbf{r}$, and its sign, $\mathrm{sgn}(\mathbf{x}) = (\mathrm{sgn}(x_1), \ldots, \mathrm{sgn}(x_p))^\top = (u_1, \ldots, u_p)^\top = \mathbf{u}$ as $\mathbf{x} = \mathbf{u} \circ \mathbf{r} = \mathbf{D}_u \mathbf{r}$, where $\mathbf{D}_u = \mathrm{diag}(u_1, \ldots, u_p)$, i.e., as the point-wise multiplication (or Hadamard product; see Exercise A.2) of $\mathbf{u}$ and $\mathbf{r}$. This fact yields the following characterization of a sign-invariant symmetric random vector $\mathbf{X} \in \mathbb{R}^p$, extending the radial-uniform representation presented in Section 2.1 for a univariate random variable:

$$\mathbf{X} \stackrel{d}{=} \mathbf{D}\mathbf{X}, \quad \forall \mathbf{D} \in \mathcal{D}_p \Longleftrightarrow \mathbf{X} \stackrel{d}{=} \mathbf{U} \circ \mathbf{R} = \mathbf{D}_U \mathbf{R}, \text{ with } \mathbf{R} \stackrel{d}{=} |\mathbf{X}| \perp\!\!\!\perp \mathbf{U} \text{ and } \mathbf{U} \sim \mathcal{U}_p(\{-1, +1\}^p),$$

where $\mathbf{D}_U = \mathrm{diag}(U_1, \ldots, U_p)$ and $\mathbf{U} = (U_1, \ldots, U_p)^\top \sim \mathcal{U}_p(\{-1, +1\}^p)$ denotes the discrete uniform random vector on $\{-1, +1\}^p$, which is such that $\mathrm{P}(\mathbf{U} = \mathbf{u}) = 2^{-p}$ for all $\mathbf{u} \in \{-1, +1\}^p$. In turn, this means that $U_1, \ldots, U_p$ are iid discrete uniform random variables on $\{-1, +1\}$. Consequently, if $\mathbf{X}$ has a pdf f, then

$$\begin{aligned} \mathbf{X} \stackrel{d}{=} \mathbf{D}\mathbf{X}, \quad \forall \mathbf{D} \in \mathcal{D}_p \quad &\Longleftrightarrow \quad f(\mathbf{x}) = f(\mathbf{D}\mathbf{x}), \quad \forall \mathbf{x} \in \mathbb{R}^p, \forall \mathbf{D} \in \mathcal{D}_p, \\ &\Longleftrightarrow \quad f(\mathbf{x}) = h(|\mathbf{x}|), \quad \forall \mathbf{x} \in \mathbb{R}^p, \end{aligned}$$

for some nonnegative scalar-valued function h.

A fundamental property is that sign-invariant random variables are conditionally independent given their absolute values. A multivariate sign-invariant symmetric distribution is also closed by marginalization, conditioning, and disjoint linear combinations. The joint distribution of any set of p symmetrically distributed (at zero) random variables is a multivariate sign-invariant symmetric distribution. This includes the multivariate spherical distributions. However, there are multivariate SIS distributions that are not spherical, as the following example shows.

Example 20 *As a simple example, consider the multivariate uniform distribution on the hypercube* $[-1,1]^p$*, with pdf of the form* $f(\mathbf{x}) = 2^{-p} I_{[-1,1]^p}(\mathbf{x}) = 2^{-p} I_{[0,1]^p}(|\mathbf{x}|)$*. This multivariate distribution is SIS but not spherical.*

Another important property that is satisfied by every partitioned sign-invariant symmetric random vector as $\mathbf{X} = (\mathbf{X}_0^\top, \mathbf{X}_1^\top)^\top$ is that

$$\mathbf{X}_1 \perp\!\!\!\perp \operatorname{sgn}(\mathbf{X}_0) \mid |\mathbf{X}_0|. \tag{2.4}$$

In turn, for any random vector satisfying (2.4), we find that the following two equivalent properties related to the selection of random vectors (see Chapter 9) hold:

$$\left(\begin{array}{c} \mathbf{X}_0 \\ \mathbf{X}_1 \end{array} \right) \Bigg| \mathbf{X}_0 > \mathbf{0}_{p_0} \stackrel{d}{=} \left(\begin{array}{c} |\mathbf{X}_0| \\ \mathbf{X}_1 \end{array} \right), \tag{2.5}$$

$$g(\mathbf{X}_0, \mathbf{X}_1) \mid \mathbf{X}_0 > \mathbf{0}_{p_0} \stackrel{d}{=} g(|\mathbf{X}_0|, \mathbf{X}_1), \text{ for any measurable function } g, \tag{2.6}$$

where the condition $\mathbf{X}_0 > \mathbf{0}_{p_0}$ is understood to be componentwise and p_0 is the dimension of $\mathbf{X}_0$. The properties (2.5) and (2.6) are very relevant to the construction of a multivariate skewed distribution starting from a multivariate sign-invariant symmetric distribution.

The sign-invariant symmetry concept based on $\mathcal{D}_p$ leads to the *$\mathcal{C}$-class* of multivariate distributions studied in Arellano-Valle et al. (2002b) and renamed as *$\mathcal{SI}$-class* (from sign-invariant) in Arellano-Valle and del Pino (2004); see also the early paper by Berman (1965).

A rich and tractable family of multivariate SIS distributions is the so-called *L_q-spherical* (or L_q-norm spherical or L_q-symmetric) distributions; see Chapter 4 for spherical distributions obtained with $q = 2$ and Section 4.5 for general q. These distributions are extensions from L_2- to L_q-spherically symmetric distributions based on replacing the L_2-norm by the L_q-norm $\|\mathbf{x}\|_q = (\sum_{i=1}^p |x_i|^q)^{1/q}$, $\mathbf{x} \in \mathbb{R}^p$, with $q > 0$ in the definition of their pdf; that is, the pdf of L_q-spherically symmetric distributions can always be written in the form $f(\mathbf{x}) = h(\|\mathbf{x}\|_q)$ for some nonnegative scalar-valued function h (Osiewalski and Steel, 1993). As was shown later by Gupta and Song (1997), they also correspond to a stochastic representation in terms of a radial variable and a q-generalized uniform vector. This is similar to the spherical case of $q = 2$, but with the radial variable corresponding to the L_q-norm of the underlying random vector; see also Arellano-Valle and Richter (2012). Here, we adopt the naming convention of Gupta and Song (1997) and call $\|\mathbf{x}\|_q$ an L_q-norm, although the triangle inequality only holds for $q \geq 1$. The L_q-spherically symmetric distributions with $q \neq 2$ are no longer invariant with respect to rotations (transformations from $\mathcal{O}_p$). Instead, they are only invariant under permutations of the coordinate axes. Further extensions and other interesting concepts related to this family of distributions can be found in Fernández et al. (1995) and Sinz and Bethge (2010); see also Section 4.5. Arellano-Valle and Richter (2012) considered skewed multivariate distributions based on the L_q-norm.

2.2.4 *Central symmetry*

A multivariate distribution of a random vector $\mathbf{X} \in \mathbb{R}^p$ is *centrally symmetric (CS)* if

$$\mathbf{X} \stackrel{d}{=} -\mathbf{X}. \tag{2.7}$$

If the pdf, f, of $\mathbf{X}$ exists, then

$$\mathbf{X} \stackrel{d}{=} -\mathbf{X} \Longleftrightarrow f(\mathbf{x}) = f(-\mathbf{x}), \quad \forall \mathbf{x} \in \mathbb{R}^p.$$

Clearly, definition (2.7) represents the most direct multivariate extension of the univariate symmetry. This is equivalent to all of the projections onto the lines through the origin having symmetric univariate distributions.

Central symmetry corresponds to $\mathcal{G}_0$-symmetry, where the action of $\mathbf{G} \in \mathcal{G}_0$ consists of the identity $\mathbf{G}(\mathbf{x}) = \mathbf{x}$ or the reflection about the origin, i.e., $\mathbf{G}(\mathbf{x}) = -\mathbf{x}$, so that $\mathcal{G}_0$ is the subgroup of $\mathcal{O}_p$ consisting of just the identity transformation and its negative, i.e., $\mathcal{G}_0 = \{\mathbf{I}_p, -\mathbf{I}_p\}$. In terms of probabilities of halfspaces (Beran and Millar, 1997), it is characterized by

$$\mathbf{X} \stackrel{d}{=} -\mathbf{X} \Longleftrightarrow \mathrm{P}(\mathbf{s}^\top \mathbf{X} \le x) = \mathrm{P}(\mathbf{s}^\top \mathbf{X} \ge -x), \quad \forall x \in \mathbb{R}, \text{ and } \mathbf{s} \in \mathbb{S}^{(p-1)},$$

which is equivalent to

$$\mathbf{X} \stackrel{d}{=} -\mathbf{X} \Longleftrightarrow \mathbf{s}^\top \mathbf{X} \stackrel{d}{=} -\mathbf{s}^\top \mathbf{X}, \quad \forall \mathbf{s} \in \mathbb{S}^{(p-1)}.$$

Central symmetry is perhaps the crudest useful notion of symmetry in $\mathbb{R}^p$; all other notions of symmetry presented reduce to this case on $\mathbb{R}$. Alternative names are polar, radial, reflective, diagonal, antipodal, or simple symmetry. However, a weaker notion of symmetry for the multivariate distribution of a random vector $\mathbf{X} \in \mathbb{R}^p$ is provided by the *angular or directional symmetry (AS or DS)*, namely

$$\frac{\mathbf{X}}{\|\mathbf{X}\|} \stackrel{d}{=} -\frac{\mathbf{X}}{\|\mathbf{X}\|}, \tag{2.8}$$

which is equivalent to central symmetry for the distribution of $\mathbf{X}/\|\mathbf{X}\|$. Here, we adopt the convention that $\mathbf{x}/\|\mathbf{x}\| = \mathbf{0}_p$ if $\mathbf{x} = \mathbf{0}_p$. It is immediate that (2.7) implies (2.8). Counter-examples to the converse can be found.

Zuo and Serfling (2000) discussed several other concepts related to central and angular symmetry, such as the notion of multidimensional location measure of median type. They generalized these notions of multidimensional symmetry by means of the so-called halfspace symmetry, and argued that it provides: a) a natural basis for requiring a center locating property in any reasonable notion of a multivariate location measure of median type, and b) an important widening of the class of actual distributions, which might be usefully approximated by symmetric distributions.

Halfspace and halfspace symmetry: For any unit length vector $\mathbf{s} \in \mathbb{R}^p$ and $t \in \mathbb{R}$, the set of points $H_{\mathbf{s},t} = \{\mathbf{x} \in \mathbb{R}^p : \mathbf{s}^\top \mathbf{x} \le t\}$ defines a *closed halfspace* in $\mathbb{R}^p$. Its interior is an *open halfspace*, and the boundary $\{\mathbf{x} \in \mathbb{R}^p : \mathbf{s}^\top \mathbf{x} = t\}$ defines a *hyperplane*. A random vector $\mathbf{X} \in \mathbb{R}^p$ has a *halfspace symmetric (HS)* distribution about $\boldsymbol{\xi} \in \mathbb{R}^p$ if $\mathrm{P}(\mathbf{X} \in H) \ge 1/2$ for each closed halfspace H with $\boldsymbol{\xi}$ in the boundary, or simply if $\mathrm{P}(\mathbf{X} \in H) \ge 1/2$ for any halfspace H containing $\boldsymbol{\xi}$. As shown in Zuo and Serfling (2000), *CS* $\Rightarrow$ *AS* $\Rightarrow$ *HS*. Counter-examples of the converses are easily found; see Exercise 2.5.

The following examples show that a distribution can have central symmetry, without being sign-invariant symmetric. Figure 2.1 depicts the geometry of the density contours for various notions of symmetry of bivariate distributions ($p = 2$).

Example 21 *Let $f_1(x)$ and $f_2(x)$ be symmetric pdfs with respect to the origin. Then, the pdf given by*

$$f(x_1,x_2) = f_1(x_1)f_2(x_2)[1+\theta\{2F_1(x_1)-1\}\{2F_2(x_2)-1\}], \quad -1 \le \theta \le 1, \tag{2.9}$$

is centrally symmetric, hence also AS and HS, but not sign-invariant symmetric and not spherically symmetric when $\theta \ne 0$. When $\theta = 0$, this pdf is centrally symmetric and sign-invariant symmetric, but not spherically symmetric unless $f_1(x) = f_2(x) = \phi(x)$, the standard univariate normal pdf; see Figure 2.1.

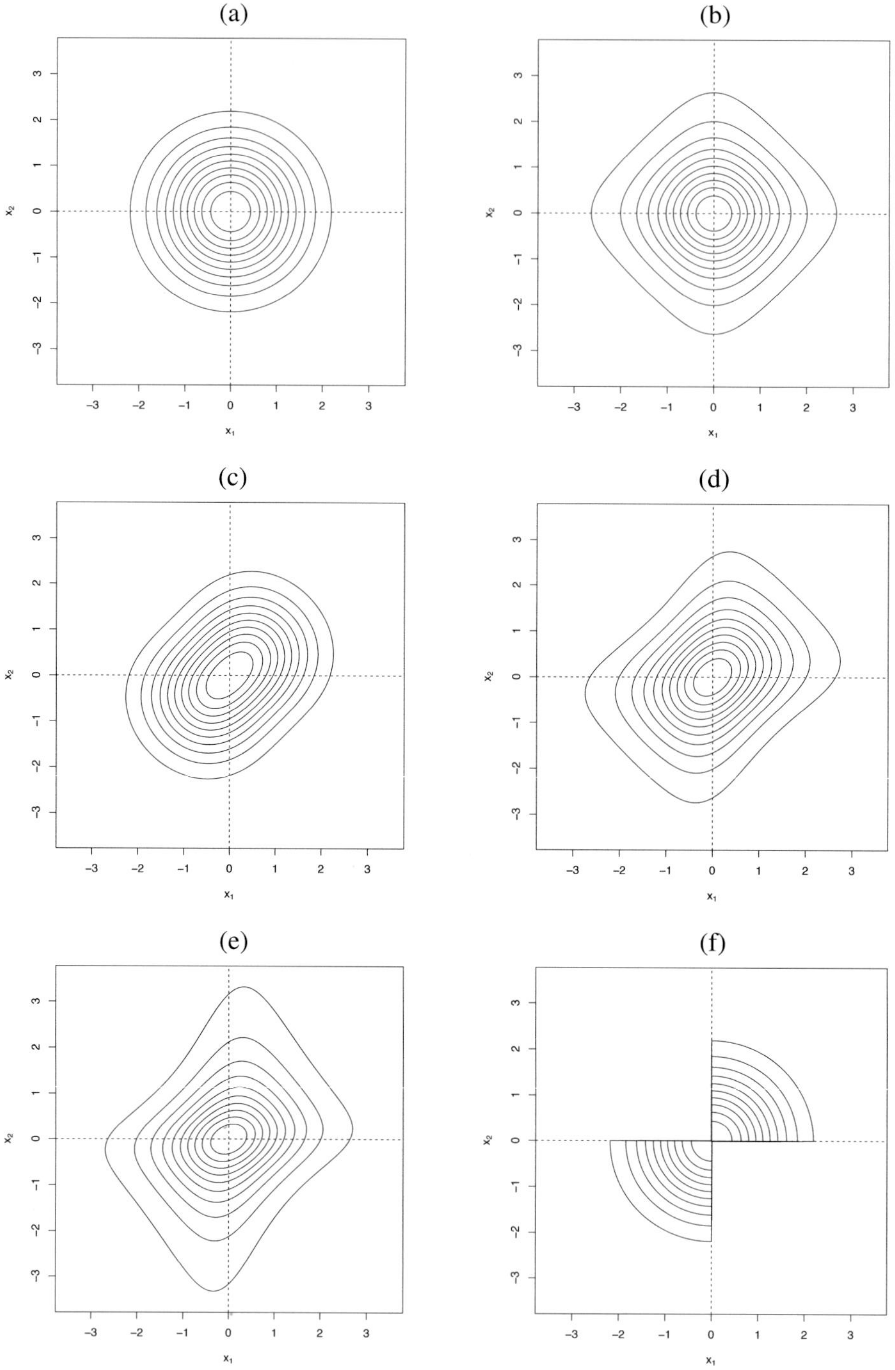

Figure 2.1 *Geometry of the density contours for various notions of symmetry of bivariate distributions ($p = 2$): (a) (2.9) with $f_1(x) = f_2(x) = \phi(x)$ and $\theta = 0$; (b) (2.9) with $f_1(x) = f_2(x) = t_3(x)$ and $\theta = 0$; (c) same as (a) with $\theta = 0.9$; (d) same as (b) with $\theta = 0.9$; (e) (2.9) with $f_1(x_1) = t_3(x_1)$, $f_2(x_2) = t_1(x_2)$, and $\theta = 0.9$; (f) (2.10) with $f_1(x) = f_2(x) = \phi(x)$. If $\theta < 0$ then the main direction for (c),(d),(e) is rotated by 90 degrees. Note that (a) is SS, (b) is SIS, and (c),(d),(e),(f) are CS. Moreover, all cases except (e) are symmetric with respect to the diagonal and to the anti-diagonal. Recall: CS $\supset$ SIS $\supset$ SS.*

Example 22 *Let $f_1(x)$ and $f_2(x)$ be symmetric pdfs with respect to the origin. Then, the pdf given by*

$$f(x_1,x_2) = \begin{cases} 2f_1(x_1)f_2(x_2), & x_1 > 0,\ x_2 > 0 \ \textit{or}\ x_1 < 0,\ x_2 < 0, \\ 0, & \textit{otherwise}, \end{cases} \tag{2.10}$$

is centrally symmetric, but not sign-invariant symmetric and not spherically symmetric. Further properties and generalizations of this distribution can be found in Dutta and Genton (2014); see Figure 2.1. Moreover, the discrete multivariate case has been explored by Ghosh et al. (2017).

In one dimension, the subgroups associated with *CS*, *SIS* and *SS* are identical, and therefore these classes of symmetric distributions are equivalent. In higher dimensions ($p \geq 2$), the following relations hold for these subgroups:

$$\mathcal{G}_{CS} \subset \mathcal{G}_{SIS} \quad \text{and} \quad \mathcal{G}_{CS} \subset \mathcal{G}_{SS},$$

but $\mathcal{G}_{SIS}$ and $\mathcal{G}_{SS}$ cannot be compared by set theoretic inclusion. On the other hand, the respective collections of all multivariate $\mathcal{G}_0$-*symmetric* distributions on $\mathbb{R}^p$ associated with these subgroups have the following inclusion relation:

$$CS \supset SIS \supset SS,$$

implying that the *SS* symmetry condition is the strongest one, while the *CS* symmetry condition is the weakest one.

2.3 Symmetry as Permutation Invariance

Another important subgroup of the orthogonal group of dimension p is the *permutation group* denoted by

$$\mathcal{P}_p = \{\mathbf{P} \in \mathbb{R}^{p\times p} : \mathbf{P}\mathbf{P}^\top = \mathbf{P}^\top\mathbf{P} = \mathbf{I}_p \text{ and } \mathbf{P}\mathbf{1}_p = \mathbf{1}_p\}.$$

Specifically, any permutation matrix in $\mathcal{P}_p$ is obtained by permuting the columns of the identity matrix of dimension p. Hence, under this group, the action $\mathbf{G}(\mathbf{x}) = \mathbf{P}\mathbf{x}$ permutes the components of the vector $\mathbf{x}$.

A random vector $\mathbf{X} \in \mathbb{R}^p$ (or its multivariate distribution) satisfying

$$\mathbf{X} \stackrel{d}{=} \mathbf{P}\mathbf{X}, \quad \forall \mathbf{P} \in \mathcal{P}_p,$$

is called *permutationally invariant* or *permutation symmetric* or *(finitely) exchangeable*; see Bilodeau and Brenner (1999), Fang et al. (1990), and Tong (1990). When F is the cdf of a random vector $\mathbf{X} \in \mathbb{R}^p$, then the (finite) exchangeability of $\mathbf{X}$ is equivalent to $F(\mathbf{x}) = F(\mathbf{P}\mathbf{x})$ for all $\mathbf{x} \in \mathbb{R}^p$ and for all $\mathbf{P} \in \mathcal{P}_p$. That is,

$$F(\mathbf{x}) \equiv F(x_1,\ldots,x_p) = F(x_{\pi_1},\ldots,x_{\pi_p}) \equiv F(\mathbf{x}_\pi),$$

for any permutation $\pi = (\pi_1,\ldots,\pi_p)$ of $(1,\ldots,p)$. This condition establishes the *symmetry of the cdf*. In particular, if

$$F(x_1,\ldots,x_p) = \prod_{i=1}^{p} F_1(x_i),$$

then $\mathbf{X}$ has iid components with cdf F_1 and is exchangeable. The iid case is a classical type of symmetry of the cdf.

Any subvector of a permutationally invariant random vector $\mathbf{X} \in \mathbb{R}^p$ is also permutationally invariant, and all its subvectors of the same dimension are identically distributed. For example, its components $X_1,\ldots,X_p$ are identically distributed random variables. Hence, if a permutationally invariant random vector $\mathbf{X}$ has a mean vector $\boldsymbol{\mu}$ and a variance matrix $\boldsymbol{\Sigma}$, then

$$\boldsymbol{\mu} = \mu \mathbf{1}_p \quad \text{and} \quad \boldsymbol{\Sigma} = \sigma^2\{(1-\rho)\mathbf{I}_p + \rho \mathbf{1}_p \mathbf{1}_p^\top\},$$

where $\mu = \mathrm{E}(X_1)$, $\sigma^2 = \mathrm{var}(X_1)$ and $\rho = \mathrm{corr}(X_1,X_2) \geq -1/(p-1)$. The latter condition must be met for the eigenvalues of $\boldsymbol{\Sigma}$ to be positive. However, in statistical applications, the assumption of (finite) exchangeability is made independently of the number of observations (here denoted by p and interpreted as the sample size), forcing the condition $\rho \geq 0$. This form of $\boldsymbol{\Sigma}$ is called the *equicorrelation matrix*; see Section A.3.3.

Clearly, (finite) exchangeability is also a prerequisite to having spherical symmetry. In fact, exchangeability, jointly with componentwise reflection symmetry, implies spherical symmetry. In particular, the components $X_1,\ldots,X_p$ of a spherical random vector $\mathbf{X}$ satisfy

$$X_i \stackrel{d}{=} -X_i \quad \forall i, \quad X_i \stackrel{d}{=} X_j \quad \forall i \neq j \quad \text{and} \quad X_i X_j \stackrel{d}{=} -X_i X_j \quad \forall i \neq j.$$

Consequently, if $\mathbf{X}$ has a multivariate spherical distribution with finite second moments, then all its components have zero mean, the same variance, σ^2 say, and are uncorrelated, i.e.:

$$\mathrm{E}(\mathbf{X}) = \mathbf{0}_p \quad \text{and} \quad \mathrm{var}(\mathbf{X}) = \sigma^2 \mathbf{I}_p.$$

Further results: The application of invariance to statistics is covered in several books, e.g., Fang and Zhang (1990), Lehmann and Casella (1998), and Eaton (1983, 1989). For a discussion on how some groups produce asymmetric distributions starting from symmetric ones, see Arellano-Valle and del Pino (2004). Sakhanenko (2009) proposed and studied a general class of tests for group symmetry, which encompasses different types of symmetry, such as ellipsoidal and permutation symmetries, among others. For a detailed study of symmetry measures and the use of symmetry in various scientific fields, see Petitjean (2003). Huang and Sen (2023) presented a unified framework for distribution-free testing of multivariate symmetry (including central symmetry, sign-invariant symmetry, spherical symmetry, among others) based on the theory of optimal transport.

2.4 Exercises

Exercise 2.1 *Let F be a cdf with symmetric pdf f. Show that $\int_{-a}^{a} F(x)\mathrm{d}x = a$ for all a and that $\int_{-\infty}^{+\infty} F(\alpha x) f(x)\mathrm{d}x = 1/2$ for all α.*

Exercise 2.2 *Show that $X_1,\ldots,X_p$ are sign-invariant symmetric random variables iff their cf has the form $\psi(t_1,\ldots,t_p) = \mathrm{E}\{\prod_{j=1}^{p}\cos(t_j X_j)\}$. (Hint: For the proof of necessity, take $p = 2$, and then use $\cos|x| = \cos x$ and $\cos(x+y) = \cos x \cos y - \sin x \sin y$. The proof for arbitrary p is similarly obtained by successive application of these arguments.)*

Exercise 2.3 *Show that $X_1,\ldots,X_p$ are sign-invariant iff they are conditionally independent given $|X_1|,\ldots,|X_p|$ with the conditional joint characteristic function $\prod_{j=1}^{p}\cos(t_j X_j)$.*

Exercise 2.4 *Suppose that a random vector $\mathbf{X}$ in $\mathbb{R}^p$ has a centrally symmetric distribution about $\boldsymbol{\xi} \in \mathbb{R}^p$, that is, $\mathbf{X} - \boldsymbol{\xi} \stackrel{d}{=} \boldsymbol{\xi} - \mathbf{X}$. Show that this representation is equivalent to each of the following representations:*

(a) $\mathbf{s}^\top(\mathbf{X}-\boldsymbol{\xi}) \stackrel{d}{=} \mathbf{s}^\top(\boldsymbol{\xi}-\mathbf{X})$, *for each unit length vector* $\mathbf{s}$ *in* $\mathbb{R}^p$;

(b) $\mathrm{P}(\mathbf{X}-\boldsymbol{\xi} \in H) = \mathrm{P}(\mathbf{X}-\boldsymbol{\xi} \in -H)$, *for each closed halfspace* $H \subset \mathbb{R}^p$.

Exercise 2.5 *In the following examples, check whether the random vector* $\mathbf{X} = (X_1, X_2)^\top$ *is centrally symmetric, angularly symmetric, or halfspace symmetric around any point in* $\mathbb{R}^2$*:*

(a) $\mathrm{P}\{(X_1, X_2) = (\pm 1, 0)\} = 1/3$, *and* $\mathrm{P}\{(X_1, X_2) = (0, 1)\} = 1/3$;

(b) $\mathrm{P}\{(X_1, X_2) = (0, 0)\} = 2/5$, $\mathrm{P}\{(X_1, X_2) = (1, \pm 1)\} = 1/5$, *and* $\mathrm{P}\{(X_1, X_2) = (-1, \pm 1)\} = 1/10$;

(c) $\mathrm{P}\{(X_1, X_2) = (0.2, 0.1)\} = 2/5$, $\mathrm{P}\{(X_1, X_2) = (1, \pm 1)\} = 1/10$, *and* $\mathrm{P}\{(X_1, X_2) = (-1, \pm 1)\} = 1/5$.

Chapter 3

The Multivariate Normal Distribution

The multivariate normal distribution is arguably the most important multivariate distribution. It has played a fundamental role in classical multivariate statistics because of the tractability and explicitness of the mathematics associated with its properties. Moreover, the multivariate normal distribution is the cornerstone of the non-normal multivariate models to be developed in the subsequent chapters. Therefore, we study this multivariate distribution in detail.

3.1 Preliminaries

In order to properly define the multivariate normal distribution and study its distributional properties more efficiently, we begin with a review of some basic facts concerning the variance matrix of a p-dimensional random vector; see also the Appendix A.

A $p \times p$ symmetric matrix $\mathbf{\Sigma}$ is said to be *positive definite (pd)* if $\mathbf{a}^\top \mathbf{\Sigma} \mathbf{a} \geq 0$ holds for all real vectors $\mathbf{a}$, and equality holds only for $\mathbf{a} = \mathbf{0}_p$. It is said to be *positive semi-definite (psd)* if $\mathbf{a}^\top \mathbf{\Sigma} \mathbf{a} \geq 0$ holds for all real vectors $\mathbf{a}$, and equality holds for some $\mathbf{a} \neq \mathbf{0}_p$. It is known that if $\mathbf{\Sigma}$ is pd (psd), then $|\mathbf{\Sigma}| > 0$ ($|\mathbf{\Sigma}| = 0$) or equivalently, the rank of $\mathbf{\Sigma}$ is p (is less than p).

Now, let $\mathbf{X}$ be a $p \times 1$ random vector with mean vector $\boldsymbol{\mu}$ and variance matrix $\mathbf{\Sigma}$. The distribution of $\mathbf{X}$ is said to be *singular* if there exists a vector $\mathbf{a} \neq \mathbf{0}_p$ such that $Y = \mathbf{a}^\top \mathbf{X}$ is singular (that is, $\mathrm{P}(Y = \mu_Y) = 1$, where $\mu_Y = \mathbf{a}^\top \boldsymbol{\mu}$). The variance of Y is $\sigma_Y^2 = \mathbf{a}^\top \mathbf{\Sigma} \mathbf{a}$ and Y is singular if and only if $\sigma_Y^2 = 0$. Thus, we find that a variance matrix $\mathbf{\Sigma}$ is either pd or psd. Furthermore,

$$
\begin{aligned}
\mathbf{\Sigma} \text{ is psd} &\Leftrightarrow |\mathbf{\Sigma}| = 0, \\
&\Leftrightarrow \text{the rank of } \mathbf{\Sigma} \text{ is less than } p, \\
&\Leftrightarrow \text{the corresponding distribution is singular.}
\end{aligned}
$$

We shall say that the distribution of $\mathbf{X}$ is *nonsingular* if it is not singular. Furthermore, for convenience in notation, we write $\mathbf{\Sigma} > 0$ ($\mathbf{\Sigma} \geq 0$) instead of $|\mathbf{\Sigma}| > 0$ ($|\mathbf{\Sigma}| = 0$) when $\mathbf{\Sigma}$ is pd (psd).

In general, if $\mathbf{A}$ is a symmetric $p \times p$ pd matrix, then:

[1] $\mathbf{A}$ is nonsingular;

[2] The eigenvalues of $\mathbf{A}$ are all positives;

[3] $\mathbf{A}$ can be written as $\mathbf{B}\mathbf{B}^\top$ where $\mathbf{B}$ is nonsingular, and the converse also holds true;

[4] If $\mathbf{C}$ is an $r \times p$ matrix of rank $r \leq p$, then $\mathbf{C}\mathbf{A}\mathbf{C}^\top$ is also pd;

[5] $\mathbf{A}^{-1}$ is also pd and symmetric.

3.2 Construction of a Multivariate Normal Distribution

There are several ways to introduce a multivariate normal distribution. Next, we present one of the most convenient and fruitful ways. Starting with the univariate normal distribution with mean $\mu \in \mathbb{R}$ and variance $\sigma^2 > 0$, namely $X \sim \mathcal{N}(\mu, \sigma^2)$, we know that it may be constructed as the distribution of the location-scale random variable

$$X = \mu + \sigma Z, \tag{3.1}$$

where $Z \sim \mathcal{N}(0,1)$ is a standard normal random variable with cf and pdf given, respectively, by

$$\text{Standard normal cf:} \quad \psi_s(t) = \mathrm{E}(e^{\mathrm{i}tZ}) = e^{-t^2/2}, \quad t \in \mathbb{R},$$

$$\text{Standard normal pdf:} \quad f_s(z) = \frac{1}{\sqrt{2\pi}} e^{-z^2/2} \equiv \phi(z), \quad z \in \mathbb{R}.$$

From the cf of Z it is straightforward to see that $\mathrm{E}(Z^{2k-1}) = 0$ (odd moments) while $\mathrm{E}(Z^{2k}) = (2k)!/(2^k k!)$ (even moments) for $k = 1, 2, \ldots$. In particular, we have $\mathrm{E}(Z) = 0$ and $\mathrm{var}(Z) = \mathrm{E}(Z^2) = 1$, implying from (3.1) that $\mathrm{E}(X) = \mu$ and $\mathrm{var}(X) = \sigma^2$. In addition, $\gamma_1 = 0$ (skewness) and $\gamma_2 = 0$ (excess kurtosis). It is also straightforward to show that the cf and pdf of X are, respectively, given by

$$\text{Normal cf:} \quad \psi(t) = \mathrm{E}(e^{\mathrm{i}tX}) = e^{\mathrm{i}t\mu}\mathrm{E}(e^{\mathrm{i}t\sigma Z}) = e^{\mathrm{i}t\mu - \sigma^2 t^2/2}, \quad t \in \mathbb{R};$$

$$\text{Normal pdf:} \quad f(x) = \frac{1}{\sigma}\phi\left(\frac{x-\mu}{\sigma}\right) = \frac{1}{\sqrt{2\pi\sigma^2}} e^{-(x-\mu)^2/(2\sigma^2)}, \quad x \in \mathbb{R}.$$

In such a case, we say that $X \sim \mathcal{N}(\mu, \sigma^2)$, which is equivalent to $Z = \sigma^{-1}(X - \mu) \sim \mathcal{N}(0,1)$. Clearly, the univariate normal distribution is uniquely determined by its mean μ and variance σ^2. This affirmation is reinforced by the well-known fact that

$$\mathrm{P}(a < X \leq b) = \Phi\left(\frac{b-\mu}{\sigma}\right) - \Phi\left(\frac{a-\mu}{\sigma}\right),$$

for all $a, b \in \mathbb{R}$ with $a < b$, where $\Phi(z) = \int_{-\infty}^{z} \phi(u)\mathrm{d}u$, $z \in \mathbb{R}$, is the standard normal cdf. For further properties of the univariate normal distribution, see, e.g., Patel and Read (1982).

Now let $Z_1, \ldots, Z_q$ be iid $\mathcal{N}(0,1)$ random variables, so that they all have the same cf and pdf as $Z \sim \mathcal{N}(0,1)$. If $\psi_s(\mathbf{t}) = \psi(t_1, \ldots, t_q) = \mathrm{E}(e^{\mathrm{i}\sum_{j=1}^{q} t_j Z_j})$ and $f_s(\mathbf{z}) = f_s(z_1, \ldots, z_q)$ are, respectively, the cf and pdf of the random vector $\mathbf{Z} = (Z_1, \ldots, Z_q)^\top$, then the independence assumption implies that:

$$q\text{-variate standard normal cf:} \quad \psi_s(\mathbf{t}) = \prod_{j=1}^{q} \psi_s(t_j) = \prod_{j=1}^{q} e^{-t_j^2/2} = e^{-\sum_{j=1}^{q} t_j^2/2} = e^{-\mathbf{t}^\top\mathbf{t}/2}, \quad \forall \mathbf{t} \in \mathbb{R}^q;$$

$$q\text{-variate standard normal pdf:} \quad f_s(\mathbf{z}) = \prod_{j=1}^{q} f_s(z_j) = \prod_{j=1}^{q} \{(2\pi)^{-1/2} e^{-z_j^2/2}\} = (2\pi)^{-q/2} e^{-\sum_{j=1}^{q} z_j^2/2}$$
$$= (2\pi)^{-q/2} e^{-\mathbf{z}^\top\mathbf{z}/2} \equiv \phi_q(\mathbf{z}), \quad \forall \mathbf{z} \in \mathbb{R}^q.$$

Of course, $\mathbf{Z} \sim \mathcal{N}_q(\mathbf{0}_q, \mathbf{I}_q)$ is the q-variate standard normal distribution which corresponds to the product of q iid $\mathcal{N}(0,1)$ pdfs and belongs to the class of multivariate spherical distributions, for which clearly $\mathrm{E}(\mathbf{Z}) = \mathbf{0}_q$, $\mathrm{var}(\mathbf{Z}) = \mathbf{I}_q$, and the multivariate distribution of $\mathbf{Z}$ is fully specified.

Finally, consider a random vector $\mathbf{X} \in \mathbb{R}^p$ obtained as the affine transformation of $\mathbf{Z} \sim \mathcal{N}_q(\mathbf{0}_q, \mathbf{I}_q)$ given by

$$\mathbf{X} = \boldsymbol{\mu} + \mathbf{A}\mathbf{Z}, \tag{3.2}$$

where $\boldsymbol{\mu} \in \mathbb{R}^p$ and $\mathbf{A} \in \mathbb{R}^{p\times q}$ are fixed arbitrarily. An immediate and clear consequence of (3.2) is that $\boldsymbol{\mu} = \mathrm{E}(\mathbf{X})$ and $\mathbf{A}\mathbf{A}^\top = \mathrm{var}(\mathbf{X})$, that is, $\boldsymbol{\mu}$ and $\mathbf{A}\mathbf{A}^\top$ are just the mean vector and variance matrix of $\mathbf{X}$, respectively, and completely determine the multivariate distributions of $\mathbf{X}$ as the multivariate distribution of $\mathbf{Z}$ does not depend on any additional parameter.

As formally defined in the next section, the multivariate distribution of a random vector $\mathbf{X}$ defined as (3.2) is called a multivariate normal distribution and is denoted by $\mathbf{X} \sim \mathcal{N}_p(\boldsymbol{\mu}, \boldsymbol{\Sigma})$, where $\boldsymbol{\Sigma} = \mathbf{A}\mathbf{A}^\top$. In fact, from (3.2) it follows that the cf of $\mathbf{X}$ is

$$\psi(\mathbf{t}) = \mathrm{E}(e^{\mathrm{i}\mathbf{t}^\top\mathbf{X}}) = \mathrm{E}(e^{\mathrm{i}\mathbf{t}^\top\boldsymbol{\mu} + \mathrm{i}\mathbf{t}^\top\mathbf{A}\mathbf{Z}}) = e^{\mathrm{i}\mathbf{t}^\top\boldsymbol{\mu}}\psi_s(\mathbf{A}^\top\mathbf{t}) = e^{\mathrm{i}\mathbf{t}^\top\boldsymbol{\mu} - \mathbf{t}^\top\boldsymbol{\Sigma}\mathbf{t}/2}, \quad \forall \mathbf{t} \in \mathbb{R}^p,$$

which clearly preserves the form of the univariate normal cf. One important point here is the fact that $\mathbf{\Sigma} \geq 0$, so that it might be either singular ($|\mathbf{\Sigma}| = 0$) or nonsingular ($|\mathbf{\Sigma}| > 0$), in which case the multivariate normal distribution is called singular or nonsingular, respectively. In addition, for each $\mathbf{\Sigma} \geq 0$, the factorization of $\mathbf{\Sigma} = \mathbf{A}\mathbf{A}^\top$ is generally not unique. In particular, we can always take a $p \times p$ square matrix as $\mathbf{A}$ with the corresponding $p \times 1$ random vector $\mathbf{Z} \sim \mathcal{N}_p(\mathbf{0}_p, \mathbf{I}_p)$ for which $\boldsymbol{\mu} + \mathbf{AZ}$ has the same multivariate normal distribution as $\mathbf{X}$ defined in (3.2). The main consequence of the singular case with $\text{rank}(\mathbf{\Sigma}) = r < p$ is that $p - r$ of the components of $\mathbf{X}$ must correspond to linear combinations of the remaining r components, and therefore the complete components $X_1, \ldots, X_p$ of $\mathbf{X}$ cannot have a joint pdf on $\mathbb{R}^p$. Hence, in the singular case, the multivariate normal distribution $\mathbf{X} \sim \mathcal{N}_p(\boldsymbol{\mu}, \mathbf{\Sigma})$ is not absolutely continuous. In fact, in this case, only r of the components of $\mathbf{X}$ are linearly independent, for which there is a joint pdf defined on $\mathbb{R}^r$. For the nonsingular case of $\mathbf{\Sigma} > 0$, we can always choose in (3.2) a nonsingular $p \times p$ matrix as $\mathbf{A}$ (which can be chosen as the unique symmetric square root of $\mathbf{\Sigma}$, $\mathbf{A} = \mathbf{\Sigma}^{1/2}$ say) with the corresponding multivariate normal random vector $\mathbf{Z} \sim \mathcal{N}_p(\mathbf{0}_p, \mathbf{I}_p)$. This is equivalent to considering the one-to-one transformation that follows from (3.2) when $p = q$ and $\mathbf{A}$ is a nonsingular square matrix. Hence, for the nonsingular case, we obtain, by applying the Jacobian method, the following pdf for $\mathbf{X}$:

$$f(\mathbf{x}) = \text{abs}(|\mathbf{A}|)^{-1} f_s\{\mathbf{A}^{-1}(\mathbf{x} - \boldsymbol{\mu})\} = |\mathbf{\Sigma}|^{-1/2}(2\pi)^{-p/2} e^{-\frac{1}{2}(\mathbf{x}-\boldsymbol{\mu})^\top \mathbf{\Sigma}^{-1}(\mathbf{x}-\boldsymbol{\mu})}, \quad \mathbf{x} \in \mathbb{R}^p.$$

In this case, we say that $\mathbf{X}$ has a nonsingular (or absolutely continuous) multivariate normal distribution, which clearly is a proper subclass of the entire family of multivariate normal distributions induced by the transformation (3.2).

Obviously, from the multivariate normal case, we can always come back to the univariate normal case. In fact, if $\mathbf{X} \sim \mathcal{N}_p(\boldsymbol{\mu}, \mathbf{\Sigma})$, then for each constant vector $\mathbf{a} \in \mathbb{R}^p$, we have

$$\psi_{\mathbf{a}^\top \mathbf{X}}(s) = \text{E}(e^{\text{i}s\mathbf{a}^\top \mathbf{X}}) = \psi_{\mathbf{X}}(s\mathbf{a}) = e^{\text{i}s(\mathbf{a}^\top \boldsymbol{\mu}) - \frac{1}{2}s^2(\mathbf{a}^\top \mathbf{\Sigma} \mathbf{a})}, \quad \forall s \in \mathbb{R},$$

implying for any linear combination that $\mathbf{a}^\top \mathbf{X} \sim \mathcal{N}(\mathbf{a}^\top \boldsymbol{\mu}, \mathbf{a}^\top \mathbf{\Sigma} \mathbf{a})$. Reciprocally, we know from the Cramér-Wold theorem that the distribution of any random vector $\mathbf{X}$ is completely determined by (and completely determines) the distribution of $\mathbf{a}^\top \mathbf{X}$ for every $\mathbf{a} \in \mathbb{R}^p$. Therefore, we obtain the following characterization of the multivariate normal distribution (which also gives a convenient alternative route for its definition):

$$\mathbf{X} \sim \mathcal{N}_p(\boldsymbol{\mu}, \mathbf{\Sigma}) \Longleftrightarrow \mathbf{a}^\top \mathbf{X} \sim \mathcal{N}(\mathbf{a}^\top \boldsymbol{\mu}, \mathbf{a}^\top \mathbf{\Sigma} \mathbf{a}), \quad \forall \mathbf{a} \in \mathbb{R}^p.$$

All of these results are formalized in the following sections.

3.3 Formal Definition and First Elementary Properties

In this section, we formalize all the previous results and explore further characterizations and the basic properties of the multivariate normal distribution.

Definition 1 [Multivariate normal distribution] *A random vector $\mathbf{X} \in \mathbb{R}^p$ is said to have a multivariate normal distribution with mean vector $\boldsymbol{\mu} \in \mathbb{R}^p$ and $p \times p$ variance matrix $\mathbf{\Sigma}$, in brief $\mathbf{X} \sim \mathcal{N}_p(\boldsymbol{\mu}, \mathbf{\Sigma})$, iff:*

$$\mathbf{X} \stackrel{d}{=} \boldsymbol{\mu} + \mathbf{AZ}, \quad \textit{with} \quad \mathbf{Z} \sim \mathcal{N}_q(\mathbf{0}_q, \mathbf{I}_q),$$

for any matrix $\mathbf{A} \in \mathbb{R}^{p \times q}$ such that $\mathbf{A}\mathbf{A}^\top = \mathbf{\Sigma}$.

Note that, from the multivariate normal distribution $\mathcal{N}_p(\boldsymbol{\mu}, \mathbf{\Sigma})$, with $\mathbf{\Sigma} > 0$, we can always return to the p-variate standard normal case simply by setting $\boldsymbol{\mu} = \mathbf{0}_p$ and $\mathbf{\Sigma} = \mathbf{I}_p$. Also, in the nonsingular case with $\mathbf{\Sigma} > 0$, the unique symmetric square root of $\mathbf{\Sigma}$ is defined by $\mathbf{\Sigma}^{1/2} = \mathbf{\Gamma}\mathbf{D}_\lambda^{1/2}\mathbf{\Gamma}^\top$

where $\mathbf{D}_\lambda = \text{diag}(\lambda_1,\ldots,\lambda_p)$ is the diagonal matrix of the eigenvalues $\lambda_1 \geq \cdots \geq \lambda_p > 0$ of $\boldsymbol{\Sigma}$ and the columns of $\boldsymbol{\Gamma} \in \mathbb{R}^{p\times p}$ are the corresponding eigenvectors. We use the notation $\mathbf{D}_\lambda^{1/2} = \text{diag}(\lambda_1^{1/2},\ldots,\lambda_p^{1/2})$. The matrix $\boldsymbol{\Gamma}$ is orthogonal, i.e., $\boldsymbol{\Gamma}^\top\boldsymbol{\Gamma} = \boldsymbol{\Gamma}\boldsymbol{\Gamma}^\top = \mathbf{I}_p$, and hence $\boldsymbol{\Gamma}^{-1} = \boldsymbol{\Gamma}^\top$. Consequently $\boldsymbol{\Sigma}^{1/2}\boldsymbol{\Sigma}^{1/2} = \boldsymbol{\Gamma}\mathbf{D}_\lambda^{1/2}\boldsymbol{\Gamma}^\top\boldsymbol{\Gamma}\mathbf{D}_\lambda^{1/2}\boldsymbol{\Gamma}^\top = \boldsymbol{\Sigma}$ and $(\boldsymbol{\Sigma}^{1/2})^\top = \boldsymbol{\Sigma}^{1/2}$. In this way, the following relationship becomes natural:

$$\mathbf{X} \overset{d}{=} \boldsymbol{\mu} + \boldsymbol{\Sigma}^{1/2}\mathbf{Z} \sim \mathcal{N}_p(\boldsymbol{\mu},\boldsymbol{\Sigma}) \Longleftrightarrow \mathbf{Z} \overset{d}{=} \boldsymbol{\Sigma}^{-1/2}(\mathbf{X}-\boldsymbol{\mu}) \sim \mathcal{N}_p(\mathbf{0}_p,\mathbf{I}_p). \tag{3.3}$$

From the standard normal random vector $\mathbf{Z} \sim \mathcal{N}_p(\mathbf{0}_p,\mathbf{I}_p)$ we can define the radial random variable R and the uniform random vector $\mathbf{U}^{(p)}$ as

$$R = \|\mathbf{Z}\| = (\mathbf{Z}^\top\mathbf{Z})^{1/2} \quad \text{and} \quad \mathbf{U}^{(p)} = \frac{\mathbf{Z}}{\|\mathbf{Z}\|},$$

which are well known to be independent, with $R^2 \sim \chi_p^2$ and $\mathbf{U}^{(p)}$ being uniformly distributed on the unit sphere $\mathbb{S}^{(p-1)} = \{\mathbf{x} \in \mathbb{R}^p : \|\mathbf{x}\| = 1\}$. Thus, the standard normal random vector $\mathbf{Z} \sim \mathcal{N}_p(\mathbf{0}_p,\mathbf{I}_p)$ can be stochastically represented as

$$\mathbf{Z} = \|\mathbf{Z}\|\frac{\mathbf{Z}}{\|\mathbf{Z}\|} = R\mathbf{U}^{(p)}, \quad \text{with } R \sim \chi_p, \ \mathbf{U}^{(p)} \sim \mathcal{U}(\mathbb{S}^{(p-1)}) \text{ and } R \perp\!\!\!\perp \mathbf{U}^{(p)}.$$

Some immediate characterizations of the multivariate normal distribution $\mathcal{N}_p(\boldsymbol{\mu},\boldsymbol{\Sigma})$ are presented in the following theorem. There, we assume the general case with $\boldsymbol{\Sigma} \geq 0$ and denote by $\boldsymbol{\Sigma}^-$ a *generalized inverse* of $\boldsymbol{\Sigma}$, i.e., a matrix satisfying $\boldsymbol{\Sigma}\boldsymbol{\Sigma}^-\boldsymbol{\Sigma} = \boldsymbol{\Sigma}$ (see also the Appendix A.2.5).

Theorem 1 *Assume that $\mathbf{X} \sim \mathcal{N}_p(\boldsymbol{\mu},\boldsymbol{\Sigma})$, with $\boldsymbol{\Sigma}$ of rank q. Then, the following statements are equivalent:*

(i) $\psi(\mathbf{t}) = \text{E}(e^{i\mathbf{t}^\top\mathbf{X}}) = e^{i\mathbf{t}^\top\boldsymbol{\mu} - \mathbf{t}^\top\boldsymbol{\Sigma}\mathbf{t}/2}$, $\mathbf{t} \in \mathbb{R}^p$;

(ii) $\mathbf{a}^\top\mathbf{X} \sim \mathcal{N}_1(\mathbf{a}^\top\boldsymbol{\mu},\mathbf{a}^\top\boldsymbol{\Sigma}\mathbf{a})$, $\forall\, \mathbf{a} \in \mathbb{R}^p$;

(iii) $\mathbf{X} \overset{d}{=} \boldsymbol{\mu} + R\mathbf{A}\mathbf{U}^{(q)}$, *with* $\mathbf{A} \in \mathbb{R}^{p\times q}$ *such that* $\mathbf{A}\mathbf{A}^\top = \boldsymbol{\Sigma}$, $R^2 \overset{d}{=} (\mathbf{X}-\boldsymbol{\mu})^\top\boldsymbol{\Sigma}^-(\mathbf{X}-\boldsymbol{\mu}) \sim \chi_q^2$, $\mathbf{U}^{(q)} \sim \mathcal{U}(\mathbb{S}^{(q-1)})$ *and* $R \perp\!\!\!\perp \mathbf{U}^{(q)}$.

Proof 1 *The equivalence of statements (i) and (ii) was proved in Section 3.2. The equivalence of (iii) with the rest of the statements is straightforward since* $\mathbf{Z} = R\mathbf{U}^{(q)} \sim \mathcal{N}_q(\mathbf{0}_q,\mathbf{I}_q)$ *and therefore* $\mathbf{X} \overset{d}{=} \boldsymbol{\mu} + R\mathbf{A}\mathbf{U}^{(q)} = \boldsymbol{\mu} + \mathbf{A}\mathbf{Z} \sim \mathcal{N}_p(\boldsymbol{\mu},\boldsymbol{\Sigma})$ *by definition.* □

The Mardia measures of multivariate skewness ($\beta_{1,p}$) and kurtosis ($\beta_{2,p}$) are also simple to obtain for the multivariate normal distribution, as shown in the following theorem.

Theorem 2 *Let $\mathbf{X} \sim \mathcal{N}_p(\boldsymbol{\mu},\boldsymbol{\Sigma})$, with $\boldsymbol{\Sigma} > 0$. Then, $R^2 = (\mathbf{X}-\boldsymbol{\mu})^\top\boldsymbol{\Sigma}^{-1}(\mathbf{X}-\boldsymbol{\mu}) \sim \chi_p^2$, and*

$$\text{E}(\mathbf{X}) = \boldsymbol{\mu}, \quad \text{var}(\mathbf{X}) = \boldsymbol{\Sigma}, \quad \beta_{1,p} = 0 \quad \textit{and} \quad \beta_{2,p} = p(p+2). \tag{3.4}$$

Proof 2 *From (3.3), we have* $R^2 = \mathbf{Z}^\top\mathbf{Z} = \sum_{i=1}^p Z_i^2$, *where* $Z_1,\ldots,Z_p$ *are iid* $\mathcal{N}(0,1)$. *Hence,* $R^2 \sim \chi_p^2$. *Thus, we have* $\beta_{2,p} = \text{E}(R^4) = \text{var}(R^2) + \{\text{E}(R^2)\}^2 = 2p + p^2 = p(p+2)$. *Finally, as we know,* $\beta_{1,p} = \text{E}(V^3)$, *where* $V = (\mathbf{X}-\boldsymbol{\mu})^\top\boldsymbol{\Sigma}^{-1}(\mathbf{Y}-\boldsymbol{\mu})$, *with* $\mathbf{X}$ *and* $\mathbf{Y}$ *being iid from* $\mathcal{N}_p(\boldsymbol{\mu},\boldsymbol{\Sigma})$. *Since* $V \overset{d}{=} -V$, *i.e.,* V *is symmetric around zero, then* $\beta_{1,p} = \text{E}(V^3) = 0$. □

Dutta and Genton (2014) have constructed a non-normal p-dimensional distribution for which any proper subset of its components has a normal distribution. Moreover, they showed that for any such subset: $\beta_{1,q} = 0$ and $\beta_{2,q} = q(q+2)$, $1 \leq q \leq p$. Hence, the Mardia measures of skewness and kurtosis in Equation (3.4) do not characterize the multivariate normal distribution. For $p = 2$, the pdf of this distribution is given in Equation (2.10) and is depicted in Figure 2.1(f).

3.4 Affine Transformations, Marginals, Independence of the Normal

Theorem 3 *Let* $\mathbf{X} \sim \mathcal{N}_p(\boldsymbol{\mu}, \boldsymbol{\Sigma})$. *If* $\mathbf{Y} = \mathbf{A}\mathbf{X} + \mathbf{b}$ *with nonsingular matrix* $\mathbf{A} \in \mathbb{R}^{p \times p}$ *and vector* $\mathbf{b} \in \mathbb{R}^p$, *then* $\mathbf{Y} \sim \mathcal{N}_p(\mathbf{A}\boldsymbol{\mu} + \mathbf{b}, \mathbf{A}\boldsymbol{\Sigma}\mathbf{A}^\top)$.

Proof 3 *Using the stochastic representation of* $\mathbf{X}$ *given by (3.3), we have:*

$$\mathbf{Y} \stackrel{d}{=} \mathbf{A}(\boldsymbol{\Sigma}^{1/2}\mathbf{Z} + \boldsymbol{\mu}) + \mathbf{b} = \mathbf{A}\boldsymbol{\Sigma}^{1/2}\mathbf{Z} + (\mathbf{A}\boldsymbol{\mu} + \mathbf{b}).$$

Hence, $\mathbf{Y}$ *has a multivariate normal distribution with mean vector* $\mathbf{A}\boldsymbol{\mu} + \mathbf{b}$ *and variance matrix* $\mathbf{A}\boldsymbol{\Sigma}^{1/2}\boldsymbol{\Sigma}^{1/2}\mathbf{A}^\top$. □

Theorem 4 *Let* $\mathbf{X} \sim \mathcal{N}_p(\boldsymbol{\mu}, \boldsymbol{\Sigma})$ *with* $\mathbf{X} = (\mathbf{X}_1^\top, \mathbf{X}_2^\top)^\top$, $\mathbf{X}_1 \in \mathbb{R}^{p_1}$, $\mathbf{X}_2 \in \mathbb{R}^{p_2}$, $p_1 + p_2 = p$, *and the induced partitions* $\boldsymbol{\mu} = (\boldsymbol{\mu}_1^\top, \boldsymbol{\mu}_2^\top)^\top$ *and* $\boldsymbol{\Sigma} = (\boldsymbol{\Sigma}_{ij})_{i,j=1,2}$. *Then:*

(a) $\mathbf{X}_1 \sim \mathcal{N}_{p_1}(\boldsymbol{\mu}_1, \boldsymbol{\Sigma}_{11})$ *and* $\mathbf{X}_2 \sim \mathcal{N}_{p_1}(\boldsymbol{\mu}_2, \boldsymbol{\Sigma}_{22})$;

(b) $\mathbf{X}_1 \perp\!\!\!\perp \mathbf{X}_2$ *iff* $\mathbf{X}_1$ *and* $\mathbf{X}_2$ *are uncorrelated, i.e.,* $\boldsymbol{\Sigma}_{12} = \mathrm{cov}(\mathbf{X}_1, \mathbf{X}_2) = \mathbf{O}$.

Proof 4 *The marginal distributions follow from the linear transformations* $\mathbf{X}_i = \mathbf{A}_i\mathbf{X}$ *for* $i = 1, 2$ *with* $\mathbf{A}_1 = (\mathbf{I}_{p_1}, \mathbf{O})$ *and* $\mathbf{A}_2 = (\mathbf{O}, \mathbf{I}_{p_2})$. *The cf of the vector* $\mathbf{X}$ *factorizes into* $\varphi_{\mathbf{X}}(\mathbf{t}) = \varphi_1(\mathbf{t}_1)\varphi_2(\mathbf{t}_2)$ *for all* $\mathbf{t} = (\mathbf{t}_1^\top, \mathbf{t}_2^\top)^\top \in \mathbb{R}^p$ *iff* $\boldsymbol{\Sigma}_{12} = \mathbf{O}$. □

Theorem 5 *Let* $\mathbf{X} \sim \mathcal{N}_p(\boldsymbol{\mu}, \boldsymbol{\Sigma})$. *Then* $\mathbf{A}\mathbf{X} \perp\!\!\!\perp \mathbf{B}\mathbf{X}$ *iff* $\mathbf{A}\boldsymbol{\Sigma}\mathbf{B}^\top = \mathbf{O}$.

Proof 5 *From the previous result, we only need* $\mathrm{cov}(\mathbf{A}\mathbf{X}, \mathbf{B}\mathbf{X}) = \mathbf{A}\boldsymbol{\Sigma}\mathbf{B}^\top = \mathbf{O}$. □

Example 23 *Let* $\mathbf{X} \sim \mathcal{N}_p(\boldsymbol{\mu}, \sigma^2\mathbf{I}_p)$ *and* $\mathbf{G} \in \mathbb{R}^{q \times p}$ *such that* $\mathbf{G}\mathbf{G}^\top = \mathbf{I}_q$. *Then* $\mathbf{G}\mathbf{X} \sim \mathcal{N}_q(\mathbf{G}\boldsymbol{\mu}, \sigma^2\mathbf{I}_q)$. *Moreover,* $\mathbf{G}\mathbf{X} \perp\!\!\!\perp (\mathbf{I}_p - \mathbf{G}^\top\mathbf{G})\mathbf{X}$ *because* $\mathbf{G}\sigma^2\mathbf{I}_p(\mathbf{I}_p - \mathbf{G}^\top\mathbf{G})^\top = \sigma^2\mathbf{G} - \sigma^2\mathbf{G}\mathbf{G}^\top\mathbf{G} = \mathbf{O}$.

3.5 Conditional Distributions of the Multivariate Normal

Theorem 6 *Let* $\mathbf{X} \sim \mathcal{N}_p(\boldsymbol{\mu}, \boldsymbol{\Sigma})$ *with* $\mathbf{X} = (\mathbf{X}_1^\top, \mathbf{X}_2^\top)^\top$, $\mathbf{X}_1 \in \mathbb{R}^{p_1}$, $\mathbf{X}_2 \in \mathbb{R}^{p_2}$, $p_1 + p_2 = p$, *and the induced partitions* $\boldsymbol{\mu} = (\boldsymbol{\mu}_1^\top, \boldsymbol{\mu}_2^\top)^\top$ *and* $\boldsymbol{\Sigma} = (\boldsymbol{\Sigma}_{ij})_{i,j=1,2}$. *Then:*

(a) $\mathbf{X}_{2\cdot 1} = \mathbf{X}_2 - \boldsymbol{\Sigma}_{21}\boldsymbol{\Sigma}_{11}^{-1}\mathbf{X}_1 \sim \mathcal{N}_{p_2}(\boldsymbol{\mu}_2 - \boldsymbol{\Sigma}_{21}\boldsymbol{\Sigma}_{11}^{-1}\boldsymbol{\mu}_1, \boldsymbol{\Sigma}_{22} - \boldsymbol{\Sigma}_{21}\boldsymbol{\Sigma}_{11}^{-1}\boldsymbol{\Sigma}_{12})$, *and* $\mathbf{X}_{2\cdot 1} \perp\!\!\!\perp \mathbf{X}_1$;

(b) $(\mathbf{X}_2 | \mathbf{X}_1 = \mathbf{x}_1) \sim \mathcal{N}_{p_2}(\boldsymbol{\mu}_2 + \boldsymbol{\Sigma}_{21}\boldsymbol{\Sigma}_{11}^{-1}(\mathbf{x}_1 - \boldsymbol{\mu}_1), \boldsymbol{\Sigma}_{22} - \boldsymbol{\Sigma}_{21}\boldsymbol{\Sigma}_{11}^{-1}\boldsymbol{\Sigma}_{12})$.

Proof 6 *The random vector $\mathbf{X}_{2\cdot 1}$ can be obtained from the linear transformation $\mathbf{X}_{2\cdot 1} = \mathbf{B}\mathbf{X}$ with $\mathbf{B} = (-\boldsymbol{\Sigma}_{21}\boldsymbol{\Sigma}_{11}^{-1}, \mathbf{I}_{p_2})$. From Theorem 3, $\mathbf{X}_1$ and $\mathbf{X}_{2\cdot 1}$ have a multivariate normal joint distribution with mean vectors $\mathbf{A}\boldsymbol{\mu}$ and $\mathbf{B}\boldsymbol{\mu}$, variance matrices $\mathbf{A}\boldsymbol{\Sigma}\mathbf{A}^\top$ and $\mathbf{B}\boldsymbol{\Sigma}\mathbf{B}^\top$, and covariance matrix $\mathbf{A}\boldsymbol{\Sigma}\mathbf{B}^\top$, which simplify to the expressions in the statement above. The independence between $\mathbf{X}_1$ and $\mathbf{X}_{2\cdot 1}$ is obtained from Theorem 5 by checking that $\mathbf{A}\boldsymbol{\Sigma}\mathbf{B}^\top = \mathbf{O}$. Therefore, the conditional distribution $(\mathbf{X}_{2\cdot 1}|\mathbf{X}_1 = \mathbf{x}_1)$ is independent of $\mathbf{x}_1$ and thus is the same as the distribution of $\mathbf{X}_{2\cdot 1}$, given in (b). Writing $\mathbf{X}_2 = \mathbf{X}_{2\cdot 1} + \boldsymbol{\Sigma}_{21}\boldsymbol{\Sigma}_{11}^{-1}\mathbf{X}_1$, for which the second term is constant when $\mathbf{X}_1 = \mathbf{x}_1$, we see that $(\mathbf{X}_2|\mathbf{X}_1 = \mathbf{x}_1)$ has a normal distribution with conditional mean $\boldsymbol{\mu}_{2\cdot 1} = \mathrm{E}(\mathbf{X}_2|\mathbf{X}_1 = \mathbf{x}_1) = \boldsymbol{\mu}_2 - \boldsymbol{\Sigma}_{21}\boldsymbol{\Sigma}_{11}^{-1}\boldsymbol{\mu}_1 + \boldsymbol{\Sigma}_{21}\boldsymbol{\Sigma}_{11}^{-1}\mathbf{x}_1 = \boldsymbol{\mu}_2 + \boldsymbol{\Sigma}_{21}\boldsymbol{\Sigma}_{11}^{-1}(\mathbf{x}_1 - \boldsymbol{\mu}_1)$ and the same conditional variance matrix as $\mathbf{X}_{2\cdot 1}$, namely $\boldsymbol{\Sigma}_{22\cdot 1} = \boldsymbol{\Sigma}_{22} - \boldsymbol{\Sigma}_{21}\boldsymbol{\Sigma}_{11}^{-1}\boldsymbol{\Sigma}_{12}$.* □

Example 24 *Let $\mathbf{X} = (X_1, X_2)^\top \sim \mathcal{N}_2(\boldsymbol{\mu}, \boldsymbol{\Sigma})$, with mean vector $\boldsymbol{\mu} = (\mu_1, \mu_2)^\top$ and variance matrix:*

$$\boldsymbol{\Sigma} = \begin{pmatrix} \sigma_1^2 & \rho\sigma_1\sigma_2 \\ \rho\sigma_1\sigma_2 & \sigma_2^2 \end{pmatrix},$$

where ρ is the correlation coefficient. Then $(X_2|X_1 = x_1) \sim \mathcal{N}_1(\mu_2 + \rho\frac{\sigma_2}{\sigma_1}(x_1 - \mu_1), \sigma_2^2(1-\rho^2))$, which is nonsingular if and only if $|\rho| < 1$.

Example 25 *Let $\mathbf{X} = (\mathbf{X}_1^\top, \mathbf{X}_2^\top)^\top \sim \mathcal{N}_{p_1+p_2}(\boldsymbol{\mu}, \boldsymbol{\Sigma})$, with $\boldsymbol{\mu} = (\boldsymbol{\mu}_1^\top, \boldsymbol{\mu}_2^\top)^\top$ and equicorrelation matrix $\boldsymbol{\Sigma} = (1-\rho)\mathbf{I}_p + \rho\mathbf{1}_p\mathbf{1}_p^\top$, with $-(p-1)^{-1} < \rho < 1$; see Section A.3.3. Then we have $\boldsymbol{\Sigma}^{-1} = (1-\rho)^{-1}[\mathbf{I}_p - \rho\{1+(p-1)\rho\}^{-1}\mathbf{1}_p\mathbf{1}_p^\top]$ and $|\boldsymbol{\Sigma}| = (1-\rho)^{p-1}\{1+\rho(p-1)\}$. Straightforward calculations yield a conditional multivariate normal distribution with conditional mean vector $\boldsymbol{\mu}_{2\cdot 1} = \boldsymbol{\mu}_2 + \alpha\mathbf{1}_{p_2}\mathbf{1}_{p_1}^\top(\mathbf{x}_1 - \boldsymbol{\mu}_1)$ and conditional variance matrix $\boldsymbol{\Sigma}_{22\cdot 1} = (1-\rho)\mathbf{I}_{p_2} + \rho(1-p_1\alpha)\mathbf{1}_{p_2}\mathbf{1}_{p_2}^\top$ where $\alpha = \rho/\{1+(p_1-1)\rho\}$.*

3.6 Cdf and Pdf of the Multivariate Normal

Theorem 7 *If $\mathbf{X} \sim \mathcal{N}_p(\boldsymbol{\mu}, \boldsymbol{\Sigma})$ is nonsingular, i.e., $\boldsymbol{\Sigma} > 0$, then its pdf is given by*

$$\phi_p(\mathbf{x}; \boldsymbol{\mu}, \boldsymbol{\Sigma}) = (2\pi)^{-p/2}|\boldsymbol{\Sigma}|^{-1/2}\exp\left\{-\frac{1}{2}(\mathbf{x}-\boldsymbol{\mu})^\top\boldsymbol{\Sigma}^{-1}(\mathbf{x}-\boldsymbol{\mu})\right\}, \qquad \mathbf{x} \in \mathbb{R}^p. \tag{3.5}$$

Proof 7 *It was proved in Section 3.2.* □

Theorem 7 is usually used as the definition of the multivariate normal distribution because in practice, we work exclusively with the nonsingular case. Thus, we can observe again that for $p = 1$, we have $\boldsymbol{\Sigma} = \sigma^2 > 0$ and we obtain the usual univariate normal pdf $\phi_1(x; \mu, \sigma^2) = (2\pi)^{-1/2}\sigma^{-1}\exp\{-(x-\mu)^2/(2\sigma^2)\}$, $x \in \mathbb{R}$. The case of $\boldsymbol{\mu} = \mathbf{0}_p$ and $\boldsymbol{\Sigma} = \mathbf{I}_p$ is the standard multivariate normal distribution and is denoted by $\mathbf{X} \sim \mathcal{N}_p(\mathbf{0}_p, \mathbf{I}_p)$. The cdf of a random vector with a multivariate normal distribution $\mathcal{N}_p(\boldsymbol{\mu}, \boldsymbol{\Sigma})$ is:

$$\Phi_p(\mathbf{x}; \boldsymbol{\mu}, \boldsymbol{\Sigma}) = \int_{-\infty}^{\mathbf{x}} \phi_p(\mathbf{y}; \boldsymbol{\mu}, \boldsymbol{\Sigma})\mathrm{d}\mathbf{y}, \quad \mathbf{x} \in \mathbb{R}^p, \tag{3.6}$$

but, in general, it does not have a closed form, except for some special cases such as, e.g., $\Phi_p(\mathbf{0}_p;\mathbf{0}_p,\mathbf{I}_p) = 1/2^p$. The cdf in (3.6) can be numerically evaluated by means of the method proposed by Genz (1992) and Genz and Bretz (2009); see also Genton et al. (2018), Cao et al. (2019, 2021, 2022b), and Zhang et al. (2024) for the computation of Φ_p when p is very large for applications in spatial statistics.

Example 26 (Orthant probability) *Assuming that* $\mathbf{X} \sim \mathcal{N}_p(\mathbf{0}_p,\mathbf{R})$, *where* $\mathbf{R} = (\rho_{ij}) > 0$ *is the correlation matrix of* $\mathbf{X}$, *several authors (see, e.g., Bacon (1963)) have been interested in obtaining the orthant probability* $\Phi_p = \mathrm{P}(\mathbf{X} \geq \mathbf{0}_p) = \mathrm{P}(X_1 \geq 0,\ldots,X_p \geq 0)$; *well-known examples are:*

$$\begin{aligned}
\Phi_1 &= \frac{1}{2},\\
\Phi_2 &= \frac{1}{4} + \frac{1}{2\pi}\arcsin(\rho_{12}),\\
\Phi_3 &= \frac{1}{8} + \frac{1}{4\pi}\{\arcsin(\rho_{12}) + \arcsin(\rho_{13}) + \arcsin(\rho_{23})\}.
\end{aligned}$$

An important fact is that if $\mathbf{X} \sim \mathcal{N}_p(\mathbf{0}_p,\mathbf{R})$ *and if* $\mathbf{Y} \sim \mathcal{E}l_p(\mathbf{0}_p,\mathbf{R},\varphi)$ *denotes an elliptical distribution (see Chapter 4) such that* $\mathrm{P}(\mathbf{Y} = \mathbf{0}_p) = 0$, *where* $\mathbf{R} = (\rho_{ij}) > 0$ *is the same correlation matrix, then* $\mathrm{P}(\mathbf{Y} \geq \mathbf{0}_p) = \mathrm{P}(\mathbf{X} \geq \mathbf{0}_p)$.

For simplicity, we will often use the notation $\phi_p(\mathbf{x})$ and $\Phi_p(\mathbf{x})$ for the standard pdf and cdf, respectively, of $\mathbf{X} \sim \mathcal{N}_p(\mathbf{0}_p,\mathbf{I}_p)$.

3.7 Geometry of the Multivariate Normal

The normal pdf (3.5) is constant on contours with an ellipsoidal shape centered at $\boldsymbol{\mu}$ and defined by those $\mathbf{x} \in \mathbb{R}^p$ that satisfy

$$(\mathbf{x}-\boldsymbol{\mu})^\top \boldsymbol{\Sigma}^{-1}(\mathbf{x}-\boldsymbol{\mu}) = c^2, \tag{3.7}$$

where c is a constant. When $\boldsymbol{\Sigma} = \mathbf{I}_p$, the contours are spheres centered at $\boldsymbol{\mu}$ with radius c. For $p = 2$, these contours are ellipses or circles in the plane. The geometry associated with the normal pdf (3.5) is best understood in terms of the *principal component transformation*

$$\mathbf{y} = \boldsymbol{\Gamma}^\top(\mathbf{x}-\boldsymbol{\mu}),$$

where $\boldsymbol{\Gamma} \in \mathbb{R}^{p\times p}$ is the matrix of eigenvectors from the spectral decomposition of $\boldsymbol{\Sigma} = \boldsymbol{\Gamma}\mathbf{D}_\lambda\boldsymbol{\Gamma}^\top$. Then

$$(\mathbf{x}-\boldsymbol{\mu})^\top \boldsymbol{\Sigma}^{-1}(\mathbf{x}-\boldsymbol{\mu}) = \mathbf{y}^\top\mathbf{D}_\lambda^{-1}\mathbf{y} = \sum_{i=1}^{p}\frac{y_i^2}{\lambda_i} = c^2,$$

where $\mathbf{D}_\lambda = \mathrm{diag}(\lambda_1,\ldots,\lambda_p)$ is the diagonal matrix of the eigenvalues of $\boldsymbol{\Sigma}$. Therefore, the components y_i of $\mathbf{y}$ represent the ellipsoid axes with corresponding semi-axes of length $\sqrt{\lambda_i}c$, for $i = 1,\ldots,p$. Because the eigenvalues $\lambda_1 \geq \cdots \geq \lambda_p > 0$ are sorted from largest to smallest, $\sqrt{\lambda_1}c$ and $\sqrt{\lambda_p}c$ are the lengths of the major and minor axes of the ellipsoid, respectively.

From Theorem 2, the probability that $\mathbf{X}$ falls within an ellipsoid determined by c is:

$$\mathrm{P}\{(\mathbf{X}-\boldsymbol{\mu})^\top \boldsymbol{\Sigma}^{-1}(\mathbf{X}-\boldsymbol{\mu}) \leq c^2\}, \tag{3.8}$$

and can be computed from the χ_p^2 distribution. Reciprocally, the constant c^2 can be determined so that a certain probability (3.8) is achieved. For example, with $p = 2$, a probability of 95% is obtained from the 95% quantile of the χ_2^2 distribution, which yields $c^2 = 5.99$. The constant c^2 depends on

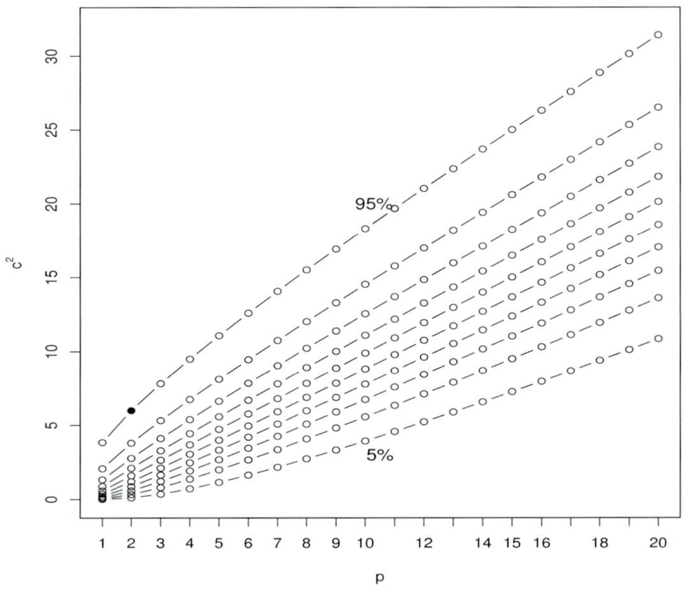

Figure 3.1 *Value of the constant c^2 determining the size of an ellipsoidal contour of the multivariate normal density in dimension p to achieve a desired probability of coverage ranging from 5% to 95% by steps of 10%. The black solid circle is for 95% in $p = 2$ dimensions.*

the dimension p and the desired probability level. Figure 3.1 depicts the value of the constant c^2 as a function of the dimension p to achieve a desired probability of coverage ranging from 5% to 95% by steps of 10%. The black solid circle is for 95% in $p = 2$ dimensions.

Example 27 *In $\mathbb{R}^2$, consider a bivariate normal distribution with mean $\boldsymbol{\mu} = \mathbf{0}_2$ and three different variance matrices:*

$$\boldsymbol{\Sigma}_1 = \begin{pmatrix} 1 & 0 \\ 0 & 1 \end{pmatrix}, \quad \boldsymbol{\Sigma}_2 = \begin{pmatrix} 2 & 0 \\ 0 & 1 \end{pmatrix}, \quad \boldsymbol{\Sigma}_3 = \begin{pmatrix} 2 & 1/2 \\ 1/2 & 1 \end{pmatrix}. \tag{3.9}$$

The eigenvalues of $\boldsymbol{\Sigma}_1$ are $\lambda_1 = \lambda_2 = 1$, and its standardized eigenvectors are $\boldsymbol{\gamma}_1 = (1,0)^\top$ and $\boldsymbol{\gamma}_2 = (0,1)^\top$. Therefore, the contours are circles centered at the origin. The eigenvalues of $\boldsymbol{\Sigma}_2$ are $\lambda_1 = 2$, $\lambda_2 = 1$, and its standardized eigenvectors are the same as for $\boldsymbol{\Sigma}_1$. Therefore, the contours are ellipses centered at the origin with major and minor axes parallel to the coordinate axes. The eigenvalues of $\boldsymbol{\Sigma}_3$ are $\lambda_1 = (3+\sqrt{2})/2$ and $\lambda_2 = (3-\sqrt{2})/2$ and its standardized eigenvectors are:

$$\begin{aligned} \boldsymbol{\gamma}_1 &= \left(1/\sqrt{2(2-\sqrt{2})}, (\sqrt{2}-1)/\sqrt{2(2-\sqrt{2})}\right)^\top, \\ \boldsymbol{\gamma}_2 &= \left(-(\sqrt{2}-1)/\sqrt{2(2-\sqrt{2})}, 1/\sqrt{2(2-\sqrt{2})}\right)^\top. \end{aligned}$$

Therefore, the contours are ellipses centered at the origin with the directions of the major and minor axes given by the eigenvectors $\boldsymbol{\gamma}_1$ and $\boldsymbol{\gamma}_2$. The contours of those three settings are depicted in Figure 3.2 with levels obtained from (3.7) in order to have associated probabilities (3.8) from 5% to 95% by steps of 10%. The last panel has 10 equispaced levels of the pdf.

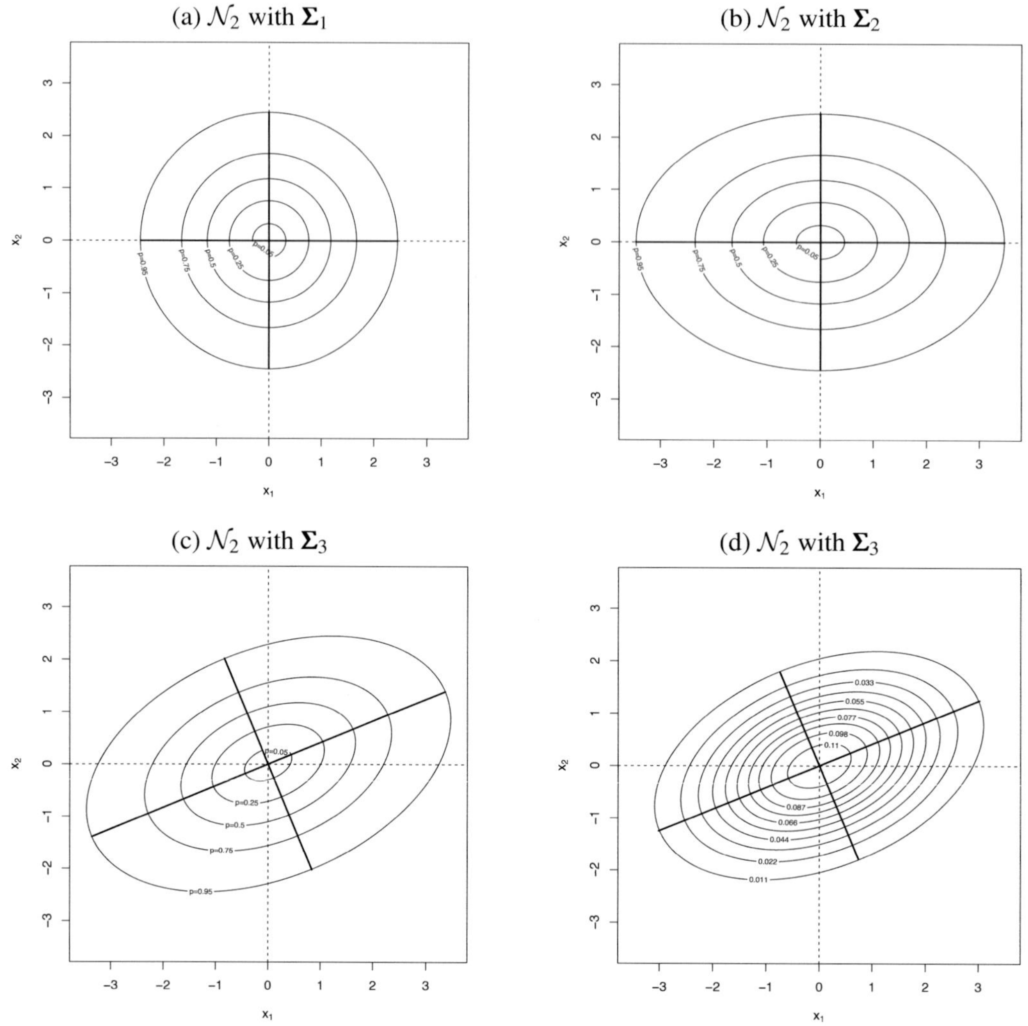

Figure 3.2 *Geometry of the multivariate normal distribution, $\mathcal{N}_p$, for $p = 2$.*

3.8 Simulations of the Multivariate Normal

In many situations, it is important to be able to simulate pseudo-realizations from a multivariate normal distribution, for instance, in order to test a new statistical method on such data. This can be easily done with the stochastic representation (3.3). First, simulate p iid pseudo-realizations from the univariate $\mathcal{N}(0,1)$ distribution and put them in a vector $\mathbf{Z}$. Based on the spectral decomposition of a positive definite matrix $\mathbf{\Sigma}$, calculate $\mathbf{\Sigma}^{1/2}$. Then, for a given $\boldsymbol{\mu}$, apply the affine transformation (3.3) to obtain $\mathbf{X} \sim \mathcal{N}_p(\boldsymbol{\mu}, \mathbf{\Sigma})$.

Alternatively, we could use the affine transformation

$$\mathbf{X} = \mathbf{\Gamma} \mathbf{D}_\lambda^{1/2} \mathbf{Z} + \boldsymbol{\mu}, \tag{3.10}$$

which also yields $\mathbf{X} \sim \mathcal{N}_p(\boldsymbol{\mu}, \mathbf{\Sigma})$. Geometrically, it introduces scales $\sqrt{\lambda_i}$ along each axis independently, rotates the axes by $\mathbf{\Gamma}$, and shifts by the vector $\boldsymbol{\mu}$. It is interesting to note that (3.10) differs from (3.3) by a rotation $\mathbf{\Gamma}^\top$.

Finally, another approach can be based on the *Choleski factorization* of a positive definite matrix $\mathbf{\Sigma} = \mathbf{L}\mathbf{L}^\top$, where $\mathbf{L}$ is a lower triangular matrix with positive diagonal entries; see Equation (A.19f). Then the vector $\mathbf{X} = \mathbf{L}\mathbf{Z} + \boldsymbol{\mu}$ will have the distribution $\mathbf{X} \sim \mathcal{N}_p(\boldsymbol{\mu}, \mathbf{\Sigma})$. From a computational point of view, the Choleski factorization is interesting because the matrix $\mathbf{L}$ is sparse. It has zeros in its upper triangular part, hence it requires less storage space, and computationally efficient procedures can be used.

3.9 Quadratic Forms in Normal Random Vectors and Related Properties

Quadratic forms play a central role in the normal theory associated with different statistical areas, in particular with linear models. In this section, we review some basic aspects of the distribution of quadratic forms in normal variables.

Given a symmetric matrix $\mathbf{A}$ of dimension $p \times p$ and a normal random vector $\mathbf{X} \sim \mathcal{N}_p(\boldsymbol{\mu}, \mathbf{\Sigma})$, the random variable defined by $Q_{\mathbf{A}}(\mathbf{X}) = \mathbf{X}^\top \mathbf{A}\mathbf{X}$ will be referred to as a *quadratic form* (in normal variables). Two more general structures are $Q_{\mathbf{A}}(\mathbf{X}-\mathbf{a}) = (\mathbf{X}-\mathbf{a})^\top \mathbf{A}(\mathbf{X}-\mathbf{a}) = Q_{\mathbf{A}}(\mathbf{X}) - 2\mathbf{a}^\top \mathbf{A}\mathbf{X} + \mathbf{a}^\top \mathbf{A}\mathbf{a}$ for some $\mathbf{a} \in \mathbb{R}^p$ and the *bilinear quadratic form* $\mathbf{X}^\top \mathbf{B}\mathbf{Y}$, where $\mathbf{X}$ is $p \times 1$, $\mathbf{Y}$ is $q \times 1$ and $\mathbf{B}$ is $p \times q$. In the last case, note that if $p = q$ and $\mathbf{B}$ is symmetric, then

$$\mathbf{X}^\top \mathbf{B}\mathbf{Y} = \mathbf{Z}^\top \mathbf{A}\mathbf{Z} \ \text{ for } \ \mathbf{Z} = \begin{pmatrix} \mathbf{X} \\ \mathbf{Y} \end{pmatrix}, \quad \mathbf{A} = \frac{1}{2}\begin{pmatrix} \mathbf{O} & \mathbf{B} \\ \mathbf{B} & \mathbf{O} \end{pmatrix}.$$

Even when the distribution of $\mathbf{X}$ is tractable, $Q_{\mathbf{A}}(\mathbf{X})$ (or $Q_{\mathbf{A}}(\mathbf{X}-\mathbf{a})$) is a very complicated function of $\mathbf{X}$, although it has a structure that can be expressed in simpler terms; this is the key to studying its distribution.

3.9.1 Moments of quadratic forms in normal random vectors

Two important and perhaps quite well-known results are related to the mean and variance of $Q_{\mathbf{A}}(\mathbf{X})$. In general, the k-th moment of $Q_{\mathbf{A}}(\mathbf{X})$ can also be obtained by using (1.8) for which we need the $2k$-th moment of $\mathbf{X}$ defined by (see also (1.7)):

$$\boldsymbol{\mu}_{2k}(\mathbf{X}) = \mathrm{E}(\underbrace{\mathbf{X}\mathbf{X}^\top \otimes \cdots \otimes \mathbf{X}\mathbf{X}^\top}_{k-\text{times}}).$$

Theorem 8 *Let $\boldsymbol{\mu}_k(\mathbf{X})$ be the k-th moment of $\mathbf{X} \sim \mathcal{N}_p(\boldsymbol{\mu}, \mathbf{\Sigma})$ as defined in (1.7). Then, for $k = 1,2,3,4$, we have:*

(i) $\boldsymbol{\mu}_1(\mathbf{X}) = \boldsymbol{\mu}$;

(ii) $\boldsymbol{\mu}_2(\mathbf{X}) = \mathbf{\Sigma} + \boldsymbol{\mu}\boldsymbol{\mu}^\top$;

(iii) $\boldsymbol{\mu}_3(\mathbf{X}) = (\mathbf{I}_{p^2} + \mathbf{K}_p)(\mathbf{\Sigma} \otimes \boldsymbol{\mu}) + \mathrm{vec}(\mathbf{\Sigma} + \boldsymbol{\mu}\boldsymbol{\mu}^\top)\boldsymbol{\mu}^\top$;

(iv) $\boldsymbol{\mu}_4(\mathbf{X}) = (\mathbf{I}_{p^2} + \mathbf{K}_p)\{(\mathbf{\Sigma} \otimes \mathbf{\Sigma}) + (\mathbf{\Sigma} \otimes \boldsymbol{\mu}\boldsymbol{\mu}^\top) + (\boldsymbol{\mu}\boldsymbol{\mu}^\top \otimes \mathbf{\Sigma})\} + \mathrm{vec}(\mathbf{\Sigma} + \boldsymbol{\mu}\boldsymbol{\mu}^\top)\mathrm{vec}(\mathbf{\Sigma} + \boldsymbol{\mu}\boldsymbol{\mu}^\top)^\top$;

(v) $\mathrm{var}(\mathbf{X} \otimes \mathbf{X}) = (\mathbf{I}_{p^2} + \mathbf{K}_p)\{(\mathbf{\Sigma} \otimes \mathbf{\Sigma}) + (\mathbf{\Sigma} \otimes \boldsymbol{\mu}\boldsymbol{\mu}^\top) + (\boldsymbol{\mu}\boldsymbol{\mu}^\top \otimes \mathbf{\Sigma})\}$,

where $\mathbf{K}_p$ is the commutation matrix, a $p^2 \times p^2$ block-matrix whose block in position (i, j) is $\mathbf{e}_j\mathbf{e}_i^\top$, with $\mathbf{e}_i$ being the i-th p-dimensional unit vector.

Proof 8 *The results for $\boldsymbol{\mu}_1(\mathbf{X})$ and $\boldsymbol{\mu}_2(\mathbf{X})$ are obvious. To prove (iii) to (v), write $\mathbf{X} = \mathbf{C}\mathbf{Z} + \boldsymbol{\mu}$, where $\mathbf{C}\mathbf{C}^\top = \mathbf{\Sigma}$ and $\mathbf{Z} \sim \mathcal{N}_p(\mathbf{0}_p, \mathbf{I}_p)$. Obviously, $\boldsymbol{\mu}_1(\mathbf{Z}) = \mathbf{0}_p$, $\boldsymbol{\mu}_2(\mathbf{Z}) = \mathrm{var}(\mathbf{Z}) = \mathbf{I}_p$ and*

$\boldsymbol{\mu}_3(\mathbf{Z}) = \mathbf{O}$ *by symmetry, and from the results in Example 17 we also find that* $\boldsymbol{\mu}_4(\mathbf{Z}) = \mathbf{I}_{p^2} + \mathbf{K}_p + \text{vec}(\mathbf{I}_\text{p})\text{vec}(\mathbf{I}_\text{p})^\top$. *Now, note that*

$$\begin{aligned}
\mathbf{X}\mathbf{X}^\top =& \mathbf{C}\mathbf{Z}\mathbf{Z}^\top\mathbf{C}^\top + \mathbf{C}\mathbf{Z}\boldsymbol{\mu}^\top + \boldsymbol{\mu}\mathbf{Z}^\top\mathbf{C}^\top + \boldsymbol{\mu}\boldsymbol{\mu}^\top \\
\implies (\mathbf{X}\otimes\mathbf{X}) =& \text{vec}(\mathbf{X}\mathbf{X}^\top) = (\mathbf{C}\otimes\mathbf{C})\text{vec}(\mathbf{Z}\mathbf{Z}^\top) + (\mathbf{I}_{\text{p}^2} + \mathbf{K}_\text{p})(\mathbf{C}\otimes\boldsymbol{\mu})\mathbf{Z} + \text{vec}(\boldsymbol{\mu}\boldsymbol{\mu}^\top) \\
\implies \text{var}\{\text{vec}(\mathbf{X}\mathbf{X}^\top)\} =& (\mathbf{C}\otimes\mathbf{C})\text{var}\{\text{vec}(\mathbf{Z}\mathbf{Z}^\top)\}(\mathbf{C}\otimes\mathbf{C})^\top + (\mathbf{I}_{\text{p}^2} + \mathbf{K}_\text{p})(\mathbf{C}\otimes\boldsymbol{\mu})\text{var}(\mathbf{Z})(\mathbf{C}\otimes\boldsymbol{\mu})^\top \\
& \times(\mathbf{I}_{p^2} + \mathbf{K}_p)^\top + (\mathbf{C}\otimes\mathbf{C})\text{cov}\{\text{vec}(\mathbf{Z}\mathbf{Z}^\top), \mathbf{Z}\}(\mathbf{C}\otimes\boldsymbol{\mu})^\top(\mathbf{I}_{\text{p}^2} + \mathbf{K}_\text{p})^\top \\
& + (\mathbf{I}_{p^2} + \mathbf{K}_p)(\mathbf{C}\otimes\boldsymbol{\mu})\text{cov}\{\mathbf{Z}, \text{vec}(\mathbf{Z}\mathbf{Z}^\top)\}(\mathbf{C}\otimes\mathbf{C})^\top \\
=& (\mathbf{C}\otimes\mathbf{C})(\mathbf{I}_{p^2} + \mathbf{K}_p)(\mathbf{C}\otimes\mathbf{C})^\top + (\mathbf{I}_{p^2} + \mathbf{K}_p)(\mathbf{C}\otimes\boldsymbol{\mu})(\mathbf{C}\otimes\boldsymbol{\mu})^\top(\mathbf{I}_{p^2} + \mathbf{K}_p) \\
=& (\mathbf{I}_{p^2} + \mathbf{K}_p)(\mathbf{C}\mathbf{C}^\top\otimes\mathbf{C}\mathbf{C}^\top) + (\mathbf{I}_{p^2} + \mathbf{K}_p)(\mathbf{C}\mathbf{C}^\top\otimes\boldsymbol{\mu}\boldsymbol{\mu}^\top)(\mathbf{I}_{p^2} + \mathbf{K}_p) \\
=& (\mathbf{I}_{p^2} + \mathbf{K}_p)\{(\boldsymbol{\Sigma}\otimes\boldsymbol{\Sigma}) + (\boldsymbol{\Sigma}\otimes\boldsymbol{\mu}\boldsymbol{\mu}^\top) + (\boldsymbol{\mu}\boldsymbol{\mu}^\top\otimes\boldsymbol{\Sigma})\},
\end{aligned}$$

since $\text{cov}\{\mathbf{Z}, \text{vec}(\mathbf{Z}\mathbf{Z}^\top)\} = \text{cov}\{\text{vec}(\mathbf{Z}\mathbf{Z}^\top), \mathbf{Z}\}^\top = \boldsymbol{\mu}_3(\mathbf{Z}) - \boldsymbol{\mu}_2(\mathbf{Z})\boldsymbol{\mu}_1(\mathbf{Z}) = \mathbf{O}$, $(\mathbf{C}\otimes\mathbf{C})\mathbf{K}_p = \mathbf{K}_p(\mathbf{C}\otimes\mathbf{C})$ *and* $(\mathbf{I}_{p^2} + \mathbf{K}_p)(\boldsymbol{\Sigma}\otimes\boldsymbol{\mu}\boldsymbol{\mu}^\top)(\mathbf{I}_{p^2} + \mathbf{K}_p) = (\mathbf{I}_{p^2} + \mathbf{K}_p)(\boldsymbol{\Sigma}\otimes\boldsymbol{\mu}\boldsymbol{\mu}^\top + \boldsymbol{\mu}\boldsymbol{\mu}^\top\otimes\boldsymbol{\Sigma})$. *This proves* (*v*). *Using* $\boldsymbol{\mu}_4(\mathbf{X}) = \text{var}\{\text{vec}(\mathbf{X}\mathbf{X}^\top)\} + \text{vec}\{\boldsymbol{\mu}_2(\mathbf{X})\}\text{vec}\{\boldsymbol{\mu}_2(\mathbf{X})\}^\top$, *we have the proof of* (*iv*). *To prove* (*iii*), *we use that*

$$\begin{aligned}
(\mathbf{X}\otimes\mathbf{X}^\top\otimes\mathbf{X}) =& (\mathbf{X}\mathbf{X}^\top\otimes\mathbf{X}) \\
=& (\mathbf{C}\otimes\mathbf{C})(\mathbf{Z}\mathbf{Z}^\top\otimes\mathbf{Z})\mathbf{C}^\top + (\mathbf{C}\otimes\boldsymbol{\mu})\mathbf{Z}\mathbf{Z}^\top\mathbf{C}^\top + (\mathbf{C}\otimes\mathbf{C})(\mathbf{Z}\otimes\mathbf{Z})\boldsymbol{\mu}^\top + (\mathbf{C}\otimes\boldsymbol{\mu})\mathbf{Z}\boldsymbol{\mu}^\top \\
& + (\boldsymbol{\mu}\otimes\mathbf{C})\mathbf{Z}\mathbf{Z}^\top\mathbf{C}^\top + (\boldsymbol{\mu}\otimes\boldsymbol{\mu})\mathbf{Z}^\top\mathbf{C}^\top + (\boldsymbol{\mu}\boldsymbol{\mu}^\top\otimes\mathbf{C}\mathbf{Z}) + (\boldsymbol{\mu}\boldsymbol{\mu}^\top\otimes\boldsymbol{\mu}) \\
\implies \boldsymbol{\mu}_3(\mathbf{X}) =& (\mathbf{C}\mathbf{C}^\top\otimes\boldsymbol{\mu}) + (\mathbf{C}\otimes\mathbf{C})\text{vec}(\mathbf{I}_\text{p})\boldsymbol{\mu}^\top + (\boldsymbol{\mu}\otimes\mathbf{C}\mathbf{C}^\top) + (\boldsymbol{\mu}\boldsymbol{\mu}^\top\otimes\boldsymbol{\mu}) \\
=& (\boldsymbol{\Sigma}\otimes\boldsymbol{\mu}) + \text{vec}(\boldsymbol{\Sigma})\boldsymbol{\mu}^\top + (\boldsymbol{\mu}\otimes\boldsymbol{\Sigma}) + (\boldsymbol{\mu}\boldsymbol{\mu}^\top\otimes\boldsymbol{\mu}) \\
=& (\mathbf{I}_{p^2} + \mathbf{K}_p)(\boldsymbol{\Sigma}\otimes\boldsymbol{\mu}) + \text{vec}(\boldsymbol{\Sigma} + \boldsymbol{\mu}\boldsymbol{\mu}^\top)\boldsymbol{\mu}^\top,
\end{aligned}$$

since $\boldsymbol{\mu}_k(\mathbf{Z}) = \mathbf{O}$ *for k odd. The theorem is proved.* □

Theorem 9 *Let* $\mathbf{X} \sim \mathcal{N}_p(\boldsymbol{\mu}, \boldsymbol{\Sigma})$. *Then:*

(*i*) $\text{E}(\mathbf{X}^\top\mathbf{A}\mathbf{X}) = \text{tr}(\mathbf{A}\boldsymbol{\Sigma}) + \boldsymbol{\mu}^\top\mathbf{A}\boldsymbol{\mu}$;

(*ii*) $\text{cov}(\mathbf{C}\mathbf{X}, \mathbf{X}^\top\mathbf{A}\mathbf{X}) = 2\mathbf{C}\boldsymbol{\Sigma}\mathbf{A}\boldsymbol{\mu}, \ \mathbf{C} \in \mathbb{R}^{k\times p}$;

(*iii*) $\text{cov}(\mathbf{X}^\top\mathbf{A}\mathbf{X}, \mathbf{X}^\top\mathbf{B}\mathbf{X}) = 2\text{tr}(\mathbf{A}\boldsymbol{\Sigma}\mathbf{B}\boldsymbol{\Sigma}) + 4\boldsymbol{\mu}^\top\mathbf{A}\boldsymbol{\Sigma}\mathbf{B}\boldsymbol{\mu}$;

(*iv*) $\text{var}(\mathbf{X}^\top\mathbf{A}\mathbf{X}) = 2\text{tr}\{(\mathbf{A}\boldsymbol{\Sigma})^2\} + 4\boldsymbol{\mu}^\top\mathbf{A}\boldsymbol{\Sigma}\mathbf{A}\boldsymbol{\mu}$.

Proof 9 *Let* $Q_\mathbf{A}(\mathbf{X}) = \mathbf{X}^\top\mathbf{A}\mathbf{X}$. *Since* $Q_\mathbf{A}(\mathbf{X}) = \text{tr}\{Q_\mathbf{A}(\mathbf{X})\} = \text{tr}(\mathbf{A}\mathbf{X}\mathbf{X}^\top)$, *then* $\text{E}\{Q_\mathbf{A}(\mathbf{X})\} = \text{E}\{\text{tr}(\mathbf{A}\mathbf{X}\mathbf{X}^\top)\} = \text{tr}\{\text{E}(\mathbf{A}\mathbf{X}\mathbf{X}^\top)\} = \text{tr}\{\mathbf{A}\text{E}(\mathbf{X}\mathbf{X}^\top)\}$, *where* $\text{E}(\mathbf{X}\mathbf{X}^\top) = \text{var}(\mathbf{X}) + \text{E}(\mathbf{X})\text{E}(\mathbf{X})^\top = \boldsymbol{\Sigma} + \boldsymbol{\mu}\boldsymbol{\mu}^\top$, *and so* $\text{E}\{Q_\mathbf{A}(\mathbf{X})\} = \text{tr}\{\mathbf{A}(\boldsymbol{\Sigma} + \boldsymbol{\mu}\boldsymbol{\mu}^\top)\} = \text{tr}(\mathbf{A}\boldsymbol{\Sigma}) + \text{tr}(\mathbf{A}\boldsymbol{\mu}\boldsymbol{\mu}^\top)$, *which proves the part* (*i*) *since* $\text{tr}(\mathbf{A}\boldsymbol{\mu}\boldsymbol{\mu}^\top) = \text{tr}(\boldsymbol{\mu}^\top\mathbf{A}\boldsymbol{\mu}) = \boldsymbol{\mu}^\top\mathbf{A}\boldsymbol{\mu}$.

To prove part (ii), we first use the covariance properties and the fact that $\mathrm{tr}(\mathbf{P}^\top\mathbf{Q}) = \mathrm{vec}(\mathbf{P})^\top\mathrm{vec}(\mathbf{Q})$ *to obtain*

$$\begin{aligned}
\mathrm{cov}(\mathbf{CX},\mathbf{X}^\top\mathbf{AX}) &= \mathrm{cov}\{\mathbf{CX},\mathrm{tr}(\mathbf{AXX}^\top)\}\\
&= \mathrm{cov}\{\mathbf{CX},\mathrm{vec}(\mathbf{A})^\top\mathrm{vec}(\mathbf{XX}^\top)\}\\
&= \mathbf{C}\mathrm{cov}\{\mathbf{X},\mathrm{vec}(\mathbf{XX}^\top)\}\mathrm{vec}(\mathbf{A})\\
&= \mathbf{C}\{\boldsymbol{\mu}_3(\mathbf{X})^\top - \boldsymbol{\mu}_1(\mathbf{X})\boldsymbol{\mu}_2(\mathbf{X})^\top\}\mathrm{vec}(\mathbf{A})\\
&= \mathbf{C}(\boldsymbol{\Sigma}\otimes\boldsymbol{\mu}^\top)(\mathbf{I}_{p^2}+\mathbf{K}_p)\mathrm{vec}(\mathbf{A})\\
&= 2\mathbf{C}\mathrm{vec}(\boldsymbol{\Sigma}\mathbf{A}\boldsymbol{\mu}) = 2\mathbf{C}\boldsymbol{\Sigma}\mathbf{A}\boldsymbol{\mu},
\end{aligned}$$

since the parts (i), (ii), and (iv) of Theorem 8 imply that $\mathrm{cov}\{\mathbf{X},\mathrm{vec}(\mathbf{XX}^\top)\} = \boldsymbol{\mu}_3(\mathbf{X})^\top - \boldsymbol{\mu}_1(\mathbf{X})\boldsymbol{\mu}_2(\mathbf{X})^\top = (\boldsymbol{\Sigma}\otimes\boldsymbol{\mu}^\top)(\mathbf{I}_{p^2}+\mathbf{K}_p)$*, and we used that* $\mathbf{K}_{pq}\mathrm{vec}(\mathbf{R}) = \mathrm{vec}(\mathbf{R}^\top)$ *for any* $p\times q$ *matrix* $\mathbf{R}$*,* $\mathrm{vec}(\mathbf{b}) = \mathrm{vec}(\mathbf{b}^\top) = \mathbf{b}$ *for any column vector* $\mathbf{b}$ *and* $\mathrm{vec}(\mathbf{PUQ}) = (\mathbf{Q}^\top\otimes\mathbf{P})\mathrm{vec}(\mathbf{U})$ *for any comformable matrices* $\mathbf{P}$*,* $\mathbf{U}$ *and* $\mathbf{Q}$*.*

Similarly to the proof of part (ii), for part (iii), we have

$$\begin{aligned}
\mathrm{cov}(\mathbf{X}^\top\mathbf{AX},\mathbf{X}^\top\mathbf{BX}) &= \mathrm{cov}\{\mathrm{tr}(\mathbf{AXX}^\top),\mathrm{tr}(\mathbf{XX}^\top\mathbf{B})\}\\
&= \mathrm{cov}\{\mathrm{vec}(\mathbf{A})^\top\mathrm{vec}(\mathbf{XX}^\top),\mathrm{vec}(\mathbf{B})^\top\mathrm{vec}(\mathbf{XX}^\top)\}\\
&= \mathrm{vec}(\mathbf{A})^\top\mathrm{var}\{\mathrm{vec}(\mathbf{XX}^\top)\}\mathrm{vec}(\mathbf{B}),
\end{aligned}$$

where $\mathrm{var}\{\mathrm{vec}(\mathbf{XX}^\top)\} = (\mathbf{I}_{p^2}+\mathbf{K}_p)\{(\boldsymbol{\Sigma}\otimes\boldsymbol{\Sigma})+(\boldsymbol{\Sigma}\otimes\boldsymbol{\mu}\boldsymbol{\mu}^\top)+(\boldsymbol{\mu}\boldsymbol{\mu}^\top\otimes\boldsymbol{\Sigma})\} = (\mathbf{I}_{p^2}+\mathbf{K}_p)\{(\boldsymbol{\Sigma}\otimes\boldsymbol{\Sigma})+(\mathbf{I}_{p^2}+\mathbf{K}_p)(\boldsymbol{\mu}\boldsymbol{\mu}^\top\otimes\boldsymbol{\Sigma})\}$ *by part (iv) of Theorem 8. So,*

$$\begin{aligned}
\mathrm{cov}(\mathbf{X}^\top\mathbf{AX},\mathbf{X}^\top\mathbf{BX}) &= \mathrm{vec}(\mathbf{A})^\top(\mathbf{I}_{p^2}+\mathbf{K}_p)\{(\boldsymbol{\Sigma}\otimes\boldsymbol{\Sigma})+(\mathbf{I}_{p^2}+\mathbf{K}_p)(\boldsymbol{\mu}\boldsymbol{\mu}^\top\otimes\boldsymbol{\Sigma})\}\mathrm{vec}(\mathbf{B})\\
&= 2\mathrm{vec}(\mathbf{A})^\top\{(\boldsymbol{\Sigma}\otimes\boldsymbol{\Sigma})\mathrm{vec}(\mathbf{B})+(\mathbf{I}_{p^2}+\mathbf{K}_p)(\boldsymbol{\mu}\boldsymbol{\mu}^\top\otimes\boldsymbol{\Sigma})\mathrm{vec}(\mathbf{B})\}\\
&= 2\mathrm{vec}(\mathbf{A})^\top\{\mathrm{vec}(\boldsymbol{\Sigma}\mathbf{B}\boldsymbol{\Sigma})+(\mathbf{I}_{p^2}+\mathbf{K}_p)\mathrm{vec}(\boldsymbol{\Sigma}\mathbf{B}\boldsymbol{\mu}\boldsymbol{\mu}^\top)\}\\
&= 2\mathrm{vec}(\mathbf{A})^\top\mathrm{vec}(\boldsymbol{\Sigma}\mathbf{B}\boldsymbol{\Sigma})+4\mathrm{vec}(\mathbf{A})^\top\mathrm{vec}(\boldsymbol{\Sigma}\mathbf{B}\boldsymbol{\mu}\boldsymbol{\mu}^\top)\\
&= 2\mathrm{tr}(\mathbf{A}^\top\boldsymbol{\Sigma}\mathbf{B}\boldsymbol{\Sigma})+4\mathrm{tr}(\mathbf{A}^\top\boldsymbol{\Sigma}\mathbf{B}\boldsymbol{\mu}\boldsymbol{\mu}^\top)\\
&= 2\mathrm{tr}(\mathbf{A}\boldsymbol{\Sigma}\mathbf{B}\boldsymbol{\Sigma})+4\boldsymbol{\mu}^\top\mathbf{A}\boldsymbol{\Sigma}\mathbf{B}\boldsymbol{\mu},
\end{aligned}$$

where we used again that $\mathrm{vec}(\mathbf{A})^\top(\mathbf{I}_{p^2}+\mathbf{K}_p) = 2\mathrm{vec}(\mathbf{A})^\top$ *and that* $\mathrm{vec}(\mathbf{C})^\top\mathrm{vec}(\mathbf{D}) = \mathrm{tr}(\mathbf{C}^\top\mathbf{D})$*.*

Finally, the proof of part (iv) is directly derived from part (iii) by letting $\mathbf{B} = \mathbf{A}$*. Therefore, the proof is completed.* □

An alternative proof of these results can also be obtained from the general facts (1.8), (1.9), and (1.10) presented in Chapter 1.

Example 28 *Let* $\mathbf{X} = (X_1,\dots,X_n)^\top$ *be* $\mathcal{N}_n(\mu\mathbf{1}_n,\sigma^2\mathbf{I}_n)$*, which is equivalent to* $X_1,\dots,X_n$ *being a random sample from* $\mathcal{N}(\mu,\sigma^2)$*. In addition,* $\bar{X} = n^{-1}\sum_{i=1}^n X_i$ *is the sample mean and* $S_X^2 = (n-1)^{-1}\sum_{i=1}^n (X_i-\bar{X})^2$ *is the sample variance. Since* $(n-1)S_X^2 = \sum_{i=1}^n (X_i-\bar{X})^2 = \mathbf{X}^\top\mathbf{QX}$*, where* $\mathbf{Q} = \mathbf{I}_n - n^{-1}\mathbf{1}_n\mathbf{1}_n^\top$ *is an* $n\times n$ *orthogonal projection matrix (i.e.,* $\mathbf{Q}$ *is symmetric and idempotent,* $\mathbf{Q}^2 = \mathbf{Q}$*) of* $\mathrm{rank}(\mathbf{Q}) = \mathrm{tr}(\mathbf{Q}) = n-1$*, then* $\mathrm{E}(S_X^2) = (n-1)^{-1}\mathrm{E}(\mathbf{X}^\top\mathbf{QX}) = (n-1)^{-1}\sigma^2\mathrm{tr}(\mathbf{Q}) = \sigma^2$ *and* $\mathrm{var}(S_X^2) = (n-1)^{-2}\mathrm{var}(\mathbf{X}^\top\mathbf{QX}) = (n-1)^{-2}\sigma^4\mathrm{tr}(\mathbf{Q}^2) = 2(n-1)^{-1}\sigma^4$ *because* $\mathbf{Q1}_n = \mathbf{0}_n$*.*

3.9.2 Non-central chi-squared distribution

Definition 2 [Univariate non-central chi-squared distribution] *A random variable Y is said to have a non-central chi-squared distribution, $Y \sim \chi^2_m(\delta)$, iff $Y \overset{d}{=} \sum_{i=1}^m X_i^2$, where $X_i \overset{ind.}{\sim} \mathcal{N}(\mu_i, 1)$, $i = 1, \ldots, m$, and $\delta = \sum_{i=1}^m \mu_i^2/2$. The central χ^2_m follows for $\delta = 0$.*

Theorem 10 *If $Y \sim \chi^2_m(\delta)$, then $\mathrm{E}(Y) = m + 2\delta$, $\mathrm{var}(Y) = 2m + 8\delta$ and its cf is $\psi_Y(t) = (1 - 2\mathrm{i}t)^{-m/2} e^{2\mathrm{i}\delta t/(1-2\mathrm{i}t)}$.*

Proof 10 *The first two results are simple using that $\mathrm{E}(Y) = \sum_{i=1}^m \mathrm{E}(X_i^2)$ and $\mathrm{var}(Y) = \sum_{i=1}^m \mathrm{var}(X_i^2)$ by independence, with $\mathrm{E}(X_i^2) = 1 + \mu_i^2$ and $\mathrm{var}(X_i^2) = 2 + 4\mu_i^2$ for $i = 1, \ldots, m$. To prove the third result, note by definition that if $Z_1, \ldots, Z_m \overset{iid}{\sim} \mathcal{N}(0, 1)$, then $Y = (Z_1 + \mu)^2 + \sum_{i=2}^m Z_i^2 \sim \chi^2_m(\delta)$, where $\delta = \mu^2/2$. Hence, $\psi_Y(t) = \psi_{(Z_1+\mu)^2}(t)\psi_{\sum_{i=2}^m Z_i^2}(t)$, where $\psi_{\sum_{i=2}^m Z_i^2}(t) = (1-2\mathrm{i}t)^{-(m-1)/2}$ since $\sum_{i=2}^m Z_i^2 \sim \chi^2_{m-1}$ and, after some simple calculations,*

$$\begin{aligned}
\psi_{(Z_1+\mu)^2}(t) &= \mathrm{E}(e^{\mathrm{i}t(Z_1^2+2\mu Z_1+\mu^2)}) \\
&= \int_{-\infty}^{\infty} e^{\mathrm{i}t(z^2+2\mu z+\mu^2)} \phi(z) \mathrm{d}z \\
&= (1-2\mathrm{i}t)^{-1/2} e^{-\frac{2\mathrm{i}t\delta}{1-2\mathrm{i}t}} \int_{-\infty}^{\infty} \phi\left(z; \frac{2\mathrm{i}t\mu}{1-2\mathrm{i}t}, \frac{1}{1-2\mathrm{i}t}\right) \mathrm{d}z,
\end{aligned}$$

where $\phi(x; \mu, \sigma^2)$ is the $\mathcal{N}(\mu, \sigma^2)$ pdf. The conclusion follows since this last integral equals one. □

Theorem 11 *Let $\mathbf{Z} \sim \mathcal{N}_p(\boldsymbol{\mu}, \mathbf{I}_p)$ and $\mathbf{Q}$ be a p-by-p orthogonal projection matrix of rank m. Then $\mathbf{Z}^\top\mathbf{Q}\mathbf{Z} \sim \chi^2_m(\delta)$, the non-central χ^2 distribution with m degrees of freedom and the non-centrality parameter $\delta = \boldsymbol{\mu}^\top\mathbf{Q}\boldsymbol{\mu}/2$.*

Proof 11 *Write $\mathbf{Q} = \mathbf{H}\mathbf{H}^\top$, where $\mathbf{H}$ is a $p \times m$ matrix of rank m such that $\mathbf{H}^\top\mathbf{H} = \mathbf{I}_m$. Then $\mathbf{Z}^\top\mathbf{Q}\mathbf{Z} = ||\mathbf{H}^\top\mathbf{Z}||^2 = ||\mathbf{Z}_0||^2$, with $\mathbf{Z}_0 = \mathbf{H}^\top\mathbf{Z} \sim \mathcal{N}_m(\mathbf{H}^\top\boldsymbol{\mu}, \mathbf{I}_m)$, and so $||\mathbf{Z}_0||^2 \sim \chi^2_m(\delta)$ with $\delta = ||\mathbf{H}^\top\boldsymbol{\mu}||^2/2$.* □

Theorem 12 *Assume that $\mathbf{X} \sim \mathcal{N}_p(\boldsymbol{\mu}, \boldsymbol{\Sigma})$ with $\boldsymbol{\Sigma} > 0$, and $\mathbf{A}$ is a p-by-p symmetric matrix such that $\mathbf{A}\boldsymbol{\Sigma}\mathbf{A} = \mathbf{A}$ and $\mathrm{rank}(\boldsymbol{\Sigma}\mathbf{A}) = m$. Then, $\mathbf{X}^\top\mathbf{A}\mathbf{X} \sim \chi^2_m(\delta)$, where $\delta = \boldsymbol{\mu}^\top\mathbf{A}\boldsymbol{\mu}/2$.*

Proof 12 *Let $\mathbf{Z} = \boldsymbol{\Sigma}^{-1/2}\mathbf{X}$ and $\mathbf{B} = \boldsymbol{\Sigma}^{1/2}\mathbf{A}\boldsymbol{\Sigma}^{1/2}$. Then, $\mathbf{X}^\top\mathbf{A}\mathbf{X} = \mathbf{Z}^\top\mathbf{B}\mathbf{Z}$, where $\mathbf{Z} \sim \mathcal{N}_p(\boldsymbol{\Sigma}^{-1/2}\boldsymbol{\mu}, \mathbf{I}_p)$, and the result follows from Theorem 11 since $\mathbf{B}$ is an orthogonal projection matrix of $\mathrm{rank}(\mathbf{B}) = \mathrm{rank}(\boldsymbol{\Sigma}\mathbf{A}) = \mathrm{rank}(\mathbf{A}) = m$ because $\boldsymbol{\Sigma}$ is nonsingular.* □

Example 29 *Under the conditions of Example 28 it follows that $(n-1)S_X^2/\sigma^2 = \mathbf{X}^\top(\sigma^{-2}\mathbf{Q})\mathbf{X} \sim \chi^2_{n-1}$ for each $n \geq 2$ since $\mathbf{Q}\mathbf{1}_n = \mathbf{0}_n$ and so $\delta = \boldsymbol{\mu}^\top(\sigma^{-2}\mathbf{Q})\boldsymbol{\mu}/2 = \mu^2\mathbf{1}_n^\top\mathbf{Q}\mathbf{1}_n/2\sigma^2 = 0$.*

3.9.3 Further properties related to quadratic forms

Theorem 13 *Assume that* $\mathbf{X} \sim \mathcal{N}_p(\boldsymbol{\mu}, \boldsymbol{\Sigma})$*. Then,* $\mathbf{X}^\top \mathbf{A}\mathbf{X} \perp\!\!\!\perp \mathbf{B}\mathbf{X}$ *iff* $\mathbf{B}\boldsymbol{\Sigma}\mathbf{A} = \mathbf{O}$.

Proof 13 *Assume that* $\mathbf{A}$ *has rank* q*. Since* $\mathbf{A}$ *is symmetric, we can always write* $\mathbf{A} = \mathbf{C}\mathbf{C}^\top$ *for some* p*-by-*q *matrix* $\mathbf{C}$ *of rank* q*. So,* $\mathbf{X}^\top\mathbf{A}\mathbf{X} = \|\mathbf{C}\mathbf{X}\|^2$*, where, by Theorem 5,* $\mathbf{C}\mathbf{X}$ *is independent of* $\mathbf{B}\mathbf{X}$ *iff* $\mathbf{B}\boldsymbol{\Sigma}\mathbf{C} = \mathbf{O}$*, which is equivalent to* $\mathbf{B}\boldsymbol{\Sigma}\mathbf{C}\mathbf{C}^\top = \mathbf{B}\boldsymbol{\Sigma}\mathbf{A} = \mathbf{O}$. □

Theorem 14 *Assume that* $\mathbf{X} \sim \mathcal{N}_p(\boldsymbol{\mu}, \boldsymbol{\Sigma})$*. Then,* $\mathbf{X}^\top \mathbf{A}\mathbf{X} \perp\!\!\!\perp \mathbf{X}^\top\mathbf{B}\mathbf{X}$ *iff* $\mathbf{A}\boldsymbol{\Sigma}\mathbf{B} = \mathbf{O}$.

Proof 14 *Since* $\mathbf{X}^\top\mathbf{A}\mathbf{X} = \|\mathbf{C}\mathbf{X}\|^2$ *and* $\mathbf{X}^\top\mathbf{D}\mathbf{X} = \|\mathbf{D}\mathbf{X}\|^2$*, where* $\mathbf{C}\mathbf{C}^\top = \mathbf{A}$ *and* $\mathbf{D}\mathbf{D}^\top = \mathbf{B}$*, the proof follows again from Theorem 5.* □

Example 30 *Assume that* $\mathbf{X} = (X_1, \ldots, X_n)^\top \sim \mathcal{N}_n(\mu\mathbf{1}_n, \sigma^2\mathbf{I}_n)$ *and consider the orthogonal projection matrices* $\mathbf{P} = \mathbf{1}_n(\mathbf{1}_n^\top\mathbf{1}_n)^{-1}\mathbf{1}_n^\top = n^{-1}\mathbf{1}_n\mathbf{1}_n^\top$ *and* $\mathbf{Q} = \mathbf{I}_n - \mathbf{P}$ *of rank* 1 *and* $n-1$*, respectively. Since* $\mathbf{P}\mathbf{1}_n = \mathbf{1}_n$ *and* $\mathbf{Q}\mathbf{1}_n = \mathbf{0}_n$*, it is clear that* $\mathbf{P}\mathbf{Q} = \mathbf{O}$*. Hence,* $\mathbf{P}\mathbf{X}$ *and* $\mathbf{Q}\mathbf{X}$ *are independent, with* $\mathbf{P}\mathbf{X} \sim \mathcal{N}_n(\mu\mathbf{1}_n, \sigma^2\mathbf{P})$ *and* $\mathbf{Q}\mathbf{X} \sim \mathcal{N}_n(\mathbf{0}_n, \sigma^2\mathbf{Q})$*. Moreover, since* $\mathbf{P}\mathbf{X}$ *and* $\mathbf{Q}\mathbf{X}$ *are projections in orthogonal complementary subspaces of* $\mathbb{R}^n$ *of dimensions* 1 *and* $n-1$*, respectively, and* $\mathbf{X} = \mathbf{P}\mathbf{X} + \mathbf{Q}\mathbf{X}$*, then by the Pythagorean theorem* $\|\mathbf{X}\|^2 = \|\mathbf{P}\mathbf{X}\|^2 + \|\mathbf{Q}\mathbf{X}\|^2 = \mathbf{X}^\top\mathbf{P}\mathbf{X} + \mathbf{X}^\top\mathbf{Q}\mathbf{X}$ *which is equivalent to* $\sum_{i=1}^n X_i^2 = n\bar{X}^2 + (n-1)S_X^2$*, where* $\bar{X}$ *and* S_X^2 *are the sample mean and sample variance of* $X_1, \ldots, X_n$*, respectively. It is again clear that* $\mathbf{X}^\top\mathbf{P}\mathbf{X} = n\bar{X}^2$ *and* $\mathbf{X}^\top\mathbf{Q}\mathbf{X} = (n-1)S_X^2$ *are independent, as are* $\bar{X} = n^{-1}\mathbf{1}_n^\top\mathbf{P}\mathbf{X}$ *and* $S_X^2 = (n-1)^{-1}\mathbf{X}^\top\mathbf{Q}\mathbf{X}$.

3.9.4 Some classical applications of quadratic forms

1. Univariate normal sampling: Considering that

$$X_1, \ldots, X_n \text{ iid } \mathcal{N}(\mu, \sigma^2) \iff \mathbf{X} = (X_1, \ldots, X_n)^\top \sim \mathcal{N}_n(\mu\mathbf{1}_n, \sigma^2\mathbf{I}_n),$$

we have for any orthogonal matrix $\mathbf{G} \in \mathcal{O}_n$ that $\mathbf{Z} = \mathbf{G}\mathbf{X} \sim \mathcal{N}_n(\mu\mathbf{G}\mathbf{1}_n, \sigma^2\mathbf{I}_n)$. In particular, if $\mathbf{G}_2 \in \mathbb{R}^{(n-1)\times n}$ is such that $\mathbf{G}_2\mathbf{1}_n = \mathbf{0}_{n-1}$ and $\mathbf{G}_2\mathbf{G}_2^\top = \mathbf{I}_{n-1}$, then:

$$\mathbf{G} = \begin{pmatrix} \frac{1}{\sqrt{n}}\mathbf{1}_n^\top \\ \mathbf{G}_2 \end{pmatrix} \in \mathcal{O}_n \Longrightarrow \mathbf{Z} = \mathbf{G}\mathbf{X} = \begin{pmatrix} \frac{1}{\sqrt{n}}\mathbf{1}_n^\top\mathbf{X} \\ \mathbf{G}_2\mathbf{X} \end{pmatrix} = \begin{pmatrix} \sqrt{n}\bar{X} \\ \mathbf{Z}_2 \end{pmatrix} \sim \mathcal{N}_n(\sqrt{n}\mu\mathbf{e}_1, \sigma^2\mathbf{I}_n),$$

where $\bar{X} = \sum_{i=1}^n X_i/n$ is the sample mean and $\mathbf{e}_1 = (1, 0, \ldots, 0)^\top$ is the first n-dimensional unit vector. Clearly,

$$Z_1 = \frac{1}{\sqrt{n}}\mathbf{1}_n^\top\mathbf{X} = \sqrt{n}\bar{X} \sim \mathcal{N}(\sqrt{n}\mu, \sigma^2), \quad \mathbf{Z}_2 = \mathbf{G}_2\mathbf{X} \sim \mathcal{N}_{n-1}(\mathbf{0}_{n-1}, \sigma^2\mathbf{I}_{n-1}), \text{ with } Z_1 \perp\!\!\!\perp \mathbf{Z}_2.$$

Also,

$$\|\mathbf{X}\|^2 = \|\mathbf{Z}\|^2 = n\bar{X}^2 + \|\mathbf{Z}_2\|^2 \Longrightarrow \|\mathbf{Z}_2\|^2 = \|\mathbf{X}\|^2 - n\bar{X}^2 = \|\mathbf{X} - \bar{X}\mathbf{1}_n\|^2 = \mathbf{X}^\top\mathbf{Q}\mathbf{X} = (n-1)S_X^2,$$

where $\mathbf{Q} = \mathbf{I}_n - \mathbf{P}$ with $\mathbf{P} = \mathbf{1}_n(\mathbf{1}_n^\top\mathbf{1}_n)^{-1}\mathbf{1}_n^\top = n^{-1}\mathbf{1}_n\mathbf{1}_n^\top$ are orthogonal projection matrices of rank $n-1$ and 1, respectively, and $S_X^2 = \sum_{i=1}^n (X_i - \bar{X})^2/(n-1)$ is the sample variance. Thus, the following sampling distributions follow:

$$\bar{X} \sim \mathcal{N}(\mu, \sigma^2/n), \ (n-1)S_X^2/\sigma^2 \sim \chi^2_{n-1}, \text{ and they are independent,}$$

which is equivalent to

$$\sqrt{n}(\bar{X} - \mu)/\sigma \sim \mathcal{N}(0,1), \ (n-1)S_X^2/\sigma^2 \sim \chi^2_{n-1}, \text{ and they are independent.}$$

At this time, we recall the following elementary definition:

Definition 3 [Univariate Student's t distribution] *A random variable T is said to have a univariate Student's t distribution, $T \sim t_\nu \equiv t(0,1,\nu)$, iff $T = V^{-1/2}Z$, where $Z \perp\!\!\!\perp V$, $Z \sim \mathcal{N}(0,1)$ and $V \sim \chi^2_\nu/\nu \equiv \mathcal{G}amma(\nu/2, \nu/2)$.*

By applying this definition to the above sampling distributions, the following pivotal function for μ follows:

$$T = \sqrt{n}(\bar{X} - \mu)/S_X \sim t_{n-1} \equiv t(0,1,n-1).$$

2. *Normal linear model (NLM):* Let $\mathbf{Y} = (Y_1, \ldots, Y_n)^\top$ be a response vector of n individuals, which is considered to be a point in the vector space $\mathbb{R}^n$. Assume that $\mathbf{Y}$ has the mean response vector $\boldsymbol{\mu} = (\mu_1, \ldots, \mu_n)^\top$ which is also a vector in $\mathbb{R}^n$. Usually, it is assumed that:

$$Y_i \overset{ind.}{\sim} \mathcal{N}(\mu_i, \sigma^2), \ i = 1, \ldots, n \iff \mathbf{Y} \sim \mathcal{N}_n(\boldsymbol{\mu}, \sigma^2 \mathbf{I}_n).$$

The general concept of a linear model for the mean vector $\boldsymbol{\mu} = \mathrm{E}(\mathbf{Y}) \in \mathbb{R}^n$ is defined as follows:

Definition 4 [Linear model] *A linear model of dimension k is a hypothesis for the mean vector $\boldsymbol{\mu} \in \mathbb{R}^n$ of the form*

$$H_0 : \boldsymbol{\mu} \in \mathcal{L}, \tag{3.11}$$

where $\mathcal{L}$ is a linear subspace of $\mathbb{R}^n$ of dimension k.

More explicitly, if we are given a basis $\mathbf{x}_1, \ldots, \mathbf{x}_k$ of $\mathcal{L}$, then the linear model (3.11) can be uniquely represented as

$$\boldsymbol{\mu} = \sum_{i=1}^{k} \mathbf{x}_j \beta_j = \mathbf{X}\boldsymbol{\beta}, \tag{3.12}$$

where $\mathbf{X} = (\mathbf{x}_1, \ldots, \mathbf{x}_k)$ is an $n \times k$ design (or model) matrix with full column rank k $(\leq n)$ and $\boldsymbol{\beta} = (\beta_1, \ldots, \beta_k)^\top$ is the $k \times 1$ vector of regression parameters. A stochastic version of (3.12) is given by

$$\mathbf{Y} = \mathbf{X}\boldsymbol{\beta} + \boldsymbol{\varepsilon}, \quad \mathrm{E}(\boldsymbol{\varepsilon}) = \mathbf{0}_n, \quad \mathrm{var}(\boldsymbol{\varepsilon}) = \sigma^2 \mathbf{I}_n,$$

where the last condition on the error vector $\boldsymbol{\varepsilon}$ means that the individual responses $Y_1, \ldots, Y_n$ are uncorrelated and homoscedastic. Finally, if we also include the normality assumption for the error vector, $\boldsymbol{\varepsilon} \sim \mathcal{N}_n(\mathbf{0}_n, \sigma^2 \mathbf{I}_n)$ say, then the classical NLM of full rank k is given by

$$\mathbf{Y} \sim \mathcal{N}_n(\mathbf{X}\boldsymbol{\beta}, \sigma^2 \mathbf{I}_n), \ \ \mathbf{X} \in \mathbb{R}^{n \times k}, \ \ \mathrm{rank}(\mathbf{X}) = k, \ \ \boldsymbol{\beta} \in \mathbb{R}^k, \ \ \sigma^2 > 0. \tag{3.13}$$

Let $\hat{\boldsymbol{\mu}} = \mathbf{P}\mathbf{Y}$ be the orthogonal projection of $\mathbf{Y}$ on $\mathcal{L} = \mathrm{span}(\mathbf{X})$, the linear subspace of $\mathbb{R}^n$ of dimension k generated by the columns $\mathbf{x}_1, \ldots, \mathbf{x}_k$ of the model matrix $\mathbf{X}$, where the orthogonal projection matrix is $\mathbf{P} = \mathbf{X}(\mathbf{X}^\top \mathbf{X})^{-1} \mathbf{X}^\top$. That is, for each observed response vector $\mathbf{y} \in \mathbb{R}^n$ of $\mathbf{Y}$, $\hat{\boldsymbol{\mu}} = \mathbf{P}\mathbf{y}$ becomes the solution to $\min_{\boldsymbol{\mu} \in \mathcal{L}} \|\mathbf{y} - \boldsymbol{\mu}\|^2$, and is called *least squares (LS)* estimate. In addition, the residual vector $\mathbf{e} = \mathbf{Y} - \hat{\boldsymbol{\mu}} = (\mathbf{I}_n - \mathbf{P})\mathbf{Y}$ becomes the orthogonal projection of $\mathbf{Y}$ on $\mathcal{L}^\perp = \mathbb{R}^n \ominus \mathcal{L}$, the orthogonal complement to $\mathcal{L}$ within $\mathbb{R}^n$, which is also a linear subspace of $\mathbb{R}^n$. Since $\hat{\boldsymbol{\mu}} = \mathbf{P}\mathbf{Y}$ is linear in $\mathbf{Y}$, then

$$\mathrm{E}(\hat{\boldsymbol{\mu}}) = \mathrm{E}(\mathbf{P}\mathbf{Y}) = \mathbf{P}\mathrm{E}(\mathbf{Y}) = \mathbf{P}\boldsymbol{\mu} = \boldsymbol{\mu} \quad (\boldsymbol{\mu} \in \mathcal{L}_k = \mathcal{L}(\mathbf{X})),$$

$$\mathrm{var}(\hat{\boldsymbol{\mu}}) = \mathrm{var}(\mathbf{P}\mathbf{Y}) = \mathbf{P}\mathrm{var}(\mathbf{Y})\mathbf{P}^\top = \sigma^2 \mathbf{P} \quad (\mathbf{P}^\top = \mathbf{P}, \ \mathbf{P}^2 = \mathbf{P}).$$

Similarly, since $\mathbf{e} = \mathbf{QY}$, where $\mathbf{Q} = \mathbf{I}_n - \mathbf{P}$ is the orthogonal projection matrix onto $\mathcal{L}^\perp$, then

$$\begin{aligned} \mathrm{E}(\mathbf{e}) &= \mathrm{E}(\mathbf{QY}) = \mathbf{Q}\mathrm{E}(\mathbf{Y}) = \mathbf{Q}\boldsymbol{\mu} = \mathbf{0}_n \quad (\boldsymbol{\mu} \perp \mathcal{L}^\perp), \\ \mathrm{var}(\mathbf{e}) &= \mathrm{var}(\mathbf{QY}) = \mathbf{Q}\mathrm{var}(\mathbf{Y})\mathbf{Q}^\top = \sigma^2\mathbf{Q} \quad (\mathbf{Q}^\top = \mathbf{Q},\ \mathbf{Q}^2 = \mathbf{Q}). \end{aligned}$$

Moreover, since $\mathbf{PQ} = \mathbf{P}(\mathbf{I}_n - \mathbf{P}) = \mathbf{P} - \mathbf{P}^2 = \mathbf{P} - \mathbf{P} = \mathbf{O}$, then

$$\mathrm{cov}(\hat{\boldsymbol{\mu}}, \mathbf{e}) = \mathrm{cov}(\mathbf{PY}, \mathbf{QY}) = \mathbf{P}\mathrm{var}(\mathbf{Y})\mathbf{Q} = \sigma^2\mathbf{PQ} = \mathbf{O}.$$

Hence, under the NLM (3.13), we have

$$\hat{\boldsymbol{\mu}} = \mathbf{PY} \sim \mathcal{N}_n(\boldsymbol{\mu}, \sigma^2\mathbf{P}), \quad \mathbf{e} = \mathbf{QY} \sim \mathcal{N}_n(\mathbf{0}_n, \sigma^2\mathbf{Q}), \text{ and they are independent.}$$

However, since $\boldsymbol{\mu} = \mathbf{X}\boldsymbol{\beta}$, where $\mathrm{rank}(\mathbf{X}) = \mathrm{rank}(\mathbf{X}^\top\mathbf{X}) = k$, then $\hat{\boldsymbol{\mu}} = \mathbf{X}\hat{\boldsymbol{\beta}}$ and we have

$$\begin{aligned} \mathbf{X}^\top\mathbf{X}\hat{\boldsymbol{\beta}} = \mathbf{X}^\top\hat{\boldsymbol{\mu}} &\Longrightarrow \hat{\boldsymbol{\beta}} = (\mathbf{X}^\top\mathbf{X})^{-1}\mathbf{X}^\top\hat{\boldsymbol{\mu}} = (\mathbf{X}^\top\mathbf{X})^{-1}\mathbf{X}^\top\mathbf{PY} \quad (\mathbf{PX} = \mathbf{X}) \\ &\Longrightarrow \hat{\boldsymbol{\beta}} = (\mathbf{X}^\top\mathbf{X})^{-1}\mathbf{X}^\top\mathbf{Y} \sim \mathcal{N}_k(\boldsymbol{\beta}, \sigma^2(\mathbf{X}^\top\mathbf{X})^{-1}), \end{aligned}$$

and it is independent of $\mathbf{e}$, and so of $\|\mathbf{e}\|^2$. In general, if $\boldsymbol{\theta} = \mathbf{A}\boldsymbol{\beta}$, where $\mathbf{A}$ is any known matrix of dimension $m \times k$, then

$$\hat{\boldsymbol{\theta}} = \mathbf{A}\hat{\boldsymbol{\beta}} \sim \mathcal{N}_m(\boldsymbol{\theta} = \mathbf{A}\boldsymbol{\beta}, \sigma^2\mathbf{A}(\mathbf{X}^\top\mathbf{X})^{-1}\mathbf{A}^\top), \text{ and it is independent of } \|\mathbf{e}\|^2.$$

On the other hand, because $\mathbf{Y} = \hat{\boldsymbol{\mu}} + \mathbf{e}$, with $\hat{\boldsymbol{\mu}} \perp \mathbf{e}$ since $\hat{\boldsymbol{\mu}} \in \mathcal{L}$ and $\mathbf{e} = \mathbf{Y} - \hat{\boldsymbol{\mu}} \in \mathcal{L}^\perp$, then by the Pythagorean theorem

$$\|\mathbf{Y}\|^2 = \|\hat{\boldsymbol{\mu}}\|^2 + \|\mathbf{e}\|^2 \Longrightarrow \|\mathbf{e}\|^2 = \|\mathbf{Y} - \hat{\boldsymbol{\mu}}\|^2 = \|\mathbf{Y}\|^2 - \|\hat{\boldsymbol{\mu}}\|^2,$$

where from the facts that $\mathrm{tr}(\mathbf{P}) = \mathrm{tr}\{\mathbf{X}(\mathbf{X}^\top\mathbf{X})^{-1}\mathbf{X}^\top\} = \mathrm{tr}(\mathbf{I}_k) = k$ and $\mathrm{tr}(\mathbf{Q}) = \mathrm{tr}(\mathbf{I}_n - \mathbf{P}) = n - k$, it follows that

$$\begin{aligned} \mathrm{E}(\|\mathbf{Y}\|^2) &= \mathrm{E}(\mathbf{Y}^\top\mathbf{Y}) = \mathrm{tr}\{\mathrm{var}(\mathbf{Y})\} + \boldsymbol{\mu}^\top\boldsymbol{\mu} \\ &= n\sigma^2 + \boldsymbol{\beta}^\top\mathbf{X}^\top\mathbf{X}\boldsymbol{\beta} \quad (\mathrm{var}(\mathbf{Y}) = \sigma^2\mathbf{I}_n,\ \boldsymbol{\mu} = \mathbf{X}\boldsymbol{\beta},\ \boldsymbol{\beta} \in \mathbb{R}^k), \\ \mathrm{E}(\|\hat{\boldsymbol{\mu}}\|^2) &= \mathrm{E}(\mathbf{Y}^\top\mathbf{PY}) = \mathrm{tr}\{\mathbf{P}\mathrm{var}(\mathbf{Y})\} + \boldsymbol{\mu}^\top\mathbf{P}\boldsymbol{\mu} \quad (\boldsymbol{\mu} \in \mathcal{L} \Longrightarrow \mathbf{P}\boldsymbol{\mu} = \boldsymbol{\mu}) \\ &= k\sigma^2 + \boldsymbol{\beta}^\top\mathbf{X}^\top\mathbf{X}\boldsymbol{\beta} \quad (\boldsymbol{\mu} = \mathbf{X}\boldsymbol{\beta}), \\ \mathrm{E}(\|\mathbf{e}\|^2) &= \mathrm{E}(\mathbf{Y}^\top\mathbf{QY}) = \mathrm{tr}\{\mathbf{Q}\mathrm{var}(\mathbf{Y})\} + \boldsymbol{\mu}^\top\mathbf{Q}\boldsymbol{\mu} \\ &= (n-k)\sigma^2 \quad (\boldsymbol{\mu} \in \mathcal{L} \Longrightarrow \mathbf{Q}\boldsymbol{\mu} = \mathbf{0}_n). \end{aligned}$$

From the LS residual vector $\mathbf{e} = \mathbf{Y} - \hat{\boldsymbol{\mu}}$, we can also obtain the unbiased estimator of σ^2 given by

$$S_e^2 = \frac{\|\mathbf{e}\|^2}{n-k} = \frac{\|\mathbf{Y} - \hat{\boldsymbol{\mu}}\|^2}{n-k} \Longrightarrow \mathrm{E}(S_e^2) = \sigma^2 \text{ and } \frac{(n-k)S_e^2}{\sigma^2} \sim \chi^2_{n-k}.$$

In summary, assuming the NLM in (3.13), we obtain the following general results:

$$\begin{aligned} &\hat{\boldsymbol{\beta}} = (\mathbf{X}^\top\mathbf{X})^{-1}\mathbf{X}^\top\mathbf{Y} \perp\!\!\!\perp S_e^2 = \frac{\|\mathbf{Y} - \mathbf{X}\hat{\boldsymbol{\beta}}\|^2}{n-k}, \\ &\hat{\boldsymbol{\beta}} \sim \mathcal{N}_k(\boldsymbol{\beta}, \sigma^2(\mathbf{X}^\top\mathbf{X})^{-1}), \quad \frac{(n-k)S_e^2}{\sigma^2} \sim \chi^2_{n-k}, \\ &\Longrightarrow \frac{\|\hat{\boldsymbol{\mu}} - \boldsymbol{\mu}\|^2}{kS_e^2} = \frac{\|\mathbf{X}(\hat{\boldsymbol{\beta}} - \boldsymbol{\beta})\|^2}{kS_e^2} \sim \mathcal{F}_{k,n-k}, \end{aligned}$$

where $\mathcal{F}_{\nu_1,\nu_2}$ denotes the central Fisher's F distribution with ν_1 and ν_2 degrees of freedom.

Definition 5 [Univariate central Fisher's F distribution] *A random variable F is said to have a central Fisher's F distribution, $F \sim \mathcal{F}_{\nu_1,\nu_2}$, iff $F \stackrel{d}{=} \dfrac{X_1/\nu_1}{X_2/\nu_2}$, with $X_1 \perp\!\!\!\perp X_2$, $X_1 \sim \chi^2_{\nu_1}$ and $X_2 \sim \chi^2_{\nu_2}$.*

Further relevant properties that follow under the NLM in (3.13) are:

[1] The LS estimators $\hat{\boldsymbol{\beta}} = (\mathbf{X}^\top\mathbf{X})^{-1}\mathbf{X}^\top\mathbf{Y}$ and $S_e^2 = \|\mathbf{Y} - \mathbf{X}\hat{\boldsymbol{\beta}}\|^2/(n-k)$ are the *best unbiased estimators* of $\boldsymbol{\beta}$ and σ^2, respectively.

[2] If $\mathrm{P}(\mathbf{Y} \in \mathcal{L}) = 0$, then $\hat{\boldsymbol{\beta}} = (\mathbf{X}^\top\mathbf{X})^{-1}\mathbf{X}^\top\mathbf{Y}$ and $\hat{\sigma}^2 = (n-k)S_e^2/n = \|\mathbf{Y} - \mathbf{X}\hat{\boldsymbol{\beta}}\|^2/n$ are the *maximum likelihood estimators (MLE)* of $\boldsymbol{\beta}$ and σ^2, respectively.

3.10 Mixtures of the Multivariate Normal

Multivariate data analysis virtually always starts with the evaluation of first- and second-order moments that are sufficient to fit a multivariate normal distribution, the maximum entropy distribution under these constraints. However, data often exhibit significant deviations from a normal distribution. In order to model these higher-order correlations, it is necessary to have more flexible distributions available. Therefore, it is an important challenge to find generalizations of the normal distribution which are more flexible but still computationally and analytically tractable.

In the literature on distribution theory, there are various formulations where a multivariate normal variable represents the basic component, but with the superposition of another random component, in the sense that the normal mean, the normal covariance matrix, or both are subject to the effect of another random variable, usually of a continuous type. We refer to these constructions as *mixtures of normal variables* or *mixtures of normal distributions*.

Presumably, the best known formulation is the scale mixtures of normal variables (Andrews and Mallows, 1974), which can be expressed as (see Example 45)

$$\mathbf{Y} = \boldsymbol{\xi} + W^{1/2}\mathbf{X}, \tag{3.14}$$

where $\mathbf{X} \sim \mathcal{N}_p(\mathbf{0}_p, \boldsymbol{\Psi})$, W is an independent random variable on $\mathbb{R}_+$, and $\boldsymbol{\xi} \in \mathbb{R}^p$ is a vector of constants. *Multivariate scale mixtures of normal (SMN) distributions* (3.14) provide a stochastic representation of a wide subset of the class of elliptically contoured distributions, often called briefly elliptical distributions. An SMN distribution simply alters the tail behavior of the normal while leaving the resultant distribution symmetric. A case of primary importance occurs when $V = 1/W \sim \chi^2_\nu/\nu$, leading to the multivariate t distribution (see Example 46).

Remark 3 *For convenience, sometimes in (3.14) the weight variable W is redefined as $W = 1/V$.*

Another very important construction is the *multivariate normal variance-mean mixtures* proposed by Barndorff-Nielsen (1977, 1978) and extensively developed by the subsequent literature, namely,

$$\mathbf{Y} = \boldsymbol{\xi} + W\boldsymbol{\eta} + W^{1/2}\mathbf{X}, \tag{3.15}$$

where $\boldsymbol{\eta} \in \mathbb{R}^p$ is a vector of constants, and W is again an independent random variable on $\mathbb{R}_+$. The best-known case is the one that assumes that W has a generalized inverse Gaussian (GIG) distribution (Jørgensen, 1982; Johnson et al., 1994), which yields the so-called *multivariate generalized hyperbolic (GH) distribution* for $\mathbf{Y}$; see also Blæsild and Jensen (1981) and Schmidt et al. (2006).

Beyond (3.14) and (3.15), numerous other constructions fall within the framework of mixtures of normals described in the introductory paragraph, such as finite normal mixtures. Many of these formulations can be encompassed in a more general expression, where both the mean and variance of the normal distribution are affected by an additional random component. This proposal is due to Arellano-Valle and Azzalini (2021), who introduced the following formulation:

$$\mathbf{Y} = \boldsymbol{\xi} + r(\mathbf{W})\boldsymbol{\eta} + s(\mathbf{W})\mathbf{X}, \quad \mathbf{W} \perp\!\!\!\perp \mathbf{X},\ \mathbf{X} \sim \mathcal{N}_p(\mathbf{0}_p, \boldsymbol{\Psi}), \tag{3.16}$$

where $r(\cdot)$ and $s(\cdot)$ are scalar functions and $\mathbf{W} \in \mathbb{R}^k$ is a continuous, discrete, or mixed random vector. An equivalent representation of (3.16) is

$$\mathbf{Y} = \boldsymbol{\xi} + R\boldsymbol{\eta} + S\mathbf{X}, \quad (R,S) \perp\!\!\!\perp \mathbf{X},\ \mathbf{X} \sim \mathcal{N}_p(\mathbf{0}_p, \boldsymbol{\Psi}), \tag{3.17}$$

where $R = r(\mathbf{W})$ and $S = s(\mathbf{W})$.

The formulation in (3.16), or its equivalent version in (3.17), is called *multivariate generalized mixtures of normal (GMN) distributions*. Clearly, it includes (3.14) and (3.15) as special cases, allowing a broader family of multivariate distributions to be covered, preserving certain properties of the multivariate normal distribution, but having sufficient flexibility to approximate any continuous distribution with a certain degree of accuracy. This family includes symmetric and asymmetric multivariate distributions according to the specifications of the functions $r(\cdot)$ and $s(\cdot)$, preserves the distribution of affine transformations and therefore their marginals, and is marginally consistent. Hierarchically, (3.16) is equivalent to

$$\begin{aligned} \mathbf{Y} \mid \mathbf{W} = \mathbf{w} &\sim \mathcal{N}_p(\boldsymbol{\xi} + r(\mathbf{w})\boldsymbol{\eta}, s(\mathbf{w})^2\boldsymbol{\Psi}), \\ \mathbf{W} &\sim G, \end{aligned}$$

where G is a cdf defined on $\mathbb{G} \subset \mathbb{R}^k$. Thus, when $\boldsymbol{\Psi} > 0$ and $s(\mathbf{w}) \neq 0$, the pdf of $\mathbf{Y}$ becomes

$$f(\mathbf{y}) = \int_{\mathbb{G}} \phi_p(\mathbf{y}; \boldsymbol{\xi} + r(\mathbf{w})\boldsymbol{\eta}, s(\mathbf{w})^2\boldsymbol{\Psi}) \mathrm{d}G(\mathbf{w}) = \int_{\mathbb{H}} \phi_p(\mathbf{y}; \boldsymbol{\xi} + r\boldsymbol{\eta}, s^2\boldsymbol{\Psi}) \mathrm{d}H(r,s), \quad \mathbf{y} \in \mathbb{R}^p,$$

where H becomes the cdf of (R,S), which is defined on $\mathbb{H} \subset \mathbb{R}^2$. Furthermore, from (3.17) it is very simple to compute $\mathrm{E}(\mathbf{Y})$ and $\mathrm{var}(\mathbf{Y})$, provided that $R = s(\mathbf{W})$ and $S = s(\mathbf{W})$ have finite first- and second-order moments, respectively, which are then given by

$$\mathrm{E}(\mathbf{Y}) = \boldsymbol{\xi} + \mathrm{E}(R)\boldsymbol{\eta}, \quad \mathrm{var}(\mathbf{Y}) = \mathrm{var}(R)\boldsymbol{\eta}\boldsymbol{\eta}^\top + \mathrm{E}(S^2)\boldsymbol{\Psi}.$$

Taking $\mathbf{W} = (U,V)^\top$, with $U \perp\!\!\!\perp V$, several well-known distributions can be generated from the GMN formulation; some examples are the following:

Skew-normal distribution (see Chapter 5):

$$\mathbf{Y} = \boldsymbol{\xi} + U\boldsymbol{\eta} + \mathbf{X}, \quad U \sim \mathcal{HN}(0,1) \perp\!\!\!\perp \mathbf{X} \sim \mathcal{N}_p(\mathbf{0}_p, \boldsymbol{\Psi}).$$

Scale mixtures of skew-normal distributions (see Section 5.11.2):

$$\mathbf{Y} = \boldsymbol{\xi} + V^{-1/2}U\boldsymbol{\eta} + V^{-1/2}\mathbf{X}, \quad U \sim \mathcal{HN}(0,1) \perp\!\!\!\perp V \sim H \perp\!\!\!\perp \mathbf{X} \sim \mathcal{N}_p(\mathbf{0}_p, \boldsymbol{\Psi}).$$

Skew-t distribution (see Chapter 7):

$$\mathbf{Y} = \boldsymbol{\xi} + U\boldsymbol{\eta} + \sqrt{\frac{\nu + U^2}{\nu + 1}} V^{-1/2}\mathbf{X}, \quad U \sim \mathcal{HT}(0,1,\nu) \perp\!\!\!\perp V \sim \frac{\chi^2_{\nu+1}}{\nu+1} \perp\!\!\!\perp \mathbf{X} \sim \mathcal{N}_p(\mathbf{0}_p, \boldsymbol{\Psi}).$$

Two-piece normal distribution (Mudholkar and Hutson, 2000; Arellano-Valle et al., 2005b):

$$\mathbf{Y} = \boldsymbol{\xi} + U\boldsymbol{\eta} + \mathbf{X}, \quad U \sim \mathcal{TPN}(0,1;\gamma) \perp\!\!\!\perp \mathbf{X} \sim \mathcal{N}_p(\mathbf{0}_p, \boldsymbol{\Psi}),$$

where $U \sim \mathcal{TPN}(0,1;\gamma)$ means that U has pdf $f(u;\gamma) = \phi\left(\frac{u}{1+\gamma}\right) I_{[0,\infty)}(u) + \phi\left(\frac{u}{1-\gamma}\right) I_{(-\infty,0)}(u)$, $|\gamma| < 1$.

Geometric skew-normal distribution (Kundu, 2017):

$$\mathbf{Y} = \boldsymbol{\xi} + U\boldsymbol{\eta} + U^{1/2}\mathbf{X}, \quad U \sim \mathcal{G}eometric(p) \perp\!\!\!\perp \mathbf{X} \sim \mathcal{N}_p(\mathbf{0}_p, \boldsymbol{\Psi}).$$

Mean mixtures of normal exponential distribution (Negarestani et al., 2019):

$$\mathbf{Y} = \boldsymbol{\xi} + U\boldsymbol{\eta} + \mathbf{X}, \quad U \sim \mathcal{E}xp(1) \perp\!\!\!\perp \mathbf{X} \sim \mathcal{N}_p(\mathbf{0}_p, \boldsymbol{\Psi}).$$

χ_ν mixing mean distribution (Arellano-Valle and Azzalini, 2021):

$$\mathbf{Y} = \boldsymbol{\xi} + U\boldsymbol{\eta} + \mathbf{X}, \quad U \sim \chi_\nu \perp\!\!\!\perp \mathbf{X} \sim \mathcal{N}_p(\mathbf{0}_p, \boldsymbol{\Psi}),$$

which includes the half-normal ($\nu = 1$) and Rayleigh ($\nu = 2$) mixing means as special cases.

In terms of pdf, a common but general form to represent the mixtures of normals becomes

$$f(\mathbf{y}) = \int \phi_p(\mathbf{y}; \boldsymbol{\mu}, \boldsymbol{\Sigma}) \mathrm{d}\Pi(\boldsymbol{\mu}, \boldsymbol{\Sigma}), \quad \mathbf{y} \in \mathbb{R}^p,$$

where the mixing distribution $\Pi(\cdot,\cdot)$ can be discrete or continuous or a mixture of both. The normal-inverse-Wishart mixing distribution is perhaps the best-known case, in which $\boldsymbol{\mu} \mid \boldsymbol{\Sigma}$ is normal and $\boldsymbol{\Sigma}$ is inverse-Wishart (i.e., $\boldsymbol{\Sigma}^{-1}$ is Wishart). It is the conjugate prior of a multivariate normal distribution with an unknown mean and variance matrix (the inverse of the precision matrix), yielding a multivariate t distribution for $\mathbf{Y}$; see p. 72 of Gelman et al. (2014). A major limitation is that the idea of a mixture of scales does not easily translate to the multivariate setting, since there are few distributions in $\boldsymbol{\Sigma}$ for which analytical results are available (basically only the inverse-Wishart distribution).

Finite mixtures of normals are another quite attractive route to obtain distributions with high flexibility. For *multivariate finite mixtures of normal (FMN) distributions*, the mixing distribution is discrete and places mass on M distinct values of $\boldsymbol{\mu}$ and $\boldsymbol{\Sigma}$. In this case, the pdf of $\mathbf{Y}$ becomes

$$f(\mathbf{y}) = \sum_{m=1}^{M} \pi_m \phi_p(\mathbf{y}; \boldsymbol{\mu}_m, \boldsymbol{\Sigma}_m), \quad \mathbf{y} \in \mathbb{R}^p,$$

where the mass points $\{\phi_p(\mathbf{y}; \boldsymbol{\mu}_m, \boldsymbol{\Sigma}_m);\ m = 1, \ldots, M\}$ are often called the *components of the mixtures*. Two reasons why this approach becomes quite attractive are: 1) it applies equally well in univariate and multivariate situations; and 2) great flexibility can be achieved with only a few components; see McLachlan and Peel (2000) for a detailed account of finite mixture models and Sections 5.11.6 and 10.9 for further applications.

3.11 Distributions Related to the Multivariate Normal

3.11.1 The normal matrix distribution

The normal matrix (variate) distribution generalizes the multivariate normal distribution to matrix-valued random variables. To introduce this distribution, let $\mathbf{X} = (X_{ij}) = (\mathbf{X}_1, \ldots, \mathbf{X}_n)^\top \in \mathbb{R}^{n\times p}$ be a random matrix whose components have finite mean and variance. As before, the mean matrix and the variance matrix of $\mathbf{X}$ are, respectively, given by

$$\mathrm{E}(\mathbf{X}) = (\mathrm{E}(X_{ij})) = (\mu_{ij}) = \mathbf{M} \in \mathbb{R}^{n\times p},$$

and

$$\mathrm{var}(\mathbf{X}) \equiv \mathrm{var}\{\mathrm{vec}(\mathbf{X}^\top)\} = (\mathrm{cov}(\mathbf{X}_i, \mathbf{X}_j)) = (\boldsymbol{\Omega}_{ij}) = \boldsymbol{\Omega} \in \mathbb{R}^{np\times np},$$

where the blocks are $\boldsymbol{\Omega}_{ij} = (\mathrm{cov}(x_{ik}, x_{jl})) \in \mathbb{R}^{p\times p}$.

Definition 6 [Normal matrix distribution] *A random matrix $\mathbf{X} \in \mathbb{R}^{n \times p}$ has a normal matrix (or matrix variate normal) distribution with mean matrix $\mathbf{M} \in \mathbb{R}^{n \times p}$ and variance matrix $\mathbf{\Omega} \in \mathbb{R}^{np \times np}$, and we write $\mathbf{X} \sim \mathcal{N}_{n,p}(\mathbf{M}, \mathbf{\Omega})$, iff* $\mathrm{vec}(\mathbf{X}^\top) \sim \mathcal{N}_{np}(\mathrm{vec}(\mathbf{M}^\top), \mathbf{\Omega})$.

In the nonsingular case with $\mathbf{\Omega} > 0$, the pdf of $\mathbf{X} \sim \mathcal{N}_{n,p}(\mathbf{M}, \mathbf{\Omega})$ is

$$f(\mathbf{X}) \equiv \phi_{n,p}(\mathbf{X}; \mathbf{M}, \mathbf{\Omega}) = (2\pi)^{-np/2} |\mathbf{\Omega}|^{-1/2} \exp\left[-\frac{1}{2}\mathrm{vec}\{(\mathbf{X}-\mathbf{M})^\top\}^\top \mathbf{\Omega}^{-1} \mathrm{vec}\{(\mathbf{X}-\mathbf{M})^\top\}\right].$$

For instance, if $\mathbf{M} = \mathbf{O}$ and $\mathbf{\Omega} = \mathbf{I}_{np} = \mathbf{I}_n \otimes \mathbf{I}_p$, then $\mathbf{X} \sim \mathcal{N}_{n,p}(\mathbf{O}, \mathbf{I}_n \otimes \mathbf{I}_p)$ is said to have a *standard normal matrix distribution*, in which case all its entries are iid $\mathcal{N}(0,1)$ random variables, and its pdf becomes

$$f(\mathbf{X}) = (2\pi)^{-np/2} \mathrm{etr}\left(-\frac{1}{2}\mathbf{X}\mathbf{X}^\top\right),$$

where $\mathrm{etr}(\cdot) \equiv \exp\{\mathrm{tr}(\cdot)\}$.

Usually, the variance matrix of $\mathbf{X}$ is represented as $\mathbf{\Omega} = \mathbf{\Psi} \otimes \mathbf{\Sigma}$, where $\mathbf{\Psi} \geq 0$ is in $\mathbb{R}^{n \times n}$ and $\mathbf{\Sigma} \geq 0$ is in $\mathbb{R}^{p \times p}$. In this case, an equivalent definition for the normal matrix distribution is

$$\mathbf{X} \sim \mathcal{N}_{n,p}(\mathbf{M}, \mathbf{\Psi} \otimes \mathbf{\Sigma}) \iff \mathrm{vec}(\mathbf{X}) \sim \mathcal{N}_{np}(\mathrm{vec}(\mathbf{M}), \mathbf{\Sigma} \otimes \mathbf{\Psi}),$$

since $\mathbf{K}_{pn}\mathrm{vec}(\mathbf{X}) = \mathrm{vec}(\mathbf{X}^\top)$, $\mathbf{K}_{pn}^\top = \mathbf{K}_{np}$ and $\mathbf{K}_{pn}(\mathbf{\Psi} \otimes \mathbf{\Sigma})\mathbf{K}_{np} = \mathbf{\Sigma} \otimes \mathbf{\Psi}$.

Further well-known properties are:

[1] If $\mathbf{X} \sim \mathcal{N}_{n,p}(\mathbf{M}, \mathbf{\Omega})$ and $\mathbf{A} \in \mathbb{R}^{m \times n}$, $\mathbf{B} \in \mathbb{R}^{p \times q}$ and $\mathbf{C} \in \mathbb{R}^{m \times q}$ are constant matrices, then $\mathbf{AXB} + \mathbf{C} \sim \mathcal{N}_{m,q}(\mathbf{AMB} + \mathbf{C}, (\mathbf{A} \otimes \mathbf{B}^\top)\mathbf{\Omega}(\mathbf{A}^\top \otimes \mathbf{B}))$ since $\mathrm{vec}(\mathbf{B}^\top\mathbf{X}\mathbf{A}^\top + \mathbf{C}^\top) = (\mathbf{A} \otimes \mathbf{B}^\top)\mathrm{vec}(\mathbf{X}^\top) + \mathrm{vec}(\mathbf{C}^\top)$. In particular, if $\mathbf{\Omega} = \mathbf{I}_n \otimes \mathbf{I}_p$, then $\mathbf{AXB} + \mathbf{C} \sim \mathcal{N}_{m,q}(\mathbf{AMB} + \mathbf{C}, \mathbf{AA}^\top \otimes \mathbf{B}^\top\mathbf{B})$.

[2] If $\mathbf{X} \sim \mathcal{N}_{n,p}(\mathbf{M}, \mathbf{\Psi} \otimes \mathbf{\Sigma})$, where $\mathbf{\Psi} > 0$ is in $\mathbb{R}^{n \times n}$ and $\mathbf{\Sigma} > 0$ is in $\mathbb{R}^{p \times p}$, then $\mathbf{X}$ has a pdf which is

$$f(\mathbf{X}) = \phi_{n,p}(\mathbf{X}; \mathbf{M}, \mathbf{\Psi} \otimes \mathbf{\Sigma}) = (2\pi)^{-np/2} |\mathbf{\Psi}|^{-p/2} |\mathbf{\Sigma}|^{-n/2} \mathrm{etr}\left\{-\frac{1}{2}\mathbf{\Sigma}^{-1}(\mathbf{X}-\mathbf{M})\mathbf{\Psi}^{-1}(\mathbf{X}-\mathbf{M})^\top\right\}.$$

This follows from $\mathbf{X} = \mathbf{\Psi}^{1/2}\mathbf{Z}\mathbf{\Sigma}^{1/2} + \mathbf{M}$, where $\mathbf{Z} \sim \mathcal{N}_{n,p}(\mathbf{0}, \mathbf{I}_n \otimes \mathbf{I}_p)$.

[3] If $\mathbf{X} = (\mathbf{X}_1, \mathbf{X}_2) \sim \mathcal{N}_{n,p_1+p_2}\left((\mathbf{M}_1, \mathbf{M}_2), \mathbf{\Psi} \otimes \begin{pmatrix} \mathbf{\Sigma}_{11} & \mathbf{\Sigma}_{12} \\ \mathbf{\Sigma}_{21} & \mathbf{\Sigma}_{22} \end{pmatrix}\right)$, with $\mathbf{\Psi} > 0$ and $\mathbf{\Sigma}_{11} > 0$. Then:

(a) $\mathbf{X}_i \sim \mathcal{N}_{n,p_i}(\mathbf{M}_i, \mathbf{\Psi} \otimes \mathbf{\Sigma}_{ii})$, $i = 1, 2$, and $\mathbf{X}_1 \perp\!\!\!\perp \mathbf{X}_2 \iff \mathbf{\Sigma}_{12} = \mathbf{O}$;

(b) $\mathbf{X}_2 \mid \mathbf{X}_1 = \mathbf{X}_1^0 \sim \mathcal{N}_{n,p_2}(\mathbf{M}_{2\cdot 1}, \mathbf{\Psi} \otimes \mathbf{\Sigma}_{22\cdot 1}), \forall \mathbf{X}_1^0 \in \mathbb{R}^{n \times p_1}$, where $\mathbf{M}_{2\cdot 1} = \mathbf{M}_2 - (\mathbf{X}_1^0 - \mathbf{M}_1)\mathbf{\Sigma}_{11}^{-1}\mathbf{\Sigma}_{12}$ and $\mathbf{\Sigma}_{22\cdot 1} = \mathbf{\Sigma}_{22} - \mathbf{\Sigma}_{21}\mathbf{\Sigma}_{11}^{-1}\mathbf{\Sigma}_{12}$.

[4] If $\mathbf{Z} \sim \mathcal{N}_{n,p}(\mathbf{O}, \mathbf{I}_{np})$, and $\mathbf{P}$ and $\mathbf{Q}$ are orthogonal projection matrices in $\mathbb{R}^{n \times n}$ such that $\mathbf{PQ} = \mathbf{O}$, then $\mathbf{PZ} \perp\!\!\!\perp \mathbf{QZ}$. To see this, note from [1] that:

$$\mathrm{var}\left\{\begin{pmatrix} \mathbf{P} \\ \mathbf{Q} \end{pmatrix} \mathbf{Z}\right\} = \left\{\begin{pmatrix} \mathbf{P} \\ \mathbf{Q} \end{pmatrix} (\mathbf{P}^\top \ \mathbf{Q}^\top)\right\} \otimes \mathbf{I}_p = \begin{pmatrix} \mathbf{P} \otimes \mathbf{I}_p & \mathbf{O} \\ \mathbf{O} & \mathbf{Q} \otimes \mathbf{I}_p \end{pmatrix}.$$

3.11.2 *Multivariate normal sampling*

Let $\mathbf{X} = (\mathbf{X}_1, \ldots, \mathbf{X}_n)^\top$ be a random matrix with row vectors $\mathbf{X}_1, \ldots, \mathbf{X}_n \overset{iid}{\sim} \mathcal{N}_p(\boldsymbol{\mu}, \boldsymbol{\Sigma})$. Then:

$$\mathrm{E}(\mathbf{X}) = \begin{pmatrix} \mathrm{E}(\mathbf{X}_1)^\top \\ \vdots \\ \mathrm{E}(\mathbf{X}_n)^\top \end{pmatrix} = \begin{pmatrix} \boldsymbol{\mu}^\top \\ \vdots \\ \boldsymbol{\mu}^\top \end{pmatrix} = \mathbf{1}_n \otimes \boldsymbol{\mu}^\top = \mathbf{1}_n \boldsymbol{\mu}^\top,$$

$$\mathrm{var}(\mathbf{X}) = \mathrm{var}\left\{ \begin{pmatrix} \mathbf{X}_1 \\ \vdots \\ \mathbf{X}_n \end{pmatrix} \right\} = (\mathrm{cov}(\mathbf{X}_i, \mathbf{X}_j)) = \begin{pmatrix} \boldsymbol{\Sigma} & \mathbf{O} & \cdots & \mathbf{O} \\ \mathbf{O} & \boldsymbol{\Sigma} & \cdots & \mathbf{O} \\ \vdots & \vdots & \ddots & \vdots \\ \mathbf{O} & \mathbf{O} & \cdots & \boldsymbol{\Sigma} \end{pmatrix} = \mathbf{I}_n \otimes \boldsymbol{\Sigma},$$

$$\Longrightarrow \mathrm{vec}(\mathbf{X}^\top) \sim \mathcal{N}_{np}(\mathbf{1}_n \otimes \boldsymbol{\mu}, \mathbf{I}_n \otimes \boldsymbol{\Sigma}) \text{ or equivalently } \mathbf{X} \sim \mathcal{N}_{n,p}(\mathbf{1}_n \boldsymbol{\mu}^\top, \mathbf{I}_n \otimes \boldsymbol{\Sigma}).$$

In other words,

$$\mathbf{X} = (\mathbf{X}_1, \ldots, \mathbf{X}_n)^\top \sim \mathcal{N}_{n,p}(\mathbf{1}_n \boldsymbol{\mu}^\top, \mathbf{I}_n \otimes \boldsymbol{\Sigma}) \Longleftrightarrow \mathbf{X}_1, \ldots, \mathbf{X}_n \overset{iid}{\sim} \mathcal{N}_p(\boldsymbol{\mu}, \boldsymbol{\Sigma}).$$

In this case, the sample matrix $\mathbf{X}$ is called the *data matrix from $\mathcal{N}_p(\boldsymbol{\mu}, \boldsymbol{\Sigma})$* or simply the *normal data matrix*. It produces the sample mean vector $\bar{\mathbf{X}}$ and the sample variance matrix $\mathbf{S}$:

$$n\bar{\mathbf{X}} = \textstyle\sum_{i=1}^n \mathbf{X}_i = \mathbf{X}^\top \mathbf{1}_n, \quad (n-1)\mathbf{S} = \sum_{i=1}^n (\mathbf{X}_i - \bar{\mathbf{X}})(\mathbf{X}_i - \bar{\mathbf{X}})^\top = \mathbf{X}^\top \mathbf{X} - n\bar{\mathbf{X}}\bar{\mathbf{X}}^\top.$$

As we know, $\bar{\mathbf{X}}$ and $\mathbf{S}$ are unbiased estimators of $\boldsymbol{\mu}$ and $\boldsymbol{\Sigma}$, respectively:

$$\mathrm{E}(\bar{\mathbf{X}}) = \frac{1}{n} \sum_{i=1}^n \mathrm{E}(\mathbf{X}_i) = \boldsymbol{\mu}, \quad \mathrm{var}(\bar{\mathbf{X}}) = \frac{1}{n^2} \sum_{i=1}^n \mathrm{var}(\mathbf{X}_i) = \frac{1}{n} \boldsymbol{\Sigma}.$$

Similarly, using $\mathrm{E}(\mathbf{X}_i \mathbf{X}_i^\top) = \boldsymbol{\Sigma} + \boldsymbol{\mu}\boldsymbol{\mu}^\top$, $\mathrm{var}(\mathbf{X}_i \otimes \mathbf{X}_i) = (\mathbf{I}_{p^2} + \mathbf{K}_p)\left(\boldsymbol{\Sigma} \otimes \boldsymbol{\Sigma} + \boldsymbol{\Sigma} \otimes \boldsymbol{\mu}\boldsymbol{\mu}^\top + \boldsymbol{\mu}\boldsymbol{\mu}^\top \otimes \boldsymbol{\Sigma}\right)$, $\mathrm{cov}(\mathbf{X}_i \otimes \mathbf{X}_i, \bar{\mathbf{X}} \otimes \bar{\mathbf{X}}) = \mathrm{cov}(\bar{\mathbf{X}} \otimes \bar{\mathbf{X}}, \mathbf{X}_i \otimes \mathbf{X}_i) = n^{-2}\mathrm{var}(\mathbf{X}_i \otimes \mathbf{X}_i)$, $i = 1, \ldots, n$, $\mathrm{E}(\bar{\mathbf{X}}\bar{\mathbf{X}}^\top) = n^{-1}\boldsymbol{\Sigma} + \boldsymbol{\mu}\boldsymbol{\mu}^\top$ and $\mathrm{var}(\bar{\mathbf{X}} \otimes \bar{\mathbf{X}}) = (\mathbf{I}_{p^2} + \mathbf{K}_p)\left(n^{-2}\boldsymbol{\Sigma} \otimes \boldsymbol{\Sigma} + n^{-1}\boldsymbol{\Sigma} \otimes \boldsymbol{\mu}\boldsymbol{\mu}^\top + n^{-1}\boldsymbol{\mu}\boldsymbol{\mu}^\top \otimes \boldsymbol{\Sigma}\right)$, we have

$$\mathrm{E}(\mathbf{S}) = \frac{1}{n-1} \sum_{i=1}^n \mathrm{E}(\mathbf{X}_i \mathbf{X}_i^\top) - \frac{n}{n-1} \mathrm{E}(\bar{\mathbf{X}}\bar{\mathbf{X}}^\top) = \frac{1}{n-1} \sum_{i=1}^n (\boldsymbol{\Sigma} + \boldsymbol{\mu}\boldsymbol{\mu}^\top) - \frac{n}{n-1}\left(\frac{1}{n}\boldsymbol{\Sigma} + \boldsymbol{\mu}\boldsymbol{\mu}^\top\right) = \boldsymbol{\Sigma},$$

$$\begin{aligned} \mathrm{var}(\mathbf{S}) =& \frac{1}{(n-1)^2} \left\{ \sum_{i=1}^n \mathrm{var}(\mathbf{X}_i \otimes \mathbf{X}_i) + n^2 \mathrm{var}(\bar{\mathbf{X}} \otimes \bar{\mathbf{X}}) - \frac{2}{n} \mathrm{var}(\mathbf{X}_i \otimes \mathbf{X}_i) \right\} \\ =& \frac{1}{(n-1)^2} (\mathbf{I}_{p^2} + \mathbf{K}_p) \left\{ n\left(\boldsymbol{\Sigma} \otimes \boldsymbol{\Sigma} + \boldsymbol{\Sigma} \otimes \boldsymbol{\mu}\boldsymbol{\mu}^\top + \boldsymbol{\mu}\boldsymbol{\mu}^\top \otimes \boldsymbol{\Sigma}\right) \right. \\ & \left. + n^2 \left(\frac{1}{n^2} \boldsymbol{\Sigma} \otimes \boldsymbol{\Sigma} + \frac{1}{n} \boldsymbol{\Sigma} \otimes \boldsymbol{\mu}\boldsymbol{\mu}^\top + \frac{1}{n} \boldsymbol{\mu}\boldsymbol{\mu}^\top \otimes \boldsymbol{\Sigma} \right) \right\} = \frac{1}{n-1} (\mathbf{I}_{p^2} + \mathbf{K}_p)(\boldsymbol{\Sigma} \otimes \boldsymbol{\Sigma}). \end{aligned}$$

Moreover, if $\boldsymbol{\Sigma} > 0$ then:

$$\begin{aligned} f(\mathbf{X}) &= (2\pi)^{-np/2} |\boldsymbol{\Sigma}|^{-n/2} \mathrm{etr}\left\{ -\frac{1}{2} \boldsymbol{\Sigma}^{-1} (\mathbf{X} - \mathbf{1}_n \boldsymbol{\mu}^\top)^\top (\mathbf{X} - \mathbf{1}_n \boldsymbol{\mu}^\top) \right\} \\ &= (2\pi)^{-np/2} |\boldsymbol{\Sigma}|^{-n/2} \mathrm{etr}\left[-\frac{1}{2} \boldsymbol{\Sigma}^{-1} \{(n-1)\mathbf{S} + n(\bar{\mathbf{X}} - \boldsymbol{\mu})(\bar{\mathbf{X}} - \boldsymbol{\mu})^\top\} \right], \end{aligned}$$

from which it can be concluded that $(\bar{\mathbf{X}}, \mathbf{S})$ are the *minimum variance unbiased estimators (MVUE)* of $(\boldsymbol{\mu}, \boldsymbol{\Sigma})$, appealing to the theorems of Rao-Blackwell and Lehmann-Scheffé; see, e.g., Arnold (1981) and Jørgensen (1993).

Further properties of $(\bar{\mathbf{X}}, \mathbf{S})$ can be studied from the representation $\mathbf{X}_i = \boldsymbol{\Sigma}^{1/2}\mathbf{Z}_i + \boldsymbol{\mu}$, $i = 1, \ldots, n$, for any $\mathbf{C}\mathbf{C}^\top = \boldsymbol{\Sigma}$, where $\mathbf{Z}_1, \ldots, \mathbf{Z}_n \overset{iid}{\sim} \mathcal{N}_p(\mathbf{0}_p, \mathbf{I}_p)$. For the normal data matrix, this means that $\mathbf{X} = \mathbf{Z}\mathbf{C}^\top + \mathbf{1}_n \boldsymbol{\mu}^\top$, where $\mathbf{Z} = (\mathbf{Z}_1, \ldots, \mathbf{Z}_n)^\top \sim \mathcal{N}_{n,p}(\mathbf{O}, \mathbf{I}_n \otimes \mathbf{I}_p)$. Therefore, it follows that

$$(\bar{\mathbf{X}}, \mathbf{S}) = (\mathbf{C}\bar{\mathbf{Z}} + \boldsymbol{\mu}, \mathbf{C}\mathbf{S}_z\mathbf{C}^\top), \quad \text{where } n\bar{\mathbf{Z}} = \mathbf{Z}^\top \mathbf{1}_n \text{ and } (n-1)\mathbf{S}_z = \mathbf{Z}^\top \mathbf{Z} - n\bar{\mathbf{Z}}\bar{\mathbf{Z}}^\top.$$

Let $\mathbf{P} = n^{-1}\mathbf{1}_n\mathbf{1}_n^\top$ and $\mathbf{Q} = \mathbf{I}_n - \mathbf{P}$, which are orthogonal projections. Since $\mathbf{PQ} = \mathbf{O}$, then $\mathbf{PZ} \perp\!\!\!\perp \mathbf{QZ}$. But, $n\bar{\mathbf{Z}} = \mathbf{Z}^\top\mathbf{1}_n = \mathbf{Z}^\top\mathbf{P}\mathbf{1}_n$ and $(n-1)\mathbf{S}_z = \mathbf{Z}^\top\mathbf{QZ} = \mathbf{Z}^\top\mathbf{QQZ}$ are functions of $\mathbf{PZ}$ and $\mathbf{QZ}$, so $\bar{\mathbf{Z}} \perp\!\!\!\perp \mathbf{S}_z$, thus implying that $\bar{\mathbf{X}} \perp\!\!\!\perp \mathbf{S}$. In summary, we can say:

$$\bar{\mathbf{X}} \sim \mathcal{N}_p(\boldsymbol{\mu}, n^{-1}\boldsymbol{\Sigma}), \quad (n-1)\mathbf{S} \sim \mathcal{W}_p(n-1, \boldsymbol{\Sigma}), \quad \bar{\mathbf{X}} \perp\!\!\!\perp \mathbf{S}.$$

Here, $\mathcal{W}_p(m, \mathbf{A})$ denotes the Wishart matrix distribution with $m > 0$ degrees of freedom and $p \times p$ scale matrix $\mathbf{A} > 0$, which is defined in the next subsection (see Definition 7 below). The above distribution of $\bar{\mathbf{X}}$ is obvious. To find the distribution of $(n-1)\mathbf{S}$, we need the distribution of $(n-1)\mathbf{S}_z$. To do this, write $\mathbf{Q} = \mathbf{HH}^\top$, where $\mathbf{H}$ gives an orthonormal basis for $\mathrm{span}(\mathbf{1}_n)^\perp$ (of dimension $n-1$). Then, $(n-1)\mathbf{S}_z = \mathbf{Z}^\top\mathbf{HH}^\top\mathbf{Z} = \mathbf{U}^\top\mathbf{U} \sim \mathcal{W}_p(n-1, \mathbf{I}_p)$, since $\mathbf{U} = \mathbf{H}^\top\mathbf{Z} \sim \mathcal{N}_{n-1,p}(\mathbf{O}, \mathbf{I}_{n-1} \otimes \mathbf{I}_p)$ by Definition 7, thus it follows that $(n-1)\mathbf{S} = (n-1)\mathbf{C}\mathbf{S}_z\mathbf{C}^\top \sim \mathcal{W}_p(n-1, \boldsymbol{\Sigma})$.

Example 31 *An interesting characterization of the normal distribution due to Carl Friedrich Gauss was revisited by Azzalini and Genton (2007) and extended to the multivariate setting. It is well known that if $\mathbf{X}_1, \ldots, \mathbf{X}_n$ is a random sample from $\mathcal{N}_p(\boldsymbol{\mu}, \boldsymbol{\Sigma})$, then the maximum likelihood estimator of $\boldsymbol{\mu}$ is the sample mean vector, $\bar{\mathbf{X}}$, even when $\boldsymbol{\Sigma}$ is unknown. The converse is less known: if the sample mean vector is the maximum likelihood estimator of the location of a given multivariate parametric location family of distributions with pdf $f(\mathbf{x} - \boldsymbol{\mu})$, can we conclude that this family is the multivariate normal, i.e., $f = \phi_p$? The answer is positive; see the aforementioned reference for further details and the proof.*

3.11.3 The Wishart matrix distribution

The Wishart matrix distribution is a matrix generalization of the univariate χ^2 distribution and has many similar properties. It occurs in many settings involving normal matrix-valued quadratic forms. As defined previously, we call $\mathbf{X} = (\mathbf{X}_1, \ldots, \mathbf{X}_n)^\top \in \mathbb{R}^{n \times p}$ with $\mathbf{X}_1, \ldots, \mathbf{X}_n$ iid from $\mathcal{N}_p(\boldsymbol{\mu}, \boldsymbol{\Sigma})$ a *normal data matrix*.

Definition 7 [Wishart matrix distribution] *A random matrix $\mathbf{W} \in \mathbb{R}^{p \times p}$ has a Wishart matrix distribution with positive definite scale matrix $\boldsymbol{\Sigma} \in \mathbb{R}^{p \times p}$ and degrees of freedom parameter m if it can be written as $\mathbf{W} = \mathbf{Z}^\top\mathbf{Z}$, where $\mathbf{Z} \in \mathbb{R}^{m \times p}$ is a normal data matrix from $\mathcal{N}_p(\mathbf{0}_p, \boldsymbol{\Sigma})$. In that case, we use the notation: $\mathbf{W} \sim \mathcal{W}_p(m, \boldsymbol{\Sigma})$.*

If $\boldsymbol{\Sigma} = \mathbf{I}_p$, then $\mathbf{W} \sim \mathcal{W}_p(m, \mathbf{I}_p)$ is said to have a standard Wishart matrix distribution. Some important properties are:

[1] $\mathbf{W} \sim \mathcal{W}_p(m, \mathbf{I}_p)$ iff $\mathbf{W} \overset{d}{=} \sum_{i=1}^m \mathbf{Z}_i\mathbf{Z}_i^\top$ with iid $\mathbf{Z}_i \sim \mathcal{N}_p(\mathbf{0}_p, \mathbf{I}_p)$;

[2] $\tilde{\mathbf{W}} \sim \mathcal{W}_p(m, \boldsymbol{\Sigma})$ iff $\tilde{\mathbf{W}} \overset{d}{=} \mathbf{CWC}^\top$ with $\boldsymbol{\Sigma} = \mathbf{CC}^\top$ and $\mathbf{W} \sim \mathcal{W}_p(m, \mathbf{I}_p)$;

[3] If $\mathbf{W} \sim \mathcal{W}_p(m, \mathbf{I}_p)$ then $\mathrm{tr}(\mathbf{W}) \sim \chi^2_{mp}$;

[4] If $\mathbf{W} \sim \mathcal{W}_p(m, \boldsymbol{\Sigma})$ and $\mathbf{A} \in \mathbb{R}^{q \times p}$, then $\mathbf{AWA}^\top \sim \mathcal{W}_q(m, \mathbf{A}\boldsymbol{\Sigma}\mathbf{A}^\top)$. Therefore, diagonal submatrices of $\mathbf{W}$ have a Wishart matrix distribution, and in particular, each diagonal element of $\mathbf{W}$ has a $\sigma_i^2\chi_m^2$ distribution. Moreover, $\boldsymbol{\Sigma}^{-1/2}\mathbf{W}\boldsymbol{\Sigma}^{-1/2} \sim \mathcal{W}_p(m, \mathbf{I}_p)$;

[5] If $\mathbf{W} \sim \mathcal{W}_p(m, \boldsymbol{\Sigma})$ and partition $\mathbf{W} = (\mathbf{W}_{ij})_{i,j=1,2}$ and $\boldsymbol{\Sigma} = (\boldsymbol{\Sigma}_{ij})_{i,j=1,2}$ with $\mathbf{W}_{ij}$ and $\boldsymbol{\Sigma}_{ij}$ of dimension $p_i \times p_j$ and $p_1 + p_2 = p$, then $\mathbf{W}_{ii} \sim \mathcal{W}_{p_i}(m, \boldsymbol{\Sigma}_{ii})$ for $i = 1, 2$;

[6] If $\mathbf{W} \sim \mathcal{W}_p(m, \boldsymbol{\Sigma})$ and $\boldsymbol{\Sigma}_{12} = \mathbf{O}$ then $\mathbf{W}_{11} \perp\!\!\!\perp \mathbf{W}_{22}$;

[7] If $\mathbf{W} \sim \mathcal{W}_p(m,\boldsymbol{\Sigma})$ with $m \geq p$, $\boldsymbol{\Sigma} > 0$, $\mathbf{W}_{11\cdot 2} = \mathbf{W}_{11} - \mathbf{W}_{12}\mathbf{W}_{22}^{-1}\mathbf{W}_{21}$, and $\boldsymbol{\Sigma}_{11\cdot 2} = \boldsymbol{\Sigma}_{11} - \boldsymbol{\Sigma}_{12}\boldsymbol{\Sigma}_{22}^{-1}\boldsymbol{\Sigma}_{21}$, then:

- $\mathbf{W}_{11\cdot 2} \sim \mathcal{W}_{p_1}(m - p_2, \boldsymbol{\Sigma}_{11\cdot 2})$,
- $(\mathbf{W}_{21}|\mathbf{W}_{22} = \mathbf{W}_{22}^0) \sim \mathcal{N}_{p_2,p_1}(\mathbf{W}_{22}^0\boldsymbol{\Sigma}_{22}^{-1}\boldsymbol{\Sigma}_{21}, \mathbf{W}_{22}^0 \otimes \boldsymbol{\Sigma}_{11\cdot 2})$,
- $\mathbf{W}_{22} \sim \mathcal{W}_{p_2}(m, \boldsymbol{\Sigma}_{22})$,
- $\mathbf{W}_{11\cdot 2} \perp\!\!\!\perp (\mathbf{W}_{21}, \mathbf{W}_{22})$;

[8] If $\mathbf{W} \sim \mathcal{W}_p(m,\boldsymbol{\Sigma})$ with $m \geq p$, $\boldsymbol{\Sigma} > 0$, then $|\mathbf{W}|/|\boldsymbol{\Sigma}| \sim \prod_{i=1}^{p} \chi^2_{m-i+1}$;

[9] If $\mathbf{W} \sim \mathcal{W}_p(m,\boldsymbol{\Sigma})$ then $\mathrm{E}(\mathbf{W}) = m\boldsymbol{\Sigma}$.

If $\mathbf{W} \sim \mathcal{W}_p(m,\boldsymbol{\Sigma})$ with $m \geq p$, $\boldsymbol{\Sigma} > 0$, then the pdf of $\mathbf{W}$ is:

$$f(\mathbf{W}) = \frac{1}{2^{mp/2}\Gamma_p(m/2)|\boldsymbol{\Sigma}|^{m/2}} |\mathbf{W}|^{(m-p-1)/2} \operatorname{etr}\left(-\frac{1}{2}\boldsymbol{\Sigma}^{-1}\mathbf{W}\right), \qquad \mathbf{W} > 0, \tag{3.18}$$

where $\Gamma_p(v) = \pi^{p(p-1)/4} \prod_{i=1}^{p} \Gamma\{v - (i-1)/2\}$, $v > (p-1)/2$ is the generalized gamma function. The pdf of $\mathbf{W} \sim \mathcal{W}_p(m, \mathbf{I}_p)$ is obtained by setting $\boldsymbol{\Sigma} = \mathbf{I}_p$ in (3.18). Moreover, if $p = 1$, then the pdf reduces to the pdf of a chi-squared random variable. The pdf (3.18) with $\boldsymbol{\Sigma} = \mathbf{I}_p$ is obtained by computing the pdf of $(W_{11\cdot 2}, \mathbf{W}_{21}, \mathbf{W}_{22})$ where, from [7] with $p_1 = 1$ and $p_2 = p - 1$, $W_{11\cdot 2} \sim \chi^2_{m-p+1}$, $(\mathbf{W}_{21}|\mathbf{W}_{22} = \mathbf{W}_{22}^0) \sim \mathcal{N}_{p-1}(\mathbf{0}_{p-1}, \mathbf{W}_{22}^0)$, and $\mathbf{W}_{22} \sim \mathcal{W}_{p-1}(m, \mathbf{I}_{p-1})$. The Jacobian of the transformation from $(W_{11\cdot 2}, \mathbf{W}_{21}, \mathbf{W}_{22})$ to $(W_{11}, \mathbf{W}_{21}, \mathbf{W}_{22})$ is equal to 1 and leads to the desired result. The general pdf (3.18) for $\boldsymbol{\Sigma} > 0$ follows from [4] with $\boldsymbol{\Sigma} = \mathbf{A}\mathbf{A}^\top$ and the Jacobian $\mathrm{abs}(|\mathbf{A}|)^{p+1}$.

3.11.4 *The Mahalanobis distance*

The notion of distance between two multivariate distributions is important in many applications. A seminal proposal is the *Mahalanobis distance* between two random vectors $\mathbf{X}_1 \in \mathbb{R}^p$ and $\mathbf{X}_2 \in \mathbb{R}^p$ in a metric given by a positive definite matrix $\boldsymbol{\Sigma} \in \mathbb{R}^{p \times p}$:

$$\Delta(\mathbf{X}_1, \mathbf{X}_2; \boldsymbol{\Sigma}) = \{(\mathbf{X}_1 - \mathbf{X}_2)^\top \boldsymbol{\Sigma}^{-1} (\mathbf{X}_1 - \mathbf{X}_2)\}^{1/2}.$$

Here, $\Delta(\mathbf{X}_1, \mathbf{X}_2; \boldsymbol{\Sigma})$ is a random variable. If $\mathbf{X}_1$ has mean vector $\boldsymbol{\mu}_1$ and covariance matrix $\boldsymbol{\Sigma}$ then $\Delta(\mathbf{X}_1, \boldsymbol{\mu}_1; \boldsymbol{\Sigma})$ is also a random variable. If $\mathbf{X}_2$ has mean vector $\boldsymbol{\mu}_2$ and covariance matrix $\boldsymbol{\Sigma}$ then $\Delta(\boldsymbol{\mu}_1, \boldsymbol{\mu}_2; \boldsymbol{\Sigma})$ is no longer a random variable: It simply measures the distance between the mean vectors of the two populations. Under an affine transformation with nonsingular matrix $\mathbf{A} \in \mathbb{R}^{p \times p}$ and shift vector $\mathbf{b} \in \mathbb{R}^p$, the Mahalanobis distance is invariant:

$$\Delta(\mathbf{A}\mathbf{X}_1 + \mathbf{b}, \mathbf{A}\mathbf{X}_2 + \mathbf{b}; \mathbf{A}\boldsymbol{\Sigma}\mathbf{A}^\top) = \Delta(\mathbf{X}_1, \mathbf{X}_2; \boldsymbol{\Sigma}).$$

3.11.5 *The Hotelling T^2 distribution*

The distribution of quadratic forms in a normal random vector and in a metric defined by a random matrix with a Wishart distribution plays a very important role in normal distribution theory and is due to Hotelling (1931).

Definition 8 [Hotelling T^2 distribution] *Let $\mathbf{Z} \sim \mathcal{N}_p(\mathbf{0}_p, \mathbf{I}_p)$ be independent of $\tilde{\mathbf{W}} \sim \mathcal{W}_p(m, \mathbf{I}_p)$. Then $H^2 = m\mathbf{Z}^\top \tilde{\mathbf{W}}^{-1}\mathbf{Z}$ has a Hotelling T^2 distribution. In that case, we use the notation: $H^2 \sim \mathcal{T}^2(p,m)$.*

Theorem 15 *Let* $\mathbf{X} \sim \mathcal{N}_p(\boldsymbol{\mu}, \boldsymbol{\Sigma})$ *be independent of* $\mathbf{W} \sim \mathcal{W}_p(m, \boldsymbol{\Sigma})$. *Then* $H^2 = m(\mathbf{X} - \boldsymbol{\mu})^\top \mathbf{W}^{-1}(\mathbf{X} - \boldsymbol{\mu}) \sim \mathcal{T}^2(p, m)$.

Proof 15 *Consider* $\mathbf{Z} = \boldsymbol{\Sigma}^{-1/2}(\mathbf{X} - \boldsymbol{\mu})$ *and* $\tilde{\mathbf{W}} = \boldsymbol{\Sigma}^{-1/2}\mathbf{W}\boldsymbol{\Sigma}^{-1/2}$ *in Definition 8.* □

More generally, H^2 is invariant under affine transformations with nonsingular matrix $\mathbf{A} \in \mathbb{R}^{p \times p}$ and shift vector $\mathbf{b} \in \mathbb{R}^p$.

The univariate case $\mathcal{T}^2(1, n)$ is the square of a Student's t random variable with n degrees of freedom, or in other words, it is a Fisher's F distribution with 1 and n degrees of freedom, $\mathcal{F}_{1,n}$. More generally, it can be shown that $\mathcal{T}^2(p, n) = \{np/(n - p + 1)\}\mathcal{F}_{1,n-p+1}$.

Example 32 *Let* $\mathbf{X} = (\mathbf{X}_1, \ldots, \mathbf{X}_n)^\top$ *be a normal data matrix from* $\mathcal{N}_p(\boldsymbol{\mu}, \boldsymbol{\Sigma})$. *As we know,* $\bar{\mathbf{X}} \perp\!\!\!\perp \mathbf{S}$ *with* $\sqrt{n}(\bar{\mathbf{X}} - \boldsymbol{\mu}) \sim \mathcal{N}_p(\mathbf{0}_p, \boldsymbol{\Sigma})$ *and* $(n-1)\mathbf{S} \sim \mathcal{W}_p(n-1, \boldsymbol{\Sigma})$. *For* $n - 1 \geq p$ *and* $|\boldsymbol{\Sigma}| \neq 0$, *we also know that* $|\mathbf{S}| \neq 0$ *with probability 1, hence:*

$$n(\bar{\mathbf{X}} - \boldsymbol{\mu})^\top \mathbf{S}^{-1}(\bar{\mathbf{X}} - \boldsymbol{\mu}) \sim \mathcal{T}^2(p, n-1),$$

which is a pivotal quantity for $\boldsymbol{\mu}$.

3.11.6 *Wishart-related distributions*

The eigenvalues of functions of Wishart matrices play an important role in various multivariate hypothesis testing procedures; hence, one can consider the following distributions.

Definition 9 *Let* $\mathbf{W}_1 \sim \mathcal{W}_p(m_1, \mathbf{I}_p)$, $m_1 \geq p$, *independent of* $\mathbf{W}_2 \sim \mathcal{W}_p(m_2, \mathbf{I}_p)$. *Then:*

[1] The joint distribution of the $\min(n_2, p)$ *eigenvalues of* $\mathbf{W}_1^{-1}\mathbf{W}_2$ *is* $\Psi(p, m_1, m_2)$.

[2] The distribution of $|\mathbf{W}_1|/|\mathbf{W}_1 + \mathbf{W}_2| = |\mathbf{I}_p + \mathbf{W}_1^{-1}\mathbf{W}_2|^{-1}$ *is* $\Lambda(p, m_1, m_2)$, *a Wilks' lambda.*

[3] The distribution of the largest eigenvalue of $(\mathbf{W}_1 + \mathbf{W}_2)^{-1}\mathbf{W}_2$ *is* $\theta(p, m_1, m_2)$.

Since $m_1 \geq p$, $\mathbf{W}_1^{-1}$ exists. Moreover, it can be shown that the distributions in Definition 9 are the same if $\mathbf{W}_i$ is replaced by $\tilde{\mathbf{W}}_i = \boldsymbol{\Sigma}^{1/2}\mathbf{W}_i\boldsymbol{\Sigma}^{1/2}$, $i = 1, 2$, in other words, if $\tilde{\mathbf{W}}_i \sim \mathcal{W}_p(n_i, \boldsymbol{\Sigma})$, $i = 1, 2$.

It can be shown that $\Psi(p, m_1, m_2)$ is identical to $\Psi(m_2, m_1 + m_2 - p, p)$, $\Lambda(p, m_1, m_2)$ is identical to $\Lambda(m_2, m_1 + m_2 - p, p)$, and $\theta(p, m_1, m_2)$ is identical to $\theta(m_2, m_1 + m_2 - p, p)$.

A generalization of Fisher's F distribution to the multivariate setting is given by:

$$|\mathbf{W}_1^{-1}\mathbf{W}_2| = \frac{|\mathbf{W}_2|}{|\mathbf{W}_1|} \sim \prod_{i=1}^{p} \frac{m_2 - i + 1}{m_1 - i + 1} F_i,$$

where, in the product, $F_i \sim \mathcal{F}_{m_2 - i + 1, m_1 - i + 1}$, and are independent. When $p = 1$, this reduces to the classical $\frac{W_2}{W_1} \sim \frac{m_2}{m_1}\mathcal{F}_{m_2, m_1}$ with $W_1 \sim \sigma^2\chi^2_{m_1}$ independent of $W_2 \sim \sigma^2\chi^2_{m_2}$ that often appears in testing procedures.

Definition 10 [Univariate beta distribution] *A random variable* B *on the unit interval is said to have a beta distribution,* $B \sim \mathcal{B}eta(a, b)$, *iff* $B \stackrel{d}{=} \dfrac{X_1}{X_1 + X_2}$, *with* $X_1 \perp\!\!\!\perp X_2$, $X_1 \sim \chi^2_{2a}$ *and* $X_2 \sim \chi^2_{2b}$.

Similarly to the above, a generalization of the beta distribution to the multivariate setting is given by:

$$\Lambda(p,m_1,m_2) \sim \prod_{i=1}^{m_2} B_i,$$

where, in the product, $B_i \sim \mathcal{B}eta((m_1+i-p)/2, p/2)$, and are independent. When $p=1$, this reduces to the classical $\frac{W_1}{W_1+W_2} \sim \mathcal{B}eta(m_1/2, m_2/2)$ with $W_1 \sim \sigma^2\chi^2_{m_1}$ independent of $W_2 \sim \sigma^2\chi^2_{m_2}$ that often appears in testing procedures. A Bartlett approximation yields $-(n_1-(p-m_2+1)/2)\log\Lambda(p,m_1,m_2) \sim \chi^2_{m_1 p}$ for large m_1.

Finally, the distribution of the largest eigenvalue of $(\mathbf{W}_1+\mathbf{W}_2)^{-1}\mathbf{W}_2$ can be related to Wilk's lambda distribution and therefore to Fisher's F distribution, e.g.:

$$\frac{\theta(p,m_1,1)}{1-\theta(p,m_1,1)} = \frac{1-\Lambda(p,m_1,1)}{\Lambda(p,m_1,1)} \sim \frac{p}{m_1-p+1}\mathcal{F}_{p,m_1-p+1}.$$

3.11.7 *Multivariate trans-normal distributions*

A simple approach to creating non-normal distributions is to apply a transformation to a normal random vector. Let $\mathbf{Z} \sim \mathcal{N}_p(\mathbf{0}_p, \mathbf{I}_p)$ and $\mathbf{r}: \mathbb{R}^p \to \mathbb{R}^p$ be a one-to-one multivariate transformation. Then

$$\mathbf{Y} = \boldsymbol{\xi} + \boldsymbol{\Omega}^{1/2}\mathbf{r}(\mathbf{Z}) \tag{3.19}$$

is said to have a *multivariate trans-normal distribution*, where $\boldsymbol{\xi} \in \mathbb{R}^p$ and $\boldsymbol{\Omega} \in \mathbb{R}^{p\times p}$ is positive definite. Based on the results for transformed random vectors, see, e.g., Section 1.6, the properties of the new random vector $\mathbf{Y}$ can be derived and studied. Moreover, the normal random vector $\mathbf{Z}$ can be recovered with the reciprocal relation $\mathbf{Z} = \mathbf{r}^{-1}\{\boldsymbol{\Omega}^{-1/2}(\mathbf{Y}-\boldsymbol{\xi})\}$. Another option is to introduce the dependence before the transformation, i.e., take $\mathbf{Z} \sim \mathcal{N}_p(\mathbf{0}_p, \bar{\mathbf{R}})$ with $\bar{\mathbf{R}} \in \mathbb{R}^{p\times p}$ a correlation matrix and $\boldsymbol{\Omega}$ to be diagonal in Equation (3.19). Next, two examples of multivariate trans-normal distributions are presented.

Example 33 (Tukey g-and-h distribution) *Let $Z \sim \mathcal{N}(0,1)$ and consider the nonlinear transformation*

$$r_{g,h}(z) = \begin{cases} z\exp(hz^2/2), & \text{if } g=0, z\in\mathbb{R}, \\ g^{-1}\{\exp(gz)-1\}\exp(hz^2/2), & \text{if } g\neq 0, z\in\mathbb{R}, \end{cases}$$

with $g \in \mathbb{R}$ and $h \geq 0$. Tukey (1977) showed that $r_{g,h}$ is a one-to-one transformation, and the random variable $Y = \xi + \omega r_{g,h}(Z)$ with $\xi \in \mathbb{R}$ and $\omega > 0$ is said to have a univariate Tukey g-and-h distribution. The parameter g controls the skewness in the distribution, and the parameter h controls its tails (heavier than the normal). Efficient maximum approximate likelihood inference for this distribution has been proposed by Xu and Genton (2015). More details can be found in Yan and Genton (2019) and references therein. The reciprocal transformation $r^{-1}_{g,h}$ has no explicit form except when $g=0$, in which case it is based on the Lambert's W function (Goerg, 2011, 2015). In the multivariate case, Field and Genton (2006) proposed using Equation (3.19) with $\mathbf{r}(\mathbf{Z}) = (r_{g_1,h_1}(Z_1),\ldots,r_{g_p,h_p}(Z_p))^\top$ where $\mathbf{Z} \sim \mathcal{N}_p(\mathbf{0}_p,\mathbf{I}_p)$, $\mathbf{g} = (g_1,\ldots,g_p)^\top \in \mathbb{R}^p$, and $\mathbf{h} = (h_1,\ldots,h_p)^\top \in \mathbb{R}^p_+$. The resulting distribution for $\mathbf{Y}$ is coined the multivariate Tukey g-and-h (TGH) distribution. The vector $\mathbf{g}$ controls the skewness and the vector $\mathbf{h}$ the kurtosis in the marginal components of $\mathbf{Y}$. If $\mathbf{g}=\mathbf{h}=\mathbf{0}_p$, then $\mathbf{Y} \sim \mathcal{N}_p(\boldsymbol{\xi},\boldsymbol{\Omega})$.

Example 34 (Sinh-arcsinh distribution) *Let $Z \sim \mathcal{N}(0,1)$ and consider the nonlinear transformation*

$$r_{e,\delta}(z) = \sinh\{\delta \operatorname{arcsinh}(z) - e\}, \quad z \in \mathbb{R},$$

with $e \in \mathbb{R}$ and $\delta \geq 0$. Jones and Pewsey (2009) showed that $r_{e,\delta}$ is a one-to-one transformation, and the random variable $Y = \xi + \omega r_{e,\delta}^{-1}(Z)$ with $\xi \in \mathbb{R}$ and $\omega > 0$ is said to have a univariate sinh-arcsinh distribution. The parameter e controls the skewness in the distribution, and the parameter δ controls its tails (heavier or lighter than the normal if $\delta < 1$ or $\delta > 1$). The reciprocal transformation has an explicit form: $r_{e,\delta}^{-1}(z) = r_{-e/\delta,1/\delta}(z)$. In the multivariate case, Jones and Pewsey (2009) proposed to use Equation (3.19) with $\mathbf{r}(\mathbf{Z}) = (r_{e_1,\delta_1}^{-1}(Z_1), \ldots, r_{e_p,\delta_p}^{-1}(Z_p))^\top$ where $\mathbf{Z} \sim \mathcal{N}_p(\mathbf{0}_p, \bar{\mathbf{R}})$ with $\bar{\mathbf{R}} \in \mathbb{R}^{p\times p}$ a correlation matrix and $\mathbf{\Omega}$ diagonal, $\mathbf{e} = (e_1, \ldots, e_p)^\top \in \mathbb{R}^p$ and $\boldsymbol{\delta} = (\delta_1, \ldots, \delta_p)^\top \in \mathbb{R}_+^p$. The resulting distribution for $\mathbf{Y}$ is coined the multivariate sinh-arcsinh (SAS) distribution. The vector $\mathbf{e}$ controls the skewness and the vector $\boldsymbol{\delta}$ the kurtosis in the marginal components of $\mathbf{Y}$. If $\mathbf{e} = \boldsymbol{\delta} = \mathbf{0}_p$, then $\mathbf{Y} \sim \mathcal{N}_p(\boldsymbol{\xi}, \mathbf{\Omega}^{1/2}\bar{\mathbf{R}}\mathbf{\Omega}^{1/2})$.

3.11.8 *Truncated multivariate normal distributions*

The truncated multivariate normal distribution (Tallis, 1961; Horrace, 2005) arises when a multivariate normal random vector is restricted to lie inside a prescribed region of the Euclidean space. Let $\mathbf{W} \sim \mathcal{N}_p(\boldsymbol{\mu}, \mathbf{\Sigma})$ and let $A \subset \mathbb{R}^p$ denote a measurable constraint set, often defined by componentwise or linear inequalities. The truncated multivariate normal distribution is the conditional distribution $\mathbf{U} \stackrel{d}{=} (\mathbf{W} \mid \mathbf{W} \in A)$ (a particular selection distribution, see Chapter 9) with pdf

$$f_{\mathbf{U}}(\mathbf{u}) = \frac{\phi_p(\mathbf{u}; \boldsymbol{\mu}, \mathbf{\Sigma}) I_A(\mathbf{u})}{\mathrm{P}(\mathbf{W} \in A)}, \quad \mathbf{u} \in \mathbb{R}^p,$$

and mgf

$$M_{\mathbf{U}}(\mathbf{t}) = M_{\mathbf{W}}(\mathbf{t}) \frac{\mathrm{P}(\mathbf{W} + \mathbf{\Sigma}^{1/2}\mathbf{t} \in A)}{\mathrm{P}(\mathbf{W} \in A)}, \quad \mathbf{t} \in \mathbb{R}^p,$$

where $M_{\mathbf{W}}(\mathbf{t}) = e^{\mathbf{t}^\top \boldsymbol{\mu} + \frac{1}{2}\mathbf{t}^\top \mathbf{\Sigma} \mathbf{t}}$ is the mgf of $\mathbf{W} \sim \mathcal{N}_p(\boldsymbol{\mu}, \mathbf{\Sigma})$. Of particular interest is the case where $A = \{\mathbf{x} \mid \mathbf{x} \geq \mathbf{c}\}$. We denote the resulting *truncated below* $\mathbf{c}$ *multivariate normal distribution* by $\mathbf{U} \sim \mathcal{TN}_p(\mathbf{c}; \boldsymbol{\mu}, \mathbf{\Sigma})$. If $\boldsymbol{\mu} = \mathbf{c} = \mathbf{0}_p$ and $\mathbf{\Sigma} = \mathbf{I}_p$, then $\mathbf{U} = |\mathbf{X}|$ and we have explicit moments:

$$\mathrm{E}(\mathbf{U}) = \sqrt{\frac{2}{\pi}} \mathbf{1}_p \quad \text{and} \quad \mathrm{var}(\mathbf{U}) = \left(1 - \frac{2}{\pi}\right) \mathbf{I}_p.$$

However, in general, the normalizing constant of the truncated multivariate normal distribution is intractable in closed form, and so efficient computation has been a major theme in the literature. The classical work of Genz (1992) and Genz and Bretz (2009) developed numerical integration methods for evaluating multivariate normal probabilities (see also Section 3.6), which form the basis of most truncated normal computations. Software implementations further support the use of truncated multivariate normal models. The R (R Core Team, 2025) package `tmvtnorm` (Wilhelm and Manjunath, 2023) provides tools for computing the pdf, cdf, moments, and random samples from the truncated multivariate normal distribution; see also Kan and Robotti (2017). Extensions to truncated multivariate elliptical distributions can be found in Morán-Vásquez and Ferrari (2021) and references therein.

3.12 Measures of Information and Geometry for the Normal

The measures of information and geometry discussed in Section 1.9 possess closed-form expressions in the case of a multivariate normal distribution. Next, we provide these quantities.

Let $(\mathbf{X}^\top, \mathbf{Y}^\top)^\top$ be jointly multivariate normal with $\mathbf{X} \sim \mathcal{N}_p(\boldsymbol{\mu}_\mathbf{X}, \boldsymbol{\Sigma}_{\mathbf{XX}})$, $\mathbf{Y} \sim \mathcal{N}_q(\boldsymbol{\mu}_\mathbf{Y}, \boldsymbol{\Sigma}_{\mathbf{YY}})$, and cross-covariance $\boldsymbol{\Sigma}_{\mathbf{XY}}$. Then the entropy of $\mathbf{X}$ does not depend on $\boldsymbol{\mu}_\mathbf{X}$, and the mutual information between $\mathbf{X}$ and $\mathbf{Y}$ does not depend on $\boldsymbol{\mu}_\mathbf{X}$ and $\boldsymbol{\mu}_\mathbf{Y}$, and they are:

$$
\begin{aligned}
H_\mathbf{X} &= \frac{1}{2}\log\{(2\pi e)^p |\boldsymbol{\Sigma}_{\mathbf{XX}}|\}, \\
I_{\mathbf{X},\mathbf{Y}} &= -\frac{1}{2}\log|\mathbf{I}_p - \boldsymbol{\Sigma}_{\mathbf{XY}}\boldsymbol{\Sigma}_{\mathbf{YY}}^{-1}\boldsymbol{\Sigma}_{\mathbf{YX}}\boldsymbol{\Sigma}_{\mathbf{XX}}^{-1}|.
\end{aligned}
$$

Note that $I_{\mathbf{X},\mathbf{Y}} \geq 0$ and it is equal to 0 iff $\boldsymbol{\Sigma}_{\mathbf{XY}} = \mathbf{O}$.

Next, let $F = \mathcal{N}_p(\boldsymbol{\mu}_0, \boldsymbol{\Sigma}_0)$ and $G = \mathcal{N}_p(\boldsymbol{\mu}_1, \boldsymbol{\Sigma}_1)$. Then the KL and J divergences between F and G are:

$$
\begin{aligned}
D_{\mathrm{KL}}(F \,\|\, G) &= \frac{1}{2}\left[\mathrm{tr}(\boldsymbol{\Sigma}_1^{-1}\boldsymbol{\Sigma}_0) + (\boldsymbol{\mu}_1 - \boldsymbol{\mu}_0)^\top \boldsymbol{\Sigma}_1^{-1}(\boldsymbol{\mu}_1 - \boldsymbol{\mu}_0) - p + \log(|\boldsymbol{\Sigma}_1|/|\boldsymbol{\Sigma}_0|)\right], \\
J(F,G) &= \frac{1}{2}\left[\mathrm{tr}(\boldsymbol{\Sigma}_1^{-1}\boldsymbol{\Sigma}_0 + \boldsymbol{\Sigma}_0^{-1}\boldsymbol{\Sigma}_1) - 2p + (\boldsymbol{\mu}_1 - \boldsymbol{\mu}_0)^\top(\boldsymbol{\Sigma}_0^{-1} + \boldsymbol{\Sigma}_1^{-1})(\boldsymbol{\mu}_1 - \boldsymbol{\mu}_0)\right].
\end{aligned}
$$

Note that $D_{\mathrm{KL}}(F\|G) \geq 0$ and $J(F,G) \geq 0$, and they are equal to 0 iff $\boldsymbol{\mu}_0 = \boldsymbol{\mu}_1$ and $\boldsymbol{\Sigma}_0 = \boldsymbol{\Sigma}_1$. Moreover, the Wasserstein distance between F and G for $q = 2$ is:

$$
W_2(F,G) = \left[\|\boldsymbol{\mu}_0 - \boldsymbol{\mu}_1\|^2 + \mathrm{tr}\left\{\boldsymbol{\Sigma}_0 + \boldsymbol{\Sigma}_1 - 2(\boldsymbol{\Sigma}_1^{1/2}\boldsymbol{\Sigma}_0\boldsymbol{\Sigma}_1^{1/2})^{1/2}\right\}\right]^{1/2}.
$$

3.13 Exercises

Exercise 3.1 *Assume* $\mathbf{X} \sim \mathcal{N}_n(\mu\mathbf{1}_n, \sigma^2\{(1-\rho)\mathbf{I}_n + \rho\mathbf{1}_n\mathbf{1}_n^\top\})$, *where* $-1/(n-1) \leq \rho \leq 1$. *Let* $T = \sqrt{n}(\bar{X} - \mu)/S_X$, *where* $\bar{X} = \sum_{i=1}^n X_i/n$ *and* $S_X^2 = \sum_{i=1}^n (X_i - \bar{X})^2/(n-1)$. *Show that* $cT \sim t_{n-1}$, *where*

$$
c = \left\{\frac{(1-\rho)}{(1-\rho)+n\rho}\right\}^{1/2}.
$$

Hint: Use the orthogonal transformation $\mathbf{Y} = \mathbf{G}\mathbf{X}$, *where* $\mathbf{G} = \begin{pmatrix} \frac{1}{\sqrt{n}}\mathbf{1}_n^\top \\ \mathbf{H} \end{pmatrix} \in \mathcal{O}_n$, *for some* $\mathbf{H} \in \mathbb{R}^{(n-1)\times n}$ *such that* $\mathbf{H}\mathbf{H}^\top = \mathbf{I}_{n-1}$ *and* $\mathbf{H}\mathbf{1}_n = \mathbf{0}_{n-1}$.

Exercise 3.2 *Assume* $\mathbf{X} \sim \mathcal{N}_n(\mu\mathbf{1}_n, \boldsymbol{\Sigma})$, *where* $\boldsymbol{\Sigma} = (\sigma_{ij})$ *with* $\sigma_{ii} = \sigma^2$ *and* $\sigma_{ij} = \sigma^2(1-\rho^2)$ *for* $i \neq j$. *Show that* $\bar{X} = n^{-1}\sum_{i=1}^n X_i$ *and* $T = \sum_{i=1}^n (X_i - \bar{X})^2$ *are independent. Hint: Use Theorem 13.*

Exercise 3.3 *Assume* $\mathbf{Y} \sim \mathcal{N}_n(\mathbf{X}\boldsymbol{\beta}, \sigma^2\mathbf{I}_n)$, *where, as usual,* $\mathbf{X}$ *is an* $n \times k$ *model matrix of rank* k *(i.e., its columns are linearly independent). Show, for any matrix* $\mathbf{A} \in \mathbb{R}^{m\times k}$ *of rank* m *and vector* $\mathbf{c} \in \mathbb{R}^m$, *that*

$$
\frac{(\mathbf{A}\hat{\boldsymbol{\beta}} - \mathbf{c})^\top\{\mathbf{A}(\mathbf{X}^\top\mathbf{X})^{-1}\mathbf{A}^\top\}^{-1}(\mathbf{A}\hat{\boldsymbol{\beta}} - \mathbf{c})}{mS_e^2} \sim \mathcal{F}_{m,n-k}(\delta),
$$

where $\delta = (\mathbf{A}\hat{\boldsymbol{\beta}} - \mathbf{c})^\top\{\mathbf{A}(\mathbf{X}^\top\mathbf{X})^{-1}\mathbf{A}^\top\}^{-1}(\mathbf{A}\hat{\boldsymbol{\beta}} - \mathbf{c})/2\sigma^2$ *and* $\mathcal{F}_{\nu_1,\nu_2}(\delta)$ *is the univariate non-central* F*-distribution defined as the distribution of* $F = \frac{Z_1/\nu_1}{Z_2/\nu_2}$, *with* $Z_1 \sim \chi^2_{\nu_1}(\delta)$ *and* $Z_2 \sim \chi^2_{\nu_2}$ *being independent.*

Exercise 3.4 *Let* $\mathbf{X} = (X_1, X_2)^\top \sim \mathcal{N}_2(\mathbf{0}_2, \mathbf{\Sigma})$, $\mathbf{\Sigma} > 0$.

(a) Find the distribution of $|X_1| - |X_2|$;

(b) Assume that $\mathbf{\Sigma}$ *is a correlation matrix with correlation coefficient* $|\rho| < 1$. *Changing to the polar coordinates:* $X_1 = r\cos(\theta)$ *and* $X_2 = r\sin(\theta)$ $(r > 0,\ 0 < \theta < 2\pi)$, *show that:*

(i) The marginal pdf of θ *is*

$$g(\theta) = \frac{\sqrt{1-\rho^2}}{2\pi\{1 - 2\rho\cos(\theta)\sin(\theta)\}}, \qquad 0 < \theta < 2\pi;$$

(ii) $\mathrm{P}(X_1 > 0, X_2 > 0) = \frac{1}{2} - \frac{1}{2\pi}\arccos(\rho)$;

(iii) $\mathrm{P}(X_1 X_2 > 0) = \frac{1}{2} + \frac{1}{\pi}\arcsin(\rho)$;

(iv) $\mathrm{P}(X_1 X_2 < 0) = \frac{1}{\pi}\arccos(\rho)$.

Exercise 3.5 *Prove the properties [1]-[4] from Section 3.11.1.*

Exercise 3.6 *Prove the following statements:*

(a) If $\mathbf{X} \sim \mathcal{N}_{n,p}(\mathbf{M}, \mathbf{\Omega})$, *then* $\mathbf{AXB} \sim \mathcal{N}_{m,q}(\mathbf{AMB}, (\mathbf{A} \otimes \mathbf{B}^\top)\mathbf{\Omega}(\mathbf{A}^\top \otimes \mathbf{B}))$, *where* $\mathbf{A} \in \mathbb{R}^{m\times n}$ *and* $\mathbf{B} \in \mathbb{R}^{p\times q}$ *are constant matrices;*

(b) If $\mathbf{X} \sim \mathcal{N}_{n,p}(\mathbf{M}, \mathbf{\Psi} \otimes \mathbf{\Sigma})$, $\mathbf{\Psi} \geq 0$ *in* $\mathbb{R}^{n\times n}$ *and* $\mathbf{\Sigma} \geq 0$ *in* $\mathbb{R}^{p\times p}$, *then* $\mathrm{E}(\mathbf{X}\mathbf{X}^\top) = \mathbf{\Psi}\mathrm{tr}(\mathbf{\Sigma}) + \mathbf{M}\mathbf{M}^\top$.

Exercise 3.7 *Show that* $\mathbf{X} \sim \mathcal{N}_{n,p}(\mathbf{M}, \mathbf{\Psi} \otimes \mathbf{\Sigma})$, *with* $\mathbf{\Psi} > 0$ *and* $\mathbf{\Sigma} > 0$, *iff* $\mathbf{X} = \mathbf{\Psi}^{1/2}\mathbf{Z}\mathbf{\Sigma}^{1/2} + \mathbf{M}$ *where* $\mathbf{Z} \sim \mathcal{N}_{n,p}(\mathbf{O}, \mathbf{I}_n \otimes \mathbf{I}_p)$.

Exercise 3.8 *Let* $\mathbf{X} = (\mathbf{X}_1, \ldots, \mathbf{X}_n)^\top \sim \mathcal{N}_{n,p}(\mathbf{O}, \mathbf{I}_n \otimes \mathbf{\Sigma})$ *and* $\mathbf{P} \in \mathbb{R}^{n\times n}$ *be an orthogonal projection matrix of rank m. Show that* $\mathbf{X}^\top\mathbf{P}\mathbf{X} \sim \mathcal{W}_p(m, \mathbf{\Sigma})$. *Hint: Write* $\mathbf{P} = \mathbf{H}\mathbf{H}^\top$, *where* $\mathbf{H} \in \mathbb{R}^{n\times m}$ *is an orthonormal basis, to show that* $\mathbf{X}^\top\mathbf{P}\mathbf{X} = \mathbf{Z}^\top\mathbf{Z}$, *where* $\mathbf{Z} = \mathbf{H}^\top\mathbf{X}$.

Exercise 3.9 *Let* $\mathbf{W} \sim \mathcal{W}_p(m, \mathbf{\Sigma})$. *Show that* $\mathrm{E}(\mathbf{W}) = m\mathbf{\Sigma}$ *and* $\mathrm{var}(\mathbf{W}) = m(\mathbf{I}_{p^2} + \mathbf{K}_p)(\mathbf{\Sigma} \otimes \mathbf{\Sigma})$, *where* $\mathbf{K}_p$ *is the commutation matrix.*

Exercise 3.10 *Prove that if* $\mathbf{W} \sim \mathcal{W}_p(m, \mathbf{I}_p)$ *with* $m \geq p$ *then* $\mathbf{W}$ *is nonsingular with probability one. Moreover, if* $\mathbf{W} \sim \mathcal{W}_p(m, \mathbf{\Sigma})$ *with* $m \geq p$ *and* $|\mathbf{\Sigma}| \neq 0$ *then* $|\mathbf{W}| \neq 0$ *with probability one.*

Exercise 3.11 *Prove that if the random matrix* $\mathbf{Z} = (Z_{ij}) \in \mathbb{R}^{n\times n}$ *has iid entries* $Z_{ij} \sim \mathcal{N}(0,1)$ *then* $\mathrm{P}(|\mathbf{Z}| = 0) = 0$. *Moreover,* $\mathrm{P}(|\mathbf{Z}| = c) = 0$ *for all* $c \in \mathbb{R}$.

Exercise 3.12 *Let* $\mathbf{W} \sim \mathcal{W}_p(m, \mathbf{I}_p)$ *and* $\mathbf{A} \geq 0$. *Show that* $\mathrm{E}(\mathbf{WAW}) = m(m+1)\mathbf{A} + m\{\mathrm{tr}(\mathbf{A})\}\mathbf{I}_p$.

Exercise 3.13 *Prove the properties [1]-[9] from Section 3.11.3.*

Exercise 3.14 *Let* $\psi_{\mathbf{W}}(\mathbf{T}) = \mathrm{E}[\exp\{\mathrm{i}\,\mathrm{tr}(\mathbf{TW})\}]$, $\mathbf{T}$ *symmetric, be the cf of* $\mathbf{W} \sim \mathcal{W}_p(m, \boldsymbol{\Sigma})$. *Show that* $\psi_{\mathbf{W}}(\mathbf{T}) = |\mathbf{I}_p - 2\mathrm{i}\mathbf{T}\boldsymbol{\Sigma}|^{-m/2}$.

Exercise 3.15 *If* $\mathbf{W}_i \overset{indep}{\sim} \mathcal{W}_p(m_i, \boldsymbol{\Sigma})$, $i = 1, \ldots, n$, *then show that* $\sum_{i=1}^n \mathbf{W}_i \sim \mathcal{W}_p(\sum_{i=1}^n m_i, \boldsymbol{\Sigma})$.

Exercise 3.16 (Inverse-Wishart matrix distribution) *Let* $\mathbf{W} \sim \mathcal{W}_p(m, \boldsymbol{\Sigma})$, $m \geq p$, $\boldsymbol{\Sigma} > 0$. *Using (3.18), show that* $\mathbf{V} = \mathbf{W}^{-1}$ *has pdf given by*

$$g(\mathbf{V}) = \frac{1}{2^{mp/2}\Gamma_p(m/2)|\boldsymbol{\Sigma}|^{m/2}} |\mathbf{V}|^{-(m+p+1)/2} \operatorname{etr}\left(-\frac{1}{2}\boldsymbol{\Sigma}^{-1}\mathbf{V}^{-1}\right), \quad \mathbf{V} > 0.$$

Exercise 3.17 (Wishart correlation matrix distribution) *Let* $\mathbf{W} = (W_{ij}) \sim \mathcal{W}_p(m, \mathbf{I}_p)$ *and* $\mathbf{R} = (R_{ij})$, *with* $R_{ij} = W_{ij}/\sqrt{W_{ii}W_{jj}}$. *Show that* $\mathbf{R}$ *has pdf given by*

$$h(\mathbf{R}) = \frac{\{\Gamma(m/2)\}^p}{\Gamma_p(m/2)} |\mathbf{R}|^{(m-p-1)/2}.$$

Hint: Consider the transformation $\mathbf{W} \to (W_{11}, \ldots, W_{pp}, \mathbf{R})$.

Exercise 3.18 *Compare Equation (3.19) with the case when the dependence is introduced before the transformation and list the pros and cons of each approach.*

Exercise 3.19 *Consider the multivariate Tukey g-and-h distribution in Example 33. Derive its pdf and moments.*

Exercise 3.20 *Consider the multivariate sinh-arcsinh distribution in Example 34. Derive its pdf and moments.*

Exercise 3.21 *Derive the entropy, mutual information, KL divergence, J divergence, and Wasserstein distance for two multivariate normal distributions with diagonal variance matrices and for two univariate* ($p = 1$) *normal distributions.*

Chapter 4

Multivariate Spherical and Elliptical Distributions

The multivariate normal distribution has occupied a central place in statistical analysis, particularly in the analysis of multivariate data. However, in various fields of statistics, the hypothesis of normality has been strongly questioned, essentially due to the low flexibility of the normal distribution to model data with complex structures. In recent decades, many statisticians have been interested in the so-called multivariate spherical (or spherically contoured) distributions and their extension to the multivariate elliptical (or elliptically contoured) distributions, because they generalize the multivariate normal distribution while preserving its symmetrical structure and most of its main properties. In addition, it also includes long- and short-tailed multivariate distributions (relative to the normal), which become important tools in robust inference. The theory of multivariate spherical and elliptical distributions was first formalized by Kelker (1970) in a nonsingular setting and then generalized by Cambanis et al. (1981) in an unrestricted setting. Although there is a short description on this topic in the book of Muirhead (1982), perhaps it was due to the books of Fang et al. (1990) and Fang and Zhang (1990) that the interest in this class of multivariate distributions had a significant increase in the statistical literature. The material presented in this book is essentially based on these references, as well as on some unpublished results presented in the Ph.D. thesis of Arellano-Valle (1994). For an elegant, clear, and concise discussion of parameter estimation and other inferential aspects, such as robustness in elliptical models, see the book of Bilodeau and Brenner (1999).

We begin this chapter with a more formal review of multivariate spherical distributions, because in the construction of the multivariate elliptical family, they play the same role as the standard multivariate normal distribution $\mathcal{N}_p(\mathbf{0}_p, \mathbf{I}_p)$ in the construction of the multivariate normal distribution $\mathcal{N}_p(\boldsymbol{\mu}, \boldsymbol{\Sigma})$. As the multivariate normal distribution is a location-scale extension of the standard multivariate normal distribution, each multivariate elliptical distribution becomes a location-scale extension of a given multivariate spherical distribution. Since $\mathcal{N}_p(\mathbf{0}_p, \mathbf{I}_p)$ is also multivariate spherical, $\mathcal{N}_p(\boldsymbol{\mu}, \Sigma)$ becomes multivariate elliptical, and in this sense, the multivariate elliptical distributions can be seen as an extension of the multivariate normal distribution.

There are several ways to define multivariate spherical and elliptical distributions. A form that seems natural and that covers both the singular and nonsingular cases is to start by defining the multivariate spherical distributions in an intuitively appealing way (see Chapter 2, Section 2.2.1) and then extend them to multivariate elliptical distributions by means of affine transformations. Following this approach, we start by defining multivariate spherical distributions and then characterizing them in different ways.

4.1 Multivariate Spherical Distributions

4.1.1 Construction

As established in Section 2.2.1 of Chapter 2, the multivariate spherical (or spherically contoured) distributions are those that remain invariant with respect to orthogonal transformations. This means, for example, that they are unaffected by rotations or reflections. Consequently, every multivariate spherical distribution is necessarily symmetric. This is formally established in the following definition.

Definition 11 [Multivariate spherical distributions] *A random vector* $\mathbf{X} \in \mathbb{R}^p$ *is said to have a multivariate spherical distribution iff* $\boldsymbol{\Gamma}\mathbf{X}$ *and* $\mathbf{X}$ *are equally distributed for every orthogonal* $p \times p$*-matrix* $\boldsymbol{\Gamma}$*, and we write:*

$$\boldsymbol{\Gamma}\mathbf{X} \stackrel{d}{=} \mathbf{X}, \quad \forall \boldsymbol{\Gamma} \in \mathcal{O}_p. \tag{4.1}$$

Definition 11 states that any orthogonally invariant random vector $\mathbf{X}$ is spherically distributed and vice versa. However, there are several equivalent ways to determine a multivariate spherical distribution. The following theorem is important in the sequel. It reveals that the multivariate distribution of any orthogonally invariant random vector is completely determined by the distribution of its first coordinate. A similar formulation of this fact was presented in Bilodeau and Brenner (1999).

Theorem 16 *A random vector* $\mathbf{X} = (X_1, \ldots, X_p)^\top$ *is orthogonally invariant (spherically distributed) iff:*

$$\mathbf{t}^\top \mathbf{X} \stackrel{d}{=} X_1, \quad \forall \mathbf{t} \in \mathbb{S}^{(p-1)},$$

where $\mathbb{S}^{(p-1)} = \{\mathbf{u} \in \mathbb{R}^p : \|\mathbf{u}\| = 1\}$ *is the unit sphere in* $\mathbb{R}^p$.

Proof 16 ($\Longrightarrow$) *Let* $\mathbf{t}^\top$ *be the first row of* $\boldsymbol{\Gamma} \in \mathcal{O}_p$*. Since* $\mathbf{t}^\top\boldsymbol{\Gamma} = \mathbf{e}_1^\top$ *where* $\mathbf{e}_1 = (1, 0, \ldots, 0)^\top \in \mathbb{R}^p$*, then*

$$\boldsymbol{\Gamma}\mathbf{X} \stackrel{d}{=} \mathbf{X}, \quad \forall \boldsymbol{\Gamma} \in \mathcal{O}_p \Longrightarrow \mathbf{t}^\top\boldsymbol{\Gamma}\mathbf{X} = X_1 \stackrel{d}{=} \mathbf{t}^\top\mathbf{X}, \quad \forall \mathbf{t} \in \mathbb{S}^{(p-1)}.$$

($\Longleftarrow$) *Let* $\boldsymbol{\Gamma} \in \mathcal{O}_p$*. Since* $\mathbf{t} \in \mathbb{S}^{(p-1)} \Longrightarrow \mathbf{t}^\top\boldsymbol{\Gamma} \in \mathbb{S}^{(p-1)}$*, it is immediate that*

$$X_1 \stackrel{d}{=} \mathbf{t}^\top\mathbf{X} \stackrel{d}{=} \mathbf{t}^\top\boldsymbol{\Gamma}\mathbf{X}, \quad \forall \mathbf{t} \in \mathbb{S}^{(p-1)}.$$

The conclusion follows from the Cramér-Wold theorem (property [6](c), Chapter 1, Section 1.7). □

4.1.2 Cf and stochastic representation

The following theorem characterizes a multivariate spherical distribution through its cf.

Theorem 17 *Let* $\mathbf{X} \in \mathbb{R}^p$ *be a spherical random vector with cf* $\psi(\mathbf{t}) = \mathrm{E}(e^{\mathrm{i}\mathbf{t}^\top\mathbf{X}})$, $\mathbf{t} \in \mathbb{R}^p$*. Then, the following statements are equivalent:*

[1] $\psi(\boldsymbol{\Gamma}^\top\mathbf{t}) = \psi(\mathbf{t})$, $\forall \mathbf{t} \in \mathbb{R}^p$ *and* $\forall \boldsymbol{\Gamma} \in \mathcal{O}_p$*;*

[2] There is a function $\varphi : [0, +\infty) \to \mathbb{R}$ *such that* $\psi(\mathbf{t}) = \varphi(\|\mathbf{t}\|^2)$, $\forall \mathbf{t} \in \mathbb{R}^p$.

Proof 17 *Let* $\mathbf{H} \in \mathbb{R}^{p\times p}$*. The cf of* $\mathbf{HX}$ *is* $\mathrm{E}(e^{\mathrm{i}\mathbf{t}^\top\mathbf{HX}}) = \psi(\mathbf{H}^\top\mathbf{t})$, $\forall \mathbf{t} \in \mathbb{R}^p$*. Thus, the equivalence of [1] with Definition 11 is immediate from the uniqueness theorem for cfs. Now [1]$\Longrightarrow$[2] because [1] means that* $\psi(\mathbf{t})$ *is invariant with respect to the left multiplication of its argument by an orthogonal matrix. Thus, the invariance property of the orthogonal group* $\mathcal{O}_p$ *(see Section 2.2.2) implies that* $\psi(\mathbf{t})$ *must depend on the argument* $\mathbf{t}$ *through its squared length* $\|\mathbf{t}\|^2$*. Finally, [2]$\Longrightarrow$[1] since* $\psi(\boldsymbol{\Gamma}^\top\mathbf{t}) = \varphi(\|\boldsymbol{\Gamma}\mathbf{t}\|^2) = \varphi(\|\mathbf{t}\|^2) = \psi(\mathbf{t})$ *for all* $\mathbf{t} \in \mathbb{R}^p$ *and* $\boldsymbol{\Gamma} \in \mathcal{O}_p$. □

Note by statement [2] in Theorem 17 that $|\varphi(u)| \leq 1$ for all $u \geq 0$.

We now know from part [2] of Theorem 17 that a random vector $\mathbf{X}$ has a multivariate spherical distribution iff its cf has the form $\varphi(\|\mathbf{t}\|^2)$ for some function φ called *characteristic generator function (cgf)*. Since the cf completely determines the multivariate distribution of a random vector $\mathbf{X}$, we can then use the notation $\mathbf{X} \sim \mathcal{S}_p(\varphi)$ to say that $\mathbf{X}$ has a multivariate spherical distribution that is uniquely determined by the cgf φ. In this sense, the cgf φ represents an infinite-dimensional parameter that identifies a unique distribution within the spherical class.

Example 35 *In the class of multivariate spherical distributions, perhaps the most popular one is the standard multivariate normal distribution; see Chapter 3. Let* $\mathbf{X} \sim \mathcal{N}_p(\mathbf{0}_p, \mathbf{I}_p)$. *As we know,* $\mathbf{X}$ *has pdf* $\phi_p(\mathbf{x}) = (2\pi)^{-p/2} e^{-\|\mathbf{x}\|^2/2}$ *for* $\mathbf{x} \in \mathbb{R}^p$ *and cf* $\psi(\mathbf{t}) = \int_{\mathbb{R}^p} e^{i\mathbf{t}^\top \mathbf{x}} \phi_p(\mathbf{x}) \mathrm{d}\mathbf{x} = e^{-\|\mathbf{t}\|^2/2}$ *for* $\mathbf{t} \in \mathbb{R}^p$. *Also, for a matrix* $\mathbf{G} \in \mathbb{R}^{p \times p}$ *we know that* $\mathbf{G}\mathbf{X} \sim \mathcal{N}_p(\mathbf{0}_p, \mathbf{G}\mathbf{G}^\top)$. *Consequently, for any orthogonal matrix* $\mathbf{\Gamma} \in \mathcal{O}_p$ *we find* $\mathbf{\Gamma}\mathbf{X} \stackrel{d}{=} \mathbf{X} \sim \mathcal{N}_p(\mathbf{0}_p, \mathbf{I}_p)$. *Thus,* $\mathbf{X} \sim \mathcal{S}_p(\varphi)$ *with cgf* $\varphi(u) = e^{-u/2}$, $u \geq 0$.

The following example presents a convenient way to construct multivariate spherical distributions by random rescaling of the standard normal.

Example 36 (Scale mixtures of normals) *Let* $\mathbf{X} = V^{-1/2}\mathbf{Z}$, *where* V *has cdf* G *on* $[0, \infty)$ *that is not related to the dimension* p *of* $\mathbf{X}$, *and* V *is independent of* $\mathbf{Z}$ *which has the distribution* $\mathcal{N}_p(\mathbf{0}_p, \mathbf{I}_p)$. *The random vector* $\mathbf{X}$ *is clearly orthogonally invariant and therefore spherically distributed since* $\mathbf{Z}$ *is (see Example 35). Also, since* $(\mathbf{X} \mid V = v) = v^{-1/2}\mathbf{Z} \sim \mathcal{N}_p(\mathbf{0}_p, v^{-1}\mathbf{I}_p)$, *we have that* $\mathbf{X} \sim \mathcal{S}_p(\varphi)$ *with cgf*

$$\varphi(u) = \int_0^\infty \varphi_{\mathcal{N}}(u/v) \mathrm{d}G(v), \quad u \geq 0,$$

where $\varphi_{\mathcal{N}}(u) = e^{-u/2}$, $u \geq 0$, *is the normal cgf which does not depend on* p. *This means that conditionally on* $V = v$, *the components* $X_1, \ldots, X_p$ *of* $\mathbf{X}$ *are iid* $\mathcal{N}_1(0, v^{-1})$ *random variables. That is, for this class we have for every integer* $p \geq 1$ *that the sequence* $(X_1, \ldots, X_p)$ *has the same distribution of* $(X_{\pi(1)}, \ldots, X_{\pi(p)})$ *for each permutation* $(\pi(1), \ldots, \pi(p))$ *of the integer numbers* $\{1, \ldots, p\}$ *and hence the infinite sequence* $X_1, X_2, \ldots$ *is exchangeable.*

In the theory of multivariate spherical distributions, an important role is played by the multivariate uniform distribution on the shell of the unit ball in $\mathbb{R}^p$, say $\mathbb{S}^{(p-1)} = \{\mathbf{x} \in \mathbb{R}^p : \|\mathbf{x}\| = 1\}$. As shown by Fang and Zhang (1990), this distribution is orthogonally invariant and therefore belongs to the class of multivariate spherical distributions (see Example 19 in Section 2.2.2). In addition, it is the only orthogonal invariant distribution on $\mathbb{S}^{(p-1)}$; see Bilodeau and Brenner (1999) for a proof. Let $\mathbf{U}^{(p)} \sim \mathcal{U}(\mathbb{S}^{(p-1)})$ denote a uniform random vector on the unit sphere $\mathbb{S}^{(p-1)}$. In particular, the cf of $\mathbf{U}^{(p)}$ was proved in Fang and Zhang (1990) (see also Fang et al. (1990)) to be

$$\psi_{\mathcal{U},p}(\mathbf{t}) = \varphi_{\mathcal{U},p}(\|\mathbf{t}\|^2) = \frac{1}{B\{(p-1)/2, 1/2\}} \int_0^\pi e^{i\|\mathbf{t}\| \cos\theta} \sin^{p-2}\theta \mathrm{d}\theta, \quad \mathbf{t} \in \mathbb{R}^p, \tag{4.2}$$

where $B(a,b) = \Gamma(a)\Gamma(b)/\Gamma(a+b)$ is the beta function. This result was originally due to Schoenberg (1938). However, it also follows easily from the fact that $\mathbf{t}^\top \mathbf{U}^{(p)} \stackrel{d}{=} \|\mathbf{t}\| U_1$ (see Theorem 16), where U_1 is the first component of $\mathbf{U}^{(p)}$ and represents the cosine of the angle between $\mathbf{t}$ and $\mathbf{U}^{(p)}$. Thus, since $\psi_{\mathcal{U},p}(\mathbf{t}) = \mathrm{E}(e^{i\mathbf{t}^\top \mathbf{U}^{(p)}}) = \mathrm{E}(e^{i\|\mathbf{t}\| U_1})$ and the marginal pdf of U_1 is (see Example 42)

$$\frac{1}{B\{(p-1), 1/2\}} (1 - u_1^2)^{(p-1)/2-1}, \quad |u_1| < 1,$$

therefore, from the change of variable $u_1 = \cos\theta$, we have

$$\psi_{\mathcal{U},p}(\mathbf{t}) = \varphi_{\mathcal{U},p}(\|\mathbf{t}\|^2) = \mathrm{E}(e^{\mathrm{i}\|\mathbf{t}\|U_1}) = \int_{-1}^{1} e^{\mathrm{i}\|\mathbf{t}\|u_1} \frac{1}{B\{(p-1),1/2\}}(1-u_1^2)^{(p-1)/2-1}du_1$$
$$= \frac{1}{B\{(p-1)/2,1/2\}}\int_0^{\pi} e^{\mathrm{i}\|\mathbf{t}\|\cos\theta}\sin^{p-2}\theta d\theta.$$

For two equivalent and more explicit forms of $\varphi_{\mathcal{U},p}(\|\mathbf{t}\|^2)$, see Exercise 4.1.

Now we characterize the class of cgfs $\varphi : [0,+\infty) \to \mathbb{R}$ of p-variate spherical distributions. Denote this class by

$$\Psi_p = \{\varphi(\cdot) : \varphi(\|\mathbf{t}\|^2), \mathbf{t} \in \mathbb{R}^p, \text{ is a cf}\}, \quad p \geq 1.$$

Schoenberg (1938) showed that

$$\varphi \in \Psi_p \Longleftrightarrow \varphi(u) = \int_0^{\infty} \varphi_{\mathcal{U},p}(r^2u)\mathrm{d}F(r), \quad u \geq 0, \tag{4.3}$$

where F is a cdf over $[0,\infty)$ and $\varphi_{\mathcal{U},p}(\|\mathbf{t}\|^2) = \mathrm{E}(e^{\mathrm{i}\mathbf{t}^\top \mathbf{U}^{(p)}})$ is the cf of the uniform random vector $\mathbf{U}^{(p)} \sim \mathcal{U}(\mathbb{S}^{(p-1)})$ defined by (4.2). This means that any orthogonal invariant distribution is a mixture of uniform distributions on spheres of varying radius $r \geq 0$. A proof of this result can also be found in Fang and Zhang (1990). Moreover, since all marginal distributions of $\mathcal{S}_p(\varphi)$ are also spherical with the same cf, we have $\varphi \in \Psi_p \Longrightarrow \varphi \in \Psi_q$ for every $1 \leq q \leq p$. Thus, since $\Psi_1 \supset \Psi_2 \supset \Psi_3 \supset \cdots$ then $\Psi_p \downarrow \Psi_\infty = \cap_{q=1}^{\infty}\Psi_q$ as $p \to \infty$ and

$$\varphi \in \Psi_\infty \Longleftrightarrow \varphi(u) = \int_0^{\infty} e^{-r^2u/2}\mathrm{d}F_\infty(r), \quad u \geq 0, \tag{4.4}$$

for some cdf F_∞ on $[0,\infty)$ which does not depend on the dimension p of the multivariate spherical distribution. Note by (4.4) that $\mathbf{X} \sim \mathcal{S}_p(\varphi)$ with $\varphi \in \Psi_\infty$ iff $\mathbf{X}$ is a scale mixtures of multivariate normal distributions, i.e., $\mathbf{X} \overset{d}{=} R_\infty \mathbf{Z}$, with $R_\infty \sim F_\infty \perp\!\!\!\perp \mathbf{Z} \sim \mathcal{N}_p(\mathbf{0}_p, \mathbf{I}_p)$; see also Example 36.

A major consequence of (4.3) is the stochastic representation presented in the following theorem, where a spherical random vector $\mathbf{X} \in \mathbb{R}^p$ is represented as the product of a *random direction* $\mathbf{U}^{(p)} \in \mathbb{S}^{(p-1)}$, which is uniformly distributed on $\mathbb{S}^{(p-1)}$, and a *radial or generating random variable* R, which has a cdf F on $[0,\infty)$ related to φ according to (4.3), and is independent of $\mathbf{U}^{(p)}$. In this sense, $\mathcal{U}(\mathbb{S}^{(p-1)})$ is referred to as *unit multivariate spherical distribution*. This fundamental representation of orthogonal random vectors was originally proved in the seminal paper of Cambanis et al. (1981) from the result in (4.3). In that paper, it was also shown that there is a one-to-one relationship between the cgf φ and the cdf F of R. Here, we reproduce an alternative proof presented in Bilodeau and Brenner (1999).

Theorem 18 *Let* $\mathbf{X} \in \mathbb{R}^p$ *be a random vector. Then,*

$$\mathbf{X} \sim \mathcal{S}_p(\varphi) \Longleftrightarrow \mathbf{X} \overset{d}{=} R\mathbf{U}^{(p)},$$

where $R \overset{d}{=} \|\mathbf{X}\|$ *has cdf* F *related to* φ *as in (4.3),* $\mathbf{U}^{(p)} \sim \mathcal{U}(\mathbb{S}^{(p-1)})$ *and* $R \perp\!\!\!\perp \mathbf{U}^{(p)}$.

Proof 18 ($\Longleftarrow$) *Since* $R \perp\!\!\!\perp \mathbf{U}^{(p)}$ *then* $R \perp\!\!\!\perp \mathbf{H}\mathbf{U}^{(p)}$ *for all* $\mathbf{H} \in \mathbb{R}^{p\times p}$. *Because* $\mathbf{U}^{(p)}$ *is orthogonally invariant, then for all* $\mathbf{\Gamma} \in \mathcal{O}_p$ *we have* $(R, \mathbf{\Gamma}\mathbf{U}^{(p)}) \overset{d}{=} (R, \mathbf{U}^{(p)})$ *and therefore* $R\mathbf{\Gamma}\mathbf{U}^{(p)} \overset{d}{=} R\mathbf{U}^{(p)}$. *In turn, this implies* $\mathbf{\Gamma}\mathbf{X} \overset{d}{=} R\mathbf{\Gamma}\mathbf{U}^{(p)} \overset{d}{=} R\mathbf{U}^{(p)} \overset{d}{=} \mathbf{X}$, $\forall \mathbf{\Gamma} \in \mathcal{O}_p$.

($\Longrightarrow$) *Assume* $R \perp\!\!\!\perp \mathbf{U}^{(p)}$ *with* $R \stackrel{d}{=} \|\mathbf{X}\|$ *and* $\mathbf{U}^{(p)} \sim \mathcal{U}(\mathbb{S}^{(p-1)})$. *Also, consider a uniform random vector* $\mathbf{Z} \sim \mathcal{U}(\mathbb{S}^{(p-1)})$ *such that* $\mathbf{Z} \perp\!\!\!\perp \mathbf{X}$, *and let* X_1, Z_1 *and* U_1 *be the first components of* $\mathbf{X}$, $\mathbf{Z}$ *and* $\mathbf{U}^{(p)}$, *respectively. Then, from the following three facts:*

$$\begin{aligned}
(R, \mathbf{U}^{(p)}) &\stackrel{d}{=} (\|\mathbf{X}\|, \mathbf{Z}) \Longrightarrow R\mathbf{U}^{(p)} \stackrel{d}{=} \|\mathbf{X}\|\mathbf{Z} \Longrightarrow RU_1 \stackrel{d}{=} \|\mathbf{X}\|Z_1, \\
\mathbf{X}^\top \mathbf{Z} \mid \mathbf{X} &\stackrel{d}{=} \|\mathbf{X}\|Z_1 \mid \mathbf{X} \Longrightarrow \mathbf{Z}^\top \mathbf{X} \stackrel{d}{=} \|\mathbf{X}\|Z_1, \\
\mathbf{Z}^\top \mathbf{X} \mid \mathbf{Z} &\stackrel{d}{=} X_1 \mid \mathbf{Z} \Longrightarrow X_1 \stackrel{d}{=} \mathbf{Z}^\top \mathbf{X},
\end{aligned}$$

we obtain $X_1 \stackrel{d}{=} RU_1$. *Also, since* $\mathbf{X}$ *and* $\mathbf{U}^{(p)}$ *are both orthogonal invariant, then for any* $\mathbf{s} \in \mathbb{S}^{(p-1)}$, *we have*

$$\mathbf{s}^\top \mathbf{X} \stackrel{d}{=} X_1 \stackrel{d}{=} RU_1 \stackrel{d}{=} \mathbf{s}^\top R\mathbf{U}^{(p)},$$

and by the Cramér-Wold theorem (property [6](c), Chapter 1, Section 1.7) we conclude that $\mathbf{X} \stackrel{d}{=} R\mathbf{U}^{(p)}$ *as stated.* □

Suppose now that $\mathbf{X} \sim \mathcal{S}_p(\varphi)$ has two stochastic representations $\mathbf{X} \stackrel{d}{=} R\mathbf{U}^{(p)}$ and $\mathbf{X} \stackrel{d}{=} R^* \mathbf{U}^{(p)}$, where $R \sim F$ and $R^* \sim F^*$ are defined over $[0, \infty)$ and are independent of $\mathbf{U}^{(p)}$. Since $R \stackrel{d}{=} \|\mathbf{X}\|$ and $R^* \stackrel{d}{=} \|\mathbf{X}\|$, we find $R \stackrel{d}{=} R^*$ or, equivalently, $F(r) = F^*(r)$ for all $r \geq 0$. This effectively confirms that the relationship (4.3) between the cgf φ and the cdf (or generator distribution) F of the radial variable R is one-to-one. As Fang et al. (1990) have said, "the number" of spherical distributions is essentially equal to "the number" of distributions for nonnegative random variables. In other words, the members of the class of spherical distributions in $\mathbb{R}^p$ are in one-to-one correspondence with the members of generator distributions, which are the one-dimensional probability distributions of nonnegative random variables.

According to Theorem 18, the multivariate distribution of any spherical random vector $\mathbf{X}$ is completely determined by the distribution of the norm $\|\mathbf{X}\|$ of $\mathbf{X}$. Furthermore, if one assumes that $\mathrm{P}(\mathbf{X} = \mathbf{0}_p) = 0$, then the direction $\mathbf{X}/\|\mathbf{X}\|$ of $\mathbf{X}$ is $\mathcal{U}(\mathbb{S}^{(p-1)})$ and is distributed independently of the norm $\|\mathbf{X}\|$.

Theorem 19 *Let* $\mathbf{X} \in \mathbb{R}^p$ *be a random vector such that* $\mathrm{P}(\mathbf{X} = \mathbf{0}_p) = 0$. *Then,*

$$\mathbf{X} \sim \mathcal{S}_p(\varphi) \Longleftrightarrow \|\mathbf{X}\| \perp\!\!\!\perp \frac{\mathbf{X}}{\|\mathbf{X}\|} \text{ and } \frac{\mathbf{X}}{\|\mathbf{X}\|} \sim \mathcal{U}(\mathbb{S}^{(p-1)}).$$

Proof 19 ($\Longleftarrow$) *Since* $\mathbf{X} = \|\mathbf{X}\|(\mathbf{X}/\|\mathbf{X}\|)$, *the result is direct from Theorem 18.*

($\Longrightarrow$) *Since* $\mathbf{X} \sim \mathcal{S}_p(\varphi)$, *then* $\mathbf{X} \stackrel{d}{=} R\mathbf{U}^{(p)}$, *with* $R \stackrel{d}{=} \|\mathbf{X}\|$, $\mathbf{U}^{(p)} \sim \mathcal{U}(\mathbb{S}^{(p-1)})$ *and* $R \perp\!\!\!\perp \mathbf{U}^{(p)}$. *Thus,* $\mathrm{P}(R = 0) = \mathrm{P}(\|\mathbf{X}\| = 0) = 0$ *because* $\mathrm{P}(\mathbf{X} = \mathbf{0}_p) = 0$, *and thus we have* $(\|\mathbf{X}\|, \mathbf{X}/\|\mathbf{X}\|) \stackrel{d}{=} (\|R\mathbf{U}^{(p)}\|, R\mathbf{U}^{(p)}/\|R\mathbf{U}^{(p)}\|) = (R, \mathbf{U}^{(p)})$. □

4.1.3 Moments

The notation $\mathbf{X} \sim \mathcal{S}_p^+(\varphi)$ is used by Fang and Zhang (1990) to indicate that $\mathbf{X} \sim \mathcal{S}_p(\varphi)$ and $\mathrm{P}(\mathbf{X} = \mathbf{0}_p) = 0$. As was stated in Theorem 19, within this class, the distribution of the "direction" vector $\mathbf{U}^{(p)} = \mathbf{X}/\|\mathbf{X}\|$ is the same for each of the elements of the class. Therefore, in order to study the other properties in the definition of $\mathbf{U}^{(p)} = \mathbf{X}/\|\mathbf{X}\|$, it is very convenient to assume $\mathbf{X} \sim \mathcal{N}_p(\mathbf{0}_p, \mathbf{I}_p)$ under which $\mathrm{P}(\mathbf{X} = \mathbf{0}_p) = 0$. In particular, as shown below, provided that the required moments exist, the components of a spherical random vector are uncorrelated random variables with zero mean.

Theorem 20 *Assume* $\mathbf{U}^{(p)} = (U_1,\ldots,U_p)^\top \sim \mathcal{U}(\mathbb{S}^{(p-1)})$. *Then,*

$$\mathrm{E}(\mathbf{U}^{(p)}) = \mathbf{0}_p \quad \textit{and} \quad \mathrm{var}(\mathbf{U}^{(p)}) = \mathrm{E}(\mathbf{U}^{(p)}\mathbf{U}^{(p)\top}) = \frac{1}{p}\mathbf{I}_p.$$

Moreover, let $(k_1,\ldots,k_p)$ *be a sequence of nonnegative integers. If at least one of them is odd, then* $\mathrm{E}(\prod_{i=1}^p U_i^{k_i}) = 0$, *and if all the* k_i *are even, then*

$$\mathrm{E}\left(\prod_{i=1}^{p} U_i^{k_i}\right) = \frac{1}{2^{k/2}\prod_{i=1}^{k/2}(\frac{p}{2}+i-1)}\prod_{i=1}^{p}\frac{k_i!}{2^{k_i/2}(k_i/2)!}, \qquad k=\sum_{i=1}^{p} k_i. \tag{4.5}$$

Proof 20 *Let* $\mathbf{X} = (X_1,\ldots,X_p)^\top \sim \mathcal{N}_p(\mathbf{0}_p,\mathbf{I}_p)$. *By Theorem 19, we find* $\mathbf{X} \stackrel{d}{=} \|\mathbf{X}\|\mathbf{U}^{(p)}$, *with* $\|\mathbf{X}\| \perp\!\!\!\perp \mathbf{U}^{(p)}$. *Since* $\|\mathbf{U}^{(p)}\| = 1$ *and* $\|\mathbf{X}\| \sim \chi_p$ *their first and second moments are finite. Thus, the proof follows from the identities* $\mathbf{0}_p = \mathrm{E}(\mathbf{X}) = \mathrm{E}(\|\mathbf{X}\|\mathbf{U}^{(p)}) = \mathrm{E}(\|\mathbf{X}\|)\mathrm{E}(\mathbf{U}^{(p)})$ *and* $\mathbf{I}_p = \mathrm{E}(\mathbf{X}\mathbf{X}^\top) = \mathrm{E}(\|\mathbf{X}\|^2\mathbf{U}^{(p)}\mathbf{U}^{(p)\top}) = \mathrm{E}(\|\mathbf{X}\|^2)\mathrm{E}(\mathbf{U}^{(p)}\mathbf{U}^{(p)\top})$, *and by using that*

$$\mathrm{E}(\|\mathbf{X}\|^k) = \frac{2^{k/2}\Gamma\{(p+k)/2\}}{\Gamma(p/2)}.$$

The proof of (4.5) is similar since $\mathrm{E}(\|\mathbf{X}\|^k)\mathrm{E}(\prod_{i=1}^p U_i^{k_i}) = \mathrm{E}(\prod_{i=1}^p X_i^{k_i})$ *and* $X_1,\ldots,X_p$ *are iid* $\mathcal{N}(0,1)$ *random variables, so that* $\mathrm{E}(\prod_{i=1}^p X_i^{k_i}) = \prod_{i=1}^p \mathrm{E}(X_i^{k_i})$ *with*

$$\mathrm{E}(X_i^{k_i}) = \begin{cases} \frac{k_i!}{2^{k_i/2}(k_i/2)!}, & if\ k_i\ is\ even,\\ 0, & if\ k_i\ is\ odd, \end{cases}$$

for all $i = 1,\ldots,p$. □

Example 37 *Let* $\mathbf{U}^{(p)} = (U_1,\ldots,U_p)^\top \sim \mathcal{U}(\mathbb{S}^{(p-1)})$. *Then:*

$$\mathrm{E}(U_1) = 0, \quad \mathrm{var}(U_1) = \mathrm{E}(U_1^2) = \tfrac{1}{p}, \quad \mathrm{E}(U_1^3) = 0, \quad \mathrm{E}(U_1^4) = \tfrac{3}{p(p+2)},$$
$$\mathrm{cov}(U_1,U_2) = \mathrm{E}(U_1U_2) = 0, \quad \mathrm{E}(U_1^2U_2) = \mathrm{E}(U_1U_2^2) = 0, \quad \mathrm{E}(U_1^2U_2^2) = \tfrac{1}{p(p+2)}.$$

The same occurs with the rest of the components since they all have the same distribution.

From Theorems 18 and 19, it is easy to verify for any spherical random vector $\mathbf{X} \sim \mathcal{S}_p(\varphi)$ that $\mathrm{E}(\mathbf{X})$ exists iff $\mathrm{E}(R) < \infty$ and $\mathrm{var}(\mathbf{X})$ exists iff $\mathrm{E}(R^2) < \infty$. For example, by the spherical symmetry, it is clear that $-\mathbf{X} \stackrel{d}{=} \mathbf{X} \Longrightarrow \mathrm{E}(-\mathbf{X}) = \mathrm{E}(\mathbf{X}) \Longrightarrow \mathrm{E}(\mathbf{X}) = \mathbf{0}_p$. The same happens for all the odd moments, provided that they exist. Furthermore, this result extends to any odd integrable vector-valued function of $\mathbf{X}$, since $-\mathbf{X} \stackrel{d}{=} \mathbf{X} \Longrightarrow \mathbf{g}(-\mathbf{X}) = -\mathbf{g}(\mathbf{X}) \stackrel{d}{=} \mathbf{g}(\mathbf{X}) \Longrightarrow \mathrm{E}\{\mathbf{g}(\mathbf{X})\} = \mathbf{0}_p$.

Theorem 21 *Assume* $\mathbf{X} = (X_1,\ldots,X_p)^\top \stackrel{d}{=} R\mathbf{U}^{(p)} \sim \mathcal{S}_p(\varphi)$ *and* $\mathrm{E}(R^2) < \infty$. *Then,*

$$\mathrm{E}(\mathbf{X}) = \mathbf{0}_p \quad \textit{and} \quad \mathrm{var}(\mathbf{X}) = \mathrm{E}(\mathbf{X}\mathbf{X}^\top) = \frac{1}{p}\mathrm{E}(R^2)\mathbf{I}_p.$$

Moreover, let $(k_1,\ldots,k_p)$ *be a sequence of nonnegative integers such that* $\mathrm{E}(R^k)=\mathrm{E}(\|\mathbf{X}\|^k)<\infty$, *where* $k=\sum_{i=1}^p k_i$. *Then* $\mathrm{E}(\prod_{i=1}^p X_i^{k_i})=0$ *if at least one of the* k_i *is odd, and*

$$\mathrm{E}\left(\prod_{i=1}^p X_i^{k_i}\right)=\frac{\mathrm{E}(R^k)}{2^{k/2}\prod_{i=1}^{k/2}(\frac{p}{2}+i-1)}\prod_{i=1}^p\frac{k_i!}{2^{k_i/2}(k_i/2)!}, \tag{4.6}$$

if all the k_i *are even.*

Proof 21 *Since* $R \perp\!\!\!\perp \mathbf{U}^{(p)}$ *and by assumption* $\mathrm{E}(R)$ *and* $\mathrm{E}(R^2)$ *are both finite, then by Theorem 20, we have* $\mathrm{E}(\mathbf{X})=\mathrm{E}(R\mathbf{U}^{(p)})=\mathrm{E}(R)\mathrm{E}(\mathbf{U}^{(p)})=\mathbf{0}_p$ *and* $\mathrm{var}(\mathbf{X})=\mathrm{E}(\mathbf{X}\mathbf{X}^\top)=\mathrm{E}(R^2\mathbf{U}^{(p)}\mathbf{U}^{(p)\top})=\mathrm{E}(R^2)\mathrm{E}(\mathbf{U}^{(p)}\mathbf{U}^{(p)\top})=\mathrm{E}(R^2)(1/p)\mathbf{I}_p$. *The proof of (4.6) is similar using (4.5).* □

Alternatively, we may use the fact that $\mathbf{X}$ has cf $\psi(\mathbf{t})=\varphi(\|\mathbf{t}\|^2)$ to obtain its moments. In particular, it is easy to find that

$$\frac{\partial\psi(\mathbf{t})}{\partial\mathbf{t}}=2\varphi'(\|\mathbf{t}\|^2)\mathbf{t}\Longrightarrow\mathrm{E}(\mathbf{X})=\frac{\partial\psi(\mathbf{t})}{\partial\mathbf{t}}\Big|_{\mathbf{t}=\mathbf{0}_p}=\mathbf{0}_p,$$

$$\frac{\partial^2\psi(\mathbf{t})}{\partial\mathbf{t}\partial\mathbf{t}^\top}=4\varphi''(\|\mathbf{t}\|^2)\mathbf{t}\mathbf{t}^\top+2\varphi'(\|\mathbf{t}\|^2)\mathbf{I}_p\Longrightarrow\mathrm{E}(\mathbf{X}\mathbf{X}^\top)=-\frac{\partial^2\psi(\mathbf{t})}{\partial\mathbf{t}\partial\mathbf{t}^\top}\Big|_{\mathbf{t}=\mathbf{0}_p}=-2\varphi'(0)\mathbf{I}_p,$$

where $\varphi'(u)$ and $\varphi''(u)$ are the first and second derivatives of $\varphi(u)$. Hence $\mathrm{var}(\mathbf{X})=\mathrm{E}(\mathbf{X}\mathbf{X}^\top)=-2\varphi'(0)\mathbf{I}_p$, from which the relation $\mathrm{E}(R^2)=-2p\varphi'(0)$ follows.

Example 38 *Let* $\mathbf{X}=(X_1,\ldots,X_p)^\top\stackrel{d}{=}R\mathbf{U}^{(p)}\sim\mathcal{S}_p(\varphi)$. *If the required moments of the radial variable* R *exist, then:*

$$\mathrm{E}(X_1)=0,\quad \mathrm{E}(X_1^2)=\frac{\mathrm{E}(R^2)}{p},\quad \mathrm{E}(X_1^3)=0,\quad \mathrm{E}(X_1^4)=\frac{3\mathrm{E}(R^4)}{p(p+2)},$$
$$\mathrm{E}(X_1X_2)=0,\quad \mathrm{E}(X_1^2X_2)=\mathrm{E}(X_1X_2^2)=0,\quad \mathrm{E}(X_1^2X_2^2)=\frac{\mathrm{E}(R^4)}{p(p+2)}.$$

The same occurs with the rest of the components since they all have the same distribution.

As $R\stackrel{d}{=}\|\mathbf{X}\|=(\sum_{i=1}^p X_i^2)^{1/2}$, it is worth mentioning here that $\mathrm{E}(R^k)=\mathrm{E}(\|\mathbf{X}\|^k)=\mathrm{E}\{(\sum_{i=1}^p X_i^2)^{k/2}\}$ for all k, provided that these moments exist. In particular, since $\mathrm{E}(R^2)=\mathrm{E}(\|\mathbf{X}\|^2)=\mathrm{E}(\sum_{i=1}^p X_i^2)=p\mathrm{E}(X_1^2)$, according to Theorem 16, we have $\mathrm{E}(R^2)<\infty$ iff $\mathrm{E}(X_1^2)<\infty$ iff the right-hand derivative of $\varphi(u)$ at $u=0$ ($\varphi'(0)$) exists and is finite. Thus, since in this case $\mathrm{E}(X_1)=0$, we can conclude that $-2\varphi'(0)=\mathrm{E}(X_1^2)=\mathrm{var}(X_1)$. Similarly, $\mathrm{E}(R^4)=\mathrm{E}(\|\mathbf{X}\|^4)=p\mathrm{E}(X_1^4)+p(p-1)\mathrm{E}(X_1^2X_2^2)$, and applying (4.6) to $\mathrm{E}(X_1^2X_2^2)$ when $p\geq 2$ we find the relation $\mathrm{E}(R^4)=p(p+2)\mathrm{E}(X_1^4)/3$, with $\mathrm{E}(X_1^4)=12\varphi''(0)$.

Example 39 *Let* $\mathbf{X}\stackrel{d}{=}R\mathbf{U}^{(p)}\sim\mathcal{S}_p(\varphi)$. *Then,* $\mathbf{X}\sim\mathcal{N}_p(\mathbf{0}_p,\mathbf{I}_p)\Longleftrightarrow\varphi(u)=e^{-u/2}\Longleftrightarrow R\stackrel{d}{=}\|\mathbf{X}\|\sim\chi_p=\sqrt{\chi_p^2}$. *In this case,* $\mathrm{E}(R)=\sqrt{2}\Gamma\{(p+1)/2\}/\Gamma(p/2)$ *and* $\mathrm{E}(R^2)=p\Longrightarrow-2\varphi'(0)=1$ *and so* $\mathrm{var}(\mathbf{X})=\mathbf{I}_p$.

Example 40 (Scale mixtures of normals) *Let* $\mathbf{X} = V^{-1/2}\mathbf{Z}$*, where* V *has cdf* G *in* $[0,\infty)$ *and is independent of* $\mathbf{Z}$ *which has the distribution* $\mathcal{N}_p(\mathbf{0}_p, \mathbf{I}_p)$*. Then,* $R \stackrel{d}{=} \|\mathbf{X}\| \stackrel{d}{=} V^{-1/2}\|\mathbf{Z}\|$ *with* $V \perp\!\!\!\perp \|\mathbf{Z}\| \sim \chi_p$*. In this case,* $\mathrm{E}(R) = \sqrt{2}\Gamma\{(p+1)/2\}\mathrm{E}(V^{-1/2})/\Gamma(p/2)$ *provided that* $\mathrm{E}(V^{-1/2}) < \infty$ *and* $\mathrm{E}(R^2) = p\mathrm{E}(V^{-1})$ *provided that* $\mathrm{E}(V^{-1}) < \infty \Longrightarrow \kappa_2 = -2\varphi'(0) = \mathrm{E}(V^{-1})$*, thus* $\mathrm{E}(\mathbf{X}) = \mathbf{0}_p$ *and* $\mathrm{var}(\mathbf{X}) = \mathrm{E}(V^{-1})\mathbf{I}_p$*. In general, for this class, we have* $\mathrm{E}(\|\mathbf{X}\|^k) < \infty \Longleftrightarrow \mathrm{E}(V^{-k}) < \infty$*,* $\forall k > 0$.

As shown previously, to calculate the moments of a spherical random vector $\mathbf{X} \stackrel{d}{=} R\mathbf{U}^{(p)}$, it is necessary to know the respective moments of the uniform vector $\mathbf{U}^{(p)}$ and the moments of a certain order of the radial variable R. In turn, the moments of $\mathbf{U}^{(p)}$ are obtained through the standard normal vector $\mathbf{Z} \sim \mathcal{N}_p(\mathbf{0}_p, \mathbf{I}_p)$. This procedure can be generalized for certain types of integrable functions as indicated in the following theorem.

Theorem 22 *Assume* $\mathbf{X} \stackrel{d}{=} R\mathbf{U}^{(p)} \sim \mathcal{S}_p(\varphi)$ *and* $\mathbf{X}_0 \stackrel{d}{=} R_0\mathbf{U}^{(p)} \sim \mathcal{N}_p(\mathbf{0}_p, \mathbf{I}_p)$*. Then, for any integrable vector-valued function of the form* $\mathbf{g}(\mathbf{x}) = g_1(\|\mathbf{x}\|)\mathbf{g}_2(\mathbf{x}/\|\mathbf{x}\|)$*,* $\mathbf{x} \in \mathbb{R}^p$*, where* $g_1 : \mathbb{R}^p \to \mathbb{R}$ *and* $\mathbf{g}_2 : \mathbb{R}^p \to \mathbb{R}^p$*, we have*

$$\mathrm{E}\{\mathbf{g}(\mathbf{X})\} = \frac{\mathrm{E}\{g_1(R)\}}{\mathrm{E}\{g_1(R_0)\}}\mathrm{E}\{\mathbf{g}(\mathbf{X}_0)\},$$

where $R \stackrel{d}{=} \|\mathbf{X}\|$ *and* $R_0 \stackrel{d}{=} \|\mathbf{X}_0\| \sim \chi_p$*.*

Proof 22 *Since* $\mathbf{X} \stackrel{d}{=} R\mathbf{U}^{(p)}$*, with* $R \stackrel{d}{=} \|\mathbf{X}\| \perp\!\!\!\perp \mathbf{U}^{(p)} \stackrel{d}{=} \frac{\mathbf{X}}{\|\mathbf{X}\|}$*, then* $\mathrm{E}\{\mathbf{g}(\mathbf{X})\} = \mathrm{E}\{g_1(R)\}\mathrm{E}\{\mathbf{g}_2(\mathbf{U}^{(p)})\}$*. Similarly,* $\mathbf{X}_0 \stackrel{d}{=} R_0\mathbf{U}^{(p)}$*, with* $R_0 \stackrel{d}{=} \|\mathbf{X}_0\| \perp\!\!\!\perp \mathbf{U}^{(p)} \stackrel{d}{=} \frac{\mathbf{X}_0}{\|\mathbf{X}_0\|} \Longrightarrow \mathrm{E}\{\mathbf{g}(\mathbf{X}_0)\} = \mathrm{E}\{g_1(R_0)\}\mathrm{E}\{\mathbf{g}_2(\mathbf{U}^{(p)})\}$ $\Longrightarrow \mathrm{E}\{\mathbf{g}_2(\mathbf{U}^{(p)})\} = [1/\mathrm{E}\{g_1(R_0)\}]\mathrm{E}\{\mathbf{g}(\mathbf{X}_0)\}$*, from which the result follows.* □

An immediate application of this result is to compute the k-th multivariate moment of a spherical random vector as defined in (1.7).

Theorem 23 *Assume* $\mathbf{X} \stackrel{d}{=} R\mathbf{U}^{(p)} \sim \mathcal{S}_p(\varphi)$ *and* $\mathbf{X}_0 \stackrel{d}{=} R_0\mathbf{U}^{(p)} \sim \mathcal{N}_p(\mathbf{0}_p, \mathbf{I}_p)$*. If* $\mathrm{E}(R^k) < \infty$*, then*

$$\boldsymbol{\mu}_k(\mathbf{X}) = \kappa_k\,\boldsymbol{\mu}_k(\mathbf{X}_0), \quad \kappa_k = \frac{\mu_k(R)}{\mu_k(R_0)},$$

where $\mu_k(R) = \mathrm{E}(R^k)$*,* $\mu_k(R_0) = \mathrm{E}(R_0^k) = 2^{k/2}\Gamma\{(p+k)/2\}/\Gamma(p/2)$ *for* $k = 1, 2, \ldots$*, and*

$$\boldsymbol{\mu}_k(\mathbf{X}_0) = \begin{cases} \mathrm{E}(\underbrace{\mathbf{X}_0\mathbf{X}_0^\top \otimes \cdots \otimes \mathbf{X}_0\mathbf{X}_0^\top}_{k\ times}), & k\ even, \\ \mathbf{O}, & k\ odd, \end{cases}$$

becomes the k-th multivariate moment of $\mathbf{X}_0 \sim \mathcal{N}_p(\mathbf{0}_p, \mathbf{I}_p)$*. For instance, for the second and fourth moments, we have*

$$\begin{aligned} \boldsymbol{\mu}_2(\mathbf{X}) &= \mathrm{E}(\mathbf{X}\mathbf{X}^\top) = \kappa_2\,\mathbf{I}_p \Longrightarrow \mathrm{E}(\mathbf{X}\otimes\mathbf{X}) = \kappa_2\,\mathrm{vec}(\mathbf{I}_p), \\ \boldsymbol{\mu}_4(\mathbf{X}) &= \mathrm{E}(\mathbf{X}\mathbf{X}^\top \otimes \mathbf{X}\mathbf{X}^\top) = \kappa_4\{\mathbf{K}_p + \mathbf{I}_{p^2} + \mathrm{vec}(\mathbf{I}_p)\mathrm{vec}(\mathbf{I}_p)^\top\} \\ &\Longrightarrow \mathrm{var}(\mathbf{X}\otimes\mathbf{X}) = \kappa_4(\mathbf{K}_p + \mathbf{I}_{p^2}) + (\kappa_4 - \kappa_2^2)\mathrm{vec}(\mathbf{I}_p)\mathrm{vec}(\mathbf{I}_p)^\top, \end{aligned}$$

with $\kappa_2 = \mathrm{E}(\|\mathbf{X}\|^2)/p = \mathrm{E}(X_1^2) = \mathrm{var}(X_1) = -2\varphi'(0)$ *and* $\kappa_4 = \mathrm{E}(\|\mathbf{X}\|^4)/\{p(p+2)\} = \mathrm{E}(X_1^4)/3 = 4\varphi''(0)$.

Proof 23 *The proof is direct from Theorem 22 since* $\boldsymbol{\mu}_k(\mathbf{X}) = \boldsymbol{\mu}_k(R\mathbf{U}^{(p)}) = \mu_k(R)\boldsymbol{\mu}_k(\mathbf{U}^{(p)})$ *and* $\boldsymbol{\mu}_k(\mathbf{X}_0) = \boldsymbol{\mu}_k(R_0\mathbf{U}^{(p)}) = \mu_k(R_0)\boldsymbol{\mu}_k(\mathbf{U}^{(p)})$, *where* $\mathbf{X}_0 \sim \mathcal{N}_p(\mathbf{0}_p, \mathbf{I}_p)$ *and so* $R_0 \overset{d}{=} \|\mathbf{X}_0\| \sim \chi_p$. *For instance (see Theorem 8 in Chapter 3),* $\boldsymbol{\mu}_2(\mathbf{X}_0) = \mathrm{E}(\mathbf{X}_0\mathbf{X}_0^\top) = \mathbf{I}_p$ *and* $\boldsymbol{\mu}_4(\mathbf{X}_0) = \mathrm{E}(\mathbf{X}_0\mathbf{X}_0^\top \otimes \mathbf{X}_0\mathbf{X}_0^\top) = \mathbf{K}_p + \mathbf{I}_{p^2} + \mathrm{vec}(\mathbf{I}_p)\mathrm{vec}(\mathbf{I}_p)^\top$. □

Note in Theorem 23 that if we set $\kappa = \kappa_4/\kappa_2^2 - 1$, then the variance matrix $\mathrm{var}(\mathbf{X}\otimes\mathbf{X})$ can be written as

$$\mathrm{var}(\mathbf{X}\otimes\mathbf{X}) = (\kappa+1)\kappa_2^2(\mathbf{K}_p + \mathbf{I}_{p^2}) + \kappa\kappa_2^2\mathrm{vec}(\mathbf{I}_p)\mathrm{vec}(\mathbf{I}_p)^\top.$$

4.1.4 Marginal distributions

When $\mathbf{X} \sim \mathcal{S}_p(\varphi)$, all its marginal distributions are spherical and belong to the same multivariate spherical class of $\mathbf{X}$. To show this, partition $\mathbf{X} = (\mathbf{X}_1^\top, \mathbf{X}_2^\top)^\top$ with $\mathbf{X}_1 \in \mathbb{R}^{p_1}$ and $\mathbf{X}_2 \in \mathbb{R}^{p_2}$. We wish to find the marginal distribution of $\mathbf{X}_1$ and of $\mathbf{X}_2$. As can be seen from the proof of the following theorem, the result for the marginal distributions is quite obvious.

Theorem 24 *If* $\mathbf{X} = (\mathbf{X}_1^\top, \mathbf{X}_2^\top)^\top \sim \mathcal{S}_p(\varphi)$, *with* $\mathbf{X}_1 \in \mathbb{R}^{p_1}$, $\mathbf{X}_2 \in \mathbb{R}^{p_2}$, $p_1 + p_2 = p$, *then* $\mathbf{X}_1 \sim \mathcal{S}_{p_1}(\varphi)$ *and* $\mathbf{X}_2 \sim \mathcal{S}_{p_2}(\varphi)$.

Proof 24 *Let* $\psi(\mathbf{t}) = \psi(\mathbf{t}_1, \mathbf{t}_2)$ *for* $\mathbf{t} = (\mathbf{t}_1^\top, \mathbf{t}_2^\top)^\top \in \mathbb{R}^p$, *with* $\mathbf{t}_1 \in \mathbb{R}^{p_1}$, *be the joint cf of* $\mathbf{X} = (\mathbf{X}_1^\top, \mathbf{X}_2^\top)^\top$. *Since* $\mathbf{X} \sim \mathcal{S}_p(\varphi)$, *then* $\psi(\mathbf{t}_1, \mathbf{t}_2) = \varphi(\|\mathbf{t}\|^2) = \varphi(\|\mathbf{t}_1\|^2 + \|\mathbf{t}_2\|^2)$ *by Theorem 17. Clearly, the marginal cf of* $\mathbf{X}_1$ *is* $\psi(\mathbf{t}_1, \mathbf{0}_{p_2}) = \varphi(\|\mathbf{t}_1\|^2)$, $\forall \mathbf{t}_1 \in \mathbb{R}^{p_1}$ *and the marginal cf of* $\mathbf{X}_2$ *is* $\psi(\mathbf{0}_{p_1}, \mathbf{t}_2) = \varphi(\|\mathbf{t}_2\|^2)$, $\forall \mathbf{t}_2 \in \mathbb{R}^{p_2}$. *Consequently,* $\mathbf{X}_1 \sim \mathcal{S}_{p_1}(\varphi)$ *and* $\mathbf{X}_2 \sim \mathcal{S}_{p_2}(\varphi)$ *by Theorem 17.* □

It is evident from this result that all the marginals of a multivariate spherical distribution with cgf φ are also spherical with the same cgf φ. A stochastic version of this result is presented below in a more general way. For a better understanding of this result and further results related to the marginal of a multivariate spherical distribution, we need to introduce a result on the uniform random vector $\mathbf{U}^{(p)}$ stated in the following theorem.

Theorem 25 *Let* $\mathbf{U}^{(p)}$ *be partitioned as* $\mathbf{U}^{(p)} = (\mathbf{U}_1^\top, \ldots, \mathbf{U}_k^\top)^\top$ *with* $\mathbf{U}_i \in \mathbb{R}^{p_i}$, $i = 1, \ldots, k$, *and* $\sum_{i=1}^k p_i = p$. *Then,*

$$\begin{pmatrix} \mathbf{U}_1 \\ \vdots \\ \mathbf{U}_k \end{pmatrix} \overset{d}{=} \begin{pmatrix} \sqrt{V_1}\,\mathbf{U}^{(p_1)} \\ \vdots \\ \sqrt{V_k}\,\mathbf{U}^{(p_k)} \end{pmatrix},$$

where $\mathbf{U}^{(p_i)} \sim \mathcal{U}(\mathbb{S}^{(p_i-1)})$, $i = 1, \ldots, k$, *and* $\mathbf{V} = (V_1, \ldots, V_k)^\top \sim \mathcal{D}_k(p_1/2, \ldots, p_k/2)$, *the multivariate Dirichlet distribution, and they are mutually independent.*

Proof 25 *Let* $\mathbf{X} \sim \mathcal{N}_p(\mathbf{0}_p, \mathbf{I}_p)$ *and similarly to* $\mathbf{U}^{(p)}$ *partition it as* $\mathbf{X} = (\mathbf{X}_1^\top, \ldots, \mathbf{X}_k^\top)^\top$. *Since* $\mathbf{X}_i \sim \mathcal{N}_{p_i}(\mathbf{0}_{p_i}, \mathbf{I}_{p_i})$, $i = 1, \ldots, k$, *are all independent, we have that* $(\|\mathbf{X}_i\|, \mathbf{X}_i/\|\mathbf{X}_i\|)$, $i = 1, \ldots, k$, *are also mutually independent. Moreover, by Theorem 19*

$$\|\mathbf{X}_i\| \perp\!\!\!\perp \frac{\mathbf{X}_i}{\|\mathbf{X}_i\|}, \text{ with } \|\mathbf{X}_i\| \sim \chi_{p_i} \text{ and } \frac{\mathbf{X}_i}{\|\mathbf{X}_i\|} \stackrel{d}{=} \mathbf{U}^{(p_i-1)} \sim \mathcal{U}(\mathbb{S}^{(p_i-1)}),\ i = 1, \ldots, k.$$

This also means that $(\mathbf{X}_1/\|\mathbf{X}_1\|, \ldots, \mathbf{X}_k/\|\mathbf{X}_k\|)$ *is independent of* $\|\mathbf{X}\| = \sqrt{\sum_{i=1}^k \|\mathbf{X}_i\|^2}$ *and therefore of* $(\|\mathbf{X}_1\|/\|\mathbf{X}\|, \ldots, \|\mathbf{X}_k\|/\|\mathbf{X}\|) \stackrel{d}{=} (\sqrt{V_1}, \ldots, \sqrt{V_k})$, *where* $(V_1, \ldots, V_k)^\top \sim \mathcal{D}_k(p_1/2, \ldots, p_k/2)$, *the multivariate Dirichlet distribution; see Example 42 below. Finally, by defining* $\mathbf{U}^{(p)} = (\mathbf{U}_1^\top, \ldots, \mathbf{U}_k^\top)^\top$ *as* $\mathbf{X}/\|\mathbf{X}\| = (\|\mathbf{X}_1\|/\|\mathbf{X}\|, \ldots, \|\mathbf{X}_k\|/\|\mathbf{X}\|)^\top$ *we obtain*

$$\begin{pmatrix} \mathbf{U}_1 \\ \vdots \\ \mathbf{U}_k \end{pmatrix} = \begin{pmatrix} \frac{\|\mathbf{X}_1\|}{\|\mathbf{X}\|} \frac{\mathbf{X}_1}{\|\mathbf{X}_1\|} \\ \vdots \\ \frac{\|\mathbf{X}_k\|}{\|\mathbf{X}\|} \frac{\mathbf{X}_k}{\|\mathbf{X}_k\|} \end{pmatrix} \stackrel{d}{=} \begin{pmatrix} \sqrt{V_1}\,\mathbf{U}^{(p_1)} \\ \vdots \\ \sqrt{V_k}\,\mathbf{U}^{(p_k)} \end{pmatrix},$$

where $\mathbf{U}^{(p_i)} \sim \mathcal{U}(\mathbb{S}^{(p_i-1)})$, $i = 1, \ldots, k$, $(V_1, \ldots, V_k)^\top \sim \mathcal{D}_k(p_1/2, \ldots, p_k/2)$ *which are all independent.* □

Now, let $\mathbf{X} = (\mathbf{X}_1^\top, \ldots, \mathbf{X}_k^\top)^\top \sim \mathcal{S}_p(\varphi)$. As a direct consequence of this last result, we have

$$\begin{pmatrix} \mathbf{X}_1 \\ \vdots \\ \mathbf{X}_k \end{pmatrix} \stackrel{d}{=} \begin{pmatrix} R_1 \mathbf{U}^{(p_1)} \\ \vdots \\ R_k \mathbf{U}^{(p_k)} \end{pmatrix} \stackrel{d}{=} \begin{pmatrix} R\sqrt{V_1}\,\mathbf{U}^{(p_1)} \\ \vdots \\ R\sqrt{V_k}\,\mathbf{U}^{(p_k)} \end{pmatrix}, \tag{4.7}$$

where $(R_1, \ldots, R_k) \stackrel{d}{=} (\|\mathbf{X}_1\|, \ldots, \|\mathbf{X}_k\|) \stackrel{d}{=} (R\sqrt{V_1}, \ldots, R\sqrt{V_k})$, $R = \sqrt{\sum_{i=1}^k R_i^2}$ and, as before, $R_i \perp\!\!\!\perp \mathbf{U}^{(p_i)} \sim \mathcal{U}(\mathbb{S}^{(p_i-1)})$ for each $i = 1, \ldots, k$, and they are all independent of $(V_1, \ldots, V_k)^\top \sim \mathcal{D}_k(p_1/2, \ldots, p_k/2)$.

Example 41 *Let* $\mathbf{X} = (X_1, \ldots, X_p)^\top \sim \mathcal{S}_p^+(\varphi)$ *and let* R *be the length of* $\mathbf{X}$. *Define also the random vectors* $\mathbf{X}_1 = (X_1, \ldots, X_{p_1})^\top$ *and* $\mathbf{X}_2 = (X_{p_1+1}, \ldots, X_{p_1+p_2})^\top$ *and let* R_1 *and* R_2 *be their respective lengths. Clearly,* $\mathbf{X}_1$ *and* $\mathbf{X}_2$ *are spherical (see Theorem 24) with* $p_1 + p_2 = p$, *and their distributions are fully determined by those of* R_1 *and* R_2, *respectively. Furthermore,*

$$\frac{R_1}{R} \perp\!\!\!\perp R \text{ and } V = \frac{R_1^2}{R^2} \sim \mathcal{B}eta\left(\frac{p_1}{2}, \frac{p_2}{2}\right),$$

$$\frac{p_2}{p_1}\frac{R_1^2}{R_2^2} \stackrel{d}{=} \frac{p_2}{p_1}\frac{V}{1-V} \sim \mathcal{F}_{p_1,p_2}.$$

A surprising fact is that these distributions do not depend on the cgf φ.

In turn, these results mean that if $\varphi \in \Psi_p$ then for each $1 \leq p_1 \leq p$, we have $\varphi \in \Psi_{p_1}$. As the relationship (4.3) between φ and F (or R) is one-to-one, there exist two radial cdfs F and F_{p_1} such that

$$\varphi(u) = \int_0^\infty \varphi_{\mathcal{U},p}(r^2 u)\mathrm{d}F(r) \text{ and } \varphi(u) = \int_0^\infty \varphi_{\mathcal{U},p_1}(r^2 u)\mathrm{d}F_{p_1}(r) \text{ for } u \geq 0.$$

To clarify the specific relationship between F and F_{p_1}, we have the following result.

Theorem 26 *If $R \sim F$ and $R_{p_1} \sim F_{p_1}$, $1 \le p_1 \le p$, then*

$$R_{p_1} \stackrel{d}{=} R\sqrt{V_{p,p_1}}, \tag{4.8}$$

where $R \perp\!\!\!\perp V_{p,p_1}$ and $V_{p,p_1} \sim \mathcal{B}eta(p_1/2,(p-p_1)/2)$.

Proof 26 *Partition $\mathbf{X} = (\mathbf{X}_1^\top, \mathbf{X}_2^\top)^\top \sim \mathcal{S}_p(\varphi)$ and $\mathbf{U}^{(p)} = (\mathbf{U}_1^\top, \mathbf{U}_2^\top)^\top$, with $\mathbf{X}_1, \mathbf{U}_1 \in \mathbb{R}^{p_1}$. Since $\mathbf{X} \stackrel{d}{=} R\mathbf{U}^{(p)}$, then $\mathbf{X}_1 \stackrel{d}{=} R\mathbf{U}_1$ and, by (4.7), $\mathbf{X}_1 \stackrel{d}{=} R\sqrt{V_{p,p_1}}\mathbf{U}^{(p_1)}$. But $\mathbf{X}_1 \sim \mathcal{S}_{p_1}(\varphi)$ and by Theorem 18 $\mathbf{X}_1 \stackrel{d}{=} R_{p_1}\mathbf{U}^{(p_1)}$ which completes the proof.* □

By conditioning in (4.8) on $R = r$, we find $(R_{p_1} \mid R = r) \stackrel{d}{=} r\sqrt{V_{p,p_1}}$ for each $r > 0$. It follows that R_{p_1} has a pdf given by

$$g_{p_1}(z) = \frac{2\Gamma(p/2)}{\Gamma(p_1/2)\Gamma\{(p-p_1)/2\}} z^{p_1-1} \int_z^\infty r^{-(p-2)}(r^2-z^2)^{(p-p_1)/2-1} \mathrm{d}F(r), \; z > 0. \tag{4.9}$$

4.1.5 *Conditional distributions*

Now we explore the conditional distribution of $\mathbf{X}_1 \in \mathbb{R}^{p_1}$ given $\mathbf{X}_2 = \mathbf{x}_2$ for each $\mathbf{x}_2 \in \mathbb{R}^{p_2}$ when $\mathbf{X}_1$ and $\mathbf{X}_2$ correspond to a partition of a spherically distributed random vector $\mathbf{X}$ of dimension p.

Theorem 27 *Let $\mathbf{X} = (\mathbf{X}_1^\top, \mathbf{X}_2^\top)^\top \sim \mathcal{S}_p(\varphi)$, with $\mathbf{X}_1 \in \mathbb{R}^{p_1}$, $\mathbf{X}_2 \in \mathbb{R}^{p_2}$, $p_1 + p_2 = p$. Then $\mathbf{X}_1 \perp\!\!\!\perp \mathbf{X}_2$ conditionally on $\|\mathbf{X}_2\|$ and*

$$(\mathbf{X}_1 \mid \mathbf{X}_2 = \mathbf{x}_2) \stackrel{d}{=} (\mathbf{X}_1 \mid \|\mathbf{X}_2\| = \|\mathbf{x}_2\|) \sim \mathcal{S}_{p_1}(\varphi_{\|\mathbf{x}_2\|^2}), \tag{4.10}$$

and with stochastic representation

$$(\mathbf{X}_1 \mid \|\mathbf{X}_2\| = \|\mathbf{x}_2\|) \stackrel{d}{=} R_{\|\mathbf{x}_2\|^2}\mathbf{U}^{(p_1)}, \tag{4.11}$$

where for each $a \ge 0$, the conditional cgf φ_{a^2} is given by (4.3) with p replaced by p_1 and F by F_{a^2}, the cdf of R_{a^2} which is independent of $\mathbf{U}^{(p_1)}$ and

$$R_{\|\mathbf{x}_2\|^2} \stackrel{d}{=} \left(\sqrt{R^2 - \|\mathbf{x}_2\|^2} \mid \|\mathbf{X}_2\| = \|\mathbf{x}_2\|\right). \tag{4.12}$$

Proof 27 *By (4.7) we have $(\mathbf{X}_1, \mathbf{X}_2) \stackrel{d}{=} (R_1\mathbf{U}^{(p_1)}, R_2\mathbf{U}^{(p_2)})$, where $(R_1, R_2) = (\sqrt{R^2 - R_2^2}, R_2) \stackrel{d}{=} (\|\mathbf{X}_1\|, \|\mathbf{X}_2\|)$ which together with $\mathbf{U}^{(p_1)}$ and $\mathbf{U}^{(p_2)}$ are all independent. It is then clear that $R_1\mathbf{U}^{(p_1)} \perp\!\!\!\perp R_2\mathbf{U}^{(p_2)} \mid R_2$ and therefore $\mathbf{X}_1 \perp\!\!\!\perp \mathbf{X}_2 \mid \|\mathbf{X}_2\|$. The conditional cf of $\mathbf{X}_1$ given $\mathbf{X}_2$ is then given by $\mathrm{E}(e^{\mathrm{i}\mathbf{t}_1^\top\mathbf{X}_1} \mid \mathbf{X}_2) = \mathrm{E}\{\mathrm{E}(e^{\mathrm{i}\mathbf{t}_1^\top\mathbf{X}_1} \mid \mathbf{X}_2) \mid \|\mathbf{X}_2\|\} = \mathrm{E}(e^{\mathrm{i}\mathbf{t}_1^\top\mathbf{X}_1} \mid \|\mathbf{X}_2\|)$. Now we easily find that $\mathrm{E}(e^{\mathrm{i}\mathbf{t}_1^\top\mathbf{X}_1} \mid \|\mathbf{X}_2\| = a) = \mathrm{E}(e^{\mathrm{i}\mathbf{t}_1^\top R_1\mathbf{U}^{(p_1)}} \mid R_2 = a) = \mathrm{E}(e^{\mathrm{i}\mathbf{t}_1^\top R_{a^2}\mathbf{U}^{(p_1)}}) = \varphi_{a^2}(\|\mathbf{t}_1\|^2)$, where $R_{a^2} = \sqrt{R^2 - a^2}$ which for each $a \ge 0$ is clearly independent of $\mathbf{U}^{(p_1)}$ and has the same distribution as $\sqrt{R^2 - R_2^2} \mid R_2 = a$. This shows (4.10) and (4.11). For a comprehensive proof of (4.12), see Cambanis et al. (1981).* □

As shown by Cambanis et al. (1981), the cdf F_{a^2} of R_{a^2} associated with φ_{a^2} is such that

$$F_{a^2}(0) = \mathrm{P}(R_{a^2} = 0) = 1 \quad \text{if } a = 0 \text{ or } F(a) = 1,$$
$$F_{a^2}(r) = \frac{\int_{(a,\sqrt{a^2+r^2}]}(s^2 - r^2)^{q/2-1}s^{-(p-2)}\mathrm{d}F(s)}{\int_{(a,\infty)}(s^2 - r^2)^{q/2-1}s^{-(p-2)}\mathrm{d}F(s)}, \quad r \geq 0, \ a > 0 \text{ and } F(a) < 1. \tag{4.13}$$

The result in (4.10) means that for each $\mathbf{x}_2 \in \mathbb{R}^{p_2}$ the conditional distribution of $\mathbf{X}_1 \mid \mathbf{X}_2 = \mathbf{x}_2$ depends on $\mathbf{x}_2$ only through its length $\|\mathbf{x}_2\|$ and it is also multivariate spherical with conditional cgf $\varphi_{\|\mathbf{x}_2\|^2}$. The computation of conditional moments (provided that they exist) is also easy from (4.11). In particular, if $\mathrm{E}(R_{\|\mathbf{x}_2\|^2}) < \infty$ then $\mathrm{E}(\mathbf{X}_1 \mid \mathbf{x}_2) = \mathrm{E}(\mathbf{X}_1 \mid \|\mathbf{x}_2\|) = \mathbf{0}_{p_1}$ and if $\mathrm{E}(R^2_{\|\mathbf{x}_2\|^2}) = -2p_1\varphi'_{\|\mathbf{x}_2\|^2}(0) < \infty$, then $\mathrm{var}(\mathbf{X}_1 \mid \mathbf{x}_2) = \mathrm{var}(\mathbf{X}_1 \mid \|\mathbf{x}_2\|) = (1/p_1)\mathrm{E}(R^2_{\|\mathbf{x}_2\|^2})\mathbf{I}_{p_1}$.

4.1.6 *Cdf and pdf*

A random vector $\mathbf{X}$ with multivariate spherical distribution does not necessarily have a pdf. However, if the pdf $f(\mathbf{x})$ of $\mathbf{X}$ exists, then due to the orthogonal invariance condition (4.1), $f(\mathbf{x})$ must depend on the argument $\mathbf{x}$ through the value $\|\mathbf{x}\|^2 = \mathbf{x}^\top\mathbf{x}$ as the following theorem shows.

Theorem 28 *Let $\mathbf{X} \in \mathbb{R}^p$ be an absolutely continuous random vector. Then, $\mathbf{X}$ is spherically distributed iff its pdf has the form*

$$f(\mathbf{x}) = h(\|\mathbf{x}\|^2), \quad \forall \mathbf{x} \in \mathbb{R}^p, \tag{4.14}$$

for some function $h : [0,\infty) \to [0,\infty)$ called density generator function (dgf).

Proof 28 ($\Longrightarrow$) *Suppose that $\mathbf{X}$ has a multivariate spherical distribution. If $f(\mathbf{x})$ is the pdf of $\mathbf{X}$, we want to show that it has the form (4.14). Let $\mathbf{Y} = \mathbf{HX}$ be a nonsingular linear transformation, i.e., $\mathbf{H} \in \mathbb{R}^{p\times p}$ and $|\mathbf{H}| \neq 0$. Clearly, the distribution of $\mathbf{Y}$ is also absolutely continuous. Moreover, by the Jacobian method, the pdf of $\mathbf{Y}$ is $\mathrm{abs}(|\mathbf{H}|)^{-1}f(\mathbf{H}^{-1}\mathbf{y})$, $\mathbf{y} \in \mathbb{R}^p$. When $\mathbf{H} = \mathbf{\Gamma}^\top$, with $\mathbf{\Gamma} \in \mathcal{O}_p$, this pdf becomes $f(\mathbf{\Gamma y})$, $\mathbf{y} \in \mathbb{R}^p$ since $\mathbf{\Gamma}^{-1} = \mathbf{\Gamma}^\top$ and $|\mathbf{\Gamma}^\top| = \pm 1$. However, by assumption $\mathbf{X}$ is spherically distributed, and hence by (4.1) the distributions of $\mathbf{Y} = \mathbf{\Gamma X}$ and $\mathbf{X}$, and therefore their respective pdfs, must be equal for all $\mathbf{\Gamma} \in \mathcal{O}_p$. That is, $f(\mathbf{\Gamma x}) = f(\mathbf{x})$, $\forall \mathbf{x} \in \mathbb{R}^p$ and $\forall \mathbf{\Gamma} \in \mathcal{O}_p$, and hence $f(\mathbf{x}) = h(\|\mathbf{x}\|^2)$.*
($\Longleftarrow$) *Now suppose $f(\mathbf{x}) = h(\|\mathbf{x}\|^2)$ for some dgf h. The cf of $\mathbf{X}$ is $\psi(\mathbf{t}) = \int_{\mathbb{R}^p} e^{\mathrm{i}\mathbf{t}^\top\mathbf{x}}h(\|\mathbf{x}\|^2)\mathrm{d}\mathbf{x}$, for $\mathbf{t} \in \mathbb{R}^p$. Let $\mathbf{t}_1$ and $\mathbf{t}_2$ be two points of $\mathbb{R}^p$. We prove that if $\|\mathbf{t}_1\| = \|\mathbf{t}_2\| \Longrightarrow \psi(\mathbf{t}_1) = \psi(\mathbf{t}_2)$. But there is $\mathbf{\Gamma} \in \mathcal{O}_p$ such that $\mathbf{t}_2 = \mathbf{\Gamma}^\top\mathbf{t}_1$ and so $\|\mathbf{t}_2\| = \|\mathbf{\Gamma t}_1\| = \|\mathbf{t}_1\|$. This result, jointly with the change of variables $\mathbf{y} = \mathbf{\Gamma x}$, implies that $\psi(\mathbf{t}_2) = \int_{\mathbb{R}^p} e^{\mathrm{i}\mathbf{t}_2^\top\mathbf{x}}h(\|\mathbf{x}\|^2)\mathrm{d}\mathbf{x} = \int_{\mathbb{R}^p} e^{\mathrm{i}\mathbf{t}_1^\top\mathbf{\Gamma x}}h(\|\mathbf{x}\|^2)\mathrm{d}\mathbf{x} = \int_{\mathbb{R}^p} e^{\mathrm{i}\mathbf{t}_1^\top\mathbf{y}}h(\|\mathbf{y}\|^2)\mathrm{d}\mathbf{y} = \psi(\mathbf{t}_1)$. This means that $\psi(\mathbf{t}) = \psi(\mathbf{\Gamma t})$ for all $\mathbf{t} \in \mathbb{R}^p$ and $\mathbf{\Gamma} \in \mathcal{O}_p$, and therefore $\psi(\mathbf{t}) = \varphi(\|\mathbf{t}\|^2)$ for some cgf φ, which by Theorem 17 concludes the proof.* □

As noted earlier by Kelker (1970), the expression $r^{p-1}h(r^2)$ or, equivalently, the expression $r^{p/2-1}h(r)$, must be integrable over $[0,\infty)$ in order for the spherical function $h(\|\mathbf{x}\|^2)$ to be effectively a pdf on $\mathbb{R}^p$. Specifically, the change of variables formula obtained by using polar coordinates, see, e.g., equation (1.32) in Fang et al. (1990),

$$\int_{\mathbb{R}^p} g(\|\mathbf{x}\|^2)\mathrm{d}\mathbf{x} = \frac{2\pi^{p/2}}{\Gamma(p/2)}\int_0^\infty r^{p-1}g(r^2)\mathrm{d}r, \tag{4.15}$$

for any nonnegative measurable function g, shows that any dgf h must satisfy

$$\int_0^\infty \frac{2\pi^{p/2}}{\Gamma(p/2)} r^{p-1} h(r^2) \mathrm{d}r = \int_0^\infty \frac{\pi^{p/2}}{\Gamma(p/2)} r^{p/2-1} h(r) \mathrm{d}r = 1. \tag{4.16}$$

If the pdf of $\mathbf{X}$ exists and is of the form (4.14), then we write $\mathbf{X} \sim \mathcal{S}_p(h)$ instead of $\mathbf{X} \sim \mathcal{S}_p(\varphi)$ and say that $\mathbf{X}$ has a multivariate spherical distribution with dgf h. It is quite simple to show that any other function of the form $\tilde{h}(u) = c^{p/2} h(cu)$ also verifies (4.16) for all $c > 0$, and therefore it is also a multivariate spherical dgf.

The following example shows that although there is no pdf for $\mathbf{U}^{(p)} \sim \mathcal{U}(\mathbb{S}^{(p-1)})$ in $\mathbb{R}^p$, all its marginal distributions have a pdf but they are not necessarily uniform.

Example 42 (Relation with the Dirichlet distribution) *Let $\mathbf{U}^{(p)} = (U_1, \ldots, U_p)^\top$ be the uniform random vector on $\mathbb{S}^{(p-1)}$ in $\mathbb{R}^p$. As $\|\mathbf{U}^{(p)}\| = \sqrt{\sum_{i=1}^p U_i^2} = 1$, there is no pdf for $\mathbf{U}^{(p)}$ in $\mathbb{R}^p$, but all its marginal distributions have a pdf which can be obtained by means of the multivariate Dirichlet distribution. In fact, since $\mathbf{U}^{(p)} \stackrel{d}{=} \mathbf{X}/\|\mathbf{X}\|$ with $\mathbf{X} = (X_1, \ldots, X_p)^\top \sim \mathcal{N}_p(\mathbf{0}_p, \mathbf{I}_p)$ then*

$$(U_1^2, \ldots, U_p^2) \stackrel{d}{=} \left(\frac{X_1^2}{\sum_{i=1}^p X_i^2}, \ldots, \frac{X_p^2}{\sum_{i=1}^p X_i^2} \right),$$

where $X_1^2, \ldots, X_p^2$ are iid χ_1^2 random variables and therefore $\sum_{i=1}^p X_i^2 \sim \chi_p^2$. Hence, we find a Dirichlet joint distribution with parameters $(\alpha_1, \ldots, \alpha_{p-1}, \alpha_p)^\top = (1/2, \ldots, 1/2, 1/2)^\top$ for the random variables $V_i = X_i^2 / \sum_{i=1}^p X_i^2$, $i = 1, \ldots, p$, which is denoted by

$$(V_1, \ldots, V_{p-1}, V_p)^\top = \left(V_1, \ldots, V_{p-1}, 1 - \sum_{i=1}^{p-1} V_i \right)^\top \sim \mathcal{D}_p\left(\alpha_1 = 1/2, \ldots, \alpha_{p-1} = 1/2, \alpha_p = 1/2\right).$$

Moreover, the random vector

$$\left(V_1, \ldots, V_q, 1 - \sum_{i=1}^{q} V_i \right)^\top \sim \mathcal{D}_{q+1}\left(1/2, \ldots, 1/2, (p-q)/2\right), \quad q \in \{1, \ldots, p-1\},$$

and the random subvector $(V_1, \ldots, V_q)^\top$ has pdf

$$\tilde{f}(v_1, \ldots, v_q) = \begin{cases} \frac{\Gamma(p/2)}{\Gamma(\frac{p-q}{2})\pi^{q/2}} \prod_{i=1}^q v_i^{\frac{1}{2}-1} \left(1 - \sum_{i=1}^q v_i\right)^{\frac{p-q}{2}-1}, & v_i \geq 0, i = 1, \ldots, q, \sum_{i=1}^q v_i < 1, \\ 0, & \textit{otherwise}. \end{cases}$$

This shows that an alternative way to obtain a uniform random vector $\mathbf{U} = (U_1, \ldots, U_p)^\top$ is from the transformation

$$(U_1, \ldots, U_p) = (S_1\sqrt{V_1}, \ldots, S_p\sqrt{V_p}), \ (S_1, \ldots, S_p) \perp\!\!\!\perp (V_1, \ldots, V_p),$$

where $(S_1, \ldots, S_p)^\top = (\mathrm{sgn}(U_1), \ldots, \mathrm{sgn}(U_p))^\top \sim \mathcal{U}(\{-1, +1\}^p)$ and $(V_1, \ldots, V_p)^\top \sim \mathcal{D}_p(1/2, \ldots, 1/2)$. Hence, by conditioning on the sign we obtain for $q \in \{1, \ldots, p-1\}$ that the random subvector $(U_1, \ldots, U_q)^\top = (S_1\sqrt{V_1}, \ldots, S_q\sqrt{V_q})^\top$ has a pdf given by

$$f(u_1, \ldots, u_q) = \sum_{(s_1, \ldots, s_q) \in \{-1, +1\}^q} \mathrm{abs}(J) \tilde{f}(u_1^2, \ldots, u_q^2) \mathrm{P}(S_1 = s_1, \ldots, S_q = s_q),$$

where $\text{abs}(J) = \prod_{i=1}^{q} 2|u_i|$ *is the absolute Jacobian (determinant) of the transformation from* $\mathbf{v}$ *to* $\mathbf{u}$*, defined by* $u_i = s_i\sqrt{u_i}$*,* $i = 1,\ldots,q$*, with* $(s_1,\ldots,s_q) \in \{-1,+1\}^q$*. Since* $\mathrm{P}(S_1 = s_1,\ldots,S_q = s_q) = 2^{-q}$*,* $\forall (s_1,\ldots,s_q) \in \{-1,+1\}^q$*, we finally obtain for* $q \in \{1,\ldots,p-1\}$ *that*

$$f(u_1,\ldots,u_q) = \begin{cases} \frac{\Gamma(p/2)}{\Gamma(\frac{p-q}{2})\pi^{q/2}} \left(1 - \sum_{i=1}^{q} u_i^2\right)^{\frac{p-q}{2}-1}, & \sum_{i=1}^{q} u_i^2 < 1, \\ 0, & otherwise. \end{cases} \tag{4.17}$$

Note from Equation (4.17) of Example 42 that if $\mathbf{X} = (X_1,\ldots,X_p)^\top \sim \mathcal{S}_p(h)$, then the marginal joint cdf of $X_1,\ldots,X_q$ is

$$F(x_1,\ldots,x_q) = \frac{\Gamma(p/2)}{\Gamma(\frac{p-q}{2})\pi^{q/2}} \int_0^\infty \mathrm{d}F(r) \int_{-1}^{x_1/r} \cdots \int_{-1}^{x_q/r} \left(1 - \sum_{i=1}^{q} u_i^2\right)^{\frac{p-q}{2}-1} I_{\{\sum_{i=1}^{q} u_i^2 < 1\}} \mathrm{d}u_1 \cdots \mathrm{d}u_q, \tag{4.18}$$

where I_A denotes the indicator of A.

As established by the following theorem, a necessary and sufficient condition for a spherical random vector $\mathbf{X} \stackrel{d}{=} R\mathbf{U}^{(p)}$ to have a pdf is that the radial random variable R has a pdf.

Theorem 29 *A spherical random vector* $\mathbf{X} \sim \mathcal{S}_p(h)$ *has a pdf* $f(\mathbf{x}) = h(\|\mathbf{x}\|^2)$*,* $\mathbf{x} \in \mathbb{R}^p$ *iff the radial random variable* $R \stackrel{d}{=} \|\mathbf{X}\|$ *has a pdf* $g(r)$*,* $r > 0$*, which is related to the dgf* h *as*

$$g(r) = \frac{2\pi^{p/2}}{\Gamma(p/2)} r^{p-1} h(r^2), \quad r > 0. \tag{4.19}$$

Proof 29 *Suppose that* $\mathbf{X}$ *has pdf* $h(\|\mathbf{x}\|^2)$ *on* $\mathbb{R}^p$*. Since* $\mathbf{X}$ *is (absolutely) continuous, then* $R \stackrel{d}{=} \|\mathbf{X}\|$ *is also (absolutely) continuous. Thus, there is a pdf* g *such that* $\mathrm{P}(R \le r) = \int_0^r g(u)\mathrm{d}u$ *for all* $r > 0$*. But* $\mathrm{P}(R \le r) = \mathrm{P}(\|\mathbf{X}\| \le r) = \int_{\mathbb{B}_r} h(\|\mathbf{x}\|^2)\mathrm{d}\mathbf{x}$*, where* $\mathbb{B}_r = \{\mathbf{x} \in \mathbb{R}^p : \|\mathbf{x}\| \le r\}$*, and a similar change of variables formula as (4.15) yields*

$$\int_{\mathbb{B}_r} h(\|\mathbf{x}\|^2)\mathrm{d}\mathbf{x} = \int_0^r \frac{2\pi^{p/2}}{\Gamma(p/2)} u^{p-1} h(u^2)\mathrm{d}u,$$

which concludes the proof. The converse is straightforward from the transformation $(r,\mathbf{u}) \to \mathbf{x} = r\mathbf{u}$ *for* $r > 0$ *and* $\mathbf{u} \in \mathbb{S}^{(p-1)}$*, since* $R \perp\!\!\!\perp \mathbf{U}^{(p)}$ *and therefore their joint pdf always exists provided that* R *has a pdf as given by (4.19); see Equation (4.17) in Example 42.* □

Example 43 *In many applications we need the distribution of the squared-radial random variable* $W = R^2$*, which in the absolutely continuous case has pdf*

$$\tilde{g}(w) = \frac{\pi^{p/2}}{\Gamma(p/2)} w^{p/2-1} h(w), \quad w > 0. \tag{4.20}$$

In particular, if $\mathbf{X} \sim \mathcal{S}_p(h)$ *then* $W = R^2 \stackrel{d}{=} \|\mathbf{X}\|^2 \sim \chi_p^2 \Longleftrightarrow \mathbf{X} \sim \mathcal{N}_p(\mathbf{0}_p, \mathbf{I}_p)$*, and (4.20) becomes*

$$\tilde{g}(w) = \frac{(1/2)^{p/2}}{\Gamma(p/2)} w^{p/2-1} e^{-w/2}, \quad w > 0.$$

As has been remarked by Kelker (1970), Cambanis et al. (1981), and Fang et al. (1990), an important property of the multivariate spherical distributions is that all the marginal distributions are absolutely continuous unless the original distribution has an atom with positive mass at zero. Even in the latter case, the marginal distributions are absolutely continuous outside the zero point; see Cambanis et al. (1981).

Theorem 30 *If* $\mathbf{X} \sim \mathcal{S}_p^+(\varphi)$, *i.e.,* $\mathrm{P}(\mathbf{X} = \mathbf{0}_p) = 0$, *then all its marginal distributions have pdfs. In particular, the marginal pdf of* $\mathbf{X}_1 = (X_1, \ldots, X_q)^\top$, $q \in \{1, \ldots, p-1\}$, *is*

$$f_1(\mathbf{x}_1) = \frac{\Gamma(p/2)}{\Gamma\{(p-q)/2\}\pi^{q/2}} \int_{\|\mathbf{x}_1\|}^{\infty} r^{-(p-2)} \left(r^2 - \|\mathbf{x}_1\|^2\right)^{(p-q)/2} \mathrm{d}F(r), \tag{4.21}$$

where F *is the cdf of* $R \stackrel{d}{=} \|\mathbf{X}\|$.

Proof 30 *The first statement is obvious since* $\mathrm{P}(\mathbf{X}_1 = \mathbf{0}_q) = \mathrm{P}(\mathbf{X} = \mathbf{0}_p) = 0$. *To prove (4.21), first note by (4.7) that* $\mathbf{X}_1 \stackrel{d}{=} R_q \mathbf{U}^{(q)}$, *where* $R_q = R_p \sqrt{V_{p,q}}$ *and has pdf given by (4.9). Since* R_q *is independent of* $\mathbf{U}^{(q)}$ *which has pdf given by (4.17) the result is easily obtained. Alternatively, we can obtain (4.21) by differentiating directly (4.18) with respect to* $\mathbf{x}_1 = (x_1, \ldots, x_q)^\top$. □

In the absolutely continuous case of $\mathbf{X} \sim \mathcal{S}_p(h)$, then all the marginal pdfs exist. In particular, for $q \in \{1, \ldots, p-1\}$ we have $\mathbf{X}_1 = (X_1, \ldots, X_q)^\top \sim \mathcal{S}_q(h^{(q)})$ with marginal pdf $f_1(\mathbf{x}_q) = h^{(q)}(\|\mathbf{x}_q\|^2)$, $\mathbf{x}_q \in \mathbb{R}^q$, where $h^{(q)}$ denotes the *marginal dgf* induced by $h = h^{(p)}$ and is given by

$$h^{(q)}(u) = \frac{\pi^{(p-q)/2}}{\Gamma\{(p-q)/2\}} \int_u^{\infty} (v-u)^{(p-q)/2} h(v)\mathrm{d}v = \frac{\pi^{(p-q)/2}}{\Gamma\{(p-q)/2\}} \int_0^{\infty} v^{(p-q)/2} h(u+v)\mathrm{d}v, \quad u > 0. \tag{4.22}$$

Moreover, from the partition $\mathbf{X} = (\mathbf{X}_1^\top, \mathbf{X}_2^\top)^\top \sim \mathcal{S}_p(h)$ we find that the conditional pdf of $\mathbf{X}_1 \mid \mathbf{X}_2 = \mathbf{x}_2$ is then

$$f_1(\mathbf{x}_1 \mid \mathbf{x}_2) = \frac{h(\|\mathbf{x}_1\|^2 + \|\mathbf{x}_2\|^2)}{h^{(p-q)}(\|\mathbf{x}_2\|^2)} = h^{(q)}_{\|\mathbf{x}_2\|^2}(\|\mathbf{x}_1\|^2), \quad \mathbf{x}_1 \in \mathbb{R}^q.$$

In this case, we use the notation $\mathbf{X}_1 \mid \mathbf{x}_2 \sim \mathcal{S}_q(h^{(q)}_{\|\mathbf{x}_2\|^2})$ and call the function given by

$$h_v^{(q)}(u) = \frac{h(u+v)}{h^{(p-q)}(v)} = \frac{\Gamma(q/2)h(u+v)}{\pi^{q/2}\int_0^{\infty} w^{q/2-1} h(v+w)\mathrm{d}w}, \quad u, v > 0, \tag{4.23}$$

the *conditional dgf*. Further properties of the marginal and conditional dgfs and the relations between them are given in Fang et al. (1990), Berkane and Bentler (1986), Kano (1994), and also in Arellano-Valle et al. (2006b). Here we have some examples.

Example 44 *Let* $\mathbf{X} \sim \mathcal{N}_p(\mathbf{0}_p, \mathbf{I}_p)$ *be partitioned as* $\mathbf{X} = (\mathbf{X}_1^\top, \mathbf{X}_2^\top)^\top$ *with* $\mathbf{X}_1 \in \mathbb{R}^{p_1}$, $\mathbf{X}_2 \in \mathbb{R}^{p_2}$, *and* $p_1 + p_2 = p$. *Since* $\mathbf{X}_1 \sim \mathcal{N}_{p_1}(\mathbf{0}_{p_1}, \mathbf{I}_{p_1}) \perp\!\!\!\perp \mathbf{X}_2 \sim \mathcal{N}_{p_2}(\mathbf{0}_{p_2}, \mathbf{I}_{p_2})$, *then* $(\mathbf{X}_1 \mid \mathbf{X}_2 = \mathbf{x}_2) \stackrel{d}{=} \mathbf{X}_1 \sim \mathcal{N}_{p_1}(\mathbf{0}_{p_1}, \mathbf{I}_{p_1})$, $\forall \mathbf{x}_2 \in \mathbb{R}^{p_2}$ *and therefore* $\mathbf{X}$, $\mathbf{X}_1$, $\mathbf{X}_2$, *and* $\mathbf{X}_1 \stackrel{d}{=} (\mathbf{X}_1 \mid \mathbf{X}_2 = \mathbf{x}_2)$ *are spherically distributed with the same cgf* $\varphi(u) = e^{-u/2}$ *and dgf* $h^{(k)}(u) = (2\pi)^{-k/2} e^{-u/2}$ *for* $k = p, p_1, p_2$, *respectively.*

As indicated in the following example, a rich class of multivariate spherical pdfs with heavy tails can be obtained from the scale mixtures of multivariate normal distributions. In this class, by the characterization of $\varphi \in \Psi_\infty$ in (4.4), a necessary and sufficient condition for $\mathbf{X} \sim \mathcal{S}_p(\varphi)$ to have a pdf, which should be given by (4.14) with dgf $h(u) = \int_0^\infty (2\pi r^2)^{-p/2} e^{-r^2 u/2} \mathrm{d}F_\infty(r)$, is that $F_\infty(0) = 0$. Otherwise, $\mathbf{X}$ places a mass $F_\infty(0)$ at the origin and except at this point has a pdf given by (4.14) with the dgf previously defined. As in the previous examples, in the definition of this class, we redefine the mixing scale variable as $R_\infty = V^{-1/2}$, with $V \sim G$ such that $G(0) = 0$.

Example 45 (Scale mixtures of normals) *If* $\mathbf{X} \mid V = v \sim \mathcal{N}_p(\mathbf{0}_p, v^{-1}\mathbf{I}_p)$ *and* $V \sim G$ *with* $G(0) = 0$, *then* $\mathbf{X} \sim \mathcal{S}_p(h)$ *with pdf*

$$f(\mathbf{x}) = h(\|\mathbf{x}\|^2) = \int_0^\infty v^{p/2} \phi_p(\sqrt{v}\|\mathbf{x}\|^2) \mathrm{d}G(v), \quad \mathbf{x} \in \mathbb{R}^p.$$

From the partition $\mathbf{X} = (\mathbf{X}_1^\top, \mathbf{X}_2^\top)^\top \in \mathbb{R}^{p_1+p_2}$ *we find that* $\mathbf{X}_1 \perp\!\!\!\perp \mathbf{X}_2 \mid V$, *with* $(\mathbf{X}_1 \mid V = v) \sim \mathcal{N}_{p_1}(\mathbf{0}_{p_1}, v^{-1}\mathbf{I}_{p_1})$ *and* $(\mathbf{X}_2 \mid V = v) \sim \mathcal{N}_{p_2}(\mathbf{0}_{p_2}, v^{-1}\mathbf{I}_{p_2})$. *Hence,* $\mathbf{X}_1 \overset{d}{=} V^{-1/2}\mathbf{Z}_1$, *with* $V \perp\!\!\!\perp \mathbf{Z}_1 \sim \mathcal{N}_{p_1}(\mathbf{0}_{p_1}, \mathbf{I}_{p_1})$, *has marginal pdf given by*

$$f_1(\mathbf{x}_1) = h^{(p_1)}(\|\mathbf{x}_1\|^2) = \int_0^\infty v^{p_1/2} \phi_{p_1}(\sqrt{v}\|\mathbf{x}_1\|) \mathrm{d}G(v), \quad \mathbf{x}_1 \in \mathbb{R}^{p_1}.$$

Similar results are found for $\mathbf{X}_2$. *For the conditional pdf of* $(\mathbf{X}_1 \mid \mathbf{X}_2 = \mathbf{x}_2)$, *from the fact* $(\mathbf{X}_1 \mid \mathbf{X}_2 = \mathbf{x}_2, V = v) \overset{d}{=} (\mathbf{X}_1 \mid V = v)$ *we easily find*

$$f_1(\mathbf{x}_1 \mid \mathbf{x}_2) = h^{(p_1)}_{\|\mathbf{x}_2\|^2}(\|\mathbf{x}_1\|^2) = \int_0^\infty v^{p_1/2} \phi_{p_1}(\sqrt{v}\|\mathbf{x}_1\|) \mathrm{d}G_{\|\mathbf{x}_2\|^2}(v), \quad \mathbf{x}_1 \in \mathbb{R}^{p_1},$$

where $G_{\|\mathbf{x}_2\|^2}$ *is the conditional cdf of* $V_{\|\mathbf{x}_2\|^2} \overset{d}{=} (V \mid \mathbf{X}_2 = \mathbf{x}_2)$ *and is such that*

$$\mathrm{d}G_{\|\mathbf{x}_2\|^2}(v) = \frac{v^{p_2/2}\phi_{p_2}(\sqrt{v}\|\mathbf{x}_2\|)\,\mathrm{d}G(v)}{\int_0^\infty v^{p_2/2}\phi_{p_2}(\sqrt{v}\|\mathbf{x}_2\|)\,\mathrm{d}G(v)}. \tag{4.24}$$

This means that $(\mathbf{X}_1 \mid \mathbf{X}_2 = \mathbf{x}_2) \overset{d}{=} V^{-1/2}_{\|\mathbf{x}_2\|^2}\mathbf{Z}_1$ *with* $V_{\|\mathbf{x}_2\|^2} \perp\!\!\!\perp \mathbf{Z}_1 \sim \mathcal{N}_{p_1}(\mathbf{0}_{p_1}, \mathbf{I}_{p_1})$. *Consequently,*

$$\mathrm{E}(\mathbf{X}_1 \mid \mathbf{X}_2 = \mathbf{x}_2) = \mathbf{0}_{p_1} \iff \mathrm{E}(V^{-1/2}_{\|\mathbf{x}_2\|^2}) = \mathrm{E}(V^{-1/2} \mid \mathbf{X}_2 = \mathbf{x}_2) < \infty$$

and

$$\mathrm{var}(\mathbf{X}_1 \mid \mathbf{X}_2 = \mathbf{x}_2) = \kappa_2(\|\mathbf{x}_2\|^2)\mathbf{I}_{p_1} \iff \kappa_2(\|\mathbf{x}_2\|^2) = \mathrm{E}(V^{-1}_{\|\mathbf{x}_2\|^2}) = \mathrm{E}(V^{-1} \mid \mathbf{X}_2 = \mathbf{x}_2) < \infty,$$

where according to (4.24) (see also Example 40)

$$\kappa_2(u) = \frac{\int_0^\infty v^{(p_2-2)/2}\phi_{p_2}(\sqrt{vu})\,\mathrm{d}G(v)}{\int_0^\infty v^{p_2/2}\phi_{p_2}(\sqrt{vu})\,\mathrm{d}G(v)}.$$

For any fixed $u > 0$, *the integrand in the numerator of* $\kappa_2(u)$ *is a continuous function of* v *on the interval* $(0, \infty)$. *As* $v \to 0$ *or* $v \to \infty$, *this function tends to* 0 *provided that* $p_2 \geq 2$. *In the case where* $p_2 = 1$, *the integrand is bounded by* $v^{-1/2}$; *consequently,* $\kappa_2(u)$ *is finite if and only if* $\mathrm{E}(V^{-1/2}) = \int_0^\infty v^{-1/2}\mathrm{d}G(v) < \infty$.

The following example illustrates the construction of the multivariate t distribution, studied extensively among others by Arellano-Valle (1994) and Arellano-Valle and Bolfarine (1995).

Example 46 (Multivariate spherical t distribution) *If in Example 45 we let $V \sim \mathcal{G}amma(\nu/2, \nu/2)$, $\nu > 0$, then we have the "standard" (spherical) version of the multivariate t distribution with ν degrees of freedom denoted by $\mathbf{X} \sim \mathcal{T}_p(\mathbf{0}_p, \mathbf{I}_p, \nu)$, and with pdf*

$$f(\mathbf{x}) = h(\|\mathbf{x}\|^2) = \frac{\Gamma\{(\nu+p)/2\}}{\Gamma(\nu/2)(\pi\nu)^{p/2}}\left(1+\frac{\|\mathbf{x}\|^2}{\nu}\right)^{-(\nu+p)/2}, \quad \mathbf{x} \in \mathbb{R}^p.$$

Since $\|\mathbf{X}\|^2 \stackrel{d}{=} V^{-1}\|\mathbf{Z}\|^2$ with $\|\mathbf{Z}\|^2 \sim \chi^2_p$ and is independent of V, it follows that $\|\mathbf{X}\|^2 \sim p\mathcal{F}_{p,\nu}$, $\mathrm{E}(\mathbf{X}) = \mathbf{0}_p$ if $\nu > 1$ and $\mathrm{var}(\mathbf{X}) = \nu(\nu-2)^{-1}\mathbf{I}_p$ if $\nu > 2$. It is also easy to find that $\mathbf{X}_1 \sim \mathcal{T}_{p_1}(\mathbf{0}_{p_1}, \mathbf{I}_{p_1}, \nu)$ and $\mathbf{X}_2 \sim \mathcal{T}_{p_2}(\mathbf{0}_{p_2}, \mathbf{I}_{p_2}, \nu)$, and by (4.24) $V_{\|\mathbf{x}_2\|^2} \sim \mathcal{G}amma\{(\nu+p_2)/2, (\nu+\|\mathbf{x}_2\|^2)/2\}$. Note that $V_{\|\mathbf{x}_2\|^2} \stackrel{d}{=} \left(\frac{\nu+p_2}{\nu+\|\mathbf{x}_2\|^2}\right)V_*$, where $V_* \sim \mathcal{G}amma\{(\nu+p_2)/2, (\nu+p_2)/2\}$ and is independent of $\mathbf{X}_2$. Therefore, $(\mathbf{X}_1 \mid \mathbf{X}_2 = \mathbf{x}_2) \stackrel{d}{=} V_{\|\mathbf{x}_2\|^2}^{-1/2}\mathbf{Z}_1 \stackrel{d}{=} \left(\frac{\nu+p_2}{\nu+\|\mathbf{x}_2\|^2}\right)^{-1/2} V_*^{-1/2}\mathbf{Z}_1$, where $V_{\|\mathbf{x}_2\|^2}$ and therefore V_*, is independent of $\mathbf{Z}_1 \sim \mathcal{N}_{p_1}(\mathbf{0}_{p_1}, \mathbf{I}_{p_1})$. Let $\mathbf{X}_{1*} = V_*^{-1/2}\mathbf{Z}_1$ and note $\mathbf{X}_{1*} \sim \mathcal{T}_{p_1}(\mathbf{0}_{p_1}, \mathbf{I}_{p_1}, \nu+p_2)$. Now we finally find $(\mathbf{X}_1 \mid \mathbf{X}_2 = \mathbf{x}_2) \stackrel{d}{=} \left(\frac{\nu+\|\mathbf{x}_2\|^2}{\nu+p_2}\right)^{1/2}\mathbf{X}_{1*} \sim \mathcal{T}_{p_1}\left(\mathbf{0}_{p_1}, \left(\frac{\nu+\|\mathbf{x}_2\|^2}{\nu+p_2}\right)\mathbf{I}_{p_1}, \nu+p_2\right)$, with conditional pdf given by

$$\begin{aligned} f_1(\mathbf{x}_1 \mid \mathbf{x}_2) &= h^{(p_1)}_{\|\mathbf{x}_2\|^2}(\|\mathbf{x}_1\|^2) \\ &= \frac{\Gamma\{(\nu+p_2+p_1)/2\}}{\Gamma\{(\nu+p_2)/2\}\{\pi(\nu+p_2)\}^{p_1/2}} \\ &\quad \times \left(\frac{\nu+p_2}{\nu+\|\mathbf{x}_2\|^2}\right)^{p_1/2}\left\{1+\left(\frac{\nu+p_2}{\nu+\|\mathbf{x}_2\|^2}\right)\frac{\|\mathbf{x}_1\|^2}{\nu+p_2}\right\}^{-(\nu+p_2+p_1)/2}, \quad \mathbf{x}_1 \in \mathbb{R}^{p_1}. \end{aligned}$$

As these results show, $\mathbf{X} = (\mathbf{X}_1^\top, \mathbf{X}_2^\top)^\top \sim \mathcal{S}_p(h)$ with $\|\mathbf{X}\|^2 \sim p\mathcal{F}_{p,\nu}$ and dgf $h(u) = h^{(p)}(u) = \frac{\Gamma\{(\nu+p)/2\}}{\Gamma(\nu/2)(\pi\nu)^{p/2}}(1+u/\nu)^{-(\nu+p)/2}$, and so $\mathbf{X}_1 \sim \mathcal{S}_{p_1}(h^{(p_1)})$ and $\mathbf{X}_2 \sim \mathcal{S}_{p_2}(h^{(p_2)})$, with $\|\mathbf{X}_1\|^2 \sim p_1\mathcal{F}_{p_1,\nu}$ and $\|\mathbf{X}_2\|^2 \sim p_2\mathcal{F}_{p_2,\nu}$, and marginal dgf $h^{(k)} = h^{(p)}\mid_{p=k}$, for $k = p_1, p_2$, and $\mathbf{X}_1 \mid \mathbf{X}_2 = \mathbf{x}_2 \sim \mathcal{S}_{p_1}(h^{(p_1)}_{\|\mathbf{x}_2\|^2})$, with conditional dgf $h^{(p_1)}_w(u) = \left(\frac{\nu+p_2}{\nu+w}\right)^{p_1/2} h^{(p_1)}\left\{\left(\frac{\nu+p_2}{\nu+w}\right)u\right\}$. Note that the conditional dgf has the form $c_w^{p/2}h^{(p)}(c_w u)$ with p replaced by p_1 and $c_w = \frac{\nu+p_2}{\nu+w}$. In addition, we should note that $\mathrm{E}(\mathbf{X}_1 \mid \mathbf{X}_2 = \mathbf{x}_2) = \mathbf{0}_{p_1}$ provided that $p_2 \geq 1$ and also that $\mathrm{var}(\mathbf{X}_1 \mid \mathbf{X}_2 = \mathbf{x}_2) = \left(\frac{\nu+\|\mathbf{x}_2\|^2}{\nu+p_2-2}\right)\mathbf{I}_{p_1}$ provided that $p_2 \geq 2$, or $p_2 = 1$ and $\nu > 1$.

The next example is useful for generating multivariate spherical components by Monte Carlo.

Example 47 (Transformation to spherical coordinates) *Let $\mathbf{X} \sim \mathcal{S}_p(h)$ for $p \geq 2$. Consider the transformation to spherical coordinates for $\mathbf{X} = (X_1, \ldots, X_p)^\top$,*

$$X_i = R\left\{\prod_{j=1}^{i-1}\sin(\theta_j)\right\}\cos(\theta_i), \quad i = 1, \ldots, p-1, \quad \text{and} \quad X_p = R\left\{\prod_{j=1}^{p-2}\sin(\theta_j)\right\}\sin(\theta_{p-1}),$$

where $R \geq 0$ and $0 \leq \theta_i < \pi$, $i = 1, \ldots, p-2$, and $0 \leq \theta_{p-1} < 2\pi$. Then $R, \theta_1, \ldots, \theta_{p-1}$ are all independent and have pdf given by (4.19) for R and

$$\begin{aligned} g_i(\theta_i) &= \frac{\sin(\theta_i)^{p-i-1}}{B\{1/2,(2p-i)/2\}}, \quad 0 \leq \theta_1 < \pi, \quad i = 1, \ldots, p-2, \\ g_{p-1}(\theta_{p-1}) &= \frac{1}{2\pi}, \quad 0 \leq \theta_{p-1} < 2\pi, \end{aligned}$$

where $B(a,b)$ denotes the beta function.

4.2 Multivariate Elliptical Distributions

4.2.1 Construction

The discussion of spherical distributions can obviously be much more extensive, as can be seen, for example, in the books of Fang et al. (1990) and Fang and Zhang (1990). Here, the aim of introducing them was essentially to be able to define the more general family of multivariate elliptical distributions. Therefore, we now state the main definition of this chapter related to this family.

Definition 12 [Multivariate elliptical distributions] *A random vector $\mathbf{X} \in \mathbb{R}^p$ is said to have a multivariate elliptical distribution with location vector $\boldsymbol{\xi} \in \mathbb{R}^p$, dispersion matrix $\boldsymbol{\Omega} \in \mathbb{R}^{p \times p}$, $\boldsymbol{\Omega} \geq 0$, and cgf φ, in symbols $\mathbf{X} \sim \mathcal{E}l_p(\boldsymbol{\xi}, \boldsymbol{\Omega}, \varphi)$, iff:*

$$\mathbf{X} \stackrel{d}{=} \boldsymbol{\xi} + \mathbf{AZ}, \quad \textit{with} \quad \mathbf{Z} \sim \mathcal{S}_q(\varphi), \tag{4.25}$$

for some matrix $\mathbf{A} \in \mathbb{R}^{p \times q}$ of rank q and such that $\mathbf{AA}^\top = \boldsymbol{\Omega}$.

According to Definition 12, we can obtain an elliptically distributed random vector $\mathbf{X}$ by applying a linear map $\mathbf{A}$ to a spherically distributed random vector $\mathbf{Z}$ and then shifting it by a location vector $\boldsymbol{\xi}$. Obviously, if $\boldsymbol{\xi} = \mathbf{0}_p$ and $\boldsymbol{\Omega} = \mathbf{I}_p$ then $\mathcal{S}_p(\varphi) = \mathcal{E}l_p(\mathbf{0}_p, \mathbf{I}_p, \varphi)$, thus recovering in this way the multivariate spherical distribution as a standardized version of the multivariate elliptical one. When $\boldsymbol{\Omega}$ has full rank p this fact can be emphasized by conveniently choosing $\mathbf{A} = \boldsymbol{\Omega}^{1/2}$, i.e., a square root of $\boldsymbol{\Omega}$, and therefore the following relation becomes natural:

$$\mathbf{X} \stackrel{d}{=} \boldsymbol{\xi} + \boldsymbol{\Omega}^{1/2}\mathbf{Z} \sim \mathcal{E}l_p(\boldsymbol{\xi}, \boldsymbol{\Omega}, \varphi) \Longleftrightarrow \mathbf{Z} \stackrel{d}{=} \boldsymbol{\Omega}^{-1/2}(\mathbf{X} - \boldsymbol{\xi}) \sim \mathcal{S}_p(\varphi). \tag{4.26}$$

The nomenclature used to represent these families must also be clear, since the image of a sphere under an affine map is an ellipsoid.

However, we should note that from (4.25) the distribution of $\mathbf{X}$ cannot be uniquely determined by $\boldsymbol{\Omega} = \mathbf{AA}^\top$ and φ for a simple reason: If $\mathbf{X} \stackrel{d}{=} \boldsymbol{\xi} + \mathbf{AZ} \sim \mathcal{E}l_p(\boldsymbol{\xi}, \boldsymbol{\Omega}, \varphi)$ is not degenerate, then for an arbitrary $c > 0$, we have

$$\mathbf{X} \stackrel{d}{=} \boldsymbol{\xi} + c\mathbf{A}(c^{-1}\mathbf{Z}) = \boldsymbol{\xi} + \tilde{\mathbf{A}}\tilde{\mathbf{Z}},$$

where $\tilde{\mathbf{A}} = c\mathbf{A}$ and $\tilde{\mathbf{Z}} = c^{-1}\mathbf{Z} \stackrel{d}{=} c^{-1}R\mathbf{U}^{(q)} = \tilde{R}\mathbf{U}^{(q)} \sim \mathcal{S}_q(\tilde{\varphi})$, where $\tilde{R} = c^{-1}R \sim \tilde{F}(r) = F(cr)$ and $\tilde{\varphi}(u) = \varphi(u/c^2)$.

4.2.2 Cf, mgf, and moments

Now we can provide an alternative representation of the multivariate elliptical distributions, similar to that given in Theorem 18 for multivariate spherical distributions.

Theorem 31 *If* $\mathbf{X} \sim \mathcal{E}l_p(\boldsymbol{\xi}, \boldsymbol{\Omega}, \varphi)$ *with* $\operatorname{rank}(\boldsymbol{\Omega}) = q$, *then:*

[1] $\mathbf{X}$ *has an alternative stochastic representation*

$$\mathbf{X} \stackrel{d}{=} \boldsymbol{\xi} + R\mathbf{A}\mathbf{U}^{(q)}, \quad \text{with} \quad R^2 \stackrel{d}{=} (\mathbf{X}-\boldsymbol{\xi})^\top \boldsymbol{\Omega}^- (\mathbf{X}-\boldsymbol{\xi}) \perp\!\!\!\perp \mathbf{U}^{(q)} \sim \mathcal{U}_q(\mathbb{S}^{(q-1)}), \tag{4.27}$$

where $\mathbf{A} \in \mathbb{R}^{p\times q}$ *is such that* $\mathbf{A}\mathbf{A}^\top = \boldsymbol{\Omega}$ *and* $\boldsymbol{\Omega}^-$ *is a generalized inverse of* $\boldsymbol{\Omega}$, *i.e., a matrix satisfying* $\boldsymbol{\Omega}\boldsymbol{\Omega}^-\boldsymbol{\Omega} = \boldsymbol{\Omega}$;

[2] The cf of $\mathbf{X}$, $\psi(\mathbf{t}) = \mathrm{E}(e^{\mathrm{i}\mathbf{t}^\top \mathbf{X}})$, *is of the form*

$$\psi(\mathbf{t}) = e^{\mathrm{i}\mathbf{t}^\top \boldsymbol{\xi}} \varphi(\mathbf{t}^\top \boldsymbol{\Omega} \mathbf{t}), \quad \mathbf{t} \in \mathbb{R}^p.$$

Proof 31 *By Definition 12,* $\mathbf{X} \stackrel{d}{=} \boldsymbol{\xi} + \mathbf{A}\mathbf{Z}$ *with* $\mathbf{Z} \sim \mathcal{S}_q(\varphi)$ *and* $\mathbf{A}\mathbf{A}^\top = \boldsymbol{\Omega}$. *Due to Theorem 18, there exist* R *and* $\mathbf{U}^{(q)}$ *with the required independence such that* $\mathbf{Z} \stackrel{d}{=} R\mathbf{U}^{(q)}$. *Hence,* $\mathbf{X} \stackrel{d}{=} \boldsymbol{\xi} + \mathbf{A}\mathbf{Z} \stackrel{d}{=} \boldsymbol{\xi} + R\mathbf{A}\mathbf{U}^{(q)}$. *Moreover, since* $\mathbf{A}^\top(\mathbf{A}\mathbf{A}^\top)^-\mathbf{A}$ *is a* $q\times q$ *projection matrix of rank* q, *i.e., a symmetric idempotent matrix, we have* $(\mathbf{X}-\boldsymbol{\xi})^\top\boldsymbol{\Omega}^-(\mathbf{X}-\boldsymbol{\xi}) \stackrel{d}{=} R^2\mathbf{U}^{(q)\top}\mathbf{A}^\top(\mathbf{A}\mathbf{A}^\top)^-\mathbf{A}\mathbf{U}^{(q)} = R^2\mathbf{U}^{(q)\top}\mathbf{U}^{(q)} = R^2$. *This proves [1]. To prove [2], let* $\mathbf{X} \stackrel{d}{=} \boldsymbol{\xi} + \mathbf{A}\mathbf{Z}$ *with* $\mathbf{Z} \sim \mathcal{S}_q(\varphi)$ *and* $\mathbf{A}\mathbf{A}^\top = \boldsymbol{\Omega}$. *Thus, according to part [2] of Theorem 17, we have*

$$\psi(\mathbf{t}) = \mathrm{E}(e^{\mathrm{i}\mathbf{t}^\top\mathbf{X}}) = \mathrm{E}\{e^{\mathrm{i}\mathbf{t}^\top(\boldsymbol{\xi}+\mathbf{A}\mathbf{Z})}\} = e^{\mathrm{i}\mathbf{t}^\top\boldsymbol{\xi}}\mathrm{E}\{e^{\mathrm{i}(\mathbf{A}^\top\mathbf{t})^\top\mathbf{Z}}\} = e^{\mathrm{i}\mathbf{t}^\top\boldsymbol{\xi}}\varphi(\mathbf{t}^\top\boldsymbol{\Omega}\mathbf{t}),$$

which proves [2] above. □

Theorem 31 is of great importance and will be used quite frequently in the sequel. This result is a demonstration of how to construct an elliptically distributed random vector. First, a random direction is determined by $\mathbf{U}^{(q)}$, then it is linearly transformed by a matrix $\mathbf{A} \in \mathbb{R}^{p\times q}$ of $\operatorname{rank}(\mathbf{A}) = q$ with $\mathbf{A}\mathbf{A}^\top = \boldsymbol{\Omega}$. Finally, the random scaling factor R is applied, and then the result is shifted by $\boldsymbol{\xi}$. However, as established by Theorem 32 below, the stochastic representation (4.27) is not unique due to arbitrariness in the choice of $\mathbf{A}$. For instance, if $\mathbf{X} \sim \mathcal{E}l_p(\boldsymbol{\xi}, \boldsymbol{\Omega}, \varphi)$, then clearly it follows that the representation (4.27) can be replaced by

$$\mathbf{X} \stackrel{d}{=} \boldsymbol{\xi} + R\bar{\mathbf{A}}\mathbf{U}^{(q)} \quad \text{with} \quad \bar{\mathbf{A}} = \mathbf{A}\mathbf{Q} \quad \text{since} \quad \mathbf{U}^{(q)} \stackrel{d}{=} \mathbf{Q}\mathbf{U}^{(q)}, \quad \forall \mathbf{Q} \in \mathcal{O}_q.$$

Theorem 32 *Let* $\mathbf{X} \stackrel{d}{=} \boldsymbol{\xi} + \mathbf{A}\mathbf{Z}$, *where* $\mathbf{Z} \sim \mathcal{S}_q(\varphi)$, $\boldsymbol{\xi} \in \mathbb{R}^p$ *and* $\mathbf{A} \in \mathbb{R}^{p\times q}$ *with* $\operatorname{rank}(\mathbf{A}) = q$ *and* $\mathbf{A}\mathbf{A}^\top = \boldsymbol{\Omega}$. *Then, for any other matrix* $\tilde{\mathbf{A}}$ *with* $\operatorname{rank}(\tilde{\mathbf{A}}) = q$ *and* $\tilde{\mathbf{A}}\tilde{\mathbf{A}}^\top = \boldsymbol{\Omega}$ *we have* $\mathbf{X} \stackrel{d}{=} \boldsymbol{\xi} + \tilde{\mathbf{A}}\mathbf{Z}$.

Proof 32 *By Theorem 31 the cf of* $\mathbf{X} \stackrel{d}{=} \boldsymbol{\xi} + \mathbf{A}\mathbf{Z}$ *is* $\psi(\mathbf{t}) = e^{\mathrm{i}\mathbf{t}^\top\boldsymbol{\xi}}\mathrm{E}\{e^{\mathrm{i}(\mathbf{A}^\top\mathbf{t})^\top\mathbf{Z}}\} = e^{\mathrm{i}\mathbf{t}^\top\boldsymbol{\xi}}\varphi(\mathbf{t}^\top\boldsymbol{\Omega}\mathbf{t})$ *since* $\mathbf{A}\mathbf{A}^\top = \boldsymbol{\Omega}$. *Let* $\tilde{\mathbf{X}} = \boldsymbol{\xi} + \tilde{\mathbf{A}}\mathbf{Z}$. *By the same theorem, the cf of* $\tilde{\mathbf{X}}$ *becomes* $\tilde{\psi}(\mathbf{t}) = e^{\mathrm{i}\mathbf{t}^\top\boldsymbol{\xi}}\mathrm{E}\{e^{\mathrm{i}(\tilde{\mathbf{A}}^\top\mathbf{t})^\top\mathbf{Z}}\} = e^{\mathrm{i}\mathbf{t}^\top\boldsymbol{\xi}}\varphi(\mathbf{t}^\top\boldsymbol{\Omega}\mathbf{t})$ *since* $\tilde{\mathbf{A}}\tilde{\mathbf{A}}^\top = \boldsymbol{\Omega}$. *Because* $\psi(\mathbf{t}) = e^{\mathrm{i}\mathbf{t}^\top\boldsymbol{\xi}}\varphi(\mathbf{t}^\top\boldsymbol{\Omega}\mathbf{t}) = \tilde{\psi}(\mathbf{t})$, $\forall \mathbf{t} \in \mathbb{R}^p$, *the uniqueness of the cf implies* $\mathbf{X} \stackrel{d}{=} \tilde{\mathbf{X}}$, *thus proving the assertion.* □

From Theorems 18 and 21, it is easy to verify for any elliptical random vector $\mathbf{X} \sim \mathcal{E}l_p(\boldsymbol{\xi}, \boldsymbol{\Omega}, \varphi)$ that $\mathrm{E}(\mathbf{X})$ exists iff $\mathrm{E}(R) < \infty$ and $\operatorname{var}(\mathbf{X})$ exists iff $\mathrm{E}(R^2) < \infty$.

Theorem 33 *Assume* $\mathbf{X} \overset{d}{=} \boldsymbol{\xi} + R\mathbf{A}\mathbf{U}^{(q)} \sim \mathcal{E}l_p(\boldsymbol{\xi}, \boldsymbol{\Omega}, \varphi)$, *where* $\mathbf{A}\mathbf{A}^\top = \boldsymbol{\Omega}$, $\text{rank}(\mathbf{A}) = q$ *and* $\text{E}(R^2) < \infty$. *Then,*

$$\text{E}(\mathbf{X}) = \boldsymbol{\xi} \quad \textit{and} \quad \text{var}(\mathbf{X}) = \frac{1}{q}\text{E}(R^2)\,\boldsymbol{\Omega} = -2\varphi'(0)\,\boldsymbol{\Omega}.$$

Proof 33 *Since* $\mathbf{X} \overset{d}{=} \boldsymbol{\xi} + R\mathbf{A}\mathbf{U}^{(q)}$, *where* $R \perp\!\!\!\perp \mathbf{U}^{(q)}$, *then by using Theorem 20 with p replaced by q, we have* $\text{E}(\mathbf{X}) = \boldsymbol{\xi} + \text{E}(R)\text{E}(\mathbf{U}^{(q)}) = \boldsymbol{\xi}$ *and* $\text{var}(\mathbf{X}) = \mathbf{A}\text{var}(R\mathbf{U}^{(q)})\mathbf{A}^\top = \mathbf{A}\text{E}\{R^2\mathbf{U}^{(q)}(\mathbf{U}^{(q)})^\top\}\mathbf{A}^\top = (1/q)\text{E}(R^2)\mathbf{A}\mathbf{A}^\top = (1/q)\text{E}(R^2)\boldsymbol{\Omega} = -2\varphi'(0)\boldsymbol{\Omega}$ *since* $\text{E}(R^2) = -2q\varphi'(0)$. □

An interesting consequence of Theorem 33 is that the correlation matrix is the same for the entire elliptical family with finite second moment. When assuming that $\boldsymbol{\Omega} > 0$, it is given by $\bar{\boldsymbol{\Omega}} = \boldsymbol{\omega}^{-1}\boldsymbol{\Omega}\boldsymbol{\omega}^{-1}$, where $\boldsymbol{\omega} = \text{diag}(\boldsymbol{\Omega})^{1/2}$.

Since $\mathbf{X} - \boldsymbol{\xi} \overset{d}{=} \mathbf{A}\mathbf{Z} \sim \mathcal{E}l_p(\mathbf{0}_p, \boldsymbol{\Omega}, \varphi)$, where $\mathbf{Z} \sim \mathcal{S}_q(\varphi)$ and $\mathbf{A}\mathbf{A}^\top = \boldsymbol{\Omega}$, expressions for higher-order centered moments of $\mathbf{X} - \boldsymbol{\xi}$ can also be obtained from the results presented in Section 4.1.3, provided that they exist. Evidently, odd-order moments of that type are zero by symmetry, while those of even order correspond to moments of the squared-radial variable $R^2 \overset{d}{=} (\mathbf{X} - \boldsymbol{\xi})^\top\boldsymbol{\Omega}^-(\mathbf{X} - \boldsymbol{\xi})$. For example, when assuming that $\boldsymbol{\Omega}$ is nonsingular and $\text{E}(R^2) < \infty$, then $\mathbf{X} \sim \mathcal{E}l_p(\boldsymbol{\xi}, \boldsymbol{\Omega}, \varphi)$ has a variance matrix $\text{var}(\mathbf{X}) = \text{E}(R^2/p)\boldsymbol{\Omega}$, which is also nonsingular. If in addition $\text{E}(R^4) < \infty$, then Mardia's measure of multivariate kurtosis of $\mathbf{X} \sim \mathcal{E}l_p(\boldsymbol{\xi}, \boldsymbol{\Omega}, \varphi)$ becomes

$$\beta_{2,p} = \text{E}[\{(\mathbf{X} - \boldsymbol{\xi})^\top\text{var}(\mathbf{X})^{-1}(\mathbf{X} - \boldsymbol{\xi})\}^2] = p(p+2)\frac{\text{E}\{R^4/p(p+2)\}}{\{\text{E}(R^2)/p\}^2} = p(p+2)\frac{\varphi''(0)}{\{\varphi'(0)\}^2},$$

since $\text{E}(R^2/p) = -2\varphi'(0)$ and $\text{E}[R^4/\{p(p+2)\}] = 4\varphi''(0)$. Since $\beta_{2,p}/\{p(p+2)\} = \beta_{2,1}/3 = \varphi''(0)/\{\varphi'(0)\}^2$, Muirhead (1982) defined the *kurtosis parameter* given by

$$\kappa = \frac{\varphi''(0)}{\{\varphi'(0)\}^2} - 1,$$

which becomes an excess of kurtosis of the marginal elliptical distributions of $X_1, \ldots, X_p$ of $\mathbf{X}$. Later, Berkane and Bentler (1986) introduced the general *moment parameters* given by

$$\kappa(m) = \frac{\varphi^{(m)}(0)}{\{\varphi'(0)\}^m} - 1, \quad m = 1, 2 \ldots.$$

They also showed that

$$\text{E}(X_{i_1}X_{i_2}\cdots X_{i_{2m}}) = (\kappa(m) + 1)\sum \text{E}(X_{i_1}X_{i_2})\text{E}(X_{i_3}X_{i_4})\cdots\text{E}(X_{i_{2m-1}}X_{i_{2m}}),$$

where the summation is for all the $(2m)!/(2^m m!)$ possible pairs of indices $\{i_j, i_{j'}\}$ obtained from the set $\{i_1, i_2, \ldots, i_{2m}\}$. For example, since $\kappa(2) = \kappa$, we have

$$\text{E}(X_iX_jX_kX_l) = (\kappa + 1)\{\text{E}(X_iX_j)\text{E}(X_kX_l) + \text{E}(X_iX_k)\text{E}(X_jX_l) + \text{E}(X_iX_l)\text{E}(X_jX_k)\},$$

which are the entries of $\boldsymbol{\mu}_4(\mathbf{X}) = \text{E}(\mathbf{X}\mathbf{X}^\top \otimes \mathbf{X}\mathbf{X}^\top)$ when $\mathbf{X} \sim \mathcal{E}l_p(\mathbf{0}_p, \boldsymbol{\Omega}, \varphi)$; see Theorem 23. In this case, it also follows that

$$\text{var}(\mathbf{X} \otimes \mathbf{X}) = (1 + \kappa)(\mathbf{I}_{p^2} + \mathbf{K}_p)(\boldsymbol{\Sigma} \otimes \boldsymbol{\Sigma}) + \kappa\text{vec}(\boldsymbol{\Sigma})\text{vec}(\boldsymbol{\Sigma})^\top,$$

where $\boldsymbol{\Sigma} = \text{var}(\mathbf{X}) = \kappa_2\boldsymbol{\Omega}$ and $\kappa_2 = \text{var}(X_1) = -2\varphi'(0)$.

4.2.3 *Affine transformations, marginals, and independence*

By construction, multivariate elliptical distributions are closed under affine transformations and then by marginalization, as shown below.

Theorem 34 *Let $\mathbf{X} \sim \mathcal{E}l_p(\boldsymbol{\xi}, \boldsymbol{\Omega}, \varphi)$. Then, for any vector $\mathbf{b} \in \mathbb{R}^q$ and matrix $\mathbf{B} \in \mathbb{R}^{q\times p}$, we have*

$$\mathbf{B}\mathbf{X} + \mathbf{b} \sim \mathcal{E}l_q(\mathbf{B}\boldsymbol{\xi} + \mathbf{b}, \mathbf{B}\boldsymbol{\Omega}\mathbf{B}^\top, \varphi).$$

Proof 34 *By Theorem 31, the cf of $\mathbf{B}\mathbf{X} + \mathbf{b}$ is $\tilde{\psi}(\mathbf{t}) = \mathrm{E}\{e^{i\mathbf{t}^\top(\mathbf{B}\mathbf{X}+\mathbf{b})}\} = e^{i\mathbf{t}^\top(\mathbf{B}\boldsymbol{\xi}+\mathbf{b})}\varphi(\mathbf{t}^\top\mathbf{B}\boldsymbol{\Omega}\mathbf{B}^\top\mathbf{t})$, $\forall \mathbf{t} \in \mathbb{R}^q$, which proves the theorem.* □

As a direct consequence of Theorem 34, we find that the marginal distributions of an elliptically distributed random vector are also elliptical. To establish this result, partition $\mathbf{X}$, $\boldsymbol{\xi}$, and $\boldsymbol{\Omega}$ into

$$\mathbf{X} = \begin{pmatrix} \mathbf{X}_1 \\ \mathbf{X}_2 \end{pmatrix}, \quad \boldsymbol{\xi} = \begin{pmatrix} \boldsymbol{\xi}_1 \\ \boldsymbol{\xi}_2 \end{pmatrix}, \quad \boldsymbol{\Omega} = \begin{pmatrix} \boldsymbol{\Omega}_{11} & \boldsymbol{\Omega}_{12} \\ \boldsymbol{\Omega}_{21} & \boldsymbol{\Omega}_{22} \end{pmatrix}, \tag{4.28}$$

where $\mathbf{X}_1, \boldsymbol{\xi}_1 \in \mathbb{R}^{p_1}$, $\mathbf{X}_2, \boldsymbol{\xi}_2 \in \mathbb{R}^{p_2}$, $\boldsymbol{\Omega}_{11} \in \mathbb{R}^{p_1\times p_1}$, $\boldsymbol{\Omega}_{22} \in \mathbb{R}^{p_2\times p_2}$, and $\boldsymbol{\Omega}_{12} = \boldsymbol{\Omega}_{21}^\top \in \mathbb{R}^{p_1\times p_2}$, with $p_1 + p_2 = p$.

Theorem 35 *Let $\mathbf{X} \sim \mathcal{E}l_p(\boldsymbol{\xi}, \boldsymbol{\Omega}, \varphi)$, and consider the partition (4.28). Then $\mathbf{X}_1 \sim \mathcal{E}l_{p_1}(\boldsymbol{\xi}_1, \boldsymbol{\Omega}_{11}, \varphi)$ and $\mathbf{X}_2 \sim \mathcal{E}l_{p_2}(\boldsymbol{\xi}_2, \boldsymbol{\Omega}_{22}, \varphi)$.*

Proof 35 *The proof is direct from Theorem 34 by taking $\mathbf{b} = \mathbf{0}_{p_1}$ and $\mathbf{B} = (\mathbf{I}_{p_1}, \mathbf{O})$ for $\mathbf{X}_1$, and $\mathbf{b} = \mathbf{0}_{p_2}$ and $\mathbf{B} = (\mathbf{O}, \mathbf{I}_{p_2})$ for $\mathbf{X}_2$.* □

4.2.4 *Conditional distributions*

As in the spherical case, the conditional distributions from a multivariate elliptical distribution are again elliptical but may have different generators from the original one. We show this by considering an elliptical random vector $\mathbf{X} \sim \mathcal{E}l_p(\boldsymbol{\xi}, \boldsymbol{\Omega}, \varphi)$ partitioned as (4.28), and we derive the conditional distribution of $(\mathbf{X}_1 \mid \mathbf{X}_2 = \mathbf{x}_2)$ assuming that $\boldsymbol{\Omega}$ is nonsingular. A similar assumption is considered in the books of Fang et al. (1990) and Fang and Zhang (1990). The result when $\boldsymbol{\Omega}$ is singular can be found in the seminal paper of Cambanis et al. (1981) or in a more detailed way in the book by Gupta et al. (2013b).

Theorem 36 *Let $\mathbf{X} \overset{d}{=} \boldsymbol{\xi} + R\mathbf{A}\mathbf{U}^{(p)} \sim \mathcal{E}l_p(\boldsymbol{\xi}, \boldsymbol{\Omega}, \varphi)$, with $\mathbf{A}\mathbf{A}^\top = \boldsymbol{\Omega} > 0$, be partitioned as (4.28). Then,*

$$(\mathbf{X}_1 \mid \mathbf{X}_2 = \mathbf{x}_2) \overset{d}{=} \boldsymbol{\xi}_{1\cdot 2} + R_{Q(\mathbf{x}_2)}\mathbf{A}_{1\cdot 2}\mathbf{U}^{(p_1)} \sim \mathcal{E}l_{p_1}(\boldsymbol{\xi}_{1\cdot 2}, \boldsymbol{\Omega}_{11\cdot 2}, \varphi_{Q(\mathbf{x}_2)}), \tag{4.29}$$

where for each $a \geq 0$, $R_{a^2} \overset{d}{=} \left(\sqrt{R^2 - a^2} \mid Q(\mathbf{X}_2) = a^2\right)$ and is independent of $\mathbf{U}^{(p_1)}$,

$$\begin{aligned} \boldsymbol{\xi}_{1\cdot 2} &= \boldsymbol{\xi}_1 + \boldsymbol{\Omega}_{12}\boldsymbol{\Omega}_{22}^{-1}(\mathbf{x}_2 - \boldsymbol{\xi}_2), \\ \boldsymbol{\Omega}_{11\cdot 2} &= \boldsymbol{\Omega}_{11} - \boldsymbol{\Omega}_{12}\boldsymbol{\Omega}_{22}^{-1}\boldsymbol{\Omega}_{21}, \\ \mathbf{A}_{1\cdot 2}\mathbf{A}_{1\cdot 2}^\top &= \boldsymbol{\Omega}_{11\cdot 2}, \quad \textit{with } \operatorname{rank}(\mathbf{A}_{1\cdot 2}) = p_1, \\ Q(\mathbf{x}_2) &= (\mathbf{x}_2 - \boldsymbol{\xi}_2)^\top\boldsymbol{\Omega}_{22}^{-1}(\mathbf{x}_2 - \boldsymbol{\xi}_2), \end{aligned}$$

and the cgf φ_{a^2} is given by (4.3) with p replaced by p_1 and F by F_{a^2}, the cdf of R_{a^2} defined by (4.13). If, in addition, $\mathrm{E}(R^2_{Q(\mathbf{x}_2)}) < \infty$, then

$$\begin{aligned} \mathrm{E}(\mathbf{X}_1 \mid \mathbf{x}_2) &= \boldsymbol{\xi}_1 + \boldsymbol{\Omega}_{12}\boldsymbol{\Omega}_{22}^{-1}(\mathbf{x}_2 - \boldsymbol{\xi}_2); \\ \mathrm{var}(\mathbf{X}_1 \mid \mathbf{x}_2) &= \kappa_2\{Q(\mathbf{x}_2)\}(\boldsymbol{\Omega}_{11} - \boldsymbol{\Omega}_{12}\boldsymbol{\Omega}_{22}^{-1}\boldsymbol{\Omega}_{21}), \end{aligned}$$

where $\kappa_2\{Q(\mathbf{x}_2)\} = \frac{1}{p_1}\mathrm{E}(R^2_{Q(\mathbf{x}_2)}) = -2\varphi'_{Q(\mathbf{x}_2)}(0)$.

Proof 36 *Let $\mathbf{A}_2$ be a square nonsingular matrix such that $\mathbf{A}_2\mathbf{A}_2^\top = \boldsymbol{\Omega}_{22}$. It is easy to see that*

$$\boldsymbol{\Omega} = \mathbf{A}\mathbf{A}^\top = \begin{pmatrix} \boldsymbol{\Omega}_{11} & \boldsymbol{\Omega}_{12} \\ \boldsymbol{\Omega}_{21} & \boldsymbol{\Omega}_{22} \end{pmatrix} = \begin{pmatrix} \mathbf{I}_{p_1} & \boldsymbol{\Omega}_{12}\boldsymbol{\Omega}_{22}^{-1} \\ \mathbf{O} & \mathbf{I}_{p_2} \end{pmatrix} \begin{pmatrix} \boldsymbol{\Omega}_{11\cdot 2} & \mathbf{O} \\ \mathbf{O} & \boldsymbol{\Omega}_{22} \end{pmatrix} \begin{pmatrix} \mathbf{I}_{p_1} & \mathbf{O} \\ \boldsymbol{\Omega}_{22}^{-1}\boldsymbol{\Omega}_{21} & \mathbf{I}_{p_2} \end{pmatrix} = \mathbf{C}\mathbf{C}^\top,$$

with

$$\mathbf{C} = \begin{pmatrix} \mathbf{A}_{1\cdot 2} & \mathbf{O} \\ \mathbf{O} & \mathbf{A}_2 \end{pmatrix} \begin{pmatrix} \mathbf{I}_{p_1} & \mathbf{O} \\ \boldsymbol{\Omega}_{22}^{-1}\boldsymbol{\Omega}_{21} & \mathbf{I}_{p_2} \end{pmatrix} = \begin{pmatrix} \mathbf{A}_{1\cdot 2} & \mathbf{O} \\ \mathbf{A}_2\boldsymbol{\Omega}_{22}^{-1}\boldsymbol{\Omega}_{21} & \mathbf{A}_2 \end{pmatrix}.$$

Also, let $\mathbf{Z} = (\mathbf{Z}_1^\top, \mathbf{Z}_2^\top)^\top \sim \mathcal{S}_p(\varphi) \equiv \mathcal{E}l_p(\mathbf{0}_p, \mathbf{I}_p, \varphi)$, where $\mathbf{Z}_1 \in \mathbb{R}^{p_1}$ and $\mathbf{Z}_2 \in \mathbb{R}^{p_2}$. By Definition 12, $\mathbf{X} \stackrel{d}{=} \boldsymbol{\xi} + \mathbf{A}\mathbf{Z}$, i.e.,

$$\begin{pmatrix} \mathbf{X}_1 \\ \mathbf{X}_2 \end{pmatrix} \stackrel{d}{=} \begin{pmatrix} \boldsymbol{\xi}_1 + \mathbf{A}_{11\cdot 2}\mathbf{Z}_1 + \boldsymbol{\Omega}_{12}\boldsymbol{\Omega}_{22}^{-1}\mathbf{Z}_2 \\ \boldsymbol{\xi}_2 + \mathbf{A}_2\mathbf{Z}_2 \end{pmatrix},$$

from where we have

$$\begin{aligned} (\mathbf{X}_1 \mid \mathbf{X}_2 = \mathbf{x}_2) &\stackrel{d}{=} \left(\boldsymbol{\xi}_1 + \mathbf{A}_{11\cdot 2}\mathbf{Z}_1 + \boldsymbol{\Omega}_{12}\boldsymbol{\Omega}_{22}^{-1}\mathbf{A}_2\mathbf{Z}_2 \mid \boldsymbol{\xi}_2 + \mathbf{A}_2\mathbf{Z}_2 = \mathbf{x}_2\right) \\ &\stackrel{d}{=} \left(\boldsymbol{\xi}_1 + \mathbf{A}_{11\cdot 2}\mathbf{Z}_1 + \boldsymbol{\Omega}_{12}\boldsymbol{\Omega}_{22}^{-1}\mathbf{A}_2\mathbf{Z}_2 \mid \mathbf{Z}_2 = \mathbf{A}_2^{-1}(\mathbf{x}_2 - \boldsymbol{\xi}_2)\right) \\ &\stackrel{d}{=} \left(\boldsymbol{\xi}_1 + \mathbf{A}_{11\cdot 2}\mathbf{Z}_1 + \boldsymbol{\Omega}_{12}\boldsymbol{\Omega}_{22}^{-1}(\mathbf{x}_2 - \boldsymbol{\xi}_2) \mid \mathbf{Z}_2 = \mathbf{A}_2^{-1}(\mathbf{x}_2 - \boldsymbol{\xi}_2)\right) \\ &\stackrel{d}{=} \boldsymbol{\xi}_{1\cdot 2} + \mathbf{A}_{11\cdot 2}\left(\mathbf{Z}_1 \mid \mathbf{Z}_2 = \mathbf{A}_2^{-1}(\mathbf{x}_2 - \boldsymbol{\xi}_2)\right), \end{aligned}$$

and by Theorem 27

$$\left(\mathbf{Z}_1 \mid \mathbf{Z}_2 = \mathbf{A}_2^{-1}(\mathbf{x}_2 - \boldsymbol{\xi}_2)\right) \stackrel{d}{=} \left(\mathbf{Z}_1 \mid \|\mathbf{Z}_2\| = \|\mathbf{A}_2^{-1}(\mathbf{x}_2 - \boldsymbol{\xi}_2)\|\right) \stackrel{d}{=} R_{Q(\mathbf{x}_2)}\mathbf{U}^{(p_1)} \sim \mathcal{S}_{p_1}(\varphi_{Q(\mathbf{x}_2)}).$$

Hence, by Definition 12, we finally have $(\mathbf{X}_1 \mid \mathbf{X}_2 = \mathbf{x}_2) \sim \mathcal{E}l_{p_1}(\boldsymbol{\xi}_1 + \boldsymbol{\Omega}_{12}\boldsymbol{\Omega}_{22}^{-1}(\mathbf{x}_2 - \boldsymbol{\xi}_2), \boldsymbol{\Omega}_{11} - \boldsymbol{\Omega}_{12}\boldsymbol{\Omega}_{22}^{-1}\boldsymbol{\Omega}_{21}, \varphi_{Q(\mathbf{x}_2)})$. The proof of the conditional mean and variance matrix is straightforward following the proof of Theorem 33. □

A similar proof can be found in the books of Fang et al. (1990) and Fang and Zhang (1990). In the singular case, if $|\boldsymbol{\Omega}_{22}| = 0$, we can replace in (4.29) the conditional dispersion matrix by $\boldsymbol{\Omega}_{11\cdot 2} = \boldsymbol{\Omega}_{11} - \boldsymbol{\Omega}_{12}\boldsymbol{\Omega}_{22}^{-}\boldsymbol{\Omega}_{21}$, where $\boldsymbol{\Omega}_{22}^{-}$ is a generalized inverse of $\boldsymbol{\Omega}_{22}$, i.e., a matrix satisfying $\boldsymbol{\Omega}_{22}\boldsymbol{\Omega}_{22}^{-}\boldsymbol{\Omega}_{22} = \boldsymbol{\Omega}_{22}$; but in this case, the proof is somewhat more tedious. One fact that can help in this situation is the existence of a matrix $\mathbf{B} \in \mathbb{R}^{p_1 \times p_2}$ such that $\boldsymbol{\Omega}_{12} = \mathbf{B}\boldsymbol{\Omega}_{22}$ and therefore $\boldsymbol{\Omega}_{12}\boldsymbol{\Omega}_{22}^{-}\boldsymbol{\Omega}_{22} = \mathbf{B}\boldsymbol{\Omega}_{22}\boldsymbol{\Omega}_{22}^{-}\boldsymbol{\Omega}_{22} = \mathbf{B}\boldsymbol{\Omega}_{22} = \boldsymbol{\Omega}_{12}$; see, e.g., Lemma 1.2.10 in Muirhead (1982).

As can be seen in Theorem 36, in the whole multivariate elliptical distribution class with finite second moment, the regression function of $\mathbf{X}_1$ on $\mathbf{X}_2 = \mathbf{x}_2$ defined by the conditional mean vector $\mathrm{E}(\mathbf{X}_1 \mid \mathbf{x}_2)$ is linear in $\mathbf{x}_2$, but, except in the normal case, it becomes a heteroscedastic linear model, because the conditional variance matrix $\mathrm{var}(\mathbf{X}_1 \mid \mathbf{x}_2)$ depends on $\mathbf{x}_2$ through the quadratic form $Q(\mathbf{x}_2) = (\mathbf{x}_2 - \boldsymbol{\xi}_2)^\top\boldsymbol{\Omega}_{22}^{-1}(\mathbf{x}_2 - \boldsymbol{\xi}_2)$.

4.2.5 Cdf and pdf

An elliptical random vector $\mathbf{X} \sim \mathcal{E}l_p(\boldsymbol{\xi}, \boldsymbol{\Omega}, \varphi)$ does not necessarily have a pdf. Obviously, a necessary condition for $\mathbf{X}$ to have a pdf is that $\text{rank}(\boldsymbol{\Omega}) = p$, and, by (4.25), $\mathbf{X} \stackrel{d}{=} \boldsymbol{\xi} + \mathbf{A}\mathbf{Z}$, but now $\mathbf{A}$ is a nonsingular matrix that can be conveniently chosen as $\mathbf{A} = \boldsymbol{\Omega}^{1/2}$ and $\mathbf{Z} \stackrel{d}{=} R\mathbf{U}^{(p)} \sim \mathcal{S}_p(\varphi)$. In view of these relations, when $\boldsymbol{\Omega} > 0$, it is clear that $\mathbf{X}$ has a pdf iff $\mathbf{Z}$ also does, which by Theorem 29 holds iff R has a pdf. In this case, since by inversion $\mathbf{Z} \stackrel{d}{=} \boldsymbol{\Omega}^{-1/2}(\mathbf{X} - \boldsymbol{\xi}) \sim \mathcal{S}_p(h)$ which by (4.14) has a pdf $h(\mathbf{z}^\top \mathbf{z})$, $\mathbf{z} \in \mathbb{R}^p$, the pdf of $\mathbf{X}$ becomes

$$f(\mathbf{x}; \boldsymbol{\xi}, \boldsymbol{\Omega}, h) = |\boldsymbol{\Omega}|^{-1/2} h\{(\mathbf{x} - \boldsymbol{\xi})^\top \boldsymbol{\Omega}^{-1} (\mathbf{x} - \boldsymbol{\xi})\}, \quad \mathbf{x} \in \mathbb{R}^p, \tag{4.30}$$

which we write as $\mathbf{X} \sim \mathcal{E}l_p(\boldsymbol{\xi}, \boldsymbol{\Omega}, h)$, and therefore now we can replace in relation (4.26) the φ by h. In this case, it is also clear that the squared-radial random variable $R^2 \stackrel{d}{=} Q(\mathbf{X}) = (\mathbf{X} - \boldsymbol{\xi})^\top \boldsymbol{\Omega}^{-1} (\mathbf{X} - \boldsymbol{\xi})$ has a pdf given by (4.20).

In the nonsingular case, with

$$\mathbf{X} \stackrel{d}{=} \boldsymbol{\xi} + R\boldsymbol{\Omega}^{1/2}\mathbf{U}^{(p)} \sim \mathcal{E}l_p(\boldsymbol{\xi}, \boldsymbol{\Omega}, h), \quad R^2 \stackrel{d}{=} Q(\mathbf{X}) = (\mathbf{X} - \boldsymbol{\xi})^\top \boldsymbol{\Omega}^{-1}(\mathbf{X} - \boldsymbol{\xi}) \perp\!\!\!\perp \mathbf{U}^{(p)} \sim \mathcal{U}(\mathbb{S}^{(p-1)}),$$

and pdf given by (4.30), which is partitioned as in (4.28), the previous results can be summarized as follows:

[1] The marginal distributions are

$$\begin{aligned}
\mathbf{X}_1 &\stackrel{d}{=} \boldsymbol{\xi}_1 + R_1 \boldsymbol{\Omega}_{11}^{1/2} \mathbf{U}^{(p_1)} \sim \mathcal{E}l_{p_1}(\boldsymbol{\xi}_1, \boldsymbol{\Omega}_{11}, h^{(p_1)}), \\
\mathbf{X}_2 &\stackrel{d}{=} \boldsymbol{\xi}_2 + R_2 \boldsymbol{\Omega}_{22}^{1/2} \mathbf{U}^{(p_2)} \sim \mathcal{E}l_{p_2}(\boldsymbol{\xi}_2, \boldsymbol{\Omega}_{22}, h^{(p_2)}), \\
R_1 &\stackrel{d}{=} R\sqrt{V_{p,p_1}}, \quad R_2 \stackrel{d}{=} R\sqrt{1 - V_{p,p_1}}, \quad R \perp\!\!\!\perp V_{p,p_1} \sim \mathcal{B}eta(p_1/2, p_2/2),
\end{aligned}$$

with marginal dgf

$$h^{(k)}(u) = \frac{\pi^{(p-k)/2}}{\Gamma\{(p-k)/2\}} \int_0^\infty v^{(p-k)/2-1} h(u+v)\mathrm{d}v, \quad k = p_1, p_2.$$

The respective marginal pdfs are

$$\begin{aligned}
f_1(\mathbf{x}_1; \boldsymbol{\xi}_1, \boldsymbol{\Omega}_{11}, h) &= |\boldsymbol{\Omega}_{11}|^{-1/2} h^{(p_1)}\{(\mathbf{x}_1 - \boldsymbol{\xi}_1)^\top \boldsymbol{\Omega}_{11}^{-1} (\mathbf{x}_1 - \boldsymbol{\xi}_1)\}, \\
f_2(\mathbf{x}_2; \boldsymbol{\xi}_2, \boldsymbol{\Omega}_{22}, h) &= |\boldsymbol{\Omega}_{22}|^{-1/2} h^{(p_2)}\{(\mathbf{x}_2 - \boldsymbol{\xi}_2)^\top \boldsymbol{\Omega}_{22}^{-1} (\mathbf{x}_2 - \boldsymbol{\xi}_2)\},
\end{aligned}$$

and the radial pdfs of R, R_1 and R_2 are, respectively,

$$g(r) = s_p r^{p-1} h(r^2), \quad g_1(r_1) = s_{p_1} r_1^{p_1-1} h^{(p_1)}(r_1^2), \quad g_2(r_2) = s_{p_2} r_2^{p_2-1} h^{(p_2)}(r_2^2), \quad r, r_1, r_2 > 0,$$

where $s_k = 2\pi^{k/2}/\Gamma(k/2)$ is the surface of the unit sphere in $\mathbb{R}^k$ denoted here by $\mathbb{S}^{(k-1)}$.

[2] The marginal mean vectors $\mathrm{E}(\mathbf{X}_1)$ and $\mathrm{E}(\mathbf{X}_2)$ exist iff $\mathrm{E}(\mathbf{X})$ exists, or equivalently, if $\mathrm{E}(R)$ exists; in this case,

$$\mathrm{E}(\mathbf{X}) = \boldsymbol{\xi}, \quad \mathrm{E}(\mathbf{X}_1) = \boldsymbol{\xi}_1, \quad \mathrm{E}(\mathbf{X}_2) = \boldsymbol{\xi}_2.$$

[3] The marginal variance matrices $\mathrm{var}(\mathbf{X}_1)$ and $\mathrm{var}(\mathbf{X}_2)$ exist iff $\mathrm{var}(\mathbf{X})$ exists, or equivalently, if $\mathrm{E}(R^2)$ exists; in this case

$$\mathrm{var}(\mathbf{X}) = \kappa_2 \boldsymbol{\Omega}, \quad \mathrm{var}(\mathbf{X}_1) = \kappa_2 \boldsymbol{\Omega}_{11}, \quad \mathrm{var}(\mathbf{X}_2) = \kappa_2 \boldsymbol{\Omega}_{22},$$

$$\kappa_2 = \frac{1}{p}\mathrm{E}(R^2) = \frac{1}{p_1}\mathrm{E}(R_1^2) = \frac{1}{p_2}\mathrm{E}(R_2^2) = -2\varphi'(0).$$

[4] The conditional distribution of $\mathbf{X}_1$ given $\mathbf{X}_2 = \mathbf{x}_2$ is

$$(\mathbf{X}_1 \mid \mathbf{X}_2 = \mathbf{x}_2) \stackrel{d}{=} \boldsymbol{\xi}_{1\cdot 2} + R_{Q(\mathbf{x}_2)} \boldsymbol{\Omega}_{11\cdot 2}^{1/2} \mathbf{U}^{(p_1)} \sim \mathcal{E}l_{p_1}(\boldsymbol{\xi}_{1\cdot 2}, \boldsymbol{\Omega}_{11\cdot 2}, h_{Q(\mathbf{x}_2)}^{(p_1)}),$$

$$R_v^2 \stackrel{d}{=} (Q(\mathbf{X}_1) \mid Q(\mathbf{X}_2) = v) \perp\!\!\!\perp \mathbf{U}^{(p_1)}, \quad v > 0,$$

with $Q(\mathbf{x}_i) = (\mathbf{x}_i - \boldsymbol{\xi}_i)^\top \boldsymbol{\Omega}_{ii}^{-1} (\mathbf{x}_i - \boldsymbol{\xi}_i)$, $i = 1, 2$, and conditional dgf

$$h_v^{(p_1)}(u) = \frac{h(u+v)}{h^{(p_2)}(v)} = \frac{\Gamma(p_1/2) h(u+v)}{\pi^{p_1/2} \int_0^\infty w^{p_1/2-1} h(v+w) \mathrm{d}w}.$$

The corresponding conditional pdf of $(\mathbf{X}_1 \mid \mathbf{X}_2 = \mathbf{x}_2)$ is

$$f_1(\mathbf{x}_1 \mid \mathbf{x}_2; \boldsymbol{\xi}_{1\cdot 2}, \boldsymbol{\Omega}_{11\cdot 2}, h) = |\boldsymbol{\Omega}_{11\cdot 2}|^{-1/2} h_{Q(\mathbf{x}_2)}^{(p_1)} \{(\mathbf{x}_1 - \boldsymbol{\xi}_{1\cdot 2})^\top \boldsymbol{\Omega}_{11\cdot 2}^{-1} (\mathbf{x}_1 - \boldsymbol{\xi}_{1\cdot 2})\},$$

and the pdf of R_v is

$$g_v(r) = s_{p_1} r^{p_1 - 1} h_v^{(p_1)}(r^2), \quad r > 0.$$

Here, it is worth noting that, for almost every $v \geq 0$, the moment $\mathrm{E}(R_v^k)$ is always finite when $p_2 \geq k$, and is given by

$$\mathrm{E}(R_v^k) = \int_0^\infty s_{p_1} r^{p_1+k-1} h_v^{(p_1)}(r^2) \mathrm{d}r = \frac{\int_0^\infty w^{(p_1+k)/2-1} h(v+w) \mathrm{d}w}{\int_0^\infty w^{p_1/2-1} h(v+w) \mathrm{d}w}.$$

This fact is due to the assumption that $\int_0^\infty w^{m/2-1} h(v+w) \mathrm{d}w < \infty$ for each $m = 1, \ldots, p-1$ and almost every $v \geq 0$.

[5] The conditional mean vector $\mathrm{E}(\mathbf{X}_1 \mid \mathbf{X}_2 = \mathbf{x}_2)$ always exists when $p_2 \geq 1$; it is given by

$$\mathrm{E}(\mathbf{X}_1 \mid \mathbf{X}_2 = \mathbf{x}_2) = \boldsymbol{\xi}_{1\cdot 2}.$$

[6] The conditional variance matrix $\mathrm{var}(\mathbf{X}_1 \mid \mathbf{X}_2 = \mathbf{x}_2)$ always exists when $p_2 \geq 2$ and iff $\mathrm{E}(R) < \infty$ when $p_2 = 1$; it is given by

$$\mathrm{var}(\mathbf{X}_1 \mid \mathbf{X}_2 = \mathbf{x}_2) = \kappa_2\{Q(\mathbf{x}_2)\} \boldsymbol{\Omega}_{11\cdot 2},$$

with

$$\kappa_2(v) = \frac{1}{p_1} \mathrm{E}(R_v^2) = \frac{1}{p_1} \frac{\int_0^\infty w^{p_1/2} h(v+w) \mathrm{d}w}{\int_0^\infty w^{p_1/2-1} h(v+w) \mathrm{d}w}.$$

In the following, we give two tables related to multivariate spherical (elliptical) distributions. Table 4.1 shows the dgf of several multivariate spherical (elliptical) pdfs. It also indicates the respective squared-radial distribution associated with the random variable variable $R^2 \stackrel{d}{=} \|\mathbf{Z}\|^2$, with $\mathbf{Z} = \boldsymbol{\Omega}^{-1/2}(\mathbf{X} - \boldsymbol{\xi}) \sim \mathcal{S}_p(h)$, which is denoted there by $\mathcal{SR}(p, h)$ or by $\mathcal{SR}_p(h)$. Table 4.2 provides their means and variances. When $p = 1$, the elliptical class $\mathcal{E}l_1(\xi, \sigma^2, \varphi)$ corresponds to the univariate symmetric distributions about ξ, i.e., $X \sim \mathcal{E}l_1(\xi, \sigma^2, \varphi)$ iff $\mathrm{P}(X \leq x - \xi) = \mathrm{P}(X \geq x + \xi)$ for all $x \in \mathbb{R}$.

4.2.6 Geometry

Some authors prefer the name "elliptically contoured distribution" for (4.30) because the contours of constant pdf of a multivariate elliptical distribution are ellipsoids, consisting of the set of points $\mathbf{x} \in \mathbb{R}^p$ satisfying

$$Q(\mathbf{x}) = (\mathbf{x} - \boldsymbol{\xi})^\top \boldsymbol{\Omega}^{-1} (\mathbf{x} - \boldsymbol{\xi}) = c^2,$$

Table 4.1 *Some subclasses of p-dimensional spherical (elliptical) distributions with their corresponding squared-radial distributions* ($u = \|\mathbf{z}\|^2$, $\mathbf{z} \in \mathbb{R}^p$). *Below,* c_p *is an appropriate normalizing constant;* $T \sim \mathcal{G}^{\frac{1}{s}}(\alpha,\lambda)$ *means that* $T^s \sim \mathcal{G}amma(\alpha,\lambda)$; $\phi^{(p)}(t) = (2\pi)^{-\frac{p}{2}} e^{-\frac{t}{2}}$ *is the generator of the p-dimensional normal distribution.*

Spherical distribution $\mathcal{S}_p(h)$	Density generator $h(u),\ u \geq 0$	Squared-radial distribution $\mathcal{SR}(p,h)$
Normal	$\phi^{(p)}(u) = c_p \exp\{-\frac{1}{2}u\}$	χ^2_p
Contaminated Normal	$(1-\varepsilon)\,\phi^{(p)}(u) + \varepsilon\,\lambda^p \phi^{(p)}(\lambda u)$ $0 < \varepsilon < 1,\ \lambda > 0$	$(1-\varepsilon)\,\chi^2_p + \varepsilon\,\lambda\,\chi^2_p$
Cauchy	$c_p\{1+u\}^{-\frac{p+1}{2}}$	$p\mathcal{F}_{p,1}$
t	$c_p\{\nu+u\}^{-\frac{p+\nu}{2}},\ \nu > 0$	$p\mathcal{F}_{p,\nu}$
Generalized-t	$c_p\{\lambda+u\}^{-\frac{\nu+p}{2}},\ \nu,\lambda > 0$	$\frac{p\lambda}{\nu}\mathcal{F}_{p,\nu}$
Exponential Power	$c_p \exp\{-\frac{1}{2}u^s\},\ s > 0$	$\mathcal{G}^{\frac{1}{s}}(\frac{p}{2s},\frac{1}{2})$
Kotz Type	$c_p u^q \exp\{-\frac{r}{2}u^s\},\ r,s > 0$ $2q+p > 0$	$\mathcal{G}^{\frac{1}{s}}(\frac{2q+p}{2s},\frac{r}{2})$
Pearson Type II	$c_p\{1-u\}^{\frac{\alpha}{2}-1},\ u < 1,\ \alpha > 0$	$\mathcal{B}eta\,(\frac{p}{2},\frac{\alpha}{2})$
Normal Scale Mixtures	$c_p \int_0^\infty v^{\frac{p}{2}}\phi^{(p)}(\sqrt{v}u)\mathrm{d}F(v)$ F a cdf with $F(0) = 0$	$R^2 = V^{-1}R_0^2$ $V \sim F \perp\!\!\!\perp R_0^2 \sim \chi^2_p$

Table 4.2 *Means and variances of squared-radial distributions* ($R^2 \sim \mathcal{SR}(p,h)$).

$\mathcal{S}_p(h)$	Mean	Variance
Normal	p	$2p$
Contaminated Normal	$1-\varepsilon+\varepsilon\lambda$	$\left\{\frac{p+2}{p}\frac{1-\varepsilon+\varepsilon\lambda^2}{(1-\varepsilon+\varepsilon\lambda)^2} - 1\right\}\{1-\varepsilon+\varepsilon\lambda\}^2$
Cauchy	–	–
t	$\frac{p\nu}{\nu-2},\ \nu > 2$	$\left\{\frac{p+2}{p}\frac{\nu-2}{\nu-4} - 1\right\}\left\{\frac{p\nu}{\nu-2}\right\}^2,\ \nu > 4$
Generalized-t	$\frac{\lambda}{\nu-2}\frac{p}{\nu},\ \nu > 2$	$\left\{\frac{p+2}{p}\frac{\nu-2}{\nu-4} - 1\right\}\left\{\frac{\lambda}{\nu-2}\frac{p}{\nu}\right\}^2,\ \nu > 2$
Exponential Power	$\frac{2^{\frac{1}{s}}\Gamma(\frac{p+2}{2s})}{\Gamma(\frac{p}{2s})}$	$\frac{2^{\frac{2}{s}}\{\Gamma(\frac{p+4}{2s})\Gamma(\frac{p}{2s})-\Gamma^2(\frac{p+2}{2s})\}}{\Gamma^2(\frac{p}{2s})}$
Kotz Type	$\frac{(\frac{2}{r})^{\frac{1}{s}}\Gamma(\frac{2q+p+2}{2s})}{\Gamma(\frac{2q+p}{2s})}$	$\frac{(\frac{2}{r})^{\frac{2}{s}}\{\Gamma(\frac{2q+p+4}{2s})\Gamma(\frac{2q+p}{2s})-\Gamma^2(\frac{2q+p+2}{2s})\}}{\Gamma^2(\frac{2q+p}{2s})}$
Pearson Type II	$\frac{p}{\alpha+p}$	$\frac{2\alpha p}{(\alpha+p)^2(\alpha+p+2)}$
Normal Scale Mixtures	$p\mathrm{E}(V^{-1})$ (if $\mathrm{E}(V^{-1}) < \infty$)	$p(p+2)\mathrm{E}(V^{-2}) - p^2\mathrm{E}(V^{-1})^2$ (if $\mathrm{E}(V^{-2}) < \infty$)

for any constant $c > 0$. In fact, by making the transformation $\mathbf{z} = \boldsymbol{\Gamma}^\top\mathbf{x}$, $\boldsymbol{\nu} = \boldsymbol{\Gamma}^\top\boldsymbol{\xi}$, where $\boldsymbol{\Gamma} \in \mathcal{O}_p$ diagonalizes $\boldsymbol{\Omega}$, i.e., $\boldsymbol{\Gamma}^\top\boldsymbol{\Omega}\boldsymbol{\Gamma} = \mathbf{D}_\lambda$, the diagonal matrix of eigenvalues of $\boldsymbol{\Omega}$, the contours are then the ellipsoids $\sum_{i=1}^p (z_i - \nu_i)^2/\lambda_i = c^2$ which are centered at $\boldsymbol{\nu}$ and have principal axes of half-length $c\sqrt{\lambda_i}$ supported by the eigenvectors given by the columns $\boldsymbol{\gamma}_1,\ldots,\boldsymbol{\gamma}_p$ of $\boldsymbol{\Gamma}$.

Example 48 (Nonsingular multivariate normal distribution) *In this case, the pdf of* $\mathbf{X} \sim \mathcal{N}_p(\boldsymbol{\xi},\boldsymbol{\Omega})$ *with* $\boldsymbol{\Omega} > 0$ *is*

$$f(\mathbf{x};\boldsymbol{\xi},\boldsymbol{\Omega}) \equiv \phi_p(\mathbf{x};\boldsymbol{\xi},\boldsymbol{\Omega}) = |\boldsymbol{\Omega}|^{-1/2}(2\pi)^{-p/2}e^{-\frac{1}{2}Q(\mathbf{x})}, \quad \mathbf{x} \in \mathbb{R}^p,$$

with $Q(\mathbf{x}) = (\mathbf{x}-\boldsymbol{\xi})^\top \boldsymbol{\Omega}^{-1}(\mathbf{x}-\boldsymbol{\xi})$. *As is well known, here* $Q(\mathbf{X}) \sim \chi^2_p$, $\mathrm{E}(\mathbf{X}) = \boldsymbol{\xi}$ *and* $var(\mathbf{X}) = \boldsymbol{\Omega}$. *Also, if* $p = 2$, *with* $|\rho| < 1$,

$$\boldsymbol{\xi} = \begin{pmatrix} \xi_1 \\ \xi_2 \end{pmatrix}, \quad \boldsymbol{\Omega} = \begin{pmatrix} \omega_1^2 & \rho\omega_1\omega_2 \\ \rho\omega_1\omega_2 & \omega_2^2 \end{pmatrix}, \quad \boldsymbol{\Omega}^{-1} = \frac{1}{1-\rho^2}\begin{pmatrix} 1/\omega_1^2 & -\rho/(\omega_1\omega_2) \\ -\rho/(\omega_1\omega_2) & 1/\omega_2^2 \end{pmatrix},$$

and therefore

$$f(x_1,x_2;\xi_1,\xi_2,\omega_1,\omega_2,\rho) = \frac{1}{2\pi\omega_1\omega_2(1-\rho^2)^{1/2}} e^{-\frac{1}{2}Q(x_1,x_2)}, \quad x_1,x_2 \in \mathbb{R},$$

with

$$Q(x_1,x_2) = \frac{1}{1-\rho^2}\left\{\left(\frac{x_1-\xi_1}{\omega_1}\right)^2 - 2\rho\left(\frac{x_1-\xi_1}{\omega_1}\right)\left(\frac{x_2-\xi_2}{\omega_2}\right) + \left(\frac{x_2-\xi_2}{\omega_2}\right)^2\right\},$$

define the nonsingular bivariate normal distribution. In terms of the z coordinates, the contours of the bivariate normal pdf are in parametric form given by

$$\begin{pmatrix} z_1 \\ z_2 \end{pmatrix} = \begin{pmatrix} v_1 \\ v_2 \end{pmatrix} + c\begin{pmatrix} \sqrt{\lambda_1}\sin\theta \\ \sqrt{\lambda_2}\cos\theta \end{pmatrix}, \quad 0 < \theta \le 2\pi.$$

In terms of the original x coordinates these contours become

$$\begin{pmatrix} x_1 \\ x_2 \end{pmatrix} = \begin{pmatrix} v_1 \\ v_2 \end{pmatrix} + c\begin{pmatrix} \gamma_{11}\sqrt{\lambda_1}\sin\theta + \gamma_{12}\sqrt{\lambda_1}\cos\theta \\ \gamma_{21}\sqrt{\lambda_2}\sin\theta + \gamma_{22}\sqrt{\lambda_1}\cos\theta \end{pmatrix}, \quad 0 < \theta \le 2\pi.$$

For instance, when $\omega_1 = \omega_2 = 1$ *it is well known that* $\mathbf{D}_\lambda = \mathrm{diag}(1+\rho, 1-\rho)$, $\boldsymbol{\gamma}_1 = (1/\sqrt{2}, 1/\sqrt{2})^\top$, *and* $\boldsymbol{\gamma}_2 = (1/\sqrt{2}, -1/\sqrt{2})^\top$.

Example 49 (Nonsingular multivariate *t* distribution) *A random vector* $\mathbf{X} \in \mathbb{R}^p$ *has a nonsingular multivariate t distribution with location vector* $\boldsymbol{\xi} \in \mathbb{R}^p$, $p \times p$ *dispersion matrix* $\boldsymbol{\Omega} > 0$, *and* $\nu > 0$ *degrees of freedom, denoted by* $\mathbf{X} \sim \mathcal{T}_p(\boldsymbol{\xi}, \boldsymbol{\Omega}, \nu)$, *iff* $\mathbf{X}$ *has a pdf given by*

$$f(\mathbf{x};\boldsymbol{\xi},\boldsymbol{\Omega},\nu) \equiv t_p(\mathbf{x};\boldsymbol{\xi},\boldsymbol{\Omega},\nu) = |\boldsymbol{\Omega}|^{-1/2}\frac{\Gamma\{(\nu+p)/2\}}{\Gamma(\nu/2)(\pi\nu)^{p/2}}\left\{1 + \frac{1}{\nu}Q(\mathbf{x})\right\}^{-(\nu+p)/2}, \quad \mathbf{x} \in \mathbb{R}^p,$$

where as before $Q(\mathbf{x}) = (\mathbf{x}-\boldsymbol{\xi})^\top\boldsymbol{\Omega}^{-1}(\mathbf{x}-\boldsymbol{\xi})$. *By construction,* $\mathbf{X} \overset{d}{=} \boldsymbol{\xi} + \boldsymbol{\Omega}^{1/2}\mathbf{Z}$, *with* $\mathbf{Z} \sim \mathcal{T}_p(\mathbf{0}_p, \mathbf{I}_p, \nu)$ *and, by Example 46,* $Q(\mathbf{X}) \sim p\mathcal{F}_{p,\nu}$, $\mathrm{E}(\mathbf{X}) = \boldsymbol{\xi}$ *if* $\nu > 1$ *and* $var(\mathbf{X}) = \frac{\nu}{\nu-2}\boldsymbol{\Omega}$ *if* $\nu > 2$. *When* $\nu = 1$, $\mathbf{X}$ *is said to have a multivariate Cauchy distribution. Similar to Example 48, if* $|\rho| < 1$ *then the bivariate t distribution has pdf given by*

$$f(x_1,x_2;\xi_1,\xi_2,\omega_1,\omega_2,\rho,\nu) = \frac{1}{2\pi\omega_1\omega_2(1-\rho^2)^{1/2}}\left\{1 + \frac{1}{\nu}Q(x_1,x_2)\right\}^{-(\nu+2)/2}, \quad x_1,x_2 \in \mathbb{R},$$

with

$$Q(x_1,x_2) = \frac{1}{1-\rho^2}\left\{\left(\frac{x_1-\xi_1}{\omega_1}\right)^2 - 2\rho\left(\frac{x_1-\xi_1}{\omega_1}\right)\left(\frac{x_2-\xi_2}{\omega_2}\right) + \left(\frac{x_2-\xi_2}{\omega_2}\right)^2\right\},$$

and with the same elliptically-shaped contours as the bivariate normal case presented in the Example 48 albeit with different regions of probability given by $Q(\mathbf{X}) \sim \chi^2_2$ for the bivariate normal and $Q(\mathbf{X}) \sim 2\mathcal{F}_{2,\nu}$ for the bivariate t; see Figure 4.1 with $\boldsymbol{\Omega} = \mathbf{I}_2$ and $\boldsymbol{\Omega} = \boldsymbol{\Sigma}_3$ from Equation (3.9), as well as $\nu = \infty$ and $\nu = 3$.

Among the various applications of the multivariate t model, starting with the early works in regression analysis of Zellner (1976), Lange et al. (1989), and Ferrari and Arellano-Valle (1996), we find those of Osorio et al. (2023), who addressed non-normality in multivariate analysis using the reparameterized t distribution in terms of its mean vector and variance matrix, and Arellano-Valle et al. (2010a) who used the t distribution for estimating the systematic risk of the Chilean stock market.

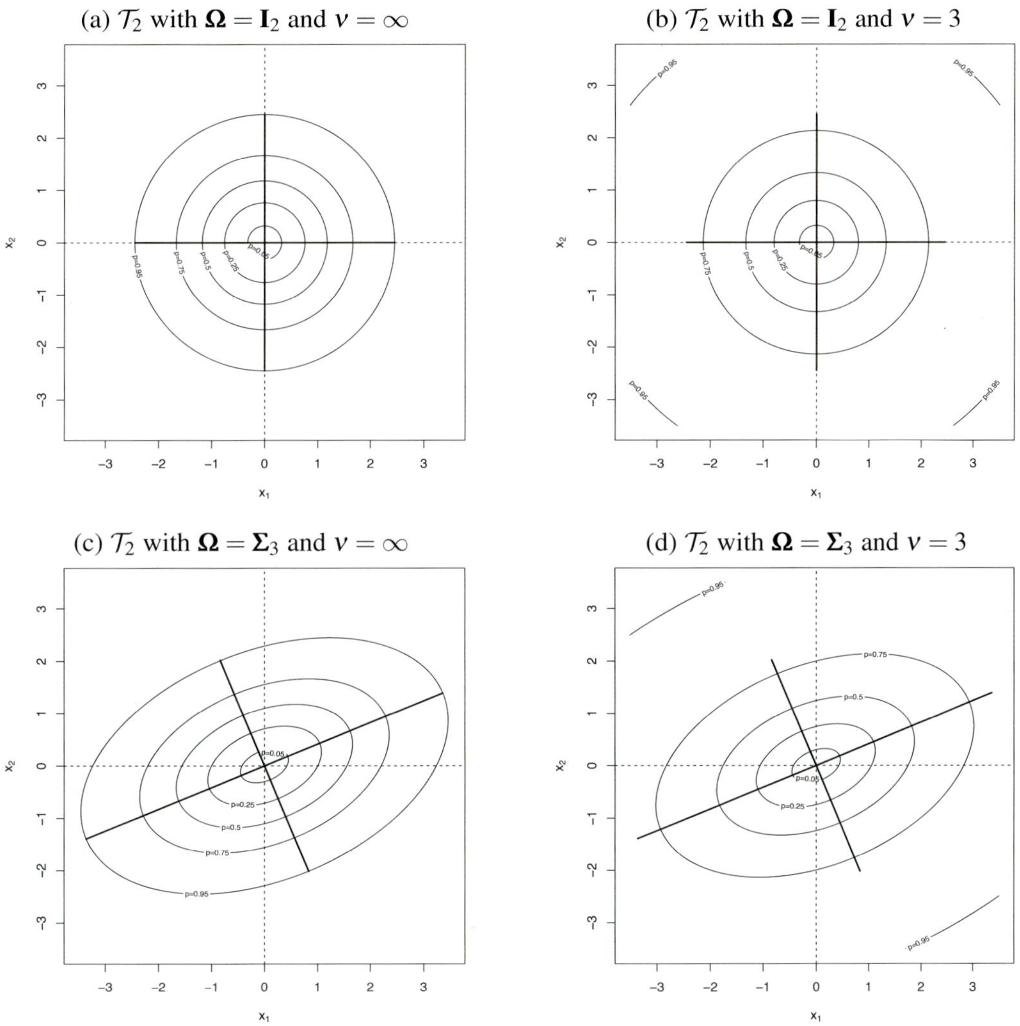

Figure 4.1 *Geometry of the multivariate t distribution, $\mathcal{T}_p$, for $p = 2$: spherical (top); elliptical (bottom); $\nu = \infty$ (left); $\nu = 3$ (right).*

4.3 Characterizations of Normality

The multivariate normal distribution has been characterized within the class of elliptical distributions in several different ways using some of its properties. Important references in this direction are Kelker (1970), Cambanis et al. (1981), Anderson and Fang (1987), and Khatri and Mukerjee (1987). Most of these results are properly documented in the books by Fang et al. (1990) and Fang and Zhang (1990). In this section, we only intend to mention those characterizations that, in our opinion, appear to be the most relevant. In particular, the one that states that assuming joint independence and spherical symmetry in the joint distribution of n random variables is equivalent to assuming normality or that the independence of the sample mean and sample variance characterizes the spherical normal distribution in the class of spherically symmetric distributions. In addition to Kelker (1970) and Cambanis et al. (1981), other early references addressing these ideas include Ali (1980) and Arnold and Lynch (1982).

As we know, the distribution $\mathcal{N}_p(\boldsymbol{\xi},\boldsymbol{\Omega})$ preserves its marginals. If, in addition, the variance matrix $\boldsymbol{\Omega}$ is diagonal, then independence is achieved. Perhaps motivated by these results, Kelker (1970) established the following two characterizations.

Theorem 37 *If* $\mathbf{X} \sim \mathcal{E}l_p(\boldsymbol{\xi},\boldsymbol{\Omega},\varphi)$*, then any marginal distribution is normal if and only if* $\mathbf{X} \sim \mathcal{N}_p(\boldsymbol{\xi},\boldsymbol{\Omega})$.

Proof 37 *Let* $\mathbf{X}_1 = \mathbf{A}_1(\mathbf{X}-\boldsymbol{\xi})$*, where* $\mathbf{A}_1 = (\mathbf{I}_{p_1},\mathbf{O})$*. Since* $\mathbf{X}_1 \sim \mathcal{E}l_{p_1}(\mathbf{0}_{p_1},\boldsymbol{\Omega}_{11},\varphi)$*, where* $\boldsymbol{\Omega}_{11} = \mathbf{A}_1\boldsymbol{\Omega}\mathbf{A}_1^\top$*, its cf is* $\varphi(\mathbf{t}_1^\top\boldsymbol{\Omega}_{11}\mathbf{t}_1)$, $\mathbf{t}_1 \in \mathbb{R}^{p_1}$*. Now, if we assume that* $\mathbf{X}_1 \sim \mathcal{N}_{p_1}(\mathbf{0}_{p_1},\boldsymbol{\Omega}_{11})$*, then its cf is* $e^{\mathbf{t}_1^\top\boldsymbol{\Omega}_{11}\mathbf{t}_1/2}$, $\mathbf{t}_1 \in \mathbb{R}^{p_1}$*, which means that* $\varphi(u) = e^{-u/2}$, $u \geq 0$*, and therefore* $\mathbf{X} \sim \mathcal{N}_p(\boldsymbol{\xi},\boldsymbol{\Omega})$*. The proof in the other direction is immediate from the properties of the multivariate normal distribution (see Chapter 3).* □

In fact, as demonstrated in Arellano-Valle (2001), within the family of elliptical distributions, each specific member can be characterized by its marginals. For example, Arellano-Valle and Bolfarine (1995) established this result for the multivariate t distribution, while Liang and Bentler (1998) did so for the Kotz and Pearson Type II distributions. Obviously, outside the elliptical family these results are not valid, as can be shown by several counterexamples; see Exercise 4.14.

Theorem 38 *Assume that* $\mathbf{X} = (X_1,\ldots,X_p)^\top \sim \mathcal{E}l_p(\boldsymbol{\xi},\boldsymbol{\Omega},\varphi)$*, where* $\boldsymbol{\Omega} = \mathrm{diag}(\omega_{11},\ldots,\omega_{pp})$*. Then,* $X_1,\ldots,X_p$ *are independent if and only if* $\mathbf{X} \sim \mathcal{N}_p(\boldsymbol{\xi},\boldsymbol{\Omega})$.

Proof 38 *Without loss of generality, let* $\boldsymbol{\xi} = \mathbf{0}_p$*. Since* $\boldsymbol{\Omega}$ *is diagonal, the joint cf of* $X_1,\ldots,X_p$ *is* $\varphi(\mathbf{t}^\top\boldsymbol{\Omega}\mathbf{t}) = \varphi(\sum_{i=1}^p \omega_{ii}t_i^2)$, $\mathbf{t} = (t_1,\ldots,t_p)^\top \in \mathbb{R}^p$*, and the marginal cf of* X_i *is* $\varphi(\omega_{ii}t_i^2)$, $t_i \in \mathbb{R}$, $i = 1,\ldots,p$*. If, in addition,* $X_1,\ldots,X_p$ *are independent, then their join cf becomes* $\prod_{i=1}^p \varphi(\omega_{ii}t_i^2)$, $(t_1,\ldots,t_p)^\top \in \mathbb{R}^p$*. Then, by setting* $u_i = \omega_{ii}t_i^2$, $i = 1,\ldots,p$*, we obtain the identity*

$$\varphi\left(\sum_{i=1}^{p} u_i^2\right) = \prod_{i=1}^{p} \varphi(u_i^2).$$

This equation is known as Hamel's equation, whose solution is $\varphi(u) = e^{-cu/2}$*, for some real constant c. Since* $\varphi(\mathbf{t}^\top\boldsymbol{\Omega}\mathbf{t}) = e^{-c\mathbf{t}^\top\boldsymbol{\Omega}\mathbf{t}/2}$, $\mathbf{t} \in \mathbb{R}^p$*, is a cf (*$\varphi$ *is a continuous function, with* $\varphi(0) = 1$ *and* $|\varphi(u)| \leq 1$*), then* $c \geq 0$*, i.e.,* $\mathbf{X} \sim \mathcal{N}_p(\mathbf{0}_p, c\boldsymbol{\Omega})$*. This proves the "only if" part. The proof of the other part is trivial from the normal theory (see Chapter 3).* □

Remark 4 *Although $\varphi(u)=1$ is also a solution of the above Hamel equation, the corresponding distribution is singular in this case.*

Now suppose that

$$\mathbf{X}=\begin{pmatrix}\mathbf{X}_1\\ \mathbf{X}_2\end{pmatrix}\sim \mathcal{E}l_{p_1+p_2}\left\{\boldsymbol{\xi}=\begin{pmatrix}\boldsymbol{\xi}_1\\ \boldsymbol{\xi}_2\end{pmatrix},\boldsymbol{\Omega}=\begin{pmatrix}\boldsymbol{\Omega}_{11} & \boldsymbol{\Omega}_{12}\\ \boldsymbol{\Omega}_{21} & \boldsymbol{\Omega}_{22}\end{pmatrix},\varphi\right\},\quad \boldsymbol{\Omega}>0. \tag{4.31}$$

According to Theorem 37, under the assumption (4.31), $\mathbf{X}_1$ (or $\mathbf{X}_2$) is normal if only if $\mathbf{X}$ is normal. Also, when $\boldsymbol{\Omega}_{12}=\mathbf{O}$, we know that $\mathbf{X}_1$ and $\mathbf{X}_2$ are uncorrelated but not independent. According to Theorem 38, $\boldsymbol{\Omega}_{12}=\mathbf{O}$ implies independence if and only if $\mathbf{X}$ is normally distributed. Furthermore, assuming the existence of the respective moments, we know that $\mathrm{E}(\mathbf{X}_1\mid \mathbf{x}_2)=\boldsymbol{\xi}_{1\cdot 2}=\boldsymbol{\xi}_1+\boldsymbol{\Omega}_{12}\boldsymbol{\Omega}_{22}^{-1}(\mathbf{x}_2-\boldsymbol{\xi}_2)$, is linear in $\mathbf{x}_2$, while $\mathrm{var}(\mathbf{X}_1\mid\mathbf{x}_2)=\kappa_2\{Q_2(\mathbf{x}_2)\}\boldsymbol{\Omega}_{11\cdot 2}$ depends on $\mathbf{x}_2$ through $Q_2(\mathbf{x}_2)=(\mathbf{x}_2-\boldsymbol{\xi}_2)^\top\boldsymbol{\Omega}_{22}^{-1}(\mathbf{x}_2-\boldsymbol{\xi}_2)$, where $\kappa_2\{Q_2(\mathbf{x}_2)\}=-2\varphi'_{Q_2(\mathbf{x}_2)}(0)$ and $\boldsymbol{\Omega}_{11\cdot 2}=\boldsymbol{\Omega}_{11}-\boldsymbol{\Omega}_{12}\boldsymbol{\Omega}_{22}^{-1}\boldsymbol{\Omega}_{21}$; that is, the regression model induced by (4.31) is linear but heteroscedastic. Kelker (1970) showed that under (4.31), the assumption of homoscedasticity characterizes normality within the class of elliptical distributions. The characterizations presented in the following theorem are due to Cambanis et al. (1981). The full proof can be found in their work, for which it is useful to remember from (4.31) that

$$(\mathbf{X}_1\mid\mathbf{X}_2=\mathbf{x}_2)\overset{d}{=}\boldsymbol{\xi}_{1\cdot 2}+R_{Q_2(\mathbf{x}_2)}\boldsymbol{\Omega}_{11\cdot 2}^{1/2}\mathbf{U}^{(p_1)}\sim\mathcal{E}l_p(\boldsymbol{\xi}_{1\cdot 2},\boldsymbol{\Omega}_{11\cdot 2},\varphi_{Q_2(\mathbf{x}_2)}),$$
$$R^2_{Q_2(\mathbf{x}_2)}\overset{d}{=}\{Q_1(\mathbf{X}_1)\mid Q_2(\mathbf{x}_2)\}\perp\!\!\!\perp\mathbf{U}^{(p_1)},$$

with $Q_i(\mathbf{x}_i)=(\mathbf{x}_i-\boldsymbol{\xi}_i)^\top\boldsymbol{\Omega}_{ii}^{-1}(\mathbf{x}_i-\boldsymbol{\xi}_i)$ for $i=1,2$.

Theorem 39 *If we assume (4.31), then:*

(a) $\varphi_{Q_2(\mathbf{x}_2)}$ does not depend on $\mathbf{x}_2$ if and only if $\mathbf{X}\sim\mathcal{N}_p(\boldsymbol{\xi},\boldsymbol{\Omega})$;

(b) $(\mathbf{X}_1\mid\mathbf{X}_2)$ is normally distributed with probability 1 if and only if $\mathbf{X}\sim\mathcal{N}_p(\boldsymbol{\xi},\boldsymbol{\Omega})$;

(c) For any fixed positive integer k, $\mathrm{E}(R^k_{Q_2(\mathbf{X}_2)}\mid\mathbf{X}_2)$ is finite and degenerate (i.e., it does not depend on $\mathbf{X}_2$) if and only if $\mathbf{X}$ is normally distributed.

Proof 39 *We will prove only part (a), for the rest see Cambanis et al. (1981). First, note that $\varphi_{Q_2(\mathbf{x}_2)}(\mathbf{t}_1^\top\boldsymbol{\Omega}_{11\cdot 2}\mathbf{t}_1)=\mathrm{E}\{e^{\mathrm{i}\mathbf{t}_1^\top(\mathbf{X}_1-\boldsymbol{\xi}_{1\cdot 2})}\mid\mathbf{X}_2\}$, $\mathbf{t}_1\in\mathbb{R}^{p_1}$. Also, for all $\mathbf{t}=(\mathbf{t}_1^\top,\mathbf{t}_2^\top)^\top\in\mathbb{R}^{p_1+p_2}$, we have*

$$\begin{aligned}\varphi(\mathbf{t}^\top\boldsymbol{\Omega}\mathbf{t})&=\mathrm{E}\{e^{\mathrm{i}\mathbf{t}^\top(\mathbf{X}-\boldsymbol{\xi})}\}\\&=\mathrm{E}[e^{\mathrm{i}\mathbf{t}_2^\top(\mathbf{X}_2-\boldsymbol{\xi}_2)}\mathrm{E}\{e^{\mathrm{i}\mathbf{t}_1^\top(\mathbf{X}_1-\boldsymbol{\xi}_1)}\mid\mathbf{X}_2\}]\\&=\mathrm{E}[e^{\mathrm{i}\mathbf{t}_2^\top(\mathbf{X}_2-\boldsymbol{\xi}_2)+\mathrm{i}\mathbf{t}_1^\top(\boldsymbol{\xi}_{1\cdot 2}-\boldsymbol{\xi}_1)}\mathrm{E}\{e^{\mathrm{i}\mathbf{t}_1^\top(\mathbf{X}_1-\boldsymbol{\xi}_{1\cdot 2})}\mid\mathbf{X}_2\}]\\&=\mathrm{E}\{e^{\mathrm{i}(\mathbf{t}_2+\boldsymbol{\Omega}_{22}^{-1}\boldsymbol{\Omega}_{21}\mathbf{t}_1)^\top(\mathbf{X}_2-\boldsymbol{\xi}_2)}\varphi_{Q_2(\mathbf{X}_2)}(\mathbf{t}_1^\top\boldsymbol{\Omega}_{11\cdot 2}\mathbf{t}_1)\}.\end{aligned}$$

Thus, if for each $u\geq 0$, $\varphi_{Q_2(\mathbf{X}_2)}(u)=\tilde{\varphi}(u)$ with probability 1, then the above result becomes

$$\begin{aligned}\varphi(\mathbf{t}^\top\boldsymbol{\Omega}\mathbf{t})&=\tilde{\varphi}(\mathbf{t}_1^\top\boldsymbol{\Omega}_{11\cdot 2}\mathbf{t}_1)\mathrm{E}\{e^{\mathrm{i}(\mathbf{t}_2+\boldsymbol{\Omega}_{22}^{-1}\boldsymbol{\Omega}_{21}\mathbf{t}_1)^\top(\mathbf{X}_2-\boldsymbol{\xi}_2)}\}\\&=\tilde{\varphi}(\mathbf{t}_1^\top\boldsymbol{\Omega}_{11\cdot 2}\mathbf{t}_1)\varphi\{(\mathbf{t}_2+\boldsymbol{\Omega}_{22}^{-1}\boldsymbol{\Omega}_{21}\mathbf{t}_1)^\top\boldsymbol{\Omega}_{22}(\mathbf{t}_2+\boldsymbol{\Omega}_{22}^{-1}\boldsymbol{\Omega}_{21}\mathbf{t}_1)\}.\end{aligned}$$

Now, from the identity $\mathbf{t}^\top \boldsymbol{\Omega}\mathbf{t} = \mathbf{t}_1^\top \boldsymbol{\Omega}_{11\cdot 2}\mathbf{t}_1 + (\mathbf{t}_2 + \boldsymbol{\Omega}_{22}^{-1}\boldsymbol{\Omega}_{21}\mathbf{t}_1)^\top \boldsymbol{\Omega}_{22}(\mathbf{t}_2 + \boldsymbol{\Omega}_{22}^{-1}\boldsymbol{\Omega}_{21}\mathbf{t}_1)$, *we find the equation (assuming that* $p_i \geq 1$, $i = 1,2$*)* $\varphi(u+v) = \tilde{\varphi}(u)\varphi(v)$, $u, v \geq 0$. *By letting* $v = 0$, *it follows that* $\tilde{\varphi} = \varphi$ *and so that* $\varphi(u+v) = \varphi(u)\varphi(v)$, $u, v \geq 0$. *Since* φ *is a cf, then* $\varphi(u) = e^{-cu/2}$, $u \geq 0$, *for some* $c \geq 0$, *and therefore* $\mathbf{X} \sim \mathcal{N}_p(\boldsymbol{\xi}, c\boldsymbol{\Omega})$. *The reverse implication is immediate.* □

Finally, as established by Anderson and Fang (1987) and Khatri and Mukerjee (1987), assuming $\boldsymbol{\xi} = \mathbf{0}_p$ and $\boldsymbol{\Omega} = \mathbf{I}_p$ with $p = p_1 + p_2$ in (4.31), normality can also be characterized from quadratic forms.

Theorem 40 *Assume* $\boldsymbol{\xi} = \mathbf{0}_p$ *and* $\boldsymbol{\Omega} = \mathbf{I}_p$ *in (4.31). Then:*

(a) $\|\mathbf{X}_1\| \perp\!\!\!\perp \|\mathbf{X}_2\|$ *if and only if* $\mathbf{X} \sim \mathcal{N}_p(\mathbf{0}_p, c\mathbf{I}_p)$ *for some constant* $c \geq 0$;

(b) $\mathbf{X}^\top \mathbf{A}\mathbf{X} \sim \chi_k^2$ *with* $\mathbf{A} = \mathbf{A}^\top$ *if and only if* $\mathbf{X} \sim \mathcal{N}_p(\mathbf{0}_p, c\mathbf{I}_p)$ *for some constant* $c > 0$, $\mathbf{A}^2 = c^{-1}\mathbf{A}$ *and* $\operatorname{rank}(\mathbf{A}) = k$.

Proof 40 *For the proof of part* (*a*)*, see Fang et al. (1990) or Fang and Zhang (1990). The proof of part* (*b*) *can be found in Khatri and Mukerjee (1987), or in Anderson and Fang (1987) under the condition* $\mathrm{P}(\mathbf{X} = \mathbf{0}_p) = 0$. *In fact, under this condition, the fact in* (*b*) *becomes a consequence of the general result in (4.37) presented below, from which it can be extended to each spherical distribution; see, e.g., Arellano-Valle (2001).* □

Under assumptions similar to those presented here, Arellano-Valle and Bolfarine (1995) characterized the multivariate t distribution within the class of elliptical/spherical distributions based on its conditionals and quadratic forms. They also characterized the t distribution within the subclass of scale mixtures of normals by assuming a linearity condition of the variance of the univariate conditional distributions. A predictive approach of this result can be found in Arellano-Valle et al. (1994). Some of these characterizations come from the research work of Arellano-Valle (1994), who also extended several inference and related results of the normal linear model to the dependent and independent elliptical linear model, with and without errors in the variables, and with emphasis on the linear t model.

The following example illustrates a highly practical application of Theorem 40, which is also of historical significance. Different ways of proving this result can be found in Ali (1980) and Arnold and Lynch (1982).

Example 50 *Let* $\mathbf{X} = (X_1, \ldots, X_n)^\top \sim \mathcal{S}_n(\varphi)$ *be nonsingular and denote by* $n\bar{X} = \sum_{i=1}^n X_i$ *and* $(n-1)S^2 = \sum_{i=1}^n (X_i - \bar{X})^2$. *Then,* $\bar{X} \perp\!\!\!\perp S^2$ *if and only if* $\mathbf{X} \sim \mathcal{N}_n(\mathbf{0}_n, \sigma^2\mathbf{I}_n)$ *for some* $\sigma > 0$. *To see this, let* $\mathbf{Y} = \boldsymbol{\Gamma}\mathbf{X}$, *where* $\boldsymbol{\Gamma} \in \mathcal{O}_n$ *whose first row is* $n^{-1/2}\mathbf{1}_n^\top$. *Clearly,* $\mathbf{Y} \stackrel{d}{=} \mathbf{X}$ *and therefore* $\mathbf{Y} = (Y_1, Y_2, \ldots, Y_n)^\top = (Y_1, \mathbf{Y}_2^\top)^\top \sim \mathcal{S}_n(\varphi)$. *Note that* $\sqrt{n}\bar{X} = Y_1$ *and* $(n-1)S^2 = \sum_{i=2}^n Y_i^2 = \|\mathbf{Y}_2\|^2$; *so* $\bar{X} \perp\!\!\!\perp S^2 \Longrightarrow Y_1^2 \perp\!\!\!\perp \|\mathbf{Y}_2\|^2$, *which in turn implies that* $\mathbf{Y}$ *is normally distributed by Theorem 40. So* $\mathbf{X}$ *is also normal since* $\mathbf{X} \stackrel{d}{=} \mathbf{Y}$.

4.4 Scale-Invariant Functions and Quadratic Forms

4.4.1 Distribution of homogeneous functions

A real-valued function $t(\mathbf{x})$, $\mathbf{x} \in \mathbb{R}^p$, is *positively homogeneous of degree k* iff $t(c\mathbf{x}) = c^k t(\mathbf{x})$ for all $\mathbf{x} \in \mathbb{R}^p$ and $c > 0$. In statistics, there are a large number of functions (statistics) with this property, particularly when $k=0$. In the context of spherical distributions, its importance lies in the fact that if $\mathbf{X} \stackrel{d}{=} R\mathbf{U}^{(p)} \sim \mathcal{S}_p^+(\varphi)$, that is, $\mathbf{X} \sim \mathcal{E}l_p(\mathbf{0}_p, \mathbf{I}_p, \varphi)$ and $\mathrm{P}(\mathbf{X}=\mathbf{0}_p)=0$, then

$$t(\mathbf{X}) \stackrel{d}{=} t(R\mathbf{U}^{(p)}) \stackrel{d}{=} R^k t(\mathbf{U}^{(p)}), \tag{4.32}$$

where the distribution of $t(\mathbf{U}^{(p)})$ is completely determined by the distribution of the uniform vector $\mathbf{U}^{(p)}$, which remains invariant whatever $\mathbf{X} \sim \mathcal{S}_p^+(\varphi)$. Furthermore, R^k and $t(\mathbf{U}^{(p)})$ are independent, since R and $\mathbf{U}^{(p)}$ are. A formal proof of (4.32) follows from the uniqueness of the characteristic function by noting that

$$\mathrm{E}(e^{\mathrm{i}st(\mathbf{X})}) = \mathrm{E}(e^{\mathrm{i}st(R\mathbf{U}^{(p)})}) = \mathrm{E}\{\mathrm{E}(e^{\mathrm{i}st(R\mathbf{U}^{(p)})}|R)\} = \mathrm{E}\{\mathrm{E}(e^{\mathrm{i}sR^k t(\mathbf{U}^{(p)})}|R)\} = \mathrm{E}(e^{\mathrm{i}sR^k t(\mathbf{U}^{(p)})}), \quad \forall s.$$

As the spherical normal distribution $\mathcal{N}_p(\mathbf{0}_p, \mathbf{I}_p)$ is $\mathcal{S}_p^+(\varphi)$, with $\varphi(u)=e^{-u/2}$, applying this property to the spherical normal vector $\mathbf{X}_0 \sim \mathcal{N}_p(\mathbf{0}_p, \mathbf{I}_p)$ allows us to obtain the moments of $t(\mathbf{U}^{(p)})$ and therefore the moments of $t(\mathbf{X})$ from (4.32), provided that they exist. In fact, since $\mathbf{X}_0 \stackrel{d}{=} R_0\mathbf{U}^{(p)}$ with $R_0 \sim \chi_p$ and $\mathrm{P}(\mathbf{X}_0=\mathbf{0}_p)=0$, we have

$$\begin{aligned}
t(\mathbf{X}_0) \stackrel{d}{=} R_0^k t(\mathbf{U}^{(p)}) &\Longrightarrow \mathrm{E}\{t(\mathbf{X}_0)^m\} = \mathrm{E}(R_0^{km})\mathrm{E}\{t(\mathbf{U}^{(p)})^m\} \\
&\Longrightarrow \mathrm{E}\{t(\mathbf{U}^{(p)})^m\} = \frac{1}{\mathrm{E}(R_0^{km})}\mathrm{E}\{t(\mathbf{X}_0)^m\} \\
&\Longrightarrow \mathrm{E}\{t(\mathbf{X})^m\} = \frac{\mathrm{E}(R^{km})}{\mathrm{E}(R_0^{km})}\mathrm{E}\{t(\mathbf{X}_0)^m\}.
\end{aligned}$$

A more general variant of the latter result is found in Theorem 22. Finally, for $k=0$, $t(c\mathbf{X})=t(\mathbf{X})$ for all $c>0$, so that $t(\mathbf{X})$ is invariant under scalar multiplication. In this case, from (4.32) we obtain the following distributional invariance of $t(\mathbf{X})$ whatever $\mathbf{X} \sim \mathcal{S}_p^+(\varphi)$:

$$t(\mathbf{X}) \stackrel{d}{=} t(\mathbf{U}^{(p)}) \stackrel{d}{=} t(\mathbf{X}_0), \quad \mathbf{X}_0 \sim \mathcal{N}_p(\mathbf{0}_p, \mathbf{I}_p). \tag{4.33}$$

Two well-known functions that satisfy this property are t- and F-statistics. In the following, we present two other examples that are quite well known in the literature of spherical distributions. For other related and well-known examples, see Example 41.

Example 51 *Suppose that* $\mathbf{X} \sim \mathcal{E}l_p(\mathbf{0}_p, \mathbf{I}_p, \varphi)$ *with* $\mathrm{P}(\mathbf{X}=\mathbf{0}_p)=0$. *Then, we have (Kariya and Eaton, 1977):*
(a) If $\mathbf{A} \in \mathbb{R}^{p\times p}$ *is a symmetric matrix such that* $\mathbf{A}^2=\mathbf{A}$ *and* $\mathrm{rank}(\mathbf{A})=q$, *then*

$$\frac{\mathbf{X}^\top\mathbf{A}\mathbf{X}}{\|\mathbf{X}\|^2} \sim \mathcal{B}eta\{q/2,(p-q)/2\}.$$

(b) If $Z=\mathbf{a}^\top\mathbf{X}/\|\mathbf{X}\|$, *with* $\mathbf{a} \in \mathbb{S}^{(p-1)}$, *then*

$$\frac{\sqrt{p-1}Z}{\sqrt{1-Z^2}} \sim t_{p-1},$$

where $t_\nu \equiv t(0,1,\nu)$ *denotes the univariate standard Student's t distribution with* $\nu>0$ *degrees of freedom. In fact, since* $\mathbf{U}^{(p)}$ *and* $\mathbf{X}/\|\mathbf{X}\|$ *have the same distribution regardless of the spherical distribution of* $\mathbf{X}$ *with* $\mathrm{P}(\mathbf{X}=\mathbf{0}_p)=0$, *to prove the above results it suffices to assume that* $\mathbf{X} \sim \mathcal{N}_p(\mathbf{0}_p, \mathbf{I}_p)$ *since in both (a) and (b) it is clear that the function* $t(\mathbf{X})$ *involved is homogeneous of degree zero and therefore* $t(\mathbf{X}) \overset{d}{=} t(\mathbf{U}^{(p)})$. *So, to obtain (a), we use the fact that if* $\mathbf{A} \in \mathbb{R}^{p\times p}$ *is a symmetric and idempotent matrix of* $\operatorname{rank}(\mathbf{A}) = q$, *then there exists* $\mathbf{\Gamma} \in \mathcal{O}_p$ *such that*

$$\mathbf{\Gamma}^\top \mathbf{A}\mathbf{\Gamma} = \mathbf{\Lambda} = \begin{pmatrix} \mathbf{I}_q & \mathbf{O} \\ \mathbf{O} & \mathbf{O} \end{pmatrix}.$$

Thus, from the partition $\mathbf{X} = (\mathbf{X}_1^\top, \mathbf{X}_2^\top)^\top$ *with* $\mathbf{X}_1 \in \mathbb{R}^q$ *we have* $\mathbf{X}^\top \mathbf{A}\mathbf{X} \overset{d}{=} \mathbf{X}^\top \mathbf{\Lambda}\mathbf{X} = \|\mathbf{X}_1\|^2$. *With this fact and the assumption that* $\mathbf{X} \sim \mathcal{N}_p(\mathbf{0}_p, \mathbf{I}_p)$, *we see that* $\|\mathbf{X}_1\|^2 \overset{d}{\sim} \chi^2_q$ *and* $\|\mathbf{X}_2\|^2 \overset{d}{\sim} \chi^2_{p-q}$ *are independent. So, since* $\|\mathbf{X}\|^2 = \|\mathbf{X}_1\|^2 + \|\mathbf{X}_2\|^2 \sim \chi^2_p$, *we have (see also Example 41)*

$$\frac{\mathbf{X}^\top \mathbf{A}\mathbf{X}}{\|\mathbf{X}\|^2} \overset{d}{=} \frac{\|\mathbf{X}_1\|^2}{\|\mathbf{X}_1\|^2 + \|\mathbf{X}_2\|^2} \overset{d}{=} V \sim \mathcal{B}eta\{q/2, (p-q)/2\}.$$

Considering again the assumption of normality, to prove (b) it is sufficient to take (without loss of generality) $\mathbf{a} = (1, 0, \ldots, 0)^\top$ *and note that*

$$\frac{\sqrt{p-1}Z}{\sqrt{1-Z^2}} = \frac{\sqrt{p-1}X_1}{\sqrt{\sum_{i=2}^p X_i^2}} \sim t_{p-1}.$$

Finally, it should be mentioned that the extension of (4.32) and (4.33) when $\mathbf{t}(\mathbf{X})$ is a vector-valued function is immediate when all its components are homogeneous functions, each of the same degree.

4.4.2 *Distributions of quadratic forms and further related functions*

Quadratic forms are homogeneous functions of degree $k = 2$, and their distributions can be studied using the results obtained in the previous subsection. In the literature, we can find several results on this topic. For example, considering the partition $\mathbf{X} = (\mathbf{X}_1^\top, \mathbf{X}_2^\top)^\top$, with $\mathbf{X}_1 \in \mathbb{R}^{p_1}$, $\mathbf{X}_2 \in \mathbb{R}^{p_2}$, $p_1 + p_2 = p$, Kelker (1970) obtained the distributions of $\|\mathbf{X}\|$, $\|\mathbf{X}\|^2$, and $(\|\mathbf{X}_1\|, \|\mathbf{X}_2\|)^\top$ assuming the existence of a pdf for $\mathbf{X}$. He also extended Cochran's theorem to the elliptical case assuming the existence of a pdf with finite fourth moment. These and other results are summarized in Muirhead (1982). Meanwhile, Anderson and Fang (1987), following Cambanis et al. (1981), managed to extend Cochran's theorem to the elliptical case with fewer restrictions than Kelker (1970). A comprehensive review of these and other properties can be found in Fang and Zhang (1990). Some results of this type are given in the previous sections, and many of them are a direct consequence of (4.7).

Recall first that if $\mathbf{X} \sim \mathcal{S}_p^+(\varphi)$ is partitioned as $\mathbf{X} = (\mathbf{X}_1^\top, \ldots, \mathbf{X}_k^\top)^\top$, with $\mathbf{X}_1, \ldots, \mathbf{X}_k$ having $p_1, \ldots, p_k$ components ($\sum_{i=1}^k p_i = p$), respectively, then (4.7) states that

$$(\mathbf{X}_1^\top, \ldots, \mathbf{X}_k^\top)^\top \overset{d}{=} (R\sqrt{V_1}\mathbf{U}_1^{(p_1)\top}, \ldots, R\sqrt{V_k}\mathbf{U}_k^{(p_k)\top})^\top,$$

where $R \overset{d}{=} \|\mathbf{X}\|$, $(V_1, \ldots, V_k)^\top \sim \mathcal{D}_k(p_1/2, \ldots, p_k/2)$ and $\mathbf{U}^{(p_i)} \overset{d}{=} \mathbf{X}_i/\|\mathbf{X}_i\|$ for $i = 1, \ldots, k$, all mutually independent. Consequently, if $\mathbf{g}$ is an arbitrary measurable function, then (see Section 1.5)

$$\mathbf{g}(\mathbf{X}_1, \ldots, \mathbf{X}_k) \overset{d}{=} \mathbf{g}(R\sqrt{V_1}\mathbf{U}_1^{(p_1)}, \ldots, R\sqrt{V_k}\mathbf{U}_k^{(p_k)}).$$

In particular, for the vector of squared lengths of $\mathbf{X}_1, \ldots, \mathbf{X}_k$, we have

$$(\|\mathbf{X}_1\|^2, \ldots, \|\mathbf{X}_k\|^2)^\top = (\mathbf{X}_1^\top\mathbf{X}_1, \ldots, \mathbf{X}_k^\top\mathbf{X}_k)^\top \overset{d}{=} R^2(V_1, \ldots, V_k)^\top. \tag{4.34}$$

Also, by (4.34) it is clear that

$$(\|\mathbf{X}_1\|^2,\dots,\|\mathbf{X}_m\|^2)^\top \stackrel{d}{=} R^2(V_1,\dots,V_m)^\top, \quad m\in\{1,\dots,k-1\},$$

where $V_1,\dots,V_m$ have Dirichlet joint pdf

$$f_{V_1,\dots,V_m}(v_1,\dots,v_m) = \frac{\Gamma(p/2)}{\Gamma(\sum_{i=m+1}^k p_i/2)\prod_{i=1}^m \Gamma(p_i/2)}\prod_{i=1}^m v_i^{\frac{p_i}{2}-1}\left(1-\sum_{i=1}^m v_i\right)^{\frac{1}{2}\sum_{i=m+1}^k p_i - 1},$$

$$v_i>0,\ i=1,\dots,m,\ \sum_{i=1}^m v_i<1,\ m\in\{1,\dots,k-1\}.$$

Moreover, since by assumption $\mathrm{P}(\mathbf{X}=\mathbf{0}_p)=0$ and $(\|\mathbf{X}_1\|^2,\dots,\|\mathbf{X}_m\|^2)^\top \mid R=r \stackrel{d}{=} r^2(V_1,\dots,V_m)^\top$ for each $m\in\{1,\dots,k-1\}$, it follows that the joint pdf of the squared lengths $\|\mathbf{X}_1\|^2,\dots,\|\mathbf{X}_m\|^2$ is

$$\begin{aligned} f_{\|\mathbf{X}_1\|^2,\dots,\|\mathbf{X}_m\|^2}(s_1,\dots,s_m) &= \int_{r>(\sum_{i=1}^m s_i)^{1/2}} r^{-2m} f_{V_1,\dots,V_m}(s_1/r^2,\dots,s_m/r^2)\mathrm{d}F(r)\\ &= \frac{\Gamma(p/2)}{\Gamma(q/2)\prod_{i=1}^m\Gamma(p_i/2)}\prod_{i=1}^m s_i^{\frac{p_i}{2}-1}\int_{s^{1/2}}^{\infty} r^{2-p}(r^2-s)^{\frac{q}{2}-1}\mathrm{d}F(r), \end{aligned}$$

where $q=\sum_{i=m+1}^k p_i$, $s=\sum_{i=1}^m s_i$ and $F(r)$ is the cdf of $R \stackrel{d}{=} \|\mathbf{X}\| = (\sum_{i=1}^k \|\mathbf{X}_i\|^2)^{1/2}$. If in addition, $\mathbf{X}=(\mathbf{X}_1^\top,\dots,\mathbf{X}_k^\top)^\top \sim \mathcal{S}_p(h)$ has a spherical pdf $h(\|\mathbf{x}\|^2)=h(\sum_{i=1}^k\|\mathbf{x}_i\|^2)$, then the joint pdf of $\|\mathbf{X}_1\|^2,\dots,\|\mathbf{X}_k\|^2$ becomes

$$f_{\|\mathbf{X}_1\|^2,\dots,\|\mathbf{X}_k\|^2}(s_1,\dots,s_k) = \frac{\pi^{p/2}}{\prod_{i=1}^k\Gamma(p_i/2)}\prod_{i=1}^k s_i^{\frac{p_i}{2}-1} h\left(\sum_{i=1}^k s_i\right),\ s_i>0,\ i=1,\dots,k. \tag{4.35}$$

Two alternative proofs of (4.35) can be found in the books of Fang and Zhang (1990) and Gupta and Varga (1993). Here, we prove (4.35) from the multiplicative rule:

$$\begin{aligned} f_{\|\mathbf{X}_1\|^2,\dots,\|\mathbf{X}_k\|^2}(s_1,\dots,s_k) &= f_{\|\mathbf{X}_1\|^2}(s_1)\times f_{\|\mathbf{X}_2\|^2\mid\|\mathbf{X}_1\|^2=s_1}(s_2)\times\cdots\\ &\quad\times f_{\|\mathbf{X}_k\|^2\mid\|\mathbf{X}_1\|^2=s_1,\dots,\|\mathbf{X}_{k-1}\|^2=s_{k-1}}(s_k)\\ &= f_{R_1^2}(s_1)\times f_{R_{s_1}^2}(s_2)\times\cdots\times f_{R^2_{\sum_{i=1}^{k-1}s_i}}(s_k)\\ &= \frac{\pi^{p_1/2}}{\Gamma(p_1/2)} s_1^{\frac{p_1}{2}-1} h^{(p_1)}(s_1)\times\frac{\pi^{p_2/2}}{\Gamma(p_2/2)} s_2^{\frac{p_2}{2}-1}\frac{h^{(p_1+p_2)}(s_1+s_2)}{h^{(p_1)}(s_1)}\times\cdots\\ &\quad\times\frac{\pi^{p_k/2}}{\Gamma(p_k/2)} s_k^{\frac{p_k}{2}-1}\frac{h^{(\sum_{i=1}^k p_i)}(\sum_{i=1}^k s_i)}{h^{(\sum_{i=1}^{k-1} p_i)}(\sum_{i=1}^{k-1} s_i)}, \end{aligned}$$

since $R_1^2 \stackrel{d}{=} \|\mathbf{X}_1\|^2$ has marginal pdf (see [1] in Section 4.2.5)

$$f_{R_1^2}(s_1) = \frac{\pi^{p_1/2}}{\Gamma(p_1/2)} s_1^{\frac{p_1}{2}-1} h^{(p_1)}(s_1),\quad s_1>0,$$

and, for each $m=2,\dots,k$,

$$\begin{aligned} R^2_{\sum_{i=1}^{m-1}\|\mathbf{x}_i\|^2} &\stackrel{d}{=} \|\mathbf{X}_m\|^2 \mid (\mathbf{X}_1=\mathbf{x}_1,\dots,\mathbf{X}_{m-1}=\mathbf{x}_{m-1})\\ &\stackrel{d}{=} \|\mathbf{X}_m\|^2 \mid (\|\mathbf{X}_1\|^2=\|\mathbf{x}_1\|^2,\dots,\|\mathbf{X}_{m-1}\|^2=\|\mathbf{x}_{m-1}\|^2) \end{aligned}$$

has conditional pdf (see [4] in Section 4.2.5)

$$f_{R^2_{\sum_{i=1}^{m-1}s_i}}(s_m) = \frac{\pi^{p_m/2}}{\Gamma(p_m/2)} s_m^{\frac{p_m}{2}-1} h^{(p_m)}_{\sum_{i=1}^{m-1}s_i}(s_1) = \frac{\pi^{p_m/2}}{\Gamma(p_m/2)} s_m^{\frac{p_m}{2}-1} \frac{h^{(\sum_{i=1}^{m}p_i)}(\sum_{i=1}^{m}s_i)}{h^{(\sum_{i=1}^{m-1}p_i)}(\sum_{i=1}^{m-1}s_i)}, \quad s_m > 0,$$

where $s_i = \|\mathbf{x}_i\|^2$, $i = 1,\dots,m$, $h^{(p_i)}$ and $h^{(\sum_{i=1}^{m}p_i)}$ are marginal dgf induced by $h^{(\sum_{i=1}^{k}p_i)} = h^{(p)} = h$.

Example 52 *Suppose that* $\mathbf{X} = (X_1,\dots,X_n)^\top \sim \mathcal{S}_n(h)$ *is an absolutely continuous spherical random vector with pdf* $h(\|\mathbf{x}\|^2)$. *Let* $\mathbf{Y} = \mathbf{\Gamma}\mathbf{X}$, *where* $\mathbf{\Gamma}$ *is an orthogonal matrix whose first row is* $n^{-1/2}\mathbf{1}_n^\top$. *Clearly* $\mathbf{Y} \stackrel{d}{=} \mathbf{X}$ *is also spherical with pdf* $h(\|\mathbf{y}\|^2)$. *Partition* $\mathbf{Y} = (Y_1,\mathbf{Y}_2^\top)^\top$ *with* $\mathbf{Y}_2 = (Y_2,\dots,Y_n)^\top$. *Then, by (4.35) the joint pdf of* Y_1^2 *and* $\|\mathbf{Y}_2\|^2 = \sum_{i=2}^{n} Y_i^2$ *becomes*

$$f_{Y_1^2,\|\mathbf{Y}_2\|^2}(s,t) = \frac{\pi^{(n-1)/2}}{\Gamma\{(n-1)/2\}} s^{\frac{1}{2}-1} t^{\frac{n-1}{2}-1} h(s+t).$$

Now, let $\bar{X} = n^{-1}\sum_{i=1}^{n} X_i = n^{-1}\mathbf{1}_n^\top\mathbf{X}$ *and* $(n-1)S_x^2 = \sum_{i=1}^{n}(X_i-\bar{X})^2 = \mathbf{X}^\top\mathbf{Q}\mathbf{X}$, *where* $\mathbf{Q} = \mathbf{I}_n - n^{-1}\mathbf{1}_n\mathbf{1}_n^\top$. *Since* $\mathbf{\Gamma}\mathbf{1}_n = n^{1/2}\mathbf{e}_1$ *and* $\mathbf{\Gamma}^\top\mathbf{Q}\mathbf{\Gamma} = \mathbf{I}_n - \mathbf{e}_1\mathbf{e}_1^\top$, *where* $\mathbf{e}_1 = (1,\mathbf{0}_{n-1}^\top)^\top$, *it is clear that*

$$\bar{X} = n^{-1/2}Y_1, \quad T = \sum_{i=2}^{n} Y_i^2 = \|\mathbf{Y}_2\|^2.$$

Then, the joint pdf of $\bar{X}$ *and* $T = (n-1)S_x^2$ *is*

$$f_{\bar{X},T}(x,t) = \frac{n^{1/2}\pi^{(n-1)/2}}{\Gamma\{(n-1)/2\}} t^{\frac{n-1}{2}-1} h(nx^2+t).$$

Furthermore, by (4.34) we have $(Y_1^2,\|\mathbf{Y}_2\|^2) \stackrel{d}{=} R^2(V,1-V)$ *with* $V \sim \mathcal{B}eta\{1/2,(n-1)/2\} \perp\!\!\!\perp R^2$. *Hence,*

$$\left(\frac{\sqrt{n}\bar{X}}{S_x}\right)^2 = \frac{(n-1)Y_1^2}{\sum_{i=2}^{n}Y_i^2} \stackrel{d}{=} \frac{(n-1)V}{1-V} \sim \mathcal{F}_{1,n-1} \perp\!\!\!\perp Y_1^2 + \sum_{i=2}^{n} Y_i^2 = \|\mathbf{X}\|^2 \stackrel{d}{=} R^2$$
$$\Longrightarrow \frac{\sqrt{n}\bar{X}}{S_x} \sim t_{n-1} \perp\!\!\!\perp \sum_{i=1}^{n} X_i^2 = n\bar{X}^2 + (n-1)S_x^2,$$

where $t_\nu \equiv t(0,1,\nu)$ *denotes the univariate standard Student's t distribution with* $\nu > 0$ *degrees of freedom. The last two facts above do not depend on the density generator h. However,* $\bar{X} \perp\!\!\!\perp S_x^2$ *if and only if* $X_1,\dots,X_n$ *have a spherical normal joint distribution, which in turn is equivalent to* $\sum_{i=1}^{n} X_i^2 = \|\mathbf{X}\|^2 \stackrel{d}{=} R^2 \sim \chi_n^2$ *(see Example 50).*

As stated in the book by Fang and Zhang (1990), we can transform the distribution theory relating to the quadratic forms of a spherical random vector $\mathbf{X}$ into that of $\mathbf{X}/\|\mathbf{X}\|$. In fact, suppose $\mathbf{X} \stackrel{d}{=} R\mathbf{U}^{(p)} \sim \mathcal{S}_p^+(\varphi)$ and let $\mathbf{A}$ be a $p \times p$ symmetric matrix with eigenvalues $\lambda_1,\dots,\lambda_p$. Since there exists $\mathbf{\Gamma} \in \mathcal{O}_p$ such that $\mathbf{\Gamma}^\top\mathbf{A}\mathbf{\Gamma} = \mathbf{\Lambda} = \mathrm{diag}(\lambda_1,\dots,\lambda_p)$, it is clear that

$$\mathbf{\Gamma}\mathbf{X} \stackrel{d}{=} \mathbf{X} \stackrel{d}{=} R\mathbf{U}^{(p)} \Longrightarrow \mathbf{X}^\top\mathbf{A}\mathbf{X} \stackrel{d}{=} \mathbf{X}^\top\mathbf{\Lambda}\mathbf{X} \stackrel{d}{=} R^2\mathbf{U}^{(p)\top}\mathbf{\Lambda}\mathbf{U}^{(p)}$$
$$\Longrightarrow \sum_{i=1}^{p} \lambda_i X_i^2 \stackrel{d}{=} R^2 \sum_{i=1}^{p} \lambda_i U_i^2.$$

Similarly, since $\mathrm{P}(\|\mathbf{X}\| = 0) = \mathrm{P}(\mathbf{X} = \mathbf{0}_p) = 0$, it is also clear that

$$\frac{\mathbf{X}}{\|\mathbf{X}\|} \overset{d}{=} \mathbf{U}^{(p)} \implies \frac{\mathbf{X}^\top \mathbf{A}\mathbf{X}}{\|\mathbf{X}\|^2} \overset{d}{=} \frac{\mathbf{X}^\top \mathbf{\Lambda}\mathbf{X}}{\|\mathbf{X}\|^2} \overset{d}{=} \mathbf{U}^{(p)\top}\mathbf{\Lambda}\mathbf{U}^{(p)}$$
$$\implies \sum_{i=1}^{p} \lambda_i \frac{X_i^2}{\|\mathbf{X}\|^2} \overset{d}{=} \sum_{i=1}^{p} \lambda_i U_i^2.$$

Here, it is also worth noting that $(X_1^2/\|\mathbf{X}\|^2, \ldots, X_p^2/\|\mathbf{X}\|^2)^\top \overset{d}{=} (U_1^2, \ldots, U_p^2)^\top \sim \mathcal{D}_p(1/2, \ldots, 1/2)$ (see Example 42). Finally, we will show that these two results are equivalent. We shall assume normality for ease of explanation.

Theorem 41 *If $\mathbf{X} = (X_1, \ldots, X_p)^\top \sim \mathcal{N}_p(\mathbf{0}_p, \mathbf{I}_p)$ and $\mathbf{A} \in \mathbb{R}^{p\times p}$ is a symmetric matrix with eigenvalues $\lambda_1, \ldots, \lambda_p$, then:*

$$\mathbf{X}^\top \mathbf{A}\mathbf{X} \overset{d}{=} \sum_{i=1}^{p} \lambda_i X_i^2 \iff \frac{\mathbf{X}^\top \mathbf{A}\mathbf{X}}{\|\mathbf{X}\|^2} \overset{d}{=} \sum_{i=1}^{p} \lambda_i V_i, \tag{4.36}$$

where $(V_1, \ldots, V_p)^\top \sim \mathcal{D}_p(1/2, \ldots, 1/2)$.

Proof 41 ($\Longrightarrow$) *As $\mathbf{X}^\top \mathbf{A}\mathbf{X} \overset{d}{=} \sum_{i=1}^{p} \lambda_i X_i^2$, then*

$$\|\mathbf{X}\|^2 \frac{\mathbf{X}^\top \mathbf{A}\mathbf{X}}{\|\mathbf{X}\|^2} = \mathbf{X}^\top \mathbf{A}\mathbf{X} \overset{d}{=} \sum_{i=1}^{p} \lambda_i X_i^2 = \|\mathbf{X}\|^2 \sum_{i=1}^{p} \lambda_i \frac{X_i^2}{\|\mathbf{X}\|^2}.$$

Since $\|\mathbf{X}\| \perp\!\!\!\perp \mathbf{X}/\|\mathbf{X}\|$, then $\|\mathbf{X}\|^2 \perp\!\!\!\perp \mathbf{X}^\top \mathbf{A}\mathbf{X}/\|\mathbf{X}\|^2$ and therefore $\|\mathbf{X}\|^2 \perp\!\!\!\perp \sum_{i=1}^{p} \lambda_i X_i^2/\|\mathbf{X}\|^2 \overset{d}{=} \sum_{i=1}^{p} \lambda_i V_i$ since $(X_1^2/\|\mathbf{X}\|^2, \ldots, X_p^2/\|\mathbf{X}\|^2)^\top \overset{d}{=} (V_1, \ldots, V_p)^\top \sim \mathcal{D}_p(1/2, \ldots, 1/2)$ (see Example 42). Since $\|\mathbf{X}\|^2 \sim \chi_p^2$, the cf of $\log\|\mathbf{X}\|^2$ is different from zero almost everywhere, so the right-hand side of (4.36) follows from the result [$\mathcal{E}$2](ii) given at the end of Section 1.7. ($\Longleftarrow$) As $\sum_{i=1}^{p} \lambda_i V_i \overset{d}{=} \sum_{i=1}^{p} \lambda_i X_i^2/\|\mathbf{X}\|^2 \perp\!\!\!\perp \|\mathbf{X}\|^2$, then $\|\mathbf{X}\|^2 \sum_{i=1}^{p} \lambda_i X_i^2/\|\mathbf{X}\|^2 = \sum_{i=1}^{p} \lambda_i X_i^2 = \mathbf{X}^\top \mathbf{\Lambda}\mathbf{X} \overset{d}{=} \mathbf{X}^\top \mathbf{A}\mathbf{X}$. □

Theorem 42 *Let $\mathbf{X} \overset{d}{=} R\mathbf{U}^{(p)} \sim \mathcal{S}_p^+(\varphi)$, $V \sim \mathcal{B}eta\{p/2, (p-q)/2\}$ be independent of R, and $\mathbf{A}$ be a p-by-p symmetric matrix. Then,*

$$\mathbf{X}^\top \mathbf{A}\mathbf{X} \overset{d}{=} R^2 V \tag{4.37}$$

if and only if $\mathbf{A}^2 = \mathbf{A}$ and $\mathrm{rank}(\mathbf{A}) = q$ $(1 \le q < p)$.

Proof 42 ($\Longleftarrow$) *Partition $\mathbf{X} = (\mathbf{X}_1^\top, \mathbf{X}_2^\top)^\top$ with $\mathbf{X}_1 \in \mathbb{R}^q$. By (4.7), $(\mathbf{X}_1^\top, \mathbf{X}_2^\top)^\top \overset{d}{=} R(\sqrt{V}, \sqrt{1-V})^\top$, where $V \sim \mathcal{B}eta\{q/2, (p-q)/2\}$ and is independent of R. So, if $\mathbf{A}^2 = \mathbf{A}$ and $\mathrm{rank}(\mathbf{A}) = q$, then $\mathbf{X}^\top \mathbf{A}\mathbf{X} \overset{d}{=} \|\mathbf{X}_1\|^2 \overset{d}{=} R^2 V$. ($\Longrightarrow$) Assume $\mathbf{X}^\top \mathbf{A}\mathbf{X} \overset{d}{=} R^2 V$. Then $\|\mathbf{X}\|^2(\mathbf{X}^\top \mathbf{A}\mathbf{X}/\|\mathbf{X}\|^2) = \mathbf{X}^\top \mathbf{A}\mathbf{X} \overset{d}{=} R^2 V$, where $V \sim \mathcal{B}eta\{q/2, (p-q)/2\}$ is independent of $\|\mathbf{X}\|^2 \overset{d}{=} R^2$. Clearly, $\mathrm{P}(V > 0) = 1$ and $\mathrm{P}(\mathbf{X}^\top \mathbf{A}\mathbf{X}/\|\mathbf{X}\|^2 > 0) = \mathrm{P}(R^2 > 0) = \mathrm{P}(\mathbf{X} \neq \mathbf{0}_p) = 1$. Since $\mathrm{P}(0 < V < 1) = 1$, then $\psi_{\log Z} \in \mathcal{E}$ (see Section 1.7). Thus, assertion [$\mathcal{E}$2](i) in Section 1.7 implies $\mathbf{X}^\top \mathbf{A}\mathbf{X}/\|\mathbf{X}\|^2 \overset{d}{=} V$.* □

A natural extension of this result, which constitutes an adaptation of Cochran's theorem to the spherical case, is as follows.

Theorem 43 *Let* $\mathbf{X} \stackrel{d}{=} R\mathbf{U}^{(p)} \sim \mathcal{S}_p^+(\varphi)$, $\mathbf{V} = (V_1,\ldots,V_k)^\top \sim \mathcal{D}_k(p_1/2,\ldots,p_k/2)$, *with* $\sum_{i=1}^k p_i = p$, *be independent of* R, *and* $\mathbf{A}_1,\ldots,\mathbf{A}_k$ *be symmetric matrices of dimensions* $p \times p$. *Then,*

$$(\mathbf{X}^\top\mathbf{A}_1\mathbf{X},\ldots,\mathbf{X}^\top\mathbf{A}_k\mathbf{X})^\top \stackrel{d}{=} R^2\mathbf{V}$$

if and only if $\mathbf{A}_i\mathbf{A}_j = \delta_{ij}\mathbf{A}_i$ *and* $\operatorname{rank}(\mathbf{A}_i) = p_i$, *where* $\delta_{ii} = 1$ *and* $\delta_{ij} = 0$ *for* $i \neq j$.

Proof 43 *We will prove the result only for* $k = 3$ *to make the exposition easier.* ($\Longrightarrow$) *Suppose* $(\mathbf{X}^\top\mathbf{A}_1\mathbf{X},\mathbf{X}^\top\mathbf{A}_2\mathbf{X},\mathbf{X}^\top\mathbf{A}_3\mathbf{X}) \stackrel{d}{=} R^2(V_1,V_2,V_3)$, *where* $R \perp\!\!\!\perp (V_1,V_2,V_3)^\top \sim \mathcal{D}_3(p_1/2,p_2/2,p_3/2)$. *Clearly, for* $i = 1,2,3$, *we have* $\mathbf{X}^\top\mathbf{A}_i\mathbf{X} \stackrel{d}{=} R^2V_i$, *with* $R \perp\!\!\!\perp V_i \sim \mathcal{B}eta\{p_i/2,(p-p_i)/2\}$ *and therefore* $\mathbf{A}_i^2 = \mathbf{A}_i$ *and* $\operatorname{rank}(\mathbf{A}_i) = p_i$ *by Theorem 42. Furthermore, for* $i \neq j$, *it follows that* $\mathbf{X}^\top(\mathbf{A}_i+\mathbf{A}_j)\mathbf{X} \stackrel{d}{=} R^2V_{ij}$ *with* $R \perp\!\!\!\perp V_{ij} = V_i + V_j \sim \mathcal{B}eta\{(p_i+p_j)/2,(p-p_i-p_j)/2\}$ *and according again to Theorem 42 we must have* $(\mathbf{A}_i+\mathbf{A}_j)^2 = \mathbf{A}_i+\mathbf{A}_j$, *which implies that* $\mathbf{A}_i\mathbf{A}_j = \mathbf{O}$. ($\Longleftarrow$) *The proof of this part is simple starting from the partition* $\mathbf{X} = (\mathbf{X}_1^\top,\ldots,\mathbf{X}_k^\top)^\top$ *with* $\mathbf{X}_i \in \mathbb{R}^{p_i}$ *for* $i = 1,\ldots,k$ *since if* $\mathbf{A}_1,\ldots,\mathbf{A}_k$ *are idempotent symmetric matrices of ranks* $p_1,\ldots,p_k$, *then* $(\mathbf{X}^\top\mathbf{A}_1\mathbf{X},\ldots,\mathbf{X}^\top\mathbf{A}_k\mathbf{X}) \stackrel{d}{=} (\|\mathbf{X}_1\|^2,\ldots,\|\mathbf{X}_k\|^2)$ *and the assertion follows from (4.34).* □

Following Arellano-Valle (1994, 2001) and Arellano-Valle et al. (2006b), when $\mathbf{X} \sim \mathcal{S}_p(\varphi)$ we call the distribution of $R^2 = \|\mathbf{X}\|^2$ a *squared-radial (SR) distribution* and denote it by $R^2 \sim \mathcal{SR}(p,\varphi)$ or $R^2 \sim \mathcal{SR}(p,h)$ in the absolutely continuous case. Next, a multivariate extension of the squared-radial distribution is given. In the literature, such distributions are also referred to as *generalized gamma*. For this, a fundamental result is that given in Equation (4.7).

Definition 13 [Multivariate squared-radial distribution] *Let* $\mathbf{X} = (X_1,\ldots,X_p)^\top \sim \mathcal{S}_p(\varphi)$, *let* $(\mathcal{V}_1,\ldots,\mathcal{V}_k)$ *be an orthogonal decomposition of* $\mathbb{R}^p$, *with* $\dim(\mathcal{V}_j) = p_j$, *and let* $\mathbf{P}_j\mathbf{X}$ *be the orthogonal projection of* $\mathbf{X}$ *onto* $\mathcal{V}_j$. *We call the distribution of*

$$\mathbf{W} = (W_1,\ldots,W_k)^\top = (\|\mathbf{P}_1\mathbf{X}\|^2,\ldots,\|\mathbf{P}_k\mathbf{X}\|^2)^\top$$

a multivariate squared-radial distribution and write $\mathbf{W} \sim \mathcal{SR}_p(p_1,\ldots,p_k,\varphi)$.

This is a valid definition because the spherical symmetry of $\mathbf{X}$ implies that the distribution of $\mathbf{W}$ depends on the subspaces only through their dimensions. A canonical form is obtained choosing $\mathcal{V}_j$ as the subspace spanned by a set of unit vectors. Let $\mathbf{e}_i$ be the i-th unit vector of $\mathbb{R}^p$ and define $N_1 = \{1,\ldots,p_1\}, N_j = \{p_1+\cdots+p_{j-1}+1,\ldots,p_1+\cdots+p_j\}, j = 2,\ldots,k$. Then the i-th component of $\mathbf{P}_j\mathbf{X}$ equals X_i for $i \in N_j$ and it is equal to 0 otherwise. Thus,

$$\mathbf{W} = (W_1,\ldots,W_k)^\top \stackrel{d}{=} \left(\sum_{i\in N_1} X_i^2,\ldots,\sum_{i\in N_k} X_i^2\right)^\top.$$

A different name and notation for this distribution is used by Fang and Zhang (1990), who showed that if $\mathrm{P}(\mathbf{X} = \mathbf{0}_p) = 0$ and $p_k \geq 1$, then $(W_1,\ldots,W_m)^\top$, $1 \leq m < k$, has a pdf. In the absolutely continuous case, Fang and Zhang (1990) also showed that if $h(\|\mathbf{x}\|^2)$ is the pdf of $\mathbf{X}$, with $h = h^{(p)}$, then $\mathbf{W} = (W_1,\ldots,W_k)^\top$ has pdf given by

$$\frac{\pi^{p/2}}{\prod_{i=1}^k \Gamma(p_i/2)}\left(\prod_{i=1}^k w_i^{p_i/2-1}\right) h\left(\sum_{i=1}^k w_i\right).$$

In this case, we write $\mathbf{W} \sim \mathcal{SR}_p(p_1,\ldots,p_k,h)$. A summary of some important properties of the family $\mathcal{SR}_p(p_1,\ldots,p_k,h)$ is as follows:

[SR1] *Closure under aggregation:* When the components are aggregated, the new distribution is still squared-radial, with the corresponding degrees of freedom being added. For example, $(W_1,W_2,W_3,W_4,W_5) \sim \mathcal{SR}_p(p_1,p_2,p_3,p_4,p_5,h) \Longrightarrow (W_1+W_2,W_3+W_4,W_5) \sim \mathcal{SR}_p(p_1+p_2,p_3+p_4,p_5,h)$.

[SR2] *Invariance for homogeneous functions:* If $T(a\mathbf{W}) = T(\mathbf{W})$ for any positive a, then $T(\mathbf{W})$ has the same distribution, say G, for all h and G may be obtained assuming $\mathbf{X}$ to be a spherical normal random vector. For instance, the distribution of the ratio of two components does not depend on the generator.

[SR3] *Independence of total and ratios:* $\frac{1}{\sum_{j=1}^k W_j}\mathbf{W} \perp\!\!\!\perp \sum_{j=1}^k W_j$.

[SR4] *Distribution of ratios:* Since the distribution does not depend on h, it may be computed under normality. In particular, $\frac{1}{\sum_{j=1}^k W_j}\mathbf{W} \sim \mathcal{D}_k\left(\frac{p_1}{2},\ldots,\frac{p_k}{2}\right)$.

[SR5] *Invariance under conditioning:* $(W_1,W_2,\ldots,W_q|\sum_{j=q+1}^k W_j = s) \sim \mathcal{SR}_p(p_1,\ldots,p_q,h_s)$, where $h_s(t) = h_s^{(q)}(t) = h^{(p)}(t+s)/h^{(p-q)}(s)$ is the corresponding conditional dgf (see [4] in Section 4.2.5).

From [SR1]-[SR5] several general properties of spherical distributions may be proved. For instance, taking $W_1 = R_q^2$ and $W_2 = R_{(p-q)}^2$, [SR3] and [SR4] become $\frac{R_q^2}{R^2} \perp\!\!\!\perp R^2$ and $V = \frac{R_q^2}{R^2} \sim \mathcal{B}eta\left(\frac{q}{2},\frac{p-q}{2}\right)$, respectively; while [SR2] becomes $\frac{p-q}{q}\frac{R_q^2}{R_{(p-q)}^2} \overset{d}{=} \frac{p-q}{q}\frac{V}{1-V} \sim \mathcal{F}_{q,p-q}$, regardless of the generator h. Under sphericity, the joint distribution of $(\mathbf{X}_q,\mathbf{X}_{(p-q)})$ is determined by that of $(R_q^2,R_{(p-q)}^2)$ and the same holds for $(\mathbf{X}_q,R_{(p-q)}^2)$. This translates into the following key relationship:

$$(R_q^2,R_{(p-q)}^2) \sim \mathcal{SR}_p(q,p-q,h) \Longleftrightarrow \mathbf{X}_q|R_{(p-q)}^2 = t \sim \mathcal{S}_q(h_t) \text{ and } R_{(p-q)}^2 \sim \mathcal{SR}_p(p-q,h).$$

4.5 Distributions Related to the Spherical and Elliptical

4.5.1 L_q-spherical normal and L_q-spherical uniform distributions

Spherical distributions are based on the L_2-norm, which corresponds to the length $\|\mathbf{x}\|$ of a vector $\mathbf{x} \in \mathbb{R}^p$. In fact, if $\mathbf{X}$ is spherically distributed, then its distribution is completely determined by the distribution of the radial variable $R = \|\mathbf{X}\|$. The direction in which $\mathbf{X}$ moves is determined by the uniform vector $\mathbf{U}^{(p)}$ independently of R through the stochastic representation $\mathbf{X} = R\mathbf{U}^{(p)}$. Since $\mathbf{U}^{(p)}$ is common to every p-dimensional spherical distribution, it can be generated from the standard p-dimensional normal distribution. Thus, the construction of spherical distributions can be seen as follows:

a) From $\mathbf{Z} \sim \mathcal{N}_p(\mathbf{0}_p,\mathbf{I}_p)$, let $\mathbf{U}^{(p)} = \frac{\mathbf{Z}}{\|\mathbf{Z}\|}$ be the uniform vector on $\mathbb{S}^{(p-1)} = \{\mathbf{x} \in \mathbb{R}^p : \|\mathbf{x}\| = 1\}$;

b) Next, define $\mathbf{X} = R\mathbf{U}^{(p)}$ for some nonnegative random variable R, with $R \perp\!\!\!\perp \mathbf{U}^{(p)}$.

Since $\|\mathbf{U}^{(p)}\| = 1$, from b) it is clear that $R = \|\mathbf{X}\| = (\sum_{i=1}^p X_i^2)^{1/2}$ becomes the L_2-norm of $\mathbf{X}$, due to which spherical distributions are also called L_2-spherical (symmetric) distributions. From a) it is also worth noting that $\mathbf{Z} = (Z_1,\ldots,Z_p)^\top \sim \mathcal{N}_p(\mathbf{0}_p,\mathbf{I}_p) \Longleftrightarrow Z_1,\ldots,Z_p \overset{iid}{\sim} \mathcal{N}(0,1)$, so the multivariate spherical normal distribution can be obtained from independent univariate normal distributions (not necessarily with the same mean, but with the same variance).

On the other hand, a natural extension of the L_2-norm $\|\mathbf{x}\|_2 = (\sum_{i=1}^p |x_i|^2)^{1/2}$ of a vector $\mathbf{x} \in \mathbb{R}^p$ is $\|\mathbf{x}\|_q = (\sum_{i=1}^p |x_i|^q)^{1/q}$ which is a norm if $q \geq 1$ and an anti-norm (i.e., not satisfying the triangular inequality) if $0 < q < 1$. Having made this clarification, we shall henceforth call $\|\mathbf{x}\|_q = (\sum_{i=1}^p |x_i|^q)^{1/q}$ the L_q-norm of $\mathbf{x} \in \mathbb{R}^p$ for any value of $q > 0$.

According to the above, L_2-spherical distributions can be extended to L_q-spherical distributions as follows. We begin by introducing the L_q-spherical normal (or q-generalized normal) distribution as the natural reference for this family, which for $q = 2$ becomes the multivariate standard normal.

Definition 14 [Multivariate L_q-spherical normal distributions] *A random vector $\mathbf{X} \in \mathbb{R}^p$ follows a multivariate L_q-spherical (or L_q-norm or q-generalized) normal distribution, denoted by $\mathbf{X} \sim \mathcal{L}_q\mathcal{N}_p(\mathbf{0}_p, \mathbf{I}_p)$, if its pdf is defined by*

$$f_p(\mathbf{x}) = c_q^p e^{-\frac{1}{q}\|\mathbf{x}\|_q^q}, \quad \mathbf{x} = (x_1, \ldots, x_p)^\top \in \mathbb{R}^p, \tag{4.38}$$

where $\|\mathbf{x}\|_q^q = \sum_{i=1}^p |x_i|^q$ and $c_q = q^{1-1/q}/\{2\Gamma(1/q)\}$, $q > 0$.

Clearly, assuming that $\mathbf{X} = (X_1, \ldots, X_p)^\top \sim \mathcal{L}_q\mathcal{N}_p(\mathbf{0}_p, \mathbf{I}_p)$ is equivalent to $X_1, \ldots, X_p$ are iid, with exponential power pdf $f_1(x) = c_q e^{-\frac{1}{q}|x|^q}$, $x \in \mathbb{R}$, which has mean $\mu = 0$ and variance $\sigma_q^2 = \frac{q^{2/q}\Gamma(3/q)}{\Gamma(1/q)}$. It is also clear that, under this distribution, we have:

[1] $\mathrm{E}(\mathbf{X}) = \mathbf{0}_p$ and $\mathrm{var}(\mathbf{X}) = \sigma_q^2 \mathbf{I}_p$, with $\sigma_q^2 = \frac{q^{2/q}\Gamma(3/q)}{\Gamma(1/q)}$ $(\sigma_2^2 = 1)$;

[2] $|X_1|, \ldots, |X_p| \overset{iid}{\sim} 2f_1(t) = 2c_q e^{-\frac{1}{q}t^q}$ for $t > 0$, and $|X_1|^q, \ldots, |X_p|^q \overset{iid}{\sim} \mathcal{G}amma(1/q, 1/q)$;

[3] If $R_q = \|\mathbf{X}\|_q$ is the L_q-norm of $\mathbf{X}$, then $R_q^q = \|\mathbf{X}\|_q^q = \sum_{i=1}^p |X_i|^q \sim \mathcal{G}amma(p/q, 1/q)$;

[4] $(|X_1|^q/\|\mathbf{X}\|_q^q, \ldots, |X_p|^q/\|\mathbf{X}\|_q^q)^\top \sim \mathcal{D}_p(1/q, \ldots, 1/q)$;

[5] $\mathbf{X} \overset{d}{=} \mathbf{D}(\mathbf{S})\mathbf{T}$, where $\mathbf{S} \perp\!\!\!\perp \mathbf{T}$, with $\mathbf{S} \overset{d}{=} \mathrm{sgn}(\mathbf{X}) = (\mathrm{sgn}(X_1), \ldots, \mathrm{sgn}(X_p))^\top \sim \mathcal{U}(\{-1, +1\}^p)$ and $\mathbf{T} \overset{d}{=} |\mathbf{X}| = (|X_1|, \ldots, |X_p|)^\top \sim 2^p f_q(\mathbf{t}) = (2c_q)^p e^{-\frac{1}{q}\sum_{i=1}^p t_i^q}$ for $\mathbf{t} = (t_1, \ldots, t_p)^\top \in \mathbb{R}_+^p$, and for each $\mathbf{s} = (s_1, \ldots, s_p)^\top \in \{-1, +1\}^p$, $\mathbf{D}(\mathbf{s}) = \mathrm{diag}(s_1, \ldots, s_p)$.

Note by [3] that the pdf of $R_q = \|\mathbf{X}\|_q$ is given by

$$g(r) = \frac{(1/q)^{(p/q)-1}}{\Gamma(p/q)} r^{p-1} e^{-r^q/q}, \quad r > 0, \tag{4.39}$$

with k-th moment

$$\mathrm{E}(R_q^k) = \frac{q^{k/q}\Gamma\{(p+k)/q\}}{\Gamma(p/q)}, \quad k = 1, 2, \ldots.$$

It is also immediate that if $\mathbf{X} \sim \mathcal{L}_q\mathcal{N}_p(\mathbf{0}_p, \mathbf{I}_p)$, then for any location vector $\boldsymbol{\xi} \in \mathbb{R}^p$ and scalar scale parameter ω, the location-scale transformation $\mathbf{Y} = \boldsymbol{\xi} + \omega\mathbf{X} \sim \mathcal{L}_q\mathcal{N}_p(\boldsymbol{\xi}, \omega^2\mathbf{I}_p)$, with L_q-norm symmetric pdf

$$f_p(\mathbf{y}; \boldsymbol{\xi}, \omega) = \omega^{-p} c_q^p e^{-\frac{1}{q}\|\mathbf{x}\|_q^q}, \quad \mathbf{x} = \omega^{-1}(\mathbf{y} - \boldsymbol{\xi}), \quad \mathbf{y} \in \mathbb{R}^p.$$

Now, from the L_q-spherical normal random vector $\mathbf{X} = (X_1, \ldots, X_p)^\top \sim \mathcal{L}_q\mathcal{N}_p(\mathbf{0}_p, \mathbf{I}_p)$ we can introduce the L_q-spherical uniform random vector $\mathbf{U}_q^{(p)} = \mathbf{X}/\|\mathbf{X}\|_q$ with L_q-norm $\|\mathbf{U}_q^{(p)}\|_q = 1$, which is uniformly distributed on the surface of the (L_q-norm) unit sphere $\mathbb{S}_q^{(p-1)} = \{\mathbf{x} \in \mathbb{R}^p : \|\mathbf{x}\|_q = 1\}$ (also called the q-generalized p-dimensional unit sphere) and is independent of the L_q-norm $R_q = \|\mathbf{X}\|_q$. To see this, consider the transformation

$$u_i = x_i/\|\mathbf{x}\|_q, \; i = 1, \ldots, p-1, \quad r = \|\mathbf{x}\|_q,$$

which is one-to-one in each of the regions

$$\mathbb{S}_1 = \{\mathbf{x} = (x_1, \ldots, x_p)^\top \in \mathbb{R}^p : x_p > 0\} \quad \text{and} \quad \mathbb{S}_2 = \{\mathbf{x} = (x_1, \ldots, x_p)^\top \in \mathbb{R}^p : x_p < 0\}, \tag{4.40}$$

with the same absolute Jacobian (determinant) of the transformation given by the expression $(1 - \sum_{i=1}^{p-1} |u_i|^q)^{(1-q)/q} r^{p-1}$. Thus, from the joint pdf of $X_1, \ldots, X_p$ given by (4.38) we get the joint pdf of $U_i = X_i/\|\mathbf{X}\|_q$, $i = 1, \ldots, p-1$, and $R_q = \|\mathbf{X}\|_q$ as

$$g(u_1, \ldots, u_{p-1}, r) = 2c_q^p r^{p-1} e^{-r^q/q} \left(1 - \sum_{i=1}^{p-1} |u_i|^q\right)^{(1-q)/q}, \ |u_i| < 1, \ \sum_{i=1}^{p-1} |u_i|^q < 1, \ r > 0. \quad (4.41)$$

From (4.41) it is clear that $(U_1, \ldots, U_{p-1}) \perp\!\!\!\perp R_q$ and therefore $\mathbf{U}_q^{(p)} = \mathbf{X}/\|\mathbf{X}\|_q \perp\!\!\!\perp R_q = \|\mathbf{X}\|_q$. Moreover, from (4.41) it also follows that the joint pdf of $U_1, \ldots, U_{p-1}$ is

$$g(u_1, \ldots, u_{p-1}) = \left(\frac{q}{2}\right)^{p-1} \frac{\Gamma(p/q)}{\{\Gamma(1/q)\}^p} \left(1 - \sum_{i=1}^{p-1} |u_i|^q\right)^{(1-q)/q}, \ |u_i| < 1, \ \sum_{i=1}^{p-1} |u_i|^q < 1. \quad (4.42)$$

This means that $\mathbf{U}_q^{(p)} = \mathbf{X}/\|\mathbf{X}\|_q \sim \mathcal{U}_p(\mathbb{S}_q^{(p-1)})$. From (4.41) it is also easy to verify that the pdf of $R_q = \|\mathbf{X}\|_q$ is that given in (4.39) and therefore that the joint pdf of $U_1, \ldots, U_{p-1}$ is that in (4.42).

We can now introduce a formal definition of the L_q-spherical uniform vector $\mathbf{U}_q^{(p)}$. For further details and properties, see Song and Gupta (1997), Gupta and Song (1997), and Szabłowski (1998).

Definition 15 [Multivariate L_q-spherical uniform distributions] *Let the random vector $\mathbf{U}_q^{(p)} = (U_1, \ldots, U_{p-1}, U_p)^\top \in \mathbb{S}_q^{(p-1)}$, that is, with $\|\mathbf{U}_q^{(p)}\|_q^q = \sum_{i=1}^p |U_i|^q = 1$ for any $q > 0$. We say that $\mathbf{U}_q^{(p)}$ has a multivariate L_q-spherical (or L_q-norm or q-generalized) uniform distribution, denoted by $\mathcal{L}_q\mathcal{U}_p$, if $U_1, \ldots, U_{p-1}$ have joint pdf given by (4.42).*

It is quite evident that the L_q-spherical uniform vector $\mathbf{U}_q^{(p)}$ is sign and permutationally invariant for any $q > 0$, but it is not rotationally invariant for $q \neq 2$. Furthermore, since $(|U_1|^q, \ldots, |U_p|^q)^\top \sim \mathcal{D}_p(1/q, \ldots, 1/q)$ then, for each $k = 1, \ldots, p-1$, $(|U_1|^q, \ldots, |U_k|^q, 1 - \sum_{i=1}^k |U_i|^q)^\top \sim \mathcal{D}_{k+1}(1/q, \ldots, 1/q, (p-k)/2)$ so that the subvector $(V_1, \ldots, V_k)^\top = (|U_1|^q, \ldots, |U_k|^q)^\top$ has pdf

$$p_k(v_1, \ldots, v_k) = \frac{\Gamma(p/q)}{\{\Gamma(1/q)\}^k \Gamma\{(p-k)/q\}} \prod_{i=1}^k v_i^{\frac{1}{q}-1} \left(1 - \sum_{i=1}^k v_i\right)^{\frac{p-k}{q}-1}, \quad v_i > 0, \ \sum_{i=1}^k v_i < 1.$$

In particular, $V_i = |U_i|^q \sim \mathcal{B}eta\{1/q, (p-1)/q\}$, $i = 1, \ldots, p$. These results can be obtained from (4.41), which for each $k = 1, \ldots, p-1$ implies that the joint pdf of $U_1, \ldots, U_k$ is given by

$$g_k(u_1, \ldots, u_k) = \frac{q^k \Gamma(p/q)}{2^k \{\Gamma(1/q)\}^k \Gamma\{(p-k)/q\}} \left(1 - \sum_{i=1}^k |u_i|^q\right)^{\frac{p-k}{q}-1}, \ |u_i| < 1, \ \sum_{i=1}^k |u_i|^q < 1. \quad (4.43)$$

Moreover, for each $k = 1, \ldots, p-1$, the joint pdf of $Z_1, \ldots, Z_k$, where $Z_i = V_i^{1/q} = |U_i|$, $i = 1, \ldots, k$, is

$$h_k(z_1, \ldots, z_k) = \frac{q^k \Gamma(p/q)}{\Gamma(1/q)^k \Gamma\{(p-k)/q\}} \left(1 - \sum_{i=1}^k z_i^q\right)^{\frac{p-k}{q}-1}, \quad z_i > 0, \ \sum_{i=1}^k z_i^q < 1.$$

4.5.2 L_q-spherical distributions

Next, we introduce the general concept of L_q-spherical distributions.

Definition 16 [Multivariate L_q-spherical distributions] *A random vector $\mathbf{X} \in \mathbb{R}^p$ is said to have a multivariate L_q-spherical (or L_q-norm spherical or q-generalized spherical) distribution, denoted by $\mathbf{X} \sim \mathcal{L}_q\mathcal{S}_p$, iff $\mathbf{X} \stackrel{d}{=} R_q \mathbf{U}_q^{(p)}$, where $R_q \perp\!\!\!\perp \mathbf{U}_q^{(p)}$, $R_q \sim F_q$ is a univariate nonnegative random variable and $\mathbf{U}_q^{(p)} \sim \mathcal{L}_q\mathcal{U}_p$.*

As in the spherical case with $q = 2$, if $\mathrm{P}(\mathbf{X} = \mathbf{0}_p) = 0$ then the L_q-spherical distribution $\mathcal{L}_q\mathcal{S}_p$ will be denoted $\mathcal{L}_q\mathcal{S}_p^+$. In this case, its relationship with $\mathcal{L}_q\mathcal{U}_p$ is established as follows.

Theorem 44 *Assume $\mathbf{X} \stackrel{d}{=} R_q \mathbf{U}_q^{(p)} \sim \mathcal{L}_q\mathcal{S}_p^+$. Then, $\|\mathbf{X}\|_q \stackrel{d}{=} R_q$, $\mathbf{X}/\|\mathbf{X}\|_q \stackrel{d}{=} \mathbf{U}_q^{(p)}$ and $\|\mathbf{X}\|_q \perp\!\!\!\perp \mathbf{X}/\|\mathbf{X}\|_q$.*

Proof 44 *Let $g_1(\mathbf{x}) = \|\mathbf{x}\|_q$ and $g_2(\mathbf{x}) = \mathbf{x}/\|\mathbf{x}\|_q$ for $\mathbf{x} \in \mathbb{R}^p$. Note that g_1 and g_2 are measurable functions such that $g_1(r\mathbf{x}) = r g_1(\mathbf{x})$ and $g_2(r\mathbf{x}) = g_2(\mathbf{x})$ for all $r > 0$. Since $\mathbf{X} \stackrel{d}{=} R_q \mathbf{U}_q^{(p)}$, with $\mathrm{P}(\mathbf{X} = \mathbf{0}_p) = 0$ and $\mathrm{P}(\|\mathbf{U}_q^{(p)}\|_q = 1) = 1$, then*

$$\begin{pmatrix} \|\mathbf{X}\|_q \\ \mathbf{X}/\|\mathbf{X}\|_q \end{pmatrix} = \begin{pmatrix} g_1(\mathbf{X}) \\ g_2(\mathbf{X}) \end{pmatrix} \stackrel{d}{=} \begin{pmatrix} g_1(R_q \mathbf{U}_q^{(p)}) \\ g_2(R_q \mathbf{U}_q^{(p)}) \end{pmatrix} = \begin{pmatrix} R_q \\ \mathbf{U}_q^{(p)} \end{pmatrix}.$$

Finally, since $R_q \perp\!\!\!\perp \mathbf{U}_q^{(p)}$ then $\|\mathbf{X}\|_q \perp\!\!\!\perp \mathbf{X}/\|\mathbf{X}\|_q$. □

From this, many of the properties studied in the L_2-spherical case can be extended to the L_q-spherical case. For example, we can note by construction that $\mathbf{X} \stackrel{d}{=} R_q \mathbf{U}_q^{(p)} \sim \mathcal{L}_q\mathcal{N}_p(\mathbf{0}_p, \mathbf{I}_p)$ iff $R_q \stackrel{d}{=} \|\mathbf{X}\|_q \sim \{\mathcal{G}amma(p/q, 1/q)\}^{1/q}$, from which it follows that

$$\mathrm{E}(\mathbf{U}_q^{(p)}) = \mathbf{0}_p \quad \text{and} \quad \mathrm{var}(\mathbf{U}_q^{(p)}) = \frac{\Gamma(3/q)\Gamma(p/q)}{\Gamma(1/q)\Gamma\{(p+2)/q\}} \mathbf{I}_p. \tag{4.44}$$

Furthermore, when partitioning $\mathbf{X} \sim \mathcal{L}_q\mathcal{N}_p(\mathbf{0}_p, \mathbf{I}_p)$ into two pieces $\mathbf{X}_1 \in \mathbb{R}^{p_1}$ and $\mathbf{X}_2 \in \mathbb{R}^{p_2}$ with $p_1 + p_2 = p$, the corresponding pieces $\mathbf{U}_1$ and $\mathbf{U}_2$ of $\mathbf{U}_q^{(p)}$ are such that

$$\begin{pmatrix} \mathbf{U}_1 \\ \mathbf{U}_2 \end{pmatrix} = \begin{pmatrix} \dfrac{\mathbf{X}_1}{\|\mathbf{X}\|_q} \\ \dfrac{\mathbf{X}_2}{\|\mathbf{X}\|_q} \end{pmatrix} = \begin{pmatrix} \dfrac{\|\mathbf{X}_1\|_q}{\|\mathbf{X}\|_q} \dfrac{\mathbf{X}_1}{\|\mathbf{X}_1\|_q} \\ \dfrac{\|\mathbf{X}_2\|_q}{\|\mathbf{X}\|_q} \dfrac{\mathbf{X}_2}{\|\mathbf{X}_2\|_q} \end{pmatrix} = \begin{pmatrix} V_{q1} \mathbf{U}_q^{(p_1)} \\ V_{q2} \mathbf{U}_q^{(p_2)} \end{pmatrix}, \tag{4.45}$$

where $V_{qi} = \|\mathbf{X}_i\|_q / \|\mathbf{X}\|_q$ and $\mathbf{U}_q^{(p_i)} = \mathbf{X}_i / \|\mathbf{X}_i\|_q$ for $i = 1, 2$. Clearly, $V_{q2} = (1 - V_{q1}^q)^{1/q}$ and $V_{q1} \perp\!\!\!\perp \mathbf{U}_q^{(p_1)} \perp\!\!\!\perp \mathbf{U}_q^{(p_2)}$, with $V_{q1}^q \sim \mathcal{B}eta\{p_1/q, (p - p_1)/q\}$ and $\mathbf{U}_q^{(p_i)} \sim \mathcal{L}_q\mathcal{N}_{p_i}$, $i = 1, 2$. From this fact, we can establish the following more general result for the L_q-spherical family.

Theorem 45 *If $\mathbf{X} \stackrel{d}{=} R_q \mathbf{U}_q^{(p)} \sim \mathcal{L}_q\mathcal{S}_p$ is partitioned in two different pieces $\mathbf{X}_i \in \mathbb{R}^{p_i}$, $i = 1, 2$, with $p_1 + p_2 = p$, then the following statements are true:*

(i) $\begin{pmatrix}\mathbf{X}_1\\ \mathbf{X}_2\end{pmatrix} \stackrel{d}{=} \begin{pmatrix}R_{q1}\mathbf{U}_q^{(p_1)}\\ R_{q2}\mathbf{U}_q^{(p_2)}\end{pmatrix}$, *where* $\begin{pmatrix}R_{q1}\\ R_{q2}\end{pmatrix} = \begin{pmatrix}R_qV_{q1}\\ R_qV_{q2}\end{pmatrix} \stackrel{d}{=} \begin{pmatrix}\|\mathbf{X}_1\|_q\\ \|\mathbf{X}_2\|_q\end{pmatrix}$, *with* $R_q = (R_{q1}^q + R_{q2}^q)^{1/q}$ *and* $V_{qi} \sim \mathcal{B}eta\{p_i/q, (p-p_i)/q\}$, $i = 1,2$, *and where* $R_q \perp\!\!\!\perp (V_{q1}, V_{q2}) \perp\!\!\!\perp \mathbf{U}_q^{(p_1)} \perp\!\!\!\perp \mathbf{U}_q^{(p_2)}$;

(ii) $\mathbf{X}_1 \perp\!\!\!\perp \mathbf{X}_2 \mid \|\mathbf{X}_2\|_q$;

(iii) $(\mathbf{X}_1 \mid \mathbf{X}_2 = \mathbf{x}_2) \stackrel{d}{=} R_{q,\|\mathbf{x}_2\|_q^q}\mathbf{U}_q^{(p_1)}$, *where for each* $a \geq 0$, $R_{q,a^q} \stackrel{d}{=} (R_q^q - a^q)^{1/q} \perp\!\!\!\perp \mathbf{U}_q^{(p_1)}$.

Proof 45 *The proof of (i) is direct from (4.45), since*

$$\begin{pmatrix}\mathbf{X}_1\\ \mathbf{X}_2\end{pmatrix} \stackrel{d}{=} R_q\mathbf{U}_q^{(p)} = R_q\begin{pmatrix}\mathbf{U}_1\\ \mathbf{U}_2\end{pmatrix} \stackrel{d}{=} R_q\begin{pmatrix}V_{q1}\mathbf{U}_q^{(p_1)}\\ V_{q2}\mathbf{U}_q^{(p_2)}\end{pmatrix} = \begin{pmatrix}R_qV_{q1}\mathbf{U}_q^{(p_1)}\\ R_qV_{q2}\mathbf{U}_q^{(p_2)}\end{pmatrix} = \begin{pmatrix}R_{q1}\mathbf{U}_q^{(p_1)}\\ R_{q2}\mathbf{U}_q^{(p_2)}\end{pmatrix},$$

where R_q, (V_{q1}, V_{q2}), $\mathbf{U}_q^{(p_1)}$ *and* $\mathbf{U}_q^{(p_2)}$ *are mutually independent.*

Now, by part (i) we have $(\mathbf{X}_1, \mathbf{X}_2) \stackrel{d}{=} (R_{q1}\,\mathbf{U}_q^{(p_1)}, R_{q2}\,\mathbf{U}_q^{(p_2)})$, *where* $(R_{q1}, R_{q2}) \stackrel{d}{=} (\|\mathbf{X}_1\|_q, \|\mathbf{X}_2\|_q)$ *which jointly with* $\mathbf{U}_q^{(p_1)}$ *and* $\mathbf{U}_q^{(p_2)}$ *are all independent. It is then clear that* $R_{q1}\,\mathbf{U}^{(p_1)} \perp\!\!\!\perp R_{q2}\,\mathbf{U}^{(p_2)} \mid R_{q2}$ *and therefore* $\mathbf{X}_1 \perp\!\!\!\perp \mathbf{X}_2 \mid \|\mathbf{X}_2\|_q$.

Finally, from parts (i) and (ii) it follows for each $\mathbf{x}_2 \in \mathbb{R}^{p_2}$ *that* $(\mathbf{X}_1 \mid \mathbf{X}_2 = \mathbf{x}_2) \stackrel{d}{=} (\mathbf{X}_1 \mid \|\mathbf{X}_2\|_q = \|\mathbf{x}_2\|_q) \stackrel{d}{=} (R_{q1}\mathbf{U}^{(p_1)} \mid R_{q2} = \|\mathbf{x}_2\|_q) \stackrel{d}{=} (R_{q1} \mid R_{q2} = \|\mathbf{x}_2\|_q)\mathbf{U}^{(p_1)}$ *since* $(R_{q1}, R_{q2}) \perp\!\!\!\perp \mathbf{U}^{(p_1)}$. *But, since* $R_{q1}^q + R_{q2}^q = R_q^q$, *then* $R_{q1}^q = R_q^q - R_{q2}^q$ *and* $(R_{q1}^q \mid R_{q2}^q = \|\mathbf{x}_2\|_q^q) = R_q^q - \|\mathbf{x}_2\|_q^q$, *i.e.,* $(R_{q1} \mid R_{q2} = \|\mathbf{x}_2\|_q) = (R_q^q - \|\mathbf{x}_2\|_q^q)^{1/q}$ *and therefore* $(\mathbf{X}_1 \mid \mathbf{X}_2 = \mathbf{x}_2) \stackrel{d}{=} R_{q,\|\mathbf{x}_2\|_q^q}\mathbf{U}_q^{(p_1)}$ *where* $R_{q,\|\mathbf{x}_2\|_q^q} \stackrel{d}{=} \{(R_q^q - \|\mathbf{x}_2\|_q^q)^{1/q} \mid \mathbf{X}_2 = \mathbf{x}_2\}$ *and is independent of* $\mathbf{U}_q^{(p_1)}$. □

Osiewalski and Steel (1993) stated, in the absolutely continuous case, that the symmetry of the L_q-spherical distributions is imposed through the pdf. As in Gupta and Song (1997), starting from Definition 16, the form of the pdf of $\mathcal{L}_q\mathcal{S}_p^+$ is established to have the form $h(\|\mathbf{x}\|_q^q)$, with $\|\mathbf{x}\|_q^q = \sum_{i=1}^p |x_i|^q$, $-\infty < x_i < \infty$, $i = 1, \ldots, p$, where $h : \mathbb{R}_+ \to \mathbb{R}_+$ is a measurable function. For this, we need the following preliminary result.

Theorem 46 *Suppose that a random vector* $\mathbf{X} = (X_1, \ldots, X_p)^\top$, *with* $\mathrm{P}(\mathbf{X} = \mathbf{0}_p) = 0$, *has a pdf of the form* $h(\|\mathbf{x}\|_q^q)$, *where* $\|\mathbf{x}\|_q^q = \sum_{i=1}^p |x_i|^q$, $-\infty < x_i < \infty$, $i = 1, \ldots, p$, $q > 0$ *and* $h : \mathbb{R}_+ \to \mathbb{R}_+$ *is a measurable function such that* $\int_0^\infty r^{p-1}h(r^q)\mathrm{d}r < \infty$. *Then,* $\mathbf{X}/\|\mathbf{X}\|_q \stackrel{d}{=} \mathbf{U}_q^{(p)}$ *and is independent of* $R = \|\mathbf{X}\|_q$, *which has a pdf related to h given by*

$$g(r) = \frac{2^p\{\Gamma(1/q)\}^p}{q^{p-1}\Gamma(p/q)} r^{p-1}h(r^q), \quad r > 0. \tag{4.46}$$

Proof 46 *The proof is analogous to that developed above for the* L_q*-spherical normal case. In fact, first recall that the transformation* $w_i = x_i/\|\mathbf{x}\|_q$, $i = 1, \ldots, p-1$, *and* $r = \|\mathbf{x}\|_q$ *is one-to-one in each of the regions* $\mathbb{S}_i$, $i = 1,2$, *defined in (4.40) with equal absolute Jacobian (determinant) given by* $(1 - \sum_{i=1}^{p-1} |w_i|^q)^{(1-q)/q}r^{p-1}$. *Now, recall that by assumption the components* $X_1, \ldots, X_p$ *of* $\mathbf{X}$ *have joint pdf* $f_p(x_1, \ldots, x_p) = h(\sum_{i=1}^p |x_i|^q)$, $-\infty < x_i < \infty$, $i = 1, \ldots, p$. *Then, the joint pdf of the new* *variables* $W_i = X_i/\|\mathbf{X}\|_q$, $i = 1,\ldots,p-1$ *and* $R = \|\mathbf{X}\|_q$ *is given by*

$$g(w_1,\ldots,w_{p-1},r) = 2\left(1-\sum_{i=1}^{p-1}|w_i|^q\right)^{(1-q)/q} r^{p-1}h(r^q),\ |w_i|<1,\ \sum_{i=1}^{p-1}|w_i|^q<1,\ r>0. \quad (4.47)$$

From (4.47) it is clear that $W_1,\ldots,W_{p-1}$ *are independent of* R *and therefore that* $\mathbf{X}/\|\mathbf{X}\|_q \perp\!\!\!\perp \|\mathbf{X}\|_q$. *It also follows from (4.47) that the joint pdf of* $W_1 = X_1/\|\mathbf{X}\|_q,\ldots,W_{p-1} = X_{p-1}/\|\mathbf{X}\|_q$ *is the same as in (4.42), which means that* $\mathbf{W} = \mathbf{X}/\|\mathbf{X}\|_q \overset{d}{=} \mathbf{U}_q^{(p)}$. □

Theorem 47 *A random vector* $\mathbf{X} = (X_1,\ldots,X_p)^\top$ *with* $\mathrm{P}(\mathbf{X} = \mathbf{0}_p) = 0$ *has a pdf of the form* $h(\|\mathbf{x}\|_q^q)$, *with* $\|\mathbf{x}\|_q^q = \sum_{i=1}^p |x_i|^q$, $-\infty < x_i < \infty$, $i = 1,\ldots,p$, *if and only if* $\mathbf{X} \overset{d}{=} R_q\mathbf{U}_q^{(p)} \sim \mathcal{L}_q\mathcal{S}_p^+$, *where* $R_q \perp\!\!\!\perp \mathbf{U}_q^{(p)}$, *and* $R_q \overset{d}{=} \|\mathbf{X}\|_q$ *has pdf* $g(\cdot)$ *related to* $h(\cdot)$ *as in (4.46).*

Proof 47 $(\Longleftarrow)$ *Suppose that* $\mathbf{X} \overset{d}{=} R_q\mathbf{U}_q^{(p)}$, *where* $R_q \perp\!\!\!\perp \mathbf{U}_q^{(p)} = (U_1,\ldots,U_{p-1},U_p)^\top$, *and* R_q *has pdf* $g(r)$. *By (4.42), the joint pdf of* $U_1,\ldots,U_{p-1},R_q$ *is then*

$$\begin{aligned} g(u_1,\ldots,u_{p-1},r) =& g(u_1,\ldots,u_{p-1})g(r) \\ =& \left(\frac{q}{2}\right)^{p-1}\frac{\Gamma(p/q)}{\{\Gamma(1/q)\}^p}\left(1-\sum_{i=1}^{p-1}|u_i|^q\right)^{(1-q)/q} g(r), \\ & |u_i|<1,\ i=1,\ldots,p-1,\ \sum_{i=1}^{p-1}|u_i|^q<1,\ r>0. \end{aligned}$$

Now, consider the transformation $x_i = ru_i$, $i = 1,\ldots,p-1$ *and* $|x_p| = r(1-\sum_{i=1}^{p-1}|u_i|^q)^{1/q}$, *that is,* $u_i = x_i/\|\mathbf{x}\|_q$, $i = 1,\ldots,p-1$ *and* $r = \|\mathbf{x}\|_q$, *which is one-to-one in each of the regions* $\mathbb{S}_1$ *and* $\mathbb{S}_2$ *defined in (4.40), with the same absolute Jacobian (determinant) of the transformation given by* $r^{1-p}(1-\sum_{i=1}^{p-1}|u_i|^q)^{-(1-q)/q}$. *Therefore, the pdf of* $\mathbf{X} = (X_1,\ldots,X_p)^\top$ *becomes*

$$\begin{aligned} f_p(x_1,\ldots,x_p) =& \frac{q^{p-1}\Gamma(p/q)}{2^p\{\Gamma(1/q)\}^p}(\|\mathbf{x}\|_q^q)^{(1-p)/q}g\{(\|\mathbf{x}\|_q^q)^{1/q}\} \\ =& h(\|\mathbf{x}\|_q^q) = h\left(\sum_{i=1}^p |x_i|^q\right),\quad -\infty<x_i<\infty,\ i=1,\ldots,p. \end{aligned}$$

$(\Longrightarrow)$ *If* $\mathbf{X}$, *with* $\mathrm{P}(\mathbf{X} = \mathbf{0}_p) = 0$, *has a pdf of the form* $h(\|\mathbf{x}\|_q^q)$, $\mathbf{x} \in \mathbb{R}^p$, *then by Theorem 46 it follows that*

$$\mathbf{X} = \|\mathbf{X}\|_q\frac{\mathbf{X}}{\|\mathbf{X}\|_q} \overset{d}{=} R_q\mathbf{U}_q^{(p)},\quad R_q \perp\!\!\!\perp \mathbf{U}_q^{(p)},$$

where R_q *has the pdf in (4.46).* □

As in the spherical case of $q = 2$, the function h is called the *density generator* of the $\mathcal{L}_q\mathcal{S}_p^+$ distribution, and the pdf g in (4.46) of the L_q-norm radial variable R_q is called the *generating density* of the distribution. As was proved in Gupta and Song (1997), a function $h(\cdot)$ on $(0,\infty)$ is a density generator function of an $\mathcal{L}_q\mathcal{S}_p$ distribution if and only if

$$\int_0^\infty u^{p/q-1}h(u)\mathrm{d}u = \left(\frac{q}{2}\right)^p\frac{\Gamma(p/q)}{\{\Gamma(1/q)\}^p}.$$

In such a case, the pdf of $\mathbf{X} \sim \mathcal{L}_q\mathcal{S}_p^+$ is $h(\|\mathbf{x}\|_q^q)$, $\mathbf{x} \in \mathbb{R}^p$, and therefore we can use the notation $\mathbf{X} \sim \mathcal{L}_q\mathcal{S}_p(h)$. This fact is due to the key relationship

$$\int_{\mathbb{R}^p} h\left(\sum_{i=1}^{p} |x_i|^q\right) \mathrm{d}x_1 \cdots \mathrm{d}x_p = \left(\frac{2}{q}\right)^p \frac{\{\Gamma(1/q)\}^p}{\Gamma(p/q)} \int_0^\infty r^{p/q-1} h(r)\mathrm{d}r.$$

Furthermore, if we define $\mathbf{Y} = \boldsymbol{\xi} + \omega\mathbf{X}$, where $\boldsymbol{\xi} \in \mathbb{R}^p$ is a location vector, $\omega > 0$ is a scalar scale parameter, and $\mathbf{X} \sim \mathcal{L}_q\mathcal{S}_p(h) \equiv \mathcal{L}_q\mathcal{S}_p(\mathbf{0}_p, \mathbf{I}_p, h)$, then it is immediate that $\mathbf{Y} \sim \mathcal{L}_q\mathcal{S}_p(\boldsymbol{\xi}, \omega^2\mathbf{I}_p, h)$ with pdf preserving the same L_q-norm symmetry given by

$$f_p(\mathbf{y}; \boldsymbol{\xi}, \omega) = \omega^{-p} h\{\|\omega^{-1}(\mathbf{y} - \boldsymbol{\xi})\|_q^q\}, \quad \mathbf{y} \in \mathbb{R}^p.$$

The extension from the more general affine transformation $\mathbf{Y} = \boldsymbol{\xi} + \mathbf{A}\mathbf{X}$, where $\mathbf{A} \in \mathbb{R}^{p\times p}$ is non-singular, yields the *multivariate L_q-elliptical distributions* with pdf given by

$$f_p(\mathbf{y}; \boldsymbol{\xi}, \mathbf{A}) = |\mathbf{A}|^{-1} h\{\|\mathbf{A}^{-1}(\mathbf{y} - \boldsymbol{\xi})\|_q^q\}, \quad \mathbf{y} \in \mathbb{R}^p,$$

which we denote as $\mathbf{Y} \sim \mathcal{L}_q\mathcal{E}l_p(\boldsymbol{\xi}, \boldsymbol{\Omega}, h)$, where $\boldsymbol{\Omega} = \mathbf{A}\mathbf{A}^\top$. For more details and other features of this approach, see Osiewalski and Steel (1993).

4.5.3 *Marginal and conditional pdf*

By Theorem 45, the marginal and conditional distributions of L_q-spherical distributions also belong to the same L_q-spherical class. Furthermore, even if the distribution $\mathcal{L}_q\mathcal{S}_p$ does not have a pdf, all its marginals do have a pdf, as established in the following theorem.

Theorem 48 *If $\mathbf{X} = (X_1, \ldots, X_p)^\top \sim \mathcal{L}_q\mathcal{S}_p^+$, then for each $k = 1, \ldots, p-1$, $\mathbf{X}_1 = (X_1, \ldots, X_k)^\top \sim \mathcal{L}_q\mathcal{S}_k^+$ with pdf*

$$f_k(x_1, \ldots, x_k) = \frac{q^k \Gamma(p/q)}{2^k \{\Gamma(1/q)\}^k \Gamma\{(p-k)/q\}} \int_{(\sum_{i=1}^k |x_i|^q)^{1/q}}^\infty \frac{(r^q - \sum_{i=1}^k |x_i|^q)^{(p-k)/q-1}}{r^{p-q}} \mathrm{d}G(r),$$
$$-\infty < x_i < \infty, \quad i = 1, \ldots, k, \tag{4.48}$$

where G is the cdf of the L_q-norm radial (or generating) variable $R_q \stackrel{d}{=} \|\mathbf{X}\|_q$. That is, $\mathbf{X}_1 = (X_1, \ldots, X_k)^\top \sim \mathcal{L}_q\mathcal{S}_k(h^{(k)})$, where the marginal density generator $h^{(k)}$ is given by

$$h^{(k)}(u) = \frac{q^k \Gamma(p/q)}{2^k \{\Gamma(1/q)\}^k \Gamma\{(p-k)/q\}} \int_{u^{1/q}}^\infty \frac{(r^q - u)^{(p-k)/q-1}}{r^{p-q}} \mathrm{d}G(r), \quad u \geq 0. \tag{4.49}$$

Moreover, the generating pdf, i.e., the pdf of $R_{q1} \stackrel{d}{=} \|\mathbf{X}_1\|_q = (\sum_{i=1}^k |X_i|^q)^{1/q}$ is given by

$$g_k(r) = \frac{2^k \{\Gamma(1/q)\}^k}{q^{k-1} \Gamma(k/q)} r^{k-1} h^{(k)}(r^q), \quad r > 0. \tag{4.50}$$

Proof 48 *Since $\mathbf{X} \stackrel{d}{=} R_q \mathbf{U}_q^{(p)} \sim \mathcal{L}_q\mathcal{S}_p^+$, then by Theorem 45 $\mathbf{X}_1 \stackrel{d}{=} R_q \mathbf{U}_{q1} \sim \mathcal{L}_q\mathcal{S}_k^+$, where $\mathbf{U}_{q1} = (U_1, \ldots, U_k)^\top \perp\!\!\!\perp R_q$ has pdf given by (4.43). Also, $\mathrm{P}(R_q = 0) = \mathrm{P}(\|\mathbf{X}\|_q = 0) = \mathrm{P}(\mathbf{X} = \mathbf{0}_p) = 0$, and given $R_q = r$, $\mathbf{X}_1 \stackrel{d}{=} r\mathbf{U}_{q1}$. So, under the conditional distribution of $\mathbf{X}_1$ given $R_q = r$, we can consider the one-to-one transformation $x_i = ru_i$, i.e., $u_i = x_i/r$, $i = 1, \ldots, k$, with Jacobian matrix $r^{-k}\mathbf{I}_k$,* *which must be such that $|x_i| < r$ for $i = 1,\ldots,k$ and $\sum_{i=1}^k |x_i|^q < r^q$ since $|u_i| < 1$ for $i = 1,\ldots,k$ and $\sum_{i=1}^k |u_i|^q < 1$. Thus, by (4.43) we have*

$$\begin{aligned}
f_k(x_1,\ldots,x_k) &= \int_{\{r\in\mathbb{R}_+ : r^q > \sum_{i=1}^k |x_i|^q\}} g_k(x_1/r,\ldots,x_k/r) r^{-k} \mathrm{d}G(r) \\
&= \int_{(\sum_{i=1}^k |x_i|^q)^{1/q}}^{\infty} \frac{q^k \Gamma(p/q)}{2^k \{\Gamma(1/q)\}^k \Gamma\{(p-k)/q\}} \left(1 - \sum_{i=1}^k \frac{|x_i|^q}{r^q}\right)^{\frac{p-k}{q}-1} r^{-k} \mathrm{d}G(r) \\
&= \frac{q^k \Gamma(p/q)}{2^k \{\Gamma(1/q)\}^k \Gamma\{(p-k)/q\}} \int_{(\sum_{i=1}^k |x_i|^q)^{1/q}}^{\infty} \left(r^q - \sum_{i=1}^k |x_i|^q\right)^{\frac{p-k}{q}-1} r^{-(p-q)} \mathrm{d}G(r), \\
&\quad -\infty < x_i < \infty, \; i = 1,\ldots,k,
\end{aligned}$$

which proves (4.48). This means that $\mathbf{X}_1 \sim \mathcal{L}_q\mathcal{S}_k(h_1)$ with marginal density generator function $h_1 = h^{(k)}$ given by (4.49) and by Theorem 47 that $\mathbf{X}_1 \stackrel{d}{=} R_{q1}\mathbf{U}_q^{(k)}$, where $R_{q1} \stackrel{d}{=} \|\mathbf{X}_1\|_q \perp\!\!\!\perp \mathbf{U}_q^{(k)} \stackrel{d}{=} \mathbf{X}_1/\|\mathbf{X}_1\|$ with generating pdf given by (4.50). □

As shown earlier, the conditional distributions from L_q-spherical distributions are also L_q-spherical, i.e., if $\mathbf{X} = (\mathbf{X}_1^\top, \mathbf{X}_2^\top)^\top \sim \mathcal{L}_q\mathcal{S}_p$, with $\mathbf{X}_1 \in \mathbb{R}^{p_1}$ and $\mathbf{X}_2 \in \mathbb{R}^{p_2}$, then $(\mathbf{X}_1 \mid \mathbf{X}_2 = \mathbf{x}_2) \sim \mathcal{L}_q\mathcal{S}_{p_1}$. In particular, if $\mathbf{X} \sim \mathcal{L}_q\mathcal{S}_p(h)$, where $h \equiv h^{(p)}$ is the density generator, i.e., with pdf $f_p(\mathbf{x}) = h(\|\mathbf{x}\|_q^q) = h^{(p_1+p_2)}(\|\mathbf{x}_1\|_q^q + \|\mathbf{x}_2\|_q^q)$, $\mathbf{x} = (\mathbf{x}_1^\top, \mathbf{x}_2^\top)^\top \in \mathbb{R}^p$, then $(\mathbf{X}_1 \mid \mathbf{X}_2 = \mathbf{x}_2) \sim \mathcal{L}_q\mathcal{S}_{p_1}(h_{\|\mathbf{x}_2\|_q^q})$, where $h_v(u) \equiv h_v^{(p_1)}(u) = h^{(p)}(u+v)/h^{(p_2)}(v)$ is the conditional density generator, and therefore with conditional pdf given by

$$f_k(\mathbf{x}_1 \mid \mathbf{x}_2) = h_{\|\mathbf{x}_2\|_q^q}^{(p_1)}(\|\mathbf{x}_1\|_q^q) = \frac{h^{(p_1+p_2)}(\|\mathbf{x}_1\|_q^q + \|\mathbf{x}_2\|_q^q)}{h^{(p_2)}(\|\mathbf{x}_2\|_q^q)},$$

where the marginal density generator $h^{(p_2)}$ is defined as (4.49).

4.5.4 *Moments*

The stochastic representation $\mathbf{X} = R_q\mathbf{U}^{(p)}$ of $\mathbf{X} \sim \mathcal{L}_q\mathcal{S}_p$ makes the calculation of the moments straightforward (provided that they exist) since $R_q \perp\!\!\!\perp \mathbf{U}^{(p)}$ and the required moments of $\mathbf{U}^{(p)}$ can be obtained from the respective moments of the L_q-spherical normal distribution. In particular, by (4.44):

$$\mathrm{E}(\mathbf{X}) = \mathbf{0}_p \text{ if } \mathrm{E}(R_q) < \infty, \quad \mathrm{var}(\mathbf{X}) = \frac{\Gamma(3/q)\Gamma(p/q)}{\Gamma(1/q)\Gamma\{(p+2)/q\}} \mathrm{E}(R_q^2)\mathbf{I}_p \text{ if } \mathrm{E}(R_q^2) < \infty.$$

See Exercises 4.16 to 4.19 for further results.

4.5.5 *Examples of L_q-spherical distributions*

Two examples of L_q-spherical distributions are the L_q-spherical normal and L_q-spherical uniform distributions presented earlier (see Definitions 14 and 15, respectively). We present further examples below, assuming the existence of pdfs. For more examples, see Gupta and Song (1997).

Example 53 (Multivariate L_q-spherical Pearson Type II distributions) *A random vector $\mathbf{X} \in \mathbb{R}^p$ follows a multivariate L_q-spherical Pearson Type II distribution if its density generator is*

$$h(u) = \frac{q^p \Gamma(p/q + \alpha)}{2^p \{\Gamma(1/q)\}^p \Gamma(\alpha)} (1-u)^{\alpha-1}, \quad 0 < u < 1, \; \alpha > 0.$$

Then, the pdf of $\mathbf{X}$ *is*

$$f_p(\mathbf{x}) = \frac{q^p\Gamma(p/q+\alpha)}{2^p\{\Gamma(1/q)\}^p\Gamma(\alpha)}(1-\|\mathbf{x}\|_q^q)^{\alpha-1}, \quad \|\mathbf{x}\|_q^q = \sum_{i=1}^p |x_i|^q < 1, \ \alpha > 0.$$

In this case, $R_q^q \sim \mathcal{B}eta(p/q,\alpha)$ *and therefore* $\mathrm{E}(R_q^k) < \infty$ *for all* $k = 1,2,\ldots$ *So,*

$$\mathrm{E}(\mathbf{X}) = \mathrm{E}(R_q)\mathrm{E}(\mathbf{U}^{(p)}) = \mathbf{0}_p, \quad \mathrm{var}(\mathbf{X}) = \mathrm{E}(R_q^2)\mathrm{E}(\mathbf{U}^{(p)}\mathbf{U}^{(p)\top}) + \frac{\Gamma(3/q)\Gamma(p/q)}{\Gamma(1/q)\Gamma\{(p+2)/q\}}\mathrm{E}(R_q^2)\,\mathbf{I}_p.$$

Example 54 (Multivariate L_q-spherical t distributions) *A random vector* $\mathbf{X} \in \mathbb{R}^p$ *follows a multivariate L_q-spherical t distribution if its density generator is*

$$h(u) = \frac{(q/2)^p\Gamma\{(\nu+p)/2\}}{\Gamma(\nu/q)\{\Gamma(1/q)\}^p\nu^{p/q}}(1+u/\nu)^{-(\nu+p)/2}, \quad u>0, \ \nu>0.$$

Then, the pdf of $\mathbf{X}$ *is*

$$f_p(\mathbf{x}) = \frac{(q/2)^p\Gamma\{(\nu+p)/2\}}{\Gamma(\nu/q)\{\Gamma(1/q)\}^p\nu^{p/q}}(1+\|\mathbf{x}\|_q^q/\nu)^{-(\nu+p)/2}, \quad \mathbf{x}\in\mathbb{R}^p, \ \nu>0.$$

In this case, the generating variable $R_q \overset{d}{=} V^{-1/q}\bar{R}_q$*, where* $V \perp\!\!\!\perp \bar{R}_q$*, with* $V \sim \mathcal{G}amma(\nu/q,\nu/q)$ *and* $\bar{R}_q^q \sim \mathcal{G}amma(p/q,1/q)$*, thus implying that*

$$\mathrm{E}(R_q^k) = \mathrm{E}(V^{-k/q})\mathrm{E}(\bar{R}_q^k) = \frac{\nu^{k/q}\Gamma\{(\nu-k)/q\}\Gamma\{(p+k)/q\}}{\Gamma(\nu/q)\Gamma(p/q)}, \quad \nu>k.$$

So, we have

$$\mathrm{E}(\mathbf{X}) = \mathbf{0}_p \ \ (\nu>1), \quad \mathrm{var}(\mathbf{X}) = \frac{\nu^{2/q}\Gamma\{(\nu-2)/q\}\Gamma\{(p+2)/q\}}{\Gamma(\nu/q)\Gamma(p/q)}\mathbf{I}_p \ \ (\nu>2).$$

For further extensions of the L_q-spherical family, see Hashorva et al. (2007), who introduced the L_q-norm generalized symmetrized Dirichlet distributions, and Arellano-Valle and Richter (2012), who studied a class of skewed continuous L_q-symmetric distributions.

4.5.6 Meta-elliptical distributions

This family was introduced by Fang et al. (2002). They assumed (without loss of generality) that $\mathbf{Z} \sim \mathcal{E}l_p(\mathbf{0}_p, \bar{\mathbf{\Omega}}, h)$, where $\bar{\mathbf{\Omega}}$ is a proper correlation matrix. In this case, all the (univariate) marginal distributions of $\mathbf{Z}$ are identical with pdf

$$f_h(z) = \frac{\pi^{(p-1)/2}}{\Gamma\{(p-1)/2\}}\int_{z^2}^{\infty}(u-z^2)^{(p-1)/2}h(u)\mathrm{d}u, \quad z\in\mathbb{R},$$

and cdf

$$F_h(z) = \frac{1}{2} + \frac{\pi^{(p-1)/2}}{\Gamma\{(p-1)/2\}}\int_0^z\int_{v^2}^{\infty}(u-v^2)^{(p-1)/2}h(u)\mathrm{d}u\mathrm{d}v, \quad z\in\mathbb{R}.$$

Let $\mathbf{X} = (X_1, \ldots, X_p)^\top$ be a random vector with each component X_i having a given (but arbitrary) continuous pdf g_i and cdf G_i. Note that $G_i(X_i) \sim \mathcal{U}(0,1)$ for $i = 1, \ldots, p$. Consider the one-to-one transformation given by

$$Z_i = F_h^{-1}\{G_i(X_i)\}, \quad i = 1, \ldots, p.$$

The Jacobian (determinant) of the transformation is

$$J_{(z_1,\ldots,z_p)\to(x_1,\ldots,x_p)} = \prod_{i=1}^{p} \frac{g_i(x_i)}{f_h[F_h^{-1}\{G_i(x_i)\}]}.$$

Therefore, the pdf of $\mathbf{X}$ becomes

$$g_p(x_1, \ldots, x_p) = \pi[F_h^{-1}\{G_1(x_1)\}, \ldots, F_h^{-1}\{G_p(x_p)\}] \prod_{i=1}^{p} g_i(x_i), \tag{4.51}$$

where

$$\pi(z_1, \ldots, z_p) = \frac{|\bar{\mathbf{\Omega}}|^{-1/2} h(\mathbf{z}^\top \bar{\mathbf{\Omega}}^{-1} \mathbf{z})}{\prod_{i=1}^{p} f_h(z_i)}, \quad \mathbf{z} = (z_1, \ldots, z_p)^\top \in \mathbb{R}^p.$$

In this case, a p-dimensional random vector $\mathbf{X}$ with pdf (4.51) is said to have a *multivariate meta-elliptical distribution*, with *density weighting function* $\pi(z_1, \ldots, z_p)$, and denoted by $\mathbf{X} \sim \mathcal{MEl}_p(\mathbf{0}_p, \bar{\mathbf{\Omega}}, h; G_1, \ldots, G_p)$.

The meta-elliptical family includes various symmetric distributions, such as the elliptical distributions themselves, if we set $G_i = F_h$ and $g_i = f_h$ for all $i = 1, \ldots, p$. It also includes asymmetric multivariate elliptical distributions, such as the multivariate asymmetric t-distributions obtained when $F_h = T_\nu$ and $f_h = t_\nu$, where T_ν and t_ν denote the Student's t cdf and pdf, respectively, with $\nu > 0$ degrees of freedom. In this case, if $G_i = T_{\nu_i}$ and $g_i = t_{\nu_i}$ for $i = 1, \ldots, p$, then the joint pdf of $X_1, \ldots, X_p$ is

$$g_p(x_1, \ldots, x_p) = \pi\{T_\nu^{-1}(T_{\nu_1}), \ldots, T_\nu^{-1}(T_{\nu_p})\} \prod_{i=1}^{p} t_{\nu_i}(x_i),$$

where

$$\pi(z_1, \ldots, z_p) = \frac{\Gamma\{(\nu+p)/2\}\{\Gamma(\nu/2)\}^{p-1}}{[\Gamma\{(\nu+1)/2\}]^p |\bar{\mathbf{\Omega}}|^{1/2}} \left(1 + \frac{\mathbf{z}^\top \bar{\mathbf{\Omega}}^{-1} \mathbf{z}}{\nu}\right)^{-(\nu+p)/2} \prod_{i=1}^{p} \left(1 + \frac{z_i^2}{\nu}\right)^{(\nu+1)/2},$$

with $\mathbf{z} = (z_1, \ldots, z_p)^\top$ and $\bar{\mathbf{\Omega}}$ being a p-by-p correlation matrix as established before. In this case, we denote the distribution of $\mathbf{X} = (X_1, \ldots, X_p)^\top$ by $\mathbf{X} \sim \mathcal{MT}_p(\mathbf{0}_p, \bar{\mathbf{\Omega}}, \nu; \nu_1, \ldots, \nu_p)$, the *multivariate meta-elliptical t distribution*, that has Student's t marginals with different degrees of freedom. Also, if $\nu_i = \nu$ for all $i = 1, \ldots, p$, then the distribution of $\mathbf{X}$ becomes the p-variate $\mathcal{T}_p(\mathbf{0}_p, \bar{\mathbf{\Omega}}, \nu)$. Another interesting fact is that $\mathcal{MT}_p(\mathbf{0}_p, \bar{\mathbf{\Omega}}, \nu; \nu_1, \ldots, \nu_p)$ and $\mathcal{T}_p(\mathbf{0}_p, \bar{\mathbf{\Omega}}, \nu)$ have the same copula (dependence structure; see Equation (10.3)).

4.5.7 Dispersion elliptical squared-radial distributions

Arellano-Valle et al. (2006b) introduced a general class of distributions which can be viewed as an extension of the well-known normal-gamma (or normal-gamma inverse) family used in the conjugate Bayesian analysis of the normal linear model. Given a random vector $\mathbf{X} \in \mathbb{R}^p$, a positive random variable T and a density generator function $h = h^{(p+q)}$ such that

$$\mathbf{X} \mid T = t \sim \mathcal{El}_p\left(\boldsymbol{\xi}, \frac{1}{bt^r}\mathbf{\Omega}, h_{at}\right) \quad \text{and} \quad T \sim \frac{1}{a}\mathcal{SR}_q(h),$$

where $a,b,r>0$ and $h_u = h_u^{(q)}$ is a conditional dgf (see [4] in Section 4.2.5), i.e., $h_u^{(q)}(v) = h^{(p+q)}(u+v)/h^{(p)}(u)$, $u,v>0$, then $(\mathbf{X},T)$ is said to have a *multivariate dispersion elliptical squared-radial (DESR) distribution* and is denoted by $\mathcal{DEl}_p\mathcal{SR}_q(\boldsymbol{\xi},\boldsymbol{\Omega},r,a,b,h)$. By ellipticity, the pdf of $(\mathbf{X},T)$ depends on $\mathbf{x}$ through $Q(\mathbf{x}) = (\mathbf{x}-\boldsymbol{\xi})^\top\boldsymbol{\Omega}^{-1}(\mathbf{x}-\boldsymbol{\xi})$; thus, straightforward computations lead to

$$f(\mathbf{x},t) = \frac{(a\pi)^{q/2}b^{p/2}}{\Gamma(q/2)}|\boldsymbol{\Omega}|^{-1/2}t^{(rp+q)/2-1}h\{bt^rQ(\mathbf{x})+at\}, \quad \mathbf{x}\in\mathbb{R}^p,\ t>0.$$

Some special cases and/or basic properties of this family are:

[1] $\frac{1}{b}$ is a scale parameter since $\mathcal{DEl}_p\mathcal{SR}_q(\boldsymbol{\xi},\boldsymbol{\Omega},r,a,b,h) = \mathcal{DEl}_p\mathcal{SR}_q(\boldsymbol{\xi},\frac{1}{b}\boldsymbol{\Omega},r,a,1,h)$.

[2] The parameter r allows the conditional variance of $\mathbf{X}\mid T=t$ to depend on t; for $r=0$ and $b=1$, the dispersion is fully determined by $\boldsymbol{\Omega}$.

[3] The *multivariate dispersion spherical squared-radial (DSSR) distribution*, $\mathcal{DS}_p\mathcal{SR}_q(r,a,b,h)$ is just the special case $\mathcal{DEl}_p\mathcal{SR}_q(\mathbf{0}_p,\mathbf{I}_p,r,a,b,h)$. On the other hand, starting from $(\mathbf{Z},T)$, where $\mathbf{Z}=\boldsymbol{\Omega}^{-1/2}(\mathbf{X}-\boldsymbol{\xi})$, we find

$$(\mathbf{X},T)\sim\mathcal{DEl}_p\mathcal{SR}_q(\boldsymbol{\xi},\boldsymbol{\Omega},r,a,b,h)\Longleftrightarrow(\mathbf{Z},T)\sim\mathcal{DS}_p\mathcal{SR}_q(r,a,b,h).$$

[4] The *multivariate spherical squared-radial (SSR) distribution*, denoted by $\mathcal{S}_p\mathcal{SR}_q(h)$, is the particular case $\mathcal{DS}_p\mathcal{SR}_q(0,1,1,h)$, which, directly from the definition, means

$$(\mathbf{Z},T)\sim\mathcal{S}_p\mathcal{SR}_q(h)\Longleftrightarrow\mathbf{Z}\mid T=t\sim\mathcal{S}_p(h) \quad\text{and}\quad T\sim\mathcal{SR}_q(h).$$

When the conditional distributions of $\mathbf{Z}\mid T=t$ are all spherical, they are determined by those of $\|\mathbf{Z}\|^2\mid T=t$, which in turn are determined by the distribution of $(\mathbf{Z},T)$. Therefore, the SSR and SR families are linked by

$$(\mathbf{Z},T)\sim\mathcal{S}_p\mathcal{SR}_q(h)\Longleftrightarrow(\|\mathbf{Z}\|^2,T)\sim\mathcal{SR}_{p+q}(h).$$

Additional properties of the general DESR family are the following.

Theorem 49 *If* $(\mathbf{X},T)\sim\mathcal{DEl}_p\mathcal{SR}_q(\boldsymbol{\xi},\boldsymbol{\Omega},r,a,b,h)$ *and* $Q(\mathbf{X}) = (\mathbf{X}-\boldsymbol{\xi})^\top\boldsymbol{\Omega}^{-1}(\mathbf{X}-\boldsymbol{\xi})$, *then* $(bT^rQ(\mathbf{X}),aT)\sim\mathcal{SR}_{p+q}(h)$. *Moreover,*

(i) $T^rQ(\mathbf{X})\sim\frac{1}{b}\mathcal{SR}_p(h)$, $\quad T\sim\frac{1}{a}\mathcal{SR}_q(h)$, $\quad Q(\mathbf{X})\mid T=t\sim\frac{1}{bt^r}\mathcal{SR}_p(h_{at})$;

(ii) $T^{r-1}Q(\mathbf{X})\sim\frac{ap}{bq}\mathcal{F}_{p,q}$;

(iii) For $r=1$, $\mathbf{X}\sim\mathcal{T}_p(\boldsymbol{\xi},\frac{a}{b}\boldsymbol{\Omega},q)$, *which does not depend on* h;

(iv) For $r=0$, $\mathbf{X}\sim\mathcal{El}_p(\boldsymbol{\xi},\frac{1}{b}\boldsymbol{\Omega},h)$ *and* $T\mid\mathbf{X}=\mathbf{x}\sim\frac{1}{a}\mathcal{SR}_q(h_{Q(\mathbf{x})})$.

Proof 49 *The proof of these results is obtained by applying the properties [SR1]-[SR5] given in Section 4.4.2.* □

Example 55 (Multivariate normal-gamma distributions) *The multivariate normal-gamma is a well-known family of distributions in the Bayesian literature. Now, it can be obtained as a special case of the DESR family by letting* $b=r=1$ *and taking the* $(p+q)$*-variate normal generator* $h(u)=(2\pi)^{-(p+q)/2}e^{-u/2}$. *In this case,*

$$\mathbf{X}\mid T=t\sim\mathcal{N}_p\left(\boldsymbol{\xi},\frac{1}{t}\boldsymbol{\Omega}\right) \quad\text{and}\quad T\sim\mathcal{G}amma(q/2,a/2),$$

since in the normal case, any conditional dgf h_u does not depend on u, i.e., $h_u^{(q)} = h^{(q)}$. Also, the same case but with $r = 0$ yields independent normal and gamma distributions for $\mathbf{X}$ *and T, respectively. This is the only situation within the DESR family where we have independence for any value of a.*

Example 56 *The DESR family induced by the $(p+q)$-variate t density generator given by $h(u) = C(p+q,\nu)(1+u/\nu)^{-(\nu+p+q)/2}$, where $c(k,\nu) = \Gamma\{(\nu+k)/2\}/\Gamma(\nu/2)(\pi\nu)^{k/2}$, is defined by*

$$\mathbf{X} \mid T = t \sim \mathcal{T}_p\left(\xi, \frac{\nu+at}{bt^r}\boldsymbol{\Omega}, \nu+q\right) \quad \text{and} \quad T \sim \frac{q}{a}\mathcal{F}_{p,q}.$$

In this case, we know that (see Example 46) if $h^{(n)}(u)$, $u > 0$, is the dgf of n-variate t distribution, then the k-variate marginal and conditional dgfs are such that $h^{(k)}(u) = h^{(n)}(u)\big|_{n=k}$ and $h_w^{(k)}(u) = \left(\frac{\nu+k}{\nu+u}\right)^{k/2} h^{(k)}\left\{\left(\frac{\nu+k}{\nu+u}\right)u\right\}$, respectively.

An extension and/or variation of the previous idea is to partition a spherical vector $\mathbf{X} \sim \mathcal{S}_p(h)$ into $m+1$ subvectors $\mathbf{X}_0, \mathbf{X}_1, \ldots, \mathbf{X}_m$ of lengths $p_0, p_1, \ldots, p_m$, respectively, with $\sum_{i=0}^m p_i = p$. Then study, for example, the joint distribution of the transformations $Y_0 = \|\mathbf{X}_0\|$ and $\mathbf{Y}_i = \mathbf{A}_i\mathbf{X}_i$, $i = 1,\ldots,m$, where $\mathbf{A}_i$ is a fixed nonsingular matrix of dimension $p_i \times p_i$ for each $i = 1,\ldots,m$, from which we can also study, e.g., the joint distribution of $\mathbf{Z}_i = g_i(Y_0)\mathbf{Y}_i$, $i = 1,\ldots,m$, for some functions $g_i(u)$, $u > 0$, $i = 1,\ldots,m$; see Exercise 4.5.

For further results on elliptical models in Bayesian analysis, see Arellano-Valle et al. (2000).

4.6 Exercises

Exercise 4.1 *Let $\varphi_{\mathcal{U},p}(\|\mathbf{t}\|^2)$, $\mathbf{t} \in \mathbb{R}^p$ be the cf of $\mathbf{U}^{(p)} \sim \mathcal{U}(\mathbb{S}^{(p-1)})$. Show that $\varphi_{\mathcal{U},p}(\|\mathbf{t}\|^2)$ can be expressed in the following equivalent forms:*

$$\varphi_{\mathcal{U},p}(\|\mathbf{t}\|^2) = \frac{\Gamma(p/2)}{\sqrt{\pi}\,\Gamma\{(p-1)/2\}}\sum_{k=0}^{\infty}\frac{(-1)^k\|\mathbf{t}\|^{2k}}{(2k)!}\frac{\Gamma\{(p-1)/2\}\Gamma\{(2k+1)/2\}}{\Gamma\{(p+2k)/2\}},$$

$$\varphi_{\mathcal{U},p}(\|\mathbf{t}\|^2) = \sum_{k=0}^{\infty}\frac{(\|\mathbf{t}\|^2/4)^k}{(p/2)_k k!} \equiv {}_0F_1(p/2; \|\mathbf{t}\|^2/4),$$

where ${}_0F_1(a;z)$ is the so-called generalized hypergeometric function defined by

$${}_0F_1(a;z) = \sum_{k=0}^{\infty}\frac{z^k}{(a)_k k!},$$

with $(a)_k$, the ascending factorial or Pochammer polynomial, defined by

$$(a)_k = \frac{\Gamma(a+k)}{\Gamma(a)} = a(a+1)\cdots(a+k-1).$$

Exercise 4.2 *Let f be a nonnegative function such that $f(\mathbf{x}^\top\mathbf{x})$ is a pdf in $\mathbb{R}^p$ with variance matrix $\mathbf{I}_p$. Show that f satisfies the following normalizing conditions:*

$$\int_0^\infty r^{p/2-1}f(r)\mathrm{d}r = \Gamma(p/2)/\pi^{p/2}, \qquad \int_0^\infty r^{p/2}f(r)\mathrm{d}r = 2\Gamma(p/2+1)/\pi^{p/2}.$$

Exercise 4.3 *Let $\mathbf{X} = (\mathbf{X}_1^\top, \mathbf{X}_2^\top)^\top \in \mathbb{R}^p$ be distributed according to a uniform distribution in the unit sphere in $\mathbb{R}^p$. Compute the conditional distribution of $\mathbf{X}_1|\mathbf{X}_2 = \mathbf{x}_2$.*

Exercise 4.4 *Let $\mathbf{X} = (X_1, X_2, X_3)^\top \in \mathbb{R}^3$ be distributed according to a uniform distribution in the ellipsoid in $\mathbb{R}^3$, centered at the origin with principal axes of half-length equal to 1, 2, 3, respectively, and parallel to the coordinate axes. Compute the conditional distribution of $X_1|X_2 = x_2, X_3 = x_3$.*

Exercise 4.5 *Find the distribution of $\mathbf{Y}_1 = (V + \|\mathbf{X}_1\|^2)^{-1/2}\mathbf{X}_1$, where $V \sim \chi^2_\nu$ and $\mathbf{X}_1 \sim \mathcal{N}_p(\mathbf{0}_p, \mathbf{I}_p)$ are independent. Also, find the distribution of $\mathbf{Y}_2 = (\nu + \|\mathbf{X}_2\|^2)^{-1/2}\mathbf{X}_2$, where $\mathbf{X}_2 \sim \mathcal{T}_p(\mathbf{0}_p, \mathbf{I}_p, \nu)$. Is it true that $\mathbf{Y}_1$ and $\mathbf{Y}_2$ have the same distribution? What is that distribution?*

Exercise 4.6 **(Multivariate normal scale mixtures distribution)** *Let $\mathbf{X} = \boldsymbol{\xi} + V^{-1/2}\mathbf{Z}$, where $V \sim G$ is independent of $\mathbf{Z} \sim \mathcal{N}_p(\mathbf{0}_p, \boldsymbol{\Omega})$, G is the mixing distribution on $[0,\infty)$, and $\boldsymbol{\Omega} > 0$.*

(a) Verify that the elliptical pdf of $\mathbf{X}$ is given by

$$f(\mathbf{x}; \boldsymbol{\xi}, \boldsymbol{\Omega}, G) = \int_0^\infty \phi_p(\mathbf{x}; \boldsymbol{\xi}, v^{-1}\boldsymbol{\Omega})\mathrm{d}G(v), \quad \mathbf{x} \in \mathbb{R}^p,$$

where $\phi_p(\mathbf{x}; \boldsymbol{\mu}, \boldsymbol{\Sigma})$ is the pdf of $\mathcal{N}_p(\boldsymbol{\mu}, \boldsymbol{\Sigma})$.

(b) Let $R = \{(\mathbf{X} - \boldsymbol{\xi})^\top\boldsymbol{\Omega}^{-1}(\mathbf{X} - \boldsymbol{\xi})\}^{1/2}$ be the corresponding radial variable. Verify that $R = V^{-1/2}R_0$, where $R_0 = (\mathbf{Z}^\top\boldsymbol{\Omega}^{-1}\mathbf{Z})^{1/2} \sim \chi_p$ and is independent of V. Give an expression for the pdf of R.

(c) Partition $\mathbf{X} = (\mathbf{X}_1^\top, \mathbf{X}_2^\top)^\top$ with $\mathbf{X}_i \in \mathbb{R}^{p_i}$, $i = 1, 2$, $p_1 + p_2 = p$ and partition $\boldsymbol{\xi}$ and $\boldsymbol{\Omega}$ in accordance with $\boldsymbol{\xi} = (\boldsymbol{\xi}_1^\top, \boldsymbol{\xi}_2^\top)^\top$ and

$$\boldsymbol{\Omega} = \begin{pmatrix} \boldsymbol{\Omega}_{11} & \boldsymbol{\Omega}_{12} \\ \boldsymbol{\Omega}_{21} & \boldsymbol{\Omega}_{22} \end{pmatrix}.$$

Find the marginal pdf of $\mathbf{X}_1$ and the conditional pdf of $\mathbf{X}_2 \mid \mathbf{X}_1 = \mathbf{x}_1$.

(d) **(Multivariate *t* distribution)** *Assume $V \sim \chi^2_\nu/\nu$. Verify that*

$$f(\mathbf{x}; \boldsymbol{\xi}, \boldsymbol{\Omega}, \nu) = c_{p,\nu}\{1 + (\mathbf{x} - \boldsymbol{\xi})^\top\boldsymbol{\Omega}^{-1}(\mathbf{x} - \boldsymbol{\xi})/\nu\}^{-(\nu+p)/2}, \quad \mathbf{x} \in \mathbb{R}^p,$$

where $c_{p,\nu} = \Gamma\{(\nu+p)/2\}/\{\Gamma(\nu/2)(\pi\nu)^{p/2}\}$. Also, find the marginal pdf of $\mathbf{X}_1$ and the conditional pdf of $\mathbf{X}_2 \mid \mathbf{X}_1 = \mathbf{x}_1$, and verify that

$$(\mathbf{X} - \boldsymbol{\xi})^\top\boldsymbol{\Omega}^{-1}(\mathbf{X} - \boldsymbol{\xi})/p \sim \mathcal{F}_{p,\nu}.$$

(e) **(Multivariate contaminated normal distribution)** *Assume the two point mixing distribution*

$$\mathrm{P}(V = 1) = \varepsilon,$$
$$\mathrm{P}(V = 1/\eta) = 1 - \varepsilon,$$

for some contamination proportion $0 < \varepsilon \leq 1$ *and contamination parameter* $0 < \eta < 1$. *Find the pdf of* $\mathbf{X}$, *the marginal pdf of* $\mathbf{X}_1$, *the conditional pdf of* $\mathbf{X}_2 \mid \mathbf{X}_1 = \mathbf{x}_1$ *and the squared radial pdf of* $(\mathbf{X}-\boldsymbol{\xi})^\top \boldsymbol{\Omega}^{-1}(\mathbf{X}-\boldsymbol{\xi})$.

Exercise 4.7 (Multivariate exponential power distribution) *Let* $\mathbf{X}$ *be distributed according to a multivariate exponential power distribution with pdf*

$$f(\mathbf{x};\boldsymbol{\xi},\boldsymbol{\Omega},\nu) = c_{p,\nu}|\boldsymbol{\Omega}|^{-1/2}\exp\left[-\frac{1}{2}\{(\mathbf{x}-\boldsymbol{\xi})^\top\boldsymbol{\Omega}^{-1}(\mathbf{x}-\boldsymbol{\xi})\}^\nu\right], \quad \mathbf{x}\in\mathbb{R}^p, \tag{4.52}$$

where $\nu > 0$.

(a) Prove that the normalizing constant is

$$c_{p,\nu} = \frac{\nu\Gamma(\nu/2)}{2^{p/(2\nu)}\pi^{p/2}\Gamma(p/2\nu)}.$$

(b) In the partition $\mathbf{X} = (\mathbf{X}_1^\top,\mathbf{X}_2^\top)^\top \in \mathbb{R}^{p_1+p_2}$, *find the marginal pdf of* $\mathbf{X}_1$ *and the conditional pdf of* $\mathbf{X}_2 \mid \mathbf{X}_1 = \mathbf{x}_1$.

(c) For $p_1 = 1$ *compare the marginal pdf of* X_1 *with the corresponding univariate pdf obtained from (4.52). In particular, taking* $\nu = 1/2$ *show that*

$$\mathrm{E}\{(X_1-\xi_1)^2\} = 4(p+1)\Omega_{11},$$

which depends on p, while the corresponding quantity in the univariate version of (4.52) with $\Omega = \Omega_{11}$ *is*

$$\mathrm{E}\{(X-\xi)^2\} = 8\Omega_{11}.$$

Conclude that this distribution cannot satisfy the consistency properties defined in Kano (1994): An elliptical family is consistent iff it is a normal mixtures family.

(d) Show that the variance matrix of $\mathbf{X}$ *is*

$$\mathrm{var}(\mathbf{X}) = \frac{2^{1/\nu}\Gamma\{(p+2)/2\nu\}}{p\Gamma(p/2\nu)}\boldsymbol{\Omega}.$$

Exercise 4.8 *Let* $\mathbf{X}_1$ *and* $\mathbf{X}_2$ *be two* $p\times 1$ *random vectors. Find the cf of* $\mathbf{X}_1+\mathbf{X}_2$ *under the following assumptions:*

(a) $\mathbf{X}_i \overset{d}{=} \boldsymbol{\xi}_i + R_i\mathbf{A}\mathbf{U}_i^{(p)}$, *with* $\boldsymbol{\Omega} = \mathbf{A}\mathbf{A}^\top$, $i = 1,2$, *where* (R_1,R_2), $\mathbf{U}_1^{(p)}$ *and* $\mathbf{U}_2^{(p)}$ *are mutually independent, whereas the generating variables* R_1 *and* R_2 *may depend on each other. Is* $\mathbf{X}_1+\mathbf{X}_2$ *elliptically distributed with location vector* $\boldsymbol{\xi}_1+\boldsymbol{\xi}_2$ *and dispersion matrix* $\boldsymbol{\Omega}$ *in this case?*

(b) $\mathbf{X}_1,\mathbf{X}_2 \overset{iid}{\sim} \mathcal{E}l_p(\boldsymbol{\xi},\boldsymbol{\Omega},\varphi)$. *Is the distribution of* $\mathbf{X}_1+\mathbf{X}_2$ *elliptical in this case?*

Exercise 4.9 *Let* $\mathbf{X} \sim \mathcal{E}l_p(\boldsymbol{\xi}, \boldsymbol{\Omega}, \varphi)$ *be nondegenerate. If the following moments exist, then prove:*

$$\begin{aligned}
&(i)\ \boldsymbol{\mu}_1(\mathbf{X}) = \boldsymbol{\xi};\\
&(ii)\ \boldsymbol{\mu}_2(\mathbf{X}) = \boldsymbol{\xi}\boldsymbol{\xi}^\top + \kappa_2\boldsymbol{\Omega};\\
&(iii)\ \boldsymbol{\mu}_3(\mathbf{X}) = \mathrm{vec}(\boldsymbol{\xi}\boldsymbol{\xi}^\top)\boldsymbol{\xi}^\top + \kappa_2\{\boldsymbol{\xi} \otimes \boldsymbol{\Omega} + \boldsymbol{\Omega} \otimes \boldsymbol{\xi} + \mathrm{vec}(\boldsymbol{\Omega})\boldsymbol{\xi}^\top\};\\
&(iv)\ \boldsymbol{\mu}_4(\mathbf{X}) = \mathrm{vec}(\boldsymbol{\xi}\boldsymbol{\xi}^\top + \kappa_2\boldsymbol{\Omega})\mathrm{vec}(\boldsymbol{\xi}\boldsymbol{\xi}^\top + \kappa_2\boldsymbol{\Omega})^\top + \mathrm{var}\{\mathrm{vec}(\mathbf{X}\mathbf{X}^\top)\};
\end{aligned}$$

where $\kappa_2 = -2\varphi'(0)$, $\kappa_4 = 4\varphi''(0)$ *and*

$$\mathrm{var}\{\mathrm{vec}(\mathbf{X}\mathbf{X}^\top)\} = 2\kappa_4\mathbf{N}_p(\boldsymbol{\Omega} \otimes \boldsymbol{\Omega}) + 4\kappa_2\mathbf{N}_p(\boldsymbol{\Omega} \otimes \boldsymbol{\xi}\boldsymbol{\xi}^\top)\mathbf{N}_p + (\kappa_4 - \kappa_2^2)\mathrm{vec}(\boldsymbol{\Omega})\mathrm{vec}(\boldsymbol{\Omega})^\top,$$

with $\mathbf{N}_p = \frac{1}{2}(\mathbf{I}_{p^2} + \mathbf{K}_p)$ *being the symmetrizer matrix (see Equation (A.28) for more details). Hint: Using* $\mathbf{X} = \boldsymbol{\Omega}^{1/2}\mathbf{Z} + \boldsymbol{\xi}$, *where* $\mathbf{Z} \sim \mathcal{S}_p(\varphi)$, *then* $\mathrm{vec}(\mathbf{X}\mathbf{X}^\top) = (\boldsymbol{\Omega}^{1/2} \otimes \boldsymbol{\Omega}^{1/2})\mathrm{vec}(\mathbf{Z}\mathbf{Z}^\top) + 2\mathbf{N}_p(\boldsymbol{\Omega}^{1/2} \otimes \boldsymbol{\xi}^\top)\mathbf{Z} + \mathrm{vec}(\boldsymbol{\xi}\boldsymbol{\xi}^\top)$.

Exercise 4.10 *Let* $\mathbf{X} \sim \mathcal{E}l_p(\boldsymbol{\xi}, \boldsymbol{\Omega}, \varphi)$ *be nondegenerate. Compute the quantities* $\kappa_2 = -2\varphi'(0)$ *and* $\kappa_4 = 4\varphi''(0)$ *for:*

(a) The multivariate normal distribution;

(b) The multivariate t distribution;

(c) The multivariate generalized t distribution;

(d) The multivariate exponential power distribution;

(e) The multivariate contaminated normal distribution;

(f) The multivariate normal scale mixtures distribution.

Then provide explicit expressions for $\boldsymbol{\mu}_i(\mathbf{X})$, $i = 1, 2, 3, 4$, *the mean vector, the variance matrix, and the Mardia measures of multivariate skewness and kurtosis of each of the aforementioned distributions.*

Exercise 4.11 *Show that* $\mathrm{E}\{\mathbf{U}^{(p)}(\mathbf{U}^{(p)})^\top\} = p^{-1}\mathbf{I}_p$ *and*

$$\mathrm{var}[\mathrm{vec}\{\mathbf{U}^{(p)}(\mathbf{U}^{(p)})^\top\}] = \frac{2}{p(p+2)}\mathbf{N}_p - \frac{4}{p^2(p+2)}\mathrm{vec}(\mathbf{I}_p)\mathrm{vec}(\mathbf{I}_p)^\top,$$

where $\mathbf{U}^{(p)} \sim \mathcal{U}(\mathbb{S}^{(p-1)})$ *and* $\mathbf{N}_p = \frac{1}{2}(\mathbf{I}_{p^2} + \mathbf{K}_p)$.

Exercise 4.12 *Let* $\varphi(t_1^2 + t_2^2)$ *be the cf of a bivariate rotationally invariant random vector. Verify by differentiation of* $\log\varphi(t_1^2 + t_2^2)$ *that the cumulants satisfy:*

$$\begin{aligned}
\kappa_{[2,0]} &= \kappa_{[0,2]} = -2\varphi'(0),\\
\kappa_{[4,0]} &= \kappa_{[0,4]} = 12\{\varphi''(0) - \varphi'(0)^2\},\\
\kappa_{[2,2]} &= 4\{\varphi''(0) - \varphi'(0)^2\}.
\end{aligned}$$

Conclude for the marginal kurtosis $\beta_2 = \beta_{2,1}$ *that*

$$\beta_2 = \frac{\kappa_{[4,0]}}{\kappa_{[2,0]}^2} = \frac{\kappa_{[0,4]}}{\kappa_{[0,2]}^2} = 3\left(\frac{\varphi''(0)}{\varphi'(0)^2} - 1\right) = 3\kappa,$$

where $\kappa = \varphi''(0)/\varphi'(0)^2 - 1$ *represents the kurtosis parameter, so that* $\kappa_{[4,0]} = \kappa_{[0,4]} = 3\kappa\kappa_2$ *and* $\kappa_{[2,2]} = \kappa\kappa_2$*, where* $\kappa_2 = -2\varphi'(0)$ *is the respective marginal variance. Finally, verify that*

$$\mathrm{var}(\mathbf{Z}\mathbf{Z}^\top) = \kappa_2^2(1+\kappa)(\mathbf{I}_{p^2} + \mathbf{K}_p) + \kappa_2^2\kappa\,\mathrm{vec}(\mathbf{I}_p)\mathrm{vec}(\mathbf{I}_p)^\top,$$

where $\mathbf{Z} \sim \mathcal{S}_p(\varphi)$ *and* $\mathrm{var}(\mathbf{Z}\mathbf{Z}^\top)$ *denotes the variance matrix of* $\mathbf{Z}\otimes\mathbf{Z} = \mathrm{vec}(\mathbf{Z}\mathbf{Z}^\top)$*. Hint: Consider the Taylor series*

$$\log\varphi(t_1^2+t_2^2) = \kappa_{[2,0]}\frac{(it_1)^2}{2!} + \kappa_{[0,2]}\frac{(it_2)^2}{2!} + \kappa_{[4,0]}\frac{(it_1)^4}{4!} + \kappa_{[0,4]}\frac{(it_2)^4}{4!} + \kappa_{[2,2]}\frac{(it_1)^2(it_1)^2}{(2!)^2} + o(\|\mathbf{t}\|^4),$$

where $\mathbf{t} = (t_1, t_2)^\top$.

Exercise 4.13 *In Exercise 4.9, check that* $\kappa_4 = (\kappa+1)\kappa_2^2$ *and indicate the value of* κ *for each of the distributions listed in Exercise 4.10.*

Exercise 4.14 *Within the family of elliptical distributions, each member can be characterized by its marginals. Find counterexamples to show that this is not the case outside the elliptical family.*

Exercise 4.15 *Explore the form of the entropy, mutual information, KL divergence, J divergence, and Wasserstein distance for two multivariate elliptical distributions. Apply these results to multivariate t distributions.*

Exercise 4.16 *If the pdf of X is* $f(x) = \{q^{(q-1)/q}/2\Gamma(1/q)\}e^{-|x|^q/q}$, $-\infty < x < \infty$, $q > 0$, *show that*

$$\mathrm{E}(X^k) = \begin{cases} \frac{q^{k/q}\Gamma\{(k+1)/q\}}{\Gamma(1/q)}, & k \text{ an even integer,} \\ 0, & k \text{ an odd integer.} \end{cases}$$

Exercise 4.17 *If Z is a random variable such that* $Z^q \sim \mathcal{G}amma(p/q, q)$*, show that*

$$\mathrm{E}(Z) = \frac{q^{1/q}\Gamma\{(p+1)/q\}}{\Gamma(p/q)}, \quad \mathrm{E}(Z^2) = \frac{q^{2/q}\Gamma\{(p+2)/q\}}{\Gamma(p/q)}.$$

Exercise 4.18 *Let* $\mathbf{U}_q^{(p)} = (U_1,\ldots,U_p)^\top \in \mathbb{S}_q^{(p-1)} = \{\mathbf{x} = (x_1,\ldots,x_p)^\top \in \mathbb{R}^p : \sum_{i=1}^p |x_i|^q = 1\}$ *be the q-generalized uniform vector. Show that*

$$\mathrm{E}\left(\prod_{i=1}^p U_i^{k_i}\right) = \begin{cases} \frac{\Gamma(p/q)\prod_{i=1}^p\{(k_i+1)/q\}}{\Gamma\{(p+\sum_{i=1}^p k_i)/q\}\{\Gamma(1/q)\}^p}, & k_i \text{ an even integer for all } i = 1,\ldots,p, \\ 0, & k_i \text{ an odd integer for some } i = 1,\ldots,p. \end{cases}$$

If $\mathbf{X} = (X_1,\ldots,X_p)^\top$ *has* L_q*-spherical distribution, show that* $\mathrm{E}(\prod_{i=1}^p X_i^{k_i}) = \mathrm{E}(R_q^k)\mathrm{E}(\prod_{i=1}^p U_i^{k_i})$ *provided* $\mathrm{E}(R_q^k) < \infty$ *for* $k = \sum_{i=1}^p k_i$*, where* $R_q \stackrel{d}{=} (\sum_{i=1}^p |X_i|^q)^{1/q}$.

Exercise 4.19 *Let* $\mathbf{Y} = V^{-1/q}\mathbf{X}$*,* $V \perp\!\!\!\perp \mathbf{X}$*, where* V *is a nonnegative mixing variable with cdf* G_V *and* $\mathbf{X} \sim \mathcal{L}_q\mathcal{N}_p$ *is an* L_q*-spherical normal random vector. Show that* $\mathbf{Y} \overset{d}{=} R_q\mathbf{U}_q^{(p)}$*,* $R_q \perp\!\!\!\perp \mathbf{U}_q^{(p)}$*, with* $R_q = V^{-1/q}\bar{R}_q$*,* $V \perp\!\!\!\perp \bar{R}_q$ *and* $\bar{R}_q^q \sim \mathcal{G}amma(p/q, q)$*. Also, show in this case that* $\mathbf{Y} \overset{d}{=} R_q\mathbf{U}_q^{(p)} \sim \mathcal{L}_q\mathcal{S}_p(h)$*, with generating pdf of* R_q *given by*

$$g(r) = \frac{r^{p-1}}{q^{p/q-1}\Gamma(p/q)} \int_0^\infty v^{p/q} e^{-r^q v/q} \mathrm{d}G_V(v), \quad r > 0,$$

and so with density generator function given by

$$h(u) = \frac{(q/2)^p}{q^{p/q}\{\Gamma(1/q)\}^p} \int_0^\infty v^{p/q} e^{-uv/q} \mathrm{d}G_V(v), \quad u > 0.$$

Exercise 4.20 (Multivariate log-elliptical distributions) *Let* $\mathbf{V} = (V_1, \ldots, V_p)^\top \in \mathbb{R}_+^p$*, where* $\mathbb{R}_+^p$ *represents the positive part of* $\mathbb{R}^p$*. Define* $\log\mathbf{V} = (\log V_1, \ldots, \log V_p)^\top$*. If* $\log\mathbf{V} \sim \mathcal{E}l_p(\boldsymbol{\xi}, \boldsymbol{\Omega}, \varphi)$*, it is said that* $\mathbf{V}$ *has a multivariate log-elliptical distribution, and we denote it by* $\mathbf{V} \sim \log\mathcal{E}l_p(\boldsymbol{\xi}, \boldsymbol{\Omega}, \varphi)$ *and by* $\mathbf{V} \sim \log\mathcal{S}_p(\varphi)$ *in the spherical case with* $\boldsymbol{\xi} = \mathbf{0}_p$ *and* $\boldsymbol{\Omega} = \mathbf{I}_p$*.*

(a) If $\mathbf{V} \sim \log\mathcal{E}l_p(\boldsymbol{\xi}, \boldsymbol{\Omega}, \varphi)$ *with* $\boldsymbol{\Omega} > 0$ *possesses a pdf with density generator* h*, show that the pdf of* $\mathbf{V}$ *is*

$$|\boldsymbol{\Omega}|^{-1/2} h\{(\log\mathbf{v} - \boldsymbol{\xi})^\top \boldsymbol{\Omega}^{-1} (\log\mathbf{v} - \boldsymbol{\xi})\} \prod_{i=1}^p v_i^{-1}.$$

In this case, we write $\mathbf{V} \sim \log\mathcal{E}l_p(\boldsymbol{\xi}, \boldsymbol{\Omega}, h)$*.*

(b) Partition

$$\mathbf{V} = \begin{pmatrix} \mathbf{V}_1 \\ \mathbf{V}_2 \end{pmatrix}, \quad \boldsymbol{\xi} = \begin{pmatrix} \boldsymbol{\xi}_1 \\ \boldsymbol{\xi}_2 \end{pmatrix}, \quad \boldsymbol{\Omega} = \begin{pmatrix} \boldsymbol{\Omega}_{11} & \boldsymbol{\Omega}_{12} \\ \boldsymbol{\Omega}_{21} & \boldsymbol{\Omega}_{22} \end{pmatrix},$$

where $\mathbf{V}_i$ *and* $\boldsymbol{\xi}_i$ *are* $p_i \times 1$ *vectors and* $\boldsymbol{\Omega}_{ij}$ *is a* $p_i \times p_j$ *matrix,* $i, j = 1, 2$*, with* $p_1 + p_2 = p$*. Show that* $\mathbf{V}_i \sim \log\mathcal{E}l_{p_i}(\boldsymbol{\xi}_i, \boldsymbol{\Omega}_{ii}, \varphi)$*,* $i = 1, 2$*, and* $(\mathbf{V}_1 \mid \mathbf{V}_2 = \mathbf{v}_2) \sim \log\mathcal{E}l_{p_1}(\boldsymbol{\xi}_{1\cdot 2}, \boldsymbol{\Omega}_{11\cdot 2}, \varphi_{Q(e^{\mathbf{v}_2})})$*, where* $\boldsymbol{\xi}_{1\cdot 2}$*,* $\boldsymbol{\Omega}_{11\cdot 2}$ *and* $\varphi_{Q(\mathbf{x})}$ *are defined as in Theorem 36, and for* $\mathbf{v} = (v_1, \ldots, v_k)^\top$*,* $e^{\mathbf{v}} = (e^{v_1}, \ldots, e^{v_k})^\top$*.*

(c) Show that $\mathrm{E}\left(\prod_{i=1}^p V_i^{k_i}\right) = e^{\mathbf{k}^\top \boldsymbol{\xi}} \varphi(\mathbf{k}^\top \boldsymbol{\Omega} \mathbf{k})$*, where* $\mathbf{k} = (k_1, \ldots, k_p)^\top$*.*

Exercise 4.21 *Show that* $\mathcal{MT}_p(\mathbf{0}_p, \bar{\boldsymbol{\Omega}}, \nu; \nu_1, \ldots, \nu_p)$ *and* $\mathcal{T}_p(\mathbf{0}_p, \bar{\boldsymbol{\Omega}}, \nu)$ *have the same copula (dependence structure); see Section 4.5.6 and Equation (10.3).*

4.7 Open Research Problems

Open Research Problem 1 *Develop the theory and inference for multivariate* L_q*-elliptical distributions.*

Open Research Problem 2 *Generalize the multivariate trans-normal distributions presented in Section 3.11.7 to multivariate trans-elliptical distributions. Develop the theory and inference for this family. Study the special case of Box-Cox transformations of elliptical distributions (Morán-Vásquez and Ferrari, 2019).*

Chapter 5

The Multivariate Skew-Normal Distribution

The multivariate skew-normal distribution introduced by Azzalini and Dalla Valle (1996) and studied in detail by Azzalini and Capitanio (1999) is one of the most important asymmetric extensions of the multivariate normal distribution. It is also one of the simplest examples of multivariate selection distributions (see Chapter 9) and, therefore, shares many properties with the multivariate normal distribution. This chapter describes its definition from a different perspective, its properties, and various extensions.

5.1 Construction

There are many ways to introduce the multivariate skew-normal distribution, particularly because it can be formulated in terms of different but equivalent settings/parameterizations. Here, we adopt an approach that allows the simultaneous introduction of both cases, the singular and the nonsingular multivariate skew-normal distribution.

5.1.1 A brief look at Azzalini's univariate skew-normal

According to Azzalini (1985), a random variable Z has a standard skew-normal (SN) distribution with skewness parameter $\alpha \in \mathbb{R}$, denoted by $Z \sim \mathcal{ASN}(\alpha) \equiv \mathcal{ASN}(0,1,\alpha)$, if its pdf and cf are, respectively, given by

$$
\begin{aligned}
f_s(z) &= 2\phi(z)\Phi(\alpha z), \quad z \in \mathbb{R}, \\
\psi_s(t) &= 2e^{-\frac{1}{2}t^2}\Phi\left(\frac{\mathrm{i}\alpha t}{\sqrt{1+\alpha^2}}\right), \quad t \in \mathbb{R},
\end{aligned}
$$

where ϕ and Φ are the pdf and cdf of the standard normal distribution, respectively. Here, $\mathcal{ASN}$ stands for "Azzalini's $\mathcal{SN}$" and the index s for "standard". It is quite clear that for $\alpha = 0$ the standard normal distribution is recovered while for $\alpha > 0$ ($\alpha < 0$) the distribution of Z has a skewness to the right (left). This is illustrated graphically in the plots in Figure 5.1; note that $f_s(0) = 1/\sqrt{2\pi}$ for any $\alpha \in \mathbb{R}$.

Basic properties: If $Z \equiv Z_\alpha \sim \mathcal{ASN}(\alpha)$, then: $Z_0 \sim \mathcal{N}(0,1)$, $-Z_\alpha \overset{d}{=} Z_{-\alpha} \sim \mathcal{ASN}(-\alpha)$, and $|Z_\alpha| \overset{d}{=} |Z_0| \sim \mathcal{HN}(0,1)$ for all α and $Z_{\pm\infty} \overset{d}{=} \pm|Z_0|$, where $\mathcal{HN}$ stands for half-normal.

Dual parameterization: In terms of the well-known dual reparameterization

$$
\delta = \frac{\alpha}{\sqrt{1+\alpha^2}}, \quad \alpha \in \mathbb{R} \quad \longleftrightarrow \quad \alpha = \frac{\delta}{\sqrt{1-\delta^2}}, \quad \delta \in (-1,1),
$$

the pdf and cf of the skew-normal random variable Z can be rewritten as

$$
\begin{aligned}
f_s(z) &= 2\phi(z)\Phi\left(\frac{\delta z}{\sqrt{1-\delta^2}}\right), \quad z \in \mathbb{R}, \\
\psi_s(t) &= 2e^{-\frac{1}{2}t^2}\Phi(\mathrm{i}\delta t), \quad t \in \mathbb{R}.
\end{aligned}
$$

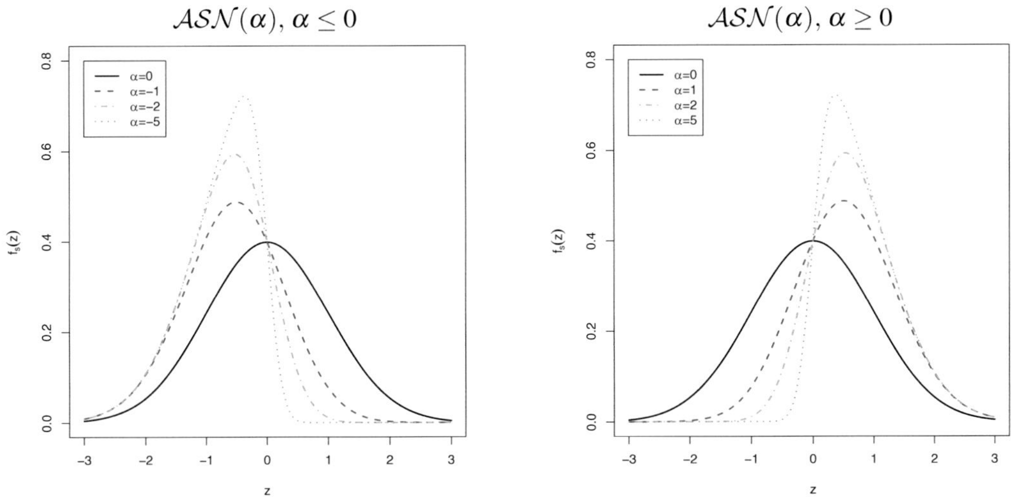

Figure 5.1 *Geometry of Azzalini's univariate standard skew-normal distribution for various values of the skewness parameter α. The solid black curve is the standard normal pdf.*

Therefore, an equivalent and perhaps more suitable notation to represent this distribution could be $Z_\delta \sim \mathcal{ASN}(\delta)$, with $|\delta| < 1$.

Mgf: The $\mathcal{ASN}(\alpha)$ distribution also has a moment generating function, which can be derived from its cf as

$$M_s(t) = \psi_s(-\mathrm{i}t) = 2e^{\frac{1}{2}t^2}\Phi(\delta t), \quad t \in \mathbb{R}.$$

Cdf: As shown later, the cdf of $Z \sim \mathcal{ASN}(\alpha)$ is

$$F_s(z) = 2\Phi_2\left(\begin{pmatrix} z \\ 0 \end{pmatrix}; \begin{pmatrix} 0 \\ 0 \end{pmatrix}, \begin{pmatrix} 1 & -\delta \\ -\delta & 1 \end{pmatrix}\right), \quad z \in \mathbb{R},$$

where $\Phi_p(\cdot;\boldsymbol{\mu},\boldsymbol{\Sigma})$ is the cdf of $\mathcal{N}_p(\boldsymbol{\mu},\boldsymbol{\Sigma})$, a p-variate normal distribution with mean vector $\boldsymbol{\mu}$ and variance matrix $\boldsymbol{\Sigma}$.

Selection/conditional representation: Two equivalent selection representations of $Z \sim \mathcal{ASN}(\alpha)$ are:

$$\text{(a)}\quad Z \stackrel{d}{=} (U \mid \alpha U > U_0), \quad U_0, U \stackrel{iid}{\sim} \mathcal{N}(0,1);$$

$$\text{(b)}\quad Z \stackrel{d}{=} (V \mid V_0 > 0), \quad \begin{pmatrix} V \\ V_0 \end{pmatrix} \sim \mathcal{N}_2\left(\begin{pmatrix} 0 \\ 0 \end{pmatrix}, \begin{pmatrix} 1 & \delta \\ \delta & 1 \end{pmatrix}\right),$$

where (U_0, U) and (V, V_0) are related by

$$V_0 = \frac{\alpha U - U_0}{\sqrt{1+\alpha^2}} \quad \text{and} \quad V = U \quad \longleftrightarrow \quad U_0 = \sqrt{1+\alpha^2}\,V_0 - \alpha V \quad \text{and} \quad U = V.$$

Note that obtaining the cdf of $Z \sim \mathcal{ASN}(\alpha)$ is now straightforward from the selection representation in (b) and the symmetry of the centered bivariate normal distribution, since

$$\mathrm{P}(Z \leq z) = \mathrm{P}(V \leq z \mid V_0 > 0) = \frac{\mathrm{P}(V \leq z, V_0 > 0)}{\mathrm{P}(V_0 > 0)} = 2\mathrm{P}(V \leq z, -V_0 < 0), \quad \forall z \in \mathbb{R},$$

and

$$\begin{pmatrix} V \\ -V_0 \end{pmatrix} \sim \mathcal{N}_2\left(\begin{pmatrix} 0 \\ 0 \end{pmatrix}, \begin{pmatrix} 1 & -\delta \\ -\delta & 1 \end{pmatrix}\right).$$

Stochastic representation: Define the selection random variable $S_\alpha \stackrel{d}{=} (W \mid W_0 > 0)$, where

$$W_0 = U_0 \quad \text{and} \quad W = \frac{\alpha U_0 + U}{\sqrt{1+\alpha^2}}, \qquad U_0, U \stackrel{iid}{\sim} \mathcal{N}(0,1).$$

Note that $(W, W_0) \stackrel{d}{=} (V, V_0)$ has a bivariate normal joint distribution as defined above in (b). Hence, $S_\alpha \stackrel{d}{=} (V \mid V_0 > 0)$ and, therefore, by the above selection representation in (b), $S_\alpha \sim \mathcal{ASN}(\alpha)$.

On the other hand, since the joint distribution of (U, U_0) is normal spherical, it is easy to show that $\{(U, U_0) \mid U_0 > 0\} \stackrel{d}{=} (U, |U_0|)$, where $|U_0| \stackrel{d}{=} (U_0 \mid U_0 > 0)$ has a half-normal distribution. From this fact, it follows that

$$(W \mid W_0 > 0) = \left(\frac{\alpha U_0 + U}{\sqrt{1+\alpha^2}} \mid U_0 > 0 \right) \stackrel{d}{=} \frac{\alpha |U_0| + U}{\sqrt{1+\alpha^2}}.$$

In turn, this proves that (Henze, 1986) if $Z \sim \mathcal{ASN}(\alpha)$, then

$$Z \stackrel{d}{=} \frac{\alpha |U_0| + U}{\sqrt{1+\alpha^2}}, \qquad U_0, U \stackrel{iid}{\sim} \mathcal{N}(0,1).$$

Hierarchical representation: A direct consequence of this last stochastic representation is the following hierarchical representation. Let $Z \sim \mathcal{ASN}(\alpha) \equiv \mathcal{ASN}(\delta/\sqrt{1-\delta^2})$ and $W = |U_0|$. Then,

$$(Z \mid W = w) \sim \mathcal{N}(\delta w, 1-\delta^2) \quad \text{and} \quad W \sim \mathcal{HN}(0,1),$$

and vice versa.

Rescaling: From the above stochastic representation, it follows that $Z \sim \mathcal{ASN}(\alpha)$ can be rescaled as

$$\tilde{Z} = \sqrt{1+\alpha^2}\, Z \stackrel{d}{=} \alpha |U_0| + U, \quad U_0, U \stackrel{iid}{\sim} \mathcal{N}(0,1).$$

Straightforward computations show that the pdf and cf of this rescaled random variable $\tilde{Z}$ are

$$\begin{aligned}
\tilde{f}_s(z) &= \frac{2}{\sqrt{1+\alpha^2}} \phi\left(\frac{z}{\sqrt{1+\alpha^2}}\right) \Phi\left(\frac{\alpha z}{\sqrt{1+\alpha^2}}\right), \quad z \in \mathbb{R}, \quad \alpha \in \mathbb{R}, \\
\tilde{\psi}_s(t) &= 2e^{-\frac{1}{2}t^2} \Phi(i\alpha t), \quad t \in \mathbb{R}, \quad \alpha \in \mathbb{R}.
\end{aligned}$$

In Azzalini's location-scale notation (see below), this means that $\tilde{Z} \sim \mathcal{ASN}(0, 1+\alpha^2, \alpha)$. Moreover, the hierarchical formulation of this rescaled skew-normal distribution is simply given by

$$(\tilde{Z} \mid W = w) \sim \mathcal{N}(\alpha w, 1) \quad \text{and} \quad W \sim \mathcal{HN}(0,1).$$

Invariance under even functions: Let $Z \sim \mathcal{ASN}(\alpha)$. For any even function g, we have $g(Z) \stackrel{d}{=} g(Z_0)$, where $Z_0 \sim \mathcal{N}(0,1)$. In fact, since $g(z) = g(-z)$ for all $z \in \mathbb{R}$, which means that g depends on z through its absolute value $|z|$, it follows for the cf of $g(Z)$ that

$$\begin{aligned}
\mathrm{E}\{e^{itg(Z)}\} &= 2\int_{-\infty}^{0} e^{itg(z)} \phi(z) \Phi(\alpha z) \mathrm{d}z + 2\int_{0}^{\infty} e^{itg(z)} \phi(z) \Phi(\alpha z) \mathrm{d}z \\
&= 2\int_{0}^{\infty} e^{itg(z)} \phi(z) \{1 - \Phi(\alpha z)\} \mathrm{d}z + 2\int_{0}^{\infty} e^{itg(z)} \phi(z) \Phi(\alpha z) \mathrm{d}z \\
&= 2\int_{0}^{\infty} e^{itg(z)} \phi(z) \mathrm{d}z \\
&= \mathrm{E}\{e^{itg(Z_0)}\}, \quad \forall t \in \mathbb{R}.
\end{aligned}$$

Since any variable Z can be represented as $Z = \text{sgn}(Z)|Z|$, the above invariance property comes from the fact that $|Z| \stackrel{d}{=} |Z_0| \sim \mathcal{HN}(0,1) \equiv \chi_1 = \sqrt{\chi_1^2}$, i.e., $Z^2 \stackrel{d}{=} Z_0^2 \sim \chi_1^2$, which does not depend on the shape/skewness parameter. Consequently, the skew-normal random variable Z has the same even moments as the standard normal random variable Z_0. The only difference between them is in the odd moments.

Note that the rescaled variable $\tilde{Z} = \sqrt{1+\alpha^2}Z$ does not possess this invariance property since $\tilde{Z}^2 = (1+\alpha^2)Z^2 \sim (1+\alpha^2)\chi_1^2$ which becomes a gamma distribution with parameters $(1/2, (1+\alpha^2)/2)$. However, it is clear from this last relationship that this rescaled random variable has the following property:

$$\frac{\tilde{Z}^2}{1+\alpha^2} \sim \chi_1^2, \quad \forall \alpha \in \mathbb{R}.$$

Moments: If $Z \sim \mathcal{ASN}(\alpha)$, then for $k = 1, 2, \ldots$ its even and odd moments are, respectively:

$$\text{E}(Z^{2k}) = \frac{(2k)!}{2^k k!} \quad \text{and} \quad \text{E}(Z^{2k-1}) = \sqrt{\frac{2}{\pi}} \frac{(2k-1)!\,\alpha}{2^{k-1}(1+\alpha^2)^{(2k-1)/2}} \sum_{j=0}^{k-1} \frac{j!\,(2\alpha)^{2j}}{(2j+1)!(k-j-1)!}.$$

In particular:

$$\text{E}(Z) = \sqrt{\frac{2}{\pi}}\,\delta, \quad \text{E}(Z^2) = 1, \quad \text{E}(Z^3) = \sqrt{\frac{2}{\pi}}\,\delta(3-\delta^2), \quad \text{E}(Z^4) = 3,$$

$$\text{var}(Z) = 1 - \frac{2}{\pi}\delta^2, \quad \gamma_1 = \text{sgn}(\delta)\left(\frac{4-\pi}{2}\right)\left(\frac{2\delta^2/\pi}{1-2\delta^2/\pi}\right)^{3/2}, \quad \gamma_2 = 2(\pi-3)\left(\frac{2\delta^2/\pi}{1-2\delta^2/\pi}\right)^2,$$

where γ_1 and γ_2 are the skewness and (excess) kurtosis, respectively.

For the moments of the rescaled variable $\tilde{Z} = \sqrt{1+\alpha^2}Z$, we have

$$\text{E}(\tilde{Z}^k) = (1+\alpha^2)^{k/2}\text{E}(Z^k), \quad k = 1, 2, \ldots.$$

In particular, $\tilde{Z}$ and Z have the same skewness and kurtosis, since these measures are invariant by rescaling. Obviously, the same occurs with the standardized skew-normal variable $Z_0 = (Z - \mu_Z)/\sigma_Z$, where $\mu_Z = \text{E}(Z)$ and $\sigma_Z^2 = \text{var}(Z)$ are given above.

Location-scale extension: Similarly to the location-scale normal distribution, the skew-normal location-scale version may be introduced as the distribution of a random variable X defined as

$$X = \xi + \omega Z, \quad Z \sim \mathcal{ASN}(\alpha), \tag{5.1}$$

where $\xi \in \mathbb{R}$ and $\omega > 0$. Straightforward algebra shows that its pdf and cf are:

$$\begin{aligned} f(x) &= \omega^{-1} f_s\{\omega^{-1}(x-\xi)\} \\ &= 2\omega^{-1}\phi\{\omega^{-1}(x-\xi)\}\Phi\{\alpha\omega^{-1}(x-\xi)\}, \quad x \in \mathbb{R}, \\ \psi(t) &= e^{it\xi}\psi_s(\omega t) \\ &= 2e^{it\xi - \frac{1}{2}\omega^2 t^2}\Phi(i\omega\delta t), \quad t \in \mathbb{R}, \end{aligned}$$

where $\alpha \in \mathbb{R}$ and $\delta = \alpha/\sqrt{1+\alpha^2} \in (-1, 1)$. This distribution is denoted by $X \sim \mathcal{ASN}(\xi, \omega^2, \alpha)$. However, sometime we will use the notation $X \sim \mathcal{ASN}(\xi, \omega, \alpha)$ in order to recall that the dependence on the scale parameter is through ω and not on ω^2, as happens in the normal case. Knowing the mean and variance of $Z \sim \mathcal{ASN}(\alpha)$, it is clear then from (5.1) that

$$\text{E}(X) = \xi + \sqrt{\frac{2}{\pi}}\,\omega\delta \quad \text{and} \quad \text{var}(X) = \left(1 - \frac{2}{\pi}\delta^2\right)\omega^2. \tag{5.2}$$

Also, the mgf of $X \sim \mathcal{ASN}(\xi, \omega^2, \alpha)$ becomes

$$M(t) = \psi(-it) = 2e^{t\xi + \frac{1}{2}\omega^2 t^2}\Phi(\omega\delta t), \quad t \in \mathbb{R}.$$

Reparameterizations: Besides the dual relation between the α and δ parameters presented above, there are further well-known reparameterizations in the location-scale skew-normal setting. One of them is the so-called *centered parameterization* which represents the case of the three-parameters skewed-normal in terms of its mean, variance, and skewness measure, that is, the one obtained from the one-to-one transformation $(\xi,\omega^2,\alpha) \longrightarrow (\mu_Z,\sigma_Z^2,\gamma_1)$ with

$$\mu_Z = \sqrt{\frac{2}{\pi}}\,\delta, \quad \sigma_Z^2 = 1-\frac{2}{\pi}\delta^2, \quad \gamma_1 = \sqrt{\frac{2}{\pi}}\left(\frac{4}{\pi}-1\right)\frac{\delta^3}{(1-2\delta^2/\pi)^{3/2}}.$$

Under this parameterization, $X \sim \mathcal{ASN}(\xi,\omega^2,\alpha)$ can be conveniently represented as

$$X = \mu + \sigma\left(\frac{Z-\mu_Z}{\sigma_Z}\right),$$

with $\mu = \mathrm{E}(X)$ and $\sigma^2 = \mathrm{var}(X)$ given in (5.2) and $Z \sim \mathcal{ASN}(\alpha)$.

Another quite convenient parameterization of $\mathcal{ASN}(\xi,\omega^2,\alpha)$ is obtained from the one-to-one transformation of $(\xi,\omega^2,\alpha) \longrightarrow (\xi,\psi^2,\eta)$ with

$$\psi = \frac{\omega}{\sqrt{1+\alpha^2}} \quad \text{and} \quad \eta = \psi\alpha = \omega\delta.$$

This parameterization comes from the rescaling of $Z \sim \mathcal{ASN}(\alpha)$ as $\tilde{Z} = \sqrt{1+\alpha^2}\,Z$ and then reformulating the location-scale case as

$$\tilde{X} = \xi + \psi\tilde{Z} = \xi + \psi\alpha|U_0| + \psi U = \xi + \eta|U_0| + \psi U, \quad U_0, U \overset{iid}{\sim} \mathcal{N}(0,1),$$

which has pdf and cf given by

$$\begin{aligned}
\tilde{f}(x) &= \frac{2}{\sqrt{\psi^2+\eta^2}}\,\phi\left(\frac{x-\xi}{\sqrt{\psi^2+\eta^2}}\right)\Phi\left\{\frac{\eta(x-\xi)}{\psi\sqrt{\psi^2+\eta^2}}\right\}, \quad x \in \mathbb{R},\\
\tilde{\varphi}(t) &= 2e^{it\xi-\frac{1}{2}\psi^2t^2}\,\Phi(it\eta), \quad t \in \mathbb{R},
\end{aligned}$$

where $(\xi,\psi^2,\eta) \in \mathbb{R}\times\mathbb{R}_+\times\mathbb{R}$. In addition, it is easy to see that the new random variable $\tilde{X}$ has a mean and variance given by

$$\mathrm{E}(\tilde{X}) = \xi + \sqrt{\frac{2}{\pi}}\,\eta \quad \text{and} \quad \mathrm{var}(\tilde{X}) = \psi^2 + \left(1-\frac{2}{\pi}\right)\eta^2.$$

Note the simplification in the expression of the mean of $\tilde{X}$ which now depends only on the location ξ and the skewness η. Finally, since $(\xi,\omega^2,\alpha) = (\xi,\psi^2+\eta^2,\psi^{-1}\eta)$, we have in Azzalini's notation

$$\tilde{X} \sim \mathcal{ASN}(\xi,\psi^2+\eta^2,\psi^{-1}\eta).$$

However, despite the abuse of notation, by convenience we will represent this distribution simply as

$$\tilde{X} \sim \mathcal{SN}(\xi,\psi^2,\eta) \quad \longleftrightarrow \quad \tilde{X} = \xi + \eta|U_0| + \psi U, \quad U_0, U \overset{iid}{\sim} \mathcal{N}(0,1).$$

This representation considerably facilitates the multivariate extension of the skew-normal distribution and the study of its main properties. For this reason, we adopt this approach by presenting, when necessary, the corresponding connections with some of the other approaches considered in the literature.

A comprehensive study of the properties of the univariate skew-normal (ASN) distribution and other related distributions can be found in the first part of the book of Azzalini and Capitanio (2014) and in the edited book by Genton (2004). Historically, the univariate skew-normal distribution can be found in various forms in Birnbaum (1950), Weinstein (1964), Roberts (1966), and even traced back to a 1909 paper by Fernando de Helguero; see the account in Azzalini and Regoli (2012b).

5.1.2 *Construction of a multivariate skew-normal*

To construct a multivariate version of the univariate skew-normal distribution, we can use the following route. First, let $U_0, U_1, U_2, \ldots, U_q$ be iid $\mathcal{N}(0,1)$ random variables. Consider the transformation

$$Z_j = \alpha_j |U_0| + U_j, \quad \alpha_j \in \mathbb{R}, \quad j = 1, \ldots, q. \tag{5.3}$$

From (5.3) one can observe for each $j = 1, \ldots, q$ that if $\alpha_j \neq 0$ then the marginal distribution of Z_j is asymmetric since it comes from a convolution type of two distributions, where one of them is asymmetric (half-normal) while the other is symmetric (normal). Recalling that the pdf and cf of the half-normal distribution are, respectively, given by

$$f_{\text{HN}}(w) = 2\phi(w), \quad w \geq 0, \quad \text{and} \quad \psi_{\text{HN}}(t) = 2e^{-\frac{1}{2}t^2}\Phi(\mathrm{i}t), \quad t \in \mathbb{R},$$

it is easy to get the pdf and cf of Z_j as follows:

$$\begin{aligned}
f_j(z_j) &= \int_0^\infty 2\phi(w)\phi(z_j - \alpha_j w)\mathrm{d}w \\
&= 2\upsilon_j^{-1}\phi\left(\upsilon_j^{-1}z_j\right)\Phi\left(\alpha_j\upsilon_j^{-1}z_j\right), \quad z_j \in \mathbb{R}, \\
\psi_j(t_j) &= \mathrm{E}(e^{\mathrm{i}t\alpha_j|U_0|})\mathrm{E}(e^{\mathrm{i}tU_j}) \\
&= 2e^{-\frac{1}{2}\upsilon_j^2 t_j^2}\Phi(\mathrm{i}t_j\alpha_j), \quad t_j \in \mathbb{R},
\end{aligned}$$

where $\upsilon_j = (1+\alpha_j^2)^{1/2}$, which holds for all $j = 1, \ldots, q$. This means that the marginal skewness of Z_j is not affected by the skewness of the rest of the variables.

Another relevant aspect of (5.3) is that the transformed variables $Z_1, \ldots, Z_q$ are not independent since they all depend on the same half-normal variable. To find the joint distribution of the Z_j's, we let $\mathbf{Z} = (Z_1, \ldots, Z_q)^\top$, $\boldsymbol{\alpha} = (\alpha_1, \ldots, \alpha_q)^\top$, and $\mathbf{U} = (U_1, \ldots, U_q)^\top$. Hence, by (5.3), we obtain

$$\mathbf{Z} = \boldsymbol{\alpha}|U_0| + \mathbf{U}, \tag{5.4}$$

where $U_0 \sim \mathcal{N}(0,1)$ and $\mathbf{U} \sim \mathcal{N}_q(\mathbf{0}_q, \mathbf{I}_q)$ and they are independent. Since U_0 and $\mathbf{U}$ are absolutely continuous and independent random quantities, the random vector $\mathbf{Z}$ is also absolutely continuous. Hence, similar to the marginal case, from (5.4) the pdf and cf of $\mathbf{Z}$ can be, respectively, computed as follows:

$$\begin{aligned}
f_{\text{s}}(\mathbf{z}) &= \int_0^\infty 2\phi(w)\phi_q(\mathbf{z} - \boldsymbol{\alpha}w)\mathrm{d}w \\
&= 2|\boldsymbol{\Upsilon}|^{-1/2}\phi_q(\boldsymbol{\Upsilon}^{-1/2}\mathbf{z})\Phi(\boldsymbol{\alpha}^\top\boldsymbol{\Upsilon}^{-1}\mathbf{z}), \quad \mathbf{z} \in \mathbb{R}^q, \qquad (5.5) \\
\psi_{\text{s}}(\mathbf{t}) &= \mathrm{E}(e^{\mathrm{i}\mathbf{t}^\top\boldsymbol{\alpha}|U_0|})\mathrm{E}(e^{\mathrm{i}\mathbf{t}^\top\mathbf{U}}) \\
&= 2e^{-\frac{1}{2}\mathbf{t}^\top\boldsymbol{\Upsilon}\mathbf{t}}\Phi(\mathrm{i}\mathbf{t}^\top\boldsymbol{\alpha}), \quad \mathbf{t} \in \mathbb{R}^q, \qquad (5.6)
\end{aligned}$$

where $\boldsymbol{\alpha} \in \mathbb{R}^q$ and $\boldsymbol{\Upsilon} = \mathbf{I}_q + \boldsymbol{\alpha}\boldsymbol{\alpha}^\top$.

Remark 5 *An alternative route for the construction of* $\mathbf{Z}$ *is to start with the transformation* $Z_1 = \alpha_1|U_0| + U_1$, $\alpha_1 \in \mathbb{R}$, *and* $Z_j = U_j$ *for* $j = 2, \ldots, q$. *Since* $U_0, U_1, \ldots, U_q$ *are independent, now the transformed random variables* $Z_1, \ldots, Z_q$ *are also independent, but with only the first asymmetric. Obviously, the asymmetry can be placed in any of the individual components, and we have chosen the first one only for convenience. Let* $\mathbf{Z}_c = (Z_1, \ldots, Z_q)^\top$. *Since* $Z_1, \ldots, Z_q$ *are absolutely continuous* *and independent random variables, then $\mathbf{Z}_c$ is also absolutely continuous with pdf and cf being the product of the marginal pdfs and cfs, respectively. Specifically,*

$$\begin{aligned} f_{\mathrm{c}}(\mathbf{z}) &= 2|\boldsymbol{\Upsilon}_1|^{-1/2}\phi_q(\boldsymbol{\Upsilon}_1^{-1/2}\mathbf{z})\Phi(\boldsymbol{\alpha}_1^\top\boldsymbol{\Upsilon}_1^{-1}\mathbf{z}), \quad \mathbf{z}\in\mathbb{R}^q, \\ \psi_{\mathrm{c}}(\mathbf{t}) &= 2e^{-\frac{1}{2}\mathbf{t}^\top\boldsymbol{\Upsilon}_1\mathbf{t}}\Phi(\mathrm{i}\mathbf{t}^\top\boldsymbol{\alpha}_1), \quad \mathbf{t}\in\mathbb{R}^q, \end{aligned}$$

where $\boldsymbol{\alpha}_1 = \alpha_1\mathbf{e}_1$, with $\mathbf{e}_1 = (1,0,\ldots,0)^\top$ being the first q-dimensional unit vector, and $\boldsymbol{\Upsilon}_1 = \mathbf{I}_q + \boldsymbol{\alpha}_1\boldsymbol{\alpha}_1^\top$. Now, in order to place the skewness in an arbitrary direction, rotate $\mathbf{Z}_c$ by letting $\mathbf{Z} = \boldsymbol{\Gamma}\mathbf{Z}_c$, where $\boldsymbol{\Gamma}\in\mathbb{R}^{q\times q}$ is orthogonal, i.e., a nonsingular matrix satisfying $\boldsymbol{\Gamma}^\top\boldsymbol{\Gamma} = \boldsymbol{\Gamma}\boldsymbol{\Gamma}^\top = \mathbf{I}_q$. Clearly, the new random vector $\mathbf{Z}$ is absolutely continuous, and its pdf and cf can be obtained as $f_s(\mathbf{z}) = f_c(\boldsymbol{\Gamma}^\top\mathbf{z})$ and $\psi_s(\mathbf{t}) = \psi_c(\boldsymbol{\Gamma}\mathbf{t})$ and they become (5.5) and (5.6), respectively, with $\boldsymbol{\alpha} = \boldsymbol{\Gamma}\boldsymbol{\alpha}_1$ and $\boldsymbol{\Upsilon} = \boldsymbol{\Gamma}^\top\boldsymbol{\Upsilon}_1\boldsymbol{\Gamma}$; we note that $\boldsymbol{\alpha} = \alpha_1\boldsymbol{\gamma}_1$, with $\boldsymbol{\gamma}_1 = \boldsymbol{\Gamma}\mathbf{e}_1$ being the first column of $\boldsymbol{\Gamma}$ and has length $\|\boldsymbol{\gamma}_1\| = 1$.

Finally, for a constant vector $\boldsymbol{\xi}\in\mathbb{R}^p$ and constant matrix $\mathbf{A}\in\mathbb{R}^{p\times q}$ $(p\geq q)$, consider the affine transformation of $\mathbf{Z}$ in (5.4) given by

$$\mathbf{X} = \boldsymbol{\xi} + \mathbf{A}\mathbf{Z}. \tag{5.7}$$

Then, it is clear that

$$\mathbf{X} = \boldsymbol{\xi} + \mathbf{A}\boldsymbol{\alpha}|U_0| + \mathbf{A}\mathbf{U},$$

where $\mathbf{AU}\sim\mathcal{N}_p(\mathbf{0}_p, \mathbf{A}\mathbf{A}^\top)$ and is independent of $|U_0|\sim\mathcal{HN}(0,1)$. Thus, taking into account the half-normal and multivariate normal cfs, the cf $\psi(\mathbf{t}) = \mathrm{E}(e^{\mathrm{i}\mathbf{t}^\top\mathbf{X}})$ of $\mathbf{X}$ can be easily calculated as follows:

$$\begin{aligned} \psi(\mathbf{t}) &= \mathrm{E}\{e^{\mathrm{i}\mathbf{t}^\top(\boldsymbol{\xi}+\boldsymbol{\eta}|U_0|+\mathbf{AU})}\} \\ &= e^{\mathrm{i}\mathbf{t}^\top\boldsymbol{\xi}}\,\mathrm{E}(e^{\mathrm{i}\mathbf{t}^\top\mathbf{A}\boldsymbol{\alpha}|U_0|})\,\mathrm{E}(e^{\mathrm{i}\mathbf{t}^\top\mathbf{AU}}) \\ &= 2e^{\mathrm{i}\mathbf{t}^\top\boldsymbol{\xi}-\frac{1}{2}\mathbf{t}^\top(\boldsymbol{\Psi}+\boldsymbol{\eta}\boldsymbol{\eta}^\top)\mathbf{t}}\Phi(\mathrm{i}\mathbf{t}^\top\boldsymbol{\eta}), \quad \mathbf{t}\in\mathbb{R}^p, \end{aligned} \tag{5.8}$$

where $\boldsymbol{\xi}\in\mathbb{R}^p$ is a location vector, $\boldsymbol{\Psi} = \mathbf{A}\mathbf{A}^\top$ is a $p\times p$ dispersion matrix and $\boldsymbol{\eta} = \mathbf{A}\boldsymbol{\alpha}\in\mathbb{R}^p$ is a skewness vector, thus completely determining the distribution of $\mathbf{X}$. Alternatively, (5.8) can be obtained directly from (5.7) and (5.6) as $\psi(\mathbf{t}) = e^{\mathrm{i}\mathbf{t}^\top\boldsymbol{\xi}}\,\psi_s(\mathbf{A}^\top\mathbf{t}) = 2e^{\mathrm{i}\mathbf{t}^\top\boldsymbol{\xi}-\frac{1}{2}\mathbf{t}^\top\mathbf{A}\boldsymbol{\Upsilon}\mathbf{A}^\top\mathbf{t}}\Phi(\mathrm{i}\mathbf{t}^\top\mathbf{A}\boldsymbol{\alpha})$ and noting again that $\mathbf{A}\boldsymbol{\alpha} = \boldsymbol{\eta}$, $\mathbf{A}\mathbf{A}^\top = \boldsymbol{\Psi}$, and so $\mathbf{A}\boldsymbol{\Upsilon}\mathbf{A}^\top = \boldsymbol{\Psi}+\boldsymbol{\eta}\boldsymbol{\eta}^\top$.

The above definition of the skewness vector $\boldsymbol{\eta} = \mathbf{A}\boldsymbol{\alpha}$, where $\mathbf{A}\in\mathbb{R}^{p\times q}$ and $\boldsymbol{\alpha}\in\mathbb{R}^q$, means that $\boldsymbol{\eta}\in\mathtt{C}(\mathbf{A})$, the column subspace of $\mathbf{A}$. Hence, $\boldsymbol{\eta} = \mathbf{0}_p$ means that $\mathbf{A}\boldsymbol{\alpha} = \mathbf{0}_p$, which in turn means that normality is obtained for $\boldsymbol{\alpha}\in\mathtt{N}(\mathbf{A})$, the null space (or kernel) of $\mathbf{A}$; see the Appendix, Section A.1.

5.2 Formal Definition and First Elementary Properties

Now we can provide a general definition of the multivariate skew-normal distribution, which can be singular or nonsingular depending on whether the dispersion matrix, $\boldsymbol{\Psi}$ say, is positive semi-definite ($\boldsymbol{\Psi}\geq 0$; $|\boldsymbol{\Psi}| = 0$) or positive definite ($\boldsymbol{\Psi} > 0$; $|\boldsymbol{\Psi}| > 0$), respectively.

Definition 17 [Multivariate skew-normal distribution] *A random vector $\mathbf{X}\in\mathbb{R}^p$ is said to have a multivariate skew-normal distribution with location parameter $\boldsymbol{\xi}\in\mathbb{R}^p$, p-by-p dispersion matrix $\boldsymbol{\Psi}$ and shape/skewness parameter $\boldsymbol{\eta}\in\mathbb{R}^p$, in symbols $\mathbf{X}\sim\mathcal{SN}_p(\boldsymbol{\xi},\boldsymbol{\Psi},\boldsymbol{\eta})$, iff $\mathbf{X}$ can be expressed as*

$$\mathbf{X} = \boldsymbol{\xi} + \boldsymbol{\eta}U + \mathbf{W}, \tag{5.9}$$

where $U = |W_0|$, with $W_0\sim\mathcal{N}(0,1)$, and is independent of $\mathbf{W}\sim\mathcal{N}_p(\mathbf{0}_p,\boldsymbol{\Psi})$.

From Equation (5.9) it is clear that

$$\boldsymbol{\eta} = \mathbf{0}_p \quad \Longrightarrow \quad \mathbf{X} = \boldsymbol{\xi} + \mathbf{W} \sim \mathcal{N}_p(\boldsymbol{\xi}, \boldsymbol{\Psi}),$$

which can be singular or nonsingular according to the dispersion matrix $\boldsymbol{\Psi} \geq 0$ or $\boldsymbol{\Psi} > 0$, respectively. Therefore, Definition 17 includes both the singular and the nonsingular multivariate skew-normal distribution. The nonsingular case also requires that $\boldsymbol{\Psi} + \boldsymbol{\eta}\boldsymbol{\eta}^\top > 0$ for the pdf to be valid, which is guaranteed if $\boldsymbol{\Psi} > 0$ since $1 + \boldsymbol{\eta}^\top \boldsymbol{\Psi}^{-1} \boldsymbol{\eta} > 0$ for all $\boldsymbol{\eta} \in \mathbb{R}^p$ and so $|\boldsymbol{\Psi} + \boldsymbol{\eta}\boldsymbol{\eta}^\top| = |\boldsymbol{\Psi}|(1 + \boldsymbol{\eta}^\top \boldsymbol{\Psi}^{-1} \boldsymbol{\eta}) > 0$.

From Equation (5.4) it is clear that $\mathbf{X} \sim \mathcal{SN}_p(\boldsymbol{\xi}, \boldsymbol{\Psi}, \boldsymbol{\eta})$ can also be represented as

$$\mathbf{X} = \boldsymbol{\xi} + \mathbf{AZ}, \qquad \mathbf{Z} \sim \mathcal{SN}_q(\mathbf{0}_q, \mathbf{I}_q, \boldsymbol{\alpha}), \tag{5.10}$$

for any $\mathbf{A} \in \mathbb{R}^{p \times q}$ and $\boldsymbol{\alpha} \in \mathbb{R}^q$ such that $\mathbf{AA}^\top = \boldsymbol{\Psi}$ and $\mathbf{A}\boldsymbol{\alpha} = \boldsymbol{\eta}$.

However, an apparently problematic fact with the representation in (5.10) of a multivariate skew-normal distribution, which has practical relevance in the nonsingular case, is that there are many options to choose a matrix $\mathbf{A}$ to factorize $\boldsymbol{\Psi}$. For example, when $p = q$ some different well-known solutions for $\mathbf{A}$ (each of which is unique when $\boldsymbol{\Psi}$ is nonsingular) are:

[1] $\mathbf{A} = \boldsymbol{\Psi}^{1/2} = \boldsymbol{\Gamma}\mathbf{D}_\lambda^{1/2}\boldsymbol{\Gamma}^\top$ (symmetric square root);

[2] $\mathbf{A} = \boldsymbol{\Gamma}\mathbf{D}_\lambda^{1/2}$, where $\boldsymbol{\Gamma} \in \mathbb{R}^{p \times p}$ is orthogonal, $\mathbf{D}_\lambda^{1/2} = \mathrm{diag}(\lambda_1^{1/2}, \ldots, \lambda_p^{1/2})$ and $\lambda_1, \ldots, \lambda_p$ are the eigenvalues of $\boldsymbol{\Psi}$;

[3] $\mathbf{A} = \mathbf{L}$ (or $\mathbf{A} = \mathbf{U}$), where $\mathbf{L} = (l_{ij}) \in \mathbb{R}^{p \times p}$ is lower triangular, with $l_{ii} > 0$, $i = 1, \ldots, p$ ($\mathbf{U} = (u_{ij}) \in \mathbb{R}^{p \times p}$ is upper triangular, with $u_{ii} > 0$, $i = 1, \ldots, p$), from the Choleski factorization of $\boldsymbol{\Psi}$.

However, it is customary for $\mathbf{A}$ to be symmetric and use the unique solution given by option [1].

Example 57 (Permutable multivariate skew-normal distribution) *Let* $\mathbf{X} \sim \mathcal{SN}_p(\xi \mathbf{1}_p, \boldsymbol{\Psi}, \eta \mathbf{1}_p)$ *with* $\xi, \eta \in \mathbb{R}$, $\boldsymbol{\Psi} = \psi^2[(1-\rho)\mathbf{I}_p + \rho \mathbf{1}_p \mathbf{1}_p^\top]$ *and* $\psi > 0$. *Straightforward computations show that:*

(a) $\boldsymbol{\Psi} > 0 \Longleftrightarrow \rho \in (-1/(p-1), 1)$;

(b) $\boldsymbol{\Psi} \geq 0 \Longleftrightarrow \rho = -1/(p-1)$ *or* $\rho = 1$.

Also, here we have $\boldsymbol{\Gamma}\mathbf{X} \stackrel{d}{=} \mathbf{X}$ *for all orthogonal matrices* $\boldsymbol{\Gamma} \in \mathbb{R}^{p \times p}$ *preserving the vector of ones* $\mathbf{1}_p \in \mathbb{R}^p$, *which are permutation matrices.*

In the following, we present an alternative stochastic representation of the multivariate skew-normal distribution.

Theorem 50 *Assume* $\mathbf{X} \sim \mathcal{SN}_p(\boldsymbol{\xi}, \boldsymbol{\Psi}, \boldsymbol{\eta})$, *with* $\boldsymbol{\Psi} = \mathbf{CC}^\top > 0$ *for some p-by-q matrix* $\mathbf{C}$ *of rank q. Then,*

$$\mathbf{X} - \boldsymbol{\xi} \stackrel{d}{=} R(\boldsymbol{\eta}|U_0| + \mathbf{CU}),$$

where R is the χ_{q+1}*-radial random variable and* $\mathbf{U}^{(q+1)} = (\mathbf{U}^\top, U_0)^\top$ *is the* $(q+1)$*-dimensional orthogonally invariant uniform random vector, both being independent.*

Proof 50 *By Equation (5.9) it follows that*

$$\mathbf{X} - \boldsymbol{\xi} = \boldsymbol{\eta}|W_0| + \mathbf{W} = (\mathbf{I}_p \; \boldsymbol{\eta}) \begin{pmatrix} \mathbf{W} \\ |W_0| \end{pmatrix}, \quad \begin{pmatrix} \mathbf{W} \\ W_0 \end{pmatrix} \sim \mathcal{N}_{p+1}\left(\begin{pmatrix} \mathbf{0}_p \\ 0 \end{pmatrix}, \begin{pmatrix} \boldsymbol{\Psi} & \mathbf{0}_p \\ \mathbf{0}_p^\top & 1 \end{pmatrix} \right).$$

Since $\boldsymbol{\Psi} = \mathbf{C}\mathbf{C}^\top$, *where* $\mathbf{C} \in \mathbb{R}^{p\times q}$ *is of rank q, we have*

$$\begin{pmatrix}\mathbf{W}\\ W_0\end{pmatrix} \stackrel{d}{=} \begin{pmatrix}\mathbf{C} & \mathbf{0}_p\\ \mathbf{0}_q^\top & 1\end{pmatrix}\begin{pmatrix}\mathbf{V}\\ V_0\end{pmatrix}, \quad \begin{pmatrix}\mathbf{V}\\ V_0\end{pmatrix} \sim \mathcal{N}_{q+1}\left(\begin{pmatrix}\mathbf{0}_q\\ 0\end{pmatrix}, \begin{pmatrix}\mathbf{I}_q & \mathbf{0}_q\\ \mathbf{0}_q^\top & 1\end{pmatrix}\right).$$

But, from the fact that $\begin{pmatrix}\mathbf{V}\\ V_0\end{pmatrix} \stackrel{d}{=} R\begin{pmatrix}\mathbf{U}\\ U_0\end{pmatrix} = R\mathbf{U}^{(q+1)}$, *where*

$$R \stackrel{d}{=} \sqrt{\|\mathbf{V}\|^2 + V_0^2} \sim \chi_{q+1}, \quad \mathbf{U}^{(q+1)} = \begin{pmatrix}\mathbf{U}\\ U_0\end{pmatrix} \stackrel{d}{=} \frac{1}{\sqrt{\|\mathbf{V}\|^2 + V_0^2}}\begin{pmatrix}\mathbf{V}\\ V_0\end{pmatrix} \sim \mathcal{U}(\mathbb{S}^{(q)}), \quad R \perp\!\!\!\perp \mathbf{U}^{(q+1)},$$

we have

$$\begin{aligned}\begin{pmatrix}\mathbf{W}\\ W_0\end{pmatrix} \stackrel{d}{=} R\begin{pmatrix}\mathbf{C} & \mathbf{0}_p\\ \mathbf{0}_q^\top & 1\end{pmatrix}\begin{pmatrix}\mathbf{U}\\ U_0\end{pmatrix} &\Longrightarrow \begin{pmatrix}\mathbf{W}\\ |W_0|\end{pmatrix} \stackrel{d}{=} R\begin{pmatrix}\mathbf{C} & \mathbf{0}_p\\ \mathbf{0}_q^\top & 1\end{pmatrix}\begin{pmatrix}\mathbf{U}\\ |U_0|\end{pmatrix}\\ &\Longrightarrow \mathbf{X} - \boldsymbol{\xi} \stackrel{d}{=} R(\mathbf{I}_p\ \boldsymbol{\eta})\begin{pmatrix}\mathbf{C} & \mathbf{0}_p\\ \mathbf{0}_q^\top & 1\end{pmatrix}\begin{pmatrix}\mathbf{U}\\ |U_0|\end{pmatrix} = R(\mathbf{C}\ \boldsymbol{\eta})\begin{pmatrix}\mathbf{U}\\ |U_0|\end{pmatrix} = R(\boldsymbol{\eta}|U_0| + \mathbf{C}\mathbf{U}).\end{aligned}$$

□

Setting $\mathbf{A} = (\mathbf{C}, \boldsymbol{\eta})$, where $\mathbf{C}\mathbf{C}^\top = \boldsymbol{\Psi}$, and recalling that $\begin{pmatrix}\mathbf{U}\\ |U_0|\end{pmatrix} \stackrel{d}{=} \left\{\begin{pmatrix}\mathbf{U}\\ U_0\end{pmatrix} \mid U_0 > 0\right\}$, the above stochastic representation can also be rewritten as

$$\mathbf{X} - \boldsymbol{\xi} \stackrel{d}{=} R\mathbf{A}(\mathbf{U}^{(p+1)}|U_0 > 0).$$

5.2.1 Mean vector and variance matrix

Definition 17 is also very convenient for obtaining the elementary properties of a skew-normal random vector $\mathbf{X}$. For example, this definition considerably facilitates the calculation of its mean vector and variance matrix.

Theorem 51 *Let* $\mathbf{X} \sim \mathcal{SN}_p(\boldsymbol{\xi}, \boldsymbol{\Psi}, \boldsymbol{\eta})$. *Then,*

$$\mathrm{E}(\mathbf{X}) = \boldsymbol{\xi} + \sqrt{\frac{2}{\pi}}\boldsymbol{\eta} \quad \text{and} \quad \mathrm{var}(\mathbf{X}) = \boldsymbol{\Psi} + \left(1 - \frac{2}{\pi}\right)\boldsymbol{\eta}\boldsymbol{\eta}^\top. \tag{5.11}$$

Proof 51 *By (5.9) we have*

$$\mathrm{E}(\mathbf{X}) = \mathrm{E}(\boldsymbol{\xi} + U\boldsymbol{\eta} + \mathbf{W}) = \boldsymbol{\xi} + \mathrm{E}(U)\boldsymbol{\eta} + \mathrm{E}(\mathbf{W}),$$

and

$$\mathrm{var}(\mathbf{X}) = \mathrm{var}(\boldsymbol{\xi} + U\boldsymbol{\eta} + \mathbf{W}) = \mathrm{var}(U)\boldsymbol{\eta}\boldsymbol{\eta}^\top + \mathrm{var}(\mathbf{W}).$$

Thus, the proof follows by noting that if $U \sim \mathcal{HN}(0,1)$ *then*

$$\mathrm{E}(U) = \sqrt{\frac{2}{\pi}} \quad \text{and} \quad \mathrm{var}(U) = 1 - \frac{2}{\pi}.$$

□

5.2.2 *Hierarchical representation*

According to Definition 17, the multivariate skew-normal is a particular member of the more general class of mean mixtures of multivariate normal distributions; see Arellano-Valle and Azzalini (2021) and Section 3.10. This class is represented analogously to (5.9), but considering an arbitrary distribution for the mixing variable U, which again can be nonnegative (with probability 1) and independent of the normal random vector $\mathbf{W} \sim \mathcal{N}_p(\mathbf{0}_p, \boldsymbol{\Psi})$. For the skew-normal case, $U = |W_0|$, with $W_0 \sim \mathcal{N}(0,1)$, and hence U becomes a standardized half-normal random variable with pdf $2\phi(u)I_{(u\geq 0)}$, and is usually denoted by $U \sim \mathcal{HN}(0,1) \equiv \chi_1$ or equivalently, by $U \sim \chi_1$. This fact is clearly visualized by conditioning in (5.9) on the half-normal latent variable $U = |W_0|$. In this way, the following hierarchical representation is obtained for the multivariate skew-normal distribution.

Theorem 52 *The random vector* $\mathbf{X} \sim \mathcal{SN}_p(\boldsymbol{\xi}, \boldsymbol{\Psi}, \boldsymbol{\eta})$ *defined in (5.9) can be represented equivalently as*

$$\mathbf{X} \mid U = u \sim \mathcal{N}_p(\boldsymbol{\xi} + \boldsymbol{\eta}\, u, \boldsymbol{\Psi}), \tag{5.12}$$

$$U \sim \mathcal{HN}(0,1). \tag{5.13}$$

Proof 52 *The proof is derived directly from the stochastic representation in (5.9).* □

This representation is useful in both the classical and Bayesian inferential processes of the nonsingular multivariate skew-normal model.

5.2.3 *Selection representation*

The multivariate skew-normal distribution also has the following selection representation.

Theorem 53 *If* $\mathbf{X} \sim \mathcal{SN}_p(\boldsymbol{\xi}, \boldsymbol{\Psi}, \boldsymbol{\eta})$, *then*

$$\mathbf{X} - \boldsymbol{\xi} \stackrel{d}{=} (\mathbf{W}_* \mid W_0 > 0), \quad \textit{with} \quad \begin{pmatrix} \mathbf{W}_* \\ W_0 \end{pmatrix} \sim \mathcal{N}_{p+1}\left(\begin{pmatrix} \mathbf{0}_p \\ 0 \end{pmatrix}, \begin{pmatrix} \boldsymbol{\Psi} + \boldsymbol{\eta}\boldsymbol{\eta}^\top & \boldsymbol{\eta} \\ \boldsymbol{\eta}^\top & 1 \end{pmatrix} \right). \tag{5.14}$$

Proof 53 *By Equation (5.9),* $\mathbf{X} - \boldsymbol{\xi} = \boldsymbol{\eta}\, |W_0| + \mathbf{W}$, *where* $W_0 \sim \mathcal{N}(0,1)$ *and* $\mathbf{W} \sim \mathcal{N}_p(\mathbf{0}_p, \boldsymbol{\Psi})$ *are independent. But, by independence* $\{(\mathbf{W}, W_0) \mid W_0 > 0\} \stackrel{d}{=} (\mathbf{W}, |W_0|)$. *In turn, this fact implies that*

$$\{g(\mathbf{W}, W_0) \mid W_0 > 0\} \stackrel{d}{=} g(\mathbf{W}, |W_0|),$$

for any measurable function g. Taking $\mathbf{W}_* = g(\mathbf{W}, W_0) = \boldsymbol{\eta}\, W_0 + \mathbf{W}$ *we have*

$$(\mathbf{W}_* \mid W_0 > 0) \stackrel{d}{=} g(\mathbf{W}, |W_0|) = \boldsymbol{\eta}\, |W_0| + \mathbf{W},$$

and the proof concludes by noting that the multivariate normal joint distribution of $(\mathbf{W}_*^\top, W_0)^\top$ *comes from the fact that* $(\mathbf{W}_* \mid W_0 = w_0) \sim \mathcal{N}_p(\boldsymbol{\eta}\, w_0, \boldsymbol{\Psi})$ *and* $W_0 \sim \mathcal{N}(0,1)$. □

An important consequence of Theorem 53 is the following result, in which, for simplicity and without loss of generality, we assume $\boldsymbol{\xi} = \mathbf{0}_p$:

Theorem 54 *Let* $\mathbf{g}$ *be a measurable vector-valued function from* $\mathbb{R}^p$ *to* $\mathbb{R}^q$ *and* C *be a Borel subset in* $\mathbb{R}^q$. *If* $\mathbf{X} \sim \mathcal{SN}_p(\mathbf{0}_p, \boldsymbol{\Psi}, \boldsymbol{\eta})$, *then the probability distribution of* $\mathbf{g}(\mathbf{X})$ *is given by*

$$\mathrm{P}(\mathbf{g}(\mathbf{X}) \in C) = \mathrm{P}(\mathbf{g}(\mathbf{W}_*) \in C \mid W_0 > 0),$$

where the joint distribution of $(\mathbf{W}_*^\top, W_0)^\top$ *is defined in (5.14).*

Proof 54 *In fact, by (5.14) we have*

$$\mathrm{P}\{\mathbf{g}(\mathbf{X}) \in C\} = \mathrm{P}\{\mathbf{X} \in \mathbf{g}^{-1}(C)\} = \mathrm{P}\{\mathbf{W}_* \in \mathbf{g}^{-1}(C) \mid W_0 > 0\} = \mathrm{P}(\mathbf{g}(\mathbf{W}_*) \in C \mid W_0 > 0\}.$$

□

5.2.4 Cdf

From the representation by selection of $\mathbf{X} \sim \mathcal{SN}_p(\boldsymbol{\xi}, \boldsymbol{\Psi}, \boldsymbol{\eta})$, or immediately from Theorem 53, its cdf can be easily calculated as shown below. Later in Section 5.5.2, we will obtain this cdf in the nonsingular case by directly integrating its pdf.

Theorem 55 *The cdf of* $\mathbf{X} \sim \mathcal{SN}_p(\boldsymbol{\xi}, \boldsymbol{\Psi}, \boldsymbol{\eta})$ *is*

$$F(\mathbf{x}) = 2\Phi_{p+1}(\mathbf{x}_*; \boldsymbol{\xi}_*, \boldsymbol{\Omega}_*), \quad \mathbf{x} \in \mathbb{R}^p,$$

where $\mathbf{x}_* = (\mathbf{x}^\top, 0)^\top$, $\boldsymbol{\xi}_* = (\boldsymbol{\xi}^\top, 0)^\top$ *and*

$$\boldsymbol{\Omega}_* = \begin{pmatrix} \boldsymbol{\Psi} + \boldsymbol{\eta}\boldsymbol{\eta}^\top & -\boldsymbol{\eta} \\ -\boldsymbol{\eta}^\top & 1 \end{pmatrix}. \tag{5.15}$$

Proof 55 *From Theorem 53, we have* $\mathbf{X} - \boldsymbol{\xi} \stackrel{d}{=} (\mathbf{W}_* \mid W_0 > 0)$, *where the joint distribution of* $(\mathbf{W}_*, W_0) = (\boldsymbol{\eta} W_0 + \mathbf{W}, W_0)$ *is defined in (5.14). Hence, we obtain*

$$\begin{aligned} F(\mathbf{x}) &= \mathrm{P}(\mathbf{X} \le \mathbf{x}) \\ &= \mathrm{P}(\mathbf{W}_* \le \mathbf{x} - \boldsymbol{\xi} \mid W_0 > 0) \\ &= \frac{\mathrm{P}(\mathbf{W}_* \le \mathbf{x} - \boldsymbol{\xi}, W_0 > 0)}{\mathrm{P}(W_0 > 0)} \\ &= 2\mathrm{P}(\mathbf{W}_* \le \mathbf{x} - \boldsymbol{\xi}, -W_0 < 0), \end{aligned}$$

which proves the result since $(\mathbf{W}_*^\top, -W_0)^\top \sim \mathcal{N}_{p+1}(\mathbf{0}_{p+1}, \boldsymbol{\Omega}_*)$. □

5.3 Cf, Mgf, Linear Combinations and Moments

We may now represent a skew-normal random vector $\mathbf{X} \sim \mathcal{SN}_p(\boldsymbol{\xi}, \boldsymbol{\Psi}, \boldsymbol{\eta})$ defined according to Definition 17 by means of its cf and the distribution of any linear combination $\mathbf{a}^\top\mathbf{X}$, $\mathbf{a} \in \mathbb{R}^p$.

Theorem 56 *Let* $\mathbf{X} \sim \mathcal{SN}_p(\boldsymbol{\xi}, \boldsymbol{\Psi}, \boldsymbol{\eta})$. *Then, its cf is*

$$\psi_{\mathbf{X}}(\mathbf{t}) = 2e^{i\mathbf{t}^\top\boldsymbol{\xi} - \frac{1}{2}\mathbf{t}^\top(\boldsymbol{\Psi} + \boldsymbol{\eta}\boldsymbol{\eta}^\top)\mathbf{t}}\Phi(i\mathbf{t}^\top\boldsymbol{\eta}), \quad \mathbf{t} \in \mathbb{R}^p. \tag{5.16}$$

Proof 56 *By Definition 17,* $\mathbf{X} = \boldsymbol{\xi} + \boldsymbol{\eta} U + \mathbf{W}$, *where* $U \sim \mathcal{HN}(0,1)$ *and* $\mathbf{W} \sim \mathcal{N}_p(\mathbf{0}_p, \boldsymbol{\Psi})$ *are independent. Hence, the proof is direct from the fact that*

$$\psi_{\mathbf{X}}(\mathbf{t}) = \psi_{\boldsymbol{\xi}+\mathbf{W}}(\mathbf{t})\psi_U(\mathbf{t}^\top \boldsymbol{\eta}),$$

where $\psi_{\boldsymbol{\xi}+\mathbf{W}}(\mathbf{t}) = e^{i\mathbf{t}^\top \boldsymbol{\xi} - \frac{1}{2}\mathbf{t}^\top \boldsymbol{\Psi}\mathbf{t}}$ *and* $\psi_U(t) = 2e^{-\frac{1}{2}t^2}\Phi(it)$. □

As we know, the cf of a random vector completely determines its probabilistic distribution and vice versa. In this regard, it should be noted from (5.16) that the cf of the $\mathcal{SN}_p(\boldsymbol{\xi}, \boldsymbol{\Psi}, \boldsymbol{\eta})$ distribution becomes the product of the cf of the $\mathcal{N}_p(\boldsymbol{\xi}, \boldsymbol{\Psi})$ distribution with the cf of the univariate $\mathcal{HN}(0,1)$ distribution. In symbols, we can write this relation as

$$\psi_{\mathrm{SN}_p}(\mathbf{t}) = \psi_{\mathrm{N}_p}(\mathbf{t})\psi_{\mathrm{HN}_1}(\mathbf{t}^\top \boldsymbol{\eta}), \quad \mathbf{t} \in \mathbb{R}^p.$$

The cf of the multivariate skew-normal distribution in its ASN parameterization can be found in Kim and Genton (2011), as well as the cf for scale mixtures of multivariate skew-normal distributions.

The next characterization involves a closure property of linear combinations of the components of $\mathbf{X} \sim \mathcal{SN}_p(\boldsymbol{\xi}, \boldsymbol{\Psi}, \boldsymbol{\eta})$.

Theorem 57 $\mathbf{X} \sim \mathcal{SN}_p(\boldsymbol{\xi}, \boldsymbol{\Psi}, \boldsymbol{\eta})$ *iff* $\mathbf{a}^\top \mathbf{X} \sim \mathcal{SN}(\mathbf{a}^\top \boldsymbol{\xi}, \mathbf{a}^\top \boldsymbol{\Psi}\mathbf{a}, \mathbf{a}^\top \boldsymbol{\eta})$ *for all* $\mathbf{a} \in \mathbb{R}^p$.

Proof 57 *In fact, by Equation (5.16) in Theorem 56 one can find that*

$$\psi_{\mathbf{X}}(t\mathbf{a}) = 2e^{it\mathbf{a}^\top \boldsymbol{\xi} - \frac{1}{2}t^2\mathbf{a}^\top(\boldsymbol{\Psi}+\boldsymbol{\eta}\boldsymbol{\eta}^\top)\mathbf{a}}\Phi(it\mathbf{a}^\top \boldsymbol{\eta}) = 2e^{it\xi_a - \frac{1}{2}t^2(\psi_a^2+\eta_a^2)}\Phi(it\eta_a) = \psi_{\mathbf{a}^\top \mathbf{X}}(t),$$

$\forall t \in \mathbb{R}$, $\forall \mathbf{a} \in \mathbb{R}^p$, *where* $\xi_a = \mathbf{a}^\top \boldsymbol{\xi}$, $\psi_a^2 = \mathbf{a}^\top \boldsymbol{\Psi}\mathbf{a}$, *and* $\eta_a = \mathbf{a}^\top \boldsymbol{\eta}$. *Therefore, this proves the stated result.* □

Remark 6 *An alternative proof of Theorem 57 is by directly using the definition of the skew-normal random vector* $\mathbf{X} \sim \mathcal{SN}_p(\boldsymbol{\xi}, \boldsymbol{\Psi}, \boldsymbol{\eta})$ *to find* $X_a = \mathbf{a}^\top \mathbf{X} = \xi_a + \eta_a U + W_a$, *where* $U \sim \mathcal{HN}(0,1)$ *and* $W_a = \mathbf{a}^\top \mathbf{W} \sim \mathcal{N}(0, \psi_a^2)$ *are independent. Hence, by definition* $X_a \sim \mathcal{SN}(\xi_a, \psi_a^2, \eta_a)$ *for all* $\mathbf{a} \in \mathbb{R}^p$, *where, as before,* $\xi_a = \mathbf{a}^\top \boldsymbol{\xi}$, $\psi_a^2 = \mathbf{a}^\top \boldsymbol{\Psi}\mathbf{a}$, *and* $\eta_a = \mathbf{a}^\top \boldsymbol{\eta}$. *Moreover, from this fact it follows that* $\mathbf{X}$ *has product-moments of any order since this fact holds for* $X_a = \mathbf{a}^\top \mathbf{X}$, $\forall \mathbf{a} \in \mathbb{R}^p$. *Thus, we can also represent the multivariate skew-normal distribution from its mgf, which can be obtained from Theorem 56 as*

$$M_{\mathbf{X}}(\mathbf{t}) = \psi_{\mathbf{X}}(-i\mathbf{t}) = 2e^{\mathbf{t}^\top \boldsymbol{\xi} + \frac{1}{2}\mathbf{t}^\top(\boldsymbol{\Psi}+\boldsymbol{\eta}\boldsymbol{\eta}^\top)\mathbf{t}}\Phi(\mathbf{t}^\top \boldsymbol{\eta}), \quad \mathbf{t} \in \mathbb{R}^p.$$

As illustrated in part (a) of the following example, specializing $\mathbf{a}$ in Theorem 57 to be the i-th p-dimensional unit vector $\mathbf{e}_i$, we then immediately deduce that all univariate marginals are skew-normal.

Example 58 (Univariate SN marginals) *Let* $\mathbf{X} = (X_1, \ldots, X_p)^\top \sim \mathcal{SN}_p(\boldsymbol{\xi}, \boldsymbol{\Psi}, \boldsymbol{\eta})$, *with* $\boldsymbol{\xi} = (\xi_1, \ldots, \xi_p)^\top$, $\boldsymbol{\Psi} = (\psi_{ij})_{i,j=1,\ldots,p}$ *and* $\boldsymbol{\eta} = (\eta_1, \ldots, \eta_p)^\top$. *By Theorem 57, we have*

$$\sum_{i=1}^p a_i X_i \sim \mathcal{SN}\left(\sum_{i=1}^p a_i \xi_i, \sum_{i=1}^p \sum_{j=1}^p a_i a_j \psi_{ij}, \sum_{i=1}^p a_i \eta_i\right).$$

Hence:

(a) *Setting* $\mathbf{a} = \mathbf{e}_i = (0,\dots,0,1,0\dots,0)^\top$ *we find the i-th marginal component* $X_i = \mathbf{e}_i^\top \mathbf{X}$*:*

$$X_i \sim \mathcal{SN}(\xi_i, \psi_{ii}, \eta_i), \quad i = 1,\dots,p,$$

with $\xi_i = \mathbf{e}_i^\top \boldsymbol{\xi}$, $\psi_{ij} = \mathbf{e}_i^\top \boldsymbol{\Psi} \mathbf{e}_j$ *and* $\eta_i = \mathbf{e}_i^\top \boldsymbol{\eta}$*;*

(b) *Similarly, setting* $\mathbf{a}$ *to be a vector with* 1 *in the i-th and j-th positions,* $i \neq j$*, and* 0 *elsewhere, we obtain*

$$X_i + X_j \sim \mathcal{SN}(\xi_i + \xi_j, \psi_{ii} + \psi_{jj} + 2\psi_{ij}, \eta_i + \eta_j), \quad i,j = 1,\dots,p, \quad i \neq j.$$

Another interesting example that characterizes skew-normality is the following.

Example 59 *Let* $U \sim \mathcal{HN}(0,1)$ *and* $W \sim \mathcal{N}(0,1)$ *be independent. If* $\mathbf{X} \sim \mathcal{SN}_p(\boldsymbol{\xi}, \boldsymbol{\Psi}, \boldsymbol{\eta})$*, with* $\boldsymbol{\Psi} > 0$*, then:*

$$\frac{\boldsymbol{\eta}^\top \boldsymbol{\Psi}^{-1}(\mathbf{X} - \boldsymbol{\xi})}{\sqrt{1 + \boldsymbol{\eta}^\top \boldsymbol{\Psi}^{-1} \boldsymbol{\eta}}} \overset{d}{=} \delta Z, \qquad Z = \eta U + W \sim \mathcal{SN}(0,1,\eta),$$

where $\delta = \eta / \sqrt{1 + \eta^2}$ *and* $\eta = \sqrt{\boldsymbol{\eta}^\top \boldsymbol{\Psi}^{-1} \boldsymbol{\eta}}$.

5.4 Affine Transformations, Marginals, and Independence

In this section, we describe further distributional properties of the multivariate skew-normal family introduced in Definition 17. We show that this family inherits several good properties of the multivariate normal family.

5.4.1 Affine transformations

By construction, the multivariate skew-normal distribution is closed under affine transformations as formalized next.

Theorem 58 *Let* $\mathbf{X} \sim \mathcal{SN}_p(\boldsymbol{\xi}, \boldsymbol{\Psi}, \boldsymbol{\eta})$*, and consider the affine transformation* $\mathbf{Y} = \mathbf{A}\mathbf{X} + \mathbf{b}$*, where the matrix* $\mathbf{A} \in \mathbb{R}^{q \times p}$ *and the vector* $\mathbf{b} \in \mathbb{R}^q$ *are constant. Then,* $\mathbf{Y} \sim \mathcal{SN}_q(\mathbf{A}\boldsymbol{\xi} + \mathbf{b}, \mathbf{A}\boldsymbol{\Psi}\mathbf{A}^\top, \mathbf{A}\boldsymbol{\eta})$.

Proof 58 *By Definition 17 it follows that* $\mathbf{Y} = \mathbf{A}\mathbf{X} + \mathbf{b} = \mathbf{A}\boldsymbol{\xi} + \mathbf{b} + \mathbf{A}\boldsymbol{\eta}U + \mathbf{A}\mathbf{W}$. *Let* $\boldsymbol{\xi}_A = \mathbf{A}\boldsymbol{\xi} + \mathbf{b}$, $\boldsymbol{\eta}_A = \mathbf{A}\boldsymbol{\eta}$ *and* $\mathbf{W}_A = \mathbf{A}\mathbf{W}$. *Since* $\mathbf{W} \sim \mathcal{N}_p(\mathbf{0}_p, \boldsymbol{\Psi})$*, then* $\mathbf{W}_A \sim \mathcal{N}_q(\mathbf{0}_q, \boldsymbol{\Psi}_A)$*, with* $\boldsymbol{\Psi}_A = \mathbf{A}\boldsymbol{\Psi}\mathbf{A}^\top$*, and is independent of* $U \sim \mathcal{HN}(0,1)$. *Hence,* $\mathbf{Y} = \boldsymbol{\xi}_A + \boldsymbol{\eta}_A U + \mathbf{W}_A$*, with* $U \sim \mathcal{HN}(0,1)$ *and* $\mathbf{W}_A \sim \mathcal{N}_q(\mathbf{0}_q, \boldsymbol{\Psi}_A)$ *being independent. Thus, the proof follows by applying Definition 17 again.* □

If $\mathbf{A}$ has full row rank q, then a nonsingular multivariate skew-normal version is obtained from Theorem 58. Another interesting result that follows from Theorem 58 is when $p = q$ and $\mathbf{A} = -\mathbf{I}_p$, which transforms $\mathbf{X}$ to $-\mathbf{X}$, as formalized in the following example.

Example 60 *If* $\mathbf{X} \sim \mathcal{SN}_p(\boldsymbol{\xi}, \boldsymbol{\Psi}, \boldsymbol{\eta})$*, then* $-\mathbf{X} \sim \mathcal{SN}_p(-\boldsymbol{\xi}, \boldsymbol{\Psi}, -\boldsymbol{\eta})$*.*

According to Theorem 58, we can recover symmetry by choosing a transformation matrix $\mathbf{A}$ to be rows-orthogonal to the skewness vector $\boldsymbol{\eta}$, i.e., when $\mathbf{A}\boldsymbol{\eta} = \mathbf{0}_q$. Therefore, in this case, the affine transformation $\mathbf{Y} = \mathbf{AX} + \mathbf{b}$ has a symmetric $\mathcal{N}_q(\mathbf{A}\boldsymbol{\xi} + \mathbf{b}, \mathbf{A}\boldsymbol{\Psi}\mathbf{A}^\top)$ distribution.

5.4.2 *Marginal distributions and independence*

Now we show that any subvector of a multivariate skew-normal random vector $\mathbf{X} \sim \mathcal{SN}_p(\boldsymbol{\xi}, \boldsymbol{\Psi}, \boldsymbol{\eta})$ is also skew-normal. For fixed nonnegative integers $p_1 + p_2 = p$, consider the following partition of $\mathbf{X}$ and the induced partition on $\boldsymbol{\xi}$, $\boldsymbol{\Psi}$, and $\boldsymbol{\eta}$:

$$\mathbf{X} = \begin{pmatrix} \mathbf{X}_1 \\ \mathbf{X}_2 \end{pmatrix}, \quad \boldsymbol{\xi} = \begin{pmatrix} \boldsymbol{\xi}_1 \\ \boldsymbol{\xi}_2 \end{pmatrix}, \quad \boldsymbol{\Psi} = \begin{pmatrix} \boldsymbol{\Psi}_{11} & \boldsymbol{\Psi}_{12} \\ \boldsymbol{\Psi}_{21} & \boldsymbol{\Psi}_{22} \end{pmatrix}, \quad \boldsymbol{\eta} = \begin{pmatrix} \boldsymbol{\eta}_1 \\ \boldsymbol{\eta}_2 \end{pmatrix}, \tag{5.17}$$

where $\mathbf{X}_j, \boldsymbol{\xi}_j, \boldsymbol{\eta}_j$ are of dimension $p_j \times 1$ and $\boldsymbol{\Psi}_{jk}$ is of dimension $p_j \times p_k$, $j,k = 1,2$. The result of the following theorem is a direct consequence of Theorem 58.

Theorem 59 *Let* $\mathbf{X} \sim \mathcal{SN}_p(\boldsymbol{\xi}, \boldsymbol{\Psi}, \boldsymbol{\eta})$ *be partitioned as (5.17). Then, the partitions* $\mathbf{X}_j \sim \mathcal{SN}_{p_j}(\boldsymbol{\xi}_j, \boldsymbol{\Psi}_{jj}, \boldsymbol{\eta}_j)$ *for* $j = 1,2$*.*

Proof 59 *Since* $\mathbf{X}_j = \mathbf{A}_j\mathbf{X}$*,* $j = 1,2$*, where* $\mathbf{A}_1 = (\mathbf{I}_{p_1}, \mathbf{O})$ *and* $\mathbf{A}_2 = (\mathbf{O}, \mathbf{I}_{p_2})$*, the proof is then direct from Theorem 58.* □

From Theorem 59, an advantage of the $(\boldsymbol{\Psi}, \boldsymbol{\eta})$ parameterization of the SN is that it yields a marginally consistent distribution, i.e., the parameters of the marginal distribution depend only on the corresponding subparameters of the full distribution. The same holds for the $(\boldsymbol{\Omega}, \boldsymbol{\delta})$ parameterization of the ASN distribution. However, this is not the case with the $(\boldsymbol{\Omega}, \boldsymbol{\alpha})$ parameterization of the ASN distribution.

In the nonsingular case with $\boldsymbol{\Psi} > 0$, all the marginals are nonsingular. In the singular case with $\boldsymbol{\Psi} \geq 0$ they can be singular or nonsingular. The following theorem establishes the condition to have a nonsingular marginal in this last case.

Theorem 60 *Let* $\mathbf{X} \sim \mathcal{SN}_p(\boldsymbol{\xi}, \boldsymbol{\Psi}, \boldsymbol{\eta})$*, with* $\operatorname{rank}(\boldsymbol{\Psi}) = r$*, be partitioned as (5.17). If* $\boldsymbol{\Psi}_{11}$ *has full rank* $p_1 = p - p_2 \leq r$*, then* $\mathbf{X}_1 \sim \mathcal{SN}_{p_1}(\boldsymbol{\xi}_1, \boldsymbol{\Psi}_{11}, \boldsymbol{\eta}_1)$ *is nonsingular.*

Proof 60 *It is direct from Theorem 59 and the assumption that* $\boldsymbol{\Psi}_{11} > 0$*.* □

The following result is also direct from Theorem 58 and states that the skewness of the distribution of each subvector of $\mathbf{X}$ is only regulated by the corresponding components of $\boldsymbol{\eta}$.

Theorem 61 *Let* $\mathbf{X} \sim \mathcal{SN}_p(\boldsymbol{\xi}, \boldsymbol{\Psi}, \boldsymbol{\eta})$ *be partitioned as (5.17). Then for every fixed* $p_1 + p_2 = p$*,* $\mathbf{X}_j$ *has symmetric* $\mathcal{N}_{p_j}(\boldsymbol{\xi}_j, \boldsymbol{\Psi}_{jj})$ *distribution if and only if* $\boldsymbol{\eta}_j = \mathbf{0}_{p_j}$*,* $j = 1,2$*.*

Proof 61 *Since* $\mathbf{X}_j \sim \mathcal{SN}_{p_j}(\boldsymbol{\xi}_j, \boldsymbol{\Psi}_{jj}, \boldsymbol{\eta}_j)$*, its cf is* $\psi_j(\mathbf{t}_j) = 2\psi_{\mathrm{N},j}(\mathbf{t}_j)\Phi(i\mathbf{t}_j^\top \boldsymbol{\eta}_j)$*, where* $\psi_{\mathrm{N},j}(\mathbf{t}_j)$ *is the cf of the* $\mathcal{N}_{p_j}(\boldsymbol{\xi}_j, \boldsymbol{\Psi}_{jj})$ *distribution. Clearly,* $\psi_j(\mathbf{t}_j) = \psi_{\mathrm{N},j}(\mathbf{t}_j), \forall \mathbf{t}_j \iff \boldsymbol{\eta}_j = \mathbf{0}_{p_j}$*.* □

A relevant connection with the latest results is that a prerequisite for independence is that one of the parts be symmetric.

Theorem 62 *Let* $\mathbf{X} \sim \mathcal{SN}_p(\boldsymbol{\xi}, \boldsymbol{\Psi}, \boldsymbol{\eta})$ *be partitioned as (5.17). Then for every fixed* $p_1 + p_2 = p$, $\mathbf{X}_1$ *and* $\mathbf{X}_2$ *are independent if and only if* $\boldsymbol{\Psi}_{12} = \mathbf{O}$ *and either* $\mathbf{X}_1$ *or* $\mathbf{X}_2$ *or both are normally distributed.*

Proof 62 *By Theorem 56 the joint cf of* $\mathbf{X}_1$ *and* $\mathbf{X}_2$ *is*

$$\psi(\mathbf{t}_1, \mathbf{t}_2) = 2e^{\mathrm{i}(\mathbf{t}_1^\top \boldsymbol{\xi}_1 + \mathbf{t}_2^\top \boldsymbol{\xi}_2) - \frac{1}{2}(\mathbf{t}_1^\top \boldsymbol{\Omega}_{11}\mathbf{t}_1 + 2\mathbf{t}_1^\top \boldsymbol{\Omega}_{12}\mathbf{t}_2 + \mathbf{t}_2^\top \boldsymbol{\Omega}_{22}\mathbf{t}_2)} \Phi(\mathrm{i}\mathbf{t}_1^\top \boldsymbol{\eta}_1 + \mathrm{i}\mathbf{t}_2^\top \boldsymbol{\eta}_2),$$

where $\boldsymbol{\Omega}_{jk} = \boldsymbol{\Psi}_{jk} + \boldsymbol{\eta}_j \boldsymbol{\eta}_k^\top$ *for* $j, k = 1, 2$. *Also, their respective marginal cfs are:*

$$\psi_1(\mathbf{t}_1) = \psi(\mathbf{t}_1, \mathbf{0}_{p_2}) = 2e^{\mathrm{i}\mathbf{t}_1^\top \boldsymbol{\xi}_1 - \frac{1}{2}\mathbf{t}_1 \boldsymbol{\Omega}_{11}\mathbf{t}_1} \Phi(\mathrm{i}\mathbf{t}_1^\top \boldsymbol{\eta}_1)$$

and

$$\psi_2(\mathbf{t}_2) = \psi(\mathbf{0}_{p_1}, \mathbf{t}_2) = 2e^{\mathrm{i}\mathbf{t}_2^\top \boldsymbol{\xi}_2 - \frac{1}{2}\mathbf{t}_1^\top \boldsymbol{\Omega}_{22}\mathbf{t}_2} \Phi(\mathrm{i}\mathbf{t}_2^\top \boldsymbol{\eta}_2).$$

Thus, it clearly follows that $\psi(\mathbf{t}_1, \mathbf{t}_2) = \psi_1(\mathbf{t}_1)\psi_2(\mathbf{t}_2)$ *for all* $(\mathbf{t}_1^\top, \mathbf{t}_2^\top)^\top \in \mathbb{R}^{p_1+p_2}$ *if and only if*

$$e^{-\mathbf{t}_1^\top \boldsymbol{\Omega}_{12}\mathbf{t}_2} \Phi(\mathrm{i}\mathbf{t}_1^\top \boldsymbol{\eta}_1 + \mathrm{i}\mathbf{t}_2^\top \boldsymbol{\eta}_2) = 2\Phi(\mathrm{i}\mathbf{t}_1^\top \boldsymbol{\eta}_1)\Phi(\mathrm{i}\mathbf{t}_2^\top \boldsymbol{\eta}_2), \quad \forall (\mathbf{t}_1^\top, \mathbf{t}_2^\top)^\top \in \mathbb{R}^{p_1+p_2},$$

which can happen if and only if $\boldsymbol{\Psi}_{12} = \mathbf{O}$ *and either* $\boldsymbol{\eta}_1 = \mathbf{0}_{p_1}$ *or* $\boldsymbol{\eta}_2 = \mathbf{0}_{p_2}$. *Since* $\mathbf{X}_1$ *and* $\mathbf{X}_2$ *are independent if and only if their joint cf is the product of the marginal cfs, the proof is complete.*

□

Example 61 *Let* $\mathbf{X} = (X_1, \ldots, X_p)^\top \sim \mathcal{SN}_p(\boldsymbol{\xi}, \boldsymbol{\Psi}, \boldsymbol{\eta})$, *with parameters* $\boldsymbol{\xi} = (\xi_1, \ldots, \xi_p)^\top$, $\boldsymbol{\Psi} = \mathrm{diag}(\psi_{11}, \ldots, \psi_{pp})$ *and* $\boldsymbol{\eta} = (\eta_1, \ldots, \eta_p)^\top$. *Then the marginal distribution of* X_j *is* $\mathcal{SN}_1(\xi_j, \psi_{jj}, \eta_j)$, $j = 1, \ldots, p$, *but they cannot be mutually independent unless* $p - 1$ *of the skewness parameters are zero; for instance* $\eta_2 = \cdots = \eta_p = 0$.

Example 62 *Suppose that* $\mathbf{X} = (\mathbf{X}_1^\top, \mathbf{X}_2^\top)^\top \sim \mathcal{SN}_{p_1+p_2}(\boldsymbol{\xi}, \boldsymbol{\Psi}, \boldsymbol{\eta})$, *with* $\boldsymbol{\xi} = (\boldsymbol{\xi}_1^\top, \boldsymbol{\xi}_2^\top)^\top \in \mathbb{R}^{p_1+p_2}$, $\boldsymbol{\Psi} = \mathrm{diag}(\boldsymbol{\Psi}_{11}, \boldsymbol{\Psi}_{22})$, *and* $\boldsymbol{\eta} = (\boldsymbol{\eta}_1^\top, \mathbf{0}_{p_2}^\top)^\top \in \mathbb{R}^{p_1+p_2}$. *Then,* $\mathbf{X}_1$ *and* $\mathbf{X}_2$ *are independent with marginal distributions given by*

$$\mathbf{X}_1 \sim \mathcal{SN}_{p_1}(\boldsymbol{\xi}_1, \boldsymbol{\Psi}_{11}, \boldsymbol{\eta}_1), \quad \mathbf{X}_2 \sim \mathcal{SN}_{p_2}(\boldsymbol{\xi}_2, \boldsymbol{\Psi}_{22}, \mathbf{0}_{p_2}) \equiv \mathcal{N}_{p_2}(\boldsymbol{\xi}_2, \boldsymbol{\Psi}_{22}).$$

An immediate extension of Theorem 62 is given next.

Theorem 63 *Let* $\mathbf{X} \sim \mathcal{SN}_p(\boldsymbol{\xi}, \boldsymbol{\Psi}, \boldsymbol{\eta})$, *where at least one of the components of* $\boldsymbol{\eta}$ *is nonzero. Then for any matrices* $\mathbf{A}_j \in \mathbb{R}^{p_j \times p}$, $j = 1, 2$, *we have* $\mathbf{A}_1\mathbf{X}$ *and* $\mathbf{A}_2\mathbf{X}$ *are independent if and only if* $\mathbf{A}_1\boldsymbol{\Psi}\mathbf{A}_2^\top = \mathbf{O}$ *and either* $\mathbf{A}_1\boldsymbol{\eta} = \mathbf{0}_{p_1}$ *or* $\mathbf{A}_2\boldsymbol{\eta} = \mathbf{0}_{p_2}$.

Proof 63 *Let $\tilde{\psi}(\mathbf{t}_1,\mathbf{t}_2) = \mathrm{E}\{e^{\mathrm{i}(\mathbf{t}_1\mathbf{Y}_1+\mathbf{t}_2\mathbf{Y}_2)}\}$ be the joint cf of $\mathbf{Y}_1 = \mathbf{A}_1\mathbf{X}$ and $\mathbf{Y}_2 = \mathbf{A}_2\mathbf{X}$. Since $\tilde{\psi}(\mathbf{t}_1,\mathbf{t}_2) = \psi(\mathbf{A}_1^\top\mathbf{t}_1,\mathbf{A}_2^\top\mathbf{t}_2)$, where $\psi(\mathbf{t}_1,\mathbf{t}_2)$ is the joint cf of $\mathbf{X}_1$ and $\mathbf{X}_2$, by Theorem 56 we have*

$$\tilde{\psi}(\mathbf{t}_1,\mathbf{t}_2) = 2e^{\mathrm{i}(\mathbf{t}_1^\top\mathbf{A}_1+\mathbf{t}_2^\top\mathbf{A}_2)\boldsymbol{\xi}-\frac{1}{2}(\mathbf{t}_1^\top\mathbf{A}_1\boldsymbol{\Omega}\mathbf{A}_1^\top\mathbf{t}_1+2\mathbf{t}_1^\top\mathbf{A}_1\boldsymbol{\Omega}\mathbf{A}_2^\top\mathbf{t}_2+\mathbf{t}_2^\top\mathbf{A}_2\boldsymbol{\Omega}\mathbf{A}_2^\top\mathbf{t}_2)}\Phi(\mathrm{i}\mathbf{t}_1^\top\mathbf{A}_1\boldsymbol{\eta}+\mathrm{i}\mathbf{t}_2^\top\mathbf{A}_2\boldsymbol{\eta}),$$

where $\boldsymbol{\Omega} = \boldsymbol{\Psi}+\boldsymbol{\eta}\boldsymbol{\eta}^\top$. Also, their respective marginal cfs are:

$$\tilde{\psi}_1(\mathbf{t}_1) = \tilde{\psi}(\mathbf{t}_1,\mathbf{0}_{p_2}) = 2e^{\mathrm{i}\mathbf{t}_1^\top\mathbf{A}_1\boldsymbol{\xi}-\frac{1}{2}\mathbf{t}_1^\top\mathbf{A}_1\boldsymbol{\Omega}\mathbf{A}_1^\top\mathbf{t}_1}\Phi(\mathrm{i}\mathbf{t}_1^\top\mathbf{A}_1\boldsymbol{\eta}),$$

and

$$\tilde{\psi}_2(\mathbf{t}_2) = \tilde{\psi}(\mathbf{0}_{p_1},\mathbf{t}_2) = 2e^{\mathrm{i}\mathbf{t}_2^\top\mathbf{A}_2\boldsymbol{\xi}-\frac{1}{2}\mathbf{t}_2^\top\mathbf{A}_2\boldsymbol{\Omega}\mathbf{A}_2^\top\mathbf{t}_2}\Phi(\mathrm{i}\mathbf{t}_2^\top\mathbf{A}_2\boldsymbol{\eta}).$$

Thus, since $\mathbf{A}_1\boldsymbol{\Omega}\mathbf{A}_2^\top = \mathbf{A}_1\boldsymbol{\Psi}\mathbf{A}_2^\top + \mathbf{A}_1\boldsymbol{\eta}\boldsymbol{\eta}^\top\mathbf{A}_2^\top$, it follows that $\tilde{\psi}(\mathbf{t}_1,\mathbf{t}_2) = \tilde{\psi}_1(\mathbf{t}_1)\tilde{\psi}_2(\mathbf{t}_2)$ for all $(\mathbf{t}_1^\top,\mathbf{t}_2^\top)^\top \in \mathbb{R}^{p_1+p_2}$ if and only if

$$e^{-\mathbf{t}_1^\top\mathbf{A}_1\boldsymbol{\Omega}\mathbf{A}_2^\top\mathbf{t}_2}\Phi(\mathrm{i}\mathbf{t}_1^\top\mathbf{A}_1\boldsymbol{\eta}+\mathrm{i}\mathbf{t}_2^\top\mathbf{A}_2\boldsymbol{\eta}) = 2\Phi(\mathrm{i}\mathbf{t}_1^\top\mathbf{A}_1\boldsymbol{\eta})\Phi(\mathrm{i}\mathbf{t}_2^\top\mathbf{A}_2\boldsymbol{\eta}), \quad \forall(\mathbf{t}_1^\top,\mathbf{t}_2^\top)^\top \in \mathbb{R}^{p_1+p_2},$$

which holds if and only if $\mathbf{A}_1\boldsymbol{\Psi}\mathbf{A}_2^\top = \mathbf{O}$ and either $\mathbf{A}_1\boldsymbol{\eta} = \mathbf{0}_{p_1}$ or $\mathbf{A}_2\boldsymbol{\eta} = \mathbf{0}_{p_2}$. Since $\mathbf{Y}_1$ and $\mathbf{Y}_2$ are independent if and only if their joint cf is the product of the marginal cfs, the proof is therefore completed. □

From the proof of Theorem 63, it follows also that a necessary condition to obtain independence between two arbitrary linear transformations of $\mathbf{X} \sim \mathcal{SN}_p(\boldsymbol{\xi},\boldsymbol{\Psi},\boldsymbol{\eta})$ is that the distribution of at least one of them must be symmetric normal. An important related result with this fact is that to have additivity under independent skew-normal random vectors, only one of them can have a non-null skewness vector, as established in the following theorem.

Theorem 64 *Let $\mathbf{Y} = a_1\mathbf{X}_1 + a_2\mathbf{X}_2$, where $\mathbf{X}_1 \sim \mathcal{SN}_p(\boldsymbol{\xi}_1,\boldsymbol{\Psi}_1,\boldsymbol{\eta}_1)$ and $\mathbf{X}_2 \sim \mathcal{N}_p(\boldsymbol{\xi}_2,\boldsymbol{\Psi}_2)$ be independent random vectors, and $a_1 \in \mathbb{R}$, $a_2 \in \mathbb{R}$. Then $\mathbf{Y} \sim \mathcal{SN}_p(a_1\boldsymbol{\xi}_1+a_2\boldsymbol{\xi}_2, a_1^2\boldsymbol{\Psi}_1+a_2^2\boldsymbol{\Psi}_2, a_1\boldsymbol{\eta}_1)$.*

Proof 64 *From the assumptions, the cf of $\mathbf{Y}$ becomes*

$$\psi_{\mathbf{Y}}(\mathbf{t}) = \psi_{\mathbf{X}_1}(a_1\mathbf{t})\psi_{\mathbf{X}_2}(a_2\mathbf{t}) = 2e^{\mathrm{i}\mathbf{t}^\top(a_1\boldsymbol{\xi}_1+a_2\boldsymbol{\xi}_2)-\frac{1}{2}\mathbf{t}^\top(a_1^2\boldsymbol{\Psi}_1+a_2^2\boldsymbol{\Psi}_2+a_1^2\boldsymbol{\eta}_1\boldsymbol{\eta}_1^\top)\mathbf{t}}\Phi(\mathrm{i}a_1\mathbf{t}^\top\boldsymbol{\eta}_1).$$

The proof concludes from the uniqueness of the cf. □

From the form of the cf of the skew-normal distribution, it can be verified that the sum of two independent SN random vectors is no longer of SN type. More generally, a random vector is said to be *infinitely divisible* if its multivariate distribution can, for every positive integer n, be expressed as the distribution of the sum of n iid random vectors. Equivalently, its characteristic function can be written as the n-th power of another characteristic function for all $n \geq 1$. Although the multivariate normal distribution is known to be infinitely divisible, Domínguez-Molina and Rocha-Arteaga (2007) have shown that the multivariate skew-normal distribution is not. This result has been confirmed in the univariate case by Kozubowski and Nolan (2008), who also proved that the univariate skew-Laplace distribution (Gupta et al., 2002) is infinitely divisible.

5.4.3 Canonical form

The so-called *canonical form* for the ASN distribution has been studied by Azzalini and Capitanio (1999) and Loperfido (2010), and extended to scale mixtures of the ASN distribution by Capitanio

(2020). In the SN parameterization, the canonical representation is related to the construction described in Remark 5, i.e., the SN random vector $\mathbf{Z} = \mathbf{X} - \boldsymbol{\xi} = \boldsymbol{\eta} U + \mathbf{W} \sim \mathcal{SN}_p(\mathbf{0}_p, \boldsymbol{\Psi}, \boldsymbol{\eta})$ has the canonical form if $\boldsymbol{\Psi} = \mathbf{I}_p$ and in the partition

$$\begin{pmatrix} Z_1 \\ \mathbf{Z}_2 \end{pmatrix} = \begin{pmatrix} \eta_1 \\ \boldsymbol{\eta}_2 \end{pmatrix} U + \begin{pmatrix} W_1 \\ \mathbf{W}_2 \end{pmatrix},$$

we put $\boldsymbol{\eta}_2 = \mathbf{0}_{p-1}$, so that $Z_1 \sim \mathcal{SN}(0,1,\eta_1)$, and $\mathbf{Z}_2 \sim \mathcal{N}_{p-1}(\mathbf{0}_{p-1}, \mathbf{I}_{p-1})$, and they are independent. To move the skewness in any direction, simply rotate the components of $\mathbf{Z}$ using an orthogonal matrix $\boldsymbol{\Gamma} \in \mathcal{O}_p$. For the general case $\mathbf{X} - \boldsymbol{\xi} \sim \mathcal{SN}_p(\mathbf{0}_p, \boldsymbol{\Psi}, \boldsymbol{\eta})$ with $\boldsymbol{\Psi} > 0$ arbitrary, we can always find a nonsingular $p \times p$ matrix $\mathbf{C}$ such that $\mathbf{C}\boldsymbol{\Psi}\mathbf{C}^\top = \mathbf{I}_p$ and $\mathbf{C}\boldsymbol{\eta} = \eta_1 \mathbf{e}_1$, where $\mathbf{e}_1$ is the first p-dimensional unit vector, so that $\mathbf{Z} = \mathbf{C}(\mathbf{X} - \boldsymbol{\xi}) \sim \mathcal{SN}_p(\mathbf{0}_p, \mathbf{I}_p, \eta_1 \mathbf{e}_1)$. Moreover, since $\eta_1^2 = \boldsymbol{\eta}^\top \mathbf{C}^\top \mathbf{C} \boldsymbol{\eta}$ and $\mathbf{C}^\top \mathbf{C} = \boldsymbol{\Psi}^{-1}$, we find that the skewness parameter $\eta_1 = (\boldsymbol{\eta}^\top \boldsymbol{\Psi}^{-1} \boldsymbol{\eta})^{1/2}$.

As highlighted by Azzalini and Capitanio (1999), the canonical transformation plays a role similar to that of converting a multivariate normal random vector into a spherical-normal one. They also emphasized the role of $(\boldsymbol{\eta}^\top \boldsymbol{\Psi}^{-1} \boldsymbol{\eta})^{1/2}$ as a quantity that summarizes the shape of the distribution. In fact, as follows from Theorem 79 given in Section 5.8.3, both Mardia measures depend on the skewness vector exclusively through $(\boldsymbol{\eta}^\top \boldsymbol{\Psi}^{-1} \boldsymbol{\eta})^{1/2}$. In summary, the canonical form is characterized by one component that absorbs the departure from normality of the whole distribution.

5.5 Nonsingular Skew-Normal

5.5.1 *Pdf*

When $\mathbf{X} = \boldsymbol{\xi} + \boldsymbol{\eta} U + \mathbf{W} \sim \mathcal{SN}_p(\boldsymbol{\xi}, \boldsymbol{\Psi}, \boldsymbol{\eta})$ and $\boldsymbol{\Psi} > 0$, i.e., $|\boldsymbol{\Psi}| > 0$, we can find an explicit pdf for $\mathbf{X}$ from the hierarchical representation (5.12) and (5.13). In fact, since $(\mathbf{X} \mid U = u) \sim \mathcal{N}_p(\boldsymbol{\xi} + \boldsymbol{\eta} u, \boldsymbol{\Psi})$, with $\boldsymbol{\Psi} > 0$, is now nonsingular, then it has a multivariate normal conditional pdf given by

$$f(\mathbf{x} \mid u) = \phi_p(\mathbf{x}; \boldsymbol{\xi} + \boldsymbol{\eta} u, \boldsymbol{\Psi}) = |\boldsymbol{\Psi}|^{-1/2} \exp\left\{ -\frac{1}{2} (\mathbf{x} - \boldsymbol{\xi} - \boldsymbol{\eta} u)^\top \boldsymbol{\Psi}^{-1} (\mathbf{x} - \boldsymbol{\xi} - \boldsymbol{\eta} u) \right\}, \quad \mathbf{x} \in \mathbb{R}^p.$$

Hence, since U also has a pdf given by $g(u) = 2\phi(u) I_{(u>0)}$, taking into account that $\phi(u) = \phi(u; 0, 1)$ and the following identity related to the marginal-conditional multivariate normal pdfs:

$$\begin{aligned} \phi_p(\mathbf{x}; \boldsymbol{\xi} + \boldsymbol{\eta} u, \boldsymbol{\Psi}) \phi(u; 0, 1) &= \phi_p(\mathbf{x}; \boldsymbol{\xi}, \boldsymbol{\Psi} + \boldsymbol{\eta}\boldsymbol{\eta}^\top) \\ &\quad \times \phi\{u; \boldsymbol{\eta}^\top (\boldsymbol{\Psi} + \boldsymbol{\eta}\boldsymbol{\eta}^\top)^{-1} (\mathbf{x} - \boldsymbol{\xi}), 1 - \boldsymbol{\eta}^\top (\boldsymbol{\Psi} + \boldsymbol{\eta}\boldsymbol{\eta}^\top)^{-1} \boldsymbol{\eta}\}, \end{aligned}$$

then using the change of variable $v = -\{u - \boldsymbol{\eta}^\top (\boldsymbol{\Psi} + \boldsymbol{\eta}\boldsymbol{\eta}^\top)^{-1} (\mathbf{x} - \boldsymbol{\xi})\} / \sqrt{1 - \boldsymbol{\eta}^\top (\boldsymbol{\Psi} + \boldsymbol{\eta}\boldsymbol{\eta}^\top)^{-1} \boldsymbol{\eta}}$, we have

$$\begin{aligned} f(\mathbf{x}) &= \int_{-\infty}^{\infty} f(\mathbf{x} \mid u) g(u) \mathrm{d}u \\ &= 2 \int_0^{\infty} \phi_p(\mathbf{x}; \boldsymbol{\xi} + \boldsymbol{\eta} u, \boldsymbol{\Psi}) \phi(u) \mathrm{d}u \\ &= 2\phi_p(\mathbf{x}; \boldsymbol{\xi}, \boldsymbol{\Psi} + \boldsymbol{\eta}\boldsymbol{\eta}^\top) \int_0^{\infty} \phi(u; \boldsymbol{\eta}^\top (\boldsymbol{\Psi} + \boldsymbol{\eta}\boldsymbol{\eta}^\top)^{-1} (\mathbf{x} - \boldsymbol{\xi}), 1 - \boldsymbol{\eta}^\top (\boldsymbol{\Psi} + \boldsymbol{\eta}\boldsymbol{\eta}^\top)^{-1} \boldsymbol{\eta}) \mathrm{d}u \\ &= 2\phi_p(\mathbf{x}; \boldsymbol{\xi}, \boldsymbol{\Psi} + \boldsymbol{\eta}\boldsymbol{\eta}^\top) \int_{-\infty}^{\frac{\boldsymbol{\eta}^\top (\boldsymbol{\Psi} + \boldsymbol{\eta}\boldsymbol{\eta}^\top)^{-1} (\mathbf{x} - \boldsymbol{\xi})}{\sqrt{1 - \boldsymbol{\eta}^\top (\boldsymbol{\Psi} + \boldsymbol{\eta}\boldsymbol{\eta}^\top)^{-1} \boldsymbol{\eta}}}} \phi(v) \mathrm{d}v \\ &= 2\phi_p(\mathbf{x}; \boldsymbol{\xi}, \boldsymbol{\Psi} + \boldsymbol{\eta}\boldsymbol{\eta}^\top) \Phi\left\{ \frac{\boldsymbol{\eta}^\top (\boldsymbol{\Psi} + \boldsymbol{\eta}\boldsymbol{\eta}^\top)^{-1} (\mathbf{x} - \boldsymbol{\xi})}{\sqrt{1 - \boldsymbol{\eta}^\top (\boldsymbol{\Psi} + \boldsymbol{\eta}\boldsymbol{\eta}^\top)^{-1} \boldsymbol{\eta}}} \right\}, \quad \mathbf{x} \in \mathbb{R}^p. \end{aligned} \tag{5.18}$$

Finally, by using the following relationships:

$$(\boldsymbol{\Psi} + \boldsymbol{\eta}\boldsymbol{\eta}^\top)^{-1} = \boldsymbol{\Psi}^{-1} - \frac{\boldsymbol{\Psi}^{-1} \boldsymbol{\eta}\boldsymbol{\eta}^\top \boldsymbol{\Psi}^{-1}}{1 + \boldsymbol{\eta}^\top \boldsymbol{\Psi}^{-1} \boldsymbol{\eta}} \quad \Longrightarrow \quad \frac{\boldsymbol{\eta}^\top (\boldsymbol{\Psi} + \boldsymbol{\eta}\boldsymbol{\eta}^\top)^{-1} (\mathbf{x} - \boldsymbol{\xi})}{\sqrt{1 - \boldsymbol{\eta}^\top (\boldsymbol{\Psi} + \boldsymbol{\eta}\boldsymbol{\eta}^\top)^{-1} \boldsymbol{\eta}}} = \frac{\boldsymbol{\eta}^\top \boldsymbol{\Psi}^{-1} (\mathbf{x} - \boldsymbol{\xi})}{\sqrt{1 + \boldsymbol{\eta}^\top \boldsymbol{\Psi}^{-1} \boldsymbol{\eta}}},$$

we find that the pdf of $\mathbf{X} \sim \mathcal{SN}_p(\boldsymbol{\xi}, \boldsymbol{\Psi}, \boldsymbol{\eta})$, $\boldsymbol{\Psi} > 0$, can be rewritten as in the next theorem.

Theorem 65 *Let $\mathbf{X} \sim \mathcal{SN}_p(\boldsymbol{\xi}, \boldsymbol{\Psi}, \boldsymbol{\eta})$, $\boldsymbol{\Psi} > 0$, be a nonsingular multivariate skew-normal distribution. Then its pdf is*

$$f(\mathbf{x}) = 2\phi_p(\mathbf{x}; \boldsymbol{\xi}, \boldsymbol{\Psi} + \boldsymbol{\eta}\boldsymbol{\eta}^\top)\Phi\left\{\frac{\boldsymbol{\eta}^\top\boldsymbol{\Psi}^{-1}(\mathbf{x}-\boldsymbol{\xi})}{\sqrt{1+\boldsymbol{\eta}^\top\boldsymbol{\Psi}^{-1}\boldsymbol{\eta}}}\right\}, \quad \mathbf{x} \in \mathbb{R}^p. \tag{5.19}$$

Proof 65 *See above.* □

Adcock and Shutes (2001) and Adcock (2004) used the same parameterization as in (5.19). Azzalini and Capitanio (2014) mentioned in their equation (5.30), but did not explore, a parameterization similar to the one in (5.19), albeit with a skewness parameter equal to $\{\text{diag}(\boldsymbol{\Psi})\}^{-1/2}\boldsymbol{\eta}$ in our notation; see also Azzalini and Dalla Valle (1996) and Theorem 67 below.

Remark 7 *Note by (5.18) that in terms of the matrix $\boldsymbol{\Omega} = \boldsymbol{\Psi} + \boldsymbol{\eta}\boldsymbol{\eta}^\top$, the multivariate skew-normal pdf can be rewritten as*

$$f(\mathbf{x}) = 2\phi_p(\mathbf{x}; \boldsymbol{\xi}, \boldsymbol{\Omega})\Phi\left\{\frac{\boldsymbol{\eta}^\top\boldsymbol{\Omega}^{-1}(\mathbf{x}-\boldsymbol{\xi})}{\sqrt{1-\boldsymbol{\eta}^\top\boldsymbol{\Omega}^{-1}\boldsymbol{\eta}}}\right\}, \quad \mathbf{x} \in \mathbb{R}^p.$$

The next proposition specializes the result in Theorem 58 to nonsingular affine transformations in an absolutely continuous skew-normal random vector $\mathbf{X}$.

Theorem 66 *Let $\mathbf{X} \sim \mathcal{SN}_p(\boldsymbol{\xi}, \boldsymbol{\Psi}, \boldsymbol{\eta})$, $\boldsymbol{\Psi} > 0$, be a nonsingular multivariate skew-normal distribution, with pdf given by (5.19). Also, let $\mathbf{Y} = \mathbf{A}\mathbf{X} + \mathbf{b}$, where $\mathbf{A} \in \mathbb{R}^{p\times p}$ is a nonsingular constant matrix and $\mathbf{b} \in \mathbb{R}^p$ is a constant vector. Then, $\mathbf{Y} \sim \mathcal{SN}_p(\boldsymbol{\xi}_A, \boldsymbol{\Psi}_A, \boldsymbol{\eta}_A)$ is also nonsingular and has pdf given by*

$$h(\mathbf{y}) = 2\phi_p(\mathbf{y}; \boldsymbol{\xi}_A, \boldsymbol{\Psi}_A + \boldsymbol{\eta}_A\boldsymbol{\eta}_A^\top)\Phi\left\{\frac{\boldsymbol{\eta}_A^\top\boldsymbol{\Psi}_A^{-1}(\mathbf{y}-\boldsymbol{\xi}_A)}{\sqrt{1+\boldsymbol{\eta}_A^\top\boldsymbol{\Psi}_A^{-1}\boldsymbol{\eta}_A}}\right\}, \quad \mathbf{y} \in \mathbb{R}^p,$$

where

$$\boldsymbol{\xi}_A = \mathbf{A}\boldsymbol{\xi} + \mathbf{b}, \quad \boldsymbol{\Psi}_A = \mathbf{A}\boldsymbol{\Psi}\mathbf{A}^\top, \quad \boldsymbol{\eta}_A = \mathbf{A}\boldsymbol{\eta}. \tag{5.20}$$

Proof 66 *We first note that if $|\mathbf{A}| \neq 0$ and $\boldsymbol{\Psi} > 0$, then $\boldsymbol{\Psi}_A = \mathbf{A}\boldsymbol{\Psi}\mathbf{A}^\top > 0$ and both $\mathbf{A}^{-1}$ and $\boldsymbol{\Psi}_A^{-1} = (\mathbf{A}\boldsymbol{\Psi}\mathbf{A}^\top)^{-1} = \mathbf{A}^{-\top}\boldsymbol{\Psi}^{-1}\mathbf{A}^{-1}$ exist. Now, inverting the affine relation $\mathbf{y} = \mathbf{A}\mathbf{x} + \mathbf{b}$, we can write $\mathbf{x} = \mathbf{A}^{-1}(\mathbf{y} - \mathbf{b})$. The pdf of $\mathbf{Y}$ is then given by*

$$\begin{aligned} h(\mathbf{y}) &= f\{\mathbf{A}^{-1}(\mathbf{y}-\mathbf{b})\}\text{abs}(|\mathbf{A}|)^{-1} \\ &= 2\phi_p\{\mathbf{A}^{-1}(\mathbf{y}-\mathbf{b}); \boldsymbol{\xi}, \boldsymbol{\Psi} + \boldsymbol{\eta}\boldsymbol{\eta}^\top\}\Phi[\boldsymbol{\eta}^\top\boldsymbol{\Psi}^{-1}\{\mathbf{A}^{-1}(\mathbf{y}-\mathbf{b}) - \boldsymbol{\xi}\}]\text{abs}(|\mathbf{A}|)^{-1}, \end{aligned}$$

since the Jacobian $J_{(\mathbf{x}\to\mathbf{y})} = |\mathbf{A}^{-1}| = |\mathbf{A}|^{-1}$, *and where* f *is the skew-normal pdf defined in (5.19). It is then straightforward to verify that*

$$\phi_p\left\{\mathbf{A}^{-1}(\mathbf{y}-\mathbf{b});\boldsymbol{\xi},\boldsymbol{\Psi}\right\}\operatorname{abs}(|\mathbf{A}|)^{-1} = \phi_p(\mathbf{y};\mathbf{A}\boldsymbol{\xi}+\mathbf{b},\mathbf{A}\boldsymbol{\Psi}\mathbf{A}^\top)$$

and

$$\Phi\left[\frac{\boldsymbol{\eta}^\top\boldsymbol{\Psi}^{-1}\{\mathbf{A}^{-1}(\mathbf{y}-\mathbf{b})-\boldsymbol{\xi}\}}{\sqrt{1+\boldsymbol{\eta}^\top\boldsymbol{\Psi}^{-1}\boldsymbol{\eta}}}\right] = \Phi\left[\frac{(\mathbf{A}\boldsymbol{\eta})^\top(\mathbf{A}\boldsymbol{\Psi}\mathbf{A}^\top)^{-1}\{\mathbf{y}-(\mathbf{A}\boldsymbol{\xi}+\mathbf{b})\}}{\sqrt{1+(\mathbf{A}\boldsymbol{\eta})^\top(\mathbf{A}\boldsymbol{\Psi}\mathbf{A}^\top)^{-1}(\mathbf{A}\boldsymbol{\eta})}}\right].$$

□

We can observe from the proof of Theorem 66 that if we consider (5.19) as the pdf of $\mathbf{X}$, then we obtain a similar form for the distribution of an affine random vector $\mathbf{Y} = \mathbf{AX}+\mathbf{b}$ with the obvious modification of the new parameters as indicated in (5.20).

In particular, if $\mathbf{A} = \boldsymbol{\Psi}^{1/2}$ is the symmetric square root of $\boldsymbol{\Psi}$, then $\mathbf{A}^{-1} = \boldsymbol{\Psi}^{-1/2}$; hence, from $\mathbf{X}\sim\mathcal{SN}_p(\boldsymbol{\xi},\boldsymbol{\Psi},\boldsymbol{\eta})$ one can define the standard random vector $\mathbf{Z} = \boldsymbol{\Psi}^{-1/2}(\mathbf{X}-\boldsymbol{\xi})$ and show from Theorem 66 (or the Jacobian method) that $\mathbf{Z}\sim\mathcal{SN}_p(\mathbf{0}_p,\mathbf{I}_p,\boldsymbol{\eta}_z)$, where $\boldsymbol{\eta}_z = \boldsymbol{\Psi}^{-1/2}\boldsymbol{\eta}$, with pdf:

$$f_{\mathbf{Z}}(\mathbf{z}) = 2\phi_p(\mathbf{z};\mathbf{0}_p,\mathbf{I}_p+\boldsymbol{\eta}_z\boldsymbol{\eta}_z^\top)\Phi\left(\frac{\boldsymbol{\eta}_z^\top\mathbf{z}}{\sqrt{1+\boldsymbol{\eta}_z^\top\boldsymbol{\eta}_z}}\right), \quad \mathbf{z}\in\mathbb{R}^p.$$

Reciprocally, starting from $\mathbf{Z}$, one can define $\mathbf{X} = \boldsymbol{\Psi}^{1/2}\mathbf{Z}+\boldsymbol{\xi}$ and again use Theorem 66 (or the Jacobian method) to obtain the pdf (5.19).

Thus, in the nonsingular case, we can formalize this fact as follows.

Theorem 67 $\mathbf{X}\sim\mathcal{SN}_p(\boldsymbol{\xi},\boldsymbol{\Psi},\boldsymbol{\eta})$, $\boldsymbol{\Psi}>0$, *iff* $\mathbf{X}\overset{d}{=}\boldsymbol{\Psi}^{1/2}\mathbf{Z}+\boldsymbol{\xi}$, *where* $\mathbf{Z}\sim\mathcal{SN}_p(\mathbf{0}_p,\mathbf{I}_p,\boldsymbol{\Psi}^{-1/2}\boldsymbol{\eta})$.

Proof 67 *It is directly derived from Theorem 66.* □

However, even if the symmetric scale matrix $\boldsymbol{\Psi}\in\mathbb{R}^{p\times p}$ is positive definite, as previously claimed in Section 5.2, the decomposition $\boldsymbol{\Psi} = \mathbf{AA}^\top$ for a nonsingular matrix $\mathbf{A}\in\mathbb{R}^{p\times p}$ is also not unique.

Remark 8 (Link between $\mathcal{SN}_p$ and $\mathcal{ASN}_p$) *The multivariate skew-normal distribution studied by Azzalini and Capitanio (1999, 2003), denoted here by* $\mathcal{ASN}_p(\boldsymbol{\xi},\boldsymbol{\Omega},\boldsymbol{\alpha})$, *is completely equivalent to the nonsingular multivariate skew-normal* $\mathcal{SN}_p(\boldsymbol{\xi},\boldsymbol{\Psi},\boldsymbol{\eta})$, *except for differences in the parameterization. The relationship between both settings is given by:*

$$\boldsymbol{\Omega} = \boldsymbol{\Psi}+\boldsymbol{\eta}\boldsymbol{\eta}^\top, \qquad \boldsymbol{\alpha} = \frac{\boldsymbol{\omega}\boldsymbol{\Psi}^{-1}\boldsymbol{\eta}}{\sqrt{1+\boldsymbol{\eta}^\top\boldsymbol{\Psi}^{-1}\boldsymbol{\eta}}},$$

where $\boldsymbol{\omega} = \operatorname{diag}(\boldsymbol{\Omega})^{1/2}$ *is a diagonal matrix whose i-th diagonal element is related to the corresponding diagonal element of* $\boldsymbol{\Psi}+\boldsymbol{\eta}\boldsymbol{\eta}^\top$ *as* $\omega_i = \sqrt{\psi_{ii}+\eta_i^2}$, $i = 1,\ldots,p$. *Reciprocally:*

$$\boldsymbol{\Psi} = \boldsymbol{\omega}(\bar{\boldsymbol{\Omega}}-\boldsymbol{\delta}\boldsymbol{\delta}^\top)\boldsymbol{\omega}, \quad \boldsymbol{\eta} = \boldsymbol{\omega}\boldsymbol{\delta}, \quad \text{with} \quad \boldsymbol{\delta} = \frac{\bar{\boldsymbol{\Omega}}\boldsymbol{\alpha}}{\sqrt{1+\boldsymbol{\alpha}^\top\bar{\boldsymbol{\Omega}}\boldsymbol{\alpha}}}, \quad \bar{\boldsymbol{\Omega}} = \boldsymbol{\omega}^{-1}\boldsymbol{\Omega}\boldsymbol{\omega}^{-1}.$$

In stochastic terms, to go from the $\mathcal{ASN}_p$ distribution to the $\mathcal{SN}_p$ distribution, consider the transformation $\mathbf{X} = \boldsymbol{\xi} + \boldsymbol{\omega}\mathbf{Z}_A$, where $\mathbf{Z}_A \sim \mathcal{ASN}_p(\mathbf{0}_p, \bar{\boldsymbol{\Omega}}, \boldsymbol{\alpha})$, then $\mathbf{X} \sim \mathcal{SN}_p(\boldsymbol{\xi}, \boldsymbol{\Psi}, \boldsymbol{\eta})$. An alternative centered parameterization based on the mean vector, the covariance matrix, and the third moment was proposed in Arellano-Valle and Azzalini (2008). A recent review of the skew-normal distribution and its extensions can be found in Azzalini (2022); see also Azzalini (2005).

5.5.2 Cdf from pdf

As shown earlier in Theorem 55, the selection representation given in Theorem 53 provides a simple route to compute the cdf of the multivariate skew-normal either in the singular case or in the nonsingular case. In the nonsingular case, with $\boldsymbol{\Psi} > 0$, the cdf of $\mathbf{X} \sim \mathcal{SN}_p(\boldsymbol{\xi}, \boldsymbol{\Psi}, \boldsymbol{\eta})$ can also be computed directly by integrating (5.19).

In order to illustrate the degree of simplicity of the selection approach, we now calculate the skew-normal cdf from its pdf. To do this, let $\boldsymbol{\Omega} = \boldsymbol{\Psi} + \boldsymbol{\eta}\boldsymbol{\eta}^\top$, and write $\phi(x+a) = \phi_1(x; -a, 1)$. Thus, we can rewrite (5.19) as

$$\begin{aligned}
f(\mathbf{z}) &= 2\phi_p(\mathbf{z}; \boldsymbol{\xi}, \boldsymbol{\Omega})\Phi_1\left\{\frac{\boldsymbol{\eta}^\top\boldsymbol{\Psi}^{-1}(\mathbf{z}-\boldsymbol{\xi})}{\sqrt{1+\boldsymbol{\eta}^\top\boldsymbol{\Psi}^{-1}\boldsymbol{\eta}}}\right\} \\
&= 2\phi_p(\mathbf{z}; \boldsymbol{\xi}, \boldsymbol{\Omega})\int_{-\infty}^{\frac{\boldsymbol{\eta}^\top\boldsymbol{\Psi}^{-1}(\mathbf{z}-\boldsymbol{\xi})}{\sqrt{1+\boldsymbol{\eta}^\top\boldsymbol{\Psi}^{-1}\boldsymbol{\eta}}}} \phi(u)\,\mathrm{d}u \\
&= 2\phi_p(\mathbf{z}; \boldsymbol{\xi}, \boldsymbol{\Omega})\int_{-\infty}^{0} \phi\left(z_0 + \frac{\boldsymbol{\eta}^\top\boldsymbol{\Psi}^{-1}(\mathbf{z}-\boldsymbol{\xi})}{\sqrt{1+\boldsymbol{\eta}^\top\boldsymbol{\Psi}^{-1}\boldsymbol{\eta}}}\right)\mathrm{d}z_0 \\
&= 2\phi_p(\mathbf{z}; \boldsymbol{\xi}, \boldsymbol{\Omega})\int_{-\infty}^{0} \phi_1\left(z_0; -\frac{\boldsymbol{\eta}^\top\boldsymbol{\Psi}^{-1}(\mathbf{z}-\boldsymbol{\xi})}{\sqrt{1+\boldsymbol{\eta}^\top\boldsymbol{\Psi}^{-1}\boldsymbol{\eta}}}, 1\right)\mathrm{d}z_0,
\end{aligned}$$

where we used the change of variable $z_0 = u - \boldsymbol{\eta}^\top\boldsymbol{\Psi}^{-1}(\mathbf{x}-\boldsymbol{\xi})/\sqrt{1+\boldsymbol{\eta}^\top\boldsymbol{\Psi}^{-1}\boldsymbol{\eta}}$. Now, from the marginal-conditional factorization of a $(p+1)$-variate normal pdf we can write

$$\phi_p(\mathbf{z}; \boldsymbol{\xi}, \boldsymbol{\Omega})\phi_1\left(z_0; -\frac{\boldsymbol{\eta}^\top\boldsymbol{\Psi}^{-1}(\mathbf{z}-\boldsymbol{\xi})}{\sqrt{1+\boldsymbol{\eta}^\top\boldsymbol{\Psi}^{-1}\boldsymbol{\eta}}}, 1\right) = \phi_{p+1}(\mathbf{z}_*; \boldsymbol{\xi}_*, \boldsymbol{\Omega}_{**}),$$

where $\mathbf{z}_* = (\mathbf{z}^\top, z_0)^\top$, $\boldsymbol{\xi}_* = (\boldsymbol{\xi}^\top, 0)^\top$ and

$$\boldsymbol{\Omega}_{**} = \begin{pmatrix} \boldsymbol{\Psi} + \boldsymbol{\eta}\boldsymbol{\eta}^\top & -\sqrt{1+\boldsymbol{\eta}^\top\boldsymbol{\Psi}^{-1}\boldsymbol{\eta}}\,\boldsymbol{\eta} \\ -\sqrt{1+\boldsymbol{\eta}^\top\boldsymbol{\Psi}^{-1}\boldsymbol{\eta}}\,\boldsymbol{\eta}^\top & 1+\boldsymbol{\eta}^\top\boldsymbol{\Psi}^{-1}\boldsymbol{\eta} \end{pmatrix}.$$

Note that $\mathbf{\Omega}_{**} = \mathbf{D}_*\mathbf{\Omega}_*\mathbf{D}_*$, where $\mathbf{D}_* = \text{diag}\left(\mathbf{I}_p, \sqrt{1+\boldsymbol{\eta}^\top\mathbf{\Psi}^{-1}\boldsymbol{\eta}}\right)$ and $\mathbf{\Omega}_*$ is defined as in (5.15). Hence, we find

$$
\begin{aligned}
F(\mathbf{x}) &= 2\int_{(-\infty,\mathbf{x}]}\phi_p(\mathbf{z};\boldsymbol{\xi},\mathbf{\Omega})\Phi\left\{\frac{\boldsymbol{\eta}^\top\mathbf{\Psi}^{-1}(\mathbf{z}-\boldsymbol{\xi})}{\sqrt{1+\boldsymbol{\eta}^\top\mathbf{\Psi}^{-1}\boldsymbol{\eta}}}\right\}\mathrm{d}\mathbf{z} \\
&= 2\int_{(-\infty,\mathbf{x}]}\phi_p(\mathbf{z};\boldsymbol{\xi},\mathbf{\Omega})\int_{-\infty}^{0}\phi_1\left(z_0;-\frac{\boldsymbol{\eta}^\top\mathbf{\Psi}^{-1}(\mathbf{z}-\boldsymbol{\xi})}{\sqrt{1+\boldsymbol{\eta}^\top\mathbf{\Psi}^{-1}\boldsymbol{\eta}}},1\right)\mathrm{d}z_0\mathrm{d}\mathbf{z} \\
&= 2\int_{(-\infty,\mathbf{x}]}\int_{-\infty}^{0}\phi_p(\mathbf{z};\boldsymbol{\xi},\mathbf{\Omega})\phi_1\left(z_0;-\frac{\boldsymbol{\eta}^\top\mathbf{\Psi}^{-1}(\mathbf{z}-\boldsymbol{\xi})}{\sqrt{1+\boldsymbol{\eta}^\top\mathbf{\Psi}^{-1}\boldsymbol{\eta}}},1\right)\mathrm{d}z_0\mathrm{d}\mathbf{z} \\
&= 2\int_{(-\infty,\mathbf{x}_*]}\phi_{p+1}(\mathbf{z}_*;\boldsymbol{\xi}_*,\mathbf{\Omega}_{**})\mathrm{d}\mathbf{z}_* \\
&= 2\Phi_{p+1}(\mathbf{x}_*;\boldsymbol{\xi}_*,\mathbf{\Omega}_{**}), \quad \text{with} \quad \mathbf{x}_* = (\mathbf{x}^\top,0)^\top, \quad \mathbf{x}\in\mathbb{R}^p.
\end{aligned}
$$

But, since $\mathbf{x}_* - \boldsymbol{\xi}_* = ((\mathbf{x}-\boldsymbol{\xi})^\top,0)^\top$, then $\mathbf{D}_*^{-1}(\mathbf{x}_*-\boldsymbol{\xi}_*) = \mathbf{x}_*-\boldsymbol{\xi}_*$ and so

$$\Phi_{p+1}(\mathbf{x}_*;\boldsymbol{\xi}_*,\mathbf{\Omega}_{**}) = \Phi_{p+1}(\mathbf{x}_*;\boldsymbol{\xi}_*,\mathbf{D}_*\mathbf{\Omega}_*\mathbf{D}_*) = \Phi_{p+1}(\mathbf{D}_*^{-1}(\mathbf{x}_*-\boldsymbol{\xi}_*);\mathbf{0}_{p+1},\mathbf{\Omega}_*) = \Phi_{p+1}(\mathbf{x}_*;\boldsymbol{\xi}_*,\mathbf{\Omega}_*).$$

5.5.3 Geometry

The skew-normal pdf (5.19) is constant on contours with a skew-ellipsoidal shape centered at $\boldsymbol{\xi}$ and defined by those $\mathbf{x}\in\mathbb{R}^p$ that satisfy

$$\exp\left\{-\frac{1}{2}(\mathbf{x}-\boldsymbol{\xi})^\top(\mathbf{\Psi}+\boldsymbol{\eta}\boldsymbol{\eta}^\top)^{-1}(\mathbf{x}-\boldsymbol{\xi})\right\}\Phi\left\{\frac{\boldsymbol{\eta}^\top\mathbf{\Psi}^{-1}(\mathbf{x}-\boldsymbol{\xi})}{\sqrt{1+\boldsymbol{\eta}^\top\mathbf{\Psi}^{-1}\boldsymbol{\eta}}}\right\} = c^2,$$

where c is a constant. When $\mathbf{\Psi}+\boldsymbol{\eta}\boldsymbol{\eta}^\top = \mathbf{I}_p$, the contours are skewed spheres; see Azzalini (2001) for regions of given probability.

Example 63 *In $\mathbb{R}^2$, consider a bivariate skew-normal distribution, $\mathcal{SN}_2$, with location $\boldsymbol{\xi} = \mathbf{0}_2$ and two different dispersion matrices:*

$$\mathbf{\Psi}_1+\boldsymbol{\eta}_1\boldsymbol{\eta}_1^\top = \begin{pmatrix} 1 & 0 \\ 0 & 1 \end{pmatrix}, \qquad \mathbf{\Psi}_2+\boldsymbol{\eta}_2\boldsymbol{\eta}_2^\top = \begin{pmatrix} 2 & 1/2 \\ 1/2 & 1 \end{pmatrix}.$$

The eigenvalues of $\mathbf{\Psi}_1+\boldsymbol{\eta}_1\boldsymbol{\eta}_1^\top$ are $\lambda_1 = \lambda_2 = 1$ and its standardized eigenvectors are $\boldsymbol{\gamma}_1 = (1,0)^\top$ and $\boldsymbol{\gamma}_2 = (0,1)^\top$. Therefore, the contours are circles centered at the origin when $\boldsymbol{\eta}_1 = (0,0)^\top$. The eigenvalues of $\mathbf{\Psi}_2+\boldsymbol{\eta}_2\boldsymbol{\eta}_2^\top$ are $\lambda_1 = (3+\sqrt{2})/2$ and $\lambda_2 = (3-\sqrt{2})/2$ and its standardized eigenvectors are $\boldsymbol{\gamma}_1 = (1/\sqrt{2(2-\sqrt{2})},(\sqrt{2}-1)/\sqrt{2(2-\sqrt{2})})^\top$ and $\boldsymbol{\gamma}_2 = ((\sqrt{2}-1)/\sqrt{2(2-\sqrt{2})},-1/\sqrt{2(2-\sqrt{2})})^\top$. Therefore, the contours are ellipses centered at the origin with axes in the direction of the two eigenvectors when $\boldsymbol{\eta}_2 = (0,0)^\top$. The contours in Figure 5.2 (a) and (b) are obtained with the matrix $\mathbf{\Psi}_1+\boldsymbol{\eta}_1\boldsymbol{\eta}_1^\top$ and the shape vector $\boldsymbol{\eta}_1$ equal to $(2/\sqrt{5},0)^\top$ and $(2/\sqrt{14},3/\sqrt{14})^\top$, respectively, which correspond to the vector $\boldsymbol{\alpha}_1$ equal to $(2,0)^\top$ and $(2,3)^\top$, respectively, in the ASN parameterization. The contours in Figure 5.2 (c) and (d) are obtained with the matrix $\mathbf{\Psi}_2+\boldsymbol{\eta}_2\boldsymbol{\eta}_2^\top$ and the shape vector $\boldsymbol{\eta}_2$ equal to $(2\sqrt{2}/\sqrt{5},1/\sqrt{10})^\top$ and

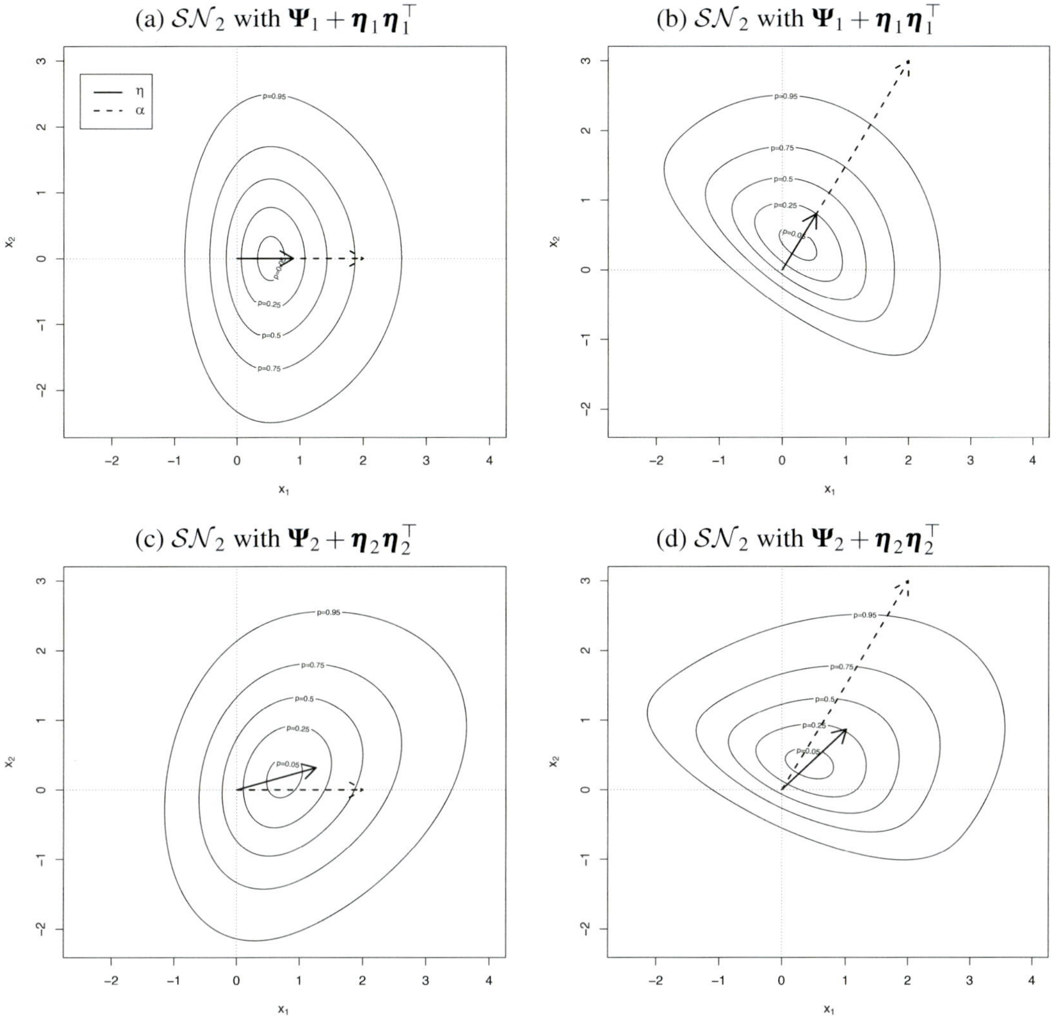

Figure 5.2 *Geometry of the multivariate skew-normal distribution,* $\mathcal{SN}_p$, *for* $p=2$.

$(2\sqrt{2}+3/2, 1/\sqrt{2}+3)^\top/\sqrt{14+6/\sqrt{2}}$, *respectively, which correspond to the vector* $\boldsymbol{\alpha}_2$ *equal to* $(2,0)^\top$ *and* $(2,3)^\top$, *respectively. In each panel, the solid vector represents* $\boldsymbol{\eta}_i$ *whereas the dashed vector represents* $\boldsymbol{\alpha}_i$ *in row* $i=1,2$. *The geometrical interpretation of* $\boldsymbol{\eta}_i$ *better indicates the direction of skewness. In fact, for the multivariate skew-normal distribution* $\mathcal{SN}_p(\boldsymbol{\xi},\mathbf{\Psi},\boldsymbol{\eta})$, *the vector* $\boldsymbol{\eta}$ *is the direction that maximizes the skewness (Loperfido, 2010). The same result applies to the multivariate skew-t distribution; see Figure 7.1(c) in Section 7.3.2 and Arevalillo and Navarro (2015).*

5.6 Singular Skew-Normal

When $\mathbf{X}\sim\mathcal{SN}_p(\boldsymbol{\xi},\mathbf{\Psi},\boldsymbol{\eta})$ is singular with $\text{rank}(\mathbf{\Psi})=q<p$ and so $|\mathbf{\Psi}|=0$, we can always find an orthogonal matrix $\mathbf{\Gamma}\in\mathcal{O}_p$ and a vector $\boldsymbol{\eta}_1\in\mathbb{R}^q$ to write

$$\mathbf{\Psi}=\mathbf{\Gamma}\mathbf{D}_\lambda\mathbf{\Gamma}^\top=(\mathbf{\Gamma}_1\;\mathbf{\Gamma}_2)\begin{pmatrix}\mathbf{D}_1 & \mathbf{O}\\ \mathbf{O} & \mathbf{O}\end{pmatrix}\begin{pmatrix}\mathbf{\Gamma}_1^\top\\ \mathbf{\Gamma}_2^\top\end{pmatrix}=\mathbf{\Gamma}_1\mathbf{D}_1\mathbf{\Gamma}_1^\top,\quad \boldsymbol{\eta}=(\mathbf{\Gamma}_1\;\mathbf{\Gamma}_2)\begin{pmatrix}\boldsymbol{\eta}_1\\ \mathbf{0}_{p-q}\end{pmatrix}=\mathbf{\Gamma}_1\boldsymbol{\eta}_1,\quad (5.21)$$

where $\mathbf{D}_1 = \text{diag}(\lambda_1, \ldots, \lambda_q)$ comprises on its diagonal the nonzero eigenvalues, $\mathbf{\Gamma}_1 = (\boldsymbol{\gamma}_1, \ldots, \boldsymbol{\gamma}_q)$ provides an orthonormal basis for the column space of $\mathbf{\Psi}$, $\mathtt{C}(\mathbf{\Psi})$, and $\mathbf{\Gamma}_2 = (\boldsymbol{\gamma}_{q+1}, \ldots, \boldsymbol{\gamma}_p)$ gives a basis for the null space of $\mathbf{\Psi}$, $\mathtt{N}(\mathbf{\Psi})$. Since $\mathtt{C}(\mathbf{\Gamma}_2) = \mathtt{C}(\mathbf{\Gamma}_1)^\perp = \mathtt{N}(\mathbf{\Psi})$ and $\mathtt{C}(\mathbf{\Gamma}_1) = \mathtt{C}(\mathbf{\Gamma}_2)^\perp = \mathtt{C}(\mathbf{\Psi})$, then $\mathbf{\Gamma}_2^\top(\mathbf{X} - \boldsymbol{\xi}) = \mathbf{0}_{p-q}$ with probability 1, which means that $\mathbf{X} - \boldsymbol{\xi} \in \mathtt{C}(\mathbf{\Gamma}_2)^\perp = \mathtt{C}(\mathbf{\Psi})$ with probability 1. Note that $\mathbf{\Gamma}_1^\top(\mathbf{X} - \boldsymbol{\xi})$ has a nonsingular $\mathcal{SN}_q(\mathbf{0}_q, \mathbf{D}_1, \boldsymbol{\eta}_1)$ distribution as established below. This is essentially equivalent to saying that $\mathbf{X} \in \boldsymbol{\xi} + \mathtt{C}(\mathbf{\Psi})$ with probability 1, and in this way, we can almost visualize $\mathbf{X}$ in this q-dimensional affine subspace of $\mathbb{R}^p$. However, despite $\mathrm{P}\{\mathbf{X} \in \boldsymbol{\xi} + \mathtt{C}(\mathbf{\Psi})\} = 1$, the distribution of $\mathbf{X}$ cannot be absolutely continuous because $\text{vol}\{\boldsymbol{\xi} + \mathtt{C}(\mathbf{\Psi})\} = 0$. The same obviously happens with the singular multivariate normal distribution, as demonstrated by Bilodeau and Brenner (1999). In fact, the development of this last discussion is carried out following the approach of these authors for the multivariate normal distribution.

To identify the distribution of $\mathbf{X}_1 = \mathbf{\Gamma}_1^\top(\mathbf{X} - \boldsymbol{\xi})$, note by (5.9) that

$$\mathbf{X}_1 = \mathbf{\Gamma}_1^\top(\boldsymbol{\eta} U + \mathbf{W}) = \boldsymbol{\eta}_1 U + \mathbf{W}_1,$$

where, as a consequence of (5.21), $\boldsymbol{\eta}_1 = \mathbf{\Gamma}_1^\top \boldsymbol{\eta} \in \mathbb{R}^q$ and $\mathbf{W}_1 = \mathbf{\Gamma}_1^\top \mathbf{W} \sim \mathcal{N}_q(\mathbf{0}_q, \mathbf{D}_1)$ is nonsingular. Since the normal random vector $\mathbf{W}_1$ is also independent of the half-normal random variable U, then $\mathbf{X}_1 \sim \mathcal{SN}_q(\mathbf{0}_q, \mathbf{D}_1, \boldsymbol{\eta}_1)$ is nonsingular with pdf

$$f_1(\mathbf{x}_1) = 2\phi_q(\mathbf{x}_1; \mathbf{0}_q, \mathbf{D}_1)\Phi\left\{\frac{\boldsymbol{\eta}_1^\top \mathbf{D}_1^{-1}\mathbf{x}_1}{\sqrt{1 + \boldsymbol{\eta}_1^\top \mathbf{D}_1^{-1}\boldsymbol{\eta}_1}}\right\}, \quad \mathbf{x}_1 \in \mathbb{R}^q.$$

This means that the p-dimensional skewness parameter associated with the multivariate skew-normal distribution of $\mathbf{X}$ can only contain a nonzero subvector of dimension q, which matches the vector $\boldsymbol{\eta}_1$. This implies that the constraint $\boldsymbol{\eta} = \mathbf{\Gamma}_1 \boldsymbol{\eta}_1$ is mathematically essential because any component of the skewness parameter $\boldsymbol{\eta}$ lying in the null space of $\mathbf{\Psi}$ would have no effect on the distribution. Since the vector $\mathbf{X} - \boldsymbol{\xi}$ belongs to the column space $\mathtt{C}(\mathbf{\Psi})$ with probability 1, its projection onto the null space is zero. Therefore, the skewness is entirely determined by the q-dimensional subvector $\boldsymbol{\eta}_1$.

5.7 Conditional Skew-Normal

Now we study the multivariate conditional skew-normal distribution. In advance, we can say that the conditional skew-normal distributions do not belong to the multivariate skew-normal family. Instead, they belong to a broader class, called the extended skew-normal distribution (ESN; see Section 6.6.3), of the more general unified skew-normal (SUN) class of multivariate distributions studied later in Chapter 6.

5.7.1 Some preliminary results

Next, we derive the conditional distribution of $\mathbf{X}_1 \mid \mathbf{X}_2 = \mathbf{x}_2$. For this, we first note from the partition of $\mathbf{\Psi}$ given in (5.17) that we can find two situations according to the submatrix $\mathbf{\Psi}_{22}$ being nonsingular ($\mathbf{\Psi}_{22} > 0$) or singular ($|\mathbf{\Psi}_{22}| = 0$). Obviously, when $\mathbf{\Psi} > 0$, the nonsingular case with $\mathbf{\Psi}_{22} > 0$ holds automatically. In this case, we also have $\mathbf{\Psi}_{11} > 0$ and $\mathbf{\Psi}_{11\cdot 2} = \mathbf{\Psi}_{11} - \mathbf{\Psi}_{12}\mathbf{\Psi}_{22}^{-1}\mathbf{\Psi}_{21} > 0$, and the relation $(\mathbf{\Psi}_{11}, \mathbf{\Psi}_{12}, \mathbf{\Psi}_{22}) \longleftrightarrow (\mathbf{\Psi}_{11\cdot 2}, \mathbf{\Psi}_{12}\mathbf{\Psi}_{22}^{-1}, \mathbf{\Psi}_{22})$ becomes one-to-one.

In any case, we can also easily verify from the partition (5.17) that for any matrix $\mathbf{C} \in \mathbb{R}^{p_1 \times p_2}$ satisfying $\mathbf{C}\mathbf{\Psi}_{22} = \mathbf{\Psi}_{12}$, it follows that

$$\underbrace{\begin{pmatrix} \mathbf{I}_{p_1} & -\mathbf{C} \\ \mathbf{O} & \mathbf{I}_{p_2} \end{pmatrix}}_{\mathbf{B}} \begin{pmatrix} \mathbf{\Psi}_{11} & \mathbf{\Psi}_{12} \\ \mathbf{\Psi}_{21} & \mathbf{\Psi}_{22} \end{pmatrix} \underbrace{\begin{pmatrix} \mathbf{I}_{p_1} & \mathbf{O} \\ -\mathbf{C}^\top & \mathbf{I}_{p_2} \end{pmatrix}}_{\mathbf{B}^\top} = \begin{pmatrix} \mathbf{\Psi}_{11} - \mathbf{C}\mathbf{\Psi}_{21} & \mathbf{O} \\ \mathbf{O} & \mathbf{\Psi}_{22} \end{pmatrix}. \tag{5.22}$$

In the special case with $\mathbf{C} = \boldsymbol{\Psi}_{12}\boldsymbol{\Psi}_{22}^{-}$, where $\boldsymbol{\Psi}_{22}^{-}$ is a generalized inverse of $\boldsymbol{\Psi}_{22}$, i.e., a matrix satisfying $\boldsymbol{\Psi}_{22}\boldsymbol{\Psi}_{22}^{-}\boldsymbol{\Psi}_{22} = \boldsymbol{\Psi}_{22}$, it holds that $\boldsymbol{\Psi}_{12}\boldsymbol{\Psi}_{22}^{-}\boldsymbol{\Psi}_{22} = \mathbf{C}\boldsymbol{\Psi}_{22}\boldsymbol{\Psi}_{22}^{-}\boldsymbol{\Psi}_{22} = \mathbf{C}\boldsymbol{\Psi}_{22} = \boldsymbol{\Psi}_{12}$, and hence $\boldsymbol{\Psi}_{11} - \mathbf{C}\boldsymbol{\Psi}_{21} = \boldsymbol{\Psi}_{11} - \boldsymbol{\Psi}_{12}\boldsymbol{\Psi}_{22}^{-}\boldsymbol{\Psi}_{21} = \boldsymbol{\Psi}_{11\cdot 2}$. Therefore, (5.22) is completely equivalent to the decomposition

$$\begin{pmatrix} \boldsymbol{\Psi}_{11} & \boldsymbol{\Psi}_{12} \\ \boldsymbol{\Psi}_{21} & \boldsymbol{\Psi}_{22} \end{pmatrix} = \underbrace{\begin{pmatrix} \mathbf{I}_{p_1} & \mathbf{C} \\ \mathbf{O} & \mathbf{I}_{p_2} \end{pmatrix}}_{\mathbf{B}^{-1}} \begin{pmatrix} \boldsymbol{\Psi}_{11\cdot 2} & \mathbf{O} \\ \mathbf{O} & \boldsymbol{\Psi}_{22} \end{pmatrix} \underbrace{\begin{pmatrix} \mathbf{I}_{p_1} & \mathbf{O} \\ \mathbf{C}^{\top} & \mathbf{I}_{p_2} \end{pmatrix}}_{\mathbf{B}^{-\top}}.$$

Now, suppose that the skew-normal random vector $\mathbf{X}$ defined by (5.9) is partitioned as

$$\begin{pmatrix} \mathbf{X}_1 \\ \mathbf{X}_2 \end{pmatrix} = \begin{pmatrix} \boldsymbol{\xi}_1 \\ \boldsymbol{\xi}_2 \end{pmatrix} + \begin{pmatrix} \boldsymbol{\eta}_1 \\ \boldsymbol{\eta}_2 \end{pmatrix} U + \begin{pmatrix} \mathbf{W}_1 \\ \mathbf{W}_2 \end{pmatrix},$$

where, as before, U and $(\mathbf{W}_1, \mathbf{W}_2)$ are independent with

$$U \sim \mathcal{HN}(0,1), \quad \begin{pmatrix} \mathbf{W}_1 \\ \mathbf{W}_2 \end{pmatrix} \sim \mathcal{N}_{p_1+p_2}\left(\begin{pmatrix} \mathbf{0}_{p_1} \\ \mathbf{0}_{p_2} \end{pmatrix}, \begin{pmatrix} \boldsymbol{\Psi}_{11} & \boldsymbol{\Psi}_{12} \\ \boldsymbol{\Psi}_{21} & \boldsymbol{\Psi}_{22} \end{pmatrix}\right),$$

and consider the nonsingular linear transformation given by

$$\tilde{\mathbf{X}} = \begin{pmatrix} \tilde{\mathbf{X}}_1 \\ \mathbf{X}_2 \end{pmatrix} = \begin{pmatrix} \mathbf{I}_{p_1} & -\mathbf{C} \\ \mathbf{O} & \mathbf{I}_{p_2} \end{pmatrix} \begin{pmatrix} \mathbf{X}_1 \\ \mathbf{X}_2 \end{pmatrix} = \begin{pmatrix} \mathbf{X}_1 - \mathbf{C}\mathbf{X}_2 \\ \mathbf{X}_2 \end{pmatrix} = \mathbf{B}\mathbf{X}. \tag{5.23}$$

Thus, we find

$$\begin{aligned} \begin{pmatrix} \tilde{\mathbf{X}}_1 \\ \mathbf{X}_2 \end{pmatrix} &= \begin{pmatrix} \boldsymbol{\xi}_1 - \mathbf{C}\boldsymbol{\xi}_2 + (\boldsymbol{\eta}_1 - \mathbf{C}\boldsymbol{\eta}_2)U + \mathbf{W}_1 - \mathbf{C}\mathbf{W}_2 \\ \boldsymbol{\xi}_2 + \boldsymbol{\eta}_2 U + \mathbf{W}_2 \end{pmatrix} \\ &= \underbrace{\begin{pmatrix} \boldsymbol{\xi}_1 - \mathbf{C}\boldsymbol{\xi}_2 \\ \boldsymbol{\xi}_2 \end{pmatrix}}_{\tilde{\boldsymbol{\xi}}} + \underbrace{\begin{pmatrix} \boldsymbol{\eta}_1 - \mathbf{C}\boldsymbol{\eta}_2 \\ \boldsymbol{\eta}_2 \end{pmatrix}}_{\tilde{\boldsymbol{\eta}}} U + \underbrace{\begin{pmatrix} \mathbf{W}_1 - \mathbf{C}\mathbf{W}_2 \\ \mathbf{W}_2 \end{pmatrix}}_{\tilde{\mathbf{W}}} \\ \Longrightarrow \tilde{\mathbf{X}} &= \tilde{\boldsymbol{\xi}} + \tilde{\boldsymbol{\eta}} U + \tilde{\mathbf{W}}, \end{aligned} \tag{5.24}$$

where again the half-normal random variable U is independent of $\tilde{\mathbf{W}}$ which, under the condition $\mathbf{C}\boldsymbol{\Psi}_{22} = \boldsymbol{\Psi}_{12}$, has a multivariate normal distribution given by

$$\tilde{\mathbf{W}} = \begin{pmatrix} \tilde{\mathbf{W}}_1 \\ \mathbf{W}_2 \end{pmatrix} = \begin{pmatrix} \mathbf{W}_1 - \mathbf{C}\mathbf{W}_2 \\ \mathbf{W}_2 \end{pmatrix} \sim \mathcal{N}_{p_1+p_2}\left(\begin{pmatrix} \mathbf{0}_{p_1} \\ \mathbf{0}_{p_2} \end{pmatrix}, \begin{pmatrix} \boldsymbol{\Psi}_{11} - \mathbf{C}\boldsymbol{\Psi}_{21} & \mathbf{O} \\ \mathbf{O} & \boldsymbol{\Psi}_{22} \end{pmatrix}\right).$$

It is clear by independence that $(\tilde{\mathbf{W}}_1 \mid \mathbf{W}_2 = \mathbf{w}_2) \stackrel{d}{=} \tilde{\mathbf{W}}_1 = \mathbf{W}_1 - \mathbf{C}\mathbf{W}_2 \sim \mathcal{N}_{p_1}(\mathbf{0}_{p_1}, \boldsymbol{\Psi}_{11} - \mathbf{C}\boldsymbol{\Psi}_{21})$ for all $\mathbf{w}_2$. Hence, since $\mathbf{W}_1 = \mathbf{C}\mathbf{W}_2 + \tilde{\mathbf{W}}_1$ we then have $(\mathbf{W}_1 \mid \mathbf{W}_2 = \mathbf{w}_2) \stackrel{d}{=} \mathbf{C}\mathbf{w}_2 + \tilde{\mathbf{W}}_1 \sim \mathcal{N}_{p_1}(\mathbf{C}\mathbf{w}_2, \boldsymbol{\Psi}_{11} - \mathbf{C}\boldsymbol{\Psi}_{21})$. In particular, if $\mathbf{C} = \boldsymbol{\Psi}_{12}\boldsymbol{\Psi}_{22}^{-}$ then $(\mathbf{W}_1 \mid \mathbf{W}_2 = \mathbf{w}_2) \sim \mathcal{N}_{p_1}(\boldsymbol{\Psi}_{12}\boldsymbol{\Psi}_{22}^{-}\mathbf{w}_2, \boldsymbol{\Psi}_{11\cdot 2})$ which is valid even if $\boldsymbol{\Psi}_{22}$ is singular; see Chapter 3 and Bilodeau and Brenner (1999).

In this way, we obtain the following theorem as a consequence of Theorem 58.

Theorem 68 *Suppose that*

$$\begin{pmatrix} \mathbf{X}_1 \\ \mathbf{X}_2 \end{pmatrix} \sim \mathcal{SN}_{p_1+p_2}\left(\begin{pmatrix} \boldsymbol{\xi}_1 \\ \boldsymbol{\xi}_2 \end{pmatrix}, \begin{pmatrix} \boldsymbol{\Psi}_{11} & \boldsymbol{\Psi}_{12} \\ \boldsymbol{\Psi}_{21} & \boldsymbol{\Psi}_{22} \end{pmatrix}, \begin{pmatrix} \boldsymbol{\eta}_1 \\ \boldsymbol{\eta}_2 \end{pmatrix}\right).$$

If we let $\mathbf{C} = \mathbf{\Psi}_{12}\mathbf{\Psi}_{22}^{-}$, *then:*

(a) $$\begin{pmatrix} \mathbf{X}_1 - \mathbf{C}\mathbf{X}_2 \\ \mathbf{X}_2 \end{pmatrix} \sim \mathcal{SN}_{p_1+p_2}\left(\begin{pmatrix} \boldsymbol{\xi}_1 - \mathbf{C}\boldsymbol{\xi}_2 \\ \boldsymbol{\xi}_2 \end{pmatrix}, \begin{pmatrix} \mathbf{\Psi}_{11\cdot 2} & \mathbf{O} \\ \mathbf{O} & \mathbf{\Psi}_{22} \end{pmatrix}, \begin{pmatrix} \boldsymbol{\eta}_1 - \mathbf{C}\boldsymbol{\eta}_2 \\ \boldsymbol{\eta}_2 \end{pmatrix}\right),$$

and $\tilde{\mathbf{X}}_1 = \mathbf{X}_1 - \mathbf{C}\mathbf{X}_2$ *and* $\mathbf{X}_2$ *are independent if* $\boldsymbol{\eta}_1 = \mathbf{C}\boldsymbol{\eta}_2$ *or* $\boldsymbol{\eta}_2 = \mathbf{0}_{p_2}$*;*

(b) $\tilde{\mathbf{X}}_1 = \mathbf{X}_1 - \mathbf{C}\mathbf{X}_2$ *and* $\mathbf{X}_2$ *are conditionally independent given* U*, with*

$$\begin{aligned} (\tilde{\mathbf{X}}_1 \mid U = u) &\sim \mathcal{N}_{p_1}(\boldsymbol{\xi}_1 - \mathbf{C}\boldsymbol{\xi}_2 + (\boldsymbol{\eta}_1 - \mathbf{C}\boldsymbol{\eta}_2)u, \mathbf{\Psi}_{11\cdot 2}), \\ (\mathbf{X}_2 \mid U = u) &\sim \mathcal{N}_{p_2}(\boldsymbol{\xi}_2 + \boldsymbol{\eta}_2 u, \mathbf{\Psi}_{22}); \end{aligned}$$

(c) $(\mathbf{X}_1 \mid \mathbf{X}_2 = \mathbf{x}_2, U = u) \sim \mathcal{N}_{p_1}(\boldsymbol{\xi}_{1\cdot 2} + \boldsymbol{\eta}_{1\cdot 2}u, \mathbf{\Psi}_{11\cdot 2})$, *with:*

$$\boldsymbol{\xi}_{1\cdot 2} = \boldsymbol{\xi}_1 + \mathbf{\Psi}_{12}\mathbf{\Psi}_{22}^{-}(\mathbf{x}_2 - \boldsymbol{\xi}_2), \quad \boldsymbol{\eta}_{1\cdot 2} = \boldsymbol{\eta}_1 - \mathbf{\Psi}_{12}\mathbf{\Psi}_{22}^{-}\boldsymbol{\eta}_2, \quad \mathbf{\Psi}_{11\cdot 2} = \mathbf{\Psi}_{11} - \mathbf{\Psi}_{12}\mathbf{\Psi}_{22}^{-}\mathbf{\Psi}_{21}. \tag{5.25}$$

Proof 68 *The proof of part (a) follows from Theorem 58. Part (b) follows from (5.24) by conditioning on* U*. Part (c) is a consequence of (b) and the properties of the conditional multivariate normal; more specifically, since* $\mathbf{X}_1 = \tilde{\mathbf{X}}_1 + \mathbf{C}\mathbf{X}_2$*, where* $\tilde{\mathbf{X}}_1$ *and* $\mathbf{X}_2$ *are conditionally independent given* U*, we have by (b) that* $(\mathbf{X}_1 \mid \mathbf{X}_2 = \mathbf{x}_2, U = u) \overset{d}{=} (\tilde{\mathbf{X}}_1 \mid U = u) + \mathbf{C}\mathbf{x}_2 \sim \mathcal{N}_{p_1}(\boldsymbol{\xi}_1 + \mathbf{C}\mathbf{x}_2 - \mathbf{C}\boldsymbol{\xi}_2 + (\boldsymbol{\eta}_1 - \mathbf{C}\boldsymbol{\eta}_2)u, \mathbf{\Psi}_{11\cdot 2})$ *which gives the proof of (c).* □

We then arrive at the following completely general result.

Theorem 69 *Under the assumptions of Theorem 68, the conditional random vector* $(\mathbf{X}_1 \mid \mathbf{X}_2 = \mathbf{x}_2)$ *has conditional cf given by*

$$\psi_{1\cdot 2}(\mathbf{t}_1) = e^{i\mathbf{t}_1^\top \boldsymbol{\xi}_{1\cdot 2} - \frac{1}{2}\mathbf{t}_1^\top \mathbf{\Psi}_{11\cdot 2}\mathbf{t}_1}\psi_{U|\mathbf{x}_2}(\boldsymbol{\eta}_{1\cdot 2}^\top \mathbf{t}_1), \tag{5.26}$$

where $\boldsymbol{\xi}_{1\cdot 2}$, $\mathbf{\Psi}_{11\cdot 2}$ *and* $\boldsymbol{\eta}_{1\cdot 2}$ *are defined in (5.25) and* $\psi_{U|\mathbf{x}_2}(t) = \mathrm{E}(e^{itU} \mid \mathbf{X}_2 = \mathbf{x}_2)$ *is the conditional cf of* $(U \mid \mathbf{X}_2 = \mathbf{x}_2)$*.*

Proof 69 *From part (c) of Theorem 68 and the properties of the conditional expectation, we have*

$$\begin{aligned} \psi_{1\cdot 2}(\mathbf{t}_1) &= \mathrm{E}(e^{i\mathbf{t}_1^\top \mathbf{X}_1} \mid \mathbf{X}_2 = \mathbf{x}_2) \\ &= \mathrm{E}\{\mathrm{E}(e^{i\mathbf{t}_1^\top \mathbf{X}_1} \mid \mathbf{X}_2 = \mathbf{x}_2, U) \mid \mathbf{X}_2 = \mathbf{x}_2\} \\ &= \mathrm{E}[\mathrm{E}\{e^{i\mathbf{t}_1^\top(\boldsymbol{\xi}_{1\cdot 2} + \boldsymbol{\eta}_{1\cdot 2}U) - \frac{1}{2}\mathbf{t}_1^\top \mathbf{\Psi}_{11\cdot 2}\mathbf{t}_1}\} \mid \mathbf{X}_2 = \mathbf{x}_2] \\ &= e^{i\mathbf{t}_1^\top \boldsymbol{\xi}_{1\cdot 2} - \frac{1}{2}\mathbf{t}_1^\top \mathbf{\Psi}_{11\cdot 2}\mathbf{t}_1}\mathrm{E}\{e^{i(\boldsymbol{\eta}_{1\cdot 2}^\top \mathbf{t}_1)U} \mid \mathbf{X}_2 = \mathbf{x}_2\}. \end{aligned}$$

□

From (5.26) it follows that the conditional distribution of $\mathbf{X}_1$ given $\mathbf{X}_2 = \mathbf{x}_2$ is completely determined by the conditional distribution of U given $\mathbf{X}_2 = \mathbf{x}_2$. To identify this last conditional distribution, first assume that $\mathbf{\Psi}_{22}$ is nonsingular. Hence, from part (b) of Theorem 68, the conditional normal distribution $\mathbf{X}_2 \mid U = u \sim \mathcal{N}_{p_2}(\boldsymbol{\xi}_2 + \boldsymbol{\eta}_2 u, \mathbf{\Psi}_{22})$ is nonsingular, and for each $u > 0$, its conditional pdf is given by

$$f_2(\mathbf{x}_2 \mid u) = \phi_{p_2}(\mathbf{x}_2; \boldsymbol{\xi}_2 + \boldsymbol{\eta}_2 u, \mathbf{\Psi}_{22}), \quad \mathbf{x}_2 \in \mathbb{R}^{p_2}.$$

Also, since $\mathbf{X}_2 \sim \mathcal{SN}_{p_2}(\boldsymbol{\xi}_2, \boldsymbol{\Psi}_{22}, \boldsymbol{\eta}_2)$ is nonsingular, it has a skew-normal marginal pdf given by

$$f_2(\mathbf{x}_2) = 2\phi_{p_2}(\mathbf{x}_2; \boldsymbol{\xi}_2, \boldsymbol{\Psi}_{22} + \boldsymbol{\eta}_2\boldsymbol{\eta}_2^\top)\Phi\left\{\frac{\boldsymbol{\eta}_2^\top\boldsymbol{\Psi}_{22}^{-1}(\mathbf{x}_2 - \boldsymbol{\xi}_2)}{\sqrt{1+\boldsymbol{\eta}_2^\top\boldsymbol{\Psi}_{22}^{-1}\boldsymbol{\eta}_2}}\right\}, \quad \mathbf{x}_2 \in \mathbb{R}^{p_2}.$$

Therefore, remembering that U has half-normal pdf $g(u) = 2\phi(u)$ for $u > 0$, from Bayes' rule we find that the conditional pdf of U given $\mathbf{X}_2 = \mathbf{x}_2$ is

$$\begin{aligned} g(u \mid \mathbf{x}_2) &= \frac{f_2(\mathbf{x}_2 \mid u)g(u)}{f_2(\mathbf{x}_2)} \\ &= \frac{2\phi_{p_2}(\mathbf{x}_2; \boldsymbol{\xi}_2 + \boldsymbol{\eta}_2 u, \boldsymbol{\Psi}_{22})\phi(u)}{2\phi_{p_2}(\mathbf{x}_2; \boldsymbol{\xi}_2, \boldsymbol{\Psi}_{22} + \boldsymbol{\eta}_2\boldsymbol{\eta}_2^\top)\Phi\left\{\frac{\boldsymbol{\eta}_2^\top\boldsymbol{\Psi}_{22}^{-1}(\mathbf{x}_2-\boldsymbol{\xi}_2)}{\sqrt{1+\boldsymbol{\eta}_2^\top\boldsymbol{\Psi}_{22}^{-1}\boldsymbol{\eta}_2}}\right\}}. \end{aligned}$$

Now, by applying the identity

$$\phi_{p_2}(\mathbf{x}_2; \boldsymbol{\xi}_2 + \boldsymbol{\eta}_2 u, \boldsymbol{\Psi}_{22})\phi(u) = \phi_{p_2}(\mathbf{x}_2; \boldsymbol{\xi}_2, \boldsymbol{\Psi}_{22} + \boldsymbol{\eta}_2\boldsymbol{\eta}_2^\top)\phi\left(u; \frac{\boldsymbol{\eta}_2^\top\boldsymbol{\Psi}_{22}^{-1}(\mathbf{x}_2 - \boldsymbol{\xi}_2)}{1+\boldsymbol{\eta}_2^\top\boldsymbol{\Psi}_{22}^{-1}\boldsymbol{\eta}_2}, \frac{1}{1+\boldsymbol{\eta}_2^\top\boldsymbol{\Psi}_{22}^{-1}\boldsymbol{\eta}_2}\right),$$

we immediately find that

$$g(u \mid \mathbf{x}_2) = \frac{\sqrt{1+\tilde{\eta}_2^2}}{\Phi(\bar{\tau}_{1\cdot 2})}\phi\left\{\sqrt{1+\tilde{\eta}_2^2}\,(u - \tau_{1\cdot 2})\right\}, \quad u > 0, \tag{5.27}$$

where

$$\tilde{\eta}_2^2 = \boldsymbol{\eta}_2^\top\boldsymbol{\Psi}_{22}^{-1}\boldsymbol{\eta}_2, \qquad \tau_{1\cdot 2} = \frac{\boldsymbol{\eta}_2^\top\boldsymbol{\Psi}_{22}^{-1}(\mathbf{x}_2 - \boldsymbol{\xi}_2)}{1+\tilde{\eta}_2^2}, \qquad \bar{\tau}_{1\cdot 2} = \sqrt{1+\tilde{\eta}_2^2}\,\tau_{1\cdot 2}.$$

Note from (5.27) that for each fixed $\mathbf{x}_2 \in \mathbb{R}^{p_2}$, the conditional distribution of $(U \mid \mathbf{X}_2 = \mathbf{x}_2)$ becomes a truncated $\mathcal{N}(\tau_{1\cdot 2}, (1+\tilde{\eta}_2^2)^{-1})$ below zero; that is, stochastically we can represent the conditional distribution of $(U \mid \mathbf{X}_2 = \mathbf{x}_2)$ as follows:

$$(U \mid \mathbf{X}_2 = \mathbf{x}_2) \stackrel{d}{=} U_{1\cdot 2}, \quad U_{1\cdot 2} \stackrel{d}{=} (Z_{1\cdot 2} \mid Z_{1\cdot 2} > 0), \quad Z_{1\cdot 2} \sim \mathcal{N}(\tau_{1\cdot 2}, (1+\tilde{\eta}_2^2)^{-1}). \tag{5.28}$$

In the singular case, with $\operatorname{rank}(\boldsymbol{\Psi}_{22}) = r < p_2$, we may always find a p_2-by-p_2 orthogonal matrix $\boldsymbol{\Gamma} = (\boldsymbol{\Gamma}_1\, \boldsymbol{\Gamma}_2)$ to write

$$\boldsymbol{\Psi}_{22} = (\boldsymbol{\Gamma}_1\, \boldsymbol{\Gamma}_2)\begin{pmatrix} \mathbf{O} & \mathbf{O} \\ \mathbf{O} & \mathbf{D}_2 \end{pmatrix}\begin{pmatrix} \boldsymbol{\Gamma}_1^\top \\ \boldsymbol{\Gamma}_2^\top \end{pmatrix} = \boldsymbol{\Gamma}_2\mathbf{D}_2\boldsymbol{\Gamma}_2^\top,$$

where $\mathbf{D}_2$ is an r-by-r diagonal matrix, and then to find the so-called pseudo-inverse of $\boldsymbol{\Psi}_{22}$:

$$\boldsymbol{\Psi}_{22}^- = (\boldsymbol{\Gamma}_1\, \boldsymbol{\Gamma}_2)\begin{pmatrix} \mathbf{O} & \mathbf{O} \\ \mathbf{O} & \mathbf{D}_2^{-1} \end{pmatrix}\begin{pmatrix} \boldsymbol{\Gamma}_1^\top \\ \boldsymbol{\Gamma}_2^\top \end{pmatrix} = \boldsymbol{\Gamma}_2\mathbf{D}_2^{-1}\boldsymbol{\Gamma}_2^\top.$$

Hence, we have $\boldsymbol{\Gamma}_1^\top(\mathbf{X}_2 - \boldsymbol{\xi}_2) = \mathbf{0}_{p_2-r}$ with probability 1 and also

$$(\boldsymbol{\Gamma}_2^\top\mathbf{X}_2 \mid U = u) \sim \mathcal{N}_r(\boldsymbol{\Gamma}_2^\top\boldsymbol{\xi}_2 + \boldsymbol{\Gamma}_2^\top\boldsymbol{\eta}_2 u, \mathbf{D}_2), \quad U \sim \mathcal{HN}(0,1).$$

Since the results in the nonsingular case now apply, with $\boldsymbol{\Psi}_{22}^{-1}$ replaced by $\boldsymbol{\Psi}_{22}^-$, we arrive at the following completely general result on the conditional multivariate skew-normal distribution.

Theorem 70 *Under the assumptions of Theorem 68, we have for each fixed* $\mathbf{x}_2 \in \mathbb{R}^{p_2}$ *that* $(\mathbf{X}_1 \mid \mathbf{X}_2 = \mathbf{x}_2) \stackrel{d}{=} \mathbf{X}_{1\cdot 2}$, *with*

$$\mathbf{X}_{1\cdot 2} = \boldsymbol{\xi}_{1\cdot 2} + \boldsymbol{\eta}_{1\cdot 2} U_{1\cdot 2} + \mathbf{W}_{1\cdot 2} = \boldsymbol{\xi}_{1\cdot 2} + \bar{\boldsymbol{\eta}}_{1\cdot 2} \bar{U}_{1\cdot 2} + \mathbf{W}_{1\cdot 2}, \tag{5.29}$$

where $\bar{U}_{1\cdot 2} = (1+\tilde{\eta}_2^2)^{1/2} U_{1\cdot 2} \sim \mathcal{TN}(0; \bar{\tau}_{1\cdot 2}, 1)$ *is independent of* $\mathbf{W}_{1\cdot 2} \sim \mathcal{N}_{p_1}(\mathbf{0}_{p_1}, \boldsymbol{\Psi}_{11\cdot 2})$, *with* $U_{1\cdot 2} \stackrel{d}{=} (U \mid \mathbf{X}_2 = \mathbf{x}_2)$, *as defined in (5.28) with* $\boldsymbol{\Psi}_{22}^{-1}$ *replaced by* $\boldsymbol{\Psi}_{22}^{-}$ *when* $\boldsymbol{\Psi}_{22}$ *is singular, which has conditional cf given by*

$$\psi_{U_{1\cdot 2}}(t) = e^{it\tau_{1\cdot 2} - \frac{1}{2}(1+\tilde{\eta}_2^2)^{-1} t^2} \frac{\Phi(\bar{\tau}_{1\cdot 2} + \mathrm{i}(1+\tilde{\eta}_2^2)^{1/2} t)}{\Phi(\bar{\tau}_{1\cdot 2})}, \quad t \in \mathbb{R}, \tag{5.30}$$

and where the conditional parameters $\boldsymbol{\xi}_{1\cdot 2}$, $\boldsymbol{\Psi}_{11\cdot 2}$ *and* $\boldsymbol{\eta}_{1\cdot 2}$ *are defined in (5.25),* $\bar{\boldsymbol{\eta}}_{1\cdot 2} = (1+\tilde{\eta}_2^2)^{-1/2} \boldsymbol{\eta}_{1\cdot 2}$, $\bar{\tau}_{1\cdot 2} = (1+\tilde{\eta}_2^2)^{1/2} \tau_{1\cdot 2}$, $\tau_{1\cdot 2} = (1+\tilde{\eta}_2^2)^{-1} \boldsymbol{\eta}_2^\top \boldsymbol{\Psi}_{22}^{-}(\mathbf{x}_2 - \boldsymbol{\xi}_2)$ *and* $\tilde{\eta}_2 = (\boldsymbol{\eta}_2^\top \boldsymbol{\Psi}_{22}^{-} \boldsymbol{\eta}_2)^{1/2}$.

Proof 70 *We first prove (5.30). Since* $U_{1\cdot 2} \stackrel{d}{=} (U \mid \mathbf{X}_2 = \mathbf{x}_2)$, *then from (5.27) (by replacing* $\boldsymbol{\Psi}_{22}^{-1}$ *by* $\boldsymbol{\Psi}_{22}^{-}$ *when* $\boldsymbol{\Psi}_{22}$ *is singular), we have, after some simple algebra, that*

$$\begin{aligned}
\psi_{U_{1\cdot 2}}(t) &= \int_0^\infty e^{\mathrm{i}tu} \frac{\sqrt{1+\tilde{\eta}_2^2}}{\Phi(\tau_{1\cdot 2})} \phi\left\{\sqrt{1+\tilde{\eta}_2^2}\,(u - \tau_{1\cdot 2})\right\} \mathrm{d}u \\
&= e^{\mathrm{i}t\tau_{1\cdot 2} - \frac{1}{2}(1+\tilde{\eta}_2^2)^{-1} t^2} \frac{\Phi(\bar{\tau}_{1\cdot 2} + \mathrm{i}(1+\tilde{\eta}_2^2)^{1/2} t)}{\Phi(\bar{\tau}_{1\cdot 2})}, \quad t \in \mathbb{R}.
\end{aligned}$$

Now, we prove (5.29). Specifically, we show that the cf of $\mathbf{X}_{1\cdot 2}$ *is (5.26) with* $\psi_{U|\mathbf{x}_2}(t) = \psi_{U_{1\cdot 2}}(t)$ *given by (5.30). In fact, since by assumption* $U_{1\cdot 2}$ *and* $\mathbf{W}_{1\cdot 2}$ *are independent, we have*

$$\begin{aligned}
\psi_{\mathbf{X}_{1\cdot 2}}(\mathbf{t}_1) &= \mathrm{E}(e^{\mathrm{i}\mathbf{t}_1^\top \mathbf{X}_{1\cdot 2}}) \\
&= \mathrm{E}\{e^{\mathrm{i}\mathbf{t}_1^\top(\boldsymbol{\xi}_{1\cdot 2} + \boldsymbol{\eta}_{1\cdot 2} U_{1\cdot 2} + \mathbf{W}_{1\cdot 2})}\} \\
&= e^{\mathrm{i}\mathbf{t}_1^\top \boldsymbol{\xi}_{1\cdot 2}} \mathrm{E}\{e^{\mathrm{i}(\boldsymbol{\eta}_{1\cdot 2}^\top \mathbf{t}_1) U_{1\cdot 2}}\} \mathrm{E}(e^{\mathrm{i}\mathbf{t}_1^\top \mathbf{W}_{1\cdot 2}}) \\
&= e^{\mathrm{i}\mathbf{t}_1^\top \boldsymbol{\xi}_{1\cdot 2} - \frac{1}{2}\mathbf{t}_1^\top \boldsymbol{\Psi}_{11\cdot 2} \mathbf{t}_1} \mathrm{E}\{e^{\mathrm{i}(\boldsymbol{\eta}_{1\cdot 2}^\top \mathbf{t}_1) U_{1\cdot 2}}\} \\
&= \psi_{\mathrm{N}_{1\cdot 2}}(\mathbf{t}_1) \psi_{U_{1\cdot 2}}(\boldsymbol{\eta}_{1\cdot 2}^\top \mathbf{t}_1).
\end{aligned}$$

The proof is concluded by the uniqueness of the cf. □

Remark 9 *Note from (5.28) that* $\sqrt{1+\tilde{\eta}_2^2}\, Z_{1\cdot 2} = \bar{\tau}_{1\cdot 2} + Z$, *where* $Z \sim \mathcal{N}(0,1)$, *and therefore the stochastic representation in (5.29) of the conditional skew-normal can also be rewritten as*

$$\mathbf{X}_{1\cdot 2} = \boldsymbol{\xi}_{1\cdot 2} + \bar{\boldsymbol{\eta}}_{1\cdot 2} \bar{\tau}_{1\cdot 2} + \bar{\boldsymbol{\eta}}_{1\cdot 2} V_{1\cdot 2} + \mathbf{W}_{1\cdot 2}, \tag{5.31}$$

where $V_{1\cdot 2} \stackrel{d}{=} (Z \mid Z + \bar{\tau}_{1\cdot 2} > 0) \sim \mathcal{TN}(-\bar{\tau}_{1\cdot 2}; 0, 1)$ *is also independent of* $\mathbf{W}_{1\cdot 2} \sim \mathcal{N}_{p_1}(\mathbf{0}_{p_1}, \boldsymbol{\Psi}_{11\cdot 2})$, *thus making explicit the additional effect generated by the conditioning on the location of the conditional distribution.*

5.7.2 *Conditional mean vector and conditional variance matrix*

From (5.29) or (5.31) it is easy to compute the conditional mean vector and the conditional variance matrix of $(\mathbf{X}_1 \mid \mathbf{X}_2 = \mathbf{x}_2)$. In fact, if we consider (5.31), then we need the first two moments of the truncated normal random variable $(Z \mid Z+\tau > 0)$, where $Z \sim \mathcal{N}(0,1)$. As we know, for any $c \in \mathbb{R}$, these moments are:

$$\mathrm{E}(Z \mid Z+c>0) = \frac{\phi(c)}{\Phi(c)}, \qquad \mathrm{E}(Z^2 \mid Z+c>0) = 1 - c\frac{\phi(c)}{\Phi(c)}. \tag{5.32}$$

Theorem 71 *Under the assumptions of Theorem 68, we have*

$$\begin{aligned}
\mathrm{E}(\mathbf{X}_1 \mid \mathbf{X}_2 = \mathbf{x}_2) &= \boldsymbol{\xi}_{1\cdot 2} + \frac{1}{\sqrt{1+\tilde{\eta}_2^2}}\left\{\bar{\tau}_{1\cdot 2} + \frac{\phi(\bar{\tau}_{1\cdot 2})}{\Phi(\bar{\tau}_{1\cdot 2})}\right\}\boldsymbol{\eta}_{1\cdot 2},\\
\mathrm{var}(\mathbf{X}_1 \mid \mathbf{X}_2 = \mathbf{x}_2) &= \boldsymbol{\Psi}_{11\cdot 2} + \frac{1}{1+\tilde{\eta}_2^2}\left[1 - \frac{\phi(\bar{\tau}_{1\cdot 2})}{\Phi(\bar{\tau}_{1\cdot 2})}\left\{\bar{\tau}_{1\cdot 2} + \frac{\phi(\bar{\tau}_{1\cdot 2})}{\Phi(\bar{\tau}_{1\cdot 2})}\right\}\right]\boldsymbol{\eta}_{1\cdot 2}\boldsymbol{\eta}_{1\cdot 2}^\top,
\end{aligned}$$

where $\boldsymbol{\xi}_{1\cdot 2}$, $\boldsymbol{\Psi}_{11\cdot 2}$, and $\boldsymbol{\eta}_{1\cdot 2}$ are given by (5.25), and, as defined earlier, $\bar{\tau}_{1\cdot 2} = (1+\tilde{\eta}_2^2)^{1/2}\tau_{1\cdot 2}$, $\tau_{1\cdot 2} = (1+\tilde{\eta}_2^2)^{-1}\boldsymbol{\eta}_2^\top\boldsymbol{\Psi}_{22}^-(\mathbf{x}_2 - \boldsymbol{\xi}_2)$ and $\tilde{\eta}_2 = (\boldsymbol{\eta}_2^\top\boldsymbol{\Psi}_{22}^-\boldsymbol{\eta}_2)^{1/2}$.

Proof 71 *From (5.31), it is clear that*

$$\mathrm{E}(\mathbf{X}_1 \mid \mathbf{X}_2 = \mathbf{x}_2) = \boldsymbol{\xi}_{1\cdot 2} + \bar{\boldsymbol{\eta}}_{1\cdot 2}\,\bar{\tau}_{1\cdot 2} + \bar{\boldsymbol{\eta}}_{1\cdot 2}\mathrm{E}(V_{1\cdot 2}), \quad \mathrm{var}(\mathbf{X}_1 \mid \mathbf{X}_2 = \mathbf{x}_2) = \boldsymbol{\Psi}_{11\cdot 2} + \mathrm{var}(V_{1\cdot 2})\bar{\boldsymbol{\eta}}_{1\cdot 2}\bar{\boldsymbol{\eta}}_{1\cdot 2}^\top.$$

Hence, the proof follows after using (5.32) to find that

$$\mathrm{E}(V_{1\cdot 2}) = \bar{\tau}_{1\cdot 2} + \frac{\phi(\bar{\tau}_{1\cdot 2})}{\Phi(\bar{\tau}_{1\cdot 2})}, \qquad \mathrm{var}(V_{1\cdot 2}) = 1 - \frac{\phi(\bar{\tau}_{1\cdot 2})}{\Phi(\bar{\tau}_{1\cdot 2})}\left\{\bar{\tau}_{1\cdot 2} + \frac{\phi(\bar{\tau}_{1\cdot 2})}{\Phi(\bar{\tau}_{1\cdot 2})}\right\}.$$

□

Since $\tau_{1\cdot 2} = \tau_{1\cdot 2}(\mathbf{x}_2)$ is a function of $\mathbf{x}_2$ and the function $\zeta(x) = \phi(x)/\Phi(x)$ is not linear, it follows from Theorem 71 that the conditional mean vector and the conditional variance matrix of $\mathbf{X}_1 \mid \mathbf{X}_2 = \mathbf{x}_2$ are generally nonlinear functions of $\mathbf{x}_2$. In the following example, two special cases are presented in which the regression of $\mathbf{X}_1$ on $\mathbf{x}_2$ is linear and homoscedastic.

Example 64 *Let $\mathbf{X} \sim \mathcal{SN}_p(\boldsymbol{\xi}, \boldsymbol{\Psi}, \boldsymbol{\eta})$ be partitioned as (5.17). We then have:*

(a) If $\boldsymbol{\eta}_{1\cdot 2} = \mathbf{0}_{p_1}$, i.e., $\boldsymbol{\eta}_1 = \boldsymbol{\Psi}_{12}\boldsymbol{\Psi}_{22}^-\boldsymbol{\eta}_2$, then $\mathbf{X}_1 \mid \mathbf{X}_2 = \mathbf{x}_2 \sim \mathcal{N}_{p_1}(\boldsymbol{\xi}_{1\cdot 2}, \boldsymbol{\Psi}_{11\cdot 2})$, with

$$\mathrm{E}(\mathbf{X}_1 \mid \mathbf{x}_2) = \boldsymbol{\xi}_{1\cdot 2} \quad \textit{and} \quad \mathrm{var}(\mathbf{X}_1 \mid \mathbf{x}_2) = \boldsymbol{\Psi}_{11\cdot 2};$$

(b) If $\boldsymbol{\eta}_2 = \mathbf{0}_{p_2}$, then $\mathbf{X}_1 \mid \mathbf{X}_2 = \mathbf{x}_2 \sim \mathcal{SN}_{p_1}(\boldsymbol{\xi}_{1\cdot 2}, \boldsymbol{\Psi}_{11\cdot 2}, \boldsymbol{\eta}_1)$, with

$$\mathrm{E}(\mathbf{X}_1 \mid \mathbf{x}_2) = \boldsymbol{\xi}_{1\cdot 2} + \sqrt{\frac{2}{\pi}}\boldsymbol{\eta}_1 \quad \textit{and} \quad \mathrm{var}(\mathbf{X}_1 \mid \mathbf{x}_2) = \boldsymbol{\Psi}_{11\cdot 2} + \left(1 - \frac{2}{\pi}\right)\boldsymbol{\eta}_1\boldsymbol{\eta}_1^\top.$$

5.7.3 Conditional pdf

As mentioned earlier, when $\boldsymbol{\Psi} > 0$, the nonsingular case with $\boldsymbol{\Psi}_{22} > 0$ holds automatically. In this case, we also have $\boldsymbol{\Psi}_{11} > 0$ and $\boldsymbol{\Psi}_{11\cdot 2} = \boldsymbol{\Psi}_{11} - \boldsymbol{\Psi}_{12}\boldsymbol{\Psi}_{22}^{-1}\boldsymbol{\Psi}_{21} > 0$, and the relation $(\boldsymbol{\Psi}_{11}, \boldsymbol{\Psi}_{12}, \boldsymbol{\Psi}_{22}) \longleftrightarrow (\boldsymbol{\Psi}_{11\cdot 2}, \boldsymbol{\Psi}_{12}\boldsymbol{\Psi}_{22}^{-1}, \boldsymbol{\Psi}_{22})$ becomes one-to-one. Obviously, under these conditions, all the marginal and conditional distributions obtained from $\mathbf{X} \sim \mathcal{SN}_p(\boldsymbol{\xi}, \boldsymbol{\Psi}, \boldsymbol{\eta})$ are nonsingular, as shown in the following theorem.

Theorem 72 *Let* $\mathbf{X} \sim \mathcal{SN}_p(\boldsymbol{\xi}, \boldsymbol{\Psi}, \boldsymbol{\eta})$, $\boldsymbol{\Psi} > 0$, *be partitioned as (5.17). Then for every fixed* $p_1 + p_2 = p$, *the conditional pdf of* $\mathbf{X}_1$ *given* $\mathbf{X}_2 = \mathbf{x}_2$, $\mathbf{x}_2 \in \mathbb{R}^{p_2}$, *is*

$$\begin{aligned} f_{1\cdot 2}(\mathbf{x}_1) &= \frac{1}{\Phi(\bar{\tau}_{1\cdot 2})}\phi_{p_1}(\mathbf{x}_1; \boldsymbol{\xi}_{1\cdot 2} + \bar{\tau}_{1\cdot 2}\bar{\boldsymbol{\eta}}_{1\cdot 2}, \boldsymbol{\Psi}_{11\cdot 2} + \bar{\boldsymbol{\eta}}_{1\cdot 2}\bar{\boldsymbol{\eta}}_{1\cdot 2}^\top) \\ &\quad \times \Phi\left\{\frac{\bar{\tau}_{1\cdot 2} + \bar{\boldsymbol{\eta}}_{1\cdot 2}^\top \boldsymbol{\Psi}_{11\cdot 2}^{-1}(\mathbf{x}_1 - \boldsymbol{\xi}_{1\cdot 2} - \bar{\tau}_{1\cdot 2}\bar{\boldsymbol{\eta}}_{1\cdot 2})}{\sqrt{1 + \bar{\boldsymbol{\eta}}_{1\cdot 2}^\top \boldsymbol{\Psi}_{11\cdot 2}^{-1}\bar{\boldsymbol{\eta}}_{1\cdot 2}}}\right\}, \quad \mathbf{x}_1 \in \mathbb{R}^{p_1}, \end{aligned} \tag{5.33}$$

where $\boldsymbol{\xi}_{1\cdot 2} = \boldsymbol{\xi}_1 + \boldsymbol{\Psi}_{12}\boldsymbol{\Psi}_{22}^{-1}(\mathbf{x}_2 - \boldsymbol{\xi}_2)$, $\boldsymbol{\Psi}_{11\cdot 2} = \boldsymbol{\Psi}_{11} - \boldsymbol{\Psi}_{12}\boldsymbol{\Psi}_{22}^{-1}\boldsymbol{\Psi}_{21}$,

$$\bar{\boldsymbol{\eta}}_{1\cdot 2} = \frac{\boldsymbol{\eta}_1 - \boldsymbol{\Psi}_{12}\boldsymbol{\Psi}_{22}^{-1}\boldsymbol{\eta}_2}{\sqrt{1 + \boldsymbol{\eta}_2^\top \boldsymbol{\Psi}_{22}^{-1}\boldsymbol{\eta}_2}} \quad \text{and} \quad \bar{\tau}_{1\cdot 2} = \frac{\boldsymbol{\eta}_2^\top \boldsymbol{\Psi}_{22}^{-1}(\mathbf{x}_2 - \boldsymbol{\xi}_2)}{\sqrt{1 + \boldsymbol{\eta}_2^\top \boldsymbol{\Psi}_{22}^{-1}\boldsymbol{\eta}_2}}.$$

Proof 72 *From the application of Theorem 66 to the linear transformation given in (5.23), or directly from the nonsingular version given in part (a) of Theorem 68, we find that the joint pdf of* $\mathbf{Y}_1 = \mathbf{X}_1 - \boldsymbol{\Psi}_{12}\boldsymbol{\Psi}_{22}^{-1}\mathbf{X}_2$ *and* $\mathbf{Y}_2 = \mathbf{X}_2$ *is*

$$\begin{aligned} h(\mathbf{y}_1, \mathbf{x}_2) &= 2\phi_{p_1}(\mathbf{y}_1; \boldsymbol{\xi}_1 - \boldsymbol{\Psi}_{12}\boldsymbol{\Psi}_{22}^{-1}\boldsymbol{\xi}_2 + \bar{\tau}_{1\cdot 2}\bar{\boldsymbol{\eta}}_{1\cdot 2}, \boldsymbol{\Psi}_{11\cdot 2} + \bar{\boldsymbol{\eta}}_{1\cdot 2}\bar{\boldsymbol{\eta}}_{1\cdot 2}^\top)\phi_{p_2}(\mathbf{x}_2; \boldsymbol{\xi}_2, \boldsymbol{\Psi}_{22} + \boldsymbol{\eta}_2\boldsymbol{\eta}_2^\top) \\ &\quad \times \Phi\left\{\frac{\boldsymbol{\eta}_{1\cdot 2}^\top \boldsymbol{\Psi}_{11\cdot 2}^{-1}(\mathbf{y}_1 - \boldsymbol{\xi}_1 + \boldsymbol{\Psi}_{12}\boldsymbol{\Psi}_{22}^{-1}\boldsymbol{\xi}_2 - \bar{\tau}_{1\cdot 2}\bar{\boldsymbol{\eta}}_{1\cdot 2}) + \boldsymbol{\eta}_2^\top \boldsymbol{\Psi}_{22}^{-1}(\mathbf{x}_2 - \boldsymbol{\xi}_2)}{\sqrt{1 + \boldsymbol{\eta}_{1\cdot 2}^\top \boldsymbol{\Psi}_{11\cdot 2}^{-1}\boldsymbol{\eta}_{1\cdot 2} + \boldsymbol{\eta}_2^\top \boldsymbol{\Psi}_{22}^{-1}\boldsymbol{\eta}_2}}\right\}. \end{aligned}$$

Also, by applying Theorem 60 to the marginal of $\mathbf{X}_2$, *we have* $\mathbf{Y}_2 = \mathbf{X}_2 \sim \mathcal{SN}_{p_2}(\boldsymbol{\xi}_2, \boldsymbol{\Psi}_{22}, \boldsymbol{\eta}_2)$, *with marginal pdf given by*

$$h_2(\mathbf{x}_2) = 2\phi_{p_2}(\mathbf{x}_2; \boldsymbol{\xi}_2, \boldsymbol{\Psi}_{22} + \boldsymbol{\eta}_2\boldsymbol{\eta}_2^\top)\Phi\left\{\frac{\boldsymbol{\eta}_2^\top \boldsymbol{\Psi}_{22}^{-1}(\mathbf{x}_2 - \boldsymbol{\xi}_2)}{\sqrt{1 + \boldsymbol{\eta}_2^\top \boldsymbol{\Psi}_{22}^{-1}\boldsymbol{\eta}_2}}\right\}.$$

Hence, the conditional pdf of $\mathbf{Y}_1$ *given* $\mathbf{Y}_2 = \mathbf{x}_2$ *is*

$$\begin{aligned} h_{1\cdot 2}(\mathbf{y}_1) = \frac{h(\mathbf{y}_1, \mathbf{x}_2)}{h_2(\mathbf{x}_2)} &= \frac{1}{\Phi(\bar{\tau}_{1\cdot 2})}\phi_{p_1}(\mathbf{y}_1; \boldsymbol{\xi}_1 - \boldsymbol{\Psi}_{12}\boldsymbol{\Psi}_{22}^{-1}\boldsymbol{\xi}_2 + \bar{\tau}_{1\cdot 2}\bar{\boldsymbol{\eta}}_{1\cdot 2}, \boldsymbol{\Psi}_{11\cdot 2} + \bar{\boldsymbol{\eta}}_{1\cdot 2}\bar{\boldsymbol{\eta}}_{1\cdot 2}^\top) \\ &\quad \times \Phi\left\{\frac{\bar{\boldsymbol{\eta}}_{1\cdot 2}^\top \boldsymbol{\Psi}_{11\cdot 2}^{-1}(\mathbf{y}_1 - \boldsymbol{\xi}_1 + \boldsymbol{\Psi}_{12}\boldsymbol{\Psi}_{22}^{-1}\boldsymbol{\xi}_2 - \bar{\tau}_{1\cdot 2}\bar{\boldsymbol{\eta}}_{1\cdot 2}) + \bar{\tau}_{1\cdot 2}}{\sqrt{1 + \bar{\boldsymbol{\eta}}_{1\cdot 2}^\top \boldsymbol{\Psi}_{11\cdot 2}^{-1}\bar{\boldsymbol{\eta}}_{1\cdot 2}}}\right\}. \end{aligned}$$

On the other hand, since $\mathbf{Y} \equiv \tilde{\mathbf{X}}$, *the transformation (5.23) implies that* $\mathbf{X} = \mathbf{B}^{-1}\mathbf{Y}$, *where* $\text{abs}(|\mathbf{B}^{-1}|) = 1$. *Hence, the Jacobian method allows us to write the joint pdf of* $(\mathbf{X}_1, \mathbf{X}_2)$ *as*

$$f(\mathbf{x}_1, \mathbf{x}_2) = h(\mathbf{x}_1 - \boldsymbol{\Psi}_{12}\boldsymbol{\Psi}_{22}^{-1}\mathbf{x}_2, \mathbf{x}_2)$$

from where the conditional pdf of $(\mathbf{X}_1 \mid \mathbf{X}_2 = \mathbf{x}_2)$ *is obtained as*

$$f_{1\cdot 2}(\mathbf{x}_1) = h_{1\cdot 2}(\mathbf{x}_1 - \boldsymbol{\Psi}_{12}\boldsymbol{\Psi}_{22}^{-1}\mathbf{x}_2).$$

This proves the result. □

Obviously, we can obtain the above result directly from (5.19) or (5.18) by using the well-known marginal-conditional factorization of the normal pdf given by

$$\phi_p(\mathbf{x};\boldsymbol{\xi},\boldsymbol{\Omega}) = \phi_{p_1}(\mathbf{x}_1;\boldsymbol{\xi}_1+\boldsymbol{\Omega}_{12}\boldsymbol{\Omega}_{22}^{-1}(\mathbf{x}_2-\boldsymbol{\xi}_2),\boldsymbol{\Omega}_{11\cdot 2})\phi_{p_2}(\mathbf{x}_2;\boldsymbol{\xi}_2,\boldsymbol{\Omega}_{22}), \quad \mathbf{x}=(\mathbf{x}_1^\top,\mathbf{x}_2^\top)^\top \in \mathbb{R}^{p_1+p_2},$$

where $\boldsymbol{\Omega} = \boldsymbol{\Psi}+\boldsymbol{\eta}\boldsymbol{\eta}^\top$ was partitioned so that $\boldsymbol{\Omega}_{ij} = \boldsymbol{\Psi}_{ij}+\boldsymbol{\eta}_i\boldsymbol{\eta}_j^\top$ for $i,j=1,2$, and showing that

$$\begin{aligned}\boldsymbol{\xi}_1+\boldsymbol{\Omega}_{12}\boldsymbol{\Omega}_{22}^{-1}(\mathbf{x}_2-\boldsymbol{\xi}_2) =& \boldsymbol{\xi}_1+\boldsymbol{\Psi}_{12}\boldsymbol{\Psi}_{22}^{-1}(\mathbf{x}_2-\boldsymbol{\xi}_2)+\frac{\boldsymbol{\eta}_2^\top\boldsymbol{\Psi}_{22}^{-1}(\mathbf{x}_2-\boldsymbol{\xi}_2)}{1+\boldsymbol{\eta}_2^\top\boldsymbol{\Psi}_{22}^{-1}\boldsymbol{\eta}_2}\boldsymbol{\eta}_{1\cdot 2}\\ =& \boldsymbol{\xi}_{11\cdot 2}+\bar{\tau}_{1\cdot 2}\bar{\boldsymbol{\eta}}_{1\cdot 2},\end{aligned}$$

and

$$\boldsymbol{\Omega}_{11\cdot 2} = \boldsymbol{\Omega}_{11}-\boldsymbol{\Omega}_{12}\boldsymbol{\Omega}_{22}^{-1}\boldsymbol{\Omega}_{21} = \boldsymbol{\Psi}_{11\cdot 2}+\bar{\boldsymbol{\eta}}_{1\cdot 2}\bar{\boldsymbol{\eta}}_{1\cdot 2}^\top.$$

Another alternative and perhaps more comprehensive proof of (5.33) is obtained from the stochastic representation (5.29) (or from its equivalent version in (5.31)). In fact, conditioning in (5.29) on the truncated normal mixing variable $U_{1\cdot 2}$, we have

$$(\mathbf{X}_{1\cdot 2} \mid U_{1\cdot 2}=u) \sim \mathcal{N}_{p_1}(\boldsymbol{\xi}_{1\cdot 2}+\boldsymbol{\eta}_{1\cdot 2}u,\boldsymbol{\Psi}_{11\cdot 2}),$$

which now is nonsingular since $\boldsymbol{\Psi}_{11\cdot 2}>0$. Therefore, using (5.27) we have

$$\begin{aligned} f_{1\cdot 2}(\mathbf{x}_1) &= \int_0^\infty \phi_{p_1}(\mathbf{x}_1;\boldsymbol{\xi}_{1\cdot 2}+\boldsymbol{\eta}_{1\cdot 2}u,\boldsymbol{\Psi}_{11\cdot 2})g(u\mid\mathbf{x}_2)\mathrm{d}u\\ &= \frac{1}{\Phi(\bar{\tau}_{1\cdot 2})}\int_0^\infty \phi_{p_1}(\mathbf{x}_1;\boldsymbol{\xi}_{1\cdot 2}+\boldsymbol{\eta}_{1\cdot 2}u,\boldsymbol{\Psi}_{11\cdot 2})\phi(u;\tau_{1\cdot 2},(1+\tilde{\eta}_2^2)^{-1})\mathrm{d}u,\end{aligned}$$

where $\tilde{\eta}_2^2 = \boldsymbol{\eta}_2^\top\boldsymbol{\Psi}_{22}^{-1}\boldsymbol{\eta}_2$, $\bar{\tau}_{1\cdot 2} = (1+\tilde{\eta}_2^2)^{-1/2}\boldsymbol{\eta}_2^\top\boldsymbol{\Psi}_{22}^{-1}(\mathbf{x}_2-\boldsymbol{\xi}_2)$ and $\tau_{1\cdot 2} = (1+\tilde{\eta}_2^2)^{-1/2}\bar{\tau}_{1\cdot 2}$. But, from the identities $\tau_{1\cdot 2}\boldsymbol{\eta}_{1\cdot 2}^\top = \bar{\tau}_{1\cdot 2}\bar{\boldsymbol{\eta}}_{1\cdot 2}^\top$ and

$$\begin{aligned}\phi_{p_1}(\mathbf{x}_1;\boldsymbol{\xi}_{1\cdot 2}+\boldsymbol{\eta}_{1\cdot 2}u,\boldsymbol{\Psi}_{11\cdot 2})\phi\left\{u;\tau_{1\cdot 2},(1+\tilde{\eta}_2^2)^{-1}\right\} &= \phi_{p_1}(\mathbf{x}_1;\boldsymbol{\xi}_{1\cdot 2}+\bar{\tau}_{1\cdot 2}\bar{\boldsymbol{\eta}}_{1\cdot 2},\boldsymbol{\Psi}_{11\cdot 2}+\bar{\boldsymbol{\eta}}_{1\cdot 2}\bar{\boldsymbol{\eta}}_{1\cdot 2}^\top)\\ &\times\phi\left[u;\tau_{1\cdot 2}+(1+\tilde{\eta}_2^2)^{-1}\boldsymbol{\eta}_{1\cdot 2}^\top(\boldsymbol{\Psi}_{11\cdot 2}+\bar{\boldsymbol{\eta}}_{1\cdot 2}\bar{\boldsymbol{\eta}}_{1\cdot 2}^\top)^{-1}(\mathbf{x}_1-\boldsymbol{\xi}_{1\cdot 2}-\bar{\tau}_{1\cdot 2}\bar{\boldsymbol{\eta}}_{1\cdot 2}),\right.\\ &\left.(1+\tilde{\eta}_2^2)^{-1}\left\{1-\bar{\boldsymbol{\eta}}_{1\cdot 2}^\top(\boldsymbol{\Psi}_{11\cdot 2}+\bar{\boldsymbol{\eta}}_{1\cdot 2}\bar{\boldsymbol{\eta}}_{1\cdot 2}^\top)^{-1}\bar{\boldsymbol{\eta}}_{1\cdot 2}\right\}\right],\end{aligned}$$

we find

$$\begin{aligned} f_{1\cdot 2}(\mathbf{x}_1) &= \frac{1}{\Phi(\bar{\tau}_{1\cdot 2})}\phi_{p_1}(\mathbf{x}_1;\boldsymbol{\xi}_{1\cdot 2}+\bar{\tau}_{1\cdot 2}\bar{\boldsymbol{\eta}}_{1\cdot 2},\boldsymbol{\Psi}_{11\cdot 2}+\bar{\boldsymbol{\eta}}_{1\cdot 2}\bar{\boldsymbol{\eta}}_{1\cdot 2}^\top)\\ &\times\int_0^\infty \phi\left\{u;\tau_{1\cdot 2}+(1+\tilde{\eta}_2^2)^{-1}\boldsymbol{\eta}_{1\cdot 2}^\top(\boldsymbol{\Psi}_{11\cdot 2}+\bar{\boldsymbol{\eta}}_{1\cdot 2}\bar{\boldsymbol{\eta}}_{1\cdot 2}^\top)^{-1}(\mathbf{x}_1-\boldsymbol{\xi}_{1\cdot 2}-\bar{\tau}_{1\cdot 2}\bar{\boldsymbol{\eta}}_{1\cdot 2}),\right.\\ &\left.(1+\tilde{\eta}_2^2)^{-1}-(1+\tilde{\eta}_2^2)^{-2}\boldsymbol{\eta}_{1\cdot 2}^\top(\boldsymbol{\Psi}_{11\cdot 2}+\bar{\boldsymbol{\eta}}_{1\cdot 2}\bar{\boldsymbol{\eta}}_{1\cdot 2}^\top)^{-1}\boldsymbol{\eta}_{1\cdot 2}\right\}\mathrm{d}u\\ &= \frac{1}{\Phi(\bar{\tau}_{1\cdot 2})}\phi_{p_1}(\mathbf{x}_1;\boldsymbol{\xi}_{1\cdot 2}+\bar{\tau}_{1\cdot 2}\bar{\boldsymbol{\eta}}_{1\cdot 2},\boldsymbol{\Psi}_{11\cdot 2}+\bar{\boldsymbol{\eta}}_{1\cdot 2}\bar{\boldsymbol{\eta}}_{1\cdot 2}^\top)\\ &\times\Phi\left\{\frac{\tau_{1\cdot 2}+(1+\tilde{\eta}_2^2)^{-1}\boldsymbol{\eta}_{1\cdot 2}^\top(\boldsymbol{\Psi}_{11\cdot 2}+\bar{\boldsymbol{\eta}}_{1\cdot 2}\bar{\boldsymbol{\eta}}_{1\cdot 2}^\top)^{-1}(\mathbf{x}_1-\boldsymbol{\xi}_{1\cdot 2}-\bar{\tau}_{1\cdot 2}\bar{\boldsymbol{\eta}}_{1\cdot 2})}{\sqrt{(1+\tilde{\eta}_2^2)^{-1}-(1+\tilde{\eta}_2^2)^{-2}\boldsymbol{\eta}_{1\cdot 2}^\top(\boldsymbol{\Psi}_{11\cdot 2}+\bar{\boldsymbol{\eta}}_{1\cdot 2}\bar{\boldsymbol{\eta}}_{1\cdot 2}^\top)^{-1}\boldsymbol{\eta}_{1\cdot 2}}}\right\}.\end{aligned}$$

Finally, from the fact that

$$(\boldsymbol{\Psi}_{11\cdot 2}+\bar{\boldsymbol{\eta}}_{1\cdot 2}\bar{\boldsymbol{\eta}}_{1\cdot 2}^\top)^{-1} = \boldsymbol{\Psi}_{11\cdot 2}^{-1}-\frac{\boldsymbol{\Psi}_{11\cdot 2}^{-1}\bar{\boldsymbol{\eta}}_{1\cdot 2}\bar{\boldsymbol{\eta}}_{1\cdot 2}^\top\boldsymbol{\Psi}_{11\cdot 2}^{-1}}{1+\bar{\boldsymbol{\eta}}_{1\cdot 2}^\top\boldsymbol{\Psi}_{11\cdot 2}^{-1}\bar{\boldsymbol{\eta}}_{1\cdot 2}},$$

we obtain

$$\begin{aligned}
&\frac{\tau_{1\cdot2}+(1+\tilde{\eta}_2^2)^{-1}\boldsymbol{\eta}_{1\cdot2}^\top(\boldsymbol{\Psi}_{11\cdot2}+\bar{\boldsymbol{\eta}}_{1\cdot2}\bar{\boldsymbol{\eta}}_{1\cdot2}^\top)^{-1}(\mathbf{x}_1-\boldsymbol{\xi}_{1\cdot2}-\bar{\tau}_{1\cdot2}\bar{\boldsymbol{\eta}}_{1\cdot2})}{\sqrt{(1+\tilde{\eta}_2^2)^{-1}-(1+\tilde{\eta}_2^2)^{-2}\boldsymbol{\eta}_{1\cdot2}^\top(\boldsymbol{\Psi}_{11\cdot2}+\bar{\boldsymbol{\eta}}_{1\cdot2}\bar{\boldsymbol{\eta}}_{1\cdot2}^\top)^{-1}\boldsymbol{\eta}_{1\cdot2}}}\\
&=\frac{\bar{\tau}_{1\cdot2}+\bar{\boldsymbol{\eta}}_{1\cdot2}^\top\boldsymbol{\Psi}_{11\cdot2}^{-1}(\mathbf{x}_1-\boldsymbol{\xi}_{1\cdot2}-\bar{\tau}_{1\cdot2}\bar{\boldsymbol{\eta}}_{1\cdot2})}{\sqrt{1+\bar{\boldsymbol{\eta}}_{1\cdot2}^\top\boldsymbol{\Psi}_{11\cdot2}^{-1}\bar{\boldsymbol{\eta}}_{1\cdot2}}}
\end{aligned}$$

and therefore

$$\begin{aligned}
f_{1\cdot2}(\mathbf{x}_1) &= \frac{1}{\Phi(\bar{\tau}_{1\cdot2})}\phi_{p_1}(\mathbf{x}_1;\boldsymbol{\xi}_{1\cdot2}+\bar{\tau}_{1\cdot2}\bar{\boldsymbol{\eta}}_{1\cdot2},\boldsymbol{\Psi}_{11\cdot2}+\bar{\boldsymbol{\eta}}_{1\cdot2}\bar{\boldsymbol{\eta}}_{1\cdot2}^\top)\\
&\quad\times\Phi\left\{\frac{\bar{\tau}_{1\cdot2}+\bar{\boldsymbol{\eta}}_{1\cdot2}^\top\boldsymbol{\Psi}_{11\cdot2}^{-1}(\mathbf{x}_1-\boldsymbol{\xi}_{1\cdot2}-\bar{\tau}_{1\cdot2}\bar{\boldsymbol{\eta}}_{1\cdot2})}{\sqrt{1+\bar{\boldsymbol{\eta}}_{1\cdot2}^\top\boldsymbol{\Psi}_{11\cdot2}^{-1}\bar{\boldsymbol{\eta}}_{1\cdot2}}}\right\},\quad \mathbf{x}_1\in\mathbb{R}^{p_1}.
\end{aligned}$$

5.8 Invariance and Quadratic Forms

5.8.1 Invariance under even functions

We next describe an invariance property of the random vector $\mathbf{Y}=\mathbf{X}-\boldsymbol{\xi}\sim\mathcal{SN}_p(\mathbf{0}_p,\boldsymbol{\Psi},\boldsymbol{\eta})$, which is very useful to compute even moments, as well as the distribution of even functions of $\mathbf{Y}$. For this, we first present an alternative stochastic representation of $\mathbf{X}$.

Theorem 73 *Let $\mathbf{Y}=\mathrm{sgn}(W_0)\,\mathbf{W}_*$, where W_0 and $\mathbf{W}_*$ have a multivariate normal joint distribution as defined in (5.14). Then, $\mathbf{Y}\sim\mathcal{SN}_p(\mathbf{0}_p,\boldsymbol{\Psi},\boldsymbol{\eta})$.*

Proof 73 *Let $S_0=\mathrm{sgn}(W_0)$. Since $W_0\overset{d}{=}-W_0$, we have*

$$\mathrm{P}(S_0=-1)=\mathrm{P}(W_0<0)=\mathrm{P}(W_0>0)=\mathrm{P}(S_0=+1)=\frac{1}{2}.$$

Now, for any Borel subset C of $\mathbb{R}^p$, note that

$$\begin{aligned}
\mathrm{P}(\mathbf{Y}\in C) &= \mathrm{P}(-\mathbf{W}_*\in C\mid S_0=-1)\mathrm{P}(S_0=-1)+\mathrm{P}(\mathbf{Y}\in C\mid S_0=+1)\mathrm{P}(S_0=+1)\\
&= \frac{1}{2}\{\mathrm{P}(-\mathbf{W}_*\in C\mid W_0<0)+\mathrm{P}(\mathbf{W}_*\in C\mid W_0>0)\}\\
&= \frac{1}{2}\{\mathrm{P}(-\mathbf{W}_*\in C\mid -W_0>0)+\mathrm{P}(\mathbf{W}_*\in C\mid W_0>0)\}.
\end{aligned}$$

However, by central symmetry $(-W_0,-\mathbf{W}_*)\overset{d}{=}(W_0,\mathbf{W}_*)$, and therefore

$$\mathrm{P}(-\mathbf{W}_*\in C\mid -W_0>0)=\mathrm{P}(\mathbf{W}_*\in C\mid W_0>0)\quad\Longrightarrow\quad \mathrm{P}(\mathbf{Y}\in C)=\mathrm{P}(\mathbf{W}_*\in C\mid W_0>0).$$

Since this holds for any Borel subset C of $\mathbb{R}^p$, then $\mathbf{Y}\overset{d}{=}(\mathbf{W}_*\mid W_0>0)$ and hence by Theorem 53, the distribution of $\mathbf{Y}$ is the same as that of $\mathbf{X}-\boldsymbol{\xi}\sim\mathcal{SN}_p(\mathbf{0}_p,\boldsymbol{\Psi},\boldsymbol{\eta})$. The result is proved. □

An immediate corollary of Theorem 73 is the following invariance property of $\mathbf{Y}\sim\mathcal{SN}_p(\mathbf{0}_p,\boldsymbol{\Psi},\boldsymbol{\eta})$.

Theorem 74 *Let* $\mathbf{Y}_0 \sim \mathcal{N}_p(\mathbf{0}_p, \boldsymbol{\Psi}+\boldsymbol{\eta}\boldsymbol{\eta}^\top)$ *and* $\mathbf{Y} \sim \mathcal{SN}_p(\mathbf{0}_p, \boldsymbol{\Psi}, \boldsymbol{\eta})$*. Then, for any measurable even vector-valued function* $\mathbf{g} : \mathbb{R}^p \to \mathbb{R}^k$*, we have*

$$\mathbf{g}(\mathbf{Y}) \stackrel{d}{=} \mathbf{g}(\mathbf{Y}_0). \tag{5.34}$$

Proof 74 *According to Theorem 73,* $\mathbf{Y} \stackrel{d}{=} S_0 \mathbf{W}_*$*, where* $S_0 = \mathrm{sgn}(W_0)$ *and* W_0 *and* $\mathbf{W}_*$ *are correlated normal random quantities, with* $W_0 \sim \mathcal{N}(0,1)$*,* $\mathbf{W}_* \sim \mathcal{N}_p(\mathbf{0}_p, \boldsymbol{\Psi}+\boldsymbol{\eta}\boldsymbol{\eta}^\top)$ *and* $\mathrm{cov}(W_0, \mathbf{W}_*) = \boldsymbol{\eta}^\top$*. Hence, since* $\mathbf{g}$ *is even, then* $\mathbf{g}(S_0\mathbf{W}_*) \stackrel{d}{=} \mathbf{g}(\mathbf{W}_*) \stackrel{d}{=} \mathbf{g}(\mathbf{Y}_0)$ *because* $\mathbf{W}_* \stackrel{d}{=} \mathbf{Y}_0$*.* □

Example 65 *If* $\mathbf{Z} \sim \mathcal{SN}_p(\mathbf{0}_p, \mathbf{I}_p, \boldsymbol{\eta})$ *and* $\mathbf{Z}_0 \sim \mathcal{N}_p(\mathbf{0}_p, \mathbf{I}_p+\boldsymbol{\eta}\boldsymbol{\eta}^\top)$*, then:*
(a) $\mathbf{Z}\mathbf{Z}^\top = (Z_iZ_j) \stackrel{d}{=} (Z_{0i}Z_{0j}) = \mathbf{Z}_0\mathbf{Z}_0^\top$*,* $i, j = 1, \ldots, p$*;*
(b) $\prod_{i=1}^p Z_i^{2k_i} \stackrel{d}{=} \prod_{i=1}^p Z_{0i}^{2k_i}$ *for any integer* k_i*,* $i = 1, \ldots, p$*;*
(c) $(\mathbf{Z}^\top\mathbf{B}_1\mathbf{Z}, \ldots, \mathbf{Z}^\top\mathbf{B}_k\mathbf{Z}) \stackrel{d}{=} (\mathbf{Z}_0^\top\mathbf{B}_1\mathbf{Z}_0, \ldots, \mathbf{Z}_0^\top\mathbf{B}_k\mathbf{Z}_0)$*, where* $\mathbf{B}_1, \ldots, \mathbf{B}_k$ *are constant symmetric matrices.*

Example 66 *Let* X *be a continuous random variable with pdf* $g(x)$*. Then* $X^2 \sim \chi_1^2$ *if and only if* X *is a generalized skew-normal random variable (Genton and Loperfido, 2005), i.e.,* $g(x) = 2\phi(x)\pi(x)$*,* $x \in \mathbb{R}$*, where the skewing function* $0 \le \pi(x) \le 1$ *satisfies* $\pi(-x) = 1 - \pi(x)$*; see Wang et al. (2004a) for further details and the proof, and also Arnold and Lin (2004) and Gupta et al. (2004b).*

Evidently, the above invariance property allows us to apply the multivariate normal theory to find the distributions and compute the moments of even functions in skew-normal random vectors; in particular, some special quadratic forms are briefly discussed in the next section.

5.8.2 Distributions of quadratic forms

In what follows, we let $\boldsymbol{\Omega}^- = (\boldsymbol{\Psi}+\boldsymbol{\eta}\boldsymbol{\eta}^\top)^-$ be a generalized inverse (or pseudo-inverse) of the matrix $\boldsymbol{\Omega} = \boldsymbol{\Psi}+\boldsymbol{\eta}\boldsymbol{\eta}^\top$. For instance, if $\boldsymbol{\Psi}^-$ is a generalized inverse (or pseudo-inverse) of $\boldsymbol{\Psi}$, we can choose $\boldsymbol{\Omega}^-$ as

$$\boldsymbol{\Omega}^- = \boldsymbol{\Psi}^- - (1+\eta^2)^{-1}\boldsymbol{\Psi}^-\boldsymbol{\eta}\boldsymbol{\eta}^\top\boldsymbol{\Psi}^-, \tag{5.35}$$

where $\eta = (\boldsymbol{\eta}^\top\boldsymbol{\Psi}^-\boldsymbol{\eta})^{1/2}$.

Theorem 75 *Let* $\mathbf{X} \sim \mathcal{SN}_p(\boldsymbol{\xi}, \boldsymbol{\Psi}, \boldsymbol{\eta})$*, and suppose that* $\boldsymbol{\Psi}$ *has rank* q*. Then, whatever the values of* $\boldsymbol{\eta} \in \mathbb{R}^p$*, we have*

$$(\mathbf{X}-\boldsymbol{\xi})^\top\boldsymbol{\Omega}^-(\mathbf{X}-\boldsymbol{\xi}) \sim \chi_q^2.$$

Proof 75 *Let* $\mathbf{y} = \mathbf{x} - \boldsymbol{\xi}$ *and note that the quadratic function* $Q(\mathbf{y}) = \mathbf{y}^\top\boldsymbol{\Omega}^-\mathbf{y}$ *is even. Then, by (5.34) we have* $Q(\mathbf{X}-\boldsymbol{\xi}) \stackrel{d}{=} Q(\mathbf{X}_0-\boldsymbol{\xi})$*, where* $\mathbf{X}_0 \sim \mathcal{N}_p(\mathbf{0}_p, \boldsymbol{\Omega})$*. Since* $\boldsymbol{\Omega} = \boldsymbol{\Psi}+\boldsymbol{\eta}\boldsymbol{\eta}^\top$*, then* $\mathrm{rank}(\boldsymbol{\Omega})=\mathrm{rank}(\boldsymbol{\Psi}) = q$ *and so* $Q(\mathbf{X}_0-\boldsymbol{\xi}) = (\mathbf{X}_0-\boldsymbol{\xi})^\top\boldsymbol{\Omega}^-(\mathbf{X}_0-\boldsymbol{\xi}) \sim \chi_q^2$*, which proves the stated result.* □

For an alternative proof, we can use (5.35) to write

$$(\mathbf{X}-\boldsymbol{\xi})^\top\boldsymbol{\Omega}^-(\mathbf{X}-\boldsymbol{\xi}) = (\mathbf{X}-\boldsymbol{\xi})^\top\boldsymbol{\Psi}^-(\mathbf{X}-\boldsymbol{\xi}) - \frac{\{\boldsymbol{\eta}^\top\boldsymbol{\Psi}^-(\mathbf{X}-\boldsymbol{\xi})\}^2}{1+\eta^2}.$$

Also, by definition, $\mathbf{X}-\boldsymbol{\xi} = \boldsymbol{\eta}U+\mathbf{W}$, where $U \sim \mathcal{HN}(0,1)$ and $\mathbf{W} \sim \mathcal{N}_p(\mathbf{0}_p, \boldsymbol{\Psi})$ are independent; and so

$$\begin{aligned}(\mathbf{X}-\boldsymbol{\xi})^\top\boldsymbol{\Omega}^-(\mathbf{X}-\boldsymbol{\xi}) &= (\boldsymbol{\eta}U+\mathbf{W})^\top\boldsymbol{\Psi}^-(\boldsymbol{\eta}U+\mathbf{W}) - \frac{\{\boldsymbol{\eta}^\top\boldsymbol{\Psi}^-(\boldsymbol{\eta}U+\mathbf{W})\}^2}{1+\eta^2}\\ &= \eta^2U^2+2U\boldsymbol{\eta}^\top\boldsymbol{\Psi}^-\mathbf{W}+\mathbf{W}^\top\boldsymbol{\Psi}^-\mathbf{W}\\ &\quad -\frac{\eta^4U^2+2\eta^2U\boldsymbol{\eta}^\top\boldsymbol{\Psi}^-\mathbf{W}+(\boldsymbol{\eta}^\top\boldsymbol{\Psi}^-\mathbf{W})^2}{1+\eta^2}\\ &= \frac{\eta^2U^2+2U\boldsymbol{\eta}^\top\boldsymbol{\Psi}^-\mathbf{W}}{1+\eta^2}+\mathbf{W}^\top\boldsymbol{\Psi}^-\mathbf{W}-\frac{(\boldsymbol{\eta}^\top\boldsymbol{\Psi}^-\mathbf{W})^2}{1+\eta^2}.\end{aligned}$$

Now, since $\boldsymbol{\Psi}$ has rank $q\,(\le p)$, we may always find an orthogonal matrix $\boldsymbol{\Gamma} = (\boldsymbol{\Gamma}_1, \boldsymbol{\Gamma}_2)$ to write

$$\boldsymbol{\Psi} = (\boldsymbol{\Gamma}_1 \quad \boldsymbol{\Gamma}_2)\begin{pmatrix}\mathbf{D}_1 & \mathbf{O}\\ \mathbf{O} & \mathbf{O}\end{pmatrix}\begin{pmatrix}\boldsymbol{\Gamma}_1^\top\\ \boldsymbol{\Gamma}_2^\top\end{pmatrix} = \boldsymbol{\Gamma}_1\mathbf{D}\boldsymbol{\Gamma}_1^\top,$$

where $\mathbf{D}_1$ is a q-by-q nonsingular diagonal matrix, and then consider the pseudo-inverse of $\boldsymbol{\Psi}$ given by

$$\boldsymbol{\Psi}^- = (\boldsymbol{\Gamma}_1 \quad \boldsymbol{\Gamma}_2)\begin{pmatrix}\mathbf{D}_1^{-1} & \mathbf{O}\\ \mathbf{O} & \mathbf{O}\end{pmatrix}\begin{pmatrix}\boldsymbol{\Gamma}_1^\top\\ \boldsymbol{\Gamma}_2^\top\end{pmatrix} = \boldsymbol{\Gamma}_1\mathbf{D}_1^{-1}\boldsymbol{\Gamma}_1^\top.$$

From this fact, we can write $\mathbf{W}^\top\boldsymbol{\Psi}^-\mathbf{W} = \mathbf{W}_0^\top\mathbf{W}_0$ (with probability 1), where $\mathbf{W}_0 \sim \mathcal{N}_q(\mathbf{0}_q, \mathbf{I}_q)$, and $\boldsymbol{\eta}^\top\boldsymbol{\Psi}^-\mathbf{W} = \eta V_0$, where $V_0 = \boldsymbol{\eta}_0^\top\mathbf{W}_0 \sim \mathcal{N}(0, \|\boldsymbol{\eta}_0\|^2)$, with $\boldsymbol{\eta}_0 \in \mathbb{R}^q$ such that $\|\boldsymbol{\eta}_0\| = 1$; and therefore

$$\begin{aligned}(\mathbf{X}-\boldsymbol{\xi})^\top\boldsymbol{\Omega}^-(\mathbf{X}-\boldsymbol{\xi}) &= \frac{(\eta U+V_0)^2}{1+\eta^2}+\|\mathbf{W}_0\|^2-V_0^2\\ &= \frac{Z^2}{1+\eta^2}+\mathbf{W}_0^\top\mathbf{M}_0\mathbf{W}_0,\end{aligned}$$

where $Z = \eta U+V_0$ and $\mathbf{M}_0 = \mathbf{I}_q - \boldsymbol{\eta}_0\boldsymbol{\eta}_0^\top$, which is a projection matrix of $\text{rank}(\mathbf{M}_0) = \text{tr}(\mathbf{M}_0) = q-1$. Here, by assumption, U is independent of $\mathbf{W}$, and therefore it is also independent of $\mathbf{W}_0$ and V_0 since they are functions of $\mathbf{W}$. But this means that $Z = \eta U + V_0 \sim \mathcal{SN}(0, 1+\eta^2, \eta)$, since $U \sim \mathcal{HN}(0,1)$ and $V_0 \sim \mathcal{N}(0,1)$, and they are independent. Also, since $\mathbf{W}_0 \sim \mathcal{N}_q(\mathbf{0}_q, \mathbf{I}_q)$ and $\mathbf{M}_0\boldsymbol{\eta}_0 = \mathbf{0}_q$, then $\mathbf{M}_0\mathbf{W}_0$ is independent of $V_0 = \boldsymbol{\eta}_0^\top\mathbf{W}_0$ and hence of Z. Finally, since

$$\frac{Z^2}{1+\eta^2} \sim \chi_1^2 \quad \text{and} \quad \mathbf{W}_0^\top\mathbf{M}_0\mathbf{W}_0 \sim \chi_{q-1}^2,$$

are independent, we conclude that $(\mathbf{X}-\boldsymbol{\xi})^\top\boldsymbol{\Omega}^-(\mathbf{X}-\boldsymbol{\xi}) \sim \chi_q^2$.

Example 67 *If* $\mathbf{X} \sim \mathcal{SN}_p(\boldsymbol{\xi}, \boldsymbol{\Psi}, \boldsymbol{\eta})$, *where* $\boldsymbol{\Psi}$ *is nonsingular, then*

$$(\mathbf{X}-\boldsymbol{\xi})^\top(\boldsymbol{\Psi}+\boldsymbol{\eta}\boldsymbol{\eta}^\top)^{-1}(\mathbf{X}-\boldsymbol{\xi}) = (\mathbf{X}-\boldsymbol{\xi})^\top\boldsymbol{\Psi}^{-1}(\mathbf{X}-\boldsymbol{\xi}) - \left\{\frac{\boldsymbol{\eta}^\top\boldsymbol{\Psi}^{-1}(\mathbf{X}-\boldsymbol{\xi})}{\sqrt{1+\boldsymbol{\eta}^\top\boldsymbol{\Psi}^{-1}\boldsymbol{\eta}}}\right\}^2 \sim \chi_p^2,$$

whatever the value of the skewness vector $\boldsymbol{\eta} \in \mathbb{R}^p$.

Theorem 75 can be generalized to an arbitrary quadratic form $(\mathbf{X}-\boldsymbol{\xi})^\top\mathbf{B}(\mathbf{X}-\boldsymbol{\xi})$ under certain conditions on the symmetric matrix $\mathbf{B}$. In fact, when $\mathbf{X}-\boldsymbol{\xi}$ is $\mathcal{SN}_p(\mathbf{0}_p,\boldsymbol{\Psi},\boldsymbol{\eta})$, then the necessary and sufficient conditions for $(\mathbf{X}-\boldsymbol{\xi})^\top\mathbf{B}(\mathbf{X}-\boldsymbol{\xi})$ to have a χ^2 distribution are

$$\boldsymbol{\Psi}\mathbf{B}\boldsymbol{\Psi}\mathbf{B}\boldsymbol{\Psi}=\boldsymbol{\Psi}\mathbf{B}\boldsymbol{\Psi} \quad \text{and} \quad \mathbf{B}\boldsymbol{\eta}=\mathbf{0}_p,$$

in which case the degrees of freedom are $r=\text{rank}(\mathbf{B}\boldsymbol{\Psi})$. When $\boldsymbol{\Psi}$ is nonsingular, the first of these conditions becomes $\mathbf{B}\boldsymbol{\Psi}\mathbf{B}\boldsymbol{\Psi}=\mathbf{B}\boldsymbol{\Psi}$, which means that $\mathbf{B}\boldsymbol{\Psi}$ is idempotent and is equivalent to $\mathbf{B}\boldsymbol{\Psi}\mathbf{B}=\mathbf{B}$. In the next theorem, we prove this by assuming that $\boldsymbol{\Psi}$ is nonsingular, but the extension of its proof to the singular case is straightforward.

Theorem 76 *If $\mathbf{X}\sim\mathcal{SN}_p(\boldsymbol{\xi},\boldsymbol{\Psi},\boldsymbol{\eta})$, with $\boldsymbol{\Psi}>0$, and $\mathbf{B}$ is a p-by-p symmetric matrix of rank r, then $(\mathbf{X}-\boldsymbol{\xi})^\top\mathbf{B}(\mathbf{X}-\boldsymbol{\xi})\sim\chi^2_r$ iff $\mathbf{B}\boldsymbol{\Psi}$ is idempotent ($\mathbf{B}\boldsymbol{\Psi}\mathbf{B}\boldsymbol{\Psi}=\mathbf{B}\boldsymbol{\Psi}$; i.e., $\mathbf{B}\boldsymbol{\Psi}\mathbf{B}=\mathbf{B}$) and $\mathbf{B}\boldsymbol{\eta}=\mathbf{0}_p$.*

Proof 76 *By Theorem 74, $(\mathbf{X}-\boldsymbol{\xi})^\top\mathbf{B}(\mathbf{X}-\boldsymbol{\xi})\overset{d}{=}(\mathbf{X}_0-\boldsymbol{\xi})^\top\mathbf{B}(\mathbf{X}_0-\boldsymbol{\xi})$, where $\mathbf{X}_0-\boldsymbol{\xi}\sim\mathcal{N}_p(\mathbf{0}_p,\boldsymbol{\Omega})$ with $\boldsymbol{\Omega}=\boldsymbol{\Psi}+\boldsymbol{\eta}\boldsymbol{\eta}^\top>0$ since $\boldsymbol{\Psi}>0$. Now, we know from the multivariate normal theory for quadratic forms (e.g., Muirhead (1982) or Chapter 3) that $(\mathbf{X}_0-\boldsymbol{\xi})^\top\mathbf{B}(\mathbf{X}_0-\boldsymbol{\xi})\sim\chi^2_r$, where $r=\text{rank}(\mathbf{B})$, if and only if $\mathbf{B}\boldsymbol{\Omega}$ is idempotent, which is equivalent to $\mathbf{B}\boldsymbol{\Omega}\mathbf{B}=\mathbf{B}$ since $\boldsymbol{\Omega}$ is nonsingular. However, since $\boldsymbol{\Omega}=\boldsymbol{\Psi}+\boldsymbol{\eta}\boldsymbol{\eta}^\top$, this last condition becomes $\mathbf{B}\boldsymbol{\Psi}\mathbf{B}+\mathbf{B}\boldsymbol{\eta}\boldsymbol{\eta}^\top\mathbf{B}=\mathbf{B}$ which in turn holds if and only if $\mathbf{B}\boldsymbol{\Psi}\mathbf{B}=\mathbf{B}$ and $\mathbf{B}\boldsymbol{\eta}=\mathbf{0}_p$. The proof ends with taking into account that $\boldsymbol{\Psi}$ is also nonsingular.* □

The non-central χ^2-distribution can also be preserved for quadratic forms of $\mathbf{X}\sim\mathcal{SN}_p(\boldsymbol{\xi},\boldsymbol{\Psi},\boldsymbol{\eta})$ of the type $\mathbf{X}^\top\mathbf{B}\mathbf{X}$ under similar conditions on the symmetric matrix $\mathbf{B}$ to those given above. To study this, we first prove the following theorem related to the cf of $\mathbf{X}^\top\mathbf{B}\mathbf{X}$ regardless of the symmetric matrix $\mathbf{B}$. This result also provides an extended version of the non-central χ^2-distribution.

Theorem 77 *Let $\mathbf{X}\sim\mathcal{SN}_p(\boldsymbol{\xi},\boldsymbol{\Psi},\boldsymbol{\eta})$, with $\boldsymbol{\Psi}>0$. If $\mathbf{B}$ is a p-by-p symmetric matrix, then the cf of $\mathbf{X}^\top\mathbf{B}\mathbf{X}$ is*

$$\begin{aligned} \mathrm{E}(e^{it\mathbf{X}^\top\mathbf{B}\mathbf{X}}) &= 2|\mathbf{I}_p-2it\mathbf{B}\boldsymbol{\Omega}|^{-1/2}e^{it\boldsymbol{\xi}^\top(\mathbf{I}_p-2it\mathbf{B}\boldsymbol{\Omega})^{-1}\mathbf{B}\boldsymbol{\xi}} \\ &\quad\times\Phi\left\{\frac{2it\boldsymbol{\eta}^\top(\mathbf{I}_p-2it\mathbf{B}\boldsymbol{\Omega})^{-1}\mathbf{B}\boldsymbol{\xi}}{\sqrt{1+\boldsymbol{\eta}^\top(\mathbf{I}_p-2it\mathbf{B}\boldsymbol{\Omega})^{-1}\boldsymbol{\Psi}^{-1}\boldsymbol{\eta}}}\right\}, \end{aligned} \tag{5.36}$$

where $\boldsymbol{\Omega}=\boldsymbol{\Psi}+\boldsymbol{\eta}\boldsymbol{\eta}^\top$.

Proof 77 *Since $\mathbf{X}\sim\mathcal{SN}_p(\boldsymbol{\xi},\boldsymbol{\Psi},\boldsymbol{\eta})$ is nonsingular, from (5.19) we have*

$$\begin{aligned} \mathrm{E}(e^{it\mathbf{X}^\top\mathbf{B}\mathbf{X}}) &= 2\int_{\mathbb{R}^p}e^{it\mathbf{x}^\top\mathbf{B}\mathbf{x}}\phi_p(\mathbf{x};\boldsymbol{\xi},\boldsymbol{\Omega})\Phi\left\{\frac{\boldsymbol{\eta}^\top\boldsymbol{\Psi}^{-1}(\mathbf{x}-\boldsymbol{\xi})}{\sqrt{1+\boldsymbol{\eta}^\top\boldsymbol{\Psi}^{-1}\boldsymbol{\eta}}}\right\}d\mathbf{x} \\ &= 2\int_{\mathbb{R}^p}e^{it(\mathbf{z}+\boldsymbol{\xi})^\top\mathbf{B}(\mathbf{z}+\boldsymbol{\xi})}\phi_p(\mathbf{z};\mathbf{0}_p,\boldsymbol{\Omega})\Phi\left(\frac{\boldsymbol{\eta}^\top\boldsymbol{\Psi}^{-1}\mathbf{z}}{\sqrt{1+\boldsymbol{\eta}^\top\boldsymbol{\Psi}^{-1}\boldsymbol{\eta}}}\right)d\mathbf{z}, \end{aligned}$$

where $\boldsymbol{\Omega} = \boldsymbol{\Psi} + \boldsymbol{\eta}\boldsymbol{\eta}^\top$*, and we have made the simple change of variables* $\mathbf{z} = \mathbf{x} - \boldsymbol{\xi}$*. Let* $\delta_{\xi\xi} = \boldsymbol{\xi}^\top\mathbf{B}\boldsymbol{\xi}$*,* $\mathbf{A} = (\boldsymbol{\Omega}^{-1} - 2\mathrm{i}t\mathbf{B})^{-1} = \boldsymbol{\Omega}(\mathbf{I}_p - 2\mathrm{i}t\mathbf{B}\boldsymbol{\Omega})^{-1}$ *and* $\mathbf{b} = \mathbf{A}\mathbf{B}\boldsymbol{\xi}$*. Then, using that*

$$\mathrm{i}t(\mathbf{z}+\boldsymbol{\xi})^\top\mathbf{B}(\mathbf{z}+\boldsymbol{\xi}) - \frac{1}{2}\mathbf{z}^\top\boldsymbol{\Omega}^{-1}\mathbf{z} = -\frac{1}{2}(\mathbf{x}-\mathbf{b})^\top\mathbf{A}^{-1}(\mathbf{x}-\mathbf{b}) + \frac{1}{2}\{2\mathrm{i}t\delta_{\xi\xi} + (2\mathrm{i}t)^2\mathbf{b}^\top\mathbf{A}^{-1}\mathbf{b}\}$$

we obtain

$$e^{\mathrm{i}t(\mathbf{z}+\boldsymbol{\xi})^\top\mathbf{B}(\mathbf{z}+\boldsymbol{\xi})}\phi_p(\mathbf{x};\mathbf{0}_p,\boldsymbol{\Omega}) = |\mathbf{I}_p - 2\mathrm{i}t\mathbf{B}\boldsymbol{\Omega}|^{-1/2}e^{\frac{1}{2}\{2\mathrm{i}t\delta_{\xi\xi}+(2\mathrm{i}t)^2\mathbf{b}^\top\mathbf{A}^{-1}\mathbf{b}\}}\phi_p(\mathbf{z};2\mathrm{i}t\mathbf{b},\mathbf{A}),$$

and, therefore, the above integral becomes

$$\begin{aligned}
\mathrm{E}(e^{\mathrm{i}t\mathbf{X}^\top\mathbf{B}\mathbf{X}}) &= 2|\mathbf{I}_p - 2\mathrm{i}t\mathbf{B}\boldsymbol{\Omega}|^{-1/2}e^{\frac{1}{2}\{2\mathrm{i}t\delta_{\xi\xi}+(2\mathrm{i}t)^2\mathbf{b}^\top\mathbf{A}^{-1}\mathbf{b}\}} \\
&\quad\times\int_{\mathbb{R}^p}\phi_p(\mathbf{z};2\mathrm{i}t\mathbf{b},\mathbf{A})\Phi\left(\frac{\boldsymbol{\eta}^\top\boldsymbol{\Psi}^{-1}\mathbf{z}}{\sqrt{1+\boldsymbol{\eta}^\top\boldsymbol{\Psi}^{-1}\boldsymbol{\eta}}}\right)\mathrm{d}\mathbf{z} \\
&= 2|\mathbf{I}_p - 2\mathrm{i}t\mathbf{B}\boldsymbol{\Omega}|^{-1/2}e^{\frac{1}{2}\{2\mathrm{i}t\delta_{\xi\xi}+(2\mathrm{i}t)^2\mathbf{b}^\top\mathbf{A}^{-1}\mathbf{b}\}}\mathrm{E}\left\{\Phi\left(\frac{\boldsymbol{\eta}^\top\boldsymbol{\Psi}^{-1}\mathbf{Z}}{\sqrt{1+\boldsymbol{\eta}^\top\boldsymbol{\Psi}^{-1}\boldsymbol{\eta}}}\right)\right\},
\end{aligned}$$

where $\mathbf{Z} \sim \mathcal{N}_p(2\mathrm{i}t\mathbf{b},\mathbf{A})$*. By Exercise 5.3, we know that*

$$\mathrm{E}\left\{\Phi\left(\frac{\boldsymbol{\eta}^\top\boldsymbol{\Psi}^{-1}\mathbf{Z}}{\sqrt{1+\boldsymbol{\eta}^\top\boldsymbol{\Psi}^{-1}\boldsymbol{\eta}}}\right)\right\} = \Phi\left(\frac{2\mathrm{i}t\boldsymbol{\eta}^\top\boldsymbol{\Psi}^{-1}\mathbf{b}}{\sqrt{1+\boldsymbol{\eta}^\top\boldsymbol{\Psi}^{-1}\boldsymbol{\eta}+\boldsymbol{\eta}^\top\boldsymbol{\Psi}^{-1}\mathbf{A}\boldsymbol{\Psi}^{-1}\boldsymbol{\eta}}}\right).$$

The proof concludes by noting that

$$\begin{aligned}
\delta_{\xi\xi} + 2\mathrm{i}t\mathbf{b}^\top\mathbf{A}^{-1}\mathbf{b} &= \boldsymbol{\xi}^\top\mathbf{B}\boldsymbol{\xi} + 2\mathrm{i}t\boldsymbol{\xi}^\top\mathbf{B}\mathbf{A}\mathbf{B}\boldsymbol{\xi} \\
&= \boldsymbol{\xi}^\top\{\mathbf{B} + 2\mathrm{i}t\mathbf{B}(\boldsymbol{\Omega}^{-1} - 2\mathrm{i}t\mathbf{B})^{-1}\mathbf{B}\}\boldsymbol{\xi} \\
&= \boldsymbol{\xi}^\top\{\boldsymbol{\Omega}^{-1} - 2\mathrm{i}t\mathbf{B} + 2\mathrm{i}t\mathbf{B}\}(\boldsymbol{\Omega}^{-1} - 2\mathrm{i}t\mathbf{B})^{-1}\mathbf{B}\boldsymbol{\xi} \\
&= \boldsymbol{\xi}^\top\boldsymbol{\Omega}^{-1}(\boldsymbol{\Omega}^{-1} - 2\mathrm{i}t\mathbf{B})^{-1}\mathbf{B}\boldsymbol{\xi} = \boldsymbol{\xi}^\top(\mathbf{I}_p - 2\mathrm{i}t\mathbf{B}\boldsymbol{\Omega})^{-1}\mathbf{B}\boldsymbol{\xi},
\end{aligned}$$

$$\begin{aligned}
\boldsymbol{\eta}^\top\boldsymbol{\Psi}^{-1}\mathbf{b} &= \boldsymbol{\eta}^\top\boldsymbol{\Psi}^{-1}\mathbf{A}\mathbf{B}\boldsymbol{\xi} \\
&= \boldsymbol{\eta}^\top\boldsymbol{\Psi}^{-1}\boldsymbol{\Omega}(\mathbf{I}_p - 2\mathrm{i}t\mathbf{B}\boldsymbol{\Omega})^{-1}\mathbf{B}\boldsymbol{\xi} \\
&= (1+\boldsymbol{\eta}^\top\boldsymbol{\Psi}^{-1}\boldsymbol{\eta})\boldsymbol{\eta}^\top(\mathbf{I}_p - 2\mathrm{i}t\mathbf{B}\boldsymbol{\Omega})^{-1}\mathbf{B}\boldsymbol{\xi},
\end{aligned}$$

and

$$\begin{aligned}
\boldsymbol{\eta}^\top\boldsymbol{\Psi}^{-1}\mathbf{A}\boldsymbol{\Psi}^{-1}\boldsymbol{\eta} &= \boldsymbol{\eta}^\top\boldsymbol{\Psi}^{-1}\boldsymbol{\Omega}(\mathbf{I}_p - 2\mathrm{i}t\mathbf{B}\boldsymbol{\Omega})^{-1}\boldsymbol{\Psi}^{-1}\boldsymbol{\eta} \\
&= (1+\boldsymbol{\eta}^\top\boldsymbol{\Psi}^{-1}\boldsymbol{\eta})\boldsymbol{\eta}^\top(\mathbf{I}_p - 2\mathrm{i}t\mathbf{B}\boldsymbol{\Omega})^{-1}\boldsymbol{\Psi}^{-1}\boldsymbol{\eta}.
\end{aligned}$$

□

When $\boldsymbol{\xi} = \mathbf{0}_p$, the result of Theorem 76 is now an immediate consequence of Theorem 77. In fact, if $\mathbf{X} \sim \mathcal{SN}_p(\mathbf{0}_p,\boldsymbol{\Psi},\boldsymbol{\eta})$ then from (5.36), we have that the cf of $\mathbf{X}^\top\mathbf{B}\mathbf{X}$ becomes

$$\mathrm{E}(e^{\mathrm{i}t\mathbf{X}^\top\mathbf{B}\mathbf{X}}) = |\mathbf{I}_p - 2\mathrm{i}t\mathbf{B}\boldsymbol{\Omega}|^{-1/2},$$

where $\mathbf{B}\boldsymbol{\Omega} = \mathbf{B}\boldsymbol{\Psi} + \mathbf{B}\boldsymbol{\eta}\boldsymbol{\eta}^\top$. However, as we know, this cf must be equal to the cf of a χ^2_r-distribution, i.e.,

$$|\mathbf{I}_p - 2\mathrm{i}t\mathbf{B}\boldsymbol{\Omega}|^{-1/2} = (1-2\mathrm{i}t)^{-r/2},$$

where $r = \text{rank}(\mathbf{B\Omega})$, which holds if and only if the symmetric matrix $\mathbf{B\Omega}$ is idempotent (i.e., a projection matrix of rank r), but this is true if and only if $\mathbf{B\Psi}$ is idempotent and $\mathbf{B}\boldsymbol{\eta} = \mathbf{0}_p$. As we prove below, the non-central χ^2-distribution can also be preserved for some quadratic forms of a skew-normal random vector.

Theorem 78 *If $\mathbf{X} \sim \mathcal{SN}_p(\boldsymbol{\xi}, \boldsymbol{\Psi}, \boldsymbol{\eta})$, where $\boldsymbol{\Psi}$ is nonsingular, and $\mathbf{B}$ is a p-by-p symmetric matrix of rank r, then $\mathbf{X}^\top \mathbf{B}\mathbf{X} \sim \chi_r^2(\delta)$, where $\delta = \boldsymbol{\xi}^\top \mathbf{B}\boldsymbol{\xi}/2$, iff $\mathbf{B\Psi}$ is idempotent ($\mathbf{B\Psi B\Psi} = \mathbf{B\Psi}$; i.e., $\mathbf{B\Psi B} = \mathbf{B}$) and $\mathbf{B}\boldsymbol{\eta} = \mathbf{0}_p$.*

Proof 78 *Since by definition $\mathbf{X} = \boldsymbol{\eta} U + (\boldsymbol{\xi} + \mathbf{W})$, where $\boldsymbol{\xi} + \mathbf{W} \sim \mathcal{N}_p(\boldsymbol{\xi}, \boldsymbol{\Psi})$ and $U \sim \mathcal{HN}(0,1)$ are independent, we have*

$$\begin{aligned} \mathbf{X}^\top \mathbf{B}\mathbf{X} &= (\boldsymbol{\xi} + \mathbf{W} + \boldsymbol{\eta} U)^\top \mathbf{B}(\boldsymbol{\xi} + \mathbf{W} + \boldsymbol{\eta} U) \\ &= (\boldsymbol{\xi} + \mathbf{W})^\top \mathbf{B}(\boldsymbol{\xi} + \mathbf{W}) + 2\boldsymbol{\eta}^\top \mathbf{B}(\boldsymbol{\xi} + \mathbf{W})U + (\boldsymbol{\eta}^\top \mathbf{B}\boldsymbol{\eta})U^2. \end{aligned} \tag{5.37}$$

If $\mathbf{B}\boldsymbol{\eta} = \mathbf{0}_p$ then it is immediate from (5.37) that

$$\mathbf{X}^\top \mathbf{B}\mathbf{X} = (\boldsymbol{\xi} + \mathbf{W})^\top \mathbf{B}(\boldsymbol{\xi} + \mathbf{W}),$$

where $\boldsymbol{\xi} + \mathbf{W} \sim \mathcal{N}_p(\boldsymbol{\xi}, \boldsymbol{\Psi})$. However, from the multivariate normal theory for quadratic forms, we know that $(\boldsymbol{\xi} + \mathbf{W})^\top \mathbf{B}(\boldsymbol{\xi} + \mathbf{W}) \sim \chi_r^2(\delta)$, where $r = \text{rank}(\mathbf{B})$ and $\delta = \boldsymbol{\xi}^\top \mathbf{B}\boldsymbol{\xi}/2$, if and only if $\mathbf{B\Psi}$ is idempotent. Therefore, if $\mathbf{B\Psi}$ is idempotent and $\mathbf{B}\boldsymbol{\eta} = \mathbf{0}_p$ then $\mathbf{X}^\top \mathbf{B}\mathbf{X} \sim \chi_r^2(\delta)$.

Reciprocally, if $\mathbf{X}^\top \mathbf{B}\mathbf{X} \sim \chi_r^2(\delta)$, where $r = \text{rank}(\mathbf{B\Psi})$ and $\delta = \boldsymbol{\xi}^\top \mathbf{B}\boldsymbol{\xi}/2$, then its cf is

$$\mathrm{E}(e^{\mathrm{i}t\mathbf{X}^\top \mathbf{B}\mathbf{X}}) = \frac{e^{\frac{2\mathrm{i}t}{1-2\mathrm{i}t}\left(\frac{1}{2}\boldsymbol{\xi}^\top \mathbf{B}\boldsymbol{\xi}\right)}}{(1-2\mathrm{i}t)^{r/2}}. \tag{5.38}$$

On the other hand, from Theorem 77, the cf of $\mathbf{X}^\top \mathbf{B}\mathbf{X}$ is given by (5.36). Therefore, Equations (5.38) and (5.36) must be the same for all t, as well as for all $\boldsymbol{\xi}$ and some values of the skewness vector $\boldsymbol{\eta}$, including $\boldsymbol{\xi} = \boldsymbol{\eta} = \mathbf{0}_p$. For instance, for $\boldsymbol{\eta} = \mathbf{0}_p$, the cf in (5.36) reduces to

$$\mathrm{E}(e^{\mathrm{i}t\mathbf{X}^\top \mathbf{B}\mathbf{X}}) = |\mathbf{I}_p - 2\mathrm{i}t\mathbf{B\Psi}|^{-1/2} e^{\mathrm{i}t\boldsymbol{\xi}^\top (\mathbf{I}_p - 2\mathrm{i}t\mathbf{B\Psi})^{-1}\mathbf{B}\boldsymbol{\xi}}, \tag{5.39}$$

where $\mathbf{X} = \boldsymbol{\xi} + \mathbf{W} \sim \mathcal{N}_p(\boldsymbol{\xi}, \boldsymbol{\Psi})$. Equating (5.38) and (5.39) for $\boldsymbol{\eta} = \mathbf{0}_p$, we see that we must have

$$\frac{|\mathbf{I}_p - 2\mathrm{i}t\mathbf{B\Psi}|^{1/2}}{(1-2\mathrm{i}t)^{r/2}} = \frac{e^{\mathrm{i}t\boldsymbol{\xi}^\top (\mathbf{I}_p - 2\mathrm{i}t\mathbf{B\Psi})^{-1}\mathbf{B}\boldsymbol{\xi}}}{e^{\frac{\mathrm{i}t}{1-2\mathrm{i}t}\boldsymbol{\xi}^\top \mathbf{B}\boldsymbol{\xi}}}.$$

As this identity must include the special case where $\boldsymbol{\xi} = \mathbf{0}_p$, we can conclude that it requires the following conditions to be met simultaneously:

$$|\mathbf{I}_p - 2\mathrm{i}t\mathbf{B\Psi}|^{1/2} = (1-2\mathrm{i}t)^{r/2}, \tag{5.40}$$

$$\boldsymbol{\xi}^\top (\mathbf{I}_p - 2\mathrm{i}t\mathbf{B\Psi})^{-1}\mathbf{B}\boldsymbol{\xi} = \frac{\boldsymbol{\xi}^\top \mathbf{B}\boldsymbol{\xi}}{1-2\mathrm{i}t}. \tag{5.41}$$

Since $\mathbf{B\Psi}$ is symmetric and has $\text{rank}(\mathbf{B\Psi}) = r$, then there is a $p \times p$ orthogonal matrix $\boldsymbol{\Gamma} = (\boldsymbol{\Gamma}_1, \boldsymbol{\Gamma}_2)$ such that

$$\mathbf{B\Psi} = \boldsymbol{\Gamma}\mathbf{D}\boldsymbol{\Gamma}^\top = (\boldsymbol{\Gamma}_1, \boldsymbol{\Gamma}_2)\begin{pmatrix} \mathbf{D}_1 & \mathbf{O} \\ \mathbf{O} & \mathbf{O} \end{pmatrix}\begin{pmatrix} \boldsymbol{\Gamma}_1^\top \\ \boldsymbol{\Gamma}_2^\top \end{pmatrix} = \boldsymbol{\Gamma}_1\mathbf{D}_1\boldsymbol{\Gamma}_1^\top,$$

where $\mathbf{D}_1 = \mathrm{diag}(\lambda_1, \ldots, \lambda_r)$ *comprises the nonzero eigenvalues of* $\mathbf{B\Psi}$. *Hence, we have*

$$\begin{aligned} |\mathbf{I}_p - 2\mathrm{i}t\mathbf{B\Psi}|^{1/2} &= |\mathbf{I}_p - 2\mathrm{i}t\boldsymbol{\Gamma}_1\mathbf{D}_1\boldsymbol{\Gamma}_1^\top| \\ &= |\mathbf{I}_r - 2\mathrm{i}t\mathbf{D}_1\boldsymbol{\Gamma}_1^\top\boldsymbol{\Gamma}_1| \\ &= |\mathbf{I}_r - 2\mathrm{i}t\mathbf{D}_1|^{1/2} = \prod_{j=1}^{r}(1-2\mathrm{i}t\lambda_j)^{1/2}, \end{aligned} \tag{5.42}$$

and

$$\begin{aligned} (\mathbf{I}_p - 2\mathrm{i}t\mathbf{B\Psi})^{-1} &= \boldsymbol{\Gamma}(\mathbf{I}_p - 2\mathrm{i}t\mathbf{D})^{-1}\boldsymbol{\Gamma}^\top \\ &= (\boldsymbol{\Gamma}_1\ \boldsymbol{\Gamma}_2)\begin{pmatrix}(\mathbf{I}_r - 2\mathrm{i}t\mathbf{D}_1)^{-1} & \mathbf{O} \\ \mathbf{O} & \mathbf{I}_{p-r}\end{pmatrix}\begin{pmatrix}\boldsymbol{\Gamma}_1^\top \\ \boldsymbol{\Gamma}_2^\top\end{pmatrix} \\ &= \boldsymbol{\Gamma}_1(\mathbf{I}_r - 2\mathrm{i}t\mathbf{D}_1)^{-1}\boldsymbol{\Gamma}_1^\top + \boldsymbol{\Gamma}_2\boldsymbol{\Gamma}_2^\top, \\ \Longrightarrow (\mathbf{I}_p - 2\mathrm{i}t\mathbf{B\Psi})^{-1}\mathbf{B\Psi} &= \{\boldsymbol{\Gamma}_1(\mathbf{I}_r - 2\mathrm{i}t\mathbf{D}_1)^{-1}\boldsymbol{\Gamma}_1^\top + \boldsymbol{\Gamma}_2\boldsymbol{\Gamma}_2^\top\}\boldsymbol{\Gamma}_1\mathbf{D}_1\boldsymbol{\Gamma}_1^\top \\ &= \boldsymbol{\Gamma}_1(\mathbf{I}_r - 2\mathrm{i}t\mathbf{D}_1)^{-1}\mathbf{D}_1\boldsymbol{\Gamma}_1^\top, \\ \Longrightarrow \boldsymbol{\xi}^\top\boldsymbol{\Gamma}_1(\mathbf{I}_p - 2\mathrm{i}t\mathbf{B\Psi})^{-1}\mathbf{B}\boldsymbol{\xi} &= \boldsymbol{\xi}^\top(\mathbf{I}_p - 2\mathrm{i}t\mathbf{B\Psi})^{-1}\mathbf{B\Psi}(\boldsymbol{\Psi}^{-1}\boldsymbol{\xi}) \\ &= \boldsymbol{\xi}^\top\boldsymbol{\Gamma}_1(\mathbf{I}_r - 2\mathrm{i}t\mathbf{D}_1)^{-1}\mathbf{D}_1\boldsymbol{\Gamma}_1^\top\boldsymbol{\Psi}^{-1}\boldsymbol{\xi}, \end{aligned}$$

since $\boldsymbol{\Gamma}_1^\top\boldsymbol{\Gamma}_1 = \mathbf{I}_r$ *and* $\boldsymbol{\Gamma}_2^\top\boldsymbol{\Gamma}_1 = \mathbf{0}_{p-r}$. *Thus, by defining the vectors* $\boldsymbol{\xi}_1 = (\xi_{11}, \ldots, \xi_{1r})^\top = \boldsymbol{\Gamma}_1^\top\boldsymbol{\xi}$ *and* $\boldsymbol{\xi}_2 = (\xi_{21}, \ldots, \xi_{2r})^\top = \boldsymbol{\Gamma}_1^\top\boldsymbol{\Psi}^{-1}\boldsymbol{\xi}$ *this last result yields*

$$\boldsymbol{\xi}^\top(\mathbf{I}_p - 2\mathrm{i}t\mathbf{B\Psi})^{-1}\mathbf{B}\boldsymbol{\xi} = \boldsymbol{\xi}_1^\top(\mathbf{I}_r - 2\mathrm{i}t\mathbf{D}_1)^{-1}\mathbf{D}_1\boldsymbol{\xi}_2 = \sum_{j=1}^{r}\frac{\lambda_j}{1-2\mathrm{i}\lambda_j t}\xi_{1j}\xi_{2j}. \tag{5.43}$$

Similarly, for the numerator of the last term in Equation (5.41), we have

$$\boldsymbol{\xi}^\top\mathbf{B}\boldsymbol{\xi} = \boldsymbol{\xi}^\top\mathbf{B\Psi}(\boldsymbol{\Psi}^{-1}\boldsymbol{\xi}) = \boldsymbol{\xi}^\top\boldsymbol{\Gamma}_1\mathbf{D}_1\boldsymbol{\Gamma}_1^\top\boldsymbol{\Psi}^{-1}\boldsymbol{\xi} = \boldsymbol{\xi}_1^\top\mathbf{D}_1\boldsymbol{\xi}_2 = \sum_{j=1}^{r}\lambda_j\xi_{1j}\xi_{2j}. \tag{5.44}$$

Therefore, from (5.42), (5.43), and (5.44), the conditions (5.40) and (5.41) become

$$\begin{aligned} \prod_{j=1}^{r}(1-2\mathrm{i}t\lambda_j)^{1/2} &= (1-2\mathrm{i}t)^{r/2}, \\ \sum_{j=1}^{r}\frac{\lambda_j}{1-2\mathrm{i}\lambda_j t}\xi_{1j}\xi_{2j} &= \sum_{j=1}^{r}\frac{\lambda_j}{1-2\mathrm{i}t}\xi_{1j}\xi_{2j}, \end{aligned}$$

from where is it clear that $\lambda_j = 1$ *for all* $j = 1, \ldots, r$ *so that* $\mathbf{D}_1 = \mathbf{I}_r$ *and so* $\mathbf{B\Psi} = \boldsymbol{\Gamma}_1\boldsymbol{\Gamma}_1^\top$ *which is idempotent. Finally, since for* $\boldsymbol{\xi} = \mathbf{0}_p$, *the cf in (5.36) reduces to*

$$\mathrm{E}(e^{\mathrm{i}t\mathbf{X}^\top\mathbf{BX}}) = |\mathbf{I}_p - 2\mathrm{i}t\mathbf{B\Omega}|^{-1/2} = |\mathbf{I}_p - 2\mathrm{i}t\mathbf{B}(\boldsymbol{\Psi} + \boldsymbol{\eta}\boldsymbol{\eta})|^{-1/2},$$

then comparing it with (5.40) leads to the condition $\mathbf{B}\boldsymbol{\eta} = \mathbf{0}_p$. □

Further results on quadratic forms in skew-normal random vectors can be found in Loperfido (2001), Gupta and Huang (2002), Wang et al. (2009), and for generalized skew-normal random vectors in Genton and Loperfido (2005) and Huang and Chen (2006).

5.8.3 *Mardia's measures of multivariate skewness and kurtosis*

The measures of multivariate skewness ($\beta_{1,p}$) and kurtosis ($\beta_{2,p}$) of the distribution of a random vector $\mathbf{X} \in \mathbb{R}^p$ with mean vector $\boldsymbol{\mu}$ and nonsingular variance matrix $\boldsymbol{\Sigma}$, were introduced earlier in Chapter 1, see also Mardia (1970, 1974), as

$$\beta_{1,p} = \mathrm{E}[\{(\mathbf{X}-\boldsymbol{\mu})^\top \boldsymbol{\Sigma}^{-1}(\mathbf{X}'-\boldsymbol{\mu})\}^3], \qquad \beta_{2,p} = \mathrm{E}[\{(\mathbf{X}-\boldsymbol{\mu})^\top \boldsymbol{\Sigma}^{-1}(\mathbf{X}-\boldsymbol{\mu})\}^2],$$

where $\mathbf{X}'$ is an independent copy of $\mathbf{X}$.

The following theorem presents these measures for the multivariate skew-normal distribution.

Theorem 79 *Let $\mathbf{X} \sim \mathcal{SN}_p(\boldsymbol{\xi}, \boldsymbol{\Psi}, \boldsymbol{\eta})$, where $\boldsymbol{\Psi}$ is nonsingular. Then,*

$$\beta_{1,p} = \gamma_1^2, \qquad \beta_{2,p} = p(p+2) + \gamma_2,$$

where $\gamma_1 = \sqrt{\beta_{1,1}}$ and $\gamma_2 = \beta_{2,1} - 3$ are the univariate skewness and excess of kurtosis, respectively, which are given by

$$\gamma_1 = \frac{\left(\frac{4-\pi}{2}\right)\left(\sqrt{\frac{2}{\pi}}\eta\right)^3}{\left\{1+\left(\frac{\pi-2}{2}\right)\left(\sqrt{\frac{2}{\pi}}\eta\right)^2\right\}^{3/2}}, \qquad \gamma_2 = \frac{2(\pi-3)\left(\sqrt{\frac{2}{\pi}}\eta\right)^4}{\left\{1+\left(\frac{\pi-2}{2}\right)\left(\sqrt{\frac{2}{\pi}}\eta\right)^2\right\}^2},$$

where $\eta = (\boldsymbol{\eta}^\top \boldsymbol{\Psi}^{-1}\boldsymbol{\eta})^{1/2}$.

Proof 79 *Taking into account the invariance of $\beta_{1,p}$ and $\beta_{2,p}$ with respect to nonsingular affine transformations, put $\mathbf{Y} = \mathbf{C}(\mathbf{X}-\boldsymbol{\mu})$ and $\mathbf{Y}' = \mathbf{C}(\mathbf{X}'-\boldsymbol{\mu})$, where $\mathbf{C}$ is a nonsingular $p \times p$ matrix such that $\mathbf{C}\boldsymbol{\Psi}\mathbf{C}^\top = \mathbf{I}_p$, we can write*

$$\beta_{1,p} = \mathrm{E}[\{\mathbf{Y}^\top \mathbf{C}^{-\top}\boldsymbol{\Sigma}^{-1}\mathbf{C}^{-1}\mathbf{Y}'\}^3], \qquad \beta_{2,p} = \mathrm{E}[\{\mathbf{Y}^\top \mathbf{C}^{-\top}\boldsymbol{\Sigma}^{-1}\mathbf{C}^{-1}\mathbf{Y}\}^2].$$

Also, from (5.9) and (5.11) we have $\mathbf{Y} = \mathbf{C}\boldsymbol{\eta}\, U_0 + \mathbf{C}\mathbf{W} = \boldsymbol{\eta}_C U_0 + \mathbf{W}_0$, where $\boldsymbol{\eta}_C = \mathbf{C}\boldsymbol{\eta} \in \mathtt{C}(\mathbf{C})$, and $U_0 = U - b$, with $b = \mathrm{E}(U) = \sqrt{\frac{2}{\pi}}$, and $\mathbf{W}_0 = \mathbf{C}\mathbf{W} \sim \mathcal{N}_p(\mathbf{0}_p, \mathbf{I}_p)$ obviously remain independent. Analogously, we obtain an independent copy of $\mathbf{Y}$ as $\mathbf{Y}' = \boldsymbol{\eta}_C U_0' + \mathbf{W}_0'$, where U_0' and $\mathbf{W}_0'$ are independent copies of U_0 and $\mathbf{W}_0$, respectively. Also, recalling the expression of $\boldsymbol{\Sigma}$ in (5.11) and the fact that $\boldsymbol{\Psi} = (\mathbf{C}^\top \mathbf{C})^{-1}$, we have

$$\begin{aligned}
\boldsymbol{\Sigma} = \boldsymbol{\Psi} + (1-b^2)\boldsymbol{\eta}\boldsymbol{\eta}^\top \quad &\Longrightarrow \quad \boldsymbol{\Sigma}^{-1} = \boldsymbol{\Psi}^{-1} - \frac{1-b^2}{1+(1-b^2)\eta^2}\boldsymbol{\Psi}^{-1}\boldsymbol{\eta}\boldsymbol{\eta}^\top\boldsymbol{\Psi}^{-1} \\
&\Longrightarrow \quad \boldsymbol{\Sigma}^{-1} = \mathbf{C}^\top \left\{\mathbf{I}_p - \frac{1-b^2}{1+(1-b^2)\eta^2}\mathbf{C}\boldsymbol{\eta}\boldsymbol{\eta}^\top\mathbf{C}^\top\right\}\mathbf{C} \\
&\Longrightarrow \quad \mathbf{C}^{-\top}\boldsymbol{\Sigma}^{-1}\mathbf{C}^{-1} = \mathbf{I}_p - \frac{1-b^2}{1+(1-b^2)\eta^2}\boldsymbol{\eta}_C\boldsymbol{\eta}_C^\top,
\end{aligned}$$

where $\eta = (\boldsymbol{\eta}^\top \boldsymbol{\Psi} \boldsymbol{\eta})^{1/2}$. Hence, we have

$$
\begin{aligned}
\beta_{1,p} &= \mathrm{E}\left[\left\{\mathbf{Y}^\top\left(\mathbf{I}_p - \frac{1-b^2}{1+(1-b^2)\eta^2}\mathbf{C}\boldsymbol{\eta}\boldsymbol{\eta}^\top\mathbf{C}^\top\right)\mathbf{Y}'\right\}^3\right] \\
&= \mathrm{E}\left[\left\{\mathbf{Y}^\top\mathbf{Y}' - \frac{1-b^2}{1+(1-b^2)\eta^2}(\boldsymbol{\eta}_C^\top\mathbf{Y})(\boldsymbol{\eta}_C^\top\mathbf{Y}')\right\}^3\right].
\end{aligned}
$$

At this point, note that $\boldsymbol{\eta}_C^\top\boldsymbol{\eta}_C = \boldsymbol{\eta}^\top\boldsymbol{\Psi}^{-1}\boldsymbol{\eta} = \eta^2$ and so $\boldsymbol{\eta}_C^\top\mathbf{W}_0 \sim \mathcal{N}(0,\eta^2)$, which allows us to write $\boldsymbol{\eta}_C^\top\mathbf{W}_0 = \eta V_0$, where $V_0 = \boldsymbol{\eta}_0^\top\mathbf{W}_0 \sim \mathcal{N}(0, \|\boldsymbol{\eta}_0\|^2)$, with $\|\boldsymbol{\eta}_0\| = 1$. From these facts, we find

$$
\begin{aligned}
\mathbf{Y}^\top\mathbf{Y}' - \frac{(1-b^2)(\boldsymbol{\eta}_C^\top\mathbf{Y})(\boldsymbol{\eta}_C^\top\mathbf{Y}')}{1+(1-b^2)\eta^2} =& (\boldsymbol{\eta}_C U_0 + \mathbf{W}_0)^\top(\boldsymbol{\eta}_C U_0' + \mathbf{W}_0') \\
& - \frac{1-b^2}{1+(1-b^2)\eta^2}\boldsymbol{\eta}_C^\top(\boldsymbol{\eta}_C U_0 + \mathbf{W}_0)\boldsymbol{\eta}_C^\top(\boldsymbol{\eta}_C U_0' + \mathbf{W}_0') \\
=& (\boldsymbol{\eta}_C^\top\boldsymbol{\eta}_C)U_0U_0' + (\boldsymbol{\eta}_C^\top\mathbf{W}_0')U_0 + (\boldsymbol{\eta}_C^\top\mathbf{W}_0)U_0' + \mathbf{W}_0^\top\mathbf{W}_0' \\
& - \frac{1-b^2}{1+(1-b^2)\eta^2}\{(\boldsymbol{\eta}_C^\top\boldsymbol{\eta}_C)^2U_0U_0' + (\boldsymbol{\eta}_C^\top\boldsymbol{\eta}_C)(\boldsymbol{\eta}_C^\top\mathbf{W}_0')U_0 \\
& + (\boldsymbol{\eta}_C^\top\boldsymbol{\eta}_C)(\boldsymbol{\eta}_C^\top\mathbf{W}_0)U_0' + (\boldsymbol{\eta}_C^\top\mathbf{W}_0)(\boldsymbol{\eta}_C^\top\mathbf{W}_0')\} \\
=& \eta^2U_0U_0' + \eta\,U_0V_0' + \eta\,U_0'V_0 + \mathbf{W}_0^\top\mathbf{W}_0' \\
& - \frac{1-b^2}{1+(1-b^2)\eta^2}\{\eta^4U_0U_0' + \eta^3U_0V_0' + \eta^3U_0'V_0 + \eta^2V_0V_0'\} \\
=& \mathbf{W}_0^\top\mathbf{W}_0' - V_0V_0' \\
& + \frac{1}{1+(1-b^2)\eta^2}\{\eta^2U_0U_0' + \eta\,U_0V_0' + \eta\,U_0'V_0 + V_0V_0'\} \\
=& \mathbf{W}_0^\top\mathbf{M}_0\mathbf{W}_0' + \frac{Z_0Z_0'}{1+(1-b^2)\eta^2},
\end{aligned}
$$

where $\mathbf{M}_0 = \mathbf{I}_p - \boldsymbol{\eta}_0\boldsymbol{\eta}_0^\top$ is a $p\times p$ projection matrix of rank $p-1$ such that $\mathbf{M}_0\boldsymbol{\eta}_0 = \mathbf{0}_p$, and $Z_0 = \eta\,U_0 + V_0$ and $Z_0' = \eta\,U_0' + V_0'$ are iid $\mathcal{SN}(-\eta\,b, 1, \eta)$ random variables. Another relevant fact here is that $\mathbf{W}_0^\top\mathbf{M}_0\mathbf{W}_0'$ and Z_0Z_0' are independent variables, since by assumption U (U') is independent of $\mathbf{W}$ ($\mathbf{W}'$) and therefore of $\mathbf{W}_0 = \mathbf{C}\mathbf{W}$ ($\mathbf{W}_0' = \mathbf{C}\mathbf{W}'$) and $V_0 = \boldsymbol{\eta}_0^\top\mathbf{W}_0$ ($V_0' = \boldsymbol{\eta}_0^\top\mathbf{W}_0'$); but also $\mathbf{M}_0\mathbf{W}_0$ ($\mathbf{M}_0\mathbf{W}_0'$) is independent of $V_0 = \boldsymbol{\eta}_0^\top\mathbf{W}_0$ ($V_0' = \boldsymbol{\eta}_0^\top\mathbf{W}_0'$) since $\mathbf{M}_0\boldsymbol{\eta}_0 = \mathbf{0}_p$. Finally, taking into account that all the required moments are finite, as well as that Z_0 and Z_0' have zero means and that the distribution of $\mathbf{W}_0$ and $\mathbf{W}_0'$ is symmetric normal around the origin, we obtain

$$
\begin{aligned}
\beta_{1,p} &= \mathrm{E}\left\{\left(\mathbf{W}_0^\top\mathbf{M}_0\mathbf{W}_0' + \frac{Z_0Z_0'}{1+(1-b^2)\eta^2}\right)^3\right\} \\
&= \mathrm{E}\{(\mathbf{W}_0^\top\mathbf{M}_0\mathbf{W}_0')^3\} + 3\mathrm{E}\{(\mathbf{W}_0^\top\mathbf{M}_0\mathbf{W}_0')^2\}\frac{\mathrm{E}\{Z_0Z_0'\}}{1+(1-b^2)\eta^2} \\
&\quad + 3\mathrm{E}\{(\mathbf{W}_0^\top\mathbf{M}_0\mathbf{W}_0')\}\frac{\mathrm{E}\{(Z_0Z_0')^2\}}{\{1+(1-b^2)\eta^2\}^2} + \frac{\mathrm{E}\{(Z_0Z_0')^3\}}{\{1+(1-b^2)\eta^2\}^3} \\
&= \frac{\{\mathrm{E}(Z_0^3)\}^2}{\{1+(1-b^2)\eta^2\}^3}.
\end{aligned}
$$

With a similar procedure, we find that

$$\mathbf{Y}^\top\mathbf{Y} - \frac{1}{1+(1-b^2)\eta^2}(\boldsymbol{\eta}_C^\top\mathbf{Y})(\boldsymbol{\eta}_C^\top\mathbf{Y}) = Q_0 + \frac{Z_0^2}{1+(1-b^2)\eta^2},$$

where $Q_0 = \mathbf{W}_0^\top\mathbf{M}_0\mathbf{W}_0 \sim \chi^2_{p-1}$ *and* $Z_0 \sim \mathcal{SN}(-\eta\, b, 1, \eta)$*, which are independent. Hence, we have*

$$\begin{aligned}
\beta_{2,p} &= \mathrm{E}\left\{\left(Q_0 + \frac{Z_0^2}{1+(1-b^2)\eta^2}\right)^2\right\}\\
&= \mathrm{E}(Q_0^2) + 2\mathrm{E}(Q_0)\frac{\mathrm{E}(Z_0^2)}{1+(1-b^2)\eta^2} + \frac{\mathrm{E}(Z_0^4)}{\{1+(1-b^2)\eta^2\}^2}\\
&= (p-1)(p+1) + 2(p-1)\frac{\mathrm{E}(Z_0^2)}{1+(1-b^2)\eta^2} + \frac{\mathrm{E}(Z_0^4)}{\{1+(1-b^2)\eta^2\}^2}.
\end{aligned}$$

The proof concludes by using the fact that the required moments of $Z_0 = \eta\, U_0 + V_0$*, where* $U_0 = U - b$*, with* $U \sim \mathcal{HN}(0,1)$ *and* $b = \sqrt{\frac{2}{\pi}}$*, and* $V_0 \sim \mathcal{N}(0,1)$ *are independent and are given by*

$$\begin{aligned}
\mathrm{E}(Z_0^2) &= \eta^2\mathrm{E}(U_0^2) + 2\eta\mathrm{E}(U_0)\mathrm{E}(V_0) + \mathrm{E}(V_0^2)\\
&= \eta^2(1-b^2) + 1,\\
\mathrm{E}(Z_0^3) &= \eta^3\mathrm{E}(U_0^3) + 3\eta^2\mathrm{E}(U_0^2)\mathrm{E}(V_0) + 3\eta\mathrm{E}(U_0)\mathrm{E}(V_0^2) + \mathrm{E}(V_0^3)\\
&= \eta^3 b(2b^2-1),\\
\mathrm{E}(Z_0^4) &= \eta^4\mathrm{E}(U_0^4) + 4\eta^3\mathrm{E}(U_0^3)\mathrm{E}(V_0) + 6\eta^2\mathrm{E}(U_0^2)\mathrm{E}(V_0^2) + 4\eta\mathrm{E}(U_0)\mathrm{E}(V_0^3) + \mathrm{E}(V_0^4)\\
&= \eta^4(3-2b^2-3b^4) + 6\eta^2(1-b^2) + 3.
\end{aligned}$$

Noting that $Z_0 \sim \mathcal{SN}(-\eta b, 1, \eta)$*, which has mean* $\mu_0 = 0$*, variance* $\sigma_0^2 = 1+\eta^2(1-b^2)$*, skewness* $\gamma_1 = \sigma_0^{-3}\mathrm{E}(Z_0^3)$ *and excess kurtosis* $\gamma_2 = \sigma_0^{-4}\mathrm{E}(Z_0^4) - 3$*, and also that* $(p-1)(p+1)+2(p-1) = p(p+2)-3$*, the proof is complete.* □

Calculations and comparisons of other measures of multivariate skewness for the multivariate skew-normal distribution have been explored by Balakrishnan and Scarpa (2012).

5.8.4 Higher-order moments of skew-normal random vectors

The k-th non-central moment of a random vector $\mathbf{X}$ (provided it exists) was defined in Equation (1.7) of Chapter 1. For instance, the first four moments of $\mathbf{X}$ are:

$$\boldsymbol{\mu}_1(\mathbf{X}) = \mathrm{E}(\mathbf{X}), \quad \boldsymbol{\mu}_2(\mathbf{X}) = \mathrm{E}(\mathbf{X}\mathbf{X}^\top), \quad \boldsymbol{\mu}_3(\mathbf{X}) = \mathrm{E}(\mathbf{X}\mathbf{X}^\top \otimes \mathbf{X}), \quad \boldsymbol{\mu}_4(\mathbf{X}) = \mathrm{E}(\mathbf{X}\mathbf{X}^\top \otimes \mathbf{X}\mathbf{X}^\top).$$

In the following, we compute these moments when $\mathbf{X} \sim \mathcal{SN}_p(\boldsymbol{\xi}, \boldsymbol{\Psi}, \boldsymbol{\eta})$; see also Genton et al. (2001), Fang (2008a), and see Kim and Mallick (2003) for skew-t distributions and Kim and Kim (2017) for scale mixtures of skew-normal distributions. Obviously, as also established in Chapter 1, an operational way to calculate these moments is by differentiating the characteristic function of $\mathbf{X}$:

$$\boldsymbol{\mu}_k(\mathbf{X}) = \begin{cases} \mathrm{i}^{-k}\dfrac{\partial^k \psi(\mathbf{t})}{\partial\mathbf{t}\partial\mathbf{t}^\top\partial\mathbf{t}\cdots\partial\mathbf{t}\partial\mathbf{t}^\top}, & \text{if } k = 2,4,\ldots \text{ (even)},\\ \mathrm{i}^{-k}\dfrac{\partial^k \psi(\mathbf{t})}{\partial\mathbf{t}\partial\mathbf{t}^\top\partial\mathbf{t}\cdots\partial\mathbf{t}^\top\partial\mathbf{t}}, & \text{if } k = 1,3,\ldots \text{ (odd)}. \end{cases}$$

Here, we calculate $\boldsymbol{\mu}_k(\mathbf{X})$, $k = 1,2,3,4$, for $\mathbf{X} \sim \mathcal{SN}_p(\boldsymbol{\xi}, \boldsymbol{\Psi}, \boldsymbol{\eta})$, starting from the respective multivariate normal moments presented in Theorem 8 of Chapter 3. To do this, we need the following preliminary result.

Theorem 80 *Let* $U \sim \mathcal{HN}(0,1)$. *Then,*

$$\mathrm{E}(U) = \sqrt{\frac{2}{\pi}} \equiv b, \quad \mathrm{E}(U^2) = 1, \quad \mathrm{E}(U^3) = 2b, \quad \mathrm{E}(U^4) = 3,$$
$$\mathrm{var}(U) = 1 - b^2, \quad \mathrm{E}\{(U-b)^3\} = b(2b^2 - 1), \quad \mathrm{E}\{(U-b)^4\} = 3 - b^2(2 + 3b^2).$$

Proof 80 *Since* U *has density* $2\phi(u)I_{(u>0)}$, *then* $\mathrm{E}(U) = 2\int_0^\infty u\phi(u)\mathrm{d}u = -2\int_0^\infty \mathrm{d}\phi(u) = 2\phi(0) = b$. *Similarly, integrating by parts yields* $\mathrm{E}(U^3) = 4\phi(0) = 2b$. *From these results and the fact that* U *has the same even moments as* $W \sim \mathcal{N}(0,1)$, *for example,* $\mathrm{E}(U^2) = 1$ *and* $\mathrm{E}(U^4) = 3$, *the proof of the other results is straightforward.* □

Theorem 81 *Let* $\boldsymbol{\mu}_k(\mathbf{X})$ *be the k-th moment of* $\mathbf{X} \sim \mathcal{SN}_p(\boldsymbol{\xi}, \boldsymbol{\Psi}, \boldsymbol{\eta})$ *as defined in (1.7). Then, for* $k = 1,2,3,4$, *we have:*

$$\boldsymbol{\mu}_1(\mathbf{X}) = \boldsymbol{\xi} + b\boldsymbol{\eta}, \quad b = \sqrt{2/\pi};$$

$$\boldsymbol{\mu}_2(\mathbf{X}) = \boldsymbol{\Psi} + \boldsymbol{\xi}\boldsymbol{\xi}^\top + b(\boldsymbol{\xi}\boldsymbol{\eta}^\top + \boldsymbol{\eta}\boldsymbol{\xi}^\top) + \boldsymbol{\eta}\boldsymbol{\eta}^\top;$$

$$\begin{aligned}\boldsymbol{\mu}_3(\mathbf{X}) =& (\mathbf{I}_{p^2} + \mathbf{K}_p)\{\boldsymbol{\Psi} \otimes (\boldsymbol{\xi} + b\boldsymbol{\eta})\} + \mathrm{vec}(\boldsymbol{\Psi} + \boldsymbol{\xi}\boldsymbol{\xi}^\top)(\boldsymbol{\xi} + b\boldsymbol{\eta})^\top \\ &+ \mathrm{vec}(\boldsymbol{\xi}\boldsymbol{\eta}^\top + \boldsymbol{\eta}\boldsymbol{\xi}^\top)(b\boldsymbol{\xi} + \boldsymbol{\eta})^\top + \mathrm{vec}(\boldsymbol{\eta}\boldsymbol{\eta}^\top)(\boldsymbol{\xi} + 2b\boldsymbol{\eta})^\top;\end{aligned}$$

$$\begin{aligned}\boldsymbol{\mu}_4(\mathbf{X}) =& (\mathbf{I}_{p^2} + \mathbf{K}_p)[\boldsymbol{\Psi} \otimes \boldsymbol{\Psi} + \boldsymbol{\Psi} \otimes \{\boldsymbol{\xi}\boldsymbol{\xi}^\top + b(\boldsymbol{\xi}\boldsymbol{\eta}^\top + \boldsymbol{\eta}\boldsymbol{\xi}^\top) + \boldsymbol{\eta}\boldsymbol{\eta}^\top\} \\ &+ \{\boldsymbol{\xi}\boldsymbol{\xi}^\top + b(\boldsymbol{\xi}\boldsymbol{\eta}^\top + \boldsymbol{\eta}\boldsymbol{\xi}^\top) + \boldsymbol{\eta}\boldsymbol{\eta}^\top\} \otimes \boldsymbol{\Psi}] + \mathrm{vec}(\boldsymbol{\eta}\boldsymbol{\eta}^\top)\mathrm{vec}\{b(\boldsymbol{\xi}\boldsymbol{\eta}^\top + \boldsymbol{\eta}\boldsymbol{\xi}^\top) + \boldsymbol{\eta}\boldsymbol{\eta}^\top\}^\top \\ &+ \mathrm{vec}\{b(\boldsymbol{\xi}\boldsymbol{\eta}^\top + \boldsymbol{\eta}\boldsymbol{\xi}^\top) + \boldsymbol{\eta}\boldsymbol{\eta}^\top\}\mathrm{vec}(\boldsymbol{\eta}\boldsymbol{\eta}^\top)^\top + (1 - b^2)\mathrm{vec}(\boldsymbol{\xi}\boldsymbol{\eta}^\top + \boldsymbol{\eta}\boldsymbol{\xi}^\top)\mathrm{vec}(\boldsymbol{\xi}\boldsymbol{\eta}^\top + \boldsymbol{\eta}\boldsymbol{\xi}^\top)^\top \\ &+ \mathrm{vec}\{\boldsymbol{\Psi} + \boldsymbol{\xi}\boldsymbol{\xi}^\top + b(\boldsymbol{\xi}\boldsymbol{\eta}^\top + \boldsymbol{\eta}\boldsymbol{\xi}^\top) + \boldsymbol{\eta}\boldsymbol{\eta}^\top\}\mathrm{vec}\{\boldsymbol{\Psi} + \boldsymbol{\xi}\boldsymbol{\xi}^\top + b(\boldsymbol{\xi}\boldsymbol{\eta}^\top + \boldsymbol{\eta}\boldsymbol{\xi}^\top) + \boldsymbol{\eta}\boldsymbol{\eta}^\top\}^\top;\end{aligned}$$

$$\begin{aligned}\mathrm{var}(\mathbf{X} \otimes \mathbf{X}) =& (\mathbf{I}_{p^2} + \mathbf{K}_p)[\boldsymbol{\Psi} \otimes \boldsymbol{\Psi} + \boldsymbol{\Psi} \otimes \{\boldsymbol{\xi}\boldsymbol{\xi}^\top + b(\boldsymbol{\xi}\boldsymbol{\eta}^\top + \boldsymbol{\eta}\boldsymbol{\xi}^\top) + \boldsymbol{\eta}\boldsymbol{\eta}^\top\} \\ &+ \{\boldsymbol{\xi}\boldsymbol{\xi}^\top + b(\boldsymbol{\xi}\boldsymbol{\eta}^\top + \boldsymbol{\eta}\boldsymbol{\xi}^\top) + \boldsymbol{\eta}\boldsymbol{\eta}^\top\} \otimes \boldsymbol{\Psi}] + \mathrm{vec}(\boldsymbol{\eta}\boldsymbol{\eta}^\top)\mathrm{vec}\{b(\boldsymbol{\xi}\boldsymbol{\eta}^\top + \boldsymbol{\eta}\boldsymbol{\xi}^\top) + \boldsymbol{\eta}\boldsymbol{\eta}^\top\}^\top \\ &+ \mathrm{vec}\{b(\boldsymbol{\xi}\boldsymbol{\eta}^\top + \boldsymbol{\eta}\boldsymbol{\xi}^\top) + \boldsymbol{\eta}\boldsymbol{\eta}^\top\}\mathrm{vec}(\boldsymbol{\eta}\boldsymbol{\eta}^\top)^\top + (1 - b^2)\mathrm{vec}(\boldsymbol{\xi}\boldsymbol{\eta}^\top + \boldsymbol{\eta}\boldsymbol{\xi}^\top)\mathrm{vec}(\boldsymbol{\xi}\boldsymbol{\eta}^\top + \boldsymbol{\eta}\boldsymbol{\xi}^\top)^\top,\end{aligned}$$

where $\mathbf{K}_p$ *is the commutation matrix, a* $p^2 \times p^2$ *block-matrix whose block in position* (i, j) *is* $\mathbf{e}_j\mathbf{e}_i^\top$, *with* $\mathbf{e}_i$ *being the i-th p-dimensional unit vector.*

Proof 81 *First, recall by Theorem 52 that* $\mathbf{X} \mid U \sim \mathcal{N}_p(\boldsymbol{\mu}(U), \boldsymbol{\Psi})$, *where* $\boldsymbol{\mu}(U) = \boldsymbol{\xi} + U\boldsymbol{\eta}$ *and* $U \sim \mathcal{HN}(0,1)$. *Hence, by Theorem 8, we find after some simple algebraic calculations that:*

$$\begin{aligned}
\boldsymbol{\mu}_1(\mathbf{X}|U) =& \boldsymbol{\mu}(U) = \boldsymbol{\xi} + U\boldsymbol{\eta};\\
\boldsymbol{\mu}_2(\mathbf{X}|U) =& \boldsymbol{\Psi} + \boldsymbol{\mu}(U)\boldsymbol{\mu}(U)^\top = \boldsymbol{\Psi} + \boldsymbol{\xi}\boldsymbol{\xi}^\top + U(\boldsymbol{\xi}\boldsymbol{\eta}^\top + \boldsymbol{\eta}\boldsymbol{\xi}^\top) + U^2\boldsymbol{\eta}\boldsymbol{\eta}^\top;\\
\boldsymbol{\mu}_3(\mathbf{X}|U) =& (\mathbf{I}_{p^2} + \mathbf{K}_p)\{\boldsymbol{\Psi} \otimes \boldsymbol{\mu}(U)\} + \mathrm{vec}\{\boldsymbol{\Psi} + \boldsymbol{\mu}(U)\boldsymbol{\mu}(U)^\top\}\boldsymbol{\mu}(U)^\top\\
=& (\mathbf{I}_{p^2} + \mathbf{K}_p)\{\boldsymbol{\Psi} \otimes \boldsymbol{\mu}(U)\} + \mathrm{vec}(\boldsymbol{\Psi} + \boldsymbol{\xi}\boldsymbol{\xi}^\top)\boldsymbol{\mu}(U)^\top + \mathrm{vec}(\boldsymbol{\xi}\boldsymbol{\eta}^\top + \boldsymbol{\eta}\boldsymbol{\xi}^\top)U\boldsymbol{\mu}(U)^\top\\
&+ \mathrm{vec}(\boldsymbol{\eta}\boldsymbol{\eta}^\top)U^2\boldsymbol{\mu}(U)^\top;\\
\boldsymbol{\mu}_4(\mathbf{X}|U) =& (\mathbf{I}_{p^2} + \mathbf{K}_p)[(\boldsymbol{\Psi} \otimes \boldsymbol{\Psi}) + \{\boldsymbol{\Psi} \otimes \boldsymbol{\mu}(U)\boldsymbol{\mu}(U)^\top\} + \{\boldsymbol{\mu}(U)\boldsymbol{\mu}(U)^\top \otimes \boldsymbol{\Psi}\}]\\
&+ \mathrm{vec}\{\boldsymbol{\Psi} + \boldsymbol{\mu}(U)\boldsymbol{\mu}(U)^\top\}\mathrm{vec}\{\boldsymbol{\Psi} + \boldsymbol{\mu}(U)\boldsymbol{\mu}(U)^\top\}^\top\\
=& (\mathbf{I}_{p^2} + \mathbf{K}_p)[(\boldsymbol{\Psi} \otimes \boldsymbol{\Psi}) + \{\boldsymbol{\Psi} \otimes \boldsymbol{\mu}(U)\boldsymbol{\mu}(U)^\top\} + \{\boldsymbol{\mu}(U)\boldsymbol{\mu}(U)^\top \otimes \boldsymbol{\Psi}\}] + \mathrm{vec}(\boldsymbol{\Psi})\mathrm{vec}(\boldsymbol{\Psi})^\top\\
&+ \mathrm{vec}(\boldsymbol{\Psi})\mathrm{vec}\{\boldsymbol{\mu}(U)\boldsymbol{\mu}(U)^\top\}^\top + \mathrm{vec}\{\boldsymbol{\mu}(U)\boldsymbol{\mu}(U)^\top\}\mathrm{vec}(\boldsymbol{\Psi})^\top\\
&+ \mathrm{vec}(\boldsymbol{\xi}\boldsymbol{\xi}^\top)\mathrm{vec}\{\boldsymbol{\mu}(U)\boldsymbol{\mu}(U)^\top\}^\top + \mathrm{vec}(\boldsymbol{\xi}\boldsymbol{\eta}^\top + \boldsymbol{\eta}\boldsymbol{\xi}^\top)\mathrm{vec}\{U\boldsymbol{\mu}(U)\boldsymbol{\mu}(U)^\top\}^\top\\
&+ \mathrm{vec}(\boldsymbol{\eta}\boldsymbol{\eta}^\top)\mathrm{vec}\{U^2\boldsymbol{\mu}(U)\boldsymbol{\mu}(U)^\top\}^\top;
\end{aligned}$$

and

$$\begin{aligned}
\mathrm{var}(\mathbf{X} \otimes \mathbf{X}|U) =& (\mathbf{I}_{p^2} + \mathbf{K}_p)[(\boldsymbol{\Psi} \otimes \boldsymbol{\Psi}) + \{\boldsymbol{\Psi} \otimes \boldsymbol{\mu}(U)\boldsymbol{\mu}(U)^\top\} + \{\boldsymbol{\mu}(U)\boldsymbol{\mu}(U)^\top \otimes \boldsymbol{\Psi}\}]\\
=& (\mathbf{I}_{p^2} + \mathbf{K}_p)((\boldsymbol{\Psi} \otimes \boldsymbol{\Psi}) + [\boldsymbol{\Psi} \otimes \{\boldsymbol{\xi}\boldsymbol{\xi}^\top + U(\boldsymbol{\xi}\boldsymbol{\eta}^\top + \boldsymbol{\eta}\boldsymbol{\xi}^\top) + U^2\boldsymbol{\eta}\boldsymbol{\eta}^\top\}]\\
&+ [\{\boldsymbol{\xi}\boldsymbol{\xi}^\top + U(\boldsymbol{\xi}\boldsymbol{\eta}^\top + \boldsymbol{\eta}\boldsymbol{\xi}^\top) + U^2\boldsymbol{\eta}\boldsymbol{\eta}^\top\} \otimes \boldsymbol{\Psi}]).
\end{aligned}$$

Since $\boldsymbol{\mu}_k(\mathbf{X}) = \mathrm{E}\{\boldsymbol{\mu}_k(\mathbf{X}|U)\}$ *for all* $k = 1, 2, \ldots$, *the proof of the first four moments concludes by applying Theorem 80, thus finding that:*

$$\begin{aligned}
\mathrm{E}\{\boldsymbol{\mu}(U)\} =& \mathrm{E}(\boldsymbol{\xi} + U\boldsymbol{\eta}) = \boldsymbol{\xi} + b\boldsymbol{\eta},\\
\mathrm{E}\{U\boldsymbol{\mu}(U)\} =& \mathrm{E}(U\boldsymbol{\xi} + U^2\boldsymbol{\eta}) = b\boldsymbol{\xi} + \boldsymbol{\eta},\\
\mathrm{E}\{U^2\boldsymbol{\mu}(U)\} =& \mathrm{E}(U^2\boldsymbol{\xi} + U^3\boldsymbol{\eta}) = \boldsymbol{\xi} + 2b\boldsymbol{\eta},\\
\mathrm{E}\{\boldsymbol{\mu}(U)\boldsymbol{\mu}(U)^\top\} =& \mathrm{E}\{\boldsymbol{\xi}\boldsymbol{\xi}^\top + U(\boldsymbol{\xi}\boldsymbol{\eta}^\top + \boldsymbol{\eta}\boldsymbol{\xi}^\top) + U^2\boldsymbol{\eta}\boldsymbol{\eta}^\top\}\\
=& \boldsymbol{\xi}\boldsymbol{\xi}^\top + b(\boldsymbol{\xi}\boldsymbol{\eta}^\top + \boldsymbol{\eta}\boldsymbol{\xi}^\top) + \boldsymbol{\eta}\boldsymbol{\eta}^\top,\\
\mathrm{E}\{U\boldsymbol{\mu}(U)\boldsymbol{\mu}(U)^\top\} =& \mathrm{E}\{U\boldsymbol{\xi}\boldsymbol{\xi}^\top + U^2(\boldsymbol{\xi}\boldsymbol{\eta}^\top + \boldsymbol{\eta}\boldsymbol{\xi}^\top) + U^3\boldsymbol{\eta}\boldsymbol{\eta}^\top\}\\
=& b\boldsymbol{\xi}\boldsymbol{\xi}^\top + \boldsymbol{\xi}\boldsymbol{\eta}^\top + \boldsymbol{\eta}\boldsymbol{\xi}^\top + 2b\boldsymbol{\eta}\boldsymbol{\eta}^\top,\\
\mathrm{E}\{U^2\boldsymbol{\mu}(U)\boldsymbol{\mu}(U)^\top\} =& \mathrm{E}\{U^2\boldsymbol{\xi}\boldsymbol{\xi}^\top + U^3(\boldsymbol{\xi}\boldsymbol{\eta}^\top + \boldsymbol{\eta}\boldsymbol{\xi}^\top) + U^4\boldsymbol{\eta}\boldsymbol{\eta}^\top\}\\
=& \boldsymbol{\xi}\boldsymbol{\xi}^\top + 2b(\boldsymbol{\xi}\boldsymbol{\eta}^\top + \boldsymbol{\eta}\boldsymbol{\xi}^\top) + 3\boldsymbol{\eta}\boldsymbol{\eta}^\top.
\end{aligned}$$

For the last result, we can use $\mathrm{var}(\mathbf{X} \otimes \mathbf{X}) = \boldsymbol{\mu}_4(\mathbf{X}) - \mathrm{vec}\{\boldsymbol{\mu}_2(\mathbf{X})\}\mathrm{vec}\{\boldsymbol{\mu}_2(\mathbf{X})\}^\top$ *or alternatively* $\mathrm{var}(\mathbf{X} \otimes \mathbf{X}) = \mathrm{E}\{\mathrm{var}(\mathbf{X} \otimes \mathbf{X}|U)\} + \mathrm{var}[\mathrm{vec}\{\boldsymbol{\mu}_2(\mathbf{X}|U)\}]$. □

Applying to the skew-normal random vector $\mathbf{X} \sim \mathcal{SN}_p(\boldsymbol{\xi}, \boldsymbol{\Psi}, \boldsymbol{\eta})$ the general results given in Equations (1.8), (1.9) and (1.10) of Chapter 1, we also find the following theorem after some algebraic operations.

Theorem 82 *Let* $\mathbf{X} \sim \mathcal{SN}_p(\boldsymbol{\xi}, \boldsymbol{\Psi}, \boldsymbol{\eta})$ *and* $\mathbf{A} = \mathbf{A}^\top$ *and* $\mathbf{B} = \mathbf{B}^\top$ *and* $\mathbf{C}$ *be constant matrices of dimensions* $p \times p$, $p \times p$, *and* $q \times p$, *respectively. Then:*

(*i*) $\mathrm{E}(\mathbf{X}^\top \mathbf{A} \mathbf{X}) = \mathrm{tr}(\mathbf{A}\boldsymbol{\Psi}) + \boldsymbol{\xi}^\top \mathbf{A} \boldsymbol{\xi} + 2b(\boldsymbol{\xi}^\top \mathbf{A} \boldsymbol{\eta}) + \boldsymbol{\eta}^\top \mathbf{A} \boldsymbol{\eta}, \quad b = \sqrt{2/\pi};$

(*ii*) $\mathrm{cov}(\mathbf{C}\mathbf{X}, \mathbf{X}^\top \mathbf{A} \mathbf{X}) = 2\mathbf{C}\boldsymbol{\Psi}\mathbf{A}(\boldsymbol{\xi} + b\boldsymbol{\eta}) + 2(1-b^2)(\boldsymbol{\xi}^\top \mathbf{A}\boldsymbol{\eta})\mathbf{C}\boldsymbol{\eta} + b(\boldsymbol{\eta}^\top \mathbf{A}\boldsymbol{\eta})\mathbf{C}\boldsymbol{\eta};$

(*iii*)
$$\begin{aligned}\mathrm{cov}(\mathbf{X}^\top \mathbf{A}\mathbf{X}, \mathbf{X}^\top \mathbf{B}\mathbf{X}) = {} & 2\mathrm{tr}(\mathbf{A}\boldsymbol{\Psi}\mathbf{B}\boldsymbol{\Psi}) + 4(\boldsymbol{\xi}^\top \mathbf{A}\boldsymbol{\Psi}\mathbf{B}\boldsymbol{\xi}) + 8b(\boldsymbol{\xi}^\top \mathbf{A}\boldsymbol{\Psi}\mathbf{B}\boldsymbol{\eta}) + 4(\boldsymbol{\eta}^\top \mathbf{A}\boldsymbol{\Psi}\mathbf{B}\boldsymbol{\eta}) \\ & + (\boldsymbol{\eta}^\top \mathbf{A}\boldsymbol{\eta})\{2b(\boldsymbol{\xi}^\top \mathbf{B}\boldsymbol{\eta}) + \boldsymbol{\eta}^\top \mathbf{B}\boldsymbol{\eta}\} + \{2b(\boldsymbol{\xi}^\top \mathbf{A}\boldsymbol{\eta}) + \boldsymbol{\eta}^\top \mathbf{A}\boldsymbol{\eta}\}(\boldsymbol{\eta}^\top \mathbf{B}\boldsymbol{\eta}) \\ & + 4(1-b^2)(\boldsymbol{\xi}^\top \mathbf{A}\boldsymbol{\eta})(\boldsymbol{\xi}^\top \mathbf{B}\boldsymbol{\eta});\end{aligned}$$

(*iv*)
$$\begin{aligned}\mathrm{var}(\mathbf{X}^\top \mathbf{A}\mathbf{X}) = {} & 2\mathrm{tr}(\mathbf{A}\boldsymbol{\Psi}\mathbf{A}\boldsymbol{\Psi}) + 4(\boldsymbol{\xi}^\top \mathbf{A}\boldsymbol{\Psi}\mathbf{A}\boldsymbol{\xi}) + 8b(\boldsymbol{\xi}^\top \mathbf{A}\boldsymbol{\Psi}\mathbf{A}\boldsymbol{\eta}) + 4(\boldsymbol{\eta}^\top \mathbf{A}\boldsymbol{\Psi}\mathbf{A}\boldsymbol{\eta}) \\ & + 4b(\boldsymbol{\xi}^\top \mathbf{A}\boldsymbol{\eta})(\boldsymbol{\eta}^\top \mathbf{A}\boldsymbol{\eta}) + 2(\boldsymbol{\eta}^\top \mathbf{A}\boldsymbol{\eta})^2 + 4(1-b^2)(\boldsymbol{\xi}^\top \mathbf{A}\boldsymbol{\eta})^2.\end{aligned}$$

Proof 82 *The proof of part* (*i*) *is straightforward. For part* (*ii*), *we have:*

$$\begin{aligned}\mathrm{E}(\mathbf{X}^\top \mathbf{A}\mathbf{X}) = {} & \mathrm{tr}\{\mathbf{A}\boldsymbol{\mu}_2(\mathbf{X})\} \\ = {} & \mathrm{tr}(\mathbf{A}\boldsymbol{\Psi}) + \boldsymbol{\xi}^\top \mathbf{A}\boldsymbol{\xi} + 2b(\boldsymbol{\xi}^\top \mathbf{A}\boldsymbol{\eta}) + \boldsymbol{\eta}^\top \mathbf{A}\boldsymbol{\eta}^\top; \\ \mathrm{E}\{(\mathbf{X}^\top \mathbf{A}\mathbf{X})\mathbf{C}\mathbf{X}\} = {} & \mathbf{C}\boldsymbol{\mu}_3(\mathbf{X})^\top \mathrm{vec}(\mathbf{A}) \\ = {} & \mathbf{C}[\{\boldsymbol{\Psi} \otimes (\boldsymbol{\xi} + b\boldsymbol{\eta})^\top\}(\mathbf{I}_{p^2} + \mathbf{K}_p) + (\boldsymbol{\xi} + b\boldsymbol{\eta})\mathrm{vec}(\boldsymbol{\Psi} + \boldsymbol{\xi}\boldsymbol{\xi}^\top)^\top \\ & + (b\boldsymbol{\xi} + \boldsymbol{\eta})\mathrm{vec}(\boldsymbol{\xi}\boldsymbol{\eta}^\top + \boldsymbol{\eta}\boldsymbol{\xi}^\top)^\top + (\boldsymbol{\xi} + 2b\boldsymbol{\eta})\mathrm{vec}(\boldsymbol{\eta}\boldsymbol{\eta}^\top)^\top]\mathrm{vec}(\mathbf{A}) \\ = {} & 2\mathbf{C}\boldsymbol{\Psi}\mathbf{A}(\boldsymbol{\xi} + b\boldsymbol{\eta}) + \{\mathrm{tr}(\mathbf{A}\boldsymbol{\Psi}) + \boldsymbol{\xi}^\top \mathbf{A}\boldsymbol{\xi}\}\mathbf{C}(\boldsymbol{\xi} + b\boldsymbol{\eta}) \\ & + 2(\boldsymbol{\xi}^\top \mathbf{A}\boldsymbol{\eta})\mathbf{C}(b\boldsymbol{\xi} + \boldsymbol{\eta}) + (\boldsymbol{\eta}^\top \mathbf{A}\boldsymbol{\eta})\mathbf{C}(\boldsymbol{\xi} + 2b\boldsymbol{\eta}) \\ \Longrightarrow \mathrm{cov}(\mathbf{C}\mathbf{X}, \mathbf{X}^\top \mathbf{A}\mathbf{X}) = {} & 2\mathbf{C}\boldsymbol{\Psi}\mathbf{A}(\boldsymbol{\xi} + b\boldsymbol{\eta}) + 2(1-b^2)(\boldsymbol{\xi}^\top \mathbf{A}\boldsymbol{\eta})\mathbf{C}\boldsymbol{\eta} + b(\boldsymbol{\eta}^\top \mathbf{A}\boldsymbol{\eta})\mathbf{C}\boldsymbol{\eta},\end{aligned}$$

where we used the relations $(\mathbf{I}_{p^2} + \mathbf{K}_p)\mathrm{vec}(\mathbf{A}) = 2\mathrm{vec}(\mathbf{A})$, $\mathrm{vec}(\mathbf{B})^\top \mathrm{vec}(\mathbf{D}) = \mathrm{tr}(\mathbf{B}^\top \mathbf{D})$ *and* $(\mathbf{D}^\top \otimes \mathbf{B})\mathrm{vec}(\mathbf{W}) = \mathrm{vec}(\mathbf{B}\mathbf{W}\mathbf{D})$; *see Section A.2.9.*

Now, in a similar form to the proof of part (iii) of Theorem 9 but using the last result in Theorem 81, for part (iii) we have:

$$\begin{aligned}
\operatorname{cov}(\mathbf{X}^\top\mathbf{A}\mathbf{X},\mathbf{X}^\top\mathbf{B}\mathbf{X}) =&\operatorname{vec}(\mathbf{A})^\top\operatorname{var}\{\operatorname{vec}(\mathbf{X}\mathbf{X}^\top)\}\operatorname{vec}(\mathbf{B})\\
=&\operatorname{vec}(\mathbf{A})^\top[(\mathbf{I}_{p^2}+\mathbf{K}_p)\{\boldsymbol{\Psi}\otimes\boldsymbol{\Psi}+\boldsymbol{\Psi}\otimes(\boldsymbol{\xi}\boldsymbol{\xi}^\top+b(\boldsymbol{\xi}\boldsymbol{\eta}^\top+\boldsymbol{\eta}\boldsymbol{\xi}^\top)+\boldsymbol{\eta}\boldsymbol{\eta}^\top)\\
&+(\boldsymbol{\xi}\boldsymbol{\xi}^\top+b(\boldsymbol{\xi}\boldsymbol{\eta}^\top+\boldsymbol{\eta}\boldsymbol{\xi}^\top)+\boldsymbol{\eta}\boldsymbol{\eta}^\top)\otimes\boldsymbol{\Psi}\}\\
&+\operatorname{vec}(\boldsymbol{\eta}\boldsymbol{\eta}^\top)\operatorname{vec}(b(\boldsymbol{\xi}\boldsymbol{\eta}^\top+\boldsymbol{\eta}\boldsymbol{\xi}^\top)+\boldsymbol{\eta}\boldsymbol{\eta}^\top)^\top\\
&+\operatorname{vec}(b(\boldsymbol{\xi}\boldsymbol{\eta}^\top+\boldsymbol{\eta}\boldsymbol{\xi}^\top)+\boldsymbol{\eta}\boldsymbol{\eta}^\top)\operatorname{vec}(\boldsymbol{\eta}\boldsymbol{\eta}^\top)^\top\\
&+(1-b^2)\operatorname{vec}(\boldsymbol{\xi}\boldsymbol{\eta}^\top+\boldsymbol{\eta}\boldsymbol{\xi}^\top)\operatorname{vec}(\boldsymbol{\xi}\boldsymbol{\eta}^\top+\boldsymbol{\eta}\boldsymbol{\xi}^\top)^\top]\operatorname{vec}(\mathbf{B})\\
=&2\operatorname{vec}(\mathbf{A})^\top\{\boldsymbol{\Psi}\otimes\boldsymbol{\Psi}+\boldsymbol{\Psi}\otimes(\boldsymbol{\xi}\boldsymbol{\xi}^\top+b(\boldsymbol{\xi}\boldsymbol{\eta}^\top+\boldsymbol{\eta}\boldsymbol{\xi}^\top)+\boldsymbol{\eta}\boldsymbol{\eta}^\top)\\
&+(\boldsymbol{\xi}\boldsymbol{\xi}^\top+b(\boldsymbol{\xi}\boldsymbol{\eta}^\top+\boldsymbol{\eta}\boldsymbol{\xi}^\top)+\boldsymbol{\eta}\boldsymbol{\eta}^\top)\otimes\boldsymbol{\Psi}\}\operatorname{vec}(\mathbf{B})\\
&+\operatorname{vec}(\mathbf{A})^\top\operatorname{vec}(\boldsymbol{\eta}\boldsymbol{\eta}^\top)\operatorname{vec}(b(\boldsymbol{\xi}\boldsymbol{\eta}^\top+\boldsymbol{\eta}\boldsymbol{\xi}^\top)+\boldsymbol{\eta}\boldsymbol{\eta}^\top)^\top\operatorname{vec}(\mathbf{B})\\
&+\operatorname{vec}(\mathbf{A})^\top\operatorname{vec}(b(\boldsymbol{\xi}\boldsymbol{\eta}^\top+\boldsymbol{\eta}\boldsymbol{\xi}^\top)+\boldsymbol{\eta}\boldsymbol{\eta}^\top)\operatorname{vec}(\boldsymbol{\eta}\boldsymbol{\eta}^\top)^\top\operatorname{vec}(\mathbf{B})\\
&+(1-b^2)\operatorname{vec}(\mathbf{A})^\top\operatorname{vec}(\boldsymbol{\xi}\boldsymbol{\eta}^\top+\boldsymbol{\eta}\boldsymbol{\xi}^\top)\operatorname{vec}(\boldsymbol{\xi}\boldsymbol{\eta}^\top+\boldsymbol{\eta}\boldsymbol{\xi}^\top)^\top\operatorname{vec}(\mathbf{B})\\
=&2\operatorname{tr}(\mathbf{A}\boldsymbol{\Psi}\mathbf{B}\boldsymbol{\Psi})+4(\boldsymbol{\xi}^\top\mathbf{A}\boldsymbol{\Psi}\mathbf{B}\boldsymbol{\xi}+2b\boldsymbol{\xi}^\top\mathbf{A}\boldsymbol{\Psi}\mathbf{B}\boldsymbol{\eta}+\boldsymbol{\eta}^\top\mathbf{A}\boldsymbol{\Psi}\mathbf{B}\boldsymbol{\eta})\\
&+(\boldsymbol{\eta}^\top\mathbf{A}\boldsymbol{\eta})\{2b(\boldsymbol{\xi}^\top\mathbf{B}\boldsymbol{\eta})+\boldsymbol{\eta}^\top\mathbf{B}\boldsymbol{\eta}\}+\{2b(\boldsymbol{\xi}^\top\mathbf{A}\boldsymbol{\eta})+\boldsymbol{\eta}^\top\mathbf{A}\boldsymbol{\eta}\}(\boldsymbol{\eta}^\top\mathbf{B}\boldsymbol{\eta})\\
&+4(1-b^2)(\boldsymbol{\xi}^\top\mathbf{A}\boldsymbol{\eta})(\boldsymbol{\xi}^\top\mathbf{B}\boldsymbol{\eta}).
\end{aligned}$$

Finally, letting $\mathbf{A}=\mathbf{B}$, *we have the proof of (iv).* □

Note that letting $\boldsymbol{\eta}=\mathbf{0}_p$ in Theorem 82 yields Theorem 9 for the multivariate normal distribution.

Setting $\boldsymbol{\mu}=\mathrm{E}(\mathbf{X})$ and $\boldsymbol{\Sigma}=\operatorname{var}(\mathbf{X})$, and assuming that $\boldsymbol{\Sigma}>0$, all previous results can be applied to the standardized skew-normal vector

$$\mathbf{Z}=\boldsymbol{\Sigma}^{-1/2}(\mathbf{X}-\boldsymbol{\mu})\sim\mathcal{SN}_p(-b\boldsymbol{\Sigma}^{-1/2}\boldsymbol{\eta},\boldsymbol{\Sigma}^{-1/2}\boldsymbol{\Psi}\boldsymbol{\Sigma}^{-1/2},\boldsymbol{\Sigma}^{-1/2}\boldsymbol{\eta}).$$

Obviously, $\boldsymbol{\mu}_1(\mathbf{Z})=\mathbf{0}$ and $\operatorname{var}(\mathbf{Z})=\boldsymbol{\mu}_2(\mathbf{Z})=\mathbf{I}_p$, and for the remaining results, it suffices to replace $\boldsymbol{\xi}$, $\boldsymbol{\Psi}$, and $\boldsymbol{\eta}$ by $-b\boldsymbol{\Sigma}^{-1/2}\boldsymbol{\eta}$, $\boldsymbol{\Sigma}^{-1/2}\boldsymbol{\Psi}\boldsymbol{\Sigma}^{-1/2}$, and $\boldsymbol{\Sigma}^{-1/2}\boldsymbol{\eta}$, respectively, in the corresponding results obtained for $\mathbf{X}$. For instance, for the third and fourth moments of $\mathbf{Z}$ we have:

$$\begin{aligned}
\boldsymbol{\mu}_3(\mathbf{Z})=&b(2b^2-1)(\boldsymbol{\Sigma}^{-1/2}\otimes\boldsymbol{\Sigma}^{-1/2})\operatorname{vec}(\boldsymbol{\eta}\boldsymbol{\eta}^\top)\boldsymbol{\eta}^\top\boldsymbol{\Sigma}^{-1/2},\\
\boldsymbol{\mu}_4(\mathbf{Z})=&(\boldsymbol{\Sigma}^{-1/2}\otimes\boldsymbol{\Sigma}^{-1/2})[(\mathbf{I}_{p^2}+\mathbf{K}_p)\{(\boldsymbol{\Psi}\otimes\boldsymbol{\Psi})+(1-b^2)(\boldsymbol{\Psi}\otimes\boldsymbol{\eta}\boldsymbol{\eta}^\top)+(1-b^2)(\boldsymbol{\eta}\boldsymbol{\eta}^\top\otimes\boldsymbol{\Psi})\}\\
&+2(1-2b^2)\operatorname{vec}(\boldsymbol{\eta}\boldsymbol{\eta}^\top)\operatorname{vec}(\boldsymbol{\eta}\boldsymbol{\eta}^\top)^\top+4b^2(1-b^2)\operatorname{vec}(\boldsymbol{\eta}\boldsymbol{\eta}^\top)\operatorname{vec}(\boldsymbol{\eta}\boldsymbol{\eta}^\top)^\top\\
&+\operatorname{vec}(\boldsymbol{\Psi}+(1-b^2)\boldsymbol{\eta}\boldsymbol{\eta}^\top)\operatorname{vec}(\boldsymbol{\Psi}+(1-b^2)\boldsymbol{\eta}\boldsymbol{\eta}^\top)^\top](\boldsymbol{\Sigma}^{-1/2}\otimes\boldsymbol{\Sigma}^{-1/2}).
\end{aligned}$$

From these results, Mardia's measures of multivariate skewness and kurtosis for the multivariate skew-normal distribution can also be computed, respectively, as:

$$\beta_{1,p}=\operatorname{tr}\{\boldsymbol{\mu}_3(\mathbf{Z})^\top\boldsymbol{\mu}_3(\mathbf{Z})\}=b^2(2b^2-1)^2(\boldsymbol{\eta}^\top\boldsymbol{\Sigma}^{-1}\boldsymbol{\eta})^3, \tag{5.45}$$

$$\beta_{2,p}=\operatorname{tr}\{\boldsymbol{\mu}_4(\mathbf{Z})\}=p(p+2)+2b^2(2-3b^2)(\boldsymbol{\eta}^\top\boldsymbol{\Sigma}^{-1}\boldsymbol{\eta})^2, \tag{5.46}$$

where

$$\boldsymbol{\eta}^\top \boldsymbol{\Sigma}^{-1} \boldsymbol{\eta} = \frac{\boldsymbol{\eta}^\top \boldsymbol{\Psi}^{-1} \boldsymbol{\eta}}{1 + (1 - b^2)(\boldsymbol{\eta}^\top \boldsymbol{\Psi}^{-1} \boldsymbol{\eta})},$$

and we used that

$$\begin{aligned}
\boldsymbol{\eta}^\top \boldsymbol{\Sigma}^{-1} \boldsymbol{\Psi} \boldsymbol{\Sigma}^{-1} \boldsymbol{\eta} &= \frac{\boldsymbol{\eta}^\top \boldsymbol{\Sigma}^{-1} \boldsymbol{\eta}}{1 + (1 - b^2)(\boldsymbol{\eta}^\top \boldsymbol{\Psi}^{-1} \boldsymbol{\eta})}, \\
\boldsymbol{\Sigma}^{-1} \boldsymbol{\Psi} &= \mathbf{I}_p - \frac{1 - b^2}{1 + (1 - b^2)\boldsymbol{\eta}^\top \boldsymbol{\Psi}^{-1} \boldsymbol{\eta}} \boldsymbol{\Psi}^{-1} \boldsymbol{\eta} \boldsymbol{\eta}^\top, \\
\operatorname{tr}(\boldsymbol{\Sigma}^{-1} \boldsymbol{\Psi}) &= p - (1 - b^2)(\boldsymbol{\eta}^\top \boldsymbol{\Sigma}^{-1} \boldsymbol{\eta}), \\
\operatorname{tr}\{(\boldsymbol{\Sigma}^{-1} \boldsymbol{\Psi})^2\} &= p - 2(1 - b^2)(\boldsymbol{\eta}^\top \boldsymbol{\Sigma}^{-1} \boldsymbol{\eta}) + (1 - b^2)^2 (\boldsymbol{\eta}^\top \boldsymbol{\Sigma}^{-1} \boldsymbol{\eta})^2.
\end{aligned}$$

5.9 Sampling from the Multivariate Skew-Normal Distribution

Let

$$\bar{\mathbf{X}} = \frac{1}{n} \sum_{i=1}^{n} \mathbf{X}_i \quad \text{and} \quad \mathbf{S} = \frac{1}{n-1} \sum_{i=1}^{n} (\mathbf{X}_i - \bar{\mathbf{X}})(\mathbf{X}_i - \bar{\mathbf{X}})^\top$$

be the sample mean vector and the sample variance matrix, respectively, of a random sample $\mathbf{X}_1, \ldots, \mathbf{X}_n$ from $\mathbf{X} \sim \mathcal{SN}_p(\boldsymbol{\xi}, \boldsymbol{\Psi}, \boldsymbol{\eta})$ with nonsingular dispersion matrix $\boldsymbol{\Psi}$. Sometimes, we consider the biased sample variance matrix given by

$$\mathbf{S_X} = \frac{1}{n} \sum_{i=1}^{n} (\mathbf{X}_i - \bar{\mathbf{X}})(\mathbf{X}_i - \bar{\mathbf{X}})^\top.$$

Then, by definition, we have

$$\mathbf{X}_i = \boldsymbol{\xi} + \boldsymbol{\eta} U_i + \mathbf{W}_i, \quad i = 1, \ldots, n,$$

where the U_is are iid $\mathcal{HN}(0,1)$ random variables, while the $\mathbf{W}_i$s are iid $\mathcal{N}_p(\mathbf{0}_p, \boldsymbol{\Psi})$ random vectors, and they are mutually independent. Hence, for the sample mean vector and sample variance matrix, we have

$$\bar{\mathbf{X}} = \boldsymbol{\xi} + \boldsymbol{\eta} \bar{U} + \bar{\mathbf{W}} \tag{5.47}$$

where

$$\bar{U} = \frac{1}{n} \sum_{i=1}^{n} U_i, \quad \bar{\mathbf{W}} = \frac{1}{n} \sum_{i=1}^{n} \mathbf{W}_i,$$

and

$$\mathbf{S_X} = S_U^2 \boldsymbol{\eta} \boldsymbol{\eta}^\top + \mathbf{S}_{U\mathbf{W}} \boldsymbol{\eta}^\top + \boldsymbol{\eta} \mathbf{S}_{U\mathbf{W}}^\top + \mathbf{S_W}, \tag{5.48}$$

where

$$S_U^2 = \frac{1}{n} \sum_{i=1}^{n} (U_i - \bar{U})^2, \quad \mathbf{S}_{U\mathbf{W}} = \frac{1}{n} \sum_{i=1}^{n} (U_i - \bar{U})(\mathbf{W}_i - \bar{\mathbf{W}}), \quad \mathbf{S_W} = \frac{1}{n} \sum_{i=1}^{n} (\mathbf{W}_i - \bar{\mathbf{W}})(\mathbf{W}_i - \bar{\mathbf{W}})^\top.$$

Since by assumption $\mathbf{U} = (U_1, \ldots, U_n)^\top$ and $\mathbf{W} = (\mathbf{W}_1, \ldots, \mathbf{W}_n)^\top$ are independent, we find that $(\bar{U}, S_U^2)$ and $(\bar{\mathbf{W}}, \mathbf{S_W})$ are also independent, and $\mathrm{E}(\mathbf{S}_{U\mathbf{W}}) = \mathbf{0}_p$. Moreover, we know that

$$\bar{\mathbf{W}} \sim \mathcal{N}_p\left(\mathbf{0}_p, \frac{1}{n}\boldsymbol{\Psi}\right), \qquad n\mathbf{S_W} = \sum_{i=1}^{n} (\mathbf{W}_i - \bar{\mathbf{W}})(\mathbf{W}_i - \bar{\mathbf{W}})^\top \sim \mathcal{W}_p(n-1, \boldsymbol{\Psi}),$$

which are also independent. From these results, we have

$$\mathrm{E}(\bar{\mathbf{W}}) = \mathbf{0}_p, \quad \operatorname{var}(\bar{\mathbf{W}}) = \frac{1}{n} \boldsymbol{\Psi},$$

and

$$\mathrm{E}(\mathbf{S_W}) = \frac{n-1}{n}\boldsymbol{\Psi}, \quad \mathrm{var}(\mathbf{S_W}) = \frac{n-1}{n^2}(\mathbf{I}_{p^2}+\mathbf{K}_p)(\boldsymbol{\Psi}\otimes\boldsymbol{\Psi}),$$

where $\mathbf{K}_p$ is the commutation matrix, which is a $p^2 \times p^2$ block-matrix whose block in position (i,j) is $\mathbf{e}_j\mathbf{e}_i^\top$ with $\mathbf{e}_i$ being the i-th p-dimensional unit vector.

The following result is related to the joint distribution of $(\bar{U}, S_U^2)$ and will be used later jointly with Theorem 80.

Theorem 83 *Let* $\mathbf{U} = (U_1,\dots,U_n)^\top \sim \mathcal{HN}_n(\mathbf{0}_n,\mathbf{I}_n)$, *i.e.,* $U_i \overset{iid}{\sim} \mathcal{HN}(0,1)$, $i=1,\dots,n$. *Then,*

$$\mathrm{E}(S_U^2) = \frac{n-1}{n}\left(1-\frac{2}{\pi}\right), \qquad \mathrm{var}(S_U^2) = \frac{2(n-1)}{n^2}\left(1-\frac{2}{\pi}\right)^2 + \frac{(n-1)^2}{n^3}\frac{4}{\pi}\left(2-\frac{6}{\pi}\right),$$

and

$$\mathrm{E}(\bar{U}S_U^2) = \frac{n-1}{n}\sqrt{\frac{2}{\pi}}\left(1-\frac{2}{\pi}\right) + \frac{n-1}{n^2}\sqrt{\frac{2}{\pi}}\left(\frac{4}{\pi}-1\right).$$

Proof 83 *Let* $U \sim \mathcal{HN}(0,1)$ *and* $b = \mathrm{E}(U) = \sqrt{\frac{2}{\pi}}$. *We know that* $\mathrm{E}(\mathbf{U}) = b\mathbf{1}_n$ *and* $\mathrm{var}(\mathbf{U}) = (1-b^2)\mathbf{I}_n$. *Also consider the orthogonal projections* $\mathbf{P} = \frac{1}{n}\mathbf{1}_n\mathbf{1}_n^\top$ *and* $\mathbf{M} = \mathbf{I}_n - \frac{1}{n}\mathbf{1}_n\mathbf{1}_n^\top$. *It is clear that* $\mathbf{P}\mathbf{1}_n = \mathbf{1}_n$ *and* $\mathbf{M}\mathbf{1}_n = \mathbf{0}_n$. *It is also clear that* $n\bar{U} = \mathbf{1}_n^\top\mathbf{U} = \mathbf{1}_n^\top\mathbf{U}_0 + bn$ *and* $nS_U^2 = \mathbf{U}^\top\mathbf{M}\mathbf{U} = \mathbf{U}_0^\top\mathbf{M}\mathbf{U}_0$, *where* $\mathbf{U}_0 = \mathbf{U} - \mathrm{E}(\mathbf{U}) = \mathbf{U} - b\mathbf{1}_n$. *Thus, for* $\mathrm{E}(S_U^2)$ *we have*

$$\mathrm{E}(S_U^2) = \frac{1}{n}\mathrm{E}(\mathbf{U}_0^\top\mathbf{M}\mathbf{U}_0) = \frac{1}{n}\mathrm{tr}\{\mathbf{M}\mathrm{E}(\mathbf{U}_0\mathbf{U}_0^\top)\} = \frac{1}{n}\mathrm{tr}\{\mathbf{M}(1-b^2)\} = \frac{n-1}{n}(1-b^2),$$

since $\mathrm{tr}(\mathbf{M}) = \mathrm{tr}\left(\mathbf{I}_n - \frac{1}{n}\mathbf{1}_n\mathbf{1}_n^\top\right) = n-1$. *To compute* $\mathrm{var}(S_U^2)$, *we also know that (see Exercise 5.8)*

$$\begin{aligned}
\mathrm{var}(S_U^2) &= \frac{1}{n^2}\mathrm{var}(\mathbf{U}_0^\top\mathbf{M}\mathbf{U}_0) \\
&= \frac{1}{n^2}\mathrm{vec}(\mathbf{M})^\top\mathrm{var}(\mathbf{U}_0\otimes\mathbf{U}_0)\mathrm{vec}(\mathbf{M}) \\
&= \frac{1}{n^2}\mathrm{vec}(\mathbf{M})^\top\left\{(1-b^2)^2(\mathbf{I}_{n^2}+\mathbf{K}_n) + (1-b^2)^2\gamma_2(U)\sum_{i=1}^n(\mathbf{E}_{ii}\otimes\mathbf{E}_{ii})\right\}\mathrm{vec}(\mathbf{M}) \\
&= \frac{1}{n^2}\left\{2(1-b^2)^2\mathrm{tr}(\mathbf{M}) + (1-b^2)^2\gamma_2(U)\sum_{i=1}^n m_{ii}^2\right\} \\
&= \frac{1}{n^2}\left\{2(n-1)(1-b^2)^2 + \frac{(n-1)^2}{n}(1-b^2)^2\gamma_2(U)\right\},
\end{aligned}$$

since $m_{ii} = 1-\frac{1}{n}$, $i=1,\dots,n$, $\mathrm{tr}(\mathbf{M}) = n-1$ *and* $\sum_{i=1}^n m_{ii}^2 = n\left(n-\frac{1}{n}\right)^2$, *and by Theorem 80,*

$$\gamma_2(U) = \mathrm{kurtosis}(U) - 3 = \frac{\mathrm{E}\{(U-b)^4\}}{\mathrm{var}(U)^2} - 3 = \frac{3-2b^2-3b^4-3(1-b^2)^2}{(1-b^2)^2} = \frac{2b^2(2-3b^2)}{(1-b^2)^2}.$$

Finally,

$$
\begin{aligned}
E(\bar{U}S_U^2) &= \frac{1}{n^2}E(\mathbf{1}_n^\top \mathbf{U}\mathbf{U}^\top \mathbf{M}\mathbf{U}) \\
&= \frac{1}{n^2}E(\mathbf{1}_n^\top \mathbf{U}_0\mathbf{U}_0^\top \mathbf{M}\mathbf{U}_0) + \frac{1}{n^2}E(b\mathbf{1}_n^\top \mathbf{1}_n\mathbf{U}_0^\top \mathbf{M}\mathbf{U}_0) \\
&= \frac{1}{n^2}\mathbf{1}_n^\top E\{\mathbf{U}_0 \mathrm{vec}(\mathbf{U}_0\mathbf{U}_0)\}\mathrm{vec}(\mathbf{M}) + bE(S_U^2) \\
&= \frac{1}{n^2}(1-b^2)^{3/2}\gamma_1(U)\mathbf{1}_n^\top \sum_{i=1}^n \mathbf{e}_i(\mathbf{e}_i \otimes \mathbf{e}_i)^\top \mathrm{vec}(\mathbf{M}) + bE(S_U^2) \\
&= \frac{1}{n^2}(1-b^2)^{3/2}\gamma_1(U)\sum_{i=1}^n \mathbf{1}_n^\top \mathbf{e}_i\mathbf{e}_i^\top \mathbf{M}\mathbf{e}_i + bE(S_U^2) \\
&= \frac{n-1}{n^2}(1-b^2)^{3/2}\gamma_1(U) + bE(S_U^2),
\end{aligned}
$$

since $\mathbf{e}_i^\top \mathbf{M}\mathbf{e}_i = m_{ii} = 1 - \frac{1}{n}$, $i = 1, \ldots, n$, *and by Theorem 80,*

$$
\gamma_1(U) = \mathrm{skewness}(U) = \frac{E\{(U-b)^3\}}{(1-b^2)^{3/2}} = \frac{b(2b^2-1)}{(1-b^2)^{3/2}}.
$$

The proof is complete. □

Using Equations (5.47) and (5.48), we have the following result.

Theorem 84 *Let* $\mathbf{X}_1, \ldots, \mathbf{X}_n$ *be a random sample from* $\mathbf{X} \sim \mathcal{SN}_p(\boldsymbol{\xi}, \boldsymbol{\Psi}, \boldsymbol{\eta})$. *Then,*

$$
E(\bar{\mathbf{X}}) = \boldsymbol{\xi} + \sqrt{\frac{2}{\pi}}\boldsymbol{\eta}, \quad \mathrm{var}(\bar{\mathbf{X}}) = \frac{1}{n}\left\{\boldsymbol{\Psi} + \left(1 - \frac{2}{\pi}\right)\boldsymbol{\eta}\boldsymbol{\eta}^\top\right\},
$$

$$
E(\mathbf{S}) = \boldsymbol{\Psi} + \left(1 - \frac{2}{\pi}\right)\boldsymbol{\eta}\boldsymbol{\eta}^\top,
$$

$$
\begin{aligned}
\mathrm{var}(\mathbf{S}) &= \frac{1}{n-1}(\mathbf{I}_{p^2} + \mathbf{K}_p)\left[(\boldsymbol{\Psi}\otimes\boldsymbol{\Psi}) + \left(1 - \frac{2}{\pi}\right)\{(\boldsymbol{\Psi}\otimes\boldsymbol{\eta}\boldsymbol{\eta}^\top) + (\boldsymbol{\eta}\boldsymbol{\eta}^\top\otimes\boldsymbol{\Psi})\}\right] \\
&\quad + \frac{1}{n-1}\left\{2\left(1 - \frac{2}{\pi}\right)^2 + \frac{n-1}{n}\frac{4}{\pi}\left(2 - \frac{6}{\pi}\right)\right\}\mathrm{vec}(\boldsymbol{\eta}\boldsymbol{\eta}^\top)\mathrm{vec}(\boldsymbol{\eta}\boldsymbol{\eta}^\top)^\top,
\end{aligned}
$$

and

$$
\mathrm{cov}(\bar{\mathbf{X}}, \mathbf{S}) = \frac{1}{n}\sqrt{\frac{2}{\pi}}\left(\frac{4}{\pi} - 1\right)\boldsymbol{\eta}\,\mathrm{vec}(\boldsymbol{\eta}\boldsymbol{\eta}^\top)^\top.
$$

Proof 84 *The result for* $E(\bar{\mathbf{X}})$, $\mathrm{var}(\bar{\mathbf{X}})$ *and* $E(\mathbf{S}) = \frac{n}{n-1}E(\mathbf{S}_\mathbf{X})$ *are direct from (5.47) and (5.48), respectively. Now we calculate* $\mathrm{var}(\mathbf{S}_\mathbf{X}) \equiv \mathrm{var}\{\mathrm{vec}(\mathbf{S}_\mathbf{X})\}$ *from which we have* $\mathrm{var}(\mathbf{S}) = \left(\frac{n}{n-1}\right)^2 \mathrm{var}(\mathbf{S}_\mathbf{X})$. *To do this, we know that*

$$
\mathrm{var}(\mathbf{S}_\mathbf{X}) = \mathrm{var}\{E(\mathbf{S}_\mathbf{X} \mid \mathbf{U})\} + E\{\mathrm{var}(\mathbf{S}_\mathbf{X} \mid \mathbf{U})\},
$$

where by (5.48),

$$\begin{aligned}
\mathrm{E}(\mathbf{S_X} \mid \mathbf{U}) &= S_U^2 \boldsymbol{\eta}\boldsymbol{\eta}^\top + \mathrm{E}(\mathbf{S_W}) = S_U^2 \boldsymbol{\eta}\boldsymbol{\eta}^\top + \frac{n-1}{n}\boldsymbol{\Psi} \\
\Longrightarrow &\quad \mathrm{var}\{\mathrm{E}(\mathbf{S_X} \mid \mathbf{U})\} = \mathrm{var}(S_U^2)\mathrm{vec}(\boldsymbol{\eta}\boldsymbol{\eta}^\top)\mathrm{vec}(\boldsymbol{\eta}\boldsymbol{\eta}^\top)^\top,
\end{aligned}$$

and

$$\mathrm{var}(\mathbf{S_X} \mid \mathbf{U}) = \mathrm{E}\{\mathrm{vec}(\mathbf{S_X})\mathrm{vec}(\mathbf{S_X})^\top \mid \mathbf{U}\} - \mathrm{E}\{\mathrm{vec}(\mathbf{S_X}) \mid \mathbf{U}\}\mathrm{E}\{\mathrm{vec}(\mathbf{S_X}) \mid \mathbf{U}\}^\top.$$

For the computation of this conditional variance matrix, note first by (5.48) that

$$\begin{aligned}
\mathrm{E}\{\mathrm{vec}(\mathbf{S_X})\mathrm{vec}(\mathbf{S_X})^\top \mid \mathbf{U}\} &= S_U^4 \mathrm{vec}(\boldsymbol{\eta}\boldsymbol{\eta}^\top)\mathrm{vec}(\boldsymbol{\eta}\boldsymbol{\eta}^\top)^\top + S_U^2 \mathrm{vec}(\boldsymbol{\eta}\boldsymbol{\eta}^\top)\mathrm{E}(\mathbf{S}_{U\mathbf{W}} \mid \mathbf{U})^\top \mathbf{A}^\top \\
&\quad + S_U^2 \mathrm{vec}(\boldsymbol{\eta}\boldsymbol{\eta}^\top)\mathrm{E}\{\mathrm{vec}(\mathbf{S_W}) \mid \mathbf{U}\}^\top + S_U^2 \mathbf{A}\mathrm{E}(\mathbf{S}_{U\mathbf{W}} \mid \mathbf{U})\mathrm{vec}(\boldsymbol{\eta}\boldsymbol{\eta}^\top)^\top \\
&\quad + \mathbf{A}\mathrm{E}(\mathbf{S}_{U\mathbf{W}}\mathbf{S}_{U\mathbf{W}}^\top \mid \mathbf{U})\mathbf{A}^\top + \mathbf{A}\mathrm{E}\{\mathbf{S}_{U\mathbf{W}}\mathrm{vec}(\mathbf{S_W})^\top \mid \mathbf{U}\} \\
&\quad + S_U^2 \mathrm{E}\{\mathrm{vec}(\mathbf{S_W}) \mid \mathbf{U}\}\mathrm{vec}(\boldsymbol{\eta}\boldsymbol{\eta}^\top)^\top + \mathrm{E}\{\mathbf{S}_{U\mathbf{W}}\mathrm{vec}(\mathbf{S_W}) \mid \mathbf{U}\}^\top \mathbf{A}^\top \\
&\quad + \mathrm{E}\{\mathrm{vec}(\mathbf{S_W})\mathrm{vec}(\mathbf{S_W})^\top \mid \mathbf{U}\}^\top,
\end{aligned}$$

where $\mathbf{A} = (\boldsymbol{\eta} \otimes \mathbf{I}_p) + (\mathbf{I}_p \otimes \boldsymbol{\eta})$ *and, recalling that* $\mathbf{W}$ *is independent of* $\mathbf{U}$*,*

$$\begin{aligned}
\mathrm{E}(\mathbf{S}_{U\mathbf{W}} \mid \mathbf{U}) &= \frac{1}{n}\sum_{i=1}^n (U_i - \bar{U})\mathrm{E}(\mathbf{W}_i) = \mathbf{0}_p, \\
\mathrm{E}(\mathbf{S}_{U\mathbf{W}}\mathbf{S}_{U\mathbf{W}}^\top \mid \mathbf{U}) &= \mathrm{var}(\mathbf{S}_{U\mathbf{W}} \mid \mathbf{U}) = \frac{1}{n^2}\sum_{i=1}^n (U_i - \bar{U})^2 \mathrm{var}(\mathbf{W}_i) = \frac{1}{n} S_U^2 \boldsymbol{\Psi}, \\
\mathrm{E}\{\mathrm{vec}(\mathbf{S_W}) \mid \mathbf{U}\} &= \mathrm{E}\{\mathrm{vec}(\mathbf{S_W})\} = \frac{n-1}{n}\mathrm{vec}(\boldsymbol{\Psi}), \\
\mathrm{E}\{\mathbf{S}_{U\mathbf{W}}\mathrm{vec}(\mathbf{S_W})^\top \mid \mathbf{U}\} &= \mathrm{E}\{\mathbf{S}_{U\mathbf{W}} \mid \mathbf{U}\}\mathrm{E}\{\mathrm{vec}(\mathbf{S_W})\}^\top = \mathbf{O}, \\
\mathrm{E}\{\mathrm{vec}(\mathbf{S_W})\mathrm{vec}(\mathbf{S_W})^\top \mid \mathbf{U}\}^\top &= \mathrm{E}\{\mathrm{vec}(\mathbf{S_W})\mathrm{vec}(\mathbf{S_W})^\top\}^\top \\
&= \mathrm{var}\{\mathrm{vec}(\mathbf{S_W})\} + \mathrm{E}\{\mathrm{vec}(\mathbf{S_W})\}\mathrm{E}\{\mathrm{vec}(\mathbf{S_W})\}^\top \\
&= \frac{n-1}{n^2}(\mathbf{I}_{p^2} + \mathbf{K}_p)(\boldsymbol{\Psi} \otimes \boldsymbol{\Psi}) + \left(\frac{n-1}{n}\right)^2 \mathrm{vec}(\boldsymbol{\Psi})\mathrm{vec}(\boldsymbol{\Psi})^\top;
\end{aligned}$$

and so

$$\begin{aligned}
\mathrm{E}\{\mathrm{vec}(\mathbf{S_X})\mathrm{vec}(\mathbf{S_X})^\top \mid \mathbf{U}\} &= S_U^4 \mathrm{vec}(\boldsymbol{\eta}\boldsymbol{\eta}^\top)\mathrm{vec}(\boldsymbol{\eta}\boldsymbol{\eta}^\top)^\top + \frac{1}{n} S_U^2 \mathbf{A}\boldsymbol{\Psi}\mathbf{A}^\top \\
&\quad + \frac{n-1}{n} S_U^2 \mathrm{vec}(\boldsymbol{\eta}\boldsymbol{\eta}^\top)\mathrm{vec}(\boldsymbol{\Psi})^\top + \frac{n-1}{n} S_U^2 \mathrm{vec}(\boldsymbol{\Psi})\mathrm{vec}(\boldsymbol{\eta}\boldsymbol{\eta}^\top)^\top \\
&\quad + \frac{n-1}{n^2}(\mathbf{I}_{p^2} + \mathbf{K}_p)(\boldsymbol{\Psi} \otimes \boldsymbol{\Psi}) + \left(\frac{n-1}{n}\right)^2 \mathrm{vec}(\boldsymbol{\Psi})\mathrm{vec}(\boldsymbol{\Psi})^\top.
\end{aligned}$$

Second, by using again (5.48) we have

$$\begin{aligned}
&\mathrm{E}\{\mathrm{vec}(\mathbf{S_X}) \mid \mathbf{U}\}\mathrm{E}\{\mathrm{vec}(\mathbf{S_X}) \mid \mathbf{U}\}^\top \\
&= [S_U^2 \mathrm{vec}(\boldsymbol{\eta}\boldsymbol{\eta}^\top) + \mathrm{E}\{\mathrm{vec}(\mathbf{S_W})\}][S_U^2 \mathrm{vec}(\boldsymbol{\eta}\boldsymbol{\eta}^\top)^\top + \mathrm{E}\{\mathrm{vec}(\mathbf{S_W})\}^\top] \\
&= \left\{S_U^2 \mathrm{vec}(\boldsymbol{\eta}\boldsymbol{\eta}^\top) + \frac{n-1}{n}\mathrm{vec}(\boldsymbol{\Psi})\right\}\left\{S_U^2 \mathrm{vec}(\boldsymbol{\eta}\boldsymbol{\eta}^\top) + \frac{n-1}{n}\mathrm{vec}(\boldsymbol{\Psi})^\top\right\}.
\end{aligned}$$

The conditional variance of $\mathbf{S_X}$ *given* $\mathbf{U}$ *is then*

$$\begin{aligned}
\mathrm{var}(\mathbf{S_X} \mid \mathbf{U}) &= \frac{n-1}{n^2}(\mathbf{I}_{p^2}+\mathbf{K}_p)(\boldsymbol{\Psi}\otimes\boldsymbol{\Psi}) + \frac{1}{n}S_U^2\mathbf{A}\boldsymbol{\Psi}\mathbf{A}^\top \\
&= \frac{n-1}{n^2}(\mathbf{I}_{p^2}+\mathbf{K}_p)(\boldsymbol{\Psi}\otimes\boldsymbol{\Psi}) + \frac{1}{n}S_U^2(\mathbf{I}_{p^2}+\mathbf{K}_p)\{(\boldsymbol{\Psi}\otimes\boldsymbol{\eta}\boldsymbol{\eta}^\top)+(\boldsymbol{\eta}\boldsymbol{\eta}^\top\otimes\boldsymbol{\Psi})\},
\end{aligned}$$

since

$$\begin{aligned}
\mathbf{A}\boldsymbol{\Psi}\mathbf{A}^\top &= \{(\boldsymbol{\eta}\otimes\mathbf{I}_p)+(\mathbf{I}_p\otimes\boldsymbol{\eta})\}\boldsymbol{\Psi}\{(\boldsymbol{\eta}^\top\otimes\mathbf{I}_p)+(\mathbf{I}_p\otimes\boldsymbol{\eta}^\top)\} \\
&= \{(\boldsymbol{\eta}\otimes\boldsymbol{\Psi})+(\boldsymbol{\Psi}\otimes\boldsymbol{\eta})\}\{(\boldsymbol{\eta}^\top\otimes\mathbf{I}_p)+(\mathbf{I}_p\otimes\boldsymbol{\eta}^\top)\} \\
&= (\boldsymbol{\eta}\otimes\boldsymbol{\Psi})(\boldsymbol{\eta}^\top\otimes\mathbf{I}_p)+(\boldsymbol{\Psi}\otimes\boldsymbol{\eta})(\boldsymbol{\eta}^\top\otimes\mathbf{I}_p)+(\boldsymbol{\eta}\otimes\boldsymbol{\Psi})(\mathbf{I}_p\otimes\boldsymbol{\eta}^\top)+(\boldsymbol{\Psi}\otimes\boldsymbol{\eta})(\mathbf{I}_p\otimes\boldsymbol{\eta}^\top) \\
&= (\boldsymbol{\eta}\boldsymbol{\eta}^\top\otimes\boldsymbol{\Psi})+\mathbf{K}_p(\boldsymbol{\eta}\boldsymbol{\eta}^\top\otimes\boldsymbol{\Psi})+\mathbf{K}_p(\boldsymbol{\Psi}\otimes\boldsymbol{\eta}\boldsymbol{\eta}^\top)+(\boldsymbol{\Psi}\otimes\boldsymbol{\eta}\boldsymbol{\eta}^\top) \\
&= (\mathbf{I}_{p^2}+\mathbf{K}_p)\{(\boldsymbol{\Psi}\otimes\boldsymbol{\eta}\boldsymbol{\eta}^\top)+(\boldsymbol{\eta}\boldsymbol{\eta}^\top\otimes\boldsymbol{\Psi})\}.
\end{aligned}$$

Hence:

$$\begin{aligned}
\mathrm{var}(\mathbf{S_X}) &= \mathrm{var}(S_U^2)\mathrm{vec}(\boldsymbol{\eta}\boldsymbol{\eta}^\top)\mathrm{vec}(\boldsymbol{\eta}\boldsymbol{\eta}^\top)^\top \\
&\quad + \frac{n-1}{n^2}(\mathbf{I}_{p^2}+\mathbf{K}_p)(\boldsymbol{\Psi}\otimes\boldsymbol{\Psi}) + \frac{1}{n}\mathrm{E}(S_U^2)(\mathbf{I}_{p^2}+\mathbf{K}_p)\left\{(\boldsymbol{\Psi}\otimes\boldsymbol{\eta}\boldsymbol{\eta}^\top)+(\boldsymbol{\eta}\boldsymbol{\eta}^\top\otimes\boldsymbol{\Psi})\right\}.
\end{aligned}$$

Finally, to find the covariance between $\bar{\mathbf{X}}$ *and* $\mathbf{S}$*, we use that* $\mathrm{cov}(\bar{\mathbf{X}},\mathbf{S}) = \frac{n}{n-1}\mathrm{cov}(\bar{\mathbf{X}},\mathbf{S_X})$*, with*

$$\mathrm{cov}(\bar{\mathbf{X}},\mathbf{S_X}) \equiv \mathrm{cov}\{\bar{\mathbf{X}},\mathrm{vec}(\mathbf{S_X})\} = \mathrm{E}\{\bar{\mathbf{X}}\mathrm{vec}(\mathbf{S_X})^\top\} - \mathrm{E}(\bar{\mathbf{X}})\mathrm{E}\{\mathrm{vec}(\mathbf{S_X})\}^\top,$$

where as before $\mathrm{E}(\bar{\mathbf{X}}) = \boldsymbol{\xi}+\mathrm{E}(\bar{U})\boldsymbol{\eta}$ *and* $\mathrm{E}\{\mathrm{vec}(\mathbf{S_X})\} = \mathrm{E}(S_U^2)\mathrm{vec}(\boldsymbol{\eta}\boldsymbol{\eta}^\top)+\frac{n-1}{n}\mathrm{vec}(\boldsymbol{\Psi})$*. But, by (5.47) and (5.48), we have*

$$\begin{aligned}
\mathrm{E}\{\bar{\mathbf{X}}\mathrm{vec}(\mathbf{S_X})^\top\} &= \mathrm{E}[\mathrm{E}\{\bar{\mathbf{X}}\mathrm{vec}(\mathbf{S_X})^\top \mid \mathbf{U}\}] \\
&= \mathrm{E}(\mathrm{E}[(\boldsymbol{\xi}+\bar{U}\boldsymbol{\eta}+\bar{\mathbf{W}})\{S_U^2\mathrm{vec}(\boldsymbol{\eta}\boldsymbol{\eta}^\top)+\mathbf{A}\mathbf{S}_{U\mathbf{W}}+\mathrm{vec}(\mathbf{S_W})\}^\top \mid \mathbf{U}]) \\
&= \mathrm{E}(\{\boldsymbol{\xi}+\bar{U}\boldsymbol{\eta}+\mathrm{E}(\bar{\mathbf{W}})\}[S_U^2\mathrm{vec}(\boldsymbol{\eta}\boldsymbol{\eta}^\top)+\mathbf{A}\mathrm{E}(\mathbf{S}_{U\mathbf{W}} \mid \mathbf{U})+\mathrm{E}\{\mathrm{vec}(\mathbf{S_W})\}^\top]) \\
&= \mathrm{E}[(\boldsymbol{\xi}+\bar{U}\boldsymbol{\eta})\{S_U^2\mathrm{vec}(\boldsymbol{\eta}\boldsymbol{\eta}^\top)+\frac{n-1}{n}\mathrm{vec}(\boldsymbol{\Psi})\}^\top] \\
&= \{\mathrm{E}(S_U^2)\boldsymbol{\xi}+\mathrm{E}(\bar{U}S_U^2)\boldsymbol{\eta}\}\mathrm{vec}(\boldsymbol{\eta}\boldsymbol{\eta}^\top)^\top + \frac{n-1}{n}\{\boldsymbol{\xi}+\mathrm{E}(\bar{U})\boldsymbol{\eta}\}\mathrm{vec}(\boldsymbol{\Psi})^\top.
\end{aligned}$$

Hence:

$$\mathrm{cov}(\bar{\mathbf{X}},\mathbf{S_X}) = \{\mathrm{E}(\bar{U}S_U^2)-\mathrm{E}(\bar{U})\mathrm{E}(S_U^2)\}\boldsymbol{\eta}\mathrm{vec}(\boldsymbol{\eta}\boldsymbol{\eta}^\top)^\top.$$

The proof concludes from some straightforward algebra after applying the results for $\mathrm{E}(S_U^2)$*,* $\mathrm{var}(S_U^2)$ *and* $\mathrm{E}(\bar{U}S_U^2)$ *given in Theorem 83.* □

Let $\boldsymbol{\mu} = \mathrm{E}(\mathbf{X})$ and $\boldsymbol{\Sigma} = \mathrm{var}(\mathbf{X})$, where $\mathbf{X} \sim \mathcal{SN}_p(\boldsymbol{\xi},\boldsymbol{\Psi},\boldsymbol{\eta})$. Theorem 84 corroborates for the multivariate skew-normal model the well-known fact that the sample mean $\bar{\mathbf{X}}$ and the sample variance $\mathbf{S}$ are unbiased and consistent estimators of $\boldsymbol{\mu} = \boldsymbol{\xi}+b\boldsymbol{\eta}$ and $\boldsymbol{\Sigma} = \boldsymbol{\Psi}+(1-b^2)\boldsymbol{\eta}\boldsymbol{\eta}^\top$ since

$$\mathrm{E}(\bar{\mathbf{X}}) = \boldsymbol{\mu}, \quad \mathrm{E}(\mathbf{S}) = \boldsymbol{\Sigma} \quad \text{and} \quad \mathrm{var}(\bar{\mathbf{X}}) \to \mathbf{0}, \quad \mathrm{var}(\mathbf{S}) \to \mathbf{0}, \quad \text{as} \quad n \to \infty.$$

The theorem given below confirms that they are also asymptotically normal, for which we need the following preliminary result.

Theorem 85 *Let* $\mathbf{X}$ *be a random vector with mean vector* $\boldsymbol{\mu}$ *and variance matrix* $\boldsymbol{\Sigma}$, *and let* $\mathbf{T} = (\mathbf{X}-\boldsymbol{\mu})(\mathbf{X}-\boldsymbol{\mu})^\top$. *If* $\mathbf{X} \sim \mathcal{SN}_p(\boldsymbol{\xi}, \boldsymbol{\Psi}, \boldsymbol{\eta})$, *then*

$$\mathrm{E}\begin{pmatrix}\mathbf{X}\\ \mathbf{T}\end{pmatrix} = \begin{pmatrix}\boldsymbol{\mu}\\ \boldsymbol{\Sigma}\end{pmatrix}, \quad \mathrm{var}\begin{pmatrix}\mathbf{X}\\ \mathbf{T}\end{pmatrix} \equiv \mathrm{var}\begin{pmatrix}\mathbf{X}\\ \mathrm{vec}(\mathbf{T})\end{pmatrix} = \begin{pmatrix}\boldsymbol{\Sigma} & \boldsymbol{\Delta}^\top\\ \boldsymbol{\Delta} & \boldsymbol{\Lambda}\end{pmatrix},$$

where, as computed earlier $\boldsymbol{\mu} = \boldsymbol{\xi} + \sqrt{\frac{2}{\pi}}\boldsymbol{\eta}$ *and* $\boldsymbol{\Sigma} = \boldsymbol{\Psi} + \left(1-\frac{2}{\pi}\right)\boldsymbol{\eta}\boldsymbol{\eta}^\top$, *and*

$$\begin{aligned}
\boldsymbol{\Delta} &= \mathrm{cov}\{\mathrm{vec}(\mathbf{T}),\mathbf{X}\} = \sqrt{\frac{2}{\pi}}\left(\frac{4}{\pi}-1\right)\mathrm{vec}(\boldsymbol{\eta}\boldsymbol{\eta}^\top)\boldsymbol{\eta}^\top,\\
\boldsymbol{\Lambda} &= \mathrm{var}\{\mathrm{vec}(\mathbf{T})\} = (\mathbf{I}_{p^2}+\mathbf{K}_p)\left[(\boldsymbol{\Psi}\otimes\boldsymbol{\Psi}) + \left(1-\frac{2}{\pi}\right)\{(\boldsymbol{\Psi}\otimes\boldsymbol{\eta}\boldsymbol{\eta}^\top)+(\boldsymbol{\eta}\boldsymbol{\eta}^\top\otimes\boldsymbol{\Psi})\}\right]\\
&\quad + 2\left(1-\frac{8}{\pi^2}\right)\mathrm{vec}(\boldsymbol{\eta}\boldsymbol{\eta}^\top)\mathrm{vec}(\boldsymbol{\eta}\boldsymbol{\eta}^\top)^\top.
\end{aligned}$$

Proof 85 *Let* $U_0 = U - b$ *and* $\mathbf{X}_0 = \mathbf{X} - \boldsymbol{\mu}$. *By definition,* $\mathbf{X} = \boldsymbol{\xi} + U\boldsymbol{\eta} + \mathbf{W}$, *where* $U \sim \mathcal{HN}(0,1)$ *and* $\mathbf{W} \sim \mathcal{N}_p(\mathbf{0}_p, \boldsymbol{\Psi})$ *are independent. Then,* $\mathbf{X}_0 = U_0\boldsymbol{\eta} + \mathbf{W}$, *with* U_0 *and* $\mathbf{W}$ *being independent random quantities with mean zero, and*

$$\begin{aligned}
&\mathbf{T} = \mathbf{X}_0\mathbf{X}_0^\top = U_0^2\boldsymbol{\eta}\boldsymbol{\eta}^\top + U_0(\boldsymbol{\eta}\mathbf{W}^\top + \mathbf{W}\boldsymbol{\eta}^\top) + \mathbf{W}\mathbf{W}^\top\\
&\Longrightarrow \mathrm{vec}(\mathbf{T}) = U_0^2\mathrm{vec}(\boldsymbol{\eta}\boldsymbol{\eta}^\top) + U_0\mathbf{A}\mathbf{W} + \mathrm{vec}(\mathbf{W}\mathbf{W}^\top),
\end{aligned}$$

where $\mathbf{A} = (\mathbf{I}_p \otimes \boldsymbol{\eta}) + (\boldsymbol{\eta} \otimes \mathbf{I}_p)$. *Hence,*

$$\begin{aligned}
\mathrm{vec}(\mathbf{T})\mathbf{X}_0^\top &= U_0^3\mathrm{vec}(\boldsymbol{\eta}\boldsymbol{\eta}^\top)\boldsymbol{\eta}^\top + U_0^2\mathbf{A}\mathbf{W}\boldsymbol{\eta}^\top + U_0\mathrm{vec}(\mathbf{W}\mathbf{W}^\top)\boldsymbol{\eta}^\top\\
&\quad + U_0^2\mathrm{vec}(\boldsymbol{\eta}\boldsymbol{\eta}^\top)\mathbf{W}^\top + U_0\mathbf{A}\mathbf{W}\mathbf{W}^\top + \mathrm{vec}(\mathbf{W}\mathbf{W}^\top)\mathbf{W}^\top,\\
\Longrightarrow \mathrm{E}\{\mathrm{vec}(\mathbf{T})\mathbf{X}_0^\top\} &= \mathrm{E}(U_0^3)\mathrm{vec}(\boldsymbol{\eta}\boldsymbol{\eta}^\top)\boldsymbol{\eta}^\top + \mathrm{E}(U_0^2)\mathbf{A}\mathrm{E}(\mathbf{W})\boldsymbol{\eta}^\top + \mathrm{E}(U_0)\mathrm{E}\{\mathrm{vec}(\mathbf{W}\mathbf{W}^\top)\}\boldsymbol{\eta}^\top\\
&\quad + \mathrm{E}(U_0^2)\mathrm{vec}(\boldsymbol{\eta}\boldsymbol{\eta}^\top)\mathrm{E}(\mathbf{W}^\top) + \mathrm{E}(U_0)\mathbf{A}\mathrm{E}(\mathbf{W}\mathbf{W}^\top) + \mathrm{E}\{\mathrm{vec}(\mathbf{W}\mathbf{W}^\top)\mathbf{W}^\top\}\\
&= \mathrm{E}(U_0^3)\mathrm{vec}(\boldsymbol{\eta}\boldsymbol{\eta}^\top)\}\boldsymbol{\eta}^\top,
\end{aligned}$$

since $\mathrm{E}(U_0) = 0$, $\mathrm{E}(\mathbf{W}) = \mathbf{0}_p$ *and* $\mathrm{E}\{\mathrm{vec}(\mathbf{W}\mathbf{W}^\top)\mathbf{W}^\top\} = \mathbf{O}$, *and where* $\mathrm{E}(U_0^3)$ *is given in Theorem 80. Since* $\boldsymbol{\Delta} = \mathrm{cov}\{\mathrm{vec}(\mathbf{T}),\mathbf{X}\} = \mathrm{cov}\{\mathrm{vec}(\mathbf{T}),\mathbf{X}_0\} = \mathrm{E}\{\mathrm{vec}(\mathbf{T})\mathbf{X}_0^\top\}$, *the result for this covariance matrix is proved.*

To prove the result for the variance matrix $\boldsymbol{\Lambda} = \mathrm{var}\{\mathrm{vec}(\mathbf{T})\}$, *note first that*

$$\begin{aligned}
\mathrm{vec}(\mathbf{T})\mathrm{vec}(\mathbf{T})^\top =&\, U_0^4\mathrm{vec}(\boldsymbol{\eta}\boldsymbol{\eta}^\top)\mathrm{vec}(\boldsymbol{\eta}\boldsymbol{\eta}^\top)^\top + U_0^3\mathbf{A}\mathbf{W}\mathrm{vec}(\boldsymbol{\eta}\boldsymbol{\eta}^\top)^\top + U_0^2\mathrm{vec}(\mathbf{W}\mathbf{W}^\top)\mathrm{vec}(\boldsymbol{\eta}\boldsymbol{\eta}^\top)^\top\\
&+ U_0^3\mathrm{vec}(\boldsymbol{\eta}\boldsymbol{\eta}^\top)\mathbf{W}^\top\mathbf{A}^\top + U_0^2\mathbf{A}\mathbf{W}\mathbf{W}^\top\mathbf{A}^\top + U_0\mathrm{vec}(\mathbf{W}\mathbf{W}^\top)\mathbf{W}^\top\mathbf{A}^\top\\
&+ U_0^2\mathrm{vec}(\boldsymbol{\eta}\boldsymbol{\eta}^\top)\mathrm{vec}(\mathbf{W}\mathbf{W}^\top)^\top + U_0\mathbf{A}\mathbf{W}\mathrm{vec}(\mathbf{W}\mathbf{W}^\top)^\top\\
&+ \mathrm{vec}(\mathbf{W}\mathbf{W}^\top)\mathrm{vec}(\mathbf{W}\mathbf{W}^\top)^\top.
\end{aligned}$$

Therefore,

$$\begin{aligned}
\mathrm{E}\{\mathrm{vec}(\mathbf{T})\mathrm{vec}(\mathbf{T})^\top\} =& \{\mathrm{E}(U_0^4)\mathrm{vec}(\boldsymbol{\eta}\boldsymbol{\eta}^\top)+\mathrm{E}(U_0^2)\mathrm{vec}(\boldsymbol{\Psi})\}\mathrm{vec}(\boldsymbol{\eta}\boldsymbol{\eta}^\top)^\top+\mathrm{E}(U_0^2)\mathbf{A}\boldsymbol{\Psi}\mathbf{A}^\top \\
&+\mathrm{E}(U_0^2)\mathrm{vec}(\boldsymbol{\eta}\boldsymbol{\eta}^\top)\mathrm{vec}(\boldsymbol{\Psi})^\top+(\mathbf{I}_{p^2}+\mathbf{K}_p)(\boldsymbol{\Psi}\otimes\boldsymbol{\Psi})+\mathrm{vec}(\boldsymbol{\Psi})\mathrm{vec}(\boldsymbol{\Psi})^\top \\
=& (\mathbf{I}_{p^2}+\mathbf{K}_p)(\boldsymbol{\Psi}\otimes\boldsymbol{\Psi})+\mathrm{E}(U_0^2)\mathbf{A}\boldsymbol{\Psi}\mathbf{A}^\top+\mathrm{E}(U_0^4)\mathrm{vec}(\boldsymbol{\eta}\boldsymbol{\eta}^\top)\mathrm{vec}(\boldsymbol{\eta}\boldsymbol{\eta}^\top)^\top \\
&+\mathrm{E}(U_0^2)\{\mathrm{vec}(\boldsymbol{\eta}\boldsymbol{\eta}^\top)\mathrm{vec}(\boldsymbol{\Psi})^\top+\mathrm{vec}(\boldsymbol{\Psi})\mathrm{vec}(\boldsymbol{\eta}\boldsymbol{\eta}^\top)^\top\}+\mathrm{vec}(\boldsymbol{\Psi})\mathrm{vec}(\boldsymbol{\Psi})^\top,
\end{aligned}$$

where we use that (see Exercise 5.8)

$$\mathrm{E}\{\mathrm{vec}(\mathbf{W}\mathbf{W}^\top)\mathrm{vec}(\mathbf{W}\mathbf{W}^\top)^\top\} = (\mathbf{I}_{p^2}+\mathbf{K}_p)(\boldsymbol{\Psi}\otimes\boldsymbol{\Psi})+\mathrm{vec}(\boldsymbol{\Psi})\mathrm{vec}(\boldsymbol{\Psi})^\top,$$

and as before $\mathbf{A}\boldsymbol{\Psi}\mathbf{A}^\top = (\mathbf{I}_{p^2}+\mathbf{K}_p)\{(\boldsymbol{\Psi}\otimes\boldsymbol{\eta}\boldsymbol{\eta}^\top)+(\boldsymbol{\eta}\boldsymbol{\eta}^\top\otimes\boldsymbol{\Psi})\}$. *Finally, since*

$$\mathrm{var}\{\mathrm{vec}(\mathbf{T})\} = \mathrm{E}\{\mathrm{vec}(\mathbf{T})\mathrm{vec}(\mathbf{T})^\top\}-\mathrm{E}\{\mathrm{vec}(\mathbf{T})\}\mathrm{E}\{\mathrm{vec}(\mathbf{T})^\top\},$$

where $\mathrm{E}\{\mathrm{vec}(\mathbf{T})\} = \mathrm{E}(U_0^2)\mathrm{vec}(\boldsymbol{\eta}\boldsymbol{\eta}^\top)+\mathrm{E}\{\mathrm{vec}(\mathbf{W}\mathbf{W}^\top)\} = \mathrm{E}(U_0^2)\mathrm{vec}(\boldsymbol{\eta}\boldsymbol{\eta}^\top)+\mathrm{vec}(\boldsymbol{\Psi})$, *we have*

$$\boldsymbol{\Lambda} = (\mathbf{I}_{p^2}+\mathbf{K}_p)(\boldsymbol{\Psi}\otimes\boldsymbol{\Psi})+\mathrm{E}(U_0^2)\mathbf{A}\boldsymbol{\Psi}\mathbf{A}^\top+\{\mathrm{E}(U_0^4)-\mathrm{E}(U_0^2)^2\}\mathrm{vec}(\boldsymbol{\eta}\boldsymbol{\eta}^\top)\mathrm{vec}(\boldsymbol{\eta}\boldsymbol{\eta}^\top)^\top,$$

and the proof concludes by replacing the moments of $U_0 = U-b$ *given in Theorem 80.* □

Of course, from Theorems 84 and 85 we can verify that

$$n\,\mathrm{var}\begin{pmatrix}\bar{\mathbf{X}}\\ \mathbf{S}\end{pmatrix} \to \mathrm{var}\begin{pmatrix}\mathbf{X}\\ \mathbf{T}\end{pmatrix} = \begin{pmatrix}\boldsymbol{\Sigma} & \boldsymbol{\Delta}^\top\\ \boldsymbol{\Delta} & \boldsymbol{\Lambda}\end{pmatrix} \quad \text{as} \quad n\to\infty.$$

Theorem 86 *Let* $\mathbf{X}_1,\ldots,\mathbf{X}_n$ *be a random sample from* $\mathbf{X}\sim\mathcal{SN}_p(\boldsymbol{\xi},\boldsymbol{\Psi},\boldsymbol{\eta})$. *Then:*

$$\sqrt{n}\begin{pmatrix}\bar{\mathbf{X}}-\boldsymbol{\mu}\\ \mathrm{vec}(\mathbf{S}-\boldsymbol{\Sigma})\end{pmatrix} \xrightarrow{d} \mathcal{N}_{p(p+1)}\left(\begin{pmatrix}\mathbf{0}_p\\ \mathrm{vec}(\mathbf{O})\end{pmatrix},\begin{pmatrix}\boldsymbol{\Sigma} & \boldsymbol{\Delta}^\top\\ \boldsymbol{\Delta} & \boldsymbol{\Lambda}\end{pmatrix}\right),$$

where, as before $\boldsymbol{\mu} = \boldsymbol{\xi}+\sqrt{\frac{2}{\pi}}\,\boldsymbol{\eta}$ *and* $\boldsymbol{\Sigma} = \boldsymbol{\Psi}+(1-\frac{2}{\pi})\,\boldsymbol{\eta}\boldsymbol{\eta}^\top$, *and*

$$\begin{aligned}
\boldsymbol{\Delta} &= \mathrm{cov}\{\mathrm{vec}(\mathbf{T}),\mathbf{X}\} = \sqrt{\frac{2}{\pi}}\left(\frac{4}{\pi}-1\right)\mathrm{vec}(\boldsymbol{\eta}\boldsymbol{\eta}^\top)\boldsymbol{\eta}^\top,\\
\boldsymbol{\Lambda} &= \mathrm{var}\{\mathrm{vec}(\mathbf{T})\} = (\mathbf{I}_{p^2}+\mathbf{K}_p)\left[(\boldsymbol{\Psi}\otimes\boldsymbol{\Psi})+\left(1-\frac{2}{\pi}\right)\{(\boldsymbol{\Psi}\otimes\boldsymbol{\eta}\boldsymbol{\eta}^\top)+(\boldsymbol{\eta}\boldsymbol{\eta}^\top\otimes\boldsymbol{\Psi})\}\right]\\
&\quad +2\left(1-\frac{8}{\pi^2}\right)\mathrm{vec}(\boldsymbol{\eta}\boldsymbol{\eta}^\top)\mathrm{vec}(\boldsymbol{\eta}\boldsymbol{\eta}^\top)^\top.
\end{aligned}$$

Proof 86 *Since* $\mathbf{X}_1,\ldots,\mathbf{X}_n$ *are iid (copies of)* $\mathbf{X}\sim\mathcal{SN}_p(\boldsymbol{\xi},\boldsymbol{\Psi},\boldsymbol{\eta})$, *then the random matrices* $\mathbf{T}_i = (\mathbf{X}_i-\boldsymbol{\mu})(\mathbf{X}_i-\boldsymbol{\mu})^\top$, $i=1,\ldots,n$, *become iid (copies of)* $\mathbf{T} = (\mathbf{X}-\boldsymbol{\mu})(\mathbf{X}-\boldsymbol{\mu})^\top$. *Hence*

$$\begin{pmatrix}\mathbf{X}_1\\ \mathbf{T}_1\end{pmatrix},\ldots,\begin{pmatrix}\mathbf{X}_n\\ \mathbf{T}_n\end{pmatrix} \quad \text{are iid (copies of)} \quad \begin{pmatrix}\mathbf{X}\\ \mathbf{T}\end{pmatrix},$$

which, by Theorem 85, has a finite mean

$$\mathrm{E}\begin{pmatrix}\mathbf{X}\\ \mathbf{T}\end{pmatrix}=\begin{pmatrix}\boldsymbol{\mu}\\ \boldsymbol{\Sigma}\end{pmatrix},$$

and finite variance

$$\operatorname{var}\begin{pmatrix}\mathbf{X}\\ \mathbf{T}\end{pmatrix}=\begin{pmatrix}\operatorname{var}(\mathbf{X}) & \operatorname{cov}\{\mathbf{X},\operatorname{vec}(\mathbf{T})\}\\ \operatorname{cov}\{\operatorname{vec}(\mathbf{T}),\mathbf{X}\} & \operatorname{var}\{\operatorname{vec}(\mathbf{T})\}\end{pmatrix}=\begin{pmatrix}\boldsymbol{\Sigma} & \boldsymbol{\Delta}^\top\\ \boldsymbol{\Delta} & \boldsymbol{\Lambda}\end{pmatrix}.$$

By the classical Central Limit Theorem (CLT) for iid random vectors, we know that

$$\sqrt{n}\begin{pmatrix}\bar{\mathbf{X}}-\boldsymbol{\mu}\\ \operatorname{vec}(\bar{\mathbf{T}}-\boldsymbol{\Sigma})\end{pmatrix}\xrightarrow{d}\mathcal{N}_{p(p+1)}\left(\begin{pmatrix}\mathbf{0}_p\\ \operatorname{vec}(\mathbf{O})\end{pmatrix},\begin{pmatrix}\boldsymbol{\Sigma} & \boldsymbol{\Delta}^\top\\ \boldsymbol{\Delta} & \boldsymbol{\Lambda}\end{pmatrix}\right).$$

Also, since $\sqrt{n}(\bar{\mathbf{X}}-\boldsymbol{\mu})=O_p(1)$ *as* $n\to\infty$, *then* $n\operatorname{vec}\{(\bar{\mathbf{X}}-\boldsymbol{\mu})(\bar{\mathbf{X}}-\boldsymbol{\mu})^\top\}\xrightarrow{P}\mathbf{0}_{p^2}$ *as* $n\to\infty$. *Hence, from the well-known identity* $\bar{\mathbf{T}}=\mathbf{S}_{\mathbf{X}}+(\bar{\mathbf{X}}-\boldsymbol{\mu})(\bar{\mathbf{X}}-\boldsymbol{\mu})^\top$ *and the multivariate Slutsky's theorem, we have*

$$\begin{aligned}\sqrt{n}\begin{pmatrix}\bar{\mathbf{X}}-\boldsymbol{\mu}\\ \operatorname{vec}(\mathbf{S}_{\mathbf{X}}-\boldsymbol{\Sigma})\end{pmatrix} &= \sqrt{n}\begin{pmatrix}\bar{\mathbf{X}}-\boldsymbol{\mu}\\ \operatorname{vec}(\bar{\mathbf{T}}-\boldsymbol{\Sigma})\end{pmatrix}-\frac{1}{\sqrt{n}}\begin{pmatrix}\mathbf{0}_p\\ n\operatorname{vec}\{(\bar{\mathbf{X}}-\boldsymbol{\mu})(\bar{\mathbf{X}}-\boldsymbol{\mu})^\top\}\end{pmatrix}\\ &\xrightarrow{d}\mathcal{N}_{p(p+1)}\left(\begin{pmatrix}\mathbf{0}_p\\ \operatorname{vec}(\mathbf{O})\end{pmatrix},\begin{pmatrix}\boldsymbol{\Sigma} & \boldsymbol{\Delta}^\top\\ \boldsymbol{\Delta} & \boldsymbol{\Lambda}\end{pmatrix}\right).\end{aligned}$$

The proof concludes from the asymptotic equivalence of $(\bar{\mathbf{X}},\mathbf{S})$ *and* $(\bar{\mathbf{X}},\mathbf{S}_{\mathbf{X}})$ *due to the identity* $\mathbf{S}=\frac{n}{n-1}\mathbf{S}_{\mathbf{X}}$. □

Further results on the asymptotic joint distribution of multivariate sample moments can be found in Arellano-Valle et al. (2021).

5.10 Parameter Estimation

5.10.1 Behavior of the likelihood function

For a random sample $\mathbf{X}_1,\ldots,\mathbf{X}_n$ from the $\mathcal{SN}_p(\boldsymbol{\xi},\boldsymbol{\Psi},\boldsymbol{\eta})$ distribution, the likelihood function for $\boldsymbol{\Theta}=(\boldsymbol{\xi},\boldsymbol{\Psi},\boldsymbol{\eta})\in\mathbb{R}^p\times\mathbb{R}^{p\times p}\times\mathbb{R}^p$, where $\boldsymbol{\Psi}>0$, is

$$L(\boldsymbol{\Theta})=\prod_{i=1}^n 2\phi_p(\mathbf{x}_i;\boldsymbol{\xi},\boldsymbol{\Psi}+\boldsymbol{\eta}\boldsymbol{\eta}^\top)\Phi\left\{\frac{\boldsymbol{\eta}^\top\boldsymbol{\Psi}^{-1}(\mathbf{x}_i-\boldsymbol{\xi})}{\sqrt{1+\boldsymbol{\eta}^\top\boldsymbol{\Psi}^{-1}\boldsymbol{\eta}}}\right\},$$

where we note that

$$\begin{aligned}\phi_p(\mathbf{x}_i;\boldsymbol{\xi},\boldsymbol{\Psi}+\boldsymbol{\eta}\boldsymbol{\eta}^\top) &= \frac{1}{(2\pi)^{p/2}|\boldsymbol{\Psi}|^{1/2}}\frac{e^{-\frac{1}{2}(\mathbf{x}_i-\boldsymbol{\xi})^\top(\boldsymbol{\Psi}+\boldsymbol{\eta}\boldsymbol{\eta}^\top)^{-1}(\mathbf{x}-\boldsymbol{\xi})}}{\sqrt{1+\boldsymbol{\eta}^\top\boldsymbol{\Psi}^{-1}\boldsymbol{\eta}}}\\ &= \frac{e^{-\frac{1}{2}(\mathbf{x}_i-\boldsymbol{\xi})^\top\boldsymbol{\Psi}^{-1}(\mathbf{x}_i-\boldsymbol{\xi})}}{(2\pi)^{p/2}|\boldsymbol{\Psi}|^{1/2}}\frac{e^{\frac{1}{2}\left\{\frac{\boldsymbol{\eta}^\top\boldsymbol{\Psi}^{-1}(\mathbf{x}_i-\boldsymbol{\xi})}{\sqrt{1+\boldsymbol{\eta}^\top\boldsymbol{\Psi}^{-1}\boldsymbol{\eta}}}\right\}^2}}{\sqrt{1+\boldsymbol{\eta}^\top\boldsymbol{\Psi}^{-1}\boldsymbol{\eta}}},\end{aligned}$$

$i = 1, \ldots, n$. Also, from the Cauchy-Schwarz inequality, we have that

$$\begin{aligned}\frac{|\boldsymbol{\eta}^\top \boldsymbol{\Psi}^{-1}(\mathbf{x}_i - \boldsymbol{\xi})|}{\sqrt{1+\boldsymbol{\eta}^\top \boldsymbol{\Psi}^{-1}\boldsymbol{\eta}}} &\leq \left(\frac{\boldsymbol{\eta}^\top \boldsymbol{\Psi}^{-1}\boldsymbol{\eta}}{1+\boldsymbol{\eta}^\top \boldsymbol{\Psi}^{-1}\boldsymbol{\eta}}\right)^{1/2} \{(\mathbf{x}_i - \boldsymbol{\xi})^\top \boldsymbol{\Psi}^{-1}(\mathbf{x}_i - \boldsymbol{\xi})\}^{1/2} \\ &\leq \{(\mathbf{x}_i - \boldsymbol{\xi})^\top \boldsymbol{\Psi}^{-1}(\mathbf{x}_i - \boldsymbol{\xi})\}^{1/2},\end{aligned}$$

for $i = 1, \ldots, n$. From this inequality, we find for each $i = 1, \ldots, n$ that

$$\begin{aligned}\phi_p(\mathbf{x}_i; \boldsymbol{\xi}, \boldsymbol{\Psi} + \boldsymbol{\eta}\boldsymbol{\eta}^\top) &= \frac{e^{-\frac{1}{2}(\mathbf{x}_i - \boldsymbol{\xi})^\top \boldsymbol{\Psi}^{-1}(\mathbf{x}_i - \boldsymbol{\xi})}}{|2\pi\boldsymbol{\Psi}|^{1/2}} \frac{e^{\frac{1}{2}\left\{\frac{\boldsymbol{\eta}^\top \boldsymbol{\Psi}^{-1}(\mathbf{x}_i - \boldsymbol{\xi})}{\sqrt{1+\boldsymbol{\eta}^\top \boldsymbol{\Psi}^{-1}\boldsymbol{\eta}}}\right\}^2}}{\sqrt{1+\boldsymbol{\eta}^\top \boldsymbol{\Psi}^{-1}\boldsymbol{\eta}}} \\ &\leq \frac{1}{|2\pi\boldsymbol{\Psi}|^{1/2}\sqrt{1+\boldsymbol{\eta}^\top \boldsymbol{\Psi}^{-1}\boldsymbol{\eta}}},\end{aligned}$$

and

$$\begin{aligned}0 \leq \Phi\left\{-\sqrt{(\mathbf{x}_i - \boldsymbol{\xi})^\top \boldsymbol{\Psi}^{-1}(\mathbf{x}_i - \boldsymbol{\xi})}\right\} &\leq \Phi\left\{\frac{\boldsymbol{\eta}^\top \boldsymbol{\Psi}^{-1}(\mathbf{x}_i - \boldsymbol{\xi})}{\sqrt{1+\boldsymbol{\eta}^\top \boldsymbol{\Psi}^{-1}\boldsymbol{\eta}}}\right\} \\ &\leq \Phi\left\{\sqrt{(\mathbf{x}_i - \boldsymbol{\xi})^\top \boldsymbol{\Psi}^{-1}(\mathbf{x}_i - \boldsymbol{\xi})}\right\} \leq 1.\end{aligned}$$

Noting also that if some or all of the components of $\boldsymbol{\eta}$ tend to $\pm\infty$, then $\boldsymbol{\eta}^\top \boldsymbol{\Psi}^{-1}\boldsymbol{\eta} \to \infty$, and it clearly follows in this setting that $\phi_p(\mathbf{x}_i; \boldsymbol{\xi}, \boldsymbol{\Psi} + \boldsymbol{\eta}\boldsymbol{\eta}^\top) \to 0$ for any fixed value of $(\boldsymbol{\xi}, \boldsymbol{\Psi})$. Thus, these results show that the profile likelihood function of $\boldsymbol{\eta}$ tends to zero whenever some or all of the components of $\boldsymbol{\eta}$ tend to $\pm\infty$. This result leads us to the conclusion that the profile likelihood function of $\boldsymbol{\eta}$ is not a monotonically increasing or decreasing function of any of the components of $\boldsymbol{\eta}$. Consequently, the profile likelihood of the skewness parameter $\boldsymbol{\eta}$ is always maximized at a finite point for the $\mathcal{SN}_p(\boldsymbol{\xi}, \boldsymbol{\Psi}, \boldsymbol{\eta})$ model. Note that this is in contrast to the skew-normal distribution with ASN parameterization for which infinite values of the maximum likelihood estimator of the skewness parameter $\boldsymbol{\alpha}$ can occur. This happens, for example, if the arguments of $\Phi_n(\boldsymbol{\alpha}^\top \mathbf{z}_1, \ldots, \boldsymbol{\alpha}^\top \mathbf{z}_n; \mathbf{0}_n, \mathbf{I}_n) = \prod_{i=1}^n \Phi(\boldsymbol{\alpha}^\top \mathbf{z}_i)$ all have the same sign in the likelihood, where $\mathbf{z}_i = \boldsymbol{\omega}^{-1}(\mathbf{x}_i - \boldsymbol{\xi})$, $i = 1, \ldots, n$; see, e.g., Liseo (1990), Sartori (2006), and Azzalini and Arellano-Valle (2013).

5.10.2 Maximum likelihood estimation, score function, and information matrix

The corresponding log-likelihood function for $\boldsymbol{\Theta} = (\boldsymbol{\xi}, \boldsymbol{\Psi}, \boldsymbol{\eta})$ is given by

$$\begin{aligned}\ell(\boldsymbol{\Theta}) = &-\frac{np}{2}\log(2\pi) - \frac{n}{2}\log\{|\boldsymbol{\Psi} + \boldsymbol{\eta}\boldsymbol{\eta}^\top|\} - \frac{1}{2}\sum_{i=1}^n (\mathbf{x}_i - \boldsymbol{\xi})^\top (\boldsymbol{\Psi} + \boldsymbol{\eta}\boldsymbol{\eta}^\top)^{-1}(\mathbf{x}_i - \boldsymbol{\xi}) \\ &+ \sum_{i=1}^n \log\left[2\Phi\left\{\frac{\boldsymbol{\eta}^\top \boldsymbol{\Psi}^{-1}(\mathbf{x}_i - \boldsymbol{\xi})}{\sqrt{1+\boldsymbol{\eta}^\top \boldsymbol{\Psi}^{-1}\boldsymbol{\eta}}}\right\}\right].\end{aligned}$$

From the approach of Magnus and Neudecker (1999) related to the differential of a vector and/or matrix function (see also Section 1.6.2), we find for the first differentials of $\ell(\boldsymbol{\Theta})$ with respect to $\boldsymbol{\xi}$, $\boldsymbol{\Psi}$, and $\boldsymbol{\eta}$ that:

$$
\begin{aligned}
\mathrm{d}_{\boldsymbol{\xi}}\ell(\boldsymbol{\Theta}) &= \sum_{i=1}^{n}(\mathrm{d}\boldsymbol{\xi})^{\top}(\boldsymbol{\Psi}+\boldsymbol{\eta}\boldsymbol{\eta}^{\top})^{-1}(\mathbf{x}_i-\boldsymbol{\xi}) - \sum_{i=1}^{n}\zeta\left\{\frac{\boldsymbol{\eta}^{\top}\boldsymbol{\Psi}^{-1}(\mathbf{x}_i-\boldsymbol{\xi})}{\sqrt{1+\boldsymbol{\eta}^{\top}\boldsymbol{\Psi}^{-1}\boldsymbol{\eta}}}\right\}\frac{(\mathrm{d}\boldsymbol{\xi})^{\top}\boldsymbol{\Psi}^{-1}\boldsymbol{\eta}}{\sqrt{1+\boldsymbol{\eta}^{\top}\boldsymbol{\Psi}^{-1}\boldsymbol{\eta}}}, \\
\mathrm{d}_{\boldsymbol{\Psi}}\ell(\boldsymbol{\Theta}) &= -\frac{n}{2}\mathrm{tr}\{(\boldsymbol{\Psi}+\boldsymbol{\eta}\boldsymbol{\eta}^{\top})^{-1}(\mathrm{d}\boldsymbol{\Psi})\} + \frac{1}{2}\sum_{i=1}^{n}(\mathbf{x}_i-\boldsymbol{\xi})^{\top}(\boldsymbol{\Psi}+\boldsymbol{\eta}\boldsymbol{\eta}^{\top})^{-1}(\mathrm{d}\boldsymbol{\Psi})(\boldsymbol{\Psi}+\boldsymbol{\eta}\boldsymbol{\eta}^{\top})^{-1}(\mathbf{x}_i-\boldsymbol{\xi}) \\
&\quad + \sum_{i=1}^{n}\zeta\left\{\frac{\boldsymbol{\eta}^{\top}\boldsymbol{\Psi}^{-1}(\mathbf{x}_i-\boldsymbol{\xi})}{\sqrt{1+\boldsymbol{\eta}^{\top}\boldsymbol{\Psi}^{-1}\boldsymbol{\eta}}}\right\} \\
&\quad \times\left\{-\frac{\boldsymbol{\eta}^{\top}\boldsymbol{\Psi}^{-1}(\mathrm{d}\boldsymbol{\Psi})\boldsymbol{\Psi}^{-1}(\mathbf{x}_i-\boldsymbol{\xi})}{(1+\boldsymbol{\eta}^{\top}\boldsymbol{\Psi}^{-1}\boldsymbol{\eta})^{1/2}} + \frac{1}{2}\frac{\boldsymbol{\eta}^{\top}\boldsymbol{\Psi}^{-1}(\mathrm{d}\boldsymbol{\Psi})\boldsymbol{\Psi}^{-1}\boldsymbol{\eta}\boldsymbol{\eta}^{\top}\boldsymbol{\Psi}^{-1}(\mathbf{x}_i-\boldsymbol{\xi})}{(1+\boldsymbol{\eta}^{\top}\boldsymbol{\Psi}^{-1}\boldsymbol{\eta})^{3/2}}\right\}, \\
\mathrm{d}_{\boldsymbol{\eta}}\ell(\boldsymbol{\Theta}) &= -\frac{n}{2}\mathrm{tr}[(\boldsymbol{\Psi}+\boldsymbol{\eta}\boldsymbol{\eta}^{\top})^{-1}\{(\mathrm{d}\boldsymbol{\eta})\boldsymbol{\eta}^{\top}+\boldsymbol{\eta}(\mathrm{d}\boldsymbol{\eta})^{\top}\}] \\
&\quad + \frac{1}{2}\sum_{i=1}^{n}(\mathbf{x}_i-\boldsymbol{\xi})^{\top}(\boldsymbol{\Psi}+\boldsymbol{\eta}\boldsymbol{\eta}^{\top})^{-1}\{(\mathrm{d}\boldsymbol{\eta})\boldsymbol{\eta}^{\top}+\boldsymbol{\eta}(\mathrm{d}\boldsymbol{\eta})^{\top}\}(\boldsymbol{\Psi}+\boldsymbol{\eta}\boldsymbol{\eta}^{\top})^{-1}(\mathbf{x}_i-\boldsymbol{\xi}) \\
&\quad + \sum_{i=1}^{n}\zeta\left\{\frac{\boldsymbol{\eta}^{\top}\boldsymbol{\Psi}^{-1}(\mathbf{x}_i-\boldsymbol{\xi})}{\sqrt{1+\boldsymbol{\eta}^{\top}\boldsymbol{\Psi}^{-1}\boldsymbol{\eta}}}\right\}\left\{\frac{(\mathrm{d}\boldsymbol{\eta})^{\top}\boldsymbol{\Psi}^{-1}(\mathbf{x}_i-\boldsymbol{\xi})}{(1+\boldsymbol{\eta}^{\top}\boldsymbol{\Psi}^{-1}\boldsymbol{\eta})^{1/2}} - \frac{(\mathrm{d}\boldsymbol{\eta})^{\top}\boldsymbol{\Psi}^{-1}\boldsymbol{\eta}\boldsymbol{\eta}^{\top}\boldsymbol{\Psi}^{-1}(\mathbf{x}_i-\boldsymbol{\xi})}{(1+\boldsymbol{\eta}^{\top}\boldsymbol{\Psi}^{-1}\boldsymbol{\eta})^{3/2}}\right\},
\end{aligned}
$$

where $\zeta(x) = \phi(x)/\Phi(x)$ denotes the inverse Mills ratio. Hence, by using the properties of the trace of matrix products, from these last expressions we find, after some computations, that the components of the score function $\mathbf{U}(\boldsymbol{\Theta}) = \partial\ell(\boldsymbol{\Theta})/\partial\boldsymbol{\theta}^{\top}$, where $\boldsymbol{\theta} = (\boldsymbol{\xi}^{\top}, \mathrm{vech}(\boldsymbol{\Psi})^{\top}, \boldsymbol{\eta}^{\top})^{\top}$ are given by

$$
\begin{aligned}
\mathbf{U}_{\boldsymbol{\xi}}(\boldsymbol{\Theta}) &= n(\boldsymbol{\Psi}+\boldsymbol{\eta}\boldsymbol{\eta}^{\top})^{-1}(\bar{\mathbf{x}}-\boldsymbol{\xi}) - \frac{1}{\sqrt{1+\boldsymbol{\eta}^{\top}\boldsymbol{\Psi}^{-1}\boldsymbol{\eta}}}\sum_{i=1}^{n}\zeta\left\{\frac{\boldsymbol{\eta}^{\top}\boldsymbol{\Psi}^{-1}(\mathbf{x}_i-\boldsymbol{\xi})}{\sqrt{1+\boldsymbol{\eta}^{\top}\boldsymbol{\Psi}^{-1}\boldsymbol{\eta}}}\right\}\boldsymbol{\Psi}^{-1}\boldsymbol{\eta}, \\
\mathbf{U}_{\mathrm{vech}(\boldsymbol{\Psi})}(\boldsymbol{\Theta}) &= \frac{n}{2}\mathbf{D}_p^{\top}\{(\boldsymbol{\Psi}+\boldsymbol{\eta}\boldsymbol{\eta}^{\top})^{-1}\otimes(\boldsymbol{\Psi}+\boldsymbol{\eta}\boldsymbol{\eta}^{\top})^{-1}\}\mathrm{vec}\{\mathbf{S}_{\mathbf{x}}+(\bar{\mathbf{x}}-\boldsymbol{\xi})(\bar{\mathbf{x}}-\boldsymbol{\xi})^{\top}-(\boldsymbol{\Psi}+\boldsymbol{\eta}\boldsymbol{\eta})\} \\
&\quad - \frac{1}{\sqrt{1+\boldsymbol{\eta}^{\top}\boldsymbol{\Psi}^{-1}\boldsymbol{\eta}}}\mathbf{D}_p^{\top}\left\{(\boldsymbol{\Psi}^{-1}\otimes\boldsymbol{\Psi}^{-1}) - \frac{1}{2}\frac{1}{1+\boldsymbol{\eta}^{\top}\boldsymbol{\Psi}^{-1}\boldsymbol{\eta}}(\boldsymbol{\Psi}^{-1}\otimes\boldsymbol{\Psi}^{-1}\boldsymbol{\eta}\boldsymbol{\eta}^{\top}\boldsymbol{\Psi}^{-1})\right\} \\
&\quad \times\sum_{i=1}^{n}\zeta\left\{\frac{\boldsymbol{\eta}^{\top}\boldsymbol{\Psi}^{-1}(\mathbf{x}_i-\boldsymbol{\xi})}{\sqrt{1+\boldsymbol{\eta}^{\top}\boldsymbol{\Psi}^{-1}\boldsymbol{\eta}}}\right\}\mathrm{vec}\{(\mathbf{x}_i-\boldsymbol{\xi})\boldsymbol{\eta}^{\top}\}, \\
\mathbf{U}_{\boldsymbol{\eta}}(\boldsymbol{\Theta}) &= n(\boldsymbol{\Psi}+\boldsymbol{\eta}\boldsymbol{\eta}^{\top})^{-1}\{\mathbf{S}_{\mathbf{x}}+(\bar{\mathbf{x}}-\boldsymbol{\xi})^{\top}(\bar{\mathbf{x}}-\boldsymbol{\xi})^{\top}-(\boldsymbol{\Psi}+\boldsymbol{\eta}\boldsymbol{\eta}^{\top})\}(\boldsymbol{\Psi}+\boldsymbol{\eta}\boldsymbol{\eta}^{\top})^{-1}\boldsymbol{\eta} \\
&\quad + \frac{1}{\sqrt{1+\boldsymbol{\eta}^{\top}\boldsymbol{\Psi}^{-1}\boldsymbol{\eta}}}(\boldsymbol{\Psi}+\boldsymbol{\eta}\boldsymbol{\eta}^{\top})^{-1}\sum_{i=1}^{n}\zeta\left\{\frac{\boldsymbol{\eta}^{\top}\boldsymbol{\Psi}^{-1}(\mathbf{x}_i-\boldsymbol{\xi})}{\sqrt{1+\boldsymbol{\eta}^{\top}\boldsymbol{\Psi}^{-1}\boldsymbol{\eta}}}\right\}(\mathbf{x}_i-\boldsymbol{\xi}),
\end{aligned}
$$

where $\mathbf{D}_p$ is the $p^2 \times p(p+1)/2$ duplication matrix such that $\mathbf{D}_p\text{vech}(\mathbf{A}) = \text{vec}(\mathbf{A})$ for any symmetric matrix $\mathbf{A}$ and $\text{vech}(\mathbf{A})$ is such that only the part on or below the diagonal (vector-half) of $\mathbf{A}$ is vectorized. Equating these expressions to zero gives the likelihood equations, which must be solved numerically to obtain the maximum likelihood estimates. However, at $\boldsymbol{\Theta}_0 = (\boldsymbol{\xi}, \boldsymbol{\Psi}, \mathbf{0}_p)$, for which $\boldsymbol{\eta} = \mathbf{0}_p$, these expressions are reduced to

$$\begin{aligned}
\mathbf{U}_{\boldsymbol{\xi}}(\boldsymbol{\Theta}_0) &= n\boldsymbol{\Psi}^{-1}(\bar{\mathbf{x}} - \boldsymbol{\xi}),\\
\mathbf{U}_{\text{vech}(\boldsymbol{\Psi})}(\boldsymbol{\Theta}_0) &= \frac{n}{2}\mathbf{D}_p^\top(\boldsymbol{\Psi}^{-1} \otimes \boldsymbol{\Psi}^{-1})\text{vec}\{\mathbf{S}_\mathbf{x} - \boldsymbol{\Psi} + (\bar{\mathbf{x}} - \boldsymbol{\xi})(\bar{\mathbf{x}} - \boldsymbol{\xi})^\top\},\\
\mathbf{U}_{\boldsymbol{\eta}}(\boldsymbol{\Theta}_0) &= n\zeta(0)\boldsymbol{\Psi}^{-1}(\bar{\mathbf{x}} - \boldsymbol{\xi}),
\end{aligned}$$

where $\zeta(0) = 2\phi(0) = \sqrt{2/\pi} = b$. As an immediate consequence of these expressions, we may notice the following:

[1] The value $(\bar{\mathbf{x}}, \mathbf{S}_\mathbf{x}, \mathbf{0}_p)$ of $(\boldsymbol{\xi}, \boldsymbol{\Psi}, \boldsymbol{\eta})$ is a solution of the likelihood equations;

[2] The components $\mathbf{U}_{\boldsymbol{\xi}}$ and $\mathbf{U}_{\boldsymbol{\eta}}$ of the score vector are linearly dependent at $\boldsymbol{\eta} = \mathbf{0}_p$; therefore, the associated information matrix is singular at $\boldsymbol{\eta} = \mathbf{0}_p$. In fact, since at $\boldsymbol{\eta} = \mathbf{0}_p$ we have $\bar{\mathbf{X}} - \boldsymbol{\xi} \sim \mathcal{N}_p(\mathbf{0}_p, \frac{1}{n}\boldsymbol{\Psi})$ and $n\mathbf{S}_\mathbf{X} \sim \mathcal{W}_p(n-1, \boldsymbol{\Psi})$, the Wishart matrix distribution, which are independent, and therefore $n\{\mathbf{S}_\mathbf{X} + (\bar{\mathbf{X}} - \boldsymbol{\xi})(\bar{\mathbf{X}} - \boldsymbol{\xi})^\top\} = \sum_{i=1}^n (\mathbf{X}_i - \boldsymbol{\xi})(\mathbf{X}_i - \boldsymbol{\xi})^\top \sim \mathcal{W}_p(n, \boldsymbol{\Psi})$, we can easily find that

$$\begin{aligned}
\text{var}_{\boldsymbol{\Theta}_0}\left\{\frac{1}{\sqrt{n}}\begin{pmatrix}\mathbf{U}_{\boldsymbol{\xi}}(\boldsymbol{\Theta}_0)\\ \mathbf{U}_{\text{vech}(\boldsymbol{\Psi})}(\boldsymbol{\Theta}_0)\\ \mathbf{U}_{\boldsymbol{\eta}}(\boldsymbol{\Theta}_0)\end{pmatrix}\right\} &= n\text{var}_{\boldsymbol{\Theta}_0}\begin{pmatrix}\boldsymbol{\Psi}^{-1}(\bar{\mathbf{X}} - \boldsymbol{\xi})\\ \frac{1}{2}\mathbf{D}_p^\top(\boldsymbol{\Psi}^{-1} \otimes \boldsymbol{\Psi}^{-1})\text{vec}\{\mathbf{S}_\mathbf{X} - \boldsymbol{\Psi} + (\bar{\mathbf{X}} - \boldsymbol{\xi})(\bar{\mathbf{X}} - \boldsymbol{\xi})^\top\}\\ \zeta(0)\boldsymbol{\Psi}^{-1}(\bar{\mathbf{X}} - \boldsymbol{\xi})\end{pmatrix}\\
&= \begin{pmatrix}\boldsymbol{\Psi}^{-1} & \mathbf{O} & \zeta(0)\boldsymbol{\Psi}^{-1}\\ \mathbf{O} & \frac{1}{2}\mathbf{D}_p^\top(\boldsymbol{\Psi}^{-1} \otimes \boldsymbol{\Psi}^{-1})\mathbf{D}_p & \mathbf{O}\\ \zeta(0)\boldsymbol{\Psi}^{-1} & \mathbf{O} & \zeta(0)^2\boldsymbol{\Psi}^{-1}\end{pmatrix},
\end{aligned}$$

since

$$\begin{aligned}
&\text{var}_{\boldsymbol{\Theta}_0}\{\boldsymbol{\Psi}^{-1}(\bar{\mathbf{X}} - \boldsymbol{\xi})\} = \boldsymbol{\Psi}^{-1}\text{var}_{\boldsymbol{\Theta}_0}(\bar{\mathbf{X}} - \boldsymbol{\xi})\boldsymbol{\Psi}^{-1} = \frac{1}{n}\boldsymbol{\Psi}^{-1},\\
&\text{var}_{\boldsymbol{\Theta}_0}[\mathbf{D}_p^\top(\boldsymbol{\Psi}^{-1} \otimes \boldsymbol{\Psi}^{-1})\text{vec}\{\mathbf{S}_\mathbf{X} - \boldsymbol{\Psi} + (\bar{\mathbf{X}} - \boldsymbol{\xi})(\bar{\mathbf{X}} - \boldsymbol{\xi})^\top\}]\\
&= \mathbf{D}_p^\top(\boldsymbol{\Psi}^{-1} \otimes \boldsymbol{\Psi}^{-1})\text{var}_{\boldsymbol{\Theta}_0}[\text{vec}\{\mathbf{S}_\mathbf{X} + (\bar{\mathbf{X}} - \boldsymbol{\xi})(\bar{\mathbf{X}} - \boldsymbol{\xi})^\top\}](\boldsymbol{\Psi}^{-1} \otimes \boldsymbol{\Psi}^{-1})\mathbf{D}_p\\
&= \frac{2}{n}\mathbf{D}_p^\top(\boldsymbol{\Psi}^{-1} \otimes \boldsymbol{\Psi}^{-1})\mathbf{D}_p,
\end{aligned}$$

and

$$\begin{aligned}
&\text{cov}_{\boldsymbol{\Theta}_0}[\boldsymbol{\Psi}^{-1}(\bar{\mathbf{X}} - \boldsymbol{\xi}), \mathbf{D}_p^\top(\boldsymbol{\Psi}^{-1} \otimes \boldsymbol{\Psi}^{-1})\text{vec}\{\mathbf{S}_\mathbf{X} - \boldsymbol{\Psi} + (\bar{\mathbf{X}} - \boldsymbol{\xi})(\bar{\mathbf{X}} - \boldsymbol{\xi})^\top\}]\\
&= \boldsymbol{\Psi}^{-1}\text{cov}[\bar{\mathbf{X}} - \boldsymbol{\xi}, \text{vec}\{\mathbf{S}_\mathbf{X} + (\bar{\mathbf{X}} - \boldsymbol{\xi})(\bar{\mathbf{X}} - \boldsymbol{\xi})^\top\}](\boldsymbol{\Psi}^{-1} \otimes \boldsymbol{\Psi}^{-1})\mathbf{D}_p\\
&= \boldsymbol{\Psi}^{-1}\text{E}[(\bar{\mathbf{X}} - \boldsymbol{\xi})\text{vec}\{\mathbf{S}_\mathbf{X} + (\bar{\mathbf{X}} - \boldsymbol{\xi})(\bar{\mathbf{X}} - \boldsymbol{\xi})^\top\}^\top](\boldsymbol{\Psi}^{-1} \otimes \boldsymbol{\Psi}^{-1})\mathbf{D}_p = \mathbf{O}.
\end{aligned}$$

The singularity of the information matrix of the skew-normal distribution in its ASN parameterization at $\boldsymbol{\alpha} = \mathbf{0}_p$ has been thoroughly investigated by Arellano-Valle and Azzalini (2008) and Hallin and Ley (2014); see Chiogna (2005) for the univariate case. Further discussion of this issue can be found in Azzalini and Genton (2008) for flexible skew-normal distributions, in Ley and Paindaveine (2010a) for generalized skew-elliptical distributions, and in Ley and Paindaveine (2010b) and Hallin and Ley (2012) for skew-symmetric distributions; see Section 5.11.1 for details about these distributions.

5.10.3 *The Fisher information matrix*

The full expected information matrix can be obtained from $-\mathrm{E}\{\mathrm{d}^2\ell(\boldsymbol{\Theta})\}$, where $\mathrm{d}^2\ell(\boldsymbol{\Theta})$ is the second differential of the log-likelihood function $\ell(\boldsymbol{\Theta})$. Thus, we can start by differentiating $\mathrm{d}_{\boldsymbol{\xi}}\ell(\boldsymbol{\Theta})$ with respect to $\boldsymbol{\xi}$, $\boldsymbol{\Psi}$, and $\boldsymbol{\eta}$, then $\mathrm{d}_{\boldsymbol{\Psi}}\ell(\boldsymbol{\Theta})$ with respect to $\boldsymbol{\Psi}$ and $\boldsymbol{\eta}$, and, finally, $\mathrm{d}_{\boldsymbol{\eta}}\ell(\boldsymbol{\eta})$ with respect to $\boldsymbol{\eta}$. From this way, we obtain the submatrices $\mathbf{I}_{\boldsymbol{\xi}\boldsymbol{\xi}}(\boldsymbol{\Theta})$, $\mathbf{I}_{\boldsymbol{\xi}\boldsymbol{\Psi}}(\boldsymbol{\Theta})$, $\mathbf{I}_{\boldsymbol{\xi}\boldsymbol{\eta}}(\boldsymbol{\Theta})$, $\mathbf{I}_{\boldsymbol{\Psi}\boldsymbol{\Psi}}(\boldsymbol{\Theta})$, $\mathbf{I}_{\boldsymbol{\Psi}\boldsymbol{\eta}}(\boldsymbol{\Theta})$, and $\mathbf{I}_{\boldsymbol{\eta}\boldsymbol{\eta}}(\boldsymbol{\Theta})$, respectively, of the Fisher information matrix $\mathbf{I}(\boldsymbol{\Theta})$.

Since $\mathbf{X}_1, \dots, \mathbf{X}_n$ are iid $\mathcal{SN}_p(\boldsymbol{\xi}, \boldsymbol{\Psi}, \boldsymbol{\eta})$, the expected information matrix becomes $\mathbf{I}(\boldsymbol{\Theta}) = n\mathbf{I}_1(\boldsymbol{\Theta})$, where $\mathbf{I}_1(\boldsymbol{\Theta})$ is the expected information matrix based on $\mathbf{X}_1$, i.e., the first observation. Let

$$a_1 = \mathrm{E}\left[\zeta\left\{\frac{\boldsymbol{\eta}^\top\boldsymbol{\Psi}^{-1}(\mathbf{X}_1-\boldsymbol{\xi})}{\sqrt{1+\boldsymbol{\eta}^\top\boldsymbol{\Psi}^{-1}\boldsymbol{\eta}}}\right\}\right], \quad a_2 = -\mathrm{E}\left[\zeta'\left\{\frac{\boldsymbol{\eta}^\top\boldsymbol{\Psi}^{-1}(\mathbf{X}_1-\boldsymbol{\xi})}{\sqrt{1+\boldsymbol{\eta}^\top\boldsymbol{\Psi}^{-1}\boldsymbol{\eta}}}\right\}\right],$$

and

$$\mathbf{a}_1 = -\mathrm{E}\left[\zeta'\left\{\frac{\boldsymbol{\eta}^\top\boldsymbol{\Psi}^{-1}(\mathbf{X}_1-\boldsymbol{\xi})}{\sqrt{1+\boldsymbol{\eta}^\top\boldsymbol{\Psi}^{-1}\boldsymbol{\eta}}}\right\}(\mathbf{X}_1-\boldsymbol{\xi})\right],$$

$$\mathbf{A}_2 = -\mathrm{E}\left[\zeta'\left\{\frac{\boldsymbol{\eta}^\top\boldsymbol{\Psi}^{-1}(\mathbf{X}_1-\boldsymbol{\xi})}{\sqrt{1+\boldsymbol{\eta}^\top\boldsymbol{\Psi}^{-1}\boldsymbol{\eta}}}\right\}(\mathbf{X}_1-\boldsymbol{\xi})(\mathbf{X}_1-\boldsymbol{\xi})^\top\right].$$

Also, we observe that $\mathrm{E}(\mathbf{X}_1-\boldsymbol{\xi}) = \sqrt{\frac{2}{\pi}}\boldsymbol{\eta}$ and

$$\mathrm{E}\{(\mathbf{X}_1-\boldsymbol{\xi})(\mathbf{X}_1-\boldsymbol{\xi})^\top\} = \boldsymbol{\Psi}+\boldsymbol{\eta}\boldsymbol{\eta}^\top, \quad \mathrm{E}\left[\zeta\left\{\frac{\boldsymbol{\eta}^\top\boldsymbol{\Psi}^{-1}(\mathbf{x}_i-\boldsymbol{\xi})}{\sqrt{1+\boldsymbol{\eta}^\top\boldsymbol{\Psi}^{-1}\boldsymbol{\eta}}}\right\}(\mathbf{X}_1-\boldsymbol{\xi})\right] = \mathbf{0}_p.$$

Thus, for the second differentials mentioned above, we have:

$$\mathrm{d}^2_{\boldsymbol{\xi}\boldsymbol{\xi}}\ell(\boldsymbol{\Theta}) = -n(\mathrm{d}\boldsymbol{\xi})^\top(\boldsymbol{\Psi}+\boldsymbol{\eta}\boldsymbol{\eta}^\top)^{-1}(\mathrm{d}\boldsymbol{\xi}) + \sum_{i=1}^n \zeta'\left\{\frac{\boldsymbol{\eta}^\top\boldsymbol{\Psi}^{-1}(\mathbf{x}_i-\boldsymbol{\xi})}{\sqrt{1+\boldsymbol{\eta}^\top\boldsymbol{\Psi}^{-1}\boldsymbol{\eta}}}\right\}\frac{(\mathrm{d}\boldsymbol{\xi})^\top\boldsymbol{\Psi}^{-1}\boldsymbol{\eta}\boldsymbol{\eta}^\top\boldsymbol{\Psi}^{-1}(\mathrm{d}\boldsymbol{\xi})}{1+\boldsymbol{\eta}^\top\boldsymbol{\Psi}^{-1}\boldsymbol{\eta}},$$

$$\begin{aligned}
\mathrm{d}^2_{\boldsymbol{\xi}\boldsymbol{\Psi}}\ell(\boldsymbol{\Theta}) &= -n(\mathrm{d}\boldsymbol{\xi})^\top(\boldsymbol{\Psi}+\boldsymbol{\eta}\boldsymbol{\eta}^\top)^{-1}(\mathrm{d}\boldsymbol{\Psi})(\boldsymbol{\Psi}+\boldsymbol{\eta}\boldsymbol{\eta}^\top)^{-1}(\bar{\mathbf{x}}-\boldsymbol{\xi}) - \sum_{i=1}^n \zeta'\left\{\frac{\boldsymbol{\eta}^\top\boldsymbol{\Psi}^{-1}(\mathbf{x}_i-\boldsymbol{\xi})}{\sqrt{1+\boldsymbol{\eta}^\top\boldsymbol{\Psi}^{-1}\boldsymbol{\eta}}}\right\}\\
&\times\left\{-\frac{\boldsymbol{\eta}^\top\boldsymbol{\Psi}^{-1}(\mathrm{d}\boldsymbol{\Psi})\boldsymbol{\Psi}^{-1}(\mathbf{x}_i-\boldsymbol{\xi})}{\sqrt{1+\boldsymbol{\eta}^\top\boldsymbol{\Psi}^{-1}\boldsymbol{\eta}}} + \frac{1}{2}\frac{\boldsymbol{\eta}^\top\boldsymbol{\Psi}^{-1}(\mathrm{d}\boldsymbol{\Psi})\boldsymbol{\Psi}^{-1}\boldsymbol{\eta}\boldsymbol{\eta}^\top\boldsymbol{\Psi}^{-1}(\mathbf{x}_i-\boldsymbol{\xi})}{(1+\boldsymbol{\eta}^\top\boldsymbol{\Psi}^{-1}\boldsymbol{\eta})^{3/2}}\right\}\frac{(\mathrm{d}\boldsymbol{\xi})^\top\boldsymbol{\Psi}^{-1}\boldsymbol{\eta}}{\sqrt{1+\boldsymbol{\eta}^\top\boldsymbol{\Psi}^{-1}\boldsymbol{\eta}}}\\
&-\sum_{i=1}^n \zeta\left\{\frac{\boldsymbol{\eta}^\top\boldsymbol{\Psi}^{-1}(\mathbf{x}_i-\boldsymbol{\xi})}{\sqrt{1+\boldsymbol{\eta}^\top\boldsymbol{\Psi}^{-1}\boldsymbol{\eta}}}\right\}\left\{-\frac{(\mathrm{d}\boldsymbol{\xi})^\top\boldsymbol{\Psi}^{-1}(\mathrm{d}\boldsymbol{\Psi})\boldsymbol{\Psi}^{-1}\boldsymbol{\eta}}{\sqrt{1+\boldsymbol{\eta}^\top\boldsymbol{\Psi}^{-1}\boldsymbol{\eta}}} + \frac{1}{2}\frac{(\mathrm{d}\boldsymbol{\xi})^\top\boldsymbol{\Psi}^{-1}\boldsymbol{\eta}\boldsymbol{\eta}^\top\boldsymbol{\Psi}^{-1}(\mathrm{d}\boldsymbol{\Psi})\boldsymbol{\Psi}^{-1}\boldsymbol{\eta}}{(1+\boldsymbol{\eta}^\top\boldsymbol{\Psi}^{-1}\boldsymbol{\eta})^{3/2}}\right\},
\end{aligned}$$

$$\begin{aligned}
\mathrm{d}^2_{\boldsymbol{\xi}\boldsymbol{\eta}}\ell(\boldsymbol{\Theta}) &= -n(\mathrm{d}\boldsymbol{\xi})^\top(\boldsymbol{\Psi}+\boldsymbol{\eta}\boldsymbol{\eta}^\top)^{-1}\{(\mathrm{d}\boldsymbol{\eta})\boldsymbol{\eta}^\top+\boldsymbol{\eta}(\mathrm{d}\boldsymbol{\eta})^\top\}(\boldsymbol{\Psi}+\boldsymbol{\eta}\boldsymbol{\eta}^\top)^{-1}(\bar{\mathbf{x}}-\boldsymbol{\xi})\\
&-\sum_{i=1}^n \zeta'\left\{\frac{\boldsymbol{\eta}^\top\boldsymbol{\Psi}^{-1}(\mathbf{x}_i-\boldsymbol{\xi})}{\sqrt{1+\boldsymbol{\eta}^\top\boldsymbol{\Psi}^{-1}\boldsymbol{\eta}}}\right\}\left\{\frac{(\mathrm{d}\boldsymbol{\eta})^\top\boldsymbol{\Psi}^{-1}(\mathbf{x}_i-\boldsymbol{\xi})}{\sqrt{1+\boldsymbol{\eta}^\top\boldsymbol{\Psi}^{-1}\boldsymbol{\eta}}} - \frac{(\mathrm{d}\boldsymbol{\eta})^\top\boldsymbol{\Psi}^{-1}\boldsymbol{\eta}\boldsymbol{\eta}^\top\boldsymbol{\Psi}^{-1}(\mathbf{x}_i-\boldsymbol{\xi})}{(1+\boldsymbol{\eta}^\top\boldsymbol{\Psi}^{-1}\boldsymbol{\eta})^{3/2}}\right\}\\
&\times\frac{(\mathrm{d}\boldsymbol{\xi})^\top\boldsymbol{\Psi}^{-1}\boldsymbol{\eta}}{\sqrt{1+\boldsymbol{\eta}^\top\boldsymbol{\Psi}^{-1}\boldsymbol{\eta}}} - \sum_{i=1}^n \zeta\left\{\frac{\boldsymbol{\eta}^\top\boldsymbol{\Psi}^{-1}(\mathbf{x}_i-\boldsymbol{\xi})}{\sqrt{1+\boldsymbol{\eta}^\top\boldsymbol{\Psi}^{-1}\boldsymbol{\eta}}}\right\}\left\{\frac{(\mathrm{d}\boldsymbol{\xi})^\top\boldsymbol{\Psi}^{-1}(\mathrm{d}\boldsymbol{\eta})}{\sqrt{1+\boldsymbol{\eta}^\top\boldsymbol{\Psi}^{-1}\boldsymbol{\eta}}} - \frac{(\mathrm{d}\boldsymbol{\xi})^\top\boldsymbol{\Psi}^{-1}\boldsymbol{\eta}\boldsymbol{\eta}^\top\boldsymbol{\Psi}^{-1}(\mathrm{d}\boldsymbol{\eta})}{(1+\boldsymbol{\eta}^\top\boldsymbol{\Psi}^{-1}\boldsymbol{\eta})^{3/2}}\right\},
\end{aligned}$$

$$\begin{aligned}
\mathrm{d}^2_{\boldsymbol{\Psi}\boldsymbol{\Psi}}\ell(\boldsymbol{\Theta}) &= \frac{n}{2}\mathrm{tr}\{(\boldsymbol{\Psi}+\boldsymbol{\eta}\boldsymbol{\eta}^\top)^{-1}(\mathrm{d}\boldsymbol{\Psi})(\boldsymbol{\Psi}+\boldsymbol{\eta}\boldsymbol{\eta}^\top)^{-1}(\mathrm{d}\boldsymbol{\Psi})\} - \sum_{i=1}^n(\mathbf{x}_i-\boldsymbol{\xi})^\top(\boldsymbol{\Psi}+\boldsymbol{\eta}\boldsymbol{\eta}^\top)^{-1}(\mathrm{d}\boldsymbol{\Psi})\\
&\times(\boldsymbol{\Psi}+\boldsymbol{\eta}\boldsymbol{\eta}^\top)^{-1}(\mathrm{d}\boldsymbol{\Psi})(\boldsymbol{\Psi}+\boldsymbol{\eta}\boldsymbol{\eta}^\top)^{-1}(\mathbf{x}_i-\boldsymbol{\xi}) + \sum_{i=1}^n \zeta'\left\{\frac{\boldsymbol{\eta}^\top\boldsymbol{\Psi}^{-1}(\mathbf{x}_i-\boldsymbol{\xi})}{\sqrt{1+\boldsymbol{\eta}^\top\boldsymbol{\Psi}^{-1}\boldsymbol{\eta}}}\right\}\\
&\times\left\{-\frac{\boldsymbol{\eta}^\top\boldsymbol{\Psi}^{-1}(\mathrm{d}\boldsymbol{\Psi})\boldsymbol{\Psi}^{-1}(\mathbf{x}_i-\boldsymbol{\xi})}{(1+\boldsymbol{\eta}^\top\boldsymbol{\Psi}^{-1}\boldsymbol{\eta})^{1/2}} + \frac{1}{2}\frac{\boldsymbol{\eta}^\top\boldsymbol{\Psi}^{-1}(\mathrm{d}\boldsymbol{\Psi})\boldsymbol{\Psi}^{-1}\boldsymbol{\eta}\boldsymbol{\eta}^\top\boldsymbol{\Psi}^{-1}(\mathbf{x}_i-\boldsymbol{\xi})}{(1+\boldsymbol{\eta}^\top\boldsymbol{\Psi}^{-1}\boldsymbol{\eta})^{3/2}}\right\}^2\\
&+\sum_{i=1}^n \zeta\left\{\frac{\boldsymbol{\eta}^\top\boldsymbol{\Psi}^{-1}(\mathbf{x}_i-\boldsymbol{\xi})}{\sqrt{1+\boldsymbol{\eta}^\top\boldsymbol{\Psi}^{-1}\boldsymbol{\eta}}}\right\}\left\{2\frac{\boldsymbol{\eta}^\top\boldsymbol{\Psi}^{-1}(\mathrm{d}\boldsymbol{\Psi})\boldsymbol{\Psi}^{-1}(\mathrm{d}\boldsymbol{\Psi})\boldsymbol{\Psi}^{-1}(\mathbf{x}_i-\boldsymbol{\xi})}{(1+\boldsymbol{\eta}^\top\boldsymbol{\Psi}^{-1}\boldsymbol{\eta})^{1/2}}\right.\\
&-\frac{1}{2}\frac{\boldsymbol{\eta}^\top\boldsymbol{\Psi}^{-1}(\mathrm{d}\boldsymbol{\Psi})\boldsymbol{\Psi}^{-1}\boldsymbol{\eta}\boldsymbol{\eta}^\top\boldsymbol{\Psi}^{-1}(\mathrm{d}\boldsymbol{\Psi})\boldsymbol{\Psi}^{-1}(\mathbf{x}_i-\boldsymbol{\xi})}{(1+\boldsymbol{\eta}^\top\boldsymbol{\Psi}^{-1}\boldsymbol{\eta})^{3/2}}\\
&-\frac{\boldsymbol{\eta}^\top\boldsymbol{\Psi}^{-1}(\mathrm{d}\boldsymbol{\Psi})\boldsymbol{\Psi}^{-1}(\mathrm{d}\boldsymbol{\Psi})\boldsymbol{\Psi}^{-1}\boldsymbol{\eta}\boldsymbol{\eta}^\top\boldsymbol{\Psi}^{-1}(\mathbf{x}_i-\boldsymbol{\xi})}{(1+\boldsymbol{\eta}^\top\boldsymbol{\Psi}^{-1}\boldsymbol{\eta})^{3/2}}\\
&-\frac{1}{2}\frac{\boldsymbol{\eta}^\top\boldsymbol{\Psi}^{-1}(\mathrm{d}\boldsymbol{\Psi})\boldsymbol{\Psi}^{-1}\boldsymbol{\eta}\boldsymbol{\eta}^\top\boldsymbol{\Psi}^{-1}(\mathrm{d}\boldsymbol{\Psi})\boldsymbol{\Psi}^{-1}(\mathbf{x}_i-\boldsymbol{\xi})}{(1+\boldsymbol{\eta}^\top\boldsymbol{\Psi}^{-1}\boldsymbol{\eta})^{3/2}}\\
&\left.+\frac{3}{4}\frac{\boldsymbol{\eta}^\top\boldsymbol{\Psi}^{-1}(\mathrm{d}\boldsymbol{\Psi})\boldsymbol{\Psi}^{-1}\boldsymbol{\eta}\boldsymbol{\eta}^\top\boldsymbol{\Psi}^{-1}(\mathrm{d}\boldsymbol{\Psi})\boldsymbol{\Psi}^{-1}\boldsymbol{\eta}\boldsymbol{\eta}^\top\boldsymbol{\Psi}^{-1}(\mathbf{x}_i-\boldsymbol{\xi})}{(1+\boldsymbol{\eta}^\top\boldsymbol{\Psi}^{-1}\boldsymbol{\eta})^{5/2}}\right\},
\end{aligned}$$

$$
\begin{aligned}
\mathrm{d}^2_{\boldsymbol{\Psi}\boldsymbol{\eta}}\ell(\boldsymbol{\Theta}) =& \frac{n}{2}\mathrm{tr}\{(\boldsymbol{\Psi}+\boldsymbol{\eta}\boldsymbol{\eta}^\top)^{-1}\{(\mathrm{d}\boldsymbol{\eta})\boldsymbol{\eta}^\top+\boldsymbol{\eta}(\mathrm{d}\boldsymbol{\eta})^\top\}(\boldsymbol{\Psi}+\boldsymbol{\eta}\boldsymbol{\eta}^\top)^{-1}(\mathrm{d}\boldsymbol{\Psi})\} \\
&-\sum_{i=1}^n(\mathbf{x}_i-\boldsymbol{\xi})^\top(\boldsymbol{\Psi}+\boldsymbol{\eta}\boldsymbol{\eta}^\top)^{-1}\{(\mathrm{d}\boldsymbol{\eta})\boldsymbol{\eta}^\top+\boldsymbol{\eta}(\mathrm{d}\boldsymbol{\eta})^\top\}(\boldsymbol{\Psi}+\boldsymbol{\eta}\boldsymbol{\eta}^\top)^{-1}(\mathrm{d}\boldsymbol{\Psi})(\boldsymbol{\Psi}+\boldsymbol{\eta}\boldsymbol{\eta}^\top)^{-1}(\mathbf{x}_i-\boldsymbol{\xi}) \\
&+\sum_{i=1}^n\zeta'\left\{\frac{\boldsymbol{\eta}^\top\boldsymbol{\Psi}^{-1}(\mathbf{x}_i-\boldsymbol{\xi})}{\sqrt{1+\boldsymbol{\eta}^\top\boldsymbol{\Psi}^{-1}\boldsymbol{\eta}}}\right\}\left\{\frac{(\mathrm{d}\boldsymbol{\eta})^\top\boldsymbol{\Psi}^{-1}(\mathbf{x}_i-\boldsymbol{\xi})}{(1+\boldsymbol{\eta}^\top\boldsymbol{\Psi}^{-1}\boldsymbol{\eta})^{1/2}}-\frac{(\mathrm{d}\boldsymbol{\eta})^\top\boldsymbol{\Psi}^{-1}\boldsymbol{\eta}\boldsymbol{\eta}^\top\boldsymbol{\Psi}^{-1}(\mathbf{x}_i-\boldsymbol{\xi})}{(1+\boldsymbol{\eta}^\top\boldsymbol{\Psi}^{-1}\boldsymbol{\eta})^{3/2}}\right\} \\
&\qquad\times\left\{-\frac{\boldsymbol{\eta}^\top\boldsymbol{\Psi}^{-1}(\mathrm{d}\boldsymbol{\Psi})\boldsymbol{\Psi}^{-1}(\mathbf{x}_i-\boldsymbol{\xi})}{(1+\boldsymbol{\eta}^\top\boldsymbol{\Psi}^{-1}\boldsymbol{\eta})^{1/2}}+\frac{1}{2}\frac{\boldsymbol{\eta}^\top\boldsymbol{\Psi}^{-1}(\mathrm{d}\boldsymbol{\Psi})\boldsymbol{\Psi}^{-1}\boldsymbol{\eta}\boldsymbol{\eta}^\top\boldsymbol{\Psi}^{-1}(\mathbf{x}_i-\boldsymbol{\xi})}{(1+\boldsymbol{\eta}^\top\boldsymbol{\Psi}^{-1}\boldsymbol{\eta})^{3/2}}\right\} \\
&\qquad\sum_{i=1}^n\zeta\left\{\frac{\boldsymbol{\eta}^\top\boldsymbol{\Psi}^{-1}(\mathbf{x}_i-\boldsymbol{\xi})}{\sqrt{1+\boldsymbol{\eta}^\top\boldsymbol{\Psi}^{-1}\boldsymbol{\eta}}}\right\}\left\{-\frac{(\mathrm{d}\boldsymbol{\eta})^\top\boldsymbol{\Psi}^{-1}(\mathrm{d}\boldsymbol{\Psi})\boldsymbol{\Psi}^{-1}(\mathbf{x}_i-\boldsymbol{\xi})}{(1+\boldsymbol{\eta}^\top\boldsymbol{\Psi}^{-1}\boldsymbol{\eta})^{1/2}}\right. \\
&\qquad+\frac{(\mathrm{d}\boldsymbol{\eta})^\top\boldsymbol{\Psi}^{-1}\boldsymbol{\eta}\boldsymbol{\eta}^\top\boldsymbol{\Psi}^{-1}(\mathrm{d}\boldsymbol{\Psi})\boldsymbol{\Psi}^{-1}(\mathbf{x}_i-\boldsymbol{\xi})}{(1+\boldsymbol{\eta}^\top\boldsymbol{\Psi}^{-1}\boldsymbol{\eta})^{3/2}} \\
&\qquad+\frac{1}{2}\frac{2(\mathrm{d}\boldsymbol{\eta})^\top\boldsymbol{\Psi}^{-1}(\mathrm{d}\boldsymbol{\Psi})\boldsymbol{\Psi}^{-1}\boldsymbol{\eta}\boldsymbol{\eta}^\top\boldsymbol{\Psi}^{-1}(\mathbf{x}_i-\boldsymbol{\xi})+\boldsymbol{\eta}^\top\boldsymbol{\Psi}^{-1}(\mathrm{d}\boldsymbol{\Psi})\boldsymbol{\Psi}^{-1}\boldsymbol{\eta}(\mathrm{d}\boldsymbol{\eta})^\top\boldsymbol{\Psi}^{-1}(\mathbf{x}_i-\boldsymbol{\xi})}{(1+\boldsymbol{\eta}^\top\boldsymbol{\Psi}^{-1}\boldsymbol{\eta})^{3/2}} \\
&\qquad\left.-\frac{3}{4}\frac{(\mathrm{d}\boldsymbol{\eta})^\top\boldsymbol{\Psi}^{-1}\boldsymbol{\eta}\boldsymbol{\eta}^\top\boldsymbol{\Psi}^{-1}(\mathrm{d}\boldsymbol{\Psi})\boldsymbol{\Psi}^{-1}\boldsymbol{\eta}\boldsymbol{\eta}^\top\boldsymbol{\Psi}^{-1}(\mathbf{x}_i-\boldsymbol{\xi})}{(1+\boldsymbol{\eta}^\top\boldsymbol{\Psi}^{-1}\boldsymbol{\eta})^{5/2}}\right\},
\end{aligned}
$$

and

$$
\begin{aligned}
&\mathrm{d}^2_{\boldsymbol{\eta}\boldsymbol{\eta}}\ell(\boldsymbol{\Theta}) = \\
&n(\mathrm{d}\boldsymbol{\eta})^\top(\boldsymbol{\Psi}+\boldsymbol{\eta}\boldsymbol{\eta}^\top)^{-1}\{(\mathrm{d}\boldsymbol{\eta})\boldsymbol{\eta}^\top+\boldsymbol{\eta}(\mathrm{d}\boldsymbol{\eta})^\top\}(\boldsymbol{\Psi}+\boldsymbol{\eta}\boldsymbol{\eta}^\top)^{-1}\boldsymbol{\eta}-n(\mathrm{d}\boldsymbol{\eta})^\top(\boldsymbol{\Psi}+\boldsymbol{\eta}\boldsymbol{\eta}^\top)^{-1}(\mathrm{d}\boldsymbol{\eta}) \\
&-\sum_{i=1}^n(\mathrm{d}\boldsymbol{\eta})^\top(\boldsymbol{\Psi}+\boldsymbol{\eta}\boldsymbol{\eta}^\top)^{-1}\{(\mathrm{d}\boldsymbol{\eta})\boldsymbol{\eta}^\top+\boldsymbol{\eta}(\mathrm{d}\boldsymbol{\eta})^\top\}(\boldsymbol{\Psi}+\boldsymbol{\eta}\boldsymbol{\eta}^\top)^{-1}(\mathbf{x}_i-\boldsymbol{\xi})(\mathbf{x}_i-\boldsymbol{\xi})^\top(\boldsymbol{\Psi}+\boldsymbol{\eta}\boldsymbol{\eta}^\top)^{-1}\boldsymbol{\eta} \\
&-\sum_{i=1}^n(\mathrm{d}\boldsymbol{\eta})^\top(\boldsymbol{\Psi}+\boldsymbol{\eta}\boldsymbol{\eta}^\top)^{-1}(\mathbf{x}_i-\boldsymbol{\xi})(\mathbf{x}_i-\boldsymbol{\xi})^\top(\boldsymbol{\Psi}+\boldsymbol{\eta}\boldsymbol{\eta}^\top)^{-1}\{(\mathrm{d}\boldsymbol{\eta})\boldsymbol{\eta}^\top+\boldsymbol{\eta}(\mathrm{d}\boldsymbol{\eta})^\top\}(\boldsymbol{\Psi}+\boldsymbol{\eta}\boldsymbol{\eta}^\top)^{-1}\boldsymbol{\eta} \\
&+\sum_{i=1}^n(\mathrm{d}\boldsymbol{\eta})^\top(\boldsymbol{\Psi}+\boldsymbol{\eta}\boldsymbol{\eta}^\top)^{-1}(\mathbf{x}_i-\boldsymbol{\xi})(\mathbf{x}_i-\boldsymbol{\xi})^\top(\boldsymbol{\Psi}+\boldsymbol{\eta}\boldsymbol{\eta}^\top)^{-1}(\mathrm{d}\boldsymbol{\eta}) \\
&+\sum_{i=1}^n\zeta'\left\{\frac{\boldsymbol{\eta}^\top\boldsymbol{\Psi}^{-1}(\mathbf{x}_i-\boldsymbol{\xi})}{\sqrt{1+\boldsymbol{\eta}^\top\boldsymbol{\Psi}^{-1}\boldsymbol{\eta}}}\right\}\left\{\frac{(\mathrm{d}\boldsymbol{\eta})^\top\boldsymbol{\Psi}^{-1}(\mathbf{x}_i-\boldsymbol{\xi})}{(1+\boldsymbol{\eta}^\top\boldsymbol{\Psi}^{-1}\boldsymbol{\eta})^{1/2}}-\frac{(\mathrm{d}\boldsymbol{\eta})^\top\boldsymbol{\Psi}^{-1}\boldsymbol{\eta}\boldsymbol{\eta}^\top\boldsymbol{\Psi}^{-1}(\mathbf{x}_i-\boldsymbol{\xi})}{(1+\boldsymbol{\eta}^\top\boldsymbol{\Psi}^{-1}\boldsymbol{\eta})^{3/2}}\right\}^2 \\
&+\sum_{i=1}^n\zeta\left\{\frac{\boldsymbol{\eta}^\top\boldsymbol{\Psi}^{-1}(\mathbf{x}_i-\boldsymbol{\xi})}{\sqrt{1+\boldsymbol{\eta}^\top\boldsymbol{\Psi}^{-1}\boldsymbol{\eta}}}\right\}\left\{-\frac{(\mathrm{d}\boldsymbol{\eta})^\top\boldsymbol{\Psi}^{-1}\boldsymbol{\eta}(\mathrm{d}\boldsymbol{\eta})^\top\boldsymbol{\Psi}^{-1}(\mathbf{x}_i-\boldsymbol{\xi})}{(1+\boldsymbol{\eta}^\top\boldsymbol{\Psi}^{-1}\boldsymbol{\eta})^{3/2}}\right. \\
&\qquad-\frac{(\mathrm{d}\boldsymbol{\eta})^\top\boldsymbol{\Psi}^{-1}(\mathrm{d}\boldsymbol{\eta})\boldsymbol{\eta}^\top\boldsymbol{\Psi}^{-1}(\mathbf{x}_i-\boldsymbol{\xi})+(\mathrm{d}\boldsymbol{\eta})^\top\boldsymbol{\Psi}^{-1}\boldsymbol{\eta}(\mathrm{d}\boldsymbol{\eta})^\top\boldsymbol{\Psi}^{-1}(\mathbf{x}_i-\boldsymbol{\xi})}{(1+\boldsymbol{\eta}^\top\boldsymbol{\Psi}^{-1}\boldsymbol{\eta})^{3/2}} \\
&\qquad\left.+3\frac{(\mathrm{d}\boldsymbol{\eta})^\top\boldsymbol{\Psi}^{-1}\boldsymbol{\eta}(\mathrm{d}\boldsymbol{\eta})^\top\boldsymbol{\Psi}^{-1}\boldsymbol{\eta}\boldsymbol{\eta}^\top\boldsymbol{\Psi}^{-1}(\mathbf{x}_i-\boldsymbol{\xi})}{(1+\boldsymbol{\eta}^\top\boldsymbol{\Psi}^{-1}\boldsymbol{\eta})^{5/2}}\right\}.
\end{aligned}
$$

Therefore, by applying the expectation operator to these expressions, we obtain, after some algebraic operations, that

$$
\mathbf{I}(\boldsymbol{\Theta}) = n\begin{pmatrix}\mathbf{I}_{\boldsymbol{\xi}\boldsymbol{\xi}} & \mathbf{I}_{\boldsymbol{\xi}\mathrm{v}(\boldsymbol{\Psi})} & \mathbf{I}_{\boldsymbol{\xi}\boldsymbol{\eta}} \\ \mathbf{I}^\top_{\boldsymbol{\xi}\mathrm{v}(\boldsymbol{\Psi})} & \mathbf{I}_{\mathrm{v}(\boldsymbol{\Psi})\mathrm{v}(\boldsymbol{\Psi})} & \mathbf{I}_{\mathrm{v}(\boldsymbol{\Psi})\boldsymbol{\eta}} \\ \mathbf{I}^\top_{\boldsymbol{\xi}\boldsymbol{\eta}} & \mathbf{I}^\top_{\mathrm{v}(\boldsymbol{\Psi})\boldsymbol{\eta}} & \mathbf{I}_{\boldsymbol{\eta}\boldsymbol{\eta}}\end{pmatrix},
$$

where

$$\begin{aligned}
\mathbf{I}_{\boldsymbol{\xi}\boldsymbol{\xi}} =& (\boldsymbol{\Psi}+\boldsymbol{\eta}\boldsymbol{\eta}^\top)^{-1} + \frac{a_2}{1+\boldsymbol{\eta}^\top\boldsymbol{\Psi}^{-1}\boldsymbol{\eta}}\boldsymbol{\Psi}^{-1}\boldsymbol{\eta}\boldsymbol{\eta}^\top\boldsymbol{\Psi}^{-1},\\
\mathbf{I}_{\boldsymbol{\xi}\mathrm{v}(\boldsymbol{\Psi})} =& \Bigg[\sqrt{\frac{2}{\pi}}\{(\boldsymbol{\Psi}+\boldsymbol{\eta}\boldsymbol{\eta}^\top)^{-1}\otimes\boldsymbol{\eta}^\top(\boldsymbol{\Psi}+\boldsymbol{\eta}\boldsymbol{\eta}^\top)^{-1}\} + \frac{1}{1+\boldsymbol{\eta}^\top\boldsymbol{\Psi}^{-1}\boldsymbol{\eta}}\\
&\times\left\{(\boldsymbol{\Psi}^{-1}\boldsymbol{\eta}\mathbf{a}_1^\top\boldsymbol{\Psi}^{-1}\otimes\boldsymbol{\eta}^\top\boldsymbol{\Psi}^{-1}) - \frac{1}{2}\frac{\boldsymbol{\eta}^\top\boldsymbol{\Psi}^{-1}\mathbf{a}_1}{1+\boldsymbol{\eta}^\top\boldsymbol{\Psi}^{-1}\boldsymbol{\eta}}(\boldsymbol{\Psi}^{-1}\boldsymbol{\eta}\boldsymbol{\eta}^\top\boldsymbol{\Psi}^{-1}\otimes\boldsymbol{\eta}^\top\boldsymbol{\Psi}^{-1})\right\}\\
&-\frac{a_1}{\sqrt{1+\boldsymbol{\eta}^\top\boldsymbol{\Psi}^{-1}\boldsymbol{\eta}}}\left\{(\boldsymbol{\Psi}^{-1}\otimes\boldsymbol{\eta}^\top\boldsymbol{\Psi}^{-1}) - \frac{1}{2(1+\boldsymbol{\eta}^\top\boldsymbol{\Psi}^{-1}\boldsymbol{\eta})}(\boldsymbol{\Psi}^{-1}\boldsymbol{\eta}\boldsymbol{\eta}^\top\boldsymbol{\Psi}^{-1}\otimes\boldsymbol{\eta}^\top\boldsymbol{\Psi}^{-1})\right\}\Bigg]\mathbf{D}_p,\\
\mathbf{I}_{\boldsymbol{\xi}\boldsymbol{\eta}} =& 2\sqrt{\frac{2}{\pi}}(\boldsymbol{\Psi}+\boldsymbol{\eta}\boldsymbol{\eta}^\top)^{-1}\boldsymbol{\eta}\boldsymbol{\eta}^\top(\boldsymbol{\Psi}+\boldsymbol{\eta}\boldsymbol{\eta}^\top)^{-1} + \frac{a_1}{\sqrt{1+\boldsymbol{\eta}^\top\boldsymbol{\Psi}^{-1}\boldsymbol{\eta}}}(\boldsymbol{\Psi}+\boldsymbol{\eta}\boldsymbol{\eta}^\top)^{-1}\\
&-\frac{1}{1+\boldsymbol{\eta}^\top\boldsymbol{\Psi}^{-1}\boldsymbol{\eta}}\left(\boldsymbol{\Psi}^{-1}\boldsymbol{\eta}\mathbf{a}_1^\top\boldsymbol{\Psi}^{-1} - \frac{\mathbf{a}_1^\top\boldsymbol{\Psi}^{-1}\boldsymbol{\eta}}{1+\boldsymbol{\eta}^\top\boldsymbol{\Psi}^{-1}\boldsymbol{\eta}}\boldsymbol{\Psi}^{-1}\boldsymbol{\eta}\boldsymbol{\eta}^\top\boldsymbol{\Psi}^{-1}\right),\\
\mathbf{I}_{\mathrm{v}(\boldsymbol{\Psi})\mathrm{v}(\boldsymbol{\Psi})} =& \mathbf{D}_p^\top\Bigg\{\frac{1}{2}\{(\boldsymbol{\Psi}+\boldsymbol{\eta}\boldsymbol{\eta}^\top)^{-1}\otimes(\boldsymbol{\Psi}+\boldsymbol{\eta}\boldsymbol{\eta}^\top)^{-1}\} + \frac{1}{1+\boldsymbol{\eta}^\top\boldsymbol{\Psi}^{-1}\boldsymbol{\eta}}(\boldsymbol{\Psi}^{-1}\mathbf{A}_2\boldsymbol{\Psi}^{-1}\\
&\otimes\boldsymbol{\Psi}^{-1}\boldsymbol{\eta}\boldsymbol{\eta}^\top\boldsymbol{\Psi}^{-1}) - \frac{1}{(1+\boldsymbol{\eta}^\top\boldsymbol{\Psi}^{-1}\boldsymbol{\eta})^2}(\boldsymbol{\Psi}^{-1}\mathbf{A}_2\boldsymbol{\Psi}^{-1}\boldsymbol{\eta}\boldsymbol{\eta}^\top\boldsymbol{\Psi}^{-1}\otimes\boldsymbol{\Psi}^{-1}\boldsymbol{\eta}\boldsymbol{\eta}^\top\boldsymbol{\Psi}^{-1})\\
&+\frac{1}{4(1+\boldsymbol{\eta}^\top\boldsymbol{\Psi}^{-1}\boldsymbol{\eta})^3}(\boldsymbol{\Psi}^{-1}\boldsymbol{\eta}\boldsymbol{\eta}^\top\boldsymbol{\Psi}^{-1}\mathbf{A}_2\boldsymbol{\Psi}^{-1}\boldsymbol{\eta}\boldsymbol{\eta}^\top\boldsymbol{\Psi}^{-1}\otimes\boldsymbol{\Psi}^{-1}\boldsymbol{\eta}\boldsymbol{\eta}^\top\boldsymbol{\Psi}^{-1})\Bigg\}\mathbf{D}_p,\\
\mathbf{I}_{\mathrm{v}(\boldsymbol{\Psi})\boldsymbol{\eta}} =& \mathbf{D}_p^\top\Bigg[\{(\boldsymbol{\Psi}+\boldsymbol{\eta}\boldsymbol{\eta}^\top)^{-1}\boldsymbol{\eta}\otimes(\boldsymbol{\Psi}+\boldsymbol{\eta}\boldsymbol{\eta}^\top)^{-1}\} - \frac{1}{1+\boldsymbol{\eta}^\top\boldsymbol{\Psi}^{-1}\boldsymbol{\eta}}\{\boldsymbol{\Psi}^{-1}\mathbf{A}_2(\boldsymbol{\Psi}+\boldsymbol{\eta}\boldsymbol{\eta}^\top)^{-1}\\
&\otimes\boldsymbol{\Psi}^{-1}\boldsymbol{\eta}\} + \frac{1}{2(1+\boldsymbol{\eta}^\top\boldsymbol{\Psi}^{-1}\boldsymbol{\eta})}(\boldsymbol{\Psi}^{-1}\boldsymbol{\eta}\boldsymbol{\eta}^\top\boldsymbol{\Psi}^{-1}\mathbf{A}_2\boldsymbol{\Psi}^{-1}\otimes\boldsymbol{\Psi}^{-1}\boldsymbol{\eta})\\
&-\frac{1}{4(1+\boldsymbol{\eta}^\top\boldsymbol{\Psi}^{-1}\boldsymbol{\eta})^2}(\boldsymbol{\Psi}^{-1}\boldsymbol{\eta}\boldsymbol{\eta}^\top\boldsymbol{\Psi}^{-1}\mathbf{A}_2\boldsymbol{\Psi}^{-1}\boldsymbol{\eta}\boldsymbol{\eta}^\top\boldsymbol{\Psi}^{-1}\otimes\boldsymbol{\Psi}^{-1}\boldsymbol{\eta})\Bigg],
\end{aligned}$$

and

$$\begin{aligned}
\mathbf{I}_{\boldsymbol{\eta}\boldsymbol{\eta}} =& \frac{\boldsymbol{\eta}^\top\boldsymbol{\Psi}^{-1}\boldsymbol{\eta}}{1+\boldsymbol{\eta}^\top\boldsymbol{\Psi}^{-1}\boldsymbol{\eta}}(\boldsymbol{\Psi}+\boldsymbol{\eta}\boldsymbol{\eta}^\top)^{-1} + \frac{\boldsymbol{\Psi}^{-1}\boldsymbol{\eta}\boldsymbol{\eta}^\top\boldsymbol{\Psi}^{-1}}{(1+\boldsymbol{\eta}^\top\boldsymbol{\Psi}^{-1}\boldsymbol{\eta})^2} + \frac{\boldsymbol{\Psi}^{-1}\mathbf{A}_2\boldsymbol{\Psi}^{-1}}{1+\boldsymbol{\eta}^\top\boldsymbol{\Psi}^{-1}\boldsymbol{\eta}}\\
&-\frac{2}{(1+\boldsymbol{\eta}^\top\boldsymbol{\Psi}^{-1}\boldsymbol{\eta})^2}\boldsymbol{\Psi}^{-1}\boldsymbol{\eta}\boldsymbol{\eta}^\top\boldsymbol{\Psi}^{-1}\mathbf{A}_2\boldsymbol{\Psi}^{-1} + \frac{\boldsymbol{\eta}^\top\boldsymbol{\Psi}^{-1}\mathbf{A}_2\boldsymbol{\Psi}^{-1}\boldsymbol{\eta}}{(1+\boldsymbol{\eta}^\top\boldsymbol{\Psi}^{-1}\boldsymbol{\eta})^3}\boldsymbol{\Psi}^{-1}\boldsymbol{\eta}\boldsymbol{\eta}^\top\boldsymbol{\Psi}^{-1}.
\end{aligned}$$

To complete the computation of these expected matrices, we must compute the quantities a_1, a_2, $\mathbf{a}_1$, and $\mathbf{A}_2$ defined above. For this, first observe that (see Exercises 5.11 and 5.12)

$$\begin{aligned}
a_1 = \mathrm{E}\left[\zeta\left\{\frac{\boldsymbol{\eta}^\top\boldsymbol{\Psi}^{-1}(\mathbf{X}_1-\boldsymbol{\xi})}{\sqrt{1+\boldsymbol{\eta}^\top\boldsymbol{\Psi}^{-1}\boldsymbol{\eta}}}\right\}\right] =& \frac{\sqrt{2/\pi}}{\sqrt{1+\boldsymbol{\eta}^\top\boldsymbol{\Psi}^{-1}\boldsymbol{\eta}}},\\
\mathrm{E}\left[\zeta\left\{\frac{\boldsymbol{\eta}^\top\boldsymbol{\Psi}^{-1}(\mathbf{X}_1-\boldsymbol{\xi})}{\sqrt{1+\boldsymbol{\eta}^\top\boldsymbol{\Psi}^{-1}\boldsymbol{\eta}}}\right\}(\mathbf{X}_1-\boldsymbol{\xi})\right] =& \mathbf{0}_p,\\
\mathrm{E}\left[\zeta\left\{\frac{\boldsymbol{\eta}^\top\boldsymbol{\Psi}^{-1}(\mathbf{X}_1-\boldsymbol{\xi})}{\sqrt{1+\boldsymbol{\eta}^\top\boldsymbol{\Psi}^{-1}\boldsymbol{\eta}}}\right\}(\mathbf{X}_1-\boldsymbol{\xi})(\mathbf{X}_1-\boldsymbol{\xi})^\top\right] =& \frac{\sqrt{2/\pi}}{\sqrt{1+\boldsymbol{\eta}^\top\boldsymbol{\Psi}^{-1}\boldsymbol{\eta}}}\boldsymbol{\Psi}.
\end{aligned}$$

Now, using $\zeta'(x) = -\zeta(x)x - \zeta(x)^2$, we have (see again Exercises 5.11 and 5.12)

$$a_2 = \bar{a}_2, \quad \mathbf{a}_1 = \frac{\sqrt{2/\pi}}{\sqrt{1+\boldsymbol{\eta}^\top\boldsymbol{\Psi}^{-1}\boldsymbol{\eta}}}\boldsymbol{\eta} + \boldsymbol{\Psi}^{1/2}\bar{\mathbf{a}}_1, \quad \mathbf{A}_2 = \boldsymbol{\Psi}^{1/2}\bar{\mathbf{A}}_2\boldsymbol{\Psi}^{1/2},$$

where

$$\bar{a}_2 = \mathrm{E}[\{\zeta(\boldsymbol{\delta}^\top\mathbf{Z})\}^2], \quad \bar{\mathbf{a}}_1 = \mathrm{E}[\{\zeta(\boldsymbol{\delta}^\top\mathbf{Z})\}^2\mathbf{Z}], \quad \bar{\mathbf{A}}_2 = \mathrm{E}[\{\zeta(\boldsymbol{\delta}^\top\mathbf{Z})\}^2\mathbf{Z}\mathbf{Z}^\top],$$

with $\mathbf{Z} = (Z_1,\dots,Z_p)^\top \sim \mathcal{N}_p(\mathbf{0}_p, \mathbf{I}_p)$ and $\boldsymbol{\delta} = (1+\boldsymbol{\eta}^\top\boldsymbol{\Psi}^{-1}\boldsymbol{\eta})^{-1/2}\boldsymbol{\Psi}^{-1/2}\boldsymbol{\eta}$.

At $\boldsymbol{\eta} = \mathbf{0}_p$, or equivalently, $\boldsymbol{\delta} = \mathbf{0}_p$, we have

$$a_1 = \zeta(0) = \sqrt{\frac{2}{\pi}}, \quad \bar{a}_2 = \zeta(0)^2 = \frac{2}{\pi}, \quad \bar{\mathbf{a}}_1 = \mathbf{0}_p, \quad \bar{\mathbf{A}}_2 = \frac{2}{\pi}\mathbf{I}_p.$$

Also, in this situation $\boldsymbol{\Theta}$ becomes $\boldsymbol{\Theta}_0 = (\boldsymbol{\xi}, \boldsymbol{\Psi}, \mathbf{0}_p)$ and so $\mathbf{I}(\boldsymbol{\Theta})$ reduces to the singular information matrix given by

$$\mathbf{I}(\boldsymbol{\Theta}_0) = \begin{pmatrix} \boldsymbol{\Psi}^{-1} & \mathbf{O} & \sqrt{\frac{2}{\pi}}\boldsymbol{\Psi}^{-1} \\ \mathbf{O} & \frac{1}{2}\mathbf{D}_p^\top(\boldsymbol{\Psi}^{-1}\otimes\boldsymbol{\Psi}^{-1})\mathbf{D}_p & \mathbf{O} \\ \sqrt{\frac{2}{\pi}}\boldsymbol{\Psi}^{-1} & \mathbf{O} & \frac{2}{\pi}\boldsymbol{\Psi}^{-1} \end{pmatrix}.$$

For $\boldsymbol{\delta} \neq \mathbf{0}_p$, $\bar{a}_2$, $\bar{\mathbf{a}}_1$, and $\bar{\mathbf{A}}_2$ must be computed numerically. These numerical computations are simplified by noting first that $\boldsymbol{\delta}^\top\mathbf{Z} \stackrel{d}{=} \delta Z_1$, where $\delta = \|\boldsymbol{\delta}\|$ and $Z_1 \sim \mathcal{N}(0,1)$, and also that $(\mathbf{Z} \mid \boldsymbol{\delta}^\top\mathbf{Z} = \boldsymbol{\delta}^\top\mathbf{z}) \sim \mathcal{N}_p\{(\bar{\boldsymbol{\delta}}^\top\mathbf{z})\bar{\boldsymbol{\delta}}, \mathbf{M}\}$, where $\bar{\boldsymbol{\delta}} = \|\boldsymbol{\delta}\|^{-1}\boldsymbol{\delta}$, $\mathbf{M} = \mathbf{I}_p - \mathbf{P}$ and $\mathbf{P} = \bar{\boldsymbol{\delta}}\bar{\boldsymbol{\delta}}^\top$ are orthogonal projection matrices since $\|\bar{\boldsymbol{\delta}}\| = 1$ and $\mathbf{MP} = \mathbf{O}$. Hence, we find that

$$\mathrm{E}(\mathbf{Z} \mid \boldsymbol{\delta}^\top\mathbf{Z}) = (\bar{\boldsymbol{\delta}}^\top\mathbf{Z})\bar{\boldsymbol{\delta}} \stackrel{d}{=} Z_1\bar{\boldsymbol{\delta}}, \quad \mathrm{E}(\mathbf{Z}\mathbf{Z}^\top \mid \boldsymbol{\delta}^\top\mathbf{Z}) = \mathbf{M} + (\bar{\boldsymbol{\delta}}^\top\mathbf{Z})^2\bar{\boldsymbol{\delta}}\bar{\boldsymbol{\delta}}^\top \stackrel{d}{=} \mathbf{M} + Z_1^2\mathbf{P}.$$

Thus, for $\boldsymbol{\delta} \neq \mathbf{0}_p$ we need to compute numerically only the univariate moments $\mathrm{E}[\{\zeta(\delta Z_1)\}^2 Z_1^s]$ for $s = 0,1,2$, since

$$\bar{a}_2 = \mathrm{E}[\{\zeta(\delta Z_1)\}^2], \quad \bar{\mathbf{a}}_1 = \mathrm{E}[\{\zeta(\delta Z_1)\}^2 Z_1]\bar{\boldsymbol{\delta}}, \quad \bar{\mathbf{A}}_2 = \mathrm{E}[\{\zeta(\delta Z_1)\}^2]\mathbf{M} + \mathrm{E}[\{\zeta(\delta Z_1)\}^2 Z_1^2]\mathbf{P}.$$

Finally, since $\eta\boldsymbol{\Psi}^{1/2}\bar{\boldsymbol{\delta}} = \boldsymbol{\eta}$, where $\eta = \sqrt{\boldsymbol{\eta}^\top\boldsymbol{\Psi}^{-1}\boldsymbol{\eta}}$, the original quantities $\mathbf{a}_1$ and $\mathbf{A}_2$ become

$$\eta\mathbf{a}_1 = \mathrm{E}[\{\zeta(\delta Z_1)\}^2 Z_1]\boldsymbol{\eta}, \quad \eta^2\mathbf{A}_2 = \eta^2\mathrm{E}[\{\zeta(\delta Z_1)\}^2]\boldsymbol{\Psi} + (\mathrm{E}[\{\zeta(\delta Z_1)\}^2 Z_1^2] - \mathrm{E}[\{\zeta(\delta Z_1)\}^2])\boldsymbol{\eta}\boldsymbol{\eta}^\top.$$

5.10.4 The EM algorithm

To estimate the parameters of the $\mathcal{SN}(\boldsymbol{\xi}, \boldsymbol{\Psi}, \boldsymbol{\eta})$ model, we shall make use of the EM technique studied by Dempster et al. (1977). For this, we first use the hierarchical representation of the multivariate skew-normal distribution (see Theorem 52) which allows us to represent the observations $\mathbf{X}_1,\dots,\mathbf{X}_n$ from $\mathcal{SN}_p(\boldsymbol{\xi}, \boldsymbol{\Psi}, \boldsymbol{\eta})$ as

$$(\mathbf{X}_i \mid U_i = u_i) \stackrel{ind.}{\sim} \mathcal{N}_p(\boldsymbol{\xi} + \boldsymbol{\eta}u_i, \boldsymbol{\Psi}),$$
$$U_i \stackrel{iid}{\sim} \mathcal{HN}(0,1),$$

for $i = 1,\dots,n$.

As a byproduct of this representation, we have that the conditional pdf of $(U_i \mid \mathbf{X}_i = \mathbf{x}_i)$ becomes

$$\begin{aligned} g(u_i \mid \mathbf{x}_i) &\propto \phi_p(\mathbf{x}_i; \boldsymbol{\xi} + \boldsymbol{\eta}u_i, \boldsymbol{\Psi})\phi(u_i; 0, 1) \\ &= \phi_p\{\mathbf{x}_i; \boldsymbol{\xi}, \boldsymbol{\Psi} + \boldsymbol{\eta}\boldsymbol{\eta}^\top\}\phi\{u_i; \boldsymbol{\eta}^\top(\boldsymbol{\Psi} + \boldsymbol{\eta}\boldsymbol{\eta}^\top)^{-1}(\mathbf{x}_i - \boldsymbol{\xi}), 1 - \boldsymbol{\eta}^\top(\boldsymbol{\Psi} + \boldsymbol{\eta}\boldsymbol{\eta}^\top)^{-1}\boldsymbol{\eta}\} \\ &= \phi_p\{\mathbf{x}_i; \boldsymbol{\xi}, \boldsymbol{\Psi} + \boldsymbol{\eta}\boldsymbol{\eta}^\top\}\phi\{u_i; \tau_i, (1+\eta^2)^{-1}\}, \quad u_i > 0, \end{aligned}$$

$i = 1, \ldots, n$, where $\eta^2 = \boldsymbol{\eta}^\top \boldsymbol{\Psi}^{-1} \boldsymbol{\eta}$ and

$$\tau_i = \tau(\mathbf{x}_i) = \boldsymbol{\eta}^\top (\boldsymbol{\Psi} + \boldsymbol{\eta}\boldsymbol{\eta}^\top)^{-1} (\mathbf{x}_i - \boldsymbol{\xi}) = \frac{1}{1+\eta^2} \boldsymbol{\eta}^\top \boldsymbol{\Psi}^{-1} (\mathbf{x}_i - \boldsymbol{\xi}),$$

since $\boldsymbol{\eta}^\top \{\boldsymbol{\Psi} + \boldsymbol{\eta}\boldsymbol{\eta}^\top\}^{-1} = (1+\eta^2)^{-1} \boldsymbol{\eta}^\top \boldsymbol{\Psi}^{-1}$. That is, the conditional distribution of the latent variables $U_1, \ldots, U_n$ given the observable $\mathbf{X}_1, \ldots, \mathbf{X}_n$ is truncated normal:

$$(U_i \mid \mathbf{X}_i = \mathbf{x}_i) \overset{ind.}{\sim} \mathcal{TN}\{0; \tau_i, (1+\eta^2)^{-1}\}, \quad i = 1, \ldots, n.$$

Also, for the first two moments of this conditional distribution, we find from (5.32) that

$$\begin{aligned} \nu_{1i} =& \mathrm{E}(U_i \mid \mathbf{X}_i = \mathbf{x}_i) = \frac{\bar{\tau}_i + \zeta(\bar{\tau}_i)}{\sqrt{1+\eta^2}}, \\ \nu_{2i} =& \mathrm{E}(U_i^2 \mid \mathbf{X}_i = \mathbf{x}_i) = \frac{1 + \bar{\tau}_i^2 + \bar{\tau}_i \zeta(\bar{\tau}_i)}{1+\eta^2}, \end{aligned}$$

$i = 1, \ldots, n$, where, as before, $\zeta(x) = \phi(x)/\Phi(x)$ and $\bar{\tau}_i = \sqrt{1+\eta^2}\tau_i = (1+\eta^2)^{-1/2} \boldsymbol{\eta}^\top \boldsymbol{\Psi}^{-1} (\mathbf{x}_i - \boldsymbol{\xi})$.

Using again the hierarchical representation above, we find that the complete log-likelihood for $\boldsymbol{\Theta} = (\boldsymbol{\xi}, \boldsymbol{\Psi}, \boldsymbol{\eta})$ based on the observed and missing data given, respectively, by $\mathbf{x} = (\mathbf{x}_1, \ldots, \mathbf{x}_n)^\top$ and $\mathbf{u} = (u_1, \ldots, u_n)^\top$ is

$$\begin{aligned} \ell_c(\boldsymbol{\Theta} \mid \mathbf{x}, \mathbf{u}) =& -\frac{np}{2} \log(2\pi) + \frac{n}{2} \log |\boldsymbol{\Lambda}| - \frac{1}{2} \sum_{i=1}^n (\mathbf{x}_i - \boldsymbol{\xi})^\top \boldsymbol{\Lambda} (\mathbf{x}_i - \boldsymbol{\xi}) + \boldsymbol{\eta}^\top \boldsymbol{\Lambda} \sum_{i=1}^n u_i (\mathbf{x}_i - \boldsymbol{\xi}) \\ & - \frac{1}{2} \boldsymbol{\eta}^\top \boldsymbol{\Lambda} \boldsymbol{\eta} \sum_{i=1}^n u_i^2 + \frac{n}{2} \log\left(\frac{2}{\pi}\right) - \frac{1}{2} \sum_{i=1}^n u_i^2, \end{aligned}$$

where, to facilitate the estimation process, we have written $\boldsymbol{\Lambda} = \boldsymbol{\Psi}^{-1}$.

Denoting by $\mathbf{X} = (\mathbf{X}_1, \ldots, \mathbf{X}_n)^\top$ and $\mathbf{U} = (U_1, \ldots, U_n)^\top$ the observable and latent random quantities, respectively, the E-Step at the $(k+1)$-th iteration of the EM algorithm requires the calculation of the Q-function defined by $Q(\boldsymbol{\Theta} | \boldsymbol{\Theta}^{(k)}) = \mathrm{E}_{\boldsymbol{\Theta}^{(k)}} \{\ell_c(\boldsymbol{\Theta} \mid \mathbf{X}, \mathbf{U}) \mid \mathbf{X} = \mathbf{x}\}$, where $\boldsymbol{\Theta}^{(k)}$ is the estimated value of $\boldsymbol{\Theta}$ in the k-th algorithm step. To do this, in the k-th algorithm step, we also need to calculate for each $i = 1, \ldots, n$ the conditional expectations given by

$$\begin{aligned} \nu_{1i}^{(k)} =& \mathrm{E}_{\boldsymbol{\Theta}^{(k)}}(U_i \mid \mathbf{X}_i = \mathbf{x}_i) = \frac{\bar{\tau}_i^{(k)} + \zeta(\bar{\tau}_i^{(k)})}{\sqrt{1+\eta^{2(k)}}}, \\ \nu_{2i}^{(k)} =& \mathrm{E}_{\boldsymbol{\Theta}^{(k)}}(U_i^2 \mid \mathbf{X}_i = \mathbf{x}_i) = \frac{1 + \bar{\tau}_i^{2(k)} + \bar{\tau}_i^{(k)} \zeta(\bar{\tau}_i^{(k)})}{1+\eta^{2(k)}}, \end{aligned}$$

where $\bar{\tau}_i^{(k)} = (1+\eta^{2(k)})^{-1/2} \boldsymbol{\eta}^{(k)\top} \boldsymbol{\Lambda}^{(k)} (\mathbf{x}_i - \boldsymbol{\xi}^{(k)})$ and $\eta^{2(k)} = \boldsymbol{\eta}^{(k)\top} \boldsymbol{\Lambda}^{(k)} \boldsymbol{\eta}^{(k)}$, with $\boldsymbol{\Lambda}^{(k)}$ being the inverse of $\boldsymbol{\Psi}^{(k)}$. Thus, in the k-th algorithm step, the Q-function becomes

$$\begin{aligned} Q(\boldsymbol{\Theta} | \boldsymbol{\Theta}^{(k)}) =& -\frac{np}{2} \log(2\pi) + \frac{n}{2} \log |\boldsymbol{\Lambda}| - \frac{1}{2} \sum_{i=1}^n (\mathbf{x}_i - \boldsymbol{\xi})^\top \boldsymbol{\Lambda} (\mathbf{x}_i - \boldsymbol{\xi}) + \boldsymbol{\eta}^\top \boldsymbol{\Lambda} \sum_{i=1}^n \nu_{1i}^{(k)} (\mathbf{x}_i - \boldsymbol{\xi}) \\ & - \frac{1}{2} \boldsymbol{\eta}^\top \boldsymbol{\Lambda} \boldsymbol{\eta} \sum_{i=1}^n \nu_{2i}^{(k)} + \frac{n}{2} \log\left(\frac{2}{\pi}\right) - \frac{1}{2} \sum_{i=1}^n \nu_{2i}^{(k)}. \end{aligned} \tag{5.49}$$

Therefore, in the $(k+1)$-th algorithm step, the EM algorithm proceeds as follows, where the symmetry condition on the matrix $\mathbf{\Lambda}$ has been taken into account:

E-Step. Compute $Q(\mathbf{\Theta}|\mathbf{\Theta}^{(k)})$ from (5.49).

M-Step. Update $\mathbf{\Theta}^{(k)}$ by maximizing $Q(\mathbf{\Theta}|\mathbf{\Theta}^{(k)})$ with respect to $\mathbf{\Theta}$, leading to the following estimating equations:

$$
\begin{aligned}
\mathrm{d}_{\boldsymbol{\xi}} Q = \mathbf{0}_p &\Longrightarrow n\mathbf{\Lambda}(\bar{\mathbf{x}}-\boldsymbol{\xi}) - \mathbf{\Lambda}\boldsymbol{\eta}\sum_{i=1}^n v_{1i}^{(k)} = \mathbf{0}_p,\\
\mathrm{d}_{\mathbf{\Lambda}} Q = \mathbf{O} &\Longrightarrow n\mathbf{\Lambda}^{-1} - \sum_{i=1}^n (\mathbf{x}_i-\boldsymbol{\xi})(\mathbf{x}_i-\boldsymbol{\xi})^\top + \sum_{i=1}^n v_{1i}^{(k)}\{(\mathbf{x}_i-\boldsymbol{\xi})\boldsymbol{\eta}^\top + \boldsymbol{\eta}(\mathbf{x}_i-\boldsymbol{\xi})^\top\}\\
&\qquad - \boldsymbol{\eta}\boldsymbol{\eta}^\top \sum_{i=1}^n v_{2i}^{(k)} = \mathbf{O},\\
\mathrm{d}_{\boldsymbol{\eta}} Q = \mathbf{0}_p &\Longrightarrow \mathbf{\Lambda}\sum_{i=1}^n v_{1i}^{(k)}(\mathbf{x}_i-\boldsymbol{\xi}) - \mathbf{\Lambda}\boldsymbol{\eta}\sum_{i=1}^n v_{2i}^{(k)} = \mathbf{0}_p.
\end{aligned}
$$

Since $\mathbf{\Lambda}^{-1} = \mathbf{\Psi}$, these equations are equivalent to

$$
\begin{aligned}
\bar{\mathbf{x}}-\boldsymbol{\xi} =& \frac{1}{n}\sum_{i=1}^n v_{1i}^{(k)}\boldsymbol{\eta},\\
\mathbf{\Psi} =& \mathbf{S}_{\mathbf{x}} + (\bar{\mathbf{x}}-\boldsymbol{\xi})(\bar{\mathbf{x}}-\boldsymbol{\xi})^\top - \left\{\frac{1}{n}\sum_{i=1}^n v_{1i}^{(k)}(\mathbf{x}_i-\bar{\mathbf{x}}) + \frac{1}{n}\sum_{i=1}^n v_{1i}^{(k)}(\bar{\mathbf{x}}-\boldsymbol{\xi})\right\}\boldsymbol{\eta}^\top\\
&- \boldsymbol{\eta}\left\{\frac{1}{n}\sum_{i=1}^n v_{1i}^{(k)}(\mathbf{x}_i-\bar{\mathbf{x}})^\top + \frac{1}{n}\sum_{i=1}^n v_{1i}^{(k)}(\bar{\mathbf{x}}-\boldsymbol{\xi})^\top\right\} + \frac{1}{n}\sum_{i=1}^n v_{2i}^{(k)}\boldsymbol{\eta}\boldsymbol{\eta}^\top,\\
\frac{1}{n}\sum_{i=1}^n v_{1i}^{(k)}(\bar{\mathbf{x}}-\boldsymbol{\xi}) =& \frac{1}{n}\sum_{i=1}^n v_{2i}^{(k)}\boldsymbol{\eta} - \frac{1}{n}\sum_{i=1}^n v_{1i}^{(k)}(\mathbf{x}_i-\bar{\mathbf{x}}).
\end{aligned}
$$

Substituting the first of these equations into the second and third, it follows that $\boldsymbol{\eta}^{(k)}$, $\boldsymbol{\xi}^{(k)}$, and $\mathbf{\Psi}^{(k)}$ must be updated using the following closed-form expressions:

$$
\begin{aligned}
\boldsymbol{\xi}^{(k+1)} =& \bar{\mathbf{x}} - \frac{1}{n}\sum_{i=1}^n v_{1i}^{(k)}\boldsymbol{\eta}^{(k+1)}\\
=& \bar{\mathbf{x}} - \left\{\frac{1}{n}\sum_{i=1}^n v_{2i}^{(k)} - \left(\frac{1}{n}\sum_{i=1}^n v_{1i}^{(k)}\right)^2\right\}^{-1}\left(\frac{1}{n}\sum_{i=1}^n v_{1i}^{(k)}\right)\frac{1}{n}\sum_{i=1}^n v_{1i}^{(k)}(\mathbf{x}_i-\bar{\mathbf{x}}),\\
\mathbf{\Psi}^{(k+1)} =& \mathbf{S}_{\mathbf{x}} + \left\{\frac{1}{n}\sum_{i=1}^n v_{2i}^{(k)} - \left(\frac{1}{n}\sum_{i=1}^n v_{1i}^{(k)}\right)^2\right\}\boldsymbol{\eta}^{(k+1)}\boldsymbol{\eta}^{(k+1)\top}\\
&- \frac{1}{n}\sum_{i=1}^n v_{1i}^{(k)}(\mathbf{x}_i-\bar{\mathbf{x}})\boldsymbol{\eta}^{(k+1)\top} - \boldsymbol{\eta}^{(k+1)}\frac{1}{n}\sum_{i=1}^n v_{1i}^{(k)}(\mathbf{x}_i-\bar{\mathbf{x}})^\top\\
=& \mathbf{S}_{\mathbf{x}} - \left\{\frac{1}{n}\sum_{i=1}^n v_{2i}^{(k)} - \left(\frac{1}{n}\sum_{i=1}^n v_{1i}^{(k)}\right)^2\right\}^{-1}\left\{\frac{1}{n}\sum_{i=1}^n v_{1i}^{(k)}(\mathbf{x}_i-\bar{\mathbf{x}})\right\}\left\{\frac{1}{n}\sum_{i=1}^n v_{1i}^{(k)}(\mathbf{x}_i-\bar{\mathbf{x}})^\top\right\},\\
\boldsymbol{\eta}^{(k+1)} =& \left\{\frac{1}{n}\sum_{i=1}^n v_{2i}^{(k)} - \left(\frac{1}{n}\sum_{i=1}^n v_{1i}^{(k)}\right)^2\right\}^{-1}\frac{1}{n}\sum_{i=1}^n v_{1i}^{(k)}(\mathbf{x}_i-\bar{\mathbf{x}}).
\end{aligned}
$$

This process is iterated until a suitable convergence rule is satisfied, e.g., if $\|\mathbf{\Theta}^{(k+1)} - \mathbf{\Theta}^{(k)}\|$ is sufficiently small, or until some measure involving two successive evaluations of the actual log-likelihood, $\ell(\mathbf{\Theta})$, like $|\ell(\mathbf{\Theta}^{(k+1)}) - \ell(\mathbf{\Theta}^{(k)})|$ or $|\ell(\mathbf{\Theta}^{(k+1)})/\ell(\mathbf{\Theta}^{(k)}) - 1|$ is small enough. A usual criticism is that EM-type procedures tend to get stuck in local modes. A convenient way to avoid such limitations is to try several EM iterations with a variety of starting values. If there exist several modes, one can find the global mode by comparing their relative masses and log-likelihood values.

5.11 Distributions Related to the Multivariate Skew-Normal

5.11.1 *Multivariate generalized and multivariate flexible skew-normal distributions*

An approach to generalizing the skew-normal distribution is to relax the structure of its skewing function. Genton and Loperfido (2005) defined a *multivariate generalized skew-normal (GSN) distribution* by its pdf:

$$2\phi_p(\mathbf{x}-\boldsymbol{\xi};\mathbf{0}_p,\boldsymbol{\Omega})\pi(\mathbf{x}-\boldsymbol{\xi}), \quad \mathbf{x}\in\mathbb{R}^p, \tag{5.50}$$

where the skewing function $0 \leq \pi(\mathbf{x}) \leq 1$ is an odd function, i.e., it satisfies $\pi(-\mathbf{x}) = 1-\pi(\mathbf{x})$, for all $\mathbf{x}\in\mathbb{R}^p$, with $\boldsymbol{\xi}\in\mathbb{R}^p$ and $\boldsymbol{\Omega}\in\mathbb{R}^{p\times p}$ a positive-definite dispersion matrix. Obviously, these conditions are satisfied by the choice $\pi(\mathbf{x}) = \Phi(\boldsymbol{\alpha}^\top\boldsymbol{\omega}^{-1}\mathbf{x})$, $\boldsymbol{\alpha}\in\mathbb{R}^p$, $\boldsymbol{\omega} = \mathrm{diag}(\boldsymbol{\Omega})^{1/2}$, for the skew-normal distribution, but many other skewing functions are possible. For example, take $\pi(\mathbf{x}) = G(\boldsymbol{\alpha}^\top\mathbf{x})$, $\boldsymbol{\alpha}\in\mathbb{R}^p$, for any univariate cdf G of a symmetric continuous random variable. It will automatically produce an odd skewing function.

Ma and Genton (2004) defined a *multivariate flexible skew-normal (FSN) distribution* by selecting in Equation (5.50) skewing functions of the form $\pi(\mathbf{x}) = G\{P_K(\mathbf{x})\}$, where $P_K(\mathbf{x})$ is an odd polynomial in $\mathbf{x}$ of order K, $K = 1,3,5,\ldots$. That is, $P_K(\mathbf{x})$ is a linear combination of terms of the form $\prod_{i=1}^p x_i^{r_i}$ with K and $k = \sum_{i=1}^p r_i \leq K$ both odd. For example, for $K = 3$, take $\pi(\mathbf{x}) = \Phi(\boldsymbol{\alpha}^\top\boldsymbol{\omega}^{-1}\mathbf{x}+\boldsymbol{\beta}^\top\boldsymbol{\omega}^{-1}\mathbf{x}\mathbf{x}^\top\boldsymbol{\omega}^{-2}\mathbf{x})$, $\boldsymbol{\alpha}\in\mathbb{R}^p$, $\boldsymbol{\beta}\in\mathbb{R}^p$. For certain values of $\boldsymbol{\alpha}$ and $\boldsymbol{\beta}$, that skewing function may not be a strictly monotonic function of $\mathbf{x}$, thus producing a multimodal pdf (5.50). Ma and Genton (2004) have proved that the approximation of the skewing function can be made arbitrarily accurate by increasing the order of the odd polynomial. A further extension that produces trimodal distributions has recently been proposed by Bufalo and Nigri (2026).

Another important class of skewing functions is $\pi(\mathbf{x}) = G\{w(\mathbf{x})\}$ for any univariate cdf G of a symmetric continuous random variable and some odd function $w(\cdot)$. It includes the aforementioned FSN distributions and the MSN distributions presented in Section 5.11.3, and in particular the earlier univariate skew-generalized-normal (SGN) pdf of Arellano-Valle et al. (2004) given by $2\phi(x)\Phi(\alpha x/\sqrt{1+\lambda x^2})$, $\alpha\in\mathbb{R}$ and $\lambda > 0$.

Equation (5.50) can be extended to pdfs of the form $2f(\mathbf{x}-\boldsymbol{\xi})\pi(\mathbf{x}-\boldsymbol{\xi})$ where f is the pdf of an elliptical distribution (Genton and Loperfido, 2005), producing *multivariate generalized skew-elliptical (GSE) distributions*, or the pdf of a centrally symmetric distribution (Wang et al., 2004b), leading to the so-called *multivariate skew-symmetric (SS) distributions*. Further properties of skew-symmetric distributions can be found in Azzalini and Regoli (2012a). Similar constructions have been explored by Azzalini and Regoli (2018) when the skewing function is not necessarily odd. Shushi (2016) proved a conjecture of Genton and Loperfido (2005) about the form of the characteristic function of GSE distributions. Shushi (2018a) also showed that GSE distributions are closed under affine transformations.

Another approach to introduce skewness in a univariate symmetric pdf f is by transformation of scale, i.e., $2f\{s(x)\}$, with conditions on the invertible function s to yield a valid pdf given by Jones (2014); see also Jones (2016) for the bivariate case.

5.11.2 *Scale and shape mixtures of multivariate skew-normal distributions*

Scale mixtures provide a major extension of the skew-normal family, designed to introduce heavy tails while retaining skewness (Branco and Dey, 2001). Following the introduction of scale mixtures of multivariate normal distributions in Section 3.10, we can define *multivariate scale mixtures of skew-normal (SMSN) distributions* by

$$\mathbf{Y} = \boldsymbol{\xi} + V^{-1/2}\mathbf{X}, \tag{5.51}$$

where $\mathbf{X} \sim \mathcal{SN}_p(\mathbf{0}_p, \boldsymbol{\Psi}, \boldsymbol{\eta})$, V is an independent random variable on $\mathbb{R}_+$ with cdf G, and $\boldsymbol{\xi} \in \mathbb{R}^p$ is a vector of constants. The resulting pdf is

$$\begin{aligned} f_{\text{SMSN}}(\mathbf{y}) &= \int_0^\infty 2\phi_p(\mathbf{y}; \boldsymbol{\xi}, v^{1/2}\boldsymbol{\Omega})\Phi\left\{\frac{v^{1/2}\boldsymbol{\eta}^\top\boldsymbol{\Omega}^{-1}(\mathbf{y}-\boldsymbol{\xi})}{\sqrt{1-\boldsymbol{\eta}^\top\boldsymbol{\Omega}^{-1}\boldsymbol{\eta}}}\right\} \mathrm{d}G(v) \\ &= \int_0^\infty 2\phi_p(\mathbf{y}; \boldsymbol{\xi}, v^{1/2}\boldsymbol{\Omega})\Phi\left\{v^{1/2}\boldsymbol{\alpha}^\top\boldsymbol{\omega}^{-1}(\mathbf{y}-\boldsymbol{\xi})\right\} \mathrm{d}G(v), \end{aligned}$$

where we have used Remarks 7 and 8. This framework includes many prominent families:

- the skew-t distribution, obtained when $V \sim \mathcal{G}amma(\nu/2, \nu/2)$;
- the skew-slash distribution, for V following a beta-power distribution;
- the skew-contaminated normal distribution, when V has a discrete-continuous mixture.

Recently, de Freitas et al. (2024) explored scale mixtures of centered skew-normal distributions while Cabral et al. (2012) studied multivariate mixture modeling using skew-normal independent (SNI) distributions (see Section 10.11).

However, recalling that $\mathbf{X} = U\boldsymbol{\eta} + \mathbf{W}$, where $U \sim \mathcal{HN}(0,1)$ and $\mathbf{W} \sim \mathcal{N}_p(\mathbf{0}_p, \boldsymbol{\Psi})$, then by placing this SN representation in (5.51) we see that

$$\mathbf{Y} = \boldsymbol{\xi} + V^{-1/2}U\boldsymbol{\eta} + V^{-1/2}\mathbf{W},$$

where $V \sim G$, $U \sim \mathcal{HN}(0,1)$ and $\mathbf{W} \sim \mathcal{N}_p(\mathbf{0}_p, \boldsymbol{\Psi})$ are mutually independent, from which it follows that, just as the SN itself, the SN scale mixtures are of the GMN type defined in (3.16) or (3.17) with $R = V^{-1/2}U$ and $S = V^{-1/2}$.

Another idea that also arises from the classic parameterization of $(\boldsymbol{\Omega}, \boldsymbol{\alpha})$ of the SN distribution, initially proposed by Arellano-Valle et al. (2008) and Arellano-Valle et al. (2009), is to treat the shape parameter $\boldsymbol{\alpha}$ not as fixed, but as a random quantity with distribution $\Pi(\boldsymbol{\alpha})$, leading to *multivariate shape mixtures of skew-normal (SHMSN) distributions*. The shape mixtures model is defined as

$$f_{\text{SHMSN}}(\mathbf{y}) = \int_{\mathbb{R}^p} 2\phi_p(\mathbf{y}; \boldsymbol{\xi}, \boldsymbol{\Omega})\Phi\left\{\boldsymbol{\alpha}^\top\boldsymbol{\omega}^{-1}(\mathbf{y}-\boldsymbol{\xi})\right\} \mathrm{d}\Pi(\boldsymbol{\alpha}), \quad \mathbf{y} \in \mathbb{R}^p,$$

which expands the modeling flexibility by averaging over different skewness configurations. Such mixtures provide much greater flexibility than single-shape skew-normal models, allowing modulated asymmetry, multimodality, and heavier tails. This construction includes important special cases such as the skew-normal, skew-generalized normal, skew-normal-t, and skew-t-normal mixtures. Arellano-Valle et al. (2008) demonstrated that these mixtures can be efficiently used in Bayesian regression, with posterior propriety guaranteed under mild conditions.

The most powerful constructions arise by combining scale and shape mixtures (Arellano-Valle et al., 2018). This leads to the most general form of *multivariate scale-and-shape mixtures of skew-normal (SSHMSN) distributions* with pdf:

$$f_{\text{SSHMSN}}(\mathbf{y}) = \int_{\mathbb{R}^{p\times p}\times\mathbb{R}^p} 2\phi_p(\mathbf{y}; \boldsymbol{\xi}, \boldsymbol{\Omega})\Phi\left\{\boldsymbol{\lambda}^\top(\mathbf{y}-\boldsymbol{\xi})\right\} \mathrm{d}\Pi(\boldsymbol{\Omega}, \boldsymbol{\lambda}), \quad \mathbf{y} \in \mathbb{R}^p,$$

where $\boldsymbol{\lambda} = \boldsymbol{\omega}^{-1}\boldsymbol{\alpha}$. Clearly, the simplest case is obtained when $(\boldsymbol{\Omega}, \boldsymbol{\alpha})$ is replaced by $(s_1(\mathbf{Z})\boldsymbol{\Omega}, s_2(\mathbf{Z})\boldsymbol{\alpha})$, so $\boldsymbol{\lambda}$ is replaced by $s_1(\mathbf{Z})^{-1/2}s_2(\mathbf{Z})\boldsymbol{\lambda}$, where only $\mathbf{Z} \in \mathbb{R}^k$ is random, and s_1 and s_2 are real valued functions with s_1 defined in $\mathbb{R}_+$. These models unify skew-normal, skew-t, skew-generalized-normal, skew-t-normal, and many other related classes. The combined mixture structure allows for simultaneous control of skewness, heavy tails, and overall dispersion.

Rezaei et al. (2020) proposed a matrix extension of the scale and shape mixtures of multivariate extended skew-normal distributions and presented some particular cases of this new class.

5.11.3 *Multivariate modified skew-normal distributions*

As discussed in Section 5.10.3, the information matrix of the multivariate skew-normal distribution is singular at $\boldsymbol{\eta} = \mathbf{0}_p$. It is therefore of interest to explore how to modify the skew-normal in order to remove this singularity. Mondal et al. (2024) have introduced a *multivariate modified skew-normal (MSN) distribution* with pdf

$$2\phi_p(\mathbf{x};\boldsymbol{\xi},\boldsymbol{\Psi}+\boldsymbol{\eta}\boldsymbol{\eta}^\top)\Phi\left\{\frac{\boldsymbol{\eta}^\top\boldsymbol{\Psi}^{-1}(\mathbf{x}-\boldsymbol{\xi})}{\sqrt{1+\boldsymbol{\eta}^\top\boldsymbol{\Psi}^{-1}\boldsymbol{\eta}}\sqrt{1+(\mathbf{x}-\boldsymbol{\xi})^\top(\boldsymbol{\Psi}+\boldsymbol{\eta}\boldsymbol{\eta}^\top)^{-1}(\mathbf{x}-\boldsymbol{\xi})}}\right\}, \quad \mathbf{x}\in\mathbb{R}^p. \tag{5.52}$$

Compared to the pdf of the $\mathcal{SN}_p(\boldsymbol{\xi},\boldsymbol{\Psi},\boldsymbol{\eta})$ distribution given in Equation (5.19), the above pdf has a new nonlinear term in the denominator of the skewing function Φ. Mondal et al. (2024) have shown that the information matrix of this multivariate modified skew-normal distribution is nonsingular at $\boldsymbol{\eta} = \mathbf{0}_p$. Further properties and applications of this distribution can be found in the aforementioned reference. When $p = 1$, the pdf (5.52) reduces to the univariate modified skew-normal pdf introduced by Arellano-Valle et al. (2004) and mentioned in Section 5.11.1. In this case, the nonsingularity of the Fisher information matrix was shown by Arellano-Valle et al. (2013c), and a method for bias reduction of maximum likelihood estimates was proposed by Arrué et al. (2016). Furthermore, Ferreira and Arellano-Valle (2018) explored estimation and diagnostic analysis for this univariate model.

5.11.4 *Multivariate skew-slash distributions*

The univariate slash distribution has been considered by Rogers and Tukey (1972) to model symmetric heavy-tailed data distributions that are less peaked than the Cauchy distribution. It is defined by the symmetric random variable $Y = \xi + \omega Z/U^{1/q}$, $q > 0$, where $Z \sim \mathcal{N}(0,1)$, $U \sim \mathcal{U}(0,1)$ independent of Z, and $\xi \in \mathbb{R}$, $\omega > 0$. The parameter q controls the heaviness of the tails with $q = 1$ producing Cauchy-like tails and $q \to \infty$ reducing to normal tails. Wang and Genton (2006) first extended this distribution to the multivariate (elliptically contoured) case by $\mathbf{Y} = \boldsymbol{\xi} + \boldsymbol{\Omega}^{1/2}\mathbf{Z}/U^{1/q}$ with $\mathbf{Z} \sim \mathcal{N}_p(\mathbf{0}_p,\mathbf{I}_p)$, $\boldsymbol{\xi} \in \mathbb{R}^p$, and $\boldsymbol{\Omega} \in \mathbb{R}^{p\times p}$ a positive-definite dispersion matrix. Then they generalized it to a *multivariate skew-slash (SSL) distribution* letting $\mathbf{Z} \sim \mathcal{SN}_p(\mathbf{0}_p,\mathbf{I}_p,\boldsymbol{\eta})$, which allowed to introduce skewness via $\boldsymbol{\eta} \in \mathbb{R}^p$. Further properties and applications of this distribution can be found in the aforementioned reference. An alternative definition would divide each component of $\mathbf{Z}$ by its own independent uniform random variable, i.e., $Z_i/U_i^{1/q_i}$, $i = 1,\dots,p$. In that case, the kurtosis of the components of $\mathbf{Y}$ would be controlled separately by the components of the vector $\mathbf{q} = (q_1,\dots,q_p)^\top > \mathbf{0}_p$.

5.11.5 *Multivariate skew-normal-Tukey-h distributions*

In Section 3.11.7, we discussed how a multivariate trans-normal distribution could be obtained by applying a transformation to a normal random vector. This idea can be extended to applying a transformation to a skew-normal random vector. For example, Mondal and Genton (2024) defined a *multivariate skew-normal-Tukey-h (SNTH) distribution* by $\mathbf{Y} = \boldsymbol{\xi} + \boldsymbol{\omega}\mathbf{r}(\mathbf{Z})$, where $\mathbf{Z} \sim \mathcal{SN}_p(\mathbf{0}_p,\bar{\boldsymbol{\Psi}},\boldsymbol{\eta})$ with $\bar{\boldsymbol{\Psi}} \in \mathbb{R}^{p\times p}$ a correlation matrix, $\boldsymbol{\eta} \in \mathbb{R}^p$ a skewness vector, $\boldsymbol{\omega} \in \mathbb{R}^{p\times p}$ a diagonal scale matrix, and $\mathbf{r}(\mathbf{Z}) = (r_{0,h_1}(Z_1),\dots,r_{0,h_p}(Z_p))^\top$ is the Tukey-h transformation; see Example 33. Here, $\mathbf{h} = (h_1,\dots,h_p)^\top \in \mathbb{R}_+^p$ controls the kurtosis in each of the components separately, an advantage compared to the multivariate skew-t distribution introduced in Section 7.4.3. Further properties and applications of this distribution can be found in the aforementioned reference.

5.11.6 Multivariate contaminated non-Gaussian distributions

A discrete mixture of multivariate distributions is a probability distribution built by combining (or mixing) several component distributions using a discrete set of mixing weights. The two-component case is often used to model outliers in robust statistics. For example, the multivariate contaminated normal distribution is a two-component normal mixtures that contains two additional parameters with respect to the parameters of the multivariate normal distribution, one to control the proportion of outliers and the other to specify the degree of contamination; see, e.g., Wang (2023) and references therein. It is natural to extend this idea to other non-Gaussian distributions, such as the skewed distributions introduced in this book.

We define a *multivariate contaminated non-Gaussian (CNG) distribution* by the pdf

$$f(\mathbf{y}) = (1-\varepsilon)g(\mathbf{y};\boldsymbol{\xi},\boldsymbol{\Omega},\boldsymbol{\theta}) + \varepsilon g\left(\mathbf{y};\boldsymbol{\xi},\frac{1}{\tau}\boldsymbol{\Omega},\boldsymbol{\theta}\right), \quad \mathbf{y} \in \mathbb{R}^p, \tag{5.53}$$

where $0 \le \varepsilon \le 1$ denotes the proportion of outliers, $0 < \tau \le 1$ represents the degree of contamination, and $g(\mathbf{y};\boldsymbol{\xi},\boldsymbol{\Omega},\boldsymbol{\theta})$ is the pdf of a non-Gaussian distribution with location $\boldsymbol{\xi} \in \mathbb{R}^p$, dispersion matrix $\boldsymbol{\Omega} \in \mathbb{R}^{p\times p}$, and additional parameters in $\boldsymbol{\theta} \in \mathbb{R}^q$. A random vector $\mathbf{Y} \in \mathbb{R}^p$ with pdf (5.53) is denoted by $\mathbf{Y} \sim \mathcal{CNG}_p(\boldsymbol{\xi},\boldsymbol{\Omega},\varepsilon,\tau,\boldsymbol{\theta})$. When $\varepsilon \to 0$ or $\tau \to 1$, then the underlying non-Gaussian distribution is recovered, i.e., $f = g$. If $g = \phi_p$, the multivariate normal pdf, then the multivariate contaminated normal distribution is obtained.

An obvious candidate for the pdf g is the one from the SN distribution studied in this chapter, leading to the multivariate SN-CNG distribution: $\mathbf{Y} \sim \mathcal{SN}\text{-}\mathcal{CNG}_p(\boldsymbol{\xi},\boldsymbol{\Omega},\varepsilon,\tau,\boldsymbol{\eta})$. The parameter $\boldsymbol{\eta} \in \mathbb{R}^p$ controls the skewness. A further extension is to use the SUN distribution studied in Chapter 6, leading to multivariate SUN-CNG distributions. Another generalization is to allow the parameter $\boldsymbol{\theta}$ to change in the two-components mixtures (5.53).

5.12 Exercises

Exercise 5.1 *Let $U \sim \mathcal{HN}(0,1)$. Show that*

$$\mathrm{E}(U^r) = 2^{\frac{r-1}{2}}\sqrt{\frac{2}{\pi}}\Gamma\left(\frac{r+1}{2}\right), \quad r > 0.$$

Use this result to find the k-th moment of $V = U - b$, where $b = \mathrm{E}(U) = \sqrt{\frac{2}{\pi}}$.

Exercise 5.2 *Let Y be a random variable with pdf given by*

$$f_Y(y) = 2\lambda\phi(y-\mu)\Phi\{\alpha(y-\mu)\} + 2(1-\lambda)\phi(y+\mu)\Phi\{-\alpha(y+\mu)\}, \quad y \in \mathbb{R},$$

where $0 \le \lambda \le 1$ and $\alpha, \mu \in \mathbb{R}$. Find $\mathrm{E}(Y)$, $\mathrm{var}(Y)$, γ_1 (skewness), and γ_2 (excess kurtosis).

Exercise 5.3 *Prove the following lemma which has been recurrently used in the development of the multivariate skew-normal theory:*
Beautiful Lemma: Let $\Phi_q(\cdot;\boldsymbol{\mu},\boldsymbol{\Sigma})$ denote the cdf of the $\mathcal{N}_q(\boldsymbol{\mu},\boldsymbol{\Sigma})$ distribution. If $\mathbf{Z} \sim \mathcal{N}_p(\boldsymbol{\xi},\boldsymbol{\Omega})$, then

$$\mathrm{E}\{\Phi_q(\mathbf{a}+\mathbf{A}\mathbf{Z};\boldsymbol{\mu},\boldsymbol{\Sigma})\} = \Phi_q(\mathbf{a};\boldsymbol{\mu}-\mathbf{A}\boldsymbol{\xi},\boldsymbol{\Sigma}+\mathbf{A}\boldsymbol{\Omega}\mathbf{A}^\top),$$

where $\mathbf{a} \in \mathbb{R}^q$ and $\mathbf{A} \in \mathbb{R}^{q\times p}$.

Exercise 5.4 *Let*

$$\begin{pmatrix} V_0 \\ \mathbf{V} \end{pmatrix} \sim \mathcal{N}_{1+p}\left(\begin{pmatrix} 0 \\ \mathbf{0}_p \end{pmatrix}, \begin{pmatrix} 1 & \boldsymbol{\delta}^\top \\ \boldsymbol{\delta} & \mathbf{I}_p \end{pmatrix}\right),$$

where $\|\boldsymbol{\delta}\| < 1$. *Show that* $(\mathbf{V}^\top\mathbf{B}\mathbf{V} \mid V_0 > 0) \overset{d}{=} \mathbf{V}^\top\mathbf{B}\mathbf{V}$ *and that* $\mathrm{cov}(V_0, \mathbf{V}^\top\mathbf{B}\mathbf{V}) = 0$. *Is* $\mathbf{V}^\top\mathbf{B}\mathbf{V}$ *independent of* $I_{(V>0)}$*? Are* V_0 *and* $\mathbf{V}^\top\mathbf{B}\mathbf{V}$ *independent?*

Exercise 5.5 *Let* $\mathbf{U} = (U_1, \ldots, U_n)^\top \sim \mathcal{HN}_n(\mathbf{0}_n, \mathbf{I}_n)$, *i.e.,* $U_1, \ldots, U_n$ *are iid* $\mathcal{HN}(0,1)$. *Show that the joint cf of* $\bar{U} = \frac{1}{n}\sum_{i=1}^n U_i$ *and* $nS_U^2 = \sum_{i=1}^n (U_i - \bar{U})^2$ *is given by*

$$\psi_{\bar{U}, nS_U^2}(t_1, t_2) = 2^n (1 - 2\mathrm{i}t_2)^{-\frac{n-1}{2}} e^{-\frac{1}{2n}t_1^2}\, \Phi_n\left(\frac{1}{n}\mathrm{i}t_1\sqrt{1-2\mathrm{i}t_2}\,\mathbf{1}_n; \mathbf{0}_n, \mathbf{I}_n - \frac{2}{n}\mathrm{i}t_2\mathbf{1}_n\mathbf{1}_n^\top\right).$$

Also, show that the marginal cfs are:

$$\psi_{\bar{U}}(t_1) = 2^n e^{-\frac{1}{2n}t_1^2}\, \Phi_n\left(\frac{1}{n}\mathrm{i}t_1\mathbf{1}_n\right), \quad \psi_{nS_U^2}(t_2) = 2^n(1-2\mathrm{i}t_2)^{-\frac{n-1}{2}}\, \Phi_n\left(\mathbf{0}_n; \mathbf{0}_n, \mathbf{I}_n - \frac{2}{n}\mathrm{i}t_2\mathbf{1}_n\mathbf{1}_n^\top\right).$$

Exercise 5.6 *Let* $\mathbf{V} = (V_1, \ldots, V_n)^\top$, *where* $V_1, \ldots, V_n$ *are iid random variables with* $\mathrm{E}(V_i) = 0$, $\mathrm{var}(V_i) = 1$, $\mathrm{E}(V_i^3) = \gamma_1$ *(the skewness), and* $\mathrm{E}(V_i^4) = \gamma_2 + 3$ *(the kurtosis). Show that:*

(a) $\mathrm{E}\{(\mathbf{V}\otimes\mathbf{V})\} = \mathrm{vec}(\mathbf{I}_n)$;

(b) $\mathrm{E}\{(\mathbf{V}\otimes\mathbf{V})\mathbf{V}^\top\} = \gamma_1\sum_{i=1}^n(\mathbf{e}_i \otimes \mathbf{E}_{ii})$;

(c) $\mathrm{E}\{(\mathbf{V}\otimes\mathbf{V})(\mathbf{V}\otimes\mathbf{V})^\top\} = \mathbf{I}_{n^2} + \mathbf{K}_n + \mathrm{vec}(\mathbf{I}_n)\mathrm{vec}(\mathbf{I}_n)^\top + \gamma_2\sum_{i=1}^n(\mathbf{E}_{ii}\otimes\mathbf{E}_{ii})$;

(d) $\mathrm{var}(\mathbf{V}\otimes\mathbf{V}) = \mathbf{I}_{n^2} + \mathbf{K}_n + \gamma_2\sum_{i=1}^n(\mathbf{E}_{ii}\otimes\mathbf{E}_{ii})$, *where* $\mathbf{e}_i$ *is the i-th unit vector of dimension n,* $\mathbf{E}_{ii} = \mathbf{e}_i\mathbf{e}_i^\top$ *and* $\mathbf{K}_n = (\mathbf{E}_{ji})$ *is a block-matrix whose block in position* (i, j) *is* $\mathbf{E}_{ji} = \mathbf{e}_j\mathbf{e}_i^\top$.

Exercise 5.7 *Let* $\mathbf{U}$ *be an n-dimensional random vector with* $\mathrm{E}(\mathbf{U}) = \boldsymbol{\mu}$ *and* $\mathrm{var}(\mathbf{U}) = \boldsymbol{\Sigma} = \mathbf{C}\mathbf{C}^\top$. *Suppose* $\mathbf{U} = \mathbf{C}\mathbf{V} + \boldsymbol{\mu}$, *where* $\mathbf{V}$ *is an n-vector with* $\mathrm{E}(\mathbf{V}) = \mathbf{0}_n$, $\mathrm{var}(\mathbf{V}) = \mathbf{I}_n$ *and for which its third- and fourth-order multivariate moments are also finite. Show that:*

$$(\mathbf{U}\otimes\mathbf{U}) = \mathrm{vec}(\mathbf{U}\mathbf{U}^\top) = (\mathbf{C}\otimes\mathbf{C})\mathrm{vec}(\mathbf{V}\mathbf{V}^\top) + (\mathbf{I}_{n^2} + \mathbf{K}_n)(\mathbf{C}\otimes\boldsymbol{\mu})\mathbf{V} + \mathrm{vec}(\boldsymbol{\mu}\boldsymbol{\mu}^\top).$$

Using this result, show that:

(a) $\mathrm{E}\{(\mathbf{U}\otimes\mathbf{U})\} = \mathrm{vec}(\boldsymbol{\Sigma}) + \mathrm{vec}(\boldsymbol{\mu}\boldsymbol{\mu}^\top)$;

(b) $\mathrm{E}\{(\mathbf{U}\otimes\mathbf{U})\mathbf{U}^\top\} = (\mathbf{C}\otimes\mathbf{C})\mathrm{E}\{(\mathbf{V}\otimes\mathbf{V})\mathbf{V}^\top\}\mathbf{C}^\top + (\mathbf{I}_{n^2}+\mathbf{K}_n)(\boldsymbol{\Sigma}\otimes\boldsymbol{\mu}) + \{\mathrm{vec}(\boldsymbol{\Sigma}) + \mathrm{vec}(\boldsymbol{\mu}\boldsymbol{\mu}^\top)\}\boldsymbol{\mu}^\top$;

(c)
$$\begin{aligned}\mathrm{E}\{(\mathbf{U}\otimes\mathbf{U})(\mathbf{U}\otimes\mathbf{U})^\top\} = {} & (\mathbf{C}\otimes\mathbf{C})\mathrm{E}\{(\mathbf{V}\otimes\mathbf{V})(\mathbf{V}\otimes\mathbf{V})^\top\}(\mathbf{C}\otimes\mathbf{C}) \\ & + (\mathbf{C}\otimes\mathbf{C})\mathrm{E}\{(\mathbf{V}\otimes\mathbf{V})\mathbf{V}^\top\}(\mathbf{C}\otimes\boldsymbol{\mu})^\top(\mathbf{I}_{n^2}+\mathbf{K}_n) \\ & + (\mathbf{I}_{n^2}+\mathbf{K}_n)(\mathbf{C}\otimes\boldsymbol{\mu})\mathrm{E}\{(\mathbf{V}\otimes\mathbf{V})\mathbf{V}^\top\}^\top(\mathbf{C}\otimes\mathbf{C})^\top \\ & + 2(\mathbf{I}_{n^2}+\mathbf{K}_n)(\boldsymbol{\Sigma}\otimes\boldsymbol{\mu}\boldsymbol{\mu}^\top)(\mathbf{I}_{n^2}+\mathbf{K}_n) + \mathrm{vec}(\boldsymbol{\Sigma})\mathrm{vec}(\boldsymbol{\mu}\boldsymbol{\mu}^\top)^\top \\ & + \mathrm{vec}(\boldsymbol{\mu}\boldsymbol{\mu}^\top)\mathrm{vec}(\boldsymbol{\Sigma})^\top + \mathrm{vec}(\boldsymbol{\mu}\boldsymbol{\mu}^\top)\mathrm{vec}(\boldsymbol{\mu}\boldsymbol{\mu}^\top)^\top.\end{aligned}$$

Exercise 5.8 *Suppose that* $\mathbf{U}$ *is an n-dimensional random vector with finite multivariate moments of order three and four. Show that*

$$\mathrm{E}(\mathbf{a}^\top \mathbf{U}\mathbf{U}^\top \mathbf{B}\mathbf{U}) = \mathbf{a}^\top \mathrm{E}\{\mathbf{U}\mathrm{vec}(\mathbf{U}\mathbf{U}^\top)^\top\}\mathrm{vec}(\mathbf{B})^\top \quad \text{and} \quad \mathrm{var}(\mathbf{U}^\top \mathbf{B}\mathbf{U}) = \mathrm{vec}(\mathbf{B})^\top \mathrm{var}(\mathbf{U}\otimes\mathbf{U})\mathrm{vec}(\mathbf{B}).$$

Apply this result to find $\mathrm{E}(\bar{U}S_U^2) = \frac{1}{n^2}\mathrm{E}(\mathbf{1}_n^\top \mathbf{U}\mathbf{U}^\top \mathbf{M}\mathbf{U})$ *and* $\mathrm{var}(nS_U^2) = \mathrm{var}(\mathbf{U}^\top \mathbf{M}\mathbf{U})$, *where* $\mathbf{M} = \mathbf{I}_n - \frac{1}{n}\mathbf{1}_n\mathbf{1}_n^\top$ *and* $\mathbf{U} \sim \mathcal{HN}_n(\mathbf{0}_n, \mathbf{I}_n)$, *i.e.,* $\mathbf{U} = |\mathbf{Z}| = (|Z_1|, \ldots, |Z_n|)^\top$, *where* $\mathbf{Z} = (Z_1, \ldots, Z_n)^\top \sim \mathcal{N}_n(\mathbf{0}_n, \mathbf{I}_n)$.

Exercise 5.9 *Let* U_1, U_2 *be iid* $\mathcal{HN}(0,1)$ *random variables. Also, let*

$$f(z;\psi^2,\eta) = \frac{2}{\sqrt{\psi^2+\eta^2}}\phi\left(\frac{z}{\sqrt{\psi^2+\eta^2}}\right)\Phi\left(\frac{\eta z}{\psi\sqrt{\psi^2+\eta^2}}\right), \quad z \in \mathbb{R},$$

be the pdf of a random variable with $\mathcal{SN}(0,\psi^2,\eta)$ *distribution.*
(a) Show that $U_1 - U_2$ *has pdf given by*

$$g(u) = 2\left\{f(u;1,-1)I_{(u\geq 0)} + f(u;1,1)I_{(u<0)}\right\};$$

(b) Find the cdf of $U_1 - U_2$*;*
(c) Find the distribution of $(U_1 - U_2)^2$.

Exercise 5.10 *Let* $\mathbf{X}_1 = U_1\boldsymbol{\eta} + \mathbf{W}_1$ *and* $\mathbf{X}_2 = U_2\boldsymbol{\eta} + \mathbf{W}_2$, *where* U_1, U_2 *are iid* $\mathcal{HN}(0,1)$ *random variables and* $\mathbf{W}_1, \mathbf{W}_2$ *are iid* $\mathcal{N}_p(\mathbf{0}_p, \boldsymbol{\Psi})$ *nonsingular random vectors, with* $U_1, U_2, \mathbf{W}_1, \mathbf{W}_2$ *being mutually independent.*

(a) Find the distribution of $\mathbf{X}_1 - \mathbf{X}_2$*;*
(b) Find the mean and variance of $\mathbf{S} = \frac{1}{2}(\mathbf{X}_1 - \mathbf{X}_2)(\mathbf{X}_1 - \mathbf{X}_2)^\top$*;*
(c) Find the distribution of $\mathbf{S}$.

Exercise 5.11 *Let* $\zeta(z) = \frac{\phi(z)}{\Phi(z)}$ *for* $z \in \mathbb{R}$. *Assuming that* $Z \sim \mathcal{ASN}(0,1,\alpha)$, *prove for any integrable function g that*

$$\mathrm{E}\{\zeta(\delta Z)g(Z)\} = \begin{cases} \frac{\sqrt{2/\pi}}{\sqrt{1+\alpha^2}}\mathrm{E}\{g(Z_0)\}, & \text{if } g(-z) = g(z) \text{ (even)}, \\ 0, & \text{if } g(-z) = -g(z) \text{ (odd)}, \end{cases}$$

where $Z_0 \sim \mathcal{N}(0,1)$ *and* $\delta = \frac{\alpha}{\sqrt{1+\alpha^2}}$. *Also, show that*

$$\mathrm{E}\{\zeta'(\delta Z)\} = -\mathrm{E}\{\zeta(\delta Z)^2\} = -\sqrt{\frac{2}{\pi}}\frac{1}{\sqrt{1+\alpha^2}}\mathrm{E}\{\zeta(\delta Z_0)\}.$$

Exercise 5.12 *Let* $\mathbf{X}_0 \sim \mathcal{N}_p(\boldsymbol{\xi}, \boldsymbol{\Psi})$ *and* $\mathbf{X} \sim \mathcal{SN}_p(\boldsymbol{\xi}, \boldsymbol{\Psi}, \boldsymbol{\eta})$. *Also, let* $\zeta'(z)$ *be the first derivative of* $\zeta(z) = \frac{\phi(z)}{\Phi(z)}$ *and consider the function given by*

$$z(\mathbf{x}-\boldsymbol{\xi}) = \frac{\boldsymbol{\eta}^\top \boldsymbol{\Psi}^{-1}(\mathbf{x}-\boldsymbol{\xi})}{\sqrt{1+\boldsymbol{\eta}^\top \boldsymbol{\Psi}^{-1}\boldsymbol{\eta}}}, \quad \mathbf{x} \in \mathbb{R}^p.$$

Show that for any integrable vector-valued function $\mathbf{g} : \mathbb{R}^p \to \mathbb{R}^q$ *it follows that:*
(a) $\mathrm{E}[\zeta\{z(\mathbf{X}-\boldsymbol{\xi})\}\mathbf{g}(\mathbf{X}-\boldsymbol{\xi})] = \begin{cases} \sqrt{2/\pi}\mathrm{E}\{\mathbf{g}(\mathbf{X}_0-\boldsymbol{\xi})\}, & \text{if } \mathbf{g} \text{ is even,} \\ \mathbf{0}_q, & \text{if } \mathbf{g} \text{ is odd;} \end{cases}$
(b) $\mathrm{E}[\zeta'\{z(\mathbf{X}-\boldsymbol{\xi})\}\mathbf{g}(\mathbf{X}-\boldsymbol{\xi})] = -\mathrm{E}[\zeta\{z(\mathbf{X}-\boldsymbol{\xi})\}z(\mathbf{X}-\boldsymbol{\xi})\mathbf{g}(\mathbf{X}-\boldsymbol{\xi})] - \mathrm{E}[\zeta\{z(\mathbf{X}-\boldsymbol{\xi})\}^2\mathbf{g}(\mathbf{X}-\boldsymbol{\xi})]$.

Exercise 5.13 *Let* $\mathbf{X} \sim \mathcal{SN}_p(\boldsymbol{\xi}, \boldsymbol{\Psi}, \boldsymbol{\eta})$.
(a) Find the distribution of $\mathbf{X}-\boldsymbol{\mu}$, *where* $\boldsymbol{\mu} = \mathrm{E}(\mathbf{X})$;
(b) Find the first four centered moments of $\mathbf{X}$;
(c) Compute $\mathrm{var}\{(\mathbf{X}-\boldsymbol{\mu}) \otimes (\mathbf{X}-\boldsymbol{\mu})\}$.

Exercise 5.14 *Let* $\mathbf{Z} = \boldsymbol{\Sigma}^{-1/2}(\mathbf{X}-\boldsymbol{\mu})$, *where* $\boldsymbol{\mu} = \mathrm{E}(\mathbf{X})$, $\boldsymbol{\Sigma} = \mathrm{var}(\mathbf{X}) > 0$ *and* $\mathbf{X} \sim \mathcal{SN}_p(\boldsymbol{\xi}, \boldsymbol{\Psi}, \boldsymbol{\eta})$.
(a) Show that $\mathbf{Z} \sim \mathcal{SN}_p(-b\boldsymbol{\Sigma}^{-1/2}\boldsymbol{\eta}, \boldsymbol{\Sigma}^{-1/2}\boldsymbol{\Psi}\boldsymbol{\Sigma}^{-1/2}, \boldsymbol{\Sigma}^{-1/2}\boldsymbol{\eta})$, *where* $b = \sqrt{2/\pi}$;
(b) Show that $\mathbf{Z} = \boldsymbol{\Sigma}^{-1/2}\mathbf{Z}_0$, *where* $\mathbf{Z}_0 = \boldsymbol{\eta}\, U_0 + \mathbf{W}$, *with* $U_0 = U - b$, $U \sim \mathcal{HN}(0,1)$, *which is independent of* $\mathbf{W} \sim \mathcal{N}_p(\mathbf{0}_p, \boldsymbol{\Psi})$;
(c) Show that $\mathrm{vec}(\mathbf{Z}\mathbf{Z}^\top) = (\boldsymbol{\Sigma}^{-1/2} \otimes \boldsymbol{\Sigma}^{-1/2})\mathrm{vec}(\mathbf{Z}_0\mathbf{Z}_0^\top)$, *where* $\mathrm{vec}(\mathbf{Z}_0\mathbf{Z}_0^\top) = \mathrm{vec}(\boldsymbol{\eta}\boldsymbol{\eta}^\top)U_0^2 + (\mathbf{I}_{p^2} + \mathbf{K}_p)(\boldsymbol{\eta} \otimes \mathbf{I}_p)U_0\mathbf{W} + \mathrm{vec}(\mathbf{W}\mathbf{W}^\top)$;
(d) Find the first four moments of $\mathbf{Z}_0$ *and* $\mathbf{Z}$ *and their variance matrices;*
(e) Find $\mathrm{E}(\|\mathbf{Z}\|^2)$ *and* $\mathrm{var}(\|\mathbf{Z}\|^2)$;
(f) Show (5.45) and (5.46).

Exercise 5.15 *Plot and investigate the shape of the following distributions:* $\mathcal{ASN}_2(\boldsymbol{\xi}, \boldsymbol{\Omega}, \boldsymbol{\alpha})$, $\mathcal{SN}_2(\boldsymbol{\xi}, \boldsymbol{\Psi}, \boldsymbol{\eta})$, *and* $\mathcal{ESN}_2(\boldsymbol{\xi}, \boldsymbol{\Psi}, \boldsymbol{\eta}, \tau)$. *Give a geometric interpretation of all the parameters.*

Exercise 5.16 *Let* $\mathbf{X} = (\mathbf{Y} \mid Z \le \boldsymbol{\lambda}^\top \mathbf{Y})$, *where* $\mathbf{Y}$ *is a random p-vector with pdf f symmetric about the origin (i.e.,* $f(-\mathbf{y}) = f(\mathbf{y})$ *for all* $\mathbf{y} \in \mathbb{R}^p$*), Z is a random variable with absolutely continuous distribution function G such that* G' *is symmetric about* 0 *and* $\boldsymbol{\lambda}$ *is a fixed p-vector.*
(a) Assuming that $\mathbf{Y}$ *and Z are independent, show that the pdf of* $\mathbf{X}$ *is given by*

$$f(\mathbf{x};\boldsymbol{\lambda}) = 2f(\mathbf{x})G(\boldsymbol{\lambda}^\top \mathbf{x}), \quad \mathbf{x} \in \mathbb{R}^p;$$

(b) Show that $-\mathbf{X}$ *has the pdf given by* $f(\mathbf{x}; -\boldsymbol{\lambda})$;
(c) Show that $2\psi_{\mathbf{Y}}(\mathbf{t}) = \psi_{\mathbf{X}}(\mathbf{t}) + \psi_{-\mathbf{X}}(\mathbf{t})$, *where* $\psi_{\mathbf{X}}$ *and* $\psi_{\mathbf{Y}}$ *are the cfs of* $\mathbf{X}$ *and* $\mathbf{Y}$, *respectively;*
(d) Show that $\mathbf{h}(\mathbf{X}) \stackrel{d}{=} \mathbf{h}(\mathbf{Y})$ *for any measurable even function* $\mathbf{h}$;
(e) Show that the even moments (provided that they exist) of $\mathbf{X}$ *do not depend on* $\boldsymbol{\lambda}$;
(f) If $\mathbf{Y} \sim \mathcal{N}_p(\mathbf{0}_p, \boldsymbol{\Omega})$ *and* $Z \sim \mathcal{N}(0,1)$ *and are independent, show that* $\mathbf{X}\mathbf{X}^\top \sim \mathcal{W}_p(1, \boldsymbol{\Omega})$.

Exercise 5.17 (Skew-normal matrix distribution)

(a) *Let* $\mathbf{X} = (\mathbf{X}_1, \ldots, \mathbf{X}_n)^\top$, *with* $\mathbf{X}_1, \ldots, \mathbf{X}_n \overset{iid}{\sim} \mathcal{SN}_p(\boldsymbol{\xi}, \boldsymbol{\Psi}, \boldsymbol{\eta})$; *i.e.,* $\mathbf{X}$ *is a skew-normal data matrix.*

(i) *Assuming* $\boldsymbol{\Psi} > 0$, *show that* $\mathrm{vec}(\mathbf{X})$ *has pdf given by*

$$2^n \phi_{np}(\mathrm{vec}(\mathbf{X}); \boldsymbol{\xi} \otimes \mathbf{1}_n, \boldsymbol{\Omega} \otimes \mathbf{I}_n)\Phi_n[(\boldsymbol{\delta} \otimes \mathbf{I}_n)^\top \{\mathrm{vec}(\mathbf{X}) - (\boldsymbol{\xi} \otimes \mathbf{1}_n)\}],$$

which can be rewritten as the pdf of the skew-normal matrix distribution:

$$2^n \phi_{n,p}(\mathbf{X}; \mathbf{M}, \mathbf{I}_n \otimes \boldsymbol{\Omega})\Phi_n\{(\mathbf{X} - \mathbf{M})\boldsymbol{\delta}\},$$

where $\mathbf{M} = \mathbf{1}_n \boldsymbol{\xi}^\top$, $\boldsymbol{\Omega} = \boldsymbol{\Psi} + \boldsymbol{\eta}\boldsymbol{\eta}^\top, \boldsymbol{\delta} = (1 + \boldsymbol{\eta}^\top \boldsymbol{\Psi}^{-1} \boldsymbol{\eta})^{-1/2} \boldsymbol{\Psi}^{-1} \boldsymbol{\eta}$, $\phi_{n,p}(\cdot; \mathbf{M}, \boldsymbol{\Lambda} \otimes \boldsymbol{\Sigma})$ *is the pdf of the normal matrix distribution (see Section 3.11.1) and* $\Phi_n(\cdot) = \Phi_n(\cdot; \mathbf{0}_n, \mathbf{I}_n)$;

(ii) *Show that* $\mathbf{X} \overset{d}{=} \mathbf{1}_n \boldsymbol{\xi}^\top + \mathbf{U}\boldsymbol{\eta}^\top + \mathbf{W}$, *where* $\mathbf{U} \sim \mathcal{HN}_n(\mathbf{0}_n, \mathbf{I}_n)$ *is a half-normal random vector and* $\mathbf{W}$ *is an* $n \times p$ *normal data matrix such that* $\mathrm{vec}(\mathbf{W}) \sim \mathcal{N}_{np}(\mathbf{0}_{np}, \boldsymbol{\Psi} \otimes \mathbf{I}_n)$, *which are independent;*

(iii) *Find* $\mathrm{E}(\mathbf{X})$ *and* $\mathrm{var}(\mathbf{X})$;

(iv) *Let* $\mathbf{Y} = \mathbf{A}\mathbf{X}\mathbf{B}^\top$, *where* $\mathbf{A} \in \mathbb{R}^{n \times n}$ *and* $\mathbf{B} \in \mathbb{R}^{p \times p}$ *are nonsingular constant matrices. Show that the Jacobian method yields the following pdf for* $\mathrm{vec}(\mathbf{Y}) = (\mathbf{B} \otimes \mathbf{A})\mathrm{vec}(\mathbf{X})$:

$$2^n \phi_{np}(\mathrm{vec}(\mathbf{Y}); \boldsymbol{\xi}_B \otimes \mathbf{A}\mathbf{1}_n, \boldsymbol{\Omega}_B \otimes \mathbf{A}\mathbf{A}^\top)\Phi_n[(\boldsymbol{\delta}_B \otimes \mathbf{A}^{-\top})^\top \{\mathrm{vec}(\mathbf{Y}) - (\boldsymbol{\xi}_B \otimes \mathbf{A}\mathbf{1}_n)\}],$$

where $\boldsymbol{\xi}_B = \mathbf{B}\boldsymbol{\xi}$, $\boldsymbol{\Omega}_B = \boldsymbol{\Psi}_B + \boldsymbol{\eta}_B \boldsymbol{\eta}_B^\top$ *and* $\boldsymbol{\delta}_B = (1 + \boldsymbol{\eta}_B^\top \boldsymbol{\Psi}_B^{-1} \boldsymbol{\eta}_B)^{-1/2} \boldsymbol{\Psi}_B^{-1} \boldsymbol{\eta}_B$ *with* $\boldsymbol{\Psi}_B = \mathbf{B}\boldsymbol{\Psi}\mathbf{B}^\top$ *and* $\boldsymbol{\eta}_B = \mathbf{B}\boldsymbol{\eta}$;

(b) *Now, assume* $\mathbf{X} = \mathbf{M} + \mathbf{U}\mathbf{H}^\top + \mathbf{W}$, *where* $\mathbf{M} \in \mathbb{R}^{n \times p}$ *is a location matrix and* $\mathbf{H} \in \mathbb{R}^{p \times q}$ *is a skewness/shape matrix. Let* $\mathbf{U} \sim \mathcal{HN}_{n,q}(\mathbf{O}, \boldsymbol{\Lambda} \otimes \boldsymbol{\Gamma})$ *follow a half-normal matrix distribution, and let* $\mathbf{W} \sim \mathcal{N}_{n,p}(\mathbf{O}, \boldsymbol{\Lambda} \otimes \boldsymbol{\Psi})$ *be independent of* $\mathbf{U}$, *with* $\boldsymbol{\Lambda}, \boldsymbol{\Gamma}, \boldsymbol{\Psi} > 0$. *Find the pdf of vec*$(\mathbf{X})$ *under this model.*

Note: Further results on skew-normal matrix distributions can be found in Chen and Gupta (2005), Harrar and Gupta (2008), Zheng et al. (2012b), Ye et al. (2014), Ning (2015), Zheng et al. (2016), and Young et al. (2020), on extended skew-normal matrix distributions in Ning and Gupta (2012), on closed skew-normal matrix distributions in Domínguez-Molina et al. (2007), and on scale and shape mixtures of matrix variate extended skew-normal distributions in Rezaei et al. (2020). Extensions to skew-t matrix distributions can be found in Zheng et al. (2012a) and Gallaugher and McNicholas (2017), and to skew-elliptical matrix distributions in Fang (2008b); see also Gallaugher and McNicholas (2019).

Exercise 5.18 *Study the formal properties of the multivariate generalized skew-normal distribution introduced in Section 5.11.1.*

Exercise 5.19 *Study the formal properties of the multivariate flexible skew-normal distribution introduced in Section 5.11.1 with skewing function* $\pi(\mathbf{x}) = \Phi(\boldsymbol{\alpha}^\top \boldsymbol{\omega}\mathbf{x} + \boldsymbol{\beta}^\top \boldsymbol{\omega}\mathbf{x}\mathbf{x}^\top \boldsymbol{\omega}^2 \mathbf{x})$, $\mathbf{x} \in \mathbb{R}^p$ *with* $\boldsymbol{\alpha} \in \mathbb{R}^p$, $\boldsymbol{\beta} \in \mathbb{R}^p$, $\boldsymbol{\omega} = \mathrm{diag}(\boldsymbol{\Omega})^{1/2}$.

Exercise 5.20 *Study the formal properties of the multivariate modified skew-normal distribution introduced in Section 5.11.3.*

Exercise 5.21 *Study the formal properties of the multivariate skew-slash distribution introduced in Section 5.11.4. Do the same for its alternative definition. Discuss the pros and cons of these two formulations.*

Exercise 5.22 *Study the formal properties of the multivariate skew-normal-Tukey-h distribution introduced in Section 5.11.5.*

Exercise 5.23 *Study the formal properties of the multivariate SN-CNG distributions introduced in Section 5.11.6.*

Exercise 5.24 *Explore the entropy, mutual information, KL divergence, and J divergence for two multivariate SN distributions. Compare your results with those for ASN distributions in Contreras-Reyes and Arellano-Valle (2012) and Arellano-Valle et al. (2013b); see also de Queiroz et al. (2016b) for the case of log-fundamental skew-normal and related distributions.*

Exercise 5.25 *Let* $\mathbf{X} = (\mathbf{X}_1^\top, \mathbf{X}_2^\top)^\top \sim \mathcal{SN}_p(\boldsymbol{\xi}, \boldsymbol{\Psi}, \boldsymbol{\eta})$ *with* $\mathbf{X}_1 \in \mathbb{R}^{p_1}$, $\mathbf{X}_2 \in \mathbb{R}^{p_2}$, *and the corresponding partitions for* $\boldsymbol{\xi}$, $\boldsymbol{\Psi}$, *and* $\boldsymbol{\eta}$. *Show that:*

$$\mathrm{E}(\|\mathbf{X}_1 - \mathbf{X}_2\|^2) = \|\boldsymbol{\xi}_1 - \boldsymbol{\xi}_2\|^2 + \mathrm{tr}(\boldsymbol{\Psi}_{11} + \boldsymbol{\Psi}_{22} - 2\boldsymbol{\Psi}_{12}) + \left(1 - \frac{2}{\pi}\right) \|\boldsymbol{\eta}_1 - \boldsymbol{\eta}_2\|^2.$$

Derive for $q = 2$ *the Wasserstein distance between the distributions of* $\mathbf{X}_1$ *and* $\mathbf{X}_2$.

Exercise 5.26 *Let* $X_1 \in \mathbb{R}$ *and* $\mathbf{X}_2 \in \mathbb{R}^{p-1}$ *be random variables with variance matrix* $\boldsymbol{\Sigma}$ *given by*

$$\boldsymbol{\Sigma} = \begin{pmatrix} \sigma_{11} & \boldsymbol{\sigma}_{21}^\top \\ \boldsymbol{\sigma}_{21} & \boldsymbol{\Sigma}_{22} \end{pmatrix}, \quad \boldsymbol{\Sigma}_{22} > 0.$$

It is known that the maximum correlation between X_1 *and a linear combination* $\mathbf{a}^\top \mathbf{X}_2$ *of* $\mathbf{X}_2$, *also called the multiple correlation coefficient between* X_1 *and* $\mathbf{X}_2$, *is*

$$R^2 = \frac{\boldsymbol{\sigma}_{21}^\top \boldsymbol{\Sigma}_{22}^{-1} \boldsymbol{\sigma}_{21}}{\sigma_{11}}.$$

Assume that X_1 *and* $\mathbf{X}_2$ *have a skew-normal joint distribution given by*

$$\begin{pmatrix} X_1 \\ \mathbf{X}_2 \end{pmatrix} \sim \mathcal{SN}_p\left(\begin{pmatrix} 0 \\ \mathbf{0}_{p-1} \end{pmatrix}, \begin{pmatrix} \psi_{11} & \boldsymbol{\psi}_{21}^\top \\ \boldsymbol{\psi}_{21} & \boldsymbol{\Psi}_{22} \end{pmatrix}, \begin{pmatrix} \eta_1 \\ \boldsymbol{\eta}_2 \end{pmatrix} \right), \quad \boldsymbol{\Psi}_{22} > 0.$$

Find the value of R^2 *in terms of the SN parameters and explain what happens if: (a)* $\eta_1 = \boldsymbol{\psi}_{21}^\top \boldsymbol{\Psi}_{22}^{-1} \boldsymbol{\eta}_2$ *or (b)* $\boldsymbol{\eta}_2 = \mathbf{0}_{p-1}$; *can you identify the regression equations of* X_1 *on* $\mathbf{X}_2$ *in each of these cases?*

Exercise 5.27 *Let* $\mathbf{X}_1,\ldots,\mathbf{X}_n$ *be a random sample from* $\mathbf{X} = (X_1, \mathbf{X}_2^\top)^\top \sim \mathcal{SN}_p(\mathbf{0}_p, \boldsymbol{\Psi}, \boldsymbol{\eta})$. *Derive the asymptotic distribution of* $R_n^2 = \mathbf{S}_{21}^\top \mathbf{S}_{22}^{-1} \mathbf{S}_{21}/S_{11}$ *which is defined in terms of the partition of* $\mathbf{S}_n = \frac{1}{n}\sum_{i=1}^n \mathbf{X}_i \mathbf{X}_i^\top$ *as*

$$\mathbf{S}_n = \begin{pmatrix} S_{11} & \mathbf{S}_{21}^\top \\ \mathbf{S}_{21} & \mathbf{S}_{22} \end{pmatrix}.$$

5.13 Open Research Problems

Open Research Problem 3 *Develop the theory and inference for multivariate skew-L_q-spherical-normal distributions.*

Open Research Problem 4 *Develop the theory and inference for multivariate skew-symmetric distributions of the form* $2f(\mathbf{x})\pi(\mathbf{x})$ *(see Section 5.11.1) where the assumption of central symmetry on the pdf* f *is replaced by other types of symmetry from Chapter 2.*

Open Research Problem 5 *Develop the theory for multivariate skew-L_q-elliptical distributions.*

Chapter 6

The Multivariate Unified Skew-Normal Distribution

The multivariate unified skew-normal (SUN) distribution constitutes a natural extension of the multivariate skew-normal (SN) distribution, which itself is a generalization of the classical multivariate normal distribution. It preserves both the marginal and conditional distributions, and has originated from multiple independent sources, with some differences in the technical development but with a common underlying structure, as explained later. For instance, Gupta et al. (2004a) have developed the so-called closed skew-normal distribution, while Arellano-Valle and Genton (2005) introduced the family of fundamental skew-normal distributions, which also includes a second version of the closed skew-normal. The interconnections among these apparently separate formulations have been examined by Arellano-Valle and Azzalini (2006), showing their essential equivalence, as well as the presence of over-parameterizations in some cases. They introduced a unifying version which embraces the above-recalled specific proposals, removing at the same time the existing over-parameterizations. This chapter provides a definition of the multivariate SUN family from a different perspective, describes its main properties, and gives further important characteristics.

The SUN family constitutes a superset of the SN family, more specifically, of the extended skew-normal (ESN) family. Its building mechanism, based on a subset of latent variables, leads to certain properties that are not amenable to the SN and ESN distributions. An important specific fact is that of closure of the family with respect to convolution; specifically, the sum of two independent SUN random vectors is also a SUN random vector. This property has proved convenient in a number of operational formulations that employ the SUN distribution as its core stochastic component.

6.1 Construction, Definition, and First Elementary Properties

Now we can provide a general definition of the multivariate unified skew-normal distribution, which can be singular or nonsingular depending on whether the dispersion matrix, $\boldsymbol{\Psi}$ say, is positive semi-definite ($\boldsymbol{\Psi} \geq 0$; $|\boldsymbol{\Psi}| = 0$) or positive definite ($\boldsymbol{\Psi} > 0$; $|\boldsymbol{\Psi}| > 0$), respectively. In addition, this definition provides a simple way for the construction of random vectors with the SUN distribution.

Definition 18 [Multivariate unified skew-normal distribution] *A random vector $\mathbf{X} \in \mathbb{R}^p$ is said to have a multivariate unified skew-normal distribution with location parameter $\boldsymbol{\xi} \in \mathbb{R}^p$, p-by-p dispersion matrix $\boldsymbol{\Psi}$, and shape/skewness parameters $\mathbf{H} \in \mathbb{R}^{p\times q}$, $\boldsymbol{\tau} \in \mathbb{R}^q$, and $\bar{\boldsymbol{\Gamma}} > 0$, a q-by-q correlation matrix, in symbols $\mathbf{X} \sim \mathcal{SUN}_{p,q}(\boldsymbol{\xi}, \boldsymbol{\Psi}, \mathbf{H}, \boldsymbol{\tau}, \bar{\boldsymbol{\Gamma}})$, iff $\mathbf{X}$ can be expressed as*

$$\mathbf{X} = \boldsymbol{\xi} + \mathbf{H}\mathbf{U} + \mathbf{W}, \tag{6.1}$$

where $\mathbf{U} \stackrel{d}{=} (\mathbf{W}_0 \mid \mathbf{W}_0 + \boldsymbol{\tau} > \mathbf{0}_q)$, with $\mathbf{W}_0 \sim \mathcal{N}_q(\mathbf{0}_q, \bar{\boldsymbol{\Gamma}})$, is a truncated normal random vector, and $\mathbf{W} \sim \mathcal{N}_p(\mathbf{0}_p, \boldsymbol{\Psi})$ which is independent of $\mathbf{W}_0$ and therefore of $\mathbf{U}$.

When $q = 1$, with $\mathbf{H} = \boldsymbol{\eta}$, $\boldsymbol{\tau} = \tau$ and $\bar{\boldsymbol{\Gamma}} = 1$, we obtain from Equation (6.1) the so-called multivariate extended skew-normal (ESN) distribution (see Section 6.6.3), which for $\tau = 0$ reduces to the multivariate skew-normal (SN) distribution. Another simple special case arises when $\boldsymbol{\tau} = \mathbf{0}_q$ and $\bar{\boldsymbol{\Gamma}} = \mathbf{I}_q$, which will be described later in Example 68.

It is also clear from Equation (6.1) that

$$\mathbf{H} = \mathbf{O} \quad \Longrightarrow \quad \mathbf{X} = \boldsymbol{\xi} + \mathbf{W} \sim \mathcal{N}_p(\boldsymbol{\xi}, \boldsymbol{\Psi}),$$

which can be singular or nonsingular according to the dispersion matrix $\boldsymbol{\Psi} \geq 0$ or $\boldsymbol{\Psi} > 0$, respectively. Therefore, Definition 18 includes both the singular and the nonsingular multivariate unified skew-normal distribution. The nonsingular case also requires that $\boldsymbol{\Psi} + \mathbf{H}\bar{\boldsymbol{\Gamma}}\mathbf{H}^\top > 0$ for the pdf to be valid, which is guaranteed if $\boldsymbol{\Psi} > 0$ for all $\mathbf{H} \in \mathbb{R}^{p\times q}$, since by assumption, $\bar{\boldsymbol{\Gamma}} > 0$ always.

Note that apparently we could also recover the $\mathcal{N}_p(\boldsymbol{\xi}, \boldsymbol{\Psi})$ distribution for $\mathbf{X}$ for any arbitrary non-null skewness matrix $\mathbf{H}$ satisfying the conditions $\mathbf{H}\boldsymbol{\tau} = \mathbf{0}_p$ and $\mathbf{H}\bar{\boldsymbol{\Gamma}} = \mathbf{O}$. However, since by assumption the matrix $\bar{\boldsymbol{\Gamma}} > 0$ is nonsingular, the last of these conditions is equivalent to $\mathbf{H} = \mathbf{O}$.

From Equation (6.1), it is also clear that $\mathbf{X} \sim \mathcal{SUN}_{p,q}(\boldsymbol{\xi}, \boldsymbol{\Psi}, \mathbf{H}, \boldsymbol{\tau}, \bar{\boldsymbol{\Gamma}})$ can be represented as

$$\mathbf{X} = \boldsymbol{\xi} + \mathbf{AZ}, \qquad \mathbf{Z} \sim \mathcal{SUN}_{q,q}(\mathbf{0}_q, \mathbf{I}_q, \mathbf{H}_q, \boldsymbol{\tau}, \bar{\boldsymbol{\Gamma}}), \tag{6.2}$$

for some $\mathbf{A} \in \mathbb{R}^{p\times q}$ and $\mathbf{H}_q \in \mathbb{R}^{q\times q}$ such that $\mathbf{AA}^\top = \boldsymbol{\Psi}$ and $\mathbf{AH}_q = \mathbf{H}$.

Remark 10 *Since by definition* $\mathbf{U} \stackrel{d}{=} (\mathbf{W}_0 \mid \mathbf{W}_0 + \boldsymbol{\tau} > \mathbf{0}_q)$, *where* $\mathbf{W}_0 \sim \mathcal{N}_q(\mathbf{0}_q, \bar{\boldsymbol{\Gamma}})$, *the condition that* $\bar{\boldsymbol{\Gamma}}$ *is a full-rank correlation matrix is due to two reasons: a) to avoid redundancies in the selection mechanism* $\mathbf{W}_0 > \mathbf{0}_q$*; and b) to avoid over-parameterizations in the model, since in (6.1) we have for any* $q \times q$ *nonsingular matrix* $\mathbf{D}$ *that*

$$\mathbf{X} = \mathbf{HU} + \mathbf{W} + \boldsymbol{\xi} = \mathbf{HD}^{-1}\mathbf{DU} + \mathbf{W} + \boldsymbol{\xi} = \tilde{\mathbf{H}}\tilde{\mathbf{U}} + \mathbf{W} + \boldsymbol{\xi},$$

where $\tilde{\mathbf{H}} = \mathbf{HD}^{-1}$ *and* $\tilde{\mathbf{U}} = \mathbf{DU}$. *Also, for any diagonal matrix* $\mathbf{D} > 0$, *the event* $\{\tilde{\mathbf{W}}_0 + \tilde{\boldsymbol{\tau}} > \mathbf{0}_q\}$, *where* $\tilde{\mathbf{W}}_0 = \mathbf{DW}$ *and* $\tilde{\boldsymbol{\tau}} = \mathbf{D}\boldsymbol{\tau}$, *is equivalent to* $\{\mathbf{W}_0 + \boldsymbol{\tau} > \mathbf{0}_q\}$ *and so* $\tilde{\mathbf{U}} \stackrel{d}{=} (\tilde{\mathbf{W}}_0 \mid \tilde{\mathbf{W}}_0 > \mathbf{0}_q)$, *where* $\tilde{\mathbf{W}}_0 \sim \mathcal{N}_q(\tilde{\boldsymbol{\tau}}, \tilde{\boldsymbol{\Gamma}})$, *with* $\tilde{\boldsymbol{\Gamma}} = \mathbf{D}\bar{\boldsymbol{\Gamma}}\mathbf{D}$. *Thus, the representations* $\mathbf{X} \sim \mathcal{SUN}_{p,q}(\boldsymbol{\xi}, \boldsymbol{\Psi}, \mathbf{H}, \boldsymbol{\tau}, \bar{\boldsymbol{\Gamma}})$ *and* $\mathbf{X} \sim \mathcal{SUN}_{p,q}(\boldsymbol{\xi}, \boldsymbol{\Psi}, \tilde{\mathbf{H}}, \tilde{\boldsymbol{\tau}}, \tilde{\boldsymbol{\Gamma}})$ *are equivalent for any diagonal matrix* $\mathbf{D} > 0$*; in particular, when* $\mathbf{D} = \operatorname{diag}(\tilde{\boldsymbol{\Gamma}})^{1/2}$ *for any variance matrix* $\tilde{\boldsymbol{\Gamma}} > 0$.

Remark 11 *If the skewness matrix* $\mathbf{H}$ *is reparameterized as* $\mathbf{H} = \boldsymbol{\Psi}^{1/2}\boldsymbol{\Lambda}$ *with* $\boldsymbol{\Lambda} \in \mathbb{R}^{p\times q}$, *then (6.1) becomes the location-scale random vector* $\mathbf{X} = \boldsymbol{\xi} + \boldsymbol{\Psi}^{1/2}\mathbf{Z}$, *where* $\mathbf{Z} \sim \mathcal{SUN}_{p,q}(\mathbf{0}_p, \mathbf{I}_p, \boldsymbol{\Lambda}, \boldsymbol{\tau}, \bar{\boldsymbol{\Gamma}})$.

Remark 12 *Let* $\mathcal{U}_m^+$ *and* $\mathcal{L}_m^+$ *be the groups of* $m \times m$ *upper and lower triangular matrices, respectively, with positive diagonal elements; see Section A.2.7. Further representations of (6.1) based on the QR decomposition of* $\mathbf{H} \in \mathbb{R}^{p\times q}$ *are:*

(a) If $\operatorname{rank}(\mathbf{H}) = q$, *then* $\mathbf{X} = \boldsymbol{\xi} + \mathbf{QRU} + \mathbf{W}$, *for some* $\mathbf{Q} \in \mathbb{R}^{p\times q}$ *such that* $\mathbf{Q}^\top\mathbf{Q} = \mathbf{I}_q$ *and* $\mathbf{R} \in \mathcal{U}_q^+$, *which are unique. In this case, if* $\boldsymbol{\Psi} = \mathbf{I}_p$, *then* $\mathbf{Q}^\top(\mathbf{X} - \boldsymbol{\xi}) = \mathbf{RU} + \mathbf{W}_q$, *where* $\mathbf{W}_q = \mathbf{Q}^\top\mathbf{W} \sim \mathcal{N}_q(\mathbf{0}_q, \mathbf{I}_q)$*;*

(b) If $\operatorname{rank}(\mathbf{H}) = p$, *then* $\mathbf{X} = \boldsymbol{\xi} + \mathbf{TQU} + \mathbf{W}$, *for some* $\mathbf{T} \in \mathcal{L}_p^+$ *and* $\mathbf{Q} \in \mathbb{R}^{p\times q}$ *such that* $\mathbf{QQ}^\top = \mathbf{I}_p$, *which are also unique. In this case,* $\mathbf{X} = \boldsymbol{\xi} + \mathbf{TU}_p + \mathbf{W}$, *where* $\mathbf{U}_p = (\mathbf{QW}_0 \mid \mathbf{W}_0 + \boldsymbol{\tau} > \mathbf{0}_q)$ *with* $\mathbf{QW}_0 \sim \mathcal{N}_p(\mathbf{0}_p, \mathbf{Q}\bar{\boldsymbol{\Gamma}}\mathbf{Q}^\top)$.

6.1.1 Mean vector and variance matrix

Definition 18 is also very convenient for obtaining the elementary properties of a unified skew-normal random vector $\mathbf{X}$. For instance, this definition considerably facilitates the calculation of its mean vector and variance matrix.

Theorem 87 *Let* $\mathbf{X} \sim \mathcal{SUN}_{p,q}(\boldsymbol{\xi}, \boldsymbol{\Psi}, \mathbf{H}, \boldsymbol{\tau}, \bar{\boldsymbol{\Gamma}})$. *Then,*

$$\mathrm{E}(\mathbf{X}) = \boldsymbol{\xi} + \mathbf{H}\mathrm{E}(\mathbf{U}) \quad \text{and} \quad \mathrm{var}(\mathbf{X}) = \boldsymbol{\Psi} + \mathbf{H}\mathrm{var}(\mathbf{U})\mathbf{H}^\top,$$

where $\mathrm{E}(\mathbf{U}) = \mathrm{E}(\mathbf{W}_0 \mid \mathbf{W}_0 + \boldsymbol{\tau} > \mathbf{0}_q)$ *and* $\mathrm{var}(\mathbf{U}) = \mathrm{var}(\mathbf{W}_0 \mid \mathbf{W}_0 + \boldsymbol{\tau} > \mathbf{0}_q)$ *for* $\mathbf{W}_0 \sim \mathcal{N}_q(\mathbf{0}_q, \bar{\boldsymbol{\Gamma}})$.

Proof 87 *By (6.1) we have*

$$\begin{aligned} \mathrm{E}(\mathbf{X}) &= \mathrm{E}(\boldsymbol{\xi} + \mathbf{HU} + \mathbf{W}) = \boldsymbol{\xi} + \mathbf{H}\mathrm{E}(\mathbf{U}) + \mathrm{E}(\mathbf{W}), \\ \mathrm{var}(\mathbf{X}) &= \mathrm{var}(\boldsymbol{\xi} + \mathbf{HU} + \mathbf{W}) = \mathbf{H}\mathrm{var}(\mathbf{U})\mathbf{H}^\top + \mathrm{var}(\mathbf{W}), \end{aligned}$$

since $\mathbf{U}$ *and* $\mathbf{W}$ *are independent.* □

Obviously, we need the mean vector and the variance matrix of the multivariate truncated normal distribution to find these quantities for the multivariate unified skew-normal distribution. With the exception of the cases of $q = 1$ or $\bar{\boldsymbol{\Gamma}} = \mathbf{I}_q$ for $q \geq 2$, there are no simple expressions for computing these quantities, so that in general they must be computed numerically. Some explicit yet complicated-to-evaluate expressions for the expectation and variance of multivariate truncated normal random vectors can be found in Arellano-Valle and Azzalini (2022).

As mentioned earlier, a simple and relevant special case occurs when $\mathbf{U} \sim \mathcal{HN}_q(\mathbf{0}_q, \mathbf{I}_q)$, in which $\boldsymbol{\tau} = \mathbf{0}_q$ and $\bar{\boldsymbol{\Gamma}} = \mathbf{I}_q$; that is, the random vector $\mathbf{U}$ has components $U_i \overset{iid}{\sim} \mathcal{HN}(0,1)$, $i = 1, \ldots, q$. In other words, $\mathbf{U} = |\mathbf{W}_0| := (|W_{01}|, \ldots, |W_{0q}|)^\top$, where $\mathbf{W}_0 \sim \mathcal{N}_q(\mathbf{0}_q, \mathbf{I}_q)$ so that $U_i = |W_{0i}|$ for $i = 1, \ldots, q$. We illustrate this case in the following example.

Example 68 (Multivariate unified skew-normal distribution with independent latent variables) *If* $\boldsymbol{\tau} = \mathbf{0}_q$ *and* $\bar{\boldsymbol{\Gamma}} = \mathbf{I}_q$, *then* $\mathbf{U} \sim \mathcal{HN}_q(\mathbf{0}_q, \mathbf{I}_q)$ *have iid* $\mathcal{HN}(0,1)$ *random components, with mean* $\sqrt{2/\pi}$ *and variance* $1 - 2/\pi$. *So* $\mathrm{E}(\mathbf{U}) = \sqrt{2/\pi}\mathbf{1}_q$ *and* $\mathrm{var}(\mathbf{U}) = (1 - 2/\pi)\mathbf{I}_q$; *hence,* $\mathbf{X} \sim \mathcal{SUN}_{p,q}(\boldsymbol{\xi}, \boldsymbol{\Psi}, \mathbf{H}, \mathbf{0}_q, \mathbf{I}_q)$, *with mean vector and variance matrix given by*

$$\mathrm{E}(\mathbf{X}) = \boldsymbol{\xi} + \sqrt{\frac{2}{\pi}}\mathbf{H}\mathbf{1}_q \quad \textit{and} \quad \mathrm{var}(\mathbf{X}) = \boldsymbol{\Psi} + \left(1 - \frac{2}{\pi}\right)\mathbf{H}\mathbf{H}^\top.$$

In this case, if $q = p$, $\boldsymbol{\Psi} = \mathrm{diag}(\psi_1^2, \ldots, \psi_p^2)$ *and* $\mathbf{H} = \mathrm{diag}(\eta_1, \ldots, \eta_p)$, *then the components* $X_1, \ldots, X_p$ *of* $\mathbf{X}$ *are independent, with* $X_i \sim \mathcal{SN}(\xi_i, \psi_i^2, \eta_i)$ *for* $i = 1, \ldots, p$.

6.1.2 Hierarchical representation

According to Definition 18, the multivariate unified skew-normal distribution is a particular member of the more general class of mean mixtures of multivariate normals. This class is represented analogously to (6.1), but considering an arbitrary distribution for the mixing variable $\mathbf{U}$, which again is nonnegative (with probability 1) and independent of the normal random vector $\mathbf{W} \sim \mathcal{N}_p(\mathbf{0}_p, \boldsymbol{\Psi})$. For the unified skew-normal case, $\mathbf{U}$ is a truncated normal random vector below $-\boldsymbol{\tau}$, with pdf

$\phi_q(\mathbf{u};\mathbf{0}_q,\bar{\mathbf{\Gamma}})/\Phi_q(\boldsymbol{\tau};\mathbf{0}_q,\bar{\mathbf{\Gamma}})$ for $\mathbf{u}+\boldsymbol{\tau} > \mathbf{0}_q$, and usually denoted by $\mathbf{U} \sim \mathcal{TN}_q(-\boldsymbol{\tau};\mathbf{0}_q,\bar{\mathbf{\Gamma}})$; see Section 3.11.8. This fact is clearly visualized by conditioning in (6.1) on the truncated latent variable $\mathbf{U}$. In this way, the following hierarchical representation is obtained for the multivariate unified skew-normal distribution.

Theorem 88 *The random vector* $\mathbf{X} \sim \mathcal{SUN}_{p,q}(\boldsymbol{\xi},\mathbf{\Psi},\mathbf{H},\boldsymbol{\tau},\bar{\mathbf{\Gamma}})$ *defined in (6.1) can be represented equivalently as*

$$\begin{aligned} \mathbf{X} \mid \mathbf{U}=\mathbf{u} &\sim \mathcal{N}_p(\boldsymbol{\xi}+\mathbf{H}\mathbf{u},\mathbf{\Psi}), && (6.3)\\ \mathbf{U} &\sim \mathcal{TN}_q(-\boldsymbol{\tau};\mathbf{0}_q,\bar{\mathbf{\Gamma}}). && (6.4) \end{aligned}$$

Proof 88 *The proof is derived directly from the stochastic representation in (6.1).* □

This representation is useful in both the classical and Bayesian inferential approaches of the nonsingular multivariate unified skew-normal model.

6.1.3 Selection representation

The multivariate unified skew-normal distribution also has the following selection representation: The essence of this constructive mechanism is that, starting from a $(p+q)$-dimensional normal distribution, q of its components play the role of hidden/latent variables that nonlinearly modify the remaining p components.

Theorem 89 *If* $\mathbf{X} \sim \mathcal{SUN}_{p,q}(\boldsymbol{\xi},\mathbf{\Psi},\mathbf{H},\boldsymbol{\tau},\bar{\mathbf{\Gamma}})$, *then* $\mathbf{X} = \boldsymbol{\xi}+\tilde{\mathbf{X}}$, *where*

$$\tilde{\mathbf{X}} = (\mathbf{W}_* \mid \mathbf{W}_0+\boldsymbol{\tau}>\mathbf{0}_q), \quad \textit{with} \quad \begin{pmatrix} \mathbf{W}_* \\ \mathbf{W}_0 \end{pmatrix} \sim \mathcal{N}_{p+q}\left(\begin{pmatrix} \mathbf{0}_p \\ \mathbf{0}_q \end{pmatrix}, \begin{pmatrix} \mathbf{\Psi}+\mathbf{H}\bar{\mathbf{\Gamma}}\mathbf{H}^\top & \mathbf{H}\bar{\mathbf{\Gamma}} \\ \bar{\mathbf{\Gamma}}\mathbf{H}^\top & \bar{\mathbf{\Gamma}} \end{pmatrix}\right). \quad (6.5)$$

Proof 89 *By Equation (6.1),* $\mathbf{X} = \mathbf{H}\mathbf{U}+\mathbf{W}+\boldsymbol{\xi}$, *where* $\mathbf{U} \stackrel{d}{=} (\mathbf{W}_0 \mid \mathbf{W}_0+\boldsymbol{\tau} > \mathbf{0}_q)$, *with* $\mathbf{W}_0 \sim \mathcal{N}_q(\mathbf{0}_q,\bar{\mathbf{\Gamma}})$ *independent of* $\mathbf{W} \sim \mathcal{N}_p(\mathbf{0}_p,\mathbf{\Psi})$. *However, by independence,* $\{(\mathbf{W},\mathbf{W}_0) \mid \mathbf{W}_0+\boldsymbol{\tau}>\mathbf{0}_q\} \stackrel{d}{=} (\mathbf{W},\mathbf{U})$. *In turn, this fact implies that*

$$\{\mathbf{g}(\mathbf{W},\mathbf{W}_0) \mid \mathbf{W}_0+\boldsymbol{\tau}>\mathbf{0}_q\} \stackrel{d}{=} \mathbf{g}(\mathbf{W},\mathbf{U}),$$

for any measurable vector-valued function $\mathbf{g}$. *Taking* $\mathbf{W}_* = \mathbf{g}(\mathbf{W},\mathbf{U}) = \mathbf{H}\mathbf{W}_0+\mathbf{W}+\boldsymbol{\xi}$, *we have*

$$(\mathbf{W}_* \mid \mathbf{W}_0+\boldsymbol{\tau}>\mathbf{0}_q) \stackrel{d}{=} \mathbf{g}(\mathbf{W},\mathbf{U}) = \mathbf{H}\mathbf{U}+\mathbf{W}+\boldsymbol{\xi},$$

and the proof concludes by noting that the multivariate normal joint distribution of $(\mathbf{W}_*,\mathbf{W}_0)$ *comes from the fact that* $(\mathbf{W}_* \mid \mathbf{W}_0=\mathbf{w}_0) \sim \mathcal{N}_p(\mathbf{H}\mathbf{w}_0+\boldsymbol{\xi},\mathbf{\Psi})$ *and* $\mathbf{W}_0 \sim \mathcal{N}_q(\boldsymbol{\tau},\bar{\mathbf{\Gamma}})$. □

The selection representation of the SUN distribution in Theorem 89 has been used in the literature to describe some other formulations of essentially the same distribution up to some different parameterizations, some of which with over-parameterizations. We shall briefly describe them next; see Table 6.1 for a summary. We start with a remark.

Remark 13 (Link between $\mathcal{SUN}_{p,q}$ and $\mathcal{AASUN}_{p,q}$) *It is important to note that the (nonsingular) multivariate unified skew-normal distribution $\mathcal{SUN}_{p,q}(\boldsymbol{\xi}, \boldsymbol{\Psi}, \mathbf{H}, \boldsymbol{\tau}, \bar{\boldsymbol{\Gamma}})$ is completely equivalent to the multivariate unified skew-normal distribution studied by Arellano-Valle and Azzalini (2006), here denoted by $\mathcal{AASUN}_{p,q}(\boldsymbol{\xi}, \boldsymbol{\Omega}, \boldsymbol{\Delta}, \boldsymbol{\tau}, \bar{\boldsymbol{\Gamma}})$, except for differences in parameterization; see Table 6.1. Since the parameters $(\boldsymbol{\xi}, \boldsymbol{\tau}, \bar{\boldsymbol{\Gamma}})$ are the same in both settings, the link between $\mathcal{SUN}_{p,q}$ and $\mathcal{AASUN}_{p,q}$ is simply reduced to:*

$$\boldsymbol{\Omega} = \boldsymbol{\Psi} + \mathbf{H}\bar{\boldsymbol{\Gamma}}\mathbf{H}^\top, \quad \boldsymbol{\omega}\boldsymbol{\Delta} = \mathbf{H}\bar{\boldsymbol{\Gamma}}, \tag{6.6}$$

where $\boldsymbol{\omega} = \mathrm{diag}(\boldsymbol{\Omega})^{1/2}$ is a diagonal matrix whose i-th diagonal element is related to the corresponding diagonal element of $\boldsymbol{\Psi} + \mathbf{H}\bar{\boldsymbol{\Gamma}}\mathbf{H}^\top$ as $\omega_i = \sqrt{\psi_{ii} + \boldsymbol{\eta}_i^\top \bar{\boldsymbol{\Gamma}} \boldsymbol{\eta}_i}$ where $\boldsymbol{\eta}_i$ is the i-th column of $\mathbf{H}^\top$, $i = 1, \ldots, p$. Reciprocally:

$$\boldsymbol{\Psi} = \boldsymbol{\Omega} - \boldsymbol{\omega}\boldsymbol{\Delta}\bar{\boldsymbol{\Gamma}}^{-1}\boldsymbol{\Delta}^\top\boldsymbol{\omega}, \quad \mathbf{H} = \boldsymbol{\omega}\boldsymbol{\Delta}\bar{\boldsymbol{\Gamma}}^{-1}. \tag{6.7}$$

Next, in a series of papers, Gupta et al. (2004a) and González-Farías et al. (2004a,b) introduced the multivariate closed skew-normal distribution. Liseo and Loperfido (2003) defined a multivariate hierarchical skew-normal distribution, while Arellano-Valle and Genton (2005) introduced a multivariate closed skew-normal distribution of type II. All these distributions are over-parameterized as noted by Arellano-Valle and Azzalini (2006); see Table 6.1 for further details and the note by Castro et al. (2013). Moreover, Arellano-Valle and Genton (2005) defined a multivariate canonical fundamental skew-normal distribution. In summary, all the distributions listed in Table 6.1 are equivalent up to changes in parameterizations and the removal of over-parameterizations. A particular case of the SUN distribution occurs when $p = q$ and was put forward by Sahu et al. (2003); the normalizing factor in its pdf is then 2^p. This model was further discussed by Azzalini et al. (2016) in comparison with the ASN distribution.

Although the relationship between the two SUN parameterizations in (6.6) and (6.7) is one-to-one, note that

$$\boldsymbol{\Omega} = \boldsymbol{\Psi} + \mathbf{H}_\mathbf{P}\bar{\boldsymbol{\Gamma}}_\mathbf{P}\mathbf{H}_\mathbf{P}^\top, \quad \boldsymbol{\omega}\boldsymbol{\Delta}_\mathbf{P} = \mathbf{H}_\mathbf{P}\bar{\boldsymbol{\Gamma}}_\mathbf{P},$$

where $\mathbf{H}_\mathbf{P} = \mathbf{H}\mathbf{P}^\top$, $\boldsymbol{\Delta}_\mathbf{P} = \boldsymbol{\Delta}\mathbf{P}^\top$ and $\bar{\boldsymbol{\Gamma}}_\mathbf{P} = \mathbf{P}\bar{\boldsymbol{\Gamma}}\mathbf{P}^\top$ with $\mathbf{P}$ a q-by-q permutation/orthogonal matrix. The requirement that $\bar{\boldsymbol{\Gamma}}$, hence $\bar{\boldsymbol{\Gamma}}_\mathbf{P}$, be correlation matrices excludes general orthogonal matrices $\mathbf{P}$ but not permutation matrices $\mathbf{P}$. This means that in any of these two parameterizations, it holds that $\mathcal{AASUN}_{p,q}(\boldsymbol{\xi}, \boldsymbol{\Omega}, \boldsymbol{\Delta}, \boldsymbol{\tau}, \bar{\boldsymbol{\Gamma}}) \equiv \mathcal{AASUN}_{p,q}(\boldsymbol{\xi}, \boldsymbol{\Omega}, \boldsymbol{\Delta}\mathbf{P}^\top, \mathbf{P}\boldsymbol{\tau}, \mathbf{P}\bar{\boldsymbol{\Gamma}}\mathbf{P}^\top)$ and $\mathcal{SUN}_{p,q}(\boldsymbol{\xi}, \boldsymbol{\Psi}, \mathbf{H}, \boldsymbol{\tau}, \bar{\boldsymbol{\Gamma}}) \equiv \mathcal{SUN}_{p,q}(\boldsymbol{\xi}, \boldsymbol{\Psi}, \mathbf{H}\mathbf{P}^\top, \mathbf{P}\boldsymbol{\tau}, \mathbf{P}\bar{\boldsymbol{\Gamma}}\mathbf{P}^\top)$ for any $q \times q$ permutation matrix $\mathbf{P}$; see Wang et al. (2023) for more details on this non-identifiability with respect to permutations of the latent variables, as well as Section 8.1.2 for further discussions. This non-identifiability with respect to permutations of the latent variables applies to all variants listed in Table 6.1.

An important consequence of Theorem 89 is the following result.

Theorem 90 *Let $\mathbf{g}$ be a measurable vector-valued function from $\mathbb{R}^p$ to $\mathbb{R}^r$ and let C be a Borel subset in $\mathbb{R}^r$. If $\mathbf{X} \sim \mathcal{SUN}_{p,q}(\mathbf{0}_p, \boldsymbol{\Psi}, \mathbf{H}, \boldsymbol{\tau}, \bar{\boldsymbol{\Gamma}})$, then the probability distribution of $\mathbf{g}(\mathbf{X})$ is given by*

$$\mathrm{P}(\mathbf{g}(\mathbf{X}) \in C) = \mathrm{P}(\mathbf{g}(\mathbf{W}_*) \in C \mid \mathbf{W}_0 + \boldsymbol{\tau} > \mathbf{0}_q),$$

where the joint distribution of $(\mathbf{W}_*, \mathbf{W}_0)$ is defined in (6.5).

Table 6.1 *Various parameterizations of the SUN distribution based on its selection representation.*

Distribution name	Selection matrix in (6.5)	Remarks
Unified skew-normal $\mathcal{SUN}_{p,q}(\boldsymbol{\xi},\boldsymbol{\Psi},\mathbf{H},\boldsymbol{\tau},\bar{\boldsymbol{\Gamma}})$ (this book)	$\begin{pmatrix} \boldsymbol{\Psi}+\mathbf{H}\bar{\boldsymbol{\Gamma}}\mathbf{H}^\top & \mathbf{H}\bar{\boldsymbol{\Gamma}} \\ \bar{\boldsymbol{\Gamma}}\mathbf{H}^\top & \bar{\boldsymbol{\Gamma}} \end{pmatrix}$	$\boldsymbol{\Psi} \in \mathbb{R}^{p\times p}$, $\bar{\boldsymbol{\Gamma}} \in \mathbb{R}^{q\times q}$ $\mathbf{H} \in \mathbb{R}^{p\times q}$ $\boldsymbol{\Psi} > 0$, $\bar{\boldsymbol{\Gamma}} > 0$
Unified skew-normal $\mathcal{AASUN}_{p,q}(\boldsymbol{\xi},\boldsymbol{\Omega},\boldsymbol{\Delta},\boldsymbol{\tau},\bar{\boldsymbol{\Gamma}})$ (Arellano-Valle and Azzalini, 2006)	$\begin{pmatrix} \bar{\boldsymbol{\Omega}} & \boldsymbol{\Delta} \\ \boldsymbol{\Delta}^\top & \bar{\boldsymbol{\Gamma}} \end{pmatrix}$ with $\mathbf{X} = \boldsymbol{\xi} + \mathrm{diag}(\boldsymbol{\Omega})^{1/2}\tilde{\mathbf{X}}$	$\bar{\boldsymbol{\Omega}} \in \mathbb{R}^{p\times p}$, $\bar{\boldsymbol{\Gamma}} \in \mathbb{R}^{q\times q}$ $\boldsymbol{\Delta} \in \mathbb{R}^{p\times q}$, $\bar{\boldsymbol{\Omega}} > 0$, $\bar{\boldsymbol{\Gamma}} > 0$, $\bar{\boldsymbol{\Omega}} - \boldsymbol{\Delta}\bar{\boldsymbol{\Gamma}}^{-1}\boldsymbol{\Delta}^\top > 0$
Closed skew-normal $\mathcal{CSN}_{p,q}(\boldsymbol{\xi},\boldsymbol{\Sigma},\mathbf{D},\boldsymbol{\tau},\boldsymbol{\Theta})$ (González-Farías et al., 2004b)	$\begin{pmatrix} \boldsymbol{\Sigma} & \boldsymbol{\Sigma}\mathbf{D}^\top \\ \mathbf{D}\boldsymbol{\Sigma} & \boldsymbol{\Theta}+\mathbf{D}\boldsymbol{\Sigma}\mathbf{D}^\top \end{pmatrix}$	$\boldsymbol{\Sigma} \in \mathbb{R}^{p\times p}$, $\boldsymbol{\Theta} \in \mathbb{R}^{q\times q}$ $\mathbf{D} \in \mathbb{R}^{q\times p}$, $\boldsymbol{\Sigma} > 0$, $\boldsymbol{\Theta} > 0$ Over-parameterized if $\boldsymbol{\Theta} \neq \bar{\boldsymbol{\Theta}}$
Closed skew-normal II $\mathcal{CSNII}_{p,q}(\boldsymbol{\xi},\boldsymbol{\Psi},\mathbf{D},\boldsymbol{\tau},\boldsymbol{\Gamma})$ (Arellano-Valle and Genton, 2005)	$\begin{pmatrix} \boldsymbol{\Psi}+\mathbf{D}^\top\boldsymbol{\Gamma}\mathbf{D} & \mathbf{D}^\top\boldsymbol{\Gamma} \\ \boldsymbol{\Gamma}\mathbf{D} & \boldsymbol{\Gamma} \end{pmatrix}$	$\boldsymbol{\Psi} \in \mathbb{R}^{p\times p}$, $\boldsymbol{\Gamma} \in \mathbb{R}^{q\times q}$ $\mathbf{D} \in \mathbb{R}^{q\times p}$, $\boldsymbol{\Psi} > 0$, $\boldsymbol{\Gamma} > 0$ Over-parameterized if $\boldsymbol{\Gamma} \neq \bar{\boldsymbol{\Gamma}}$
Hierarchical skew-normal $\mathcal{HSN}_{p,q}(\boldsymbol{\xi},\boldsymbol{\Sigma},\mathbf{C},\boldsymbol{\tau},\boldsymbol{\Upsilon})$ (Liseo and Loperfido, 2003)	$\begin{pmatrix} \boldsymbol{\Upsilon}+\boldsymbol{\Sigma} & -\boldsymbol{\Upsilon}\mathbf{C}^\top \\ -\mathbf{C}\boldsymbol{\Upsilon} & \mathbf{C}\boldsymbol{\Upsilon}\mathbf{C}^\top \end{pmatrix}$	$\boldsymbol{\Sigma} \in \mathbb{R}^{p\times p}$, $\boldsymbol{\Upsilon} \in \mathbb{R}^{p\times p}$ $\mathbf{C} \in \mathbb{R}^{q\times p}$, $\boldsymbol{\Sigma} > 0$, $\boldsymbol{\Upsilon} > 0$ Over-parameterized if $\boldsymbol{\Upsilon} \neq \bar{\boldsymbol{\Upsilon}}$
Canonical fundamental skew-normal $\mathcal{CFUSN}_{p,q}(\boldsymbol{\xi},\boldsymbol{\Omega},\boldsymbol{\Delta})$ (Arellano-Valle and Genton, 2005)	$\begin{pmatrix} \mathbf{I}_p & \boldsymbol{\Delta} \\ \boldsymbol{\Delta}^\top & \mathbf{I}_q \end{pmatrix}$ with $\mathbf{X} = \boldsymbol{\xi} + \boldsymbol{\Omega}^{1/2}\tilde{\mathbf{X}}$	$\boldsymbol{\Delta} \in \mathbb{R}^{p\times q}$, $\mathbf{I}_p - \boldsymbol{\Delta}\boldsymbol{\Delta}^\top > 0$

Proof 90 *In fact, by (6.5), we have*

$$\mathrm{P}\{\mathbf{g}(\mathbf{X}) \in C\} = \mathrm{P}\{\mathbf{X} \in \mathbf{g}^{-1}(C)\} = \mathrm{P}\{\mathbf{W}_* \in \mathbf{g}^{-1}(C) \mid \mathbf{W}_0 + \boldsymbol{\tau} > \mathbf{0}_q\} = \mathrm{P}\{\mathbf{g}(\mathbf{W}_*) \in C \mid \mathbf{W}_0 + \boldsymbol{\tau} > \mathbf{0}_q\}.$$

□

6.1.4 Cdf

From the representation by selection of $\mathbf{X} \sim \mathcal{SUN}_{p,q}(\boldsymbol{\xi},\boldsymbol{\Psi},\mathbf{H},\boldsymbol{\tau},\bar{\boldsymbol{\Gamma}})$, or immediately from Theorem 90, its cdf can be easily calculated as shown below. Later in Section 6.4.2, we will obtain this cdf in the nonsingular case by directly integrating its pdf.

Theorem 91 *The cdf of* $\mathbf{X} \sim \mathcal{SUN}_{p,q}(\boldsymbol{\xi},\boldsymbol{\Psi},\mathbf{H},\boldsymbol{\tau},\bar{\boldsymbol{\Gamma}})$ *is*

$$F(\mathbf{x}) = \frac{1}{\Phi_q(\boldsymbol{\tau};\mathbf{0}_q,\bar{\boldsymbol{\Gamma}})}\Phi_{p+q}(\mathbf{x}_*;\boldsymbol{\xi}_*,\boldsymbol{\Omega}_*), \quad \mathbf{x} \in \mathbb{R}^p,$$

where $\mathbf{x}_* = (\mathbf{x}^\top,\boldsymbol{\tau}^\top)^\top$, $\boldsymbol{\xi}_* = (\boldsymbol{\xi}^\top,\mathbf{0}_q^\top)^\top$, *and*

$$\boldsymbol{\Omega}_* = \begin{pmatrix} \boldsymbol{\Psi}+\mathbf{H}\bar{\boldsymbol{\Gamma}}\mathbf{H}^\top & -\mathbf{H}\bar{\boldsymbol{\Gamma}} \\ -\bar{\boldsymbol{\Gamma}}\mathbf{H}^\top & \bar{\boldsymbol{\Gamma}} \end{pmatrix}. \tag{6.8}$$

Proof 91 *By Theorem 89, we have* $\mathbf{X} \stackrel{d}{=} (\mathbf{W}_* \mid \mathbf{W}_0 + \boldsymbol{\tau} > \mathbf{0}_q)$*, which is equivalent to* $\mathbf{X} \stackrel{d}{=} (\mathbf{W}_* \mid \tilde{\mathbf{W}}_0 < \boldsymbol{\tau})$*, where* $\tilde{\mathbf{W}}_0 = -\mathbf{W}_0 \sim \mathcal{N}_q(\mathbf{0}_q, \bar{\boldsymbol{\Gamma}})$ *and the multivariate joint distribution of* $\mathbf{W}_* = \mathbf{H}\mathbf{W}_0 + \mathbf{W} + \boldsymbol{\xi}$ *and* $\tilde{\mathbf{W}}_0 = -\mathbf{W}_0$ *is* $\mathcal{N}_{p+q}(\boldsymbol{\xi}_*, \boldsymbol{\Omega}_*)$*. Hence, we obtain*

$$F(\mathbf{x}) = \mathrm{P}(\mathbf{X} \le \mathbf{x}) = \mathrm{P}(\mathbf{W}_* \le \mathbf{x} \mid \tilde{\mathbf{W}}_0 < \boldsymbol{\tau}) = \frac{\mathrm{P}(\mathbf{W}_* \le \mathbf{x}, \tilde{\mathbf{W}}_0 < \boldsymbol{\tau})}{\mathrm{P}(\tilde{\mathbf{W}}_0 < \boldsymbol{\tau})},$$

which proves the result. □

6.2 Cf, Mgf, Linear Combinations and Moments

Now, we may represent a unified skew-normal random vector $\mathbf{X} \sim \mathcal{SUN}_{p,q}(\boldsymbol{\xi}, \boldsymbol{\Psi}, \mathbf{H}, \boldsymbol{\tau}, \bar{\boldsymbol{\Gamma}})$ defined according to Definition 18 by means of its cf and the distribution of any linear combination $\mathbf{a}^\top \mathbf{X}$, $\mathbf{a} \in \mathbb{R}^p$.

Theorem 92 *Let* $\mathbf{X} \sim \mathcal{SUN}_{p,q}(\boldsymbol{\xi}, \boldsymbol{\Psi}, \mathbf{H}, \boldsymbol{\tau}, \bar{\boldsymbol{\Gamma}})$*. Then, its cf is*

$$\psi_{\mathbf{X}}(\mathbf{t}) = e^{\mathrm{i}\mathbf{t}^\top \boldsymbol{\xi} - \frac{1}{2}\mathbf{t}^\top (\boldsymbol{\Psi} + \mathbf{H}\bar{\boldsymbol{\Gamma}}\mathbf{H}^\top)\mathbf{t}} \frac{\Phi_q(\boldsymbol{\tau} + \mathrm{i}\bar{\boldsymbol{\Gamma}}\mathbf{H}^\top \mathbf{t}; \mathbf{0}_q, \bar{\boldsymbol{\Gamma}})}{\Phi_q(\boldsymbol{\tau}; \mathbf{0}_q, \bar{\boldsymbol{\Gamma}})}, \quad \mathbf{t} \in \mathbb{R}^p. \tag{6.9}$$

Proof 92 *By Definition 18,* $\mathbf{X} = \mathbf{H}\mathbf{U} + \mathbf{W} + \boldsymbol{\xi}$*, where* $\mathbf{U} \sim \mathcal{TN}_q(-\boldsymbol{\tau}; \mathbf{0}_q, \bar{\boldsymbol{\Gamma}})$ *and* $\mathbf{W} + \boldsymbol{\xi} \sim \mathcal{N}_p(\boldsymbol{\xi}, \boldsymbol{\Psi})$ *are independent. Hence, the proof is direct from the fact that*

$$\psi_{\mathbf{X}}(\mathbf{t}) = \psi_{\boldsymbol{\xi}+\mathbf{W}}(\mathbf{t})\, \psi_{\mathbf{U}}(\mathbf{H}^\top \mathbf{t}),$$

where $\psi_{\boldsymbol{\xi}+\mathbf{W}}(\mathbf{t}) = e^{\mathrm{i}\mathbf{t}^\top \boldsymbol{\xi} - \frac{1}{2}\mathbf{t}^\top \boldsymbol{\Psi}\mathbf{t}}$ *and*

$$\begin{aligned}
\psi_{\mathbf{U}}(\mathbf{s}) &= \psi_{(\mathbf{W}_0 \mid \mathbf{W}_0 + \boldsymbol{\tau} > \mathbf{0}_q)}(\mathbf{s}) \quad (\mathbf{s} \in \mathbb{R}^q) \\
&= \mathrm{E}(e^{\mathrm{i}\mathbf{s}^\top \mathbf{W}_0} \mid \mathbf{W}_0 + \boldsymbol{\tau} > \mathbf{0}_q), \quad (\mathbf{W}_0 \sim \mathcal{N}_q(\mathbf{0}_q, \bar{\boldsymbol{\Gamma}})) \\
&= \int_{\mathbf{w}+\boldsymbol{\tau}>\mathbf{0}_q} e^{\mathrm{i}\mathbf{s}^\top \mathbf{w}} f_{(\mathbf{W}_0 \mid \mathbf{W}_0 + \boldsymbol{\tau} > \mathbf{0}_q)}(\mathbf{w}) \mathrm{d}\mathbf{w} \\
&= \int_{\mathbf{w}+\boldsymbol{\tau}>\mathbf{0}_q} e^{\mathrm{i}\mathbf{s}^\top \mathbf{w}} \frac{1}{\Phi_q(\boldsymbol{\tau}; \mathbf{0}_q, \bar{\boldsymbol{\Gamma}})} \phi_q(\mathbf{w}; \mathbf{0}_q, \bar{\boldsymbol{\Gamma}}) \mathrm{d}\mathbf{w} \\
&= \frac{1}{\Phi_q(\boldsymbol{\tau}; \mathbf{0}_q, \bar{\boldsymbol{\Gamma}})} \int_{\mathbf{w}+\boldsymbol{\tau}>\mathbf{0}_q} e^{\mathrm{i}\mathbf{s}^\top \mathbf{w}} \phi_q(\mathbf{w}; \mathbf{0}_q, \bar{\boldsymbol{\Gamma}}) \mathrm{d}\mathbf{w} \\
&= e^{-\frac{1}{2}\mathbf{s}^\top \bar{\boldsymbol{\Gamma}}\mathbf{s}} \frac{1}{\Phi_q(\boldsymbol{\tau}; \mathbf{0}_q, \bar{\boldsymbol{\Gamma}})} \int_{\mathbf{w}+\boldsymbol{\tau}>\mathbf{0}_q} \phi_q(\mathbf{w}; \mathrm{i}\bar{\boldsymbol{\Gamma}}\mathbf{s}, \bar{\boldsymbol{\Gamma}}) \mathrm{d}\mathbf{w} \\
&= e^{-\frac{1}{2}\mathbf{s}^\top \bar{\boldsymbol{\Gamma}}\mathbf{s}} \frac{1}{\Phi_q(\boldsymbol{\tau}; \mathbf{0}_q, \bar{\boldsymbol{\Gamma}})} \int_{\mathbf{v}+\boldsymbol{\tau}+\mathrm{i}\bar{\boldsymbol{\Gamma}}\mathbf{s}>\mathbf{0}_q} \phi_q(\mathbf{v}; \mathbf{0}_q, \bar{\boldsymbol{\Gamma}}) \mathrm{d}\mathbf{v} \quad (\mathbf{w} = \mathbf{v} + \mathrm{i}\bar{\boldsymbol{\Gamma}}\mathbf{s}) \\
&= e^{-\frac{1}{2}\mathbf{s}^\top \bar{\boldsymbol{\Gamma}}\mathbf{s}} \frac{\Phi_q(\boldsymbol{\tau} + \mathrm{i}\bar{\boldsymbol{\Gamma}}\mathbf{s}; \mathbf{0}_q, \bar{\boldsymbol{\Gamma}})}{\Phi_q(\boldsymbol{\tau}; \mathbf{0}_q, \bar{\boldsymbol{\Gamma}})} \\
\Longrightarrow \psi_{\mathbf{U}}(\mathbf{H}^\top \mathbf{t}) &= e^{-\frac{1}{2}\mathbf{t}^\top \mathbf{H}\bar{\boldsymbol{\Gamma}}\mathbf{H}^\top \mathbf{t}} \frac{\Phi_q(\boldsymbol{\tau} + \mathrm{i}\bar{\boldsymbol{\Gamma}}\mathbf{H}^\top \mathbf{t}; \mathbf{0}_q, \bar{\boldsymbol{\Gamma}})}{\Phi_q(\boldsymbol{\tau}; \mathbf{0}_q, \bar{\boldsymbol{\Gamma}})}.
\end{aligned}$$

□

As we know, the cf of a random vector completely determines its probability distribution and vice versa. In this regard, it should be noted from (6.9) that the cf of the $\mathcal{SUN}_{p,q}(\boldsymbol{\xi},\boldsymbol{\Psi},\mathbf{H},\boldsymbol{\tau},\bar{\boldsymbol{\Gamma}})$ distribution becomes the product of the cf of the $\mathcal{N}_p(\boldsymbol{\xi},\boldsymbol{\Psi}+\mathbf{H}\bar{\boldsymbol{\Gamma}}\mathbf{H}^\top)$ distribution with the cf of the $\mathcal{TN}_q(-\boldsymbol{\tau};\mathbf{0}_q,\bar{\boldsymbol{\Gamma}})$ distribution. In symbols, we can write this relation as

$$\psi_{\mathrm{SUN}_{p,q}}(\mathbf{t}) = \psi_{\mathrm{N}_p}(\mathbf{t})\,\psi_{\mathrm{TN}_q}(\mathbf{H}^\top\mathbf{t}), \quad \mathbf{t}\in\mathbb{R}^p. \tag{6.10}$$

From (6.10) or (6.9) it is now evident that if the symmetry condition $\mathbf{H}=\mathbf{O}$ holds, then the cf of $\mathbf{X}$ becomes $\psi_{\mathbf{X}}(\mathbf{t}) = e^{\mathrm{i}\mathbf{t}^\top\boldsymbol{\xi}-\frac{1}{2}\mathbf{t}^\top\boldsymbol{\Psi}\mathbf{t}}$, $\mathbf{t}\in\mathbb{R}^p$, that is, the cf of the $\mathcal{N}_p(\boldsymbol{\xi},\boldsymbol{\Psi})$ distribution.

The next characterization involves a closure property of linear combinations of the components of $\mathbf{X}\sim\mathcal{SUN}_{p,q}(\boldsymbol{\xi},\boldsymbol{\Psi},\mathbf{H},\boldsymbol{\tau},\bar{\boldsymbol{\Gamma}})$.

Theorem 93 $\mathbf{X}\sim\mathcal{SUN}_{p,q}(\boldsymbol{\xi},\boldsymbol{\Psi},\mathbf{H},\boldsymbol{\tau},\bar{\boldsymbol{\Gamma}})$ *iff* $\mathbf{a}^\top\mathbf{X}\sim\mathcal{SUN}_{1,q}(\mathbf{a}^\top\boldsymbol{\xi},\mathbf{a}^\top\boldsymbol{\Psi}\mathbf{a},\mathbf{a}^\top\mathbf{H},\boldsymbol{\tau},\bar{\boldsymbol{\Gamma}})$ *for all* $\mathbf{a}\in\mathbb{R}^p$.

Proof 93 *By Equation (6.9) in Theorem 92, one can find that*

$$\begin{aligned}
\psi_{\mathbf{X}}(t\mathbf{a}) &= e^{\mathrm{i}t\mathbf{a}^\top\boldsymbol{\xi}-\frac{1}{2}t^2\mathbf{a}^\top(\boldsymbol{\Psi}+\mathbf{H}\bar{\boldsymbol{\Gamma}}\mathbf{H}^\top)\mathbf{a}}\,\frac{\Phi_q(\boldsymbol{\tau}+\mathrm{i}t\bar{\boldsymbol{\Gamma}}\mathbf{H}^\top\mathbf{a};\mathbf{0}_q,\bar{\boldsymbol{\Gamma}})}{\Phi_q(\boldsymbol{\tau};\mathbf{0}_q,\bar{\boldsymbol{\Gamma}})}\\
&= 2e^{\mathrm{i}t\xi_a-\frac{1}{2}t^2(\psi_a^2+\boldsymbol{\eta}_a^\top\bar{\boldsymbol{\Gamma}}\boldsymbol{\eta}_a)}\,\frac{\Phi_q(\boldsymbol{\tau}+\mathrm{i}t\bar{\boldsymbol{\Gamma}}\boldsymbol{\eta}_a;\mathbf{0}_q,\bar{\boldsymbol{\Gamma}})}{\Phi_q(\boldsymbol{\tau};\mathbf{0}_q,\bar{\boldsymbol{\Gamma}})}\\
&= \psi_{\mathbf{a}^\top\mathbf{X}}(t),\ \forall t\in\mathbb{R},\ \forall\mathbf{a}\in\mathbb{R}^p,
\end{aligned}$$

where $\xi_a=\mathbf{a}^\top\boldsymbol{\xi}$, $\psi_a^2=\mathbf{a}^\top\boldsymbol{\Psi}\mathbf{a}$, *and* $\boldsymbol{\eta}_a=\mathbf{H}^\top\mathbf{a}$. *This proves the result.* □

Remark 14 *An alternative way to obtain the distribution of a linear combination of the unified skew-normal random vector* $\mathbf{X}\sim\mathcal{SUN}_{p,q}(\boldsymbol{\xi},\boldsymbol{\Psi},\mathbf{H},\boldsymbol{\tau},\bar{\boldsymbol{\Gamma}})$ *is to directly use its definition to find* $X_a=\mathbf{a}^\top\mathbf{X}=\xi_a+\boldsymbol{\eta}_a^\top\mathbf{U}+W_a$, *where* $\mathbf{U}\sim\mathcal{TN}_q(-\boldsymbol{\tau};\mathbf{0}_q,\bar{\boldsymbol{\Gamma}})$ *and* $W_a=\mathbf{a}^\top\mathbf{W}\sim\mathcal{N}(0,\psi_a^2)$ *are independent. Hence, by definition* $X_a\sim\mathcal{SUN}_{1,q}(\xi_a,\psi_a^2,\boldsymbol{\eta}_a^\top,\boldsymbol{\tau},\bar{\boldsymbol{\Gamma}})$ *for all* $\mathbf{a}\in\mathbb{R}^p$, *where as before* $\xi_a=\mathbf{a}^\top\boldsymbol{\xi}$, $\psi_a^2=\mathbf{a}^\top\boldsymbol{\Psi}\mathbf{a}$, $\boldsymbol{\eta}_a=\mathbf{H}^\top\mathbf{a}$. *Moreover, from this fact it follows that* $\mathbf{X}$ *has product-moments of any order since this holds for* $X_a=\mathbf{a}^\top\mathbf{X}$, $\forall\mathbf{a}\in\mathbb{R}^p$. *Thus, we can also represent the multivariate unified skew-normal distribution from its moment generating function (mgf), which can be obtained from Theorem 92 as*

$$M_{\mathbf{X}}(\mathbf{t}) = \psi_{\mathbf{X}}(-\mathrm{i}\mathbf{t}) = e^{\mathbf{t}^\top\boldsymbol{\xi}+\frac{1}{2}\mathbf{t}^\top(\boldsymbol{\Psi}+\mathbf{H}\bar{\boldsymbol{\Gamma}}\mathbf{H}^\top)\mathbf{t}}\,\frac{\Phi_q(\boldsymbol{\tau}+\bar{\boldsymbol{\Gamma}}\mathbf{H}^\top\mathbf{t};\mathbf{0}_q,\bar{\boldsymbol{\Gamma}})}{\Phi_q(\boldsymbol{\tau};\mathbf{0}_q,\bar{\boldsymbol{\Gamma}})}, \quad \mathbf{t}\in\mathbb{R}^p.$$

As illustrated in part (a) of the following example, by specializing $\mathbf{a}$ in Theorem 93 to be the i-th p-dimensional unit vector $\mathbf{e}_i$, we then immediately deduce that all univariate marginals are unified skew-normal.

Example 69 (Univariate marginals) *Let* $\mathbf{X}=(X_1,\ldots,X_p)^\top\sim\mathcal{SUN}_{p,q}(\boldsymbol{\xi},\boldsymbol{\Psi},\mathbf{H},\boldsymbol{\tau},\bar{\boldsymbol{\Gamma}})$, *with* $\boldsymbol{\xi}=(\xi_1,\ldots,\xi_p)^\top$, $\boldsymbol{\Psi}=(\psi_{ij})_{i,j=1,\ldots,p}$ *and* $\mathbf{H}=(\boldsymbol{\eta}_1,\ldots,\boldsymbol{\eta}_p)^\top$, *where* $\boldsymbol{\eta}_i\in\mathbb{R}^q$ *is the i-th row (column) of* $\mathbf{H}$ *(*$\mathbf{H}^\top$*). By Theorem 93, we have*

$$\sum_{i=1}^p a_iX_i\sim\mathcal{SUN}_{1,q}\left(\sum_{i=1}^p a_i\xi_i,\sum_{i=1}^p\sum_{j=1}^p a_ia_j\psi_{ij},\sum_{i=1}^p a_i\boldsymbol{\eta}_i^\top,\boldsymbol{\tau},\bar{\boldsymbol{\Gamma}}\right).$$

Hence:

(a) Setting $\mathbf{a} = \mathbf{e}_i = (0,\dots,0,1,0,\dots,0)^\top$ *we find the i-th marginal component* $X_i = \mathbf{e}_i^\top \mathbf{X}$*:*

$$X_i \sim \mathcal{SUN}_{1,q}(\xi_i, \psi_{ii}, \boldsymbol{\eta}_i^\top, \boldsymbol{\tau}, \bar{\boldsymbol{\Gamma}}), \quad i = 1,\dots,p,$$

with $\xi_i = \mathbf{e}_i^\top \boldsymbol{\xi}$, $\psi_{ij} = \mathbf{e}_i^\top \boldsymbol{\Psi} \mathbf{e}_j$ *and* $\boldsymbol{\eta}_i = \mathbf{H}^\top \mathbf{e}_i$*;*

(b) Similarly, setting $\mathbf{a}$ *to be a vector with* 1 *in the i-th and j-th positions,* $i \neq j$*, and* 0 *elsewhere, we obtain*

$$X_i + X_j \sim \mathcal{SUN}_{1,q}(\xi_i + \xi_j, \psi_{ii} + \psi_{jj} + 2\psi_{ij}, \boldsymbol{\eta}_i^\top + \boldsymbol{\eta}_j^\top, \boldsymbol{\tau}, \bar{\boldsymbol{\Gamma}}), \quad i,j = 1,\dots,p, \quad i \neq j.$$

Another interesting example that characterizes unified skew-normality is the following.

Example 70 *Let* $\mathbf{U} = (U_1,\dots,U_q)^\top \sim \mathcal{TN}_q(-\boldsymbol{\tau}; \mathbf{0}_q, \bar{\boldsymbol{\Gamma}})$ *and* $\mathbf{W} = (W_1,\dots,W_q)^\top \sim \mathcal{N}_q(\mathbf{0}_q, \mathbf{I}_q)$ *be mutually independent. Suppose that* $\mathbf{X} \sim \mathcal{SUN}_{p,q}(\boldsymbol{\xi}, \boldsymbol{\Psi}, \mathbf{H}, \boldsymbol{\tau}, \bar{\boldsymbol{\Gamma}})$*, with* $\boldsymbol{\Psi} > 0$ *and* $\mathbf{H}^\top \boldsymbol{\Psi}^{-1} \mathbf{H} = \mathrm{diag}(\tilde{\eta}_1,\dots,\tilde{\eta}_q)$*, where* $\tilde{\eta}_j = \sqrt{\boldsymbol{\eta}_j^\top \boldsymbol{\Psi}^{-1} \boldsymbol{\eta}_j}$ *and* $\boldsymbol{\eta}_j$ *is the j-th column of* $\mathbf{H}$*,* $j = 1,\dots,q$*. Then:*

$$\boldsymbol{\eta}_j^\top \boldsymbol{\Psi}^{-1}(\mathbf{X} - \boldsymbol{\xi}) \stackrel{d}{=} \tilde{\eta}_j Z_j, \quad Z_j = \tilde{\eta}_j U_j + W_j \sim \mathcal{SUN}_{1,q}(0, 1, \tilde{\eta}_j, \boldsymbol{\tau}, \bar{\boldsymbol{\Gamma}}), \quad j = 1,\dots,q.$$

Moreover, if $\bar{\boldsymbol{\Gamma}} = \mathbf{I}_q$*, then* $Z_1,\dots,Z_q$ *are independent because in this case* $U_i \stackrel{ind.}{\sim} \mathcal{TN}_1(-\tau_i; 0, 1)$*,* $i = 1,\dots,q$*.*

6.3 Affine Transformations, Marginals, and Independence

In this section, we describe further distributional properties of the multivariate unified skew-normal family introduced in Definition 18. As we will show, this family inherits several of the good properties of the multivariate normal family.

6.3.1 Affine transformations

By construction, the multivariate unified skew-normal distribution is closed under affine transformations as formalized next.

Theorem 94 *Let* $\mathbf{X} \sim \mathcal{SUN}_{p,q}(\boldsymbol{\xi}, \boldsymbol{\Psi}, \mathbf{H}, \boldsymbol{\tau}, \bar{\boldsymbol{\Gamma}})$*, and consider the affine transformation* $\mathbf{Y} = \mathbf{A}\mathbf{X} + \mathbf{b}$*, where the matrix* $\mathbf{A} \in \mathbb{R}^{r \times p}$ *and the vector* $\mathbf{b} \in \mathbb{R}^r$ *are constant. Then,* $\mathbf{Y} \sim \mathcal{SUN}_{r,q}(\mathbf{A}\boldsymbol{\xi} + \mathbf{b}, \mathbf{A}\boldsymbol{\Psi}\mathbf{A}^\top, \mathbf{A}\mathbf{H}, \boldsymbol{\tau}, \bar{\boldsymbol{\Gamma}})$.

Proof 94 *By Definition 18 it follows that* $\mathbf{Y} = \mathbf{A}\mathbf{X} + \mathbf{b} = \mathbf{A}\boldsymbol{\xi} + \mathbf{b} + \mathbf{A}\mathbf{H}\mathbf{U} + \mathbf{A}\mathbf{W}$*. Let* $\boldsymbol{\xi}_A = \mathbf{A}\boldsymbol{\xi} + \mathbf{b}$*,* $\mathbf{H}_A = \mathbf{A}\mathbf{H}$*, and* $\mathbf{W}_A = \mathbf{A}\mathbf{W}$*. Since* $\mathbf{W} \sim \mathcal{N}_p(\mathbf{0}_p, \boldsymbol{\Psi})$*, we have* $\mathbf{W}_A \sim \mathcal{N}_r(\mathbf{0}_r, \boldsymbol{\Psi}_A)$*, with* $\boldsymbol{\Psi}_A = \mathbf{A}\boldsymbol{\Psi}\mathbf{A}^\top$*, and is independent of* $\mathbf{U} \sim \mathcal{TN}_q(-\boldsymbol{\tau}; \mathbf{0}_q, \bar{\boldsymbol{\Gamma}})$*. Hence,* $\mathbf{Y} = \boldsymbol{\xi}_A + \mathbf{H}_A\mathbf{U} + \mathbf{W}_A$*, with* $\mathbf{U} \sim \mathcal{TN}_q(-\boldsymbol{\tau}; \mathbf{0}_q, \bar{\boldsymbol{\Gamma}})$ *and* $\mathbf{W}_A \sim \mathcal{N}_r(\mathbf{0}_r, \boldsymbol{\Psi}_A)$ *being independent. Thus, the proof follows by applying Definition 18 again.* □

If $\mathbf{A}$ has full row rank r, then a nonsingular multivariate unified skew-normal version is obtained from Theorem 94. Another interesting result that follows from Theorem 94 is when $p = q$ and $\mathbf{A} = -\mathbf{I}_p$, which transforms $\mathbf{X}$ to $-\mathbf{X}$, as formalized in the following example.

Example 71 *If* $\mathbf{X} \sim \mathcal{SUN}_{p,q}(\boldsymbol{\xi}, \boldsymbol{\Psi}, \mathbf{H}, \boldsymbol{\tau}, \bar{\boldsymbol{\Gamma}})$*, then* $-\mathbf{X} \sim \mathcal{SUN}_{p,q}(-\boldsymbol{\xi}, \boldsymbol{\Psi}, -\mathbf{H}, \boldsymbol{\tau}, \bar{\boldsymbol{\Gamma}})$.

According to Theorem 94 we can recover symmetry by choosing a transformation matrix $\mathbf{A}$ such that $\mathbf{AH} = \mathbf{O}$. Therefore, in this case, the affine transformation $\mathbf{Y} = \mathbf{AX} + \mathbf{b}$ has a symmetric $\mathcal{N}_p(\mathbf{A}\boldsymbol{\xi} + \mathbf{b}, \mathbf{A}\boldsymbol{\Psi}\mathbf{A}^\top)$ distribution. In the following example, we illustrate a case where the new skewness matrix $\mathbf{AH}$ is diagonal, with non-null elements in its diagonal.

Example 72 *Suppose that* $\mathbf{X} \sim \mathcal{SUN}_{p,q}(\boldsymbol{\xi}, \boldsymbol{\Psi}, \mathbf{H}, \boldsymbol{\tau}, \bar{\boldsymbol{\Gamma}})$*, with* $\boldsymbol{\Psi} > 0$ *and* $\mathbf{H}^\top \boldsymbol{\Psi}^{-1} \mathbf{H} = \mathbf{D}(\tilde{\boldsymbol{\eta}}) = \mathrm{diag}(\tilde{\eta}_1^2, \ldots, \tilde{\eta}_q^2)$*, where* $\tilde{\boldsymbol{\eta}} = (\tilde{\eta}_1, \ldots, \tilde{\eta}_q)^\top$*, with* $\tilde{\eta}_i = \sqrt{\boldsymbol{\eta}_i^\top \boldsymbol{\Psi}^{-1} \boldsymbol{\eta}_i}$ *and* $\boldsymbol{\eta}_i$ *being the i-th column of* $\mathbf{H}$*,* $i = 1, \ldots, q$*. Then:*

$$\mathbf{H}^\top \boldsymbol{\Psi}^{-1}(\mathbf{X} - \boldsymbol{\xi}) \stackrel{d}{=} \mathbf{D}(\tilde{\boldsymbol{\eta}})\mathbf{U} + \mathbf{W}_1 \sim \mathcal{SUN}_{q,q}(\mathbf{0}_q, \mathbf{D}(\tilde{\boldsymbol{\eta}}), \mathbf{D}(\tilde{\boldsymbol{\eta}}), \boldsymbol{\tau}, \bar{\boldsymbol{\Gamma}}),$$

where $\mathbf{U} \sim \mathcal{TN}_q(-\boldsymbol{\tau}; \mathbf{0}_q, \bar{\boldsymbol{\Gamma}})$ *and* $\mathbf{W}_1 \sim \mathcal{N}_q(\mathbf{0}_q, \mathbf{D}(\tilde{\boldsymbol{\eta}}))$ *are mutually independent.*

6.3.2 Marginal distributions and independence

Now we show that any subvector of a multivariate unified skew-normal random vector $\mathbf{X} \sim \mathcal{SUN}_{p,q}(\boldsymbol{\xi}, \boldsymbol{\Psi}, \mathbf{H}, \boldsymbol{\tau}, \bar{\boldsymbol{\Gamma}})$ is also unified skew-normal. For fixed nonnegative integers $p_1 + p_2 = p$, consider the following partition of $\mathbf{X}$ and the induced partition on $\boldsymbol{\xi}$, $\boldsymbol{\Psi}$, and $\mathbf{H}$:

$$\mathbf{X} = \begin{pmatrix} \mathbf{X}_1 \\ \mathbf{X}_2 \end{pmatrix}, \quad \boldsymbol{\xi} = \begin{pmatrix} \boldsymbol{\xi}_1 \\ \boldsymbol{\xi}_2 \end{pmatrix}, \quad \boldsymbol{\Psi} = \begin{pmatrix} \boldsymbol{\Psi}_{11} & \boldsymbol{\Psi}_{12} \\ \boldsymbol{\Psi}_{21} & \boldsymbol{\Psi}_{22} \end{pmatrix}, \quad \mathbf{H} = \begin{pmatrix} \mathbf{H}_1 \\ \mathbf{H}_2 \end{pmatrix}, \tag{6.11}$$

where $\mathbf{X}_j, \boldsymbol{\xi}_j$ are of dimension $p_j \times 1$, $\boldsymbol{\Psi}_{jk}$ is of dimension $p_j \times p_k$ and $\mathbf{H}_j$ is of dimension $p_j \times q$, $j,k = 1,2$. The following result is a direct consequence of Theorem 94.

Theorem 95 *Let* $\mathbf{X} \sim \mathcal{SUN}_{p,q}(\boldsymbol{\xi}, \boldsymbol{\Psi}, \mathbf{H}, \boldsymbol{\tau}, \bar{\boldsymbol{\Gamma}})$ *be partitioned as (6.11). Then, the partitions* $\mathbf{X}_j \sim \mathcal{SUN}_{p_j,q}(\boldsymbol{\xi}_j, \boldsymbol{\Psi}_{jj}, \mathbf{H}_j, \boldsymbol{\tau}, \bar{\boldsymbol{\Gamma}})$ *for* $j = 1,2$.

Proof 95 *Since* $\mathbf{X}_j = \mathbf{A}_j\mathbf{X}$*,* $j = 1,2$*, where* $\mathbf{A}_1 = (\mathbf{I}_{p_1}, \mathbf{O})$ *and* $\mathbf{A}_2 = (\mathbf{O}, \mathbf{I}_{p_2})$*, the proof is then direct from Theorem 94.* □

Obviously, considering the partition (6.11), it is straightforward from Definition 18 that the marginals are given by

$$\mathbf{X}_j = \boldsymbol{\xi}_j + \mathbf{H}_j\mathbf{U} + \mathbf{W}_j, \quad j = 1,2, \tag{6.12}$$

where $\mathbf{W}_j \sim \mathcal{N}_{p_j}(\mathbf{0}_{p_j}, \boldsymbol{\Psi}_{jj})$ which, for $j = 1,2$, is independent of $\mathbf{U} \sim \mathcal{TN}_q(-\boldsymbol{\tau}; \mathbf{0}_q, \bar{\boldsymbol{\Gamma}})$.

From Theorem 95, an advantage of the $(\boldsymbol{\Psi}, \mathbf{H})$ parameterization of the SUN is that it yields a marginally consistent distribution, i.e., the parameters of the marginal distribution depend only on the corresponding subparameters of the full distribution (treating the latent parameters as fixed).

In the nonsingular case with $\boldsymbol{\Psi} > 0$, all the marginals are nonsingular. In the singular case with $\boldsymbol{\Psi} \geq 0$, they can be singular or nonsingular. The following theorem establishes the condition for having a nonsingular marginal in the latter case.

Theorem 96 *Let* $\mathbf{X} \sim \mathcal{SUN}_{p,q}(\boldsymbol{\xi}, \boldsymbol{\Psi}, \mathbf{H}, \boldsymbol{\tau}, \bar{\boldsymbol{\Gamma}})$, *with rank*$(\boldsymbol{\Psi}) = r$, *be partitioned as (6.11). If* $\boldsymbol{\Psi}_{11}$ *has full rank* $p_1 = p - p_2 \leq r$, *then* $\mathbf{X}_1 \sim \mathcal{SUN}_{p_1,q}(\boldsymbol{\xi}_1, \boldsymbol{\Psi}_{11}, \mathbf{H}_1, \boldsymbol{\tau}, \bar{\boldsymbol{\Gamma}})$ *is nonsingular.*

Proof 96 *It is directly derived from Theorem 95 and the assumption that* $\boldsymbol{\Psi}_{11} > 0$. □

The following result is also direct from Theorem 94 and states that the skewness of the distribution of each subvector of **X** is only regulated by the corresponding components of **H**.

Theorem 97 *Let* $\mathbf{X} \sim \mathcal{SUN}_{p,q}(\boldsymbol{\xi}, \boldsymbol{\Psi}, \mathbf{H}, \boldsymbol{\tau}, \bar{\boldsymbol{\Gamma}})$ *be partitioned as (6.11). Then for every fixed* $p_1 + p_2 = p$, $\mathbf{X}_j$ *has a symmetric* $\mathcal{N}_{p_j}(\boldsymbol{\xi}_j, \boldsymbol{\Psi}_{jj})$ *distribution iff* $\mathbf{H}_j = \mathbf{O}$.

Proof 97 *By Theorem 95,* $\mathbf{X}_j \sim \mathcal{SUN}_{p_j,q}(\boldsymbol{\xi}_j, \boldsymbol{\Psi}_{jj}, \mathbf{H}_j, \boldsymbol{\tau}, \bar{\boldsymbol{\Gamma}})$. *Then, according to Theorem 92, its cf is* $\psi_j(\mathbf{t}_j) = \psi_{\mathrm{N},j}(\mathbf{t}_j)\Phi_q(\boldsymbol{\tau} + \mathrm{i}\bar{\boldsymbol{\Gamma}}\mathbf{H}_j^\top \mathbf{t}_j; \mathbf{0}_q, \bar{\boldsymbol{\Gamma}})/\Phi_q(\boldsymbol{\tau}; \mathbf{0}_q, \bar{\boldsymbol{\Gamma}})$, *where* $\psi_{\mathrm{N},j}(\mathbf{t}_j) = e^{\mathrm{i}\mathbf{t}_j^\top \boldsymbol{\xi}_j - \frac{1}{2}\mathbf{t}_j^\top(\boldsymbol{\Psi}_{jj} + \mathbf{H}_j\bar{\boldsymbol{\Gamma}}\mathbf{H}_j^\top)\mathbf{t}_j}$ *is the cf of the* $\mathcal{N}_{p_j}(\boldsymbol{\xi}_j, \boldsymbol{\Psi}_{jj} + \mathbf{H}_j\bar{\boldsymbol{\Gamma}}\mathbf{H}_j^\top)$ *distribution. Clearly,* $\psi_j(\mathbf{t}_j) = e^{\mathrm{i}\mathbf{t}_j^\top \boldsymbol{\xi}_j - \frac{1}{2}\mathbf{t}_j^\top \boldsymbol{\Psi}_{jj}\mathbf{t}_j}\ \forall \mathbf{t}_j \iff \psi_{\mathrm{N},j}(\mathbf{t}_j) = e^{\mathrm{i}\mathbf{t}_j^\top \boldsymbol{\xi}_j - \frac{1}{2}\mathbf{t}_j^\top \boldsymbol{\Psi}_{jj}\mathbf{t}_j}$ *and* $\Phi_q(\boldsymbol{\tau} + \mathrm{i}\bar{\boldsymbol{\Gamma}}\mathbf{H}_j^\top \mathbf{t}_j; \mathbf{0}_q, \bar{\boldsymbol{\Gamma}}) = \Phi_q(\boldsymbol{\tau}; \mathbf{0}_q, \bar{\boldsymbol{\Gamma}})$ *for all* $\mathbf{t}_j \iff \bar{\boldsymbol{\Gamma}}\mathbf{H}_j^\top = \mathbf{O}$ $\iff \mathbf{H}_j = \mathbf{O}$ *since by assumption* $\bar{\boldsymbol{\Gamma}} > 0$ *and so nonsingular.* □

Another relevant property with the marginals occurs when the correlation matrix $\bar{\boldsymbol{\Gamma}}$ is block-diagonal, which produces a nice simplification for marginal distributions (a SUN distribution with block-diagonal $\bar{\boldsymbol{\Gamma}}$ is called a SUNSET distribution, see Section 6.8). To see this, we first partition the parameters $(\mathbf{H}, \boldsymbol{\tau}, \bar{\boldsymbol{\Gamma}})$ as follows:

$$\mathbf{H} = \begin{pmatrix} \mathbf{H}_1 \\ \mathbf{H}_2 \end{pmatrix} = \begin{pmatrix} \mathbf{H}_{11} & \mathbf{H}_{12} \\ \mathbf{H}_{21} & \mathbf{H}_{22} \end{pmatrix}, \quad \boldsymbol{\tau} = \begin{pmatrix} \boldsymbol{\tau}_1 \\ \boldsymbol{\tau}_2 \end{pmatrix}, \quad \bar{\boldsymbol{\Gamma}} = \begin{pmatrix} \bar{\boldsymbol{\Gamma}}_{11} & \bar{\boldsymbol{\Gamma}}_{12} \\ \bar{\boldsymbol{\Gamma}}_{21} & \bar{\boldsymbol{\Gamma}}_{22} \end{pmatrix}, \tag{6.13}$$

where for $i, j = 1, 2$, the matrix $\mathbf{H}_{ij}$ is $p_i \times q_j$, the vector $\boldsymbol{\tau}_i$ is $q_i \times 1$, and the matrix $\bar{\boldsymbol{\Gamma}}_{ij}$ is $q_i \times q_j$, with $p_1 + p_2 = p$ and $q_1 + q_2 = q$.

Theorem 98 *Let* $\mathbf{X} \sim \mathcal{SUN}_{p,q}(\boldsymbol{\xi}, \boldsymbol{\Psi}, \mathbf{H}, \boldsymbol{\tau}, \bar{\boldsymbol{\Gamma}})$ *be partitioned as (6.11), and consider also the partition (6.13) of* $(\mathbf{H}, \boldsymbol{\tau}, \bar{\boldsymbol{\Gamma}})$ *with the assumption that* $\bar{\boldsymbol{\Gamma}}_{12} = \bar{\boldsymbol{\Gamma}}_{21}^\top = \mathbf{O}$, *i.e.,* $\bar{\boldsymbol{\Gamma}} = \mathrm{diag}(\bar{\boldsymbol{\Gamma}}_{11}, \bar{\boldsymbol{\Gamma}}_{22})$. *If in addition we have:*

(i) $\mathbf{H}_{jk} = \mathbf{O}$, $j \neq k$, *then* $\mathbf{X}_j \sim \mathcal{SUN}_{p_j,q_j}(\boldsymbol{\xi}_j, \boldsymbol{\Psi}_{jj}, \mathbf{H}_{jj}, \boldsymbol{\tau}_j, \bar{\boldsymbol{\Gamma}}_{jj})$ *for* $j, k = 1, 2$,

(ii) $\mathbf{H}_{jj} = \mathbf{O}$, $j \neq k$, *then* $\mathbf{X}_j \sim \mathcal{SUN}_{p_j,q_k}(\boldsymbol{\xi}_j, \boldsymbol{\Psi}_{jj}, \mathbf{H}_{jk}, \boldsymbol{\tau}_k, \bar{\boldsymbol{\Gamma}}_{kk})$ *for* $j, k = 1, 2$,

which hold for every fixed $p_1 + p_2 = p$ *and* $q_1 + q_2 = q$.

Proof 98 *By Theorem 92 the joint cf of* $\mathbf{X}_1$ *and* $\mathbf{X}_2$ *can be written as*

$$\begin{aligned} \psi(\mathbf{t}_1, \mathbf{t}_2) \quad = \quad & e^{\mathrm{i}(\mathbf{t}_1^\top \boldsymbol{\xi}_1 + \mathbf{t}_2^\top \boldsymbol{\xi}_2) - \frac{1}{2}(\mathbf{t}_1^\top \boldsymbol{\Omega}_{11}\mathbf{t}_1 + 2\mathbf{t}_1^\top \boldsymbol{\Omega}_{12}\mathbf{t}_2 + \mathbf{t}_2^\top \boldsymbol{\Omega}_{22}\mathbf{t}_2)} \\ & \times \frac{\Phi_q(\boldsymbol{\tau} + \mathrm{i}\bar{\boldsymbol{\Gamma}}\mathbf{H}_1^\top \mathbf{t}_1 + \mathrm{i}\bar{\boldsymbol{\Gamma}}\mathbf{H}_2^\top \mathbf{t}_2; \mathbf{0}_q, \bar{\boldsymbol{\Gamma}})}{\Phi_q(\boldsymbol{\tau}; \mathbf{0}_q, \bar{\boldsymbol{\Gamma}})}, \quad \mathbf{t}_1 \in \mathbb{R}^{p_1},\ \mathbf{t}_2 \in \mathbb{R}^{p_2}, \end{aligned} \tag{6.14}$$

where, from the partition (6.13),

$$\boldsymbol{\Omega}_{jk} = \boldsymbol{\Psi}_{jk} + \mathbf{H}_j\bar{\boldsymbol{\Gamma}}\mathbf{H}_k^\top = \boldsymbol{\Psi}_{jk} + \mathbf{H}_{j1}\bar{\boldsymbol{\Gamma}}_{11}\mathbf{H}_{k1}^\top + \mathbf{H}_{j1}\bar{\boldsymbol{\Gamma}}_{12}\mathbf{H}_{k2}^\top + \mathbf{H}_{j2}\bar{\boldsymbol{\Gamma}}_{21}\mathbf{H}_{k1}^\top + \mathbf{H}_{j2}\bar{\boldsymbol{\Gamma}}_{22}\mathbf{H}_{k2}^\top,$$

for $j = 1,2$, and

$$\Phi_q(\boldsymbol{\tau};\mathbf{0}_q,\bar{\boldsymbol{\Gamma}}) = \Phi_{q_1+q_2}\left\{\begin{pmatrix}\boldsymbol{\tau}_1\\ \boldsymbol{\tau}_2\end{pmatrix};\begin{pmatrix}\mathbf{0}_{q_1}\\ \mathbf{0}_{q_2}\end{pmatrix},\begin{pmatrix}\bar{\boldsymbol{\Gamma}}_{11} & \bar{\boldsymbol{\Gamma}}_{12}\\ \bar{\boldsymbol{\Gamma}}_{21} & \bar{\boldsymbol{\Gamma}}_{22}\end{pmatrix}\right\}, \tag{6.15}$$

$$\begin{aligned}&\Phi_q(\boldsymbol{\tau}+\mathrm{i}\bar{\boldsymbol{\Gamma}}\mathbf{H}_1^\top\mathbf{t}_1+\mathrm{i}\bar{\boldsymbol{\Gamma}}\mathbf{H}_2^\top\mathbf{t}_2;\mathbf{0}_q,\bar{\boldsymbol{\Gamma}})\\ &= \Phi_{q_1+q_2}\left\{\begin{pmatrix}\boldsymbol{\tau}_1+\mathrm{i}(\bar{\boldsymbol{\Gamma}}_{11}\mathbf{H}_{11}^\top+\bar{\boldsymbol{\Gamma}}_{12}\mathbf{H}_{12}^\top)\mathbf{t}_1+\mathrm{i}(\bar{\boldsymbol{\Gamma}}_{11}\mathbf{H}_{21}^\top+\bar{\boldsymbol{\Gamma}}_{12}\mathbf{H}_{22}^\top)\mathbf{t}_2\\ \boldsymbol{\tau}_2+\mathrm{i}(\bar{\boldsymbol{\Gamma}}_{21}\mathbf{H}_{11}^\top+\bar{\boldsymbol{\Gamma}}_{22}\mathbf{H}_{12}^\top)\mathbf{t}_1+\mathrm{i}(\bar{\boldsymbol{\Gamma}}_{21}\mathbf{H}_{21}^\top+\bar{\boldsymbol{\Gamma}}_{22}\mathbf{H}_{22}^\top)\mathbf{t}_2\end{pmatrix};\right.\\ &\qquad\left.\begin{pmatrix}\mathbf{0}_{q_1}\\ \mathbf{0}_{q_2}\end{pmatrix},\begin{pmatrix}\bar{\boldsymbol{\Gamma}}_{11} & \bar{\boldsymbol{\Gamma}}_{12}\\ \bar{\boldsymbol{\Gamma}}_{21} & \bar{\boldsymbol{\Gamma}}_{22}\end{pmatrix}\right\}.\end{aligned} \tag{6.16}$$

By applying the assumption that $\bar{\boldsymbol{\Gamma}}_{12} = \mathbf{O}$ in (6.15) and (6.16), the joint cf (6.14) of $\mathbf{X}_1$ and $\mathbf{X}_2$ becomes

$$\begin{aligned}\psi(\mathbf{t}_1,\mathbf{t}_2) &= e^{\mathrm{i}(\mathbf{t}_1^\top\boldsymbol{\xi}_1+\mathbf{t}_2^\top\boldsymbol{\xi}_2)-\frac{1}{2}(\mathbf{t}_1^\top\boldsymbol{\Omega}_{11}\mathbf{t}_1+2\mathbf{t}_1^\top\boldsymbol{\Omega}_{12}\mathbf{t}_2+\mathbf{t}_2^\top\boldsymbol{\Omega}_{22}\mathbf{t}_2)}\\ &\quad\times\frac{\Phi_{q_1}(\boldsymbol{\tau}_1+\mathrm{i}\bar{\boldsymbol{\Gamma}}_{11}\mathbf{H}_{11}^\top\mathbf{t}_1+\mathrm{i}\bar{\boldsymbol{\Gamma}}_{11}\mathbf{H}_{21}^\top\mathbf{t}_2;\mathbf{0}_{q_1},\bar{\boldsymbol{\Gamma}}_{11})}{\Phi_{q_1}(\boldsymbol{\tau}_1;\mathbf{0}_{q_1},\bar{\boldsymbol{\Gamma}}_{11})}\\ &\quad\times\frac{\Phi_{q_2}(\boldsymbol{\tau}_2+\mathrm{i}\bar{\boldsymbol{\Gamma}}_{22}\mathbf{H}_{12}^\top\mathbf{t}_1+\mathrm{i}\bar{\boldsymbol{\Gamma}}_{22}\mathbf{H}_{22}^\top\mathbf{t}_2;\mathbf{0}_{q_2},\bar{\boldsymbol{\Gamma}}_{22})}{\Phi_{q_2}(\boldsymbol{\tau}_2;\mathbf{0}_{q_2},\bar{\boldsymbol{\Gamma}}_{22})}, \quad \mathbf{t}_1\in\mathbb{R}^{p_1},\ \mathbf{t}_2\in\mathbb{R}^{p_2},\end{aligned} \tag{6.17}$$

with $\boldsymbol{\Omega}_{jk} = \boldsymbol{\Psi}_{jk}+\mathbf{H}_{j1}\bar{\boldsymbol{\Gamma}}_{11}\mathbf{H}_{k1}^\top+\mathbf{H}_{j2}\bar{\boldsymbol{\Gamma}}_{22}\mathbf{H}_{k2}^\top$ for $j,k = 1,2$. If, in addition to the assumption $\bar{\boldsymbol{\Gamma}}_{12} = \mathbf{O}$, we suppose that $\mathbf{H}_{jk} = \mathbf{O}$ for $k \neq j$, then from (6.17) the marginal cf of $\mathbf{X}_j$ is given by

$$\psi_j(\mathbf{t}_j) = e^{\mathrm{i}\mathbf{t}_j^\top\tilde{\boldsymbol{\xi}}_j-\frac{1}{2}\mathbf{t}_j\boldsymbol{\Omega}_{jj}\mathbf{t}_j}\frac{\Phi_{q_j}(\boldsymbol{\tau}_j+\mathrm{i}\bar{\boldsymbol{\Gamma}}_{jj}\mathbf{H}_{jj}^\top\mathbf{t}_j;\mathbf{0}_{q_j},\bar{\boldsymbol{\Gamma}}_{jj})}{\Phi_{q_j}(\boldsymbol{\tau}_j;\mathbf{0}_{q_j},\bar{\boldsymbol{\Gamma}}_{jj})}, \quad \mathbf{t}_j\in\mathbb{R}^{p_j},$$

with $\tilde{\boldsymbol{\xi}}_j = \boldsymbol{\xi}_j+\mathbf{H}_{jj}\boldsymbol{\tau}_j$ and $\boldsymbol{\Omega}_{jj} = \boldsymbol{\Psi}_{jj}+\mathbf{H}_{jj}\bar{\boldsymbol{\Gamma}}_{jj}\mathbf{H}_{jj}^\top$ for $j = 1,2$. Due to the uniqueness of the cf of $\mathbf{X}_j$, the proof of part (i) is completed by applying Theorem 92 again. The proof of part (ii) is analogous, but in this case, the additional assumption to $\bar{\boldsymbol{\Gamma}}_{12} = \mathbf{O}$ is $\mathbf{H}_{kk} = \mathbf{O}$ for $k \neq j$, and therefore, from (6.17), we have

$$\psi_j(\mathbf{t}_j) = e^{\mathrm{i}\mathbf{t}_j^\top\boldsymbol{\xi}_j-\frac{1}{2}\mathbf{t}_j\boldsymbol{\Omega}_{jj}\mathbf{t}_j}\frac{\Phi_{q_j}(\boldsymbol{\tau}_j+\mathrm{i}\bar{\boldsymbol{\Gamma}}_{jj}\mathbf{H}_{jk}^\top\mathbf{t}_j;\mathbf{0}_{q_j},\bar{\boldsymbol{\Gamma}}_{kk})}{\Phi_{q_k}(\boldsymbol{\tau}_k;\mathbf{0}_{q_k},\bar{\boldsymbol{\Gamma}}_{jk})}, \quad \mathbf{t}_j\in\mathbb{R}^{p_j},$$

with $\boldsymbol{\Omega}_{jj} = \boldsymbol{\Psi}_{jj}+\mathbf{H}_{jk}\bar{\boldsymbol{\Gamma}}_{jj}\mathbf{H}_{jk}^\top$ for $j,k = 1,2$ and $k \neq j$. □

An alternative and perhaps more straightforward proof of Theorem 98 is by applying the partition (6.13) to the stochastic definition in (6.1) to obtain

$$\mathbf{X}_j = \boldsymbol{\xi}_j+\mathbf{H}_{j1}\mathbf{U}_1+\mathbf{H}_{j2}\mathbf{U}_2+\mathbf{W}_j, \quad j = 1,2,$$

where the condition $\bar{\boldsymbol{\Gamma}}_{12} = \mathbf{O}$ implies that $\mathbf{U}_j \sim \mathcal{TN}_{q_j}(-\boldsymbol{\tau}_j;\mathbf{0}_{q_j},\bar{\boldsymbol{\Gamma}}_{jj})$ which for $j = 1,2$ are independent and also independent of $\mathbf{W}_j \sim \mathcal{N}_{p_j}(\mathbf{0}_{p_j},\boldsymbol{\Psi}_{jj})$ for each $j = 1,2$. Hence, by condition (i) and again using the stochastic definition (6.1), we have $\mathbf{X}_j = \boldsymbol{\xi}_j+\mathbf{H}_{jj}\mathbf{U}_j+\mathbf{W}_j \sim \mathcal{SUN}_{p_j,q_j}(\boldsymbol{\xi}_j,\boldsymbol{\Psi}_{jj},\mathbf{H}_{jj},\boldsymbol{\tau}_j,\bar{\boldsymbol{\Gamma}}_{jj})$ for $j = 1,2$. Similarly, condition (ii) and the stochastic definition (6.1) imply that $\mathbf{X}_j = \boldsymbol{\xi}_j+\mathbf{H}_{jk}\mathbf{U}_k+\mathbf{W}_j \sim \mathcal{SUN}_{p_j,q_k}(\boldsymbol{\xi}_j,\boldsymbol{\Psi}_{jj},\mathbf{H}_{jk},\boldsymbol{\tau}_j,\bar{\boldsymbol{\Gamma}}_{kk})$ for $j,k = 1,2$ with $j \neq k$.

Example 73 *Let $q = p$ and $\mathbf{X} \sim \mathcal{SUN}_{p,p}(\boldsymbol{\xi},\boldsymbol{\Psi},\mathbf{H},\boldsymbol{\tau},\bar{\boldsymbol{\Gamma}})$ be partitioned as (6.11), and consider also the partition (6.13) of $(\mathbf{H},\boldsymbol{\tau},\bar{\boldsymbol{\Gamma}})$. Assume that $\mathbf{H}$ is lower triangular. Then $\mathbf{H}_{11}$ and $\mathbf{H}_{22}$ are also lower triangular, and $\mathbf{H}_{12} = \mathbf{O}$, but $\mathbf{H}_{21} \neq \mathbf{O}$. Moreover, if $\bar{\boldsymbol{\Gamma}} = \mathrm{diag}(\bar{\boldsymbol{\Gamma}}_{11},\bar{\boldsymbol{\Gamma}}_{22})$ then according to*

Theorem 98 we have $\mathbf{X}_1 \sim \mathcal{SUN}_{p_1,p_1}(\boldsymbol{\xi}_1, \boldsymbol{\Psi}_{11}, \mathbf{H}_{11}, \boldsymbol{\tau}_1, \bar{\boldsymbol{\Gamma}}_{11})$*, but according to Theorem 95 we have* $\mathbf{X}_2 \sim \mathcal{SUN}_{p_2,p}(\boldsymbol{\xi}_2, \boldsymbol{\Psi}_{22}, (\mathbf{H}_{21}, \mathbf{H}_{22}), \boldsymbol{\tau}, \bar{\boldsymbol{\Gamma}})$*. This holds in particular if* $\boldsymbol{\tau} = \mathbf{0}_p$ *and* $\bar{\boldsymbol{\Gamma}} = \mathbf{I}_p$*. This means that the SUN distribution in this setting is marginally consistent for* $\mathbf{X}_1$ *but not for* $\mathbf{X}_2$*. A possible choice for* $\mathbf{H}$ *is* $\mathbf{H} = \boldsymbol{\Psi}^{\pm 1/2}$ *where* $\boldsymbol{\Psi}^{\pm 1/2}$ *is the lower triangular factor of the Choleski factorization of* $\boldsymbol{\Psi}$ *or* $\boldsymbol{\Psi}^{-1}$*.*

A connection with these results is that they are prerequisites for independence, as established in the following theorem.

Theorem 99 *Let* $\mathbf{X} \sim \mathcal{SUN}_{p,q}(\boldsymbol{\xi}, \boldsymbol{\Psi}, \mathbf{H}, \boldsymbol{\tau}, \bar{\boldsymbol{\Gamma}})$ *be partitioned as (6.11), and with* $(\mathbf{H}, \boldsymbol{\tau}, \bar{\boldsymbol{\Gamma}})$ *partitioned as (6.13). Then, for every fixed* $p_1 + p_2 = p$ *and* $q_1 + q_2 = q$*,* $\mathbf{X}_1$ *and* $\mathbf{X}_2$ *are independent if and only if* $\boldsymbol{\Psi}_{12} = \mathbf{O}$ *and either of the following conditions holds:*

(i) $\mathbf{H}_1 = \mathbf{O}$ *or* $\mathbf{H}_2 = \mathbf{O}$*, in which case one of the parts is symmetric;*

(ii) $\bar{\boldsymbol{\Gamma}}_{12} = \mathbf{O}$*, and either* $\mathbf{H}_{21} = \mathbf{O}$ *and* $\mathbf{H}_{12} = \mathbf{O}$ *or* $\mathbf{H}_{11} = \mathbf{O}$ *and* $\mathbf{H}_{22} = \mathbf{O}$*.*

Proof 99 *Using Theorem 91, we find that the joint cdf of* $\mathbf{X}_1$ *and* $\mathbf{X}_2$ *and their corresponding marginal cdfs are, respectively, given by*

$$F(\mathbf{x}_1, \mathbf{x}_2) = \frac{1}{\Phi_q(\boldsymbol{\tau}; \mathbf{0}_q, \bar{\boldsymbol{\Gamma}})} \Phi_{p_1+p_2+q}\left\{ \begin{pmatrix} \mathbf{x}_1 \\ \mathbf{x}_2 \\ \boldsymbol{\tau} \end{pmatrix}; \begin{pmatrix} \boldsymbol{\xi}_1 \\ \boldsymbol{\xi}_2 \\ \mathbf{0}_q \end{pmatrix}, \begin{pmatrix} \boldsymbol{\Omega}_{11} & \boldsymbol{\Omega}_{12} & -\mathbf{H}_1\bar{\boldsymbol{\Gamma}} \\ \boldsymbol{\Omega}_{21} & \boldsymbol{\Omega}_{22} & -\mathbf{H}_2\bar{\boldsymbol{\Gamma}} \\ -\bar{\boldsymbol{\Gamma}}\mathbf{H}_1^\top & -\bar{\boldsymbol{\Gamma}}\mathbf{H}_2^\top & \bar{\boldsymbol{\Gamma}} \end{pmatrix} \right\}, \tag{6.18}$$

$$F_1(\mathbf{x}_1) = F(\mathbf{x}_1, \infty) = \frac{1}{\Phi_q(\boldsymbol{\tau}; \mathbf{0}_q, \bar{\boldsymbol{\Gamma}})} \Phi_{p_1+q}\left\{ \begin{pmatrix} \mathbf{x}_1 \\ \boldsymbol{\tau} \end{pmatrix}; \begin{pmatrix} \boldsymbol{\xi}_1 \\ \mathbf{0}_q \end{pmatrix}, \begin{pmatrix} \boldsymbol{\Omega}_{11} & -\mathbf{H}_1\bar{\boldsymbol{\Gamma}} \\ -\bar{\boldsymbol{\Gamma}}\mathbf{H}_1^\top & \bar{\boldsymbol{\Gamma}} \end{pmatrix} \right\}, \tag{6.19}$$

$$F_2(\mathbf{x}_2) = F(\infty, \mathbf{x}_2) = \frac{1}{\Phi_q(\boldsymbol{\tau}; \mathbf{0}_q, \bar{\boldsymbol{\Gamma}})} \Phi_{p_2+q}\left\{ \begin{pmatrix} \mathbf{x}_2 \\ \boldsymbol{\tau} \end{pmatrix}; \begin{pmatrix} \boldsymbol{\xi}_2 \\ \mathbf{0}_q \end{pmatrix}, \begin{pmatrix} \boldsymbol{\Omega}_{22} & -\mathbf{H}_2\bar{\boldsymbol{\Gamma}} \\ -\bar{\boldsymbol{\Gamma}}\mathbf{H}_2^\top & \bar{\boldsymbol{\Gamma}} \end{pmatrix} \right\}, \tag{6.20}$$

for $\mathbf{x}_1 \in \mathbb{R}^{p_1}$ *and* $\mathbf{x}_2 \in \mathbb{R}^{p_2}$*, where* $\boldsymbol{\Omega}_{jk} = \boldsymbol{\Psi}_{jk} + \mathbf{H}_j\bar{\boldsymbol{\Gamma}}\mathbf{H}_k^\top$ *for* $j,k = 1,2$*.*

Also, we recall that $\mathbf{X}_1$ *and* $\mathbf{X}_2$ *are independent if and only if* $F(\mathbf{x}_1, \mathbf{x}_2) = F_1(\mathbf{x}_1)F_2(\mathbf{x}_2)$ *for all* $\mathbf{x}_1 \in \mathbb{R}^{p_1}$ *and* $\mathbf{x}_2 \in \mathbb{R}^{p_2}$*.*

First, note that the joint distribution of $\mathbf{X}_1$ *and* $\mathbf{X}_2$ *becomes multivariate normal if and only if* $\mathbf{H} = \mathbf{O}$*, which, of course, is equivalent to* $\mathbf{H}_j = \mathbf{O}$ *for* $j = 1,2$ *or to* $\mathbf{H}_{jk} = \mathbf{O}$ *for* $j,k = 1,2$*. In turn, this is equivalent to* $\boldsymbol{\Omega}_{jk} = \boldsymbol{\Psi}_{jk}$ *for* $j,k = 1,2$*. Therefore, the above joint and marginal cdfs reduce to the multivariate normal cdfs given by*

$$F(\mathbf{x}_1, \mathbf{x}_2) = \Phi_{p_1+p_2}\left\{ \begin{pmatrix} \mathbf{x}_1 \\ \mathbf{x}_2 \end{pmatrix}; \begin{pmatrix} \boldsymbol{\xi}_1 \\ \boldsymbol{\xi}_2 \end{pmatrix}, \begin{pmatrix} \boldsymbol{\Psi}_{11} & \boldsymbol{\Psi}_{12} \\ \boldsymbol{\Psi}_{21} & \boldsymbol{\Psi}_{22} \end{pmatrix} \right\}, \quad \begin{pmatrix} \mathbf{x}_1 \\ \mathbf{x}_2 \end{pmatrix} \in \mathbb{R}^{p_1+p_2},$$
$$F_1(\mathbf{x}_1) = \Phi_{p_1}(\mathbf{x}_1; \boldsymbol{\xi}_1, \boldsymbol{\Psi}_{11}), \quad \mathbf{x}_1 \in \mathbb{R}^{p_1},$$
$$F_2(\mathbf{x}_2) = \Phi_{p_2}(\mathbf{x}_2; \boldsymbol{\xi}_2, \boldsymbol{\Psi}_{22}), \quad \mathbf{x}_2 \in \mathbb{R}^{p_2}.$$

As is well known, in this setting, the independence between $\mathbf{X}_1$ *and* $\mathbf{X}_2$ *is equivalent to* $\mathrm{cov}(\mathbf{X}_1, \mathbf{X}_2) = \boldsymbol{\Psi}_{12} = \mathbf{O}$*; in other words,*

$$\Phi_{p_1+p_2}\left\{ \begin{pmatrix} \mathbf{x}_1 \\ \mathbf{x}_2 \end{pmatrix}; \begin{pmatrix} \boldsymbol{\xi}_1 \\ \boldsymbol{\xi}_2 \end{pmatrix}, \begin{pmatrix} \boldsymbol{\Psi}_{11} & \boldsymbol{\Psi}_{12} \\ \boldsymbol{\Psi}_{21} & \boldsymbol{\Psi}_{22} \end{pmatrix} \right\} = \Phi_{p_1}(\mathbf{x}_1; \boldsymbol{\xi}_1, \boldsymbol{\Psi}_{11})\Phi_{p_2}(\mathbf{x}_2; \boldsymbol{\xi}_2, \boldsymbol{\Psi}_{22}), \ \forall \begin{pmatrix} \mathbf{x}_1 \\ \mathbf{x}_2 \end{pmatrix} \in \mathbb{R}^{p_1+p_2}$$
$$\iff \boldsymbol{\Psi}_{12} = \mathbf{O}.$$

Now, we prove part (i) assuming skewness in one of the marginals. Thus, considering that the multivariate normal cdf can be factorized only if the associated variance matrix is block-diagonal, it clearly follows that the joint cdf (6.18) becomes the product of the marginal cdfs (6.19) and (6.20) for all $\mathbf{x}_1 \in \mathbb{R}^{p_1}$ and $\mathbf{x}_2 \in \mathbb{R}^{p_2}$ if and only if in the variance matrix associated with the (p_1+p_2+q)-variate normal cdf that defines (6.18) we have: $(i1)$ $\boldsymbol{\Omega}_{12}=\mathbf{O}$ and $-\mathbf{H}_1\bar{\boldsymbol{\Gamma}}=\mathbf{O}$ or $(i2)$ $\boldsymbol{\Omega}_{12}=\mathbf{O}$ and $-\mathbf{H}_2\bar{\boldsymbol{\Gamma}}=\mathbf{O}$, which are completely equivalent to the conditions in part (i) since $\boldsymbol{\Omega}_{12}=\boldsymbol{\Psi}_{12}+\mathbf{H}_1^\top\bar{\boldsymbol{\Gamma}}\mathbf{H}_2$ and $\bar{\boldsymbol{\Gamma}}>0$, and so, for each $j=1,2$, we have that $\boldsymbol{\Omega}_{12}=\mathbf{O}$ and $-\mathbf{H}_j\bar{\boldsymbol{\Gamma}}=\mathbf{O} \Longleftrightarrow \boldsymbol{\Psi}_{12}+\mathbf{H}_1^\top\bar{\boldsymbol{\Gamma}}\mathbf{H}_2=\mathbf{O}$ and $-\mathbf{H}_j=\mathbf{O} \Longleftrightarrow \boldsymbol{\Psi}_{12}=\mathbf{O}$ and $\mathbf{H}_j=\mathbf{O}$; that is, when:

$$
\begin{aligned}
(i1)\ F(\mathbf{x}_1,\mathbf{x}_2) &= \frac{1}{\Phi_q(\boldsymbol{\tau};\mathbf{0}_q,\bar{\boldsymbol{\Gamma}})}\Phi_{p_1+p_2+q}\left\{\begin{pmatrix}\mathbf{x}_1\\ \mathbf{x}_2\\ \boldsymbol{\tau}\end{pmatrix};\begin{pmatrix}\boldsymbol{\xi}_1\\ \boldsymbol{\xi}_2\\ \mathbf{0}_q\end{pmatrix},\begin{pmatrix}\boldsymbol{\Psi}_{11} & \mathbf{O} & \mathbf{O}\\ \mathbf{O} & \boldsymbol{\Omega}_{22} & -\mathbf{H}_2\bar{\boldsymbol{\Gamma}}\\ \mathbf{O} & -\bar{\boldsymbol{\Gamma}}\mathbf{H}_2^\top & \bar{\boldsymbol{\Gamma}}\end{pmatrix}\right\}\\
&= \underbrace{\Phi_{p_1}(\mathbf{x}_1;\boldsymbol{\xi}_1,\boldsymbol{\Psi}_{11})}_{\mathcal{N}_{p_1}(\boldsymbol{\xi}_1,\boldsymbol{\Psi}_{11})-\text{cdf}}\times\underbrace{\frac{1}{\Phi_q(\boldsymbol{\tau};\mathbf{0}_q,\bar{\boldsymbol{\Gamma}})}\Phi_{p_2+q}\left\{\begin{pmatrix}\mathbf{x}_2\\ \boldsymbol{\tau}\end{pmatrix};\begin{pmatrix}\boldsymbol{\xi}_2\\ \mathbf{0}_q\end{pmatrix},\begin{pmatrix}\boldsymbol{\Omega}_{22} & -\mathbf{H}_2\bar{\boldsymbol{\Gamma}}\\ -\bar{\boldsymbol{\Gamma}}\mathbf{H}_2^\top & \bar{\boldsymbol{\Gamma}}\end{pmatrix}\right\}}_{\mathcal{SUN}_{p_2,q}(\boldsymbol{\xi}_2,\boldsymbol{\Psi}_{22},\mathbf{H}_2,\boldsymbol{\tau},\bar{\boldsymbol{\Gamma}})-\text{cdf}},
\end{aligned}
$$

or

$$
\begin{aligned}
(i2)\ F(\mathbf{x}_1,\mathbf{x}_2) &= \frac{1}{\Phi_q(\boldsymbol{\tau};\mathbf{0}_q,\bar{\boldsymbol{\Gamma}})}\Phi_{p_1+p_2+q}\left\{\begin{pmatrix}\mathbf{x}_1\\ \mathbf{x}_2\\ \boldsymbol{\tau}\end{pmatrix};\begin{pmatrix}\boldsymbol{\xi}_1\\ \boldsymbol{\xi}_2\\ \mathbf{0}_q\end{pmatrix},\begin{pmatrix}\boldsymbol{\Omega}_{11} & \mathbf{O} & -\mathbf{H}_1\bar{\boldsymbol{\Gamma}}\\ \mathbf{O} & \boldsymbol{\Psi}_{22} & \mathbf{O}\\ -\bar{\boldsymbol{\Gamma}}\mathbf{H}_1^\top & -\mathbf{O} & \bar{\boldsymbol{\Gamma}}\end{pmatrix}\right\}\\
&= \underbrace{\frac{1}{\Phi_q(\boldsymbol{\tau};\mathbf{0}_q,\bar{\boldsymbol{\Gamma}})}\Phi_{p_1+q}\left\{\begin{pmatrix}\mathbf{x}_1\\ \boldsymbol{\tau}\end{pmatrix};\begin{pmatrix}\boldsymbol{\xi}_1\\ \mathbf{0}_q\end{pmatrix},\begin{pmatrix}\boldsymbol{\Omega}_{11} & -\mathbf{H}_1\bar{\boldsymbol{\Gamma}}\\ -\bar{\boldsymbol{\Gamma}}\mathbf{H}_1^\top & \bar{\boldsymbol{\Gamma}}\end{pmatrix}\right\}}_{\mathcal{SUN}_{p_1,q}(\boldsymbol{\xi}_1,\boldsymbol{\Psi}_{11},\mathbf{H}_1,\boldsymbol{\tau},\bar{\boldsymbol{\Gamma}})-\text{cdf}}\times\underbrace{\Phi_{p_2}(\mathbf{x}_2;\boldsymbol{\xi}_2,\boldsymbol{\Psi}_{22})}_{\mathcal{N}_{p_2}(\boldsymbol{\xi}_2,\boldsymbol{\Psi}_{22})-\text{cdf}}.
\end{aligned}
$$

Thus, part (i) is proved.

Now, we prove part (ii). For this, we first find by applying the partition (6.13) in (6.18) to (6.20) that the associated normal cdfs involved in those expressions can be rewritten as

$$
\Phi_q(\boldsymbol{\tau};\mathbf{0}_q,\bar{\boldsymbol{\Gamma}})=\Phi_{q_1+q_2}\left\{\begin{pmatrix}\boldsymbol{\tau}_1\\ \boldsymbol{\tau}_2\end{pmatrix};\begin{pmatrix}\mathbf{0}_{q_1}\\ \mathbf{0}_{q_2}\end{pmatrix},\begin{pmatrix}\bar{\boldsymbol{\Gamma}}_{11} & \bar{\boldsymbol{\Gamma}}_{12}\\ \bar{\boldsymbol{\Gamma}}_{21} & \bar{\boldsymbol{\Gamma}}_{22}\end{pmatrix}\right\},
$$

$$
\begin{aligned}
&\Phi_{p_1+p_2+q}\left\{\begin{pmatrix}\mathbf{x}_1\\ \mathbf{x}_2\\ \boldsymbol{\tau}\end{pmatrix};\begin{pmatrix}\boldsymbol{\xi}_1\\ \boldsymbol{\xi}_2\\ \mathbf{0}_q\end{pmatrix},\begin{pmatrix}\boldsymbol{\Omega}_{11} & \boldsymbol{\Omega}_{12} & -\mathbf{H}_1\bar{\boldsymbol{\Gamma}}\\ \boldsymbol{\Omega}_{21} & \boldsymbol{\Omega}_{22} & -\mathbf{H}_2\bar{\boldsymbol{\Gamma}}\\ -\bar{\boldsymbol{\Gamma}}\mathbf{H}_1^\top & -\bar{\boldsymbol{\Gamma}}\mathbf{H}_2^\top & \bar{\boldsymbol{\Gamma}}\end{pmatrix}\right\}=\Phi_{p_1+p_2+q_1+q_2}\left\{\begin{pmatrix}\mathbf{x}_1\\ \mathbf{x}_2\\ \boldsymbol{\tau}_1\\ \boldsymbol{\tau}_2\end{pmatrix};\begin{pmatrix}\boldsymbol{\xi}_1\\ \boldsymbol{\xi}_2\\ \mathbf{0}_{q_1}\\ \mathbf{0}_{q_2}\end{pmatrix},\right.\\
&\left.\begin{pmatrix}\boldsymbol{\Omega}_{11} & \boldsymbol{\Omega}_{12} & -\mathbf{H}_{11}\bar{\boldsymbol{\Gamma}}_{11}-\mathbf{H}_{12}\bar{\boldsymbol{\Gamma}}_{21} & -\mathbf{H}_{11}\bar{\boldsymbol{\Gamma}}_{12}-\mathbf{H}_{12}\bar{\boldsymbol{\Gamma}}_{22}\\ \boldsymbol{\Omega}_{21} & \boldsymbol{\Omega}_{22} & -\mathbf{H}_{21}\bar{\boldsymbol{\Gamma}}_{11}-\mathbf{H}_{22}\bar{\boldsymbol{\Gamma}}_{21} & -\mathbf{H}_{21}\bar{\boldsymbol{\Gamma}}_{12}-\mathbf{H}_{22}\bar{\boldsymbol{\Gamma}}_{22}\\ -\bar{\boldsymbol{\Gamma}}_{11}\mathbf{H}_{11}^\top-\bar{\boldsymbol{\Gamma}}_{12}\mathbf{H}_{12}^\top & -\bar{\boldsymbol{\Gamma}}_{11}\mathbf{H}_{21}^\top-\bar{\boldsymbol{\Gamma}}_{12}\mathbf{H}_{22}^\top & \bar{\boldsymbol{\Gamma}}_{11} & \bar{\boldsymbol{\Gamma}}_{12}\\ -\bar{\boldsymbol{\Gamma}}_{21}\mathbf{H}_{11}^\top-\bar{\boldsymbol{\Gamma}}_{22}\mathbf{H}_{12}^\top & -\bar{\boldsymbol{\Gamma}}_{21}\mathbf{H}_{21}^\top-\bar{\boldsymbol{\Gamma}}_{22}\mathbf{H}_{22}^\top & \bar{\boldsymbol{\Gamma}}_{21} & \bar{\boldsymbol{\Gamma}}_{22}\end{pmatrix}\right\},
\end{aligned}
$$

$$
\begin{aligned}
&\Phi_{p_j+q}\left\{\begin{pmatrix}\mathbf{x}_j\\ \boldsymbol{\tau}\end{pmatrix};\begin{pmatrix}\boldsymbol{\xi}_j\\ \mathbf{0}_q\end{pmatrix},\begin{pmatrix}\boldsymbol{\Omega}_{jj} & -\mathbf{H}_j\bar{\boldsymbol{\Gamma}}\\ -\bar{\boldsymbol{\Gamma}}\mathbf{H}_j^\top & \bar{\boldsymbol{\Gamma}}\end{pmatrix}\right\}=\Phi_{p_j+q_1+q_2}\left\{\begin{pmatrix}\mathbf{x}_j\\ \boldsymbol{\tau}_1\\ \boldsymbol{\tau}_2\end{pmatrix};\begin{pmatrix}\boldsymbol{\xi}_j\\ \mathbf{0}_{q_1}\\ \mathbf{0}_{q_2}\end{pmatrix},\right.\\
&\left.\begin{pmatrix}\boldsymbol{\Omega}_{jj} & -\mathbf{H}_{j1}\bar{\boldsymbol{\Gamma}}_{11}-\mathbf{H}_{j2}\bar{\boldsymbol{\Gamma}}_{21} & -\mathbf{H}_{j1}\bar{\boldsymbol{\Gamma}}_{12}-\mathbf{H}_{j2}\bar{\boldsymbol{\Gamma}}_{22}\\ -\bar{\boldsymbol{\Gamma}}_{11}\mathbf{H}_{j1}^\top-\bar{\boldsymbol{\Gamma}}_{12}\mathbf{H}_{j2}^\top & \bar{\boldsymbol{\Gamma}}_{11} & \bar{\boldsymbol{\Gamma}}_{12}\\ -\bar{\boldsymbol{\Gamma}}_{21}\mathbf{H}_{j1}^\top-\bar{\boldsymbol{\Gamma}}_{22}\mathbf{H}_{j2}^\top & \bar{\boldsymbol{\Gamma}}_{21} & \bar{\boldsymbol{\Gamma}}_{22}\end{pmatrix}\right\},\quad j=1,2.
\end{aligned}
$$

Thus, to have independence with skewness in both marginal distributions, we need that all these last expressions on the right side can be simultaneously factorized for $\mathbf{H}_j\neq\mathbf{O}$ for $j=1,2$, which holds if *and only if either:* $(ii1)$ $\bar{\mathbf{\Gamma}}_{12} = \mathbf{O}$, $\mathbf{\Omega}_{12} = \mathbf{O}$ *and* $-\mathbf{H}_{11}\bar{\mathbf{\Gamma}}_{12} - \mathbf{H}_{12}\bar{\mathbf{\Gamma}}_{22} = \mathbf{O}$ *or* $(ii2)$ $\bar{\mathbf{\Gamma}}_{12} = \mathbf{O}$, $\mathbf{\Omega}_{12} = \mathbf{O}$ *and* $-\mathbf{H}_{11}\bar{\mathbf{\Gamma}}_{11} - \mathbf{H}_{12}\bar{\mathbf{\Gamma}}_{21} = \mathbf{O}$. *The equivalence of these conditions with the conditions in part* (ii) *is straightforward by noting that* $\bar{\mathbf{\Gamma}}_{jj} > 0$ *for* $j = 1,2$, *and* $\mathbf{\Omega}_{12} = \mathbf{\Psi}_{12} + \mathbf{H}_{11}\bar{\mathbf{\Gamma}}_{11}\mathbf{H}_{21}^\top + \mathbf{H}_{12}\bar{\mathbf{\Gamma}}_{21}\mathbf{H}_{21}^\top + \mathbf{H}_{11}\bar{\mathbf{\Gamma}}_{12}\mathbf{H}_{22}^\top + \mathbf{H}_{12}\bar{\mathbf{\Gamma}}_{22}\mathbf{H}_{22}^\top$, *under which we then have*

$$(ii1)\ F(\mathbf{x}_1,\mathbf{x}_2) = \underbrace{\frac{1}{\Phi_{q_1}(\boldsymbol{\tau}_1;\mathbf{0}_{q_1},\bar{\mathbf{\Gamma}}_{11})}\Phi_{p_1+q_1}\left\{\begin{pmatrix}\mathbf{x}_1\\ \boldsymbol{\tau}_1\end{pmatrix};\begin{pmatrix}\boldsymbol{\xi}_1\\ \mathbf{0}_{q_1}\end{pmatrix},\begin{pmatrix}\mathbf{\Psi}_{11} & \mathbf{H}_{11}\bar{\mathbf{\Gamma}}_{11}\\ \bar{\mathbf{\Gamma}}_{11}\mathbf{H}_{11}^\top & \bar{\mathbf{\Gamma}}_{11}\end{pmatrix}\right\}}_{\mathcal{SUN}_{p_1,q_1}(\boldsymbol{\xi}_1,\mathbf{\Psi}_{11},\mathbf{H}_{11},\boldsymbol{\tau}_1,\bar{\mathbf{\Gamma}}_{11})-\text{cdf}}$$

$$\times\underbrace{\frac{1}{\Phi_{q_2}(\boldsymbol{\tau}_2;\mathbf{0}_{q_2},\bar{\mathbf{\Gamma}}_{22})}\Phi_{p_2+q_2}\left\{\begin{pmatrix}\mathbf{x}_2\\ \boldsymbol{\tau}_2\end{pmatrix};\begin{pmatrix}\boldsymbol{\xi}_2\\ \mathbf{0}_{q_2}\end{pmatrix},\begin{pmatrix}\mathbf{\Psi}_{22} & \mathbf{H}_{22}\bar{\mathbf{\Gamma}}_{22}\\ \bar{\mathbf{\Gamma}}_{22}\mathbf{H}_{22}^\top & \bar{\mathbf{\Gamma}}_{22}\end{pmatrix}\right\}}_{\mathcal{SUN}_{p_2,q_2}(\boldsymbol{\xi}_2,\mathbf{\Psi}_{22},\mathbf{H}_{22},\boldsymbol{\tau}_2,\bar{\mathbf{\Gamma}}_{22})-\text{cdf}},$$

or

$$(ii2)\ F(\mathbf{x}_1,\mathbf{x}_2) = \underbrace{\frac{1}{\Phi_{q_2}(\boldsymbol{\tau}_2;\mathbf{0}_{q_2},\bar{\mathbf{\Gamma}}_{22})}\Phi_{p_1+q_2}\left\{\begin{pmatrix}\mathbf{x}_1\\ \boldsymbol{\tau}_2\end{pmatrix};\begin{pmatrix}\boldsymbol{\xi}_1\\ \mathbf{0}_{q_2}\end{pmatrix},\begin{pmatrix}\mathbf{\Psi}_{11} & \mathbf{H}_{12}\bar{\mathbf{\Gamma}}_{22}\\ \bar{\mathbf{\Gamma}}_{22}\mathbf{H}_{12}^\top & \bar{\mathbf{\Gamma}}_{22}\end{pmatrix}\right\}}_{\mathcal{SUN}_{p_1,q_2}(\boldsymbol{\xi}_1,\mathbf{\Psi}_{11},\mathbf{H}_{12},\boldsymbol{\tau}_2,\bar{\mathbf{\Gamma}}_{22})-\text{cdf}}$$

$$\times\underbrace{\frac{1}{\Phi_{q_1}(\boldsymbol{\tau}_1;\mathbf{0}_{q_1},\bar{\mathbf{\Gamma}}_{11})}\Phi_{p_2+q_1}\left\{\begin{pmatrix}\mathbf{x}_2\\ \boldsymbol{\tau}_1\end{pmatrix};\begin{pmatrix}\boldsymbol{\xi}_2\\ \mathbf{0}_{q_1}\end{pmatrix},\begin{pmatrix}\mathbf{\Psi}_{22} & \mathbf{H}_{21}\bar{\mathbf{\Gamma}}_{11}\\ \bar{\mathbf{\Gamma}}_{11}\mathbf{H}_{21}^\top & \bar{\mathbf{\Gamma}}_{11}\end{pmatrix}\right\}}_{\mathcal{SUN}_{p_2,q_1}(\boldsymbol{\xi}_2,\mathbf{\Psi}_{22},\mathbf{H}_{21},\boldsymbol{\tau}_1,\bar{\mathbf{\Gamma}}_{11})-\text{cdf}}.$$

So part (ii) *is also proved.* □

For an informal but perhaps more concise proof, note from (6.12) and the partition (6.13) that

$$\begin{aligned}\mathbf{X}_1 &= \boldsymbol{\xi}_1 + \mathbf{H}_1\mathbf{U} + \mathbf{W}_1 = \boldsymbol{\xi}_1 + \mathbf{H}_{11}\mathbf{U}_1 + \mathbf{H}_{12}\mathbf{U}_2 + \mathbf{W}_1,\\ \mathbf{X}_2 &= \boldsymbol{\xi}_2 + \mathbf{H}_2\mathbf{U} + \mathbf{W}_2 = \boldsymbol{\xi}_2 + \mathbf{H}_{21}\mathbf{U}_1 + \mathbf{H}_{22}\mathbf{U}_2 + \mathbf{W}_2,\end{aligned}$$

where $\mathbf{U} = (\mathbf{U}_1^\top,\mathbf{U}_2^\top)^\top$ and $\mathbf{W} = (\mathbf{W}_1^\top,\mathbf{W}_2^\top)^\top$ are independent, with

$$\begin{aligned}\begin{pmatrix}\mathbf{U}_1\\ \mathbf{U}_2\end{pmatrix} &\sim \mathcal{TN}_{q_1+q_2}\left(\begin{pmatrix}\mathbf{0}_{q_1}\\ \mathbf{0}_{q_2}\end{pmatrix};\begin{pmatrix}\boldsymbol{\tau}_1\\ \boldsymbol{\tau}_2\end{pmatrix},\begin{pmatrix}\bar{\mathbf{\Gamma}}_{11} & \bar{\mathbf{\Gamma}}_{12}\\ \bar{\mathbf{\Gamma}}_{21} & \bar{\mathbf{\Gamma}}_{22}\end{pmatrix}\right),\\ \begin{pmatrix}\mathbf{W}_1\\ \mathbf{W}_2\end{pmatrix} &\sim \mathcal{N}_{p_1+p_2}\left(\begin{pmatrix}\mathbf{0}_{p_1}\\ \mathbf{0}_{p_2}\end{pmatrix},\begin{pmatrix}\mathbf{\Psi}_{11} & \mathbf{\Psi}_{12}\\ \mathbf{\Psi}_{21} & \mathbf{\Psi}_{22}\end{pmatrix}\right).\end{aligned}$$

Here, note that $\mathbf{U}_1$ and $\mathbf{U}_2$ are independent if and only if $\bar{\mathbf{\Gamma}}_{12} = \bar{\mathbf{\Gamma}}_{21}^\top = \mathbf{O}$, and also that $\mathbf{W}_1$ and $\mathbf{W}_2$ are independent if and only if $\mathbf{\Psi}_{12} = \mathbf{\Psi}_{21}^\top = \mathbf{O}$. Thus, it clearly follows that $\mathbf{X}_1$ and $\mathbf{X}_2$ are independent if and only if $\mathbf{W}_1$ and $\mathbf{W}_2$ are independent, equivalently $\mathbf{\Psi}_{12} = \mathbf{O}$, and $\mathbf{H}_1 = \mathbf{O}$ or $\mathbf{H}_2 = \mathbf{O}$; as well as $\mathbf{\Psi}_{12} = \mathbf{O}$, $\mathbf{U}_1$ and $\mathbf{U}_2$ are independent, equivalently $\bar{\mathbf{\Gamma}}_{12} = \mathbf{O}$, and $\mathbf{H}_{12} = \mathbf{O}$ and $\mathbf{H}_{21} = \mathbf{O}$, or $\mathbf{H}_{11} = \mathbf{O}$ and $\mathbf{H}_{22} = \mathbf{O}$. That is, if and only if $\mathbf{\Psi}_{12} = \mathbf{O}$ and, either:

$$\begin{aligned}\mathbf{X}_1 &= \boldsymbol{\xi}_1 + \mathbf{W}_1 \quad \text{and} \quad \mathbf{X}_2 = \boldsymbol{\xi}_2 + \mathbf{H}_2\mathbf{U} + \mathbf{W}_2, \text{ or}\\ \mathbf{X}_1 &= \boldsymbol{\xi}_1 + \mathbf{H}_1\mathbf{U} + \mathbf{W}_1 \quad \text{and} \quad \mathbf{X}_2 = \boldsymbol{\xi}_2 + \mathbf{W}_2, \text{ or}\\ \mathbf{X}_1 &= \boldsymbol{\xi}_1 + \mathbf{W}_1 \quad \text{and} \quad \mathbf{X}_2 = \boldsymbol{\xi}_2 + \mathbf{W}_2;\end{aligned}$$

or $\mathbf{\Psi}_{12} = \mathbf{O}$ and either:

$$\begin{aligned}\mathbf{X}_1 &= \boldsymbol{\xi}_1 + \mathbf{H}_{11}\mathbf{U}_1 + \mathbf{W}_1 \quad \text{and} \quad \mathbf{X}_2 = \boldsymbol{\xi}_2 + \mathbf{W}_2, \text{ or}\\ \mathbf{X}_1 &= \boldsymbol{\xi}_1 + \mathbf{W}_1 \quad \text{and} \quad \mathbf{X}_2 = \boldsymbol{\xi}_2 + \mathbf{H}_{22}\mathbf{U}_2 + \mathbf{W}_2, \text{ or}\\ \mathbf{X}_1 &= \boldsymbol{\xi}_1 + \mathbf{H}_{12}\mathbf{U}_2 + \mathbf{W}_1 \quad \text{and} \quad \mathbf{X}_2 = \boldsymbol{\xi}_2 + \mathbf{W}_2, \text{ or}\\ \mathbf{X}_1 &= \boldsymbol{\xi}_1 + \mathbf{W}_1 \quad \text{and} \quad \mathbf{X}_2 = \boldsymbol{\xi}_2 + \mathbf{H}_{21}\mathbf{U}_1 + \mathbf{W}_2;\end{aligned}$$

or $\boldsymbol{\Psi}_{12} = \mathbf{O}$, $\bar{\boldsymbol{\Gamma}}_{12} = \mathbf{O}$ and either:

$$\begin{aligned}\mathbf{X}_1 &= \boldsymbol{\xi}_1 + \mathbf{H}_{11}\mathbf{U}_1 + \mathbf{W}_1 \quad \text{and} \quad \mathbf{X}_2 = \boldsymbol{\xi}_2 + \mathbf{H}_{22}\mathbf{U}_2 + \mathbf{W}_2, \text{ or}\\ \mathbf{X}_1 &= \boldsymbol{\xi}_1 + \mathbf{H}_{12}\mathbf{U}_2 + \mathbf{W}_1 \quad \text{and} \quad \mathbf{X}_2 = \boldsymbol{\xi}_2 + \mathbf{H}_{21}\mathbf{U}_1 + \mathbf{W}_2.\end{aligned}$$

Example 74 *Suppose that* $\mathbf{X} = (\mathbf{X}_1^\top, \mathbf{X}_2^\top)^\top \sim \mathcal{SUN}_{p_1+p_2, q_1+q_2}(\boldsymbol{\xi}, \boldsymbol{\Psi}, \mathbf{H}, \boldsymbol{\tau}, \bar{\boldsymbol{\Gamma}})$, *with* $\boldsymbol{\xi} = (\boldsymbol{\xi}_1^\top, \boldsymbol{\xi}_2^\top)^\top \in \mathbb{R}^{p_1+p_2}$, $\boldsymbol{\Psi} = \mathrm{diag}(\boldsymbol{\Psi}_{11}, \boldsymbol{\Psi}_{22})$, *with* $\boldsymbol{\Psi}_{jj}$ *of dimension* $p_j \times p_j$ *for* $j = 1,2$, $\mathbf{H} = \mathrm{diag}(\mathbf{H}_{11}, \mathbf{H}_{22})$, *with* $\mathbf{H}_{jj}$ *of dimension* $p_j \times q_j$ *for* $j = 1,2$, $\boldsymbol{\tau} = (\boldsymbol{\tau}_1^\top, \boldsymbol{\tau}_2^\top)^\top \in \mathbb{R}^{q_1+q_2}$, *and* $\bar{\boldsymbol{\Gamma}} = \mathrm{diag}(\bar{\boldsymbol{\Gamma}}_{11}, \bar{\boldsymbol{\Gamma}}_{22})$, *with* $\bar{\boldsymbol{\Gamma}}_{jj}$ *of dimension* $q_j \times q_j$ *for* $j = 1,2$. *Then,* $\mathbf{X}_j \sim \mathcal{SUN}_{p_j, q_j}(\boldsymbol{\xi}_j, \boldsymbol{\Psi}_{jj}, \mathbf{H}_{jj}, \boldsymbol{\tau}_j, \bar{\boldsymbol{\Gamma}}_{jj})$ *for* $j = 1,2$, *and they are independent.*

Example 75 *Let* $\mathbf{X} = (X_1, \ldots, X_p)^\top \sim \mathcal{SUN}_{p,p}(\boldsymbol{\xi}, \boldsymbol{\Psi}, \mathbf{H}, \boldsymbol{\tau}, \bar{\boldsymbol{\Gamma}})$, *with* $\boldsymbol{\xi} = (\xi_1, \ldots, \xi_p)^\top$, $\boldsymbol{\Psi} = (\psi_{ij})_{i,j=1}^p$, $\mathbf{H} = \mathrm{diag}(\eta_1, \ldots, \eta_p)$, $\boldsymbol{\tau} = \mathbf{0}_p$, *and* $\bar{\boldsymbol{\Gamma}} = \mathbf{I}_p$. *Then, the marginal distribution of* X_j *is* $\mathcal{SN}(\xi_j, \psi_{jj}, \eta_j)$, $j = 1, \ldots, p$, *and they are mutually independent. However, for any* $1 < k < p$:

$$(X_1, \ldots, X_k)^\top \sim \mathcal{SUN}_{k,k}((\xi_1, \ldots, \xi_k)^\top, (\psi_{ij})_{i,j=1}^k, \mathrm{diag}(\eta_1, \ldots, \eta_k), \mathbf{0}_k, \mathbf{I}_k),$$

which is different from an $\mathcal{SN}_k$ *distribution.*

Next, a general result related to Theorem 99 is given.

Theorem 100 *Let* $\mathbf{X} \sim \mathcal{SUN}_{p,q}(\boldsymbol{\xi}, \boldsymbol{\Psi}, \mathbf{H}, \boldsymbol{\tau}, \bar{\boldsymbol{\Gamma}})$. *Then for all matrices* $\mathbf{A}_j \in \mathbb{R}^{p_j \times p}$, $j = 1,2$, *we have that* $\mathbf{A}_1\mathbf{X}$ *and* $\mathbf{A}_2\mathbf{X}$ *are independent if and only if* $\mathbf{A}_1\boldsymbol{\Psi}\mathbf{A}_2^\top = \mathbf{O}$ *and either* $\mathbf{A}_1\mathbf{H} = \mathbf{O}$ *or* $\mathbf{A}_2\mathbf{H} = \mathbf{O}$, *in which case one of the parts is symmetric.*

Proof 100 *Let* $\tilde{\psi}(\mathbf{t}_1, \mathbf{t}_2) = \mathrm{E}\{e^{\mathrm{i}(\mathbf{t}_1^\top \mathbf{Y}_1 + \mathbf{t}_2^\top \mathbf{Y}_2)}\}$ *be the joint cf of* $\mathbf{Y}_1 = \mathbf{A}_1\mathbf{X}$ *and* $\mathbf{Y}_2 = \mathbf{A}_2\mathbf{X}$. *Since* $\tilde{\psi}(\mathbf{t}_1, \mathbf{t}_2) = \psi(\mathbf{A}_1^\top \mathbf{t}_1 + \mathbf{A}_2^\top \mathbf{t}_2)$, *where* $\psi(\mathbf{t})$ *is the cf of* $\mathbf{X}$ *given in (6.9), by Theorem 92, we have*

$$\begin{aligned}\tilde{\psi}(\mathbf{t}_1, \mathbf{t}_2) &= e^{\mathrm{i}\mathbf{t}_1^\top \mathbf{A}_1 \boldsymbol{\xi} + \mathrm{i}\mathbf{t}_2^\top \mathbf{A}_2 \boldsymbol{\xi} - \frac{1}{2}(\mathbf{t}_1^\top \mathbf{A}_1 \boldsymbol{\Omega} \mathbf{A}_1^\top \mathbf{t}_1 + 2\mathbf{t}_1^\top \mathbf{A}_1 \boldsymbol{\Omega} \mathbf{A}_2^\top \mathbf{t}_2 + \mathbf{t}_2^\top \mathbf{A}_2 \boldsymbol{\Omega} \mathbf{A}_2^\top \mathbf{t}_2)}\\ &\quad \times \frac{\Phi_q(\boldsymbol{\tau} + \mathrm{i}\bar{\boldsymbol{\Gamma}}\mathbf{H}^\top \mathbf{A}_1^\top \mathbf{t}_1 + \mathrm{i}\bar{\boldsymbol{\Gamma}}\mathbf{H}^\top \mathbf{A}_2^\top \mathbf{t}_2; \mathbf{0}_q, \bar{\boldsymbol{\Gamma}})}{\Phi_q(\boldsymbol{\tau}; \mathbf{0}_q, \bar{\boldsymbol{\Gamma}})}, \quad \mathbf{t}_1 \in \mathbb{R}^{p_1}, \mathbf{t}_2 \in \mathbb{R}^{p_2},\end{aligned}$$

where $\boldsymbol{\Omega} = \boldsymbol{\Psi} + \mathbf{H}\bar{\boldsymbol{\Gamma}}\mathbf{H}^\top$ *and* $\mathbf{H} = (\mathbf{H}_1^\top, \mathbf{H}_2^\top)^\top$, *with* $\mathbf{H}_j$ *of dimension* $p_j \times q$, $j = 1,2$. *Also, their respective marginal cfs are:*

$$\tilde{\psi}_1(\mathbf{t}_1) = \tilde{\psi}(\mathbf{t}_1, \mathbf{0}_{p_2}) = e^{\mathrm{i}\mathbf{t}_1^\top \mathbf{A}_1 \boldsymbol{\xi} - \frac{1}{2}\mathbf{t}_1^\top \mathbf{A}_1 \boldsymbol{\Omega} \mathbf{A}_1^\top \mathbf{t}_1} \frac{\Phi_q(\boldsymbol{\tau} + \mathrm{i}\bar{\boldsymbol{\Gamma}}\mathbf{H}^\top \mathbf{A}_1^\top \mathbf{t}_1; \mathbf{0}_q, \bar{\boldsymbol{\Gamma}})}{\Phi_q(\boldsymbol{\tau}; \mathbf{0}_q, \bar{\boldsymbol{\Gamma}})}, \quad \mathbf{t}_1 \in \mathbb{R}^{p_1}$$

and

$$\tilde{\psi}_2(\mathbf{t}_2) = \tilde{\psi}(\mathbf{0}_{p_1}, \mathbf{t}_2) = e^{\mathrm{i}\mathbf{t}_2^\top \mathbf{A}_2 \boldsymbol{\xi} - \frac{1}{2}\mathbf{t}_2^\top \mathbf{A}_2 \boldsymbol{\Omega} \mathbf{A}_2^\top \mathbf{t}_2} \frac{\Phi_q(\boldsymbol{\tau} + \mathrm{i}\bar{\boldsymbol{\Gamma}}\mathbf{H}^\top \mathbf{A}_2^\top \mathbf{t}_2; \mathbf{0}_q, \bar{\boldsymbol{\Gamma}})}{\Phi_q(\boldsymbol{\tau}; \mathbf{0}_q, \bar{\boldsymbol{\Gamma}})}, \quad \mathbf{t}_2 \in \mathbb{R}^{p_2}.$$

Thus, it clearly follows that $\tilde{\psi}(\mathbf{t}_1, \mathbf{t}_2) = \tilde{\psi}_1(\mathbf{t}_1)\tilde{\psi}_2(\mathbf{t}_2)$ *for all* $(\mathbf{t}_1^\top, \mathbf{t}_2^\top)^\top \in \mathbb{R}^{p_1+p_2}$ *if and only if* $\mathbf{t}_1^\top \mathbf{A}_1 \boldsymbol{\Omega} \mathbf{A}_2^\top \mathbf{t}_2 = 0$ *and either* $\bar{\boldsymbol{\Gamma}}\mathbf{H}^\top \mathbf{A}_1^\top \mathbf{t}_1 = \mathbf{0}_q$ *or* $\bar{\boldsymbol{\Gamma}}\mathbf{H}^\top \mathbf{A}_2^\top \mathbf{t}_2 = \mathbf{0}_q$ *for all* $(\mathbf{t}_1^\top, \mathbf{t}_2^\top)^\top \in \mathbb{R}^{p_1+p_2}$ *which hold if and only if* $\mathbf{A}_1 \boldsymbol{\Omega} \mathbf{A}_2^\top = \mathbf{O}$ *and either* $\bar{\boldsymbol{\Gamma}}\mathbf{H}^\top \mathbf{A}_1^\top = \mathbf{O}$ *or* $\bar{\boldsymbol{\Gamma}}\mathbf{H}^\top \mathbf{A}_2^\top = \mathbf{O}$. *However, since* $\mathbf{A}_1 \boldsymbol{\Omega} \mathbf{A}_2^\top = \mathbf{A}_1 \boldsymbol{\Psi} \mathbf{A}_2^\top + \mathbf{A}_1 \mathbf{H}\bar{\boldsymbol{\Gamma}}\mathbf{H}^\top \mathbf{A}_2^\top$ *and* $\bar{\boldsymbol{\Gamma}} > 0$, *these last conditions are equivalent to* $\mathbf{A}_1 \boldsymbol{\Psi} \mathbf{A}_2^\top = \mathbf{O}$ *and either* $\mathbf{H}^\top \mathbf{A}_1^\top = \mathbf{O}$ *or* $\mathbf{H}^\top \mathbf{A}_2^\top = \mathbf{O}$, *which prove the result.* □

An important specific fact related to the last independence result is that of closure of the SUN family with respect to convolution; specifically, the sum of two independent SUN variables of type $\mathcal{SUN}_{p,q_1}$ and $\mathcal{SUN}_{p,q_2}$ is of type $\mathcal{SUN}_{p,q_1+q_2}$ as established below. Gupta et al. (2013a) provided related results in the AASUN parameterization.

Theorem 101 *Let* $\mathbf{Y} = a_1\mathbf{X}_1 + a_2\mathbf{X}_2$, *where* a_1 *and* a_2 *are non-null real numbers, and* $\mathbf{X}_1 \sim \mathcal{SUN}_{p,q_1}(\boldsymbol{\xi}_1, \boldsymbol{\Psi}_1, \mathbf{H}_1, \boldsymbol{\tau}_1, \bar{\boldsymbol{\Gamma}}_1)$ *and* $\mathbf{X}_2 \sim \mathcal{SUN}_{p,q_2}(\boldsymbol{\xi}_2, \boldsymbol{\Psi}_2, \mathbf{H}_2, \boldsymbol{\tau}_2, \bar{\boldsymbol{\Gamma}}_2)$ *are independent random vectors. Then,* $\mathbf{Y} \sim \mathcal{SUN}_{p,q_1+q_2}(\boldsymbol{\xi}_a, \boldsymbol{\Psi}_a, \mathbf{H}_a, \boldsymbol{\tau}, \bar{\boldsymbol{\Gamma}})$, *where* $\boldsymbol{\xi}_a = a_1\boldsymbol{\xi}_1 + a_2\boldsymbol{\xi}_2$, $\boldsymbol{\Psi}_a = a_1^2\boldsymbol{\Psi}_1 + a_2^2\boldsymbol{\Psi}_2$, $\mathbf{H}_a = (a_1\mathbf{H}_1, a_2\mathbf{H}_2)$, $\boldsymbol{\tau} = (\boldsymbol{\tau}_1^\top, \boldsymbol{\tau}_2^\top)^\top$, *and* $\bar{\boldsymbol{\Gamma}} = \operatorname{diag}(\bar{\boldsymbol{\Gamma}}_1, \bar{\boldsymbol{\Gamma}}_2)$.

Proof 101 *From the fact that*

$$\Phi_{q_1+q_2}\left\{\begin{pmatrix}\mathbf{z}_1\\ \mathbf{z}_2\end{pmatrix}; \begin{pmatrix}\boldsymbol{\xi}_1\\ \boldsymbol{\xi}_2\end{pmatrix}, \begin{pmatrix}\bar{\boldsymbol{\Gamma}}_1 & \mathbf{O}\\ \mathbf{O} & \bar{\boldsymbol{\Gamma}}_2\end{pmatrix}\right\} = \Phi_{q_1}(\mathbf{z}_1; \boldsymbol{\xi}_1, \bar{\boldsymbol{\Gamma}}_1)\Phi_{q_2}(\mathbf{z}_2; \boldsymbol{\xi}_2, \bar{\boldsymbol{\Gamma}}_2),\ \forall \begin{pmatrix}\mathbf{x}_1\\ \mathbf{x}_2\end{pmatrix} \in \mathbb{R}^{p_1+p_2},$$

and the independence assumption, the cf of $\mathbf{Y}$ *becomes*

$$\begin{aligned}
\psi_{\mathbf{Y}}(\mathbf{t}) &= \psi_{\mathbf{X}_1}(a_1\mathbf{t})\psi_{\mathbf{X}_2}(a_2\mathbf{t})\\
&= e^{i\mathbf{t}^\top(a_1\boldsymbol{\xi}_1 + a_2\boldsymbol{\xi}_2) - \frac{1}{2}\mathbf{t}^\top(a_1^2\boldsymbol{\Psi}_1 + a_2^2\boldsymbol{\Psi}_2 + a_1^2\mathbf{H}_1\bar{\boldsymbol{\Gamma}}_1\mathbf{H}_1^\top + a_2^2\mathbf{H}_2\bar{\boldsymbol{\Gamma}}_2\mathbf{H}_2^\top)\mathbf{t}}\\
&\quad \times \frac{\Phi_{q_1}(\boldsymbol{\tau}_1 + ia_1\bar{\boldsymbol{\Gamma}}_1\mathbf{H}_1^\top\mathbf{t}; \mathbf{0}_{q_1}, \bar{\boldsymbol{\Gamma}}_1)}{\Phi_{q_1}(\boldsymbol{\tau}_1; \mathbf{0}_{q_1}, \bar{\boldsymbol{\Gamma}}_1)} \frac{\Phi_{q_2}(\boldsymbol{\tau}_2 + ia_2\bar{\boldsymbol{\Gamma}}_2\mathbf{H}_2^\top\mathbf{t}; \mathbf{0}_{q_2}, \bar{\boldsymbol{\Gamma}}_2)}{\Phi_{q_2}(\boldsymbol{\tau}_2; \mathbf{0}_{q_2}, \bar{\boldsymbol{\Gamma}}_2)}\\
&= e^{i\mathbf{t}^\top\boldsymbol{\xi}_a - \frac{1}{2}\mathbf{t}^\top(\boldsymbol{\Psi}_a + \mathbf{H}_a\bar{\boldsymbol{\Gamma}}\mathbf{H}_a^\top)\mathbf{t}} \frac{\Phi_q(\boldsymbol{\tau} + i\bar{\boldsymbol{\Gamma}}\mathbf{H}_a^\top\mathbf{t}; \mathbf{0}_q, \bar{\boldsymbol{\Gamma}})}{\Phi_q(\boldsymbol{\tau}; \mathbf{0}_q, \bar{\boldsymbol{\Gamma}})}, \quad \mathbf{t} \in \mathbb{R}^p.
\end{aligned}$$

The proof concludes with the uniqueness of the cf. □

Example 76 *Let* $\mathbf{X}_1, \dots, \mathbf{X}_n$ *be a random sample from* $\mathbf{X} \sim \mathcal{SUN}_{p,q}(\boldsymbol{\xi}, \boldsymbol{\Psi}, \mathbf{H}, \boldsymbol{\tau}, \bar{\boldsymbol{\Gamma}})$. *Then, the sample mean vector* $\bar{\mathbf{X}} = \frac{1}{n}\sum_{i=1}^n \mathbf{X}_i$ *has a SUN distribution given by*

$$\bar{\mathbf{X}} \sim \mathcal{SUN}_{p,nq}\left(\boldsymbol{\xi}, \frac{1}{n}\boldsymbol{\Psi}, \frac{1}{n}\mathbf{1}_n \otimes \mathbf{H}, \mathbf{1}_n \otimes \boldsymbol{\tau}, \mathbf{I}_n \otimes \bar{\boldsymbol{\Gamma}}\right).$$

6.3.3 Canonical form

For the SUN distribution, the *canonical form* is related to the transformation given in Equation (6.2). To discuss this, we only consider the case where $\mathbf{X} - \boldsymbol{\xi} = \mathbf{HU} + \mathbf{W} \sim \mathcal{SUN}_{p,q}(\mathbf{0}_p, \boldsymbol{\Psi}, \mathbf{H}, \boldsymbol{\tau}, \bar{\boldsymbol{\Gamma}})$ is nonsingular, i.e., $\boldsymbol{\Psi} > 0$ so that $\operatorname{rank}(\boldsymbol{\Psi}) = p$. The idea is to find an affine transformation $\mathbf{Z} = \mathbf{C}(\mathbf{X} - \boldsymbol{\xi})$ that eliminates redundant components of the latent vector $\mathbf{HU}$ and standardizes the normal vector $\mathbf{W}$. In this sense, we say that the random vector $\mathbf{Z} \sim \mathcal{SUN}_{p,q}(\mathbf{0}_p, \mathbf{C}\boldsymbol{\Psi}\mathbf{C}^\top, \mathbf{CH}, \boldsymbol{\tau}, \bar{\boldsymbol{\Gamma}})$ has a canonical form of rank/size $r \leq \min\{p, q\}$ if $\mathbf{H}$ has rank r and the nonsingular matrix $\mathbf{C} = (\mathbf{C}_1^\top, \mathbf{C}_2^\top)^\top \in \mathbb{R}^{p\times p}$, where $\mathbf{C}_1$ is $r \times p$ and $\mathbf{C}_2$ is $(p-r) \times p$, is such that $\mathbf{C}_1\boldsymbol{\Psi}\mathbf{C}_1^\top = \mathbf{I}_r$, $\mathbf{C}_1\boldsymbol{\Psi}\mathbf{C}_2^\top = \mathbf{O}$ and $\mathbf{C}_2\boldsymbol{\Psi}\mathbf{C}_2^\top = \mathbf{I}_{p-r}$, and also allows the partition of $\mathbf{Z}$ as

$$\begin{pmatrix}\mathbf{Z}_1\\ \mathbf{Z}_2\end{pmatrix} = \begin{pmatrix}\mathbf{H}_1\mathbf{U} + \mathbf{W}_1\\ \mathbf{H}_2\mathbf{U} + \mathbf{W}_2\end{pmatrix},$$

where we have $\mathbf{H}_1 = \mathbf{C}_1\mathbf{H} \neq \mathbf{O}$, $\mathbf{H}_2 = \mathbf{C}_2\mathbf{H} = \mathbf{O}$, $\mathbf{W}_1 = \mathbf{C}_1\mathbf{W} \sim \mathcal{N}_r(\mathbf{0}_r, \mathbf{I}_r)$, and $\mathbf{W}_2 = \mathbf{C}_2\mathbf{W} \sim \mathcal{N}_{p-r}(\mathbf{0}_{p-r}, \mathbf{I}_{p-r})$, so that $\mathbf{Z}_1 = \mathbf{H}_1\mathbf{U} + \mathbf{W}_1 \sim \mathcal{SUN}_{r,q}(\mathbf{0}_r, \mathbf{I}_r, \mathbf{H}_1, \boldsymbol{\tau}, \bar{\boldsymbol{\Gamma}})$ and $\mathbf{Z}_2 = \mathbf{W}_2 \sim \mathcal{N}_{p-r}(\mathbf{0}_{p-r}, \mathbf{I}_{p-r})$ which are also independent.

Note that the above canonical transformation matrix $\mathbf{C}$ requires that the column space of $\mathbf{H}$ be a subspace of the null space of $\mathbf{C}_2$. Hence, for $2 \le q \le p$, the canonical transformation does not always exist. This result is different from the SN family, for which the canonical form always exists, and which, setting $q = 1$, is equivalent to $\mathcal{SUN}_{p,1}$. The guaranteed existence of the canonical transformation for the SN family is due to the fact that it is always possible to construct a transformation matrix with a one-dimensional null space. In the SUN case ($q \ge 2$), the matrix $\mathbf{H} \in \mathbb{R}^{p\times q}$, hence $\text{rank}(\mathbf{H}) \le \min\{p,q\}$. The null space for $\mathbf{C}_2 \in \mathbb{R}^{(p-r)\times p}$ is at most $(p-r-1)$-dimensional. Hence, we need $q \le p-r-1$ in order for the canonical transformation to exist. Thus, for example, there is no canonical form of rank $r = 1$ for $p = 3$ and $q = 2$.

6.4 Nonsingular Unified Skew-Normal

6.4.1 Pdf

If $\mathbf{X} = \boldsymbol{\xi} + \mathbf{HU} + \mathbf{W} \sim \mathcal{SUN}_{p,q}(\boldsymbol{\xi}, \boldsymbol{\Psi}, \mathbf{H}, \boldsymbol{\tau}, \bar{\boldsymbol{\Gamma}})$ is nonsingular, i.e., with $\boldsymbol{\Psi} > 0$ so that $|\boldsymbol{\Psi}| > 0$, then the pdf of $\mathbf{X}$ may be found in an explicit form from its hierarchical representation given by (6.3) and (6.4). In fact, since $\boldsymbol{\Psi} > 0$ and therefore the multivariate normal conditional distribution $(\mathbf{X} \mid \mathbf{U} = \mathbf{u}) \sim \mathcal{N}_p(\boldsymbol{\xi} + \mathbf{Hu}, \boldsymbol{\Psi})$ is also nonsingular, its conditional pdf is, for $\mathbf{x} \in \mathbb{R}^p$, given by

$$f(\mathbf{x} \mid \mathbf{u}) = \phi_p(\mathbf{x}; \boldsymbol{\xi} + \mathbf{Hu}, \boldsymbol{\Psi}) = (2\pi)^{-p/2}|\boldsymbol{\Psi}|^{-1/2} \exp\left\{-\frac{1}{2}(\mathbf{x} - \boldsymbol{\xi} - \mathbf{Hu})^\top \boldsymbol{\Psi}^{-1}(\mathbf{x} - \boldsymbol{\xi} - \mathbf{Hu})\right\}.$$

Similarly, the assumption that $\bar{\boldsymbol{\Gamma}} > 0$ means that the truncated normal random vector $\mathbf{U} \sim \mathcal{TN}_q(-\boldsymbol{\tau}; \mathbf{0}_q, \bar{\boldsymbol{\Gamma}})$ is also nonsingular with pdf $g(\mathbf{u}) = \phi_q(\mathbf{u}; \boldsymbol{\tau}, \bar{\boldsymbol{\Gamma}})/\Phi_q(\boldsymbol{\tau}; \mathbf{0}_q, \bar{\boldsymbol{\Gamma}})$ for $\mathbf{u} + \boldsymbol{\tau} > \mathbf{0}_q$. Therefore, for the pdf of $\mathbf{X}$, we have

$$\begin{aligned} f(\mathbf{x}) &= \int_{\mathbb{R}^q} f(\mathbf{x} \mid \mathbf{u}) g(\mathbf{u}) d\mathbf{u} \\ &= \frac{1}{\Phi_q(\boldsymbol{\tau}; \mathbf{0}_q, \bar{\boldsymbol{\Gamma}})} \int_{\mathbf{u}+\boldsymbol{\tau}>\mathbf{0}_q} \phi_p(\mathbf{x}; \boldsymbol{\xi} + \mathbf{Hu}, \boldsymbol{\Psi}) \phi_q(\mathbf{u}; \mathbf{0}_q, \bar{\boldsymbol{\Gamma}}) d\mathbf{u}. \end{aligned}$$

Here, the marginal-conditional representation of a $(p+q)$-variate normal pdf yields the following identity:

$$\begin{aligned} &\phi_p(\mathbf{x}; \boldsymbol{\xi} + \mathbf{Hu}, \boldsymbol{\Psi}) \phi_q(\mathbf{u}; \mathbf{0}_q, \bar{\boldsymbol{\Gamma}}) \\ &= \phi_p(\mathbf{x}; \boldsymbol{\xi}, \boldsymbol{\Psi} + \mathbf{H}\bar{\boldsymbol{\Gamma}}\mathbf{H}^\top) \phi_q\{\mathbf{u}; \bar{\boldsymbol{\Gamma}}\mathbf{H}^\top(\boldsymbol{\Psi} + \mathbf{H}\bar{\boldsymbol{\Gamma}}\mathbf{H}^\top)^{-1}(\mathbf{x} - \boldsymbol{\xi}), \bar{\boldsymbol{\Gamma}} - \bar{\boldsymbol{\Gamma}}\mathbf{H}^\top(\boldsymbol{\Psi} + \mathbf{H}\bar{\boldsymbol{\Gamma}}\mathbf{H}^\top)^{-1}\mathbf{H}\bar{\boldsymbol{\Gamma}}\} \\ &= \phi_p(\mathbf{x}; \boldsymbol{\xi}, \boldsymbol{\Psi} + \mathbf{H}\bar{\boldsymbol{\Gamma}}\mathbf{H}^\top) \phi_q\{\mathbf{u}; \tilde{\boldsymbol{\Gamma}}\mathbf{H}^\top\boldsymbol{\Psi}^{-1}(\mathbf{x} - \boldsymbol{\xi}), \tilde{\boldsymbol{\Gamma}}\}, \end{aligned} \tag{6.21}$$

where $\tilde{\boldsymbol{\Gamma}} = (\bar{\boldsymbol{\Gamma}}^{-1} + \mathbf{H}^\top\boldsymbol{\Psi}^{-1}\mathbf{H})^{-1}$, and for the last equality we used the following matrix identities:

$$\bar{\boldsymbol{\Gamma}}\mathbf{H}^\top(\boldsymbol{\Psi} + \mathbf{H}\bar{\boldsymbol{\Gamma}}\mathbf{H}^\top)^{-1} = (\bar{\boldsymbol{\Gamma}}^{-1} + \mathbf{H}^\top\boldsymbol{\Psi}^{-1}\mathbf{H})^{-1}\mathbf{H}^\top\boldsymbol{\Psi}^{-1}, \tag{6.22}$$

$$\bar{\boldsymbol{\Gamma}} - \bar{\boldsymbol{\Gamma}}\mathbf{H}^\top(\boldsymbol{\Psi} + \mathbf{H}\bar{\boldsymbol{\Gamma}}\mathbf{H}^\top)^{-1}\mathbf{H}\bar{\boldsymbol{\Gamma}} = (\bar{\boldsymbol{\Gamma}}^{-1} + \mathbf{H}^\top\boldsymbol{\Psi}^{-1}\mathbf{H})^{-1}. \tag{6.23}$$

Using (6.21) and the change of variables $\mathbf{u} = \mathbf{v} + \mathbf{h}$, where $\mathbf{h} = \tilde{\boldsymbol{\Gamma}}\mathbf{H}^\top\boldsymbol{\Psi}^{-1}(\mathbf{x} - \boldsymbol{\xi})$, we have

$$\begin{aligned} f(\mathbf{x}) &= \frac{1}{\Phi_q(\boldsymbol{\tau}; \mathbf{0}_q, \bar{\boldsymbol{\Gamma}})} \phi_p(\mathbf{x}; \boldsymbol{\xi}, \boldsymbol{\Psi} + \mathbf{H}\bar{\boldsymbol{\Gamma}}\mathbf{H}^\top) \int_{\mathbf{u}+\boldsymbol{\tau}>\mathbf{0}_q} \phi_q\{\mathbf{u}; \tilde{\boldsymbol{\Gamma}}\mathbf{H}^\top\boldsymbol{\Psi}^{-1}(\mathbf{x} - \boldsymbol{\xi}), \tilde{\boldsymbol{\Gamma}}\} d\mathbf{u} \\ &= \frac{1}{\Phi_q(\boldsymbol{\tau}; \mathbf{0}_q, \bar{\boldsymbol{\Gamma}})} \phi_p(\mathbf{x}; \boldsymbol{\xi}, \boldsymbol{\Psi} + \mathbf{H}\bar{\boldsymbol{\Gamma}}\mathbf{H}^\top) \int_{\mathbf{v}+\boldsymbol{\tau}+\mathbf{h}>\mathbf{0}_q} \phi_q(\mathbf{v}; \mathbf{0}_q, \tilde{\boldsymbol{\Gamma}}) d\mathbf{v}, \end{aligned}$$

where by symmetry

$$\int_{\mathbf{v}+\boldsymbol{\tau}+\mathbf{h}>\mathbf{0}_q} \phi_q(\mathbf{v}; \mathbf{0}_q, \tilde{\boldsymbol{\Gamma}}) d\mathbf{v} = \int_{\mathbf{v}-\boldsymbol{\tau}-\mathbf{h}<\mathbf{0}_q} \phi_q(\mathbf{v}; \mathbf{0}_q, \tilde{\boldsymbol{\Gamma}}) d\mathbf{v} = \Phi_q\{\boldsymbol{\tau} + \tilde{\boldsymbol{\Gamma}}\mathbf{H}^\top\boldsymbol{\Psi}^{-1}(\mathbf{x} - \boldsymbol{\xi}); \mathbf{0}_q, \tilde{\boldsymbol{\Gamma}}\}.$$

Thus, the pdf of $\mathbf{X} \sim \mathcal{SUN}_{p,q}(\boldsymbol{\xi}, \boldsymbol{\Psi}, \mathbf{H}, \boldsymbol{\tau}, \bar{\boldsymbol{\Gamma}})$ when $\boldsymbol{\Psi} > 0$ becomes as stated in the next theorem.

Theorem 102 *Let* $\mathbf{X} \sim \mathcal{SUN}_{p,q}(\boldsymbol{\xi}, \boldsymbol{\Psi}, \mathbf{H}, \boldsymbol{\tau}, \bar{\boldsymbol{\Gamma}})$, $\boldsymbol{\Psi} > 0$, *be a nonsingular multivariate unified skew-normal distribution. Then its pdf is:*

$$f(\mathbf{x}) = \frac{1}{\Phi_q(\boldsymbol{\tau}; \mathbf{0}_q, \bar{\boldsymbol{\Gamma}})} \phi_p(\mathbf{x}; \boldsymbol{\xi}, \boldsymbol{\Psi} + \mathbf{H}\bar{\boldsymbol{\Gamma}}\mathbf{H}^\top) \Phi_q\{\boldsymbol{\tau} + \tilde{\boldsymbol{\Gamma}}\mathbf{H}^\top\boldsymbol{\Psi}^{-1}(\mathbf{x} - \boldsymbol{\xi}); \mathbf{0}_q, \tilde{\boldsymbol{\Gamma}}\}, \quad \mathbf{x} \in \mathbb{R}^p, \tag{6.24}$$

where $\tilde{\boldsymbol{\Gamma}} = \bar{\boldsymbol{\Gamma}} - \bar{\boldsymbol{\Gamma}}\mathbf{H}^\top(\boldsymbol{\Psi} + \mathbf{H}\bar{\boldsymbol{\Gamma}}\mathbf{H}^\top)^{-1}\mathbf{H}\bar{\boldsymbol{\Gamma}} = (\bar{\boldsymbol{\Gamma}}^{-1} + \mathbf{H}^\top\boldsymbol{\Psi}^{-1}\mathbf{H})^{-1}$.

Proof 102 *See above.* □

Note from (6.24) and the matrix identities (6.22) and (6.23) that, in terms of the matrix $\boldsymbol{\Omega} = \boldsymbol{\Psi} + \mathbf{H}\bar{\boldsymbol{\Gamma}}\mathbf{H}^\top$, the SUN pdf can be written as

$$f(\mathbf{x}) = \frac{1}{\Phi_q(\boldsymbol{\tau}; \mathbf{0}_q, \bar{\boldsymbol{\Gamma}})} \phi_p(\mathbf{x}; \boldsymbol{\xi}, \boldsymbol{\Omega}) \Phi_q\{\boldsymbol{\tau} + \bar{\boldsymbol{\Gamma}}\mathbf{H}^\top\boldsymbol{\Omega}^{-1}(\mathbf{x} - \boldsymbol{\xi}); \mathbf{0}_q, \bar{\boldsymbol{\Gamma}} - \bar{\boldsymbol{\Gamma}}\mathbf{H}^\top\boldsymbol{\Omega}^{-1}\mathbf{H}\bar{\boldsymbol{\Gamma}}\}, \quad \mathbf{x} \in \mathbb{R}^p. \tag{6.25}$$

Clearly, if $\mathbf{H} = \mathbf{O}$ then $f(\mathbf{x}) = \phi_p(\mathbf{x}; \boldsymbol{\xi}, \boldsymbol{\Psi})$ for all $\mathbf{x} \in \mathbb{R}^p$. Note that from (6.24) or (6.25), the distribution remains skewed if $\boldsymbol{\tau} = \mathbf{0}_q$ unless also $\mathbf{H} = \mathbf{O}$.

Example 77 (Tractable SUN pdf) *Let* $\tilde{\boldsymbol{\Gamma}} = \bar{\boldsymbol{\Gamma}} - \bar{\boldsymbol{\Gamma}}\mathbf{H}^\top\boldsymbol{\Omega}^{-1}\mathbf{H}\bar{\boldsymbol{\Gamma}}$ *be the variance matrix associated with the last term in (6.24). Note that* $\tilde{\boldsymbol{\Gamma}} = \bar{\boldsymbol{\Gamma}} - \boldsymbol{\Upsilon}$, *where* $\boldsymbol{\Upsilon} = \bar{\boldsymbol{\Gamma}}\mathbf{H}^\top\boldsymbol{\Omega}^{-1}\mathbf{H}\bar{\boldsymbol{\Gamma}}$ *with* $\boldsymbol{\Omega} = \boldsymbol{\Psi} + \mathbf{H}\bar{\boldsymbol{\Gamma}}\mathbf{H}^\top$. *Write* $\bar{\boldsymbol{\Gamma}} = (\bar{\gamma}_{ij})$ *and* $\boldsymbol{\Upsilon} = (\upsilon_{ij})$. *Then* $\upsilon_{ij} = \bar{\boldsymbol{\gamma}}_i^\top\mathbf{H}^\top\boldsymbol{\Omega}^{-1}\mathbf{H}\bar{\boldsymbol{\gamma}}_j$, $i, j = 1, \ldots, q$, *where* $\bar{\boldsymbol{\gamma}}_1, \ldots, \bar{\boldsymbol{\gamma}}_q$ *are the column vectors of* $\bar{\boldsymbol{\Gamma}}$. *Suppose that* $\bar{\gamma}_{ij} = \upsilon_{ij}$ *for all* $i \neq j$. *This condition means that in (6.24) the variance matrix* $\tilde{\boldsymbol{\Gamma}} \equiv \mathbf{D}(\tilde{\boldsymbol{\Gamma}})$ *becomes diagonal, that is,* $\mathbf{D}(\tilde{\boldsymbol{\Gamma}}) = \mathbf{I}_q - \mathbf{D}(\boldsymbol{\Upsilon})$, *where for any matrix* $\mathbf{A} = (a_{ij}) \in \mathbb{R}^{q \times q}$, $\mathbf{D}(\mathbf{A}) = \mathrm{diag}(a_{11}, \ldots, a_{qq})$. *Write* $\mathbf{D}(\tilde{\boldsymbol{\Gamma}}) = \mathbf{I}_q - \mathbf{D}(\boldsymbol{\Upsilon}) = \mathrm{diag}(1 - \upsilon_{11}, \ldots, 1 - \upsilon_{qq})$ *and let* $\tilde{\tau}_i = (1 - \upsilon_{ii})^{-1/2}\tau_i$ *and* $\bar{\boldsymbol{\eta}}_i = (1 - \upsilon_{ii})^{1/2}\boldsymbol{\eta}_i$, $i = 1, \ldots, q$, *where* $\tau_1, \ldots, \tau_q$ *are the components of the vector* $\boldsymbol{\tau}$ *and* $\boldsymbol{\eta}_1, \ldots, \boldsymbol{\eta}_q$ *are the column vectors of the matrix* $\mathbf{H}$. *So, under the conditions* $\bar{\gamma}_{ij} = \bar{\boldsymbol{\gamma}}_i^\top\mathbf{H}^\top\boldsymbol{\Omega}^{-1}\mathbf{H}\bar{\boldsymbol{\gamma}}_j$ *for all* $i \neq j$, *the SUN pdf given in (6.24) simplifies to*

$$f(\mathbf{x}) = \frac{1}{\Phi_q(\boldsymbol{\tau}; \mathbf{0}_q, \bar{\boldsymbol{\Gamma}})} \phi_p(\mathbf{x}; \boldsymbol{\xi}, \boldsymbol{\Psi} + \mathbf{H}\bar{\boldsymbol{\Gamma}}\mathbf{H}^\top) \prod_{i=1}^{q} \Phi_1\{\tilde{\tau}_i + \bar{\boldsymbol{\eta}}_i^\top\boldsymbol{\Psi}^{-1}(\mathbf{x} - \boldsymbol{\xi})\}, \quad \mathbf{x} \in \mathbb{R}^p. \tag{6.26}$$

In the special case of $\bar{\boldsymbol{\Gamma}} = \mathbf{I}_q$, *we have* $\bar{\boldsymbol{\gamma}}_i = \mathbf{e}_i$, *the i-th q-dimensional unit vector, and therefore* $\mathbf{H}\bar{\boldsymbol{\gamma}}_i = \mathbf{H}\mathbf{e}_i = \boldsymbol{\eta}_i$ *for* $i = 1, \ldots, q$. *So, in this situation, the previous diagonalization condition is reduced to* $(\mathbf{H}^\top\boldsymbol{\Omega}^{-1}\mathbf{H})_{ij} = \boldsymbol{\eta}_i^\top\boldsymbol{\Omega}^{-1}\boldsymbol{\eta}_j = 0$ *for all* $i \neq j$, *and the SUN pdf to*

$$f(\mathbf{x}) = \phi_p(\mathbf{x}; \boldsymbol{\xi}, \boldsymbol{\Psi} + \mathbf{H}\mathbf{H}^\top) \prod_{i=1}^{q} \frac{\Phi_1\{\tilde{\tau}_i + \bar{\boldsymbol{\eta}}_i^\top\boldsymbol{\Psi}^{-1}(\mathbf{x} - \boldsymbol{\xi})\}}{\Phi_1(\tau_i)}, \quad \mathbf{x} \in \mathbb{R}^p.$$

Moreover, if $\boldsymbol{\tau} = \mathbf{0}_q$ *then* $1/\Phi_q(\mathbf{0}_q; \mathbf{0}_q, \bar{\boldsymbol{\Gamma}}) = \{1/\Phi_1(0)\}^q = 2^q$. *Finally, note from the selection representation given in Theorem 89 that* $\tilde{\boldsymbol{\Gamma}} = \mathrm{var}(\mathbf{W}_0 | \mathbf{W}_* = \mathbf{w})$ *where* $\mathbf{W}_0$ *and* $\mathbf{W}_*$ *have a multivariate normal joint distribution as in (6.5). So, while the matrix* $\tilde{\boldsymbol{\Gamma}} = \mathrm{var}(\mathbf{W}_0 | \mathbf{W}_* = \mathbf{w})$ *becomes diagonal if and only if for all* $i \neq j$ *the components* W_{0i} *and* W_{0j} *are conditionally independent given the random vector* $\mathbf{W}_*$, *the additional diagonal condition on the correlation matrix* $\bar{\boldsymbol{\Gamma}} = \mathrm{var}(\mathbf{W}_0)$ *means that these components are also marginally independent. From these ideas, the SUN pdf can be simplified to intermediate situations by assuming that* $\tilde{\boldsymbol{\Gamma}} = \mathrm{var}(\mathbf{W}_0 | \mathbf{W}_* = \mathbf{w})$ *and/or* $\bar{\boldsymbol{\Gamma}} = \mathrm{var}(\mathbf{W}_0)$ *are block-diagonal; see, e.g., the SUNSET, BSUNSET, and SUNNY distributions in Section 6.8.*

The next proposition specializes the result in Theorem 94 to a nonsingular affine transformation in an absolutely continuous unified skew-normal random vector $\mathbf{X}$.

Theorem 103 *Let* $\mathbf{X} \sim \mathcal{SUN}_{p,q}(\boldsymbol{\xi}, \boldsymbol{\Psi}, \mathbf{H}, \boldsymbol{\tau}, \bar{\boldsymbol{\Gamma}})$, $\boldsymbol{\Psi} > 0$, *be a nonsingular multivariate unified skew-normal distribution, with pdf given by (6.24). Also, let* $\mathbf{Y} = \mathbf{A}\mathbf{X} + \mathbf{b}$, *where* $\mathbf{A} \in \mathbb{R}^{p\times p}$ *is a nonsingular constant matrix and* $\mathbf{b} \in \mathbb{R}^p$ *is a constant vector. Then,* $\mathbf{Y} \sim \mathcal{SUN}_{p,q}(\boldsymbol{\xi}_A, \boldsymbol{\Psi}_A, \mathbf{H}_A, \boldsymbol{\tau}, \bar{\boldsymbol{\Gamma}})$ *is also nonsingular and has pdf given by*

$$h(\mathbf{y}) = \frac{1}{\Phi_q(\boldsymbol{\tau}; \mathbf{0}_q, \bar{\boldsymbol{\Gamma}})} \phi_p(\mathbf{y}; \boldsymbol{\xi}_A, \boldsymbol{\Psi}_A + \mathbf{H}_A \bar{\boldsymbol{\Gamma}} \mathbf{H}_A^\top) \Phi_q\{\boldsymbol{\tau} + \tilde{\boldsymbol{\Gamma}}_A \mathbf{H}_A^\top \boldsymbol{\Psi}_A^{-1}(\mathbf{y} - \boldsymbol{\xi}_A); \mathbf{0}_q, \tilde{\boldsymbol{\Gamma}}_A\}, \quad \mathbf{y} \in \mathbb{R}^p, \tag{6.27}$$

where

$$\boldsymbol{\xi}_A = \mathbf{A}\boldsymbol{\xi} + \mathbf{b}, \quad \boldsymbol{\Psi}_A = \mathbf{A}\boldsymbol{\Psi}\mathbf{A}^\top, \quad \mathbf{H}_A = \mathbf{A}\mathbf{H}, \quad \tilde{\boldsymbol{\Gamma}}_A = (\bar{\boldsymbol{\Gamma}}^{-1} + \mathbf{H}_A^\top \boldsymbol{\Psi}_A^{-1} \mathbf{H}_A)^{-1} = \tilde{\boldsymbol{\Gamma}}. \tag{6.28}$$

Proof 103 *We first note that if* $|\mathbf{A}| \neq 0$ *and* $\boldsymbol{\Psi} > 0$, *then* $\boldsymbol{\Psi}_A = \mathbf{A}\boldsymbol{\Psi}\mathbf{A}^\top > 0$ *and both* $\mathbf{A}^{-1}$ *and* $\boldsymbol{\Psi}_A^{-1} = (\mathbf{A}\boldsymbol{\Psi}\mathbf{A}^\top)^{-1} = \mathbf{A}^{-\top}\boldsymbol{\Psi}^{-1}\mathbf{A}^{-1}$ *exist. Now, inverting the affine relation* $\mathbf{y} = \mathbf{A}\mathbf{x} + \mathbf{b}$, *we can write* $\mathbf{x} = \mathbf{A}^{-1}(\mathbf{y} - \mathbf{b})$. *Thus, the pdf of* $\mathbf{Y}$ *is*

$$\begin{aligned} h(\mathbf{y}) &= f\{\mathbf{A}^{-1}(\mathbf{y} - \mathbf{b})\}\text{abs}(|\mathbf{A}|)^{-1} \\ &= \frac{1}{\Phi_q(\boldsymbol{\tau}; \mathbf{0}_q, \bar{\boldsymbol{\Gamma}})} \phi_p\{\mathbf{A}^{-1}(\mathbf{y} - \mathbf{b}); \boldsymbol{\xi}, \boldsymbol{\Psi} + \mathbf{H}\bar{\boldsymbol{\Gamma}}\mathbf{H}^\top\} \\ &\quad \times \Phi_q[\boldsymbol{\tau} + \tilde{\boldsymbol{\Gamma}}\mathbf{H}^\top\boldsymbol{\Psi}^{-1}\{\mathbf{A}^{-1}(\mathbf{y} - \mathbf{b}) - \boldsymbol{\xi}\}; \mathbf{0}_q, \tilde{\boldsymbol{\Gamma}}]\text{abs}(|\mathbf{A}|)^{-1}, \end{aligned}$$

since the Jacobian $J_{(\mathbf{x}\to\mathbf{y})} = |\mathbf{A}^{-1}| = |\mathbf{A}|^{-1}$, *and where* f *is the unified skew-normal pdf defined in (6.24). It is then straightforward to verify that*

$$\phi_p\{\mathbf{A}^{-1}(\mathbf{y} - \mathbf{b}); \boldsymbol{\xi}, \boldsymbol{\Psi} + \mathbf{H}\bar{\boldsymbol{\Gamma}}\mathbf{H}^\top\}\text{abs}(|\mathbf{A}|)^{-1} = \phi_p(\mathbf{y}; \mathbf{A}\boldsymbol{\xi} + \mathbf{b}, \mathbf{A}\boldsymbol{\Psi}\mathbf{A}^\top + \mathbf{A}\mathbf{H}\bar{\boldsymbol{\Gamma}}\mathbf{H}^\top\mathbf{A}^\top)$$

and

$$\Phi_q[\boldsymbol{\tau} + \tilde{\boldsymbol{\Gamma}}\mathbf{H}^\top\boldsymbol{\Psi}^{-1}\{\mathbf{A}^{-1}(\mathbf{y} - \mathbf{b}) - \boldsymbol{\xi}\}; \mathbf{0}_q, \tilde{\boldsymbol{\Gamma}}] = \Phi_q[\boldsymbol{\tau} + \tilde{\boldsymbol{\Gamma}}\mathbf{H}^\top\mathbf{A}^\top(\mathbf{A}\boldsymbol{\Psi}\mathbf{A}^\top)^{-1}\{\mathbf{y} - (\mathbf{A}\boldsymbol{\xi} + \mathbf{b})\}; \mathbf{0}_q, \tilde{\boldsymbol{\Gamma}}],$$

and the proof concludes by noting that

$$\tilde{\boldsymbol{\Gamma}} = (\bar{\boldsymbol{\Gamma}}^{-1} + \mathbf{H}^\top\boldsymbol{\Psi}^{-1}\mathbf{H})^{-1} = \{\bar{\boldsymbol{\Gamma}}^{-1} + \mathbf{H}^\top\mathbf{A}^\top(\mathbf{A}\boldsymbol{\Psi}\mathbf{A}^\top)^{-1}\mathbf{A}\mathbf{H}\}^{-1} = \tilde{\boldsymbol{\Gamma}}_A.$$

□

We can observe from Theorem 103 that if we consider (6.24) as the pdf of $\mathbf{X}$, then we obtain a similar form for the pdf of a nonsingular affine random vector $\mathbf{Y} = \mathbf{A}\mathbf{X} + \mathbf{b}$ with the obvious modification of the new parameters as indicated in (6.28). The same result is also valid for any matrix $\mathbf{A}$ of dimension $r \times p$ but with full row-rank r, since in this case the associated dispersion matrix $\boldsymbol{\Psi}_A = \mathbf{A}\boldsymbol{\Psi}\mathbf{A}^\top$ is also nonsingular. That is, (6.27) also becomes the pdf for any affine random vector $\mathbf{Y} = \mathbf{A}\mathbf{X} + \mathbf{b}$ when $\mathbf{A} \in \mathbb{R}^{r\times p}$ is of rank r. The proof follows by considering a $(p-r)$-dimensional auxiliary random vector $\mathbf{Z} = \mathbf{C}\mathbf{X} + \mathbf{d}$ such that $\mathbf{C} \in \mathbb{R}^{(p-r)\times p}$ has full row-rank $p - r$, and then applying the Jacobian procedure to the extended affine transformation of full rank p given by

$$\tilde{\mathbf{Y}} = \begin{pmatrix} \mathbf{Y} \\ \mathbf{Z} \end{pmatrix} = \begin{pmatrix} \mathbf{A}\mathbf{X} + \mathbf{b} \\ \mathbf{C}\mathbf{X} + \mathbf{c} \end{pmatrix} = \underbrace{\begin{pmatrix} \mathbf{A} \\ \mathbf{C} \end{pmatrix}}_{\tilde{\mathbf{A}}} \mathbf{X} + \underbrace{\begin{pmatrix} \mathbf{b} \\ \mathbf{c} \end{pmatrix}}_{\tilde{\mathbf{b}}} = \tilde{\mathbf{A}}\mathbf{X} + \tilde{\mathbf{b}},$$

to obtain the joint pdf $\tilde{h}(\mathbf{y},\mathbf{z}) = \tilde{h}(\tilde{\mathbf{y}})$. The marginal pdf of $\mathbf{Y}$ is then obtained as $h(\mathbf{y}) = \int_{\mathbb{R}^{p-r}} \tilde{h}(\mathbf{y},\mathbf{z})d\mathbf{z}$.

In particular, all p_j-dimensional marginal distributions of a nonsingular unified skew-normal random vector $\mathbf{X} \sim \mathcal{SUN}_{p,q}(\boldsymbol{\xi},\boldsymbol{\Psi},\mathbf{H},\boldsymbol{\tau},\bar{\boldsymbol{\Gamma}})$, $\boldsymbol{\Psi} > 0$, are also nonsingular. More specifically, since in the partition (6.11) we see that if $\boldsymbol{\Psi} > 0$ then $\boldsymbol{\Psi}_{jj} > 0$ for $j = 1,2$, and then the marginal $\mathbf{X}_j \sim \mathcal{SUN}_{p_j,q}(\boldsymbol{\xi}_j,\boldsymbol{\Psi}_{jj},\mathbf{H}_j,\boldsymbol{\tau},\bar{\boldsymbol{\Gamma}})$ has pdf given by

$$\begin{aligned} f_j(\mathbf{x}_j) =& \frac{1}{\Phi_q(\boldsymbol{\tau};\mathbf{0}_q,\bar{\boldsymbol{\Gamma}})} \phi_{p_j}(\mathbf{x}_j;\boldsymbol{\xi}_j,\boldsymbol{\Psi}_{jj}+\mathbf{H}_j\bar{\boldsymbol{\Gamma}}\mathbf{H}_j^\top) \\ & \times \Phi_q\{\boldsymbol{\tau}+\tilde{\boldsymbol{\Gamma}}_j\mathbf{H}_j^\top\boldsymbol{\Psi}_{jj}^{-1}(\mathbf{x}_j-\boldsymbol{\xi}_j);\mathbf{0}_q,\tilde{\boldsymbol{\Gamma}}_j\}, \quad \mathbf{x}_j \in \mathbb{R}^{p_j},\ j=1,2, \end{aligned}$$

where $\tilde{\boldsymbol{\Gamma}}_j = (\bar{\boldsymbol{\Gamma}}^{-1} + \mathbf{H}_j^\top\boldsymbol{\Psi}_{jj}^{-1}\mathbf{H}_j)^{-1}$ with $\mathbf{H}_j$ defined as in (6.13). Equivalently, according to (6.25), this marginal pdf can be rewritten as

$$\begin{aligned} f_j(\mathbf{x}_j) =& \frac{1}{\Phi_q(\boldsymbol{\tau};\mathbf{0}_q,\bar{\boldsymbol{\Gamma}})} \phi_{p_j}(\mathbf{x}_j;\boldsymbol{\xi}_j,\boldsymbol{\Omega}_{jj}) \\ & \times \Phi_q\{\boldsymbol{\tau}+\bar{\boldsymbol{\Gamma}}\mathbf{H}_j^\top\boldsymbol{\Omega}_{jj}^{-1}(\mathbf{x}_j-\boldsymbol{\xi}_j);\mathbf{0}_q,\bar{\boldsymbol{\Gamma}}-\bar{\boldsymbol{\Gamma}}\mathbf{H}_j^\top\boldsymbol{\Omega}_{jj}^{-1}\mathbf{H}_j\bar{\boldsymbol{\Gamma}}\}, \quad \mathbf{x}_j \in \mathbb{R}^{p_j},\ j=1,2, \end{aligned}$$

with $\boldsymbol{\Omega}_{jj} = \boldsymbol{\Psi}_{jj} + \mathbf{H}_j\bar{\boldsymbol{\Gamma}}\mathbf{H}_j^\top$. Furthermore, also recalling the partition in (6.13) for $\boldsymbol{\tau}$, $\mathbf{H}$ and $\bar{\boldsymbol{\Gamma}}$, we can note that $\boldsymbol{\Omega}_{jj} = \boldsymbol{\Psi}_{jj} + \mathbf{H}_{jj}\bar{\boldsymbol{\Gamma}}_{jj}\mathbf{H}_{jj}^\top$ if and only if $\mathbf{H}_{jk} = \mathbf{O}$ for $j \neq k$, in which case $\mathbf{H} = \text{diag}(\mathbf{H}_{11},\mathbf{H}_{22})$, and thus it follows that $\mathbf{X}_j \sim \mathcal{SUN}_{p_j,q}(\boldsymbol{\xi}_j,\boldsymbol{\Psi}_{jj},\mathbf{H}_{jj},\boldsymbol{\tau},\bar{\boldsymbol{\Gamma}})$ for $j = 1,2$. If in addition to the condition $\mathbf{H} = \text{diag}(\mathbf{H}_{11},\mathbf{H}_{22})$, we assume that $\bar{\boldsymbol{\Gamma}} = \text{diag}(\bar{\boldsymbol{\Gamma}}_{11},\bar{\boldsymbol{\Gamma}}_{22})$, that is, $\bar{\boldsymbol{\Gamma}}_{12} = \mathbf{O}$, then (see Theorem 98) $\mathbf{X}_j \sim \mathcal{SUN}_{p_j,q_j}(\boldsymbol{\xi}_j,\boldsymbol{\Psi}_{jj},\mathbf{H}_{jj},\boldsymbol{\tau}_j,\bar{\boldsymbol{\Gamma}}_{jj})$, $j = 1,2$, and its marginal pdf becomes

$$\begin{aligned} f_j(\mathbf{x}_j) =& \frac{1}{\Phi_{q_j}(\boldsymbol{\tau}_j;\mathbf{0}_{q_j},\bar{\boldsymbol{\Gamma}}_{jj})} \phi_{p_j}(\mathbf{x}_j;\boldsymbol{\xi}_j,\boldsymbol{\Omega}_{jj}) \\ & \times \Phi_{q_j}\{\boldsymbol{\tau}_j+\bar{\boldsymbol{\Gamma}}_{jj}\mathbf{H}_{jj}^\top\boldsymbol{\Omega}_{jj}^{-1}(\mathbf{x}_j-\boldsymbol{\xi}_j);\mathbf{0}_{q_j},\bar{\boldsymbol{\Gamma}}_{jj}-\bar{\boldsymbol{\Gamma}}_{jj}\mathbf{H}_{jj}^\top\boldsymbol{\Omega}_{jj}^{-1}\mathbf{H}_{jj}\bar{\boldsymbol{\Gamma}}_{jj}\}, \quad \mathbf{x}_j \in \mathbb{R}^{p_j}, \end{aligned} \tag{6.29}$$

with $\boldsymbol{\Omega}_{jj} = \boldsymbol{\Psi}_{jj} + \mathbf{H}_{jj}\bar{\boldsymbol{\Gamma}}_{jj}\mathbf{H}_{jj}^\top$. In particular, if $\boldsymbol{\tau}_j = \mathbf{0}_{q_j}$ and $\bar{\boldsymbol{\Gamma}}_{jj} = \mathbf{I}_{q_j}$, then this last marginal pdf is

$$\begin{aligned} f_j(\mathbf{x}_j) =& 2^{q_j}\phi_{p_j}(\mathbf{x}_j;\boldsymbol{\xi}_j,\boldsymbol{\Psi}_{jj}+\mathbf{H}_{jj}\mathbf{H}_{jj}^\top)\Phi_{q_j}\{\tilde{\boldsymbol{\Gamma}}_{jj}\mathbf{H}_{jj}^\top\boldsymbol{\Psi}_{jj}^{-1}(\mathbf{x}_j-\boldsymbol{\xi}_j);\mathbf{0}_{q_j},\tilde{\boldsymbol{\Gamma}}_{jj}\} \\ =& 2^{q_j}\phi_{p_j}(\mathbf{x}_j;\boldsymbol{\xi}_j,\boldsymbol{\Omega}_{jj})\Phi_{q_j}\{\mathbf{H}_{jj}^\top\boldsymbol{\Omega}_{jj}^{-1}(\mathbf{x}_j-\boldsymbol{\xi}_j);\mathbf{0}_{q_j},\mathbf{I}_{q_j}-\mathbf{H}_{jj}^\top\boldsymbol{\Omega}_{jj}^{-1}\mathbf{H}_{jj}\}, \quad \mathbf{x}_j \in \mathbb{R}^{p_j}, \end{aligned}$$

with $\tilde{\boldsymbol{\Gamma}}_{jj} = (\mathbf{I}_{q_j} + \mathbf{H}_{jj}^\top\boldsymbol{\Psi}_{jj}^{-1}\mathbf{H}_{jj})^{-1} = \mathbf{I}_{q_j} - \mathbf{H}_{jj}^\top\boldsymbol{\Omega}_{jj}^{-1}\mathbf{H}_{jj}$ and $\boldsymbol{\Omega}_{jj} = \boldsymbol{\Psi}_{jj} + \mathbf{H}_{jj}\mathbf{H}_{jj}^\top$.

Example 78 *Let* $\mathbf{X} = (\mathbf{X}_1^\top,\mathbf{X}_2^\top)^\top \sim \mathcal{SUN}_{p_1+p_2,q_1+q_2}(\boldsymbol{\xi},\boldsymbol{\Psi},\mathbf{H},\boldsymbol{\tau},\bar{\boldsymbol{\Gamma}})$*, where* $\boldsymbol{\xi} = (\boldsymbol{\xi}_1^\top,\boldsymbol{\xi}_2^\top)^\top \in \mathbb{R}^{p_1+p_2}$, $\boldsymbol{\Psi} = \text{diag}(\boldsymbol{\Psi}_{11},\boldsymbol{\Psi}_{22})$*, with* $\boldsymbol{\Psi}_{jj} > 0$ *of dimension* $p_j \times p_j$ *for* $j = 1,2$, $\mathbf{H} = \text{diag}(\mathbf{H}_{11},\mathbf{H}_{22})$*, with* $\mathbf{H}_{jj}$ *of dimension* $p_j \times q_j$ *for* $j = 1,2$, $\boldsymbol{\tau} = (\boldsymbol{\tau}_1^\top,\boldsymbol{\tau}_2^\top)^\top \in \mathbb{R}^{q_1+q_2}$*, and* $\bar{\boldsymbol{\Gamma}} = \text{diag}(\bar{\boldsymbol{\Gamma}}_{11},\bar{\boldsymbol{\Gamma}}_{22})$*, with* $\bar{\boldsymbol{\Gamma}}_{jj} > 0$ *of dimension* $q_j \times q_j$ *for* $j = 1,2$*. Then,* $\mathbf{X}_j \overset{ind.}{\sim} \mathcal{SUN}_{p_j,q_j}(\boldsymbol{\xi}_j,\boldsymbol{\Psi}_{jj},\mathbf{H}_{jj},\boldsymbol{\tau}_j,\bar{\boldsymbol{\Gamma}}_{jj})$, $j = 1,2$*, with marginal pdf given by (6.29). In particular, if* $\mathbf{X} = (X_1,\dots,X_p)^\top \sim \mathcal{SUN}_{p,p}(\boldsymbol{\xi},\boldsymbol{\Psi},\mathbf{H},\boldsymbol{\tau},\bar{\boldsymbol{\Gamma}})$*, with* $\boldsymbol{\xi} = (\xi_1,\dots,\xi_p)^\top$, $\boldsymbol{\Psi} = \text{diag}(\psi_1^2,\dots,\psi_p^2)$, $\mathbf{H} = \text{diag}(\eta_1,\dots,\eta_p)$, $\boldsymbol{\tau} = \mathbf{0}_p$*, and* $\bar{\boldsymbol{\Gamma}} = \mathbf{I}_p$*, then* $X_j \overset{ind.}{\sim} \mathcal{SN}(\xi_j,\psi_j^2,\eta_j)$, $j = 1,\dots,p$*, with joint pdf*

$$f(x_1,\dots,x_p) = \prod_{j=1}^{p} \frac{2}{\sqrt{\psi_j^2+\eta_j^2}}\phi_1\left(\frac{x_j-\xi_j}{\sqrt{\psi_j^2+\eta_j^2}}\right)\Phi_1\left(\frac{\eta_j(x_j-\xi_j)}{\psi_j\sqrt{\psi_j^2+\eta_j^2}}\right), \quad (x_1,\dots,x_p)^\top \in \mathbb{R}^p.$$

6.4.2 *Cdf*

As shown earlier in Theorem 91, the selection representation given in Theorem 89 provides a simple route to compute the cdf of the multivariate unified skew-normal either in the singular case or in the nonsingular case. In the nonsingular case, with $\boldsymbol{\Psi} > 0$, the cdf of $\mathbf{X} \sim \mathcal{SUN}_{p,q}(\boldsymbol{\xi}, \boldsymbol{\Psi}, \mathbf{H}, \boldsymbol{\tau}, \bar{\boldsymbol{\Gamma}})$ can also be computed directly by integrating (6.24).

In order to illustrate the degree of simplicity of the selection approach, we now calculate the unified skew-normal cdf from its pdf. To do this, we rewrite (6.24) as

$$\begin{aligned} f(\mathbf{u}) &= \frac{1}{\Phi_q(\boldsymbol{\tau}; \mathbf{0}_q, \bar{\boldsymbol{\Gamma}})} \phi_p(\mathbf{u}; \boldsymbol{\xi}, \boldsymbol{\Psi} + \mathbf{H}\bar{\boldsymbol{\Gamma}}\mathbf{H}^\top) \Phi_q\{\boldsymbol{\tau} + \tilde{\boldsymbol{\Gamma}}\mathbf{H}^\top \boldsymbol{\Psi}^{-1}(\mathbf{u} - \boldsymbol{\xi}); \mathbf{0}_q, \tilde{\boldsymbol{\Gamma}}\} \\ &= \frac{1}{\Phi_q(\boldsymbol{\tau}; \mathbf{0}_q, \bar{\boldsymbol{\Gamma}})} \phi_p(\mathbf{u}; \boldsymbol{\xi}, \boldsymbol{\Psi} + \mathbf{H}\bar{\boldsymbol{\Gamma}}\mathbf{H}^\top) \int_{\mathbf{u} \le \boldsymbol{\tau}} \phi_q\{\mathbf{v}; -\tilde{\boldsymbol{\Gamma}}\mathbf{H}^\top \boldsymbol{\Psi}^{-1}(\mathbf{u} - \boldsymbol{\xi}), \tilde{\boldsymbol{\Gamma}}\} d\mathbf{v}, \end{aligned}$$

where, as before $\tilde{\boldsymbol{\Gamma}} = (\bar{\boldsymbol{\Gamma}}^{-1} + \mathbf{H}^\top \boldsymbol{\Psi}^{-1}\mathbf{H})^{-1}$. Hence, we find

$$\begin{aligned} F(\mathbf{x}) &= \int_{\mathbf{u} \le \mathbf{x}} \frac{1}{\Phi_q(\boldsymbol{\tau}; \mathbf{0}_q, \bar{\boldsymbol{\Gamma}})} \phi_p(\mathbf{u}; \boldsymbol{\xi}, \boldsymbol{\Psi} + \mathbf{H}\bar{\boldsymbol{\Gamma}}\mathbf{H}^\top) \Phi_q\{\boldsymbol{\tau} + \tilde{\boldsymbol{\Gamma}}\mathbf{H}^\top \boldsymbol{\Psi}^{-1}(\mathbf{u} - \boldsymbol{\xi}); \mathbf{0}_q, \tilde{\boldsymbol{\Gamma}}\} d\mathbf{u} \\ &= \frac{1}{\Phi_q(\boldsymbol{\tau}; \mathbf{0}_q, \bar{\boldsymbol{\Gamma}})} \int_{\mathbf{u} \le \mathbf{x}} \phi_p(\mathbf{u}; \boldsymbol{\xi}, \boldsymbol{\Psi} + \mathbf{H}\bar{\boldsymbol{\Gamma}}\mathbf{H}^\top) \int_{\mathbf{v} \le \boldsymbol{\tau}} \phi_q\{\mathbf{v}; -\tilde{\boldsymbol{\Gamma}}\mathbf{H}^\top \boldsymbol{\Psi}^{-1}(\mathbf{u} - \boldsymbol{\xi}), \tilde{\boldsymbol{\Gamma}}\} d\mathbf{v} d\mathbf{u} \\ &= \frac{1}{\Phi_q(\boldsymbol{\tau}; \mathbf{0}_q, \bar{\boldsymbol{\Gamma}})} \int_{\mathbf{u} \le \mathbf{x}} \int_{\mathbf{v} \le \boldsymbol{\tau}} \phi_p(\mathbf{u}; \boldsymbol{\xi}, \boldsymbol{\Psi} + \mathbf{H}\bar{\boldsymbol{\Gamma}}\mathbf{H}^\top) \phi_q\{\mathbf{v}; -\tilde{\boldsymbol{\Gamma}}\mathbf{H}^\top \boldsymbol{\Psi}^{-1}(\mathbf{u} - \boldsymbol{\xi}), \tilde{\boldsymbol{\Gamma}}\} d\mathbf{v} d\mathbf{u}. \end{aligned}$$

But, similar to the marginal-conditional factorization of a $(p+q)$-variate normal pdf given in (6.21), we have

$$\begin{aligned} \phi_p(\mathbf{u}; \boldsymbol{\xi}, \boldsymbol{\Psi} + \mathbf{H}\bar{\boldsymbol{\Gamma}}\mathbf{H}^\top) \phi_q(\mathbf{v}; -\tilde{\boldsymbol{\Gamma}}\mathbf{H}^\top \boldsymbol{\Psi}^{-1}(\mathbf{u} - \boldsymbol{\xi}), \tilde{\boldsymbol{\Gamma}}) &= \phi_p(\mathbf{u}; \boldsymbol{\xi} - \mathbf{H}\mathbf{v}, \boldsymbol{\Psi}) \phi_q(\mathbf{v}; \mathbf{0}_q, \bar{\boldsymbol{\Gamma}}) \\ &= \phi_{p+q}(\mathbf{u}_*; \boldsymbol{\xi}_*, \boldsymbol{\Omega}_*), \end{aligned}$$

where $\mathbf{u}_* = (\mathbf{u}^\top, \mathbf{v}^\top)^\top$, $\boldsymbol{\xi}_* = (\boldsymbol{\xi}^\top, \mathbf{0}_q^\top)^\top$ and, as defined in (6.8),

$$\boldsymbol{\Omega}_* = \begin{pmatrix} \boldsymbol{\Psi} + \mathbf{H}\bar{\boldsymbol{\Gamma}}\mathbf{H}^\top & -\mathbf{H}\bar{\boldsymbol{\Gamma}} \\ -\bar{\boldsymbol{\Gamma}}\mathbf{H}^\top & \bar{\boldsymbol{\Gamma}} \end{pmatrix}.$$

Therefore, in the nonsingular case, the cdf of $\mathbf{X} \sim \mathcal{SUN}_{p,q}(\boldsymbol{\xi}, \boldsymbol{\Psi}, \mathbf{H}, \boldsymbol{\tau}, \bar{\boldsymbol{\Gamma}})$ is also given by

$$F(\mathbf{x}) = \frac{1}{\Phi_q(\boldsymbol{\tau}; \mathbf{0}_q, \bar{\boldsymbol{\Gamma}})} \int_{\mathbf{u}_* \le \mathbf{x}_*} \phi_{q+p}(\mathbf{u}_*; \boldsymbol{\xi}_*, \boldsymbol{\Omega}_*) d\mathbf{u}_* = \frac{1}{\Phi_q(\boldsymbol{\tau}; \mathbf{0}_q, \bar{\boldsymbol{\Gamma}})} \Phi_{q+p}(\mathbf{x}_*; \boldsymbol{\xi}_*, \boldsymbol{\Omega}_*), \quad \mathbf{x} \in \mathbb{R}^p,$$

where $\mathbf{x}_* = (\mathbf{x}^\top, \boldsymbol{\tau}^\top)^\top$.

6.4.3 *Geometry*

The unified skew-normal pdf (6.24) is constant on contours with a skew-ellipsoidal shape centered at $\boldsymbol{\xi}$ and defined by those $\mathbf{x} \in \mathbb{R}^p$ that satisfy

$$\exp\left\{-\frac{1}{2}(\mathbf{x} - \boldsymbol{\xi})^\top (\boldsymbol{\Psi} + \mathbf{H}\bar{\boldsymbol{\Gamma}}\mathbf{H}^\top)^{-1}(\mathbf{x} - \boldsymbol{\xi})\right\} \Phi_q\{\boldsymbol{\tau} + \tilde{\boldsymbol{\Gamma}}\mathbf{H}^\top \boldsymbol{\Psi}^{-1}(\mathbf{x} - \boldsymbol{\xi}); \mathbf{0}_q, \tilde{\boldsymbol{\Gamma}}\} = c^2, \qquad (6.30)$$

where $\tilde{\boldsymbol{\Gamma}} = (\bar{\boldsymbol{\Gamma}}^{-1} + \mathbf{H}^\top \boldsymbol{\Psi}^{-1}\mathbf{H})^{-1}$ and c is a constant. When $\boldsymbol{\Psi} + \mathbf{H}\bar{\boldsymbol{\Gamma}}\mathbf{H}^\top = \mathbf{I}_p$, the contours are skew-spheres.

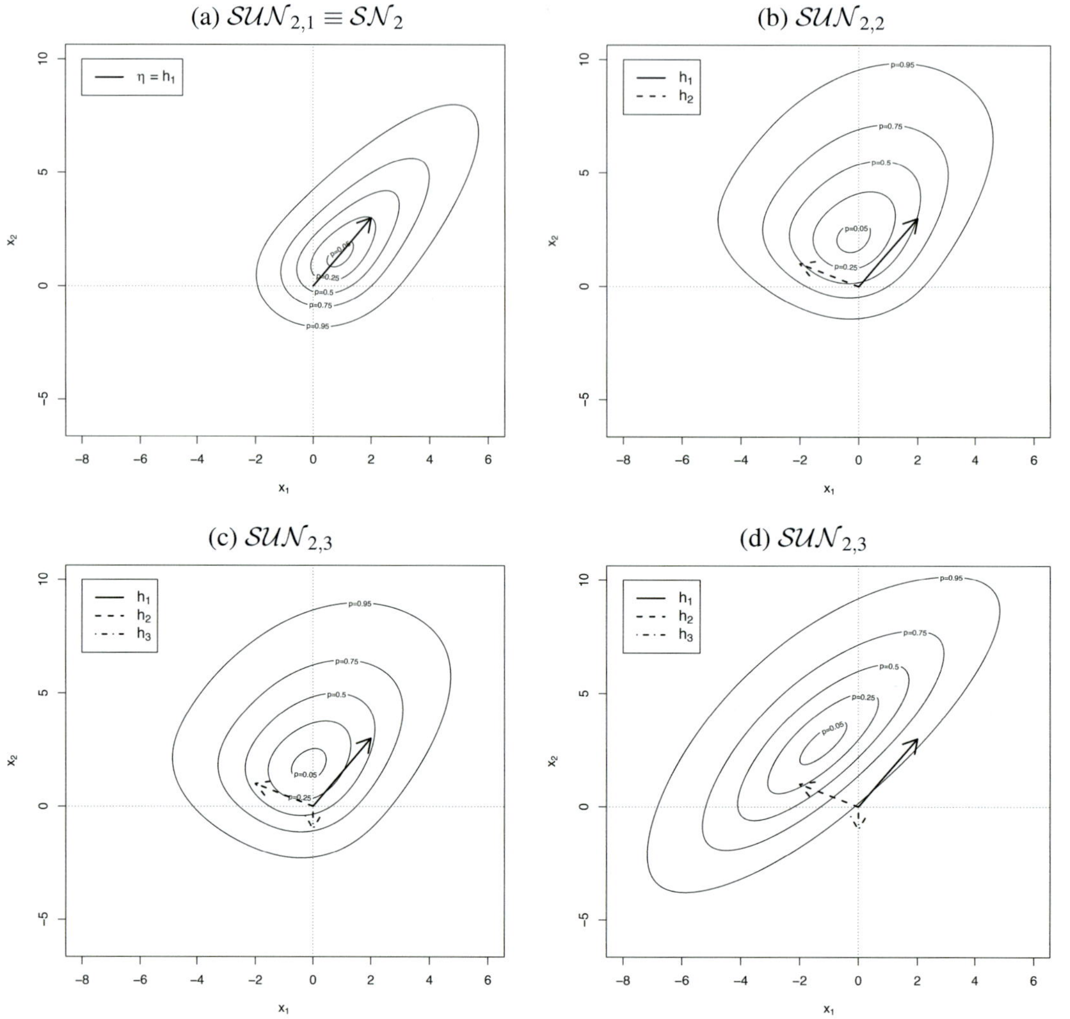

Figure 6.1 *Geometry of the multivariate unified skew-normal distribution, $\mathcal{SUN}_{p,q}$, for $p = 2$ and $q = 1, 2, 3$.*

Example 79 *In $\mathbb{R}^2$, consider a bivariate unified skew-normal distribution, $\mathcal{SUN}_{2,q}(\boldsymbol{\xi}, \boldsymbol{\Psi}, \mathbf{H}, \boldsymbol{\tau}, \bar{\boldsymbol{\Gamma}})$, with location $\boldsymbol{\xi} = \mathbf{0}_2$, latent correlation matrix $\bar{\boldsymbol{\Gamma}} = \frac{1}{2}\mathbf{I}_q + \frac{1}{2}\mathbf{1}_q\mathbf{1}_q^\top$ and:*

(a) $q = 1$, $\tau = 0$, $\boldsymbol{\Psi} = \mathbf{I}_2$, $\mathbf{H} = (2,3)^\top$;

(b) $q = 2$, $\boldsymbol{\tau} = \mathbf{0}_2$, $\boldsymbol{\Psi} = \mathbf{I}_2$, $\mathbf{H} = \begin{pmatrix} 2 & -2 \\ 3 & 1 \end{pmatrix}$;

(c) $q = 3$, $\boldsymbol{\tau} = \mathbf{0}_3$, $\boldsymbol{\Psi} = \mathbf{I}_2$, $\mathbf{H} = \begin{pmatrix} 2 & -2 & 0 \\ 3 & 1 & -1 \end{pmatrix}$;

(d) $q = 3$, $\boldsymbol{\tau} = (1, -1, 0)^\top$, $\boldsymbol{\Psi} = \begin{pmatrix} 3 & 2 \\ 2 & 2 \end{pmatrix}$, $\mathbf{H} = \begin{pmatrix} 2 & -2 & 0 \\ 3 & 1 & -1 \end{pmatrix}$.

The contours (6.30) of the pdf of the associated $\mathcal{SUN}$ distributions are plotted in Figure 6.1. In each panel, the vectors represent the columns $\mathbf{h}_i$ $(i = 1, \ldots, q)$ of the matrix $\mathbf{H}$. The geometrical interpretation of $\mathbf{h}_i$ $(i = 1, \ldots, q)$ is that they indicate the directions of skewness.

6.5 Singular Unified Skew-Normal

When $\mathbf{X} \sim \mathcal{SUN}_{p,q}(\boldsymbol{\xi}, \boldsymbol{\Psi}, \mathbf{H}, \boldsymbol{\tau}, \bar{\boldsymbol{\Gamma}})$ is singular, i.e., the dispersion matrix $\boldsymbol{\Psi}$ has rank $r < p$ and so $|\boldsymbol{\Psi}| = 0$, we can always find an orthogonal matrix $\mathbf{G} = (\mathbf{G}_1, \mathbf{G}_2) \in \mathcal{O}_p$ and an $r \times q$ matrix $\mathbf{H}_1$ to write

$$\boldsymbol{\Psi} = \mathbf{G}\mathbf{D}_\lambda\mathbf{G}^\top = (\mathbf{G}_1, \mathbf{G}_2)\begin{pmatrix} \mathbf{D}_1 & \mathbf{O} \\ \mathbf{O} & \mathbf{O} \end{pmatrix}\begin{pmatrix} \mathbf{G}_1^\top \\ \mathbf{G}_2^\top \end{pmatrix} = \mathbf{G}_1\mathbf{D}_1\mathbf{G}_1^\top, \quad \mathbf{H} = (\mathbf{G}_1, \mathbf{G}_2)\begin{pmatrix} \mathbf{H}_1 \\ \mathbf{O} \end{pmatrix} = \mathbf{G}_1\mathbf{H}_1, \tag{6.31}$$

where $\mathbf{D}_1 = \mathrm{diag}(\lambda_1, \ldots, \lambda_r)$ comprises on its diagonal the nonzero eigenvalues, $\mathbf{G}_1 = (\mathbf{g}_1, \ldots, \mathbf{g}_r)$ provides an orthonormal basis for the column space of $\boldsymbol{\Psi}$, $\mathtt{C}(\boldsymbol{\Psi})$, and $\mathbf{G}_2 = (\mathbf{g}_{r+1}, \ldots, \mathbf{g}_p)$ gives a basis for the null space (or kernel) of $\boldsymbol{\Psi}$, $\mathtt{N}(\boldsymbol{\Psi})$. Since $\mathtt{C}(\mathbf{G}_2) = \mathtt{C}(\mathbf{G}_1)^\perp = \mathtt{N}(\boldsymbol{\Psi})$, $\mathtt{C}(\mathbf{G}_1) = \mathtt{C}(\mathbf{G}_2)^\perp = \mathtt{C}(\boldsymbol{\Psi})$ and $\mathbf{G}_2^\top\mathbf{H} = \mathbf{O}$, then $\mathbf{G}_2^\top(\mathbf{X} - \boldsymbol{\xi}) = \mathbf{0}_{p-r}$ with probability 1 which means that $\mathbf{X} - \boldsymbol{\xi} \in \mathtt{C}(\mathbf{G}_2)^\perp = \mathtt{C}(\boldsymbol{\Psi})$ with probability 1, whereas $\mathbf{G}_1^\top(\mathbf{X} - \boldsymbol{\xi})$ has a nonsingular $\mathcal{SUN}_{r,q}(\mathbf{0}_r, \mathbf{D}_1, \mathbf{H}_1, \boldsymbol{\tau}, \bar{\boldsymbol{\Gamma}})$ as established below. This is essentially equivalent to saying that $\mathbf{X} \in \boldsymbol{\xi} + \mathtt{C}(\boldsymbol{\Psi})$ with probability 1, and in this way we can almost visualize $\mathbf{X}$ in this r-dimensional affine subspace of $\mathbb{R}^p$. However, despite $\mathrm{P}\{\mathbf{X} \in \boldsymbol{\xi} + \mathtt{C}(\boldsymbol{\Psi})\} = 1$ the distribution of $\mathbf{X}$ cannot be absolutely continuous because $\mathrm{vol}\{\boldsymbol{\xi} + \mathtt{C}(\boldsymbol{\Psi})\} = 0$. The same obviously happens with the singular multivariate normal distribution as demonstrated by Bilodeau and Brenner (1999). In fact, the development of this last discussion is carried out following that of the said authors for the multivariate normal distribution.

To identify the distribution of $\mathbf{X}_1 = \mathbf{G}_1^\top(\mathbf{X} - \boldsymbol{\xi})$, note by (6.1) that

$$\mathbf{X}_1 = \mathbf{G}_1^\top(\mathbf{H}\mathbf{U} + \mathbf{W}) = \mathbf{H}_1\mathbf{U} + \mathbf{W}_1,$$

where, as a consequence of (6.31), $\mathbf{H}_1 = \mathbf{G}_1^\top\mathbf{H}$ and $\mathbf{W}_1 = \mathbf{G}_1^\top\mathbf{W} \sim \mathcal{N}_r(\mathbf{0}_r, \mathbf{D}_1)$ is nonsingular. Since the normal random vector $\mathbf{W}_1$ is also independent of the truncated normal random vector $\mathbf{U}$, then $\mathbf{X}_1 \sim \mathcal{SUN}_{r,q}(\mathbf{0}_r, \mathbf{D}_1, \mathbf{H}_1, \boldsymbol{\tau}, \bar{\boldsymbol{\Gamma}})$ is nonsingular with pdf

$$f_1(\mathbf{x}_1) = \frac{1}{\Phi_q(\boldsymbol{\tau}; \mathbf{0}_q, \bar{\boldsymbol{\Gamma}})}\phi_r(\mathbf{x}_1; \mathbf{0}_r, \mathbf{D}_1 + \mathbf{H}_1\bar{\boldsymbol{\Gamma}}\mathbf{H}_1^\top)\Phi_q(\tilde{\boldsymbol{\Gamma}}_1\mathbf{H}_1^\top\mathbf{D}_1^{-1}\mathbf{x}_1; \mathbf{0}_q, \tilde{\boldsymbol{\Gamma}}_1), \quad \mathbf{x}_1 \in \mathbb{R}^r,$$

where $\tilde{\boldsymbol{\Gamma}}_1 = (\bar{\boldsymbol{\Gamma}}^{-1} + \mathbf{H}_1^\top\mathbf{D}_1^{-1}\mathbf{H}_1)^{-1}$. This means that the $p \times q$ skewness matrix $\mathbf{H}$ associated with the multivariate unified skew-normal distribution of $\mathbf{X}$ can only contain a non-null submatrix of dimension $r \times q$, which matches the matrix $\mathbf{H}_1$.

6.6 Conditional Unified Skew-Normal

Assuming the partition (6.11) of $\mathbf{X} \sim \mathcal{SUN}_{p,q}(\boldsymbol{\xi}, \boldsymbol{\Psi}, \mathbf{H}, \boldsymbol{\tau}, \bar{\boldsymbol{\Gamma}})$, we derive next the conditional distribution of $\mathbf{X}_1 \mid \mathbf{X}_2 = \mathbf{x}_2$. As in the skew-normal case, we want to include both situations, the one where the dispersion matrix $\boldsymbol{\Psi}_{22}$ associated with $\mathbf{X}_2$ is nonsingular ($\boldsymbol{\Psi}_{22} > 0$) and the one where this matrix is singular ($|\boldsymbol{\Psi}_{22}| = 0$). To do this, first consider that the matrix $\boldsymbol{\Psi}$ can be decomposed as

$$\begin{pmatrix} \boldsymbol{\Psi}_{11} & \boldsymbol{\Psi}_{12} \\ \boldsymbol{\Psi}_{21} & \boldsymbol{\Psi}_{22} \end{pmatrix} = \underbrace{\begin{pmatrix} \mathbf{I}_{p_1} & \mathbf{C} \\ \mathbf{O} & \mathbf{I}_{p_2} \end{pmatrix}}_{\mathbf{B}^{-1}}\begin{pmatrix} \boldsymbol{\Psi}_{11\cdot 2} & \mathbf{O} \\ \mathbf{O} & \boldsymbol{\Psi}_{22} \end{pmatrix}\underbrace{\begin{pmatrix} \mathbf{I}_{p_1} & \mathbf{O} \\ \mathbf{C}^\top & \mathbf{I}_{p_2} \end{pmatrix}}_{\mathbf{B}^{-\top}}, \tag{6.32}$$

where $\mathbf{C} = \boldsymbol{\Psi}_{12}\boldsymbol{\Psi}_{22}^-$ and $\boldsymbol{\Psi}_{22}^-$ is a generalized inverse of $\boldsymbol{\Psi}_{22}$, i.e., a matrix satisfying $\boldsymbol{\Psi}_{22}\boldsymbol{\Psi}_{22}^-\boldsymbol{\Psi}_{22} = \boldsymbol{\Psi}_{22}$. Now, partition the unified skew-normal random vector $\mathbf{X}$ defined by (6.1) as

$$\begin{pmatrix} \mathbf{X}_1 \\ \mathbf{X}_2 \end{pmatrix} = \begin{pmatrix} \boldsymbol{\xi}_1 \\ \boldsymbol{\xi}_2 \end{pmatrix} + \begin{pmatrix} \mathbf{H}_1 \\ \mathbf{H}_2 \end{pmatrix}\mathbf{U} + \begin{pmatrix} \mathbf{W}_1 \\ \mathbf{W}_2 \end{pmatrix},$$

where by assumption $\mathbf{U}$, $\mathbf{W}_1$ and $\mathbf{W}_2$ are independent, with

$$\mathbf{U} \sim \mathcal{TN}_q(\mathbf{0}_q; \boldsymbol{\tau}, \bar{\boldsymbol{\Gamma}}), \quad \begin{pmatrix} \mathbf{W}_1 \\ \mathbf{W}_2 \end{pmatrix} \sim \mathcal{N}_{p_1+p_2}\left(\begin{pmatrix} \mathbf{0}_{p_1} \\ \mathbf{0}_{p_2} \end{pmatrix}, \begin{pmatrix} \boldsymbol{\Psi}_{11} & \boldsymbol{\Psi}_{12} \\ \boldsymbol{\Psi}_{21} & \boldsymbol{\Psi}_{22} \end{pmatrix}\right), \tag{6.33}$$

and consider the nonsingular linear transformation given by

$$\begin{pmatrix} \tilde{\mathbf{X}}_1 \\ \tilde{\mathbf{X}}_2 \end{pmatrix} = \begin{pmatrix} \mathbf{I}_{p_1} & -\mathbf{C} \\ \mathbf{O} & \mathbf{I}_{p_2} \end{pmatrix} \begin{pmatrix} \mathbf{X}_1 \\ \mathbf{X}_2 \end{pmatrix} = \begin{pmatrix} \mathbf{X}_1 - \mathbf{C}\mathbf{X}_2 \\ \mathbf{X}_2 \end{pmatrix} = \mathbf{B}\mathbf{X},$$

where, as before $\mathbf{C} = \boldsymbol{\Psi}_{12}\boldsymbol{\Psi}_{22}^{-}$. Thus, we find

$$\begin{aligned}\begin{pmatrix} \tilde{\mathbf{X}}_1 \\ \tilde{\mathbf{X}}_2 \end{pmatrix} &= \begin{pmatrix} \boldsymbol{\xi}_1 - \mathbf{C}\boldsymbol{\xi}_2 + (\mathbf{H}_1 - \mathbf{C}\mathbf{H}_2)\mathbf{U} + \mathbf{W}_1 - \mathbf{C}\mathbf{W}_2 \\ \boldsymbol{\xi}_2 + \mathbf{H}_2\mathbf{U} + \mathbf{W}_2 \end{pmatrix} \\ &= \begin{pmatrix} \boldsymbol{\xi}_1 - \mathbf{C}\boldsymbol{\xi}_2 \\ \boldsymbol{\xi}_2 \end{pmatrix} + \begin{pmatrix} \mathbf{H}_1 - \mathbf{C}\mathbf{H}_2 \\ \mathbf{H}_2 \end{pmatrix} + \begin{pmatrix} \mathbf{W}_1 - \mathbf{C}\mathbf{W}_2 \\ \mathbf{W}_2 \end{pmatrix}, \end{aligned} \tag{6.34}$$

where again the truncated normal random vector $\mathbf{U}$ is independent of $\tilde{\mathbf{W}}$, which for $\mathbf{C} = \boldsymbol{\Psi}_{12}\boldsymbol{\Psi}_{22}^{-}$ has a multivariate normal distribution given by

$$\begin{pmatrix} \tilde{\mathbf{W}}_1 \\ \tilde{\mathbf{W}}_2 \end{pmatrix} = \begin{pmatrix} \mathbf{W}_1 - \mathbf{C}\mathbf{W}_2 \\ \mathbf{W}_2 \end{pmatrix} \sim \mathcal{N}_{p_1+p_2}\left(\begin{pmatrix} \mathbf{0}_{p_1} \\ \mathbf{0}_{p_2} \end{pmatrix}, \begin{pmatrix} \boldsymbol{\Psi}_{11\cdot 2} & \mathbf{O} \\ \mathbf{O} & \boldsymbol{\Psi}_{22} \end{pmatrix} \right), \tag{6.35}$$

where $\boldsymbol{\Psi}_{11\cdot 2} = \boldsymbol{\Psi}_{11} - \boldsymbol{\Psi}_{12}\boldsymbol{\Psi}_{22}^{-}\boldsymbol{\Psi}_{21}$. Clearly, the marginal $\tilde{\mathbf{W}}_1 = \mathbf{W}_1 - \mathbf{C}\mathbf{W}_2$ and $\tilde{\mathbf{W}}_2 = \mathbf{W}_2$ are independent. In turn, this means that $(\tilde{\mathbf{W}}_1 \mid \mathbf{W}_2 = \mathbf{w}_2) \overset{d}{=} \tilde{\mathbf{W}}_1 = \mathbf{W}_1 - \mathbf{C}\mathbf{W}_2 \sim \mathcal{N}_{p_1}(\mathbf{0}_{p_1}, \boldsymbol{\Psi}_{11\cdot 2})$ for all $\mathbf{w}_2$. Hence, since $\mathbf{W}_1 = \mathbf{C}\mathbf{W}_2 + \tilde{\mathbf{W}}_1$, with $\mathbf{C} = \boldsymbol{\Psi}_{12}\boldsymbol{\Psi}_{22}^{-}$, then we have

$$(\mathbf{W}_1 \mid \mathbf{W}_2 = \mathbf{w}_2) \overset{d}{=} \mathbf{C}\mathbf{w}_2 + \tilde{\mathbf{W}}_1 \sim \mathcal{N}_{p_1}(\mathbf{C}\mathbf{w}_2, \boldsymbol{\Psi}_{11\cdot 2}), \tag{6.36}$$

which is valid even if $\boldsymbol{\Psi}_{22}$ is singular; see Chapter 4 and Bilodeau and Brenner (1999).

In this way, we arrive at the following theorem as a consequence of Theorem 94.

Theorem 104 *Suppose that*

$$\begin{pmatrix} \mathbf{X}_1 \\ \mathbf{X}_2 \end{pmatrix} \sim \mathcal{SUN}_{p_1+p_2,q}\left(\begin{pmatrix} \boldsymbol{\xi}_1 \\ \boldsymbol{\xi}_2 \end{pmatrix}, \begin{pmatrix} \boldsymbol{\Psi}_{11} & \boldsymbol{\Psi}_{12} \\ \boldsymbol{\Psi}_{21} & \boldsymbol{\Psi}_{22} \end{pmatrix}, \begin{pmatrix} \mathbf{H}_1 \\ \mathbf{H}_2 \end{pmatrix}, \boldsymbol{\tau}, \bar{\boldsymbol{\Gamma}} \right).$$

If we let $\mathbf{C} = \boldsymbol{\Psi}_{12}\boldsymbol{\Psi}_{22}^{-}$*, then:*

(a) $$\begin{pmatrix} \mathbf{X}_1 - \mathbf{C}\mathbf{X}_2 \\ \mathbf{X}_2 \end{pmatrix} \sim \mathcal{SUN}_{p_1+p_2,q}\left(\begin{pmatrix} \boldsymbol{\xi}_1 - \mathbf{C}\boldsymbol{\xi}_2 \\ \boldsymbol{\xi}_2 \end{pmatrix}, \begin{pmatrix} \boldsymbol{\Psi}_{11\cdot 2} & \mathbf{O} \\ \mathbf{O} & \boldsymbol{\Psi}_{22} \end{pmatrix}, \begin{pmatrix} \mathbf{H}_1 - \mathbf{C}\mathbf{H}_2 \\ \mathbf{H}_2 \end{pmatrix}, \boldsymbol{\tau}, \bar{\boldsymbol{\Gamma}} \right),$$

and $\tilde{\mathbf{X}}_1 = \mathbf{X}_1 - \mathbf{C}\mathbf{X}_2$ *and* $\mathbf{X}_2$ *are independent if* $\mathbf{H}_1 = \mathbf{C}\mathbf{H}_2$ *or* $\mathbf{H}_2 = \mathbf{O}$*;*

(b) $\tilde{\mathbf{X}}_1 = \mathbf{X}_1 - \mathbf{C}\mathbf{X}_2$ *and* $\mathbf{X}_2$ *are conditionally independent given* $\mathbf{U}$*, with*

$$\begin{aligned} (\tilde{\mathbf{X}}_1 \mid \mathbf{U} = \mathbf{u}) &\sim \mathcal{N}_{p_1}(\boldsymbol{\xi}_1 - \mathbf{C}\boldsymbol{\xi}_2 + (\mathbf{H}_1 - \mathbf{C}\mathbf{H}_2)\mathbf{u}, \boldsymbol{\Psi}_{11\cdot 2}), \\ (\mathbf{X}_2 \mid \mathbf{U} = \mathbf{u}) &\sim \mathcal{N}_{p_2}(\boldsymbol{\xi}_2 + \mathbf{H}_2\mathbf{u}, \boldsymbol{\Psi}_{22}); \end{aligned}$$

(c) $(\mathbf{X}_1 \mid \mathbf{X}_2 = \mathbf{x}_2, \mathbf{U} = \mathbf{u}) \sim \mathcal{N}_{p_1}(\boldsymbol{\xi}_{1\cdot 2} + \mathbf{H}_{1\cdot 2}\mathbf{u}, \boldsymbol{\Psi}_{11\cdot 2})$*, with:*

$$\boldsymbol{\xi}_{1\cdot 2} = \boldsymbol{\xi}_1 + \boldsymbol{\Psi}_{12}\boldsymbol{\Psi}_{22}^{-}(\mathbf{x}_2 - \boldsymbol{\xi}_2), \quad \mathbf{H}_{1\cdot 2} = \mathbf{H}_1 - \boldsymbol{\Psi}_{12}\boldsymbol{\Psi}_{22}^{-}\mathbf{H}_2, \quad \boldsymbol{\Psi}_{11\cdot 2} = \boldsymbol{\Psi}_{11} - \boldsymbol{\Psi}_{12}\boldsymbol{\Psi}_{22}^{-}\boldsymbol{\Psi}_{21}. \tag{6.37}$$

Proof 104 *The proof of part (a) follows from Theorem 94. Part (b) follows from (6.34) and (6.35) by conditioning on* $\mathbf{U}=\mathbf{u}$. *Part (c) is a consequence of (b). More specifically, since* $\mathbf{X}_j=\boldsymbol{\xi}_j+\mathbf{H}_j\mathbf{U}+\mathbf{W}_j$ *for* $j=1,2$, *where* $\mathbf{U}$ *and* $(\mathbf{W}_1,\mathbf{W}_2)$ *are independently distributed as in (6.33), then by using (6.36), we have*

$$\begin{aligned}(\mathbf{X}_1\mid\mathbf{X}_2=\mathbf{x}_2,\mathbf{U}=\mathbf{u}) &= (\boldsymbol{\xi}_1+\mathbf{H}_1\mathbf{U}+\mathbf{W}_1\mid\boldsymbol{\xi}_2+\mathbf{H}_2\mathbf{U}+\mathbf{W}_2=\mathbf{x}_2,\mathbf{U}=\mathbf{u})\\ &= \boldsymbol{\xi}_1+\mathbf{H}_1\mathbf{u}+(\mathbf{W}_1\mid\mathbf{W}_2=\mathbf{x}_2-\boldsymbol{\xi}_2-\mathbf{H}_1\mathbf{u},\mathbf{U}=\mathbf{u})\\ &\sim \mathcal{N}_{p_1}(\boldsymbol{\xi}_1+\mathbf{H}_1\mathbf{u}+\mathbf{C}(\mathbf{x}_2-\boldsymbol{\xi}_2-\mathbf{H}_1\mathbf{u}),\boldsymbol{\Psi}_{11\cdot 2}),\end{aligned}$$

which proves part (c) since for $\mathbf{C}=\boldsymbol{\Psi}_{12}\boldsymbol{\Psi}_{22}^{-}$, *we have*

$$\boldsymbol{\xi}_1+\mathbf{H}_1\mathbf{u}+\mathbf{C}(\mathbf{x}_2-\boldsymbol{\xi}_2-\mathbf{H}_1\mathbf{u})=\boldsymbol{\xi}_1+\mathbf{C}(\mathbf{x}_2-\boldsymbol{\xi}_2)+(\mathbf{H}_1-\mathbf{C}\mathbf{H}_2)\mathbf{u}=\boldsymbol{\xi}_{1\cdot 2}+\mathbf{H}_{1\cdot 2}\mathbf{u}.$$

□

Taking into account part (c) of the previous theorem, we arrive at the following completely general result.

Theorem 105 *Under the assumptions of Theorem 104, the conditional random vector* $(\mathbf{X}_1\mid\mathbf{X}_2=\mathbf{x}_2)$ *has conditional cf given by*

$$\psi_{1\cdot 2}(\mathbf{t}_1)=e^{i\mathbf{t}_1^\top\boldsymbol{\xi}_{1\cdot 2}-\frac{1}{2}\mathbf{t}_1^\top\boldsymbol{\Psi}_{11\cdot 2}\mathbf{t}_1}\psi_{\mathbf{U}|\mathbf{x}_2}(\mathbf{H}_{1\cdot 2}^\top\mathbf{t}_1),\quad \mathbf{t}_1\in\mathbb{R}^{p_1},\tag{6.38}$$

where $\boldsymbol{\xi}_{1\cdot 2}$, $\mathbf{H}_{1\cdot 2}$ *and* $\boldsymbol{\Psi}_{11\cdot 2}$ *are given in (6.37) and* $\psi_{\mathbf{U}|\mathbf{x}_2}(\mathbf{t})=\mathrm{E}(e^{i\mathbf{t}^\top\mathbf{U}}\mid\mathbf{X}_2=\mathbf{x}_2)$, $\mathbf{t}\in\mathbb{R}^q$, *is the conditional cf of* $(\mathbf{U}\mid\mathbf{X}_2=\mathbf{x}_2)$.

Proof 105 *From part (c) of Theorem 104 and the properties of the conditional expectation, we have*

$$\begin{aligned}\psi_{1\cdot 2}(\mathbf{t}_1) &= \mathrm{E}(e^{i\mathbf{t}_1^\top\mathbf{X}_1}\mid\mathbf{X}_2=\mathbf{x}_2)\\ &= \mathrm{E}\{\mathrm{E}(e^{i\mathbf{t}_1^\top\mathbf{X}_1}\mid\mathbf{X}_2=\mathbf{x}_2,\mathbf{U})\mid\mathbf{X}_2=\mathbf{x}_2\}\\ &= \mathrm{E}[\mathrm{E}\{e^{i\mathbf{t}_1^\top(\boldsymbol{\xi}_{1\cdot 2}+\mathbf{H}_{1\cdot 2}\mathbf{U})-\frac{1}{2}\mathbf{t}_1^\top\boldsymbol{\Psi}_{11\cdot 2}\mathbf{t}_1}\}\mid\mathbf{X}_2=\mathbf{x}_2]\\ &= e^{i\mathbf{t}_1^\top\boldsymbol{\xi}_{1\cdot 2}-\frac{1}{2}\mathbf{t}_1^\top\boldsymbol{\Psi}_{11\cdot 2}\mathbf{t}_1}\mathrm{E}\{e^{i(\mathbf{H}_{1\cdot 2}^\top\mathbf{t}_1)^\top\mathbf{U}}\mid\mathbf{X}_2=\mathbf{x}_2\}.\end{aligned}$$

□

From (6.38) it follows that the conditional distribution of $\mathbf{X}_1$ given $\mathbf{X}_2=\mathbf{x}_2$ is completely determined by the conditional distribution of the latent random vector $\mathbf{U}$ given $\mathbf{X}_2=\mathbf{x}_2$. To identify this last conditional distribution, first assume that $\boldsymbol{\Psi}_{22}$ is nonsingular. Hence, from part (b) of Theorem 104, the conditional multivariate normal distribution $\mathbf{X}_2\mid\mathbf{U}=\mathbf{u}\sim\mathcal{N}_{p_2}(\boldsymbol{\xi}_2+\mathbf{H}_2\mathbf{u},\boldsymbol{\Psi}_{22})$ is nonsingular, and for each $\mathbf{u}+\boldsymbol{\tau}>\mathbf{0}_q$, its conditional pdf is given by

$$f_2(\mathbf{x}_2\mid\mathbf{u})=\phi_{p_2}(\mathbf{x}_2;\boldsymbol{\xi}_2+\mathbf{H}_2\mathbf{u},\boldsymbol{\Psi}_{22}),\quad \mathbf{x}_2\in\mathbb{R}^{p_2}.$$

Also, since $\mathbf{X}_2\sim\mathcal{SUN}_{p_2,q}(\boldsymbol{\xi}_2,\boldsymbol{\Psi}_{22},\mathbf{H}_2,\boldsymbol{\tau},\bar{\boldsymbol{\Gamma}})$ is nonsingular, then its pdf is given by

$$\begin{aligned}f_2(\mathbf{x}_2) &= \frac{1}{\Phi_q(\boldsymbol{\tau};\mathbf{0}_q,\bar{\boldsymbol{\Gamma}})}\phi_{p_2}(\mathbf{x}_2;\boldsymbol{\xi}_2,\boldsymbol{\Psi}_{22}+\mathbf{H}_2\bar{\boldsymbol{\Gamma}}\mathbf{H}_2^\top)\\ &\quad\times\Phi_q\{\boldsymbol{\tau}+\tilde{\boldsymbol{\Gamma}}_2\mathbf{H}_2^\top\boldsymbol{\Psi}_{22}^{-1}(\mathbf{x}_2-\boldsymbol{\xi}_2);\mathbf{0}_q,\tilde{\boldsymbol{\Gamma}}_2\},\quad \mathbf{x}_2\in\mathbb{R}^{p_2},\end{aligned}$$

where $\tilde{\mathbf{\Gamma}}_2 = (\bar{\mathbf{\Gamma}}^{-1} + \mathbf{H}_2^\top \mathbf{\Psi}_{22}^{-1} \mathbf{H}_2)^{-1}$. Thus, remembering that $\mathbf{U}$ has a truncated normal pdf $g(\mathbf{u}) = \phi_q(\mathbf{u}; \mathbf{0}_q, \bar{\mathbf{\Gamma}})/\Phi_q(\boldsymbol{\tau}; \mathbf{0}_q, \bar{\mathbf{\Gamma}})$ for $\mathbf{u} + \boldsymbol{\tau} > \mathbf{0}_q$, we obtain for the conditional pdf of $\mathbf{U}$ given $\mathbf{X}_2 = \mathbf{x}_2$ that

$$\begin{aligned} g_2(\mathbf{u} \mid \mathbf{x}_2) &= \frac{f_2(\mathbf{x}_2 \mid \mathbf{u}) g(\mathbf{u})}{f_2(\mathbf{x}_2)} \\ &= \frac{\phi_{p_2}(\mathbf{x}_2; \boldsymbol{\xi}_2 + \mathbf{H}_2\mathbf{u}, \mathbf{\Psi}_{22}) \phi_q(\mathbf{u}; \mathbf{0}_q, \bar{\mathbf{\Gamma}})}{\phi_{p_2}(\mathbf{x}_2; \boldsymbol{\xi}_2, \mathbf{\Psi}_{22} + \mathbf{H}_2 \bar{\mathbf{\Gamma}} \mathbf{H}_2^\top) \Phi_q\{\boldsymbol{\tau} + \tilde{\mathbf{\Gamma}}_2 \mathbf{H}_2^\top \mathbf{\Psi}_{22}^{-1}(\mathbf{x}_2 - \boldsymbol{\xi}_2); \mathbf{0}_q, \tilde{\mathbf{\Gamma}}_2\}} \\ &= \frac{\phi_q\{\mathbf{u}; \tilde{\mathbf{\Gamma}}_2 \mathbf{H}_2^\top \mathbf{\Psi}_{22}^{-1}(\mathbf{x}_2 - \boldsymbol{\xi}_2), \tilde{\mathbf{\Gamma}}_2\}}{\Phi_q\{\boldsymbol{\tau} + \tilde{\mathbf{\Gamma}}_2 \mathbf{H}_2^\top \mathbf{\Psi}_{22}^{-1}(\mathbf{x}_2 - \boldsymbol{\xi}_2); \mathbf{0}_q, \tilde{\mathbf{\Gamma}}_2\}}, \quad \mathbf{u} + \boldsymbol{\tau} > \mathbf{0}_q, \end{aligned}$$

where, as in (6.21), we have used the identity

$$\phi_{p_2}(\mathbf{x}_2; \boldsymbol{\xi}_2 + \mathbf{H}_2\mathbf{u}, \mathbf{\Psi}_{22}) \phi_q(\mathbf{u}; \mathbf{0}_q, \bar{\mathbf{\Gamma}}) = \phi_{p_2}(\mathbf{x}_2; \boldsymbol{\xi}_2, \mathbf{\Psi}_{22} + \mathbf{H}_2 \bar{\mathbf{\Gamma}} \mathbf{H}_2^\top) \phi_q\{\mathbf{u}; \tilde{\mathbf{\Gamma}}_2 \mathbf{H}_2^\top \mathbf{\Psi}_{22}^{-1}(\mathbf{x}_2 - \boldsymbol{\xi}_2), \tilde{\mathbf{\Gamma}}_2\}.$$

Let

$$\boldsymbol{\tau}_{1\cdot 2} = \boldsymbol{\tau} + \tilde{\mathbf{\Gamma}}_2 \mathbf{H}_2^\top \mathbf{\Psi}_{22}^{-1}(\mathbf{x}_2 - \boldsymbol{\xi}_2), \quad \bar{\mathbf{\Gamma}}_{1\cdot 2} \equiv \tilde{\mathbf{\Gamma}}_2 = (\bar{\mathbf{\Gamma}}^{-1} + \mathbf{H}_2^\top \mathbf{\Psi}_{22}^{-1} \mathbf{H}_2)^{-1}. \tag{6.39}$$

Then, a simpler and more comprehensive expression for the conditional pdf of $(\mathbf{U} \mid \mathbf{X}_2 = \mathbf{x}_2)$ is

$$g_2(\mathbf{u} \mid \mathbf{x}_2) = \frac{\phi_q(\mathbf{u}; \boldsymbol{\tau}_{1\cdot 2} - \boldsymbol{\tau}, \bar{\mathbf{\Gamma}}_{1\cdot 2})}{\Phi_q(\boldsymbol{\tau}_{1\cdot 2}; \mathbf{0}_q, \bar{\mathbf{\Gamma}}_{1\cdot 2})}, \quad \mathbf{u} + \boldsymbol{\tau} > \mathbf{0}_q, \tag{6.40}$$

where $\boldsymbol{\tau}_{1\cdot 2} - \boldsymbol{\tau} = \tilde{\mathbf{\Gamma}}_2 \mathbf{H}_2^\top \mathbf{\Psi}_{22}^{-1}(\mathbf{x}_2 - \boldsymbol{\xi}_2)$. Thus, for each fixed $\mathbf{x}_2 \in \mathbb{R}^{p_2}$, it follows from (6.40) that the conditional distribution of $(\mathbf{U} \mid \mathbf{X}_2 = \mathbf{x}_2)$ becomes a truncated $\mathcal{N}_q(\boldsymbol{\tau}_{1\cdot 2} - \boldsymbol{\tau}, \bar{\mathbf{\Gamma}}_{1\cdot 2})$ below $-\boldsymbol{\tau}$. Stochastically, it is straightforward to show, for each fixed $\mathbf{x}_2 \in \mathbb{R}^{p_2}$, that

$$\boldsymbol{\tau} + (\mathbf{U} \mid \mathbf{X}_2 = \mathbf{x}_2) \stackrel{d}{=} \boldsymbol{\tau}_{1\cdot 2} + \mathbf{U}_{1\cdot 2}, \quad \mathbf{U}_{1\cdot 2} \stackrel{d}{=} (\mathbf{Z}_{1\cdot 2} \mid \mathbf{Z}_{1\cdot 2} + \boldsymbol{\tau}_{1\cdot 2} > \mathbf{0}_q), \quad \mathbf{Z}_{1\cdot 2} \sim \mathcal{N}_q(\mathbf{0}_q, \bar{\mathbf{\Gamma}}_{1\cdot 2}). \tag{6.41}$$

Moreover, the conditional cf of $(\mathbf{U} \mid \mathbf{X}_2 = \mathbf{x}_2)$ is

$$\begin{aligned} \psi_{\mathbf{U}|\mathbf{x}_2}(\mathbf{t}) &= \int_{\mathbf{u}+\boldsymbol{\tau}>\mathbf{0}_q} \frac{e^{i\mathbf{t}^\top \mathbf{u}} \phi_q(\mathbf{u}; \boldsymbol{\tau}_{1\cdot 2} - \boldsymbol{\tau}, \bar{\mathbf{\Gamma}}_{1\cdot 2})}{\Phi_q(\boldsymbol{\tau}_{1\cdot 2}; \mathbf{0}_q, \bar{\mathbf{\Gamma}}_{1\cdot 2})} d\mathbf{u}, \quad (\boldsymbol{\tau}_{1\cdot 2} - \boldsymbol{\tau} = \bar{\mathbf{\Gamma}}_{1\cdot 2} \mathbf{H}_2 \mathbf{\Psi}_{22}^{-1}(\mathbf{x}_2 - \boldsymbol{\xi}_2)) \\ &= \int_{\mathbf{u}+\boldsymbol{\tau}_{1\cdot 2}>\mathbf{0}_q} \frac{e^{i\mathbf{t}^\top \mathbf{u}} \phi_q(\mathbf{u}; \mathbf{0}_q, \bar{\mathbf{\Gamma}}_{1\cdot 2})}{\Phi_q(\boldsymbol{\tau}_{1\cdot 2}; \mathbf{0}_q, \bar{\mathbf{\Gamma}}_{1\cdot 2})} d\mathbf{u} \\ &= e^{i\mathbf{t}^\top(\boldsymbol{\tau}_{1\cdot 2} - \boldsymbol{\tau}) - \frac{1}{2}\mathbf{t}^\top \bar{\mathbf{\Gamma}}_{1\cdot 2}\mathbf{t}} \frac{\Phi_q(\boldsymbol{\tau}_{1\cdot 2} + i\bar{\mathbf{\Gamma}}_{1\cdot 2}\mathbf{t}; \mathbf{0}_q, \bar{\mathbf{\Gamma}}_{1\cdot 2})}{\Phi_q(\boldsymbol{\tau}_{1\cdot 2}; \mathbf{0}_q, \bar{\mathbf{\Gamma}}_{1\cdot 2})}, \quad \mathbf{t} \in \mathbb{R}^q \end{aligned} \tag{6.42}$$

$$\Rightarrow \psi_{\mathbf{U}_{1\cdot 2}}(\mathbf{t}) = e^{-\frac{1}{2}\mathbf{t}^\top \bar{\mathbf{\Gamma}}_{1\cdot 2}\mathbf{t}} \frac{\Phi_q(\boldsymbol{\tau}_{1\cdot 2} + i\bar{\mathbf{\Gamma}}_{1\cdot 2}\mathbf{t}; \mathbf{0}_q, \bar{\mathbf{\Gamma}}_{1\cdot 2})}{\Phi_q(\boldsymbol{\tau}_{1\cdot 2}; \mathbf{0}_q, \bar{\mathbf{\Gamma}}_{1\cdot 2})}, \quad \mathbf{t} \in \mathbb{R}^q. \tag{6.43}$$

In the singular case, with $\text{rank}(\mathbf{\Psi}_{22}) = r < p_2$, we may always find a p_2-by-p_2 orthogonal matrix $\mathbf{G} = (\mathbf{G}_1, \mathbf{G}_2)$ to write

$$\mathbf{\Psi}_{22} = (\mathbf{G}_1, \mathbf{G}_2) \begin{pmatrix} \mathbf{O} & \mathbf{O} \\ \mathbf{O} & \mathbf{D}_2 \end{pmatrix} \begin{pmatrix} \mathbf{G}_1^\top \\ \mathbf{G}_2^\top \end{pmatrix} = \mathbf{G}_2 \mathbf{D}_2 \mathbf{G}_2^\top,$$

where $\mathbf{D}_2$ is an r-by-r diagonal matrix, and then find the so-called pseudo-inverse of $\mathbf{\Psi}_{22}$:

$$\mathbf{\Psi}_{22}^- = (\mathbf{G}_1, \mathbf{G}_2) \begin{pmatrix} \mathbf{O} & \mathbf{O} \\ \mathbf{O} & \mathbf{D}_2^{-1} \end{pmatrix} \begin{pmatrix} \mathbf{G}_1^\top \\ \mathbf{G}_2^\top \end{pmatrix} = \mathbf{G}_2 \mathbf{D}_2^{-1} \mathbf{G}_2^\top.$$

Hence, we have $\mathbf{G}_1^\top(\mathbf{X}_2 - \boldsymbol{\xi}_2) = \mathbf{0}_{p_2 - r}$ (with probability 1) and also

$$(\mathbf{G}_2^\top \mathbf{X}_2 \mid \mathbf{U} = \mathbf{u}) \sim \mathcal{N}_r(\mathbf{G}_2^\top \boldsymbol{\xi}_2 + \mathbf{G}_2^\top \mathbf{H}_2 \mathbf{u}, \mathbf{D}_2), \quad \mathbf{U} \sim \mathcal{TN}_q(-\boldsymbol{\tau}; \mathbf{0}_q, \bar{\mathbf{\Gamma}}).$$

Since the results in the nonsingular case now apply, we arrive at the completely general result on the conditional multivariate unified skew-normal distribution given below in Theorem 106, where $\mathbf{\Psi}_{22}^{-1}$ can be replaced by $\mathbf{\Psi}_{22}^-$ in the singular case.

Theorem 106 *Under the assumptions of Theorem 104, we have for each fixed* $\mathbf{x}_2 \in \mathbb{R}^{p_2}$ *that* $(\mathbf{X}_1 \mid \mathbf{X}_2 = \mathbf{x}_2) \sim \mathcal{SUN}_{p_1,q}(\boldsymbol{\xi}_{1\cdot 2} + \bar{\mathbf{H}}_{1\cdot 2}\bar{\bar{\boldsymbol{\Gamma}}}_{1\cdot 2}\bar{\mathbf{H}}_2^\top \boldsymbol{\Psi}_{22}^{-1}(\mathbf{x}_2 - \boldsymbol{\xi}_2), \boldsymbol{\Psi}_{11\cdot 2}, \bar{\mathbf{H}}_{1\cdot 2}, \bar{\boldsymbol{\tau}}_{1\cdot 2}, \bar{\bar{\boldsymbol{\Gamma}}}_{1\cdot 2})$*, where* $\boldsymbol{\xi}_{1\cdot 2}$ *and* $\boldsymbol{\Psi}_{11\cdot 2}$ *are defined in (6.37) with* $\boldsymbol{\Psi}_{22}^{-}$ *replaced by* $\boldsymbol{\Psi}_{22}^{-1}$*,*

$$\bar{\mathbf{H}}_{1\cdot 2} = \mathbf{H}_{1\cdot 2}\bar{\boldsymbol{\gamma}}_{1\cdot 2}, \quad \bar{\mathbf{H}}_2 = \mathbf{H}_2\bar{\boldsymbol{\gamma}}_{1\cdot 2}, \quad \bar{\boldsymbol{\tau}}_{1\cdot 2} = \bar{\boldsymbol{\gamma}}_{1\cdot 2}^{-1}\boldsymbol{\tau}_{1\cdot 2} = \bar{\boldsymbol{\tau}} + \bar{\bar{\boldsymbol{\Gamma}}}_{1\cdot 2}\bar{\mathbf{H}}_2^\top \boldsymbol{\Psi}_{22}^{-1}(\mathbf{x}_2 - \boldsymbol{\xi}_2),$$
$$\bar{\boldsymbol{\tau}} = \bar{\boldsymbol{\gamma}}_{1\cdot 2}^{-1}\boldsymbol{\tau}, \quad \bar{\bar{\boldsymbol{\Gamma}}}_{1\cdot 2} = \bar{\boldsymbol{\gamma}}_{1\cdot 2}^{-1}\bar{\boldsymbol{\Gamma}}_{1\cdot 2}\bar{\boldsymbol{\gamma}}_{1\cdot 2}^{-1} = (\tilde{\boldsymbol{\Gamma}}^{-1} + \bar{\mathbf{H}}_2^\top \boldsymbol{\Psi}_{22}^{-1}\bar{\mathbf{H}}_2)^{-1}, \quad \bar{\boldsymbol{\gamma}}_{1\cdot 2} = \operatorname{diag}(\bar{\boldsymbol{\Gamma}}_{1\cdot 2})^{1/2},$$

with $\boldsymbol{\tau}_{1\cdot 2}$ *and* $\bar{\boldsymbol{\Gamma}}_{1\cdot 2}$ *defined as in (6.39),* $\mathbf{H}_{1\cdot 2} = \mathbf{H}_1 - \boldsymbol{\Psi}_{12}\boldsymbol{\Psi}_{22}^{-1}\mathbf{H}_2$ *and* $\tilde{\boldsymbol{\Gamma}} = \bar{\boldsymbol{\gamma}}_{1\cdot 2}^{-1}\bar{\boldsymbol{\Gamma}}\bar{\boldsymbol{\gamma}}_{1\cdot 2}^{-1}$*. That is, stochastically* $(\mathbf{X}_1 \mid \mathbf{X}_2 = \mathbf{x}_2) \stackrel{d}{=} \mathbf{X}_{1\cdot 2}$*, with*

$$\mathbf{X}_{1\cdot 2} = \boldsymbol{\xi}_{1\cdot 2} + \bar{\mathbf{H}}_{1\cdot 2}\bar{\bar{\boldsymbol{\Gamma}}}_{1\cdot 2}\bar{\mathbf{H}}_2^\top \boldsymbol{\Psi}_{22}^{-1}(\mathbf{x}_2 - \boldsymbol{\xi}_2) + \bar{\mathbf{H}}_{1\cdot 2}\bar{\mathbf{U}}_{1\cdot 2} + \mathbf{W}_{1\cdot 2}, \tag{6.44}$$

where $\bar{\mathbf{U}}_{1\cdot 2} \sim \mathcal{TN}_q(-\bar{\boldsymbol{\tau}}_{1\cdot 2}; \mathbf{0}_q, \bar{\bar{\boldsymbol{\Gamma}}}_{1\cdot 2})$ *and is independent of* $\mathbf{W}_{1\cdot 2} \sim \mathcal{N}_{p_1}(\mathbf{0}_{p_1}, \boldsymbol{\Psi}_{11\cdot 2})$*.*

Proof 106 *Since* $\mathcal{SUN}_{p_1,q}(\boldsymbol{\xi}_{1\cdot 2} + \bar{\mathbf{H}}_{1\cdot 2}\bar{\bar{\boldsymbol{\Gamma}}}_{1\cdot 2}\bar{\mathbf{H}}_2^\top \boldsymbol{\Psi}_{22}^{-1}(\mathbf{x}_2 - \boldsymbol{\xi}_2), \boldsymbol{\Psi}_{11\cdot 2}, \bar{\mathbf{H}}_{1\cdot 2}, \bar{\boldsymbol{\tau}}_{1\cdot 2}, \bar{\bar{\boldsymbol{\Gamma}}}_{1\cdot 2})$ *is equivalent to* $\mathcal{SUN}_{p_1,q}(\boldsymbol{\xi}_{1\cdot 2} + \mathbf{H}_{1\cdot 2}\bar{\boldsymbol{\Gamma}}_{1\cdot 2}\mathbf{H}_2^\top \boldsymbol{\Psi}_{22}^{-1}(\mathbf{x}_2 - \boldsymbol{\xi}_2), \boldsymbol{\Psi}_{11\cdot 2}, \mathbf{H}_{1\cdot 2}, \boldsymbol{\tau}_{1\cdot 2}, \bar{\boldsymbol{\Gamma}}_{1\cdot 2})$*, for simplicity of notation, we will first prove that* $(\mathbf{X}_1 \mid \mathbf{X}_2 = \mathbf{x}_2) \sim \mathcal{SUN}_{p_1,q}(\boldsymbol{\xi}_{1\cdot 2}, \boldsymbol{\Psi}_{11\cdot 2}, \mathbf{H}_{1\cdot 2}, \boldsymbol{\tau}_{1\cdot 2}, \bar{\boldsymbol{\Gamma}}_{1\cdot 2})$ *and then that*

$$(\mathbf{X}_1 \mid \mathbf{X}_2 = \mathbf{x}_2) \stackrel{d}{=} \mathbf{X}_{1\cdot 2} = \boldsymbol{\xi}_{1\cdot 2} + \mathbf{H}_{1\cdot 2}\bar{\boldsymbol{\Gamma}}_{1\cdot 2}\mathbf{H}_2^\top \boldsymbol{\Psi}_{22}^{-1}(\mathbf{x}_2 - \boldsymbol{\xi}_2) + \mathbf{H}_{1\cdot 2}\mathbf{U}_{1\cdot 2} + \mathbf{W}_{1\cdot 2}, \tag{6.45}$$

where $\mathbf{U}_{1\cdot 2} = \{\mathbf{U} - \bar{\boldsymbol{\Gamma}}_{1\cdot 2}\mathbf{H}_2^\top \boldsymbol{\Psi}_{22}^{-1}(\mathbf{X}_2 - \boldsymbol{\xi}_2)\} \mid \mathbf{X}_2 = \mathbf{x}_2 \sim \mathcal{TN}_q(-\boldsymbol{\tau}_{1\cdot 2}; \mathbf{0}_q, \bar{\boldsymbol{\Gamma}}_{1\cdot 2})$ *is also independent of* $\mathbf{W}_{1\cdot 2} \sim \mathcal{N}_{p_1}(\mathbf{0}_{p_1}, \boldsymbol{\Psi}_{11\cdot 2})$*. To do this, first note that for each fixed* $\mathbf{x}_2 \in \mathbb{R}^{p_2}$*, according to Theorem 105, the cf of* $(\mathbf{X}_1 \mid \mathbf{X}_2 = \mathbf{x}_2)$ *is obtained by applying (6.42) in (6.38), which for* $\mathbf{t} \in \mathbb{R}^q$ *is given by*

$$\psi_{1\cdot 2}(\mathbf{t}) = e^{\mathrm{i}\mathbf{t}^\top\{\boldsymbol{\xi}_{1\cdot 2} + \mathbf{H}_{1\cdot 2}\bar{\boldsymbol{\Gamma}}_{1\cdot 2}\mathbf{H}_2^\top \boldsymbol{\Psi}_{22}^{-1}(\mathbf{x}_2 - \boldsymbol{\xi}_2)\} - \frac{1}{2}\mathbf{t}^\top(\boldsymbol{\Psi}_{11\cdot 2} + \mathbf{H}_{1\cdot 2}\bar{\boldsymbol{\Gamma}}_{1\cdot 2}\mathbf{H}_{1\cdot 2}^\top)\mathbf{t}}$$
$$\times \frac{\Phi_q(\boldsymbol{\tau}_{1\cdot 2} + \mathrm{i}\mathbf{H}_{1\cdot 2}\bar{\boldsymbol{\Gamma}}_{1\cdot 2}\mathbf{t}; \mathbf{0}_q, \bar{\boldsymbol{\Gamma}}_{1\cdot 2})}{\Phi_q(\boldsymbol{\tau}_{1\cdot 2}; \mathbf{0}_q, \bar{\boldsymbol{\Gamma}}_{1\cdot 2})}, \quad \mathbf{t} \in \mathbb{R}^q. \tag{6.46}$$

By Theorem 92, this means that $(\mathbf{X}_1 \mid \mathbf{X}_2 = \mathbf{x}_2) \sim \mathcal{SUN}_{p_1,q}(\boldsymbol{\xi}_{1\cdot 2}, \boldsymbol{\Psi}_{11\cdot 2}, \mathbf{H}_{1\cdot 2}, \boldsymbol{\tau}_{1\cdot 2}, \bar{\boldsymbol{\Gamma}}_{1\cdot 2})$*. Now, we show that the cf of* $\mathbf{X}_{1\cdot 2}$ *in (6.45) also becomes (6.46), which is precisely the conditional cf of* $(\mathbf{X}_1 \mid \mathbf{X}_2 = \mathbf{x}_2)$*. In fact, since by assumption in (6.45) the random vectors* $\mathbf{U}_{1\cdot 2}$ *and* $\mathbf{W}_{1\cdot 2}$ *are independent, we have*

$$\begin{aligned}
\psi_{\mathbf{X}_{1\cdot 2}}(\mathbf{t}_1) &= \mathrm{E}(e^{\mathrm{i}\mathbf{t}_1^\top \mathbf{X}_{1\cdot 2}}) \\
&= \mathrm{E}\{e^{\mathrm{i}\mathbf{t}_1^\top\{\boldsymbol{\xi}_{1\cdot 2} + \mathbf{H}_{1\cdot 2}\bar{\boldsymbol{\Gamma}}_{1\cdot 2}\mathbf{H}_2^\top \boldsymbol{\Psi}_{22}^{-1}(\mathbf{x}_2 - \boldsymbol{\xi}_2) + \mathbf{H}_{1\cdot 2}\mathbf{U}_{1\cdot 2} + \mathbf{W}_{1\cdot 2}\}}\} \\
&= e^{\mathrm{i}\mathbf{t}_1^\top\{\boldsymbol{\xi}_{1\cdot 2} + \mathbf{H}_{1\cdot 2}\bar{\boldsymbol{\Gamma}}_{1\cdot 2}\mathbf{H}_2^\top \boldsymbol{\Psi}_{22}^{-1}(\mathbf{x}_2 - \boldsymbol{\xi}_2)\}}\mathrm{E}\{e^{\mathrm{i}(\mathbf{H}_{1\cdot 2}^\top\mathbf{t}_1)^\top \mathbf{U}_{1\cdot 2}}\}\mathrm{E}(e^{\mathrm{i}\mathbf{t}_1^\top \mathbf{W}_{1\cdot 2}}) \\
&= e^{\mathrm{i}\mathbf{t}_1^\top\{\boldsymbol{\xi}_{1\cdot 2} + \mathbf{H}_{1\cdot 2}\bar{\boldsymbol{\Gamma}}_{1\cdot 2}\mathbf{H}_2^\top \boldsymbol{\Psi}_{22}^{-1}(\mathbf{x}_2 - \boldsymbol{\xi}_2)\} - \frac{1}{2}\mathbf{t}_1^\top \boldsymbol{\Psi}_{11\cdot 2}\mathbf{t}_1}\mathrm{E}\{e^{\mathrm{i}(\mathbf{H}_{1\cdot 2}^\top\mathbf{t}_1)^\top \mathbf{U}_{1\cdot 2}}\} \\
&= e^{\mathrm{i}\mathbf{t}_1^\top\{\boldsymbol{\xi}_{1\cdot 2} + \mathbf{H}_{1\cdot 2}\bar{\boldsymbol{\Gamma}}_{1\cdot 2}\mathbf{H}_2^\top \boldsymbol{\Psi}_{22}^{-1}(\mathbf{x}_2 - \boldsymbol{\xi}_2)\} - \frac{1}{2}\mathbf{t}_1^\top \boldsymbol{\Psi}_{11\cdot 2}\mathbf{t}_1}\psi_{\mathbf{U}_{1\cdot 2}}(\mathbf{H}_{1\cdot 2}^\top\mathbf{t}_1),
\end{aligned}$$

which, by (6.43), also becomes (6.46). The proof is concluded by the uniqueness of the cf. □

Remark 15 *Note that the stochastic representation in (6.44) makes explicit the additional effect* $\mathbf{H}_{1\cdot 2}\bar{\boldsymbol{\Gamma}}_{1\cdot 2}\mathbf{H}_2^\top \boldsymbol{\Psi}_{22}^{-1}(\mathbf{x}_2 - \boldsymbol{\xi}_2)$ *generated by conditioning on the location of the conditional unified skew-normal distribution. In other words, (6.44) can be rewritten as* $\mathbf{X}_{1\cdot 2} = \tilde{\boldsymbol{\xi}}_{1\cdot 2} + \bar{\mathbf{H}}_{1\cdot 2}\bar{\mathbf{U}}_{1\cdot 2} +$

$\mathbf{W}_{1\cdot 2}$, *where the conditional location vector becomes* $\tilde{\boldsymbol{\xi}}_{1\cdot 2} = \boldsymbol{\xi}_1 + \boldsymbol{\Omega}_{12}\boldsymbol{\Omega}_{22}^{-1}(\mathbf{x}_2 - \boldsymbol{\xi}_2) = \boldsymbol{\xi}_{1\cdot 2} + \bar{\mathbf{H}}_{1\cdot 2}\bar{\bar{\boldsymbol{\Gamma}}}_{1\cdot 2}\bar{\mathbf{H}}_2^\top\boldsymbol{\Psi}_{22}^{-1}(\mathbf{x}_2 - \boldsymbol{\xi}_2)$, *since* $\boldsymbol{\Omega}_{ij} = \boldsymbol{\Psi}_{ij} + \mathbf{H}_i\bar{\boldsymbol{\Gamma}}\mathbf{H}_j^\top$, $i,j = 1,2$, *and as proved below in Section 6.6.2,* $\boldsymbol{\Omega}_{12}\boldsymbol{\Omega}_{22}^{-1} = \boldsymbol{\Psi}_{12}\boldsymbol{\Psi}_{22}^{-1} + \mathbf{H}_{1\cdot 2}\bar{\boldsymbol{\Gamma}}_{1\cdot 2}\mathbf{H}_2^\top\boldsymbol{\Psi}_{22}^{-1} = \boldsymbol{\Psi}_{12}\boldsymbol{\Psi}_{22}^{-1} + \bar{\mathbf{H}}_{1\cdot 2}\bar{\bar{\boldsymbol{\Gamma}}}_{1\cdot 2}\bar{\mathbf{H}}_2^\top\boldsymbol{\Psi}_{22}^{-1}$.

Remark 16 *Note that the conditional distribution presented in Theorem 106 as* $(\mathbf{X}_1 \mid \mathbf{X}_2 = \mathbf{x}_2) \sim \mathcal{SUN}_{p_1,q}(\boldsymbol{\xi}_{1\cdot 2} + \bar{\mathbf{H}}_{1\cdot 2}\bar{\bar{\boldsymbol{\Gamma}}}_{1\cdot 2}\bar{\mathbf{H}}_2^\top\boldsymbol{\Psi}_{22}^{-1}(\mathbf{x}_2 - \boldsymbol{\xi}_2), \boldsymbol{\Psi}_{11\cdot 2}, \bar{\mathbf{H}}_{1\cdot 2}, \bar{\boldsymbol{\tau}}_{1\cdot 2}, \bar{\bar{\boldsymbol{\Gamma}}}_{1\cdot 2})$ *is due solely to the formal requirement that* $\bar{\bar{\boldsymbol{\Gamma}}}_{1\cdot 2} = \bar{\boldsymbol{\gamma}}_{1\cdot 2}^{-1}\bar{\boldsymbol{\Gamma}}_{1\cdot 2}\bar{\boldsymbol{\gamma}}_{1\cdot 2}^{-1}$ *be precisely the correlation matrix induced by* $\bar{\boldsymbol{\Gamma}}_{1\cdot 2}$, *since* $\bar{\boldsymbol{\gamma}}_{1\cdot 2}$ *is the diagonal matrix formed by the square root of the corresponding diagonal entries of* $\bar{\boldsymbol{\Gamma}}_{1\cdot 2}$.

6.6.1 *Conditional mean vector and conditional variance matrix*

From (6.44) it is easy to compute the conditional mean vector and the conditional variance matrix of $(\mathbf{X}_1 \mid \mathbf{X}_2 = \mathbf{x}_2)$ as

$$\begin{aligned}
\mathrm{E}(\mathbf{X}_1 \mid \mathbf{X}_2 = \mathbf{x}_2) &= \boldsymbol{\xi}_{1\cdot 2} + \mathbf{H}_{1\cdot 2}\bar{\boldsymbol{\Gamma}}_{1\cdot 2}\mathbf{H}_2^\top\boldsymbol{\Psi}_{22}^{-1}(\mathbf{x}_2 - \boldsymbol{\xi}_2) + \mathbf{H}_{1\cdot 2}\mathrm{E}(\mathbf{U}_{1\cdot 2}),\\
\mathrm{var}(\mathbf{X}_1 \mid \mathbf{X}_2 = \mathbf{x}_2) &= \boldsymbol{\Psi}_{11\cdot 2} + \mathbf{H}_{1\cdot 2}\mathrm{var}(\mathbf{U}_{1\cdot 2})\mathbf{H}_{1\cdot 2}^\top.
\end{aligned}$$

However, as in the unconditional case, to do this, we need, for each fixed $\mathbf{x}_2 \in \mathbb{R}^{p_2}$, the first two moments of the truncated normal random vector $\mathbf{U}_{1\cdot 2} \overset{d}{=} (\mathbf{Z}_{1\cdot 2} \mid \mathbf{Z}_{1\cdot 2} + \boldsymbol{\tau}_{1\cdot 2} > \mathbf{0}_q)$, where $\mathbf{Z}_{1\cdot 2} \sim \mathcal{N}_q(-\boldsymbol{\tau}_{1\cdot 2}; \mathbf{0}_q, \bar{\boldsymbol{\Gamma}}_{1\cdot 2})$.

A relevant fact is that the conditional mean vector and the conditional variance matrix of $\mathbf{X}_1 \mid \mathbf{X}_2 = \mathbf{x}_2$ are, in general, nonlinear functions in $\mathbf{x}_2$. In the following example, two special cases are presented in which the regression of $\mathbf{X}_1$ on $\mathbf{x}_2$ is linear and homoscedastic.

Example 80 *Let* $\mathbf{X} \sim \mathcal{SUN}_{p,q}(\boldsymbol{\xi}, \boldsymbol{\Psi}, \mathbf{H}, \boldsymbol{\tau}, \bar{\boldsymbol{\Gamma}})$ *be partitioned as (6.11). We then have:*

i) If $\mathbf{H}_{1\cdot 2} = \mathbf{O}$, *then* $\mathbf{X}_1 \mid \mathbf{X}_2 = \mathbf{x}_2 \sim \mathcal{N}_{p_1}(\boldsymbol{\xi}_{1\cdot 2}, \boldsymbol{\Psi}_{11\cdot 2})$, *with*

$$\mathrm{E}(\mathbf{X}_1 \mid \mathbf{x}_2) = \boldsymbol{\xi}_{1\cdot 2} \quad \textit{and} \quad \mathrm{var}(\mathbf{X}_1 \mid \mathbf{x}_2) = \boldsymbol{\Psi}_{11\cdot 2};$$

ii) If $\mathbf{H}_2 = \mathbf{O}$, *then* $\mathbf{X}_1 \mid \mathbf{X}_2 = \mathbf{x}_2 \sim \mathcal{SUN}_{p_1,q}(\boldsymbol{\xi}_{1\cdot 2}, \boldsymbol{\Psi}_{11\cdot 2}, \mathbf{H}_1, \boldsymbol{\tau}, \bar{\boldsymbol{\Gamma}})$, *with*

$$\mathrm{E}(\mathbf{X}_1 \mid \mathbf{x}_2) = \boldsymbol{\xi}_{1\cdot 2} + \sqrt{\frac{2}{\pi}}\mathbf{H}_1\mathrm{E}(\mathbf{U}) \quad \textit{and} \quad \mathrm{var}(\mathbf{X}_1 \mid \mathbf{x}_2) = \boldsymbol{\Psi}_{11\cdot 2} - \frac{2}{\pi}\mathbf{H}_1\mathrm{var}(\mathbf{U})\mathbf{H}_1^\top,$$

where $\mathbf{U} \sim \mathcal{TN}_q(-\boldsymbol{\tau}; \mathbf{0}_q, \bar{\boldsymbol{\Gamma}})$.

6.6.2 *Conditional unified skew-normal pdf*

As mentioned earlier, when $\boldsymbol{\Psi} > 0$, the nonsingular case with $\boldsymbol{\Psi}_{22} > 0$ holds automatically. In this case, we also have that $\boldsymbol{\Psi}_{11} > 0$ and $\boldsymbol{\Psi}_{11\cdot 2} = \boldsymbol{\Psi}_{11} - \boldsymbol{\Psi}_{12}\boldsymbol{\Psi}_{22}^{-1}\boldsymbol{\Psi}_{21} > 0$, and the relation $(\boldsymbol{\Psi}_{11}, \boldsymbol{\Psi}_{12}, \boldsymbol{\Psi}_{22}) \longleftrightarrow (\boldsymbol{\Psi}_{11\cdot 2}, \boldsymbol{\Psi}_{12}\boldsymbol{\Psi}_{22}^{-1}, \boldsymbol{\Psi}_{22})$ becomes one-to-one. Obviously, under these conditions, all the marginal and conditional distributions obtained from $\mathbf{X} \sim \mathcal{SUN}_{p,q}(\boldsymbol{\xi}, \boldsymbol{\Psi}, \mathbf{H}, \boldsymbol{\tau}, \bar{\boldsymbol{\Gamma}})$ are nonsingular. The following theorem shows the conditional pdf of $\mathbf{X}_1$ given $\mathbf{X}_2 = \mathbf{x}_2$ according to the partition (6.11).

Theorem 107 *Let* $\mathbf{X} \sim \mathcal{SUN}_{p,q}(\boldsymbol{\xi}, \boldsymbol{\Psi}, \mathbf{H}, \boldsymbol{\tau}, \bar{\boldsymbol{\Gamma}})$, $\boldsymbol{\Psi} > 0$, *be partitioned as (6.11). Then for every fixed* $p_1 + p_2 = p$, *the conditional pdf of* $\mathbf{X}_1$ *given* $\mathbf{X}_2 = \mathbf{x}_2$, $\mathbf{x}_2 \in \mathbb{R}^{p_2}$, *is given by*

$$f_{1\cdot 2}(\mathbf{x}_1 \mid \mathbf{x}_2) = \frac{1}{\Phi_q(\boldsymbol{\tau}_{1\cdot 2}; \mathbf{0}_q, \bar{\boldsymbol{\Gamma}}_{1\cdot 2})} \phi_{p_1}\{\mathbf{x}_1; \boldsymbol{\xi}_{1\cdot 2} + \mathbf{H}_{1\cdot 2}\bar{\boldsymbol{\Gamma}}_{1\cdot 2}\mathbf{H}_2^\top \boldsymbol{\Psi}_{22}^{-1}(\mathbf{x}_2 - \boldsymbol{\xi}_2), \boldsymbol{\Psi}_{11\cdot 2} + \mathbf{H}_{1\cdot 2}\bar{\boldsymbol{\Gamma}}_{1\cdot 2}\mathbf{H}_{1\cdot 2}^\top\}$$
$$\times \Phi_q\left[\boldsymbol{\tau}_{1\cdot 2} + \tilde{\boldsymbol{\Gamma}}_{1\cdot 2}\mathbf{H}_{1\cdot 2}^\top \boldsymbol{\Psi}_{11\cdot 2}^{-1}\{\mathbf{x}_1 - \boldsymbol{\xi}_{1\cdot 2} - \mathbf{H}_{1\cdot 2}(\boldsymbol{\tau}_{1\cdot 2} - \boldsymbol{\tau})\}; \mathbf{0}_q, \tilde{\boldsymbol{\Gamma}}_{1\cdot 2}\right], \quad \mathbf{x}_1 \in \mathbb{R}^{p_1}, \tag{6.47}$$

where $\boldsymbol{\xi}_{1\cdot 2} = \boldsymbol{\xi}_1 + \boldsymbol{\Psi}_{12}\boldsymbol{\Psi}_{22}^{-1}(\mathbf{x}_2 - \boldsymbol{\xi}_2)$, $\boldsymbol{\Psi}_{11\cdot 2} = \boldsymbol{\Psi}_{11} - \boldsymbol{\Psi}_{12}\boldsymbol{\Psi}_{22}^{-1}\boldsymbol{\Psi}_{21}$, $\mathbf{H}_{1\cdot 2} = \mathbf{H}_1 - \boldsymbol{\Psi}_{12}\boldsymbol{\Psi}_{22}^{-1}\mathbf{H}_2$, $\boldsymbol{\tau}_{1\cdot 2} = \boldsymbol{\tau} + \bar{\boldsymbol{\Gamma}}_{1\cdot 2}\mathbf{H}_2^\top\boldsymbol{\Psi}_{22}^{-1}(\mathbf{x}_2 - \boldsymbol{\xi}_2)$, $\tilde{\boldsymbol{\Gamma}}_{1\cdot 2} = (\bar{\boldsymbol{\Gamma}}_{1\cdot 2}^{-1} + \mathbf{H}_{1\cdot 2}^\top\boldsymbol{\Psi}_{11\cdot 2}^{-1}\mathbf{H}_{1\cdot 2})^{-1}$, $\bar{\boldsymbol{\Gamma}}_{1\cdot 2} = (\bar{\boldsymbol{\Gamma}}^{-1} + \mathbf{H}_2^\top\boldsymbol{\Psi}_{22}^{-1}\mathbf{H}_2)^{-1}$.

Proof 107 *Let* $\mathbf{Y} = (\mathbf{Y}_1^\top, \mathbf{Y}_2^\top)^\top$ *be a linear transformation of* $\mathbf{X} = (\mathbf{X}_1^\top, \mathbf{X}_2^\top)^\top$ *such that* $\mathbf{Y}_1 = \mathbf{A}_1\mathbf{X} = \mathbf{X}_1 - \boldsymbol{\Psi}_{12}\boldsymbol{\Psi}_{22}^{-1}\mathbf{X}_2$ *and* $\mathbf{Y}_2 = \mathbf{A}_2\mathbf{X} = \mathbf{X}_2$, *i.e.,* $\mathbf{Y} = \mathbf{A}\mathbf{X}$, *with* $\mathbf{A} = (\mathbf{A}_1^\top\, \mathbf{A}_2^\top)^\top$, *where* $\mathbf{A}_1 = (\mathbf{I}_{p_1}, -\boldsymbol{\Psi}_{22}^{-1}\boldsymbol{\Psi}_{21})$ *and* $\mathbf{A}_2 = (\mathbf{O}, \mathbf{I}_{p_2})$. *By Theorem 94, we have* $\mathbf{Y} \sim \mathcal{SUN}_{p_1+p_2,q}(\boldsymbol{\xi}_A, \boldsymbol{\Psi}_A, \mathbf{H}_A, \boldsymbol{\tau}, \bar{\boldsymbol{\Gamma}})$, *where*

$$\boldsymbol{\xi}_A = \mathbf{A}\boldsymbol{\xi} = \begin{pmatrix} \boldsymbol{\xi}_1 - \boldsymbol{\Psi}_{12}\boldsymbol{\Psi}_{22}^{-1}\boldsymbol{\xi}_2 \\ \boldsymbol{\xi}_2 \end{pmatrix}, \quad \boldsymbol{\Psi}_A = \mathbf{A}\boldsymbol{\Psi}\mathbf{A}^\top = \begin{pmatrix} \boldsymbol{\Psi}_{11\cdot 2} & \mathbf{O} \\ \mathbf{O} & \boldsymbol{\Psi}_{22} \end{pmatrix}, \quad \mathbf{H}_A = \mathbf{A}\mathbf{H} = \begin{pmatrix} \mathbf{H}_{1\cdot 2} \\ \mathbf{H}_2 \end{pmatrix}.$$

Now, since $\mathbf{A}$ *is a nonsingular* $(p_1 + p_2) \times (p_1 + p_2)$ *matrix and by assumption* $\boldsymbol{\Psi} > 0$, *and so* $\mathbf{X} \sim \mathcal{SUN}_{p_1+p_2,q}(\boldsymbol{\xi}, \boldsymbol{\Psi}, \mathbf{H}, \boldsymbol{\tau}, \bar{\boldsymbol{\Gamma}})$ *is nonsingular, then* $\boldsymbol{\Psi}_A = \mathrm{diag}(\boldsymbol{\Psi}_{11\cdot 2}, \boldsymbol{\Psi}_{22})$ *is also nonsingular, i.e., both* $\boldsymbol{\Psi}_{11\cdot 2} > 0$ *and* $\boldsymbol{\Psi}_{22} > 0$. *Hence, by Theorem 103, we can obtain the pdf of* $\mathbf{Y} = \mathbf{A}\mathbf{X}$ *from (6.27), which also becomes the joint pdf of* $(\mathbf{Y}_1^\top, \mathbf{Y}_2^\top)^\top$, *and is given by*

$$h(\mathbf{y}_1, \mathbf{x}_2) = \frac{1}{\Phi_q(\boldsymbol{\tau}; \mathbf{0}_q, \bar{\boldsymbol{\Gamma}})}$$
$$\times \phi_{p_1+p_2}\left\{\begin{pmatrix} \mathbf{y}_1 \\ \mathbf{x}_2 \end{pmatrix}; \begin{pmatrix} \boldsymbol{\xi}_1 - \boldsymbol{\Psi}_{12}\boldsymbol{\Psi}_{22}^{-1}\boldsymbol{\xi}_2 \\ \boldsymbol{\xi}_2 \end{pmatrix}, \begin{pmatrix} \boldsymbol{\Psi}_{11\cdot 2} & \mathbf{O} \\ \mathbf{O} & \boldsymbol{\Psi}_{22} \end{pmatrix} + \begin{pmatrix} \mathbf{H}_{1\cdot 2} \\ \mathbf{H}_2 \end{pmatrix} \bar{\boldsymbol{\Gamma}} \begin{pmatrix} \mathbf{H}_{1\cdot 2}^\top \mathbf{H}_2^\top \end{pmatrix}\right\}$$
$$\times \Phi_q\left\{\boldsymbol{\tau} + \bar{\boldsymbol{\Gamma}}_A \begin{pmatrix} \mathbf{H}_{1\cdot 2}^\top \mathbf{H}_2^\top \end{pmatrix} \begin{pmatrix} \boldsymbol{\Psi}_{11\cdot 2}^{-1} & \mathbf{O} \\ \mathbf{O}^\top & \boldsymbol{\Psi}_{22}^{-1} \end{pmatrix} \begin{pmatrix} \mathbf{y}_1 - \boldsymbol{\xi}_1 + \boldsymbol{\Psi}_{12}\boldsymbol{\Psi}_{22}^{-1}\boldsymbol{\xi}_2 \\ \mathbf{x}_2 - \boldsymbol{\xi}_2 \end{pmatrix}; \mathbf{0}_q, \bar{\boldsymbol{\Gamma}}_A\right\}$$
$$= \frac{1}{\Phi_q(\boldsymbol{\tau}; \mathbf{0}_q, \bar{\boldsymbol{\Gamma}})}$$
$$\times \phi_{p_1+p_2}\left\{\begin{pmatrix} \mathbf{y}_1 \\ \mathbf{x}_2 \end{pmatrix}; \begin{pmatrix} \boldsymbol{\xi}_1 - \boldsymbol{\Psi}_{12}\boldsymbol{\Psi}_{22}^{-1}\boldsymbol{\xi}_2 \\ \boldsymbol{\xi}_2 \end{pmatrix}, \begin{pmatrix} \boldsymbol{\Psi}_{1\cdot 2} + \mathbf{H}_{1\cdot 2}\bar{\boldsymbol{\Gamma}}\mathbf{H}_{1\cdot 2}^\top & \mathbf{H}_{1\cdot 2}\bar{\boldsymbol{\Gamma}}\mathbf{H}_2^\top \\ \mathbf{H}_2\bar{\boldsymbol{\Gamma}}\mathbf{H}_{1\cdot 2}^\top & \boldsymbol{\Psi}_{22} + \mathbf{H}_2\bar{\boldsymbol{\Gamma}}\mathbf{H}_2^\top \end{pmatrix}\right\}$$
$$\times \Phi_q\{\boldsymbol{\tau} + \bar{\boldsymbol{\Gamma}}_A\mathbf{H}_{1\cdot 2}^\top\boldsymbol{\Psi}_{11\cdot 2}^{-1}(\mathbf{y}_1 - \boldsymbol{\xi}_1 + \boldsymbol{\Psi}_{12}\boldsymbol{\Psi}_{22}^{-1}\boldsymbol{\xi}_2) + \bar{\boldsymbol{\Gamma}}_A\mathbf{H}_2^\top\boldsymbol{\Psi}_{22}^{-1}(\mathbf{x}_2 - \boldsymbol{\xi}_2); \mathbf{0}_q, \bar{\boldsymbol{\Gamma}}_A\}.$$

Also, the marginal-conditional representation of the normal pdf involved in the above joint pdf allows us to rewrite this joint pdf as

$$\begin{aligned}
h(\mathbf{y}_1,\mathbf{x}_2) =& \frac{1}{\Phi_q(\boldsymbol{\tau};\mathbf{0}_q,\bar{\boldsymbol{\Gamma}})} \\
&\times \phi_{p_1}\{\mathbf{y}_1;\boldsymbol{\xi}_1-\boldsymbol{\Psi}_{12}\boldsymbol{\Psi}_{22}^{-1}\boldsymbol{\xi}_2+\mathbf{H}_{1\cdot2}\bar{\boldsymbol{\Gamma}}\mathbf{H}_2^\top(\boldsymbol{\Psi}_{22}+\mathbf{H}_2\bar{\boldsymbol{\Gamma}}\mathbf{H}_2^\top)^{-1}(\mathbf{x}_2-\boldsymbol{\xi}_2),\\
&\qquad \boldsymbol{\Psi}_{11\cdot2}+\mathbf{H}_{1\cdot2}\bar{\boldsymbol{\Gamma}}\mathbf{H}_{1\cdot2}^\top-\mathbf{H}_{1\cdot2}\bar{\boldsymbol{\Gamma}}\mathbf{H}_2^\top(\boldsymbol{\Psi}_{22}+\mathbf{H}_2\bar{\boldsymbol{\Gamma}}\mathbf{H}_2^\top)^{-1}\mathbf{H}_2\bar{\boldsymbol{\Gamma}}\mathbf{H}_{1\cdot2}^\top\}\\
&\times \phi_{p_2}(\mathbf{x}_2;\boldsymbol{\xi}_2,\boldsymbol{\Psi}_{22}+\mathbf{H}_2\bar{\boldsymbol{\Gamma}}\mathbf{H}_2^\top)\\
&\times \Phi_q\{\boldsymbol{\tau}+\bar{\boldsymbol{\Gamma}}_A\mathbf{H}_{1\cdot2}^\top\boldsymbol{\Psi}_{11\cdot2}^{-1}(\mathbf{y}_1-\boldsymbol{\xi}_1+\boldsymbol{\Psi}_{12}\boldsymbol{\Psi}_{22}^{-1}\boldsymbol{\xi}_2)+\bar{\boldsymbol{\Gamma}}_A\mathbf{H}_2^\top\boldsymbol{\Psi}_{22}^{-1}(\mathbf{x}_2-\boldsymbol{\xi}_2);\mathbf{0}_q,\bar{\boldsymbol{\Gamma}}_A\}\\
=& \frac{1}{\Phi_q(\boldsymbol{\tau};\mathbf{0}_q,\bar{\boldsymbol{\Gamma}})}\phi_{p_1}\{\mathbf{y}_1;\boldsymbol{\xi}_1-\boldsymbol{\Psi}_{12}\boldsymbol{\Psi}_{22}^{-1}\boldsymbol{\xi}_2+\mathbf{H}_{1\cdot2}\bar{\boldsymbol{\Gamma}}_{1\cdot2}\mathbf{H}_2^\top\boldsymbol{\Psi}_{22}^{-1}(\mathbf{x}_2-\boldsymbol{\xi}_2),\\
&\qquad \boldsymbol{\Psi}_{11\cdot2}+\mathbf{H}_{1\cdot2}\bar{\boldsymbol{\Gamma}}_{1\cdot2}\mathbf{H}_{1\cdot2}^\top\}\times\phi_{p_2}(\mathbf{x}_2;\boldsymbol{\xi}_2,\boldsymbol{\Psi}_{22}+\mathbf{H}_2\bar{\boldsymbol{\Gamma}}\mathbf{H}_2^\top)\\
&\times \Phi_q[\boldsymbol{\tau}_{1\cdot2}+\bar{\boldsymbol{\Gamma}}_A\mathbf{H}_{1\cdot2}^\top\boldsymbol{\Psi}_{11\cdot2}^{-1}\{\mathbf{y}_1-\boldsymbol{\xi}_1+\boldsymbol{\Psi}_{12}\boldsymbol{\Psi}_{22}^{-1}\boldsymbol{\xi}_2-\mathbf{H}_{1\cdot2}\bar{\boldsymbol{\Gamma}}_{1\cdot2}\mathbf{H}_2^\top\boldsymbol{\Psi}_{22}^{-1}(\mathbf{x}_2-\boldsymbol{\xi}_2)\};\mathbf{0}_q,\bar{\boldsymbol{\Gamma}}_A],
\end{aligned}$$

since by definition $\boldsymbol{\tau}_{1\cdot2}=\boldsymbol{\tau}+\bar{\boldsymbol{\Gamma}}_{1\cdot2}\mathbf{H}_2^\top\boldsymbol{\Psi}_{22}^{-1}(\mathbf{x}_2-\boldsymbol{\xi}_2)$ *and we used the matrix identities*

$$\begin{aligned}
\bar{\boldsymbol{\Gamma}}\mathbf{H}_2^\top(\boldsymbol{\Psi}_{22}+\mathbf{H}_2\bar{\boldsymbol{\Gamma}}\mathbf{H}_2^\top)^{-1}&=\bar{\boldsymbol{\Gamma}}_{1\cdot2}\mathbf{H}_2^\top\boldsymbol{\Psi}_{22}^{-1},\\
\bar{\boldsymbol{\Gamma}}-\bar{\boldsymbol{\Gamma}}\mathbf{H}_2^\top(\boldsymbol{\Psi}_{22}+\mathbf{H}_2\bar{\boldsymbol{\Gamma}}\mathbf{H}_2^\top)^{-1}\mathbf{H}_2\bar{\boldsymbol{\Gamma}}&=(\bar{\boldsymbol{\Gamma}}^{-1}+\mathbf{H}_2^\top\boldsymbol{\Psi}_{22}^{-1}\mathbf{H}_2)^{-1}=\bar{\boldsymbol{\Gamma}}_{1\cdot2}\\
\Rightarrow \mathbf{H}_{1\cdot2}\bar{\boldsymbol{\Gamma}}\mathbf{H}_{1\cdot2}^\top-\mathbf{H}_{1\cdot2}\bar{\boldsymbol{\Gamma}}\mathbf{H}_2^\top(\boldsymbol{\Psi}_{22}+\mathbf{H}_2\bar{\boldsymbol{\Gamma}}\mathbf{H}_2^\top)^{-1}\mathbf{H}_2\bar{\boldsymbol{\Gamma}}\mathbf{H}_{1\cdot2}^\top&=\mathbf{H}_{1\cdot2}\bar{\boldsymbol{\Gamma}}_{1\cdot2}\mathbf{H}_{1\cdot2}^\top,
\end{aligned}$$

and the fact that the term involved in the skewing factor becomes

$$\begin{aligned}
\boldsymbol{\tau}&+\bar{\boldsymbol{\Gamma}}_A\mathbf{H}_{1\cdot2}^\top\boldsymbol{\Psi}_{11\cdot2}^{-1}(\mathbf{y}_1-\boldsymbol{\xi}_1+\boldsymbol{\Psi}_{12}\boldsymbol{\Psi}_{22}^{-1}\boldsymbol{\xi}_2)+\bar{\boldsymbol{\Gamma}}_A\mathbf{H}_2^\top\boldsymbol{\Psi}_{22}^{-1}(\mathbf{x}_2-\boldsymbol{\xi}_2)\\
&=\boldsymbol{\tau}+\bar{\boldsymbol{\Gamma}}_A\mathbf{H}_{1\cdot2}^\top\boldsymbol{\Psi}_{11\cdot2}^{-1}\{\mathbf{y}_1-\boldsymbol{\xi}_1+\boldsymbol{\Psi}_{12}\boldsymbol{\Psi}_{22}^{-1}\boldsymbol{\xi}_2-\mathbf{H}_{1\cdot2}\bar{\boldsymbol{\Gamma}}_{1\cdot2}\mathbf{H}_2^\top\boldsymbol{\Psi}_{22}^{-1}(\mathbf{x}_2-\boldsymbol{\xi}_2)\}\\
&\quad+\bar{\boldsymbol{\Gamma}}_A\mathbf{H}_{1\cdot2}^\top\boldsymbol{\Psi}_{11\cdot2}^{-1}\mathbf{H}_{1\cdot2}\bar{\boldsymbol{\Gamma}}_{1\cdot2}\mathbf{H}_2^\top\boldsymbol{\Psi}_{22}^{-1}(\mathbf{x}_2-\boldsymbol{\xi}_2)+\bar{\boldsymbol{\Gamma}}_A\mathbf{H}_2^\top\boldsymbol{\Psi}_{22}^{-1}(\mathbf{x}_2-\boldsymbol{\xi}_2)\\
&=\boldsymbol{\tau}+\bar{\boldsymbol{\Gamma}}_A\mathbf{H}_{1\cdot2}^\top\boldsymbol{\Psi}_{11\cdot2}^{-1}\{\mathbf{y}_1-\boldsymbol{\xi}_1+\boldsymbol{\Psi}_{12}\boldsymbol{\Psi}_{22}^{-1}\boldsymbol{\xi}_2-\mathbf{H}_{1\cdot2}\bar{\boldsymbol{\Gamma}}_{1\cdot2}\mathbf{H}_2^\top\boldsymbol{\Psi}_{22}^{-1}(\mathbf{x}_2-\boldsymbol{\xi}_2)\}\\
&\quad+\bar{\boldsymbol{\Gamma}}_A(\mathbf{H}_{1\cdot2}^\top\boldsymbol{\Psi}_{11\cdot2}^{-1}\mathbf{H}_{1\cdot2}\bar{\boldsymbol{\Gamma}}_{1\cdot2}+\mathbf{I}_q)\mathbf{H}_2^\top\boldsymbol{\Psi}_{22}^{-1}(\mathbf{x}_2-\boldsymbol{\xi}_2)\\
&=\boldsymbol{\tau}+\bar{\boldsymbol{\Gamma}}_A\mathbf{H}_{1\cdot2}^\top\boldsymbol{\Psi}_{11\cdot2}^{-1}\{\mathbf{y}_1-\boldsymbol{\xi}_1+\boldsymbol{\Psi}_{12}\boldsymbol{\Psi}_{22}^{-1}\boldsymbol{\xi}_2-\mathbf{H}_{1\cdot2}\bar{\boldsymbol{\Gamma}}_{1\cdot2}\mathbf{H}_2^\top\boldsymbol{\Psi}_{22}^{-1}(\mathbf{x}_2-\boldsymbol{\xi}_2)\}\\
&\quad+\bar{\boldsymbol{\Gamma}}_A(\mathbf{H}_{1\cdot2}^\top\boldsymbol{\Psi}_{11\cdot2}^{-1}\mathbf{H}_{1\cdot2}+\bar{\boldsymbol{\Gamma}}_{1\cdot2}^{-1})\bar{\boldsymbol{\Gamma}}_{1\cdot2}\mathbf{H}_2^\top\boldsymbol{\Psi}_{22}^{-1}(\mathbf{x}_2-\boldsymbol{\xi}_2)\\
&=\boldsymbol{\tau}+\bar{\boldsymbol{\Gamma}}_A\mathbf{H}_{1\cdot2}^\top\boldsymbol{\Psi}_{11\cdot2}^{-1}\{\mathbf{y}_1-\boldsymbol{\xi}_1+\boldsymbol{\Psi}_{12}\boldsymbol{\Psi}_{22}^{-1}\boldsymbol{\xi}_2-\mathbf{H}_{1\cdot2}\bar{\boldsymbol{\Gamma}}_{1\cdot2}\mathbf{H}_2^\top\boldsymbol{\Psi}_{22}^{-1}(\mathbf{x}_2-\boldsymbol{\xi}_2)\}\\
&\quad+\bar{\boldsymbol{\Gamma}}_A\bar{\boldsymbol{\Gamma}}_A^{-1}\bar{\boldsymbol{\Gamma}}_{1\cdot2}\mathbf{H}_2^\top\boldsymbol{\Psi}_{22}^{-1}(\mathbf{x}_2-\boldsymbol{\xi}_2)\\
&=\boldsymbol{\tau}_{1\cdot2}+\bar{\boldsymbol{\Gamma}}_A\mathbf{H}_{1\cdot2}^\top\boldsymbol{\Psi}_{11\cdot2}^{-1}\{\mathbf{y}_1-\boldsymbol{\xi}_1+\boldsymbol{\Psi}_{12}\boldsymbol{\Psi}_{22}^{-1}\boldsymbol{\xi}_2-\mathbf{H}_{1\cdot2}\bar{\boldsymbol{\Gamma}}_{1\cdot2}\mathbf{H}_2^\top\boldsymbol{\Psi}_{22}^{-1}(\mathbf{x}_2-\boldsymbol{\xi}_2)\}.
\end{aligned}$$

On the other hand, from the application of Theorem 96 to the marginal of $\mathbf{X}_2$*, we have* $\mathbf{Y}_2=\mathbf{X}_2\sim\mathcal{SUN}_{p_2,q}(\boldsymbol{\xi}_2,\boldsymbol{\Psi}_{22},\mathbf{H}_2,\boldsymbol{\tau},\bar{\boldsymbol{\Gamma}})$*, with marginal pdf given by*

$$\begin{aligned}
h_2(\mathbf{x}_2)=f_2(\mathbf{x}_2)=&\frac{1}{\Phi_q(\boldsymbol{\tau};\mathbf{0}_q,\bar{\boldsymbol{\Gamma}})}\phi_{p_2}(\mathbf{x}_2;\boldsymbol{\xi}_2,\boldsymbol{\Psi}_{22}+\mathbf{H}_2\bar{\boldsymbol{\Gamma}}\mathbf{H}_2^\top)\Phi_q\{\boldsymbol{\tau}+\tilde{\boldsymbol{\Gamma}}_2\mathbf{H}_2^\top\boldsymbol{\Psi}_{22}^{-1}(\mathbf{x}_2-\boldsymbol{\xi}_2);\mathbf{0}_q,\tilde{\boldsymbol{\Gamma}}_2\}\\
=&\frac{1}{\Phi_q(\boldsymbol{\tau};\mathbf{0}_q,\bar{\boldsymbol{\Gamma}})}\phi_{p_2}(\mathbf{x}_2;\boldsymbol{\xi}_2,\boldsymbol{\Psi}_{22}+\mathbf{H}_2\bar{\boldsymbol{\Gamma}}\mathbf{H}_2^\top)\Phi_q\{\boldsymbol{\tau}_{1\cdot2};\mathbf{0}_q,\bar{\boldsymbol{\Gamma}}_{1\cdot2}\},
\end{aligned}$$

since $\bar{\boldsymbol{\Gamma}}_{1\cdot 2} \equiv \tilde{\boldsymbol{\Gamma}}_2 = (\bar{\boldsymbol{\Gamma}}^{-1} + \mathbf{H}_2\boldsymbol{\Psi}_{22}^{-1}\mathbf{H}_2)^{-1}$*. Hence, the conditional pdf of* $\mathbf{Y}_1$ *given* $\mathbf{Y}_2 = \mathbf{x}_2$ *is*

$$\begin{aligned} h_{1\cdot 2}(\mathbf{y}_1 \mid \mathbf{x}_2) &= \frac{h(\mathbf{y}_1, \mathbf{x}_2)}{h_2(\mathbf{x}_2)} \\ &= \frac{1}{\Phi_q\{\boldsymbol{\tau}_{1\cdot 2}; \mathbf{0}_q, \bar{\boldsymbol{\Gamma}}_{1\cdot 2}\}} \\ &\quad \times \phi_{p_1}\{\mathbf{y}_1; \boldsymbol{\xi}_1 - \boldsymbol{\Psi}_{12}\boldsymbol{\Psi}_{22}^{-1}\boldsymbol{\xi}_2 + \mathbf{H}_{1\cdot 2}\bar{\boldsymbol{\Gamma}}_{1\cdot 2}\mathbf{H}_2^\top\boldsymbol{\Psi}_{22}^{-1}(\mathbf{x}_2 - \boldsymbol{\xi}_2), \boldsymbol{\Psi}_{11\cdot 2} + \mathbf{H}_{1\cdot 2}\bar{\boldsymbol{\Gamma}}_{1\cdot 2}\mathbf{H}_{1\cdot 2}^\top\} \\ &\quad \times \Phi_q[\boldsymbol{\tau}_{1\cdot 2} + \bar{\boldsymbol{\Gamma}}_A\mathbf{H}_{1\cdot 2}^\top\boldsymbol{\Psi}_{11\cdot 2}^{-1}\{\mathbf{y}_1 - \boldsymbol{\xi}_1 + \boldsymbol{\Psi}_{12}\boldsymbol{\Psi}_{22}^{-1}\boldsymbol{\xi}_2 - \mathbf{H}_{1\cdot 2}\bar{\boldsymbol{\Gamma}}_{1\cdot 2}\mathbf{H}_2^\top\boldsymbol{\Psi}_{22}^{-1}(\mathbf{x}_2 - \boldsymbol{\xi}_2)\}; \mathbf{0}_q, \bar{\boldsymbol{\Gamma}}_A], \end{aligned}$$

where $\bar{\boldsymbol{\Gamma}}_A = \tilde{\boldsymbol{\Gamma}} = (\bar{\boldsymbol{\Gamma}}^{-1} + \mathbf{H}^\top\boldsymbol{\Psi}^{-1}\mathbf{H})^{-1} = (\bar{\boldsymbol{\Gamma}}^{-1} + \mathbf{H}_A^\top\boldsymbol{\Psi}_A^{-1}\mathbf{H}_A)^{-1} = \bar{\boldsymbol{\Gamma}}_{1\cdot 2}$.

Finally, since $\mathbf{X} = \mathbf{A}^{-1}\mathbf{Y}$*, where* $\mathrm{abs}(|\mathbf{A}^{-1}|) = \mathrm{abs}(|\mathbf{A}|^{-1}) = \mathrm{abs}(|\mathbf{A}|)^{-1} = 1$*, then the Jacobian method allows us to write the joint pdf of* $(\mathbf{X}_1, \mathbf{X}_2)$ *as*

$$f(\mathbf{x}_1, \mathbf{x}_2) = h(\mathbf{x}_1 - \boldsymbol{\Psi}_{12}\boldsymbol{\Psi}_{22}^{-1}\mathbf{x}_2, \mathbf{x}_2)$$

from where the conditional pdf of $(\mathbf{X}_1 \mid \mathbf{X}_2 = \mathbf{x}_2)$ *is obtained as*

$$f_{1\cdot 2}(\mathbf{x}_1 \mid \mathbf{x}_2) = h_{1\cdot 2}(\mathbf{x}_1 - \boldsymbol{\Psi}_{12}\boldsymbol{\Psi}_{22}^{-1}\mathbf{x}_2 \mid \mathbf{x}_2).$$

This proves the result. □

Obviously, we can also obtain the conditional pdf $f_{1\cdot 2}(\mathbf{x}_1 \mid \mathbf{x}_2)$ directly from (6.24) which becomes the joint pdf of $\mathbf{X}_1$ and $\mathbf{X}_2$. To see this, we will use the alternative expression of this joint pdf presented in (6.25), in which we first notice that the normal pdf $\phi_p(\mathbf{x}; \boldsymbol{\xi}, \boldsymbol{\Omega})$, where $\boldsymbol{\Omega} = \boldsymbol{\Psi} + \mathbf{H}\bar{\boldsymbol{\Gamma}}\mathbf{H}^\top$, can be factorized from the marginal-conditional representation of the multivariate normal distribution, as

$$\phi_p(\mathbf{x}; \boldsymbol{\xi}, \boldsymbol{\Omega}) = \phi_{p_1}\{\mathbf{x}_1; \boldsymbol{\xi}_1 + \boldsymbol{\Omega}_{12}\boldsymbol{\Omega}_{22}^{-1}(\mathbf{x}_2 - \boldsymbol{\xi}_2), \boldsymbol{\Omega}_{11\cdot 2}\}\phi_{p_2}(\mathbf{x}_2; \boldsymbol{\xi}_2, \boldsymbol{\Omega}_{22}),$$

where $\boldsymbol{\xi}$ is partitioned into the marginals $\boldsymbol{\xi}_j$, $j = 1, 2$, and $\boldsymbol{\Omega} = \boldsymbol{\Psi} + \mathbf{H}\bar{\boldsymbol{\Gamma}}\mathbf{H}^\top$ is partitioned to have submatrices $\boldsymbol{\Omega}_{ij} = \boldsymbol{\Psi}_{ij} + \mathbf{H}_i\bar{\boldsymbol{\Gamma}}\mathbf{H}_j^\top$, $i, j = 1, 2$. Also, by applying the decomposition (6.32) to the matrix $\boldsymbol{\Omega}$ in the nonsingular case, i.e., where

$$\mathbf{B} = \begin{pmatrix} \mathbf{I}_{p_1} & -\mathbf{C} \\ \mathbf{O} & \mathbf{I}_{p_2} \end{pmatrix}, \quad \text{with} \quad \mathbf{C} = \boldsymbol{\Omega}_{12}\boldsymbol{\Omega}_{22}^{-1},$$

we have for the argument in the skewing factor $\Phi_q\{\boldsymbol{\tau} + \bar{\boldsymbol{\Gamma}}\mathbf{H}^\top\boldsymbol{\Omega}^{-1}(\mathbf{x} - \boldsymbol{\xi}); \mathbf{0}_q, \bar{\boldsymbol{\Gamma}} - \bar{\boldsymbol{\Gamma}}\mathbf{H}^\top\boldsymbol{\Omega}^{-1}\mathbf{H}\bar{\boldsymbol{\Gamma}}\}$ of (6.25) that

$$\begin{aligned} \boldsymbol{\tau} + \bar{\boldsymbol{\Gamma}}\mathbf{H}^\top\boldsymbol{\Omega}^{-1}(\mathbf{x} - \boldsymbol{\xi}) &= (\mathbf{B}\mathbf{H})^\top \begin{pmatrix} \boldsymbol{\Omega}_{11\cdot 2}^{-1} & \mathbf{O} \\ \mathbf{O} & \boldsymbol{\Omega}_{22}^{-1} \end{pmatrix} \mathbf{B}(\mathbf{x} - \boldsymbol{\xi}) \\ &= \boldsymbol{\tau} + \bar{\boldsymbol{\Gamma}}\mathbf{H}_2\boldsymbol{\Omega}_{22}^{-1}(\mathbf{x}_2 - \boldsymbol{\xi}_2) \\ &\quad + \bar{\boldsymbol{\Gamma}}(\mathbf{H}_1 - \boldsymbol{\Omega}_{12}\boldsymbol{\Omega}_{22}^{-1}\mathbf{H}_2)\boldsymbol{\Omega}_{11\cdot 2}^{-1}\{\mathbf{x}_1 - \boldsymbol{\xi}_1 - \boldsymbol{\Omega}_{12}\boldsymbol{\Omega}_{22}^{-1}(\mathbf{x}_2 - \boldsymbol{\xi}_2)\}, \end{aligned}$$

where $\boldsymbol{\Omega}_{11\cdot 2}^{-1} = \boldsymbol{\Omega}_{12} - \boldsymbol{\Omega}_{12}\boldsymbol{\Omega}_{22}^{-1}\boldsymbol{\Omega}_{21}$. Furthermore, from (6.25), the marginal pdf of $\mathbf{X}_2$ becomes

$$f_2(\mathbf{x}_2) = \frac{1}{\Phi_q(\boldsymbol{\tau}; \mathbf{0}_q, \bar{\boldsymbol{\Gamma}})}\phi_{p_2}(\mathbf{x}_2; \boldsymbol{\xi}_2, \boldsymbol{\Omega}_{22})\Phi_q\{\boldsymbol{\tau} + \bar{\boldsymbol{\Gamma}}\mathbf{H}_2^\top\boldsymbol{\Omega}_{22}^{-1}(\mathbf{x}_2 - \boldsymbol{\xi}_2); \mathbf{0}_q, \bar{\boldsymbol{\Gamma}} - \bar{\boldsymbol{\Gamma}}\mathbf{H}^\top\boldsymbol{\Omega}^{-1}\mathbf{H}\bar{\boldsymbol{\Gamma}}\}, \ \mathbf{x}_2 \in \mathbb{R}^{p_2}.$$

Therefore, for each fixed $\mathbf{x}_2 \in \mathbb{R}^{p_2}$, the conditional pdf of $(\mathbf{X}_1 \mid \mathbf{X}_2 = \mathbf{x}_2)$ is

$$\begin{aligned} f_{1\cdot 2}(\mathbf{x}_1 \mid \mathbf{x}_2) &= \frac{1}{\Phi_q\{\boldsymbol{\tau} + \bar{\boldsymbol{\Gamma}}\mathbf{H}_2^\top\boldsymbol{\Omega}_{22}^{-1}(\mathbf{x}_2 - \boldsymbol{\xi}_2); \mathbf{0}_q, \bar{\boldsymbol{\Gamma}} - \bar{\boldsymbol{\Gamma}}\mathbf{H}^\top\boldsymbol{\Omega}^{-1}\mathbf{H}\bar{\boldsymbol{\Gamma}}\}} \\ &\quad \times \phi_{p_1}\{\mathbf{x}_1; \boldsymbol{\xi}_1 + \boldsymbol{\Omega}_{12}\boldsymbol{\Omega}_{22}^{-1}(\mathbf{x}_2 - \boldsymbol{\xi}_2), \boldsymbol{\Omega}_{11\cdot 2}\}\Phi_q[\boldsymbol{\tau} + \bar{\boldsymbol{\Gamma}}\mathbf{H}_2^\top\boldsymbol{\Omega}_{22}^{-1}(\mathbf{x}_2 - \boldsymbol{\xi}_2) \\ &\quad + \bar{\boldsymbol{\Gamma}}(\mathbf{H}_1 - \boldsymbol{\Omega}_{12}\boldsymbol{\Omega}_{22}^{-1}\mathbf{H}_2)^\top\boldsymbol{\Omega}_{11\cdot 2}^{-1}\{\mathbf{x}_1 - \boldsymbol{\xi}_1 - \boldsymbol{\Omega}_{12}\boldsymbol{\Omega}_{22}^{-1}(\mathbf{x}_2 - \boldsymbol{\xi}_2)\}; \mathbf{0}_q, \bar{\boldsymbol{\Gamma}} - \bar{\boldsymbol{\Gamma}}\mathbf{H}^\top\boldsymbol{\Omega}^{-1}\mathbf{H}\bar{\boldsymbol{\Gamma}}], \end{aligned} \tag{6.48}$$

with $\mathbf{x}_1 \in \mathbb{R}^{p_1}$.

To show that expressions (6.47) and (6.48) are in agreement, we need some additional matrix calculations. For this, first recall the matrix relations:

$$\bar{\boldsymbol{\Gamma}}-\bar{\boldsymbol{\Gamma}}\mathbf{H}^\top\boldsymbol{\Omega}^{-1}\mathbf{H}\bar{\boldsymbol{\Gamma}}=(\bar{\boldsymbol{\Gamma}}^{-1}+\mathbf{H}^\top\boldsymbol{\Psi}^{-1}\mathbf{H})^{-1}=(\bar{\boldsymbol{\Gamma}}_{1\cdot2}^{-1}+\mathbf{H}_{1\cdot2}^\top\boldsymbol{\Psi}_{11\cdot2}^{-1}\mathbf{H}_{1\cdot2})^{-1}=\tilde{\boldsymbol{\Gamma}}_{1\cdot2},$$

where:

$$\begin{aligned}\mathbf{H}_{1\cdot2}=\mathbf{H}_1-\boldsymbol{\Psi}_{12}\boldsymbol{\Psi}_{22}^{-1}\mathbf{H}_2,\quad \boldsymbol{\Psi}_{11\cdot2}&=\boldsymbol{\Psi}_{11}-\boldsymbol{\Psi}_{12}\boldsymbol{\Psi}_{22}^{-1}\boldsymbol{\Psi}_{21},\\ \bar{\boldsymbol{\Gamma}}_{1\cdot2}=(\bar{\boldsymbol{\Gamma}}^{-1}+\mathbf{H}_2^\top\boldsymbol{\Psi}_{22}^{-1}\mathbf{H}_2)^{-1}\ \Rightarrow\ \bar{\boldsymbol{\Gamma}}_{1\cdot2}^{-1}&-\mathbf{H}_2^\top\boldsymbol{\Psi}_{22}^{-1}\mathbf{H}_2=\bar{\boldsymbol{\Gamma}}^{-1}.\end{aligned}$$

Thus, we have

$$\begin{aligned}\boldsymbol{\Omega}_{22}^{-1}&=(\boldsymbol{\Psi}_{22}+\mathbf{H}_2\bar{\boldsymbol{\Gamma}}\mathbf{H}_2^\top)^{-1}=\boldsymbol{\Psi}_{22}^{-1}-\boldsymbol{\Psi}_{22}^{-1}\mathbf{H}_2\bar{\boldsymbol{\Gamma}}_{1\cdot2}\mathbf{H}_2^\top\boldsymbol{\Psi}_{22}^{-1},\\ \Rightarrow\ \boldsymbol{\Omega}_{12}\boldsymbol{\Omega}_{22}^{-1}&=(\boldsymbol{\Psi}_{12}+\mathbf{H}_1\bar{\boldsymbol{\Gamma}}\mathbf{H}_2^\top)(\boldsymbol{\Psi}_{22}^{-1}-\boldsymbol{\Psi}_{22}^{-1}\mathbf{H}_2\bar{\boldsymbol{\Gamma}}_{1\cdot2}\mathbf{H}_2^\top\boldsymbol{\Psi}_{22}^{-1})\\ &=\boldsymbol{\Psi}_{12}(\boldsymbol{\Psi}_{22}^{-1}-\boldsymbol{\Psi}_{22}^{-1}\mathbf{H}_2\bar{\boldsymbol{\Gamma}}_{1\cdot2}\mathbf{H}_2^\top\boldsymbol{\Psi}_{22}^{-1})+\mathbf{H}_1\bar{\boldsymbol{\Gamma}}\mathbf{H}_2^\top(\boldsymbol{\Psi}_{22}^{-1}-\boldsymbol{\Psi}_{22}^{-1}\mathbf{H}_2\bar{\boldsymbol{\Gamma}}_{1\cdot2}\mathbf{H}_2^\top\boldsymbol{\Psi}_{22}^{-1})\\ &=\boldsymbol{\Psi}_{12}\boldsymbol{\Psi}_{22}^{-1}-\boldsymbol{\Psi}_{12}\boldsymbol{\Psi}_{22}^{-1}\mathbf{H}_2\bar{\boldsymbol{\Gamma}}_{1\cdot2}\mathbf{H}_2^\top\boldsymbol{\Psi}_{22}^{-1}+\mathbf{H}_1\bar{\boldsymbol{\Gamma}}(\mathbf{H}_2^\top\boldsymbol{\Psi}_{22}^{-1}-\mathbf{H}_2^\top\boldsymbol{\Psi}_{22}^{-1}\mathbf{H}_2\bar{\boldsymbol{\Gamma}}_{1\cdot2}\mathbf{H}_2^\top\boldsymbol{\Psi}_{22}^{-1})\\ &=\boldsymbol{\Psi}_{12}\boldsymbol{\Psi}_{22}^{-1}-\boldsymbol{\Psi}_{12}\boldsymbol{\Psi}_{22}^{-1}\mathbf{H}_2\bar{\boldsymbol{\Gamma}}_{1\cdot2}\mathbf{H}_2^\top\boldsymbol{\Psi}_{22}^{-1}+\mathbf{H}_1\bar{\boldsymbol{\Gamma}}(\bar{\boldsymbol{\Gamma}}_{1\cdot2}^{-1}-\mathbf{H}_2^\top\boldsymbol{\Psi}_{22}^{-1}\mathbf{H}_2)\bar{\boldsymbol{\Gamma}}_{1\cdot2}\mathbf{H}_2^\top\boldsymbol{\Psi}_{22}^{-1}\\ &=\boldsymbol{\Psi}_{12}\boldsymbol{\Psi}_{22}^{-1}-\boldsymbol{\Psi}_{12}\boldsymbol{\Psi}_{22}^{-1}\mathbf{H}_2\bar{\boldsymbol{\Gamma}}_{1\cdot2}\mathbf{H}_2^\top\boldsymbol{\Psi}_{22}^{-1}+\mathbf{H}_1\bar{\boldsymbol{\Gamma}}(\bar{\boldsymbol{\Gamma}}^{-1})\bar{\boldsymbol{\Gamma}}_{1\cdot2}\mathbf{H}_2^\top\boldsymbol{\Psi}_{22}^{-1}\\ &=\boldsymbol{\Psi}_{12}\boldsymbol{\Psi}_{22}^{-1}+(\mathbf{H}_1-\boldsymbol{\Psi}_{12}\boldsymbol{\Psi}_{22}^{-1}\mathbf{H}_2)\bar{\boldsymbol{\Gamma}}_{1\cdot2}\mathbf{H}_2^\top\boldsymbol{\Psi}_{22}^{-1}\\ &=\boldsymbol{\Psi}_{12}\boldsymbol{\Psi}_{22}^{-1}+\mathbf{H}_{1\cdot2}\bar{\boldsymbol{\Gamma}}_{1\cdot2}\mathbf{H}_2^\top\boldsymbol{\Psi}_{22}^{-1}.\end{aligned}$$

Similarly, we find

$$\begin{aligned}\boldsymbol{\Omega}_{11\cdot2}=\boldsymbol{\Psi}_{11\cdot2}+\mathbf{H}_{1\cdot2}\bar{\boldsymbol{\Gamma}}_{1\cdot2}\mathbf{H}_{1\cdot2}^\top&=\boldsymbol{\Psi}_{11\cdot2}+\bar{\mathbf{H}}_{1\cdot2}\bar{\bar{\boldsymbol{\Gamma}}}_{1\cdot2}\bar{\mathbf{H}}_{1\cdot2}^\top,\\ \bar{\boldsymbol{\Gamma}}\mathbf{H}_2^\top\boldsymbol{\Omega}_{22}^{-1}=\bar{\boldsymbol{\Gamma}}_{1\cdot2}\mathbf{H}_2^\top\boldsymbol{\Psi}_{22}^{-1},\quad \mathbf{H}_1-\boldsymbol{\Omega}_{12}\boldsymbol{\Omega}_{22}^{-1}\mathbf{H}_2&=\mathbf{H}_{1\cdot2}\bar{\boldsymbol{\Gamma}}_{1\cdot2}\bar{\boldsymbol{\Gamma}}^{-1}\\ \Rightarrow\ \bar{\boldsymbol{\Gamma}}(\mathbf{H}_1-\boldsymbol{\Omega}_{12}\boldsymbol{\Omega}_{22}^{-1}\mathbf{H}_2)^\top&=\bar{\boldsymbol{\Gamma}}_{1\cdot2}\mathbf{H}_{1\cdot2}^\top\\ \Rightarrow\ \bar{\boldsymbol{\Gamma}}(\mathbf{H}_1-\boldsymbol{\Omega}_{12}\boldsymbol{\Omega}_{22}^{-1}\mathbf{H}_2)^\top\boldsymbol{\Omega}_{11\cdot2}^{-1}&=\tilde{\boldsymbol{\Gamma}}_{1\cdot2}\mathbf{H}_{1\cdot2}^\top\boldsymbol{\Psi}_{11\cdot2}^{-1}.\end{aligned}$$

From the application of these relations to each of the ingredients of (6.25), it is obtained that

$$\begin{aligned}&\Phi_q\{\boldsymbol{\tau}+\bar{\boldsymbol{\Gamma}}\mathbf{H}_2^\top\boldsymbol{\Omega}_{22}^{-1}(\mathbf{x}_2-\boldsymbol{\xi}_2);\mathbf{0}_q,\bar{\boldsymbol{\Gamma}}-\bar{\boldsymbol{\Gamma}}\mathbf{H}^\top\boldsymbol{\Omega}^{-1}\mathbf{H}\bar{\boldsymbol{\Gamma}}\}=\Phi_q\{\boldsymbol{\tau}+\bar{\boldsymbol{\Gamma}}_{1\cdot2}\mathbf{H}_2^\top\boldsymbol{\Psi}_{22}^{-1}(\mathbf{x}_2-\boldsymbol{\xi}_2);\mathbf{0}_q,\tilde{\boldsymbol{\Gamma}}_{1\cdot2}\},\\ &\phi_{p_1}\{\mathbf{x}_1;\boldsymbol{\xi}_1+\boldsymbol{\Omega}_{12}\boldsymbol{\Omega}_{22}^{-1}(\mathbf{x}_2-\boldsymbol{\xi}_2),\boldsymbol{\Omega}_{11\cdot2}\}=\phi_{p_1}\{\mathbf{x}_1;\boldsymbol{\xi}_1+\boldsymbol{\Psi}_{12}\boldsymbol{\Psi}_{22}^{-1}(\mathbf{x}_2-\boldsymbol{\xi}_2)\\ &\qquad\qquad+\mathbf{H}_{1\cdot2}\bar{\boldsymbol{\Gamma}}_{1\cdot2}\mathbf{H}_2^\top\boldsymbol{\Psi}_{22}^{-1}(\mathbf{x}_2-\boldsymbol{\xi}_2),\boldsymbol{\Psi}_{11\cdot2}+\mathbf{H}_{1\cdot2}\bar{\boldsymbol{\Gamma}}_{1\cdot2}\mathbf{H}_{1\cdot2}^\top\},\end{aligned}$$

and

$$\begin{aligned}&\Phi_q[\boldsymbol{\tau}+\bar{\boldsymbol{\Gamma}}\mathbf{H}_2^\top\boldsymbol{\Omega}_{22}^{-1}(\mathbf{x}_2-\boldsymbol{\xi}_2)+\bar{\boldsymbol{\Gamma}}(\mathbf{H}_1-\boldsymbol{\Omega}_{12}\boldsymbol{\Omega}_{22}^{-1}\mathbf{H}_2)^\top\boldsymbol{\Omega}_{11\cdot2}^{-1}\{\mathbf{x}_1-\boldsymbol{\xi}_1-\boldsymbol{\Omega}_{12}\boldsymbol{\Omega}_{22}^{-1}(\mathbf{x}_2-\boldsymbol{\xi}_2)\};\\ &\qquad\qquad\mathbf{0}_q,\bar{\boldsymbol{\Gamma}}-\bar{\boldsymbol{\Gamma}}\mathbf{H}^\top\boldsymbol{\Omega}^{-1}\mathbf{H}\bar{\boldsymbol{\Gamma}}]\\ &=\Phi_q[\boldsymbol{\tau}+\bar{\boldsymbol{\Gamma}}_{1\cdot2}\mathbf{H}_2^\top\boldsymbol{\Psi}_{22}^{-1}(\mathbf{x}_2-\boldsymbol{\xi}_2)+\tilde{\boldsymbol{\Gamma}}_{1\cdot2}\mathbf{H}_{1\cdot2}^\top\boldsymbol{\Psi}_{11\cdot2}^{-1}\{\mathbf{x}_1-\boldsymbol{\xi}_1-\boldsymbol{\Omega}_{12}\boldsymbol{\Omega}_{22}^{-1}(\mathbf{x}_2-\boldsymbol{\xi}_2)\};\mathbf{0}_q,\tilde{\boldsymbol{\Gamma}}_{1\cdot2}].\end{aligned}$$

This shows that (6.24) and (6.25) coincide.

Another alternative and perhaps more comprehensive proof of (6.47) is obtained from the stochastic representation (6.44); that is, considering, for each fixed $\mathbf{x}_2\in\mathbb{R}^{p_2}$, the hierarchical representation

$$\begin{aligned}(\mathbf{X}_{1\cdot2}\mid\mathbf{U}_{1\cdot2}=\mathbf{u})&\sim\mathcal{N}_{p_1}(\boldsymbol{\xi}_{1\cdot2}+\mathbf{H}_{1\cdot2}\bar{\boldsymbol{\Gamma}}_{1\cdot2}\mathbf{H}_2^\top\boldsymbol{\Psi}_{22}^{-1}(\mathbf{x}_2-\boldsymbol{\xi}_2)+\mathbf{H}_{1\cdot2}\mathbf{u},\boldsymbol{\Psi}_{11\cdot2}),\\ \mathbf{U}_{1\cdot2}&\sim\mathcal{TN}_q(-\boldsymbol{\tau}_{1\cdot2};\mathbf{0}_q,\bar{\boldsymbol{\Gamma}}_{1\cdot2}),\end{aligned}$$

which now is nonsingular since $\boldsymbol{\Psi}_{11\cdot 2} > 0$. Also, from (6.41), the pdf of $\mathbf{U}_{1\cdot 2}$ becomes

$$g_{1\cdot 2}(\mathbf{u}|\mathbf{x}_2) = \frac{\phi_q(\mathbf{u}; \mathbf{0}_q, \bar{\boldsymbol{\Gamma}}_{1\cdot 2})}{\Phi_q(\boldsymbol{\tau}_{1\cdot 2}; \mathbf{0}_q, \bar{\boldsymbol{\Gamma}}_{1\cdot 2})}, \quad \mathbf{u} + \boldsymbol{\tau}_{1\cdot 2} > \mathbf{0}_q.$$

Therefore, the pdf of $\mathbf{X}_{1\cdot 2}$ becomes

$$\begin{aligned} f_{1\cdot 2}(\mathbf{x}_1 \mid \mathbf{x}_2) &= \int_{\mathbf{u}+\boldsymbol{\tau}_{1\cdot 2}>\mathbf{0}_q} \phi_{p_1}(\mathbf{x}_1; \boldsymbol{\xi}_{1\cdot 2} + \mathbf{H}_{1\cdot 2}\bar{\boldsymbol{\Gamma}}_{1\cdot 2}\mathbf{H}_2^\top \boldsymbol{\Psi}_{22}^{-1}(\mathbf{x}_2 - \boldsymbol{\xi}_2) + \mathbf{H}_{1\cdot 2}\mathbf{u}, \boldsymbol{\Psi}_{11\cdot 2}) g_{1\cdot 2}(\mathbf{u}|\mathbf{x}_2) d\mathbf{u} \\ &= \frac{1}{\Phi_q(\boldsymbol{\tau}_{1\cdot 2}; \mathbf{0}_q, \bar{\boldsymbol{\Gamma}}_{1\cdot 2})} \int_{\mathbf{u}+\boldsymbol{\tau}_{1\cdot 2}>\mathbf{0}_q} \phi_{p_1}(\mathbf{x}_1; \boldsymbol{\xi}_{1\cdot 2} + \mathbf{H}_{1\cdot 2}\mathbf{u}, \boldsymbol{\Psi}_{11\cdot 2}) \phi_q(\mathbf{u}; \mathbf{0}_q, \bar{\boldsymbol{\Gamma}}_{1\cdot 2}) d\mathbf{u}, \end{aligned}$$

where

$$\begin{aligned} &\phi_{p_1}(\mathbf{x}_1; \boldsymbol{\xi}_{1\cdot 2} + \mathbf{H}_{1\cdot 2}\bar{\boldsymbol{\Gamma}}_{1\cdot 2}\mathbf{H}_2^\top \boldsymbol{\Psi}_{22}^{-1}(\mathbf{x}_2 - \boldsymbol{\xi}_2) + \mathbf{H}_{1\cdot 2}\mathbf{u}, \boldsymbol{\Psi}_{11\cdot 2}) \phi_q(\mathbf{u}; \mathbf{0}_q, \bar{\boldsymbol{\Gamma}}_{1\cdot 2}) \\ &\quad = \phi_{p_1}(\mathbf{x}_1; \boldsymbol{\xi}_{1\cdot 2} + \mathbf{H}_{1\cdot 2}\bar{\boldsymbol{\Gamma}}_{1\cdot 2}\mathbf{H}_2^\top \boldsymbol{\Psi}_{22}^{-1}(\mathbf{x}_2 - \boldsymbol{\xi}_2), \boldsymbol{\Psi}_{11\cdot 2} + \mathbf{H}_{1\cdot 2}\bar{\boldsymbol{\Gamma}}_{1\cdot 2}\mathbf{H}_{1\cdot 2}^\top) \\ &\qquad \times \phi_q\{\mathbf{u}; \bar{\boldsymbol{\Gamma}}_{1\cdot 2}\mathbf{H}_{1\cdot 2}^\top(\boldsymbol{\Psi}_{11\cdot 2} + \mathbf{H}_{1\cdot 2}\bar{\boldsymbol{\Gamma}}_{1\cdot 2}\mathbf{H}_{1\cdot 2}^\top)^{-1}(\mathbf{x}_1 - \boldsymbol{\xi}_{1\cdot 2} - \mathbf{H}_{1\cdot 2}\bar{\boldsymbol{\Gamma}}_{1\cdot 2}\mathbf{H}_2^\top \boldsymbol{\Psi}_{22}^{-1}(\mathbf{x}_2 - \boldsymbol{\xi}_2)), \\ &\qquad\qquad \bar{\boldsymbol{\Gamma}}_{1\cdot 2} - \bar{\boldsymbol{\Gamma}}_{1\cdot 2}\mathbf{H}_{1\cdot 2}^\top(\boldsymbol{\Psi}_{11\cdot 2} + \mathbf{H}_{1\cdot 2}\bar{\boldsymbol{\Gamma}}_{1\cdot 2}\mathbf{H}_{1\cdot 2}^\top)^{-1}\mathbf{H}_{1\cdot 2}\bar{\boldsymbol{\Gamma}}_{1\cdot 2}\} \\ &\quad = \phi_{p_1}(\mathbf{x}_1; \boldsymbol{\xi}_{1\cdot 2} + \mathbf{H}_{1\cdot 2}\bar{\boldsymbol{\Gamma}}_{1\cdot 2}\mathbf{H}_2^\top \boldsymbol{\Psi}_{22}^{-1}(\mathbf{x}_2 - \boldsymbol{\xi}_2), \boldsymbol{\Psi}_{11\cdot 2} + \mathbf{H}_{1\cdot 2}\bar{\boldsymbol{\Gamma}}_{1\cdot 2}\mathbf{H}_{1\cdot 2}^\top) \\ &\qquad \times \phi_q\{\mathbf{u}; \boldsymbol{\tau}_{1\cdot 2} + \tilde{\boldsymbol{\Gamma}}_{1\cdot 2}\mathbf{H}_{1\cdot 2}^\top \boldsymbol{\Psi}_{11\cdot 2}^{-1}(\mathbf{x}_1 - \boldsymbol{\xi}_{1\cdot 2} - \mathbf{H}_{1\cdot 2}\bar{\boldsymbol{\tau}}_{1\cdot 2}), \tilde{\boldsymbol{\Gamma}}_{1\cdot 2}\}, \end{aligned}$$

since $\bar{\boldsymbol{\Gamma}}_{1\cdot 2} - \bar{\boldsymbol{\Gamma}}_{1\cdot 2}\mathbf{H}_{1\cdot 2}^\top(\boldsymbol{\Psi}_{11\cdot 2} + \mathbf{H}_{1\cdot 2}\bar{\boldsymbol{\Gamma}}_{1\cdot 2}\mathbf{H}_{1\cdot 2}^\top)^{-1}\mathbf{H}_{1\cdot 2}\bar{\boldsymbol{\Gamma}}_{1\cdot 2} = (\bar{\boldsymbol{\Gamma}}_{1\cdot 2}^{-1} + \mathbf{H}_{1\cdot 2}^\top \boldsymbol{\Psi}_{1\cdot 2}\mathbf{H}_{1\cdot 2})^{-1} = \tilde{\boldsymbol{\Gamma}}_{1\cdot 2}$ and $\bar{\boldsymbol{\Gamma}}_{1\cdot 2}\mathbf{H}_{1\cdot 2}^\top(\boldsymbol{\Psi}_{11\cdot 2} + \mathbf{H}_{1\cdot 2}\bar{\boldsymbol{\Gamma}}_{1\cdot 2}\mathbf{H}_{1\cdot 2}^\top)^{-1} = \tilde{\boldsymbol{\Gamma}}_{1\cdot 2}\mathbf{H}_{1\cdot 2}^\top \boldsymbol{\Psi}_{11\cdot 2}^{-1}$. Hence, by using this last factorization in the integral above, the desired result is obtained, that is,

$$\begin{aligned} f_{1\cdot 2}(\mathbf{x}_1 \mid \mathbf{x}_2) =& \frac{1}{\Phi_q(\boldsymbol{\tau}_{1\cdot 2}; \mathbf{0}_q, \bar{\boldsymbol{\Gamma}}_{1\cdot 2})} \phi_{p_1}(\mathbf{x}_1; \boldsymbol{\xi}_{1\cdot 2} + \mathbf{H}_{1\cdot 2}\bar{\boldsymbol{\Gamma}}_{1\cdot 2}\mathbf{H}_2^\top \boldsymbol{\Psi}_{22}^{-1}(\mathbf{x}_2 - \boldsymbol{\xi}_2), \boldsymbol{\Psi}_{11\cdot 2} + \bar{\mathbf{H}}_{1\cdot 2}\bar{\bar{\boldsymbol{\Gamma}}}_{1\cdot 2}\bar{\mathbf{H}}_{1\cdot 2}^\top) \\ &\times \int_{\mathbf{u}+\boldsymbol{\tau}_{1\cdot 2}>\mathbf{0}_q} \phi_q\{\mathbf{u}; \tilde{\boldsymbol{\Gamma}}_{1\cdot 2}\mathbf{H}_{1\cdot 2}^\top \boldsymbol{\Psi}_{11\cdot 2}^{-1}(\mathbf{x}_1 - \boldsymbol{\xi}_{1\cdot 2} - \mathbf{H}_{1\cdot 2}\bar{\boldsymbol{\Gamma}}_{1\cdot 2}\mathbf{H}_2^\top \boldsymbol{\Psi}_{22}^{-1}(\mathbf{x}_2 - \boldsymbol{\xi}_2)), \tilde{\boldsymbol{\Gamma}}_{1\cdot 2}\} d\mathbf{u} \\ =& \frac{1}{\Phi_q(\boldsymbol{\tau}_{1\cdot 2}; \mathbf{0}_q, \bar{\boldsymbol{\Gamma}}_{1\cdot 2})} \phi_{p_1}(\mathbf{x}_1; \boldsymbol{\xi}_{1\cdot 2} + \mathbf{H}_{1\cdot 2}\bar{\boldsymbol{\Gamma}}_{1\cdot 2}\mathbf{H}_2^\top \boldsymbol{\Psi}_{22}^{-1}(\mathbf{x}_2 - \boldsymbol{\xi}_2), \boldsymbol{\Psi}_{11\cdot 2} + \mathbf{H}_{1\cdot 2}\bar{\bar{\boldsymbol{\Gamma}}}_{1\cdot 2}\mathbf{H}_{1\cdot 2}^\top) \\ &\times \Phi_q\{\boldsymbol{\tau}_{1\cdot 2} + \tilde{\boldsymbol{\Gamma}}_{1\cdot 2}\mathbf{H}_{1\cdot 2}^\top \boldsymbol{\Psi}_{11\cdot 2}^{-1}(\mathbf{x}_1 - \boldsymbol{\xi}_{1\cdot 2} - \mathbf{H}_{1\cdot 2}\bar{\boldsymbol{\Gamma}}_{1\cdot 2}\mathbf{H}_2^\top \boldsymbol{\Psi}_{22}^{-1}(\mathbf{x}_2 - \boldsymbol{\xi}_2)); \mathbf{0}_q, \tilde{\boldsymbol{\Gamma}}_{1\cdot 2}\}, \ \mathbf{x}_1 \in \mathbb{R}^{p_1}. \end{aligned}$$

6.6.3 *Multivariate extended skew-normal distribution*

When $q = 1$, we have $\bar{\boldsymbol{\Gamma}} = 1$ and the multivariate unified skew-normal distribution becomes the so-called *multivariate extended skew-normal (ESN) distribution*, and we write $\mathbf{X} \sim \mathcal{ESN}_p(\boldsymbol{\xi}, \boldsymbol{\Psi}, \boldsymbol{\eta}, \tau)$; stochastically, this distribution can be represented as

$$\mathbf{X} = \boldsymbol{\xi} + \boldsymbol{\eta} U + \mathbf{W}, \tag{6.49}$$

where U and $\mathbf{W}$ are independent random quantities, with $U \stackrel{d}{=} (Z \mid Z + \tau > 0)$, with $Z \sim \mathcal{N}(0, 1)$, and $\mathbf{W} \sim \mathcal{N}_p(\mathbf{0}_p, \boldsymbol{\Psi})$.

The cf of $\mathbf{X} \sim \mathcal{ESN}_p(\boldsymbol{\xi}, \boldsymbol{\Psi}, \boldsymbol{\eta}, \tau)$ is given by

$$\psi_{\mathrm{ESN}}(\mathbf{t}) = \frac{1}{\Phi(\tau)} e^{\mathrm{i}\mathbf{t}^\top \boldsymbol{\xi} - \frac{1}{2}\mathbf{t}^\top(\boldsymbol{\Psi} + \boldsymbol{\eta}\boldsymbol{\eta}^\top)\mathbf{t}} \Phi(\tau + \mathrm{i}\boldsymbol{\eta}^\top \mathbf{t}), \quad \mathbf{t} \in \mathbb{R}^p,$$

from where it is straightforward to show that

$$\mathrm{E}(\mathbf{X}) = \boldsymbol{\xi} + \zeta(\tau)\boldsymbol{\eta}, \qquad \mathrm{var}(\mathbf{X}) = \boldsymbol{\Psi} + [1 - \zeta(\tau)\{\tau + \zeta(\tau)\}]\boldsymbol{\eta}\boldsymbol{\eta}^\top,$$

where $\zeta(z) = \phi(z)/\Phi(z)$. Another way to obtain these results is to use (6.49) and to consider the first and second moments of $U \sim \mathcal{TN}(-\tau;0,1)$.

In the nonsingular case with $\boldsymbol{\Psi} > 0$, the pdf of $\mathbf{X} \sim \mathcal{ESN}_p(\boldsymbol{\xi}, \boldsymbol{\Psi}, \boldsymbol{\eta}, \tau)$ is given by

$$f_{\text{ESN}}(\mathbf{x}) = \frac{1}{\Phi(\tau)} \phi_p(\mathbf{x}; \boldsymbol{\xi}, \boldsymbol{\Psi} + \boldsymbol{\eta}\boldsymbol{\eta}^\top)\Phi\left\{\frac{\tau + \boldsymbol{\eta}^\top\boldsymbol{\Psi}^{-1}(\mathbf{x}-\boldsymbol{\xi})}{\sqrt{1+\boldsymbol{\eta}^\top\boldsymbol{\Psi}^{-1}\boldsymbol{\eta}}}\right\}, \quad \mathbf{x} \in \mathbb{R}^p. \tag{6.50}$$

In the univariate case, statistical aspects of the extended skew-normal distribution can be found in Canale (2011, 2015).

6.7 Quadratic Forms and Some Multivariate Moments

6.7.1 Distribution of quadratic forms

In this section, we explore the distribution of quadratic forms of unified skew-normal random vectors. In particular, we give the conditions under which these quadratic forms preserve the chi-squared distribution. Some related results can be found in Arellano-Valle and Azzalini (2006), Arellano-Valle and Genton (2010a), and Aziz and Gupta (2013). We start with the following general result.

Theorem 108 *Suppose $\mathbf{X} \sim \mathcal{SUN}_{p,q}(\boldsymbol{\xi}, \boldsymbol{\Psi}, \mathbf{H}, \boldsymbol{\tau}, \bar{\boldsymbol{\Gamma}})$ is nonsingular ($\boldsymbol{\Psi} > 0$). Then, the characteristic function of the quadratic form $Q_A(\mathbf{X}-\mathbf{a}) = (\mathbf{X}-\mathbf{a})^\top\mathbf{A}(\mathbf{X}-\mathbf{a})$, where $\mathbf{A}$ is a fixed $p \times p$ symmetric matrix and $\mathbf{a}$ is a fixed $p \times 1$ vector, is, $\forall t \in \mathbb{R}$:*

$$\begin{aligned}\text{E}\{e^{\text{i}tQ_A(\mathbf{X}-\mathbf{a})}\} =& |\mathbf{I}_p - 2\text{i}t\mathbf{A}\boldsymbol{\Omega}|^{-1/2} e^{\text{i}t\boldsymbol{\xi}_a^\top(\mathbf{I}_p - 2\text{i}t\mathbf{A}\boldsymbol{\Omega})^{-1}\mathbf{A}\boldsymbol{\xi}_a} \\ &\times \frac{\Phi_q\{\boldsymbol{\tau} + 2\text{i}t\bar{\boldsymbol{\Gamma}}\mathbf{H}^\top(\mathbf{I}_p - 2\text{i}t\mathbf{A}\boldsymbol{\Omega})^{-1}\mathbf{A}\boldsymbol{\xi}_a; \mathbf{0}_q, \bar{\boldsymbol{\Gamma}} + 2\text{i}t\bar{\boldsymbol{\Gamma}}\mathbf{H}^\top(\mathbf{I}_p - 2\text{i}t\mathbf{A}\boldsymbol{\Omega})^{-1}\mathbf{A}\mathbf{H}\bar{\boldsymbol{\Gamma}}\}}{\Phi_q(\boldsymbol{\tau}; \mathbf{0}_q, \bar{\boldsymbol{\Gamma}})},\end{aligned} \tag{6.51}$$

where $\boldsymbol{\xi}_a = \boldsymbol{\xi} - \mathbf{a}$ and $\boldsymbol{\Omega} = \boldsymbol{\Psi} + \mathbf{H}\bar{\boldsymbol{\Gamma}}\mathbf{H}^\top$.

Proof 108 *Since $\mathbf{X}$ has a pdf given by (6.24), then from the change of variables $\mathbf{z} = \mathbf{x} - \mathbf{a}$, we have*

$$\begin{aligned}\text{E}\{e^{\text{i}t(\mathbf{X}-\mathbf{a})^\top\mathbf{A}(\mathbf{X}-\mathbf{a})}\} =& \frac{1}{\Phi_q(\boldsymbol{\tau}; \mathbf{0}_q, \bar{\boldsymbol{\Gamma}})} \int_{\mathbb{R}^p} e^{\text{i}t(\mathbf{x}-\mathbf{a})^\top\mathbf{A}(\mathbf{x}-\mathbf{a})} \phi_p(\mathbf{x}; \boldsymbol{\xi}, \boldsymbol{\Omega})\Phi_q\{\boldsymbol{\tau} + \tilde{\boldsymbol{\Gamma}}\mathbf{H}^\top\boldsymbol{\Psi}^{-1}(\mathbf{x}-\boldsymbol{\xi}); \mathbf{0}_q, \tilde{\boldsymbol{\Gamma}}\}\text{d}\mathbf{x} \\ =& \frac{1}{\Phi_q(\boldsymbol{\tau}; \mathbf{0}_q, \bar{\boldsymbol{\Gamma}})} \int_{\mathbb{R}^p} e^{\text{i}t\mathbf{z}^\top\mathbf{A}\mathbf{z}} \phi_p(\mathbf{z}; \boldsymbol{\xi}_a, \boldsymbol{\Omega})\Phi_q\{\boldsymbol{\tau} + \tilde{\boldsymbol{\Gamma}}\mathbf{H}^\top\boldsymbol{\Psi}^{-1}(\mathbf{z}-\boldsymbol{\xi}_a); \mathbf{0}_q, \tilde{\boldsymbol{\Gamma}}\}\text{d}\mathbf{z}\end{aligned}$$

where, as before, $\boldsymbol{\Omega} = \boldsymbol{\Psi} + \mathbf{H}\bar{\boldsymbol{\Gamma}}\mathbf{H}^\top$, $\tilde{\boldsymbol{\Gamma}} = (\bar{\boldsymbol{\Gamma}}^{-1} + \mathbf{H}^\top\boldsymbol{\Psi}^{-1}\mathbf{H})^{-1}$, and $\boldsymbol{\xi}_a = \boldsymbol{\xi} - \mathbf{a}$. Also, after some algebra, we obtain the relation:

$$\begin{aligned}e^{\text{i}t\mathbf{z}^\top\mathbf{A}\mathbf{z}}\phi_p(\mathbf{z}; \boldsymbol{\xi}_a, \boldsymbol{\Omega}) =& e^{-\frac{1}{2}\boldsymbol{\xi}_a^\top\boldsymbol{\Lambda}(t)\boldsymbol{\xi}_a}|\boldsymbol{\Omega}|^{-1/2}|\boldsymbol{\Omega}^{-1} - 2\text{i}t\mathbf{A}|^{-1/2} \\ &\times \phi_p\{\mathbf{z}; (\boldsymbol{\Omega}^{-1} - 2\text{i}t\mathbf{A})^{-1}\boldsymbol{\Omega}^{-1}\boldsymbol{\xi}_a, (\boldsymbol{\Omega}^{-1} - 2\text{i}t\mathbf{A})^{-1}\},\end{aligned}$$

where $\boldsymbol{\Lambda}(t) = \boldsymbol{\Omega}^{-1} - \boldsymbol{\Omega}^{-1}(\boldsymbol{\Omega}^{-1} - 2\text{i}t\mathbf{A})^{-1}\boldsymbol{\Omega}^{-1}$. Using this last result in the above integral and recalling the relations $\bar{\boldsymbol{\Gamma}}\mathbf{H}^\top\boldsymbol{\Psi}^{-1} = \bar{\boldsymbol{\Gamma}}\mathbf{H}^\top\boldsymbol{\Omega}^{-1}$ and $(\bar{\boldsymbol{\Gamma}}^{-1} + \mathbf{H}^\top\boldsymbol{\Psi}^{-1}\mathbf{H})^{-1} = \bar{\boldsymbol{\Gamma}} - \bar{\boldsymbol{\Gamma}}\mathbf{H}^\top\boldsymbol{\Omega}^{-1}\mathbf{H}\bar{\boldsymbol{\Gamma}}$ and *also Theorem 90, we find that*

$$\begin{aligned}
\mathrm{E}\{e^{it(\mathbf{X}-\mathbf{a})^\top \mathbf{A}(\mathbf{X}-\mathbf{a})}\} =& e^{-\frac{1}{2}\boldsymbol{\xi}_a^\top \boldsymbol{\Lambda}(t)\boldsymbol{\xi}_a} |\boldsymbol{\Omega}|^{-1/2}|\boldsymbol{\Omega}^{-1} - 2it\mathbf{A}|^{-1/2} \\
&\times \frac{1}{\Phi_q(\boldsymbol{\tau};\mathbf{0}_q,\bar{\boldsymbol{\Gamma}})} \int_{\mathbb{R}^p} \phi_p\{\mathbf{z};(\boldsymbol{\Omega}^{-1}-2it\mathbf{A})^{-1}\boldsymbol{\Omega}^{-1}\boldsymbol{\xi}_a,(\boldsymbol{\Omega}^{-1}-2it\mathbf{A})^{-1}\} \\
&\times \Phi_q\{\boldsymbol{\tau}+\bar{\boldsymbol{\Gamma}}\mathbf{H}^\top\boldsymbol{\Omega}^{-1}(\mathbf{z}-\boldsymbol{\xi}_a);\mathbf{0}_q,\bar{\boldsymbol{\Gamma}}-\bar{\boldsymbol{\Gamma}}\mathbf{H}^\top\boldsymbol{\Omega}^{-1}\mathbf{H}\bar{\boldsymbol{\Gamma}}\}d\mathbf{z} \\
=& |\mathbf{I}_p - 2it\mathbf{A}\boldsymbol{\Omega}|^{-1/2} e^{-\frac{1}{2}\boldsymbol{\xi}_a^\top \boldsymbol{\Lambda}(t)\boldsymbol{\xi}_a} \frac{\Phi_q\{\boldsymbol{\tau}-\bar{\boldsymbol{\Gamma}}\mathbf{H}^\top\boldsymbol{\Lambda}(t)\boldsymbol{\xi}_a;\mathbf{0}_q,\bar{\boldsymbol{\Gamma}}-\bar{\boldsymbol{\Gamma}}\mathbf{H}^\top\boldsymbol{\Lambda}(t)\mathbf{H}\bar{\boldsymbol{\Gamma}}\}}{\Phi_q(\boldsymbol{\tau};\mathbf{0}_q,\bar{\boldsymbol{\Gamma}})}.
\end{aligned}$$

Thus, the proof follows by noting that $\boldsymbol{\Lambda}(t) = \boldsymbol{\Omega}^{-1} - (\boldsymbol{\Omega} - 2it\boldsymbol{\Omega}\mathbf{A}\boldsymbol{\Omega})^{-1} = -2it(\mathbf{I}_p - 2it\mathbf{A}\boldsymbol{\Omega})^{-1}\mathbf{A}$. □

Some relevant special cases are as follows:

[1] If $\boldsymbol{\xi}_a = \mathbf{0}_p$, then (6.51) reduces to

$$\mathrm{E}\{e^{itQ_A(\mathbf{X}-\mathbf{a})}\} = |\mathbf{I}_p - 2it\mathbf{A}\boldsymbol{\Omega}|^{-1/2} \frac{\Phi_q\{\boldsymbol{\tau};\mathbf{0}_q,\bar{\boldsymbol{\Gamma}}+2it\bar{\boldsymbol{\Gamma}}\mathbf{H}^\top(\mathbf{I}_p-2it\mathbf{A}\boldsymbol{\Omega})^{-1}\mathbf{A}\mathbf{H}\bar{\boldsymbol{\Gamma}}\}}{\Phi_q(\boldsymbol{\tau};\mathbf{0}_q,\bar{\boldsymbol{\Gamma}})},$$

which agrees with the result obtained in Arellano-Valle and Azzalini (2006) for the moment generating function of the unified skew-normal distribution.

[2] If $\mathbf{a} = \boldsymbol{\xi}$ and $\mathbf{A}\mathbf{H} = \mathbf{O}$, then (6.51) becomes

$$\mathrm{E}\{e^{itQ_A(\mathbf{X}-\boldsymbol{\xi})}\} = |\mathbf{I}_p - 2it\mathbf{A}\boldsymbol{\Psi}|^{-1/2}.$$

If, in addition, $\mathbf{A}\boldsymbol{\Psi}$ is an idempotent matrix of rank $r \le p$ (which is equivalent to $\mathbf{A}\boldsymbol{\Psi}\mathbf{A} = \mathbf{A}$ and $\mathrm{rank}(\mathbf{A}) = r$ since $\boldsymbol{\Psi}$ is nonsingular), then $|\mathbf{I}_p - 2it\mathbf{A}\boldsymbol{\Psi}|^{-1/2} = (1-2it)^{-r/2}$ and so

$$Q_A(\mathbf{X}-\boldsymbol{\xi}) = (\mathbf{X}-\boldsymbol{\xi})^\top\mathbf{A}(\mathbf{X}-\boldsymbol{\xi}) \sim \chi^2_r.$$

[3] If $\mathbf{a} = \boldsymbol{\xi}$, $\mathbf{A} = \boldsymbol{\Omega}^{-1} = (\boldsymbol{\Psi} + \mathbf{H}\bar{\boldsymbol{\Gamma}}\mathbf{H}^\top)^{-1}$, $\boldsymbol{\tau} = \mathbf{0}_q$, $\bar{\boldsymbol{\Gamma}} = \mathbf{I}_p$ and $\mathbf{H}^\top\boldsymbol{\Psi}^{-1}\mathbf{H}$ is a diagonal matrix, then $\mathrm{E}\{e^{itQ_A(\mathbf{X}-\boldsymbol{\xi})}\} = (1-2it)^{-p/2}$, so that

$$Q_A(\mathbf{X}-\boldsymbol{\xi}) = (\mathbf{X}-\boldsymbol{\xi})^\top(\boldsymbol{\Psi}+\mathbf{H}\mathbf{H}^\top)^{-1}(\mathbf{X}-\boldsymbol{\xi}) \sim \chi^2_p.$$

Example 81 *Let* $\mathbf{X} \sim \mathcal{SUN}_{p,q}(\boldsymbol{\xi},\mathbf{I}_p,\mathbf{H},\boldsymbol{\tau},\bar{\boldsymbol{\Gamma}})$. *Suppose that the skewness matrix* $\mathbf{H}$ *has rank* $q \le p$ *and let* $\mathbf{P}$ *be a* $p \times p$ *projection matrix of rank* q *such that* $\mathbf{P}\mathbf{H} = \mathbf{H}$. *Then, for* $\mathbf{A} = \mathbf{I}_p - \mathbf{P}$ *we have* $Q_A(\mathbf{X}-\boldsymbol{\xi}) = (\mathbf{X}-\boldsymbol{\xi})^\top(\mathbf{I}_p-\mathbf{P})(\mathbf{X}-\boldsymbol{\xi}) = \|\mathbf{X}-\boldsymbol{\xi}\|^2 - Q_P(\mathbf{X}-\boldsymbol{\xi}) \sim \chi^2_{p-q}$, *since* $\mathbf{A}\mathbf{H} = (\mathbf{I}_p-\mathbf{P})\mathbf{H} = \mathbf{H}-\mathbf{H} = \mathbf{O}$ *and* $\mathbf{A}\boldsymbol{\Psi} = \mathbf{I}_p - \mathbf{P}$ *is also a projection matrix of rank* $p-q$, *and so the matrix* $\mathbf{A}\boldsymbol{\Psi}$ *is idempotent and has rank* $r = p-q$.

6.7.2 *Higher-order moments*

Although in principle successive differentiations of the cf could be considered to compute higher-order moments of $\mathbf{X} \sim \mathcal{SUN}_{p,q}(\boldsymbol{\xi},\boldsymbol{\Psi},\mathbf{H},\boldsymbol{\tau},\bar{\boldsymbol{\Gamma}})$, this process becomes algebraically cumbersome. We therefore follow another route, based on the additive representation (6.1) used to define the unified skew-normal random vector $\mathbf{X}$. For this, we need some additional definitions and preliminary results. The main results of this section are based on Arellano-Valle and Azzalini (2022).

For a p-dimensional random vector $\mathbf{X}$, define its moments up to the fourth order as

$$\begin{aligned}
\boldsymbol{\mu}_1(\mathbf{X}) &= \mathrm{E}(\mathbf{X}),\\
\boldsymbol{\mu}_2(\mathbf{X}) &= \mathrm{E}(\mathbf{X}\otimes\mathbf{X}^\top) = \mathrm{E}(\mathbf{X}\mathbf{X}^\top),\\
\boldsymbol{\mu}_3(\mathbf{X}) &= \mathrm{E}(\mathbf{X}\otimes\mathbf{X}^\top\otimes\mathbf{X}) = \mathrm{E}(\mathbf{X}\otimes\mathbf{X}\mathbf{X}^\top) = \mathrm{E}(\mathbf{X}\mathbf{X}^\top\otimes\mathbf{X}) = \mathrm{E}\{\mathrm{vec}(\mathbf{X}\mathbf{X}^\top)\mathbf{X}^\top\},\\
\boldsymbol{\mu}_4(\mathbf{X}) &= \mathrm{E}(\mathbf{X}\otimes\mathbf{X}^\top\otimes\mathbf{X}\otimes\mathbf{X}^\top) = \mathrm{E}(\mathbf{X}\mathbf{X}^\top\otimes\mathbf{X}\mathbf{X}^\top) = \mathrm{E}\{\mathrm{vec}(\mathbf{X}\mathbf{X}^\top)\mathrm{vec}(\mathbf{X}\mathbf{X}^\top)^\top\},
\end{aligned}$$

provided that the expected values involved exist. The equivalence of the various expressions for a given moment follows from standard properties of the Kronecker product; see Chapter 1 and Appendix A.2.9.

In addition, the following notation will be used, adopted from Magnus and Neudecker (1999). Let $\mathbf{e}_{i:s}$ be the i-th s-dimensional unit vector, and define $\mathbf{E}_{ij} = \mathbf{e}_{i:p}\mathbf{e}_{j:q}^\top$; when $p = q$, we will remove the index indicating dimensions from these unit vectors. Furthermore, denote by

$$\mathbf{K}_{pq} = \sum_{i=1}^{p}\sum_{j=1}^{q}\mathbf{E}_{ij}\otimes\mathbf{E}_{ij}^\top$$

the pq-by-pq square commutation matrix and let $\mathbf{K}_r = \mathbf{K}_{rr}$. Note that $\mathbf{K}_{pq}^\top = \mathbf{K}_{qp} = \mathbf{K}_{pq}^{-1}$.

Now, consider the following preliminary results.

Theorem 109 *If* $\mathbf{W}_0 \sim \mathcal{N}_p(\mathbf{0}_p, \mathbf{I}_p)$, *then:*

(i) $\mathrm{E}(\mathbf{W}_0\otimes\mathbf{W}_0^\top) = \mathrm{E}(\mathbf{W}_0\mathbf{W}_0^\top) = \mathbf{I}_p$,

(ii) $\mathrm{E}(\mathbf{W}_0\mathbf{W}_0^\top\otimes\mathbf{W}_0) = \mathrm{E}(\mathbf{W}_0\otimes\mathbf{W}_0\mathbf{W}_0^\top) = \mathbf{O}$,

(iii) $\mathrm{E}(\mathbf{W}_0\mathbf{W}_0^\top\otimes\mathbf{W}_0\mathbf{W}_0^\top) = \mathbf{I}_{p^2} + \mathbf{K}_p + \mathrm{vec}(\mathbf{I}_p)\mathrm{vec}(\mathbf{I}_p)^\top$,

(iv) $\mathrm{var}\{\mathrm{vec}(\mathbf{W}_0\mathbf{W}_0^\top)\} = \mathbf{I}_{p^2} + \mathbf{K}_p$.

Proof 109 *The proof of statements (i) and (ii) is direct. For (iii) and (iv), see, for instance, Theorem 4.1 (i) and Lemma 4.1 (ii) of Magnus and Neudecker (1979).* □

The following theorem is related to the moments of a linear combination of two independent random vectors.

Theorem 110 *Let* $\mathbf{X} = \mathbf{A}\mathbf{U} + \mathbf{B}\mathbf{V}$, *where* $\mathbf{A}\in\mathbb{R}^{p\times q}$ *and* $\mathbf{B}\in\mathbb{R}^{p\times r}$ *are constant matrices, and* $\mathbf{U}\in\mathbb{R}^q$ *and* $\mathbf{V}\in\mathbb{R}^r$ *are independent random vectors. If the required moments exist, then*

$$\begin{aligned}
\boldsymbol{\mu}_1(\mathbf{X}) &= \mathbf{A}\boldsymbol{\mu}_1(\mathbf{U}) + \mathbf{B}\boldsymbol{\mu}_1(\mathbf{V}),\\
\boldsymbol{\mu}_2(\mathbf{X}) &= \mathbf{A}\boldsymbol{\mu}_2(\mathbf{U})\mathbf{A}^\top + \mathbf{A}\boldsymbol{\mu}_1(\mathbf{U})\boldsymbol{\mu}_1(\mathbf{V})^\top\mathbf{B}^\top + \mathbf{B}\boldsymbol{\mu}_1(\mathbf{V})\boldsymbol{\mu}_1(\mathbf{U})^\top\mathbf{A}^\top + \mathbf{B}\boldsymbol{\mu}_2(\mathbf{V})\mathbf{B}^\top,\\
\boldsymbol{\mu}_3(\mathbf{X}) &= (\mathbf{A}\otimes\mathbf{A})\boldsymbol{\mu}_3(\mathbf{U})\mathbf{A}^\top + (\mathbf{A}\otimes\mathbf{A})\mathrm{vec}\{\boldsymbol{\mu}_2(\mathbf{U})\}\boldsymbol{\mu}_1(\mathbf{V})^\top\mathbf{B}^\top\\
&\quad + (\mathbf{I}_{p^2} + \mathbf{K}_p)(\mathbf{A}\otimes\mathbf{B})\left\{(\boldsymbol{\mu}_2(\mathbf{U})\otimes\boldsymbol{\mu}_1(\mathbf{V}))\mathbf{A}^\top + (\boldsymbol{\mu}_1(\mathbf{U})\otimes\boldsymbol{\mu}_2(\mathbf{V}))\mathbf{B}^\top\right\}\\
&\quad + (\mathbf{B}\otimes\mathbf{B})\mathrm{vec}\{\boldsymbol{\mu}_2(\mathbf{V})\}\boldsymbol{\mu}_1(\mathbf{U})^\top\mathbf{A}^\top + (\mathbf{B}\otimes\mathbf{B})\boldsymbol{\mu}_3(\mathbf{V})\mathbf{B}^\top,
\end{aligned}$$

and

$$\begin{aligned}
\boldsymbol{\mu}_4(\mathbf{X}) &= (\mathbf{A}\otimes\mathbf{A})\boldsymbol{\mu}_4(\mathbf{U})(\mathbf{A}\otimes\mathbf{A})^\top \\
&+(\mathbf{A}\otimes\mathbf{A})\boldsymbol{\mu}_3(\mathbf{U})(\mathbf{I}_q\otimes\boldsymbol{\mu}_1(\mathbf{V}))^\top(\mathbf{A}\otimes\mathbf{B})^\top \\
&+\mathbf{K}_p(\mathbf{A}\otimes\mathbf{A})\boldsymbol{\mu}_3(\mathbf{U})(\mathbf{I}_q\otimes\boldsymbol{\mu}_1(\mathbf{V}))^\top(\mathbf{A}\otimes\mathbf{B})^\top\mathbf{K}_p \\
&+(\mathbf{A}\otimes\mathbf{B})(\mathbf{I}_q\otimes\boldsymbol{\mu}_1(\mathbf{V}))\boldsymbol{\mu}_3(\mathbf{U})^\top(\mathbf{A}\otimes\mathbf{A})^\top \\
&+\mathbf{K}_p(\mathbf{A}\otimes\mathbf{B})(\mathbf{I}_q\otimes\boldsymbol{\mu}_1(\mathbf{V}))\boldsymbol{\mu}_3(\mathbf{U})^\top(\mathbf{A}\otimes\mathbf{A})^\top\mathbf{K}_p \\
&+(\mathbf{A}\otimes\mathbf{B})(\boldsymbol{\mu}_2(\mathbf{U})\otimes\boldsymbol{\mu}_2(\mathbf{V}))(\mathbf{A}\otimes\mathbf{B})^\top \\
&+(\mathbf{A}\otimes\mathbf{B})(\boldsymbol{\mu}_2(\mathbf{U})\otimes\boldsymbol{\mu}_2(\mathbf{V}))(\mathbf{A}\otimes\mathbf{B})^\top\mathbf{K}_p \\
&+\mathbf{K}_p(\mathbf{A}\otimes\mathbf{B})(\boldsymbol{\mu}_2(\mathbf{U})\otimes\boldsymbol{\mu}_2(\mathbf{V}))(\mathbf{A}\otimes\mathbf{B})^\top \\
&+\mathbf{K}_p(\mathbf{A}\otimes\mathbf{B})(\boldsymbol{\mu}_2(\mathbf{U})\otimes\boldsymbol{\mu}_2(\mathbf{V}))(\mathbf{A}\otimes\mathbf{B})^\top\mathbf{K}_p \\
&+(\mathbf{A}\otimes\mathbf{A})\mathrm{vec}\{\boldsymbol{\mu}_2(\mathbf{U})\}\mathrm{vec}\{\boldsymbol{\mu}_2(\mathbf{V})\}^\top(\mathbf{B}\otimes\mathbf{B})^\top \\
&+(\mathbf{B}\otimes\mathbf{B})\mathrm{vec}\{\boldsymbol{\mu}_2(\mathbf{V})\}\mathrm{vec}\{\boldsymbol{\mu}_2(\mathbf{U})\}^\top(\mathbf{A}\otimes\mathbf{A})^\top \\
&+(\mathbf{A}\otimes\mathbf{B})(\boldsymbol{\mu}_1(\mathbf{U})\otimes\mathbf{I}_r)\boldsymbol{\mu}_3(\mathbf{V})^\top(\mathbf{B}\otimes\mathbf{B})^\top \\
&+\mathbf{K}_p(\mathbf{A}\otimes\mathbf{B})(\boldsymbol{\mu}_1(\mathbf{U})\otimes\mathbf{I}_r)\boldsymbol{\mu}_3(\mathbf{V})^\top(\mathbf{B}\otimes\mathbf{B})^\top\mathbf{K}_p \\
&+(\mathbf{B}\otimes\mathbf{B})\boldsymbol{\mu}_3(\mathbf{V})(\boldsymbol{\mu}_1(\mathbf{U})\otimes\mathbf{I}_r)^\top(\mathbf{A}\otimes\mathbf{B})^\top \\
&+\mathbf{K}_p(\mathbf{B}\otimes\mathbf{B})\boldsymbol{\mu}_3(\mathbf{V})(\boldsymbol{\mu}_1(\mathbf{U})\otimes\mathbf{I}_r)^\top(\mathbf{A}\otimes\mathbf{B})^\top\mathbf{K}_p \\
&+(\mathbf{B}\otimes\mathbf{B})\boldsymbol{\mu}_4(\mathbf{V})(\mathbf{B}\otimes\mathbf{B})^\top.
\end{aligned}$$

In particular, if $\boldsymbol{\mu}_1(\mathbf{V})$ and $\boldsymbol{\mu}_3(\mathbf{V})$ are zero, then

$$\begin{aligned}
\boldsymbol{\mu}_1(\mathbf{X}) &= \mathbf{A}\boldsymbol{\mu}_1(\mathbf{U}), \\
\boldsymbol{\mu}_2(\mathbf{X}) &= \mathbf{A}\boldsymbol{\mu}_2(\mathbf{U})\mathbf{A}^\top+\mathbf{B}\boldsymbol{\mu}_2(\mathbf{V})\mathbf{B}^\top, \\
\boldsymbol{\mu}_3(\mathbf{X}) &= (\mathbf{A}\otimes\mathbf{A})\boldsymbol{\mu}_3(\mathbf{U})\mathbf{A}^\top+(\mathbf{I}_{p^2}+\mathbf{K}_p)(\mathbf{A}\otimes\mathbf{B})(\boldsymbol{\mu}_1(\mathbf{U})\otimes\boldsymbol{\mu}_2(\mathbf{V}))\mathbf{B}^\top \\
&+(\mathbf{B}\otimes\mathbf{B})\mathrm{vec}\{\boldsymbol{\mu}_2(\mathbf{V})\}\boldsymbol{\mu}_1(\mathbf{U})^\top\mathbf{A}^\top, \\
\boldsymbol{\mu}_4(\mathbf{X}) &= (\mathbf{A}\otimes\mathbf{A})\boldsymbol{\mu}_4(\mathbf{U})(\mathbf{A}\otimes\mathbf{A})^\top+(\mathbf{I}_{p^2}+\mathbf{K}_p)\{(\mathbf{A}\otimes\mathbf{B})(\boldsymbol{\mu}_2(\mathbf{U})\otimes\boldsymbol{\mu}_2(\mathbf{V}))(\mathbf{A}\otimes\mathbf{B})^\top \\
&+(\mathbf{B}\otimes\mathbf{A})(\boldsymbol{\mu}_2(\mathbf{V})\otimes\boldsymbol{\mu}_2(\mathbf{U}))(\mathbf{B}\otimes\mathbf{A})^\top\}+(\mathbf{A}\otimes\mathbf{A})\mathrm{vec}\{\boldsymbol{\mu}_2(\mathbf{U})\}\mathrm{vec}\{\boldsymbol{\mu}_2(\mathbf{V})\}^\top(\mathbf{B}\otimes\mathbf{B})^\top \\
&+(\mathbf{B}\otimes\mathbf{B})\mathrm{vec}\{\boldsymbol{\mu}_2(\mathbf{V})\}\mathrm{vec}\{\boldsymbol{\mu}_2(\mathbf{U})\}^\top(\mathbf{A}\otimes\mathbf{A})^\top+(\mathbf{B}\otimes\mathbf{B})\boldsymbol{\mu}_4(\mathbf{V})(\mathbf{B}\otimes\mathbf{B})^\top.
\end{aligned}$$

Moreover,

$$\begin{aligned}
\mathrm{var}(\mathbf{X}) &= \boldsymbol{\mu}_2(\mathbf{X})-\boldsymbol{\mu}_1(\mathbf{X})\boldsymbol{\mu}_1(\mathbf{X})^\top \\
&= \mathbf{A}\,\mathrm{var}(\mathbf{U})\mathbf{A}^\top+\mathbf{B}\,\mathrm{var}(\mathbf{V})\mathbf{B}^\top, \\
\mathrm{cov}(\mathbf{X}\otimes\mathbf{X},\mathbf{X}) &= \boldsymbol{\mu}_3(\mathbf{X})-\mathrm{vec}\{\boldsymbol{\mu}_2(\mathbf{X})\}\boldsymbol{\mu}_1(\mathbf{X})^\top \\
&= (\mathbf{A}\otimes\mathbf{A})\mathrm{cov}(\mathbf{U}\otimes\mathbf{U},\mathbf{U})\mathbf{A}^\top+(\mathbf{I}_{p^2}+\mathbf{K}_p)(\mathbf{A}\otimes\mathbf{B})(\boldsymbol{\mu}_1(\mathbf{U})\otimes\boldsymbol{\mu}_2(\mathbf{V}))\mathbf{B}^\top, \\
\mathrm{var}(\mathbf{X}\otimes\mathbf{X}) &= \boldsymbol{\mu}_4(\mathbf{X})-\mathrm{vec}\{\boldsymbol{\mu}_2(\mathbf{X})\}\mathrm{vec}\{\boldsymbol{\mu}_2(\mathbf{X})\}^\top \\
&= (\mathbf{A}\otimes\mathbf{A})\mathrm{var}(\mathbf{U}\otimes\mathbf{U})(\mathbf{A}\otimes\mathbf{A})^\top+(\mathbf{B}\otimes\mathbf{B})\mathrm{var}(\mathbf{V}\otimes\mathbf{V})(\mathbf{B}\otimes\mathbf{B})^\top \\
&+(\mathbf{I}_{p^2}+\mathbf{K}_p)\{(\mathbf{A}\otimes\mathbf{B})(\boldsymbol{\mu}_2(\mathbf{U})\otimes\boldsymbol{\mu}_2(\mathbf{V}))(\mathbf{A}\otimes\mathbf{B})^\top \\
&+(\mathbf{B}\otimes\mathbf{A})(\boldsymbol{\mu}_2(\mathbf{V})\otimes\boldsymbol{\mu}_2(\mathbf{U}))(\mathbf{B}\otimes\mathbf{A})^\top\}.
\end{aligned}$$

Proof 110 *See Exercise 6.13.* □

Now, we apply the above results to a unified skew-normal $\mathbf{Z} \sim \mathcal{SUN}_{p,q}(\mathbf{0}_p, \boldsymbol{\Psi}, \mathbf{H}, \boldsymbol{\tau}, \bar{\boldsymbol{\Gamma}})$. For this, we denote by $\boldsymbol{\Psi}^{1/2}$ the unique symmetric square root of $\boldsymbol{\Psi}$; however, it would not make any difference if another square root of $\boldsymbol{\Psi}$ were considered.

Theorem 111 *Let* $\mathbf{Z} = \mathbf{H}\mathbf{U} + \boldsymbol{\Psi}^{1/2}\mathbf{W}_0 \sim \mathcal{SUN}_{p,q}(\mathbf{0}_p, \boldsymbol{\Psi}, \mathbf{H}, \boldsymbol{\tau}, \bar{\boldsymbol{\Gamma}})$*, where according to (6.1),* $\mathbf{U} \sim \mathcal{TN}_q(-\boldsymbol{\tau}; \mathbf{0}_q, \bar{\boldsymbol{\Gamma}})$*,* $\mathbf{W}_0 \sim \mathcal{N}_p(\mathbf{0}_p, \mathbf{I}_p)$ *and they are independent. Then:*

$$\begin{aligned}
\boldsymbol{\mu}_1(\mathbf{Z}) =& \mathbf{H}\boldsymbol{\mu}_1(\mathbf{U}),\\
\boldsymbol{\mu}_2(\mathbf{Z}) =& \mathbf{H}\boldsymbol{\mu}_2(\mathbf{U})\mathbf{H}^\top + \boldsymbol{\Psi},\\
\boldsymbol{\mu}_3(\mathbf{Z}) =& (\mathbf{H}\otimes\mathbf{H})\boldsymbol{\mu}_3(\mathbf{U})\mathbf{H}^\top + (\mathbf{I}_{p^2} + \mathbf{K}_p)(\mathbf{H}\boldsymbol{\mu}_1(\mathbf{U})\otimes\boldsymbol{\Psi}) + \mathrm{vec}(\boldsymbol{\Psi})\boldsymbol{\mu}_1(\mathbf{U})^\top\mathbf{H}^\top,\\
\boldsymbol{\mu}_4(\mathbf{Z}) =& (\mathbf{H}\otimes\mathbf{H})\boldsymbol{\mu}_4(\mathbf{U})(\mathbf{H}\otimes\mathbf{H})^\top + (\mathbf{I}_{p^2} + \mathbf{K}_p)\{(\mathbf{H}\boldsymbol{\mu}_2(\mathbf{U})\mathbf{H}^\top\otimes\boldsymbol{\Psi}) + (\boldsymbol{\Psi}\otimes\mathbf{H}\boldsymbol{\mu}_2(\mathbf{U})\mathbf{H}^\top) + \boldsymbol{\Psi}\otimes\boldsymbol{\Psi}\}\\
& + \mathrm{vec}(\mathbf{H}\boldsymbol{\mu}_2(\mathbf{U})\mathbf{H}^\top)\mathrm{vec}(\boldsymbol{\Psi})^\top + \mathrm{vec}(\boldsymbol{\Psi})\mathrm{vec}(\mathbf{H}\boldsymbol{\mu}_2(\mathbf{U})\mathbf{H}^\top)^\top + \mathrm{vec}(\boldsymbol{\Psi})\mathrm{vec}(\boldsymbol{\Psi})^\top\\
=& (\mathbf{I}_{p^2} + \mathbf{K}_p)\{(\mathbf{H}\boldsymbol{\mu}_2(\mathbf{U})\mathbf{H}^\top\otimes\boldsymbol{\Psi}) + (\boldsymbol{\Psi}\otimes\mathbf{H}\boldsymbol{\mu}_2(\mathbf{U})\mathbf{H}^\top) + \boldsymbol{\Psi}\otimes\boldsymbol{\Psi}\}\\
& + \mathrm{vec}(\mathbf{H}\boldsymbol{\mu}_2(\mathbf{U})\mathbf{H}^\top + \boldsymbol{\Psi})\mathrm{vec}(\mathbf{H}\boldsymbol{\mu}_2(\mathbf{U})\mathbf{H}^\top + \boldsymbol{\Psi})^\top + (\mathbf{H}\otimes\mathbf{H})\mathrm{var}(\mathbf{U}\otimes\mathbf{U})(\mathbf{H}\otimes\mathbf{H})^\top.
\end{aligned}$$

Moreover, the variance matrix of $\mathbf{Z}$ *is*

$$\mathrm{var}(\mathbf{Z}) = \mathbf{H}\boldsymbol{\Sigma}_U\mathbf{H}^\top + \boldsymbol{\Psi} = \boldsymbol{\Sigma},$$

say, having set $\boldsymbol{\Sigma}_U = \mathrm{var}(\mathbf{U}) = \boldsymbol{\mu}_2(\mathbf{U}) - \boldsymbol{\mu}_1(\mathbf{U})\boldsymbol{\mu}_1(\mathbf{U})^\top$.

Proof 111 *Make use of Theorem 110, combined with expressions in Theorem 109 for the moments of* $\mathbf{W}_0 \sim \mathcal{N}_p(\mathbf{0}_p, \mathbf{I}_p)$*. After some algebraic simplifications, we arrive at the stated expressions.* □

The actual usage of the expressions provided in Theorem 111 requires knowledge of the moments of the truncated normal component, $\boldsymbol{\mu}_k(\mathbf{U})$. In general, these moments are not amenable to explicit treatment, and one must resort to numerical computations. In recent years, two different methods for computing truncated normal moments have been presented by Arismendi (2013) and Kan and Robotti (2017). For the latter approach, there exist publicly available computing routines written by the authors in the Matlab language. Of these routines, a corresponding version is available in the R computing environment within its package `mnormt` (Azzalini and Genz, 2020); see also the discussions in Sections 3.11.8 and 9.4.

The expressions $\boldsymbol{\mu}_k(\mathbf{Z})$ given in Theorem 111 refer to a unified skew-normal random vector $\mathbf{Z}$ with location parameter $\boldsymbol{\xi} = \mathbf{0}_p$. For the general case with an arbitrary $\boldsymbol{\xi}$, consider the shifted variable $\mathbf{X} = \boldsymbol{\xi} + \mathbf{Z}$ and use the expressions given below:

$$\begin{aligned}
\boldsymbol{\mu}_1(\mathbf{X}) &= \boldsymbol{\xi} + \boldsymbol{\mu}_1(\mathbf{Z}),\\
\boldsymbol{\mu}_2(\mathbf{X}) &= \boldsymbol{\xi}\boldsymbol{\xi}^\top + \boldsymbol{\xi}\boldsymbol{\mu}_1(\mathbf{Z})^\top + \boldsymbol{\mu}_1(\mathbf{Z})\boldsymbol{\xi}^\top + \boldsymbol{\mu}_1(\mathbf{Z})\boldsymbol{\mu}_1(\mathbf{Z})^\top,\\
\boldsymbol{\mu}_3(\mathbf{X}) &= \boldsymbol{\xi}\otimes\boldsymbol{\xi}\boldsymbol{\xi}^\top + (\boldsymbol{\xi}\otimes\boldsymbol{\xi})\boldsymbol{\mu}_1(\mathbf{Z})^\top + (\mathbf{I}_{p^2} + \mathbf{K}_p)\{\boldsymbol{\xi}\boldsymbol{\xi}^\top\otimes\boldsymbol{\mu}_1(\mathbf{Z}) + \boldsymbol{\xi}\otimes\boldsymbol{\mu}_2(\mathbf{Z})\}\\
&\quad + \mathrm{vec}\{\boldsymbol{\mu}_2(\mathbf{Z})\}\boldsymbol{\xi}^\top + \boldsymbol{\mu}_3(\mathbf{Z}),\\
\boldsymbol{\mu}_4(\mathbf{X}) &= \boldsymbol{\xi}\boldsymbol{\xi}\otimes\boldsymbol{\xi}\boldsymbol{\xi}^\top + (\mathbf{I}_{p^2} + \mathbf{K}_p)\{\boldsymbol{\xi}\boldsymbol{\xi}^\top\otimes\boldsymbol{\xi}\boldsymbol{\mu}_1(\mathbf{Z})^\top + \boldsymbol{\xi}\boldsymbol{\xi}^\top\otimes\boldsymbol{\mu}_1(\mathbf{Z})\boldsymbol{\xi}^\top\\
&\quad + \boldsymbol{\xi}\boldsymbol{\xi}^\top\otimes\boldsymbol{\mu}_2(\mathbf{Z}) + \boldsymbol{\mu}_2(\mathbf{Z})\otimes\boldsymbol{\xi}\boldsymbol{\xi}^\top + (\boldsymbol{\xi}\otimes\mathbf{I}_p)\boldsymbol{\mu}_3(\mathbf{Z})^\top + \boldsymbol{\mu}_3(\mathbf{Z})(\boldsymbol{\xi}^\top\otimes\mathbf{I}_p)\}\\
&\quad + \mathrm{vec}\{\boldsymbol{\mu}_2(\mathbf{Z})\}(\boldsymbol{\xi}\otimes\boldsymbol{\xi})^\top + (\boldsymbol{\xi}\otimes\boldsymbol{\xi})\mathrm{vec}\{\boldsymbol{\mu}_2(\mathbf{Z})\}^\top + \boldsymbol{\mu}_4(\mathbf{Z}).
\end{aligned}$$

6.7.3 *Mardia's measures of multivariate skewness and kurtosis*

We now want to obtain expressions for Mardia's measures of multivariate skewness and kurtosis, denoted $\beta_{1,p}$ and $\beta_{2,p}$ in the original publications of Mardia (1970, 1974). To simplify the algebraic derivations, it is convenient to work with a suitably transformed variable, exploiting the invariance properties of Mardia's measures with respect to nonsingular affine transformations. For a random vector $\mathbf{X} \sim \mathcal{SUN}_{p,q}(\boldsymbol{\xi}, \boldsymbol{\Psi}, \mathbf{H}, \boldsymbol{\tau}, \bar{\boldsymbol{\Gamma}})$, consider again its representation (6.1), and introduce additionally:

$$\boldsymbol{\mu}_0 = \mathbf{H}\boldsymbol{\mu}_1(\mathbf{U}), \quad \boldsymbol{\mu} = \mathrm{E}(\mathbf{X}) = \boldsymbol{\xi} + \boldsymbol{\mu}_0, \quad \mathbf{X} = \boldsymbol{\mu} + \mathbf{X}_0, \quad \mathbf{X}_0 = \mathbf{X} - \boldsymbol{\mu}_0, \quad \mathbf{U}_0 = \mathbf{U} - \boldsymbol{\mu}_0 .$$

Leaving out degenerate cases, $\boldsymbol{\Sigma} = \mathrm{var}(\mathbf{X}) = \mathrm{var}(\mathbf{X}_0)$ is nonsingular. Hence, consider any nonsingular $p \times p$ matrix $\mathbf{C}$ such that $\boldsymbol{\Sigma} = \mathbf{C}\mathbf{C}^\top$ (e.g., we can always choose $\mathbf{C} = \boldsymbol{\Sigma}^{1/2}$), and introduce the standardized variable:

$$\tilde{\mathbf{X}}_0 = \mathbf{C}^{-1}(\mathbf{X} - \boldsymbol{\mu}) = \mathbf{C}^{-1}\mathbf{H}\mathbf{U}_0 + \mathbf{C}^{-1}\boldsymbol{\Psi}^{1/2}\mathbf{W}_0 \sim \mathcal{SUN}_{p,q}(-\mathbf{C}^{-1}\boldsymbol{\mu}_0, \mathbf{C}^{-1}\boldsymbol{\Psi}(\mathbf{C}^{-1})^\top, \mathbf{C}^{-1}\mathbf{H}, \boldsymbol{\tau}, \bar{\boldsymbol{\Gamma}}) \tag{6.52}$$

such that

$$\mathrm{E}(\tilde{\mathbf{X}}_0) = \mathbf{0}_p, \qquad \mathrm{var}(\tilde{\mathbf{X}}_0) = \mathbf{I}_p .$$

On setting $\tilde{\mathbf{H}} = \mathbf{C}^{-1}\mathbf{H}$ and $\tilde{\boldsymbol{\Psi}} = \mathbf{C}^{-1}\boldsymbol{\Psi}(\mathbf{C}^{-1})^\top$, and a matching definition of $\tilde{\boldsymbol{\Psi}}^{1/2}$, it is algebraically equivalent to write:

$$\tilde{\mathbf{X}}_0 = \tilde{\mathbf{H}}\mathbf{U}_0 + \tilde{\boldsymbol{\Psi}}^{1/2}\mathbf{W}_0 . \tag{6.53}$$

The reason for considering the standardized random vector $\tilde{\mathbf{X}}_0$, having zero mean vector and identity variance matrix, is that Mardia's measures can be conveniently computed using the general expressions given by Kollo and Srivastava (2005), namely (see Exercise 6.9):

$$\beta_{1,p} = \mathrm{tr}\{\boldsymbol{\mu}_3(\tilde{\mathbf{X}}_0)^\top \boldsymbol{\mu}_3(\tilde{\mathbf{X}}_0)\} = \mathrm{vec}\{\boldsymbol{\mu}_3(\tilde{\mathbf{X}}_0)\}^\top \mathrm{vec}\{\boldsymbol{\mu}_3(\tilde{\mathbf{X}}_0)\}, \qquad \beta_{2,p} = \mathrm{tr}\{\boldsymbol{\mu}_4(\tilde{\mathbf{X}}_0)\} .$$

The next statement presents the evaluation of these expressions for the unified skew-normal random vector $\tilde{\mathbf{X}}_0$.

Theorem 112 *For a random vector $\tilde{\mathbf{X}}_0$ specified as in (6.52), or equivalently as in (6.53), the following moments hold:*

$$\begin{aligned}
\boldsymbol{\mu}_3(\tilde{\mathbf{X}}_0) &= (\tilde{\mathbf{H}} \otimes \tilde{\mathbf{H}})\boldsymbol{\mu}_3(\mathbf{U}_0)\tilde{\mathbf{H}}^\top, \\
\boldsymbol{\mu}_4(\tilde{\mathbf{X}}_0) &= (\tilde{\mathbf{H}} \otimes \tilde{\mathbf{H}})\boldsymbol{\mu}_4(\mathbf{U}_0)(\tilde{\mathbf{H}} \otimes \tilde{\mathbf{H}})^\top \\
&\quad + (\mathbf{I}_{p^2} + \mathbf{K}_p)(\tilde{\mathbf{H}}\boldsymbol{\Sigma}_U\tilde{\mathbf{H}}^\top \otimes \tilde{\boldsymbol{\Psi}} + \tilde{\boldsymbol{\Psi}} \otimes \tilde{\mathbf{H}}\boldsymbol{\Sigma}_U\tilde{\mathbf{H}}^\top + \tilde{\boldsymbol{\Psi}} \otimes \tilde{\boldsymbol{\Psi}}) \\
&\quad + \mathrm{vec}(\tilde{\mathbf{H}}\boldsymbol{\Sigma}_U\tilde{\mathbf{H}}^\top)\mathrm{vec}(\tilde{\boldsymbol{\Psi}})^\top + \mathrm{vec}(\tilde{\boldsymbol{\Psi}})\mathrm{vec}(\tilde{\mathbf{H}}\boldsymbol{\Sigma}_U\tilde{\mathbf{H}}^\top)^\top + \mathrm{vec}(\tilde{\boldsymbol{\Psi}})\mathrm{vec}(\tilde{\boldsymbol{\Psi}})^\top,
\end{aligned}$$

where $\boldsymbol{\Sigma}_U = \boldsymbol{\mu}_2(\mathbf{U}_0) = \mathrm{var}(\mathbf{U})$. Moreover, Mardia's measures of multivariate skewness and kurtosis are:

$$\begin{aligned}
\beta_{1,p} &= \mathrm{tr}\{(\mathbf{H}^\top\boldsymbol{\Sigma}^{-1}\mathbf{H} \otimes \mathbf{H}^\top\boldsymbol{\Sigma}^{-1}\mathbf{H})\boldsymbol{\mu}_3(\mathbf{U}_0)\mathbf{H}^\top\boldsymbol{\Sigma}^{-1}\mathbf{H}\boldsymbol{\mu}_3(\mathbf{U}_0)^\top\} \\
&= \mathrm{vec}\{\boldsymbol{\mu}_3(\mathbf{U}_0)\}^\top(\tilde{\mathbf{H}}^\top\tilde{\mathbf{H}} \otimes \tilde{\mathbf{H}}^\top\tilde{\mathbf{H}} \otimes \tilde{\mathbf{H}}^\top\tilde{\mathbf{H}})\mathrm{vec}\{\boldsymbol{\mu}_3(\mathbf{U}_0)\}, \\
\beta_{2,p} &= \mathrm{tr}\{(\mathbf{H}^\top\boldsymbol{\Sigma}^{-1}\mathbf{H} \otimes \mathbf{H}^\top\boldsymbol{\Sigma}^{-1}\mathbf{H})\boldsymbol{\mu}_4(\mathbf{U}_0)\} + 2\mathrm{tr}(\boldsymbol{\Sigma}_U\mathbf{H}^\top\boldsymbol{\Sigma}^{-1}\mathbf{H})\mathrm{tr}(\boldsymbol{\Psi}\boldsymbol{\Sigma}^{-1}) \\
&\quad + \mathrm{tr}(\boldsymbol{\Psi}\boldsymbol{\Sigma}^{-1})^2 + 4\mathrm{tr}(\boldsymbol{\Sigma}_U\mathbf{H}^\top\boldsymbol{\Sigma}^{-1}\boldsymbol{\Psi}\boldsymbol{\Sigma}^{-1}\mathbf{H}) + 2\mathrm{tr}(\boldsymbol{\Psi}\boldsymbol{\Sigma}^{-1}\boldsymbol{\Psi}\boldsymbol{\Sigma}^{-1}).
\end{aligned}$$

Proof 112 *The expressions of* $\boldsymbol{\mu}_3(\tilde{\mathbf{X}}_0)$ *and* $\boldsymbol{\mu}_4(\tilde{\mathbf{X}}_0)$ *follow directly from Theorem 111. Therefore, we concentrate on the proof for* $\beta_{1,p}$ *and* $\beta_{2,p}$ *only. Let* $\mathbf{X}' = \boldsymbol{\mu} + \mathbf{C}\tilde{\mathbf{X}}_0'$ *be an independent replicate of* $\mathbf{X} = \boldsymbol{\mu} + \mathbf{C}\tilde{\mathbf{X}}_0$. *Thus, Mardia's measure of multivariate skewness can be expressed as:*

$$\begin{aligned}\beta_{1,p} &= \mathrm{E}[(\mathbf{X}-\boldsymbol{\mu})^\top \boldsymbol{\Sigma}^{-1}(\mathbf{X}'-\boldsymbol{\mu})]^3 \\ &= \mathrm{E}(\tilde{\mathbf{X}}_0^\top \tilde{\mathbf{X}}_0')^3,\end{aligned}$$

where by introducing the matrices $\mathbf{M}_{00} = \mathbf{H}^\top\boldsymbol{\Sigma}^{-1}\mathbf{H}$, $\mathbf{M}_{01} = \mathbf{H}^\top\boldsymbol{\Sigma}^{-1}\boldsymbol{\Psi}^{1/2}$, $\mathbf{M}_{10} = \boldsymbol{\Psi}^{1/2}\boldsymbol{\Sigma}^{-1}\mathbf{H} = \mathbf{M}_{01}^\top$, *and* $\mathbf{M}_{11} = \boldsymbol{\Psi}^{1/2}\boldsymbol{\Sigma}^{-1}\boldsymbol{\Psi}^{1/2}$, *we have the expansion:*

$$\begin{aligned}(\tilde{\mathbf{X}}_0^\top \tilde{\mathbf{X}}_0')^3 &= [\{\mathbf{U}_0^\top \mathbf{H}^\top(\mathbf{C}^{-1})^\top + \mathbf{W}_0^\top\boldsymbol{\Psi}^{1/2}(\mathbf{C}^{-1})^\top\}(\mathbf{C}^{-1}\mathbf{H}\mathbf{U}_0' + \mathbf{C}^{-1}\boldsymbol{\Psi}^{1/2}\mathbf{W}_0')]^3 \\ &= (\mathbf{U}_0^\top\mathbf{M}_{00}\mathbf{U}_0' + \mathbf{U}_0^\top\mathbf{M}_{01}\mathbf{W}_0' + \mathbf{W}_0^\top\mathbf{M}_{01}^\top\mathbf{U}_0' + \mathbf{W}_0^\top\mathbf{M}_{11}\mathbf{W}_0')^3 \\ &= (\mathbf{U}_0^\top\mathbf{M}_{00}\mathbf{U}_0' + \mathbf{U}_0^\top\mathbf{M}_{01}\mathbf{W}_0')^3 \\ &\quad +3(\mathbf{U}_0^\top\mathbf{M}_{00}\mathbf{U}_0' + \mathbf{U}_0^\top\mathbf{M}_{01}\mathbf{W}_0')^2(\mathbf{W}_0^\top\mathbf{M}_{01}^\top\mathbf{U}_0' + \mathbf{W}_0^\top\mathbf{M}_{11}\mathbf{W}_0') \\ &\quad +3(\mathbf{U}_0^\top\mathbf{M}_{00}\mathbf{U}_0' + \mathbf{U}_0^\top\mathbf{M}_{01}\mathbf{W}_0')(\mathbf{W}_0^\top\mathbf{M}_{01}^\top\mathbf{U}_0' + \mathbf{W}_0^\top\mathbf{M}_{11}\mathbf{W}_0')^2 \\ &\quad +(\mathbf{W}_0^\top\mathbf{M}_{01}^\top\mathbf{U}_0' + \mathbf{W}_0^\top\mathbf{M}_{11}\mathbf{W}_0')^3 \\ &= (\mathbf{U}_0^\top\mathbf{M}_{00}\mathbf{U}_0')^3 + 3(\mathbf{U}_0^\top\mathbf{M}_{00}\mathbf{U}_0')^2(\mathbf{U}_0^\top\mathbf{M}_{01}\mathbf{W}_0') \\ &\quad +3(\mathbf{U}_0^\top\mathbf{M}_{00}\mathbf{U}_0')(\mathbf{U}_0^\top\mathbf{M}_{01}\mathbf{W}_0')^2 + (\mathbf{U}_0^\top\mathbf{M}_{01}\mathbf{W}_0')^3 \\ &\quad +3[\{(\mathbf{U}_0^\top\mathbf{M}_{00}\mathbf{U}_0')^2 + 2\mathbf{U}_0^\top\mathbf{M}_{00}\mathbf{U}_0'\mathbf{U}_0^\top\mathbf{M}_{01}\mathbf{W}_0' + (\mathbf{U}_0^\top\mathbf{M}_{01}\mathbf{W}_0')^2\}\mathbf{W}_0^\top\mathbf{M}_{01}^\top\mathbf{U}_0' \\ &\quad +\{(\mathbf{U}_0^\top\mathbf{M}_{00}\mathbf{U}_0')^2 + 2\mathbf{U}_0^\top\mathbf{M}_{00}\mathbf{U}_0'\mathbf{U}_0^\top\mathbf{M}_{01}\mathbf{W}_0' + (\mathbf{U}_0^\top\mathbf{M}_{01}\mathbf{W}_0')^2\}\mathbf{W}_0^\top\mathbf{M}_{11}\mathbf{W}_0'] \\ &\quad +3[\mathbf{U}_0^\top\mathbf{M}_{00}\mathbf{U}_0'\{(\mathbf{W}_0^\top\mathbf{M}_{01}^\top\mathbf{U}_0')^2 + 2\mathbf{W}_0^\top\mathbf{M}_{01}^\top\mathbf{U}_0'\mathbf{W}_0^\top\mathbf{M}_{11}\mathbf{W}_0' + (\mathbf{W}_0^\top\mathbf{M}_{11}\mathbf{W}_0')^2\} \\ &\quad +\mathbf{U}_0^\top\mathbf{M}_{01}\mathbf{W}_0'\{(\mathbf{W}_0^\top\mathbf{M}_{01}^\top\mathbf{U}_0')^2 + 2(\mathbf{W}_0^\top\mathbf{M}_{01}^\top\mathbf{U}_0'\mathbf{W}_0^\top\mathbf{M}_{11}\mathbf{W}_0' + (\mathbf{W}_0^\top\mathbf{M}_{11}\mathbf{W}_0')^2\}] \\ &\quad +(\mathbf{W}_0^\top\mathbf{M}_{01}^\top\mathbf{U}_0')^3 + 3(\mathbf{W}_0^\top\mathbf{M}_{01}^\top\mathbf{U}_0')^2(\mathbf{W}_0^\top\mathbf{M}_{11}\mathbf{W}_0') \\ &\quad +3(\mathbf{W}_0^\top\mathbf{M}_{01}^\top\mathbf{U}_0')(\mathbf{W}_0^\top\mathbf{M}_{11}\mathbf{W}_0')^2 + (\mathbf{W}_0^\top\mathbf{M}_{11}\mathbf{W}_0')^3.\end{aligned}$$

Taking into account that $\mathbf{U}_0$ *and* $\mathbf{U}_0'$ *are iid random vectors with mean zero, as well as that* $\mathbf{W}_0$ *and* $\mathbf{W}_0'$ *are iid random vectors with a spherical normal distribution, all the terms involving odd functions of* $\mathbf{W}_0$ *or* $\mathbf{W}_0'$ *have zero expectation. Consider also that* $\mathbf{U}_0$, $\mathbf{U}_0'$, $\mathbf{W}_0$, *and* $\mathbf{W}_0'$ *are mutually independent. Then, by taking the expectation and removing the terms with zero mean, we have:*

$$\begin{aligned}\mathrm{E}\{(\tilde{\mathbf{X}}_0^\top\tilde{\mathbf{X}}_0')^3\} &= \mathrm{E}\{(\mathbf{U}_0^\top\mathbf{M}_{00}\mathbf{U}_0')^3\} \\ &= \mathrm{E}(\mathbf{U}_0^\top\mathbf{M}_{00}\mathbf{U}_0'\mathbf{U}_0^\top\mathbf{M}_{00}\mathbf{U}_0'\mathbf{U}_0^\top\mathbf{M}_{00}\mathbf{U}_0') \\ &= \mathrm{E}\{(\mathbf{U}_0')^\top\mathbf{M}_{00}\mathbf{U}_0\mathbf{U}_0^\top\mathbf{M}_{00}\mathbf{U}_0'(\mathbf{U}_0')^\top\mathbf{M}_{00}\mathbf{U}_0\} \\ &= \mathrm{E}[\mathrm{tr}\{\mathbf{M}_{00}\mathbf{U}_0\mathbf{U}_0^\top\mathbf{M}_{00}\mathbf{U}_0'(\mathbf{U}_0')^\top\mathbf{M}_{00}\mathbf{U}_0(\mathbf{U}_0')^\top\}].\end{aligned}$$

Now use the property $\mathrm{tr}(\mathbf{NPQR}) = \mathrm{vec}(\mathbf{R}^\top)^\top(\mathbf{Q}^\top\otimes\mathbf{N})\mathrm{vec}(\mathbf{P})$ *(see Equation (A.24g)) with* $\mathbf{N} = \mathbf{M}_{00}\mathbf{U}_0\mathbf{U}_0^\top\mathbf{M}_{00}$, $\mathbf{P} = \mathbf{U}_0'(\mathbf{U}_0')^\top$, $\mathbf{Q} = \mathbf{M}_{00}\mathbf{U}_0$ *and* $\mathbf{R} = (\mathbf{U}_0')^\top$, *and write:*

$$\begin{aligned}\beta_{1,p} = \mathrm{E}\{(\tilde{\mathbf{X}}_0^\top\tilde{\mathbf{X}}_0')^3\} &= \mathrm{E}\{\mathrm{vec}(\mathbf{U}_0')^\top(\mathbf{U}_0^\top\mathbf{M}_{00}\otimes\mathbf{M}_{00}\mathbf{U}_0\mathbf{U}_0^\top\mathbf{M}_{00})\mathrm{vec}(\mathbf{U}_0'(\mathbf{U}_0')^\top)\} \\ &= \mathrm{E}\{(\mathbf{U}_0')^\top\mathbf{M}_{00}(\mathbf{U}_0^\top\otimes\mathbf{U}_0\mathbf{U}_0^\top)(\mathbf{M}_{00}\otimes\mathbf{M}_{00})(\mathbf{U}_0'\otimes\mathbf{U}_0')\} \\ &= \mathrm{E}[\mathrm{tr}\{(\mathbf{M}_{00}\otimes\mathbf{M}_{00})(\mathbf{U}_0'(\mathbf{U}_0')^\top\otimes\mathbf{U}_0')\mathbf{M}_{00}(\mathbf{U}_0^\top\otimes\mathbf{U}_0\mathbf{U}_0^\top)\}] \\ &= \mathrm{tr}\left\{(\mathbf{M}_{00}\otimes\mathbf{M}_{00})\mathrm{E}\{\mathbf{U}_0'(\mathbf{U}_0')^\top\otimes\mathbf{U}_0'\}\mathbf{M}_{00}\mathrm{E}(\mathbf{U}_0^\top\otimes\mathbf{U}_0\mathbf{U}_0^\top)\right\} \\ &= \mathrm{tr}\{(\mathbf{M}_{00}\otimes\mathbf{M}_{00})\boldsymbol{\mu}_3(\mathbf{U}_0')\mathbf{M}_{00}\boldsymbol{\mu}_3(\mathbf{U}_0)^\top\},\end{aligned}$$

where $\boldsymbol{\mu}_3(\mathbf{U}_0') = \boldsymbol{\mu}_3(\mathbf{U}_0)$ *since* $\mathbf{U}_0$ *and* $\mathbf{U}_0'$ *are iid variables.*

Proceeding in a similar way for the measure of kurtosis, we have:

$$\begin{aligned}
\beta_{2,p} &= \mathrm{E}[\{(\mathbf{X}-\boldsymbol{\mu})^\top \boldsymbol{\Sigma}^{-1}(\mathbf{X}-\boldsymbol{\mu})\}^2] \\
&= \mathrm{E}\{(\tilde{\mathbf{X}}_0^\top \tilde{\mathbf{X}}_0)^2\} \\
&= \mathrm{E}\left(\left[\{\mathbf{U}_0^\top \mathbf{H}^\top (\mathbf{C}^{-1})^\top + \mathbf{W}_0^\top \boldsymbol{\Psi}^{1/2}(\mathbf{C}^{-1})^\top\}(\mathbf{C}^{-1}\mathbf{H}\mathbf{U}_0 + \mathbf{C}^{-1}\boldsymbol{\Psi}^{1/2}\mathbf{W}_0)\right]^2\right) \\
&= \mathrm{E}(\mathbf{U}_0^\top \mathbf{M}_{00}\mathbf{U}_0 + \mathbf{U}_0^\top \mathbf{M}_{01}\mathbf{W}_0 + \mathbf{W}_0^\top \mathbf{M}_{01}^\top \mathbf{U}_0 + \mathbf{W}_0^\top \mathbf{M}_{11}\mathbf{W}_0)^2 \\
&= \mathrm{E}(\mathbf{U}_0^\top \mathbf{M}_{00}\mathbf{U}_0 + \mathbf{U}_0^\top \mathbf{M}_{01}\mathbf{W}_0)^2 \\
&\quad +2\mathrm{E}\{(\mathbf{U}_0^\top \mathbf{M}_{00}\mathbf{U}_0 + \mathbf{U}_0^\top \mathbf{M}_{01}\mathbf{W}_0)(\mathbf{W}_0^\top \mathbf{M}_{01}^\top \mathbf{U}_0 + \mathbf{W}_0^\top \mathbf{M}_{11}\mathbf{W}_0)\} \\
&\quad +\mathrm{E}\{(\mathbf{W}_0^\top \mathbf{M}_{01}^\top \mathbf{U}_0 + \mathbf{W}_0^\top \mathbf{M}_{11}\mathbf{W}_0)^2\} \\
&= \mathrm{E}\{(\mathbf{U}_0^\top \mathbf{M}_{00}\mathbf{U}_0)^2\} + 2\mathrm{E}\{(\mathbf{U}_0^\top \mathbf{M}_{00}\mathbf{U}_0)(\mathbf{U}_0^\top \mathbf{M}_{01}\mathbf{W}_0)\} + \mathrm{E}\{(\mathbf{U}_0^\top \mathbf{M}_{01}\mathbf{W}_0)^2\} \\
&\quad +2\mathrm{E}\{(\mathbf{U}_0^\top \mathbf{M}_{00}\mathbf{U}_0)(\mathbf{W}_0^\top \mathbf{M}_{01}^\top \mathbf{U}_0 + \mathbf{W}_0^\top \mathbf{M}_{11}\mathbf{W}_0)\} \\
&\quad +2\mathrm{E}\{(\mathbf{U}_0^\top \mathbf{M}_{01}\mathbf{W}_0)(\mathbf{W}_0^\top \mathbf{M}_{01}^\top \mathbf{U}_0 + \mathbf{W}_0^\top \mathbf{M}_{11}\mathbf{W}_0)\} \\
&\quad +\mathrm{E}\{(\mathbf{W}_0^\top \mathbf{M}_{01}^\top \mathbf{U}_0)^2\} + 2\mathrm{E}\{(\mathbf{W}_0^\top \mathbf{M}_{01}^\top \mathbf{U}_0)(\mathbf{W}_0^\top \mathbf{M}_{11}\mathbf{W}_0)\} + \mathrm{E}\{(\mathbf{W}_0^\top \mathbf{M}_{11}\mathbf{W}_0)^2\} \\
&= \mathrm{E}\{(\mathbf{U}_0^\top \mathbf{M}_{00}\mathbf{U}_0)^2\} + \mathrm{E}\{(\mathbf{U}_0^\top \mathbf{M}_{01}\mathbf{W}_0)^2\} \\
&\quad +2\mathrm{E}\{(\mathbf{U}_0^\top \mathbf{M}_{00}\mathbf{U}_0)(\mathbf{W}_0^\top \mathbf{M}_{11}\mathbf{W}_0)\} + 2\mathrm{E}\{(\mathbf{U}_0^\top \mathbf{M}_{01}\mathbf{W}_0)(\mathbf{W}_0^\top \mathbf{M}_{01}^\top \mathbf{U}_0)\} \\
&\quad +\mathrm{E}\{(\mathbf{W}_0^\top \mathbf{M}_{01}^\top \mathbf{U}_0)^2\} + \mathrm{E}\{(\mathbf{W}_0^\top \mathbf{M}_{11}\mathbf{W}_0)^2\},
\end{aligned}$$

where the terms with zero expectation have been removed, namely those associated with odd functions of $\mathbf{W}_0$. *The remaining expected values can be worked out, recalling that the powers of quadratic forms can be expressed as the trace of matrix products, combined with properties of the trace of products of matrices, specifically that* $\mathrm{tr}\{\mathbf{K}_{pq}(\mathbf{P}^\top \otimes \mathbf{Q})\} = \mathrm{tr}(\mathbf{P}^\top \mathbf{Q})$ *as stated in Equation (A.24h) and, in case* $p = q$, $\mathrm{tr}(\mathbf{P}\otimes\mathbf{Q}) = \mathrm{tr}(\mathbf{P})\mathrm{tr}(\mathbf{Q})$, $\mathrm{tr}(\mathbf{P}^\top \mathbf{Q}) = \mathrm{vec}(\mathbf{P})^\top \mathrm{vec}(\mathbf{Q})$. *We then obtain:*

$$\begin{aligned}
\beta_{2,p} =& \mathrm{E}\{\mathrm{tr}(\mathbf{M}_{00}\mathbf{U}_0\mathbf{U}_0^\top \mathbf{M}_{00}\mathbf{U}_0\mathbf{U}_0^\top)\} + 2\,\mathrm{E}\{\mathrm{tr}(\mathbf{M}_{01}\mathbf{W}_0\mathbf{W}_0^\top \mathbf{M}_{01}^\top \mathbf{U}_0\mathbf{U}_0^\top)\} \\
&+ 2\,\mathrm{E}\{\mathrm{tr}(\mathbf{M}_{00}\mathbf{U}_0\mathbf{U}_0^\top)\mathrm{tr}(\mathbf{M}_{11}\mathbf{W}_0\mathbf{W}_0^\top)\} + 2\,\mathrm{E}\{\mathrm{tr}(\mathbf{M}_{01}\mathbf{W}_0\mathbf{W}_0^\top \mathbf{M}_{01}^\top \mathbf{U}_0\mathbf{U}_0^\top)\} \\
&+ \mathrm{E}\{\mathrm{tr}(\mathbf{M}_{11}\mathbf{W}_0\mathbf{W}_0^\top \mathbf{M}_{11}\mathbf{W}_0\mathbf{W}_0^\top)\} \\
=& \mathrm{E}\{\mathrm{vec}(\mathbf{U}_0\mathbf{U}_0^\top)^\top (\mathbf{M}_{00}\otimes\mathbf{M}_{00})\mathrm{vec}(\mathbf{U}_0\mathbf{U}_0^\top)\} + 2\,\mathrm{E}\{\mathrm{vec}(\mathbf{U}_0\mathbf{U}_0^\top)^\top (\mathbf{M}_{01}\otimes\mathbf{M}_{01})\mathrm{vec}(\mathbf{W}_0\mathbf{W}_0^\top)\} \\
&+ 2\,\mathrm{E}\{\mathrm{tr}(\mathbf{M}_{00}\mathbf{U}_0\mathbf{U}_0^\top)\mathrm{tr}(\mathbf{M}_{11}\mathbf{W}_0\mathbf{W}_0^\top)\} + 2\,\mathrm{E}\{\mathrm{tr}(\mathbf{M}_{01}\mathbf{W}_0\mathbf{W}_0^\top \mathbf{M}_{01}^\top \mathbf{U}_0\mathbf{U}_0^\top)\} \\
&+ \mathrm{E}\{\mathrm{vec}(\mathbf{W}_0\mathbf{W}_0^\top)^\top (\mathbf{M}_{11}\otimes\mathbf{M}_{11})\mathrm{vec}(\mathbf{W}_0\mathbf{W}_0^\top)\} \\
=& \mathrm{tr}[(\mathbf{M}_{00}\otimes\mathbf{M}_{00})\mathrm{E}\{\mathrm{vec}(\mathbf{U}_0\mathbf{U}_0^\top)\mathrm{vec}(\mathbf{U}_0\mathbf{U}_0^\top)^\top\}] \\
&+ 2\,\mathrm{E}\{\mathrm{vec}(\mathbf{U}_0\mathbf{U}_0^\top)^\top\}(\mathbf{M}_{01}\otimes\mathbf{M}_{01})\mathrm{E}\{\mathrm{vec}(\mathbf{W}_0\mathbf{W}_0^\top)\} \\
&+ 2\,\mathrm{tr}\{\mathbf{M}_{00}\mathrm{E}(\mathbf{U}_0\mathbf{U}_0^\top)\}\mathrm{tr}\{\mathbf{M}_{11}\mathrm{E}\mathbf{W}_0\mathbf{W}_0^\top\} + 2\,\mathrm{tr}\{\mathbf{M}_{01}\mathrm{E}(\mathbf{W}_0\mathbf{W}_0^\top)\mathbf{M}_{01}^\top \mathrm{E}(\mathbf{U}_0\mathbf{U}_0^\top)\} \\
&+ \mathrm{tr}[(\mathbf{M}_{11}\otimes\mathbf{M}_{11})\mathrm{E}\{\mathrm{vec}(\mathbf{W}_0\mathbf{W}_0^\top)\mathrm{vec}(\mathbf{W}_0\mathbf{W}_0^\top)^\top\}] \\
=& \mathrm{tr}\{(\mathbf{M}_{00}\otimes\mathbf{M}_{00})\boldsymbol{\mu}_4(\mathbf{U}_0)\} + 2\mathrm{vec}\{\boldsymbol{\mu}_2(\mathbf{U}_0)\}^\top (\mathbf{M}_{01}\otimes\mathbf{M}_{01})\mathrm{vec}\{\boldsymbol{\mu}_2(\mathbf{W}_0)\} \\
&+ 2\,\mathrm{tr}\{\mathbf{M}_{00}\boldsymbol{\mu}_2(\mathbf{U}_0)\}\mathrm{tr}\{\mathbf{M}_{11}\boldsymbol{\mu}_2(\mathbf{W}_0)\} + 2\,\mathrm{tr}\{\mathbf{M}_{01}\boldsymbol{\mu}_2(\mathbf{W}_0)\mathbf{M}_{01}^\top \boldsymbol{\mu}_2(\mathbf{U}_0)\} \\
&+ \mathrm{tr}[\{\mathbf{M}_{11}\otimes\mathbf{M}_{11}\}\boldsymbol{\mu}_4(\mathbf{W}_0)].
\end{aligned}$$

Taking into account Theorem 109, we can substitute $\boldsymbol{\mu}_2(\mathbf{W}_0) = \mathbf{I}_p$, $\mathrm{vec}\{\boldsymbol{\mu}_2(\mathbf{W}_0)\} = \mathrm{vec}(\mathbf{I}_p)$, $\boldsymbol{\mu}_4(\mathbf{W}_0) = \mathbf{I}_{p^2} + \mathbf{K}_p + \mathrm{vec}(\mathbf{I}_p)\mathrm{vec}(\mathbf{I}_p)^\top$, *and* $\boldsymbol{\mu}_2(\mathbf{U}_0) = \mathrm{var}(\mathbf{U}_0)$, *leading to:*

$$\begin{aligned}
\beta_{2,p} &= \mathrm{tr}\left[(\mathbf{M}_{00}\otimes\mathbf{M}_{00})\boldsymbol{\mu}_4(\mathbf{U}_0)\right] + 2\mathrm{vec}\{\boldsymbol{\mu}_2(\mathbf{U}_0)\}^\top(\mathbf{M}_{01}\otimes\mathbf{M}_{01})\mathrm{vec}(\mathbf{I}_d)\\
&\quad +2\,\mathrm{tr}\left(\mathbf{M}_{00}\boldsymbol{\mu}_2(\mathbf{U}_0)\right)\mathrm{tr}\left(\mathbf{M}_{11}\right) + 2\,\mathrm{tr}(\mathbf{M}_{01}\mathbf{M}_{01}^\top)\\
&\quad +\mathrm{tr}\left\{(\mathbf{M}_{11}\otimes\mathbf{M}_{11})(\mathbf{I}_{p^2}+\mathbf{K}_p)\right\} + \mathrm{tr}\left\{(\mathbf{M}_{11}\otimes\mathbf{M}_{11})\mathrm{vec}(\mathbf{I}_p)\mathrm{vec}(\mathbf{I}_p)^\top\right\}\\
&= \mathrm{tr}\left\{\boldsymbol{\mu}_4(\mathbf{U}_0)(\mathbf{M}_{00}\otimes\mathbf{M}_{00})\right\} + 4\,\mathrm{tr}(\boldsymbol{\mu}_2(\mathbf{U}_0)\mathbf{M}_{01}\mathbf{M}_{01}^\top)\\
&\quad +2\,\mathrm{tr}(\boldsymbol{\mu}_2(\mathbf{U}_0)\mathbf{M}_{00})\mathrm{tr}(\mathbf{M}_{11}) + \mathrm{tr}(\mathbf{M}_{11})^2 + 2\,\mathrm{tr}(\mathbf{M}_{11}^2),
\end{aligned}$$

where $\mathbf{M}_{01}\mathbf{M}_{01}^\top = \mathbf{H}^\top\boldsymbol{\Sigma}^{-1}\boldsymbol{\Psi}\boldsymbol{\Sigma}^{-1}\mathbf{H}$, $\mathrm{tr}(\mathbf{M}_{11}) = \mathrm{tr}(\boldsymbol{\Psi}\boldsymbol{\Sigma}^{-1})$ *and* $\mathrm{tr}(\mathbf{M}_{11}^2) = \mathrm{tr}(\boldsymbol{\Psi}\boldsymbol{\Sigma}^{-1}\boldsymbol{\Psi}\boldsymbol{\Sigma}^{-1})$. □

Example 82 *Let* $\mathbf{Z} = \mathbf{H}\mathbf{U} + \boldsymbol{\Psi}^{-1/2}\mathbf{W}_0$, *where* $\mathbf{U} \sim \mathcal{TN}_q(\mathbf{0}_q;\mathbf{0}_q,\mathbf{I}_q)$, $\mathbf{W}_0 \sim \mathcal{N}_p(\mathbf{0}_p,\mathbf{I}_p)$ *and they are independent. In this case, the components* $U_1,\ldots,U_q$ *of* $\mathbf{U}$ *are iid* $\mathcal{HN}(0,1)$ *random variables, with* $\mathrm{E}(U_i) = b \equiv \sqrt{2/\pi}$, $\mathrm{E}(U_i^2) = 1$, $\mathrm{var}(U_i) = 1-b^2$, $\mathrm{E}(U_i^3) = 2b$, $\mathrm{E}(U_i^4) = 3$, $\gamma_1(U) = b(2b^2-1)/(1-b^2)^{3/2}$ *(skewness) and* $\gamma_2(U) = 2b^2(2-3b^2)/(1-b^2)^2$ *(excess kurtosis), and therefore the first four multivariate moments of* $\mathbf{U}_0 = \mathbf{U}-\mathrm{E}(\mathbf{U}) = \mathbf{U}-b\mathbf{1}_q$ *are (see Theorem 80 and Exercise 5.6 in Chapter 5):*

$$\begin{aligned}
&\boldsymbol{\mu}_1(\mathbf{U}_0) = \mathbf{0}_q, \quad \boldsymbol{\mu}_2(\mathbf{U}_0) = \mathrm{var}(\mathbf{U}_0) = \mathrm{var}(\mathbf{U}) = (1-b^2)\mathbf{I}_q = \boldsymbol{\Sigma}_U,\\
&\boldsymbol{\mu}_3(\mathbf{U}_0) = (1-b^2)^{3/2}\gamma_1(U)\sum_{i=1}^q(\mathbf{e}_i\otimes\mathbf{E}_{ii}),\\
&\boldsymbol{\mu}_4(\mathbf{U}_0) = (1-b^2)^2\{\mathbf{I}_{q^2}+\mathbf{K}_q+\mathrm{vec}(\mathbf{I}_q)\mathrm{vec}(\mathbf{I}_q)^\top + \gamma_2\sum_{i=1}^q(\mathbf{E}_{ii}\otimes\mathbf{E}_{ii})\}.
\end{aligned}$$

Therefore, for the multivariate measures of skewness and kurtosis of $\mathbf{Z} \sim \mathcal{SUN}_{p,q}(\boldsymbol{\xi},\boldsymbol{\Psi},\mathbf{H},\mathbf{0}_q,\mathbf{I}_q)$, *we have after some algebra that:*

$$\begin{aligned}
\beta_{1,p} &= (1-b^2)^3\gamma_1(U)^2\mathrm{tr}\{\sum_{i=1}^q(\mathbf{H}^\top\boldsymbol{\Sigma}^{-1}\mathbf{H}\mathbf{e}_i\otimes\mathbf{H}^\top\boldsymbol{\Sigma}^{-1}\mathbf{H}\mathbf{E}_{ii})\sum_{j=1}^q(\mathbf{e}_j^\top\otimes\mathbf{H}^\top\boldsymbol{\Sigma}^{-1}\mathbf{H}\mathbf{E}_{jj})\}\\
&= (1-b^2)^3\gamma_1(U)^2\sum_{i=1}^q\sum_{j=1}^q\mathrm{tr}\{(\mathbf{H}^\top\boldsymbol{\Sigma}^{-1}\mathbf{H}\mathbf{E}_{ij}\otimes\mathbf{H}^\top\boldsymbol{\Sigma}^{-1}\mathbf{H}\mathbf{E}_{ii}\mathbf{H}^\top\boldsymbol{\Sigma}^{-1}\mathbf{H}\mathbf{E}_{jj})\}\\
&= (1-b^2)^3\gamma_1(U)^2\sum_{i=1}^q\sum_{j=1}^q\mathrm{tr}\{(\mathbf{H}^\top\boldsymbol{\Sigma}^{-1}\mathbf{H}\mathbf{E}_{ij}\}\mathrm{tr}\{\mathbf{H}^\top\boldsymbol{\Sigma}^{-1}\mathbf{H}\mathbf{E}_{ii}\mathbf{H}^\top\boldsymbol{\Sigma}^{-1}\mathbf{H}\mathbf{E}_{jj})\}\\
&= (1-b^2)^3\gamma_1(U)^2\sum_{i=1}^q\sum_{j=1}^q(\mathbf{e}_i^\top\mathbf{H}^\top\boldsymbol{\Sigma}^{-1}\mathbf{H}\mathbf{e}_j)^3,\\
\beta_{2,p} &= (1-b^2)^2[\mathrm{tr}(\mathbf{H}^\top\boldsymbol{\Sigma}^{-1}\mathbf{H}\otimes\mathbf{H}^\top\boldsymbol{\Sigma}^{-1}\mathbf{H}) + \mathrm{tr}\{\mathbf{K}_q(\mathbf{H}^\top\boldsymbol{\Sigma}^{-1}\mathbf{H}\otimes\mathbf{H}^\top\boldsymbol{\Sigma}^{-1}\mathbf{H})\}\\
&\quad +\mathrm{vec}(\mathbf{H}^\top\boldsymbol{\Sigma}^{-1}\mathbf{H})^\top\mathrm{vec}(\mathbf{H}^\top\boldsymbol{\Sigma}^{-1}\mathbf{H}) + \gamma_2(U)\sum_{i=1}^q\mathrm{tr}(\mathbf{H}^\top\boldsymbol{\Sigma}^{-1}\mathbf{H}\mathbf{E}_{ii}\otimes\mathbf{H}^\top\boldsymbol{\Sigma}^{-1}\mathbf{H}\mathbf{E}_{ii})]\\
&\quad +2(1-b^2)\mathrm{tr}(\mathbf{H}^\top\boldsymbol{\Sigma}^{-1}\mathbf{H})\mathrm{tr}(\boldsymbol{\Psi}\boldsymbol{\Sigma}^{-1}) + \mathrm{tr}(\boldsymbol{\Psi}\boldsymbol{\Sigma}^{-1})^2 + 2\mathrm{tr}\{(\boldsymbol{\Psi}\boldsymbol{\Sigma}^{-1})^2\}\\
&\quad +4(1-b^2)\mathrm{tr}(\mathbf{H}^\top\boldsymbol{\Sigma}^{-1}\boldsymbol{\Psi}\boldsymbol{\Sigma}^{-1}\mathbf{H})\\
&= (1-b^2)^2[\mathrm{tr}(\mathbf{H}^\top\boldsymbol{\Sigma}^{-1}\mathbf{H})^2 + 2\mathrm{tr}\{(\mathbf{H}^\top\boldsymbol{\Sigma}^{-1}\mathbf{H})^2\} + \gamma_2(U)\sum_{i=1}^q(\mathbf{e}_i^\top\mathbf{H}^\top\boldsymbol{\Sigma}^{-1}\mathbf{H}\mathbf{e}_i)^2]\\
&\quad +2(1-b^2)\mathrm{tr}(\mathbf{H}^\top\boldsymbol{\Sigma}^{-1}\mathbf{H})\mathrm{tr}(\boldsymbol{\Psi}\boldsymbol{\Sigma}^{-1}) + \mathrm{tr}(\boldsymbol{\Psi}\boldsymbol{\Sigma}^{-1})^2 + 2\mathrm{tr}\{(\boldsymbol{\Psi}\boldsymbol{\Sigma}^{-1})^2\}\\
&\quad +4(1-b^2)\mathrm{tr}(\mathbf{H}^\top\boldsymbol{\Sigma}^{-1}\boldsymbol{\Psi}\boldsymbol{\Sigma}^{-1}\mathbf{H}),
\end{aligned}$$

where $\boldsymbol{\Sigma}^{-1} = \{\boldsymbol{\Psi} + (1-b^2)\mathbf{H}\mathbf{H}^\top\}^{-1} = \boldsymbol{\Psi}^{-1} - (1-b^2)\boldsymbol{\Psi}^{-1}\mathbf{H}\{\mathbf{I}_q + (1-b^2)\mathbf{H}^\top\boldsymbol{\Psi}^{-1}\mathbf{H}\}^{-1}\mathbf{H}^\top\boldsymbol{\Psi}^{-1}$. *Two special cases are the following:*

(i) If $\boldsymbol{\Psi} = \mathbf{I}_p$ *and* $\mathbf{H}^\top\boldsymbol{\Psi}^{-1}\mathbf{H} = \mathbf{H}^\top\mathbf{H} = \mathbf{I}_q$, *then* $\boldsymbol{\Sigma}^{-1} = \mathbf{I}_p - (1-b^2)(2-b^2)^{-1}\mathbf{H}\mathbf{H}^\top$, $\mathbf{H}^\top\boldsymbol{\Sigma}^{-1}\mathbf{H} = (2-b^2)^{-1}\mathbf{I}_q$ *and* $\mathbf{H}^\top\boldsymbol{\Sigma}^{-1}\boldsymbol{\Psi}\boldsymbol{\Sigma}^{-1}\mathbf{H} = (2-b^2)^{-2}\mathbf{I}_q$, *so that*

$$
\begin{aligned}
\beta_{1,p} =& (1-b^2)^3(2-b^2)^{-3}\gamma_1(U)^2, \\
\beta_{2,p} =& (1-b^2)^2(2-b^2)^{-2}\{q^2+2q+\gamma_2(U)q\} + 2(1-b^2)(2-b^2)^{-1}q\{p-(1-b^2)(2-b^2)^{-1}q\} \\
&+ \{p-(1-b^2)(2-b^2)^{-1}q\}^2 + 2\{p-2(1-b^2)(2-b^2)^{-1}q + (1-b^2)^2(2-b^2)^{-2}q\} \\
&+ 4(1-b^2)(2-b^2)^{-2}q \\
=& 2b^2(2-3b^2)(2-b^2)^{-2}q + p(p+2).
\end{aligned}
$$

We can see that in this case, the dimension q *has no effect on* $\beta_{1,p}$ *and a linear effect on* $\beta_{2,p}$.

(ii) If $q = 1$, *with* $\mathbf{H} = \boldsymbol{\eta}$, $\eta^2 = \boldsymbol{\eta}^\top\boldsymbol{\Psi}^{-1}\boldsymbol{\eta}$ *and* $\theta^2 = (1-b^2)\boldsymbol{\eta}^\top\boldsymbol{\Sigma}^{-1}\boldsymbol{\eta} = (1-b^2)\eta^2\{1+(1-b^2)\eta^2\}^{-1}$, *then*

$$
\begin{aligned}
\beta_{1,p} &= \gamma_1(U)^2\theta^6 = b^2(2b^2-1)^2\eta^6\{1+(1-b^2)\eta^2\}^{-3}, \\
\beta_{2,p} &= 3\theta^4 + \gamma_2(U)\theta^4 + 2\theta^2(p-\theta^2) + (p-\theta^2)^2 + 2(p-2\theta^2+\theta^4) + 4\theta^2(1-\theta^2) \\
&= (1-b^2)\gamma_2(U)\theta + p(p+2) = 2b^2(2-3b^2)\eta^4\{1+(1-b^2)\eta^2\}^{-2} + p(p+2).
\end{aligned}
$$

These expressions are equal to those for the SN distribution obtained in Theorem 79.

6.8 Distributions Related to the Unified Skew-Normal

6.8.1 Multivariate SUNSET distribution

An interesting particular SUN distribution is presented next.

Definition 19 [Multivariate unified skew-normal with independent sets of latent variables distribution] *A random vector* $\mathbf{X} \in \mathbb{R}^p$ *is said to have a multivariate unified skew-normal with S independent sets of latent variables distribution, or SUNSET in brief, with location parameter* $\boldsymbol{\xi} \in \mathbb{R}^p$, *p-by-p dispersion matrix* $\boldsymbol{\Psi} > 0$ *and shape/skewness parameters* $\mathbf{H} = (\mathbf{H}_1, \ldots, \mathbf{H}_S) \in \mathbb{R}^{p\times q}$, $\mathbf{H}_s \in \mathbb{R}^{p\times q_s}$, $\boldsymbol{\tau} = (\boldsymbol{\tau}_1^\top, \ldots, \boldsymbol{\tau}_S^\top)^\top \in \mathbb{R}^q$, $\boldsymbol{\tau}_s \in \mathbb{R}^{q_s}$, *and* $\bar{\boldsymbol{\Gamma}} = \mathrm{diag}(\bar{\boldsymbol{\Gamma}}_1, \ldots, \bar{\boldsymbol{\Gamma}}_S) > 0$, *a q-by-q block-diagonal correlation matrix with* $\bar{\boldsymbol{\Gamma}}_s \in \mathbb{R}^{q_s\times q_s}$, $s = 1, \ldots, S$, *with* $\sum_{s=1}^S q_s = q$, *in symbols* $\mathbf{X} \sim \mathcal{SUNSET}_{p,q_1,\ldots,q_S}(\boldsymbol{\xi}, \boldsymbol{\Psi}, \mathbf{H}, \boldsymbol{\tau}, \bar{\boldsymbol{\Gamma}})$, *if* $\mathbf{X}$ *can be expressed as*

$$
\mathbf{X} = \boldsymbol{\xi} + \sum_{s=1}^{S} \mathbf{H}_s\mathbf{U}_s + \mathbf{W},
$$

where $\mathbf{U}_s \stackrel{d}{=} (\mathbf{W}_{0s} \mid \mathbf{W}_{0s} + \boldsymbol{\tau}_s > \mathbf{0}_{q_s})$, *with* $\mathbf{W}_{0s} \sim \mathcal{N}_{q_s}(\mathbf{0}_{q_s}, \bar{\boldsymbol{\Gamma}}_s)$, $s = 1, \ldots, S$, *are independent truncated normal random vectors, and* $\mathbf{W} \sim \mathcal{N}_p(\mathbf{0}_p, \boldsymbol{\Psi})$ *which is independent of* $\mathbf{W}_{0s}$ *and therefore of* $\mathbf{U}_s$.

The SUNSET distribution is said to be fading if $q < p$, lasting if $q > p$, and perfect if $q = p$. Moreover, the SUNSET distribution is clear if $\bar{\boldsymbol{\Gamma}} = \mathbf{I}_q$.

The SUNSET distribution has many appealing properties that can be derived from those of the SUN distribution provided in this chapter; see Wang et al. (2026) for further details. For example, if $\mathbf{X} \sim \mathcal{SUNSET}_{p,q_1,\ldots,q_S}(\boldsymbol{\xi}, \boldsymbol{\Psi}, \mathbf{H}, \boldsymbol{\tau}, \bar{\boldsymbol{\Gamma}})$ then its pdf is

$$f_{\mathbf{X}}(\mathbf{x}) = \phi_p(\mathbf{x}; \boldsymbol{\xi}, \boldsymbol{\Psi} + \mathbf{H}\bar{\boldsymbol{\Gamma}}\mathbf{H}^\top) \frac{\Phi_q\{\boldsymbol{\tau} + \tilde{\boldsymbol{\Gamma}}\mathbf{H}^\top\boldsymbol{\Psi}^{-1}(\mathbf{x} - \boldsymbol{\xi}); \mathbf{0}_q, \tilde{\boldsymbol{\Gamma}}\}}{\prod_{s=1}^S \Phi_{q_s}(\boldsymbol{\tau}_s; \mathbf{0}_{q_s}, \bar{\boldsymbol{\Gamma}}_s)}, \quad \mathbf{x} \in \mathbb{R}^p, \tag{6.54}$$

where $\tilde{\boldsymbol{\Gamma}} = (\bar{\boldsymbol{\Gamma}}^{-1} + \mathbf{H}^\top\boldsymbol{\Psi}^{-1}\mathbf{H})^{-1}$. An interesting aspect of the SUNSET distribution is that it has fewer parameters than the SUN distribution. The number of parameters can be further reduced by appropriate choices of the blocks $\bar{\boldsymbol{\Gamma}}_s$, e.g., equicorrelation $\bar{\boldsymbol{\Gamma}}_s = (1 - \rho_s)\mathbf{I}_{q_s} + \rho_s\mathbf{1}_{q_s}\mathbf{1}_{q_s}^\top$. Moreover, the block-diagonal structure of $\bar{\boldsymbol{\Gamma}}$ leads to simplifications of the marginal distributions as described in Theorem 98. The SUNSET distribution also arises as the distribution of the sum of two independent SUN random vectors; see Theorem 101. To improve the interpretation of the skewness, some structure in the matrix $\mathbf{H}$ can be introduced, as shown in the following example.

Example 83 (Multivariate beautiful SUNSET distribution) *Consider a random vector* $\mathbf{X} \sim \mathcal{SUNSET}_{p,q_1,\ldots,q_S}(\boldsymbol{\xi}, \boldsymbol{\Psi}, \mathbf{H}, \boldsymbol{\tau}, \bar{\boldsymbol{\Gamma}})$ *with* $q \le p$ *and the spectral decomposition* $\boldsymbol{\Psi} = \sum_{i=1}^p \lambda_i \boldsymbol{\psi}_i \boldsymbol{\psi}_i^\top$. *Define* $\mathbf{H} = (h_1\boldsymbol{\psi}_1, \ldots, h_q\boldsymbol{\psi}_q)$, *where* $\mathbf{h} = (h_1, \ldots, h_q)^\top \in \mathbb{R}^q$ *controls the magnitude of the skewness in the directions given by the first q eigenvectors of* $\boldsymbol{\Psi}$, *hence providing a clear geometrical interpretation. If* $q > p$, *then we use all the p eigenvectors of* $\boldsymbol{\Psi}$ *and add* $q - p$ *vectors to form a basis of* $\mathbb{R}^q$. *We call such a distribution a multivariate beautiful SUNSET (BSUNSET) distribution and denote it by* $\mathbf{X} \sim \mathcal{BSUNSET}_{p,q_1,\ldots,q_S}(\boldsymbol{\xi}, \boldsymbol{\Psi}, \mathbf{h}, \boldsymbol{\tau}, \bar{\boldsymbol{\Gamma}})$. *Hence, the matrix* $\mathbf{H}$ *has only q unknown parameters from* $\mathbf{h}$ *rather than pq parameters when* $\mathbf{H}$ *is unstructured. Note that the above structure of* $\mathbf{H}$ *can also be used in SUN distributions to improve interpretability. Moreover, the pdf of* $\mathbf{X}$ *further simplifies to*

$$f_{\mathbf{X}}(\mathbf{x}) = \phi_p(\mathbf{x}; \boldsymbol{\xi}, \boldsymbol{\Psi} + \mathbf{H}\bar{\boldsymbol{\Gamma}}\mathbf{H}^\top) \prod_{s=1}^S \frac{\Phi_{q_s}\{\boldsymbol{\tau}_s + \tilde{\boldsymbol{\Gamma}}_s\mathbf{H}_s^\top\boldsymbol{\Psi}^{-1}(\mathbf{x} - \boldsymbol{\xi}); \mathbf{0}_{q_s}, \tilde{\boldsymbol{\Gamma}}_s\}}{\Phi_{q_s}(\boldsymbol{\tau}_s; \mathbf{0}_{q_s}, \bar{\boldsymbol{\Gamma}}_s)}, \quad \mathbf{x} \in \mathbb{R}^p, \tag{6.55}$$

where $\tilde{\boldsymbol{\Gamma}}_s = (\bar{\boldsymbol{\Gamma}}_s^{-1} + \mathbf{D}_s)^{-1}$ *and* $\mathbf{D}_s = \mathrm{diag}(h_i^2/\lambda_i) \in \mathbb{R}^{q_s \times q_s}$. *Therefore, evaluating the pdf (6.55) is computationally simpler than evaluating the pdf (6.54) because only normal cdfs* Φ_{q_s} *rather than* Φ_q, $q_s < q$, *need to be computed.*

6.8.2 *Multivariate SUNNY distribution*

We explore another distribution related to the SUN family. We define a *multivariate unified skew-normal with a new youth (SUNNY) distribution* by the following relationship:

$$\mathcal{SUNNY}_p(\boldsymbol{\xi}, \boldsymbol{\Omega}, \boldsymbol{\alpha}, q) \equiv \mathcal{AASUN}_{p,q}(\boldsymbol{\xi}, \boldsymbol{\Omega}, \boldsymbol{\Delta}, \mathbf{0}_q, \bar{\boldsymbol{\Gamma}}),$$

with

$$\boldsymbol{\Delta} = \bar{\boldsymbol{\Omega}} \frac{\boldsymbol{\alpha}}{\sqrt{1 + \boldsymbol{\alpha}^\top\bar{\boldsymbol{\Omega}}\boldsymbol{\alpha}}} \mathbf{1}_q^\top, \quad \bar{\boldsymbol{\Gamma}} = \frac{1}{1 + \boldsymbol{\alpha}^\top\bar{\boldsymbol{\Omega}}\boldsymbol{\alpha}} \mathbf{I}_q + \frac{\boldsymbol{\alpha}^\top\bar{\boldsymbol{\Omega}}\boldsymbol{\alpha}}{1 + \boldsymbol{\alpha}^\top\bar{\boldsymbol{\Omega}}\boldsymbol{\alpha}} \mathbf{1}_q\mathbf{1}_q^\top,$$

where $\boldsymbol{\xi} \in \mathbb{R}^p$, $\boldsymbol{\Omega} \in \mathbb{R}^{p \times p}$, and $\boldsymbol{\alpha} \in \mathbb{R}^p$. Straightforward calculations lead to:

$$\boldsymbol{\Delta}^\top\bar{\boldsymbol{\Omega}}^{-1} = \frac{\mathbf{1}_q\boldsymbol{\alpha}^\top}{\sqrt{1 + \boldsymbol{\alpha}^\top\bar{\boldsymbol{\Omega}}\boldsymbol{\alpha}}}, \tag{6.56}$$

$$\bar{\boldsymbol{\Gamma}} - \boldsymbol{\Delta}^\top\bar{\boldsymbol{\Omega}}^{-1}\boldsymbol{\Delta} = \frac{1}{1 + \boldsymbol{\alpha}^\top\bar{\boldsymbol{\Omega}}\boldsymbol{\alpha}} \mathbf{I}_q. \tag{6.57}$$

Therefore, from the pdf of the AASUN in Arellano-Valle and Azzalini (2006) and Equations (6.56) and (6.57), the pdf of the SUNNY distribution becomes:

$$f(\mathbf{x}) = \frac{1}{\Phi_q(\mathbf{0}_q;\mathbf{0}_q,\bar{\mathbf{\Gamma}})}\phi_p(\mathbf{x};\boldsymbol{\xi},\mathbf{\Omega})\left[\Phi\{\boldsymbol{\alpha}^\top\boldsymbol{\omega}^{-1}(\mathbf{x}-\boldsymbol{\xi})\}\right]^q, \quad \mathbf{x}\in\mathbb{R}^p, \tag{6.58}$$

where $\Phi_q(\mathbf{0}_q;\mathbf{0}_q,\bar{\mathbf{\Gamma}}) = K_q\left(\sqrt{\boldsymbol{\alpha}^\top\bar{\mathbf{\Omega}}\boldsymbol{\alpha}}\right)$ with $K_q(\theta) = \int_{-\infty}^{+\infty}\{\Phi(\theta t)\}^q\phi(t)\mathrm{d}t = \mathrm{E}[\{\Phi(\theta X)\}^q]$ and $X\sim\mathcal{N}(0,1)$. When $q=1$, $K_1(\theta)=1/2$ and the SUNNY distribution reduces to the skew-normal (ASN) distribution. We can derive the expressions $K_2(\theta)=\pi^{-1}\mathrm{atan}(\sqrt{1+2\theta^2})$ and $K_3(\theta) = 3/(2\pi)\mathrm{atan}(\sqrt{1+2\theta^2})-1/4$. When $q=2,3,\ldots$, the pdf of the SUNNY distribution becomes more peaked/skewed. When $p=1$, the pdf (6.58) was studied by Balakrishnan (2002) and Gupta and Gupta (2004). The case $p=2$ was explored by Sharafi and Behboodian (2008); see also Yadegari et al. (2008) for a generalization. The extension of the case $p=1$ to a skew-t version can be found in Shafiei and Doostparast (2014).

An alternative parameterization of the SUNNY distribution can be obtained from Example 77 on tractable SUN pdfs. Indeed, in Equation (6.26), setting $\boldsymbol{\eta}_1=\cdots=\boldsymbol{\eta}_q=\boldsymbol{\eta}$, one obtains a pdf similar to (6.58) but in the $(\mathbf{\Psi},\boldsymbol{\eta})$ parameterization instead of the $(\mathbf{\Omega},\boldsymbol{\alpha})$ parameterization. The formal properties of multivariate SUNNY distributions can be derived from those of the multivariate AASUN/SUN distributions; see Open Research Problems 8 and 16.

In our parameterization, the SUNNY distribution becomes a $\mathcal{SUN}_{p,q}(\boldsymbol{\xi},\mathbf{\Psi},\mathbf{H},\boldsymbol{\tau},\bar{\mathbf{\Gamma}})$ with

$$\mathbf{\Psi}=\mathbf{\Omega}-q\beta_q^2\boldsymbol{\eta}\boldsymbol{\eta}^\top,\quad \mathbf{H}=\beta_q^2\boldsymbol{\eta}\mathbf{1}_q^\top,\quad \boldsymbol{\tau}=\mathbf{0}_q,\quad \bar{\mathbf{\Gamma}}=(1-\delta^2)\mathbf{I}_q+\delta^2\mathbf{1}_q\mathbf{1}_q^\top,$$

where

$$\beta_q^2=\{1+(q-1)\delta^2\}^{-1},\quad \boldsymbol{\eta}=\boldsymbol{\omega}\boldsymbol{\delta},\quad \boldsymbol{\delta}=(1+\boldsymbol{\alpha}^\top\bar{\mathbf{\Omega}}\boldsymbol{\alpha})^{-1/2}\bar{\mathbf{\Omega}}\boldsymbol{\alpha},\quad \delta^2=\boldsymbol{\delta}^\top\bar{\mathbf{\Omega}}^{-1}\boldsymbol{\delta}.$$

That is, if $\mathbf{U}\sim\mathcal{TN}_q(\mathbf{0}_q;\mathbf{0}_q,(1-\delta^2)\mathbf{I}_q+\delta^2\mathbf{1}_q\mathbf{1}_q^\top)$ and $\mathbf{W}\sim\mathcal{N}_p(\mathbf{0}_p,\mathbf{\Omega}-q\beta_q^2\boldsymbol{\eta}\boldsymbol{\eta}^\top)$ are independent, then

$$\begin{aligned}\mathbf{X}&=\boldsymbol{\xi}+\beta_q^2\boldsymbol{\eta}\mathbf{1}_q^\top\mathbf{U}+\mathbf{W}\\&=\boldsymbol{\xi}+\beta_q^2\boldsymbol{\eta}U+\mathbf{W},\end{aligned}$$

where $U=\mathbf{1}_q^\top\mathbf{U}=(\mathbf{1}_q^\top\mathbf{U}_0\mid\mathbf{U}_0>\mathbf{0}_q)$, with $\mathbf{U}_0\sim\mathcal{N}_q(\mathbf{0}_q,(1-\delta^2)\mathbf{I}_q+\delta^2\mathbf{1}_q\mathbf{1}_q^\top)$ and so $\mathbf{1}_q^\top\mathbf{U}_0\sim\mathcal{N}(0,q/\beta_q^2)$. Let $\mu_U=\mathrm{E}(U)$ and $\sigma_U^2=\mathrm{var}(U)$. Then,

$$\boldsymbol{\mu}=\mathrm{E}(\mathbf{X})=\boldsymbol{\xi}+\beta_q^2\boldsymbol{\eta}\mu_U,\quad \mathbf{\Sigma}=\mathrm{var}(\mathbf{X})=\mathbf{\Omega}-a_q\boldsymbol{\eta}\boldsymbol{\eta}^\top,\quad a_q=\beta_q^2(q-\beta_q^2\sigma_U^2).$$

Let $\mathbf{Y}=\mathbf{C}(\mathbf{X}-\boldsymbol{\mu})$ and $\mathbf{Y}'=\mathbf{C}(\mathbf{X}'-\boldsymbol{\mu})$, where $\mathbf{X}'$ is an independent replicate of $\mathbf{X}$ and $\mathbf{C}$ is a $p\times p$ nonsingular matrix. We have $\beta_{1,p}=\mathrm{E}[\{\mathbf{Y}^\top(\mathbf{C}\mathbf{\Sigma}\mathbf{C}^\top)^{-1}\mathbf{Y}'\}^3]$ and $\beta_{2,p}=\mathrm{E}[\{\mathbf{Y}^\top(\mathbf{C}\mathbf{\Sigma}\mathbf{C}^\top)^{-1}\mathbf{Y}\}^2]$. Choosing $\mathbf{C}$ such that $\mathbf{C}\mathbf{\Omega}\mathbf{C}^\top=\mathbf{I}_p$ and making $\boldsymbol{\eta}_c=\mathbf{C}\boldsymbol{\eta}$, we have

$$\mathbf{C}\mathbf{\Sigma}\mathbf{C}^\top=\mathbf{I}_p-a_q\boldsymbol{\eta}_c\boldsymbol{\eta}_c^\top \implies (\mathbf{C}\mathbf{\Sigma}\mathbf{C}^\top)^{-1}=\mathbf{I}_p+b_q\boldsymbol{\eta}_c\boldsymbol{\eta}_c^\top,\quad b_q=\frac{a_q}{1-a_q\delta^2},$$

since $\boldsymbol{\eta}_c^\top\boldsymbol{\eta}_c=\delta^2$. Thus,

$$\beta_{1,p}=\mathrm{E}[\{\mathbf{Y}^\top\mathbf{Y}'+b_q(\boldsymbol{\eta}_c^\top\mathbf{Y})(\boldsymbol{\eta}_c^\top\mathbf{Y}')\}^3],\quad \beta_{2,p}=\mathrm{E}[\{\|\mathbf{Y}\|^2+b_q(\boldsymbol{\eta}_c^\top\mathbf{Y})^2\}^2],$$

where $\mathbf{Y}=\beta_q^2\boldsymbol{\eta}_c(U-\mu_U)+\mathbf{W}_c$, with $\mathbf{W}_c=\mathbf{C}\mathbf{W}\sim\mathcal{N}_p(\mathbf{0}_p,\mathbf{I}_p-q\beta_q^2\boldsymbol{\eta}_c\boldsymbol{\eta}_c^\top)$. Let $U_0=U-\mu_U$ and note that $\boldsymbol{\eta}_c^\top\mathbf{W}_c=\beta_q\delta\sqrt{1-\delta^2}\,V_0$, where $V_0=\mathbf{e}_1^\top\mathbf{W}_0\sim\mathcal{N}(0,1)$ with $\mathbf{W}_0\sim\mathcal{N}_p(\mathbf{0}_p,\mathbf{I}_p)$ which is independent of U_0. Thus, we have $\boldsymbol{\eta}_c^\top\mathbf{Y}=\beta_q^2\delta^2U_0+\beta_q\delta\sqrt{1-\delta^2}\,V_0$, where U_0 and V_0 are independent. Similarly, $\boldsymbol{\eta}_c^\top\mathbf{Y}'=\beta_q^2\delta^2U_0'+\beta_q\delta\sqrt{1-\delta^2}\,V_0'$, where $U_0'=U'-\mu_U$, $V_0'=\boldsymbol{\eta}_c^\top\mathbf{W}_0'$ and $\mathbf{W}_0'$ are independent replicates of U_0, V_0 and $\mathbf{W}_0$, respectively. Also, let $\mathbf{M}_0 = \mathbf{I}_p - q\beta_q^2 \boldsymbol{\eta}_c \boldsymbol{\eta}_c^\top$, and write $\mathbf{W}_c^\top \mathbf{W}_c' = \mathbf{W}_0^\top \mathbf{M}_0 \mathbf{W}_0' = \mathbf{W}_0^\top \mathbf{W}_0' - q\beta_q^2 \delta^2 V_0 V_0'$, where $\mathbf{W}_0, \mathbf{W}_0'$ are iid $\mathcal{N}_p(\mathbf{0}_p, \mathbf{I}_p)$. Thus, after some algebra, we find

$$\begin{aligned}\mathbf{Y}^\top \mathbf{Y}' + b_q(\boldsymbol{\eta}_c^\top \mathbf{Y})(\boldsymbol{\eta}_c^\top \mathbf{Y}') &= Z_0 Z_0' + \mathbf{W}_0^\top \mathbf{W}_0' - V_0 V_0' + b_q \delta^2 Z_0 Z_0' \\ &= (1 + b_q \delta^2) Z_0 Z_0' + \mathbf{W}_0^\top \mathbf{P}_0 \mathbf{W}_0',\end{aligned}$$

where $Z_0 = \beta_q^2 \delta U_0 + \beta_q \sqrt{1-\delta^2}\, V_0$, $Z_0' = \beta_q^2 \delta U_0' + \beta_q \sqrt{1-\delta^2}\, V_0'$, and $\mathbf{P}_0 = \mathbf{I}_p - \mathbf{e}_1 \mathbf{e}_1^\top$ is an orthogonal projection matrix of rank $p-1$. Note that U_0, U_0', V_0, V_0', $\mathbf{P}_0 \mathbf{W}_0$ and $\mathbf{P}_0 \mathbf{W}_0'$ are all mutually independent; Z_0, Z_0', $\mathbf{P}_0 \mathbf{W}_0$ and $\mathbf{P}_0 \mathbf{W}_0'$ are also mutually independent. Furthermore, note that

$$\begin{aligned}&\mu_Z \equiv \mathrm{E}(Z_0) = 0, \quad \sigma_Z^2 \equiv \mathrm{var}(Z_0) = \beta_q^4 \delta^2 \sigma_U^2 + \beta_q^2 (1-\delta^2) \\ \Longrightarrow\ &\mathrm{E}(Z_0^3) = \sigma_Z^3 \beta_{1,1}(Z_0), \quad \mathrm{E}(Z_0^4) = \sigma_Z^4 \beta_{2,1}(Z_0).\end{aligned}$$

Thus, omitting those terms with zero expectation, we have for Mardia's measures of multivariate skewness and kurtosis:

$$\begin{aligned}\beta_{1,p} &= \mathrm{E}[\{\mathbf{Y}^\top \mathbf{Y}' + b_q(\boldsymbol{\eta}_c^\top \mathbf{Y})(\boldsymbol{\eta}_c^\top \mathbf{Y}')\}^3] \\ &= \mathrm{E}[\{(1+b_q\delta^2) Z_0 Z_0' + \mathbf{W}_0^\top \mathbf{P}_0 \mathbf{W}_0'\}^3] \\ &= (1+b_q\delta^2)^3 \mathrm{E}(Z_0^3)^2 \\ &= (1+b_q\delta^2)^3 \sigma_Z^6 \beta_{1,1}(Z_0)^2, \\ \beta_{2,p} &= \mathrm{E}[\{\|\mathbf{Y}^\top\|^2 + b_q(\boldsymbol{\eta}_c^\top \mathbf{Y})^2\}^2] \\ &= \mathrm{E}[\{(1+b_q\delta^2) Z_0^2 + \mathbf{W}_0^\top \mathbf{P}_0 \mathbf{W}_0\}^2] \\ &= (1+b_q\delta^2)^2 \mathrm{E}(Z_0^4) + 2(1+b_q\delta^2)\mathrm{E}(Z_0^2)\mathrm{E}(\mathbf{W}_0^\top \mathbf{P}_0 \mathbf{W}_0) + \mathrm{E}\{(\mathbf{W}_0^\top \mathbf{P}_0 \mathbf{W}_0)^2\} \\ &= (1+b_q\delta^2)^2 \sigma_Z^4 \beta_{1,2}(Z_0) + 2(p-1)(1+b_q\delta^2)\sigma_Z^2 + p^2 - 1.\end{aligned}$$

Note here that $Z_0 = \beta_q^2 \delta U_0 + \beta_q \sqrt{1-\delta^2}\, V_0 = -\mu_U + \beta_q^2 \delta U + \beta_q \sqrt{1-\delta^2}\, V_0$ is not exactly an SN variable, since $U = (\mathbf{1}_q^\top \mathbf{U}_0 \mid \mathbf{U}_0 > \mathbf{0}_q)$, where $\mathbf{1}_q^\top \mathbf{U}_0 \sim \mathcal{N}(0, q/\beta_q^2)$, is not a truncated variable, but rather a selection variable (see Chapter 9). Its pdf is (see Exercise 6.6):

$$\tilde{g}(u) = \frac{1}{\Phi_q(\mathbf{0}_q; \mathbf{0}_q, \bar{\boldsymbol{\Gamma}})} \phi(u; 0, q\beta_q^{-2}) \mathrm{P}(\mathbf{U}_0 > \mathbf{0}_q \mid \mathbf{1}_q^\top \mathbf{U}_0 = u), \quad u > 0.$$

For $q = 1$, $\tilde{g}(u)$ becomes the $\mathcal{HN}(0,1)$-pdf. In this case, Z_0 has an ASN/SN distribution. In fact, for $q = 1$, the SUNNY distribution becomes the multivariate ASN/SN distribution. Thus, it is easy to verify that the above measures reduce to $\beta_{1,p} = \beta_{1,1}(Z_0)^2$ and $\beta_{2,p} = \beta_{2,1}(Z_0) - 3 + p(p+2)$, where $Z_0 \sim \mathcal{ASN}(-\sqrt{2/\pi}\,\delta, 1, \delta/\sqrt{1-\delta^2})$ or equivalently, $Z_0 \sim \mathcal{SN}(-\sqrt{2/\pi}\,\delta, 1-\delta^2, \delta)$.

6.9 Exercises

Exercise 6.1 *Prove Theorem 94 using the cf.*

Exercise 6.2 *Prove Theorem 99 using the cf.*

Exercise 6.3 *Prove Theorem 100 using the definition of the SUN distribution.*

Exercise 6.4 *Prove Theorem 101 using the definition of the SUN distribution.*

Exercise 6.5 *Prove Theorem 101 using the selection representation.*

Exercise 6.6 *Let* $\mathbf{U}_0 = \sqrt{1-\delta^2}\,\mathbf{U}_1 + \delta U_2 \mathbf{1}_q$, *where* $|\delta| \le 1$, $\mathbf{U}_1 \sim \mathcal{N}_q(\mathbf{0}_q, \mathbf{I}_q)$ *and* $U_2 \sim \mathcal{N}(0,1)$ *which are independent.*

(a) Show that $\mathrm{P}(\mathbf{U}_0 > \mathbf{0}_q) = \mathrm{E}[\{\Phi(\alpha Z)\}^q]$, *where* $\alpha = \delta/\sqrt{1-\delta^2}$ *and* $Z \sim \mathcal{N}(0,1)$;

(b) Find the pdf, cdf and cf of $\mathbf{U} = (\mathbf{U}_0 \mid \mathbf{U}_0 > \mathbf{0}_q)$;

(c) Study the conditional distribution of $(\mathbf{U}_0 \mid \mathbf{1}_q^\top \mathbf{U}_0 = u)$ *and then the distribution of* $U = \mathbf{1}_q^\top \mathbf{U}$. *What can be said about the moments of* U*?*

Exercise 6.7 *Let* $\mathbf{X} \sim \mathcal{SUN}_{p,q}(\boldsymbol{\xi}, \boldsymbol{\Psi}, \mathbf{H}, \boldsymbol{\tau}, \bar{\boldsymbol{\Gamma}})$. *Show that there exists an orthogonal matrix* $\mathbf{P} \in \mathcal{O}_p$ *such that* $\mathbf{PX} \sim \mathcal{SUN}_{p,q}(\mathbf{P}\boldsymbol{\xi}, \mathbf{D}, \mathbf{PH}, \boldsymbol{\tau}, \bar{\boldsymbol{\Gamma}})$ *where* $\mathbf{D} > 0$ *is a p-by-p diagonal matrix.*

Exercise 6.8 *Let* $\mathbf{X}_1, \ldots, \mathbf{X}_n$ *be a random sample from* $\mathbf{X} \sim \mathcal{SUN}_{p,q}(\boldsymbol{\xi}, \boldsymbol{\Psi}, \mathbf{H}, \boldsymbol{\tau}, \bar{\boldsymbol{\Gamma}})$. *Compute the following expectations and variances:* $\mathrm{E}(\bar{\mathbf{X}})$, $\mathrm{var}(\bar{\mathbf{X}})$, $\mathrm{E}(\mathbf{S})$, $\mathrm{var}(\mathbf{S})$.

Exercise 6.9 *Let* $\mathbf{Y}$ *and* $\mathbf{Y}'$ *be iid random p-vectors with mean vector* $\mathbf{0}_p$, *variance matrix* $\mathbf{I}_p$, *and with third and fourth moments* $\boldsymbol{\mu}_3(\mathbf{Y})$ *and* $\boldsymbol{\mu}_4(\mathbf{Y})$, *respectively. Prove that*

$$\mathrm{tr}\{\boldsymbol{\mu}_3(\mathbf{Y})^\top \boldsymbol{\mu}_3(\mathbf{Y})\} = \mathrm{E}\{(\mathbf{Y}^\top \mathbf{Y}')^3\} \quad \text{and} \quad \mathrm{tr}\{\boldsymbol{\mu}_4(\mathbf{Y})\} = \mathrm{E}\{(\mathbf{Y}^\top \mathbf{Y})^2\}.$$

Exercise 6.10 *Prove that the unified skew-normal pdf is log-concave; see Arellano-Valle and Azzalini (2022).*

Exercise 6.11 *Prove Theorem 109.*

Exercise 6.12 *Let* $\mathbf{X} \sim \mathcal{SUN}_{p,q}(\boldsymbol{\xi}, \boldsymbol{\Psi}, \mathbf{H}, \boldsymbol{\tau}, \bar{\boldsymbol{\Gamma}})$. *Find an expression for* $\mathrm{E}\{e^{\mathrm{i}tQ_A(\mathbf{X}-\mathbf{a})}\}$ *assuming that the dispersion matrix* $\boldsymbol{\Psi}$ *is singular, i.e., when* $\boldsymbol{\Psi} \ge 0$.

Exercise 6.13 *Prove Theorem 110; see Arellano-Valle and Azzalini (2022).*

Exercise 6.14 *Let* $\mathbf{X} = (\mathbf{Y} \mid \mathbf{Z} - \boldsymbol{\Lambda}\mathbf{Y} \le \boldsymbol{\tau})$, *where* $\mathbf{Y} \sim \mathcal{N}_p(\mathbf{0}_p, \boldsymbol{\Omega})$ *and* $\mathbf{Z} \sim \mathcal{N}_q(\mathbf{0}_q, \boldsymbol{\Gamma})$ *are nonsingular and independent normal random vectors and the vector* $\boldsymbol{\tau} \in \mathbb{R}^q$ *and the matrix* $\boldsymbol{\Lambda} \in \mathbb{R}^{q \times p}$ *are fixed. Show that* $\mathbf{X}$ *has a pdf given by*

$$f_{\mathbf{X}}(\mathbf{x}) = \frac{1}{\Phi_q(\boldsymbol{\tau}; \mathbf{0}_q, \boldsymbol{\Gamma} + \boldsymbol{\Lambda}\boldsymbol{\Omega}\boldsymbol{\Lambda}^\top)} \phi_p(\mathbf{x}; \mathbf{0}_q, \boldsymbol{\Omega}) \Phi_q(\boldsymbol{\tau} + \boldsymbol{\Lambda}\mathbf{x}; \mathbf{0}_q, \boldsymbol{\Gamma}), \quad \mathbf{x} \in \mathbb{R}^p.$$

Exercise 6.15 *Let* $\mathbf{X} \in \mathbb{R}^p$ *and* $\mathbf{Y} \in \mathbb{R}^q$ *be two absolutely continuous random vectors with some joint distribution such that* $\mathbf{X}$ *has pdf* $f_{\mathbf{X}}$. *Using that* $\mathrm{P}(\mathbf{Y} \in B) = \int_B \mathrm{P}(\mathbf{Y} \in B \mid \mathbf{X} = \mathbf{x}) f_{\mathbf{X}}(\mathbf{x}) \mathrm{d}\mathbf{x}$ *for each Borel subset B of* $\mathbb{R}^q$, *show that the selection random vector* $\mathbf{X}_{\mathrm{s}} = (\mathbf{X} \mid \mathbf{Y} \in B)$ *has pdf given by*

$$f_{\mathbf{X}_{\mathrm{s}}}(\mathbf{x}) = f_{\mathbf{X}}(\mathbf{x}) \frac{\mathrm{P}(\mathbf{Y} \in B \mid \mathbf{X} = \mathbf{x})}{\mathrm{P}(\mathbf{Y} \in B)}.$$

Apply this result to the case in which the joint distribution of $\mathbf{X}$ *and* $\mathbf{Y}$ *is multivariate normal with* $\mathrm{E}(\mathbf{X}) = \boldsymbol{\mu}_{\mathbf{X}}$, $\mathrm{E}(\mathbf{Y}) = \boldsymbol{\mu}_{\mathbf{Y}}$, $\mathrm{var}(\mathbf{X}) = \boldsymbol{\Sigma}_{\mathbf{XX}}$, $\mathrm{var}(\mathbf{Y}) = \boldsymbol{\Sigma}_{\mathbf{YY}}$ *and* $\mathrm{cov}(\mathbf{X}, \mathbf{Y}) = \boldsymbol{\Sigma}_{\mathbf{XY}}$.

Exercise 6.16 *Consider a q-dimensional random vector* $\mathbf{V}$ *and an arbitrary Borel set* $A \subset \mathbb{R}^q$. *Assume that the variance matrix* $\mathrm{var}(\mathbf{V})$ *exists and is positive semi-definite* ($\mathrm{var}(\mathbf{V}) \geq 0$). *Show that*

$$\begin{aligned} \mathrm{var}(\mathbf{V}) =& p_A \mathrm{var}(\mathbf{V} \mid \mathbf{V} \in A) + (1 - p_A) \mathrm{var}(\mathbf{V} \mid \mathbf{V} \in A^c) \\ &+ p_A(1 - p_A)\{\mathrm{E}(\mathbf{V}) - \mathrm{E}(\mathbf{V} \mid \mathbf{V} \in A^c)\}\{\mathrm{E}(\mathbf{V}) - \mathrm{E}(\mathbf{V} \mid \mathbf{V} \in A^c)\}^\top, \end{aligned}$$

where $p_A = \mathrm{P}(\mathbf{V} \in A)$.

Exercise 6.17 *Give an expression for the first 4 multivariate moments and Mardia's measures of multivariate skewness and kurtosis of* $\mathbf{X} \sim \mathcal{SUN}_{p,q}(\boldsymbol{\xi}, \boldsymbol{\Psi}, \mathbf{H}, \mathbf{0}_q, \mathbf{I}_q)$.

Exercise 6.18 *Give an expression for the first 4 multivariate moments and Mardia's measures of multivariate skewness and kurtosis of* $\mathbf{X} \sim \mathcal{SUN}_{p,q}(\boldsymbol{\xi}, \boldsymbol{\Psi}, \boldsymbol{\eta}\mathbf{1}_q^\top, \tau\mathbf{1}_q, (1-\rho)\mathbf{I}_q + \rho\mathbf{1}_q\mathbf{1}_q^\top)$, *with* $\boldsymbol{\eta} \in \mathbb{R}^p$, $\tau \in \mathbb{R}$ *and* $\rho \in (-(q-1)^{-1}, 1)$ *for* $q > 1$.

Exercise 6.19 *In the Exercises 6.17 and 6.18 find* $\mathrm{var}(\mathbf{X}\mathbf{X}^\top)$.

Exercise 6.20 *In the Exercises 6.17 and 6.18 partition* $\mathbf{X} = (\mathbf{X}_1^\top, \mathbf{X}_2^\top)^\top$ *and accordingly partition the respective parameters. Find the marginal distribution of* $\mathbf{X}_1$ *and the conditional distribution* $\mathbf{X}_2 \mid \mathbf{X}_1 = \mathbf{x}_1$.

Exercise 6.21 *In the Exercises 6.17 and 6.18 study the marginal and joint asymptotic distributions of the sample mean vector* $\bar{\mathbf{X}}$ *and sample covariance matrix* $\mathbf{S}$ *of a random sample of size n.*

6.10 Open Research Problems

Open Research Problem 6 *Explore whether the representations given in Remark 12 yield identifiable SUN distributions.*

Open Research Problem 7 *Develop the theory and inference for multivariate BSUNSET distributions introduced in Section 6.8.1.*

Open Research Problem 8 *Consider the multivariate SUNNY distribution introduced in Section 6.8.2:*

(a) For $p = 2$, plot the contours of the SUNNY pdf for various values of q and investigate the effect of q on the shape of the pdf;

(b) Study the formal properties of the multivariate SUNNY distribution from those of the multivariate AASUN/SUN distributions. In particular, derive the first 4 multivariate moments;

(c) Implement likelihood inference for the SUNNY distribution and propose a method to select q;

(d) Fit the multivariate SUNNY distribution to some datasets.

Open Research Problem 9 *Extend the multivariate skew-slash distribution introduced in Section 5.11.4 to a multivariate unified skew-normal-slash (SUNSL) distribution and develop its theory and inference. Compare its properties with the multivariate SUT distribution introduced in Chapter 7.*

Open Research Problem 10 *Explore the entropy, mutual information, KL divergence, J divergence, and Wasserstein distance for two multivariate SUN distributions.*

Open Research Problem 11 *Develop the theory of sampling from the multivariate SUN distribution. Then extend the results in Exercise 5.27 to the multivariate SUN distribution.*

Open Research Problem 12 *Develop maximum likelihood inference for the multivariate SUN distribution based on the EM algorithm. Pay attention to identifiability issues and consider submodels of interest.*

Open Research Problem 13 *Develop inference for the multivariate SUN distribution based on the factor analysis model (see Section 10.8) where the latent dimension q corresponds to the number of underlying factors.*

Chapter 7

The Multivariate Unified Skew-t Distribution

The multivariate unified skew-t (SUT) family provides a very flexible and analytically tractable distribution and includes the multivariate unified skew-normal (SUN) class as a limit case. It is indexed by the same set of parameters as the SUN family to model location and dispersion, $(\boldsymbol{\xi}, \boldsymbol{\Psi})$ say, with $\boldsymbol{\xi} \in \mathbb{R}^p$ and $\boldsymbol{\Psi} > 0\,(\boldsymbol{\Psi} \geq 0)$, a positive (semi-)definite matrix of dimension $p \times p$, and to regulate skewness/shape, $(\mathbf{H}, \boldsymbol{\tau}, \bar{\boldsymbol{\Gamma}})$ say, with $\mathbf{H} \in \mathbb{R}^{p \times q}$, $\boldsymbol{\tau} \in \mathbb{R}^q$ and $\bar{\boldsymbol{\Gamma}} > 0$, a proper $q \times q$ correlation matrix. However, the SUT family also includes an additional parameter of degrees of freedom, $\nu > 0$, to regulate the kurtosis, thus allowing situations with heavy tails, such as its symmetric parent, the multivariate t distribution. Although the multivariate unified skew-t distribution can also be singular or nonsingular depending on whether the dispersion matrix $\boldsymbol{\Psi}$ is positive semi-definite ($\boldsymbol{\Psi} \geq 0$; $|\boldsymbol{\Psi}| = 0$) or positive definite ($\boldsymbol{\Psi} > 0$; $|\boldsymbol{\Psi}| > 0$), respectively, throughout this chapter we will pay particular attention to the nonsingular case.

7.1 Construction, Definition, and First Elementary Properties

7.1.1 Construction

Similarly to the multivariate symmetric t distribution, independence between its marginals is not possible within the multivariate unified skew-t family. This fact forces us to construct this family in a slightly different way from the multivariate unified skew-normal family.

We start by considering the symmetric $(p+q)$-dimensional random vector given by

$$\begin{pmatrix} \mathbf{W} \\ \mathbf{W}_0 \end{pmatrix} \sim \mathcal{T}_{p+q}\left(\begin{pmatrix} \mathbf{0}_p \\ \mathbf{0}_q \end{pmatrix}, \begin{pmatrix} \boldsymbol{\Psi} & \mathbf{O} \\ \mathbf{O} & \bar{\boldsymbol{\Gamma}} \end{pmatrix}, \nu \right), \tag{7.1}$$

which has uncorrelated (but not independent) marginals $\mathbf{W} \sim \mathcal{T}_p(\mathbf{0}_p, \boldsymbol{\Psi}, \nu)$ and $\mathbf{W}_0 \sim \mathcal{T}_q(\mathbf{0}_q, \bar{\boldsymbol{\Gamma}}, \nu)$, where $\mathcal{T}_k(\boldsymbol{\mu}, \boldsymbol{\Sigma}, \nu)$ denotes the multivariate t distribution with $k \times 1$ location vector $\boldsymbol{\mu}$, $k \times k$ dispersion matrix $\boldsymbol{\Sigma}$, and $\nu > 0$ degrees of freedom. As in the SUN case, we focus our analysis on the case where $\bar{\boldsymbol{\Gamma}}$ is a positive definite correlation matrix (which avoids redundant latent variables and/or identifiability problems) so that the marginal $\mathbf{W}_0 \sim \mathcal{T}_q(\mathbf{0}_q, \bar{\boldsymbol{\Gamma}}, \nu)$ will always be nonsingular, while $\mathbf{W} \sim \mathcal{T}_p(\mathbf{0}_p, \boldsymbol{\Psi}, \nu)$ could be singular or nonsingular depending on whether we assume $\boldsymbol{\Psi} \geq 0$ or $\boldsymbol{\Psi} > 0$, respectively.

Now, we consider some preliminary key results, for which we need some random functions of $\mathbf{W}_0 \sim \mathcal{T}_q(\mathbf{0}_q, \bar{\boldsymbol{\Gamma}}, \nu)$, with $\bar{\boldsymbol{\Gamma}} > 0$, such as the quadratic form $Q(\mathbf{W}_0) = Q(\mathbf{W}_0; \mathbf{0}_q, \bar{\boldsymbol{\Gamma}})$ given by

$$Q(\mathbf{W}_0) = \mathbf{W}_0^\top \bar{\boldsymbol{\Gamma}}^{-1} \mathbf{W}_0 \Longrightarrow \frac{1}{q} Q(\mathbf{W}_0) \sim \mathcal{F}_{q,\nu},$$

where $\mathcal{F}_{q,\nu}$ is the Fisher distribution with $q = \text{rank}(\bar{\boldsymbol{\Gamma}})$ and ν degrees of freedom, and the truncated random vector defined by

$$\mathbf{U} \stackrel{d}{=} (\mathbf{W}_0 \mid \mathbf{W}_0 + \boldsymbol{\tau} > \mathbf{0}_q) \sim \mathcal{TT}_q(-\boldsymbol{\tau}; \mathbf{0}_q, \bar{\boldsymbol{\Gamma}}, \nu),$$

where $\mathcal{TT}_q(\mathbf{c};\boldsymbol{\mu},\boldsymbol{\Sigma},\nu)$ denotes the *truncated below* **c** *multivariate t distribution* $\mathcal{T}_q(\boldsymbol{\mu},\boldsymbol{\Sigma},\nu)$. Another important consequence of the condition $\bar{\boldsymbol{\Gamma}} > 0$ is that $\mathbf{U} \sim \mathcal{TT}_q(-\boldsymbol{\tau};\mathbf{0}_q,\bar{\boldsymbol{\Gamma}},\nu)$ is absolutely continuous, with pdf given by

$$g(\mathbf{u}) = \frac{1}{T_q(\boldsymbol{\tau};\mathbf{0}_q,\bar{\boldsymbol{\Gamma}},\nu)} t_q(\mathbf{u};\mathbf{0}_q,\bar{\boldsymbol{\Gamma}},\nu), \quad \mathbf{u}+\boldsymbol{\tau} > \mathbf{0}_q, \tag{7.2}$$

where $t_k(\mathbf{z};\boldsymbol{\mu},\boldsymbol{\Sigma},\nu)$ (with $\boldsymbol{\Sigma} > 0$) and $T_k(\mathbf{z};\boldsymbol{\mu},\boldsymbol{\Sigma},\nu)$ are the pdf and cdf of $\mathbf{Z} \sim \mathcal{T}_k(\boldsymbol{\mu},\boldsymbol{\Sigma},\nu)$, respectively; i.e.,

$$t_k(\mathbf{z};\boldsymbol{\mu},\boldsymbol{\Sigma},\nu) = \frac{\Gamma\{(\nu+k)/2\}}{\Gamma(\nu/2)(\pi\nu)^{k/2}} \left\{1 + \frac{1}{\nu}(\mathbf{z}-\boldsymbol{\mu})^\top \boldsymbol{\Sigma}^{-1}(\mathbf{z}-\boldsymbol{\mu})\right\}^{-(\nu+k)/2}, \quad \mathbf{z} \in \mathbb{R}^k$$

and

$$T_k(\mathbf{z};\boldsymbol{\mu},\boldsymbol{\Sigma},\nu) = \mathrm{P}(\mathbf{Z} \le \mathbf{z}) = \int_{\mathbf{v}\le\mathbf{z}} t_k(\mathbf{v};\boldsymbol{\mu},\boldsymbol{\Sigma},\nu)\mathrm{d}\mathbf{v}.$$

With these facts, we can establish the following key result.

Theorem 113 *Under the multivariate t joint distribution in (7.1), we have:*

(i) If we let $\mathbf{W}_* = \sqrt{\frac{\nu+q}{\nu+Q(\mathbf{W}_0)}}\mathbf{W}$, *then* $\mathbf{W}_* \sim \mathcal{T}_p(\mathbf{0}_p,\boldsymbol{\Psi},\nu+q)$ *and is independent of* $\mathbf{W}_0 \sim \mathcal{T}_q(\mathbf{0}_q,\bar{\boldsymbol{\Gamma}},\nu)$;

(ii) For any fixed matrix $\mathbf{H} \in \mathbb{R}^{p\times q}$,

$$\mathbf{H}\mathbf{W}_0 + \sqrt{\frac{\nu+Q(\mathbf{W}_0)}{\nu+q}}\mathbf{W}_* \overset{d}{=} \mathbf{H}\mathbf{W}_0 + \mathbf{W} \sim \mathcal{T}_p(\mathbf{0}_p, \mathbf{H}\bar{\boldsymbol{\Gamma}}\mathbf{H}^\top + \boldsymbol{\Psi}, \nu); \tag{7.3}$$

(iii) If we let $\mathbf{U} \overset{d}{=} (\mathbf{W}_0 \mid \mathbf{W}_0 + \boldsymbol{\tau} > \mathbf{0}_q)$, *then* $Q(\mathbf{U}) \overset{d}{=} (Q(\mathbf{W}_0) \mid \mathbf{W}_0 + \boldsymbol{\tau} > \mathbf{0}_q)$ *and, for any fixed matrix* $\mathbf{H} \in \mathbb{R}^{p\times q}$,

$$\mathbf{H}\mathbf{U} + \sqrt{\frac{\nu+Q(\mathbf{U})}{\nu+q}}\mathbf{W}_* \overset{d}{=} \left(\mathbf{H}\mathbf{W}_0 + \sqrt{\frac{\nu+Q(\mathbf{W}_0)}{\nu+q}}\mathbf{W}_*\right) \Big| \mathbf{W}_0 + \boldsymbol{\tau} > \mathbf{0}_q \tag{7.4}$$

$$= (\mathbf{H}\mathbf{W}_0 + \mathbf{W}) \,\big|\, \mathbf{W}_0 + \boldsymbol{\tau} > \mathbf{0}_q, \tag{7.5}$$

where **U** *and* $\mathbf{W}_*$ *are independent;*

(iv) If $\boldsymbol{\tau} = \mathbf{0}_q$, *then* $Q(\mathbf{U}) \overset{d}{=} Q(\mathbf{W}_0)$.

Proof 113 *From (7.1) and the fact that the multivariate t distribution is closed by conditioning, it follows that the conditional distribution of* **W** *given* $\mathbf{W}_0 = \mathbf{w}_0$ *is*

$$\begin{aligned}
(\mathbf{W} \mid \mathbf{W}_0 = \mathbf{w}_0) &\sim \mathcal{T}_p\left(\mathbf{0}_p, \frac{\nu+Q(\mathbf{w}_0)}{\nu+q}\boldsymbol{\Psi}, \nu+q\right) \\
\Longrightarrow &\left(\sqrt{\frac{\nu+q}{\nu+Q(\mathbf{w}_0)}}\mathbf{W} \,\Big|\, \mathbf{W}_0 = \mathbf{w}_0\right) \sim \mathcal{T}_p(\mathbf{0}_p,\boldsymbol{\Psi},\nu+q), \quad \forall \mathbf{w}_0 \in \mathbb{R}^q \\
\Longrightarrow &\sqrt{\frac{\nu+q}{\nu+Q(\mathbf{W}_0)}}\mathbf{W} \sim \mathcal{T}_p(\mathbf{0}_p,\boldsymbol{\Psi},\nu+q),
\end{aligned}$$

which is independent of $\mathbf{W}_0$ *since for any* $\mathbf{w}_0 \in \mathbb{R}^q$ *its conditional distribution given* $\mathbf{W}_0 = \mathbf{w}_0$ *does not depend on* $\mathbf{w}_0$*. Hence,* $\mathbf{W}_* = \sqrt{\frac{\nu+q}{\nu+Q(\mathbf{W}_0)}}\mathbf{W}$ *is independent of any Borel function of* $\mathbf{W}_0$*, in particular of* $Q(\mathbf{W}_0)$ *and of* $\mathbf{U} \overset{d}{=} (\mathbf{W}_0 \mid \mathbf{W}_0 + \boldsymbol{\tau} > \mathbf{0}_q)$*. This proves part* (i) *as well as the statement in* (iii) *that the random vectors* $\mathbf{U}$ *and* $\mathbf{W}_*$ *are independent.*

The proof of (ii) *is straightforward since by definition* $\mathbf{W}_* = \sqrt{\frac{\nu+q}{\nu+Q(\mathbf{W}_0)}}\mathbf{W}$*. The distribution of* $\mathbf{H}\mathbf{W}_0 + \mathbf{W}$ *in (7.3) is also straightforward from (7.1) and the fact that the multivariate t distribution is closed under affine transformations. To prove the first result of part* (iii)*, note that for any Borel function* $T(\mathbf{z})$*,* $\mathbf{z} \in \mathbb{R}^q$*, and Borel subset* C *of* $\mathbb{R}$*, we have*

$$\begin{aligned} \mathrm{P}\{T(\mathbf{U}) \in C\} &= \mathrm{P}\{\mathbf{U} \in T^{-1}(C)\} = \mathrm{P}\{\mathbf{W}_0 \in T^{-1}(C) \mid \mathbf{W}_0 + \boldsymbol{\tau} > \mathbf{0}_q\} \\ &= \mathrm{P}\{T(\mathbf{W}_0) \in C \mid \mathbf{W}_0 + \boldsymbol{\tau} > \mathbf{0}_q\}, \end{aligned}$$

where $T^{-1}(C)$ *is the inverse image of* C *under the function* T*. This proves that*

$$\mathbf{U} \overset{d}{=} (\mathbf{W}_0 \mid \mathbf{W}_0 + \boldsymbol{\tau} > \mathbf{0}_q) \Longrightarrow T(\mathbf{U}) \overset{d}{=} (T(\mathbf{W}_0) \mid \mathbf{W}_0 + \boldsymbol{\tau} > \mathbf{0}_q).$$

Thus, using this general fact with $T(\mathbf{u}) = Q(\mathbf{u}) = \mathbf{u}^\top \bar{\boldsymbol{\Gamma}}^{-1}\mathbf{u}$*, we have the proof of the first result of part* (iii)*. Now, using similar arguments and the fact that* $\mathbf{W}_*$ *is independent of* $\mathbf{W}_0$*, we have* $(\mathbf{W}_*, \mathbf{U}) \overset{d}{=} \{(\mathbf{W}_*, \mathbf{W}_0) \mid \mathbf{W}_0 + \boldsymbol{\tau} > \mathbf{0}_q\}$*, and therefore* $S(\mathbf{W}_*, \mathbf{U}) \overset{d}{=} \{S(\mathbf{W}_*, \mathbf{W}_0) \mid \mathbf{W}_0 + \boldsymbol{\tau} > \mathbf{0}_q\}$ *for any Borel function* $S(\mathbf{w}_*, \mathbf{u})$*,* $(\mathbf{w}_*, \mathbf{u}) \in \mathbb{R}^{p+q}$*. Thus, taking* $S(\mathbf{w}_*, \mathbf{u}) = \mathbf{H}\mathbf{u} + \sqrt{\frac{\nu+Q(\mathbf{u})}{\nu+q}}\mathbf{w}_*$*, the proof of (7.4) follows. The proof of (7.5) is straightforward from* (ii)*.*

Finally, to prove (iv), first note that we can write $\mathbf{W}_0 = R_0 \bar{\boldsymbol{\Gamma}}^{1/2}\mathbf{U}^{(q)}$*, where* $R_0 \overset{d}{=} Q(\mathbf{W}_0)^{1/2} = (\mathbf{W}_0^\top \bar{\boldsymbol{\Gamma}}^{-1}\mathbf{W}_0)^{1/2}$ *and is independent of* $\mathbf{U}^{(q)}$*, the uniform vector on the unit q-dimensional sphere (see Chapter 4). Also, since the radial variable* $R_0 > 0$ *(with probability 1) and* $\|\mathbf{U}^{(q)}\| = \{(\mathbf{U}^{(q)})^\top \mathbf{U}^{(q)}\}^{1/2} = 1$ *(with probability 1), then for* $\boldsymbol{\tau} = \mathbf{0}_q$ *we have*

$$Q(\mathbf{U}) \overset{d}{=} (Q(\mathbf{W}_0) \mid \mathbf{W}_0 > \mathbf{0}_q) = (R_0^2 (\mathbf{U}^{(q)})^\top \mathbf{U}^{(q)} \mid R_0 \bar{\boldsymbol{\Gamma}}^{1/2}\mathbf{U}^{(q)} > \mathbf{0}_q) \overset{d}{=} (R_0^2 \mid \bar{\boldsymbol{\Gamma}}^{1/2}\mathbf{U}^{(q)} > \mathbf{0}_q) \overset{d}{=} R_0^2,$$

thus ending the proof of part (iv)*.* □

The converse of part (iii) of Theorem 113 is also true; that is, $Q(\mathbf{U}) \overset{d}{=} Q(\mathbf{W}_0) \Longleftrightarrow \boldsymbol{\tau} = \mathbf{0}_q$ (see Exercise 7.1).

With these elements, we can define/construct a SUT random vector as shown next.

7.1.2 Definition

Definition 20 [Multivariate unified skew-*t* distribution] *A random vector* $\mathbf{X} \in \mathbb{R}^p$ *is said to have a multivariate unified skew-t distribution with location parameter* $\boldsymbol{\xi} \in \mathbb{R}^p$*, p-by-p dispersion matrix* $\boldsymbol{\Psi}$*, shape/skewness parameters* $\mathbf{H} \in \mathbb{R}^{p\times q}$*,* $\boldsymbol{\tau} \in \mathbb{R}^q$ *and* $\bar{\boldsymbol{\Gamma}} > 0$*, a q-by-q positive definite correlation matrix, and* $\nu > 0$ *degrees of freedom, in symbols* $\mathbf{X} \sim \mathcal{SUT}_{p,q}(\boldsymbol{\xi}, \boldsymbol{\Psi}, \mathbf{H}, \boldsymbol{\tau}, \bar{\boldsymbol{\Gamma}}, \nu)$*, iff* $\mathbf{X}$ *can be represented as follow:*

$$\mathbf{X} = \boldsymbol{\xi} + \mathbf{H}\mathbf{U} + \sqrt{\frac{\nu + Q(\mathbf{U})}{\nu + q}}\mathbf{W}_*, \tag{7.6}$$

where $Q(\mathbf{U}) = \mathbf{U}^\top \bar{\boldsymbol{\Gamma}}^{-1}\mathbf{U}$*,* $\mathbf{U}$ *and* $\mathbf{W}_*$ *are independent, with* $\mathbf{U} \sim \mathcal{TT}_q(-\boldsymbol{\tau}; \mathbf{0}_q, \bar{\boldsymbol{\Gamma}}, \nu)$*, a truncated t random vector, and* $\mathbf{W}_* \sim \mathcal{T}_p(\mathbf{0}_p, \boldsymbol{\Psi}, \nu + q)$*.*

It is important not to lose sight of the fact that for the construction of (7.6), we have used that $\mathbf{U} \overset{d}{=} (\mathbf{W}_0 \mid \mathbf{W}_0 + \boldsymbol{\tau} > \mathbf{0}_q)$ and $\mathbf{W}_* \overset{d}{=} \sqrt{\frac{\nu+q}{\nu+Q(\mathbf{W}_0)}}\mathbf{W}$, with $Q(\mathbf{W}_0) = \mathbf{W}_0^\top \bar{\boldsymbol{\Gamma}}^{-1}\mathbf{W}_0$, which is independent of $\mathbf{W}_0$, and where $\mathbf{W}_0$ and $\mathbf{W}$ have a multivariate t joint distribution as given by (7.1). Furthermore, if we set $\mathbf{W}_{Q(\mathbf{U})} = \sqrt{\frac{\nu+Q(\mathbf{U})}{\nu+q}}\mathbf{W}_*$, then we can observe that $\mathbf{W}_{Q(\mathbf{U})}|\mathbf{U} \sim \mathcal{T}_p(\mathbf{0}_p, \frac{\nu+Q(\mathbf{U})}{\nu+q}\boldsymbol{\Psi}, \nu+q)$ and therefore $\mathbf{W}_{Q(\mathbf{U})} \perp\!\!\!\perp \mathbf{U}|Q(\mathbf{U})$. Thus, we can rewrite (7.6) as

$$\mathbf{X} = \boldsymbol{\xi} + \mathbf{H}\mathbf{U} + \mathbf{W}_{Q(\mathbf{U})}.$$

The SUT distribution can also be defined as a mixture of location and scale of the normal distribution. In fact, since $\mathbf{W}_* \overset{d}{=} V_*^{-1/2}\mathbf{Z}_*$ where $V_* \sim \chi^2_{\nu+q}/(\nu+q)$ and $\mathbf{Z} \sim \mathcal{N}_p(\mathbf{0}_p, \boldsymbol{\Psi})$, which are mutually independent and also independent of $\mathbf{U}$, then (7.6) can be rewritten as

$$\mathbf{X} = \boldsymbol{\xi} + \mathbf{H}\mathbf{U} + \sqrt{\frac{\nu+Q(\mathbf{U})}{\nu+q}}V_*^{-1/2}\mathbf{Z}_*. \tag{7.7}$$

Furthermore, setting $V_{Q(\mathbf{U})} = \frac{\nu+q}{\nu+Q(\mathbf{U})}V_*$, where $V_* \perp\!\!\!\perp \mathbf{U}$, we have $V_{Q(\mathbf{U})} \perp\!\!\!\perp \mathbf{Z}_* \sim \mathcal{N}_p(\mathbf{0}_p, \boldsymbol{\Psi})$ and $V_{Q(\mathbf{U})}|\mathbf{U} = \mathbf{u} \sim \mathcal{G}amma\{(\nu+q)/2, (\nu+Q(\mathbf{u}))/2\}$; so in (7.7) we can replace the term $\sqrt{\frac{\nu+Q(\mathbf{U})}{\nu+q}}V_*^{-1/2}$ by $V_{Q(\mathbf{U})}^{-1/2}$.

Remark 17 (Link between $\mathcal{SUT}_{p,q}$ and $\mathcal{AASUT}_{p,q}$) *It is important to note that the (nonsingular) multivariate unified skew-t distribution $\mathcal{SUT}_{p,q}(\boldsymbol{\xi}, \boldsymbol{\Psi}, \mathbf{H}, \boldsymbol{\tau}, \bar{\boldsymbol{\Gamma}}, \nu)$ is completely equivalent to the multivariate unified skew-t distribution arising from the parameterization described by Arellano-Valle and Azzalini (2006), here denoted by $\mathcal{AASUT}_{p,q}(\boldsymbol{\xi}, \boldsymbol{\Omega}, \boldsymbol{\Delta}, \boldsymbol{\tau}, \bar{\boldsymbol{\Gamma}}, \nu)$, except for differences in parameterization. Moreover, since both parameterizations use the same ν, the link between $\mathcal{SUT}_{p,q}$ and $\mathcal{AASUT}_{p,q}$ is the same as the one given in Remark 13.*

When $q = 1$, the SUT distribution is called an extended skew-t (EST) distribution; see, e.g., Arellano-Valle and Genton (2010b) and Adcock (2010). Recall that a similar fact occurs with the SUN distribution, which becomes the extended skew-normal (ESN) distributions when $q = 1$. In fact, the SUN distribution is a limit particular case of the SUT distribution as $\nu \to \infty$ and, similarly, the ESN distribution can be obtained from the EST one by letting $\nu \to \infty$. If, in addition to $q = 1$, we also set $\tau = 0$, then we obtain the skew-t (ST) distribution; see, e.g., Branco and Dey (2001) and Azzalini and Capitanio (2003), from which the skew-normal (SN) distribution follows as $\nu \to \infty$; see, e.g., Azzalini and Capitanio (1999) and Azzalini and Dalla Valle (1996). An alternative form of skew-t distribution in the univariate case has been put forward by Jones and Faddy (2003), with a multivariate extension in Jones and Larsen (2004). Finally, when $\mathbf{H} = \mathbf{O}$ and $\boldsymbol{\tau} = \mathbf{0}_q$, the multivariate symmetric t distribution is recovered, from which the multivariate normal distribution is obtained by making $\nu \to \infty$. Note that in the SUN case, the condition $\boldsymbol{\tau} = \mathbf{0}_q$ is not necessary to recover the symmetric normal parent distribution.

The special case with $\boldsymbol{\tau} = \mathbf{0}_q$ substantially simplifies the SUT representation (7.6) and incorporates further important properties into the SUT distribution. In fact, when $\boldsymbol{\tau} = \mathbf{0}_q$ the SUT distribution becomes a scale mixtures of the SUN distribution, and therefore a location-scale (or mean-variance) mixtures of the multivariate normal distribution. To see this, first we recall that the symmetric random vector in (7.1) can be represented as scale mixtures of normals as follows (see Example 46). Let

$$\begin{pmatrix} \tilde{\mathbf{W}} \\ \tilde{\mathbf{W}}_0 \end{pmatrix} \sim \mathcal{N}_{p+q}\left(\begin{pmatrix} \mathbf{0}_p \\ \mathbf{0}_q \end{pmatrix}, \begin{pmatrix} \boldsymbol{\Psi} & \mathbf{O} \\ \mathbf{O} & \bar{\boldsymbol{\Gamma}} \end{pmatrix}\right) \quad \text{and} \quad V \sim \chi^2_\nu/\nu$$

be independent random quantities, where $\mathcal{N}_k(\boldsymbol{\mu},\boldsymbol{\Sigma})$ denotes the multivariate normal distribution with $k\times 1$ mean vector $\boldsymbol{\mu}$ and $k\times k$ variance matrix $\boldsymbol{\Sigma}$. Note that the marginals $\tilde{\mathbf{W}}\sim\mathcal{N}_p(\mathbf{0}_p,\boldsymbol{\Psi})$ and $\tilde{\mathbf{W}}_0\sim\mathcal{N}_q(\mathbf{0}_q,\bar{\boldsymbol{\Gamma}})$ are independent. Moreover, they are independent of V. Thus, in (7.1) we have:

$$\begin{aligned}\begin{pmatrix}\mathbf{W}\\ \mathbf{W}_0\end{pmatrix} &= V^{-1/2}\begin{pmatrix}\tilde{\mathbf{W}}\\ \tilde{\mathbf{W}}_0\end{pmatrix}\sim\mathcal{T}_{p+q}\left(\begin{pmatrix}\mathbf{0}_p\\ \mathbf{0}_q\end{pmatrix},\begin{pmatrix}\boldsymbol{\Psi} & \mathbf{O}\\ \mathbf{O} & \bar{\boldsymbol{\Gamma}}\end{pmatrix},\nu\right)\\ \Longrightarrow\begin{pmatrix}\mathbf{W}\\ \mathbf{W}_0\end{pmatrix} &\mid \mathbf{W}_0+\boldsymbol{\tau}>\mathbf{0}_q = V^{-1/2}\begin{pmatrix}\tilde{\mathbf{W}}\\ \tilde{\mathbf{W}}_0\end{pmatrix}\mid V^{-1/2}\tilde{\mathbf{W}}_0+\boldsymbol{\tau}>\mathbf{0}_q.\end{aligned}$$

Thus, since $\tilde{\mathbf{W}}\perp\!\!\!\perp\tilde{\mathbf{W}}_0\perp\!\!\!\perp V$ and $V>0$ (with probability 1) if $\boldsymbol{\tau}=\mathbf{0}_q$, then

$$\begin{aligned}\begin{pmatrix}\mathbf{W}\\ \mathbf{W}_0\end{pmatrix}\mid\mathbf{W}_0>\mathbf{0}_q &= V^{-1/2}\begin{pmatrix}\tilde{\mathbf{W}}\\ \tilde{\mathbf{W}}_0\end{pmatrix}\mid\tilde{\mathbf{W}}_0>\mathbf{0}_q = V^{-1/2}\begin{pmatrix}\tilde{\mathbf{W}}\\ \tilde{\mathbf{U}}\end{pmatrix}\\ \Longrightarrow(\mathbf{H}\mathbf{W}_0+\mathbf{W})|\mathbf{W}_0>\mathbf{0}_q &= V^{-1/2}(\mathbf{H}\tilde{\mathbf{U}}+\tilde{\mathbf{W}}),\end{aligned}\tag{7.8}$$

for any $\mathbf{H}\in\mathbb{R}^{p\times q}$, where $\tilde{\mathbf{W}}\sim\mathcal{N}_p(\mathbf{0}_p,\boldsymbol{\Psi})$, $\tilde{\mathbf{W}}_0\sim\mathcal{N}_p(\mathbf{0}_p,\bar{\boldsymbol{\Gamma}})$, $\tilde{\mathbf{U}}\overset{d}{=}(\tilde{\mathbf{W}}_0\mid\tilde{\mathbf{W}}_0>\mathbf{0}_q)\sim\mathcal{TN}_q(\mathbf{0}_q;\mathbf{0}_q,\bar{\boldsymbol{\Gamma}})$ and $V\sim\chi^2_\nu/\nu$, all mutually independent. Consequently, if $\boldsymbol{\tau}=\mathbf{0}_q$, then according to (7.8), part (*iii*) of Theorem 113, and (7.6), we have

$$\begin{aligned}\mathbf{X}-\boldsymbol{\xi} &\overset{d}{=} V^{-1/2}(\mathbf{H}\tilde{\mathbf{U}}+\tilde{\mathbf{W}})\\ &= V^{-1/2}\tilde{\mathbf{Z}},\end{aligned}$$

where $\tilde{\mathbf{Z}}=\mathbf{H}\tilde{\mathbf{U}}+\tilde{\mathbf{W}}$, which by definition (see Chapter 6) is $\tilde{\mathbf{Z}}\sim\mathcal{SUN}_{p,q}(\mathbf{0}_p,\boldsymbol{\Psi},\mathbf{H},\mathbf{0}_q,\bar{\boldsymbol{\Gamma}})$, since $\tilde{\mathbf{U}}\sim\mathcal{TN}_q(\mathbf{0}_q;\mathbf{0}_q,\bar{\boldsymbol{\Gamma}})$ and $\tilde{\mathbf{W}}\sim\mathcal{N}_p(\mathbf{0}_p,\boldsymbol{\Psi})$ and are independent. This yields the following formal fact.

Theorem 114 *A random vector* $\mathbf{X}\sim\mathcal{SUT}_{p,q}(\boldsymbol{\xi},\boldsymbol{\Psi},\mathbf{H},\boldsymbol{\tau},\bar{\boldsymbol{\Gamma}},\nu)$ *with* $\boldsymbol{\tau}=\mathbf{0}_q$ *can be equivalently formulated as*

$$\mathbf{X}=\boldsymbol{\xi}+V^{-1/2}\tilde{\mathbf{Z}},\tag{7.9}$$

where $V\sim\chi^2_\nu/\nu\equiv\mathcal{G}amma(\nu/2,\nu/2)$ *is independent of* $\tilde{\mathbf{Z}}=\mathbf{H}\tilde{\mathbf{U}}+\tilde{\mathbf{W}}\sim\mathcal{SUN}_{p,q}(\mathbf{0}_p,\boldsymbol{\Psi},\mathbf{H},\mathbf{0}_q,\bar{\boldsymbol{\Gamma}})$, *in which* $\tilde{\mathbf{U}}\sim\mathcal{TN}_q(\mathbf{0}_q;\mathbf{0}_q,\bar{\boldsymbol{\Gamma}})$ *and* $\tilde{\mathbf{W}}\sim\mathcal{N}_p(\mathbf{0}_p,\boldsymbol{\Psi})$ *are independent.*

Proof 114 *It was done above.* □

When $q=1$, we have $\boldsymbol{\tau}=\tau$, $\mathbf{H}=\boldsymbol{\eta}$, and $\bar{\boldsymbol{\Gamma}}=1$, so if $\tau=0$, then Equation (7.9) represents the multivariate skew-t distribution. Also, since $V\sim\chi^2_\nu/\nu$, then $V\overset{P}{\to}1$ as $\nu\to\infty$, and therefore from Equation (7.9) clearly follows that

$$\mathbf{X}=\boldsymbol{\xi}+V^{-1/2}\tilde{\mathbf{Z}}\ \overset{d}{\to}\ \tilde{\mathbf{X}}=\boldsymbol{\xi}+\tilde{\mathbf{Z}}\sim\mathcal{SUN}_{p,q}(\boldsymbol{\xi},\boldsymbol{\Psi},\mathbf{H},\mathbf{0}_q,\bar{\boldsymbol{\Gamma}})\ \text{ as }\ \nu\to\infty,$$

which, for $q=1$, becomes the skew-normal distribution. Moreover, from Equation (7.9), it is also clear that

$$\mathbf{H}=\mathbf{O}\Longrightarrow\mathbf{X}=\boldsymbol{\xi}+V^{-1/2}\tilde{\mathbf{W}}\sim\mathcal{T}_p(\boldsymbol{\xi},\boldsymbol{\Psi},\nu)\ \overset{d}{\to}\ \tilde{\mathbf{X}}=\boldsymbol{\xi}+\tilde{\mathbf{W}}\sim\mathcal{N}_p(\boldsymbol{\xi},\boldsymbol{\Psi})\ \text{ as }\ \nu\to\infty.$$

That is, as mentioned earlier, when $\boldsymbol{\tau}=\mathbf{0}_q$ and $\mathbf{H}=\mathbf{O}$, the distribution of $\mathbf{X}$ becomes the multivariate symmetric t distribution with location vector $\boldsymbol{\xi}$, dispersion matrix $\boldsymbol{\Psi}$, and ν degrees of freedom, which includes the multivariate normal distribution as a limit case as $\nu\to\infty$, and can be singular or nonsingular according to the dispersion matrix $\boldsymbol{\Psi}\geq 0$ or $\boldsymbol{\Psi}>0$, respectively.

In summary, Definition 20 and its equivalent representations include both the singular and the nonsingular multivariate unified skew-t distribution. The nonsingular case also requires that

$\boldsymbol{\Psi} + \mathbf{H}\bar{\boldsymbol{\Gamma}}\mathbf{H}^\top > 0$ for all $\mathbf{H} \in \mathbb{R}^{p\times q}$ for the pdf to be valid, which is guaranteed if $\boldsymbol{\Psi} > 0$ since by assumption $\bar{\boldsymbol{\Gamma}} > 0$.

7.1.3 *Mean vector and variance matrix*

The Definition 20 and its equivalent representations for the special case when $\boldsymbol{\tau} = \mathbf{0}_q$ are also very convenient to obtain the elementary properties of the SUT family. For instance, this definition considerably facilitates the calculation of the mean vector and the variance matrix of the SUT random vector $\mathbf{X} \sim \mathcal{SUT}_{p,q}(\boldsymbol{\xi}, \boldsymbol{\Psi}, \mathbf{H}, \boldsymbol{\tau}, \bar{\boldsymbol{\Gamma}}, \nu)$. However, to study this, we need some auxiliary results.

Theorem 115 *Let* $Q(\mathbf{U}) = \mathbf{U}^\top\bar{\boldsymbol{\Gamma}}^{-1}\mathbf{U}$*, where* $\mathbf{U} \sim \mathcal{TT}_q(-\boldsymbol{\tau}; \mathbf{0}_q, \bar{\boldsymbol{\Gamma}}, \nu)$*. Then,*

$$\mathrm{E}\left\{\left(\frac{\nu+Q(\mathbf{U})}{\nu+q}\right)^{k/2} h(\mathbf{U})\right\} = \frac{T_q\left(\sqrt{\frac{\nu-k}{\nu}}\,\boldsymbol{\tau}; \mathbf{0}_q, \bar{\boldsymbol{\Gamma}}, \nu-k\right)}{T_q(\boldsymbol{\tau}; \mathbf{0}_q, \bar{\boldsymbol{\Gamma}}, \nu)} \left(\frac{\nu}{\nu-k}\right)^{q/2} \left(\frac{\nu}{\nu+q}\right)^{k/2} \frac{c(\nu,q)}{c(\nu-k,q)}$$
$$\times \mathrm{E}\left\{h\left(\sqrt{\frac{\nu}{\nu-k}}\,\mathbf{U}_k\right)\right\},$$

where $\mathbf{U}_k \sim \mathcal{TT}_q\left(-\sqrt{\frac{\nu-k}{\nu}}\,\boldsymbol{\tau}; \mathbf{0}_q, \bar{\boldsymbol{\Gamma}}, \nu-k\right)$ *for* $0 < k < \nu$*, and*

$$c(\upsilon, m) = \frac{\Gamma\{(\upsilon+m)/2\}}{\Gamma(\upsilon/2)(\pi\upsilon)^{m/2}}, \quad \upsilon, m > 0,$$

$$\mathrm{E}\left\{\left(\frac{\nu+Q(\mathbf{U})}{\nu+q}\right)^{k/2}\right\} = \frac{T_q\left(\sqrt{\frac{\nu-k}{\nu}}\,\boldsymbol{\tau}; \mathbf{0}_q, \bar{\boldsymbol{\Gamma}}, \nu-k\right)}{T_q(\boldsymbol{\tau}; \mathbf{0}_q, \bar{\boldsymbol{\Gamma}}, \nu)} \left(\frac{\nu}{\nu-k}\right)^{q/2} \left(\frac{\nu}{\nu+q}\right)^{k/2} \frac{c(\nu,q)}{c(\nu-k,q)},$$

for $0 < k < \nu$*. In particular, if* $k = 2$ *and* $\nu > 2$ *then:*

$$\mathrm{E}\left\{\left(\frac{\nu+Q(\mathbf{U})}{\nu+q}\right)\right\} = \frac{T_q\left(\sqrt{\frac{\nu-k}{\nu}}\,\boldsymbol{\tau}; \mathbf{0}_q, \bar{\boldsymbol{\Gamma}}, \nu-2\right)}{T_q(\boldsymbol{\tau}; \mathbf{0}_q, \bar{\boldsymbol{\Gamma}}, \nu)} \left(\frac{\nu+q-2}{\nu+q}\right)\left(\frac{\nu}{\nu-2}\right);$$

$$\mathrm{E}\left\{\left(\frac{\nu+Q(\mathbf{U})}{\nu+q}\right)\mathbf{U}\right\} = \frac{T_q\left(\sqrt{\frac{\nu-k}{\nu}}\,\boldsymbol{\tau}; \mathbf{0}_q, \bar{\boldsymbol{\Gamma}}, \nu-2\right)}{T_q(\boldsymbol{\tau}; \mathbf{0}_q, \bar{\boldsymbol{\Gamma}}, \nu)} \left(\frac{\nu+q-2}{\nu+q}\right)\left(\frac{\nu}{\nu-2}\right)^{3/2} \mathrm{E}(\mathbf{U}_2);$$

$$\mathrm{E}\left\{\left(\frac{\nu+Q(\mathbf{U})}{\nu+q}\right)\mathbf{U}\mathbf{U}^\top\right\} = \frac{T_q\left(\sqrt{\frac{\nu-k}{\nu}}\,\boldsymbol{\tau}; \mathbf{0}_q, \bar{\boldsymbol{\Gamma}}, \nu-2\right)}{T_q(\boldsymbol{\tau}; \mathbf{0}_q, \bar{\boldsymbol{\Gamma}}, \nu)} \left(\frac{\nu+q-2}{\nu+q}\right)\left(\frac{\nu}{\nu-2}\right)^{2} \mathrm{E}(\mathbf{U}_2\mathbf{U}_2^\top),$$

with $\mathbf{U}_2 \sim \mathcal{TT}_q\left(-\sqrt{\frac{\nu-2}{\nu}}\,\boldsymbol{\tau}; \mathbf{0}_q, \bar{\boldsymbol{\Gamma}}, \nu-2\right)$*.*

Proof 115 *First of all, we use the relation that*

$$\left(\frac{\nu+Q(\mathbf{u})}{\nu+q}\right)^{k/2} t_q(\mathbf{u}; \mathbf{0}_q, \bar{\boldsymbol{\Gamma}}, \nu) = \left(\frac{\nu}{\nu+q}\right)^{k/2} \frac{c(\nu,q)}{c(\nu-k,q)}\, t_q\left(\sqrt{\frac{\nu-k}{\nu}}\,\mathbf{u}; \mathbf{0}_q, \bar{\boldsymbol{\Gamma}}, \nu-k\right),$$

where $\nu > k > 0$. *Next, because* $\mathbf{U}$ *has pdf* $g(\mathbf{u}) = t_q(\mathbf{u};\mathbf{0}_q,\bar{\mathbf{\Gamma}},\nu)/T_q(\boldsymbol{\tau};\mathbf{0}_q,\bar{\mathbf{\Gamma}},\nu)$ *for* $\mathbf{u}+\boldsymbol{\tau} > \mathbf{0}_q$, *we have*

$$\begin{aligned}
\mathrm{E}\left\{\left(\frac{\nu+Q(\mathbf{U})}{\nu+q}\right)^{k/2} h(\mathbf{U})\right\} &= \int_{\mathbf{u}+\boldsymbol{\tau}>\mathbf{0}_q} \left(\frac{\nu+Q(\mathbf{u})}{\nu+q}\right)^{k/2} h(\mathbf{u})g(\mathbf{u})d\mathbf{u} \\
&= \frac{1}{T_q(\boldsymbol{\tau};\mathbf{0}_q,\bar{\mathbf{\Gamma}},\nu)} \int_{\mathbf{u}+\boldsymbol{\tau}>\mathbf{0}_q} \left(\frac{\nu+Q(\mathbf{u})}{\nu+q}\right)^{k/2} h(\mathbf{u})t_q(\mathbf{u};\mathbf{0}_q,\bar{\mathbf{\Gamma}},\nu)d\mathbf{u} \\
&= \frac{1}{T_q(\boldsymbol{\tau};\mathbf{0}_q,\bar{\mathbf{\Gamma}},\nu)} \left(\frac{\nu}{\nu+q}\right)^{k/2} \frac{c(\nu,q)}{c(\nu-k,q)} \\
&\quad \times \int_{\mathbf{u}+\boldsymbol{\tau}>\mathbf{0}_q} h(\mathbf{u})t_q\left(\sqrt{\frac{\nu-k}{\nu}}\mathbf{u};\mathbf{0}_q,\bar{\mathbf{\Gamma}},\nu-k\right) d\mathbf{u}.
\end{aligned}$$

Thus, the proof is made by the change of variables $\mathbf{v} = \sqrt{\frac{\nu-k}{\nu}}\mathbf{u}$ *and using the fact that the truncated vector* $\mathbf{U}_k$ *has pdf* $g_k(\mathbf{u}) = t_q(\mathbf{u};\mathbf{0}_q,\bar{\mathbf{\Gamma}},\nu-k)/T_q\left(\sqrt{\frac{\nu-k}{\nu}}\boldsymbol{\tau};\mathbf{0}_q,\bar{\mathbf{\Gamma}},\nu-k\right)$ *for* $\mathbf{u}+\sqrt{\frac{\nu-k}{\nu}}\boldsymbol{\tau} > \mathbf{0}_q$ *and* $\nu-k>0$. □

Theorem 116 *Let* $\mathbf{X} \sim \mathcal{SUT}_{p,q}(\boldsymbol{\xi},\mathbf{\Psi},\mathbf{H},\boldsymbol{\tau},\bar{\mathbf{\Gamma}},\nu)$. *Then:*

$$\mathrm{E}(\mathbf{X}) = \boldsymbol{\xi} + \mathbf{H}\mathrm{E}(\mathbf{U}) \quad (\nu > 1), \tag{7.10}$$

$$\mathrm{var}(\mathbf{X}) = \mathbf{H}\mathrm{var}(\mathbf{U})\mathbf{H}^\top + \frac{T_q\left(\sqrt{\frac{\nu-2}{\nu}}\boldsymbol{\tau};\mathbf{0}_q,\bar{\mathbf{\Gamma}},\nu-2\right)}{T_q(\boldsymbol{\tau};\mathbf{0}_q,\bar{\mathbf{\Gamma}},\nu)}\left(\frac{\nu}{\nu-2}\right)\mathbf{\Psi} \quad (\nu>2), \tag{7.11}$$

where $\mathrm{E}(\mathbf{U}) = \mathrm{E}(\mathbf{W}_0 \mid \mathbf{W}_0+\boldsymbol{\tau} > \mathbf{0}_q)$ *and* $\mathrm{var}(\mathbf{U}) = \mathrm{var}(\mathbf{W}_0 \mid \mathbf{W}_0+\boldsymbol{\tau}>\mathbf{0}_q)$ *for* $\mathbf{W}_0 \sim \mathcal{T}_q(\boldsymbol{\tau},\bar{\mathbf{\Gamma}},\nu)$.

Proof 116 *By Equation (7.6) and the fact that* $\mathbf{U}$, *and therefore* $\sqrt{\frac{\nu+Q(\mathbf{U})}{\nu+q}}$, *are independent of* $\mathbf{W}_*$, *all with finite expectation for* $\nu>1$, *we have*

$$\mathrm{E}(\mathbf{X}) = \boldsymbol{\xi} + \mathbf{H}\mathrm{E}(\mathbf{U}) + \mathrm{E}\left(\sqrt{\frac{\nu+Q(\mathbf{U})}{\nu+q}}\right)\mathrm{E}(\mathbf{W}_*) = \boldsymbol{\xi}+\mathbf{H}\mathrm{E}(\mathbf{U}),$$

since $\mathrm{E}(\mathbf{W}_*) = \mathbf{0}_p$ *for* $\nu+q>1$. *This proves (7.10).*

To compute $\mathrm{var}(\mathbf{X})$ *we use the well-known fact that, for* $\nu>2$,

$$\mathrm{var}(\mathbf{X}) = \mathrm{var}\{\mathrm{E}(\mathbf{X}\mid\mathbf{U})\} + \mathrm{E}\{\mathrm{var}(\mathbf{X}\mid\mathbf{U})\},$$

where from (7.6),

$$\begin{aligned}
\mathrm{E}(\mathbf{X}\mid\mathbf{U}) &= \boldsymbol{\xi}+\mathbf{H}\mathbf{U}+\sqrt{\frac{\nu+Q(\mathbf{U})}{\nu+q}}\mathrm{E}(\mathbf{W}_*) = \boldsymbol{\xi}+\mathbf{H}\mathbf{U} \quad (\nu+q>1) \\
&\Longrightarrow \mathrm{var}\{\mathrm{E}(\mathbf{X}\mid\mathbf{U})\} = \mathrm{var}(\mathbf{H}\mathbf{U}) = \mathbf{H}\mathrm{var}(\mathbf{U})\mathbf{H}^\top, \\
\mathrm{var}(\mathbf{X}\mid\mathbf{U}) &= \mathrm{var}\left(\sqrt{\frac{\nu+Q(\mathbf{U})}{\nu+q}}\mathbf{W}_*\middle|\mathbf{U}\right) = \left(\frac{\nu+Q(\mathbf{U})}{\nu+q}\right)\mathrm{var}(\mathbf{W}_*\mid\mathbf{U}) \\
&= \left(\frac{\nu+Q(\mathbf{U})}{\nu+q}\right)\mathrm{var}(\mathbf{W}_*) = \left(\frac{\nu+Q(\mathbf{U})}{\nu+q}\right)\left(\frac{\nu+q}{\nu+q-2}\right)\mathbf{\Psi} \quad (\nu+q>2) \\
&\Longrightarrow \mathrm{E}\{\mathrm{var}(\mathbf{X}\mid\mathbf{U})\} = \mathrm{E}\left(\frac{\nu+Q(\mathbf{U})}{\nu+q}\right)\left(\frac{\nu+q}{\nu+q-2}\right)\mathbf{\Psi}.
\end{aligned}$$

Thus, the proof of (7.11) follows by applying the results of Theorem 115. □

Obviously, we need the mean vector and the variance matrix of the multivariate truncated *t* distribution to find these quantities for the multivariate unified skew-*t* distribution.

When $\boldsymbol{\tau} = \mathbf{0}_q$ we have further simplifications, since in this case $T_q(\mathbf{0}_q;\mathbf{0}_q,\bar{\boldsymbol{\Gamma}},\nu-2) = T_q(\mathbf{0}_q;\mathbf{0}_q,\bar{\boldsymbol{\Gamma}},\nu) = \Phi_q(\mathbf{0}_q;\mathbf{0}_q,\bar{\boldsymbol{\Gamma}})$ for all $\nu > 2$ and $\mathbf{U} = V^{-1/2}\tilde{\mathbf{U}}$, where as before $V \sim \mathcal{G}amma(\nu/2,\nu/2)$ and $\tilde{\mathbf{U}} \sim \mathcal{TN}_q(\mathbf{0}_q;\mathbf{0}_q,\bar{\boldsymbol{\Gamma}})$, which are independent. Hence, we have

$$\begin{aligned}
\mathrm{E}(\mathbf{U}) &= \mathrm{E}(V^{-1/2}\tilde{\mathbf{U}}) = \mathrm{E}(V^{-1/2})\mathrm{E}(\tilde{\mathbf{U}}) \quad (\nu > 1), \\
\mathrm{var}(\mathbf{U}) &= \mathrm{var}(V^{-1/2}\tilde{\mathbf{U}}) = \mathrm{E}(V^{-1})\mathrm{E}(\tilde{\mathbf{U}}\tilde{\mathbf{U}}^\top) - \mathrm{E}(V^{-1/2})^2\mathrm{E}(\tilde{\mathbf{U}})\mathrm{E}(\tilde{\mathbf{U}})^\top \\
&= \mathrm{E}(V^{-1})\mathrm{var}(\tilde{\mathbf{U}}) + \{\mathrm{E}(V^{-1}) - \mathrm{E}(V^{-1/2})^2\}\mathrm{E}(\tilde{\mathbf{U}})\mathrm{E}(\tilde{\mathbf{U}})^\top \quad (\nu > 2),
\end{aligned}$$

where

$$\mathrm{E}(V^{-1/2}) = \int_0^\infty \frac{(\nu/2)^{\nu/2}}{\Gamma(\nu/2)} x^{(\nu-1)/2-1} e^{-\nu x/2} \mathrm{d}x = \sqrt{\frac{\nu}{2}} \frac{\Gamma\{(\nu-1)/2\}}{\Gamma(\nu/2)} \equiv b_1(\nu) \quad (\nu > 1), \tag{7.12}$$

$$\mathrm{E}(V^{-1}) = \frac{\nu}{\nu-2} \quad (\nu > 2).$$

Note that $b_1(\nu) \to 1$ as $\nu \to \infty$. The notation $b_k(\nu)$, with $0 < k < \nu$, will be used for the expectation of $V^{-k/2}$ throughout this chapter.

Example 84 *Let* $\mathbf{U} = V^{-1/2}\tilde{\mathbf{U}} \sim \mathcal{TT}_q(\mathbf{0}_q;\mathbf{0}_q,\mathbf{I}_q)$ *where* $\tilde{\mathbf{U}} \sim \mathcal{HN}_q(\mathbf{0}_q,\mathbf{I}_q)$ *is independent of* $V \sim \mathcal{G}amma(\nu/2,\nu/2)$. *Then:*

$$\begin{aligned}
\mathrm{E}(\mathbf{U}) &= \mathrm{E}(V^{-1/2})\mathrm{E}(\tilde{\mathbf{U}}) = b_1(\nu)\sqrt{\frac{2}{\pi}}\mathbf{1}_q, \quad (\nu > 1) \\
\mathrm{var}(\mathbf{U}) &= \mathrm{E}(V^{-1})\mathrm{E}(\tilde{\mathbf{U}}\tilde{\mathbf{U}}^\top) - \mathrm{E}(V^{-1/2})^2\mathrm{E}(\tilde{\mathbf{U}})\mathrm{E}(\tilde{\mathbf{U}})^\top \\
&= \frac{\nu}{\nu-2}\left(1-\frac{2}{\pi}\right)\mathbf{I}_q + \frac{2}{\pi}\left\{\frac{\nu}{\nu-2} - b_1(\nu)^2\right\}\mathbf{1}_q\mathbf{1}_q^\top, \quad (\nu > 2).
\end{aligned}$$

Theorem 117 *Let* $\mathbf{X} \sim \mathcal{SUT}_{p,q}(\boldsymbol{\xi},\boldsymbol{\Psi},\mathbf{H},\boldsymbol{\tau},\bar{\boldsymbol{\Gamma}},\nu)$. *If* $\boldsymbol{\tau} = \mathbf{0}_q$, *then:*

$$\begin{aligned}
\mathrm{E}(\mathbf{X}) &= \boldsymbol{\xi} + b_1(\nu)\mathbf{H}\mathrm{E}(\tilde{\mathbf{U}}) \quad (\nu > 1), \\
\mathrm{var}(\mathbf{X}) &= \left(\frac{\nu}{\nu-2}\right)\{\mathbf{H}\mathrm{var}(\tilde{\mathbf{U}})\mathbf{H}^\top + \boldsymbol{\Psi}\} + \left(\frac{\nu}{\nu-2} - b_1(\nu)^2\right)\mathbf{H}\mathrm{E}(\tilde{\mathbf{U}})\mathrm{E}(\tilde{\mathbf{U}})^\top\mathbf{H}^\top \quad (\nu > 2),
\end{aligned}$$

where $b_1(\nu)$ *is defined in (7.12),* $\mathrm{E}(\tilde{\mathbf{U}}) = \mathrm{E}(\tilde{\mathbf{W}}_0 \mid \tilde{\mathbf{W}}_0 > \mathbf{0}_q)$ *and* $\mathrm{var}(\tilde{\mathbf{U}}) = \mathrm{var}(\tilde{\mathbf{W}}_0 \mid \tilde{\mathbf{W}}_0 > \mathbf{0}_q)$ *for* $\tilde{\mathbf{W}}_0 \sim \mathcal{N}_q(\mathbf{0}_q,\bar{\boldsymbol{\Gamma}})$.

Proof 117 *Done above.* □

When $\boldsymbol{\tau} = \mathbf{0}_q$, we only need the mean vector and the variance matrix of the multivariate truncated normal distribution to find these quantities for the multivariate unified skew-*t* distribution. However, again, with the exception of the cases of $q = 1$ or $\bar{\boldsymbol{\Gamma}} = \mathbf{I}_q$, for $q \geq 2$, there are no simple explicit expressions to calculate these quantities, so that in general they must be computed numerically.

As mentioned earlier, a simple but relevant special case occurs when $\boldsymbol{\tau} = \mathbf{0}_q$ and $\bar{\boldsymbol{\Gamma}} = \mathbf{I}_q$, so that $\mathbf{U} = V^{-1/2}\tilde{\mathbf{U}}$ with $\tilde{\mathbf{U}} \sim \mathcal{HN}_q(\mathbf{0}_q,\mathbf{I}_q)$; that is, $\tilde{\mathbf{U}} = |\tilde{\mathbf{W}}_0| := (|\tilde{W}_{01}|,\ldots,|\tilde{W}_{0q}|)^\top$, with $\tilde{\mathbf{W}}_0 \sim \mathcal{N}_q(\mathbf{0}_q,\mathbf{I}_q)$

being independent of $V \sim \mathcal{G}amma(\nu/2, \nu/2)$, and therefore the components $\tilde{U}_1 = |\tilde{W}_{01}|, \ldots, \tilde{U}_q = |\tilde{W}_{0q}|$ of $\tilde{\mathbf{U}}$ are iid $\mathcal{HN}(0,1)$ random variables that are also independent of the random scale factor V. We illustrate this case in the following example.

Example 85 (Multivariate unified skew-*t* distribution with uncorrelated latent variables) *If $\boldsymbol{\tau} = \mathbf{0}_q$ and $\bar{\boldsymbol{\Gamma}} = \mathbf{I}_q$, then $\mathbf{X} = \boldsymbol{\xi} + V^{-1/2}(\mathbf{H}\tilde{\mathbf{U}} + \tilde{\mathbf{W}}) \sim \mathcal{SUT}_{p,q}(\boldsymbol{\xi}, \boldsymbol{\Psi}, \mathbf{H}, \mathbf{0}_q, \mathbf{I}_q, \nu)$, with $\tilde{\mathbf{U}} \sim \mathcal{HN}_q(\mathbf{0}_q, \mathbf{I}_q)$ having iid $\mathcal{HN}(0,1)$ random components, with mean $b \equiv \sqrt{2/\pi}$ and variance $1-b^2$. So $\mathrm{E}(\tilde{\mathbf{U}}) = b\mathbf{1}_q$, $\mathrm{E}(\tilde{\mathbf{U}}\tilde{\mathbf{U}}^\top) = (1-b^2)\mathbf{I}_q + b^2\mathbf{1}_q\mathbf{1}_q^\top$ and so $\mathrm{var}(\tilde{\mathbf{U}}) = (1-b^2)\mathbf{I}_q$; hence the SUT random vector $\mathbf{X} \sim \mathcal{SUT}_{p,q}(\boldsymbol{\xi}, \boldsymbol{\Psi}, \mathbf{H}, \mathbf{0}_q, \mathbf{I}_q, \nu)$ has mean vector and variance matrix given by*

$$\mathrm{E}(\mathbf{X}) = \boldsymbol{\xi} + b b_1(\nu)\mathbf{H}\mathbf{1}_q, \quad \mathrm{var}(\mathbf{X}) = \frac{\nu}{\nu-2}\{(1-b^2)\mathbf{H}\mathbf{H}^\top + \boldsymbol{\Psi}\} + \left(\frac{\nu}{\nu-2} - b_1(\nu)^2\right) b^2\mathbf{H}\mathbf{1}_q\mathbf{1}_q^\top\mathbf{H}^\top,$$

where $b_1(\nu)$ is defined in (7.12); see also Example 84.

7.1.4 *Hierarchical representation*

According to Definition 20, the multivariate unified skew-*t* is a location-scale mixture of a multivariate symmetric *t* distribution. This fact follows directly from (7.6) by conditioning on the truncated random vector $\mathbf{U}$. When $\boldsymbol{\tau} = \mathbf{0}_q$, the SUT distribution can also be represented as a mean-variance mixture of the multivariate normal distribution, which is clearly seen from (7.6) by conditioning simultaneously on the latent scale factor V and on the truncated vector $\mathbf{U}$. In this way, the following hierarchical representation is obtained for the multivariate unified skew-*t* distribution.

Theorem 118 *The random vector $\mathbf{X} \sim \mathcal{SUT}_{p,q}(\boldsymbol{\xi}, \boldsymbol{\Psi}, \mathbf{H}, \boldsymbol{\tau}, \bar{\boldsymbol{\Gamma}}, \nu)$ defined in (7.6) can be represented equivalently as*

$$\mathbf{X} \mid \mathbf{U} = \mathbf{u} \sim \mathcal{T}_p\left(\boldsymbol{\xi} + \mathbf{H}\mathbf{u}, \frac{\nu + Q(\mathbf{u})}{\nu + q}\boldsymbol{\Psi}, \nu + q\right), \tag{7.13}$$

$$\mathbf{U} \sim \mathcal{TT}_q(-\boldsymbol{\tau}; \mathbf{0}_q, \bar{\boldsymbol{\Gamma}}, \nu). \tag{7.14}$$

Moreover, if $\boldsymbol{\tau} = \mathbf{0}_q$, then:

$$\begin{aligned} \mathbf{X} \mid \mathbf{U} = \mathbf{u}, V = \upsilon &\sim \mathcal{N}_p(\boldsymbol{\xi} + \mathbf{H}\mathbf{u}, \upsilon^{-1}\boldsymbol{\Psi}), \\ \mathbf{U} \mid V = \upsilon &\sim \mathcal{TN}_q(\mathbf{0}_q; \mathbf{0}_q, \upsilon^{-1}\bar{\boldsymbol{\Gamma}}), \\ V &\sim \mathcal{G}amma(\nu/2, \nu/2), \end{aligned}$$

where the mixing variables V and $\mathbf{U}$ are independent.

Proof 118 *The proof is derived directly from the stochastic representations in (7.6) and (7.9), respectively.* □

These representations provide several useful tools, both in the classical and Bayesian inferential paradigms, for the nonsingular multivariate unified skew-*t* model.

7.1.5 *Selection representation*

The multivariate unified skew-*t* distribution also has the following selection representation: The essence of this constructive mechanism is that, starting from a $(p+q)$-dimensional *t* distribution, *q* of its components play the role of hidden variables that nonlinearly modify the remaining p components.

Theorem 119 *If* $\mathbf{X} \sim \mathcal{SUT}_{p,q}(\boldsymbol{\xi}, \boldsymbol{\Psi}, \mathbf{H}, \boldsymbol{\tau}, \bar{\boldsymbol{\Gamma}}, \nu)$, *then*

$$\mathbf{X} \stackrel{d}{=} \boldsymbol{\xi} + (\bar{\mathbf{W}} \mid \mathbf{W}_0 + \boldsymbol{\tau} > \mathbf{0}_q), \text{ with } \begin{pmatrix} \bar{\mathbf{W}} \\ \mathbf{W}_0 \end{pmatrix} \sim \mathcal{T}_{p+q}\left\{ \begin{pmatrix} \mathbf{0}_p \\ \mathbf{0}_q \end{pmatrix}, \begin{pmatrix} \boldsymbol{\Psi} + \mathbf{H}\bar{\boldsymbol{\Gamma}}\mathbf{H}^\top & \mathbf{H}\bar{\boldsymbol{\Gamma}} \\ \bar{\boldsymbol{\Gamma}}\mathbf{H}^\top & \bar{\boldsymbol{\Gamma}} \end{pmatrix}, \nu \right\}. \tag{7.15}$$

Proof 119 *Using (7.6) and (7.4), we have*

$$\mathbf{X} = \boldsymbol{\xi} + \mathbf{H}\mathbf{U} + \sqrt{\frac{\nu + Q(\mathbf{U})}{\nu + q}}\mathbf{W}_* \stackrel{d}{=} (\boldsymbol{\xi} + \mathbf{H}\mathbf{W}_0 + \mathbf{W}) \mid \mathbf{W}_0 + \boldsymbol{\tau} > \mathbf{0}_q,$$

where $\mathbf{W}$ *and* $\mathbf{W}_0$ *have a joint multivariate t distribution given by (7.1). Thus, the proof follows by using that*

$$\begin{aligned}\begin{pmatrix} \bar{\mathbf{W}} \\ \mathbf{W}_0 \end{pmatrix} &= \begin{pmatrix} \mathbf{H}\mathbf{W}_0 + \mathbf{W} \\ \mathbf{W}_0 \end{pmatrix} = \begin{pmatrix} \mathbf{I}_p & \mathbf{H} \\ \mathbf{O} & \mathbf{I}_q \end{pmatrix} \begin{pmatrix} \mathbf{W} \\ \mathbf{W}_0 \end{pmatrix} \\ &\sim \mathcal{T}_{p+q}\left\{ \begin{pmatrix} \mathbf{I}_p & \mathbf{H} \\ \mathbf{O} & \mathbf{I}_q \end{pmatrix} \begin{pmatrix} \mathbf{0}_p \\ \mathbf{0}_q \end{pmatrix}, \begin{pmatrix} \mathbf{I}_p & \mathbf{H} \\ \mathbf{O} & \mathbf{I}_q \end{pmatrix} \begin{pmatrix} \boldsymbol{\Psi} & \mathbf{O} \\ \mathbf{O} & \bar{\boldsymbol{\Gamma}} \end{pmatrix} \begin{pmatrix} \mathbf{I}_p & \mathbf{O} \\ \mathbf{H}^\top & \mathbf{I}_q \end{pmatrix}, \nu \right\} \\ &= \mathcal{T}_{p+q}\left\{ \begin{pmatrix} \mathbf{0}_p \\ \mathbf{0}_q \end{pmatrix}, \begin{pmatrix} \boldsymbol{\Psi} + \mathbf{H}\bar{\boldsymbol{\Gamma}}\mathbf{H}^\top & \mathbf{H}\bar{\boldsymbol{\Gamma}} \\ \bar{\boldsymbol{\Gamma}}\mathbf{H}^\top & \bar{\boldsymbol{\Gamma}} \end{pmatrix}, \nu \right\}.\end{aligned}$$

□

An important consequence of Theorem 119 is the following result.

Theorem 120 *Let* $\mathbf{g}$ *be a measurable vector-valued function from* $\mathbb{R}^p$ *to* $\mathbb{R}^r$ *and* C *be a Borel subset in* $\mathbb{R}^r$. *If* $\mathbf{X} \sim \mathcal{SUT}_{p,q}(\boldsymbol{\xi}, \boldsymbol{\Psi}, \mathbf{H}, \boldsymbol{\tau}, \bar{\boldsymbol{\Gamma}}, \nu)$, *then the probability distribution of* $\mathbf{g}(\mathbf{X})$ *is given by*

$$\mathrm{P}(\mathbf{g}(\mathbf{X}) \in C) = \mathrm{P}(\mathbf{g}(\bar{\mathbf{W}}) \in C \mid \mathbf{W}_0 > \mathbf{0}_q),$$

where the joint distribution of $(\bar{\mathbf{W}}, \mathbf{W}_0)$ *is defined in (7.15).*

Proof 120 *In fact, by (7.15), we have*

$$\mathrm{P}\{\mathbf{g}(\mathbf{X}) \in C\} = \mathrm{P}\{\mathbf{X} \in \mathbf{g}^{-1}(C)\} = \mathrm{P}\{\bar{\mathbf{W}} \in \mathbf{g}^{-1}(C) \mid \mathbf{W}_0 + \boldsymbol{\tau} > \mathbf{0}_q\} = \mathrm{P}(\mathbf{g}(\bar{\mathbf{W}}) \in C \mid \mathbf{W}_0 + \boldsymbol{\tau} > \mathbf{0}_q\}.$$

□

7.1.6 Cdf

From the representation by selection of $\mathbf{X} \sim \mathcal{SUT}_{p,q}(\boldsymbol{\xi}, \boldsymbol{\Psi}, \mathbf{H}, \boldsymbol{\tau}, \bar{\boldsymbol{\Gamma}}, \nu)$, or immediately from Theorem 120, its cdf can be easily calculated as shown below.

Theorem 121 *The cdf of* $\mathbf{X} \sim \mathcal{SUT}_{p,q}(\boldsymbol{\xi}, \boldsymbol{\Psi}, \mathbf{H}, \boldsymbol{\tau}, \bar{\boldsymbol{\Gamma}}, \nu)$ *is*

$$F(\mathbf{x}) = \frac{1}{T_q(\boldsymbol{\tau}; \mathbf{0}_q, \bar{\boldsymbol{\Gamma}}, \nu)} T_{p+q}(\mathbf{x}_*; \boldsymbol{\xi}_*, \boldsymbol{\Omega}_*, \nu), \quad \mathbf{x} \in \mathbb{R}^p,$$

where $\mathbf{x}_* = (\mathbf{x}^\top, \boldsymbol{\tau}^\top)^\top$, $\boldsymbol{\xi}_* = (\boldsymbol{\xi}^\top, \mathbf{0}_q^\top)^\top$ *and*

$$\boldsymbol{\Omega}_* = \begin{pmatrix} \boldsymbol{\Psi} + \mathbf{H}\bar{\boldsymbol{\Gamma}}\mathbf{H}^\top & -\mathbf{H}\bar{\boldsymbol{\Gamma}} \\ -\bar{\boldsymbol{\Gamma}}\mathbf{H}^\top & \bar{\boldsymbol{\Gamma}} \end{pmatrix}.$$

Proof 121 *By Theorem 119, we have* $\mathbf{X} \stackrel{d}{=} (\bar{\mathbf{W}} \mid \mathbf{W}_0 + \boldsymbol{\tau} > \mathbf{0}_q)$ *which is equivalent to* $\mathbf{X} \stackrel{d}{=} (\bar{\mathbf{W}} \mid \tilde{\mathbf{W}}_0 < \boldsymbol{\tau})$, *where* $\bar{\mathbf{W}}_0 = -\mathbf{W}_0 \sim \mathcal{T}_q(\mathbf{0}_q, \bar{\boldsymbol{\Gamma}}, \nu)$ *and the multivariate joint distribution of* $(\bar{\mathbf{W}}, \bar{\mathbf{W}}_0) = (\mathbf{H}\mathbf{W}_0 + \mathbf{W} + \boldsymbol{\xi}, -\mathbf{W}_0)$ *is* $\mathcal{T}_{p+q}(\boldsymbol{\xi}_*, \boldsymbol{\Omega}_*, \nu)$. *Hence, we obtain*

$$F(\mathbf{x}) = \mathrm{P}(\mathbf{X} \le \mathbf{x}) = \mathrm{P}(\bar{\mathbf{W}} \le \mathbf{x} \mid \bar{\mathbf{W}}_0 < \boldsymbol{\tau}) = \frac{\mathrm{P}(\bar{\mathbf{W}} \le \mathbf{x}, \bar{\mathbf{W}}_0 < \boldsymbol{\tau})}{\mathrm{P}(\bar{\mathbf{W}}_0 < \boldsymbol{\tau})},$$

which proves the result. □

7.2 Affine Transformations and Marginals

In this section, we describe further distributional properties of the multivariate unified skew-*t* family introduced in Definition 20. As we show, this family inherits several of the good properties of the multivariate *t* family.

7.2.1 *Affine transformations*

By construction, the multivariate unified skew-*t* distribution is closed under affine transformations as formalized next.

Theorem 122 *Let* $\mathbf{X} \sim \mathcal{SUT}_{p,q}(\boldsymbol{\xi}, \boldsymbol{\Psi}, \mathbf{H}, \boldsymbol{\tau}, \bar{\boldsymbol{\Gamma}}, \nu)$, *and consider the affine transformation* $\mathbf{Y} = \mathbf{A}\mathbf{X} + \mathbf{b}$, *where the matrix* $\mathbf{A} \in \mathbb{R}^{r \times p}$ *and the vector* $\mathbf{b} \in \mathbb{R}^r$ *are constant. Then,* $\mathbf{Y} \sim \mathcal{SUT}_{r,q}(\mathbf{A}\boldsymbol{\xi} + \mathbf{b}, \mathbf{A}\boldsymbol{\Psi}\mathbf{A}^\top, \mathbf{A}\mathbf{H}, \boldsymbol{\tau}, \bar{\boldsymbol{\Gamma}}, \nu)$.

Proof 122 *By Equation (7.6) of Definition 20, we have* $\mathbf{Y} = \mathbf{A}\mathbf{X} + \mathbf{b} = \mathbf{A}\boldsymbol{\xi} + \mathbf{b} + \mathbf{A}\mathbf{H}\mathbf{U} + \sqrt{\frac{\nu + Q(\mathbf{U})}{\nu + q}}\mathbf{A}\mathbf{W}_*$. *Let* $\boldsymbol{\xi}_A = \mathbf{A}\boldsymbol{\xi} + \mathbf{b}$, $\mathbf{H}_A = \mathbf{A}\mathbf{H}$, *and* $\mathbf{W}_{*A} = \mathbf{A}\mathbf{W}_*$. *Since* $\mathbf{W}_* \sim \mathcal{T}_q(\mathbf{0}_q, \boldsymbol{\Psi}, \nu + q)$, *then* $\mathbf{W}_{*A} \sim \mathcal{T}_q(\mathbf{0}_q, \boldsymbol{\Psi}_A, \nu + q)$, *with* $\boldsymbol{\Psi}_A = \mathbf{A}\boldsymbol{\Psi}\mathbf{A}^\top$, *which is independent of* $\mathbf{U} \sim \mathcal{TT}_q(-\boldsymbol{\tau}; \mathbf{0}_q, \bar{\boldsymbol{\Gamma}}, \nu)$. *Hence,* $\mathbf{Y} = \boldsymbol{\xi}_A + \mathbf{H}_A\mathbf{U} + \sqrt{\frac{\nu + Q(\mathbf{U})}{\nu + q}}\mathbf{W}_{*A}$, *with* $\mathbf{U} \sim \mathcal{TT}_q(-\boldsymbol{\tau}; \mathbf{0}_q, \bar{\boldsymbol{\Gamma}}, \nu)$ *and* $\mathbf{W}_{*A} \sim \mathcal{T}_q(\mathbf{0}_q, \boldsymbol{\Psi}_A, \nu + q)$ *being independent. Thus, the proof follows by applying Definition 20 again.* □

Note that if $\mathbf{A}$ has full row rank r, then a nonsingular multivariate unified skew-*t* version is obtained from Theorem 122, which will be formalized later in Section 7.2.2.

The next example describes the closure property of linear combinations of the components of $\mathbf{X} \sim \mathcal{SUT}_{p,q}(\boldsymbol{\xi}, \boldsymbol{\Psi}, \mathbf{H}, \boldsymbol{\tau}, \bar{\boldsymbol{\Gamma}}, \nu)$.

Example 86 (Linear combinations for SUT) *For a fixed vector* $\mathbf{a} \in \mathbb{R}^p$, *let* $X_a = \mathbf{a}^\top \mathbf{X}$, $\xi_a = \mathbf{a}^\top \boldsymbol{\xi}$, $\boldsymbol{\eta}_a = \mathbf{H}^\top \mathbf{a}$ *and* $W_{*a} = \mathbf{a}^\top \mathbf{W}_*$, *where* $\mathbf{W}_* \sim \mathcal{T}_p(\mathbf{0}_p, \boldsymbol{\Psi}, \nu + q)$. *Then, from Equation (7.6) given in Definition 20, we have*

$$X_a = \mathbf{a}^\top \boldsymbol{\xi} + \mathbf{a}^\top \mathbf{H}\mathbf{U} + \sqrt{\frac{\nu + Q(\mathbf{U})}{\nu + q}}\, \mathbf{a}^\top \mathbf{W}_* = \xi_a + \boldsymbol{\eta}_a^\top \mathbf{U} + \sqrt{\frac{\nu + Q(\mathbf{U})}{\nu + q}}\, W_{*a},$$

where by definition $\mathbf{U} \sim \mathcal{TT}_q(-\boldsymbol{\tau}; \mathbf{0}_q, \bar{\boldsymbol{\Gamma}}, \nu)$ *and is independent of* $\mathbf{W}_*$, *and therefore of* $W_{*a} = \mathbf{a}^\top \mathbf{W}_* \sim \mathcal{T}_1(0, \mathbf{a}^\top \boldsymbol{\Psi} \mathbf{a}, \nu + q)$. *Hence, by applying again the Definition 20, but now with* $p = 1$, *we have that* $X_a \sim \mathcal{SUT}_{1,q}(\xi_a, \psi_a^2, \boldsymbol{\eta}_a^\top, \boldsymbol{\tau}, \bar{\boldsymbol{\Gamma}}, \nu)$ *for all* $\mathbf{a} \in \mathbb{R}^p$, *where* $\xi_a = \mathbf{a}^\top \boldsymbol{\xi}$, $\psi_a^2 = \mathbf{a}^\top \boldsymbol{\Psi} \mathbf{a}$, $\boldsymbol{\eta}_a = \mathbf{H}^\top \mathbf{a}$. *This result is now direct from Theorem 122 by taking* $\mathbf{A} = \mathbf{a}^\top$, *a p-dimensional row vector.*

An interesting application of Example 86 is presented in the following example.

Example 87 *Let* $\mathbf{U} = (U_1, \ldots, U_q)^\top \sim \mathcal{TT}_q(-\boldsymbol{\tau}; \mathbf{0}_q, \bar{\boldsymbol{\Gamma}}, \nu)$ *and* $\mathbf{W}_* = (W_{*1}, \ldots, W_{*q})^\top \sim \mathcal{T}_q(\mathbf{0}_q, \mathbf{I}_q, \nu)$ *be mutually independent. Suppose that* $\mathbf{X} \sim \mathcal{SUT}_{p,q}(\boldsymbol{\xi}, \boldsymbol{\Psi}, \mathbf{H}, \boldsymbol{\tau}, \bar{\boldsymbol{\Gamma}}, \nu)$, *with* $\boldsymbol{\Psi} > 0$ *and* $\mathbf{H}^\top \boldsymbol{\Psi}^{-1} \mathbf{H} = \mathrm{diag}(\tilde{\eta}_1^2, \ldots, \tilde{\eta}_q^2)$, *where* $\tilde{\eta}_j = \sqrt{\boldsymbol{\eta}_j^\top \boldsymbol{\Psi}^{-1} \boldsymbol{\eta}_j}$ *and* $\boldsymbol{\eta}_j$ *is the j-th column of* $\mathbf{H}$, $j = 1, \ldots, q$. *Then:*

$$\boldsymbol{\eta}_j^\top \boldsymbol{\Psi}^{-1}(\mathbf{X} - \boldsymbol{\xi}) \stackrel{d}{=} \tilde{\eta}_j Z_j, \quad Z_j = \tilde{\eta}_j U_j + \sqrt{\frac{\nu + Q(\mathbf{U})}{\nu + q}}\, W_{*j} \sim \mathcal{SUT}_{1,q}(0, 1, \tilde{\eta}_j, \boldsymbol{\tau}, \bar{\boldsymbol{\Gamma}}, \nu), \quad j = 1, \ldots, q.$$

Another interesting result that follows from Theorem 122 is when $p = q$ and $\mathbf{A} = -\mathbf{I}_p$, which transforms $\mathbf{X}$ to $-\mathbf{X}$, as formalized in the following example.

Example 88 *If* $\mathbf{X} \sim \mathcal{SUT}_{p,q}(\boldsymbol{\xi}, \boldsymbol{\Psi}, \mathbf{H}, \boldsymbol{\tau}, \bar{\boldsymbol{\Gamma}}, \nu)$, *then* $-\mathbf{X} \sim \mathcal{SUT}_{p,q}(-\boldsymbol{\xi}, \boldsymbol{\Psi}, -\mathbf{H}, \boldsymbol{\tau}, \bar{\boldsymbol{\Gamma}}, \nu)$.

According to Theorem 122, we can recover symmetry by letting $\boldsymbol{\tau} = \mathbf{0}_q$ and choosing a transformation matrix $\mathbf{A}$ such that $\mathbf{AH} = \mathbf{O}$. Therefore, in this case, the affine transformation $\mathbf{Y} = \mathbf{AX} + \mathbf{b}$ has a symmetric $\mathcal{T}_p(\mathbf{A}\boldsymbol{\xi} + \mathbf{b}, \mathbf{A}\boldsymbol{\Psi}\mathbf{A}^\top, \nu)$ distribution. In the following example, we illustrate a case where the new skewness matrix $\mathbf{AH}$ is diagonal, with non-null elements in its diagonal.

Example 89 *Let* $\mathbf{U} \sim \mathcal{TT}_q(\mathbf{0}_q; \mathbf{0}_q, \mathbf{I}_q, \nu)$ *and* $\mathbf{W}_* \sim \mathcal{T}_q(\mathbf{0}_q, \mathbf{I}_q, \nu + q)$ *be mutually independent. Suppose that* $\mathbf{X} \sim \mathcal{SUT}_{p,q}(\boldsymbol{\xi}, \boldsymbol{\Psi}, \mathbf{H}, \boldsymbol{\tau}, \bar{\boldsymbol{\Gamma}}, \nu)$, *with* $\boldsymbol{\Psi} > 0$ *and* $\mathbf{H}^\top \boldsymbol{\Psi}^{-1} \mathbf{H} = \mathbf{D}(\tilde{\boldsymbol{\eta}}) = \mathrm{diag}(\tilde{\eta}_1^2, \ldots, \tilde{\eta}_q^2)$, *where* $\tilde{\boldsymbol{\eta}} = (\tilde{\eta}_1, \ldots, \tilde{\eta}_q)^\top$, *with* $\tilde{\eta}_i = \sqrt{\boldsymbol{\eta}_i^\top \boldsymbol{\Psi}^{-1} \boldsymbol{\eta}_i}$ *and* $\boldsymbol{\eta}_i$ *being the i-th column of* $\mathbf{H}$, $i = 1, \ldots, q$. *Then:*

$$\mathbf{H}^\top \boldsymbol{\Psi}^{-1}(\mathbf{X} - \boldsymbol{\xi}) \stackrel{d}{=} \mathbf{D}(\tilde{\boldsymbol{\eta}})^{1/2} \mathbf{Z}, \; \mathbf{Z} = \mathbf{D}(\tilde{\boldsymbol{\eta}})^{1/2} \mathbf{U} + \sqrt{\frac{\nu + Q(\mathbf{U})}{\nu + q}}\, \mathbf{W}_* \sim \mathcal{SUT}_{q,q}(\mathbf{0}_q, \mathbf{I}_q, \mathbf{D}(\tilde{\boldsymbol{\eta}})^{1/2}, \boldsymbol{\tau}, \bar{\boldsymbol{\Gamma}}, \nu).$$

7.2.2 Marginal distributions

From Theorem 122, we can also show that the multivariate unified skew-t family is closed by marginalization, with all its marginals having the same degrees of freedom. However, as mentioned earlier, similar to the symmetric case, this family does not allow independence between its marginals.

Let $\mathbf{X} \sim \mathcal{SUT}_{p,q}(\boldsymbol{\xi}, \boldsymbol{\Psi}, \mathbf{H}, \boldsymbol{\tau}, \bar{\boldsymbol{\Gamma}}, \nu)$ and, for fixed nonnegative integers $p_1 + p_2 = p$, consider the following partitions:

$$\mathbf{X} = \begin{pmatrix} \mathbf{X}_1 \\ \mathbf{X}_2 \end{pmatrix}, \quad \boldsymbol{\xi} = \begin{pmatrix} \boldsymbol{\xi}_1 \\ \boldsymbol{\xi}_2 \end{pmatrix}, \quad \boldsymbol{\Psi} = \begin{pmatrix} \boldsymbol{\Psi}_{11} & \boldsymbol{\Psi}_{12} \\ \boldsymbol{\Psi}_{21} & \boldsymbol{\Psi}_{22} \end{pmatrix}, \quad \mathbf{H} = \begin{pmatrix} \mathbf{H}_1 \\ \mathbf{H}_2 \end{pmatrix}, \tag{7.16}$$

where $\mathbf{X}_j, \boldsymbol{\xi}_j$ are of dimension $p_j \times 1$, $\boldsymbol{\Psi}_{jk}$ is of dimension $p_j \times p_k$ and $\mathbf{H}_j$ is of dimension $p_j \times q$, $j,k = 1,2$. Also, denote by $\mathbf{W}_j$ and $\mathbf{W}_{*j}$ for $j = 1,2$ the induced partitions on the random vectors $\mathbf{W}$ and $\mathbf{W}_*$ considered in the definition of the SUT random vector $\mathbf{X}$. Thus, the following result is a direct consequence of Theorem 122.

Theorem 123 *Let $\mathbf{X} \sim \mathcal{SUT}_{p,q}(\boldsymbol{\xi}, \boldsymbol{\Psi}, \mathbf{H}, \boldsymbol{\tau}, \bar{\boldsymbol{\Gamma}}, \nu)$ be partitioned as (7.16). Then, the partitions $\mathbf{X}_j \sim \mathcal{SUT}_{p_j,q}(\boldsymbol{\xi}_j, \boldsymbol{\Psi}_{jj}, \mathbf{H}_j, \boldsymbol{\tau}, \bar{\boldsymbol{\Gamma}}, \nu)$ for $j = 1,2$.*

Proof 123 *Since $\mathbf{X}_j = \mathbf{A}_j \mathbf{X}$, $j = 1,2$, where $\mathbf{A}_1 = (\mathbf{I}_{p_1}, \mathbf{O})$ and $\mathbf{A}_2 = (\mathbf{O}, \mathbf{I}_{p_2})$, the proof is then direct from Theorem 122.* □

Obviously, using the partition (7.16) in Equation (7.6), it is straightforward to see that the marginals are given by

$$\mathbf{X}_j = \boldsymbol{\xi}_j + \mathbf{H}_j \mathbf{U} + \sqrt{\frac{\nu + Q(\mathbf{U})}{\nu + q}}\, \mathbf{W}_{*j}, \quad j = 1,2, \tag{7.17}$$

where $\mathbf{W}_{*j} \sim \mathcal{T}_{p_j}(\mathbf{0}_{p_j}, \boldsymbol{\Psi}_{jj}, \nu + q)$, which, for $j = 1,2$, is independent of $\mathbf{U} \sim \mathcal{TT}_q(-\boldsymbol{\tau}; \mathbf{0}_q, \bar{\boldsymbol{\Gamma}}, \nu)$.

From Theorem 123, an advantage of the $(\boldsymbol{\Psi}, \mathbf{H})$ parameterization of the SUT is again that it yields a marginally consistent distribution, i.e., the parameters of the marginal distribution depend only on the corresponding subparameters of the full distribution (treating the latent parameters as fixed).

As illustrated in part (a) of the following example, by specializing the vector $\mathbf{a}$ in Example 86 to be the i-th p-dimensional unit vector $\mathbf{e}_i$, we then immediately deduce that all the univariate marginals are unified skew-t.

Example 90 (Univariate marginals) *Let $\mathbf{X} = (X_1, \ldots, X_p)^\top \sim \mathcal{SUT}_{p,q}(\boldsymbol{\xi}, \boldsymbol{\Psi}, \mathbf{H}, \boldsymbol{\tau}, \bar{\boldsymbol{\Gamma}}, \nu)$, with $\boldsymbol{\xi} = (\xi_1, \ldots, \xi_p)^\top$, $\boldsymbol{\Psi} = (\psi_{ij})_{i,j=1,\ldots,p}$, and $\mathbf{H} = (\boldsymbol{\eta}_1, \ldots, \boldsymbol{\eta}_p)^\top$, where $\boldsymbol{\eta}_i \in \mathbb{R}^q$ is the i-th row (column) of $\mathbf{H}$ ($\mathbf{H}^\top$). By Example 86, we have*

$$\sum_{i=1}^{p} a_i X_i \sim \mathcal{SUT}_{1,q}\left(\sum_{i=1}^{p} a_i \xi_i, \sum_{i=1}^{p} \sum_{j=1}^{p} a_i a_j \psi_{ij}, \sum_{i=1}^{p} a_i \boldsymbol{\eta}_i^\top, \boldsymbol{\tau}, \bar{\boldsymbol{\Gamma}}, \nu \right).$$

Hence:

(a) Setting $\mathbf{a} = \mathbf{e}_i = (0, \ldots, 0, 1, 0, \ldots, 0)^\top$ we find the i-th marginal component $X_i = \mathbf{e}_i^\top \mathbf{X}$:

$$X_i \sim \mathcal{SUT}_{1,q}(\xi_i, \psi_{ii}, \boldsymbol{\eta}_i^\top, \boldsymbol{\tau}, \bar{\boldsymbol{\Gamma}}, \nu), \quad i = 1, \ldots, p,$$

with $\xi_i = \mathbf{e}_i^\top \boldsymbol{\xi}$, $\psi_{ij} = \mathbf{e}_i^\top \boldsymbol{\Psi} \mathbf{e}_j$ and $\boldsymbol{\eta}_i = \mathbf{H}^\top \mathbf{e}_i$;

(b) *Similarly, setting* **a** *to be a vector with* 1 *in the i-th and j-th positions,* $i \neq j$*, and* 0 *elsewhere, we obtain*

$$X_i + X_j \sim \mathcal{SUT}_{1,q}(\xi_i + \xi_j, \psi_{ii} + \psi_{jj} + 2\psi_{ij}, \boldsymbol{\eta}_i^\top + \boldsymbol{\eta}_j^\top, \boldsymbol{\tau}, \bar{\boldsymbol{\Gamma}}, \nu), \quad i,j = 1,\ldots,p, \quad i \neq j.$$

In the nonsingular case with $\boldsymbol{\Psi} > 0$, all the marginals are nonsingular. In the singular case with $\boldsymbol{\Psi} \geq 0$, they can be singular or nonsingular. The following theorem establishes the condition to have a nonsingular marginal in the latter case.

Theorem 124 *Let* $\mathbf{X} \sim \mathcal{SUT}_{p,q}(\boldsymbol{\xi}, \boldsymbol{\Psi}, \mathbf{H}, \boldsymbol{\tau}, \bar{\boldsymbol{\Gamma}}, \nu)$*, with rank*$(\boldsymbol{\Psi}) = r$*, be partitioned as (7.16). If* $\boldsymbol{\Psi}_{11}$ *has full rank* $p_1 = p - p_2 \leq r$*, then* $\mathbf{X}_1 \sim \mathcal{SUT}_{p_1,q}(\boldsymbol{\xi}_1, \boldsymbol{\Psi}_{11}, \mathbf{H}_1, \boldsymbol{\tau}, \bar{\boldsymbol{\Gamma}}, \nu)$ *is nonsingular.*

Proof 124 *It is directly derived from Theorem 123 and the assumption that* $\boldsymbol{\Psi}_{11} > 0$. □

The following result is also direct from Theorem 122 and states, when $\boldsymbol{\tau} = \mathbf{0}_q$, that the skewness of the distribution of each subvector of $\mathbf{X}$ is regulated by the corresponding components of $\mathbf{H}$ only.

Theorem 125 *Let* $\mathbf{X} \sim \mathcal{SUT}_{p,q}(\boldsymbol{\xi}, \boldsymbol{\Psi}, \mathbf{H}, \boldsymbol{\tau}, \bar{\boldsymbol{\Gamma}}, \nu)$ *be partitioned as (7.16). Then for every fixed* $p_1 + p_2 = p$*,* $\mathbf{X}_j$ *has symmetric* $\mathcal{T}_{p_j}(\boldsymbol{\xi}_j, \boldsymbol{\Psi}_{jj}, \nu)$ *distribution if and only if* $\boldsymbol{\tau} = \mathbf{0}_q$ *and* $\mathbf{H}_j = \mathbf{O}$.

Proof 125 *By Theorem 123,* $\mathbf{X}_j \sim \mathcal{SUT}_{p_j,q}(\boldsymbol{\xi}_j, \boldsymbol{\Psi}_{jj}, \mathbf{H}_j, \boldsymbol{\tau}, \bar{\boldsymbol{\Gamma}}, \nu)$*. Hence, by definition,* $\mathbf{X}_j$ *is stochastically represented by (7.17), with* $\mathbf{U} \stackrel{d}{=} (\mathbf{W}_0 \mid \mathbf{W}_0 + \boldsymbol{\tau} > \mathbf{0}_q)$ *and* $\mathbf{W}_{*j} = \sqrt{\frac{\nu+q}{\nu+Q(\mathbf{W}_0)}}\mathbf{W}_j$ *being independent, where, according to (7.1),*

$$\begin{pmatrix} \mathbf{W}_j \\ \mathbf{W}_0 \end{pmatrix} \sim \mathcal{T}_{p_j+q}\left(\begin{pmatrix} \mathbf{0}_{p_j} \\ \mathbf{0}_q \end{pmatrix}, \begin{pmatrix} \boldsymbol{\Psi}_{jj} & \mathbf{O} \\ \mathbf{O} & \bar{\boldsymbol{\Gamma}} \end{pmatrix}, \nu\right),$$

with marginals $\mathbf{W}_0 \sim \mathcal{T}_q(\mathbf{0}_q, \bar{\boldsymbol{\Gamma}}, \nu)$ *and* $\mathbf{W}_j \sim \mathcal{T}_{p_j}(\mathbf{0}_{p_j}, \boldsymbol{\Psi}_{jj}, \nu)$*. Therefore, since* $\boldsymbol{\xi}_j + \mathbf{W}_j \sim \mathcal{T}_{p_j}(\boldsymbol{\xi}_j, \boldsymbol{\Psi}_{jj}, \nu)$*, we have that:*

$$\begin{aligned}
\mathbf{X}_j =& \boldsymbol{\xi}_j + \mathbf{H}_j\mathbf{U} + \sqrt{\frac{\nu + Q(\mathbf{U})}{\nu + q}}\mathbf{W}_{*j} \sim \mathcal{T}_{p_j}(\boldsymbol{\xi}_j, \boldsymbol{\Psi}_{jj}, \nu) \\
\iff & \boldsymbol{\xi}_j + \mathbf{H}_j\mathbf{U} + \sqrt{\frac{\nu + Q(\mathbf{U})}{\nu + q}}\mathbf{W}_{*j} \stackrel{d}{=} \boldsymbol{\xi}_j + \mathbf{W}_j \\
\iff & \mathbf{H}_j = \mathbf{O} \text{ and } \sqrt{\frac{\nu + Q(\mathbf{U})}{\nu + q}}\mathbf{W}_{*j} \stackrel{d}{=} \mathbf{W}_j \\
\iff & \mathbf{H}_j = \mathbf{O} \text{ and } \sqrt{\frac{\nu + Q(\mathbf{U})}{\nu + q}}\sqrt{\frac{\nu + q}{\nu + Q(\mathbf{W}_0)}}\mathbf{W}_j \stackrel{d}{=} \mathbf{W}_j \\
\iff & \mathbf{H}_j = \mathbf{O} \text{ and } Q(\mathbf{U}) \stackrel{d}{=} Q(\mathbf{W}_0) \\
\iff & \mathbf{H}_j = \mathbf{O} \text{ and } \boldsymbol{\tau} = \mathbf{0}_q,
\end{aligned}$$

where the last conclusion follows from part (iii) of Theorem 113. □

Another relevant property of the marginals occurs when the correlation matrix $\bar{\mathbf{\Gamma}}$ is block-diagonal, which produces a nice simplification for the marginal distributions. To see this, we first partition the parameters $(\mathbf{H}, \boldsymbol{\tau}, \bar{\mathbf{\Gamma}})$ as follows:

$$\mathbf{H} = \begin{pmatrix} \mathbf{H}_1 \\ \mathbf{H}_2 \end{pmatrix} = \begin{pmatrix} \mathbf{H}_{11} & \mathbf{H}_{12} \\ \mathbf{H}_{21} & \mathbf{H}_{22} \end{pmatrix}, \quad \boldsymbol{\tau} = \begin{pmatrix} \boldsymbol{\tau}_1 \\ \boldsymbol{\tau}_2 \end{pmatrix}, \quad \bar{\mathbf{\Gamma}} = \begin{pmatrix} \bar{\mathbf{\Gamma}}_{11} & \bar{\mathbf{\Gamma}}_{12} \\ \bar{\mathbf{\Gamma}}_{21} & \bar{\mathbf{\Gamma}}_{22} \end{pmatrix}, \tag{7.18}$$

where for $i, j = 1,2$, the matrix $\mathbf{H}_{ij}$ is $p_i \times q_j$, the vector $\boldsymbol{\tau}_i$ is $q_i \times 1$, and the matrix $\bar{\mathbf{\Gamma}}_{ij}$ is $q_i \times q_j$, with $p_1 + p_2 = p$ and $q_1 + q_2 = q$. Now, we consider the following auxiliary fact.

Theorem 126 *The (q_1+q_2)-dimensional truncated t distribution given by*

$$\mathbf{U} = \begin{pmatrix} \mathbf{U}_1 \\ \mathbf{U}_2 \end{pmatrix} \sim \mathcal{TT}_{q_1+q_2}\left(\begin{pmatrix} -\boldsymbol{\tau}_1 \\ -\boldsymbol{\tau}_2 \end{pmatrix}; \begin{pmatrix} \mathbf{0}_{q_1} \\ \mathbf{0}_{q_2} \end{pmatrix}, \begin{pmatrix} \bar{\mathbf{\Gamma}}_{11} & \bar{\mathbf{\Gamma}}_{12} \\ \bar{\mathbf{\Gamma}}_{21} & \bar{\mathbf{\Gamma}}_{22} \end{pmatrix}, \nu\right)$$

has marginals given by

$$g_j(\mathbf{u}_j) = \frac{T_{q_k}\left(\sqrt{\frac{\nu+q_j}{\nu+Q_j(\mathbf{u}_j)}}\boldsymbol{\tau}_{k\cdot j}; \mathbf{0}_{q_k}, \bar{\mathbf{\Gamma}}_{kk\cdot j}, \nu+q_j\right)}{T_{q_1+q_2}\left(\begin{pmatrix} \boldsymbol{\tau}_1 \\ \boldsymbol{\tau}_2 \end{pmatrix}; \begin{pmatrix} \mathbf{0}_{q_1} \\ \mathbf{0}_{q_2} \end{pmatrix}, \begin{pmatrix} \bar{\mathbf{\Gamma}}_{11} & \bar{\mathbf{\Gamma}}_{12} \\ \bar{\mathbf{\Gamma}}_{21} & \bar{\mathbf{\Gamma}}_{22} \end{pmatrix}, \nu\right)} t_{q_j}(\mathbf{u}_j; \mathbf{0}_{q_j}, \bar{\mathbf{\Gamma}}_{jj}, \nu), \quad \mathbf{u}_j + \boldsymbol{\tau}_j > \mathbf{0}_{q_j},$$

$j,k = 1,2$, $j \neq k$, where $Q_j(\mathbf{u}_j) = \mathbf{u}_j^\top \bar{\mathbf{\Gamma}}_{jj}^{-1}\mathbf{u}_j$, $\boldsymbol{\tau}_{k\cdot j} = \boldsymbol{\tau}_k + \bar{\mathbf{\Gamma}}_{kj}\bar{\mathbf{\Gamma}}_{jj}^{-1}\mathbf{u}_j$ and $\bar{\mathbf{\Gamma}}_{kk\cdot j} = \bar{\mathbf{\Gamma}}_{kk} - \bar{\mathbf{\Gamma}}_{kj}\bar{\mathbf{\Gamma}}_{jj}^{-1}\bar{\mathbf{\Gamma}}_{jk}$.
Moreover, if $\boldsymbol{\tau}_k = \mathbf{0}_{q_k}$ for each $k \neq j$ and $\bar{\mathbf{\Gamma}}_{12} = \mathbf{O}$, then

$$g_j(\mathbf{u}_j) = \frac{1}{T_{q_j}(\boldsymbol{\tau}_j; \mathbf{0}_{q_j}, \bar{\mathbf{\Gamma}}_{jj}, \nu)} t_{q_j}(\mathbf{u}_j; \mathbf{0}_{q_j}, \bar{\mathbf{\Gamma}}_{jj}, \nu), \quad \mathbf{u}_j + \boldsymbol{\tau}_j > \mathbf{0}_{q_j};$$

that is, $\mathbf{U}_j \sim \mathcal{TT}_{q_j}(\mathbf{0}_{q_j}; \mathbf{0}_{q_j}, \bar{\mathbf{\Gamma}}_{jj}, \nu)$, $j = 1,2$.

Proof 126 *We prove the result for the marginal pdf of $\mathbf{U}_1$ only. The proof for the pdf of $\mathbf{U}_2$ is analogous. By (7.2), the joint pdf of $\mathbf{U}_1$ and $\mathbf{U}_2$ is*

$$\begin{aligned} g(\mathbf{u}_1, \mathbf{u}_2) &= \frac{1}{T_{q_1+q_2}\left(\begin{pmatrix} \boldsymbol{\tau}_1 \\ \boldsymbol{\tau}_2 \end{pmatrix}; \begin{pmatrix} \mathbf{0}_{q_1} \\ \mathbf{0}_{q_2} \end{pmatrix}, \begin{pmatrix} \bar{\mathbf{\Gamma}}_{11} & \bar{\mathbf{\Gamma}}_{12} \\ \bar{\mathbf{\Gamma}}_{21} & \bar{\mathbf{\Gamma}}_{22} \end{pmatrix}, \nu\right)} t_{q_1+q_2}\left(\begin{pmatrix} \mathbf{u}_1 \\ \mathbf{u}_2 \end{pmatrix}; \begin{pmatrix} \mathbf{0}_{q_1} \\ \mathbf{0}_{q_2} \end{pmatrix}, \begin{pmatrix} \bar{\mathbf{\Gamma}}_{11} & \bar{\mathbf{\Gamma}}_{12} \\ \bar{\mathbf{\Gamma}}_{21} & \bar{\mathbf{\Gamma}}_{22} \end{pmatrix}, \nu\right) \\ &= \frac{1}{T_{q_1+q_2}\left(\begin{pmatrix} \boldsymbol{\tau}_1 \\ \boldsymbol{\tau}_2 \end{pmatrix}; \begin{pmatrix} \mathbf{0}_{q_1} \\ \mathbf{0}_{q_2} \end{pmatrix}, \begin{pmatrix} \bar{\mathbf{\Gamma}}_{11} & \bar{\mathbf{\Gamma}}_{12} \\ \bar{\mathbf{\Gamma}}_{21} & \bar{\mathbf{\Gamma}}_{22} \end{pmatrix}, \nu\right)} \\ &\quad \times t_{q_1}(\mathbf{u}_1; \mathbf{0}_{q_1}, \bar{\mathbf{\Gamma}}_{11}, \nu) t_{q_2}\left(\mathbf{u}_2; \bar{\mathbf{\Gamma}}_{21}\bar{\mathbf{\Gamma}}_{11}^{-1}\mathbf{u}_1, \sqrt{\frac{\nu+Q_1(\mathbf{u}_1)}{\nu+q_1}}\bar{\mathbf{\Gamma}}_{22\cdot 1}, \nu+q_1\right), \quad \mathbf{u}_j + \boldsymbol{\tau}_j > \mathbf{0}_{q_j}, \end{aligned}$$

$j = 1,2$, where the marginal-conditional factorization of the joint (q_1+q_2)-variate t pdf was used. Hence, considering the change of variables $\mathbf{v} = \sqrt{\frac{\nu+q_1}{\nu+Q_1(\mathbf{u}_1)}}(\mathbf{u}_2 - \bar{\mathbf{\Gamma}}_{21}\bar{\mathbf{\Gamma}}_{11}^{-1}\mathbf{u}_1)$, the marginal pdf of

$\mathbf{U}_1$ *becomes*

$$
\begin{aligned}
g_1(\mathbf{u}_1) &= \int_{\mathbf{u}_2+\boldsymbol{\tau}_2>\mathbf{0}_{q_2}} g(\mathbf{u}_1,\mathbf{u}_2)d\mathbf{u}_2 \\
&= \frac{1}{T_{q_1+q_2}\left(\begin{pmatrix}\boldsymbol{\tau}_1\\ \boldsymbol{\tau}_2\end{pmatrix};\begin{pmatrix}\mathbf{0}_{q_1}\\ \mathbf{0}_{q_2}\end{pmatrix},\begin{pmatrix}\bar{\boldsymbol{\Gamma}}_{11} & \bar{\boldsymbol{\Gamma}}_{12}\\ \bar{\boldsymbol{\Gamma}}_{21} & \bar{\boldsymbol{\Gamma}}_{22}\end{pmatrix},\nu\right)} \\
&\quad \times t_{q_1}(\mathbf{u}_1;\mathbf{0}_{q_1},\bar{\boldsymbol{\Gamma}}_{11},\nu)\int_{\mathbf{u}_2+\boldsymbol{\tau}_2>\mathbf{0}_{q_2}} t_{q_2}\left(\mathbf{u}_2;\bar{\boldsymbol{\Gamma}}_{21}\bar{\boldsymbol{\Gamma}}_{11}^{-1}\mathbf{u}_1,\sqrt{\frac{\nu+Q_1(\mathbf{u}_1)}{\nu+q_1}}\bar{\boldsymbol{\Gamma}}_{22\cdot 1},\nu+q_1\right)d\mathbf{u}_2 \\
&= \frac{1}{T_{q_1+q_2}\left(\begin{pmatrix}\boldsymbol{\tau}_1\\ \boldsymbol{\tau}_2\end{pmatrix};\begin{pmatrix}\mathbf{0}_{q_1}\\ \mathbf{0}_{q_2}\end{pmatrix},\begin{pmatrix}\bar{\boldsymbol{\Gamma}}_{11} & \bar{\boldsymbol{\Gamma}}_{12}\\ \bar{\boldsymbol{\Gamma}}_{21} & \bar{\boldsymbol{\Gamma}}_{22}\end{pmatrix},\nu\right)} \\
&\quad \times t_{q_1}(\mathbf{u}_1;\mathbf{0}_{q_1},\bar{\boldsymbol{\Gamma}}_{11},\nu)\int_{\mathbf{v}+\sqrt{\frac{\nu+q_1}{\nu+Q_1(\mathbf{u}_1)}}\boldsymbol{\tau}_{2\cdot 1}>\mathbf{0}_{q_2}} t_{q_2}\left(\mathbf{v};\mathbf{0}_{q_2},\bar{\boldsymbol{\Gamma}}_{22\cdot 1},\nu+q_1\right)d\mathbf{v},
\end{aligned}
$$

and the proof for the marginal of $\mathbf{U}_1$ *follows by noting by symmetry that*

$$
\begin{aligned}
\int_{\mathbf{v}>-\sqrt{\frac{\nu+q_1}{\nu+Q_1(\mathbf{u}_1)}}\boldsymbol{\tau}_{2\cdot 1}} t_{q_2}\left(\mathbf{v};\mathbf{0}_{q_2},\bar{\boldsymbol{\Gamma}}_{22\cdot 1},\nu+q_1\right)d\mathbf{v} &= \int_{\mathbf{v}\le\sqrt{\frac{\nu+q_1}{\nu+Q_1(\mathbf{u}_1)}}\boldsymbol{\tau}_{2\cdot 1}} t_{q_2}\left(\mathbf{v};\mathbf{0}_{q_2},\bar{\boldsymbol{\Gamma}}_{22\cdot 1},\nu+q_1\right)d\mathbf{v} \\
&= T_{q_2}\left(\sqrt{\frac{\nu+q_1}{\nu+Q_1(\mathbf{u}_1)}}\boldsymbol{\tau}_{2\cdot 1};\mathbf{0}_{q_2},\bar{\boldsymbol{\Gamma}}_{22\cdot 1},\nu+q_1\right).
\end{aligned}
$$

Finally, if $\boldsymbol{\tau}_2=\mathbf{0}_{q_2}$ *and* $\bar{\boldsymbol{\Gamma}}_{12}=\mathbf{O}$, *then* $\boldsymbol{\tau}_{2\cdot 1}=\mathbf{0}_{q_2}$ *and* $\bar{\boldsymbol{\Gamma}}_{22\cdot 1}=\bar{\boldsymbol{\Gamma}}_{22}$. *Hence, for the probability ratio in the marginal pdf* $g_1(\mathbf{u}_1)$, *we have (see Exercise 7.2)*

$$
\begin{aligned}
\frac{T_{q_2}\left(\sqrt{\frac{\nu+q_1}{\nu+Q_1(\mathbf{u}_1)}}\boldsymbol{\tau}_{2\cdot 1};\mathbf{0}_{q_2},\bar{\boldsymbol{\Gamma}}_{22},\nu+q_1\right)}{T_{q_1+q_2}\left(\begin{pmatrix}\boldsymbol{\tau}_1\\ \boldsymbol{\tau}_2\end{pmatrix};\begin{pmatrix}\mathbf{0}_{q_1}\\ \mathbf{0}_{q_2}\end{pmatrix},\begin{pmatrix}\bar{\boldsymbol{\Gamma}}_{11} & \bar{\boldsymbol{\Gamma}}_{12}\\ \bar{\boldsymbol{\Gamma}}_{21} & \bar{\boldsymbol{\Gamma}}_{22}\end{pmatrix},\nu\right)} &= \frac{T_{q_2}\left(\mathbf{0}_{q_2};\mathbf{0}_{q_2},\bar{\boldsymbol{\Gamma}}_{22},\nu+q_1\right)}{T_{q_1+q_2}\left(\begin{pmatrix}\boldsymbol{\tau}_1\\ \mathbf{0}_{q_2}\end{pmatrix};\begin{pmatrix}\mathbf{0}_{q_1}\\ \mathbf{0}_{q_2}\end{pmatrix},\begin{pmatrix}\bar{\boldsymbol{\Gamma}}_{11} & \mathbf{O}\\ \mathbf{O} & \bar{\boldsymbol{\Gamma}}_{22}\end{pmatrix},\nu\right)} \\
&= \frac{T_{q_2}\left(\mathbf{0}_{q_2};\mathbf{0}_{q_2},\bar{\boldsymbol{\Gamma}}_{22},\nu+q_1\right)}{T_{q_1}\left(\boldsymbol{\tau}_1;\mathbf{0}_{q_1},\bar{\boldsymbol{\Gamma}}_{11},\nu\right)T_{q_2}\left(\mathbf{0}_{q_2};\mathbf{0}_{q_2},\bar{\boldsymbol{\Gamma}}_{22},\nu\right)} \\
&= \frac{1}{T_{q_1}\left(\boldsymbol{\tau}_1;\mathbf{0}_{q_1},\bar{\boldsymbol{\Gamma}}_{11},\nu\right)},
\end{aligned}
$$

since $T_{q_2}\left(\mathbf{0}_{q_2};\mathbf{0}_{q_2},\bar{\boldsymbol{\Gamma}}_{22},\nu+q_1\right)=T_{q_2}\left(\mathbf{0}_{q_2};\mathbf{0}_{q_2},\bar{\boldsymbol{\Gamma}}_{22},\nu\right)=\Phi_{q_2}(\mathbf{0}_{q_2};\mathbf{0}_{q_2},\bar{\boldsymbol{\Gamma}}_{22})$ *for all* $\nu>0$. *The proof is complete.* □

Theorem 127 *Let* $\mathbf{X}\sim\mathcal{SUT}_{p,q}(\boldsymbol{\xi},\boldsymbol{\Psi},\mathbf{H},\boldsymbol{\tau},\bar{\boldsymbol{\Gamma}},\nu)$ *be partitioned as (7.16). Also, consider the induced partition (7.18) of* $(\mathbf{H},\boldsymbol{\tau},\bar{\boldsymbol{\Gamma}})$ *with the assumption* $\bar{\boldsymbol{\Gamma}}_{12}=\bar{\boldsymbol{\Gamma}}_{21}^{\top}=\mathbf{O}$, *i.e.,* $\bar{\boldsymbol{\Gamma}}=\mathrm{diag}(\bar{\boldsymbol{\Gamma}}_{11},\bar{\boldsymbol{\Gamma}}_{22})$. *Then, we have:*

(i) if $\mathbf{H}_{jk}=\mathbf{O}$ *and* $\boldsymbol{\tau}_k=\mathbf{0}_{q_k}$, $j\neq k$, *then* $\mathbf{X}_j\sim\mathcal{SUT}_{p_j,q_j}(\boldsymbol{\xi}_j,\boldsymbol{\Psi}_{jj},\mathbf{H}_{jj},\boldsymbol{\tau}_j,\bar{\boldsymbol{\Gamma}}_{jj},\nu)$ *for* $j,k=1,2$,

(ii) if $\mathbf{H}_{jj}=\mathbf{O}$, *and* $\boldsymbol{\tau}_j=\mathbf{0}_{q_j}$ $j\neq k$, *then* $\mathbf{X}_j\sim\mathcal{SUT}_{p_j,q_k}(\boldsymbol{\xi}_j,\boldsymbol{\Psi}_{jj},\mathbf{H}_{jk},\boldsymbol{\tau}_k,\bar{\boldsymbol{\Gamma}}_{kk},\nu)$ *for* $j,k=1,2$,

which hold for all fixed $p_1+p_2=p$ *and* $q_1+q_2=q$.

Proof 127 *We prove the results for $j=1$ and $k=2$ only. The proof of the opposite case is analogous. By Theorem 123, $\mathbf{X}_1 \sim \mathcal{SUT}_{p_1,q}(\boldsymbol{\xi}_1, \boldsymbol{\Psi}_{11}, \mathbf{H}_1, \boldsymbol{\tau}, \bar{\boldsymbol{\Gamma}}, \nu)$, where $\mathbf{H}_1 = (\mathbf{H}_{11}, \mathbf{H}_{12})$, $\boldsymbol{\tau} = (\boldsymbol{\tau}_1^\top, \boldsymbol{\tau}_2^\top)^\top$ and $\bar{\boldsymbol{\Gamma}} = (\bar{\boldsymbol{\Gamma}}_{ij})_{i,j=1,2}$. So, by Theorem 121, its cdf is*

$$F_1(\mathbf{x}_1) = \frac{1}{T_q(\boldsymbol{\tau}; \mathbf{0}_q, \bar{\boldsymbol{\Gamma}}, \nu)} T_{p_1+q}\left\{\begin{pmatrix}\mathbf{x}_1\\ \boldsymbol{\tau}\end{pmatrix}; \begin{pmatrix}\boldsymbol{\xi}_1\\ \mathbf{0}_q\end{pmatrix}, \begin{pmatrix}\boldsymbol{\Omega}_{11} & -\mathbf{H}_1\bar{\boldsymbol{\Gamma}}\\ -\bar{\boldsymbol{\Gamma}}\mathbf{H}_1^\top & \bar{\boldsymbol{\Gamma}}\end{pmatrix}, \nu\right\}, \quad \mathbf{x}_1 \in \mathbb{R}^{p_1}, \tag{7.19}$$

where $q = q_1 + q_2$, and

$$T_q(\boldsymbol{\tau}; \mathbf{0}_q, \bar{\boldsymbol{\Gamma}}, \nu) = T_{q_1+q_2}\left\{\begin{pmatrix}\boldsymbol{\tau}_1\\ \boldsymbol{\tau}_2\end{pmatrix}; \begin{pmatrix}\mathbf{0}_{q_1}\\ \mathbf{0}_{q_2}\end{pmatrix}, \begin{pmatrix}\bar{\boldsymbol{\Gamma}}_{11} & \bar{\boldsymbol{\Gamma}}_{12}\\ \bar{\boldsymbol{\Gamma}}_{21} & \bar{\boldsymbol{\Gamma}}_{22}\end{pmatrix}, \nu\right\}$$

so for the blocks $(1,2)$ and $(1,1)$ in the dispersion matrix of the numerator of (7.19), we have

$$\begin{aligned}-\mathbf{H}_1\bar{\boldsymbol{\Gamma}} &= -\left(\mathbf{H}_{11}\bar{\boldsymbol{\Gamma}}_{11} + \mathbf{H}_{12}\bar{\boldsymbol{\Gamma}}_{21}, \mathbf{H}_{11}\bar{\boldsymbol{\Gamma}}_{12} + \mathbf{H}_{12}\bar{\boldsymbol{\Gamma}}_{22}\right),\\ \boldsymbol{\Omega}_{11} = \boldsymbol{\Psi}_{11} + \mathbf{H}_1\bar{\boldsymbol{\Gamma}}\mathbf{H}_1^\top &= \boldsymbol{\Psi}_{11} + \mathbf{H}_{11}\bar{\boldsymbol{\Gamma}}_{11}\mathbf{H}_{11}^\top + \mathbf{H}_{12}\bar{\boldsymbol{\Gamma}}_{21}\mathbf{H}_{11}^\top + \mathbf{H}_{11}\bar{\boldsymbol{\Gamma}}_{12}\mathbf{H}_{12}^\top + \mathbf{H}_{12}\bar{\boldsymbol{\Gamma}}_{22}\mathbf{H}_{12}^\top.\end{aligned}$$

Now, the assumptions that $\boldsymbol{\tau}_2 = \mathbf{0}_{q_2}$ and $\bar{\boldsymbol{\Gamma}}_{12} = \bar{\boldsymbol{\Gamma}}_{21}^\top = \mathbf{O}$ (see Exercise 7.2) imply

$$\begin{aligned}T_q(\boldsymbol{\tau}; \mathbf{0}_q, \bar{\boldsymbol{\Gamma}}, \nu) &= T_{q_1+q_2}\left\{\begin{pmatrix}\boldsymbol{\tau}_1\\ \mathbf{0}_{q_2}\end{pmatrix}; \begin{pmatrix}\mathbf{0}_{q_1}\\ \mathbf{0}_{q_2}\end{pmatrix}, \begin{pmatrix}\bar{\boldsymbol{\Gamma}}_{11} & \mathbf{O}\\ \mathbf{O} & \bar{\boldsymbol{\Gamma}}_{22}\end{pmatrix}, \nu\right\}\\ &= T_{q_1}(\boldsymbol{\tau}_1; \mathbf{0}_{q_1}, \bar{\boldsymbol{\Gamma}}_{11}, \nu) T_{q_2}(\mathbf{0}_{q_2}; \mathbf{0}_{q_2}, \bar{\boldsymbol{\Gamma}}_{22}, \nu),\end{aligned} \tag{7.20}$$

while the assumptions that $\bar{\boldsymbol{\Gamma}}_{12} = \bar{\boldsymbol{\Gamma}}_{21}^\top = \mathbf{O}$ and $\mathbf{H}_{12} = \mathbf{O}$ imply

$$\begin{aligned}-\mathbf{H}_1\bar{\boldsymbol{\Gamma}} &= -\left(\mathbf{H}_{11}\bar{\boldsymbol{\Gamma}}_{11}, \mathbf{H}_{12}\bar{\boldsymbol{\Gamma}}_{22}\right) = -(\mathbf{H}_{11}\bar{\boldsymbol{\Gamma}}_{11}, \mathbf{O}),\\ \boldsymbol{\Omega}_{11} = \boldsymbol{\Psi}_{11} + \mathbf{H}_1\bar{\boldsymbol{\Gamma}}\mathbf{H}_1^\top &= \boldsymbol{\Psi}_{11} + \mathbf{H}_{11}\bar{\boldsymbol{\Gamma}}_{11}\mathbf{H}_{11}^\top + \mathbf{H}_{12}\bar{\boldsymbol{\Gamma}}_{22}\mathbf{H}_{12}^\top = \boldsymbol{\Psi}_{11} + \mathbf{H}_{11}\bar{\boldsymbol{\Gamma}}_{11}\mathbf{H}_{11}^\top,\end{aligned}$$

which in turn imply that

$$\begin{aligned}&T_{p_1+q}\left\{\begin{pmatrix}\mathbf{x}_1\\ \boldsymbol{\tau}\end{pmatrix}; \begin{pmatrix}\boldsymbol{\xi}_1\\ \mathbf{0}_q\end{pmatrix}, \begin{pmatrix}\boldsymbol{\Psi}_{11} + \mathbf{H}_1\bar{\boldsymbol{\Gamma}}\mathbf{H}_1^\top & -\mathbf{H}_1\bar{\boldsymbol{\Gamma}}\\ -\bar{\boldsymbol{\Gamma}}\mathbf{H}_1^\top & \bar{\boldsymbol{\Gamma}}\end{pmatrix}, \nu\right\}\\ &\quad= T_{p_1+q_1+q_2}\left\{\begin{pmatrix}\mathbf{x}_1\\ \boldsymbol{\tau}_1\\ \mathbf{0}_{q_2}\end{pmatrix}; \begin{pmatrix}\boldsymbol{\xi}_1\\ \mathbf{0}_{q_1}\\ \mathbf{0}_{q_2}\end{pmatrix}, \begin{pmatrix}\boldsymbol{\Psi}_{11} + \mathbf{H}_{11}\bar{\boldsymbol{\Gamma}}_{11}\mathbf{H}_{11}^\top & -\mathbf{H}_{11}\bar{\boldsymbol{\Gamma}}_{11} & \mathbf{O}\\ -\bar{\boldsymbol{\Gamma}}_{11}\mathbf{H}_{11}^\top & \bar{\boldsymbol{\Gamma}}_{11} & \mathbf{O}\\ \mathbf{O} & \mathbf{O} & \bar{\boldsymbol{\Gamma}}_{22}\end{pmatrix}, \nu\right\}\\ &\quad= T_{p_1+q_1}\left\{\begin{pmatrix}\mathbf{x}_1\\ \boldsymbol{\tau}_1\end{pmatrix}; \begin{pmatrix}\boldsymbol{\xi}_1\\ \mathbf{0}_{q_1}\end{pmatrix}, \begin{pmatrix}\boldsymbol{\Psi}_{11} + \mathbf{H}_{11}\bar{\boldsymbol{\Gamma}}_{11}\mathbf{H}_{11}^\top & -\mathbf{H}_{11}\bar{\boldsymbol{\Gamma}}_{11}\\ -\bar{\boldsymbol{\Gamma}}_{11}\mathbf{H}_{11}^\top & \bar{\boldsymbol{\Gamma}}_{11}\end{pmatrix}, \nu\right\} T_{q_2}(\mathbf{0}_{q_2}; \mathbf{0}_{q_2}, \bar{\boldsymbol{\Gamma}}_{22}, \nu).\end{aligned} \tag{7.21}$$

The incorporation of (7.20) and (7.21) in (7.19) proves part (i).

Now, we prove part (ii). Again, we will prove this for $j=1$ and $k=2$ only, since the opposite case is analogous. In this case, for the denominator of (7.19), we have

$$\boldsymbol{\tau}_1 = \mathbf{0}_{q_1},\ \bar{\boldsymbol{\Gamma}}_{12} = \mathbf{O} \implies T_q(\boldsymbol{\tau}; \mathbf{0}_q, \bar{\boldsymbol{\Gamma}}, \nu) = T_{q_1}(\mathbf{0}_{q_1}; \mathbf{0}_{q_1}, \bar{\boldsymbol{\Gamma}}_{11}, \nu) T_{q_2}(\boldsymbol{\tau}_2; \mathbf{0}_{q_2}, \bar{\boldsymbol{\Gamma}}_{22}, \nu),$$

while the conditions

$$\begin{aligned}&\mathbf{H}_{11} = \mathbf{O},\ \bar{\boldsymbol{\Gamma}}_{12} = \mathbf{O} \implies -\mathbf{H}_1\bar{\boldsymbol{\Gamma}} = -\mathbf{H}_{12}\bar{\boldsymbol{\Gamma}}_{22},\ \boldsymbol{\Omega}_{11} = \boldsymbol{\Psi}_{11} + \mathbf{H}_{12}\bar{\boldsymbol{\Gamma}}_{22}\mathbf{H}_{12}^\top\\ &\quad\implies T_{p_1+q}\left\{\begin{pmatrix}\mathbf{x}_1\\ \boldsymbol{\tau}\end{pmatrix}; \begin{pmatrix}\boldsymbol{\xi}_1\\ \mathbf{0}_q\end{pmatrix}, \begin{pmatrix}\boldsymbol{\Omega}_{11} & -\mathbf{H}_1\bar{\boldsymbol{\Gamma}}\\ -\bar{\boldsymbol{\Gamma}}\mathbf{H}_1^\top & \bar{\boldsymbol{\Gamma}}\end{pmatrix}, \nu\right\} = T_{q_1}(\mathbf{0}_{q_1}; \mathbf{0}_{q_1}, \bar{\boldsymbol{\Gamma}}_{11}, \nu)\\ &\qquad\times T_{p_1+q_2}\left\{\begin{pmatrix}\mathbf{x}_1\\ \boldsymbol{\tau}_2\end{pmatrix}; \begin{pmatrix}\boldsymbol{\xi}_1\\ \mathbf{0}_{q_2}\end{pmatrix}, \begin{pmatrix}\boldsymbol{\Psi}_{11} + \mathbf{H}_{12}\bar{\boldsymbol{\Gamma}}_{22}\mathbf{H}_{12}^\top & -\mathbf{H}_{12}\bar{\boldsymbol{\Gamma}}_{22}\\ -\bar{\boldsymbol{\Gamma}}_{22}\mathbf{H}_{12}^\top & \bar{\boldsymbol{\Gamma}}_{22}\end{pmatrix}, \nu\right\}.\end{aligned}$$

With these last results in (7.19), the proof of (ii) when $j=1$ and $k=2$ follows. □

Remark 18 *Note that in the special case where both $\boldsymbol{\tau}_1=\mathbf{0}_{q_1}$ and $\boldsymbol{\tau}_2=\mathbf{0}_{q_2}$, i.e., $\boldsymbol{\tau}=\mathbf{0}_q$, the proof of Theorem 127 simplifies substantially. In fact, recall by Equation (7.9) in Theorem 114 that if $\boldsymbol{\tau}=\mathbf{0}_q$ or equivalently $\boldsymbol{\tau}_j=\mathbf{0}_{q_j}$ for $j=1,2$, then $\mathbf{X}=\boldsymbol{\xi}+V^{-1/2}\tilde{\mathbf{Z}}$, where $\tilde{\mathbf{Z}}=\mathbf{H}\tilde{\mathbf{U}}+\tilde{\mathbf{W}}\sim \mathcal{SUN}_{p,q}(\mathbf{0}_p,\boldsymbol{\Psi},\mathbf{H},\mathbf{0}_q,\bar{\boldsymbol{\Gamma}})$ and $V\sim \mathcal{G}amma(\nu/2,\nu/2)$ are independent; hence, by using the partitions (7.16) and (7.18), we obtain*

$$\mathbf{X}_j=\boldsymbol{\xi}_j+V^{-1/2}\tilde{\mathbf{Z}}_j,\quad j=1,2, \tag{7.22}$$

where by Theorem 95, $\tilde{\mathbf{Z}}_j=\mathbf{H}_{j1}\tilde{\mathbf{U}}_1+\mathbf{H}_{j2}\tilde{\mathbf{U}}_2+\tilde{\mathbf{W}}_j\sim \mathcal{SUN}_{p_j,q}(\mathbf{0}_{p_j},\boldsymbol{\Psi}_{jj},(\mathbf{H}_{j1},\mathbf{H}_{j2}),\mathbf{0}_q,\bar{\boldsymbol{\Gamma}})$, which is independent of $V\sim \mathcal{G}amma(\nu/2,\nu/2)$ for each $j=1,2$. Thus, if in addition we consider the condition (i) or (ii), we have the following marginal representations:

(i) $\mathbf{H}_{jk}=\mathbf{O},\ j\neq k\Longrightarrow \tilde{\mathbf{Z}}_j=\boldsymbol{\xi}_j+\mathbf{H}_{jj}\tilde{\mathbf{U}}_j+\tilde{\mathbf{W}}_j\sim \mathcal{SUT}_{p_j,q_j}(\mathbf{0}_{p_j},\boldsymbol{\Psi}_{jj},\mathbf{H}_{jj},\mathbf{0}_{q_j},\bar{\boldsymbol{\Gamma}}_{jj})$,

(ii) $\mathbf{H}_{jj}=\mathbf{O},\ j\neq k\Longrightarrow \tilde{\mathbf{Z}}_j=\boldsymbol{\xi}_j+\mathbf{H}_{jk}\tilde{\mathbf{U}}_k+\tilde{\mathbf{W}}_j\sim \mathcal{SUT}_{p_j,q_k}(\mathbf{0}_{p_j},\boldsymbol{\Psi}_{jj},\mathbf{H}_{jk},\mathbf{0}_{q_k},\bar{\boldsymbol{\Gamma}}_{kk})$.

Using these facts in (7.22) the proof follows by using again Theorem 114.

Example 91 *Let $q=p$ and $\mathbf{X}\sim \mathcal{SUT}_{p,p}(\boldsymbol{\xi},\boldsymbol{\Psi},\mathbf{H},\boldsymbol{\tau},\bar{\boldsymbol{\Gamma}},\nu)$ be partitioned as (7.16), and consider also the partition (7.18) of $(\mathbf{H},\boldsymbol{\tau},\bar{\boldsymbol{\Gamma}})$. Assume that $\mathbf{H}$ is lower triangular. Then $\mathbf{H}_{11}$ and $\mathbf{H}_{22}$ are also lower triangular, and $\mathbf{H}_{12}=\mathbf{O}$, but $\mathbf{H}_{21}\neq\mathbf{O}$. Moreover, if $\bar{\boldsymbol{\Gamma}}=\mathrm{diag}(\bar{\boldsymbol{\Gamma}}_{11},\bar{\boldsymbol{\Gamma}}_{22})$ and $\boldsymbol{\tau}_2=\mathbf{0}_{p_2}$ then according to Theorem 127 we have $\mathbf{X}_1\sim \mathcal{SUT}_{p_1,p_1}(\boldsymbol{\xi}_1,\boldsymbol{\Psi}_{11},\mathbf{H}_{11},\boldsymbol{\tau}_1,\bar{\boldsymbol{\Gamma}}_{11},\nu)$, but according to Theorem 123 we have $\mathbf{X}_2\sim \mathcal{SUT}_{p_2,p}(\boldsymbol{\xi}_2,\boldsymbol{\Psi}_{22},(\mathbf{H}_{21},\mathbf{H}_{22}),\boldsymbol{\tau},\bar{\boldsymbol{\Gamma}},\nu)$. This holds in particular if $\boldsymbol{\tau}=\mathbf{0}_p$ and $\bar{\boldsymbol{\Gamma}}=\mathbf{I}_p$. This means that the SUT distribution in this setting is marginally consistent for $\mathbf{X}_1$ but not for $\mathbf{X}_2$. A possible choice for $\mathbf{H}$ is $\mathbf{H}=\boldsymbol{\Psi}^{\pm1/2}$ where $\boldsymbol{\Psi}^{\pm1/2}$ is the lower triangular factor of the Choleski factorization of $\boldsymbol{\Psi}$ or $\boldsymbol{\Psi}^{-1}$.*

Example 92 *Suppose that $\mathbf{X}=(\mathbf{X}_1^\top,\mathbf{X}_2^\top)^\top\sim \mathcal{SUT}_{p_1+p_2,q_1+q_2}(\boldsymbol{\xi},\boldsymbol{\Psi},\mathbf{H},\mathbf{0}_{q_1+q_2},\bar{\boldsymbol{\Gamma}},\nu)$, with $\boldsymbol{\xi}=(\boldsymbol{\xi}_1^\top,\boldsymbol{\xi}_2^\top)^\top\in\mathbb{R}^{p_1+p_2}$, $\boldsymbol{\Psi}=\mathrm{diag}(\boldsymbol{\Psi}_{11},\boldsymbol{\Psi}_{22})$, with $\boldsymbol{\Psi}_{jj}$ of dimension $p_j\times p_j$ for $j=1,2$, $\mathbf{H}=\mathrm{diag}(\mathbf{H}_{11},\mathbf{H}_{22})$, with $\mathbf{H}_{jj}$ of dimension $p_j\times q_j$ for $j=1,2$ and $\bar{\boldsymbol{\Gamma}}=\mathrm{diag}(\bar{\boldsymbol{\Gamma}}_{11},\bar{\boldsymbol{\Gamma}}_{22})$, with $\bar{\boldsymbol{\Gamma}}_{jj}$ of dimension $q_j\times q_j$ for $j=1,2$. Then, $\mathbf{X}_j\sim \mathcal{SUT}_{p_j,q_j}(\boldsymbol{\xi}_j,\boldsymbol{\Psi}_{jj},\mathbf{H}_{jj},\mathbf{0}_{q_j},\bar{\boldsymbol{\Gamma}}_{jj},\nu)$ for $j=1,2$, and they are uncorrelated but not independent.*

Example 93 *Let $\mathbf{X}=(X_1,\dots,X_p)^\top\sim \mathcal{SUT}_{p,p}(\boldsymbol{\xi},\boldsymbol{\Psi},\mathbf{H},\mathbf{0}_p,\mathbf{I}_p,\nu)$, with $\boldsymbol{\xi}=(\xi_1,\dots,\xi_p)^\top$, $\boldsymbol{\Psi}=(\psi_{ij})_{i,j=1}^p$ and $\mathbf{H}=\mathrm{diag}(\eta_1,\dots,\eta_p)$. Then, the marginal distribution of X_j is $\mathcal{ST}(\xi_j,\psi_{jj},\eta_j,\nu)$, $j=1,\dots,p$, and all of them are uncorrelated but not independent. However, for any $1<k<p$:*

$$(X_1,\dots,X_k)^\top\sim \mathcal{SUT}_{k,k}((\xi_1,\dots,\xi_k)^\top,(\psi_{ij})_{i,j=1}^k,\mathrm{diag}(\eta_1,\dots,\eta_k),\mathbf{0}_k,\mathbf{I}_k,\nu),$$

which is different from a $\mathcal{ST}_k$ distribution.

7.3 Nonsingular Unified Skew-t

7.3.1 Pdf

When $\mathbf{X} = \boldsymbol{\xi} + \mathbf{HU} + \sqrt{\frac{\nu+Q(\mathbf{U})}{\nu+q}}\mathbf{W} \sim \mathcal{SUT}_{p,q}(\boldsymbol{\xi}, \boldsymbol{\Psi}, \mathbf{H}, \boldsymbol{\tau}, \bar{\boldsymbol{\Gamma}}, \nu)$ is nonsingular, i.e., with $\boldsymbol{\Psi} > 0$ so that $|\boldsymbol{\Psi}| > 0$, then the pdf of $\mathbf{X}$ may be found in an explicit form from its definition, or more specifically from its hierarchical representation given by (7.13) and (7.14).

Theorem 128 *A random vector* $\mathbf{X} \sim \mathcal{SUT}_{p,q}(\boldsymbol{\xi}, \boldsymbol{\Psi}, \mathbf{H}, \boldsymbol{\tau}, \bar{\boldsymbol{\Gamma}}, \nu)$*, with* $\boldsymbol{\Psi} > 0$*, has pdf given by*

$$f_{\mathbf{X}}(\mathbf{x}) = \frac{1}{T_q(\boldsymbol{\tau}; \mathbf{0}_q, \bar{\boldsymbol{\Gamma}}, \nu)} t_p(\mathbf{x}; \boldsymbol{\xi}, \boldsymbol{\Psi} + \mathbf{H}\bar{\boldsymbol{\Gamma}}\mathbf{H}^\top, \nu) \\ \times T_q\left[\sqrt{\frac{\nu+p}{\nu+\tilde{Q}(\mathbf{x})}}\left\{\boldsymbol{\tau} + \tilde{\boldsymbol{\Gamma}}\mathbf{H}^\top\boldsymbol{\Psi}^{-1}(\mathbf{x}-\boldsymbol{\xi})\right\}; \mathbf{0}_q, \tilde{\boldsymbol{\Gamma}}, \nu+p\right], \quad \mathbf{x} \in \mathbb{R}^p, \tag{7.23}$$

where $\tilde{Q}(\mathbf{x}) = (\mathbf{x}-\boldsymbol{\xi})^\top(\boldsymbol{\Psi} + \mathbf{H}\bar{\boldsymbol{\Gamma}}\mathbf{H}^\top)^{-1}(\mathbf{x}-\boldsymbol{\xi})$*, and* $\tilde{\boldsymbol{\Gamma}} = (\bar{\boldsymbol{\Gamma}}^{-1} + \mathbf{H}^\top\boldsymbol{\Psi}^{-1}\mathbf{H})^{-1}$.

Proof 128 *Since* $\boldsymbol{\Psi} > 0$*, by (7.13) and (7.14), we have* $(\mathbf{X} \mid \mathbf{U} = \mathbf{u}) \sim \mathcal{T}_p\left(\boldsymbol{\xi} + \mathbf{Hu}, \sqrt{\frac{\nu+Q(\mathbf{u})}{\nu+q}}\boldsymbol{\Psi}, \nu+q\right)$ *is also nonsingular; moreover, for each* $\mathbf{u} > \mathbf{0}_q$*, its conditional pdf is*

$$f(\mathbf{x} \mid \mathbf{u}) = t_p\left(\mathbf{x}; \boldsymbol{\xi} + \mathbf{Hu}, \sqrt{\frac{\nu+Q(\mathbf{u})}{\nu+q}}\boldsymbol{\Psi}, \nu+q\right), \quad \mathbf{x} \in \mathbb{R}^p. \tag{7.24}$$

Similarly, since by assumption $\bar{\boldsymbol{\Gamma}} > 0$*, the truncated t random vector* $\mathbf{U} \sim \mathcal{TT}_q(-\boldsymbol{\tau}; \mathbf{0}_q, \bar{\boldsymbol{\Gamma}}, \nu)$ *is also nonsingular with pdf given by (7.2). Therefore, for the pdf of* $\mathbf{X}$*, we have that*

$$f_{\mathbf{X}}(\mathbf{x}) = \int_{\mathbf{u}+\boldsymbol{\tau}>\mathbf{0}_q} f(\mathbf{x} \mid \mathbf{u}) g(\mathbf{u}) d\mathbf{u} \\ = \frac{1}{T_q(\boldsymbol{\tau}; \mathbf{0}_q, \bar{\boldsymbol{\Gamma}}, \nu)} \int_{\mathbf{u}+\boldsymbol{\tau}>\mathbf{0}_q} t_p\left(\mathbf{x}; \boldsymbol{\xi} + \mathbf{Hu}, \sqrt{\frac{\nu+Q(\mathbf{u})}{\nu+q}}\boldsymbol{\Psi}, \nu+q\right) t_q(\mathbf{u}; \mathbf{0}_q, \bar{\boldsymbol{\Gamma}}, \nu) d\mathbf{u}.$$

Here, the marginal-conditional representations of the $(p+q)$*-variate t pdf given by*

$$t_{p+q}\left(\begin{pmatrix} \mathbf{x} \\ \mathbf{u} \end{pmatrix}; \begin{pmatrix} \boldsymbol{\xi} \\ \mathbf{0}_q \end{pmatrix}, \begin{pmatrix} \boldsymbol{\Psi} + \mathbf{H}\bar{\boldsymbol{\Gamma}}\mathbf{H}^\top & \mathbf{H}\bar{\boldsymbol{\Gamma}} \\ \bar{\boldsymbol{\Gamma}}\mathbf{H}^\top & \bar{\boldsymbol{\Gamma}} \end{pmatrix}, \nu\right)$$

yields the following identity:

$$t_p\left(\mathbf{x}; \boldsymbol{\xi} + \mathbf{Hu}, \sqrt{\frac{\nu+Q(\mathbf{u})}{\nu+q}}\boldsymbol{\Psi}, \nu+q\right) t_q(\mathbf{u}; \mathbf{0}_q, \bar{\boldsymbol{\Gamma}}, \nu) = t_p(\mathbf{x}; \boldsymbol{\xi}, \boldsymbol{\Psi} + \mathbf{H}\bar{\boldsymbol{\Gamma}}\mathbf{H}^\top, \nu) \\ \times t_q\left[\mathbf{u}; \bar{\boldsymbol{\Gamma}}\mathbf{H}^\top(\boldsymbol{\Psi} + \mathbf{H}\bar{\boldsymbol{\Gamma}}\mathbf{H}^\top)^{-1}(\mathbf{x}-\boldsymbol{\xi}), \sqrt{\frac{\nu+\tilde{Q}(\mathbf{x})}{\nu+p}}\left\{\bar{\boldsymbol{\Gamma}} - \bar{\boldsymbol{\Gamma}}\mathbf{H}^\top(\boldsymbol{\Psi} + \mathbf{H}\bar{\boldsymbol{\Gamma}}\mathbf{H}^\top)^{-1}\mathbf{H}\bar{\boldsymbol{\Gamma}}\right\}, \nu+p\right] \\ = t_p(\mathbf{x}; \boldsymbol{\xi}, \boldsymbol{\Psi} + \mathbf{H}\bar{\boldsymbol{\Gamma}}\mathbf{H}^\top, \nu) t_q\left\{\mathbf{u}; \tilde{\boldsymbol{\Gamma}}\mathbf{H}^\top\boldsymbol{\Psi}^{-1}(\mathbf{x}-\boldsymbol{\xi}), \sqrt{\frac{\nu+\tilde{Q}(\mathbf{x})}{\nu+p}}\tilde{\boldsymbol{\Gamma}}, \nu+p\right\}, \tag{7.25}$$

where $\tilde{Q}(\mathbf{x}) = (\mathbf{x}-\boldsymbol{\xi})^\top(\boldsymbol{\Psi}+\mathbf{H}\bar{\boldsymbol{\Gamma}}\mathbf{H}^\top)^{-1}(\mathbf{x}-\boldsymbol{\xi})$, $\tilde{\boldsymbol{\Gamma}} = (\bar{\boldsymbol{\Gamma}}^{-1}+\mathbf{H}^\top\boldsymbol{\Psi}^{-1}\mathbf{H})^{-1}$, *and the following matrix identities have also been used:*

$$\bar{\boldsymbol{\Gamma}}\mathbf{H}^\top(\boldsymbol{\Psi}+\mathbf{H}\bar{\boldsymbol{\Gamma}}\mathbf{H}^\top)^{-1} = (\bar{\boldsymbol{\Gamma}}^{-1}+\mathbf{H}^\top\boldsymbol{\Psi}^{-1}\mathbf{H})^{-1}\mathbf{H}^\top\boldsymbol{\Psi}^{-1}$$

and

$$\bar{\boldsymbol{\Gamma}} - \bar{\boldsymbol{\Gamma}}\mathbf{H}^\top(\boldsymbol{\Psi}+\mathbf{H}\bar{\boldsymbol{\Gamma}}\mathbf{H}^\top)^{-1}\mathbf{H}\bar{\boldsymbol{\Gamma}} = (\bar{\boldsymbol{\Gamma}}^{-1}+\mathbf{H}^\top\boldsymbol{\Psi}^{-1}\mathbf{H})^{-1}.$$

Now, using (7.25) and the change of variables $\mathbf{u} = \sqrt{\frac{\nu+\tilde{Q}(\mathbf{x})}{\nu+p}}\,\mathbf{v}+\mathbf{h}(\mathbf{x})$, *where* $\mathbf{h}(\mathbf{x}) = \tilde{\boldsymbol{\Gamma}}\mathbf{H}^\top\boldsymbol{\Psi}^{-1}(\mathbf{x}-\boldsymbol{\xi})$, *we have*

$$\begin{aligned}
f_{\mathbf{X}}(\mathbf{x}) =& \frac{1}{T_q(\boldsymbol{\tau};\mathbf{0}_q,\bar{\boldsymbol{\Gamma}},\nu)} t_p(\mathbf{x};\boldsymbol{\xi},\boldsymbol{\Psi}+\mathbf{H}\bar{\boldsymbol{\Gamma}}\mathbf{H}^\top,\nu)\\
&\times \int_{\mathbf{u}+\boldsymbol{\tau}>\mathbf{0}_q} t_q\left\{\mathbf{u};\tilde{\boldsymbol{\Gamma}}\mathbf{H}^\top\boldsymbol{\Psi}^{-1}(\mathbf{x}-\boldsymbol{\xi}),\sqrt{\frac{\nu+\tilde{Q}(\mathbf{x})}{\nu+p}}\,\tilde{\boldsymbol{\Gamma}},\nu+p\right\}d\mathbf{u}\\
=& \frac{1}{T_q(\boldsymbol{\tau};\mathbf{0}_q,\bar{\boldsymbol{\Gamma}},\nu)} t_p(\mathbf{x};\boldsymbol{\xi},\boldsymbol{\Psi}+\mathbf{H}\bar{\boldsymbol{\Gamma}}\mathbf{H}^\top,\nu)\int_{\sqrt{\frac{\nu+\tilde{Q}(\mathbf{x})}{\nu+p}}\mathbf{v}+\mathbf{h}(\mathbf{x})+\boldsymbol{\tau}>\mathbf{0}_q} t_q(\mathbf{v};\mathbf{0}_q,\tilde{\boldsymbol{\Gamma}},\nu+p)d\mathbf{v},
\end{aligned}$$

where by symmetry

$$\begin{aligned}
\int_{\sqrt{\frac{\nu+\tilde{Q}(\mathbf{x})}{\nu+p}}\mathbf{v}+\boldsymbol{\tau}+\mathbf{h}(\mathbf{x})>\mathbf{0}_q} t_q(\mathbf{v};\mathbf{0}_q,\tilde{\boldsymbol{\Gamma}},\nu+p)d\mathbf{v} &= \int_{\mathbf{v}\le\sqrt{\frac{\nu+p}{\nu+\tilde{Q}(\mathbf{x})}}\{\boldsymbol{\tau}+\mathbf{h}(\mathbf{x})\}} t_q(\mathbf{v};\mathbf{0}_q,\tilde{\boldsymbol{\Gamma}},\nu+p)d\mathbf{v}\\
&= T_q\left[\sqrt{\frac{\nu+p}{\nu+\tilde{Q}(\mathbf{x})}}\left\{\boldsymbol{\tau}+\tilde{\boldsymbol{\Gamma}}\mathbf{H}^\top\boldsymbol{\Psi}^{-1}(\mathbf{x}-\boldsymbol{\xi})\right\};\mathbf{0}_q,\tilde{\boldsymbol{\Gamma}},\nu+p\right].
\end{aligned}$$

This concludes the proof. □

Obviously, when $\mathbf{X} \sim \mathcal{SUT}_{p,q}(\boldsymbol{\xi},\boldsymbol{\Psi},\mathbf{H},\boldsymbol{\tau},\bar{\boldsymbol{\Gamma}},\nu)$ is absolutely continuous, its pdf can be obtained directly from its cdf presented in Theorem 121. This is equivalent to considering the selection representation in Theorem 119 to obtain the selection pdf (see Chapter 9) of $\mathbf{X} \overset{d}{=} (\mathbf{W} \mid \mathbf{W}_0+\boldsymbol{\tau}>\mathbf{0}_q)$ given by (Arellano-Valle et al., 2006a):

$$f_{\mathbf{X}}(\mathbf{x}) = f_{\mathbf{W}}(\mathbf{x})\frac{\mathrm{P}(\mathbf{W}_0+\boldsymbol{\tau}>\mathbf{0}_q \mid \mathbf{W}=\mathbf{x})}{\mathrm{P}(\mathbf{W}_0+\boldsymbol{\tau}>\mathbf{0}_q)}, \quad \mathbf{x}\in\mathbb{R}^p,$$

where $(\mathbf{W}_0 \mid \mathbf{W}=\mathbf{x}) \sim \mathcal{T}_q\left\{\tilde{\boldsymbol{\Gamma}}\mathbf{H}^\top\boldsymbol{\Psi}^{-1}(\mathbf{x}-\boldsymbol{\xi}),\sqrt{\frac{\nu+\tilde{Q}(\mathbf{x})}{\nu+p}}\,\tilde{\boldsymbol{\Gamma}},\nu+p\right\}$ gives $\mathrm{P}(\mathbf{W}_0+\boldsymbol{\tau}>\mathbf{0}_q \mid \mathbf{W}=\mathbf{x})$, $\mathbf{W}\sim\mathcal{T}_p(\boldsymbol{\xi}+\mathbf{H}\boldsymbol{\tau},\boldsymbol{\Psi}+\mathbf{H}\bar{\boldsymbol{\Gamma}}\mathbf{H}^\top,\nu)$ gives $f_{\mathbf{W}}(\mathbf{x})$, and $\mathbf{W}_0\sim\mathcal{T}_q(\mathbf{0}_q,\bar{\boldsymbol{\Gamma}},\nu)$ gives $\mathrm{P}(\mathbf{W}_0+\boldsymbol{\tau}>\mathbf{0}_q)$.

Remark 19 *In terms of the matrix* $\boldsymbol{\Omega} = \boldsymbol{\Psi}+\mathbf{H}\bar{\boldsymbol{\Gamma}}\mathbf{H}^\top$ *the SUT pdf (7.23) becomes*

$$\begin{aligned}
f_{\mathbf{X}}(\mathbf{x}) =& \frac{1}{T_q(\boldsymbol{\tau};\mathbf{0}_q,\bar{\boldsymbol{\Gamma}},\nu)} t_p(\mathbf{x};\boldsymbol{\xi},\boldsymbol{\Omega},\nu)\\
&\times T_q\left[\sqrt{\frac{\nu+p}{\nu+\tilde{Q}(\mathbf{x})}}\left\{\boldsymbol{\tau}+\bar{\boldsymbol{\Gamma}}\mathbf{H}^\top\boldsymbol{\Omega}^{-1}(\mathbf{x}-\boldsymbol{\xi})\right\};\mathbf{0}_q,\bar{\boldsymbol{\Gamma}}-\bar{\boldsymbol{\Gamma}}\mathbf{H}^\top\boldsymbol{\Omega}^{-1}\mathbf{H}\bar{\boldsymbol{\Gamma}},\nu+p\right], \quad \mathbf{x}\in\mathbb{R}^p,
\end{aligned} \tag{7.26}$$

where $\tilde{Q}(\mathbf{x}) = (\mathbf{x}-\boldsymbol{\xi})^\top\boldsymbol{\Omega}^{-1}(\mathbf{x}-\boldsymbol{\xi})$.

The next proposition specializes the result in Theorem 122 for a nonsingular affine transformation in an absolutely continuous unified skew-*t* random vector.

Theorem 129 *Let* $\mathbf{X} \sim \mathcal{SUT}_{p,q}(\boldsymbol{\xi}, \boldsymbol{\Psi}, \mathbf{H}, \boldsymbol{\tau}, \bar{\boldsymbol{\Gamma}}, \nu)$, $\boldsymbol{\Psi} > 0$, *be a nonsingular multivariate unified skew-t random vector with pdf given by (7.23). Also, let* $\mathbf{Y} = \mathbf{A}\mathbf{X} + \mathbf{b}$, *where* $\mathbf{A} \in \mathbb{R}^{p \times p}$ *is a nonsingular constant matrix and* $\mathbf{b} \in \mathbb{R}^p$ *is a constant vector. Then,* $\mathbf{Y} \sim \mathcal{SUT}_{p,q}(\boldsymbol{\xi}_A, \boldsymbol{\Psi}_A, \mathbf{H}_A, \boldsymbol{\tau}, \bar{\boldsymbol{\Gamma}}, \nu)$ *is also nonsingular and has pdf given by*

$$\begin{aligned} h(\mathbf{y}) =& \frac{1}{T_q(\boldsymbol{\tau}; \mathbf{0}_q, \bar{\boldsymbol{\Gamma}}, \nu)} t_p(\mathbf{y}; \boldsymbol{\xi}_A, \boldsymbol{\Psi}_A + \mathbf{H}_A \bar{\boldsymbol{\Gamma}} \mathbf{H}_A^\top, \nu) \\ & \times T_q \left[\sqrt{\frac{\nu + p}{\nu + \tilde{Q}_A(\mathbf{y})}} \left\{ \boldsymbol{\tau} + \tilde{\boldsymbol{\Gamma}}_A \mathbf{H}_A^\top \boldsymbol{\Psi}_A^{-1} (\mathbf{x} - \boldsymbol{\xi}_A) \right\}; \mathbf{0}_q, \tilde{\boldsymbol{\Gamma}}_A, \nu + p \right], \quad \mathbf{y} \in \mathbb{R}^p, \end{aligned} \tag{7.27}$$

where $\tilde{Q}_A(\mathbf{y}) = (\mathbf{y} - \boldsymbol{\xi}_A)^\top (\boldsymbol{\Psi}_A + \mathbf{H}_A \bar{\boldsymbol{\Gamma}} \mathbf{H}_A^\top)^{-1} (\mathbf{y} - \boldsymbol{\xi}_A)$, *with*

$$\boldsymbol{\xi}_A = \mathbf{A}\boldsymbol{\xi} + \mathbf{b}, \quad \boldsymbol{\Psi}_A = \mathbf{A}\boldsymbol{\Psi}\mathbf{A}^\top, \quad \mathbf{H}_A = \mathbf{A}\mathbf{H}, \quad \tilde{\boldsymbol{\Gamma}}_A = (\bar{\boldsymbol{\Gamma}}^{-1} + \mathbf{H}_A^\top \boldsymbol{\Psi}_A^{-1} \mathbf{H}_A)^{-1} = \tilde{\boldsymbol{\Gamma}}. \tag{7.28}$$

Proof 129 *We first note that if* $|\mathbf{A}| \neq 0$ *and* $\boldsymbol{\Psi} > 0$, *then* $\boldsymbol{\Psi}_A = \mathbf{A}\boldsymbol{\Psi}\mathbf{A}^\top > 0$ *and both* $\mathbf{A}^{-1}$ *and* $\boldsymbol{\Psi}_A^{-1} = (\mathbf{A}\boldsymbol{\Psi}\mathbf{A}^\top)^{-1} = \mathbf{A}^{-\top}\boldsymbol{\Psi}^{-1}\mathbf{A}^{-1}$ *exist. Now, by inverting the affine relation* $\mathbf{y} = \mathbf{A}\mathbf{x} + \mathbf{b}$ *we can write* $\mathbf{x} = \mathbf{A}^{-1}(\mathbf{y} - \mathbf{b})$. *Thus, the pdf of* $\mathbf{Y}$ *is*

$$\begin{aligned} h(\mathbf{y}) =& \text{abs}(|\mathbf{A}|)^{-1} f\{\mathbf{A}^{-1}(\mathbf{y} - \mathbf{b})\} \\ =& \frac{\text{abs}(|\mathbf{A}|)^{-1}}{T_q(\boldsymbol{\tau}; \mathbf{0}_q, \bar{\boldsymbol{\Gamma}}, \nu)} t_p\{\mathbf{A}^{-1}(\mathbf{y} - \mathbf{b}); \boldsymbol{\xi}, \boldsymbol{\Psi} + \mathbf{H}\bar{\boldsymbol{\Gamma}}\mathbf{H}^\top, \nu\} \\ & \times T_q \left(\sqrt{\frac{\nu + p}{\nu + \tilde{Q}\{\mathbf{A}^{-1}(\mathbf{y} - \mathbf{b})\}}} \left[\boldsymbol{\tau} + \tilde{\boldsymbol{\Gamma}}\mathbf{H}^\top \boldsymbol{\Psi}^{-1} \{\mathbf{A}^{-1}(\mathbf{y} - \mathbf{b}) - \boldsymbol{\xi}\} \right]; \mathbf{0}_q, \tilde{\boldsymbol{\Gamma}}, \nu + p \right), \end{aligned}$$

since the Jacobian $J_{(\mathbf{x} \to \mathbf{y})} = |\mathbf{A}^{-1}| = |\mathbf{A}|^{-1}$, *and where* $f(\mathbf{x})$ *is the unified skew-t pdf defined in (7.23). It is then straightforward to verify that*

$$\begin{aligned} & \tilde{Q}\{\mathbf{A}^{-1}(\mathbf{y} - \mathbf{b})\} = \{\mathbf{A}^{-1}(\mathbf{y} - \mathbf{b}) - \boldsymbol{\xi}\}^\top (\boldsymbol{\Psi} + \mathbf{H}\bar{\boldsymbol{\Gamma}}\mathbf{H}^\top)^{-1} \{\mathbf{A}^{-1}(\mathbf{y} - \mathbf{b}) - \boldsymbol{\xi}\} = \tilde{Q}_A(\mathbf{y}), \\ & t_p\{\mathbf{A}^{-1}(\mathbf{y} - \mathbf{b}); \boldsymbol{\xi}, \boldsymbol{\Psi} + \mathbf{H}\bar{\boldsymbol{\Gamma}}\mathbf{H}^\top, \nu\} \text{abs}(|\mathbf{A}|)^{-1} = t_p(\mathbf{y}; \mathbf{A}\boldsymbol{\xi} + \mathbf{b}, \mathbf{A}\boldsymbol{\Psi}\mathbf{A}^\top + \mathbf{A}\mathbf{H}\bar{\boldsymbol{\Gamma}}\mathbf{H}^\top\mathbf{A}^\top, \nu) \end{aligned}$$

and also that $T_q \left(\sqrt{\frac{\nu+p}{\nu+\tilde{Q}\{\mathbf{A}^{-1}(\mathbf{y}-\mathbf{b})\}}} \left[\boldsymbol{\tau} + \tilde{\boldsymbol{\Gamma}}\mathbf{H}^\top \boldsymbol{\Psi}^{-1} \{\mathbf{A}^{-1}(\mathbf{y} - \mathbf{b}) - \boldsymbol{\xi}\} \right]; \mathbf{0}_q, \tilde{\boldsymbol{\Gamma}}, \nu + p \right)$ *is equal to* $T_q \left[\sqrt{\frac{\nu+p}{\nu+\tilde{Q}_A(\mathbf{y})}} \left\{ \boldsymbol{\tau} + \tilde{\boldsymbol{\Gamma}}_A \mathbf{H}_A^\top \boldsymbol{\Psi}_A^{-1} (\mathbf{x} - \boldsymbol{\xi}_A) \right\}; \mathbf{0}_q, \tilde{\boldsymbol{\Gamma}}, \nu + p \right]$. *The proof concludes by noting that*

$$\tilde{\boldsymbol{\Gamma}} = (\bar{\boldsymbol{\Gamma}}^{-1} + \mathbf{H}^\top \boldsymbol{\Psi}^{-1} \mathbf{H})^{-1} = \{\bar{\boldsymbol{\Gamma}}^{-1} + \mathbf{H}^\top \mathbf{A}^\top (\mathbf{A}\boldsymbol{\Psi}\mathbf{A}^\top)^{-1} \mathbf{A}\mathbf{H}\}^{-1} = \tilde{\boldsymbol{\Gamma}}_A.$$

□

We can observe from Theorem 129 that if we consider (7.23) as the pdf of $\mathbf{X}$, then we obtain a similar form for the pdf of a nonsingular affine random vector $\mathbf{Y} = \mathbf{A}\mathbf{X} + \mathbf{b}$ with the obvious modification of the new set of parameters as indicated in (7.28). A similar result is also valid for any matrix $\mathbf{A}$ of dimension $r \times p$ with full row-rank r, since in this case the associated dispersion matrix $\boldsymbol{\Psi}_A = \mathbf{A}\boldsymbol{\Psi}\mathbf{A}^\top$ is also nonsingular. That is, (7.27) with $\nu + p$ replaced by $\nu + r$ also becomes the pdf of any affine random vector $\mathbf{Y} = \mathbf{A}\mathbf{X} + \mathbf{b}$ when $\mathbf{A} \in \mathbb{R}^{r \times p}$ has rank r. The proof follows by considering a $(p-r)$-dimensional auxiliary random vector $\mathbf{Z} = \mathbf{C}\mathbf{X} + \mathbf{d}$ such that $\mathbf{C} \in \mathbb{R}^{(p-r)\times p}$ has full row-rank $p-r$, and then applying the Jacobian procedure to the extended affine transformation of full rank p given by

$$\tilde{\mathbf{Y}} = \begin{pmatrix} \mathbf{Y} \\ \mathbf{Z} \end{pmatrix} = \begin{pmatrix} \mathbf{A}\mathbf{X}+\mathbf{b} \\ \mathbf{C}\mathbf{X}+\mathbf{c} \end{pmatrix} = \underbrace{\begin{pmatrix} \mathbf{A} \\ \mathbf{C} \end{pmatrix}}_{\tilde{\mathbf{A}}}\mathbf{X} + \underbrace{\begin{pmatrix} \mathbf{b} \\ \mathbf{c} \end{pmatrix}}_{\tilde{\mathbf{b}}} = \tilde{\mathbf{A}}\mathbf{X} + \tilde{\mathbf{b}}$$

to obtain the joint pdf $\tilde{h}(\mathbf{y},\mathbf{z}) = \tilde{h}(\tilde{\mathbf{y}})$. The marginal pdf of $\mathbf{Y}$ is then obtained as $h(\mathbf{y}) = \int_{\mathbb{R}^{p-r}} \tilde{h}(\mathbf{y},\mathbf{z})d\mathbf{z}$.

In particular, all the marginal distributions of a nonsingular unified skew-*t* random vector $\mathbf{X} \sim \mathcal{SUT}_{p,q}(\boldsymbol{\xi},\boldsymbol{\Psi},\mathbf{H},\boldsymbol{\tau},\bar{\boldsymbol{\Gamma}},\nu)$, $\boldsymbol{\Psi} > 0$, are also nonsingular. More specifically, since in the partition (7.16), we have that if $\boldsymbol{\Psi} > 0$, then $\boldsymbol{\Psi}_{jj} > 0$ for $j = 1,2$, so that the marginal $\mathbf{X}_j \sim \mathcal{SUT}_{p_j,q}(\boldsymbol{\xi}_j,\boldsymbol{\Psi}_{jj},\mathbf{H}_j,\boldsymbol{\tau},\bar{\boldsymbol{\Gamma}},\nu)$ (see Theorem 123) has pdf given by

$$\begin{aligned} f_j(\mathbf{x}_j) =& \frac{1}{T_q(\boldsymbol{\tau};\mathbf{0}_q,\bar{\boldsymbol{\Gamma}},\nu)} t_{p_j}(\mathbf{x}_j;\boldsymbol{\xi}_j,\boldsymbol{\Psi}_{jj}+\mathbf{H}_j\bar{\boldsymbol{\Gamma}}\mathbf{H}_j^\top,\nu) \\ &\times T_q\left[\sqrt{\frac{\nu+p_j}{\nu+\tilde{Q}_j(\mathbf{x}_j)}}\left\{\boldsymbol{\tau}+\tilde{\boldsymbol{\Gamma}}_j\mathbf{H}_j^\top\boldsymbol{\Psi}_{jj}^{-1}(\mathbf{x}_j-\boldsymbol{\xi}_j)\right\};\mathbf{0}_q,\tilde{\boldsymbol{\Gamma}}_j,\nu+p_j\right], \quad \mathbf{x}_j \in \mathbb{R}^{p_j},\ j=1,2, \end{aligned} \tag{7.29}$$

where $\tilde{Q}_j(\mathbf{x}_j) = (\mathbf{x}_j-\boldsymbol{\xi}_j)^\top(\boldsymbol{\Psi}_{jj}+\mathbf{H}_j\bar{\boldsymbol{\Gamma}}\mathbf{H}_j^\top)^{-1}(\mathbf{x}_j-\boldsymbol{\xi}_j)$ and $\tilde{\boldsymbol{\Gamma}}_j = (\bar{\boldsymbol{\Gamma}}^{-1}+\mathbf{H}_j^\top\boldsymbol{\Psi}_{jj}^{-1}\mathbf{H}_j)^{-1}$ with $\mathbf{H}_j$ defined as in (7.18). According to (7.26), in terms of $\boldsymbol{\Omega}_{jj} = \boldsymbol{\Psi}_{jj}+\mathbf{H}_j\bar{\boldsymbol{\Gamma}}\mathbf{H}_j^\top$, this marginal pdf can also be rewritten as

$$\begin{aligned} f_j(\mathbf{x}_j) =& \frac{1}{T_q(\boldsymbol{\tau};\mathbf{0}_q,\bar{\boldsymbol{\Gamma}},\nu)} t_{p_j}(\mathbf{x}_j;\boldsymbol{\xi}_j,\boldsymbol{\Omega}_{jj},\nu) \\ &\times T_q\left[\sqrt{\frac{\nu+p_j}{\nu+\tilde{Q}_j(\mathbf{x}_j)}}\left\{\boldsymbol{\tau}+\bar{\boldsymbol{\Gamma}}\mathbf{H}_j^\top\boldsymbol{\Omega}_{jj}^{-1}(\mathbf{x}_j-\boldsymbol{\xi}_j)\right\};\mathbf{0}_q,\bar{\boldsymbol{\Gamma}}-\bar{\boldsymbol{\Gamma}}\mathbf{H}_j^\top\boldsymbol{\Omega}_{jj}^{-1}\mathbf{H}_j\bar{\boldsymbol{\Gamma}},\nu+p_j\right], \mathbf{x}_j \in \mathbb{R}^{p_j}, \end{aligned} \tag{7.30}$$

where $\tilde{Q}_j(\mathbf{x}_j) = (\mathbf{x}_j-\boldsymbol{\xi}_j)^\top\boldsymbol{\Omega}_{jj}^{-1}(\mathbf{x}_j-\boldsymbol{\xi}_j)$. In particular, if $\boldsymbol{\tau} = \mathbf{0}_q$ and $\bar{\boldsymbol{\Gamma}} = \mathbf{I}_q$, then the denominator of (7.29) or (7.30) simplifies to $T_q(\mathbf{0}_q;\mathbf{0}_q,\mathbf{I}_q,\nu) = 2^{-q}$.

Example 94 *Suppose that* $\mathbf{X} = (\mathbf{X}_1^\top,\mathbf{X}_2^\top)^\top \sim \mathcal{SUT}_{p_1+p_2,q_1+q_2}(\boldsymbol{\xi},\boldsymbol{\Psi},\mathbf{H},\boldsymbol{\tau},\bar{\boldsymbol{\Gamma}},\nu)$, *with* $\boldsymbol{\xi} = (\boldsymbol{\xi}_1^\top,\boldsymbol{\xi}_2^\top)^\top \in \mathbb{R}^{p_1+p_2}$, $\boldsymbol{\Psi} = (\boldsymbol{\Psi}_{ij}) > 0$, *with* $\boldsymbol{\Psi}_{ij} > 0$ *of dimension* $p_i \times p_j$ *for* $i,j = 1,2$, $\mathbf{H} = \mathrm{diag}(\mathbf{H}_{11},\mathbf{H}_{22})$, *with* $\mathbf{H}_{jj}$ *of dimension* $p_j \times q_j$ *for* $j = 1,2$, $\boldsymbol{\tau} = (\boldsymbol{\tau}_1^\top,\boldsymbol{\tau}_2^\top)^\top \in \mathbb{R}^{q_1+q_2}$ *and* $\bar{\boldsymbol{\Gamma}} = \mathrm{diag}(\bar{\boldsymbol{\Gamma}}_{11},\bar{\boldsymbol{\Gamma}}_{22})$, *with* $\bar{\boldsymbol{\Gamma}}_{jj} > 0$ *of dimension* $q_j \times q_j$ *for* $j = 1,2$. *If in addition* $\boldsymbol{\tau}_2 = \mathbf{0}_{q_2}$, *then* $\mathbf{X}_1 \sim \mathcal{SUT}_{p_1,q_1}(\boldsymbol{\xi}_1,\boldsymbol{\Psi}_{11},\mathbf{H}_{11},\boldsymbol{\tau}_1,\bar{\boldsymbol{\Gamma}}_{11},\nu)$ *(see Example 92), with marginal pdf*

$$\begin{aligned} f_1(\mathbf{x}_1) =& \frac{1}{T_{q_1}(\boldsymbol{\tau}_1;\mathbf{0}_{q_1},\bar{\boldsymbol{\Gamma}}_{11},\nu)} t_{p_1}(\mathbf{x}_1;\boldsymbol{\xi}_1,\boldsymbol{\Psi}_{11}+\mathbf{H}_{11}\bar{\boldsymbol{\Gamma}}\mathbf{H}_{11}^\top,\nu) \\ &\times T_{q_1}\left[\sqrt{\frac{\nu+p_1}{\nu+\tilde{Q}_1(\mathbf{x}_1)}}\left\{\boldsymbol{\tau}_1+\tilde{\boldsymbol{\Gamma}}_{11}\mathbf{H}_{11}^\top\boldsymbol{\Psi}_{11}^{-1}(\mathbf{x}_1-\boldsymbol{\xi}_1)\right\};\mathbf{0}_{q_1},\tilde{\boldsymbol{\Gamma}}_{11},\nu+p_1\right], \quad \mathbf{x}_1 \in \mathbb{R}^{p_1}, \end{aligned}$$

where $\tilde{Q}_1(\mathbf{x}_1) = (\mathbf{x}_1-\boldsymbol{\xi}_1)^\top(\boldsymbol{\Psi}_{11}+\mathbf{H}_{11}\bar{\boldsymbol{\Gamma}}_{11}\mathbf{H}_{11}^\top)^{-1}(\mathbf{x}_1-\boldsymbol{\xi}_1)$ *and* $\tilde{\boldsymbol{\Gamma}}_{11} = (\bar{\boldsymbol{\Gamma}}_{11}^{-1}+\mathbf{H}_{11}^\top\boldsymbol{\Psi}_{11}^{-1}\mathbf{H}_{11})^{-1}$. *Similarly, we can obtain the marginal pdf of* $\mathbf{X}_2 \sim \mathcal{SUT}_{p_2,q_2}(\boldsymbol{\xi}_2,\boldsymbol{\Psi}_{22},\mathbf{H}_{22},\boldsymbol{\tau}_2,\bar{\boldsymbol{\Gamma}}_{22},\nu)$ *if* $\boldsymbol{\tau}_1 = \mathbf{0}_{q_1}$.

Example 95 *Let* $\mathbf{X} = (X_1,\dots,X_p)^\top \sim \mathcal{SUT}_{p,p}(\boldsymbol{\xi},\boldsymbol{\Psi},\mathbf{H},\boldsymbol{\tau},\bar{\boldsymbol{\Gamma}},\nu)$, *with* $\boldsymbol{\xi} = (\xi_1,\dots,\xi_p)^\top$, $\boldsymbol{\Psi} = \operatorname{diag}(\psi_1^2,\dots,\psi_p^2)$, $\mathbf{H} = \operatorname{diag}(\eta_1,\dots,\eta_p)$, $\boldsymbol{\tau} = \mathbf{0}_p$, *and* $\bar{\boldsymbol{\Gamma}} = \mathbf{I}_p$. *Then,* $X_j \sim \mathcal{ST}(\xi_j,\psi_j^2,\eta_j,\nu)$, $j = 1,\dots,p$ *(see Example 93), with marginal pdf*

$$f_j(x_j) = 2t_1(x_j;\xi_j,\psi_j^2+\eta_j^2,\nu) \times T_1\left\{\sqrt{\frac{\nu+1}{\nu+\tilde{Q}_j(x_j)}}\frac{\eta_j(x_j-\xi_j)}{\psi_j\sqrt{\psi_j^2+\eta_j^2}};0,1,\nu+1\right\}, \quad x_j\in\mathbb{R},\ j=1,\dots,p,$$

where $\tilde{Q}_j(x_j) = (x_j-\xi_j)^2/(\psi_j^2+\eta_j^2)$. *However, as in the symmetric case,* $X_1,\dots,X_p$ *are not mutually independent.*

7.3.2 Geometry

The unified skew-t pdf (7.23) is constant on contours with a skew-ellipsoidal shape centered at $\boldsymbol{\xi}$ and defined by those $\mathbf{x}\in\mathbb{R}^p$ that satisfy

$$t_p(\mathbf{x};\boldsymbol{\xi},\boldsymbol{\Psi}+\mathbf{H}\bar{\boldsymbol{\Gamma}}\mathbf{H}^\top,\nu)T_q\left[\sqrt{\frac{\nu+p}{\nu+\tilde{Q}(\mathbf{x})}}\left\{\boldsymbol{\tau}+\tilde{\boldsymbol{\Gamma}}\mathbf{H}^\top\boldsymbol{\Psi}^{-1}(\mathbf{x}-\boldsymbol{\xi})\right\};\mathbf{0}_q,\tilde{\boldsymbol{\Gamma}},\nu+p\right] = c^2, \tag{7.31}$$

where $\tilde{Q}(\mathbf{x}) = (\mathbf{x}-\boldsymbol{\xi})^\top(\boldsymbol{\Psi}+\mathbf{H}\bar{\boldsymbol{\Gamma}}\mathbf{H}^\top)^{-1}(\mathbf{x}-\boldsymbol{\xi})$, $\tilde{\boldsymbol{\Gamma}} = (\bar{\boldsymbol{\Gamma}}^{-1}+\mathbf{H}^\top\boldsymbol{\Psi}^{-1}\mathbf{H})^{-1}$ and c is a constant. When $\boldsymbol{\Psi}+\mathbf{H}\bar{\boldsymbol{\Gamma}}\mathbf{H}^\top = \mathbf{I}_p$, the contours are skew-spheres.

Example 96 *In* $\mathbb{R}^2$, *consider a bivariate unified skew-t distribution,* $\mathcal{SUT}_{2,q}(\boldsymbol{\xi},\boldsymbol{\Psi},\mathbf{H},\boldsymbol{\tau},\bar{\boldsymbol{\Gamma}},\nu)$, *with location* $\boldsymbol{\xi} = \mathbf{0}_2$, *latent correlation matrix* $\bar{\boldsymbol{\Gamma}} = \frac{1}{2}\mathbf{I}_q + \frac{1}{2}\mathbf{1}_q\mathbf{1}_q^\top$ *and:*

(a) $q=1$, $\tau=0$, $\boldsymbol{\Psi}=\mathbf{I}_2$, $\mathbf{H}=(2,3)^\top$, $\nu=\infty$;

(b) $q=2$, $\boldsymbol{\tau}=\mathbf{0}_2$, $\boldsymbol{\Psi}=\mathbf{I}_2$, $\mathbf{H}=\begin{pmatrix}2 & -2\\ 3 & 1\end{pmatrix}$, $\nu=\infty$;

(c) $q=1$, $\tau=0$, $\boldsymbol{\Psi}=\mathbf{I}_2$, $\mathbf{H}=(2,3)^\top$, $\nu=3$;

(d) $q=2$, $\boldsymbol{\tau}=\mathbf{0}_2$, $\boldsymbol{\Psi}=\mathbf{I}_2$, $\mathbf{H}=\begin{pmatrix}2 & -2\\ 3 & 1\end{pmatrix}$, $\nu=3$.

The contours (7.31) of the pdf of the associated $\mathcal{SUT}$ *distributions are plotted in Figure 7.1. In each panel, the vectors represent the columns* $\mathbf{h}_i$ $(i=1,\dots,q)$ *of the matrix* $\mathbf{H}$. *The geometrical interpretation of* $\mathbf{h}_i$ $(i=1,\dots,q)$ *is that they indicate the directions of skewness.*

7.4 Conditional Unified Skew-t

Assuming the partition (7.16) of $\mathbf{X}\sim\mathcal{SUT}_{p,q}(\boldsymbol{\xi},\boldsymbol{\Psi},\mathbf{H},\boldsymbol{\tau},\bar{\boldsymbol{\Gamma}},\nu)$, we derive next the conditional distribution of $\mathbf{X}_1 \mid \mathbf{X}_2 = \mathbf{x}_2$ for the nonsingular case with $\boldsymbol{\Psi} > 0$, and therefore only for $\boldsymbol{\Psi}_{22} > 0$, where $\boldsymbol{\Psi}_{22}$ is the dispersion matrix associated with $\mathbf{X}_2$. The general case where this matrix can also be singular ($|\boldsymbol{\Psi}_{22}| = 0$) can be studied in a similar way with the obvious necessary changes (i.e., with $\boldsymbol{\Psi}_{22}^{-1}$ replaced by $\boldsymbol{\Psi}_{22}^{-}$, a generalized inverse of $\boldsymbol{\Psi}_{22}$). To do this, some preliminary facts are necessary.

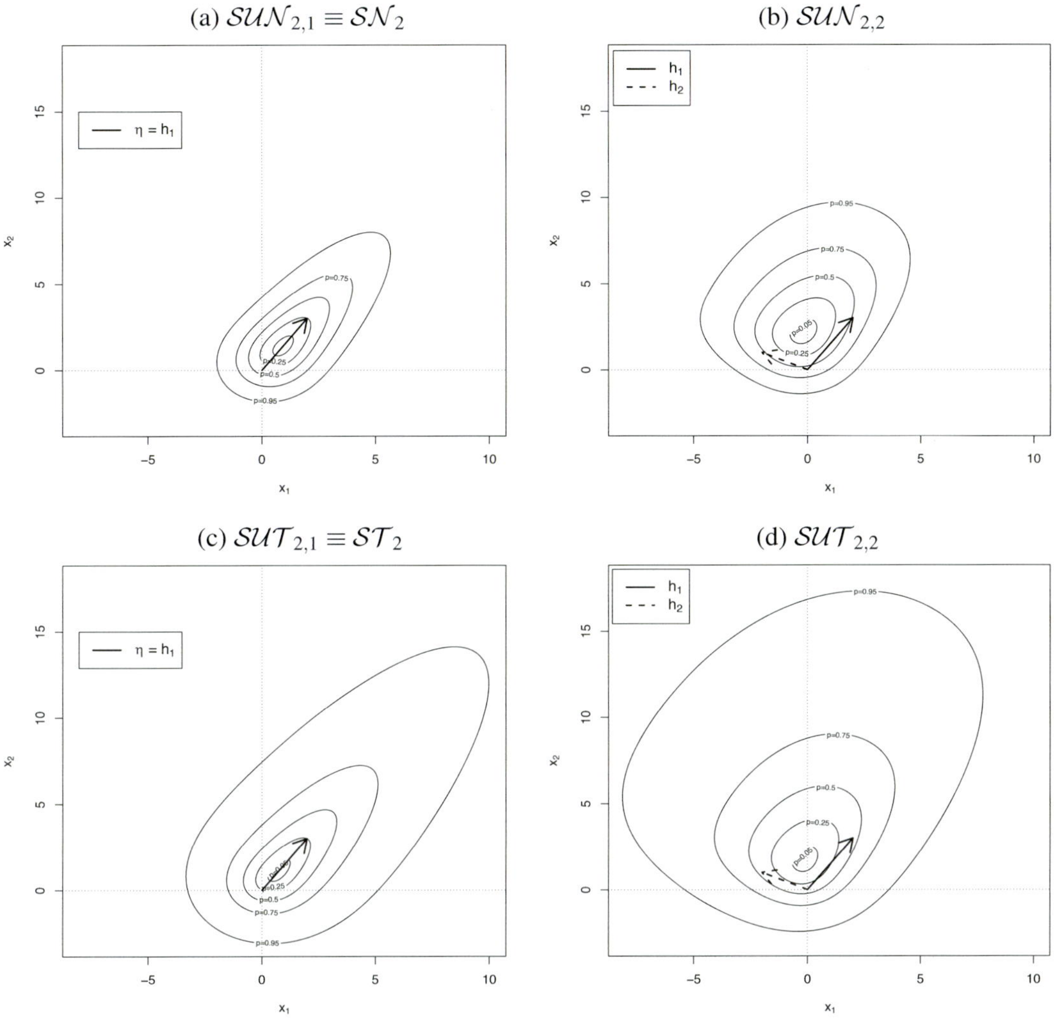

Figure 7.1 *Geometry of the multivariate unified skew-t distribution, $\mathcal{SUT}_{p,q}$, for $p = 2$ and $q = 1, 2$, and (a) & (b): $\nu = \infty$; (c) & (d): $\nu = 3$.*

Theorem 130 *Let* $\mathbf{X} = \boldsymbol{\xi} + \mathbf{H}\mathbf{U} + \sqrt{\frac{\nu + Q(\mathbf{u})}{\nu + q}}\mathbf{W}_* \sim \mathcal{SUT}_{p,q}(\boldsymbol{\xi}, \boldsymbol{\Psi}, \mathbf{H}, \boldsymbol{\tau}, \bar{\boldsymbol{\Gamma}}, \nu)$ *be nonsingular and partitioned as in (7.16). Then:*

(a) $(\mathbf{U} \mid \mathbf{X} = \mathbf{x}) \sim \mathcal{TT}_q\left(\mathbf{0}_q; \tilde{\boldsymbol{\Gamma}}\mathbf{H}^\top\boldsymbol{\Psi}^{-1}(\mathbf{x} - \boldsymbol{\xi}), \frac{\nu + \tilde{Q}(\mathbf{x})}{\nu + p}\tilde{\boldsymbol{\Gamma}}, \nu + p\right)$*, with conditional pdf:*

$$g(\mathbf{u} \mid \mathbf{x}) = \frac{1}{T_q\left[\sqrt{\frac{\nu+p}{\nu+\tilde{Q}(\mathbf{x})}}\{\boldsymbol{\tau} + \tilde{\boldsymbol{\Gamma}}\mathbf{H}^\top\boldsymbol{\Psi}^{-1}(\mathbf{x} - \boldsymbol{\xi})\}; \mathbf{0}_q, \tilde{\boldsymbol{\Gamma}}, \nu + p\right]} \times t_q\left\{\mathbf{u}; \tilde{\boldsymbol{\Gamma}}\mathbf{H}^\top\boldsymbol{\Psi}^{-1}(\mathbf{x} - \boldsymbol{\xi}), \frac{\nu + \tilde{Q}(\mathbf{x})}{\nu + p}\tilde{\boldsymbol{\Gamma}}, \nu + p\right\}, \quad \mathbf{u} + \boldsymbol{\tau} > \mathbf{0}_q,$$

where, as defined earlier, $\tilde{Q}(\mathbf{x}) = (\mathbf{x} - \boldsymbol{\xi})^\top(\boldsymbol{\Psi} + \mathbf{H}\bar{\boldsymbol{\Gamma}}\mathbf{H}^\top)^{-1}(\mathbf{x} - \boldsymbol{\xi})$ *and* $\tilde{\boldsymbol{\Gamma}} = (\bar{\boldsymbol{\Gamma}}^{-1} + \mathbf{H}^\top\boldsymbol{\Psi}^{-1}\mathbf{H})^{-1}$*;*

(b) $(\mathbf{U} \mid \mathbf{X}_2 = \mathbf{x}_2) \sim \mathcal{TT}_q\left(-\boldsymbol{\tau}; \tilde{\boldsymbol{\Gamma}}_2\mathbf{H}_2^\top\boldsymbol{\Psi}_{22}^{-1}(\mathbf{x}_2-\boldsymbol{\xi}_2), \frac{\nu+\tilde{Q}_2(\mathbf{x}_2)}{\nu+p_2}\tilde{\boldsymbol{\Gamma}}_2, \nu+p_2\right)$, *with conditional pdf:*

$$g_2(\mathbf{u} \mid \mathbf{x}_2) = \frac{1}{T_q\left[\sqrt{\frac{\nu+p_2}{\nu+\tilde{Q}_2(\mathbf{x}_2)}}\{\boldsymbol{\tau}+\tilde{\boldsymbol{\Gamma}}_2\mathbf{H}_2^\top\boldsymbol{\Psi}_{22}^{-1}(\mathbf{x}_2-\boldsymbol{\xi}_2)\};\mathbf{0}_q,\tilde{\boldsymbol{\Gamma}}_2,\nu+p_2\right]} \times t_q\left\{\mathbf{u};\tilde{\boldsymbol{\Gamma}}_2\mathbf{H}_2^\top\boldsymbol{\Psi}_{22}^{-1}(\mathbf{x}_2-\boldsymbol{\xi}_2),\frac{\nu+\tilde{Q}_2(\mathbf{x}_2)}{\nu+p_2}\tilde{\boldsymbol{\Gamma}}_2,\nu+p_2\right\}, \quad \mathbf{u}+\boldsymbol{\tau}>\mathbf{0}_q, \tag{7.32}$$

where $\tilde{Q}_2(\mathbf{x}_2) = (\mathbf{x}_2-\boldsymbol{\xi}_2)^\top(\boldsymbol{\Psi}_{22}+\mathbf{H}_2\bar{\boldsymbol{\Gamma}}\mathbf{H}_2^\top)^{-1}(\mathbf{x}_2-\boldsymbol{\xi}_2)$ *and* $\tilde{\boldsymbol{\Gamma}}_2 = (\bar{\boldsymbol{\Gamma}}^{-1}+\mathbf{H}_2^\top\boldsymbol{\Psi}_{22}^{-1}\mathbf{H}_2)^{-1}$;

(c) $\mathbf{W}_{1\cdot2}^* = \sqrt{\frac{\nu+q+p_2}{\nu+q+\mathbf{W}_{*2}^\top\boldsymbol{\Psi}_{22}^{-1}\mathbf{W}_{*2}}}(\mathbf{W}_{*1}-\boldsymbol{\Psi}_{12}\boldsymbol{\Psi}_{22}^{-1}\mathbf{W}_{*2}) \sim \mathcal{T}_{p_1}(\mathbf{0}_{p_1},\boldsymbol{\Psi}_{11\cdot2},\nu+q+p_2)$, *with* $\boldsymbol{\Psi}_{11\cdot2} = \boldsymbol{\Psi}_{11}-\boldsymbol{\Psi}_{12}\boldsymbol{\Psi}_{22}^{-1}\boldsymbol{\Psi}_{21}$, *and is independent of* $\mathbf{W}_{*2}$ *and also of* $\mathbf{U}$.

Proof 130 *From the hierarchical presentation in (7.13) and (7.14), we have* $\mathbf{X} \mid \mathbf{U}=\mathbf{u} \sim \mathcal{T}_p(\boldsymbol{\xi}+\mathbf{H}\mathbf{u},\boldsymbol{\Psi},\nu)$ *and* $\mathbf{U} \sim \mathcal{TT}_q(-\boldsymbol{\tau};\mathbf{0}_q,\bar{\boldsymbol{\Gamma}},\nu)$. *Hence, since these distributions are nonsingular, by using the Bayes formula with (7.24), (7.2), and (7.23), we have*

$$\begin{aligned} g(\mathbf{u} \mid \mathbf{x}) &= \frac{f(\mathbf{x} \mid \mathbf{u})g(\mathbf{u})}{f(\mathbf{x})} \\ &= \frac{t_p\left(\mathbf{x};\boldsymbol{\xi}+\mathbf{H}\mathbf{u},\frac{\nu+Q(\mathbf{u})}{\nu+q}\boldsymbol{\Psi},\nu+q\right)t_q(\mathbf{u};\mathbf{0}_q,\bar{\boldsymbol{\Gamma}},\nu)}{t_p(\mathbf{x};\boldsymbol{\xi},\boldsymbol{\Psi}+\mathbf{H}\bar{\boldsymbol{\Gamma}}\mathbf{H}^\top,\nu)T_q\left[\sqrt{\frac{\nu+p}{\nu+\tilde{Q}(\mathbf{x})}}\{\boldsymbol{\tau}+\tilde{\boldsymbol{\Gamma}}\mathbf{H}^\top\boldsymbol{\Psi}^{-1}(\mathbf{x}-\boldsymbol{\xi})\};\mathbf{0}_q,\tilde{\boldsymbol{\Gamma}},\nu+p\right]}, \quad \mathbf{u}+\boldsymbol{\tau}>\mathbf{0}_q, \end{aligned}$$

and the proof of part (a) *follows by again using the identity (7.25) in the numerator of this last expression.*

The proof of the result in (b) *is analogous by replacing the conditional pdf of* $\mathbf{X} \mid \mathbf{U}=\mathbf{u}$ *by the marginal-conditional pdf of* $\mathbf{X}_2 \mid \mathbf{U}=\mathbf{u}$ *and the joint pdf* $f(\mathbf{x}) = f(\mathbf{x}_1,\mathbf{x}_2)$ *by the marginal pdf* $f_2(\mathbf{x}_2)$ *given in (7.29).*

The proof of part (c) *follows from the fact that*

$$\begin{aligned} \mathbf{W}_* = \begin{pmatrix}\mathbf{W}_{*1}\\ \mathbf{W}_{*2}\end{pmatrix} &\sim \mathcal{T}_{p_1+p_2}\left(\begin{pmatrix}\mathbf{0}_{p_1}\\ \mathbf{0}_{p_2}\end{pmatrix},\begin{pmatrix}\boldsymbol{\Psi}_{11} & \boldsymbol{\Psi}_{12}\\ \boldsymbol{\Psi}_{21} & \boldsymbol{\Psi}_{22}\end{pmatrix},\nu+q\right) \\ \Longrightarrow \begin{pmatrix}\mathbf{W}_{*1}-\boldsymbol{\Psi}_{12}\boldsymbol{\Psi}_{22}^{-1}\mathbf{W}_{*2}\\ \mathbf{W}_{*2}\end{pmatrix} &\sim \mathcal{T}_{p_1+p_2}\left(\begin{pmatrix}\mathbf{0}_{p_1}\\ \mathbf{0}_{p_2}\end{pmatrix},\begin{pmatrix}\boldsymbol{\Psi}_{11\cdot2} & \mathbf{O}\\ \mathbf{O} & \boldsymbol{\Psi}_{22}\end{pmatrix},\nu+q\right) \\ \Longrightarrow (\mathbf{W}_{*1}-\boldsymbol{\Psi}_{12}\boldsymbol{\Psi}_{22}^{-1}\mathbf{W}_{*2}) \mid \mathbf{W}_{*2}=\mathbf{w}_{*2} &\sim \mathcal{T}_{p_1}\left(\mathbf{0}_{p_1},\frac{\nu+q+\mathbf{w}_{*2}^\top\boldsymbol{\Psi}_{22}^{-1}\mathbf{w}_{*2}}{\nu+q+p_2}\boldsymbol{\Psi}_{11\cdot2},\nu+q+p_2\right) \\ \Longrightarrow \sqrt{\frac{\nu+q+p_2}{\nu+q+\mathbf{w}_{*2}^\top\boldsymbol{\Psi}_{22}^{-1}\mathbf{w}_{*2}}}(\mathbf{W}_{*1}-\boldsymbol{\Psi}_{12}\boldsymbol{\Psi}_{22}^{-1}\mathbf{W}_{*2}) \mid \mathbf{W}_{*2}=\mathbf{w}_{*2} &\sim \mathcal{T}_{p_1}(\mathbf{0}_{p_1},\boldsymbol{\Psi}_{11\cdot2},\nu+q+p_2), \\ \forall \mathbf{w}_{*2} \in \mathbb{R}^{p_2} & \\ \Longrightarrow \mathbf{W}_{1\cdot2}^* = \sqrt{\frac{\nu+q+p_2}{\nu+q+\mathbf{W}_{*2}^\top\boldsymbol{\Psi}_{22}^{-1}\mathbf{W}_{*2}}}(\mathbf{W}_{*1}-\boldsymbol{\Psi}_{12}\boldsymbol{\Psi}_{22}^{-1}\mathbf{W}_{*2}) &\sim \mathcal{T}_{p_1}(\mathbf{0}_{p_1},\boldsymbol{\Psi}_{11\cdot2},\nu+q+p_2). \end{aligned}$$

Since the conditional distribution of $\mathbf{W}_{1\cdot2}^*$ *given* $\mathbf{W}_{*2}=\mathbf{w}_{*2}$ *does not depend on* $\mathbf{w}_{*2}$, *then* $\mathbf{W}_{1\cdot2}^*$ *is independent of* $\mathbf{W}_{*2}$. *Also,* $\mathbf{W}_{1\cdot2}^*$ *is independent of* $\mathbf{U}$, *since* $\mathbf{W}_* = (\mathbf{W}_{*1}^\top,\mathbf{W}_{*2}^\top)^\top$ *is also independent of* $\mathbf{U}$. □

From parts (b) and (c) of Theorem 130, we can obtain a nonsingular version of the conditional distribution of $\mathbf{X}_1 \mid \mathbf{X}_2 = \mathbf{x}_2$ under the multivariate unified skew-t distribution. For this, we need some additional notation. Let:

$$\bar{\boldsymbol{\Gamma}}_{1\cdot 2} = \boldsymbol{\gamma}_{1\cdot 2}^{-1}\boldsymbol{\Gamma}_{1\cdot 2}\boldsymbol{\gamma}_{1\cdot 2}^{-1}, \text{ with } \boldsymbol{\Gamma}_{1\cdot 2} = (\bar{\boldsymbol{\Gamma}}^{-1} + \mathbf{H}_2^\top\boldsymbol{\Psi}_{22}^{-1}\mathbf{H}_2)^{-1} = \tilde{\boldsymbol{\Gamma}}_2 \text{ and } \boldsymbol{\gamma}_{1\cdot 2} = \operatorname{diag}(\boldsymbol{\Gamma}_{1\cdot 2})^{1/2}, \tag{7.33}$$

$$\bar{\boldsymbol{\tau}}_{1\cdot 2} = \boldsymbol{\gamma}_{1\cdot 2}^{-1}\boldsymbol{\tau}_{1\cdot 2}, \text{ with } \boldsymbol{\tau}_{1\cdot 2} = \boldsymbol{\tau} + \boldsymbol{\Gamma}_2\mathbf{H}_2^\top\boldsymbol{\Psi}_{22}^{-1}(\mathbf{x}_2 - \boldsymbol{\xi}_2) = \boldsymbol{\tau} + \boldsymbol{\Gamma}_{1\cdot 2}\mathbf{H}_2^\top\boldsymbol{\Psi}_{22}^{-1}(\mathbf{x}_2 - \boldsymbol{\xi}_2), \tag{7.34}$$

$$\bar{\mathbf{H}}_{1\cdot 2} = \mathbf{H}_{1\cdot 2}\boldsymbol{\gamma}_{1\cdot 2}, \text{ with } \mathbf{H}_{1\cdot 2} = \mathbf{H}_1 - \boldsymbol{\Psi}_{12}\boldsymbol{\Psi}_{22}^{-1}\mathbf{H}_2. \tag{7.35}$$

Note that $\bar{\boldsymbol{\tau}}_{1\cdot 2} = \bar{\boldsymbol{\tau}} + \bar{\boldsymbol{\Gamma}}_{1\cdot 2}\bar{\mathbf{H}}_2^\top\boldsymbol{\Psi}_{22}^{-1}(\mathbf{x}_2 - \boldsymbol{\xi}_2)$, with $\bar{\mathbf{H}}_2 = \mathbf{H}_2\bar{\boldsymbol{\gamma}}_{1\cdot 2}$ and $\bar{\boldsymbol{\tau}} = \boldsymbol{\gamma}_{1\cdot 2}^{-1}\boldsymbol{\tau}$, and $\bar{\boldsymbol{\Gamma}}_{1\cdot 2} = (\bar{\bar{\boldsymbol{\Gamma}}}^{-1} + \bar{\mathbf{H}}_2^\top\boldsymbol{\Psi}_{22}^{-1}\bar{\mathbf{H}}_2)^{-1}$, with $\bar{\bar{\boldsymbol{\Gamma}}} = \boldsymbol{\gamma}_{1\cdot 2}^{-1}\bar{\boldsymbol{\Gamma}}\bar{\boldsymbol{\gamma}}_{1\cdot 2}^{-1}$, which is the correlation matrix induced by $\boldsymbol{\Gamma}_{1\cdot 2}$.

Furthermore, we need the following result which comes directly from part (b) of Theorem 130. Let

$$\mathbf{U}_{12} = \sqrt{\frac{\nu + p_2}{\nu + \tilde{Q}_2(\mathbf{X}_2)}}\{\mathbf{U} - \boldsymbol{\Gamma}_{1\cdot 2}\mathbf{H}_2^\top\boldsymbol{\Psi}_{22}^{-1}(\mathbf{X}_2 - \boldsymbol{\xi}_2)\},$$

where $\tilde{Q}_2(\mathbf{X}_2) = (\mathbf{X}_2 - \boldsymbol{\xi}_2)^\top(\boldsymbol{\Psi}_{22} + \mathbf{H}_2\bar{\boldsymbol{\Gamma}}\mathbf{H}_2^\top)^{-1}(\mathbf{X}_2 - \boldsymbol{\xi}_2)$. Then, for each fixed $\mathbf{x}_2 \in \mathbb{R}^{p_2}$, it is easy to see that

$$\mathbf{U}_{1\cdot 2} \stackrel{d}{=} (\mathbf{U}_{12} \mid \mathbf{X}_2 = \mathbf{x}_2) \sim \mathcal{TT}_q\left(-\sqrt{\frac{\nu + p_2}{\nu + \tilde{Q}_2(\mathbf{x}_2)}}\boldsymbol{\tau}_{1\cdot 2}; \mathbf{0}_q, \boldsymbol{\Gamma}_{1\cdot 2}, \nu + q\right) \tag{7.36}$$

$$\Longrightarrow (\mathbf{U} \mid \mathbf{X}_2 = \mathbf{x}_2) \stackrel{d}{=} \sqrt{\frac{\nu + \tilde{Q}_2(\mathbf{x}_2)}{\nu + p_2}}\mathbf{U}_{1\cdot 2} + \boldsymbol{\Gamma}_{1\cdot 2}\mathbf{H}_2^\top\boldsymbol{\Psi}_{22}^{-1}(\mathbf{x}_2 - \boldsymbol{\xi}_2).$$

Theorem 131 *Under the assumptions of Theorem 130, we have for each fixed $\mathbf{x}_2 \in \mathbb{R}^{p_2}$ that the conditional distribution of $(\mathbf{X}_1 \mid \mathbf{X}_2 = \mathbf{x}_2)$ is*

$$\mathcal{SUT}_{p_1,q}\left(\tilde{\boldsymbol{\xi}}_{1\cdot 2}, \frac{\nu + \tilde{Q}_2(\mathbf{x}_2)}{\nu + p_2}\boldsymbol{\Psi}_{11\cdot 2}, \sqrt{\frac{\nu + \tilde{Q}_2(\mathbf{x}_2)}{\nu + p_2}}\bar{\mathbf{H}}_{1\cdot 2}, \sqrt{\frac{\nu + p_2}{\nu + \tilde{Q}_2(\mathbf{x}_2)}}\bar{\boldsymbol{\tau}}_{1\cdot 2}, \bar{\boldsymbol{\Gamma}}_{1\cdot 2}, \nu_{1\cdot 2}\right),$$

where $\tilde{\boldsymbol{\xi}}_{1\cdot 2} = \boldsymbol{\xi}_{1\cdot 2} + \mathbf{H}_{1\cdot 2}\boldsymbol{\Gamma}_{1\cdot 2}\mathbf{H}_2^\top\boldsymbol{\Psi}_{22}^{-1}(\mathbf{x}_2 - \boldsymbol{\xi}_2)$, with $\boldsymbol{\xi}_{1\cdot 2} = \boldsymbol{\xi}_1 + \boldsymbol{\Psi}_{12}\boldsymbol{\Psi}_{22}^{-1}(\mathbf{x}_2 - \boldsymbol{\xi}_2)$, $\boldsymbol{\Psi}_{11\cdot 2} = \boldsymbol{\Psi}_{11} - \boldsymbol{\Psi}_{12}\boldsymbol{\Psi}_{22}^{-1}\boldsymbol{\Psi}_{21}$, $\tilde{Q}_2(\mathbf{x}_2) = (\mathbf{x}_2 - \boldsymbol{\xi}_2)^\top(\boldsymbol{\Psi}_{22} + \mathbf{H}_2\bar{\boldsymbol{\Gamma}}\mathbf{H}_2^\top)^{-1}(\mathbf{x}_2 - \boldsymbol{\xi}_2)$, $\nu_{1\cdot 2} = \nu + p_2$, and with $\bar{\boldsymbol{\Gamma}}_{1\cdot 2}$, $\bar{\boldsymbol{\tau}}_{1\cdot 2}$ and $\bar{\mathbf{H}}_{1\cdot 2}$ given in (7.33), (7.34), and (7.35), respectively. That is, for each fixed $\mathbf{x}_2 \in \mathbb{R}^{p_2}$, $(\mathbf{X}_1 \mid \mathbf{X}_2 = \mathbf{x}_2) \stackrel{d}{=} \mathbf{X}_{1\cdot 2}$, with

$$\mathbf{X}_{1\cdot 2} = \tilde{\boldsymbol{\xi}}_{1\cdot 2} + \sqrt{\frac{\nu + \tilde{Q}_2(\mathbf{x}_2)}{\nu + p_2}}\bar{\mathbf{H}}_{1\cdot 2}\bar{\mathbf{U}}_{1\cdot 2} + \sqrt{\frac{\nu_{1\cdot 2} + \bar{Q}_{1\cdot 2}(\bar{\mathbf{U}}_{1\cdot 2})}{\nu_{1\cdot 2} + q}}\sqrt{\frac{\nu + \tilde{Q}_2(\mathbf{x}_2)}{\nu + p_2}}\mathbf{W}_{1\cdot 2}^*, \tag{7.37}$$

where by (7.36) $\bar{\mathbf{U}}_{1\cdot 2} = \sqrt{\frac{\nu + \tilde{Q}_2(\mathbf{x}_2)}{\nu + p_2}}\boldsymbol{\gamma}_{1\cdot 2}^{-1}\mathbf{U}_{1\cdot 2} \sim \mathcal{TT}_q\left(-\sqrt{\frac{\nu + \tilde{Q}_2(\mathbf{x}_2)}{\nu + p_2}}\bar{\boldsymbol{\tau}}_{1\cdot 2}; \mathbf{0}_q, \bar{\boldsymbol{\Gamma}}_{1\cdot 2}, \nu_{1\cdot 2}\right)$, which is independent of $\mathbf{W}_{1\cdot 2}^* \sim \mathcal{T}_{p_1}(\mathbf{0}_{p_1}, \boldsymbol{\Psi}_{11\cdot 2}, \nu_{1\cdot 2})$, and $\bar{Q}_{1\cdot 2}(\bar{\mathbf{U}}_{1\cdot 2}) = \left(\frac{\nu + p_2}{\nu + \tilde{Q}_2(\mathbf{x}_2)}\right) Q_{1\cdot 2}(\mathbf{U}_{1\cdot 2})$, with $Q(\mathbf{U}_{1\cdot 2}) = \mathbf{U}_{1\cdot 2}^\top\bar{\boldsymbol{\Gamma}}_{1\cdot 2}^{-1}\mathbf{U}_{1\cdot 2}$.

Proof 131 *Let $\mathbf{X}_{12} = \mathbf{X}_1 - \boldsymbol{\Psi}_{12}\boldsymbol{\Psi}_{22}^{-1}\mathbf{X}_2$, where by (7.17) $\mathbf{X}_j = \boldsymbol{\xi}_j + \mathbf{H}_2\mathbf{U} + \sqrt{\frac{\nu + Q(\mathbf{U})}{\nu + q}}\mathbf{W}_{*j}$ for $j = 1, 2$. Also, let $\mathbf{W}_{1\cdot 2}^* = \sqrt{\frac{\nu + q + p_2}{\nu + q + Q_2(\mathbf{W}_{*2})}}\left(\mathbf{W}_{*1} - \boldsymbol{\Psi}_{12}\boldsymbol{\Psi}_{22}^{-1}\mathbf{W}_{*2}\right)$, where $Q_2(\mathbf{W}_{*2}) = \mathbf{W}_{*2}^\top\boldsymbol{\Psi}_{22}^{-1}\mathbf{W}_{*2}$.*

Recall in part (c) of Theorem 130 that $\mathbf{W}^*_{1\cdot 2} \sim \mathcal{T}_{p_1}(\mathbf{0}_{p_1}, \boldsymbol{\Psi}_{11\cdot 2}, \nu+q+p_2)$ *and is independent of* $\mathbf{W}_{*2}$ *and* $\mathbf{U}$*, and therefore of* $\mathbf{X}_2 = \boldsymbol{\xi}_2 + \mathbf{H}_2\mathbf{U} + \sqrt{\frac{\nu+Q(\mathbf{U})}{\nu+q}}\,\mathbf{W}_{*2}$*. Thus, we have that*

$$\mathbf{X}_{12} = \boldsymbol{\xi}_1 - \boldsymbol{\Psi}_{12}\boldsymbol{\Psi}_{22}^{-1}\boldsymbol{\xi}_2 + \mathbf{H}_{1\cdot 2}\mathbf{U} + \sqrt{\frac{\nu+Q(\mathbf{U})}{\nu+q}}\sqrt{\frac{\nu+q+Q_2(\mathbf{W}_{*2})}{\nu+q+p_2}}\,\mathbf{W}^*_{1\cdot 2}. \tag{7.38}$$

However, since $\mathbf{X}_1 = \mathbf{X}_{12} + \boldsymbol{\Psi}_{12}\boldsymbol{\Psi}_{22}^{-1}\mathbf{X}_2$*, then by (7.38) we have*

$$\mathbf{X}_1 = \boldsymbol{\xi}_1 + \boldsymbol{\Psi}_{12}\boldsymbol{\Psi}_{22}^{-1}(\mathbf{X}_2 - \boldsymbol{\xi}_2) + \mathbf{H}_{1\cdot 2}\mathbf{U} + \sqrt{\frac{\nu+Q(\mathbf{U})}{\nu+q}}\sqrt{\frac{\nu+q+Q_2(\mathbf{W}_{*2})}{\nu+q+p_2}}\,\mathbf{W}^*_{1\cdot 2}. \tag{7.39}$$

Hence, from (7.39) we have, by conditioning $\mathbf{X}_1$ *simultaneously on* $\mathbf{U}=\mathbf{u}$ *and on* $\mathbf{X}_2 = \mathbf{x}_2$*, that*

$$\begin{aligned}(\mathbf{X}_1 \mid \mathbf{U}=\mathbf{u}, \mathbf{X}_2=\mathbf{x}_2) =& \boldsymbol{\xi}_1 + \boldsymbol{\Psi}_{12}\boldsymbol{\Psi}_{22}^{-1}(\mathbf{x}_2 - \boldsymbol{\xi}_2) + \mathbf{H}_{1\cdot 2}\mathbf{u} + \sqrt{\frac{\nu+Q(\mathbf{u})}{\nu+q}} \\ &\times \left\{\sqrt{\frac{\nu+q+Q_2(\mathbf{W}_{*2})}{\nu+q+p_2}} \,\Big|\, \mathbf{U}=\mathbf{u}, \mathbf{W}_{*2} = \sqrt{\frac{\nu+q}{\nu+Q(\mathbf{u})}}(\mathbf{x}_2 - \boldsymbol{\xi}_2 - \mathbf{H}_2\mathbf{u})\right\}\mathbf{W}^*_{1\cdot 2} \\ =& \boldsymbol{\xi}_{1\cdot 2} + \mathbf{H}_{1\cdot 2}\mathbf{u} + \sqrt{\frac{\nu+Q(\mathbf{u})}{\nu+q}}\sqrt{\frac{\nu+q+\left(\frac{\nu+q}{\nu+Q(\mathbf{u})}\right)\tilde{Q}_2(\mathbf{x}_2|\mathbf{u})}{\nu+q+p_2}}\,\mathbf{W}^*_{1\cdot 2} \\ =& \boldsymbol{\xi}_{1\cdot 2} + \mathbf{H}_{1\cdot 2}\mathbf{u} + \sqrt{\frac{\nu+Q(\mathbf{u})+\tilde{Q}_2(\mathbf{x}_2|\mathbf{u})}{\nu+q+p_2}}\,\mathbf{W}^*_{1\cdot 2},\end{aligned} \tag{7.40}$$

where $\tilde{Q}_2(\mathbf{x}_2|\mathbf{u}) = (\mathbf{x}_2 - \boldsymbol{\xi}_2 - \mathbf{H}_2\mathbf{u})^\top\boldsymbol{\Psi}_{22}^{-1}(\mathbf{x}_2 - \boldsymbol{\xi}_2 - \mathbf{H}_2\mathbf{u})$*, and from some extensive but straightforward algebra, we find that*

$$\begin{aligned}Q(\mathbf{u}) + \tilde{Q}_2(\mathbf{x}_2|\mathbf{u}) &= \mathbf{u}^\top\bar{\boldsymbol{\Gamma}}^{-1}\mathbf{u} + (\mathbf{x}_2 - \boldsymbol{\xi}_2 - \mathbf{H}_2\mathbf{u})^\top\boldsymbol{\Psi}_{22}^{-1}(\mathbf{x}_2 - \boldsymbol{\xi}_2 - \mathbf{H}_2\mathbf{u}) \\ &= \{\mathbf{u} - \boldsymbol{\Gamma}_{1\cdot 2}\mathbf{H}_2^\top\boldsymbol{\Psi}_{22}^{-1}(\mathbf{x}_2 - \boldsymbol{\xi}_2)\}^\top\boldsymbol{\Gamma}_{1\cdot 2}^{-1}\{\mathbf{u} - \boldsymbol{\Gamma}_{1\cdot 2}\mathbf{H}_2^\top\boldsymbol{\Psi}_{22}^{-1}(\mathbf{x}_2 - \boldsymbol{\xi}_2)\} \\ &\quad + (\mathbf{x}_2 - \boldsymbol{\xi}_2)^\top(\boldsymbol{\Psi}_{22} + \mathbf{H}_2\bar{\boldsymbol{\Gamma}}\mathbf{H}_2^\top)^{-1}(\mathbf{x}_2 - \boldsymbol{\xi}_2) \\ &= Q_{1\cdot 2}(\mathbf{u}) + \tilde{Q}_2(\mathbf{x}_2),\end{aligned}$$

where $Q_{1\cdot 2}(\mathbf{u}) = \{\mathbf{u} - \boldsymbol{\Gamma}_{1\cdot 2}\mathbf{H}_2^\top\boldsymbol{\Psi}_{22}^{-1}(\mathbf{x}_2 - \boldsymbol{\xi}_2)\}^\top\boldsymbol{\Gamma}_{1\cdot 2}^{-1}\{\mathbf{u} - \boldsymbol{\Gamma}_{1\cdot 2}\mathbf{H}_2^\top\boldsymbol{\Psi}_{22}^{-1}(\mathbf{x}_2 - \boldsymbol{\xi}_2)\}$ *and, as defined earlier,* $\tilde{Q}_2(\mathbf{x}_2) = (\mathbf{x}_2 - \boldsymbol{\xi}_2)^\top(\boldsymbol{\Psi}_{22} + \mathbf{H}_2\bar{\boldsymbol{\Gamma}}\mathbf{H}_2^\top)^{-1}(\mathbf{x}_2 - \boldsymbol{\xi}_2)$*. Therefore, from (7.40) and part (c) of Theorem 130, we have*

$$(\mathbf{X}_1 \mid \mathbf{U}=\mathbf{u}, \mathbf{X}_2=\mathbf{x}_2) \sim \mathcal{T}_{p_1}\left(\boldsymbol{\xi}_{1\cdot 2} + \mathbf{H}_{1\cdot 2}\mathbf{u}, \frac{\nu+Q_{1\cdot 2}(\mathbf{u})+\tilde{Q}_2(\mathbf{x}_2)}{\nu+q+p_2}\boldsymbol{\Psi}_{11\cdot 2}, \nu+q+p_2\right), \tag{7.41}$$

where, by part (b) of Theorem 130,

$$(\mathbf{U} \mid \mathbf{X}_2=\mathbf{x}_2) \sim \mathcal{TT}_q\left(-\boldsymbol{\tau}; \boldsymbol{\Gamma}_{1\cdot 2}\mathbf{H}_2^\top\boldsymbol{\Psi}_{22}^{-1}(\mathbf{x}_2 - \boldsymbol{\xi}_2), \frac{\nu+\tilde{Q}_2(\mathbf{x}_2)}{\nu+p_2}\boldsymbol{\Gamma}_{1\cdot 2}, \nu+p_2\right), \tag{7.42}$$

which has a conditional pdf given by (7.32). However, since the conditional location and dispersion of (7.41) can be rewritten as

$$\boldsymbol{\xi}_{1\cdot 2} + \mathbf{H}_{1\cdot 2}\mathbf{u} = \tilde{\boldsymbol{\xi}}_{1\cdot 2} + \mathbf{H}_{1\cdot 2}\{\mathbf{u} - \boldsymbol{\Gamma}_{1\cdot 2}\mathbf{H}_2^\top\boldsymbol{\Psi}_{22}^{-1}(\mathbf{x}_2 - \boldsymbol{\xi}_2)\},$$

$$\frac{\nu+Q_{1\cdot 2}(\mathbf{u})+\tilde{Q}_2(\mathbf{x}_2)}{\nu+q+p_2}\boldsymbol{\Psi}_{11\cdot 2} = \frac{\nu+\tilde{Q}_2(\mathbf{x}_2)}{\nu+p_2}\,\frac{\nu+p_2+\left(\frac{\nu+p_2}{\nu+\tilde{Q}_2(\mathbf{x}_2)}\right)Q_{1\cdot 2}(\mathbf{u})}{\nu+p_2+q}\boldsymbol{\Psi}_{11\cdot 2},$$

where $\tilde{\boldsymbol{\xi}}_{1\cdot 2} = \boldsymbol{\xi}_{1\cdot 2} + \boldsymbol{\Gamma}_{1\cdot 2}\mathbf{H}_2^\top \boldsymbol{\Psi}_{22}^{-1}(\mathbf{x}_2 - \boldsymbol{\xi}_2)$, then (7.41) and (7.42) form a hierarchical representation of the conditional distribution of $(\mathbf{X}_1 \mid \mathbf{X}_2 = \mathbf{x}_2)$. Thus, from Theorem 118, we have

$$(\mathbf{X}_1 \mid \mathbf{X}_2 = \mathbf{x}_2) \sim \mathcal{SUT}_{p_1,q}\left(\tilde{\boldsymbol{\xi}}_{1\cdot 2}, \frac{\nu + \tilde{Q}_2(\mathbf{x}_2)}{\nu + p_2}\boldsymbol{\Psi}_{11\cdot 2}, \mathbf{H}_{1\cdot 2}, \boldsymbol{\tau}_{1\cdot 2}, \frac{\nu + \tilde{Q}_2(\mathbf{x}_2)}{\nu + p_2}\boldsymbol{\Gamma}_{1\cdot 2}, \nu + p_2\right).$$

Finally, since $\mathrm{diag}\left(\frac{\nu+\tilde{Q}_2(\mathbf{x}_2)}{\nu+p_2}\boldsymbol{\Gamma}_{1\cdot 2}\right)^{1/2} = \sqrt{\frac{\nu+\tilde{Q}_2(\mathbf{x}_2)}{\nu+p_2}}\boldsymbol{\gamma}_{1\cdot 2}$, the proof concludes by noting that this last conditional distribution is completely equivalent to the distribution of the random vector $\mathbf{X}_{1\cdot 2}$ defined by (7.37), which by definition is

$$\mathcal{SUT}_{p_1,q}\left(\tilde{\boldsymbol{\xi}}_{1\cdot 2}, \frac{\nu + \tilde{Q}_2(\mathbf{x}_2)}{\nu + p_2}\boldsymbol{\Psi}_{11\cdot 2}, \sqrt{\frac{\nu + \tilde{Q}_2(\mathbf{x}_2)}{\nu + p_2}}\mathbf{H}_{1\cdot 2}\boldsymbol{\gamma}_{1\cdot 2}, \sqrt{\frac{\nu + p_2}{\nu + \tilde{Q}_2(\mathbf{x}_2)}}\boldsymbol{\gamma}_{1\cdot 2}^{-1}\boldsymbol{\tau}_{1\cdot 2}, \boldsymbol{\gamma}_{1\cdot 2}^{-1}\boldsymbol{\Gamma}_{1\cdot 2}\boldsymbol{\gamma}_{1\cdot 2}^{-1}, \nu_{1\cdot 2}\right).$$

□

7.4.1 *Conditional mean vector and conditional variance matrix*

From (7.37), it is easy to compute the conditional mean vector and the conditional variance matrix of $(\mathbf{X}_1 \mid \mathbf{X}_2 = \mathbf{x}_2)$, which, according to Theorem 116, with the corresponding adaptation of the parameters, become

$$\mathrm{E}(\mathbf{X}_1 \mid \mathbf{X}_2 = \mathbf{x}_2) = \tilde{\boldsymbol{\xi}}_{1\cdot 2} + \sqrt{\frac{\nu + \tilde{Q}_2(\mathbf{x}_2)}{\nu + p_2}}\bar{\mathbf{H}}_{1\cdot 2}\mathrm{E}(\bar{\mathbf{U}}_{1\cdot 2}),$$

$$\begin{aligned}\mathrm{var}(\mathbf{X}_1 \mid \mathbf{X}_2 = \mathbf{x}_2) &= \left(\frac{\nu + \tilde{Q}_2(\mathbf{x}_2)}{\nu + p_2}\right)\bar{\mathbf{H}}_{1\cdot 2}\mathrm{var}(\bar{\mathbf{U}}_{1\cdot 2})\bar{\mathbf{H}}_{1\cdot 2}^\top \\ &\quad + \frac{T_q\left(\sqrt{\frac{\nu+p_2}{\nu+\tilde{Q}_2(\mathbf{x}_2)}}\bar{\boldsymbol{\tau}}_{1\cdot 2}; \mathbf{0}_q, \bar{\boldsymbol{\Gamma}}_{1\cdot 2}, \nu + p_2 - 2\right)}{T_q\left(\sqrt{\frac{\nu+p_2}{\nu+\tilde{Q}_2(\mathbf{x}_2)}}\bar{\boldsymbol{\tau}}_{1\cdot 2}; \mathbf{0}_q, \bar{\boldsymbol{\Gamma}}_{1\cdot 2}, \nu + p_2\right)}\left(\frac{\nu + p_2}{\nu + p_2 - 2}\right)\left(\frac{\nu + \tilde{Q}_2(\mathbf{x}_2)}{\nu + p_2}\right)\boldsymbol{\Psi}_{11\cdot 2}.\end{aligned}$$

However, as in the unconditional case, to do this we need, for each fixed $\mathbf{x}_2 \in \mathbb{R}^{p_2}$, the first two moments of the truncated t random vector $\bar{\mathbf{U}}_{1\cdot 2} \sim \mathcal{TT}_q\left(-\sqrt{\frac{\nu+p_2}{\nu+\tilde{Q}_2(\mathbf{x}_2)}}\bar{\boldsymbol{\tau}}_{1\cdot 2}; \mathbf{0}_q, \bar{\boldsymbol{\Gamma}}_{1\cdot 2}, \nu_{1\cdot 2}\right)$.

A relevant fact is that the conditional mean vector and the conditional variance matrix of $(\mathbf{X}_1 \mid \mathbf{X}_2 = \mathbf{x}_2)$ are, in general, nonlinear functions of $\mathbf{x}_2$. In the following example, two special cases are presented where we can note for the first of them that the regression of $\mathbf{X}_1$ on $\mathbf{x}_2$ is linear but heteroscedastic.

Example 97 *Let $\mathbf{X} \sim \mathcal{SUT}_{p,q}(\boldsymbol{\xi}, \boldsymbol{\Psi}, \mathbf{H}, \boldsymbol{\tau}, \bar{\boldsymbol{\Gamma}}, \nu)$ be partitioned as (7.16). We then have:*

(a) If $\mathbf{H}_1 = \boldsymbol{\Psi}_{12}\boldsymbol{\Psi}_{22}^{-1}\mathbf{H}_2$, then $\bar{\mathbf{H}}_{1\cdot 2} = \mathbf{O}$, and therefore $(\mathbf{X}_1 \mid \mathbf{X}_2 = \mathbf{x}_2) \sim \mathcal{SUT}_{p_1,q}\left(\boldsymbol{\xi}_{1\cdot 2}, \frac{\nu+\tilde{Q}_2(\mathbf{x}_2)}{\nu+p_2}\boldsymbol{\Psi}_{11\cdot 2}, \mathbf{O}, \sqrt{\frac{\nu+p_2}{\nu+\tilde{Q}_2(\mathbf{x}_2)}}\bar{\boldsymbol{\tau}}_{1\cdot 2}, \bar{\boldsymbol{\Gamma}}_{1\cdot 2}, \nu + p_2\right)$ and

$$\mathrm{E}(\mathbf{X}_1 \mid \mathbf{x}_2) = \boldsymbol{\xi}_{1\cdot 2},$$

$$\mathrm{var}(\mathbf{X}_1 \mid \mathbf{x}_2) = \frac{T_q\left(\sqrt{\frac{\nu+p_2}{\nu+\tilde{Q}_2(\mathbf{x}_2)}}\bar{\boldsymbol{\tau}}_{1\cdot 2}; \mathbf{0}_q, \bar{\boldsymbol{\Gamma}}_{1\cdot 2}, \nu + p_2 - 2\right)}{T_q\left(\sqrt{\frac{\nu+p_2}{\nu+\tilde{Q}_2(\mathbf{x}_2)}}\bar{\boldsymbol{\tau}}_{1\cdot 2}; \mathbf{0}_q, \bar{\boldsymbol{\Gamma}}_{1\cdot 2}, \nu + p_2\right)}\left(\frac{\nu + \tilde{Q}_2(\mathbf{x}_2)}{\nu + p_2 - 2}\right)\boldsymbol{\Psi}_{11\cdot 2};$$

(b) If $\boldsymbol{\tau} = \mathbf{0}_q$ *and* $\mathbf{H}_2 = \mathbf{O}$*, then* $\bar{\mathbf{H}}_{1\cdot 2} = \mathbf{H}_1$*,* $\bar{\boldsymbol{\Gamma}}_{1\cdot 2} = \bar{\boldsymbol{\Gamma}}$*,* $\boldsymbol{\gamma}_{1\cdot 2} = \mathbf{I}_q$*, and* $\bar{\boldsymbol{\tau}}_{1\cdot 2} = \boldsymbol{\tau} = \mathbf{0}_q$*, so that* $(\mathbf{X}_1 \mid \mathbf{X}_2 = \mathbf{x}_2) \sim \mathcal{SUT}_{p_1,q}\left(\boldsymbol{\xi}_{1\cdot 2}, \frac{\nu+\tilde{Q}_2(\mathbf{x}_2)}{\nu+p_2}\boldsymbol{\Psi}_{11\cdot 2}, \sqrt{\frac{\nu+\tilde{Q}_2(\mathbf{x}_2)}{\nu+p_2}}\mathbf{H}_1, \mathbf{0}_q, \bar{\boldsymbol{\Gamma}}, \nu+p_2\right)$*, with*

$$\mathrm{E}(\mathbf{X}_1 \mid \mathbf{x}_2) = \boldsymbol{\xi}_{1\cdot 2} + \sqrt{\frac{\nu+\tilde{Q}_2(\mathbf{x}_2)}{\nu+p_2}}\mathbf{H}_1\mathrm{E}(\bar{\mathbf{U}}_{1\cdot 2}),$$

$$\mathrm{var}(\mathbf{X}_1 \mid \mathbf{x}_2) = \left(\frac{\nu+\tilde{Q}_2(\mathbf{x}_2)}{\nu+p_2}\right)\left\{\mathbf{H}_1\mathrm{var}(\bar{\mathbf{U}}_{1\cdot 2})\mathbf{H}_1^\top + \left(\frac{\nu+p_2}{\nu+p_2-2}\right)\boldsymbol{\Psi}_{11\cdot 2}\right\},$$

where $\bar{\mathbf{U}}_{1\cdot 2} = \mathbf{U}_{1\cdot 2} \sim \mathcal{TT}_q(\mathbf{0}_q; \mathbf{0}_q, \bar{\boldsymbol{\Gamma}}, \nu+p_2)$*.*

7.4.2 *Conditional unified skew-t pdf*

As mentioned earlier, if $\boldsymbol{\Psi} > 0$, then the nonsingular case with $\boldsymbol{\Psi}_{22} > 0$ holds automatically. In addition, we have $\boldsymbol{\Psi}_{11} > 0$ and $\boldsymbol{\Psi}_{11\cdot 2} = \boldsymbol{\Psi}_{11} - \boldsymbol{\Psi}_{12}\boldsymbol{\Psi}_{22}^{-1}\boldsymbol{\Psi}_{21} > 0$, and the relation $(\boldsymbol{\Psi}_{11}, \boldsymbol{\Psi}_{12}, \boldsymbol{\Psi}_{22}) \longleftrightarrow (\boldsymbol{\Psi}_{11\cdot 2}, \boldsymbol{\Psi}_{12}\boldsymbol{\Psi}_{22}^{-1}, \boldsymbol{\Psi}_{22})$ is one-to-one. Obviously, under these conditions, all the marginal and conditional distributions obtained from $\mathbf{X} \sim \mathcal{SUT}_{p,q}(\boldsymbol{\xi}, \boldsymbol{\Psi}, \mathbf{H}, \boldsymbol{\tau}, \bar{\boldsymbol{\Gamma}}, \nu)$ are nonsingular. The following theorem describes the conditional pdf of $\mathbf{X}_1$ given $\mathbf{X}_2 = \mathbf{x}_2$ according to the partition (7.16).

Theorem 132 *Let* $\mathbf{X} \sim \mathcal{SUT}_{p,q}(\boldsymbol{\xi}, \boldsymbol{\Psi}, \mathbf{H}, \boldsymbol{\tau}, \bar{\boldsymbol{\Gamma}}, \nu)$ *be nonsingular* ($\boldsymbol{\Psi} > 0$) *and partitioned as (7.16). Then for every fixed* $p_1 + p_2 = p$*, the conditional pdf of* $\mathbf{X}_1$ *given* $\mathbf{X}_2 = \mathbf{x}_2$*,* $\mathbf{x}_2 \in \mathbb{R}^{p_2}$*, is*

$$\begin{aligned} f_{1\cdot 2}(\mathbf{x}_1 \mid \mathbf{x}_2) \quad = \quad & \frac{1}{T_q\left\{\sqrt{\frac{\nu+p_2}{\nu+\tilde{Q}_2(\mathbf{x}_2)}}\,\bar{\boldsymbol{\tau}}_{1\cdot 2}; \mathbf{0}_q, \bar{\boldsymbol{\Gamma}}_{1\cdot 2}, \nu+p_2\right\}} \\ & \times t_{p_1}\left\{\mathbf{x}_1; \tilde{\boldsymbol{\xi}}_{1\cdot 2}, \frac{\nu+\tilde{Q}_2(\mathbf{x}_2)}{\nu+p_2}(\boldsymbol{\Psi}_{11\cdot 2} + \bar{\mathbf{H}}_{1\cdot 2}^\top\bar{\boldsymbol{\Gamma}}_{1\cdot 2}\bar{\mathbf{H}}_{1\cdot 2}), \nu+p_2\right\} \\ & \times T_q\left[c_{1\cdot 2}^{-1/2}\left\{\bar{\boldsymbol{\tau}}_{1\cdot 2} + \tilde{\boldsymbol{\Gamma}}_{1\cdot 2}\bar{\mathbf{H}}_{1\cdot 2}^\top\boldsymbol{\Psi}_{11\cdot 2}^{-1}(\mathbf{x}_1 - \tilde{\boldsymbol{\xi}}_{1\cdot 2})\right\}; \mathbf{0}_q, \tilde{\boldsymbol{\Gamma}}_{1\cdot 2}, \nu+p_2+p_1\right], \\ & \qquad \mathbf{x}_1 \in \mathbb{R}^{p_1}, \end{aligned} \quad (7.43)$$

where $\tilde{\boldsymbol{\xi}}_{1\cdot 2} = \boldsymbol{\xi}_{1\cdot 2} + \mathbf{H}_{1\cdot 2}\boldsymbol{\Gamma}_{1\cdot 2}\mathbf{H}_2^\top\boldsymbol{\Psi}_{22}^{-1}(\mathbf{x}_2 - \boldsymbol{\xi}_2)$*, with* $\boldsymbol{\xi}_{1\cdot 2} = \boldsymbol{\xi}_1 + \boldsymbol{\Psi}_{12}\boldsymbol{\Psi}_{22}^{-1}(\mathbf{x}_2 - \boldsymbol{\xi}_2)$*,* $\boldsymbol{\Psi}_{11\cdot 2} = \boldsymbol{\Psi}_{11} - \boldsymbol{\Psi}_{12}\boldsymbol{\Psi}_{22}^{-1}\boldsymbol{\Psi}_{21}$*,* $\bar{\mathbf{H}}_{1\cdot 2} = (\mathbf{H}_1 - \boldsymbol{\Psi}_{12}\boldsymbol{\Psi}_{22}^{-1}\mathbf{H}_2)\boldsymbol{\gamma}_{1\cdot 2}^{-1}$*,* $\bar{\boldsymbol{\tau}}_{1\cdot 2} = \bar{\boldsymbol{\tau}} + \bar{\boldsymbol{\Gamma}}_{1\cdot 2}\bar{\mathbf{H}}_2^\top\boldsymbol{\Psi}_{22}^{-1}(\mathbf{x}_2 - \boldsymbol{\xi}_2)$*,* $\bar{\boldsymbol{\tau}} = \boldsymbol{\gamma}_{1\cdot 2}^{-1}\boldsymbol{\tau}$*,* $\tilde{\boldsymbol{\Gamma}}_{1\cdot 2} = (\bar{\boldsymbol{\Gamma}}_{1\cdot 2}^{-1} + \bar{\mathbf{H}}_{1\cdot 2}^\top\boldsymbol{\Psi}_{11\cdot 2}^{-1}\bar{\mathbf{H}}_{1\cdot 2})^{-1}$*,* $\bar{\boldsymbol{\Gamma}}_{1\cdot 2} = \boldsymbol{\gamma}_{1\cdot 2}^{-1}\boldsymbol{\Gamma}_{1\cdot 2}\boldsymbol{\gamma}_{1\cdot 2}^{-1}$*, with* $\boldsymbol{\Gamma}_{1\cdot 2} = (\bar{\boldsymbol{\Gamma}}^{-1} + \mathbf{H}_2^\top\boldsymbol{\Psi}_{22}^{-1}\mathbf{H}_2)^{-1}$ *and* $\boldsymbol{\gamma}_{1\cdot 2} = \mathrm{diag}(\boldsymbol{\Gamma}_{1\cdot 2})^{1/2}$*, and*

$$c_{1\cdot 2} = \frac{\nu+p_2+\tilde{Q}_2(\mathbf{x}_2)+\tilde{Q}_{1\cdot 2}(\mathbf{x}_1)}{\nu+p_2+p_1},$$

with the quadratic forms $\tilde{Q}_2(\mathbf{x}_2) = (\mathbf{x}_2 - \boldsymbol{\xi}_2)^\top(\boldsymbol{\Psi}_{22} + \mathbf{H}_2\bar{\boldsymbol{\Gamma}}\mathbf{H}_2^\top)^{-1}(\mathbf{x}_2 - \boldsymbol{\xi}_2)$ *and* $\tilde{Q}_{1\cdot 2}(\mathbf{x}_1) = (\mathbf{x}_1 - \tilde{\boldsymbol{\xi}}_{1\cdot 2})^\top(\boldsymbol{\Psi}_{11\cdot 2} + \bar{\mathbf{H}}_{1\cdot 2}\bar{\boldsymbol{\Gamma}}_{1\cdot 2}\bar{\mathbf{H}}_{1\cdot 2}^\top)^{-1}(\mathbf{x}_1 - \tilde{\boldsymbol{\xi}}_{1\cdot 2})$*.*

Proof 132 *We obtain (7.43) from the hierarchical representation in (7.41) and (7.42), which are nonsingular distributions. In fact, since by assumption* $\boldsymbol{\Psi} > 0$*, and so* $\boldsymbol{\Psi}_{22} > 0$ *and* $\boldsymbol{\Psi}_{11\cdot 2} > 0$*, it follows that (7.41) is nonsingular, with pdf given by*

$$f(\mathbf{x}_1 \mid \mathbf{u}, \mathbf{x}_2) = t_{p_1}\left\{\mathbf{x}_1; \boldsymbol{\xi}_{1\cdot 2} + \mathbf{H}_{1\cdot 2}\mathbf{u}, \frac{\nu+Q_{1\cdot 2}(\mathbf{u})+\tilde{Q}_2(\mathbf{x}_2)}{\nu+q+p_2}\boldsymbol{\Psi}_{11\cdot 2}, \nu+q+p_2\right\}, \quad \mathbf{x}_1 \in \mathbb{R}^{p_1},$$

where, as defined earlier, $Q_{1\cdot 2}(\mathbf{u}) = (\mathbf{u}-\mathbf{h}_2)^\top \boldsymbol{\Gamma}^{-1}(\mathbf{u}-\mathbf{h}_2)$, *with* $\mathbf{h}_2 = \boldsymbol{\Gamma}_{1\cdot 2}\mathbf{H}_2\boldsymbol{\Psi}_{22}^{-1}(\mathbf{x}_2-\boldsymbol{\xi}_2)$, *and* $\tilde{Q}_2(\mathbf{x}_2) = (\mathbf{x}_2-\boldsymbol{\xi}_2)^\top(\boldsymbol{\Psi}_{22}+\mathbf{H}_2\bar{\boldsymbol{\Gamma}}\mathbf{H}_2^\top)^{-1}(\mathbf{x}_2-\boldsymbol{\xi}_2)$. *Also, since* $\boldsymbol{\Psi}_{22}>0$ *and* $\bar{\boldsymbol{\Gamma}}>0$, *then* $\boldsymbol{\Gamma}_{1\cdot 2} = \tilde{\boldsymbol{\Gamma}}_2 = (\bar{\boldsymbol{\Gamma}}^{-1}+\mathbf{H}_2^\top\boldsymbol{\Psi}_{22}^{-1}\mathbf{H}_2)^{-1} > 0$, *which means that (7.42) is also nonsingular, with pdf given by (7.32). Hence, the conditional pdf of* $\mathbf{X}_1 \mid \mathbf{X}_2 = \mathbf{x}_2$ *becomes*

$$
\begin{aligned}
f_{1\cdot 2}(\mathbf{x}_1 \mid \mathbf{x}_2) &= \int_{\mathbb{R}^q} f(\mathbf{x}_1 \mid \mathbf{u},\mathbf{x}_2)g_2(\mathbf{u}\mid\mathbf{x}_2)d\mathbf{u}\\
&= \frac{1}{T_q\left\{\sqrt{\frac{\nu+p_2}{\nu+\tilde{Q}_2(\mathbf{x}_2)}}\boldsymbol{\tau}_{1\cdot 2};\mathbf{0}_q,\boldsymbol{\Gamma}_{1\cdot 2},\nu+p_2\right\}}\\
&\quad\times\int_{\mathbf{u}+\boldsymbol{\tau}>\mathbf{0}_q} t_{p_1}\left\{\mathbf{x}_1;\boldsymbol{\xi}_{1\cdot 2}+\mathbf{H}_{1\cdot 2}\mathbf{u},\frac{\nu+Q_{1\cdot 2}(\mathbf{u})+\tilde{Q}_2(\mathbf{x}_2)}{\nu+q+p_2}\boldsymbol{\Psi}_{11\cdot 2},\nu+q+p_2\right\}\\
&\quad\times t_q\left\{\mathbf{u};\mathbf{h}_2,\frac{\nu+\tilde{Q}_2(\mathbf{x}_2)}{\nu+p_2}\boldsymbol{\Gamma}_{1\cdot 2},\nu+p_2\right\}d\mathbf{u}\\
&= \frac{1}{T_q\left\{\sqrt{\frac{\nu+p_2}{\nu+\tilde{Q}_2(\mathbf{x}_2)}}\bar{\boldsymbol{\tau}}_{1\cdot 2};\mathbf{0}_q,\bar{\boldsymbol{\Gamma}}_{1\cdot 2},\nu+p_2\right\}}\\
&\quad\times\int_{\mathbf{v}+\bar{\boldsymbol{\tau}}_{1\cdot 2}>\mathbf{0}_q} t_{p_1}\left\{\mathbf{x}_1;\tilde{\boldsymbol{\xi}}_{1\cdot 2}+\bar{\mathbf{H}}_{1\cdot 2}\mathbf{v},\frac{\nu+\tilde{Q}_{1\cdot 2}(\mathbf{v})+\tilde{Q}_2(\mathbf{x}_2)}{\nu+q+p_2}\boldsymbol{\Psi}_{11\cdot 2},\nu+q+p_2\right\}\\
&\quad\times t_q\left\{\mathbf{v};\mathbf{0}_q,\frac{\nu+\tilde{Q}_2(\mathbf{x}_2)}{\nu+p_2}\bar{\boldsymbol{\Gamma}}_{1\cdot 2},\nu+p_2\right\}d\mathbf{v}\quad (\mathbf{v}=\boldsymbol{\gamma}_{1\cdot 2}^{-1}(\mathbf{u}-\mathbf{h}_2)),
\end{aligned}
$$

with $\tilde{Q}_{1\cdot 2}(\mathbf{v}) = \mathbf{v}^\top\boldsymbol{\Gamma}^{-1}\mathbf{v}$. *However, from the marginal/conditional representation of the multivariate* t *distribution, we have*

$$
\begin{aligned}
& t_{p_1}\left\{\mathbf{x}_1;\tilde{\boldsymbol{\xi}}_{1\cdot 2}+\bar{\mathbf{H}}_{1\cdot 2}\mathbf{v},\frac{\nu+\tilde{Q}_{1\cdot 2}(\mathbf{v})+\tilde{Q}_2(\mathbf{x}_2)}{\nu+q+p_2}\boldsymbol{\Psi}_{11\cdot 2},\nu+q+p_2\right\} t_q\left\{\mathbf{v};\mathbf{0}_q,\frac{\nu+\tilde{Q}_2(\mathbf{x}_2)}{\nu+p_2}\bar{\boldsymbol{\Gamma}}_{1\cdot 2},\nu+p_2\right\}\\
&= t_{p_1}\left\{\mathbf{x}_1;\tilde{\boldsymbol{\xi}}_{1\cdot 2}+\bar{\mathbf{H}}_{1\cdot 2}\mathbf{v},\frac{\nu+p_2+\left(\frac{\nu+p_2}{\nu+\tilde{Q}_2(\mathbf{x}_2)}\right)\tilde{Q}_{1\cdot 2}(\mathbf{v})}{\nu+p_2+q}\,\frac{\nu+\tilde{Q}_2(\mathbf{x}_2)}{\nu+p_2}\boldsymbol{\Psi}_{11\cdot 2},\nu+p_2+q\right\}\\
&\quad\times t_q\left\{\mathbf{v};\mathbf{0}_q,\frac{\nu+\tilde{Q}_2(\mathbf{x}_2)}{\nu+p_2}\bar{\boldsymbol{\Gamma}}_{1\cdot 2},\nu+p_2\right\}\\
&= t_{p_1+q}\left\{\begin{pmatrix}\mathbf{x}_1\\ \mathbf{v}\end{pmatrix};\begin{pmatrix}\tilde{\boldsymbol{\xi}}_{1\cdot 2}\\ \mathbf{0}_q\end{pmatrix},\frac{\nu+\tilde{Q}_2(\mathbf{x}_2)}{\nu+p_2}\begin{pmatrix}\boldsymbol{\Psi}_{11\cdot 2}+\bar{\mathbf{H}}_{1\cdot 2}\bar{\boldsymbol{\Gamma}}_{1\cdot 2}\bar{\mathbf{H}}_{1\cdot 2}^\top & \bar{\mathbf{H}}_{1\cdot 2}\bar{\boldsymbol{\Gamma}}_{1\cdot 2}\\ \bar{\boldsymbol{\Gamma}}_{1\cdot 2}\bar{\mathbf{H}}_{1\cdot 2}^\top & \bar{\boldsymbol{\Gamma}}_{1\cdot 2}\end{pmatrix},\nu+p_2\right\}\\
&= t_{p_1}\left\{\mathbf{x}_1;\tilde{\boldsymbol{\xi}}_{1\cdot 2},\frac{\nu+\tilde{Q}_2(\mathbf{x}_2)}{\nu+p_2}\left(\boldsymbol{\Psi}_{11\cdot 2}+\bar{\mathbf{H}}_{1\cdot 2}\bar{\boldsymbol{\Gamma}}_{1\cdot 2}\bar{\mathbf{H}}_{1\cdot 2}^\top\right),\nu+p_2\right\}\\
&\quad\times t_q\left[\mathbf{v};\bar{\boldsymbol{\Gamma}}_{1\cdot 2}\bar{\mathbf{H}}_{1\cdot 2}^\top\left(\boldsymbol{\Psi}_{11\cdot 2}+\bar{\mathbf{H}}_{1\cdot 2}\bar{\boldsymbol{\Gamma}}_{1\cdot 2}\bar{\mathbf{H}}_{1\cdot 2}^\top\right)^{-1}(\mathbf{x}_1-\tilde{\boldsymbol{\xi}}_{1\cdot 2}),\right.\\
&\quad\left.\frac{\nu+p_2+\left(\frac{\nu+p_2}{\nu+\tilde{Q}_2(\mathbf{x}_2)}\right)\tilde{Q}_{1\cdot 2}(\mathbf{x}_1)}{\nu+p_2+p_1}\,\frac{\nu+\tilde{Q}_2(\mathbf{x}_2)}{\nu+p_2}\left\{\bar{\boldsymbol{\Gamma}}_{1\cdot 2}-\bar{\boldsymbol{\Gamma}}_{1\cdot 2}\left(\boldsymbol{\Psi}_{11\cdot 2}+\bar{\mathbf{H}}_{1\cdot 2}\bar{\boldsymbol{\Gamma}}_{1\cdot 2}\bar{\mathbf{H}}_{1\cdot 2}^\top\right)^{-1}\bar{\boldsymbol{\Gamma}}_{1\cdot 2}\right\},\nu+p_2+p_1\right]\\
&= t_{p_1}\left\{\mathbf{x}_1;\tilde{\boldsymbol{\xi}}_{1\cdot 2},\frac{\nu+\tilde{Q}_2(\mathbf{x}_2)}{\nu+p_2}\left(\boldsymbol{\Psi}_{11\cdot 2}+\bar{\mathbf{H}}_{1\cdot 2}\bar{\boldsymbol{\Gamma}}_{1\cdot 2}\bar{\mathbf{H}}_{1\cdot 2}^\top\right),\nu+p_2\right\}\\
&\quad\times t_q\left\{\mathbf{v};\bar{\boldsymbol{\Gamma}}_{1\cdot 2}\bar{\mathbf{H}}_{1\cdot 2}^\top\boldsymbol{\Psi}_{11\cdot 2}^{-1}(\mathbf{x}_1-\tilde{\boldsymbol{\xi}}_{1\cdot 2}),\frac{\nu+p_2+\tilde{Q}_2(\mathbf{x}_2)+\tilde{Q}_{1\cdot 2}(\mathbf{x}_1)}{\nu+p_2+p_1}\tilde{\boldsymbol{\Gamma}}_{1\cdot 2},\nu+p_2+p_1\right\},
\end{aligned}
$$

where $\tilde{Q}_{1\cdot 2}(\mathbf{x}_1) = (\mathbf{x}_1-\tilde{\boldsymbol{\xi}}_{1\cdot 2})^\top\left(\boldsymbol{\Psi}_{11\cdot 2}+\bar{\mathbf{H}}_{1\cdot 2}\bar{\boldsymbol{\Gamma}}_{1\cdot 2}\bar{\mathbf{H}}_{1\cdot 2}^\top\right)^{-1}(\mathbf{x}_1-\tilde{\boldsymbol{\xi}}_{1\cdot 2})$, *and we used that* $\bar{\boldsymbol{\Gamma}}_{1\cdot 2} - \bar{\boldsymbol{\Gamma}}_{1\cdot 2}\bar{\mathbf{H}}_{1\cdot 2}^\top(\boldsymbol{\Psi}_{11\cdot 2}+\bar{\mathbf{H}}_{1\cdot 2}\bar{\boldsymbol{\Gamma}}_{1\cdot 2}\bar{\mathbf{H}}_{1\cdot 2}^\top)^{-1}\bar{\mathbf{H}}_{1\cdot 2}\bar{\boldsymbol{\Gamma}}_{1\cdot 2} = (\bar{\boldsymbol{\Gamma}}_{1\cdot 2}^{-1}+\bar{\mathbf{H}}_{1\cdot 2}^\top\boldsymbol{\Psi}_{11\cdot 2}^{-1}\bar{\mathbf{H}}_{1\cdot 2})^{-1} \equiv \tilde{\boldsymbol{\Gamma}}_{1\cdot 2}$ *and also that*

$\bar{\boldsymbol{\Gamma}}_{1\cdot2}\bar{\mathbf{H}}_{1\cdot2}^{\top}(\boldsymbol{\Psi}_{11\cdot2}+\bar{\mathbf{H}}_{1\cdot2}\bar{\boldsymbol{\Gamma}}_{1\cdot2}\bar{\mathbf{H}}_{1\cdot2}^{\top})^{-1}=\tilde{\boldsymbol{\Gamma}}_{1\cdot2}\bar{\mathbf{H}}_{1\cdot2}^{\top}\boldsymbol{\Psi}_{11\cdot2}^{-1}$. *Hence, we have*

$$\begin{aligned}f_{1\cdot2}(\mathbf{x}_1\mid\mathbf{x}_2)=&\frac{1}{T_q\left\{\sqrt{\frac{\nu+p_2}{\nu+\tilde{Q}_2(\mathbf{x}_2)}}\,\bar{\boldsymbol{\tau}}_{1\cdot2};\mathbf{0}_q,\bar{\boldsymbol{\Gamma}}_{1\cdot2},\nu+p_2\right\}}\\&\times t_{p_1}\left\{\mathbf{x}_1;\tilde{\boldsymbol{\xi}}_{1\cdot2},\frac{\nu+\tilde{Q}_2(\mathbf{x}_2)}{\nu+p_2}\,(\boldsymbol{\Psi}_{11\cdot2}+\bar{\mathbf{H}}_{1\cdot2}^{\top}\bar{\boldsymbol{\Gamma}}_{1\cdot2}\bar{\mathbf{H}}_{1\cdot2}),\nu+p_2\right\}\\&\times\int_{\mathbf{v}+\boldsymbol{\tau}_{1\cdot2}>\mathbf{0}_q}t_q\left\{\mathbf{v};\tilde{\boldsymbol{\Gamma}}_{1\cdot2}\bar{\mathbf{H}}_{1\cdot2}^{\top}\boldsymbol{\Psi}_{11\cdot2}^{-1}(\mathbf{x}_1-\tilde{\boldsymbol{\xi}}_{1\cdot2}),c_{1\cdot2}\tilde{\boldsymbol{\Gamma}}_{1\cdot2},\nu+p_2+p_1\right\}d\mathbf{v},\end{aligned}$$

where $c_{1\cdot2}=\frac{\nu+p_2+\tilde{Q}_2(\mathbf{x}_2)+\tilde{Q}_{1\cdot2}(\mathbf{x}_1)}{\nu+p_2+p_1}$. *The proof concludes by making the change of variables* $\mathbf{u}=c_{1\cdot2}^{-1/2}(\mathbf{v}-\mathbf{h}_{1\cdot2})$, *where* $\mathbf{h}_{1\cdot2}=\mathbf{h}_{1\cdot2}(\mathbf{x}_1)=\tilde{\boldsymbol{\Gamma}}_{1\cdot2}\bar{\mathbf{H}}_{1\cdot2}^{\top}\boldsymbol{\Psi}_{11\cdot2}^{-1}(\mathbf{x}_1-\tilde{\boldsymbol{\xi}}_{1\cdot2})$, *to obtain in the last integral that*

$$\int_{\mathbf{u}+c_{1\cdot2}^{-1/2}(\bar{\boldsymbol{\tau}}_{1\cdot2}+\mathbf{h}_{1\cdot2})>\mathbf{0}_q}t_q\left\{\mathbf{u};\mathbf{0}_q,\tilde{\boldsymbol{\Gamma}}_{1\cdot2},\nu+p_2+p_1\right\}d\mathbf{u}=T_q\left\{c_{1\cdot2}^{-1/2}(\bar{\boldsymbol{\tau}}_{1\cdot2}+\mathbf{h}_{1\cdot2});\mathbf{0}_q,\tilde{\boldsymbol{\Gamma}}_{1\cdot2},\nu+p_2+p_1\right\}.$$

□

Example 98 *Two special cases of the conditional SUT pdf are the following:*

(a) If $\mathbf{H}_{1\cdot2}=\mathbf{H}_1-\boldsymbol{\Psi}_{12}\boldsymbol{\Psi}_{22}^{-1}\mathbf{H}_2=\mathbf{O}$, *then the conditional pdf of* $(\mathbf{X}_1\mid\mathbf{X}_2=\mathbf{x}_2)\sim\mathcal{SUT}_{p_1,q}\left(\boldsymbol{\xi}_{1\cdot2},\frac{\nu+\tilde{Q}_2(\mathbf{x}_2)}{\nu+p_2}\boldsymbol{\Psi}_{11\cdot2},\mathbf{O},\sqrt{\frac{\nu+p_2}{\nu+\tilde{Q}_2(\mathbf{x}_2)}}\,\bar{\boldsymbol{\tau}}_{1\cdot2},\bar{\boldsymbol{\Gamma}}_{1\cdot2},\nu+p_2\right)$ *is given by*

$$\begin{aligned}f_{1\cdot2}(\mathbf{x}_1\mid\mathbf{x}_2)=&\frac{1}{T_q\left\{\sqrt{\frac{\nu+p_2}{\nu+\tilde{Q}_2(\mathbf{x}_2)}}\,\bar{\boldsymbol{\tau}}_{1\cdot2};\mathbf{0}_q,\bar{\boldsymbol{\Gamma}}_{1\cdot2},\nu+p_2\right\}}t_{p_1}\left\{\mathbf{x}_1;\boldsymbol{\xi}_{1\cdot2},\frac{\nu+\tilde{Q}_2(\mathbf{x}_2)}{\nu+p_2}\boldsymbol{\Psi}_{11\cdot2},\nu+p_2\right\}\\&\times T_q\left\{c_{1\cdot2}^{-1/2}\bar{\boldsymbol{\tau}}_{1\cdot2};\mathbf{0}_q,\bar{\boldsymbol{\Gamma}}_{1\cdot2},\nu+p_2+p_1\right\},\quad\mathbf{x}_1\in\mathbb{R}^{p_1};\end{aligned}$$

(b) If $\boldsymbol{\tau}=\mathbf{0}_q$ *and* $\mathbf{H}_2=\mathbf{O}$, *then the conditional pdf of* $(\mathbf{X}_1\mid\mathbf{X}_2=\mathbf{x}_2)$ *becomes*

$$\begin{aligned}f_{1\cdot2}(\mathbf{x}_1\mid\mathbf{x}_2)=&\frac{1}{T_q\left(\mathbf{0}_q;\mathbf{0}_q,\bar{\boldsymbol{\Gamma}},\nu+p_2\right)}t_{p_1}\left\{\mathbf{x}_1;\boldsymbol{\xi}_{1\cdot2},\frac{\nu+\tilde{Q}_2(\mathbf{x}_2)}{\nu+p_2}\,(\boldsymbol{\Psi}_{11\cdot2}+\mathbf{H}_1\bar{\boldsymbol{\Gamma}}\mathbf{H}_1^{\top}),\nu+p_2\right\}\\&\times T_q\left\{c_{1\cdot2}^{-1/2}\bar{\boldsymbol{\Gamma}}\mathbf{H}_1^{\top}\boldsymbol{\Psi}_{11\cdot2}^{-1}(\mathbf{x}_1-\boldsymbol{\xi}_{1\cdot2});\mathbf{0}_q,\bar{\boldsymbol{\Gamma}},\nu+p_2+p_1\right\},\quad\mathbf{x}_1\in\mathbb{R}^{p_1}.\end{aligned}$$

7.4.3 Multivariate extended skew-t and multivariate skew-t distributions

When $q=1$, and so $\bar{\boldsymbol{\Gamma}}=1$, the multivariate unified skew-*t* distribution becomes the so-called *multivariate extended skew-t (EST) distribution*; see, e.g., Arellano-Valle and Genton (2010b) and Adcock (2010). In this special case, we write $\mathbf{X}\sim\mathcal{EST}_p(\boldsymbol{\xi},\boldsymbol{\Psi},\boldsymbol{\eta},\tau,\nu)$. Stochastically, this distribution can be represented as

$$\mathbf{X}=\boldsymbol{\xi}+\boldsymbol{\eta}U+\sqrt{\frac{\nu+U^2}{\nu+1}}\mathbf{W}_*,\tag{7.44}$$

where $U\sim\mathcal{TT}_1(-\tau;0,1,\nu)$ and $\mathbf{W}_*\sim\mathcal{T}_p(\mathbf{0}_p,\boldsymbol{\Psi},\nu+1)$ which are independent. Hierarchically, (7.44) is equivalent to

$$\begin{aligned}(\mathbf{X}\mid U=u)&\sim\mathcal{T}_p\left(\boldsymbol{\xi}+\boldsymbol{\eta}u,\frac{\nu+u^2}{\nu+1}\boldsymbol{\Psi},\nu+1\right),\\U&\sim\mathcal{TT}_1(-\tau;0,1,\nu).\end{aligned}$$

In the nonsingular case with $\boldsymbol{\Psi} > 0$, it follows from this last representation (or directly from (7.23) with $q = 1$) that the pdf of $\mathbf{X} \sim \mathcal{EST}_p(\boldsymbol{\xi}, \boldsymbol{\Psi}, \nu, \boldsymbol{\eta}, \tau)$ is

$$\begin{aligned} f_{\mathrm{EST}}(\mathbf{x}) &= \frac{1}{T(\tau;0,1,\nu)} \int_{-\tau}^{\infty} t_p\left(\mathbf{x};\boldsymbol{\xi}+\boldsymbol{\eta}\,u, \frac{\nu+u^2}{\nu+1}\boldsymbol{\Psi}, \nu+1\right) t_1(u;0,1,\nu)\mathrm{d}u \\ &= \frac{1}{T(\tau;0,1,\nu)} t_p(\mathbf{x};\boldsymbol{\xi},\boldsymbol{\Psi}+\boldsymbol{\eta}\boldsymbol{\eta}^\top,\nu) \\ &\quad \times T\left[\sqrt{\frac{\nu+1}{\nu+\tilde{Q}(\mathbf{x})}} \frac{\tau+\boldsymbol{\eta}^\top\boldsymbol{\Psi}^{-1}(\mathbf{x}-\boldsymbol{\xi})}{\sqrt{1+\boldsymbol{\eta}^\top\boldsymbol{\Psi}^{-1}\boldsymbol{\eta}}}, \nu+1\right], \quad \mathbf{x} \in \mathbb{R}^p, \end{aligned} \tag{7.45}$$

where, as defined earlier, $\tilde{Q}(\mathbf{x}) = (\mathbf{x}-\boldsymbol{\xi})^\top(\boldsymbol{\Psi}+\boldsymbol{\eta}\boldsymbol{\eta}^\top)^{-1}(\mathbf{x}-\boldsymbol{\xi})$.

From (7.44) we can also obtain the mean vector and the variance matrix of $\mathbf{X} \sim \mathcal{EST}_p(\boldsymbol{\xi}, \boldsymbol{\Psi}, \nu, \boldsymbol{\eta}, \tau)$. However, these results can now be obtained directly from Theorem 116. Thus, we have

$$\mathrm{E}(\mathbf{X}) = \boldsymbol{\xi} + \mathrm{E}(U)\boldsymbol{\eta} \quad (\nu > 1), \quad \mathrm{var}(\mathbf{X}) = \mathrm{var}(U)\boldsymbol{\eta}\boldsymbol{\eta}^\top + \frac{T(\tau;0,1,\nu-2)}{T(\tau;0,1,\nu)}\left(\frac{\nu}{\nu-2}\right)\boldsymbol{\Psi} \quad (\nu > 2). \tag{7.46}$$

For a more explicit expression of these results, we need the mean and variance of the truncated random variable U. For this, we recall that $U \stackrel{d}{=} (W_0 \mid W_0 + \tau > 0)$, with $W_0 \sim \mathcal{T}(0,1,\nu)$. Thus, if we assume $\nu > k$, then the k-th moment of U is given by $\mathrm{E}(U^k) = \mathrm{E}(W_0^k \mid W_0 + \tau > 0)$. To compute the t-truncated moments, we can use the fact that the t-pdf can be represented as a scale mixtures of the normal pdf; that is, by writing $t(w;0,1,\nu) = \int_0^\infty s^{1/2}\phi(s^{1/2}w;0,1)g(s;\nu)\mathrm{d}s$, with $\phi(x;0,1) = (2\pi)^{-1/2}e^{-x^2/2}$, $x \in \mathbb{R}$, and $g(s;\nu) = \{(\nu/2)^{\nu/2}/\Gamma(\nu/2)\}s^{(\nu/2)-1}e^{-(\nu/2)s}$, $s > 0$, the $\mathcal{G}amma(\nu/2,\nu/2)$-pdf, we have that

$$\begin{aligned} \mathrm{E}(U^k) &= \mathrm{E}(W_0^k \mid W_0 + \tau > 0) \\ &= \frac{1}{T(\tau;0,1,\nu)} \int_{-\tau}^{\infty} w^k t(w;0,1,\nu)\mathrm{d}w \\ &= \frac{1}{T(\tau;0,1,\nu)} \int_{-\tau}^{\infty} w^k \int_0^\infty s^{1/2}\phi(s^{1/2}w;0,1)g(s;\nu)\mathrm{d}s\mathrm{d}w \\ &= \frac{1}{T(\tau;0,1,\nu)} \int_0^\infty \int_{-s^{1/2}\tau}^{\infty} z^k\phi(z;0,1)\mathrm{d}z\, s^{-k/2}g(s;\nu)\mathrm{d}s \quad (z = s^{1/2}w) \\ &= \frac{1}{T(\tau;0,1,\nu)} \int_0^\infty s^{-k/2}\Phi(s^{1/2}\tau;0,1)\mathrm{E}(\tilde{W}_0^k \mid \tilde{W}_0 + s^{1/2}\tau > 0, V = s)g(s;\nu)\mathrm{d}s \quad (\nu > m) \\ &= \frac{1}{T(\tau;0,1,\nu)} \mathrm{E}\left\{V^{-k/2}\Phi(V^{1/2}\tau;0,1)\mathrm{E}(\tilde{W}_0^k \mid \tilde{W}_0 + V^{1/2}\tau > 0, V)\right\} \quad (\nu > k), \end{aligned} \tag{7.47}$$

where $V \sim \mathcal{G}amma(\nu/2,\nu/2)$ and $\tilde{W}_0 \sim \mathcal{N}(0,1)$, which are independent, and $\mathrm{E}(\tilde{W}_0^k \mid \tilde{W}_0 + s^{1/2}\tau > 0, V = s)\}$, $k = 1, 2, \ldots$, are conditional truncated normal moments. These truncated normal moments can be computed from the moment generating function of $(\tilde{W}_0 \mid \tilde{W}_0 + a > 0)$, which for each $a \in \mathbb{R}$, is given by

$$M_{(\tilde{W}_0 \mid \tilde{W}_0 + a > 0)}(t) = e^{t^2/2}\frac{\Phi(a+t;0,1)}{\Phi(a;0,1)}, \quad t \in \mathbb{R},$$

where $\Phi(x;0,1)$, $x \in \mathbb{R}$, is the standard normal cdf. For instance, up to the fourth order, these truncated normal moments become

$$\begin{aligned} &\mathrm{E}(\tilde{W}_0 \mid \tilde{W}_0 + a > 0) = \zeta(a), \quad \mathrm{E}(\tilde{W}_0^2 \mid \tilde{W}_0 + a > 0) = 1 - a\zeta(a), \\ &\mathrm{E}(\tilde{W}_0^3 \mid \tilde{W}_0 + a > 0) = (2 + a^2)\zeta(a), \quad \mathrm{E}(\tilde{W}_0^4 \mid \tilde{W}_0 + a > 0) = 3 - (3a + a^2)\zeta(a), \end{aligned}$$

where $\zeta(a) = \frac{\phi(a;0,1)}{\Phi(a;0,1)}$, with $\phi(x;0,1) = \Phi'(x;0,1)$, $x \in \mathbb{R}$, the $\mathcal{N}(0,1)$-pdf. Hence, using these results in (7.47), we have for $\mathrm{E}(U^k) = \mathrm{E}(W_0^k \mid W_0 + \tau > 0)$, with $W_0 \sim \mathcal{T}(0,1,\nu)$ and $k = 1,2,3,4$, that:

$$
\begin{aligned}
\mathrm{E}(U) &= \frac{1}{T(\tau;0,1,\nu)} \mathrm{E}\left\{V^{-1/2}\Phi\left(V^{1/2}\tau;0,1\right)\zeta\left(V^{1/2}\tau\right)\right\} \\
&= \frac{1}{T(\tau;0,1,\nu)} \mathrm{E}\left\{V^{-1/2}\phi\left(V^{1/2}\tau;0,1\right)\right\}, \\
\mathrm{E}(U^2) &= \frac{1}{T(\tau;0,1,\nu)} \mathrm{E}\left[V^{-1}\Phi\left(V^{1/2}\tau;0,1\right)\left\{1 - V^{1/2}\tau\zeta\left(V^{1/2}\tau\right)\right\}\right] \\
&= \frac{1}{T(\tau;0,1,\nu)} \left[\mathrm{E}\left\{V^{-1}\Phi\left(V^{1/2}\tau;0,1\right)\right\} - \tau\mathrm{E}\left\{V^{-1}\phi\left(V^{1/2}\tau;0,1\right)\right\}\right], \\
\mathrm{E}(U^3) &= \frac{1}{T(\tau;0,1,\nu)} \mathrm{E}\left\{V^{-3/2}\Phi\left(V^{1/2}\tau;0,1\right)(2+V\tau^2)\,\zeta\left(V^{1/2}\tau\right)\right\} \\
&= \frac{1}{T(\tau;0,1,\nu)} \left[\mathrm{E}\left\{V^{-3/2}\Phi\left(V^{1/2}\tau;0,1\right)\right\} + \tau^2\mathrm{E}\left\{V^{-1/2}\phi\left(V^{1/2}\tau;0,1\right)\right\}\right], \\
\mathrm{E}(U^4) &= \frac{1}{T(\tau;0,1,\nu)} \mathrm{E}\left[V^{-2}\Phi\left(V^{1/2}\tau;0,1\right)\left\{3 - \left(3V^{1/2}\tau + V\tau^2\right)\zeta\left(V_4^{1/2}\tau\right)\right\}\right] \\
&= \frac{1}{T(\tau;0,1,\nu)} \left[3\mathrm{E}\left\{V^{-2}\Phi\left(V^{1/2}\tau;0,1\right)\right\} - 3\tau\mathrm{E}\left\{V^{-3/2}\phi\left(V^{1/2}\tau;0,1\right)\right\}\right. \\
&\quad \left. -\tau^2\mathrm{E}\left\{V^{-1}\phi\left(V^{1/2}\tau;0,1\right)\right\}\right],
\end{aligned}
$$

where

$$
\begin{aligned}
\mathrm{E}\{V^{-k/2}\phi(V^{1/2}\tau;0,1)\} &= \frac{(\nu/2)^{k/2}\Gamma\{(\nu-k)/2\}}{\Gamma(\nu/2)}\left(1+\frac{\tau^2}{\nu}\right)^{-(\nu-k)/2} \quad (\nu > k), \\
\mathrm{E}\{V^{-k/2}\Phi(V^{1/2}\tau;0,1)\} &= \frac{(\nu/2)^{k/2}\Gamma\{(\nu-k)/2\}}{\Gamma(\nu/2)}\, T\left(\sqrt{\frac{\nu-k}{\nu}}\,\tau;0,1,\nu-k\right) \quad (\nu > k).
\end{aligned}
$$

When $\tau = 0$ in addition to $q = 1$, we obtain the *multivariate skew-t (ST) distribution* denoted by $\mathbf{X} \sim \mathcal{ST}_p(\boldsymbol{\xi}, \boldsymbol{\Psi}, \boldsymbol{\eta}, \nu)$. In that case, we have $U = V^{-1/2}|\tilde{W}_0|$ and so $\mathrm{E}(U^k) = \mathrm{E}(V^{-k/2})\mathrm{E}(|\tilde{W}_0|^k)$, for $\nu > k \geq 0$, where as before $V \sim \mathcal{G}amma(\nu/2, \nu/2)$ and is independent of $|\tilde{W}_0| \sim \mathcal{HN}(0,1)$. Hence, from (7.46) we find that the mean vector and variance matrix of $\mathbf{X} \sim \mathcal{ST}_p(\boldsymbol{\xi}, \boldsymbol{\Psi}, \boldsymbol{\eta}, \nu)$ are

$$
\mathrm{E}(\mathbf{X}) = \boldsymbol{\xi} + \mathrm{E}(U)\boldsymbol{\eta} \quad (\nu > 1), \quad \mathrm{var}(\mathbf{X}) = \frac{\nu}{\nu-2}\boldsymbol{\Psi} + \mathrm{var}(U)\boldsymbol{\eta}\boldsymbol{\eta}^\top \quad (\nu > 2),
$$

with

$$
\mathrm{E}(U) = \sqrt{\frac{\nu}{\pi}}\,\frac{\Gamma\{(\nu-1)/2\}}{\Gamma(\nu/2)} \quad (\nu > 1), \quad \mathrm{var}(U) = \frac{\nu}{\nu-2} - \frac{\nu}{\pi}\,\frac{\Gamma\{(\nu-1)/2\}^2}{\Gamma(\nu/2)^2} \quad (\nu > 2).
$$

Also, in this case, the EST pdf in (7.45) reduces to the pdf of the multivariate skew-t distribution:

$$
f_{\mathrm{ST}}(\mathbf{x}) = 2t_p(\mathbf{x};\boldsymbol{\xi}, \boldsymbol{\Psi} + \boldsymbol{\eta}\boldsymbol{\eta}^\top, \nu)\,T\left\{\sqrt{\frac{\nu+1}{\nu+\tilde{Q}(\mathbf{x})}}\,\frac{\boldsymbol{\eta}^\top\boldsymbol{\Psi}^{-1}(\mathbf{x}-\boldsymbol{\xi})}{\sqrt{1+\boldsymbol{\eta}^\top\boldsymbol{\Psi}^{-1}\boldsymbol{\eta}}};0,1,\nu+1\right\}, \quad \mathbf{x} \in \mathbb{R}^p. \quad (7.48)
$$

In turn, this last distribution becomes the multivariate t, $\mathcal{T}_p(\boldsymbol{\xi}, \boldsymbol{\Psi}, \nu)$, for $\boldsymbol{\eta} = \mathbf{0}_p$, and the multivariate skew-normal, $\mathcal{SN}_p(\boldsymbol{\xi}, \boldsymbol{\Psi}, \boldsymbol{\eta})$, as $\nu \to \infty$. In terms of the parameterization $\boldsymbol{\Omega} = \boldsymbol{\Psi} + \boldsymbol{\eta}\boldsymbol{\eta}^\top$ and $\boldsymbol{\omega}\boldsymbol{\delta} =$

$\boldsymbol{\eta}$ of Azzalini and Capitanio (1999), where $\boldsymbol{\delta} = (1 - \boldsymbol{\alpha}^\top \bar{\boldsymbol{\Omega}}^{-1} \boldsymbol{\alpha})^{-1/2} \bar{\boldsymbol{\Omega}}^{-1} \boldsymbol{\alpha}$, $\boldsymbol{\omega} = \text{diag}(\boldsymbol{\Omega})^{1/2}$ and $\bar{\boldsymbol{\Omega}} = \boldsymbol{\omega}^{-1} \boldsymbol{\Omega} \boldsymbol{\omega}^{-1}$, the ST pdf in (7.48) becomes

$$f_{\text{AST}}(\mathbf{x}) = 2t_p(\mathbf{x}; \boldsymbol{\xi}, \boldsymbol{\Omega}, \nu) T \left\{ \sqrt{\frac{\nu + 1}{\nu + \tilde{Q}(\mathbf{x})}} \boldsymbol{\alpha}^\top \boldsymbol{\omega}^{-1} (\mathbf{x} - \boldsymbol{\xi}); 0, 1, \nu + 1 \right\}, \quad \mathbf{x} \in \mathbb{R}^p,$$

with $\tilde{Q}(\mathbf{x}) = (\mathbf{x} - \boldsymbol{\xi})^\top \boldsymbol{\Omega}^{-1} (\mathbf{x} - \boldsymbol{\xi})$, and denoted by $\mathbf{X} \sim \mathcal{AST}_p(\boldsymbol{\xi}, \boldsymbol{\Omega}, \boldsymbol{\alpha}, \nu)$; see also Gupta (2003). Properties and other aspects related to this distribution can be found in the book of Azzalini and Capitanio (2014); see also Alruwaili (2023) for the mode of the univariate skew-*t* distribution. An alternative centered parameterization based on the mean vector, the variance matrix and higher moments was proposed by Arellano-Valle and Azzalini (2013). When $\nu = 1$, the *multivariate skew-Cauchy (SC) distribution* is obtained; see Arnold and Beaver (2000b) and Behboodian et al. (2006), and univariate extensions by Huang and Chen (2007) and Arrué et al. (2015). Because the skew-*t* distribution allows a wide range of skewness and kurtosis, its use in real-world applications has become increasingly common in recent years; see Chapter 10. Inferential aspects of the skew-*t* distribution can be found, e.g., in Carota (2010) and DiCiccio and Monti (2011, 2018) for the univariate case, and in Azzalini and Capitanio (2003) and Azzalini and Genton (2008) for the multivariate case.

7.5 Quadratic Forms and Some Multivariate Moments

7.5.1 Distribution of quadratic forms

In general, for a fixed symmetric matrix $\mathbf{A} \in \mathbb{R}^{p \times p}$ and a vector $\mathbf{a} \in \mathbb{R}^p$, obtaining the distribution of the quadratic form $Q_A(\mathbf{X} - \mathbf{a}) = (\mathbf{X} - \mathbf{a})^\top \mathbf{A} (\mathbf{X} - \mathbf{a})$ of a random vector $\mathbf{X} \in \mathbb{R}^p$ is a somewhat complex problem. However, if the matrix $\mathbf{A}$ is nonsingular and the random vector $\mathbf{X}$ has a selection representation, namely $\mathbf{X} = (\mathbf{W}_* \mid \mathbf{W}_0 \in C)$ for some random vector $(\mathbf{W}_*^\top, \mathbf{W}_0^\top)^\top \in \mathbb{R}^{p+q}$ and Borel subset C of $\mathbb{R}^q$, then the property that $Q_A(\mathbf{X} - \mathbf{a}) = (Q_A(\mathbf{W}_* - \mathbf{a}) \mid \mathbf{W}_0 \in C)$ implies that its pdf could be obtained from the selection pdf given by (see Chapter 9)

$$f_{Q_A(\mathbf{X}-\mathbf{a})}(s) = f_{Q_A(\mathbf{W}_*-\mathbf{a})}(s) \frac{\text{P}(\mathbf{W}_0 \in C \mid Q_A(\mathbf{W}_* - \mathbf{a}) = s)}{\text{P}(\mathbf{W}_0 \in C)}, \quad s > 0.$$

However, from this route, we need to know the pdf of $Q_A(\mathbf{W}_* - \mathbf{a}) = (\mathbf{W}_* - \mathbf{a})^\top \mathbf{A} (\mathbf{W}_* - \mathbf{a})$, the conditional probability $\text{P}(\mathbf{W}_0 \in C \mid Q_A(\mathbf{W}_* - \mathbf{a}) = s)$ and the marginal probability $\text{P}(\mathbf{W}_0 \in C)$, which are not always possible to obtain in a simple way. Furthermore, unlike the SUN case where it was possible to obtain semi-explicit expressions for the characteristic function of $Q_A(\mathbf{X} - \mathbf{a})$ (see Theorem 108 in Chapter 6), in the SUT case this is not simple either. In view of this, in this section, we shall only analyze the distribution of some special quadratic forms with $\mathbf{a} = \boldsymbol{\xi}$ and $\boldsymbol{\tau} = \mathbf{0}_q$ in the nonsingular SUT case.

Theorem 133 *Let $Q_A(\mathbf{X}) = (\mathbf{X} - \boldsymbol{\xi})^\top \mathbf{A} (\mathbf{X} - \boldsymbol{\xi})$, where $\mathbf{X} \sim \mathcal{SUT}_{p,q}(\boldsymbol{\xi}, \boldsymbol{\Psi}, \mathbf{H}, \mathbf{0}_q, \bar{\boldsymbol{\Gamma}}, \nu)$ is nonsingular ($\boldsymbol{\Psi} > 0$). Then:*

(i) If $\mathbf{A} = \mathbf{C}^\top \mathbf{C}$, with $\mathbf{CH} = \mathbf{O}$, then $Q_A(\mathbf{X}) \stackrel{d}{=} V^{-1} \tilde{\mathbf{W}}^\top \mathbf{A} \tilde{\mathbf{W}}$, where $V \sim \mathcal{G}amma(\nu/2, \nu/2)$ and $\tilde{\mathbf{W}} \sim \mathcal{N}_p(\mathbf{0}_p, \boldsymbol{\Psi})$ are independent. If in addition $\mathbf{A}\boldsymbol{\Psi}$ is an idempotent matrix of rank $r \leq p$ (which is equivalent to $\mathbf{A}\boldsymbol{\Psi}\mathbf{A} = \mathbf{A}$ and $\text{rank}(\mathbf{A}) = r$ since $\boldsymbol{\Psi}$ is nonsingular), then $\tilde{\mathbf{W}}^\top \mathbf{A} \tilde{\mathbf{W}} \sim \chi_r^2$ and so $Q_A(\mathbf{X} - \boldsymbol{\xi}) = (\mathbf{X} - \boldsymbol{\xi})^\top \mathbf{A} (\mathbf{X} - \boldsymbol{\xi}) \sim r\mathcal{F}_{r,\nu}$.

(ii) If $\mathbf{A} = \boldsymbol{\Omega}^{-1}$, where $\boldsymbol{\Omega} = \boldsymbol{\Psi} + \mathbf{H}\bar{\boldsymbol{\Gamma}}\mathbf{H}^\top$, then $Q_A(\mathbf{X}) \stackrel{d}{=} V^{-1} \tilde{\mathbf{Z}}^\top \boldsymbol{\Omega}^{-1} \tilde{\mathbf{Z}}$, where $V \sim \mathcal{G}amma(\nu/2, \nu/2)$ and $\tilde{\mathbf{Z}} \sim \mathcal{SUN}_{p,q}(\mathbf{0}_p, \boldsymbol{\Psi}, \mathbf{H}, \mathbf{0}_q, \bar{\boldsymbol{\Gamma}})$ are independent. Moreover, if $\bar{\boldsymbol{\Gamma}} = \mathbf{I}_q$ and $\mathbf{H}^\top \boldsymbol{\Psi}^{-1} \mathbf{H}$ is a diagonal matrix, then $Q_A(\mathbf{X} - \boldsymbol{\xi}) = (\mathbf{X} - \boldsymbol{\xi})^\top \boldsymbol{\Omega}^{-1} (\mathbf{X} - \boldsymbol{\xi}) \sim p\mathcal{F}_{p,\nu}$.

Proof 133 *If $\boldsymbol{\tau} = \mathbf{0}_q$, then by Theorem 114, we have $\mathbf{X} - \boldsymbol{\xi} \stackrel{d}{=} V^{-1/2}\tilde{\mathbf{Z}}$, where V and $\tilde{\mathbf{Z}}$ are independent, with $V \sim \mathcal{G}amma(\nu/2, \nu/2)$ and $\tilde{\mathbf{Z}} = \mathbf{H}\tilde{\mathbf{U}} + \tilde{\mathbf{W}} \sim \mathcal{SUN}_{p,q}(\mathbf{0}_p, \boldsymbol{\Psi}, \mathbf{H}, \mathbf{0}_q, \bar{\boldsymbol{\Gamma}})$ since $\tilde{\mathbf{U}} \sim \mathcal{TN}_q(\mathbf{0}_q; \mathbf{0}_q, \bar{\boldsymbol{\Gamma}})$ and $\tilde{\mathbf{W}} \sim \mathcal{N}_p(\mathbf{0}_p, \boldsymbol{\Psi})$ and they are independent. Thus, for part (i) we have*

$$
\begin{aligned}
Q_A(\mathbf{X}-\boldsymbol{\xi}) &= (\mathbf{X}-\boldsymbol{\xi})^\top \mathbf{A}(\mathbf{X}-\boldsymbol{\xi}) \\
&\stackrel{d}{=} V^{-1}\tilde{\mathbf{Z}}^\top \mathbf{C}^\top \mathbf{C}\tilde{\mathbf{Z}} \\
&= V^{-1}(\mathbf{H}\tilde{\mathbf{U}}+\tilde{\mathbf{W}})^\top \mathbf{C}^\top \mathbf{C}(\mathbf{H}\tilde{\mathbf{U}}+\tilde{\mathbf{W}}) \\
&= V^{-1}(\mathbf{C}\mathbf{H}\tilde{\mathbf{U}}+\mathbf{C}\tilde{\mathbf{W}})^\top (\mathbf{C}\mathbf{H}\tilde{\mathbf{U}}+\mathbf{C}\tilde{\mathbf{W}}) \\
&= V^{-1}\tilde{\mathbf{W}}^\top \mathbf{A}\tilde{\mathbf{W}},
\end{aligned}
$$

since by assumption $\mathbf{CH} = \mathbf{O}$, which proves the first result of part (i). For the second result of part (i) we have by Theorem 108 of Chapter 6 (see the second special case of that theorem) under the conditions established for the matrix $\mathbf{A}$ that $\tilde{\mathbf{W}}^\top \mathbf{A}\tilde{\mathbf{W}} \sim \chi^2_r$ and so that $Q_A(\mathbf{X}) \stackrel{d}{=} V^{-1}\tilde{\mathbf{W}}^\top \mathbf{A}\tilde{\mathbf{W}} \sim r\mathcal{F}_{r,\nu}$ since V and $\tilde{\mathbf{W}}^\top \mathbf{A}\tilde{\mathbf{W}}$ are independent. The proof of the first result of part (ii) is similar, since in this case $Q_A(\mathbf{X}-\boldsymbol{\xi}) \stackrel{d}{=} V^{-1}\tilde{\mathbf{Z}}^\top \boldsymbol{\Omega}^{-1}\tilde{\mathbf{Z}}$. The second result of part (ii) is direct from Theorem 108 of Chapter 6, since under the conditions $\boldsymbol{\tau} = \mathbf{0}_q$, $\bar{\boldsymbol{\Gamma}} = \mathbf{I}_q$ and $\mathbf{H}^\top \boldsymbol{\Psi}^{-1}\mathbf{H}$ is diagonal (see the third special case of Theorem 108), then $Q_A(\tilde{\mathbf{Z}}) = \tilde{\mathbf{Z}}^\top \boldsymbol{\Omega}^{-1}\tilde{\mathbf{Z}} \sim \chi^2_p$. □

Example 99 *Let $\mathbf{X} \sim \mathcal{SUT}_{p,q}(\boldsymbol{\xi}, \mathbf{I}_p, \mathbf{H}, \mathbf{0}_q, \bar{\boldsymbol{\Gamma}}, \nu)$. Suppose that the skewness matrix $\mathbf{H}$ has rank $q \le p$ and let $\mathbf{P}$ be a $p \times p$ projection matrix of rank q such that $\mathbf{PH} = \mathbf{H}$. Then, for $\mathbf{A} = \mathbf{I}_p - \mathbf{P}$ we have $Q_A(\mathbf{X}-\boldsymbol{\xi}) = (\mathbf{X}-\boldsymbol{\xi})^\top(\mathbf{I}_p - \mathbf{P})(\mathbf{X}-\boldsymbol{\xi}) = \|\mathbf{X}-\boldsymbol{\xi}\|^2 - Q_P(\mathbf{X}-\boldsymbol{\xi}) \sim (p-q)\mathcal{F}_{p-q,\nu}$, since $\mathbf{AH} = (\mathbf{I}_p - \mathbf{P})\mathbf{H} = \mathbf{H} - \mathbf{H} = \mathbf{O}$ and $\mathbf{A}\boldsymbol{\Psi} = \mathbf{I}_p - \mathbf{P}$ is also a projection matrix of rank $p-q$ and so the matrix $\mathbf{A}\boldsymbol{\Psi}$ is idempotent and has rank $r = p-q$.*

7.5.2 Higher-order moments

As developed in Chapter 6 for the SUN case, explicit higher-order moments of the non-shifted SUT random vector $\mathbf{Z} = \mathbf{X} - \boldsymbol{\xi} \sim \mathcal{SUT}_{p,q}(\mathbf{0}_p, \boldsymbol{\Psi}, \mathbf{H}, \boldsymbol{\tau}, \bar{\boldsymbol{\Gamma}}, \nu)$ can also be computed by appropriately applying the general results in Theorem 110 presented in Chapter 6 for the SUN. For this, we will use the stochastic representation of $\mathbf{X}$ presented in Equation (7.6) and the following preliminary results related to the multivariate t distribution.

Theorem 134 *If $\mathbf{W} \sim \mathcal{T}_p(\mathbf{0}_p, \boldsymbol{\Psi}, \nu)$, then:*

$$
\begin{aligned}
&\mu_2(\mathbf{W}) = \mathrm{E}(\mathbf{W} \otimes \mathbf{W}^\top) = \frac{\nu}{\nu-2}\boldsymbol{\Psi}, \quad \nu > 2, \\
&\mu_3(\mathbf{W}) = \mathrm{E}(\mathbf{W}\mathbf{W}^\top \otimes \mathbf{W}) = \mathbf{O}, \quad \nu > 3, \\
&\mu_4(\mathbf{W}) = \mathrm{E}(\mathbf{W}\mathbf{W}^\top \otimes \mathbf{W}\mathbf{W}^\top) = \frac{\nu^2}{(\nu-2)(\nu-4)}\{(\mathbf{I}_{p^2} + \mathbf{K}_p)(\boldsymbol{\Psi} \otimes \boldsymbol{\Psi}) + \mathrm{vec}(\boldsymbol{\Psi})\mathrm{vec}(\boldsymbol{\Psi})^\top\}, \quad \nu > 4, \\
&\mathrm{var}\{\mathrm{vec}(\mathbf{W}\mathbf{W}^\top)\} = \frac{\nu^2}{(\nu-2)(\nu-4)}\{(\mathbf{I}_{p^2} + \mathbf{K}_p)(\boldsymbol{\Psi} \otimes \boldsymbol{\Psi}) + 2\mathrm{vec}(\boldsymbol{\Psi})\mathrm{vec}(\boldsymbol{\Psi})^\top\}, \quad \nu > 4.
\end{aligned}
$$

Proof 134 *Since $\mathbf{W} = V^{-1/2}\tilde{\mathbf{W}}$, where $V \sim \mathcal{G}amma(\nu/2, \nu/2)$ and $\tilde{\mathbf{W}} \sim \mathcal{N}_p(\mathbf{0}_p, \boldsymbol{\Psi})$, which are independent, the proof follows by using Theorem 109 of Chapter 6 and the fact that $\mathrm{E}(V^{-k/2}) = (\nu/2)^{k/2}\Gamma\{(\nu-k)/2\}/\Gamma(\nu/2)$ for $\nu > k \ge 0$.* □

Next, we present only the third and fourth moments of $\mathbf{Z} = \mathbf{X} - \boldsymbol{\xi}$ because the first and second can be obtained directly through the mean and variance computed in Theorem 116 of $\mathbf{X} = \boldsymbol{\xi} + \mathbf{Z}$.

Theorem 135 *Let* $\mathbf{Z} = \mathbf{H}\mathbf{U} + V_*^{-1/2}\mathbf{W}_* \sim \mathcal{SUT}_{p,q}(\mathbf{0}_p, \boldsymbol{\Psi}, \mathbf{H}, \boldsymbol{\tau}, \bar{\boldsymbol{\Gamma}}, \nu)$, *where* $\mathbf{U} \sim \mathcal{TT}_q(-\boldsymbol{\tau}; \mathbf{0}_q, \bar{\boldsymbol{\Gamma}}, \nu)$ *and* $\mathbf{W}_* \sim \mathcal{T}_p(\mathbf{0}_p, \boldsymbol{\Psi}, \nu + q)$ *are independent,* $V_* = \frac{\nu+q}{\nu+Q(\mathbf{U})}$ *and* $Q(\mathbf{U}) = \mathbf{U}^\top \bar{\boldsymbol{\Gamma}}^{-1}\mathbf{U}$. *Then:*

$$
\begin{aligned}
\boldsymbol{\mu}_3(\mathbf{Z}) =& (\mathbf{H}\otimes\mathbf{H})\boldsymbol{\mu}_3(\mathbf{U})\mathbf{H}^\top + \frac{T_q(\boldsymbol{\tau}_2; \mathbf{0}_q, \bar{\boldsymbol{\Gamma}}, \nu-2)}{T_q(\boldsymbol{\tau}; \mathbf{0}_q, \bar{\boldsymbol{\Gamma}}, \nu)}\left(\frac{\nu+q-2}{\nu+q}\right)\left(\frac{\nu}{\nu-2}\right)^{5/2}\\
&\times\left[(\mathbf{I}_{p^2}+\mathbf{K}_p)\{\mathbf{H}\boldsymbol{\mu}_1(\mathbf{U}_2)\otimes\boldsymbol{\Psi}\} + \mathrm{vec}(\boldsymbol{\Psi})\boldsymbol{\mu}_1(\mathbf{U}_2)^\top\mathbf{H}^\top\right],\\
\boldsymbol{\mu}_4(\mathbf{Z}) =& (\mathbf{H}\otimes\mathbf{H})\boldsymbol{\mu}_4(\mathbf{U})(\mathbf{H}\otimes\mathbf{H})^\top + \frac{T_q(\boldsymbol{\tau}_2; \mathbf{0}_q, \bar{\boldsymbol{\Gamma}}, \nu-2)}{T_q(\boldsymbol{\tau}; \mathbf{0}_q, \bar{\boldsymbol{\Gamma}}, \nu)}\left(\frac{\nu+q-2}{\nu+q}\right)\left(\frac{\nu}{\nu-2}\right)^{3}\\
&\times\left[\{\mathbf{H}\boldsymbol{\mu}_2(\mathbf{U})\mathbf{H}^\top\otimes\boldsymbol{\Psi}\} + \{\mathbf{H}\boldsymbol{\mu}_2(\mathbf{U}_2)\mathbf{H}^\top\otimes\boldsymbol{\Psi}\}\mathbf{K}_p + \mathbf{K}_p\{\mathbf{H}\boldsymbol{\mu}_2(\mathbf{U}_2)\mathbf{H}^\top\otimes\boldsymbol{\Psi}\}\right.\\
&\left.+\mathbf{K}_p\{\mathbf{H}\boldsymbol{\mu}_2(\mathbf{U}_2)\mathbf{H}^\top\otimes\boldsymbol{\Psi}\}\mathbf{K}_p + \mathrm{vec}\{\mathbf{H}\boldsymbol{\mu}_2(\mathbf{U}_2)\mathbf{H}^\top\}\mathrm{vec}(\boldsymbol{\Psi})^\top + \mathrm{vec}(\boldsymbol{\Psi})\mathrm{vec}\{\mathbf{H}\boldsymbol{\mu}_2(\mathbf{U}_2)\mathbf{H}^\top\}^\top\right]\\
&+\frac{T_q(\boldsymbol{\tau}_2; \mathbf{0}_q, \bar{\boldsymbol{\Gamma}}, \nu-2)}{T_q(\boldsymbol{\tau}; \mathbf{0}_q, \bar{\boldsymbol{\Gamma}}, \nu)}\left(\frac{\nu+q-2}{\nu+q}\right)\left(\frac{\nu}{\nu-2}\right)^{2}\left(\frac{\nu}{\nu-4}\right)\\
&\times\left\{(\mathbf{I}_{p^2}+\mathbf{K}_p)(\boldsymbol{\Psi}\otimes\boldsymbol{\Psi}) + \mathrm{vec}(\boldsymbol{\Psi})\mathrm{vec}(\boldsymbol{\Psi})^\top\right\},
\end{aligned}
$$

where $\boldsymbol{\mu}_k(\cdot)$ *represents the k-th multivariate moment of the indicated variable and* $\mathbf{K}_p$ *is the commutation matrix of order* $p \times p$, $\boldsymbol{\tau}_2 = \sqrt{\frac{\nu-2}{\nu}}\boldsymbol{\tau}$, *and* $\mathbf{U}_2 \sim \mathcal{TT}_q(-\boldsymbol{\tau}_2; \mathbf{0}_q, \bar{\boldsymbol{\Gamma}}, \nu-2)$.

Proof 135 *First, notice that* $\mathbf{Z}$ *can be written as* $\mathbf{Z} = \mathbf{A}\mathbf{U} + \mathbf{B}\mathbf{V}$, *with* $\mathbf{A} = \mathbf{H}$, $\mathbf{B} = \mathbf{I}_p$, *and* $\mathbf{V} = V_*^{1/2}\mathbf{W}_*$. *In this way, although in our case the vectors* $\mathbf{U}$ *and* $\mathbf{V} = V_*^{1/2}\mathbf{W}_*$ *are not independent since* V_* *depends on* $\mathbf{U}$ *through* $Q(\mathbf{U})$, *the moments up to the fourth order can be calculated by directly applying the formula obtained in Theorem 110 in Chapter 6, but replacing* $\mathbf{U}$ *with* $V_*^{1/2}\mathbf{U}$ *and* $\mathbf{V} = V_*^{1/2}\mathbf{W}_*$ *by* $\mathbf{W}_*$*; see directly the proof of Theorem 110. Thus, we find:*

$$
\begin{aligned}
\mathrm{E}(\mathbf{Z}\otimes\mathbf{Z}\mathbf{Z}^\top) =& (\mathbf{A}\otimes\mathbf{A})\boldsymbol{\mu}_3(\mathbf{U})\mathbf{A}^\top + (\mathbf{I}_{p^2}+\mathbf{K}_p)(\mathbf{A}\otimes\mathbf{B})\{\mathrm{E}(V_*\mathbf{U})\otimes\boldsymbol{\mu}_2(\mathbf{W}_*)\}\mathbf{B}^\top\\
&+(\mathbf{B}\otimes\mathbf{B})\mathrm{vec}\{\boldsymbol{\mu}_2(\mathbf{W}_*)\}\mathrm{E}(V_*\mathbf{U}^\top)\mathbf{A}^\top,\\
\mathrm{E}(\mathbf{Z}\mathbf{Z}^\top\otimes\mathbf{Z}\mathbf{Z}^\top) =& (\mathbf{A}\otimes\mathbf{A})\boldsymbol{\mu}_4(\mathbf{U})(\mathbf{A}\otimes\mathbf{A})^\top + (\mathbf{A}\otimes\mathbf{B})\{\mathrm{E}(V_*\mathbf{U}\mathbf{U}^\top)\otimes\boldsymbol{\mu}_2(\mathbf{W}_*)\}(\mathbf{A}^\top\otimes\mathbf{B}^\top)\\
&+(\mathbf{A}\otimes\mathbf{B})\{\mathrm{E}(V_*\mathbf{U}\mathbf{U}^\top)\otimes\boldsymbol{\mu}_2(\mathbf{W}_*)\}(\mathbf{A}^\top\otimes\mathbf{B}^\top)\mathbf{K}_p\\
&+\mathbf{K}_p(\mathbf{A}\otimes\mathbf{B})\{\mathrm{E}(V_*\mathbf{U}\mathbf{U}^\top)\otimes\boldsymbol{\mu}_2(\mathbf{W}_*)\}(\mathbf{A}^\top\otimes\mathbf{B}^\top)\\
&+\mathbf{K}_p(\mathbf{A}\otimes\mathbf{B})\{\mathrm{E}(V_*\mathbf{U}\mathbf{U}^\top)\otimes\boldsymbol{\mu}_2(\mathbf{W}_*)\}(\mathbf{A}^\top\otimes\mathbf{B}^\top)\mathbf{K}_p\\
&+(\mathbf{A}\otimes\mathbf{A})\mathrm{vec}\{\mathrm{E}(V_*\mathbf{U}\mathbf{U}^\top)\}\mathrm{vec}\{\boldsymbol{\mu}_2(\mathbf{W}_*)\}^\top(\mathbf{B}^\top\otimes\mathbf{B}^\top)\\
&+(\mathbf{B}\otimes\mathbf{B})\mathrm{vec}\{\boldsymbol{\mu}_2(\mathbf{W}_*)\}\mathrm{vec}\{\mathrm{E}(V_*\mathbf{U}\mathbf{U}^\top)\}^\top(\mathbf{A}^\top\otimes\mathbf{A}^\top)\\
&+(\mathbf{B}\otimes\mathbf{B})\boldsymbol{\mu}_2(V_*)\boldsymbol{\mu}_4(\mathbf{W}_*)(\mathbf{B}^\top\otimes\mathbf{B}^\top).
\end{aligned}
$$

The proof concludes by applying the results in Theorem 134 for the moments of $\mathbf{W}_*$ *and those in Theorem 115 for the moments of* $V_*^{1/2}\mathbf{U}$ *and after some algebra.* □

Example 100 *Let* $\mathbf{Z} = \mathbf{H}\mathbf{U} + \mathbf{W} \sim \mathcal{SUT}_{p,q}(\mathbf{0}_p, \boldsymbol{\Psi}, \mathbf{H}, \mathbf{0}_q, \mathbf{I}_q, \nu)$, *where* $\mathbf{U} \sim \mathcal{TT}_q(\mathbf{0}_q; \mathbf{0}_q, \mathbf{I}_q, \nu)$ *and* $\mathbf{W} \sim \mathcal{T}_p(\mathbf{0}_p, \boldsymbol{\Psi}, \nu)$ *which are uncorrelated but not independent. That is,* $(\mathbf{U}, \mathbf{W}) \stackrel{d}{=} V^{-1/2}(\tilde{\mathbf{U}}, \tilde{\mathbf{W}})$, *where* $V \sim \mathcal{G}amma(\nu/2, \nu/2)$, $\tilde{\mathbf{U}} \sim \mathcal{HN}_q(\mathbf{0}_q, \mathbf{I}_q)$, *and* $\tilde{\mathbf{W}} \sim \mathcal{N}_p(\mathbf{0}_p, \boldsymbol{\Psi})$ *which are all independent. Hence,* $\mathbf{Z} \stackrel{d}{=} V^{-1/2}\tilde{\mathbf{Z}}$, $\tilde{\mathbf{Z}} = \mathbf{H}\tilde{\mathbf{U}} + \tilde{\mathbf{W}} \sim \mathcal{SUN}_{p,q}(\mathbf{0}_p, \boldsymbol{\Psi}, \mathbf{H}, \mathbf{0}_q, \mathbf{I}_q)$ *and is independent of the scale variable* V. *In this case, by letting* $b = \sqrt{2/\pi}$ *and* $b_k(\nu) = \mathrm{E}(V^{-k/2}) = (\nu/2)^{k/2}\Gamma\{(\nu-k)/2\}/\Gamma(\nu/2)$ *for* $\nu > k \geq 0$, *then we have for* $\nu > k$ *that*

$$\boldsymbol{\mu}_k(\mathbf{U}) = b_k(\nu)\boldsymbol{\mu}_k(\tilde{\mathbf{U}}) \quad \textit{and} \quad \boldsymbol{\mu}_k(\mathbf{Z}) = b_k(\nu)\boldsymbol{\mu}_k(\tilde{\mathbf{Z}}),$$

where $\boldsymbol{\mu}_k(\tilde{\mathbf{U}})$ *and* $\boldsymbol{\mu}_k(\tilde{\mathbf{Z}})$ *can be obtained from Example 82 and Theorem 111 for* $k = 1, 2, 3, 4$.

7.5.3 *Mardia's measures of multivariate skewness and kurtosis*

Mardia's measure of multivariate skewness and kurtosis can also be computed for the SUT random vector $\mathbf{X} \sim \mathcal{SUT}_{p,q}(\boldsymbol{\xi}, \boldsymbol{\Psi}, \mathbf{H}, \boldsymbol{\tau}, \bar{\boldsymbol{\Gamma}}, \nu)$, for $\nu > 4$, in the same way as for the SUN distribution. That is, from a standardized version $\tilde{\mathbf{X}}_0$ of $\mathbf{X}$, namely

$$\tilde{\mathbf{X}}_0 = \mathbf{C}^{-1}(\mathbf{X} - \boldsymbol{\mu}) = \mathbf{C}^{-1}\mathbf{H}\mathbf{U}_0 + \sqrt{\frac{\nu + Q(\mathbf{U})}{\nu + q}}\,\mathbf{C}^{-1}\mathbf{W}_*, \tag{7.49}$$

where $\mathbf{U} \sim \mathcal{TT}_q(-\boldsymbol{\tau}; \mathbf{0}_q, \bar{\boldsymbol{\Gamma}}, \nu)$ and $\mathbf{W}_* \sim \mathcal{T}_p(\mathbf{0}_p, \boldsymbol{\Psi}, \nu + q)$ are independent, $Q(\mathbf{U}) = \mathbf{U}^\top \bar{\boldsymbol{\Gamma}}^{-1}\mathbf{U}$, $\boldsymbol{\mu} = \mathrm{E}(\mathbf{X}) = \boldsymbol{\xi} + \mathbf{H}\boldsymbol{\mu}_1(\mathbf{U})$, $\boldsymbol{\mu}_1(\mathbf{U}) = \mathrm{E}(\mathbf{U})$, $\mathbf{U}_0 = \mathbf{U} - \boldsymbol{\mu}_1(\mathbf{U})$, and $\mathbf{C}\mathbf{C}^\top = \boldsymbol{\Sigma} = \mathrm{var}(\mathbf{X})$; for instance, we can always take $\mathbf{C} = \boldsymbol{\Sigma}^{1/2}$, the symmetric square root of $\boldsymbol{\Sigma}$. Hence, since $\mathrm{E}(\tilde{\mathbf{X}}_0) = \mathbf{0}_p$ for $\nu > 1$ and $\mathrm{var}(\tilde{\mathbf{X}}_0) = \mathbf{I}_p$ for $\nu > 2$, the Mardia measures of multivariate skewness ($\beta_{1,p}$) and kurthosis ($\beta_{2,p}$) of the standardized random vector $\tilde{\mathbf{X}}_0$ can then be computed, for $\nu > 4$, from Theorem 112 in Chapter 6 by using the truncated t random vector $\mathbf{U} \sim \mathcal{TT}_q(-\boldsymbol{\tau}; \mathbf{0}_q, \bar{\boldsymbol{\Gamma}}, \nu)$ and the standardized SUT random vector $\tilde{\mathbf{X}}_0$ defined by (7.49). Thus, Mardia's measures of multivariate skewness and kurtosis are:

$$\begin{aligned}
\beta_{1,p} =& \mathrm{tr}\{(\mathbf{H}^\top\boldsymbol{\Sigma}^{-1}\mathbf{H} \otimes \mathbf{H}^\top\boldsymbol{\Sigma}^{-1}\mathbf{H})\boldsymbol{\mu}_3(\mathbf{U}_0)\mathbf{H}^\top\boldsymbol{\Sigma}^{-1}\mathbf{H}\boldsymbol{\mu}_3(\mathbf{U}_0)^\top\} \\
=& \mathrm{vec}\{\boldsymbol{\mu}_3(\mathbf{U}_0)\}^\top(\tilde{\mathbf{H}}^\top\tilde{\mathbf{H}} \otimes \tilde{\mathbf{H}}^\top\tilde{\mathbf{H}} \otimes \tilde{\mathbf{H}}^\top\tilde{\mathbf{H}})\mathrm{vec}\{\boldsymbol{\mu}_3(\mathbf{U}_0)\}, \\
\beta_{2,p} =& \mathrm{tr}\{(\mathbf{H}^\top\boldsymbol{\Sigma}^{-1}\mathbf{H} \otimes \mathbf{H}^\top\boldsymbol{\Sigma}^{-1}\mathbf{H})\boldsymbol{\mu}_4(\mathbf{U}_0)\} + 2\mathrm{tr}(\boldsymbol{\Sigma}_U\mathbf{H}^\top\boldsymbol{\Sigma}^{-1}\mathbf{H})\mathrm{tr}(\boldsymbol{\Psi}\boldsymbol{\Sigma}^{-1}) \\
&+ \mathrm{tr}(\boldsymbol{\Psi}\boldsymbol{\Sigma}^{-1})^2 + 4\mathrm{tr}(\boldsymbol{\Sigma}_U\mathbf{H}^\top\boldsymbol{\Sigma}^{-1}\boldsymbol{\Psi}\boldsymbol{\Sigma}^{-1}\mathbf{H}) + 2\mathrm{tr}(\boldsymbol{\Psi}\boldsymbol{\Sigma}^{-1}\boldsymbol{\Psi}\boldsymbol{\Sigma}^{-1}),
\end{aligned}$$

where $\boldsymbol{\Sigma} = \mathbf{H}\boldsymbol{\Sigma}_U\mathbf{H}^\top + \frac{T_q(\boldsymbol{\tau}_2; \mathbf{0}_q, \bar{\boldsymbol{\Gamma}}, \nu-2)}{T_q(\boldsymbol{\tau}; \mathbf{0}_q, \bar{\boldsymbol{\Gamma}}, \nu)}\left(\frac{\nu}{\nu-2}\right)\boldsymbol{\Psi}$, with $\boldsymbol{\Sigma}_U = \mathrm{var}(\mathbf{U})$ and $\boldsymbol{\tau}_2 = \sqrt{\frac{\nu-2}{\nu}}\,\boldsymbol{\tau}$, and $\boldsymbol{\mu}_3(\mathbf{U}_0)$ and $\boldsymbol{\mu}_4(\mathbf{U}_0)$ can be computed from $\boldsymbol{\mu}_k(\mathbf{U})$ for $k = 1, 2, 3, 4$ by developing the relationship $\boldsymbol{\mu}_k(\mathbf{U}_0) = \boldsymbol{\mu}_k(\mathbf{U} - \boldsymbol{\mu}_1(\mathbf{U}))$ for $k = 1, 2, 3, 4$.

Example 101 *Assume that* $\mathbf{X} \sim \mathcal{SUT}_{p,q}(\boldsymbol{\xi}, \boldsymbol{\Psi}, \mathbf{H}, \boldsymbol{\tau}, \bar{\boldsymbol{\Gamma}}, \nu)$ *with* $\boldsymbol{\tau} = \mathbf{0}_q$. *Then we have* $\boldsymbol{\mu}_k(\mathbf{U}) = b_k(\nu)\boldsymbol{\mu}_k(\tilde{\mathbf{U}})$, *where* $b_k(\nu) = \mathrm{E}(V^{-k/2}) = (\nu/2)^{k/2}\Gamma\{(\nu-k)/2\}/\Gamma(\nu/2)$ *for* $\nu > k \geq 0$ *and* $\tilde{\mathbf{U}} \sim \mathcal{HN}_q(\mathbf{0}_q, \bar{\boldsymbol{\Gamma}})$. *Hence,*

$$\boldsymbol{\Sigma} = \mathrm{var}(\mathbf{X}) = \mathbf{H}\boldsymbol{\Sigma}_U\mathbf{H}^\top + \frac{\nu}{\nu-2}\boldsymbol{\Psi}, \quad \textit{with } \boldsymbol{\Sigma}_U = \mathrm{var}(\mathbf{U}) = b_2(\nu)\boldsymbol{\mu}_2(\tilde{\mathbf{U}}) - b_1(\nu)^2\boldsymbol{\mu}_1(\tilde{\mathbf{U}})\boldsymbol{\mu}_1(\tilde{\mathbf{U}})^\top,$$

$$
\begin{aligned}
\boldsymbol{\mu}_3(\mathbf{U}_0) =& (\mathbf{I}_{q^2}+\mathbf{K}_q)\{b_1^3(\nu)\boldsymbol{\mu}_1(\tilde{\mathbf{U}})\otimes\boldsymbol{\mu}_1(\tilde{\mathbf{U}})\boldsymbol{\mu}_1(\tilde{\mathbf{U}})^\top - b_1(\nu)b_2(\nu)\boldsymbol{\mu}_1(\tilde{\mathbf{U}})\otimes\boldsymbol{\mu}_2(\tilde{\mathbf{U}})\} \\
& - b_1(\nu)b_2(\nu)\mathrm{vec}\{\boldsymbol{\mu}_2(\tilde{\mathbf{U}})\}\boldsymbol{\mu}_1(\tilde{\mathbf{U}})^\top + b_3(\nu)\boldsymbol{\mu}_3(\tilde{\mathbf{U}}), \\
\boldsymbol{\mu}_4(\mathbf{U}_0) =& b_1^4(\nu)\boldsymbol{\mu}_1(\tilde{\mathbf{U}})\boldsymbol{\mu}_1(\tilde{\mathbf{U}})\otimes\boldsymbol{\mu}_1(\tilde{\mathbf{U}})\boldsymbol{\mu}_1(\tilde{\mathbf{U}})^\top + (\mathbf{I}_{q^2}+\mathbf{K}_q)\{-2b_1^4(\nu)\boldsymbol{\mu}_1(\tilde{\mathbf{U}})\boldsymbol{\mu}_1(\tilde{\mathbf{U}})^\top \\
& \otimes\boldsymbol{\mu}_1(\tilde{\mathbf{U}})\boldsymbol{\mu}_1(\tilde{\mathbf{U}})^\top + b_1^2(\nu)b_2(\nu)\boldsymbol{\mu}_1(\tilde{\mathbf{U}})\boldsymbol{\mu}_1(\tilde{\mathbf{U}})^\top\otimes\boldsymbol{\mu}_2(\tilde{\mathbf{U}}) + b_1^2(\nu)b_2(\nu)\boldsymbol{\mu}_2(\tilde{\mathbf{U}})\otimes\boldsymbol{\mu}_1(\tilde{\mathbf{U}})\boldsymbol{\mu}_1(\tilde{\mathbf{U}})^\top \\
& + b_1(\nu)b_3(\nu)(-\boldsymbol{\mu}_1(\tilde{\mathbf{U}})\otimes\mathbf{I}_q)\boldsymbol{\mu}_3(\tilde{\mathbf{U}})^\top + b_1(\nu)b_3(\nu)\boldsymbol{\mu}_3(\tilde{\mathbf{U}})(-\boldsymbol{\mu}_1(\tilde{\mathbf{U}})^\top\otimes\mathbf{I}_q)\} \\
& + b_1^2(\nu)b_2(\nu)\mathrm{vec}\{\boldsymbol{\mu}_2(\tilde{\mathbf{U}})\}(\boldsymbol{\mu}_1(\tilde{\mathbf{U}})\otimes\boldsymbol{\mu}_1(\tilde{\mathbf{U}}))^\top + b_1(\nu)^2b_2(\nu)(\boldsymbol{\mu}_1(\tilde{\mathbf{U}})\otimes\boldsymbol{\mu}_1(\tilde{\mathbf{U}}))\mathrm{vec}\{\boldsymbol{\mu}_2(\tilde{\mathbf{U}})\}^\top \\
& + b_4(\nu)\boldsymbol{\mu}_4(\tilde{\mathbf{U}}).
\end{aligned}
$$

Further properties of the unified skew-*t* distribution with the AASUT parameterization can be found in Wang et al. (2024).

7.6 Exercises

Exercise 7.1 *Prove that the converse of part (iv) of Theorem 113 is also true; i.e.,* $Q(\mathbf{U}) \overset{d}{=} Q(\mathbf{W}_0) \Longleftrightarrow \boldsymbol{\tau} = \mathbf{0}_q$.

Exercise 7.2 *Let* $T_q(\mathbf{z};\mathbf{0}_q,\boldsymbol{\Gamma},\nu)$ *be the cdf of* $\mathbf{Z} \sim \mathcal{T}_q(\mathbf{0}_q,\boldsymbol{\Gamma},\nu)$. *If* $\boldsymbol{\Gamma} = \mathrm{diag}(\boldsymbol{\Gamma}_{11},\boldsymbol{\Gamma}_{22})$, *show that*

$$
T_{q_1+q_2}\left\{\begin{pmatrix}\mathbf{z}_1\\ \mathbf{0}_{q_2}\end{pmatrix};\begin{pmatrix}\mathbf{0}_{q_1}\\ \mathbf{0}_{q_2}\end{pmatrix},\begin{pmatrix}\boldsymbol{\Gamma}_{11} & \mathbf{O}\\ \mathbf{O} & \boldsymbol{\Gamma}_{22}\end{pmatrix},\nu\right\} = T_{q_1}(\mathbf{z}_1;\mathbf{0}_{q_1},\boldsymbol{\Gamma}_{11},\nu)T_{q_2}(\mathbf{0}_{q_2};\mathbf{0}_{q_2},\boldsymbol{\Gamma}_{22},\nu),\ \forall\mathbf{z}_1\in\mathbb{R}^{q_1}.
$$

Exercise 7.3 *In Theorem 126 suppose* $\bar{\boldsymbol{\Gamma}}_{12} = \mathbf{O}$. *Find the marginal pdf of* $\mathbf{U}_1$ *when* $\boldsymbol{\tau}_1 \neq \mathbf{0}_{q_1}$ *and* $\boldsymbol{\tau}_2 = \mathbf{0}_{q_2}$, *and the marginal pdf of* $\mathbf{U}_2$ *when* $\boldsymbol{\tau}_1 = \mathbf{0}_{q_1}$ *and* $\boldsymbol{\tau}_2 \neq \mathbf{0}_{q_2}$.

Exercise 7.4 *Assume*

$$
\begin{pmatrix}\mathbf{X}_1\\ \mathbf{X}_2\end{pmatrix} \sim \mathcal{SUT}_{p_1+p_2,q}\left\{\begin{pmatrix}\boldsymbol{\xi}_1\\ \boldsymbol{\xi}_2\end{pmatrix},\begin{pmatrix}\boldsymbol{\Psi}_{11} & \boldsymbol{\Psi}_{12}\\ \boldsymbol{\Psi}_{21} & \boldsymbol{\Psi}_{22}\end{pmatrix},\begin{pmatrix}\mathbf{H}_1\\ \mathbf{H}_2\end{pmatrix},\boldsymbol{\tau},\bar{\boldsymbol{\Gamma}},\nu\right\}
$$

to be nonsingular. Obtain the marginal pdf of $\mathbf{X}_1$ *and the conditional pdf of* $\mathbf{X}_2|\mathbf{X}_1 = \mathbf{x}_1$ *directly from (7.29) or (7.30).*

Exercise 7.5 *Find the cf of the multivariate SUT distribution. Hint: First find the cf of the multivariate symmetric t distribution; see Sutradhar (1986).*

Exercise 7.6 *Prove Theorem 122 using the cf.*

Exercise 7.7 *Prove Theorem 127 using the cf.*

Exercise 7.8 *Obtain the pdf of* $\mathbf{X} \sim \mathcal{SUT}_{p,q}(\boldsymbol{\xi}, \boldsymbol{\Psi}, \mathbf{H}, \boldsymbol{\tau}, \bar{\boldsymbol{\Gamma}}, \nu)$, $\boldsymbol{\Psi} > 0$, *from the selection stochastic representation.*

Exercise 7.9 *Obtain the cdf of* $\mathbf{X} \sim \mathcal{SUT}_{p,q}(\boldsymbol{\xi}, \boldsymbol{\Psi}, \mathbf{H}, \boldsymbol{\tau}, \bar{\boldsymbol{\Gamma}}, \nu)$, $\boldsymbol{\Psi} > 0$ *by directly integrating its pdf.*

Exercise 7.10 *Prove or disprove that the unified skew-t pdf is log-concave; see Exercise 6.10 for the unified skew-normal case.*

Exercise 7.11 *Let* $\mathbf{X}_1, \ldots, \mathbf{X}_n$ *be a random sample from* $\mathbf{X} \sim \mathcal{SUT}_{p,q}(\boldsymbol{\xi}, \boldsymbol{\Psi}, \mathbf{H}, \boldsymbol{\tau}, \bar{\boldsymbol{\Gamma}}, \nu)$. *Compute the following expectations and variances:* $\mathrm{E}(\bar{\mathbf{X}})$, $\mathrm{var}(\bar{\mathbf{X}})$, $\mathrm{E}(\mathbf{S})$, $\mathrm{var}(\mathbf{S})$.

Exercise 7.12 *Give an expression for the first 4 multivariate moments and Mardia's measures of multivariate skewness and kurtosis of* $\mathbf{X} \sim \mathcal{SUT}_{p,q}(\boldsymbol{\xi}, \boldsymbol{\Psi}, \mathbf{H}, \mathbf{0}_p, \mathbf{I}_q, \nu)$.

Exercise 7.13 *Give an expression for the first 4 multivariate moments and Mardia's measures of multivariate skewness and kurtosis of* $\mathbf{X} \sim \mathcal{SUT}_{p,q}(\boldsymbol{\xi}, \boldsymbol{\Psi}, \boldsymbol{\eta}\mathbf{1}_q^\top, \tau\mathbf{1}_q, (1-\rho)\mathbf{I}_q + \rho\mathbf{1}_q\mathbf{1}_q^\top, \nu)$, *with* $\boldsymbol{\eta} \in \mathbb{R}^p$, $\tau \in \mathbb{R}$, *and* $\rho \in (-(q-1)^{-1}, 1)$ *for* $q > 1$.

Exercise 7.14 *In Exercises 7.12 and 7.13 find* $\mathrm{var}(\mathbf{X}\mathbf{X}^\top)$.

Exercise 7.15 *In Exercises 7.12 and 7.13 partition* $\mathbf{X} = (\mathbf{X}_1^\top, \mathbf{X}_2^\top)^\top$ *and accordingly partition the respective parameters. Find the marginal distribution of* $\mathbf{X}_1$ *and the conditional distribution* $\mathbf{X}_2 \mid \mathbf{X}_1 = \mathbf{x}_1$.

Exercise 7.16 *Assume*

$$\begin{pmatrix}\mathbf{X}_1\\ \mathbf{X}_2\end{pmatrix} \sim \mathcal{SUT}_{p_1+p_2, q_1+q_2}\left\{\begin{pmatrix}\boldsymbol{\xi}_1\\ \boldsymbol{\xi}_2\end{pmatrix}, \begin{pmatrix}\boldsymbol{\Psi}_{11} & \mathbf{O}\\ \mathbf{O} & \boldsymbol{\Psi}_{22}\end{pmatrix}, \begin{pmatrix}\mathbf{H}_{11} & \mathbf{O}\\ \mathbf{O} & \mathbf{H}_{22}\end{pmatrix}, \begin{pmatrix}\boldsymbol{\tau}_1\\ \boldsymbol{\tau}_2\end{pmatrix}, \begin{pmatrix}\bar{\boldsymbol{\Gamma}}_{11} & \mathbf{O}\\ \mathbf{O} & \bar{\boldsymbol{\Gamma}}_{22}\end{pmatrix}, \nu\right\},$$

with $\boldsymbol{\Psi}_{jj} > 0$ *and* $\bar{\boldsymbol{\Gamma}}_{jj} > 0$ *for* $j = 1, 2$. *Find the marginal distribution of* $\mathbf{X}_1$ *and the conditional distribution* $\mathbf{X}_2 \mid \mathbf{X}_1 = \mathbf{x}_1$.

Exercise 7.17 *Let* $\mathbf{W} = S^{-1/2}\mathbf{V} \in \mathbb{R}^p$, *where the scale random variable* S *has a finite fourth moment and is independent of the components* $V_1, \ldots, V_p$ *of* $\mathbf{V}$, *which are iid random variables with mean* 0, *variance* 1, *skewness* γ_1 *and kurtosis* $\gamma_2 + 3$ *(see Exercise 5.6). Find* $\boldsymbol{\mu}_k(\mathbf{W})$ *for* $k = 1, 2, 3, 4$.

Exercise 7.18 *In Exercises 7.12 and 7.13 study the marginal and joint asymptotic distributions of the sample mean vector* $\bar{\mathbf{X}}$ *and sample variance matrix* $\mathbf{S}$ *of a random sample of size n.*

Exercise 7.19 *Extend the canonical form for the SUN distribution from Section 6.3.3 to the SUT distribution. For the case of rank $r = 1$, compare these results with those for the AASUT parameterization in Wang et al. (2024).*

7.7 Open Research Problems

Open Research Problem 14 *Let $\mathbf{X}_1, \ldots, \mathbf{X}_n$ be a random sample from $\mathbf{X} = (X_1, \mathbf{X}_2^\top)^\top \sim \mathcal{ST}_p(\mathbf{0}_p, \boldsymbol{\Psi}, \boldsymbol{\eta}, \nu)$.*

(a) Derive the asymptotic joint distribution of the sample mean vector and sample variance matrix;

(b) Derive the asymptotic distribution of $R_n^2 = \mathbf{S}_{21}^\top \mathbf{S}_{22}^{-1} \mathbf{S}_{21} / S_{11}$ which is defined in terms of the partition of $\mathbf{S}_n = \frac{1}{n} \sum_{i=1}^n \mathbf{X}_i \mathbf{X}_i^\top$ as

$$\mathbf{S}_n = \begin{pmatrix} S_{11} & \mathbf{S}_{21}^\top \\ \mathbf{S}_{21} & \mathbf{S}_{22} \end{pmatrix};$$

(c) Extend the previous results to the multivariate SUT distribution.

Open Research Problem 15 *Adapt the representations given in Remark 12 for the SUN distribution to the SUT distribution, and explore whether they yield identifiable SUT distributions.*

Open Research Problem 16 *Extend the multivariate SUNNY distribution introduced in Section 6.8.2 to a multivariate unified skew-t with new youth (SUTNY) distribution and redo Open Research Problem 8 for SUTNY. Moreover, derive the Mardia measures of multivariate skewness $\beta_{1,p}$ and multivariate kurtosis $\beta_{2,p}$.*

Open Research Problem 17 *Explore the entropy, mutual information, KL divergence, J divergence, and Wasserstein distance for two multivariate SUT distributions.*

Open Research Problem 18 *Develop the theory of sampling from the multivariate SUT distribution.*

Open Research Problem 19 *Develop maximum likelihood inference for the multivariate SUT distribution based on the EM algorithm. Pay attention to identifiability issues and consider submodels of interest.*

Open Research Problem 20 *Develop inference for the multivariate SUT distribution based on the factor analysis model (see Section 10.8) where the latent dimension q corresponds to the number of underlying factors.*

Chapter 8

Multivariate Unified Skew-Elliptical Distributions

In this chapter, we define the family of multivariate unified skew-elliptical (SUE) distributions, which constitute a rich and analytically tractable class of flexible models. Their main properties are discussed in detail. While specific members of the SUE family, such as the multivariate unified skew-normal (SUN) and skew-t (SUT), were thoroughly examined in Chapters 6 and 7, the broader SUE class is indexed by the same parameter sets for location and dispersion $(\boldsymbol{\xi}, \boldsymbol{\Psi})$, and for skewness and shape $(\mathbf{H}, \boldsymbol{\tau}, \bar{\boldsymbol{\Gamma}})$. Notably, unlike the SUN case, the SUE class regulates kurtosis through an additional parameter (or vector), allowing for both light and heavy-tailed distributions. Finally, although SUE distributions can be singular or nonsingular depending on the dispersion matrix $\boldsymbol{\Psi}$, primary attention in this chapter will be given to the nonsingular case.

8.1 Construction, Definition, and First Elementary Properties

8.1.1 Construction/definition

As is well known, within the multivariate symmetric elliptical class of distributions, the independence of the marginals characterizes the multivariate normal distribution. This means that only for the multivariate normal and hence for the induced SUN case, the independence between marginals is allowed. Since the unified skew-elliptical distributions preserve this property of their symmetric parents, we are forced to construct this family in a slightly different way than the multivariate unified skew-normal family.

We start by considering the symmetric $(p+q)$-dimensional random vector given by

$$\begin{pmatrix} \mathbf{W} \\ \mathbf{W}_0 \end{pmatrix} \sim \mathcal{E}l_{p+q}\left(\begin{pmatrix} \mathbf{0}_p \\ \mathbf{0}_q \end{pmatrix}, \begin{pmatrix} \boldsymbol{\Psi} & \mathbf{O} \\ \mathbf{O} & \bar{\boldsymbol{\Gamma}} \end{pmatrix}, \varphi\right), \tag{8.1}$$

which has uncorrelated (but not independent) marginals $\mathbf{W} \sim \mathcal{E}l_p(\mathbf{0}_p, \boldsymbol{\Psi}, \varphi)$ and $\mathbf{W}_0 \sim \mathcal{E}l_q(\mathbf{0}_q, \bar{\boldsymbol{\Gamma}}, \varphi)$, where $\mathcal{E}l_k(\boldsymbol{\mu}, \boldsymbol{\Sigma}, \varphi)$ denotes the multivariate elliptical distribution with $k \times 1$ location vector $\boldsymbol{\mu}$, $k \times k$ dispersion matrix $\boldsymbol{\Sigma}$, and characteristic generator function φ. As in the SUN case, we focus our analysis on the case where $\bar{\boldsymbol{\Gamma}} = (\bar{\gamma}_{ij})$ is a $q \times q$ symmetric and positive definite matrix, with entries $\bar{\gamma}_{ii} = 1$ for $i = 1, \ldots, q$ and $|\bar{\gamma}_{ij}| < 1$ for all $i \neq j$. That is, for a given $q \times q$ symmetric and positive definite matrix $\boldsymbol{\Gamma}$, $\bar{\boldsymbol{\Gamma}} = \boldsymbol{\gamma}^{-1}\boldsymbol{\Gamma}\boldsymbol{\gamma}^{-1}$, where $\boldsymbol{\gamma} = \mathrm{diag}(\boldsymbol{\Gamma})^{1/2}$. In other words, $\bar{\boldsymbol{\Gamma}}$ becomes a proper correlation matrix, which avoids redundant latent variables and/or identifiability problems due to scale changes in the components of $\mathbf{W}_0$. Furthermore, the condition $\bar{\boldsymbol{\Gamma}} > 0$ is necessary for $\mathbf{W}_0 \sim \mathcal{E}l_q(\mathbf{0}_q, \bar{\boldsymbol{\Gamma}}, \varphi)$ to be nonsingular and to have a pdf, while $\mathbf{W} \sim \mathcal{E}l_p(\mathbf{0}_p, \boldsymbol{\Psi}, \varphi)$ could be singular or nonsingular depending on whether we assume $\boldsymbol{\Psi} \geq 0$ or $\boldsymbol{\Psi} > 0$, respectively. Also, from (8.1) and the fact that the elliptical distributions are closed under affine transformations (see Section 4.2.3 on spherical and elliptical distributions), it follows that the marginal distribution of $\mathbf{W}_* = \mathbf{H}\mathbf{W}_0 + \mathbf{W}$ and its joint distribution with $\mathbf{W}_0$ are, respectively, given by

$$\mathbf{W}_* = \mathbf{H}\mathbf{W}_0 + \mathbf{W} = \begin{pmatrix} \mathbf{I}_p & \mathbf{H} \end{pmatrix} \begin{pmatrix} \mathbf{W} \\ \mathbf{W}_0 \end{pmatrix} \sim \mathcal{E}l_p(\mathbf{0}_p, \boldsymbol{\Psi} + \mathbf{H}\bar{\boldsymbol{\Gamma}}\mathbf{H}^\top, \varphi), \tag{8.2}$$

$$\begin{pmatrix} \mathbf{W}_* \\ \mathbf{W}_0 \end{pmatrix} = \begin{pmatrix} \mathbf{I}_p & \mathbf{H} \\ \mathbf{O} & \mathbf{I}_q \end{pmatrix} \begin{pmatrix} \mathbf{W} \\ \mathbf{W}_0 \end{pmatrix} \sim \mathcal{E}l_{p+q}\left(\begin{pmatrix} \mathbf{0}_p \\ \mathbf{0}_q \end{pmatrix}, \begin{pmatrix} \boldsymbol{\Psi} + \mathbf{H}\bar{\boldsymbol{\Gamma}}\mathbf{H}^\top & \mathbf{H}\bar{\boldsymbol{\Gamma}} \\ \bar{\boldsymbol{\Gamma}}\mathbf{H}^\top & \bar{\boldsymbol{\Gamma}} \end{pmatrix}, \varphi \right). \tag{8.3}$$

Note also that if $\boldsymbol{\Psi} > 0$, then $\boldsymbol{\Omega} = \boldsymbol{\Psi} + \mathbf{H}\bar{\boldsymbol{\Gamma}}\mathbf{H}^\top > 0$ since by assumption $\bar{\boldsymbol{\Gamma}} > 0$ always. This fact becomes a necessary condition for the above marginal and joint distributions to be nonsingular. Moreover, the corresponding conditional distributions are $(\mathbf{W}_* \mid \mathbf{W}_0 = \mathbf{w}_0) \sim \mathcal{E}l_p(\mathbf{H}\mathbf{w}_0, \boldsymbol{\Psi}, \varphi_{Q(\mathbf{w}_0)})$ and $(\mathbf{W}_0 \mid \mathbf{W}_* = \mathbf{w}_*) \sim \mathcal{E}l_q(\bar{\boldsymbol{\Gamma}}\mathbf{H}^\top\boldsymbol{\Omega}^{-1}\mathbf{w}_*, \bar{\boldsymbol{\Gamma}} - \bar{\boldsymbol{\Gamma}}\bar{\mathbf{H}}^\top\boldsymbol{\Omega}^{-1}\mathbf{H}\boldsymbol{\Gamma}, \varphi_{\tilde{Q}(\mathbf{w}_*)})$, where $Q(\mathbf{w}_0) = \mathbf{w}_0^\top\bar{\boldsymbol{\Gamma}}^{-1}\mathbf{w}_0$, $\tilde{Q}(\mathbf{w}_*) = \mathbf{w}_*^\top\boldsymbol{\Omega}^{-1}\mathbf{w}_*$ and φ_{a^2} is a conditional characteristic generator (cgf).

Using (8.2) and (8.3), the construction of a multivariate SUE distribution can be obtained directly from a selection point of view (see Chapter 9) as follows.

Definition 21 [Multivariate unified skew-elliptical distributions] *A random vector $\mathbf{X} \in \mathbb{R}^p$ is said to have a multivariate unified skew-elliptical distribution with location parameter $\boldsymbol{\xi} \in \mathbb{R}^p$, p-by-p dispersion matrix $\boldsymbol{\Psi}$, characteristic generator function φ and shape/skewness parameters $\mathbf{H} \in \mathbb{R}^{p\times q}$, $\boldsymbol{\tau} \in \mathbb{R}^q$ and $\bar{\boldsymbol{\Gamma}} > 0$, a q-by-q positive definite correlation matrix, in symbols $\mathbf{X} \sim \mathcal{SUE}_{p,q}(\boldsymbol{\xi}, \boldsymbol{\Psi}, \mathbf{H}, \boldsymbol{\tau}, \bar{\boldsymbol{\Gamma}}, \varphi)$, iff $\mathbf{X}$ can be represented as follows:*

$$\mathbf{X} = \boldsymbol{\xi} + \mathbf{X}_*, \tag{8.4}$$

where $\mathbf{X}_* = (\mathbf{W}_* \mid \mathbf{W}_0 + \boldsymbol{\tau} > \mathbf{0}_q)$ is the selection version of $\mathbf{W}_* = \mathbf{H}\mathbf{W}_0 + \mathbf{W}$ under the selection condition $\mathbf{W}_0 + \boldsymbol{\tau} > \mathbf{0}_q$, with $\mathbf{W}$ and $\mathbf{W}_0$ jointly distributed according to the multivariate elliptical distribution given by (8.1).

However, since $\mathbf{U} = (\mathbf{W}_0 | \mathbf{W}_0 + \boldsymbol{\tau} > \mathbf{0}_q) \sim \mathcal{TE}l_q(-\boldsymbol{\tau}; \mathbf{0}_q, \bar{\boldsymbol{\Gamma}}, \varphi)$, where $\mathcal{TE}l_q(\mathbf{c}; \boldsymbol{\mu}, \boldsymbol{\Sigma}, \varphi)$ denotes the *truncated below* **c** *multivariate elliptical distributions* $\mathcal{E}l_q(\boldsymbol{\mu}, \boldsymbol{\Sigma}, \varphi)$, and we know that $Q(\mathbf{U}) = \mathbf{U}^\top\bar{\boldsymbol{\Gamma}}^{-1}\mathbf{U} \overset{d}{=} (Q(\mathbf{W}_0) | \mathbf{W}_0 + \boldsymbol{\tau} > \mathbf{0}_q)$ and by (8.1) $(\mathbf{W} \mid \mathbf{W}_0 = \mathbf{w}_0) \sim \mathcal{E}l_p(\mathbf{0}_p, \boldsymbol{\Psi}, \varphi_{Q(\mathbf{w}_0)})$, which means that $\mathbf{W} \perp\!\!\!\perp \mathbf{W}_0 \big| Q(\mathbf{W}_0)$, we have

$$\mathbf{X}_* = (\mathbf{W}_* \mid \mathbf{W}_0 + \boldsymbol{\tau} > \mathbf{0}_q) = \{(\mathbf{H}\mathbf{W}_0 + \mathbf{W}) | \mathbf{W}_0 + \boldsymbol{\tau} > \mathbf{0}_q\} \overset{d}{=} \mathbf{H}\mathbf{U} + \mathbf{W}_{Q(\mathbf{U})},$$

where $\mathbf{W}_{Q(\mathbf{U})} | \mathbf{U} = \mathbf{u} \sim \mathcal{E}l_p(\mathbf{0}_p, \boldsymbol{\Psi}, \varphi_{Q(\mathbf{u})})$, which is conditionally independent of $\mathbf{U}$ given $Q(\mathbf{U})$. Thus, (8.4) can also be represented as

$$\mathbf{X} \overset{d}{=} \boldsymbol{\xi} + \mathbf{H}\mathbf{U} + \mathbf{W}_{Q(\mathbf{U})}, \tag{8.5}$$

where $\mathbf{U} \perp\!\!\!\perp \mathbf{W}_{Q(\mathbf{U})} | Q(\mathbf{U})$, with $\mathbf{U} \sim \mathcal{TE}l_q\left(-\boldsymbol{\tau}; \mathbf{0}_q, \bar{\boldsymbol{\Gamma}}, \varphi\right)$ and $\mathbf{W}_{Q(\mathbf{U})} | Q(\mathbf{U}) \sim \mathcal{E}l_p(\mathbf{0}_p, \boldsymbol{\Psi}, \varphi_{Q(\mathbf{U})})$. Hierarchically, (8.5) becomes

$$\mathbf{X} | \mathbf{U} = \mathbf{u} \sim \mathcal{E}l_p\left(\boldsymbol{\xi} + \mathbf{H}\mathbf{u}, \boldsymbol{\Psi}, \varphi_{Q(\mathbf{u})}\right), \tag{8.6}$$

$$\mathbf{U} \sim \mathcal{TE}l_q\left(-\boldsymbol{\tau}; \mathbf{0}_q, \bar{\boldsymbol{\Gamma}}, \varphi\right). \tag{8.7}$$

The representations given in (8.5) and (8.6)-(8.7) can be used as equivalent definitions of the SUE family. Perhaps (8.5) is more appropriate in the sense that it is more similar to the way the SUN and SUT distributions were defined in Chapters 6 and 7.

Two important and well-known members of the SUE family are the SUN and SUT classes, which were thoroughly studied in Chapters 6 and 7, respectively. In fact, for the SUN distribution in (8.5), we have $\mathbf{W}_{Q(\mathbf{U})} = \mathbf{W} \sim \mathcal{N}_p(\mathbf{0}_p, \mathbf{\Psi}) \perp\!\!\!\perp \mathbf{U} \sim \mathcal{TN}_q(-\boldsymbol{\tau}; \mathbf{0}_q, \bar{\mathbf{\Gamma}})$, so that

$$\begin{aligned} \mathbf{X}|\mathbf{U} = \mathbf{u} &\sim \mathcal{N}_p(\boldsymbol{\xi} + \mathbf{H}\mathbf{u}, \mathbf{\Psi}), \\ \mathbf{U} &\sim \mathcal{TN}_q(-\boldsymbol{\tau}; \mathbf{0}_q, \bar{\mathbf{\Gamma}}). \end{aligned}$$

For the SUT distribution, $\mathbf{W}_{Q(\mathbf{U})} = \sqrt{\frac{\nu+Q(\mathbf{U})}{\nu+q}}\mathbf{W}_*$, where $\mathbf{W}_* \sim \mathcal{T}_p(\mathbf{0}_p, \mathbf{\Psi}, \nu+q) \perp\!\!\!\perp \mathbf{U} \sim \mathcal{TT}_q(-\boldsymbol{\tau}; \mathbf{0}_q, \bar{\mathbf{\Gamma}}, \nu)$, and thus $\mathbf{W}_{Q(\mathbf{U})}|\mathbf{U} = \mathbf{u} \sim \mathcal{T}_p\left(\mathbf{0}_p, \frac{\nu+Q(\mathbf{u})}{\nu+q}\mathbf{\Psi}, \nu+q\right)$. That is,

$$\begin{aligned} \mathbf{X}|\mathbf{U} = \mathbf{u} &\sim \mathcal{T}_p\left(\boldsymbol{\xi} + \mathbf{H}\mathbf{u}, \frac{\nu+Q(\mathbf{u})}{\nu+q}\mathbf{\Psi}, \nu+q\right), \\ \mathbf{U} &\sim \mathcal{TT}_q(-\boldsymbol{\tau}; \mathbf{0}_q, \bar{\mathbf{\Gamma}}, \nu). \end{aligned}$$

Remark 20 **(Link between $\mathcal{SUE}_{p,q}$ and $\mathcal{AASUE}_{p,q}$)** *It is important to note that the (nonsingular) multivariate unified skew-elliptical distribution $\mathcal{SUE}_{p,q}(\boldsymbol{\xi}, \mathbf{\Psi}, \mathbf{H}, \boldsymbol{\tau}, \bar{\mathbf{\Gamma}}, \varphi)$ is completely equivalent to the multivariate unified skew-elliptical distribution arising from the parameterization described by Arellano-Valle and Azzalini (2006), here denoted by $\mathcal{AASUE}_{p,q}(\boldsymbol{\xi}, \mathbf{\Omega}, \mathbf{\Delta}, \boldsymbol{\tau}, \bar{\mathbf{\Gamma}}, \varphi)$, except for differences in parameterization. Moreover, since both parameterizations use the same φ, the link between $\mathcal{SUE}_{p,q}$ and $\mathcal{AASUE}_{p,q}$ is the same as the one given in Remark 13.*

Remark 21 *Many properties of unified skew-elliptical distributions have been studied by Arellano-Valle and Genton (2010c) in a third parameterization which is different from the SUE and AASUE parameterizations, but they are all equivalent by one-to-one transformations.*

When $q = 1$, the SUE distributions are called *multivariate extended skew-elliptical (ESE) distributions* and denoted by $\mathbf{X} \sim \mathcal{ESE}_p(\boldsymbol{\xi}, \mathbf{\Psi}, \boldsymbol{\eta}, \tau, \varphi)$; see, e.g., Arellano-Valle and Genton (2010c). Recall that a similar fact occurs with the SUN and SUT distributions, from which come the extended skew-normal (ESN) and extended skew-*t* (EST) distributions, respectively, when $q = 1$.

If, in addition to $q = 1$, we also set $\tau = 0$, then we obtain the *multivariate skew-elliptical (SE) distributions* denoted by $\mathbf{X} \sim \mathcal{SE}_p(\boldsymbol{\xi}, \mathbf{\Psi}, \boldsymbol{\eta}, \varphi)$ or alternatively $\mathbf{X} \sim \mathcal{ASE}_p(\boldsymbol{\xi}, \mathbf{\Omega}, \boldsymbol{\alpha}, \varphi)$; see, e.g., Branco and Dey (2001, 2002), Azzalini and Capitanio (2003), and Liu and Dey (2004). The SE family includes the skew-normal (SN) and skew-*t* (ST) important subfamilies; see, e.g., Azzalini and Dalla Valle (1996) and Azzalini and Capitanio (1999, 2003). Various statistics and their properties have been studied under skew-elliptical distributions by Fang (2005e,c,b,a, 2006); see also Yin and Balakrishnan (2024) for recent results.

Finally, as will be formally proved later, when $\mathbf{H} = \mathbf{O}$ and $\boldsymbol{\tau} = \mathbf{0}_q$, the multivariate symmetric elliptical family is recovered, from which the multivariate normal and multivariate *t* distributions are important special cases. Note that in the SUN case, the additional condition $\boldsymbol{\tau} = \mathbf{0}_q$ is not necessary to recover the symmetric normal parent distribution.

Remark 22 *Note that if we set $\mathbf{W}_0 = \mathbf{W}_{0*} + \mathbf{\Lambda}\mathbf{W}_*$, where $\mathbf{\Lambda} = \bar{\mathbf{\Gamma}}\mathbf{H}^\top\mathbf{\Omega}^{-1}$, then it is easy to see from (8.3) that $(\mathbf{W}_*^\top, \tilde{\mathbf{W}}_{0*}^\top)^\top \sim \mathcal{E}l_{p+q}((\mathbf{0}_p^\top, \mathbf{0}_q^\top)^\top, \mathrm{diag}(\mathbf{\Omega}, \bar{\mathbf{\Gamma}} - \mathbf{\Lambda}\mathbf{\Lambda}^\top), \varphi)$ and from (8.4) we find that the SUE random vector could also be defined as*

$$\mathbf{X} - \boldsymbol{\xi} = (\mathbf{W}_* \mid \mathbf{W}_{0*} < \mathbf{\Lambda}\mathbf{W}_* + \boldsymbol{\tau}).$$

8.1.2 Identifiability

The SUE distribution is non-identifiable with respect to permutations of the components of $\mathbf{W}_0 + \boldsymbol{\tau} \sim \mathcal{E}l_q(\boldsymbol{\tau}, \bar{\boldsymbol{\Gamma}}, \varphi)$ if $q > 1$. Let $\mathcal{P}_q = \{\mathbf{P} \in \mathbb{R}^{q\times q} : \mathbf{P}\mathbf{P}^\top = \mathbf{P}^\top\mathbf{P} = \mathbf{I}_q \text{ and } \mathbf{P}\mathbf{1}_q = \mathbf{1}_q\}$ be the set of $q \times q$ permutation matrices. This lack of identifiability arises because the selection events $\{\mathbf{W}_0 + \boldsymbol{\tau} > \mathbf{0}_q\}$ and $\{\mathbf{A}\mathbf{W}_0 + \mathbf{A}\boldsymbol{\tau} > \mathbf{0}_q\}$ are equivalent for all nonsingular matrices of the form $\mathbf{A} = \mathbf{P}\mathbf{D}$, where $\mathbf{P} \in \mathcal{P}_q$ and $\mathbf{D} > 0$ is a diagonal matrix. Consequently, the selection vectors $\mathbf{X}_* = (\mathbf{W}_* \mid \mathbf{W}_0 + \boldsymbol{\tau} > \mathbf{0}_q)$ and $\mathbf{X}_{*A} = (\mathbf{W}_* \mid \mathbf{A}\mathbf{W}_0 + \mathbf{A}\boldsymbol{\tau} > \mathbf{0}_q)$ share the same distribution for any such $\mathbf{A}$. However, since $\mathbf{A}\mathbf{W}_0 + \mathbf{A}\boldsymbol{\tau} \sim \mathcal{E}l_q(\mathbf{A}\boldsymbol{\tau}, \mathbf{A}\bar{\boldsymbol{\Gamma}}\mathbf{A}^\top, \varphi)$, the requirement that $\mathbf{A}\bar{\boldsymbol{\Gamma}}\mathbf{A}^\top$ be a correlation matrix forces $\mathbf{D} = \mathbf{I}_q$, and thus $\mathbf{A} = \mathbf{P}$. In particular, the random vector $\mathbf{X} \sim \mathcal{SUE}_{p,q}(\boldsymbol{\xi}, \boldsymbol{\Psi}, \mathbf{H}, \boldsymbol{\tau}, \bar{\boldsymbol{\Gamma}}, \varphi)$ is equal in distribution to $\mathbf{X}_P \sim \mathcal{SUE}_{p,q}(\boldsymbol{\xi}, \boldsymbol{\Psi}, \mathbf{H}\mathbf{P}^\top, \mathbf{P}\boldsymbol{\tau}, \mathbf{P}\bar{\boldsymbol{\Gamma}}\mathbf{P}^\top, \varphi)$. This property is inherited by all members of the SUE family, including the SUN and SUT distributions, as first noted by Wang et al. (2023) for the SUN case. Another consequence is that if we set $\boldsymbol{\Gamma} = \boldsymbol{\gamma}\bar{\boldsymbol{\Gamma}}\boldsymbol{\gamma}$, where $\boldsymbol{\gamma} > 0$ is a $q \times q$ diagonal matrix, the SUE distribution can be represented as $\mathcal{SUE}_{p,q}(\boldsymbol{\xi}, \boldsymbol{\Psi}, \mathbf{H}, \boldsymbol{\tau}, \bar{\boldsymbol{\Gamma}}, \varphi)$ or as $\mathcal{SUE}_{p,q}(\boldsymbol{\xi}, \boldsymbol{\Psi}, \mathbf{H}\boldsymbol{\gamma}, \boldsymbol{\gamma}\boldsymbol{\tau}, \boldsymbol{\Gamma}, \varphi)$.

Such non-identifiability is problematic for parameter inference, as the resulting optimization surface may possess multiple peaks, potentially leading algorithms to erroneous stopping values and rendering the SUE family non-applicable in practice. To resolve this issue, several strategies can be employed:

1. Ordering $\boldsymbol{\tau}$: Ranking the components τ_i in a strictly increasing (or decreasing) order. Note that this does not resolve the case where $\boldsymbol{\tau} = \mathbf{0}_q$.
2. Ordering the conditional dispersion: Assuming that $\tilde{\boldsymbol{\Gamma}} = \bar{\boldsymbol{\Gamma}} - \bar{\boldsymbol{\Gamma}}\mathbf{H}^\top\boldsymbol{\Omega}^{-1}\mathbf{H}\boldsymbol{\Gamma}$ is diagonal with its elements in a fixed order.
3. Ordering the columns of $\mathbf{H}$: Following the standard convention in factor analysis (see, e.g., Jolliffe, 2002), the columns $\mathbf{h}_i$ of $\mathbf{H}$ can be ordered according to their Euclidean norms, requiring $\|\mathbf{h}_1\|_2 > \|\mathbf{h}_2\|_2 > \cdots > \|\mathbf{h}_q\|_2$. This eliminates permutation ambiguity even when $\boldsymbol{\tau} = \mathbf{0}_q$.
4. Structural constraints: Eliminating the ordering flexibility by imposing relationships such as $\mathbf{H} = \eta\boldsymbol{\Psi}^{1/2}$ for $\eta \in \mathbb{R}$, or assuming that $\bar{\boldsymbol{\Gamma}}$ is an equicorrelation matrix with $\mathbf{H}$ having constant rows.

Example 102 *The following are two examples of identifiable SUE distributions:*

(a) $\mathcal{SUE}_{p,q}(\boldsymbol{\xi}, \boldsymbol{\Psi}, \boldsymbol{\eta}\mathbf{1}_q^\top, \tau\mathbf{1}_q, (1-\rho)\mathbf{I}_q + \rho\mathbf{1}_q\mathbf{1}_q^\top, \varphi)$*, where* $\boldsymbol{\tau} = \tau\mathbf{1}_q$ $(\tau \in \mathbb{R})$, $\mathbf{H} = \boldsymbol{\eta}\mathbf{1}_q^\top$ $(\boldsymbol{\eta} \in \mathbb{R}^p)$*, and* $\bar{\boldsymbol{\Gamma}} = (1-\rho)\mathbf{I}_q + \rho\mathbf{1}_q\mathbf{1}_q^\top$ *with* $\rho \in (-(q-1)^{-1}, 1)$*. In this case, if* $\rho \approx 1$*, the model reduces toward an SE distribution.*

(b) $\mathcal{SUE}_{p,q}(\boldsymbol{\xi}, \boldsymbol{\Psi}, \mathbf{H}, \alpha\mathbf{1}_q + \beta\mathbf{j}_q, \bar{\boldsymbol{\Gamma}}, \varphi)$*, where* $\boldsymbol{\tau} = \alpha\mathbf{1}_q + \beta\mathbf{j}_q$ *for* $\alpha, \beta \in \mathbb{R}$ $(\beta \neq 0)$ *and* $\mathbf{j}_q = (1, \ldots, q)^\top$.

8.1.3 Cdf and cf

As shown below, from Definition 21 it is very easy to obtain the cdf of a SUE random vector.

Theorem 136 *The cdf of* $\mathbf{X} \sim \mathcal{SUE}_{p,q}(\boldsymbol{\xi}, \boldsymbol{\Psi}, \mathbf{H}, \boldsymbol{\tau}, \bar{\boldsymbol{\Gamma}}, \varphi)$*, with* $\operatorname{rank}(\boldsymbol{\Psi}) = p$ *and* $\operatorname{rank}(\bar{\boldsymbol{\Gamma}}) = q$*, is*

$$F(\mathbf{x}) = \frac{1}{F_q(\boldsymbol{\tau}; \mathbf{0}_q, \bar{\boldsymbol{\Gamma}}, \varphi)} F_{p+q}\left\{\begin{pmatrix} \mathbf{x} \\ \boldsymbol{\tau} \end{pmatrix}; \begin{pmatrix} \boldsymbol{\xi} \\ \mathbf{0}_q \end{pmatrix}, \begin{pmatrix} \boldsymbol{\Psi} + \mathbf{H}\bar{\boldsymbol{\Gamma}}\mathbf{H}^\top & -\mathbf{H}\bar{\boldsymbol{\Gamma}} \\ -\bar{\boldsymbol{\Gamma}}\mathbf{H}^\top & \bar{\boldsymbol{\Gamma}} \end{pmatrix}, \varphi\right\}, \quad \mathbf{x} \in \mathbb{R}^p, \quad (8.8)$$

where $F_k(\cdot; \boldsymbol{\mu}, \boldsymbol{\Sigma}, \varphi)$ *denotes the cdf of a k-variate elliptical distribution with location vector* $\boldsymbol{\mu}$*, dispersion matrix* $\boldsymbol{\Sigma}$*, and characteristic generator* φ*.*

Proof 136 *From (8.4):*

$$\begin{aligned}
F(\mathbf{x}) &= \mathrm{P}(\mathbf{X} \le \mathbf{x}) \\
&= \mathrm{P}(\mathbf{X}_* \le \mathbf{x} - \boldsymbol{\xi}) \\
&= \mathrm{P}(\mathbf{W}_* \le \mathbf{x} - \boldsymbol{\xi} \mid \mathbf{W}_0 + \boldsymbol{\tau} > \mathbf{0}_q) \\
&= \frac{\mathrm{P}\{\mathbf{W}_* \le \mathbf{x} - \boldsymbol{\xi}, -\mathbf{W}_0 < \boldsymbol{\tau}\}}{\mathrm{P}\{-\mathbf{W}_0 < \boldsymbol{\tau}\}} \\
&= \frac{1}{F_q(\boldsymbol{\tau}; \mathbf{0}_q, \bar{\boldsymbol{\Gamma}}, \varphi)} F_{p+q}\left\{ \begin{pmatrix} \mathbf{x} - \boldsymbol{\xi} \\ \boldsymbol{\tau} \end{pmatrix}; \begin{pmatrix} \mathbf{0}_p \\ \mathbf{0}_q \end{pmatrix}, \begin{pmatrix} \boldsymbol{\Psi} + \mathbf{H}\bar{\boldsymbol{\Gamma}}\mathbf{H}^\top & -\mathbf{H}\bar{\boldsymbol{\Gamma}} \\ -\bar{\boldsymbol{\Gamma}}\mathbf{H}^\top & \bar{\boldsymbol{\Gamma}} \end{pmatrix}, \varphi \right\}, \quad \mathbf{x} \in \mathbb{R}^p,
\end{aligned}$$

from which the proof follows. □

Next, we prove that if $\boldsymbol{\tau} = \mathbf{0}_q$ and all the entries in the matrix $\mathbf{H}$ are zero, $\mathbf{H} = \mathbf{O}$ say, then in Definition 21, we obtain an $\mathcal{E}l_p(\boldsymbol{\xi}, \boldsymbol{\Psi}, \varphi)$ distribution for the SUE random vector $\mathbf{X}$. To show this fact formally, we need the following preliminary result.

Theorem 137 *The joint cdf of the elliptical random vectors* $\mathbf{W}$ *and* $\mathbf{W}_0$ *defined in (8.1), with* $\operatorname{rank}(\boldsymbol{\Psi}) = p$ *and* $\operatorname{rank}(\bar{\boldsymbol{\Gamma}}) = q$, *is such that*

$$F_{p+q}\left\{ \begin{pmatrix} \mathbf{w} \\ \mathbf{0}_q \end{pmatrix}; \begin{pmatrix} \mathbf{0}_p \\ \mathbf{0}_q \end{pmatrix}, \begin{pmatrix} \boldsymbol{\Psi} & \mathbf{O} \\ \mathbf{O} & \bar{\boldsymbol{\Gamma}} \end{pmatrix}, \varphi \right\} = F_p(\mathbf{w}; \mathbf{0}_p, \boldsymbol{\Psi}, \varphi) F_q(\mathbf{0}_q; \mathbf{0}_q, \bar{\boldsymbol{\Gamma}}, \varphi), \quad \forall \mathbf{w} \in \mathbb{R}^p, \tag{8.9}$$

where $F_p(\mathbf{w}; \mathbf{0}_p, \boldsymbol{\Psi}, \varphi)$ *and* $F_q(\mathbf{w}_0; \mathbf{0}_q, \bar{\boldsymbol{\Gamma}}, \varphi)$ *are the marginal cdfs of* $\mathbf{W}$ *and of* $\mathbf{W}_0$*, respectively.*

Proof 137 *Since* $(\mathbf{W}, \mathbf{W}_0) \overset{d}{=} (R\mathbf{A}\mathbf{U}^{(p)}, R_0\mathbf{A}_0\mathbf{U}^{(q)})$*, where* $\mathbf{A}\mathbf{A}^\top = \boldsymbol{\Psi}$*, with* $\operatorname{rank}(\mathbf{A}) = p$ *and* $\mathbf{A}_0\mathbf{A}_0^\top = \bar{\boldsymbol{\Gamma}}$*, with* $\operatorname{rank}(\mathbf{A}_0) = q$*, and* (R, R_0) *are (nonnegative) radial variables independent of* $(\mathbf{U}^{(p)}, \mathbf{U}^{(q)})$*, which are independent uniform random vectors on the p-dimensional and q-dimensional unit spheres, respectively, we have that*

$$\begin{aligned}
\mathrm{P}(\mathbf{W} \le \mathbf{w}, \mathbf{W}_0 \le \mathbf{0}_q) &= \mathrm{P}(R\mathbf{A}\mathbf{U}^{(p)} \le \mathbf{w}, R_0\mathbf{A}_0\mathbf{U}^{(q)} \le \mathbf{0}_q) \\
&= \mathrm{E}\{\mathrm{P}(R\mathbf{A}\mathbf{U}^{(p)} \le \mathbf{w}, R_0\mathbf{A}_0\mathbf{U}^{(q)} \le \mathbf{0}_q \mid R_0)\} \\
&= \mathrm{E}\{\mathrm{P}(R\mathbf{A}\mathbf{U}^{(p)} \le \mathbf{w}, \mathbf{A}_0\mathbf{U}^{(q)} \le \mathbf{0}_q \mid R_0)\} \\
&= \mathrm{E}\{\mathrm{P}(R\mathbf{A}\mathbf{U}^{(p)} \le \mathbf{w} \mid R_0)\mathrm{P}(\mathbf{A}_0\mathbf{U}^{(q)} \le \mathbf{0}_q \mid R_0)\} \\
&= \mathrm{E}\{\mathrm{P}(R\mathbf{A}\mathbf{U}^{(p)} \le \mathbf{w} \mid R_0)\mathrm{P}(\mathbf{A}_0\mathbf{U}^{(q)} \le \mathbf{0}_q)\} \\
&= \mathrm{E}\{\mathrm{P}(R\mathbf{A}\mathbf{U}^{(p)} \le \mathbf{w} \mid R_0)\}\mathrm{P}(\mathbf{A}_0\mathbf{U}^{(q)} \le \mathbf{0}_q),
\end{aligned}$$

which yields the proof since $\mathrm{E}\{\mathrm{P}(R\mathbf{A}\mathbf{U}^{(p)} \le \mathbf{w} \mid R_0)\} = \mathrm{P}(R\mathbf{A}\mathbf{U}^{(p)} \le \mathbf{w}) = \mathrm{P}(\mathbf{W} \le \mathbf{w})$ *and* $\mathrm{P}(\mathbf{A}_0\mathbf{U}^{(q)} \le \mathbf{0}_q) = \mathrm{E}\{\mathrm{P}\{R_0\mathbf{A}_0\mathbf{U}^{(q)} \le \mathbf{0}_q \mid R_0)\} = \mathrm{P}(R_0\mathbf{A}_0\mathbf{U}^{(q)} \le \mathbf{0}_q) = \mathrm{P}(\mathbf{W}_0 \le \mathbf{0}_q)$. □

As a consequence of this last result, we can formalize the fact that the SUE distribution becomes its symmetric elliptical parent when both $\boldsymbol{\tau} = \mathbf{0}_q$ and $\mathbf{H} = \mathbf{O}$.

Theorem 138 *Let* $\mathbf{X} \sim \mathcal{SUE}_{p,q}(\boldsymbol{\xi}, \boldsymbol{\Psi}, \mathbf{H}, \boldsymbol{\tau}, \bar{\boldsymbol{\Gamma}}, \varphi)$*, with* $\operatorname{rank}(\boldsymbol{\Psi}) = p$ *and* $\operatorname{rank}(\bar{\boldsymbol{\Gamma}}) = q$*. If* $\boldsymbol{\tau} = \mathbf{0}_q$ *and* $\mathbf{H} = \mathbf{O}$*, then* $\mathbf{X} \sim \mathcal{E}l_p(\boldsymbol{\xi}, \boldsymbol{\Psi}, \varphi)$.

Proof 138 *Using the assumption that $\boldsymbol{\tau} = \mathbf{0}_q$ and $\mathbf{H} = \mathbf{O}$ in Equation (8.8) of Theorem 136 and then the result in Equation (8.9) of Theorem 137, we find that*

$$\begin{aligned} F(\mathbf{x}) &= \frac{1}{F_q(\mathbf{0}_q;\mathbf{0}_q,\bar{\boldsymbol{\Gamma}},\varphi)} F_{p+q}\left\{\begin{pmatrix} \mathbf{x} \\ \mathbf{0}_q \end{pmatrix};\begin{pmatrix} \boldsymbol{\xi} \\ \mathbf{0}_q \end{pmatrix},\begin{pmatrix} \boldsymbol{\Psi} & \mathbf{O} \\ \mathbf{O} & \bar{\boldsymbol{\Gamma}} \end{pmatrix},\varphi\right\} \\ &= \frac{1}{F_q(\mathbf{0}_q;\mathbf{0}_q,\bar{\boldsymbol{\Gamma}},\varphi)} F_p(\mathbf{x};\boldsymbol{\xi},\boldsymbol{\Psi},\varphi)F_q(\mathbf{0}_q;\mathbf{0}_q,\bar{\boldsymbol{\Gamma}},\varphi) \\ &= F_p(\mathbf{x};\boldsymbol{\xi},\boldsymbol{\Psi},\varphi), \quad \forall \mathbf{x} \in \mathbb{R}^p, \end{aligned}$$

which is just the cdf of $\mathbf{W} \sim \mathcal{E}l_p(\boldsymbol{\xi},\boldsymbol{\Psi},\varphi)$. Hence, for $\boldsymbol{\tau} = \mathbf{0}_q$ and $\mathbf{H} = \mathbf{O}$, $\mathbf{X}$ and $\mathbf{W}$ have the same distribution. This proves the result. □

Theorem 139 *The cf of the SUE random vector $\mathbf{X} \sim \mathcal{SUE}_{p,q}(\boldsymbol{\xi},\boldsymbol{\Psi},\mathbf{H},\boldsymbol{\tau},\bar{\boldsymbol{\Gamma}},\varphi)$ can be computed as $\psi(\mathbf{t}) = e^{\mathrm{i}\mathbf{t}^\top\boldsymbol{\xi}}\mathrm{E}\{e^{\mathrm{i}(\mathbf{H}^\top\mathbf{t})^\top\mathbf{U}}\varphi_{Q(\mathbf{U})}(\mathbf{t}^\top\boldsymbol{\Psi}\mathbf{t})\}$, where $\mathbf{U} \sim \mathcal{TE}l_p(-\boldsymbol{\tau};\mathbf{0}_q,\bar{\boldsymbol{\Gamma}},\varphi)$ and $Q(\mathbf{U}) = \mathbf{U}^\top\bar{\boldsymbol{\Gamma}}^{-1}\mathbf{U}$. Furthermore, provided that the mean vector and variance matrix of $\mathbf{X}$ exist, then*

$$\mathrm{E}(\mathbf{X}) = \boldsymbol{\xi} + \mathbf{H}\mathrm{E}(\mathbf{U}), \quad \mathrm{var}(\mathbf{X}) = \mathbf{H}\mathrm{var}(\mathbf{U})\mathbf{H}^\top + \mathrm{E}\{-2\varphi'_{Q(\mathbf{U})}(0)\}\boldsymbol{\Psi}.$$

Proof 139 *It is direct from the stochastic representation in (8.5).* □

8.2 The $\boldsymbol{\tau} = \mathbf{0}_q$ Case

The special case with $\boldsymbol{\tau} = \mathbf{0}_q$ allows us to incorporate additional important properties into the SUE distribution. Indeed, if $\boldsymbol{\tau} = \mathbf{0}_q$ then $\mathbf{U} = (\mathbf{W}_0 \mid \mathbf{W}_0 + \boldsymbol{\tau} > \mathbf{0}_q) = (\mathbf{W}_0 \mid \mathbf{W}_0 > \mathbf{0}_q) \sim \mathcal{TE}l_q(\mathbf{0}_q;\mathbf{0}_q,\bar{\boldsymbol{\Gamma}},\varphi)$, where $\mathbf{W}_0 \sim \mathcal{E}l_q(\mathbf{0}_q,\bar{\boldsymbol{\Gamma}},\varphi)$. Thus, as shown in the following theorem, an immediate consequence of this fact is $Q(\mathbf{U}) = \mathbf{U}^\top\bar{\boldsymbol{\Gamma}}^{-1}\mathbf{U} \stackrel{d}{=} \mathbf{W}_0^\top\bar{\boldsymbol{\Gamma}}^{-1}\mathbf{W}_0 = Q(\mathbf{W}_0)$, a property that actually holds for any measurable function.

Theorem 140 *If $\mathbf{U} = (\mathbf{W}_0 \mid \mathbf{W}_0 > \mathbf{0}_q) \sim \mathcal{TE}l_q(\mathbf{0}_q;\mathbf{0}_q,\bar{\boldsymbol{\Gamma}},\varphi)$, where $\mathbf{W}_0 \sim \mathcal{E}l_q(\mathbf{0}_q,\bar{\boldsymbol{\Gamma}},\varphi)$, then $Q(\mathbf{U}) = \mathbf{U}^\top\bar{\boldsymbol{\Gamma}}^{-1}\mathbf{U} \stackrel{d}{=} \mathbf{W}_0^\top\bar{\boldsymbol{\Gamma}}^{-1}\mathbf{W}_0 = Q(\mathbf{W}_0)$. Moreover, if the k-th moment of $R_0 = (\mathbf{W}_0^\top\bar{\boldsymbol{\Gamma}}^{-1}\mathbf{W}_0)^{1/2}$ exists, then the k-th multivariate moment of the truncated elliptical random vector $\mathbf{U} = (\mathbf{W}_0 \mid \mathbf{W}_0 > \mathbf{0}_q)$ is given by*

$$\boldsymbol{\mu}_k(\mathbf{U}) = \frac{\mathrm{E}(R_0^k)}{\mathrm{E}(\bar{R}_0^k)}\boldsymbol{\mu}_k(\bar{\mathbf{U}}), \quad k = 1,2,\ldots,$$

where $\boldsymbol{\mu}_k(\bar{\mathbf{U}})$ is the k-th multivariate moment of the truncated normal random vector $\bar{\mathbf{U}} = (\bar{\mathbf{W}}_0 \mid \bar{\mathbf{W}}_0 > \mathbf{0}_q) \sim \mathcal{TN}_q(\mathbf{0}_q;\mathbf{0}_q,\bar{\boldsymbol{\Gamma}},\varphi)$, with $\bar{\mathbf{W}}_0 \sim \mathcal{N}_q(\mathbf{0}_q,\bar{\boldsymbol{\Gamma}})$, and $\bar{R}_0 = (\bar{\mathbf{W}}_0^\top\bar{\boldsymbol{\Gamma}}^{-1}\bar{\mathbf{W}}_0)^{1/2} \sim \chi_q$, with $\mathrm{E}(\bar{R}_0^k) = 2^{k/2}\Gamma\{(q+k)/2\}/\Gamma(q/2)$. In particular,

$$\mathrm{E}(\mathbf{U}) = \frac{\mathrm{E}(R_0)}{\mathrm{E}(\bar{R}_0)}\boldsymbol{\mu}_1(\bar{\mathbf{U}}), \quad \mathrm{var}(\mathbf{U}) = \frac{\mathrm{E}(R_0^2)}{\mathrm{E}(\bar{R}_0^2)}\boldsymbol{\mu}_2(\bar{\mathbf{U}}) - \frac{\mathrm{E}(R_0)^2}{\mathrm{E}(\bar{R}_0)^2}\boldsymbol{\mu}_1(\bar{\mathbf{U}})\boldsymbol{\mu}_1(\bar{\mathbf{U}})^\top.$$

Proof 140 *As in the proof of Theorem 137, we have that $\mathbf{W}_0 = R_0\mathbf{A}_0\mathbf{U}^{(q)}$, where $\mathbf{A}_0\mathbf{A}_0^\top = \bar{\boldsymbol{\Gamma}}$ and we know that the variates R_0 and $\mathbf{U}^{(q)}$ are independent. Also, it is immediate that the event $\{\mathbf{W}_0 > \mathbf{0}_q\} = \{R_0\mathbf{A}_0\mathbf{U}^{(q)} > \mathbf{0}_q\}$ reduces to $\{\mathbf{A}_0\mathbf{U}^{(q)} > \mathbf{0}_q\}$ since $R_0 > 0$ with probability 1; so that $\mathbf{U} =$*

$(\mathbf{W}_0|\mathbf{W}_0 > \mathbf{0}_q) \overset{d}{=} R_0\mathbf{U}_0$, where $\mathbf{U}_0 = (\mathbf{A}_0\mathbf{U}^{(q)}|\mathbf{A}_0\mathbf{U}^{(q)} > \mathbf{0}_q)$ and is independent of R_0. Then, for every measurable vector-valued function $\mathbf{g}$, we have that $\mathbf{g}(\mathbf{U}) = (\mathbf{g}(\mathbf{W}_0)|\mathbf{W}_0 > \mathbf{0}_q) \overset{d}{=} \mathbf{g}(R_0\mathbf{U}_0)$. In particular, if $\mathbf{g}$ is homogeneous of degree k, i.e., $\mathbf{g}(a\mathbf{u}) = a^k\mathbf{g}(\mathbf{u})$ for all $a > 0$, then $\mathbf{g}(\mathbf{U}) \overset{d}{=} R_0^k\mathbf{g}(\mathbf{U}_0)$, with $\mathbf{g}(\mathbf{U}_0) = (\mathbf{g}(\mathbf{A}_0\mathbf{U}^{(q)})|\mathbf{A}_0\mathbf{U}^{(q)} > \mathbf{0}_q)$. Thus, since we can always choose $\mathbf{A}_0 = \bar{\boldsymbol{\Gamma}}^{1/2}$, for $\mathbf{g}(\mathbf{u}) = Q(\mathbf{u}) = \mathbf{u}^\top\bar{\boldsymbol{\Gamma}}^{-1}\mathbf{u}$ we have $Q(\mathbf{U}) \overset{d}{=} R_0^2 Q(\bar{\boldsymbol{\Gamma}}^{1/2}\mathbf{U}_0) = R_0^2((\mathbf{U}^{(q)})^\top\mathbf{U}^{(q)}|\bar{\boldsymbol{\Gamma}}^{1/2}\mathbf{U}^{(q)} > \mathbf{0}_q) = R_0^2$. This proves the first result. The proof of the second result is similar since $\boldsymbol{\mu}_k(\mathbf{U}) = \mathrm{E}(R_0^k)\boldsymbol{\mu}_k(\mathbf{U}_0)$ provided that $\mathrm{E}(R_0^k)$ is finite (see Theorem 23). Since this fact also holds for the multivariate normal distribution, then

$$\boldsymbol{\mu}_k(\bar{\mathbf{U}}) = \mathrm{E}(\bar{R}_0^k)\boldsymbol{\mu}_k(\mathbf{U}_0) \Longrightarrow \boldsymbol{\mu}_k(\mathbf{U}_0) = \frac{1}{\mathrm{E}(\bar{R}_0^k)}\boldsymbol{\mu}_k(\bar{\mathbf{U}}), \quad k = 1,2,\ldots,$$

from which we arrive at the desired result. The result for $\mathrm{E}(\bar{R}_0^k)$ is well known since $\bar{R}_0^2 \sim \chi_q^2$. □

Some immediate consequences of Theorems 139 and 140 are given below.

Theorem 141 *If $\mathbf{X} \sim \mathcal{SUE}_{p,q}(\boldsymbol{\xi},\boldsymbol{\Psi},\mathbf{H},\mathbf{0}_q,\bar{\boldsymbol{\Gamma}},\varphi)$, then its cf is given by $\psi(\mathbf{t}) = \mathrm{E}(e^{\mathrm{i}\mathbf{t}^\top\mathbf{X}}) = e^{\mathrm{i}\mathbf{t}^\top\boldsymbol{\xi}}\mathrm{E}\{\psi_0(R_0\mathbf{H}^\top\mathbf{t})\varphi_{R_0^2}(\mathbf{t}^\top\boldsymbol{\Psi}\mathbf{t})\}$, $\mathbf{t}\in\mathbb{R}^p$, where $\psi_0(\mathbf{s}) = \mathrm{E}(e^{\mathrm{i}\mathbf{s}^\top\mathbf{U}_0})$, $\mathbf{s}\in\mathbb{R}^q$, is the cf of $\mathbf{U}_0 = (\bar{\boldsymbol{\Gamma}}^{1/2}\mathbf{U}^{(q)} \mid \bar{\boldsymbol{\Gamma}}^{1/2}\mathbf{U}^{(q)} > \mathbf{0}_q)$ and $R_0^2 = Q(\mathbf{W}_0) = \mathbf{W}_0^\top\bar{\boldsymbol{\Gamma}}^{-1}\mathbf{W}_0$, with $\mathbf{W}_0 \sim \mathcal{E}l_q(\mathbf{0}_q,\bar{\boldsymbol{\Gamma}},\varphi)$. Moreover, provided that the second moment of the radial variable $R_0 = Q(\mathbf{W}_0)^{1/2}$ and the expectation $\mathrm{E}\{-2\varphi'_{Q(\mathbf{W}_0)}(0)\}$ are finite, then*

$$\begin{aligned}
\mathrm{E}(\mathbf{X}) =& \boldsymbol{\xi} + \frac{\mathrm{E}(R_0)}{\mathrm{E}(\bar{R}_0)}\mathbf{H}\boldsymbol{\mu}_1(\bar{\mathbf{U}}),\\
\mathrm{var}(\mathbf{X}) =& \mathbf{H}\left\{\frac{\mathrm{E}(R_0^2)}{\mathrm{E}(\bar{R}_0^2)}\boldsymbol{\mu}_2(\bar{\mathbf{U}}) - \frac{\mathrm{E}(R_0)^2}{\mathrm{E}(\bar{R}_0)^2}\boldsymbol{\mu}_1(\bar{\mathbf{U}})\boldsymbol{\mu}_1(\bar{\mathbf{U}})^\top\right\}\mathbf{H}^\top + \mathrm{E}\{-2\varphi'_{Q(\mathbf{W}_0)}(0)\}\boldsymbol{\Psi},
\end{aligned}$$

where $\bar{R}_0 \sim \chi_q$, with $\mathrm{E}(\bar{R}_0) = 2^{1/2}\Gamma\{(q+1)/2\}/\Gamma(q/2)$ and $\mathrm{E}(\bar{R}_0^2) = q$, and $\bar{\mathbf{U}} \sim \mathcal{TN}_q(\mathbf{0}_q;\mathbf{0}_q,\bar{\boldsymbol{\Gamma}})$.

Proof 141 *It is directly derived from Theorems 139 and 140.* □

On the other hand, when $\boldsymbol{\tau} = \mathbf{0}_q$ and the baseline elliptical distribution in (8.1) corresponds to the elliptical subfamily generated as scale mixtures of the multivariate normal distribution (SMN), it is possible to build an interesting SUE subfamily from a scale mixture of the SUN subclass with $\boldsymbol{\tau} = \mathbf{0}_q$. The following result illustrates this important subclass of the SUE family with $\boldsymbol{\tau} = \mathbf{0}_q$.

Theorem 142 *Let $\mathbf{X} \overset{d}{=} \boldsymbol{\xi} + \mathbf{H}\mathbf{U} + \mathbf{W}_{Q(\mathbf{W}_0)} \sim \mathcal{SUE}_{p,q}(\boldsymbol{\xi},\boldsymbol{\Psi},\mathbf{H},\mathbf{0}_q,\bar{\boldsymbol{\Gamma}},\varphi)$, where $\mathbf{U} = (\mathbf{W}_0 \mid \mathbf{W}_0 > \mathbf{0}_q) \sim \mathcal{TE}l_q(\mathbf{0}_q;\mathbf{0}_q,\bar{\boldsymbol{\Gamma}},\varphi)$, $\mathbf{W}_{Q(\mathbf{w}_0)} = (\mathbf{W} \mid \mathbf{W}_0 = \mathbf{w}_0) \sim \mathcal{E}l_p(\mathbf{0}_p,\boldsymbol{\Psi},\varphi_{Q(\mathbf{w}_0)})$. Also, assume that*

$$\begin{pmatrix}\mathbf{W}\\ \mathbf{W}_0\end{pmatrix} = V^{-1/2}\begin{pmatrix}\tilde{\mathbf{W}}\\ \tilde{\mathbf{W}}_0\end{pmatrix}, \quad \begin{pmatrix}\tilde{\mathbf{W}}\\ \tilde{\mathbf{W}}_0\end{pmatrix} \sim \mathcal{N}_{p+q}\left(\begin{pmatrix}\mathbf{0}_p\\ \mathbf{0}_q\end{pmatrix},\begin{pmatrix}\boldsymbol{\Psi} & \mathbf{O}\\ \mathbf{O} & \bar{\boldsymbol{\Gamma}}\end{pmatrix},\varphi\right),$$

where the scale random variable V ($V > 0$ with probability 1) is independent of the normal random vectors $\tilde{\mathbf{W}} \sim \mathcal{N}_p(\mathbf{0}_p,\boldsymbol{\Psi})$ and $\tilde{\mathbf{W}}_0 \sim \mathcal{N}_q(\mathbf{0}_q,\bar{\boldsymbol{\Gamma}})$, which are also independent since they are uncorrelated. Then,

$$\mathbf{X} \overset{d}{=} \boldsymbol{\xi} + V^{-1/2}\mathbf{Z}, \tag{8.10}$$

where $\mathbf{Z} \sim \mathcal{SUN}_{p,q}(\mathbf{0}_p, \boldsymbol{\Psi}, \mathbf{H}, \mathbf{0}_q, \bar{\boldsymbol{\Gamma}})$ *and is independent of the scale factor* V*. Moreover, in this case*

$$\begin{aligned} \mathrm{E}(\mathbf{X}) &= \boldsymbol{\xi} + \mathrm{E}(V^{-1/2})\mathrm{E}(\mathbf{Z}) \quad \textit{if } \mathrm{E}(V^{-1/2}) < \infty, \\ \mathrm{var}(\mathbf{X}) &= \mathrm{E}(V^{-1})\mathrm{var}(\mathbf{Z}) + \mathrm{var}(V^{-1/2})\mathrm{E}(\mathbf{Z})\mathrm{E}(\mathbf{Z})^\top \quad \textit{if } \mathrm{E}(V^{-1}) < \infty, \end{aligned}$$

where

$$\mathrm{E}(\mathbf{Z}) = \boldsymbol{\xi} + \mathbf{H}\mathrm{E}(\tilde{\mathbf{W}}_0 \mid \tilde{\mathbf{W}}_0 > \mathbf{0}_q) \quad \textit{and} \quad \mathrm{var}(\mathbf{Z}) = \boldsymbol{\Psi} + \mathbf{H}\mathrm{var}(\tilde{\mathbf{W}}_0 \mid \tilde{\mathbf{W}}_0 > \mathbf{0}_q)\mathbf{H}^\top.$$

Proof 142 *Since* $V > 0$ *(with probability 1) and is independent of* $(\bar{\mathbf{W}}, \bar{\mathbf{W}}_0)$*, then by (8.4) we have for* $\mathbf{X}_* = \mathbf{X} - \boldsymbol{\xi}$ *that*

$$\begin{aligned} \mathbf{X}_* &\stackrel{d}{=} \{(\mathbf{H}\mathbf{W}_0 + \mathbf{W}) \mid \mathbf{W}_0 > \mathbf{0}_q\} \\ &= \{V^{-1/2}(\mathbf{H}\tilde{\mathbf{W}}_0 + \tilde{\mathbf{W}}) \mid V^{-1/2}\tilde{\mathbf{W}}_0 > \mathbf{0}_q\} \\ &\stackrel{d}{=} V^{-1/2}\{(\mathbf{H}\tilde{\mathbf{W}}_0 + \tilde{\mathbf{W}}) \mid \tilde{\mathbf{W}}_0 > \mathbf{0}_q\} \\ &= V^{-1/2}\mathbf{Z}, \end{aligned}$$

where $\mathbf{Z} = \{(\mathbf{H}\tilde{\mathbf{W}}_0 + \tilde{\mathbf{W}}) \mid \tilde{\mathbf{W}}_0 > \mathbf{0}_q\} \stackrel{d}{=} \mathbf{H}\tilde{\mathbf{U}} + \tilde{\mathbf{W}}$*, with* $\tilde{\mathbf{U}} = (\tilde{\mathbf{W}}_0 \mid \tilde{\mathbf{W}}_0 > \mathbf{0}_q)$*, since* $\tilde{\mathbf{W}}$ *and* $\tilde{\mathbf{W}}_0$ *are independent. Finally, since* $\tilde{\mathbf{U}} \sim \mathcal{TN}_q(\mathbf{0}_q; \mathbf{0}_q, \bar{\boldsymbol{\Gamma}})$ *and* $\tilde{\mathbf{W}} \sim \mathcal{N}_p(\mathbf{0}_p, \boldsymbol{\Psi})$ *are also independent, then* $\mathbf{Z} \stackrel{d}{=} \mathbf{H}\tilde{\mathbf{U}} + \tilde{\mathbf{W}} \sim \mathcal{SUN}_{p,q}(\mathbf{0}_p, \boldsymbol{\Psi}, \mathbf{H}, \mathbf{0}_q, \bar{\boldsymbol{\Gamma}})$*, which proves (8.10). The results for* $\mathrm{E}(\mathbf{X})$ *and* $\mathrm{var}(\mathbf{X})$ *are direct from (8.10), and the results for* $\mathrm{E}(\mathbf{Z})$ *and* $\mathrm{var}(\mathbf{Z})$ *are obtained from Theorem 139 with* $R_0^2 \sim \chi_q^2$ *and* $-2\varphi'_{Q(\mathbf{W}_0)}(0) \equiv 1$. □

Under the conditions of the previous Theorem 142, to calculate the expectation and variance of the SUE vector $\mathbf{X}$, the first and second moments of the variable $V^{-1/2}$ are required.

8.3 Affine Transformations, Marginals, and Conditional Distributions

The next theorem points out that any linear combination of unified skew-elliptically distributed variates is still unified skew-elliptical. This means in particular that all the marginals of an SUE distribution belong to the same SUE family.

Theorem 143 *Assume that* $\mathbf{X} \sim \mathcal{SUE}_{p,q}(\boldsymbol{\xi}, \boldsymbol{\Psi}, \mathbf{H}, \boldsymbol{\tau}, \bar{\boldsymbol{\Gamma}}, \varphi)$*,* $\mathbf{A}$ *is a* $k \times p$ *constant matrix and* $\mathbf{b}$ *is a* $k \times 1$ *constant vector. Then,*

$$\mathbf{A}\mathbf{X} + \mathbf{b} \sim \mathcal{SUE}_{k,q}(\mathbf{A}\boldsymbol{\xi} + \mathbf{b}, \mathbf{A}\boldsymbol{\Psi}\mathbf{A}^\top, \mathbf{A}\mathbf{H}, \boldsymbol{\tau}, \bar{\boldsymbol{\Gamma}}, \varphi).$$

Proof 143 *By Definition 21 it follows straightforwardly that* $\mathbf{A}\mathbf{X} + \mathbf{b} = \boldsymbol{\xi}_A + \mathbf{X}_{*A}$*, where* $\boldsymbol{\xi}_A = \mathbf{A}\boldsymbol{\xi} + \mathbf{b}$ *and*

$$\mathbf{X}_{*A} = \mathbf{A}\mathbf{X}_* = (\mathbf{A}\mathbf{W}_* \mid \mathbf{W}_0 + \boldsymbol{\tau} > \mathbf{0}_q) = (\mathbf{W}_{*A} \mid \mathbf{W}_0 + \boldsymbol{\tau} > \mathbf{0}_q),$$

with $\mathbf{W}_{*A} = \mathbf{A}\mathbf{W}_* = \mathbf{A}\mathbf{H}\mathbf{W}_0 + \mathbf{A}\mathbf{W}$. *Finally, using (8.1) and the fact that the elliptical distributions are closed under affine transformations, it follows that*

$$\mathbf{W}_{*A} = \mathbf{A}\mathbf{H}\mathbf{W}_0 + \mathbf{A}\mathbf{W} = \begin{pmatrix} \mathbf{A} & \mathbf{A}\mathbf{H} \end{pmatrix} \begin{pmatrix} \mathbf{W} \\ \mathbf{W}_0 \end{pmatrix} \sim \mathcal{E}l_k\left(\mathbf{0}_k, \mathbf{\Psi}_A + \mathbf{H}_A \bar{\mathbf{\Gamma}} \mathbf{H}_A^\top, \varphi\right),$$

$$\begin{pmatrix} \mathbf{W}_{*A} \\ \mathbf{W}_0 \end{pmatrix} = \begin{pmatrix} \mathbf{A} & \mathbf{A}\mathbf{H} \\ \mathbf{O} & \mathbf{I}_q \end{pmatrix} \begin{pmatrix} \mathbf{W} \\ \mathbf{W}_0 \end{pmatrix} \sim \mathcal{E}l_{k+q}\left(\begin{pmatrix} \mathbf{0}_k \\ \mathbf{0}_q \end{pmatrix}, \begin{pmatrix} \mathbf{\Psi}_A + \mathbf{H}_A \bar{\mathbf{\Gamma}} \mathbf{H}_A^\top & \mathbf{H}_A \bar{\mathbf{\Gamma}} \\ \bar{\mathbf{\Gamma}} \mathbf{H}_A^\top & \bar{\mathbf{\Gamma}} \end{pmatrix}, \varphi \right),$$

where $\mathbf{\Psi}_A = \mathbf{A}\mathbf{\Psi}\mathbf{A}^\top$ *and* $\mathbf{H}_A = \mathbf{A}\mathbf{H}$. *Thus, according to Definition 21 the result is proved.* □

Next, we partition $\mathbf{X}$, $(\boldsymbol{\xi}, \mathbf{\Psi}, \mathbf{H})$ and the working matrix $\mathbf{\Omega} = \mathbf{\Psi} + \mathbf{H}\bar{\mathbf{\Gamma}}\mathbf{H}^\top$ into

$$\mathbf{X} = \begin{pmatrix} \mathbf{X}_1 \\ \mathbf{X}_2 \end{pmatrix}, \ \boldsymbol{\xi} = \begin{pmatrix} \boldsymbol{\xi}_1 \\ \boldsymbol{\xi}_2 \end{pmatrix}, \ \mathbf{\Psi} = \begin{pmatrix} \mathbf{\Psi}_{11} & \mathbf{\Psi}_{12} \\ \mathbf{\Psi}_{21} & \mathbf{\Psi}_{22} \end{pmatrix}, \ \mathbf{H} = \begin{pmatrix} \mathbf{H}_1 \\ \mathbf{H}_2 \end{pmatrix}, \ \mathbf{\Omega} = \begin{pmatrix} \mathbf{\Omega}_{11} & \mathbf{\Omega}_{12} \\ \mathbf{\Omega}_{21} & \mathbf{\Omega}_{22} \end{pmatrix}, \tag{8.11}$$

where for $i, j = 1, 2$, the vectors $\mathbf{X}_i$ and $\boldsymbol{\xi}_i$ are of dimension $p_i \times 1$, with $p_1 + p_2 = p$, and the matrices $\mathbf{\Psi}_{ij}$ and $\mathbf{H}_i$ have dimensions $p_i \times p_j$ and $p_i \times q$, respectively. The following theorem provides the marginal distributions and shows that they are also SUE.

Theorem 144 *If* $\mathbf{X} \sim \mathcal{SUE}_{p,q}(\boldsymbol{\xi}, \mathbf{\Psi}, \mathbf{H}, \boldsymbol{\tau}, \bar{\mathbf{\Gamma}}, \varphi)$, *then* $\mathbf{X}_j \sim \mathcal{SUE}_{p_j,q}(\boldsymbol{\xi}_j, \mathbf{\Psi}_{jj}, \mathbf{H}_j, \boldsymbol{\tau}, \bar{\mathbf{\Gamma}}, \varphi)$ *for* $j = 1, 2$.

Proof 144 *It is directly derived from Theorem 143 since* $\mathbf{X}_j = \mathbf{A}_j \mathbf{X}$, $j = 1, 2$, *with* $\mathbf{A}_1 = (\mathbf{I}_{p_1}, \mathbf{O})$ *and* $\mathbf{A}_2 = (\mathbf{O}, \mathbf{I}_{p_2})$.

□

This can also be observed by extracting the marginal cdf of $\mathbf{X}_j$ from Theorem 136 as

$$F_j(\mathbf{x}_j) = \frac{1}{F_q(\boldsymbol{\tau}; \mathbf{0}_q, \bar{\mathbf{\Gamma}}, \varphi)} F_{p_j+q}\left\{ \begin{pmatrix} \mathbf{x}_j \\ \boldsymbol{\tau} \end{pmatrix}; \begin{pmatrix} \boldsymbol{\xi}_j \\ \mathbf{0}_q \end{pmatrix}, \begin{pmatrix} \mathbf{\Psi}_{jj} + \mathbf{H}_j \bar{\mathbf{\Gamma}} \mathbf{H}_j^\top & -\mathbf{H}_j \bar{\mathbf{\Gamma}} \\ -\bar{\mathbf{\Gamma}} \mathbf{H}_j^\top & \bar{\mathbf{\Gamma}} \end{pmatrix}, \varphi \right\},$$
$$\mathbf{x}_j \in \mathbb{R}^{p_j}, \quad j = 1, 2. \tag{8.12}$$

From Theorem 144 and (8.12), we see again the advantage of the $(\mathbf{\Psi}, \mathbf{H})$ parameterization of the SUE, namely that it yields a marginally consistent distribution, i.e., the parameters of the marginal distribution depend only on the corresponding subparameters of the full distribution (treating the latent parameters as fixed).

In addition to the partition (8.11), we also assume that the matrix $\mathbf{H}$ and the parameters $(\boldsymbol{\tau}, \bar{\mathbf{\Gamma}})$ associated with the latent vector $\mathbf{U}$ are partitioned as

$$\mathbf{H} = \begin{pmatrix} \mathbf{H}_{11} & \mathbf{H}_{12} \\ \mathbf{H}_{21} & \mathbf{H}_{22} \end{pmatrix}, \quad \boldsymbol{\tau} = \begin{pmatrix} \boldsymbol{\tau}_1 \\ \boldsymbol{\tau}_2 \end{pmatrix}, \quad \bar{\mathbf{\Gamma}} = \begin{pmatrix} \bar{\mathbf{\Gamma}}_{11} & \bar{\mathbf{\Gamma}}_{12} \\ \bar{\mathbf{\Gamma}}_{21} & \bar{\mathbf{\Gamma}}_{22} \end{pmatrix}, \tag{8.13}$$

where $\boldsymbol{\tau}_i$ is a $q_i \times 1$ vector, $\mathbf{H}_{ij}$ is a $p_i \times q_j$ matrix and $\bar{\mathbf{\Gamma}}_{ij}$ is a $q_i \times q_j$ matrix, $i, j = 1, 2$, with $q_1 + q_2 = q$. Therefore, assuming also that $\mathbf{H}_{ij} = \mathbf{O}$ and $\bar{\mathbf{\Gamma}}_{ij} = \mathbf{O}$ for $i \neq j$, that is, $\mathbf{H} = \mathrm{diag}(\mathbf{H}_{11}, \mathbf{H}_{22})$

and $\bar{\mathbf{\Gamma}} = \operatorname{diag}(\bar{\mathbf{\Gamma}}_{11}, \bar{\mathbf{\Gamma}}_{22})$, then, with the support of Theorem 137, we can see that

$$\begin{aligned}
\boldsymbol{\tau}_2 = \mathbf{0}_{q_2} \implies F_1(\mathbf{x}_1) =& \frac{1}{F_{q_1}(\boldsymbol{\tau}_1; \mathbf{0}_{q_1}, \bar{\mathbf{\Gamma}}_{11}, \varphi)} \\
& \times F_{p_1+q_1}\left\{ \begin{pmatrix} \mathbf{x}_1 \\ \boldsymbol{\tau}_1 \end{pmatrix}; \begin{pmatrix} \boldsymbol{\xi}_1 \\ \mathbf{0}_{q_1} \end{pmatrix}, \begin{pmatrix} \mathbf{\Psi}_{11} + \mathbf{H}_{11}\bar{\mathbf{\Gamma}}_{11}\mathbf{H}_{11}^\top & -\mathbf{H}_{11}\bar{\mathbf{\Gamma}}_{11} \\ -\bar{\mathbf{\Gamma}}_{11}\mathbf{H}_{11}^\top & \bar{\mathbf{\Gamma}}_{11} \end{pmatrix}, \varphi \right\} \\
& \implies \mathbf{X}_1 \sim \mathcal{SUE}_{p_1,q_1}(\boldsymbol{\xi}_1, \mathbf{\Psi}_{11}, \mathbf{H}_{11}, \boldsymbol{\tau}_1, \bar{\mathbf{\Gamma}}_{11}, \varphi), \\
\boldsymbol{\tau}_1 = \mathbf{0}_{q_1} \implies F_2(\mathbf{x}_2) =& \frac{1}{F_{q_2}(\boldsymbol{\tau}_2; \mathbf{0}_{q_2}, \bar{\mathbf{\Gamma}}_{22}, \varphi)} \\
& \times F_{p_2+q_2}\left\{ \begin{pmatrix} \mathbf{x}_2 \\ \boldsymbol{\tau}_2 \end{pmatrix}; \begin{pmatrix} \boldsymbol{\xi}_2 \\ \mathbf{0}_{q_2} \end{pmatrix}, \begin{pmatrix} \mathbf{\Psi}_{22} + \mathbf{H}_{22}\bar{\mathbf{\Gamma}}_{22}\mathbf{H}_{22}^\top & -\mathbf{H}_{22}\bar{\mathbf{\Gamma}}_{22} \\ -\bar{\mathbf{\Gamma}}_{22}\mathbf{H}_{22}^\top & \bar{\mathbf{\Gamma}}_{22} \end{pmatrix}, \varphi \right\} \\
& \implies \mathbf{X}_2 \sim \mathcal{SUE}_{p_2,q_2}(\boldsymbol{\xi}_2, \mathbf{\Psi}_{22}, \mathbf{H}_{22}, \boldsymbol{\tau}_2, \bar{\mathbf{\Gamma}}_{22}, \varphi).
\end{aligned}$$

Example 103 *Let $q = p$ and $\mathbf{X} \sim \mathcal{SUE}_{p,p}(\boldsymbol{\xi}, \mathbf{\Psi}, \mathbf{H}, \boldsymbol{\tau}, \bar{\mathbf{\Gamma}}, \varphi)$ be partitioned as (8.11), and consider also the partition (8.13) of $(\mathbf{H}, \boldsymbol{\tau}, \bar{\mathbf{\Gamma}})$. Assume that $\mathbf{H}$ is lower triangular. Then $\mathbf{H}_{11}$ and $\mathbf{H}_{22}$ are also lower triangular, and $\mathbf{H}_{12} = \mathbf{O}$, but $\mathbf{H}_{21} \neq \mathbf{O}$. Moreover, if $\bar{\mathbf{\Gamma}} = \operatorname{diag}(\bar{\mathbf{\Gamma}}_{11}, \bar{\mathbf{\Gamma}}_{22})$ and $\boldsymbol{\tau}_2 = \mathbf{0}_{p_2}$ then according to the results above, we have $\mathbf{X}_1 \sim \mathcal{SUE}_{p_1,p_1}(\boldsymbol{\xi}_1, \mathbf{\Psi}_{11}, \mathbf{H}_{11}, \boldsymbol{\tau}_1, \bar{\mathbf{\Gamma}}_{11}, \varphi)$, but according to Theorem 144 we have $\mathbf{X}_2 \sim \mathcal{SUE}_{p_2,p}(\boldsymbol{\xi}_2, \mathbf{\Psi}_{22}, (\mathbf{H}_{21}, \mathbf{H}_{22}), \boldsymbol{\tau}, \bar{\mathbf{\Gamma}}, \varphi)$. This holds in particular if $\boldsymbol{\tau} = \mathbf{0}_p$ and $\bar{\mathbf{\Gamma}} = \mathbf{I}_p$. This means that the SUE distribution in this setting is marginally consistent for $\mathbf{X}_1$ but not for $\mathbf{X}_2$. A possible choice for $\mathbf{H}$ is $\mathbf{H} = \mathbf{\Psi}^{\pm 1/2}$ where $\mathbf{\Psi}^{\pm 1/2}$ is the lower triangular factor of the Choleski factorization of $\mathbf{\Psi}$ or $\mathbf{\Psi}^{-1}$.*

Example 104 *Suppose that $\mathbf{X} = (\mathbf{X}_1^\top, \mathbf{X}_2^\top)^\top \sim \mathcal{SUE}_{p_1+p_2,q_1+q_2}(\boldsymbol{\xi}, \mathbf{\Psi}, \mathbf{H}, \boldsymbol{\tau}, \bar{\mathbf{\Gamma}}, \varphi)$, with $\boldsymbol{\xi} = (\boldsymbol{\xi}_1^\top, \boldsymbol{\xi}_2^\top)^\top \in \mathbb{R}^{p_1+p_2}$, $\mathbf{\Psi} = (\mathbf{\Psi}_{ij})$, with $\mathbf{\Psi}_{ij}$ of dimension $p_i \times p_j$ for $j = 1, 2$, $\mathbf{H} = \operatorname{diag}(\mathbf{H}_{11}, \mathbf{H}_{22})$, with $\mathbf{H}_{jj}$ of dimension $p_j \times q_j$ for $j = 1, 2$, $\boldsymbol{\tau} = (\mathbf{0}_{q_1}^\top, \mathbf{0}_{q_2}^\top)^\top$ and $\bar{\mathbf{\Gamma}} = \operatorname{diag}(\bar{\mathbf{\Gamma}}_{11}, \bar{\mathbf{\Gamma}}_{22})$, with $\bar{\mathbf{\Gamma}}_{jj}$ of dimension $q_j \times q_j$ for $j = 1, 2$. Then, by using Theorem 144 we can find that*

$$\mathbf{X}_j \sim \mathcal{SUE}_{p_j,q_j}(\boldsymbol{\xi}_j, \mathbf{\Psi}_{jj}, \mathbf{H}_{jj}, \mathbf{0}_{q_j}, \bar{\mathbf{\Gamma}}_{jj}, \varphi), \quad j = 1, 2.$$

Example 105 *Let $\mathbf{X} = (X_1, \ldots, X_p)^\top \sim \mathcal{SUE}_{p,p}(\boldsymbol{\xi}, \mathbf{\Psi}, \mathbf{H}, \boldsymbol{\tau}, \bar{\mathbf{\Gamma}}, \varphi)$, with $\boldsymbol{\xi} = (\xi_1, \ldots, \xi_p)^\top$, $\mathbf{\Psi} = (\psi_{ij})_{i,j=1}^p$, $\mathbf{H} = \operatorname{diag}(\eta_1, \ldots, \eta_p)$, $\boldsymbol{\tau} = \mathbf{0}_p$, and $\bar{\mathbf{\Gamma}} = \mathbf{I}_p$. Then, the marginal distribution of X_j is $\mathcal{SE}(\xi_j, \psi_{jj}, \eta_j, \varphi)$, $j = 1, \ldots, p$, which are uncorrelated but not mutually independent. However, for any $1 < k < p$:*

$$(X_1, \ldots, X_k)^\top \sim \mathcal{SUE}_{k,k}((\xi_1, \ldots, \xi_k)^\top, (\psi_{ij})_{i,j=1}^k, \operatorname{diag}(\eta_1, \ldots, \eta_k), \mathbf{0}_k, \mathbf{I}_k, \varphi),$$

which is different from a $\mathcal{SE}_k$ distribution.

Conditional distributions also belong to the same SUE family as the original SUE random vector. This fact is established in the following theorem.

Theorem 145 *Let* $\mathbf{X} \sim \mathcal{SUE}_{p,q}(\boldsymbol{\xi}, \boldsymbol{\Psi}, \mathbf{H}, \boldsymbol{\tau}, \bar{\boldsymbol{\Gamma}}, \varphi)$ *with* $\boldsymbol{\Psi} > 0$. *Then, under the partition (8.11) we have*

$$(\mathbf{X}_1 \mid \mathbf{X}_2 = \mathbf{x}_2) \sim \mathcal{SUE}_{p_1,q}(\tilde{\boldsymbol{\xi}}_{1\cdot 2}, \boldsymbol{\Psi}_{11\cdot 2}, \bar{\mathbf{H}}_{1\cdot 2}, \bar{\boldsymbol{\tau}}_{1\cdot 2}, \bar{\boldsymbol{\Gamma}}_{1\cdot 2}, \varphi_{Q_2(\mathbf{x}_2)}),$$

where $Q_2(\mathbf{x}_2) = (\mathbf{x}_2 - \boldsymbol{\xi}_2)^\top \boldsymbol{\Omega}_{22}^{-1}(\mathbf{x}_2 - \boldsymbol{\xi}_2)$, *with* $\boldsymbol{\Omega}_{22} = \boldsymbol{\Psi}_{22} + \mathbf{H}_2\bar{\boldsymbol{\Gamma}}\mathbf{H}_2^\top$, $\varphi_{Q_2(\mathbf{x}_2)}$ *is the corresponding conditional characteristic generator, and*

$$\begin{aligned}
&\tilde{\boldsymbol{\xi}}_{1\cdot 2} = \boldsymbol{\xi} + \bar{\boldsymbol{\Gamma}}\mathbf{H}_2^\top \boldsymbol{\Omega}_{22}^{-1}(\mathbf{x}_2 - \boldsymbol{\xi}_2) = \boldsymbol{\xi}_{1\cdot 2} + \mathbf{H}_{1\cdot 2}\boldsymbol{\Gamma}_{1.2}\mathbf{H}_2^\top \boldsymbol{\Psi}_{22}^{-1}(\mathbf{x}_2 - \boldsymbol{\xi}_2), \quad \boldsymbol{\Psi}_{11\cdot 2} = \boldsymbol{\Psi}_{11} - \boldsymbol{\Psi}_{12}\boldsymbol{\Psi}_{22}^{-1}\boldsymbol{\Psi}_{21},\\
&\bar{\mathbf{H}}_{1\cdot 2} = \mathbf{H}_{1\cdot 2}\boldsymbol{\gamma}_{1\cdot 2}, \quad \mathbf{H}_{1\cdot 2} = \mathbf{H}_1 - \boldsymbol{\Psi}_{12}\boldsymbol{\Psi}_{22}^{-1}\mathbf{H}_2 = (\mathbf{H}_1 - \boldsymbol{\Omega}_{12}\boldsymbol{\Omega}_{22}^{-1}\mathbf{H}_2)\bar{\boldsymbol{\Gamma}}\boldsymbol{\Gamma}_{1\cdot 2}^{-1}, \quad \bar{\boldsymbol{\tau}}_{1\cdot 2} = \boldsymbol{\gamma}_{1\cdot 2}^{-1}\boldsymbol{\tau}_{1\cdot 2},\\
&\boldsymbol{\tau}_{1\cdot 2} = \boldsymbol{\tau} + \bar{\boldsymbol{\Gamma}}\mathbf{H}_2^\top \boldsymbol{\Omega}_{22}^{-1}(\mathbf{x}_2 - \boldsymbol{\xi}_2) = \boldsymbol{\tau} + \bar{\boldsymbol{\Gamma}}_{1\cdot 2}\mathbf{H}_2^\top \boldsymbol{\Psi}_{22}^{-1}(\mathbf{x}_2 - \boldsymbol{\xi}_2), \quad \bar{\boldsymbol{\Gamma}}_{1\cdot 2} = \boldsymbol{\gamma}_{1\cdot 2}^{-1}\boldsymbol{\Gamma}_{1\cdot 2}\boldsymbol{\gamma}_{1\cdot 2}^{-1},\\
&\boldsymbol{\Gamma}_{1\cdot 2} = \bar{\boldsymbol{\Gamma}} - \bar{\boldsymbol{\Gamma}}\mathbf{H}_2^\top \boldsymbol{\Omega}_{22}^{-1}\mathbf{H}_2\bar{\boldsymbol{\Gamma}}, \quad \boldsymbol{\gamma}_{1\cdot 2} = \mathrm{diag}(\boldsymbol{\Gamma}_{1\cdot 2})^{1/2}, \quad \boldsymbol{\Omega}_{ij} = \boldsymbol{\Psi}_{ij} + \mathbf{H}_i\bar{\boldsymbol{\Gamma}}\mathbf{H}_j^\top,\ i,j = 1,2.
\end{aligned}$$

Proof 145 *From Definition 21 and the partition (8.11), we easily find that*

$$\begin{pmatrix} \mathbf{X}_1 \\ \mathbf{X}_2 \end{pmatrix} = \begin{pmatrix} \boldsymbol{\xi}_1 \\ \boldsymbol{\xi}_2 \end{pmatrix} + \begin{pmatrix} \mathbf{X}_{*1} \\ \mathbf{X}_{*2} \end{pmatrix} = \begin{pmatrix} \boldsymbol{\xi}_1 + \mathbf{X}_{*1} \\ \boldsymbol{\xi}_2 + \mathbf{X}_{*2} \end{pmatrix}, \tag{8.14}$$

where $\mathbf{X}_{*i} = (\mathbf{W}_{*i} \mid \mathbf{W}_0 + \boldsymbol{\tau} > \mathbf{0}_q)$, *with* $\mathbf{W}_{*i} = \mathbf{H}_i\mathbf{W}_0 + \mathbf{W}_i$, $i = 1,2$, *are the induced partitions on* $\mathbf{X}_* = (\mathbf{W}_* \mid \mathbf{W}_0 + \boldsymbol{\tau} > \mathbf{0}_q)$ *and* $\mathbf{W}_* = \mathbf{H}\mathbf{W}_0 + \mathbf{W}$, *respectively. Thus, by (8.14)*

$$\begin{aligned}
(\mathbf{X}_1 \mid \mathbf{X}_2 = \mathbf{x}_2) &= \{(\boldsymbol{\xi}_1 + \mathbf{X}_{*1}) \mid \boldsymbol{\xi}_2 + \mathbf{X}_{*2} = \mathbf{x}_2\}\\
&= \boldsymbol{\xi}_1 + (\mathbf{X}_{*1} \mid \mathbf{X}_{*2} = \mathbf{x}_2 - \boldsymbol{\xi}_2)\\
&= \boldsymbol{\xi}_1 + (\mathbf{W}_{*1} \mid \mathbf{W}_0 + \boldsymbol{\tau} > \mathbf{0}_q, \mathbf{W}_{*2} = \mathbf{x}_2 - \boldsymbol{\xi}_2).
\end{aligned}$$

However, by writing $\mathbf{W}_{*1} = \mathbf{W}_{*12} + \boldsymbol{\Omega}_{12}\boldsymbol{\Omega}_{22}^{-1}\mathbf{W}_{*2}$ *and* $\mathbf{W}_0 = \mathbf{W}_{02} + \bar{\boldsymbol{\Gamma}}\mathbf{H}_2^\top \boldsymbol{\Omega}_{22}^{-1}\mathbf{W}_{*2}$, *where* $\mathbf{W}_{*12} = \mathbf{W}_{*1} - \boldsymbol{\Omega}_{12}\boldsymbol{\Omega}_{22}^{-1}\mathbf{W}_{*2}$ *and* $\mathbf{W}_{02} = \mathbf{W}_0 - \bar{\boldsymbol{\Gamma}}\mathbf{H}_2^\top \boldsymbol{\Omega}_{22}^{-1}\mathbf{W}_{*2}$ *which, as shown below, are uncorrelated with* $\mathbf{W}_{*2}$, *we have*

$$\begin{aligned}
&(\mathbf{W}_{*1} \mid \mathbf{W}_0 + \boldsymbol{\tau} > \mathbf{0}_q, \mathbf{W}_{*2} = \mathbf{x}_2 - \boldsymbol{\xi}_2)\\
&\qquad = \{(\mathbf{W}_{*12} + \boldsymbol{\Omega}_{12}\boldsymbol{\Omega}_{22}^{-1}\mathbf{W}_{*2}) \mid (\mathbf{W}_{02} + \bar{\boldsymbol{\Gamma}}\mathbf{H}_2^\top \boldsymbol{\Omega}_{22}^{-1}\mathbf{W}_{*2} + \boldsymbol{\tau} > \mathbf{0}_q, \mathbf{W}_{*2} = \mathbf{x}_2 - \boldsymbol{\xi}_2)\}\\
&\qquad \stackrel{d}{=} [\{\mathbf{W}_{*,1\cdot 2} + \boldsymbol{\Omega}_{12}\boldsymbol{\Omega}_{22}^{-1}(\mathbf{x}_2 - \boldsymbol{\xi}_2)\} \mid \{\mathbf{W}_{0\cdot 2} + \bar{\boldsymbol{\Gamma}}\mathbf{H}_2^\top \boldsymbol{\Omega}_{22}^{-1}(\mathbf{x}_2 - \boldsymbol{\xi}_2) + \boldsymbol{\tau} > \mathbf{0}_q\}]\\
&\qquad = \boldsymbol{\Omega}_{12}\boldsymbol{\Omega}_{22}^{-1}(\mathbf{x}_2 - \boldsymbol{\xi}_2) + (\mathbf{W}_{*,1\cdot 2} \mid \mathbf{W}_{0\cdot 2} + \boldsymbol{\tau}_{1\cdot 2} > \mathbf{0}_q),
\end{aligned}$$

where $\boldsymbol{\tau}_{1\cdot 2} = \boldsymbol{\tau} + \bar{\boldsymbol{\Gamma}}\mathbf{H}_2^\top \boldsymbol{\Omega}_{22}^{-1}(\mathbf{x}_2 - \boldsymbol{\xi}_2)$ *and jointly* $\mathbf{W}_{*,1\cdot 2}$ *and* $\mathbf{W}_{0\cdot 2}$ *are conditional elliptically distributed random vectors given by*

$$\begin{aligned}
\begin{pmatrix} \mathbf{W}_{*,1\cdot 2} \\ \mathbf{W}_{0\cdot 2} \end{pmatrix} &= \begin{pmatrix} \mathbf{W}_{*12} \\ \mathbf{W}_{02} \end{pmatrix} \mid (\mathbf{W}_{*2} = \mathbf{x}_2 - \boldsymbol{\xi}_2) = \begin{pmatrix} \mathbf{W}_{*1} - \boldsymbol{\Omega}_{12}\boldsymbol{\Omega}_{22}^{-1}\mathbf{W}_{*2} \\ \mathbf{W}_0 - \bar{\boldsymbol{\Gamma}}\mathbf{H}_2^\top \boldsymbol{\Omega}_{22}^{-1}\tilde{\mathbf{W}}_2 \end{pmatrix} \mid (\mathbf{W}_{*2} = \mathbf{x}_2 - \boldsymbol{\xi}_2)\\
&\sim \mathcal{E}l_{p_1+q}\left(\begin{pmatrix} \mathbf{0}_{p_1} \\ \mathbf{0}_q \end{pmatrix}, \begin{pmatrix} \boldsymbol{\Omega}_{11\cdot 2} & \mathbf{H}_{1\cdot 2}\bar{\boldsymbol{\Gamma}} \\ \bar{\boldsymbol{\Gamma}}\mathbf{H}_{1\cdot 2}^\top & \boldsymbol{\Gamma}_{1\cdot 2} \end{pmatrix}, \varphi_{Q_2(\mathbf{x}_2)}\right)
\end{aligned}$$

$$\Rightarrow \begin{cases} \mathbf{W}_{*,1\cdot 2} = (\mathbf{W}_{*1} \mid \mathbf{W}_{*2} = \mathbf{x}_2 - \boldsymbol{\xi}_2) \sim \mathcal{E}l_{p_1}(\mathbf{0}_{p_1}, \boldsymbol{\Omega}_{11\cdot 2}, \varphi_{Q_2(\mathbf{x}_2)}),\\ \mathbf{W}_{0\cdot 2} = (\mathbf{W}_0 \mid \mathbf{W}_{*2} = \mathbf{x}_2 - \boldsymbol{\xi}_2) \sim \mathcal{E}l_q(\mathbf{0}_q, \boldsymbol{\Gamma}_{1\cdot 2}, \varphi_{Q_2(\mathbf{x}_2)}),\end{cases}$$

since

$$\begin{pmatrix} \mathbf{W}_{*1} \\ \mathbf{W}_{*2} \\ \mathbf{W}_0 \end{pmatrix} = \begin{pmatrix} \mathbf{H}_1\mathbf{W}_0 + \mathbf{W}_1 \\ \mathbf{H}_2\mathbf{W}_0 + \mathbf{W}_2 \\ \mathbf{W}_0 \end{pmatrix} \sim \mathcal{E}l_{p_1+p_2+q}\left(\begin{pmatrix} \mathbf{0}_{p_1} \\ \mathbf{0}_{p_2} \\ \mathbf{0}_q \end{pmatrix}, \begin{pmatrix} \boldsymbol{\Omega}_{11} & \boldsymbol{\Omega}_{12} & \mathbf{H}_1\bar{\boldsymbol{\Gamma}} \\ \boldsymbol{\Omega}_{21} & \boldsymbol{\Omega}_{22} & \mathbf{H}_2\bar{\boldsymbol{\Gamma}} \\ \bar{\boldsymbol{\Gamma}}\mathbf{H}_1^\top & \bar{\boldsymbol{\Gamma}}\mathbf{H}_2^\top & \bar{\boldsymbol{\Gamma}} \end{pmatrix}, \varphi \right),$$

$$\Rightarrow \begin{pmatrix} \mathbf{W}_{1*} - \boldsymbol{\Omega}_{12}\boldsymbol{\Omega}_{22}^{-1}\mathbf{W}_{*2} \\ \mathbf{W}_0 - \bar{\boldsymbol{\Gamma}}\mathbf{H}_2^\top\boldsymbol{\Omega}_{22}^{-1}\mathbf{W}_{*2} \\ \mathbf{W}_{*2} \end{pmatrix} = \begin{pmatrix} \mathbf{I}_{p_1} & -\boldsymbol{\Omega}_{12}\boldsymbol{\Omega}_{22}^{-1} & \mathbf{O} \\ \mathbf{O} & -\bar{\boldsymbol{\Gamma}}\mathbf{H}_2^\top\boldsymbol{\Omega}_{22}^{-1} & \mathbf{I}_q \\ \mathbf{O} & \mathbf{I}_{p_2} & \mathbf{O} \end{pmatrix} \begin{pmatrix} \mathbf{W}_{*1} \\ \mathbf{W}_{*2} \\ \mathbf{W}_0 \end{pmatrix}$$

$$\sim \mathcal{E}l_{p_1+q+p_2}\left(\begin{pmatrix} \mathbf{0}_{p_1} \\ \mathbf{0}_q \\ \mathbf{0}_{p_2} \end{pmatrix}, \begin{pmatrix} \boldsymbol{\Omega}_{11\cdot 1} & \mathbf{H}_{1\cdot 2}\bar{\boldsymbol{\Gamma}} & \mathbf{O} \\ \bar{\boldsymbol{\Gamma}}\mathbf{H}_{1\cdot 2}^\top & \boldsymbol{\Gamma}_{1\cdot 2} & \mathbf{O} \\ \mathbf{O} & \mathbf{O} & \boldsymbol{\Omega}_{22} \end{pmatrix}, \varphi \right),$$

with $\boldsymbol{\Omega}_{ij} = \boldsymbol{\Psi}_{ij} + \mathbf{H}_i\bar{\boldsymbol{\Gamma}}\mathbf{H}_j^\top$, $i,j = 1,2$, $\boldsymbol{\Omega}_{11\cdot 2} = \boldsymbol{\Omega}_{11} - \boldsymbol{\Omega}_{12}\boldsymbol{\Omega}_{22}^{-1}\boldsymbol{\Omega}_{21}$, $\mathbf{H}_{1\cdot 2} = \mathbf{H}_1 - \boldsymbol{\Omega}_{12}\boldsymbol{\Omega}_{22}^{-1}\mathbf{H}_2$ *and* $\boldsymbol{\Gamma}_{1\cdot 2} = \bar{\boldsymbol{\Gamma}} - \bar{\boldsymbol{\Gamma}}\mathbf{H}_2^\top\boldsymbol{\Omega}_{22}^{-1}\mathbf{H}_2\bar{\boldsymbol{\Gamma}}$. *Hence, by writing* $\mathbf{X}_{1\cdot 2} = (\mathbf{X}_1 \mid \mathbf{X}_2 = \mathbf{x}_2)$ *and* $\mathbf{X}_{*,1\cdot 2} = (\mathbf{W}_{*,1\cdot 2} \mid \mathbf{W}_{0\cdot 2} + \boldsymbol{\tau}_{1\cdot 2} > \mathbf{0}_q)$, *we obtain that* $\mathbf{X}_{1\cdot 2} = \tilde{\boldsymbol{\xi}}_{1\cdot 2} + \mathbf{X}_{*,1\cdot 2}$, *with* $\tilde{\boldsymbol{\xi}}_{1\cdot 2} = \boldsymbol{\xi}_1 + \bar{\boldsymbol{\Gamma}}\mathbf{H}_2^\top\boldsymbol{\Omega}_{22}^{-1}(\mathbf{x}_2 - \boldsymbol{\xi}_2)$, *and the proof follows according to Definition 21 and considering that* $\mathbf{X}_{*,1\cdot 2}$ *can be rewritten as* $\mathbf{X}_{*,1\cdot 2} = (\mathbf{W}_{*,1\cdot 2} \mid \boldsymbol{\gamma}_{1\cdot 2}^{-1}\mathbf{W}_{0\cdot 2} + \boldsymbol{\gamma}_{1\cdot 2}^{-1}\boldsymbol{\tau}_{1\cdot 2} > \mathbf{0}_q)$, *where* $\boldsymbol{\gamma}_{1\cdot 2} = \mathrm{diag}(\boldsymbol{\Gamma}_{1\cdot 2})^{1/2}$ *so that* $\bar{\boldsymbol{\Gamma}}_{1\cdot 2} = \boldsymbol{\gamma}_{1\cdot 2}^{-1}\boldsymbol{\Gamma}_{1\cdot 2}\boldsymbol{\gamma}_{1\cdot 2}^{-1}$ *is a correlation matrix.*

□

Similarly to the SUN case (see Section 6.5 of Chapter 6), this last theorem can be extended to the singular case by replacing $\boldsymbol{\Omega}_{22}^{-1}$ with $\boldsymbol{\Omega}_{22}^{-}$, a generalized inverse of $\boldsymbol{\Omega}_{22} = \boldsymbol{\Psi}_{22} + \mathbf{H}_2\bar{\boldsymbol{\Gamma}}\mathbf{H}_2^\top$. In this context, it is convenient to recall that if $\boldsymbol{\Psi}$ and $\bar{\boldsymbol{\Gamma}}$ are nonsingular matrices, then $\boldsymbol{\Omega} = \boldsymbol{\Psi} + \mathbf{H}\bar{\boldsymbol{\Gamma}}\mathbf{H}^\top$ is also nonsingular, whatever the matrix $\mathbf{H} \in \mathbb{R}^{p\times q}$.

8.4 Pdf

Here, we must remember that, in general, a random vector $\mathbf{Y} \sim \mathcal{E}l_p(\boldsymbol{\mu}, \boldsymbol{\Sigma}, \varphi)$ does not necessarily have a pdf (see Chapter 4). A necessary condition for $\mathbf{Y} \sim \mathcal{E}l_p(\boldsymbol{\mu}, \boldsymbol{\Sigma}, \varphi)$ to have a pdf is that $\mathrm{rank}(\boldsymbol{\Sigma}) = p$. In that case, if $\mathbf{Y}$ has a pdf, it is of the form

$$f_p(\mathbf{x}; \boldsymbol{\mu}, \boldsymbol{\Sigma}, h) = |\boldsymbol{\Sigma}|^{-1/2} h\{(\mathbf{y}-\boldsymbol{\mu})^\top\boldsymbol{\Sigma}^{-1}(\mathbf{y}-\boldsymbol{\mu})\}, \quad \mathbf{y} \in \mathbb{R}^p,$$

where h is the density generator function. Also, in this case, we sometimes use the notation $\mathcal{E}l_p(\boldsymbol{\mu}, \boldsymbol{\Sigma}, h)$ instead of $\mathcal{E}l_p(\boldsymbol{\mu}, \boldsymbol{\Sigma}, \varphi)$. Moreover, to make the dimension explicit, we sometimes write $h^{(p)}$ instead of h.

So, if we assume that the elliptical joint distribution of $(\tilde{\mathbf{W}}^\top, \mathbf{W}_0^\top)^\top$ in (8.1) has a pdf, for which it is necessary that $\mathrm{rank}(\boldsymbol{\Psi}) = p$ and $\mathrm{rank}(\bar{\boldsymbol{\Gamma}}) = q$, then the elliptical joint distribution of $(\tilde{\mathbf{W}}^\top, \mathbf{W}_0^\top)^\top = ((\mathbf{H}\mathbf{W}_0 + \mathbf{W})^\top, \mathbf{W}_0^\top)^\top$ given by (8.3) also has a pdf, since this random vector comes from a nonsingular linear transformation of $(\tilde{\mathbf{W}}^\top, \mathbf{W}_0^\top)^\top$. Under these circumstances, the SUE random vector $\mathbf{X} \sim \mathcal{SUE}_{p,q}(\boldsymbol{\xi}, \boldsymbol{\Psi}, \mathbf{H}, \boldsymbol{\tau}, \bar{\boldsymbol{\Gamma}}, \varphi)$, defined as in (8.4), possesses a pdf. In this case, we write $\mathbf{X} \sim \mathcal{SUE}_{p,q}(\boldsymbol{\xi}, \boldsymbol{\Psi}, \mathbf{H}, \boldsymbol{\tau}, \bar{\boldsymbol{\Gamma}}, h)$ or $\mathbf{X} \sim \mathcal{SUE}_{p,q}(\boldsymbol{\xi}, \boldsymbol{\Psi}, \mathbf{H}, \boldsymbol{\tau}, \bar{\boldsymbol{\Gamma}}, h^{(p+q)})$, where the characteristic generator φ is replaced by the $(p+q)$-variate density generator $h = h^{(p+q)}$, from which we can obtain the p-variate and q-variate marginal density generators $h^{(p)}$ and $h^{(q)}$, respectively, according to [1] in Section 4.2.5. We formalize this in the following theorem.

Theorem 146 *Assume that* $\mathbf{X} \sim \mathcal{SUE}_{p,q}(\boldsymbol{\xi}, \boldsymbol{\Psi}, \mathbf{H}, \boldsymbol{\tau}, \bar{\boldsymbol{\Gamma}}, h^{(p+q)})$ *is defined as in (8.4), but assuming in (8.1) that* $(\mathbf{W}^\top, \mathbf{W}_0^\top)^\top \sim \mathcal{E}l_{p+q}((\mathbf{0}_p^\top, \mathbf{0}_q^\top)^\top, \mathrm{diag}(\boldsymbol{\Psi}, \bar{\boldsymbol{\Gamma}}), h^{(p+q)})$. *Then,* $\mathbf{X}$ *has a pdf given by*

$$f(\mathbf{x}) = \frac{1}{F_q(\boldsymbol{\tau}; \mathbf{0}_q, \bar{\boldsymbol{\Gamma}}, h^{(q)})} f_p(\mathbf{x}; \boldsymbol{\xi}, \boldsymbol{\Omega}, h^{(p)}) F_q\{\boldsymbol{\tau} + \bar{\boldsymbol{\Gamma}}\mathbf{H}^\top\boldsymbol{\Omega}^{-1}(\mathbf{x}-\boldsymbol{\xi}); \mathbf{0}_q, \bar{\boldsymbol{\Gamma}} - \bar{\boldsymbol{\Gamma}}\mathbf{H}^\top\boldsymbol{\Omega}^{-1}\mathbf{H}\bar{\boldsymbol{\Gamma}}; h^{(q)}_{Q(\mathbf{x})}\}, \tag{8.15}$$

for $\mathbf{x} \in \mathbb{R}^p$, *where* $\boldsymbol{\Omega} = \boldsymbol{\Psi} + \mathbf{H}\bar{\boldsymbol{\Gamma}}\mathbf{H}^\top$, $Q(\mathbf{x}) = (\mathbf{x}-\boldsymbol{\xi})^\top\boldsymbol{\Omega}^{-1}(\mathbf{x}-\boldsymbol{\xi})$,

$$f_p(\mathbf{x}; \boldsymbol{\xi}, \boldsymbol{\Omega}, h^{(p)}) = |\boldsymbol{\Omega}|^{-1/2} h^{(p)}\{(\mathbf{x}-\boldsymbol{\xi})^\top\boldsymbol{\Omega}^{-1}(\mathbf{x}-\boldsymbol{\xi})\}, \quad \mathbf{x} \in \mathbb{R}^p,$$

is an $\mathcal{E}l_p(\boldsymbol{\xi}, \boldsymbol{\Omega}, h^{(p)})$ *pdf,* $F_k(\mathbf{z}; \boldsymbol{\mu}, \boldsymbol{\Sigma}, h^{(k)})$, $\mathbf{z} \in \mathbb{R}^k$, *is an* $\mathcal{E}l_k(\boldsymbol{\mu}, \boldsymbol{\Sigma}, h^{(k)})$ *cdf evaluated at* $\mathbf{z}$, *and* $h_u^{(q)}(v) = h^{(p+q)}(u+v)/h^{(p)}(u)$, $u, v > 0$, *is a conditional density generator. In addition, an equivalent version of (8.15) is*

$$f(\mathbf{x}) = \frac{1}{F_q(\boldsymbol{\tau}; \mathbf{0}_q, \bar{\boldsymbol{\Gamma}}, h^{(q)})} f_p(\mathbf{x}; \boldsymbol{\xi}, \boldsymbol{\Psi} + \mathbf{H}\bar{\boldsymbol{\Gamma}}\mathbf{H}^\top, h^{(p)}) F_q\{\boldsymbol{\tau} + \tilde{\boldsymbol{\Gamma}}\mathbf{H}^\top\boldsymbol{\Psi}^{-1}(\mathbf{x}-\boldsymbol{\xi}); \mathbf{0}_q, \tilde{\boldsymbol{\Gamma}}; h^{(q)}_{Q(\mathbf{x})}\}, \quad \mathbf{x} \in \mathbb{R}^p, \tag{8.16}$$

where $\tilde{\boldsymbol{\Gamma}} = (\bar{\boldsymbol{\Gamma}}^{-1} + \mathbf{H}^\top\boldsymbol{\Psi}^{-1}\mathbf{H})^{-1}$.

Proof 146 *Since* $\mathbf{X} = \boldsymbol{\xi} + \tilde{\mathbf{X}}$, *then* $f(\mathbf{x}) = \tilde{f}(\mathbf{x}-\boldsymbol{\xi})$, *where* $\tilde{f}(\tilde{\mathbf{x}})$ *is the pdf of the selection vector* $\tilde{\mathbf{X}} = (\tilde{\mathbf{W}} \mid \mathbf{W}_0 + \boldsymbol{\tau} > \mathbf{0}_q)$, *where* $\tilde{\mathbf{W}} = \mathbf{H}\mathbf{W}_0 + \mathbf{W}$. *Then, we just need to find* $\tilde{f}(\tilde{\mathbf{x}})$, *the selection pdf of* $\tilde{\mathbf{X}} = (\tilde{\mathbf{W}} \mid \mathbf{W}_0 + \boldsymbol{\tau} > \mathbf{0}_q)$. *However, the cdf of* $\tilde{\mathbf{X}} = (\tilde{\mathbf{W}} \mid \mathbf{W}_0 + \boldsymbol{\tau} > \mathbf{0}_q)$ *is*

$$\begin{aligned}
\tilde{F}(\mathbf{x}) &= \mathrm{P}(\tilde{\mathbf{W}} \le \mathbf{x} \mid \mathbf{W}_0 + \boldsymbol{\tau} > \mathbf{0}_q) \\
&= \frac{1}{\mathrm{P}(\mathbf{W}_0 + \boldsymbol{\tau} > \mathbf{0}_q)} \mathrm{P}(\tilde{\mathbf{W}} \le \mathbf{x}, \mathbf{W}_0 + \boldsymbol{\tau} > \mathbf{0}_q) \\
&= \frac{1}{\mathrm{P}(\mathbf{W}_0 + \boldsymbol{\tau} > \mathbf{0}_q)} \int_{\tilde{\mathbf{w}} \le \mathbf{x}} \int_{\mathbf{w}_0 + \boldsymbol{\tau} > \mathbf{0}_q} f_{p+q}(\tilde{\mathbf{w}}, \mathbf{w}_0) \mathrm{d}\mathbf{w}_0 \mathrm{d}\tilde{\mathbf{w}},
\end{aligned}$$

where $f_{p+q}(\tilde{\mathbf{w}}, \mathbf{w}_0)$ *is the joint pdf of* $(\tilde{\mathbf{W}}^\top, \mathbf{W}_0^\top)^\top$, *which can be rewritten as* $\tilde{f}_p(\tilde{\mathbf{w}}) f_q(\mathbf{w}_0 \mid \tilde{\mathbf{w}})$, *where* $\tilde{f}_p(\tilde{\mathbf{w}})$ *is the marginal pdf of* $\tilde{\mathbf{W}}$ *and* $f_q(\mathbf{w}_0 \mid \tilde{\mathbf{w}})$ *is the conditional pdf of* $(\mathbf{W}_0 \mid \tilde{\mathbf{W}} = \tilde{\mathbf{w}})$. *Hence, we find that*

$$\begin{aligned}
\tilde{F}(\mathbf{x}) &= \frac{1}{\mathrm{P}(\mathbf{W}_0 + \boldsymbol{\tau} > \mathbf{0}_q)} \int_{\tilde{\mathbf{w}} \le \mathbf{x}} \int_{\mathbf{w}_0 + \boldsymbol{\tau} > \mathbf{0}_q} \tilde{f}_p(\tilde{\mathbf{w}}) f_q(\mathbf{w}_0 \mid \tilde{\mathbf{w}}) \mathrm{d}\mathbf{w}_0 \mathrm{d}\tilde{\mathbf{w}} \\
&= \frac{1}{\mathrm{P}(\mathbf{W}_0 + \boldsymbol{\tau} > \mathbf{0}_q)} \int_{\tilde{\mathbf{w}} \le \mathbf{x}} \tilde{f}_p(\tilde{\mathbf{w}}) \int_{\mathbf{w}_0 + \boldsymbol{\tau} > \mathbf{0}_q} f_q(\mathbf{w}_0 \mid \tilde{\mathbf{w}}) \mathrm{d}\mathbf{w}_0 \mathrm{d}\tilde{\mathbf{w}} \\
\Longrightarrow \tilde{f}(\mathbf{x}) &= \frac{1}{\mathrm{P}(\mathbf{W}_0 + \boldsymbol{\tau} > \mathbf{0}_q)} \tilde{f}_p(\mathbf{x}) \int_{\mathbf{w}_0 + \boldsymbol{\tau} > \mathbf{0}_q} f_q(\mathbf{w}_0 \mid \mathbf{x}) \mathrm{d}\mathbf{w}_0 \\
&= \frac{1}{\mathrm{P}(\mathbf{W}_0 + \boldsymbol{\tau} > \mathbf{0}_q)} \tilde{f}_p(\mathbf{x}) \mathrm{P}(\mathbf{W}_0 + \boldsymbol{\tau} > \mathbf{0}_q \mid \mathbf{x}).
\end{aligned}$$

The proof of (8.15) is finished by noting that $\tilde{f}_p(\mathbf{x}) = f_p(\mathbf{x}; \boldsymbol{\xi}, \boldsymbol{\Omega}, h^{(p)})$ *since* $\tilde{\mathbf{W}} \sim \mathcal{E}l_p(\boldsymbol{\xi}, \boldsymbol{\Omega}, h^{(p)})$, $\mathrm{P}(\mathbf{W}_0 + \boldsymbol{\tau} > \mathbf{0}_q) = \mathrm{P}(-\mathbf{W}_0 < \boldsymbol{\tau}) = F_q(\boldsymbol{\tau}; \mathbf{0}_q, \bar{\boldsymbol{\Gamma}}, h^{(q)})$ *since* $-\mathbf{W}_0 \sim \mathcal{E}l_q(\mathbf{0}_q, \bar{\boldsymbol{\Gamma}}, h^{(q)})$ *and* $\mathrm{P}(\mathbf{W}_0 + \boldsymbol{\tau} > \mathbf{0}_q \mid \tilde{\mathbf{W}} = \mathbf{x}) = \mathrm{P}(-\mathbf{W}_0 < \boldsymbol{\tau} \mid \tilde{\mathbf{W}} = \mathbf{x}) = F_q(\boldsymbol{\tau} + \bar{\boldsymbol{\Gamma}}\mathbf{H}^\top\boldsymbol{\Omega}^{-1}(\mathbf{x}-\boldsymbol{\xi}); \mathbf{0}_q, \bar{\boldsymbol{\Gamma}} - \bar{\boldsymbol{\Gamma}}\mathbf{H}^\top\boldsymbol{\Omega}^{-1}\mathbf{H}\bar{\boldsymbol{\Gamma}}, h^{(q)}_{Q(\mathbf{x})})$ *since* $(-\mathbf{W}_0 \mid \tilde{\mathbf{W}} = \mathbf{x}) \sim \mathcal{E}l_q(-\bar{\boldsymbol{\Gamma}}\mathbf{H}^\top\boldsymbol{\Omega}^{-1}(\mathbf{x}-\boldsymbol{\xi}), \bar{\boldsymbol{\Gamma}} - \bar{\boldsymbol{\Gamma}}\mathbf{H}^\top\boldsymbol{\Omega}^{-1}\mathbf{H}\bar{\boldsymbol{\Gamma}}, h^{(q)}_{Q(\mathbf{x})})$.

Finally, to prove that (8.15) is equivalent to (8.16), it suffices to consider the matrix relations:

$$\boldsymbol{\Omega} = \boldsymbol{\Psi} + \mathbf{H}\bar{\boldsymbol{\Gamma}}\mathbf{H}^\top, \quad \bar{\boldsymbol{\Gamma}} - \bar{\boldsymbol{\Gamma}}\mathbf{H}^\top\boldsymbol{\Omega}^{-1}\mathbf{H}\bar{\boldsymbol{\Gamma}} = (\bar{\boldsymbol{\Gamma}}^{-1} + \mathbf{H}^\top\boldsymbol{\Psi}^{-1}\mathbf{H})^{-1} \equiv \tilde{\boldsymbol{\Gamma}}, \quad \bar{\boldsymbol{\Gamma}}\mathbf{H}^\top\boldsymbol{\Omega}^{-1} = \tilde{\boldsymbol{\Gamma}}\mathbf{H}^\top\boldsymbol{\Psi}^{-1}.$$

□

Notice that an interesting property of the pdf and cdf of the SUE is that, unlike the SUN distribution, for which $\mathbf{H} = \mathbf{O}$ is sufficient to recover its corresponding elliptically symmetric parent distribution, the SUE requires both $\mathbf{H} = \mathbf{O}$ and $\boldsymbol{\tau} = \mathbf{0}_q$. Indeed, when $\mathbf{H} = \mathbf{O}$ and $\boldsymbol{\tau} = \mathbf{0}_q$, the SUE pdf of $\mathbf{X}$ in (8.15) becomes:

$$f(\mathbf{x}) = \frac{F_q(\mathbf{0}_q;\mathbf{0}_q,\bar{\boldsymbol{\Gamma}};h^{(q)}_{Q(\mathbf{x})})}{F_q(\mathbf{0}_q;\mathbf{0}_q,\bar{\boldsymbol{\Gamma}},h^{(q)})} f_p(\mathbf{x};\boldsymbol{\xi},\boldsymbol{\Omega},h^{(p)}) = f_p(\mathbf{x};\boldsymbol{\xi},\boldsymbol{\Omega},h^{(p)}), \ \forall\, \mathbf{x} \in \mathbb{R}^p, \tag{8.17}$$

since $F_q(\mathbf{0}_q;\mathbf{0}_q,\bar{\boldsymbol{\Gamma}};h^{(q)}_{Q(\mathbf{x})}) = F_q(\mathbf{0}_q;\mathbf{0}_q,\bar{\boldsymbol{\Gamma}},h^{(q)}) = \Phi_q(\mathbf{0}_q;\mathbf{0}_q,\bar{\boldsymbol{\Gamma}})$, and the pdf of $\mathbf{X}$ is then $\mathcal{E}l_p(\boldsymbol{\xi},\boldsymbol{\Omega},h^{(p)})$. This last identity comes from the fact that if $\mathbf{Y} = R\mathbf{C}\mathbf{U}^{(p)} \sim \mathcal{E}l_p(\mathbf{0}_p,\boldsymbol{\Sigma},h)$, where $\boldsymbol{\Sigma} = \mathbf{C}\mathbf{C}^\top$ and R and $\mathbf{U}^{(p)}$ are independent, then $\mathrm{P}(\mathbf{Y} \le \mathbf{0}_p) = \mathrm{P}(R\mathbf{C}\mathbf{U}^{(p)} \le \mathbf{0}_p) = \mathrm{P}(\mathbf{C}\mathbf{U}^{(p)} \le \mathbf{0}_p)$ regardless of the positive radial variable R.

Obviously, from (8.16) we can obtain the marginal and conditional pdfs associated with a certain partition of $\mathbf{X}$. Next, we present such results, and their proofs are left as exercises.

Theorem 147 *Consider the SUE random vector* $\mathbf{X} = (\mathbf{X}_1^\top,\mathbf{X}_2^\top)^\top \sim \mathcal{SUE}_{p_1+p_2,q}(\boldsymbol{\xi},\boldsymbol{\Psi},\mathbf{H},\boldsymbol{\tau},\bar{\boldsymbol{\Gamma}},h)$ *with the partition in (8.11). Then, according to Theorems 144 and 146,* $\mathbf{X}_j \sim \mathcal{SUE}_{p_j,q}(\boldsymbol{\xi}_j,\boldsymbol{\Psi}_{jj},\mathbf{H}_j,\boldsymbol{\tau},\bar{\boldsymbol{\Gamma}},h)$, $j = 1,2$, *with marginal pdf given by*

$$\begin{aligned} f_j(\mathbf{x}_j) =& \frac{1}{F_q(\boldsymbol{\tau};\mathbf{0}_q,\bar{\boldsymbol{\Gamma}},h^{(q)})} f_{p_j}(\mathbf{x}_j;\boldsymbol{\xi}_j,\boldsymbol{\Omega}_{jj},h^{(p_j)}) \\ &\times F_q\{\boldsymbol{\tau}+\bar{\boldsymbol{\Gamma}}\mathbf{H}_j^\top\boldsymbol{\Omega}_{jj}^{-1}(\mathbf{x}_j-\boldsymbol{\xi}_j);\mathbf{0}_q,\bar{\boldsymbol{\Gamma}}-\bar{\boldsymbol{\Gamma}}\mathbf{H}_j^\top\boldsymbol{\Omega}_{jj}^{-1}\mathbf{H}_j\bar{\boldsymbol{\Gamma}};h^{(q)}_{Q_j(\mathbf{x}_j)}\}, \quad \mathbf{x}_j \in \mathbb{R}^{p_j}, \end{aligned}$$

with $\boldsymbol{\Omega}_{jj} = \boldsymbol{\Psi}_{jj} + \mathbf{H}_j\bar{\boldsymbol{\Gamma}}\mathbf{H}_j^\top$ *and* $Q_j(\mathbf{x}_j) = (\mathbf{x}_j-\boldsymbol{\xi}_j)^\top\boldsymbol{\Omega}_{jj}^{-1}(\mathbf{x}_j-\boldsymbol{\xi}_j)$, *or by*

$$\begin{aligned} f_j(\mathbf{x}_j) =& \frac{1}{F_q(\boldsymbol{\tau};\mathbf{0}_q,\bar{\boldsymbol{\Gamma}},h^{(q)})} f_{p_j}(\mathbf{x}_j;\boldsymbol{\xi}_j,\boldsymbol{\Psi}_{jj}+\mathbf{H}_j\bar{\boldsymbol{\Gamma}}\mathbf{H}_j^\top,h^{(p_j)}) \\ &\times F_q\{\boldsymbol{\tau}+\tilde{\boldsymbol{\Gamma}}_j\mathbf{H}_j^\top\boldsymbol{\Psi}_{jj}^{-1}(\mathbf{x}_j-\boldsymbol{\xi}_j);\mathbf{0}_q,\tilde{\boldsymbol{\Gamma}}_j;h^{(q)}_{Q(\mathbf{x}_j)}\}, \quad \mathbf{x}_j \in \mathbb{R}^{p_j}, \end{aligned}$$

with $\tilde{\boldsymbol{\Gamma}}_j = (\bar{\boldsymbol{\Gamma}}^{-1} + \mathbf{H}_j^\top\boldsymbol{\Psi}_{jj}^{-1}\mathbf{H}_j)^{-1}$.

Proof 147 *See Exercise 8.5.* □

Example 106 *Suppose*

$$\begin{pmatrix}\mathbf{X}_1\\ \mathbf{X}_2\end{pmatrix} \sim \mathcal{SUE}_{p_1+p_2,q}\left\{\begin{pmatrix}\boldsymbol{\xi}_1\\ \boldsymbol{\xi}_2\end{pmatrix},\begin{pmatrix}\boldsymbol{\Psi}_{11} & \mathbf{O}\\ \mathbf{O} & \boldsymbol{\Psi}_{22}\end{pmatrix},\begin{pmatrix}\mathbf{H}_{11} & \mathbf{O}\\ \mathbf{O} & \mathbf{H}_{22}\end{pmatrix},\begin{pmatrix}\mathbf{0}_{q_1}\\ \mathbf{0}_{q_2}\end{pmatrix},\begin{pmatrix}\bar{\boldsymbol{\Gamma}}_{11} & \mathbf{O}\\ \mathbf{O} & \bar{\boldsymbol{\Gamma}}_{22}\end{pmatrix},h\right\},$$

with $\boldsymbol{\Psi}_{jj} > 0$ *of dimension* $p_j \times p_j$, $\mathbf{H}_{jj}$ *of dimension* $p_j \times q_j$ *and* $\bar{\boldsymbol{\Gamma}}_{jj} > 0$ *of dimension* $q_j \times q_j$ *for* $j = 1,2$. *Then,* $\mathbf{X}_j \sim \mathcal{SUE}_{p_j,q_j}(\boldsymbol{\xi}_j,\boldsymbol{\Psi}_{jj},\mathbf{H}_{jj},\mathbf{0}_{q_j},\bar{\boldsymbol{\Gamma}}_{jj},h)$ *for* $j = 1,2$, *which are uncorrelated but not independent unless* $h = h^{(p+q)}$ *is the normal density generator. In this case, the marginal pdf of* $\mathbf{X}_j$ *becomes*

$$\begin{aligned} f_j(\mathbf{x}_j) =& \frac{1}{F_{q_j}(\mathbf{0}_{q_j};\mathbf{0}_{q_j},\bar{\boldsymbol{\Gamma}}_{jj},h^{(q_j)})} f_{p_j}(\mathbf{x}_j;\boldsymbol{\xi}_j,\boldsymbol{\Omega}_{jj},h^{(p_j)}) \\ &\times F_{q_j}\{\bar{\boldsymbol{\Gamma}}_{jj}\mathbf{H}_{jj}^\top\boldsymbol{\Omega}_{jj}^{-1}(\mathbf{x}_j-\boldsymbol{\xi}_j);\mathbf{0}_{q_j},\bar{\boldsymbol{\Gamma}}_{jj}-\bar{\boldsymbol{\Gamma}}_{jj}\mathbf{H}_{jj}^\top\boldsymbol{\Omega}_{jj}^{-1}\mathbf{H}_{jj}\bar{\boldsymbol{\Gamma}}_{jj};h^{(q_j)}_{Q_j(\mathbf{x}_j)}\}, \quad \mathbf{x}_j \in \mathbb{R}^{p_j}, \ j = 1,2, \end{aligned}$$

with $\boldsymbol{\Omega}_{jj} = \boldsymbol{\Psi}_{jj} + \mathbf{H}_{jj}\bar{\boldsymbol{\Gamma}}_{jj}\mathbf{H}_{jj}^\top$. *Furthermore, note that* $F_{q_j}(\mathbf{0}_{q_j};\mathbf{0}_{q_j},\bar{\boldsymbol{\Gamma}}_{jj},h^{(q_j)}) = \Phi_{q_j}(\mathbf{0}_{q_j};\mathbf{0}_{q_j},\bar{\boldsymbol{\Gamma}}_{jj})$ *regardless of the density generator* $h^{(q_j)}$ *and the correlation matrix* $\bar{\boldsymbol{\Gamma}}_{jj}$. *In particular,* $F_{q_j}(\mathbf{0}_{q_j};\mathbf{0}_{q_j},\mathbf{I}_{q_j},h^{(q_j)}) = \Phi_{q_j}(\mathbf{0}_{q_j};\mathbf{0}_{q_j},\mathbf{I}_{q_j}) = 2^{-q_j}$.

Example 107 *Let* $\mathbf{X} = (X_1,\ldots,X_p)^\top \sim \mathcal{SUE}_{p,p}(\boldsymbol{\xi},\boldsymbol{\Psi},\mathbf{H},\boldsymbol{\tau},\bar{\boldsymbol{\Gamma}},h)$, *with* $\boldsymbol{\xi} = (\xi_1,\ldots,\xi_p)^\top$, $\boldsymbol{\Psi} = \mathrm{diag}(\psi_{11},\ldots,\psi_{pp})$, $\mathbf{H} = \mathrm{diag}(\eta_1,\ldots,\eta_p)$, $\boldsymbol{\tau} = \mathbf{0}_p$, *and* $\bar{\boldsymbol{\Gamma}} = \mathbf{I}_p$. *Then, the marginal distribution of* X_j *is* $\mathcal{SE}(\xi_j,\psi_{jj},\eta_j,h)$, $j = 1,\ldots,p$, *which are uncorrelated but not mutually independent unless* $h = h^{(p+q)}$ *is the normal density generator. Moreover, the marginal pdfs are*

$$f_j(x_j) = 2f(x_j;\xi_j,\psi_j^2+\eta_j^2,h^{(1)})F\left\{\frac{\eta_j(x_j-\xi_j)}{\psi_j\sqrt{\psi_j^2+\eta_j^2}};0,1,h^{(p)}_{\tilde{Q}(x_j)}\right\}, \quad x_j \in \mathbb{R},\ j=1,\ldots,p,$$

where $\psi_j = \sqrt{\psi_{jj}}$ *and* $\tilde{Q}_j(x_j) = (x_j-\xi_j)^2/(\psi_j^2+\eta_j^2)$.

Finally, when considering the partition (8.11), the pdf in (8.16) corresponds to the joint pdf of $\mathbf{X}_1$ and $\mathbf{X}_2$. Thus, since $f(\mathbf{x}) \equiv f(\mathbf{x}_1,\mathbf{x}_2) = f_{1\cdot 2}(\mathbf{x}_1 \mid \mathbf{x}_2)f_2(\mathbf{x}_2)$, then applying the decomposition (6.32) to the working matrix $\boldsymbol{\Omega} = (\boldsymbol{\Omega}_{ij})_{i,j=1,2}$, with $\boldsymbol{\Omega}_{ij} = \boldsymbol{\Psi}_{ij} + \mathbf{H}_i\bar{\boldsymbol{\Gamma}}\mathbf{H}_j^\top$ and $\mathbf{H}_i = (\mathbf{H}_{i1},\mathbf{H}_{i2})$, allows us to obtain in (8.16) that

$$\begin{aligned}
f_p(\mathbf{x};\boldsymbol{\xi},\boldsymbol{\Omega},h^{(p)}) &= f_{p_1}(\mathbf{x}_1;\tilde{\boldsymbol{\xi}}_{1\cdot 2},\boldsymbol{\Omega}_{11\cdot 2},h^{(p_1)}_{Q_2(\mathbf{x}_2)})f_{p_2}(\mathbf{x}_2;\boldsymbol{\xi}_2,\boldsymbol{\Omega}_{22},h^{(p_2)}),\\
\boldsymbol{\tau}+\bar{\boldsymbol{\Gamma}}\mathbf{H}^\top\boldsymbol{\Omega}^{-1}(\mathbf{x}-\boldsymbol{\xi}) &= \boldsymbol{\tau}+\bar{\boldsymbol{\Gamma}}\mathbf{H}_{1\cdot 2}^\top\boldsymbol{\Omega}_{11\cdot 2}^{-1}(\mathbf{x}_1-\tilde{\boldsymbol{\xi}}_{1\cdot 2})+\bar{\boldsymbol{\Gamma}}\mathbf{H}_2^\top\boldsymbol{\Omega}_{22}^{-1}(\mathbf{x}_2-\boldsymbol{\xi}_2)\\
&= \boldsymbol{\tau}_{1\cdot 2}+\bar{\boldsymbol{\Gamma}}\mathbf{H}_{1\cdot 2}^\top\boldsymbol{\Omega}_{11\cdot 2}^{-1}(\mathbf{x}_1-\tilde{\boldsymbol{\xi}}_{1\cdot 2}),\\
Q(\mathbf{x}) = (\mathbf{x}-\boldsymbol{\xi})^\top\boldsymbol{\Omega}^{-1}(\mathbf{x}-\boldsymbol{\xi}) &= (\mathbf{x}_1-\tilde{\boldsymbol{\xi}}_{1\cdot 2})^\top\boldsymbol{\Omega}_{11\cdot 2}^{-1}(\mathbf{x}_1-\tilde{\boldsymbol{\xi}}_{1\cdot 2})+(\mathbf{x}_2-\boldsymbol{\xi}_2)^\top\boldsymbol{\Omega}_{22}^{-1}(\mathbf{x}_2-\boldsymbol{\xi}_2)\\
&= Q_{1\cdot 2}(\mathbf{x}_1)+Q_2(\mathbf{x}_2),
\end{aligned}$$

where, as defined before, $\tilde{\boldsymbol{\xi}}_{1\cdot 2} = \boldsymbol{\xi}_1 + \boldsymbol{\Omega}_{12}\boldsymbol{\Omega}_{22}^{-1}(\mathbf{x}_2-\boldsymbol{\xi}_2)$, $\boldsymbol{\tau}_{1\cdot 2} = \boldsymbol{\tau}+\bar{\boldsymbol{\Gamma}}\mathbf{H}_2^\top\boldsymbol{\Omega}_{22}^{-1}(\mathbf{x}_2-\boldsymbol{\xi}_2)$ and $Q_2(\mathbf{x}_2) = (\mathbf{x}_2-\boldsymbol{\xi}_2)^\top\boldsymbol{\Omega}_{22}^{-1}(\mathbf{x}_2-\boldsymbol{\xi}_2)$ while $Q_{1\cdot 2}(\mathbf{x}_1) = (\mathbf{x}_1-\tilde{\boldsymbol{\xi}}_{1\cdot 2})^\top\boldsymbol{\Omega}_{11\cdot 2}^{-1}(\mathbf{x}_1-\tilde{\boldsymbol{\xi}}_{1\cdot 2})$ and $\boldsymbol{\Omega}_{11\cdot 2} = \boldsymbol{\Omega}_{11}-\boldsymbol{\Omega}_{12}\boldsymbol{\Omega}_{22}^{-1}\boldsymbol{\Omega}_{21}$which can be rewritten as $\boldsymbol{\Omega}_{11\cdot 2} = \boldsymbol{\Psi}_{11\cdot 2}+\mathbf{H}_{1\cdot 2}\boldsymbol{\Gamma}_{1\cdot 2}\mathbf{H}_{1\cdot 2}^\top$, with $\boldsymbol{\Psi}_{11\cdot 2}$, $\mathbf{H}_{1\cdot 2}$ and $\boldsymbol{\Gamma}_{1\cdot 2}$ defined earlier. With these ingredients, we find the conditional pdf of $(\mathbf{X}_1 \mid \mathbf{X}_2 = \mathbf{x}_2)$ presented below.

Theorem 148 *If* $\mathbf{X} \sim \mathcal{SUE}_{p,q}(\boldsymbol{\xi},\boldsymbol{\Psi},\mathbf{H},\boldsymbol{\tau},\bar{\boldsymbol{\Gamma}},h^{(p+q)})$, *with* $\boldsymbol{\Psi} > 0$, *is partitioned as in (8.11), then* $(\mathbf{X}_1 \mid \mathbf{X}_2 = \mathbf{x}_2) \sim \mathcal{SUE}_{p_1,q}(\tilde{\boldsymbol{\xi}}_{1\cdot 2},\boldsymbol{\Psi}_{11\cdot 2},\bar{\mathbf{H}}_{1\cdot 2},\bar{\boldsymbol{\tau}}_{1\cdot 2},\bar{\boldsymbol{\Gamma}}_{1\cdot 2},h^{(p_1+q)}_{Q_2(\mathbf{x}_2)})$, *with conditional pdf for* $\mathbf{x}_1 \in \mathbb{R}^{p_1}$:

$$\begin{aligned}
f_{1\cdot 2}(\mathbf{x}_1 \mid \mathbf{x}_2) = &\frac{1}{F_q(\boldsymbol{\tau}_{1\cdot 2};\mathbf{0}_q,\boldsymbol{\Gamma}_{1\cdot 2};h^{(q)}_{Q_2(\mathbf{x}_2)})}f_{p_1}(\mathbf{x}_1;\tilde{\boldsymbol{\xi}}_{1\cdot 2},\boldsymbol{\Omega}_{11\cdot 2},h^{(p_1)}_{Q_2(\mathbf{x}_2)})\\
&\times F_q\{\boldsymbol{\tau}_{1\cdot 2}+\bar{\boldsymbol{\Gamma}}\mathbf{H}_{1\cdot 2}^\top\boldsymbol{\Omega}_{11\cdot 2}^{-1}(\mathbf{x}_1-\tilde{\boldsymbol{\xi}}_{1\cdot 2});\mathbf{0}_q,\bar{\boldsymbol{\Gamma}}-\bar{\boldsymbol{\Gamma}}\mathbf{H}_1^\top\boldsymbol{\Omega}_{11}^{-1}\mathbf{H}_1\bar{\boldsymbol{\Gamma}};h^{(q)}_{Q_{1\cdot 2}(\mathbf{x}_1|\mathbf{x}_2)}\},
\end{aligned}$$

where $\tilde{\boldsymbol{\xi}}_{1\cdot 2} = \boldsymbol{\xi}_{1\cdot 2}+\mathbf{H}_{1\cdot 2}\boldsymbol{\Gamma}_{1\cdot 2}\mathbf{H}_2^\top\boldsymbol{\Psi}_{22}^{-1}(\mathbf{x}_2-\boldsymbol{\xi}_2)$, $\boldsymbol{\Omega}_{11\cdot 2} = \boldsymbol{\Psi}_{11\cdot 2}+\mathbf{H}_{1\cdot 2}\boldsymbol{\Gamma}_{1\cdot 2}\mathbf{H}_{1\cdot 2}^\top$, $\boldsymbol{\tau}_{1\cdot 2} = \boldsymbol{\tau}+\boldsymbol{\Gamma}\mathbf{H}_2^\top\boldsymbol{\Psi}_{22}^{-1}(\mathbf{x}_2-\boldsymbol{\xi}_2)$, $\boldsymbol{\Omega}_{11} = \boldsymbol{\Psi}_{11}-\mathbf{H}_1\bar{\boldsymbol{\Gamma}}\mathbf{H}_2^\top$, $Q_{1\cdot 2}(\mathbf{x}_1|\mathbf{x}_2) = Q_{1\cdot 2}(\mathbf{x}_1)+Q_2(\mathbf{x}_2)$, *with* $Q_{1\cdot 2}(\mathbf{x}_1) = (\mathbf{x}_1-\tilde{\boldsymbol{\xi}}_{1\cdot 2})^\top(\boldsymbol{\Psi}_{11\cdot 2}+\mathbf{H}_{1\cdot 2}\boldsymbol{\Gamma}_{1\cdot 2}\mathbf{H}_{1\cdot 2}^\top)^{-1}(\mathbf{x}_1-\tilde{\boldsymbol{\xi}}_{1\cdot 2})$ *and* $Q_2(\mathbf{x}_2) = (\mathbf{x}_2-\boldsymbol{\xi}_2)^\top(\boldsymbol{\Psi}_{22}+\mathbf{H}_2\bar{\boldsymbol{\Gamma}}\mathbf{H}_2^\top)^{-1}(\mathbf{x}_2-\boldsymbol{\xi}_2)$,

$h_v^{(p_1+q)}(u) = h^{(p+q)}(u+v)/h^{(p_2+q)}(v)$ *and* $h_v^{(k)}(u) = h^{(p+q)}(u+v)/h^{(p-k+q)}(v)$ *for* $k = p_1, q$ *are conditional density generators, and where* $\boldsymbol{\Psi}_{11\cdot 2} = \boldsymbol{\Psi}_{11} - \boldsymbol{\Psi}_{12}\boldsymbol{\Psi}_{22}^{-1}\boldsymbol{\Psi}_{21}$, $\mathbf{H}_{1\cdot 2} = \mathbf{H}_1 - \boldsymbol{\Psi}_{12}\boldsymbol{\Psi}_{22}^{-1}\mathbf{H}_2$ *and* $\boldsymbol{\Gamma}_{1\cdot 2} = \bar{\boldsymbol{\Gamma}} - \bar{\boldsymbol{\Gamma}}\mathbf{H}_2^\top\boldsymbol{\Omega}_{22}^{-1}\mathbf{H}_2\bar{\boldsymbol{\Gamma}}$.

Proof 148 *See Exercise 8.5.* □

8.5 Exercises

Exercise 8.1 *Study the formal properties of the following identifiable SUE distribution:* $\mathcal{SUE}_{p,q}(\boldsymbol{\xi}, \boldsymbol{\Psi}, \boldsymbol{\eta}\mathbf{1}_q^\top, \tau\mathbf{1}_q, (1-\rho)\mathbf{I}_q + \rho\mathbf{1}_q\mathbf{1}_q^\top, \varphi)$, *where* $\boldsymbol{\tau} = \tau\mathbf{1}_q$ $(\tau \in \mathbb{R})$, $\mathbf{H} = \boldsymbol{\eta}\mathbf{1}_q^\top$ $(\boldsymbol{\eta} \in \mathbb{R}^p)$ *and* $\bar{\boldsymbol{\Gamma}} = (1-\rho)\mathbf{I}_q + \rho\mathbf{1}_q\mathbf{1}_q^\top$, *with* $\rho \in (-(q-1)^{-1}, 1)$.

Exercise 8.2 *Study the formal properties of the following identifiable SUE distribution:* $\mathcal{SUE}_{p,q}(\boldsymbol{\xi}, \boldsymbol{\Psi}, \mathbf{H}, \alpha\mathbf{1}_q + \beta\mathbf{j}_q, \bar{\boldsymbol{\Gamma}}, \varphi)$, *where* $\boldsymbol{\tau} = \alpha\mathbf{1}_q + \beta\mathbf{j}_q$ $(\alpha, \beta \in \mathbb{R},\ \beta \neq 0)$ *and* $\mathbf{j}_q = (1, \ldots, q)^\top$.

Exercise 8.3 *Assuming* $\mathbf{X} \sim \mathcal{SUE}_{p,q}(\boldsymbol{\xi}, \boldsymbol{\Psi}, \mathbf{H}, \boldsymbol{\tau}, \bar{\boldsymbol{\Gamma}}, h)$ *with* $\boldsymbol{\Psi} > 0$, *obtain the pdf of* $\mathbf{X}$ *from (8.6). Also, using the partition in (8.11), obtain the conditional density of* $\mathbf{X}_1$ *given* $\mathbf{X}_2 = \mathbf{x}_2$ *and give an expression for* $\mathrm{E}(\mathbf{X}_1 \mid \mathbf{x}_2)$ *and* $\mathrm{var}(\mathbf{X}_1 \mid \mathbf{x}_2)$ *(provided that they exist).*

Exercise 8.4 *Let* $\mathbf{W}_0 = (\mathbf{W}_{01}^\top, \mathbf{W}_{02}^\top)^\top \in \mathbb{R}^{q_1+q_2}$ *and* $\mathbf{U} = (\mathbf{U}_1^\top, \mathbf{U}_2^\top)^\top \in \mathbb{R}^{q_1+q_2}$ *be two random vectors such that* $\mathbf{U} = (\mathbf{W}_0 \mid \mathbf{W}_0 > \mathbf{0}_q)$ *and* $\mathbf{W}_0 \sim \mathcal{E}l_q(\mathbf{0}_q, \mathrm{diag}(\bar{\boldsymbol{\Gamma}}_{11}, \bar{\boldsymbol{\Gamma}}_{22}), \varphi)$, *where* $q = q_1 + q_2$ *and* $\bar{\Gamma}_{jj} > 0$ *has dimension* $q_j \times q_j$, $j = 1, 2$. *Can we say that marginally,* $\mathbf{U}_j = (\mathbf{W}_{0j} \mid \mathbf{W}_{0j} > \mathbf{0}_{q_j})$ *for* $j = 1, 2$? *If so, provide a formal proof.*

Exercise 8.5 *Prove Theorems 147 and 148.*

Exercise 8.6 *Let* $Q_A(\mathbf{X}) = (\mathbf{X} - \boldsymbol{\xi})^\top\mathbf{A}(\mathbf{X} - \boldsymbol{\xi})$, *where* $\mathbf{X} \sim \mathcal{SUE}_{p,q}(\boldsymbol{\xi}, \boldsymbol{\Psi}, \mathbf{H}, \mathbf{0}_q, \bar{\boldsymbol{\Gamma}}, h)$ *is nonsingular* $(\boldsymbol{\Psi} > 0)$ *and* $\mathbf{A} \in \mathbb{R}^{p\times p}$ *is symmetric. Discuss the truthfulness of the following statements:*

(a) $\mathbf{X} \stackrel{d}{=} \boldsymbol{\xi} + \mathbf{H}\mathbf{U} + R_{Q(\mathbf{W}_0)}\boldsymbol{\Psi}^{1/2}\mathbf{U}^{(p)}$, *where* $R_{Q(\mathbf{W}_0)}$ *is the radial variable associated with the conditional distribution* $(\mathbf{W} \mid \mathbf{W}_0 = \mathbf{w}_0) \sim \mathcal{E}l_p(\mathbf{0}_p, \boldsymbol{\Psi}, \varphi_{Q(\mathbf{w}_0)})$ *induced by (8.1);*

(b) If $\mathbf{A} = \mathbf{C}^\top\mathbf{C}$, *with* $\mathbf{C}\mathbf{H} = \mathbf{O}$ *and* $\mathbf{C}\boldsymbol{\Psi}\mathbf{C}^\top = \mathbf{I}_p$, *then* $Q_A(\mathbf{X}) \stackrel{d}{=} R^2_{Q(\mathbf{W}_0)}$;

(c) If $\mathbf{A} = \boldsymbol{\Omega}^{-1}$, *where* $\boldsymbol{\Omega} = \boldsymbol{\Psi} + \mathbf{H}\bar{\boldsymbol{\Gamma}}\mathbf{H}^\top$ *and* $\mathbf{X}$ *is representable as in (8.10) then* $Q_A(\mathbf{X}) \stackrel{d}{=} V^{-1}\mathbf{Z}^\top\boldsymbol{\Omega}^{-1}\mathbf{Z}$, *where* $V > 0$ *(with probability 1) and* $\mathbf{Z} \sim \mathcal{SUN}_{p,q}(\mathbf{0}_p, \boldsymbol{\Psi}, \mathbf{H}, \mathbf{0}_q, \bar{\boldsymbol{\Gamma}})$ *are independent. What happens if* $\bar{\boldsymbol{\Gamma}} = \mathbf{I}_q$ *and* $\mathbf{H}^\top\boldsymbol{\Psi}^{-1}\mathbf{H}$ *is a diagonal matrix?*

Note: Other results on quadratic forms in skew-elliptical random vectors can be found in Fang (2003, 2005d).

8.6 Open Research Problems

Open Research Problem 21 *Let* $\mathbf{X}_1, \ldots, \mathbf{X}_n$ *be a random sample from* $\mathbf{X} = (X_1, \mathbf{X}_2^\top)^\top \sim \mathcal{SE}_p(\mathbf{0}_p, \boldsymbol{\Psi}, \boldsymbol{\eta}, \varphi)$.

(a) *Derive the asymptotic joint distribution of the sample mean vector and sample variance matrix;*

(b) *Derive the asymptotic distribution of* $R_n^2 = \mathbf{S}_{21}^\top \mathbf{S}_{22}^{-1} \mathbf{S}_{21}/S_{11}$ *which is defined in terms of the partition of* $\mathbf{S}_n = \frac{1}{n}\sum_{i=1}^n \mathbf{X}_i \mathbf{X}_i^\top$ *as*

$$\mathbf{S}_n = \begin{pmatrix} S_{11} & \mathbf{S}_{21}^\top \\ \mathbf{S}_{21} & \mathbf{S}_{22} \end{pmatrix};$$

(c) *Extend the previous results to the multivariate SUE distributions.*

Open Research Problem 22 *Adapt the representations given in Remark 12 for the SUN distribution to the SUE distribution, and explore whether they yield identifiable SUE distributions.*

Open Research Problem 23 *Explore characterizations of the multivariate SUN distribution within the multivariate SUE family; see Section 4.3 for characterizations of the multivariate normal distribution within the elliptical family.*

Open Research Problem 24 *Explore the entropy, mutual information, KL divergence, J divergence, and Wasserstein distance for two multivariate SUE distributions.*

Open Research Problem 25 *Extend the canonical form for the SUN distribution from Section 6.3.3 to the SUE distribution and to the AASUE parameterization.*

Open Research Problem 26 *Develop the higher-order moments and Mardia measures of multivariate skewness and kurtosis for the multivariate SUE distribution.*

Open Research Problem 27 *Develop the theory of sampling from the multivariate SUE distribution.*

Open Research Problem 28 *Develop maximum likelihood inference for the multivariate SUE distribution based on the EM algorithm. Pay attention to identifiability issues and consider submodels of interest.*

Open Research Problem 29 *Develop the theory and inference for multivariate skew-meta-elliptical distributions.*

Chapter 9

Weighted and Selection Multivariate Distributions

Weighted and selection multivariate distributions are very important because they are the cornerstone of many non-normal distributions presented in this book. They provide a very general approach to introducing skewed multivariate distributions. Although most of the theory considered in this chapter is developed for the absolutely continuous case, the procedure can be extended to other settings.

9.1 Weighted Distributions

Weighted multivariate distributions (Patil and Rao, 1978; Rao, 1985) are useful for the appropriate formulation of statistical models. Therefore, they can play an important role in the efficient modeling of statistical data and in prediction when standard distributions are unsuitable. In this section, a brief description of these distributions is provided. For a more detailed review of the properties and implications of these distributions, see, e.g., Saghir et al. (2017) and references therein.

As a starting point, if $f(\mathbf{x})$ is the (reference) pdf of a random vector $\mathbf{X} \in \mathbb{R}^p$, and $w(\mathbf{x};\boldsymbol{\lambda}) > 0$ is an integrable (scalar-valued) weight function (which can be indexed by a parameter vector $\boldsymbol{\lambda} \in \mathbb{R}^q$), then

$$\begin{aligned} 0 < \mathrm{E}\{w(\mathbf{X};\boldsymbol{\lambda})\} &= \int_{\mathbb{R}^p} w(\mathbf{x};\boldsymbol{\lambda}) f(\mathbf{x}) \mathrm{d}\mathbf{x} < \infty \\ \Longrightarrow f^{\mathrm{w}}(\mathbf{x};\boldsymbol{\lambda}) &= \frac{w(\mathbf{x};\boldsymbol{\lambda})}{\mathrm{E}\{w(\mathbf{X};\boldsymbol{\lambda})\}} f(\mathbf{x}) \end{aligned}$$

is also a pdf on $\mathbb{R}^p$ called a *weighted pdf*, which corresponds to the pdf of a random vector $\mathbf{X}^{\mathrm{w}}$ called the *weighted version* of $\mathbf{X}$. A natural and convenient condition on the weight function is

$$w(\mathbf{x};\mathbf{0}_q) = \mathrm{E}\{w(\mathbf{X};\boldsymbol{\lambda})\}\big|_{\boldsymbol{\lambda}=\mathbf{0}_q}, \quad \forall \mathbf{x} \in \mathbb{R}^p,$$

which implies that $f^{\mathrm{w}}(\mathbf{x};\boldsymbol{\lambda})\big|_{\boldsymbol{\lambda}=\mathbf{0}_q} = f(\mathbf{x})$, $\forall \mathbf{x} \in \mathbb{R}^p$. Below are two non-standard (unconventional) examples of weighted normal distributions.

Example 108 (Multivariate weighted normal distributions) *Let $f(\mathbf{x}) = \phi_p(\mathbf{x})$, $\mathbf{x} \in \mathbb{R}^p$, be the pdf of $\mathbf{X} \sim \mathcal{N}_p(\mathbf{0}_p, \mathbf{I}_p)$. Then:*
(a) If $w(\mathbf{x};\boldsymbol{\lambda}) = \exp(\boldsymbol{\lambda}^\top \mathbf{x})$, then

$$\begin{aligned} \mathrm{E}\{w(\mathbf{X};\boldsymbol{\lambda})\} &= \mathrm{E}\{\exp(\boldsymbol{\lambda}^\top \mathbf{X})\} = M_{\mathbf{X}}(\boldsymbol{\lambda}) = \exp(\boldsymbol{\lambda}^\top \boldsymbol{\lambda}/2), \quad \boldsymbol{\lambda} \in \mathbb{R}^p \\ \Longrightarrow f^{\mathrm{w}}(\mathbf{x};\boldsymbol{\lambda}) &= \frac{\exp(\boldsymbol{\lambda}^\top \mathbf{x})}{\exp(\boldsymbol{\lambda}^\top \boldsymbol{\lambda}/2)} \phi_p(\mathbf{x}), \quad \mathbf{x} \in \mathbb{R}^p. \end{aligned}$$

(b) If $w(\mathbf{x};\boldsymbol{\lambda}) = 1+(\boldsymbol{\lambda}^\top \mathbf{x})^2$, *then as* $\boldsymbol{\lambda}^\top \mathbf{X} \sim \mathcal{N}(0,\boldsymbol{\lambda}^\top\boldsymbol{\lambda})$ *we have*

$$\begin{aligned} \mathrm{E}\{w(\mathbf{X};\boldsymbol{\lambda})\} &= 1+\mathrm{E}\{(\boldsymbol{\lambda}^\top \mathbf{X})^2\} = 1+\boldsymbol{\lambda}^\top\boldsymbol{\lambda}, \quad \boldsymbol{\lambda} \in \mathbb{R}^p \\ \Longrightarrow f^{\mathrm{w}}(\mathbf{x};\boldsymbol{\lambda}) &= \frac{1+(\boldsymbol{\lambda}^\top \mathbf{x})^2}{1+\boldsymbol{\lambda}^\top\boldsymbol{\lambda}}\phi_p(\mathbf{x}), \quad \mathbf{x} \in \mathbb{R}^p. \end{aligned}$$

Let $\mathbf{X}^{\mathrm{w}} \sim f^{\mathrm{w}}$ *be the respective weighted version of* $\mathbf{X}$. *Then, a location and scale extension of the above weighted normal distributions are obtained by the usual affine transformation*

$$\mathbf{Y} = \boldsymbol{\mu} + \boldsymbol{\Sigma}^{1/2}\mathbf{X}^{\mathrm{w}} \Longrightarrow f(\mathbf{y};\boldsymbol{\mu},\boldsymbol{\Sigma},\boldsymbol{\lambda}) = |\boldsymbol{\Sigma}|^{-1/2} f^{\mathrm{w}}\{\boldsymbol{\Sigma}^{-1/2}(\mathbf{y}-\boldsymbol{\mu});\boldsymbol{\lambda}\}, \quad \mathbf{y} \in \mathbb{R}^p,$$

for some location vector $\boldsymbol{\mu} \in \mathbb{R}^p$ *and* $p \times p$ *dispersion matrix* $\boldsymbol{\Sigma} > 0$.

When the baseline pdf f is symmetric (e.g., $f(-\mathbf{x}) = f(\mathbf{x})$ for all $\mathbf{x}$), then the weighted pdf

$$f^{\mathrm{w}}(\mathbf{x};\boldsymbol{\lambda}) = c w(\mathbf{x};\boldsymbol{\lambda}) f(\mathbf{x}) \quad (c^{-1} = \mathrm{E}\{w(\mathbf{X};\boldsymbol{\lambda})\},\ \mathbf{X} \sim f)$$

becomes an asymmetric density in $\mathbb{R}^p$, and the weighting function $w(\mathbf{x};\boldsymbol{\lambda})$ becomes a skewing mechanism. An important particular case is:

$$w(-\mathbf{x};\boldsymbol{\lambda}) = 1 - w(\mathbf{x};\boldsymbol{\lambda}),\ \forall \mathbf{x} \Longrightarrow \mathrm{E}\{w(\mathbf{X};\boldsymbol{\lambda})\} = \frac{1}{2} \Longrightarrow c = 2.$$

Weight functions with this property are obtained, for example, by choosing $w(\mathbf{x};\boldsymbol{\lambda}) = G\{u(\mathbf{x};\boldsymbol{\lambda})\}$, where $G(u)$ is the cdf of a symmetric univariate distribution, i.e., $G(-u) = 1 - G(u)$ for all $u \in \mathbb{R}$, and $u(\mathbf{x};\boldsymbol{\lambda})$ is an odd scalar-valued function, i.e., $u(-\mathbf{x};\boldsymbol{\lambda}) = -u(\mathbf{x};\boldsymbol{\lambda})$ for all $\mathbf{x}$ and $\boldsymbol{\lambda}$. This yields the well-known *multivariate generalized skew-symmetric* pdf (Genton and Loperfido, 2005) given by

$$f^{\mathrm{w}}(\mathbf{x};\boldsymbol{\lambda}) = 2G\{u(\mathbf{x};\boldsymbol{\lambda})\} f(\mathbf{x}), \quad \mathbf{x} \in \mathbb{R}^p.$$

Example 109 *Two well-known examples of generalized skew-symmetric pdf are:*

(a) Skew-normal pdf: $f^{\mathrm{w}}(\mathbf{x};\boldsymbol{\lambda}) = 2\Phi(\boldsymbol{\lambda}^\top \mathbf{x})\phi_p(\mathbf{x}), \mathbf{x} \in \mathbb{R}^p$, *where* $f(\mathbf{x}) = \phi_p(\mathbf{x})$, $G(u) = \Phi(u)$ *and* $u(\mathbf{x};\boldsymbol{\lambda}) = \boldsymbol{\lambda}^\top \mathbf{x}$ *(linear).*

(b) Skew-t pdf: $f^{\mathrm{w}}(\mathbf{x};\boldsymbol{\lambda}) = 2T_1\left\{\sqrt{\frac{\nu+1}{\nu+\|\mathbf{x}\|^2}}\boldsymbol{\lambda}^\top \mathbf{x}; \nu+1\right\} t_p(\mathbf{x};\nu)$, *where* $f(\mathbf{x}) = t_p(\mathbf{x};\nu)$, $G(u) = T_1(u;\nu+1)$ *and* $u(\mathbf{x};\boldsymbol{\lambda}) = \sqrt{\frac{\nu+1}{\nu+\|\mathbf{x}\|^2}}\boldsymbol{\lambda}^\top \mathbf{x}$.

Weighted distributions are used when the sampling mechanism records observations according to a certain probability. For applications and other useful concepts, see, e.g., Patil and Rao (1978), Rao (1985), and more recently Copas and Li (1997) and Ma et al. (2005). Currently, these distributions have various applications for statistical analysis based on complex data.

9.2 Selection Distributions

Perhaps the basis of selection distributions dates back to the fundamental works of Azzalini (1985), Azzalini (1986), and Azzalini and Dalla Valle (1996) to construct univariate and multivariate skew-normal distributions, respectively. This was also used in Branco and Dey (2001) in the construction of multivariate skew-elliptical distributions. Later, Arellano-Valle et al. (2002b), Arellano-Valle and del Pino (2004) and Arellano-Valle and Genton (2005) extended this idea to obtain skewed distributions from arbitrary symmetric ones. In all these works, a selection mechanism of the same type was considered, which consists of truncating a latent variable at the origin. Selection mechanisms based on truncations at an arbitrary location vector were considered, e.g., by Arellano-Valle and Azzalini (2006) to obtain the SUN distribution and later by Arellano-Valle and Genton (2010c) to extend the SUN to the SUE family. However, the concept of selection distributions under arbitrary selection mechanisms, together with its extension beyond elliptical distributions and the formal study of their probabilistic properties, was developed in Arellano-Valle et al. (2006a). More recent explorations of selection distributions based on generalized symmetry settings can be found in Azzalini (2012). In parallel, similar ideas under the concept of *hidden truncation distributions* appeared in the works of Arnold et al. (1993), Arnold and Beaver (2000a, 2002), and Arnold and Beaver (2004), and under *conditionally specified multivariate skewed distributions* in Arnold et al. (2002); for a recent review in this direction, with several examples and application fields, see Arnold and Sarabia (2022) and Ghosh and Ng (2025).

As a starting point, given two random vectors $\mathbf{X} \in \mathbb{R}^p$ and $\mathbf{X}_0 \in \mathbb{R}^q$ and a Borel proper subset S_0 of $\mathbb{R}^q$, the total probability theorem states the following:

$$\mathrm{P}(\mathbf{X}_0 \in S_0) = \int_{S_0} \mathrm{P}(\mathbf{X}_0 \in S_0 \mid \mathbf{X} = \mathbf{x}) \mathrm{d}F(\mathbf{x}) \Longrightarrow 1 = \int_{S_0} \frac{\mathrm{P}(\mathbf{X}_0 \in S_0 \mid \mathbf{X} = \mathbf{x})}{\mathrm{P}(\mathbf{X}_0 \in S_0)} \mathrm{d}F(\mathbf{x}),$$

where $F(\mathbf{x})$ is the cdf of $\mathbf{X}$. Thus, when it is assumed that $F(\mathbf{x})$ is absolutely continuous with pdf $f(\mathbf{x})$ on $\mathbb{R}^p$, then it follows from this simple result that the function given by

$$f(\mathbf{x} \mid \mathbf{X}_0 \in S_0) = \frac{\mathrm{P}(\mathbf{X}_0 \in S_0 \mid \mathbf{X} = \mathbf{x})}{\mathrm{P}(\mathbf{X}_0 \in S_0)} f(\mathbf{x}),$$

becomes a pdf in $\mathbb{R}^p$. It is called a *selection pdf* and corresponds to the *selection multivariate distribution* of the *selection random vector* $\mathbf{Z} \stackrel{d}{=} (\mathbf{X} \mid \mathbf{X}_0 \in S_0)$, also called the *selection version* of $\mathbf{X}$ under the *selection mechanism* $\{\mathbf{X}_0 \in S_0\}$ (Arellano-Valle et al., 2006a). The corresponding *selection cdf* is then

$$F(\mathbf{x} \mid \mathbf{X}_0 \in S_0) = \mathrm{P}(\mathbf{X} \leq \mathbf{x} \mid \mathbf{X}_0 \in S_0) = \frac{\mathrm{P}(\mathbf{X} \leq \mathbf{x}, \mathbf{X}_0 \in S_0)}{\mathrm{P}(\mathbf{X}_0 \in S_0)}. \tag{9.1}$$

Writing the joint pdf of $\mathbf{X}$ and $\mathbf{X}_0$ as $f(\mathbf{x}, \mathbf{x}_0) = f(\mathbf{x}) f_0(\mathbf{x}_0|\mathbf{x})$, where $f(\mathbf{x})$ is the marginal pdf of $\mathbf{X}$ and $f_0(\mathbf{x}_0|\mathbf{x})$ the conditional pdf of $(\mathbf{X}_0 \mid \mathbf{X} = \mathbf{x})$, we see that the conditional joint pdf of $(\mathbf{X}, \mathbf{X}_0)$ given the event $\{\mathbf{X}_0 \in S_0\}$ is

$$f(\mathbf{x}, \mathbf{x}_0 \mid \mathbf{X}_0 \in S_0) = \frac{f(\mathbf{x}, \mathbf{x}_0)}{\mathrm{P}(\mathbf{X}_0 \in S_0)} = \frac{f(\mathbf{x}) f_0(\mathbf{x}_0|\mathbf{x})}{\mathrm{P}(\mathbf{X}_0 \in S_0)}. \tag{9.2}$$

In such a case, we can obtain the selection pdf as

$$f(\mathbf{x} \mid \mathbf{X}_0 \in S_0) = \frac{f(\mathbf{x}) \int_{S_0} f_0(\mathbf{x}_0|\mathbf{x}) \mathrm{d}\mathbf{x}_0}{\mathrm{P}(\mathbf{X}_0 \in S_0)} = \frac{f(\mathbf{x}) \mathrm{P}(\mathbf{X}_0 \in S_0 \mid \mathbf{X} = \mathbf{x})}{\mathrm{P}(\mathbf{X}_0 \in S_0)}. \tag{9.3}$$

Taking into account (9.3), it is clear that the joint pdf (9.2) can be factorized as

$$\begin{aligned} f(\mathbf{x}, \mathbf{x}_0 \mid \mathbf{X}_0 \in S_0) &= \frac{f(\mathbf{x}) \mathrm{P}(\mathbf{X}_0 \in S_0 \mid \mathbf{X} = \mathbf{x})}{\mathrm{P}(\mathbf{X}_0 \in S_0)} \times \frac{f_0(\mathbf{x}_0|\mathbf{x})}{\mathrm{P}(\mathbf{X}_0 \in S_0 \mid \mathbf{X} = \mathbf{x})} \\ &= f(\mathbf{x} \mid \mathbf{X}_0 \in S_0) \times f_0(\mathbf{x}_0 \mid \mathbf{x}, \mathbf{X}_0 \in S_0), \end{aligned}$$

where $f_0(\mathbf{x}_0 \mid \mathbf{x}, \mathbf{X}_0 \in S_0) = f_0(\mathbf{x}_0|\mathbf{x}) / \mathrm{P}(\mathbf{X}_0 \in S_0 \mid \mathbf{X} = \mathbf{x})$ is just the conditional pdf of $(\mathbf{X}_0 \mid \mathbf{X} = \mathbf{x})$ restricted by the subset S_0. In other words, any selection pdf can also be expressed as

$$f(\mathbf{x} \mid \mathbf{X}_0 \in S_0) = \frac{f(\mathbf{x}, \mathbf{x}_0 \mid \mathbf{X}_0 \in S_0)}{f(\mathbf{x}_0 \mid \mathbf{x}, \mathbf{X}_0 \in S_0)}. \tag{9.4}$$

For simplicity, we denote the cdf of $\mathbf{Z} \stackrel{d}{=} (\mathbf{X} \mid \mathbf{X}_0 \in S_0)$ by $F(\mathbf{z} \mid S_0)$ and its pdf when it exists by $f(\mathbf{z} \mid S_0)$. Thus, we have that $F(\mathbf{z} \mid S_0) = F(\mathbf{z} \mid \mathbf{X}_0 \in S_0)$ and $f(\mathbf{z} \mid S_0) = f(\mathbf{z} \mid \mathbf{X}_0 \in S_0)$ are given by (9.1) and (9.3) or (9.4), respectively. Similarly, the *selection cf* of $\mathbf{Z}$ is denoted by $\psi(\mathbf{t} \mid S_0) = \int_{\mathbb{R}^p} e^{i\mathbf{t}^\top \mathbf{z}} \mathrm{d}F(\mathbf{z} \mid S_0)$.

Example 110 (Univariate skew-normal distribution) *Let* $(X_0 \mid X = x) \sim \mathcal{N}(\rho x, 1-\rho^2)$, *where* $|\rho| < 1$, *and* $X \sim \mathcal{N}(0,1)$. *Note that* ρ *is the correlation between* X *and* X_0. *In addition, it is straightforward to see that* $X_0 \sim \mathcal{N}(0,1)$ *and that* $\tilde{X}_0 = (X_0 - \rho X)/\sqrt{1-\rho^2} \sim \mathcal{N}(0,1)$, *which is independent of* X. *Let* $\phi(x) = (2\pi)^{-1/2} e^{-x^2/2}$ *and* $\Phi(x) = \int_{-\infty}^{x} \phi(t)\mathrm{d}t$ *be the pdf and cdf of a* $\mathcal{N}(0,1)$ *distribution, respectively. By letting* $\alpha = \rho/\sqrt{1-\rho^2}$, *we have for* $S_0 = (0, +\infty)$ *that* $\mathrm{P}(X_0 > 0) = 1/2$ *and* $\mathrm{P}(X_0 > 0 \mid X = x) = \mathrm{P}(\tilde{X}_0 > -\alpha x) = \mathrm{P}(\tilde{X}_0 \leq \alpha x) = \Phi(\alpha x)$. *Therefore we find that the selection pdf of* $Z \stackrel{d}{=} (X \mid X_0 > 0)$ *is just the Azzalini skew-normal pdf (Azzalini, 1985) defined by*

$$f(z) = 2\phi(z)\Phi(\alpha z), \quad \forall z \in \mathbb{R}.$$

We write this as $Z \sim \mathcal{ASN}(\alpha)$, $\alpha \in \mathbb{R}$.

Note that the selection distributions are weighted distributions, where the weight function is $w(\mathbf{x}) = \mathrm{P}(\mathbf{X}_0 \in S_0 \mid \mathbf{X} = \mathbf{x})$ and so $\mathrm{E}\{w(\mathbf{X})\} = \mathrm{P}(\mathbf{X}_0 \in S_0)$ with $\mathbf{X} \sim f$.

Even if the baseline pdf f of $\mathbf{X}$ is symmetric, the selection pdf at (9.3) is always asymmetric, unless one of the following holds: (i) $\mathrm{P}(\mathbf{X}_0 \in S_0 \mid \mathbf{X} = \mathbf{x}) = \mathrm{P}(\mathbf{X}_0 \in S_0 \mid \mathbf{X} = -\mathbf{x})$ for all $\mathbf{x}$ or (ii) $\mathrm{P}(\mathbf{X}_0 \in S_0 \mid \mathbf{X} = \mathbf{x}) = \mathrm{P}(\mathbf{X}_0 \in S_0)$ for all $\mathbf{x}$. Condition (ii) is automatically fulfilled when $\mathbf{X} \perp\!\!\!\perp \mathbf{X}_0$, but it can also be verified without independence when these vectors are uncorrelated. In fact, the asymmetry is closely linked to the correlation matrix between $\mathbf{X}$ and $\mathbf{X}_0$.

Example 111 (Examples of selection mechanisms) *Some examples for* $(X, X_0) \in \mathbb{R}^2$ *are:*

(a) $(X \mid X_0 + \tau > 0)$, $\tau \in \mathbb{R}$;

(b) $(X \mid \tau < X_0 < \lambda)$, $\tau \in \mathbb{R}$, $\lambda \in \mathbb{R}$, $\tau < \lambda$;

(c) $(X \mid X_0 < \lambda X + \tau)$, $\tau \in \mathbb{R}$, $\lambda \in \mathbb{R}$;

(d) $(X \mid X X_0 > 0)$;

(e) $(X \mid X^2 + X_0^2 > \tau)$, $\tau > 0$;

(f) $(X \mid X_0 < \lambda |X| + \tau)$, $\tau \in \mathbb{R}$, $\lambda \in \mathbb{R}$.

9.3 Some General Properties of Selection Random Vectors

Many properties of a selection random vector $\mathbf{Z} \stackrel{d}{=} (\mathbf{X} \mid \mathbf{X}_0 \in S_0)$ can be studied directly from its definition; see Arellano-Valle et al. (2006a) and Arellano-Valle and Genton (2010a) for details. For instance, we have:

(P1) *Independence and truncated distribution:* If $\mathbf{X} \perp\!\!\!\perp \mathbf{X}_0$ then $\mathbf{Z} \stackrel{d}{=} \mathbf{X} \sim f$; and if $\mathbf{X} = \mathbf{X}_0$ then we obtain the *truncated random vector* $\mathbf{Z}_0 \stackrel{d}{=} (\mathbf{X}_0 \mid \mathbf{X}_0 \in S_0)$ and its cdf is the *truncated cdf* $F_0(\mathbf{x}_0 \mid S_0) = F_0(\mathbf{x}_0)/\mathrm{P}(\mathbf{X}_0 \in S_0)$.

(P2) *Selection cf:* The selection cf of $\mathbf{Z} \stackrel{d}{=} (\mathbf{X} \mid \mathbf{X}_0 \in S_0)$ may be computed as

$$\psi(\mathbf{t} \mid S_0) = \mathrm{E}(e^{i\mathbf{t}^\top \mathbf{X}} \mid \mathbf{X}_0 \in S_0) = \int_{\mathbf{x}_0 \in S_0} \psi(\mathbf{t}|\mathbf{x}_0) \mathrm{d}F_0(\mathbf{x}_0 \mid \mathbf{X}_0 \in S_0), \tag{9.5}$$

where $\psi(\mathbf{t}|\mathbf{x}_0)$ is the conditional cf of $(\mathbf{X} \mid \mathbf{X}_0 = \mathbf{x}_0)$.

(P3) *Transformations:* For any Borel vector-valued function $\mathbf{g}$, we have $\mathbf{g}(\mathbf{Z}) \stackrel{d}{=} \{\mathbf{g}(\mathbf{X}) \mid \mathbf{X}_0 \in S_0\}$. Moreover, if $\mathbf{g}(\mathbf{X})$ has a pdf $\tilde{f}(\mathbf{y})$, then $\mathbf{g}(\mathbf{Z})$ also has a selection pdf of the form

$$\tilde{f}(\mathbf{y} \mid S_0) = \tilde{f}(\mathbf{y}) \frac{\mathrm{P}\{\mathbf{X}_0 \in S_0 \mid \mathbf{g}(\mathbf{X}) = \mathbf{y}\}}{\mathrm{P}(\mathbf{X}_0 \in S_0)},$$

and its selection cf is

$$\tilde{\psi}(\mathbf{t} \mid S_0) = \int_{\mathbf{x}_0 \in S_0} \tilde{\psi}(\mathbf{t}|\mathbf{x}_0) \mathrm{d}F_0(\mathbf{x}_0 \mid \mathbf{X}_0 \in S_0), \tag{9.6}$$

where $\tilde{\psi}(\mathbf{t}|\mathbf{x}_0)$ is the conditional cf of $\{\mathbf{g}(\mathbf{X}) \mid \mathbf{X}_0 = \mathbf{x}_0\}$;
Two direct consequences of (P3) are:

(a) If $\mathrm{P}\{\mathbf{X}_0 \in S_0 \mid \mathbf{g}(\mathbf{X}) = \mathbf{y}\} = \mathrm{P}(\mathbf{X}_0 \in S_0)$, $\forall \mathbf{y}$, then $\mathbf{g}(\mathbf{Z}) \stackrel{d}{=} \mathbf{g}(\mathbf{X})$, which is trivially true if $\mathbf{X}_0 \perp\!\!\!\perp \mathbf{g}(\mathbf{X})$, but it can also be true if $\mathbf{X}_0$ and $\mathbf{g}(\mathbf{X})$ are uncorrelated. For example, if $S_0 = \{\mathbf{x} \in \mathbb{R}^q : \mathbf{x} > \mathbf{0}_q\} = \mathbb{R}^q_+$ and the random vectors $\mathbf{X}_0$ and $\mathbf{X}$ are (centrally) symmetric, then $\mathbf{g}(\mathbf{Z}) \stackrel{d}{=} \mathbf{g}(\mathbf{X})$ for any even Borel function $\mathbf{g}(\mathbf{x})$ since in such a case $\mathbf{X}_0$ and $\mathbf{g}(\mathbf{X})$ are uncorrelated;

(b) If $\mathbf{g}(\mathbf{X}) = \mathbf{A}\mathbf{X} + \mathbf{b}$, then $\mathbf{g}(\mathbf{Z}) \stackrel{d}{=} \mathbf{A}\mathbf{Z} + \mathbf{b}$, for any matrix $\mathbf{A} \in \mathbb{R}^{q \times p}$ and vector $\mathbf{b} \in \mathbb{R}^q$ fixed.

Note also from (b) that any selection random vector $\mathbf{Z} \stackrel{d}{=} (\mathbf{X} \mid \mathbf{X}_0 \in S_0)$ is closed under affine transformations and, in particular, under marginalizations, when the original vector $\mathbf{X}$ has such a property.

(P4) *Marginalization, conditioning, independence and additivity:* Let $\mathbf{Z} \stackrel{d}{=} (\mathbf{X} \mid \mathbf{X}_0 \in S_0)$ be partitioned as $(\mathbf{Z}_1, \mathbf{Z}_2)$, which is induced by a partition $(\mathbf{X}_1, \mathbf{X}_2)$ of $\mathbf{X}$. Suppose that $\mathbf{X} = (\mathbf{X}_1^\top, \mathbf{X}_2^\top)^\top$ has a pdf with marginals f_1 and f_2 for $\mathbf{X}_1$ and $\mathbf{X}_2$, respectively. Then:

(a) *(Marginalization)* $\mathbf{Z}_i \stackrel{d}{=} (\mathbf{X}_i \mid \mathbf{X}_0 \in S_0)$ and has a *marginal selection pdf* given by

$$f_i(\mathbf{z}_i \mid S_0) = f_i(\mathbf{x}_i) \frac{\mathrm{P}(\mathbf{X}_0 \in S_0 \mid \mathbf{X}_i = \mathbf{z}_i)}{\mathrm{P}(\mathbf{X}_0 \in S_0)}, \quad i = 1, 2;$$

(b) *(Conditioning)* $(\mathbf{Z}_1 \mid \mathbf{Z}_2 = \mathbf{z}_2) \stackrel{d}{=} (\mathbf{X}_1 \mid \mathbf{X}_2 = \mathbf{z}_2, \mathbf{X}_0 \in S_0)$ and has a *conditional selection pdf* given by

$$f(\mathbf{z}_1 \mid \mathbf{z}_2, S_0) = f(\mathbf{z}_1|\mathbf{z}_2) \frac{\mathrm{P}(\mathbf{X}_0 \in S_0 \mid \mathbf{X}_1 = \mathbf{z}_1, \mathbf{X}_2 = \mathbf{z}_2)}{\mathrm{P}(\mathbf{X}_0 \in S_0 \mid \mathbf{X}_2 = \mathbf{z}_2)};$$

(c) *(Independence)* If the selection region S_0 is such that $\{\mathbf{X}_0 \in S_0\} = \{\mathbf{X}_{01} \in S_{01}, \mathbf{X}_{02} \in S_{02}\}$ for any partition $(\mathbf{X}_{01}, \mathbf{X}_{02})$ of $\mathbf{X}_0$, which is such that $(\mathbf{X}_{01}, \mathbf{X}_1)$ and $(\mathbf{X}_{02}, \mathbf{X}_2)$ are independent, then we have $\mathbf{Z}_i \stackrel{d}{=} (\mathbf{X}_i \mid \mathbf{X}_{0i} \in S_{0i})$, $i = 1, 2$, and they are independent;

(d) *(Additivity)* Under the same conditions as in (c), we have

$$(\mathbf{Z}_1 + \mathbf{Z}_2) \stackrel{d}{=} (\mathbf{X}_1 \mid \mathbf{X}_{01} \in S_{01}) + (\mathbf{X}_2 \mid \mathbf{X}_{02} \in S_{02}) \stackrel{d}{=} (\mathbf{X}_1 + \mathbf{X}_2 \mid \mathbf{X}_0 \in S_0).$$

(P5) *Invariance of quadratic forms in selection random vectors:* Arellano-Valle and Genton (2010a) have studied the conditions for the invariance of quadratic forms in random vectors having a selection distribution with an arbitrary selection subset S_0 of $\mathbb{R}^q$. The results consider the cases where the symmetric pdf f is normal, elliptical, or a scale mixture of normals.

(P6) *Invariance by rescaling and permutations:* Let $\mathbf{Z} \overset{d}{=} (\mathbf{X} \mid \mathbf{X}_0 \in S_0)$ and $\mathbf{Z}_\mathbf{A} \overset{d}{=} (\mathbf{X} \mid \mathbf{X}_0 \in S_\mathbf{A})$, where $S_\mathbf{A} = \{\mathbf{x}_\mathbf{A} = \mathbf{A}\mathbf{x} \mid \mathbf{x} \in S_0\}$ for some matrix $\mathbf{A} \in \mathbb{R}^{q\times q}$. It is clear that if $\mathbf{A} = \mathbf{DP}$, with $\mathbf{D}$ a $q \times q$ positive diagonal matrix and $\mathbf{P} \in \mathcal{P}_q = \{\mathbf{P} \in \mathbb{R}^{q\times q} : \mathbf{P}\mathbf{P}^\top = \mathbf{P}^\top\mathbf{P} = \mathbf{I}_q \text{ and } \mathbf{P}\mathbf{1}_q = \mathbf{1}_q\}$, the group of $q \times q$ permutation matrices, then $\mathrm{P}(\mathbf{A}\mathbf{X}_0 \in S_\mathbf{A}) = \mathrm{P}(\mathbf{X}_0 \in S_0)$ and $\mathrm{P}(\mathbf{A}\mathbf{X}_0 \in S_\mathbf{A} | \mathbf{X} = \mathbf{x}) = \mathrm{P}(\mathbf{X}_0 \in S_0 | \mathbf{X} = \mathbf{x})$ for every $\mathbf{x} \in \mathbb{R}^p$. Consequently, for every matrix $\mathbf{A} = \mathbf{DP}$, we have $f(\mathbf{z} \mid S_\mathbf{A}) = f(\mathbf{z} \mid S_0)$ for all $\mathbf{z} \in \mathbb{R}^d$ and therefore $\mathbf{Z} \overset{d}{=} \mathbf{Z}_\mathbf{A}$.
In this sense, the identifiability issues discussed in Section 8.1.2 for the SUE family (which also apply to the SUN and SUT subfamilies) are direct consequences of the symmetry of the joint distribution of $\mathbf{X}_0$ and $\mathbf{X}$, and noting that $S_\mathbf{A} = (-\mathbf{A}\boldsymbol{\tau}, \infty)$, $\boldsymbol{\tau} \in \mathbb{R}^q$, so $\mathrm{P}(\mathbf{A}\mathbf{X}_0 \in S_\mathbf{A}) = \mathrm{P}(\mathbf{A}\mathbf{X}_0 > -\mathbf{A}\boldsymbol{\tau}) = \mathrm{P}(-\mathbf{A}\mathbf{X}_0 < \mathbf{A}\boldsymbol{\tau}) = F_{\mathbf{A}\mathbf{X}_0}(\mathbf{A}\boldsymbol{\tau})$, since $-\mathbf{A}\mathbf{X}_0$ has the same distribution as $\mathbf{A}\mathbf{X}_0$. Similarly, $\mathrm{P}(\mathbf{A}\mathbf{X}_0 \in S_\mathbf{A} \mid \mathbf{X} = \mathbf{z}) = F_{\mathbf{A}\mathbf{X}_0|\mathbf{X}=\mathbf{z}}\{\mathbf{A}\boldsymbol{\tau} + \mathbf{A}\boldsymbol{\mu}_{0\cdot 1}(\mathbf{z})\}$, where $\boldsymbol{\mu}_{0\cdot 1}(\mathbf{z})$ is the conditional location of $\mathbf{X}_0 | \mathbf{X} = \mathbf{z}$.

(P7) *Linear and nonlinear selection mechanisms:* By redefining $\mathbf{X}_0$ as $\mathbf{X}_0 = \mathbf{g}(\mathbf{X}, \tilde{\mathbf{X}}_0)$ for some function $\mathbf{g}$ and random vector $\tilde{\mathbf{X}}_0$ jointly distributed with $\mathbf{X}$, we can obtain linear and nonlinear selection mechanisms depending on $\mathbf{X}$. The independence (or null correlation) between $\mathbf{X}$ and $\tilde{\mathbf{X}}_0$ may be a useful and reasonable assumption in this case. In the linear case, a common example is $\mathbf{g}(\mathbf{X}, \tilde{\mathbf{X}}_0) = \boldsymbol{\Lambda}\mathbf{X} - \tilde{\mathbf{X}}_0 + \boldsymbol{\tau}$ for some fixed matrix $\boldsymbol{\Lambda} \in \mathbb{R}^{q\times p}$ and a fixed vector $\boldsymbol{\tau} \in \mathbb{R}^q$, and taking $S_0 = \mathbb{R}^q_+$. Similarly, we can redefine the random vector $\mathbf{X}$ as $\mathbf{X} = \mathbf{g}(\tilde{\mathbf{X}}, \mathbf{X}_0)$, and consider the independence assumption $\tilde{\mathbf{X}} \perp\!\!\!\perp \mathbf{X}_0$ to obtain for the selection random vector $\mathbf{Z} \overset{d}{=} \{\mathbf{g}(\tilde{\mathbf{X}}, \mathbf{X}_0) \mid \mathbf{X}_0 \in S_0\}$ the stochastic representation $\mathbf{Z} \overset{d}{=} \mathbf{g}(\tilde{\mathbf{X}}, \mathbf{Z}_0)$, where $\mathbf{Z}_0 \overset{d}{=} (\tilde{\mathbf{X}} \mid \mathbf{X}_0 \in S_0)$. An example in the linear case is $\mathbf{g}(\tilde{\mathbf{X}}, \mathbf{Z}_0) = \mathbf{H}\mathbf{X}_0 + \tilde{\mathbf{X}}$ for some fixed matrix $\mathbf{H} \in \mathbb{R}^{p\times q}$, so it follows that $\mathbf{Z} \overset{d}{=} \mathbf{H}\mathbf{Z}_0 + \tilde{\mathbf{X}}$.

(P8) *Most common case:* If $\mathbf{X} \overset{d}{=} -\mathbf{X}$ and $\mathbf{X}_0 \overset{d}{=} -\mathbf{X}_0$, i.e., they are (centrally) symmetric (around the origin) random vectors, then the selection pdf of $\mathbf{Z} \overset{d}{=} (\mathbf{X} \mid \mathbf{X}_0 + \boldsymbol{\tau} > \mathbf{0}_q)$ with $\boldsymbol{\tau} \in \mathbb{R}^q$ is

$$f(\mathbf{z} \mid S_0) = \frac{F_0(\boldsymbol{\tau} \mid \mathbf{z})}{F_0(\boldsymbol{\tau})} f(\mathbf{z}), \quad \mathbf{z} \in \mathbb{R}^p.$$

Moreover, if $\mathbf{X}_0 = \boldsymbol{\eta}(\mathbf{X}) - \tilde{\mathbf{X}}_0$, where $\boldsymbol{\eta}(-\mathbf{u}) = -\boldsymbol{\eta}(\mathbf{u})$, $\forall \mathbf{u}$, $\mathbf{X} \perp\!\!\!\perp \tilde{\mathbf{X}}_0$ and both $\mathbf{X}$ and $\tilde{\mathbf{X}}_0$ are (centrally) symmetric, then

$$f(\mathbf{z} \mid S_0) = \frac{F_0\{\boldsymbol{\tau} + \boldsymbol{\eta}(\mathbf{z})\}}{F_0(\boldsymbol{\tau})} f(\mathbf{z}), \quad \mathbf{z} \in \mathbb{R}^p.$$

In particular, if $\boldsymbol{\tau} = \mathbf{0}_q$ and $\tilde{\mathbf{X}}_0$ is sign-invariant ($\mathbf{D}\tilde{\mathbf{X}}_0 \overset{d}{=} \tilde{\mathbf{X}}_0$, $\forall \mathbf{D} = \mathrm{diag}(\pm 1, \ldots, \pm 1)$, see Section 2.2.3), then

$$F_0(\mathbf{0}_q) = 2^{-q} \Longrightarrow f(\mathbf{z} \mid S_0) = 2^q F_0\{\boldsymbol{\eta}(\mathbf{z})\} f(\mathbf{z}), \quad \mathbf{z} \in \mathbb{R}^p.$$

(P9) *A tractable selection distribution:* When assuming that the components $X_{01}, \ldots, X_{0q}$ of $\mathbf{X}_0$ are conditionally independent given $\mathbf{X} = \mathbf{x}$, then the selection pdf of $\mathbf{Z} = (\mathbf{X} \mid \mathbf{X}_0 + \boldsymbol{\tau} > \mathbf{0}_q)$ becomes

$$f(\mathbf{z} \mid S_0) = f(\mathbf{z}) \frac{\prod_{i=1}^q \mathrm{P}(X_i + \tau_i > 0 \mid \mathbf{X} = \mathbf{z})}{\mathrm{P}(\mathbf{X}_0 + \boldsymbol{\tau} > \mathbf{0}_q)}, \quad \mathbf{z} \in \mathbb{R}^p.$$

As an illustration, see Example 77 on tractable SUN pdfs.

Example 112 *In Example 110, we found that the selection random variable $Z \overset{d}{=} (X \mid X_0 > 0)$, where $(X_0 \mid X = x) \sim \mathcal{N}(\rho x, 1 - \rho^2)$ and $X \sim \mathcal{N}(0, 1)$, has an $\mathcal{ASN}(\alpha)$ distribution with $\alpha = \rho/\sqrt{1 - \rho^2}$. Taking into account (P7), two equivalent selection representations of $Z \sim \mathcal{ASN}(\alpha)$ are:*

(a) Let $\tilde{X}_0 = (\rho X - X_0)/\sqrt{1-\rho^2}$, *and note that* X *and* $\tilde{X}_0$ *are iid* $\mathcal{N}(0,1)$. *Since* $X_0 = \rho X - \sqrt{1-\rho^2}\,\tilde{X}_0$, *it is clear that* $X_0 > 0 \Longleftrightarrow \tilde{X}_0 < \alpha X$. *We then find* $Z \stackrel{d}{=} (X \mid \tilde{X}_0 < \alpha X)$;

(b) Let now $\tilde{X} = (X - \rho X_0)/\sqrt{1-\rho^2}$, *and note again that* X *and* $\tilde{X}_0$ *are iid* $\mathcal{N}(0,1)$. *Here we have* $X = \rho X_0 + \sqrt{1-\rho^2}\,\tilde{X}$. *Hence, we obtain*

$$Z \stackrel{d}{=} (X \mid X_0 > 0) = \left(\rho X_0 + \sqrt{1-\rho^2}\,\tilde{X} \mid X_0 > 0\right) \stackrel{d}{=} \rho Z_0 + \sqrt{1-\rho^2}\,\tilde{X},$$

where $Z_0 = |X_0| \stackrel{d}{=} (X_0 \mid X_0 > 0)$ *and* $Z_0 \sim 2\phi(z)I_{(0,+\infty)}(z)$, *the half-normal pdf.*

9.4 Moments of Selection Random Vectors

As indicated by (9.5) and (9.6), the moments of a selection random vector $\mathbf{Z} \stackrel{d}{=} (\mathbf{X} \mid \mathbf{X}_0 \in S_0)$ are strongly linked to the truncated moments of $(\mathbf{X}_0 \mid \mathbf{X}_0 \in S_0)$. For instance, if $\mathbf{Z} \in \mathbb{R}^p$ has a selection pdf $f(\mathbf{z} \mid S_0)$, then by (9.3) we have for any integrable scalar-valued function $g(\mathbf{z})$ that

$$\begin{aligned}
\mathrm{E}\{g(\mathbf{Z})\} &= \int_{\mathbb{R}^p} g(\mathbf{x}) f(\mathbf{x} \mid S_0)\mathrm{d}\mathbf{x} = \frac{\int_{\mathbb{R}^p} g(\mathbf{x}) f(\mathbf{x})\{\int_{S_0} f(\mathbf{x}) f_0(\mathbf{x}_0|\mathbf{x})\mathrm{d}\mathbf{x}_0\}\mathrm{d}\mathbf{x}}{\mathrm{P}(\mathbf{X}_0 \in S_0)} \\
&= \frac{\int_{S_0}\{\int_{\mathbb{R}^p} g(\mathbf{x}) f(\mathbf{x}|\mathbf{x}_0)\mathrm{d}\mathbf{x}\} f_0(\mathbf{x}_0)\mathrm{d}\mathbf{x}_0}{\mathrm{P}(\mathbf{X}_0 \in S_0)} = \int_{S_0} \mathrm{E}\{g(\mathbf{X}) \mid \mathbf{x}_0\} f_0(\mathbf{x}_0 \mid \mathbf{X}_0 \in S_0)\mathrm{d}\mathbf{x}_0 \\
&= \mathrm{E}[\mathrm{E}\{g(\mathbf{X}) \mid \mathbf{X}_0\} \mid \mathbf{X}_0 \in S_0].
\end{aligned}$$

In particular, for the *selection mean vector* $\mathrm{E}(\mathbf{Z}) = \mathrm{E}(\mathbf{X} \mid \mathbf{X}_0 \in S_0)$ and *selection variance matrix* $\mathrm{var}(\mathbf{Z}) = \mathrm{var}(\mathbf{X} \mid \mathbf{X}_0 \in S_0)$ we have

$$\mathrm{E}(\mathbf{Z}) = \mathrm{E}\{\mathrm{E}(\mathbf{X} \mid \mathbf{X}_0) \mid \mathbf{X}_0 \in S_0\} \quad \text{and} \quad \mathrm{var}(\mathbf{Z}) = \mathrm{E}\{\mathrm{E}(\mathbf{X}\mathbf{X}^\top \mid \mathbf{X}_0) \mid \mathbf{X}_0 \in S_0\} - \mathrm{E}(\mathbf{Z})\mathrm{E}(\mathbf{Z})^\top.$$

Similarly, from the partition $(\mathbf{Z}_1, \mathbf{Z}_2)$ of $\mathbf{Z}$ which is induced by the partition $(\mathbf{X}_1, \mathbf{X}_2)$ of $\mathbf{X}$, we may compute the selection covariance matrix of $\mathbf{Z}_1$ and $\mathbf{Z}_2$ as

$$\mathrm{cov}(\mathbf{Z}_1, \mathbf{Z}_2) = \mathrm{cov}(\mathbf{X}_1, \mathbf{X}_2 \mid \mathbf{X}_0 \in S_0) = \mathrm{E}\{\mathrm{E}(\mathbf{X}_1\mathbf{X}_2^\top \mid \mathbf{X}_0) \mid \mathbf{X}_0 \in S_0\} - \mathrm{E}(\mathbf{Z}_1)\mathrm{E}(\mathbf{Z}_2)^\top,$$

where, as before, $\mathrm{E}(\mathbf{Z}_i) = \mathrm{E}\{\mathrm{E}(\mathbf{X}_i \mid \mathbf{X}_0) \mid \mathbf{X}_0 \in S_0\}$, $i = 1, 2$.

Example 113 *In Examples 110 and 112, we obtained the selection* $\mathcal{ASN}(\alpha)$ *random variable* $Z_0 \stackrel{d}{=} (X \mid X_0 > 0)$, *where* $(X \mid X_0 = x_0) \sim \mathcal{N}(\rho x_0, 1-\rho^2)$ *and* $X_0 \sim \mathcal{N}(0,1)$, *and where* α *is related to* ρ *through* $\alpha = \rho/\sqrt{1-\rho^2}$, $|\rho| < 1$. *Thus, for any integrable function* g, *we have* $\mathrm{E}\{g(Z_0)\} = \mathrm{E}[\mathrm{E}\{g(X) \mid X_0\} \mid X_0 > 0]$. *In particular, since* $\mathrm{E}(X \mid x_0) = \rho x_0$ *and* $\mathrm{E}(X^2 \mid x_0) = 1 - \rho^2 + \rho^2 x_0^2$ *we have for the mean and variance of* Z_0 *that*

$$\begin{aligned}
\mathrm{E}(Z_0) &= \mathrm{E}(\rho X_0 \mid X_0 > 0) = \mathrm{E}(|X_0|)\rho = \sqrt{\frac{2}{\pi}}\rho, \\
\mathrm{var}(Z_0) &= \mathrm{E}(1 - \rho^2 + \rho^2 X_0^2 \mid X_0 > 0) - \frac{2}{\pi}\rho^2 = 1 - \frac{2}{\pi}\rho^2,
\end{aligned}$$

since $\mathrm{E}(Z_0^2) = \mathrm{E}(X_0^2 \mid X_0 > 0) = \mathrm{E}(X_0^2) = 1$

Example 114 *Let* $Y = V^{-1/2}X$ *and* $Y_0 = V^{-1/2}X_0$, *where* V *is a scalar random variable independent of* X *and of* X_0, *which are the normal random variables defined in Example 113. Now, let* $Z \stackrel{d}{=} (Y \mid Y_0 > 0)$. *Then,* $Z \stackrel{d}{=} V^{-1/2}Z_0$, *where* $Z_0 \stackrel{d}{=} (X \mid X_0 > 0)$ *and is independent of* V. *So, if* g *is an integrable function, then* $\mathrm{E}\{g(Z)\} = \mathrm{E}\{g(V^{-1/2}Z_0)\}$. *In particular, if* $\mathrm{E}(V^{-1/2}) < \infty$ *then*

$$\mathrm{E}(Z) = \mathrm{E}(V^{-1/2}Z_0) = \mathrm{E}(V^{-1/2})\mathrm{E}(Z_0) = \sqrt{\frac{2}{\pi}}\rho\mathrm{E}(V^{-1/2}).$$

Also, if $\mathrm{E}(V^{-1}) < \infty$, *then*

$$\mathrm{var}(Z) = \mathrm{var}(V^{-1/2}Z_0) = \mathrm{E}(V^{-1})\mathrm{E}(Z_0^2) - \mathrm{E}(V^{-1/2})^2\mathrm{E}(Z_0)^2 = \mathrm{E}(V^{-1}) - \frac{2}{\pi}\rho^2\mathrm{E}(V^{-1/2})^2.$$

Numerical computations of moments of selection elliptical distributions, such as folded and doubly truncated multivariate distributions, can be obtained with the R package `MomTrunc` (Galarza Morales and Lachos, 2024); see also Galarza Morales et al. (2022b) and Galarza Morales et al. (2022a) and references therein.

The results on selection distributions presented here are still quite limited, so more theory on these distributions and their connections to the various areas of application to statistics is needed.

9.5 Maximum and L-Statistics as Selections

The distribution of the maximum of a set of absolutely continuous and potentially dependent random variables beyond the classical independent case is of importance for many applications, e.g., in genetic selection problems or best performance studies. Arellano-Valle and Genton (2008) showed that the exact distribution of the maximum can be interpreted through a selection mechanism: The maximum has the same distribution as one component of the random vector conditioned on being the largest, which naturally places it within the class of fundamental skew (FUS) distributions (Arellano-Valle and Genton, 2005). Indeed, if $\mathbf{X} = (X_1, \ldots, X_p)^\top$ is an absolutely continuous random vector with joint pdf $f_{\mathbf{X}}$ on $\mathbb{R}^p$ and $X_{(p)} = \max_{1 \le i \le p} X_i$, then the pdf of $X_{(p)}$ was shown in this general dependent setting to be:

$$f_{X_{(p)}}(x) = \sum_{i=1}^{p} f_{X_i}(x)\, F_{\mathbf{X}_{-i}|X_i=x}(x\mathbf{1}_{p-1}), \quad x \in \mathbb{R}, \tag{9.7}$$

a finite mixture of densities obtained by conditioning on the event $\{X_i \le X_j,\ \forall j \ne i\}$, where f_{X_i} is the marginal pdf of X_i and $F_{\mathbf{X}_{-i}|X_i=x}$ is the conditional cdf of $\mathbf{X}_{-i} = (X_1, \ldots, X_{i-1}, X_{i+1}, \ldots, X_p)^\top$ given $X_i = x$. When $\mathbf{X}$ is exchangeable, the mixture collapses to:

$$f_{X_{(p)}}(x) = p f_{X_p}(x)\, F_{\mathbf{X}_{-p}|X_p=x}(x\mathbf{1}_{p-1}), \quad x \in \mathbb{R}. \tag{9.8}$$

Equivalently, $X_{(p)} \stackrel{d}{=} (X_p \,|\, X_1 \le \cdots \le X_p)$, so that the maximum is precisely the distribution of one coordinate selected under the event that the components are ordered. From the selection point of view, the factor $F_{\mathbf{X}_{-p}|X_p=x}(x\mathbf{1}_{p-1})$ acts as a skewing function: Even when f_{X_p} is symmetric, conditioning on an ordering event induces skewness. The pdfs in (9.7) and (9.8) can be specialized to elliptical distributions for $\mathbf{X}$, and in particular to the multivariate normal and t distributions, where the resulting formulas become fully explicit. For example, when $\mathbf{X}$ is multivariate normal and exchangeable, i.e., $\mathbf{X} \sim \mathcal{N}_p(\mu\mathbf{1}_p, \sigma^2\{(1-\rho)\mathbf{I}_p + \rho\mathbf{1}_p\mathbf{1}_p^\top\})$, $\rho \in [0,1)$, the pdf of the maximum is:

$$f_{X_{(p)}}(x) = p\phi_1(x;\mu,\sigma^2)\Phi_{p-1}\left(\sqrt{1-\rho}\{(x-\mu)/\sigma\}\mathbf{1}_{p-1}; \mathbf{0}_{p-1}, \mathbf{I}_{p-1} + \rho\mathbf{1}_{p-1}\mathbf{1}_{p-1}^\top\right), \quad x \in \mathbb{R}. \tag{9.9}$$

Note that for $p = 2$, the pdf in (9.9) is exactly skew-normal. These results show how dependence affects the shape of the maximum's distribution; see also Loperfido (2008).

Arellano-Valle and Genton (2007) extended the above results to L-statistics, i.e., linear combinations of order statistics. Denote by $\mathbf{X}_{(p)} = (X_{(1)}, \ldots, X_{(p)})^\top$ the vector of order statistics induced by $\mathbf{X}$ where $X_{(1)} \leq \cdots \leq X_{(p)}$. The aim is to find the joint k-dimensional distribution of $\mathbf{LX}_{(p)}$ where $\mathbf{L} \in \mathbb{R}^{k\times p}$. Let $\mathbf{\Delta} \in \mathbb{R}^{(p-1)\times p}$ be the difference matrix such that $\mathbf{\Delta X} = (X_2 - X_1, X_3 - X_2, \ldots, X_p - X_{p-1})^\top$ and let $S_k = \{\mathbf{y} = \mathbf{Lx} \in \mathbb{R}^k; \mathbf{\Delta x} \geq \mathbf{0}_{p-1}, \mathbf{x} \in \mathbb{R}^p\}$ be the support of $\mathbf{LX}|\mathbf{\Delta X} \geq \mathbf{0}_{p-1}$. Then the joint pdf of $\mathbf{LX}_{(p)}$ is:

$$f_{\mathbf{LX}_{(p)}}(\mathbf{y}) = \sum_{i=1}^{p!} f_{\mathbf{LX}_i}(\mathbf{y}) \mathrm{P}(\mathbf{\Delta X}_i \geq \mathbf{0}_{p-1} | \mathbf{LX}_i = \mathbf{y}), \quad \mathbf{y} \in S_k, \tag{9.10}$$

where $\mathbf{X}_i$ is one of the $p!$ different permutations of the components of $\mathbf{X}$. If $\mathbf{X}$ is exchangeable, then we have $\mathbf{LX}_{(p)} \stackrel{d}{=} (\mathbf{LX}|\mathbf{\Delta X} \geq \mathbf{0}_{p-1})$ and therefore the joint pdf of $\mathbf{LX}_{(p)}$ becomes:

$$f_{\mathbf{LX}_{(p)}}(\mathbf{y}) = p! f_{\mathbf{LX}}(\mathbf{y}) \mathrm{P}(\mathbf{\Delta X} \geq \mathbf{0}_{p-1} | \mathbf{LX} = \mathbf{y}), \quad \mathbf{y} \in S_k. \tag{9.11}$$

The pdfs in (9.10) and (9.11) can be specialized to elliptical distributions for $\mathbf{X}$, and in particular to the multivariate normal and t distributions, where the resulting formulas become fully explicit.

Jamalizadeh and Balakrishnan (2008) studied order statistics from bivariate skew-normal and skew-t distributions. Jamalizadeh et al. (2009c) explored recurrence relations for the cdf of bivariate t and extended skew-t distributions aimed at an application to order statistics from a bivariate t, while Jamalizadeh et al. (2009a) developed recurrence relations for distributions of a skew-t and a linear combination of order statistics from a bivariate t. Jamalizadeh and Balakrishnan (2009) examined order statistics from trivariate normal and t distributions in terms of generalized skew-normal and skew-t distributions. Jamalizadeh et al. (2009b) developed an exact expression for the distribution of a linear combination involving one component of a trivariate elliptical random vector and the order statistics of the other two components. After introducing a general univariate skew-elliptical distribution derived through a selection mechanism, these authors showed that any such linear combination can be expressed as a mixture of two skew-elliptical distributions, with mixture weights and parameters determined by the dependence structure of the original trivariate elliptical vector. This result holds for general elliptical families and produces explicit forms in the important normal and t cases, including closed-form expressions for the corresponding moments. Jamalizadeh and Balakrishnan (2010) studied distributions of order statistics and linear combinations of order statistics from an elliptical distribution as mixtures of unified skew-elliptical distributions, while Jamalizadeh et al. (2010) considered order statistics and linear combinations of order statistics arising from a bivariate selection normal distribution.

A general framework for deriving the exact distributions of order statistics and L-statistics when the underlying random vector follows a multivariate unified skew-elliptical (SUE) distribution was put forward by Arellano-Valle et al. (2014). Building on selection representations of skew-elliptical distributions, they showed that each order statistic can be expressed as a mixture of univariate SUE distributions and that any linear combination of order statistics is likewise a mixture of multivariate SUE distributions. These forms of mixtures provide a unified way to obtain moments, distributional properties, and conditional distributions, extending earlier results available only for symmetric elliptical distributions such as the multivariate normal or t. The framework is motivated by applications in genetic selection and vision research, where ordered dependent measurements naturally arise, and it delivers new, exact tools for analyzing such problems in the presence of skewness and dependence; see the paper for further details.

9.6 Exercises

Exercise 9.1 *Let X be a random variable with pdf f and cdf F:*

(a) *Let* X^w *be a weighted version of X for a weight function* $w(x)$*. Show that* $\mathrm{E}(X^w) > \mathrm{E}(X)$ *if* $\mathrm{cov}\{X, w(X)\} > 0$ *and* $\mathrm{E}(X^w) < \mathrm{E}(X)$ *if* $\mathrm{cov}\{X, w(X)\} < 0$*. Furthermore, if X is nonnegative, then show that* $\mathrm{E}(X^w) > \mathrm{E}(X)$ *if* $w(x)$ *is an increasing function and* $\mathrm{E}(X^w) < \mathrm{E}(X)$ *if* $w(x)$ *is a decreasing function;*

(b) *Assuming that* $X \sim f$ *is nonnegative and such that* $\mathrm{E}(X) < \infty$*, let* $X^w \sim f^w(x) = xf(x)/\mathrm{E}(X)$*. Show that* $\mathrm{E}(X^w) - \mathrm{E}(X) = \mathrm{var}(X)/\mathrm{E}(X)$ *and therefore* $\mathrm{E}(X^w) > \mathrm{E}(X)$ *for nondegenerate X;*

(c) *Let* $r(x) = w_1(x)/w_2(x)$*, where* $w_1(x)$ *and* $w_2(x)$ *are two weight functions, and let* $X^{w_i} \sim f^{w_i}(x) = w_i(x)f(x)/\mathrm{E}\{w_i(X)\}$ *for* $i = 1,2$ *be the respective weighted versions of* $X \sim f$*. Show that* $\mathrm{E}(X^{w_1}) > \mathrm{E}(X^{w_2})$ *if* $r(x)$ *is increasing and* $\mathrm{E}(X^{w_1}) < \mathrm{E}(X^{w_2})$ *if* $r(x)$ *is decreasing;*

(d) *If X is nonnegative and* $w(x;\lambda) = 1 - (1-\lambda)^x$ *for* $\lambda \in [0,1]$*, show that*

$$f^w(x) = \frac{w(x;\lambda)}{\mathrm{E}\{w(X;\lambda)\}} f(x) \to \frac{x}{\mathrm{E}(X)} f(x), \quad \text{as } \lambda \to 0;$$

(e) *If* $w(x;\lambda) = 1 - \{1 - F(x)\}^\lambda$ *for* $\lambda > 0$*, show that* $\mathrm{E}\{w(X;\lambda)\} = \lambda/(1+\lambda)$*.*

Exercise 9.2 *Study the formal properties of the distributions resulting from the selection mechanisms (a)-(f) in Example 111.*

Exercise 9.3 *In property (P4) of Section 9.3, prove parts (a) and (b) from the joint pdf of* $\mathbf{Z}_1$ *and* $\mathbf{Z}_2$*.*

Exercise 9.4 *Let* $\mathbf{Z}_0 \overset{d}{=} (\mathbf{X} \mid X_0 > 0)$*, where* $(\mathbf{X} \mid X_0 = x_0) \sim \mathcal{N}_p(\boldsymbol{\rho} x_0, \mathbf{I}_p - \boldsymbol{\rho}\boldsymbol{\rho}^\top)$ *and* $X_0 \sim \mathcal{N}(0,1)$*, where* $\boldsymbol{\rho} \in \mathbb{R}^p$ *is such that* $\|\boldsymbol{\rho}\| < 1$*. Find* $\mathrm{E}(\mathbf{Z}_0)$ *and* $\mathrm{var}(\mathbf{Z}_0)$*.*

Exercise 9.5 *Let* $\mathbf{Y} = V^{-1/2}\mathbf{X}$ *and* $Y_0 = V^{-1/2}X_0$*, where V is a scalar random variable independent of* $\mathbf{X}$ *and of* X_0*, which are the normal random variables defined in Exercise 9.4. Also, let* $\mathbf{Z} \overset{d}{=} (\mathbf{Y} \mid Y_0 > 0)$*. Find* $\mathrm{E}(\mathbf{Z})$ *and* $\mathrm{var}(\mathbf{Z})$*.*

Exercise 9.6 *Let the random vector* $\mathbf{X} \in \mathbb{R}^p$ *have a multivariate normal and exchangeable distribution. Consider the partition* $\mathbf{X} = (\mathbf{X}_1^\top, \mathbf{X}_2^\top)^\top$ *with* $\mathbf{X}_1 \in \mathbb{R}^{p_1}$*,* $\mathbf{X}_2 \in \mathbb{R}^{p_2}$*,* $p_1 + p_2 = p$*. Derive the joint pdf of* $X_{(p_1)} = \max(X_1, \ldots, X_{p_1})$ *and* $X_{(p_2)} = \max(X_{p_1+1}, \ldots, X_{p_1+p_2})$*.*

9.7 Open Research Problems

Open Research Problem 30 *Explore the entropy, mutual information, KL divergence, J divergence, and Wasserstein distance for some key weighted and selection distributions.*

Open Research Problem 31 *Develop the theory for multivariate L_q-selection/weighted distributions, i.e., selection/weighted distributions based on the L_q-norm.*

Open Research Problem 32 *Develop the theory for matrix selection/weighted distributions.*

Open Research Problem 33 *Develop efficient methods for simulations from selection/weighted distributions.*

Chapter 10

Data Applications Beyond Normality

Many data applications have traditionally employed the multivariate normal distribution due to its simple and appealing properties. In this chapter, we illustrate the practical use of multivariate distributions beyond normality to analyze datasets in various settings, and we survey and discuss many areas of application; see also Part II in the edited book by Genton (2004) for a collection of eleven applications and case studies.

10.1 Inference

10.1.1 Testing normality

The assumption of normality has been at the core of many developments in statistics, and thus a large number of statistical tests have been proposed, both in the univariate and multivariate settings. Chen and Genton (2023) presented a thorough review of many normality tests in the iid case and a discussion about testing normality for time series and spatial data; see also Kim et al. (2025) for a recent approach for dependent data based on neural networks. Therefore, any data analysis usually starts with a test of normality, and in many cases, especially if the sample size tends to be large, the null hypothesis of (multivariate) normality will be rejected. This motivates the need for non-normal multivariate models, for example, like those introduced in this book.

10.1.2 Skew-normal and skew-t distributions

Likelihood-based inferential methods for the skew-normal and skew-t distributions have been studied extensively; see the book by Azzalini and Capitanio (2014) and references therein. For the ASN/AST parameterizations, an R (R Core Team, 2025) package sn has been developed by Azzalini (2023). For the SN/ST parameterizations presented in this book, parameter estimation has been discussed in Section 5.10, including an EM algorithm. Because the change in parameterization from ASN/AST to SN/ST is one-to-one (see Remark 8), the R package sn can be used for both cases. ASN/AST distributions can also be fitted in Stata; see Marchenko and Genton (2010b).

Arellano-Valle (2010) derived and studied the information matrix for the multivariate AST distribution and showed that it is nonsingular for finite values of the degrees of freedom when the skewness is zero. Azzalini and Arellano-Valle (2013) proposed maximum penalized likelihood estimation to reduce the occurrence of infinite estimates of the skewness parameter of ASN/AST distributions. Azzalini and Salehi (2020) explored some computational aspects of maximum likelihood estimation for the AST distribution such as multiple local maxima of the log-likelihood function, and a quick and reliable initialization method for the subsequent numerical maximization of that function. Franco and Azzalini (2025) studied the information matrix of the bivariate extended skew-normal distribution, showing its singularity not only at zero skewness but also when the extension parameter becomes unbounded.

Bayesian inference for the ASN/AST distributions has been investigated by various authors on the basis of their hierarchical stochastic representation. For example, Liseo and Loperfido (2003) described the Bayesian interpretation of the multivariate ASN distribution. For objective Bayesian analysis, Liseo and Loperfido (2006) computed the Jeffreys prior for the skewness parameter of the univariate SN, and Liseo and Parisi (2013) for the multivariate ASN. Bayes and Branco (2007) considered Bayesian inference for the skewness parameter of the univariate ASN distribution, while Cabras et al. (2012) studied a matching prior in that setting. Branco et al. (2013) studied the objective Bayesian analysis of the univariate AST distribution. They examined the Jeffreys prior and its properties for the skewness parameter and showed that the resulting priors for the skewness parameter are symmetric around zero and proper. Moreover, they proposed a Student's t approximation of the Jeffreys prior. Godoi et al. (2017) derived the concentration function (a tool for comparing probability measures) for the skew-normal and skew-t distributions, with application in robust Bayesian analysis. More recently, Parisi and Liseo (2018) explored the Bayesian analysis of the multivariate AST distribution, providing a new parameterization, a set of non-informative priors, and a sampler specifically designed to explore the posterior density of the model parameters. Liseo (2004) provided a discussion of skew-elliptical distributions in Bayesian inference.

Balakrishnan et al. (2014) proposed a goodness-of-fit test to assess whether multivariate data follow a multivariate skew-normal (ASN) distribution. Their test is based on the canonical form of the multivariate skew-normal, which is obtained via a linear transformation that yields mutually independent components with all asymmetry concentrated in a single coordinate. Using this representation, they constructed a test statistic whose null distribution under multivariate skew-normality was derived, first for a single observation and then extended to the practical case of a random sample with unknown location and scale parameters. The performance of the proposed test was investigated through Monte Carlo simulations under several alternative distributions, showing good power and reasonable computational demands compared to the existing, more limited procedures. Finally, three real-data examples illustrated how the test can be applied in practice to check the adequacy of multivariate skew-normal models in applied work. Recently, Karling et al. (2023) proposed goodness-of-fit tests for certain popular families of multivariate skewed distributions based on the empirical characteristic function. They considered five classes of multivariate distributions: ASN, AST, skew-Laplace (SL), Tukey g-and-h, and α-stable distributions. They demonstrated their approach in various datasets. Goodness-of-fit tests are useful for assessing whether a given multivariate dataset can be modeled by various skewed distributions.

10.1.3 Robust models

One approach to the robust estimation problem in the presence of outliers is to introduce a more flexible parametric model that can control deviations from normality. For example, Lange et al. (1989) proposed using the Student's t distribution to regulate tail behavior and accommodate outliers in the tails symmetrically, while Lange and Sinsheimer (1993) suggested the use of normal/independent (NI) distributions (i.e., scale mixtures of normals) in robust regression; see also chapter 13 of Bilodeau and Brenner (1999) for robust modeling based on elliptical distributions. Azzalini and Genton (2008) further advocated the use of the skew-t distribution to regulate both skewness and kurtosis to accommodate outliers in the tails asymmetrically. The skew-t provides a general-purpose compromise between robustness and simplicity, as was illustrated on various wide-ranging examples with real datasets that covered regression, time series, and classification problems. Zeller et al. (2016) proposed robust regression modeling based on scale mixtures of skew-normal distributions. Further results on the skew-t distribution as a robust model can be found in Azzalini (2016); see also Harnik (2023) for recent developments.

Another extension of the aforementioned approach to robustness is to consider generalized skew-normal distributions (Genton and Loperfido, 2005) for which the base normal model is of interest, but the skewing function is treated as a nuisance parameter that models the outliers, hence producing robustness. Ma et al. (2005) derived efficient semi-parametric estimators of the location and scale parameters of the central normal part of this model. The proposed estimators are consistent and asymptotically normal even if the model for the skewing function is mis-specified, and the loss of efficiency can be computed in such cases. The method is also applicable to generalized skew-elliptical distributions. Further extensions of this approach using constrained local-likelihood estimators can be found in Ma and Hart (2007). Other methods for skew-symmetric distributions have been proposed based on their invariance properties (Azzalini et al., 2010) and their characteristic function (Potgieter and Genton, 2013).

Unlike the previous setting of generalized skew-normal distributions, Genton et al. (2012) considered the reverse case where the central symmetric model is unknown up to a location parameter and the skewing function has a parametric form, hence representing semi-parametric location estimation under non-random sampling. Efficient location estimators that are robust to model misspecifications were proposed and studied. A further extension where the skewing function is not necessarily odd (i.e., the normalizing factor is not 2) was investigated by Ma et al. (2013).

Marchenko and Genton (2012) extended the classical Heckman sample selection model from normal errors to elliptical errors and established a new link between sample selection models and extended skew-elliptical distributions. The particular case of the Student's t distribution of errors leads to a robust sample selection model that was used to analyze a medical dataset of ambulatory expenditures; see also Lachos et al. (2021) for parameter estimation via the EM algorithm. Further sample selection models with skew-normal error distributions have been investigated by Ogundimu and Hutton (2016), while the case of discrete and other non-normal distributions has been studied by Azzalini et al. (2019).

10.1.4 SUN/SUT distributions

The SUN/SUT distributions are generally difficult to fit directly because of their large number of parameters, of their need of evaluating a multivariate normal cdf in their pdf, and, more importantly, of their lack of identifiability without additional constraints and special submodels. However, some SUN/SUT submodels may be more suitable for direct inference, such as the tractable pdfs discussed in Example 77, and more generally SUNSET, BSUNSET, and SUNNY multivariate distributions (see Section 6.8). However, the SUN/SUT distributions have been found to be very useful in Bayesian inference due to their conjugacy properties; see Section 10.7.

10.2 SN/ST Data Fitting

To demonstrate the use of SN/ST distributions in data fitting, we consider bivariate wind speed data produced by Yip (2018) with the Weather Research and Forecasting (WRF) computer model. The dataset consists of $n = 156$ wind speed vectors of dimension $p = 2$ representing bi-weekly midday wind speed during the years 2009-2014 at two locations near Dumat Al Jandal, where the first wind farm of the Kingdom of Saudi Arabia is located. According to a Ljung-Box test, there is no serial dependence hence the data can be treated as a random sample from a bivariate distribution. It is important to study this bivariate wind speed distribution because understanding wind patterns, which will influence the energy production by the nearby wind farm, is crucial. It is indeed of interest to assess whether a non-normal distribution is needed rather than a common normal model.

The data are plotted in Figure 10.1(a) along with the density contours of a fitted normal distribution. The associated probability-probability (P-P) plot of the Mahalanobis distances is provided in Figure 10.1(d) and shows a moderate lack of fit. Since there is some apparent skewness in the data, we first fit by maximum likelihood an SN distribution depicted in Figure 10.1(b) with the skewness vector $\hat{\boldsymbol{\eta}} = (2.31, -0.38)^\top$. The associated P-P plot is shown in Figure 10.1(e). Both plots indicate a better fit of the SN distribution compared to the normal model. We further consider to fit by maximum likelihood an ST distribution depicted in Figure 10.1(c) with the skewness vector $\hat{\boldsymbol{\eta}} = (1.59, -0.26)^\top$ and $\hat{\nu} = 8.46$. The associated P-P plot is shown in Figure 10.1(f). Both plots and the moderate value of $\hat{\nu}$ suggest a further improvement of the ST distribution compared to the SN model. A likelihood ratio test of the null hypothesis $H_0 : \boldsymbol{\eta} = \mathbf{0}_2$ yields a p-value of 0.0043, thus rejecting symmetry. We therefore conclude that the ST distribution is a good model for describing the wind speed patterns at those two locations.

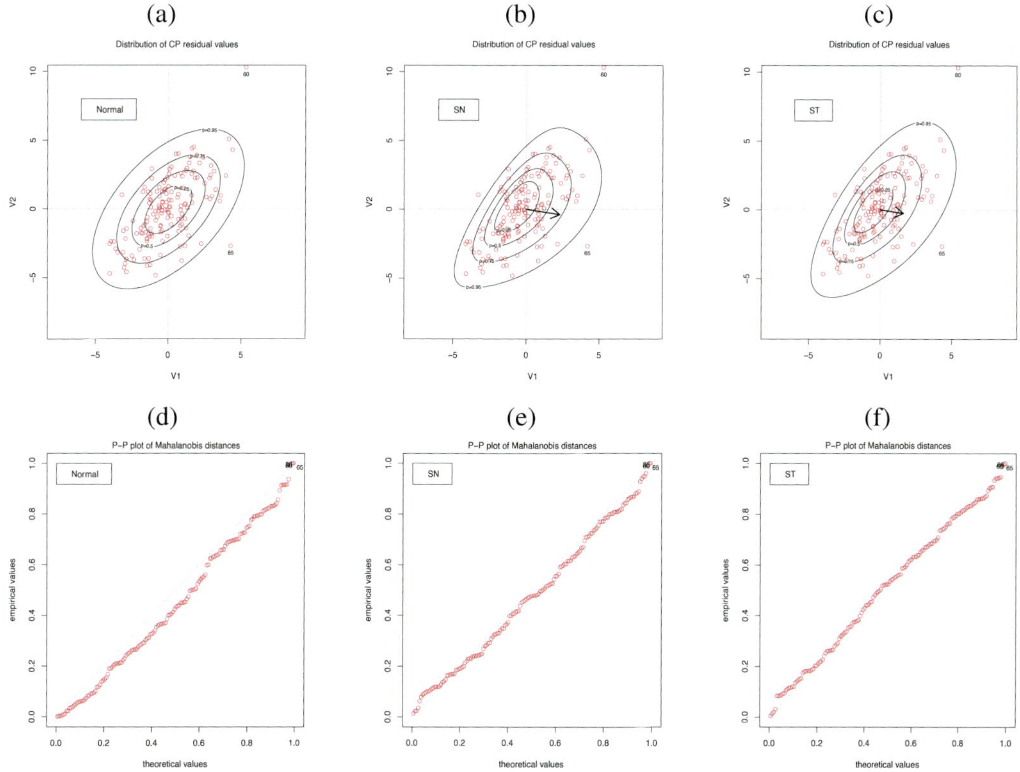

Figure 10.1 *Bivariate wind speed data from two locations near Dumat Al Jandal wind farm: (a) data and normal fit; (b) data and SN fit; (c) data and ST fit; P-P plots of Mahalanobis distances based on (d) normal, (e) SN, and (f) ST distributions. The two arrows in (b) and (c) indicate the skewness vector* $\hat{\boldsymbol{\eta}}$.

10.3 Time Series with SN/ST Errors

Models for time series data can be made more flexible by replacing their error distributions by more adaptable distributions than the normal, for instance, the SN/ST distributions. This flexibility also allows us to cope with possible outliers in the data, hence yielding more "robust" models. For example, Azzalini and Genton (2008) used an AR(1) model with AST error distributions to model a time series of monthly interest rates of an Austrian bank that included some outliers. They demonstrated that the classical least squares fit of the AR(1) model is severely affected by outliers, whereas the fit with AST errors behaves similarly to an optimal robust estimator of the AR(1) model parameters.

We consider a monthly time series of inward movement of residential telephone extensions in a certain region during January 1966 to May 1973 that contains some outliers; see Maronna et al. (2006) for a robust analysis and references therein for previous studies. This time series is plotted in Figure 10.2(a). Earlier analyses identified a seasonal integrated AR(2) model for these data. Therefore, we start by robustly removing a trend and seasonal component to obtain the remainder series depicted in Figure 10.2(b), to which we fit an AR(2) model of the form:

$$Y(t) = \beta_0 + \beta_1 Y(t-1) + \beta_2 Y(t-2) + \varepsilon(t), \tag{10.1}$$

where t represents the time in months, β_0, β_1, β_2 are unknown real parameters assumed to satisfy $|\beta_2| < 1$, $\beta_2 \pm \beta_1 < 1$ for stationarity, and $\varepsilon(t)$ models iid error components with an ST distribution. The P-P plots of scaled residuals squared of the model (10.1), fitted by maximum likelihood, are presented in Figure 10.2 for the cases of the (d) normal, (e) SN, and (f) ST error distributions. One

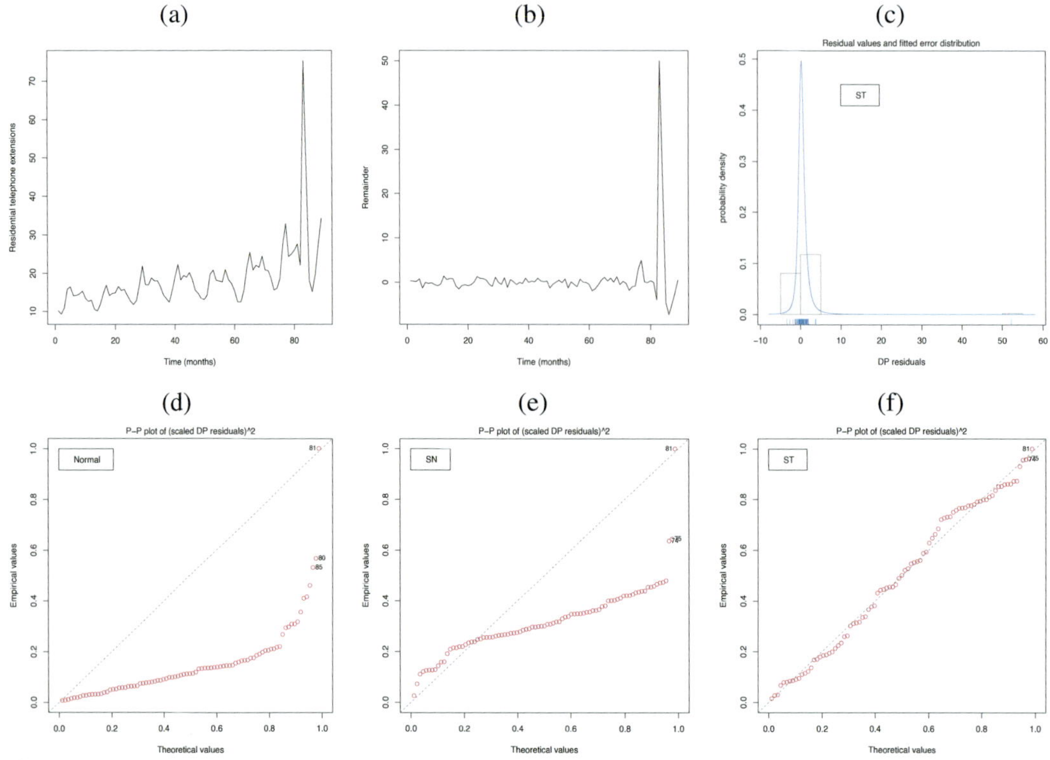

Figure 10.2 *Residential telephone extension time series: (a) data; (b) remainder after robust trend and seasonality removal; (c) fitted ST distribution to residuals of AR(2) model; P-P plots of scaled residuals squared of AR(2) model based on (d) normal, (e) SN, and (f) ST distributions.*

can appreciate how the skewness of the SN distribution improves the fit compared to the normal, and how, in addition, the heavier tails of the ST lead to a very good fitted model. The fitted ST distribution to the residuals of the AR(2) model is plotted in Figure 10.2(c) and confirms an adequate error model. For the ST fit, the estimated degrees of freedom is $\hat{\nu} = 2.0019$, with standard error 0.518, indicating very heavy tails due to the outliers. Note that to compute fitted values $\hat{Y}(t)$, the skewness of the error component must be taken into account, for example, using the median of the fitted distribution of $\varepsilon(t)$; see Azzalini and Genton (2008) for a more detailed discussion on this point.

The fitting of the model (10.1) is treated as a regression of $Y(t)$ on $Y(t-1)$ and $Y(t-2)$ with a univariate SN/ST error term. It is tempting to explore the possibility of building stationary time series models whose finite-dimensional marginal or joint distributions are SN/ST. Unfortunately, even in the simple SN case, Pourahmadi (2007) has shown that the stationary distribution of ARMA models with SN errors is not of SN type. The problem stems from the fact that the sum of two independent SN random variables is not SN anymore, but instead is SUN distributed with latent dimension increased to two. This accumulation of latent dimensions makes computations with such a model difficult.

There is a family of nonlinear autoregressive models, named self-exciting threshold autoregressive (SETAR) models, that are naturally linked to the skew-normal distribution because they essentially convolve a normal random variable with the absolute value of another independent normal random variable; see Anděl et al. (1984) and Chan and Tong (1986). Further results in this direction have been obtained by Das and Genton (2020). For example, they showed that the stationary marginal distribution of some multivariate SETAR processes of order one with multivariate normal innovations is a SUN distribution.

Another time series setting in which the SN/ST distributions have been used with some success is the state-space model framework used to model and forecast dynamical systems. The Kalman filter provides an optimal approach for estimating the parameters of such models under normality. Its extension to the skew-normal distribution was proposed by Naveau et al. (2005); see also Naveau et al. (2004). A further generalization to mixtures of skewed Kalman filters, which allows one to deal with heavy-tailed data, can be found in Kim et al. (2014); see also Rezaie and Eidsvik (2014) for Kalman filter variants using the closed skew-normal distribution. Related work for dynamic linear models (state-space models where both the observation equation and the state evolution equation are linear, and typically Gaussian noise) can be found in Arellano-Valle et al. (2019) for the skew-normal distribution and in Aliverti et al. (2023) for a flexible two-piece normal distribution.

Arellano-Valle et al. (2013a) developed a Bayesian framework for detecting a single change point in the parameters of the skew-normal distribution, focusing on changes in location, scale, and especially the skewness (shape) parameter, an aspect that was largely unexplored in previous change point literature. They derived analytical results for Bayesian inference under general proper priors for the skewness parameter, including stochastic representations and closed-form expressions for posterior means and variances, and extended these results to a mixture-based change-point model with an efficient Markov chain Monte Carlo (MCMC) algorithm. Applications to Latin American emerging market stock returns demonstrated that skew-normal change-point models provide a substantially better fit and richer interpretation of structural changes and asymmetry than classical normal models, highlighting the importance of modeling skewness dynamics in financial time series.

Some other developments of skewed models for time series are the following. Corns and Satchell (2007) developed a skewed Brownian motion for the pricing of European options. Chang and Zimmerman (2016) proposed a skew-normal antedependence model for skewed longitudinal data exhibiting serial correlation. Maleki et al. (2017) investigated robust skewed autoregressive processes in a Bayesian context. Maleki and Arellano-Valle (2017) considered autoregressive processes based on finite mixtures of scale mixtures of skew-normal distributions. Zarrin et al. (2019) studied time series models based on the unrestricted skew-normal process. Maleki et al. (2020) explored symmetric and asymmetric mixture autoregressive processes.

10.4 Non-Gaussian Spatial Random Fields

Like for time series data, building more flexible models for spatial data is of great interest in applications where Gaussian random fields have mostly been the norm. Kim and Mallick (2004) and Kim et al. (2004) introduced a skew-Gaussian model for spatial data based on the multivariate ASN distribution; see also Gualtierotti (2005). Unfortunately, a spatial random field with a finite-dimensional multivariate ASN joint distribution is not identifiable by a single realization, even with a large number of spatial locations, as shown by Genton and Zhang (2012) and Minozzo and Ferracuti (2012). Another approach, proposed by Allard and Naveau (2007), is to use a multivariate closed skew-normal (CSN) distribution. However, also for this case, Minozzo and Ferracuti (2012) have shown that there are non-identifiability issues when the number of latent variables in the CSN is smaller than the number of spatial locations. Zhang and El-Shaarawi (2010) introduced the spatial process:

$$Z(\mathbf{s}) = \delta|X_0(\mathbf{s})| + \sqrt{1-\delta^2}X(\mathbf{s}),$$

where $\mathbf{s} \in \mathbb{R}^d$ is the spatial location, $X_0(\mathbf{s})$ and $X(\mathbf{s})$ are two standard Gaussian processes, second-order stationary, with correlation functions $\rho_0(h)$ and $\rho(h)$, respectively, with h denoting the distance in space, and $\delta \in [-1,1]$. Then $Z(\mathbf{s})$ is second-order stationary with skew-normal marginals, $\mathcal{SN}(0, 1-\delta^2, \delta)$; see Section 5.1.1. Moreover, the correlation function of $Z(\mathbf{s})$ is an explicit function of $\rho_0(h)$ and $\rho(h)$. Next, if this process is observed at n spatial locations $\mathbf{s}_1, \ldots, \mathbf{s}_n$ and we let

$\mathbf{Z} = (Z(\mathbf{s}_1), \ldots, Z(\mathbf{s}_n))^\top$, then the joint pdf of $\mathbf{Z}$ is a weighted sum of 2^n multivariate normal pdfs, the evaluation of the corresponding likelihood is not tractable. Therefore, Zhang and El-Shaarawi (2010) proposed an EM algorithm for inference.

Rimstad and Omre (2014) developed and analyzed a spatial random field model based on the multivariate closed skew-normal (CSN) distribution, allowing explicit modeling of skewness in spatial data. They presented a Metropolis-Hastings algorithm for efficient simulation and a Monte Carlo maximum likelihood approach for parameter estimation. They used simulation studies to examine how spatial correlation reduces the attainable marginal skewness and to assess estimator consistency. They applied the model to seismic amplitude-versus-offset data from the Sleipner field in the North Sea, demonstrating that the CSN random field provides a better fit and improved predictions of subsurface elastic properties compared with a purely Gaussian model, albeit at higher computational cost.

Palacios and Steel (2006) introduced a Gaussian-Log-Gaussian (GLG) model for non-Gaussian Bayesian geostatistical modeling. It is defined as the ratio of a Gaussian random field and the square root of a log-Gaussian random field, independent of each other. The mixing term on the logarithmic scale calibrates the fourth-order moment of the underlying field and allows to model (symmetric) outliers. Bayesian inference was performed by MCMC. Zareifard and Khaledi (2013) extended the previous model to accommodate skewness as well. They replaced the Gaussian random field in the numerator by a SUN distribution and coined the resulting field a Unified Skew Gaussian-Log-Gaussian (SUGLG) process. They used a Stochastic Approximation Expectation Maximization (SAEM) algorithm for parameter inference.

Mahmoudian (2017) introduced a skewed and heavy-tailed latent random field model for spatial extremes. He combined the multivariate skew-normal distribution proposed by Sahu et al. (2003) with scale mixing to model more extreme observations. Bayesian prediction was performed with a Gibbs sampler. The performance of the model was evaluated by simulations and a data analysis of extreme wind speeds in Iran. Furthermore, Mahmoudian (2017) considered the consistency condition of the Kolmogorov existence theorem for random fields, which requires the finite-dimensional distributions of a random field model to be symmetric under permutations and to be identical to the ones whose non-common dimensions are integrated out (marginal consistency). In particular, he proved that a family of skew-normal distributions satisfies the marginal consistency condition if and only if it is closed under marginalization, and further, the marginal distribution of an arbitrary finite dimension only depends on its corresponding parameters and dimension. With this result, he showed that his aforementioned model corresponds to a well-defined random field.

Mahmoudian (2018) studied more closely the existence of some skew-Gaussian processes. He started with the skew-Gaussian random field proposed by Kim and Mallick (2004) based on the $\mathcal{ASN}_n(\boldsymbol{\xi}, \omega^2\bar{\boldsymbol{\Omega}}, \alpha\mathbf{1}_n)$ distribution and showed that it is not marginally consistent. However, its δ-parameterization, and therefore also its $\mathcal{SN}_n$ parameterization, are marginally consistent. Nevertheless, the associated random field is not identifiable with a single realization, as mentioned above. He then demonstrated that the skew-Gaussian random field proposed by Allard and Naveau (2007) based on the closed skew-normal distribution, is not valid but described another closed skew-normal distribution, that is marginally consistent. He then further confirmed that skew-Gaussian random fields based on the multivariate skew-normal distribution of Sahu et al. (2003) are valid because the latter is marginally consistent. Finally, he showed that the skew-Gaussian random field introduced by Zareifard and Khaledi (2013) and based on the $\mathcal{AASUN}_{n,n}(\mathbf{0}_n, \omega^2\bar{\boldsymbol{\Omega}}_{\boldsymbol{\theta}}, \omega\delta\bar{\boldsymbol{\Omega}}_{\boldsymbol{\theta}}, \mathbf{0}_n, \bar{\boldsymbol{\Omega}}_{\boldsymbol{\theta}})$ distribution, with $\omega > 0$, $\delta \in (-1,1)$, $\bar{\boldsymbol{\Omega}}_{\boldsymbol{\theta}}$ a correlation matrix parameterized by $\boldsymbol{\theta}$, is not valid because that distribution is not marginally consistent. However, this model is identifiable with replicates as shown by Wang et al. (2023). Mahmoudian (2018) then argued that an $\mathcal{AASUN}_{n,n}(\mathbf{0}_n, \omega^2\bar{\boldsymbol{\Omega}}_{\boldsymbol{\theta}}, \delta\bar{\boldsymbol{\Omega}}_{\boldsymbol{\theta}}^{1/2}, \mathbf{0}_n, \mathbf{I}_n)$ distribution is marginally consistent and leads to a valid skew-Gaussian random field. A SUN-based geostatistical factor model for multivariate spatial data was proposed by Minozzo and Bagnato (2021).

Tagle et al. (2019) developed a spatio-temporal model for daily wind speed data that captures both asymmetry and heavy tails using a multivariate skew-t distribution. Their approach decomposes daily wind speed observations into a mean structure driven by covariates, a vector-autoregressive process to capture the temporal auto- and cross-correlations, coupled with independent realizations from a multivariate skew-t distribution to model the spatial residuals. Tagle et al. (2020a) generalized the previous model by presenting a spatio-temporal non-Gaussian model that assumes a skew-t data distribution with a hierarchical representation defined as the sum of a small- and a large-scale spatial latent effect. They partitioned the spatial domain into multiple regions, each with a different stationary behavior, linked with a latent large-scale spatial effect. They provided semi-closed-form expressions for inference with the Monte Carlo EM algorithm and the conditional distribution for spatial prediction at unsampled locations. Tagle et al. (2020b) then developed a high-resolution bi-level skew-t stochastic generator to assess Saudi Arabia's wind energy resources.

Bevilacqua et al. (2021) proposed a stationary spatial process with Student's t marginals obtained via scale mixing of a Gaussian process with an inverse square root process with gamma marginals. They studied its properties and generalized their construction by replacing the Gaussian process with a generalization of the skew-Gaussian process defined by Zhang and El-Shaarawi (2010), whose likelihood is still intractable, but appears to be marginally consistent. The resulting process has skew-t marginal distributions, but inference remains difficult. Therefore, the authors focused on the t process in their simulation study and their application to maximum temperature data in Australia.

Márquez-Urbina and González-Farías (2022) introduced a parsimonious, flexible special case of the closed skew-normal distribution, called FS-CSN, for spatial modeling and prediction. They proved that FS-CSN is identifiable, closed under marginalization and conditioning, and that zero correlation implies independence, ensuring Kolmogorov marginal consistency and straightforward interpretation of mean and variance. They developed both discrete and continuous FS-CSN spatial models, proposed maximum likelihood and profiled predictive likelihood methods for parameter estimation and spatial prediction. They showed via simulations that the model provides a more reliable estimation of correlation parameters than the Gaussian counterpart when data are skewed. They used an application to a COVID-19 mortality risk index in municipalities in northeast Mexico to demonstrate that the FS-CSN model yields accurate predictions and captures spatial dependence and skewness better than a purely Gaussian model, albeit somewhat marginally, while retaining the Gaussian case as a special case. The authors did not provide comparisons with other skew-Gaussian models. We also note that this model might only be partially marginally consistent; see Example 73. Another spatial model based on the closed skew-normal distribution and with a specified covariance function was proposed by Jafari Khaledi et al. (2023).

From another perspective, Bevilacqua et al. (2022) introduced a new family of non-Gaussian random fields. Their marginals follow two-piece distributions, allowing both skewness and heavy tails. The authors provided theoretical results for second-order properties, bivariate densities, and correlation functions. They presented two key examples: two-piece Gaussian and two-piece Tukey-h random fields. A notable feature of these models is their ability to generate and model spatial outliers, a phenomenon often overlooked in geostatistical modeling. Because the full likelihood is analytically intractable, they proposed parameter estimation via weighted pairwise likelihood, supported by simulation studies showing accurate inference and good prediction performance. An application to mean-temperature data from the Middle East demonstrated that these two-piece models outperform standard Gaussian and Tukey-h random fields in capturing asymmetry, heavy tails, and local spatial anomalies.

Recently, Wang and Genton (2026) introduced a generalized unified skew-normal (GSUN) model to capture both skewness and heavy tails in non-Gaussian spatial data. It leads to a multivariate SUN distribution:

$$\mathbf{Z} \sim \mathcal{SUN}_{n,n}(\mathbf{0}_n, \mathbf{\Sigma}_{\boldsymbol{\theta}_1}, \mathbf{H}, \mathbf{0}_n, \bar{\mathbf{\Gamma}}_{\boldsymbol{\theta}_2}),$$

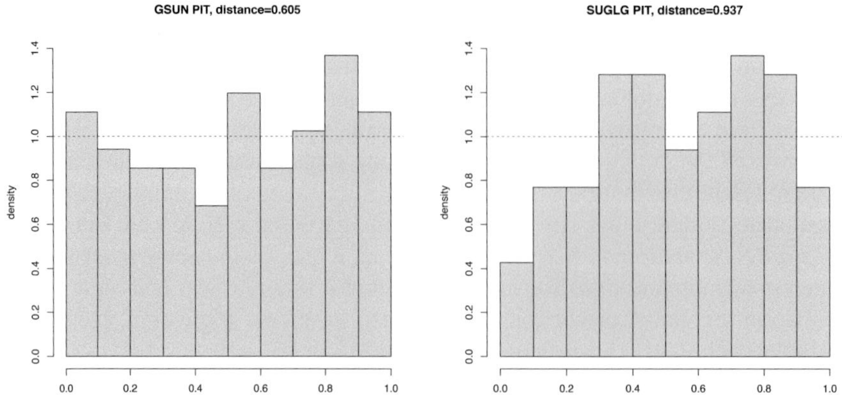

Figure 10.3 *Histograms of PIT values from GSUN (left) and SUGLG (right) models fitted to Pb-contaminated soil data in northern Iran. The horizontal dashed red line indicates a uniform density, to which the distances from the GSUN and SUGLG histograms are provided, indicating a better fit of the GSUN model.*

where $\mathbf{\Sigma}_{\boldsymbol{\theta}_1}$ and $\bar{\mathbf{\Gamma}}_{\boldsymbol{\theta}_2}$ are the variance and correlation matrices based on two sets of parameters $\boldsymbol{\theta}_1$ and $\boldsymbol{\theta}_2$ of a Matérn covariance function, and $\mathbf{H}$ is a diagonal skewness matrix linked to the eigenstructure of the two previous matrices. The GSUN model ensures vanishing correlations at large distances and allows a flexible, interpretable skewness structure through a diagonal skewness matrix. To address the computational challenges of high-dimensional inference, the authors designed a neural Bayes estimator using graph attention networks and an encoder-transformer architecture, which achieves stable and accurate parameter estimation after simulation-based training. Simulation studies showed that the GSUN model differs meaningfully from the Gaussian and Tukey g-and-h processes (Xu and Genton, 2017) and that the proposed neural estimator outperforms conventional CNN-based methods. An application to $n = 117$ Pb-contaminated soil measurements in northern Iran demonstrated the superior fit and predictive performance of the GSUN model compared to Zareifard and Khaledi (2013)'s SUGLG model, and therefore also compared to other existing skewed spatial models. Figure 10.3 presents the histograms of the probability integral transforms (PIT) values from the GSUN and SUGLG models fitted to the Pb-contaminated soil data. The horizontal dashed red line indicates a uniform density, to which the Euclidean distances from the GSUN and SUGLG histograms are 0.605 and 0.937, respectively. Although the SUGLG model is not marginally consistent, the GSUN model is well-defined. The results in Wang and Genton (2026) and in Figure 10.3 demonstrate the superiority of the GSUN as a flexible model to model non-Gaussian spatial data. Future investigations are needed to better understand the consequences of a lack of marginal consistency for non-Gaussian spatial models in practical applications.

10.5 Classification

Classification refers to the task of assigning observations to one of several predefined categories or classes based on their features. It is a type of supervised learning in which a training dataset with known class labels is used to build a model that can classify new unlabeled observations. Popular approaches include linear/quadratic discriminant analysis, logistic regression, decision trees, support vector machines, and random forests, among others. The goal is to find a rule or discriminant function that separates the classes as accurately as possible. Classification is widely applied across domains, such as medical diagnosis (disease vs healthy), finance (creditworthy vs risky), and image or text recognition (e.g., classifying handwritten digits or emails as spam vs not spam); see, e.g., Mardia et al. (1979, Chapter 11) and Izenman (2008, Chapter 8) for more details.

Linear discriminant analysis (LDA) assumes that the distribution of the features in each class c is multivariate normal with the same covariance matrix, i.e., $\mathcal{N}_p(\boldsymbol{\mu}_c, \boldsymbol{\Sigma})$, $c = 1, \ldots, C$, leading to linear discriminant functions. Quadratic discriminant analysis (QDA) assumes that the distribution of features in each class c is multivariate normal, i.e., $\mathcal{N}_p(\boldsymbol{\mu}_c, \boldsymbol{\Sigma}_c)$, $c = 1, \ldots, C$, leading to quadratic discriminant functions. Azzalini and Capitanio (1999) and Azzalini and Genton (2008) proposed to replace $\mathcal{N}_p(\boldsymbol{\mu}_c, \boldsymbol{\Sigma})$ in LDA by $\mathcal{ASN}_p(\boldsymbol{\xi}_c, \boldsymbol{\Omega}, \boldsymbol{\alpha})$ and $\mathcal{AST}_p(\boldsymbol{\xi}_c, \boldsymbol{\Omega}, \boldsymbol{\alpha}, \nu)$, respectively, leading to nonlinear decision boundaries between classes. Here, we allow the parameters of the skew-t distributions to be different across the classes, leading to the discriminant function:

$$\log f_{\mathrm{ST}}(\mathbf{x}; \boldsymbol{\xi}_c, \boldsymbol{\Psi}_c, \boldsymbol{\eta}_c, \nu_c) + \log \pi_c, \qquad \mathbf{x} \in \mathbb{R}^p, \tag{10.2}$$

where f_{ST} is the pdf in class c from an $\mathcal{ST}_p(\boldsymbol{\xi}_c, \boldsymbol{\Psi}_c, \boldsymbol{\eta}_c, \nu_c)$ distribution and π_c is the prior probability of class c. Compared to QDA which has $p + p(p+1)/2$ parameters to describe the normal distribution per class, the ST distribution has $p+1$ additional parameters per class.

We consider a dataset produced by Monte Carlo simulations using the CORSIKA program (Heck et al., 1998) that mimics the detection of high-energy gamma rays by ground-based atmospheric Cherenkov telescopes. These telescopes capture the Cherenkov light produced in the atmosphere by charged particles from gamma-induced showers. The signals are recorded as clusters of pulses on a photomultiplier camera, with each shower image typically forming an elongated cluster whose orientation and shape carry information about the primary gamma. By applying principal component analysis (see next section), the shower image is summarized into geometric descriptors (Hillas parameters) of a fitted ellipse, such as orientation, extent, and asymmetry, which help to discriminate between gamma-ray events (signal) and hadronic cosmic-ray showers (background). The simulations cover energies down to below 50 giga-electron volt, providing realistic data for testing image-based classification and reconstruction techniques.

The complete dataset can be found in Bock (2007). It has a total of 19,020 observations on 10 variables. We randomly split the data into a training set of 10,000 observations and a testing set of 9,020 observations. We consider three variables, namely the length of the major and minor axes of an ellipse fitted to the particle image and the $\log_{10}$ of the sum of the content of all pixels in the particle image. We selected these three variables because they are the simplest fundamental features of this problem. The goal is to find a discriminant function to classify new observations into $C = 2$ classes: high-energy gamma (g) particle or hadron (h) particle. We compare the classification performance of LDA, QDA, and SN/ST distributions.

We used the training data to fit a normal, SN, and ST distribution to the three variables above in each class. Figure 10.4 presents the associated P-P plots of the Mahalanobis distances. We observe improved distributional fits of the SN/ST, especially the ST, compared to the normal. This is supported by small degrees of freedom estimates of $\hat{\nu}_1 = 3.64$ in the gamma class and $\hat{\nu}_1 = 1.78$ in the hadron class.

The parameters estimated from the training data in each class are then inserted in the discriminant function (10.2) for the ST and its variants for the SN and normal densities. The observed percentages of misclassified cases on the training and testing data are given in Table 10.1 for the LDA, QDA, SN, and ST classification methods. It was expected that the use of the SN and ST distribution would lead to some improvement with respect to the normal associated with LDA and QDA, because of the nesting of these parametric classes. In practical terms, the improvement is non-negligible, e.g., on the testing data, the misclassification rate is reduced to 22% for ST compared to 27% for LDA. It is also worth noting that the addition of a single extra parameter leading from the SN to the ST produces a noticeable improvement. Table 10.1 also includes the percentages of hadron (h) particles misclassified as gamma (g) particles, since this is a worse error than the reverse. We see that the ST reduces this error to 15% compared to 25% for LDA.

Other authors have also used more sophisticated methods for classification in this dataset; see Karthick et al. (2024) and the references therein. For example, Karthick et al. (2024) used a deep learning approach based on a long-short-term memory (LSTM) model, a type of recurrent neural

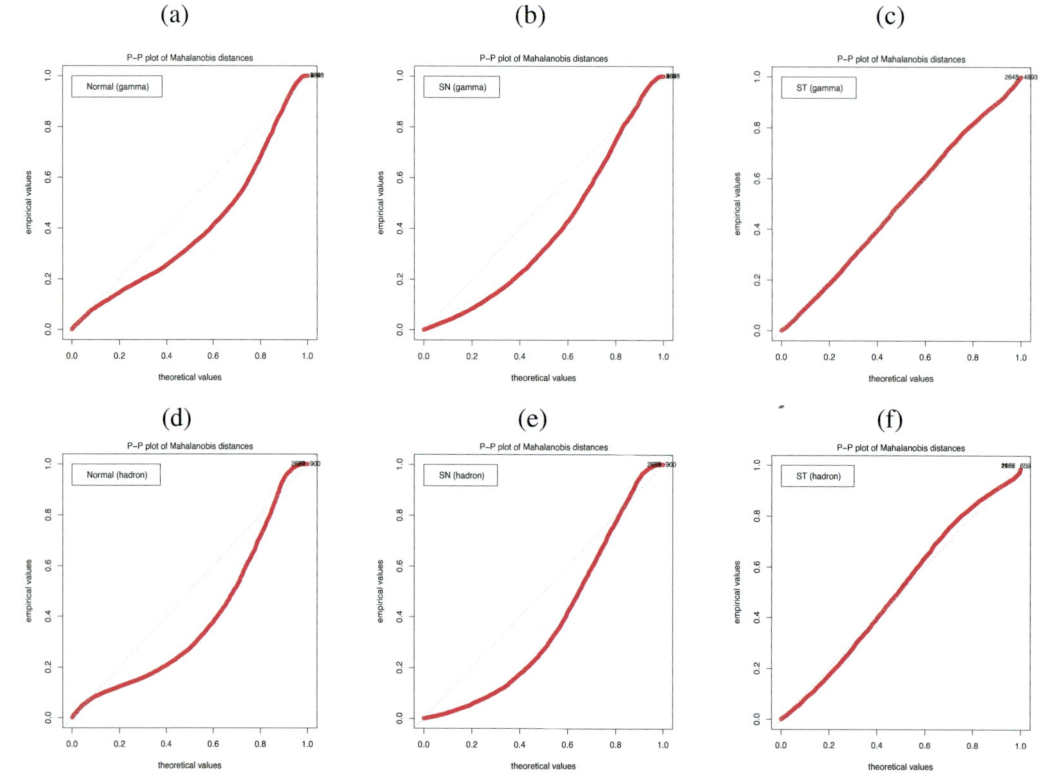

Figure 10.4 *Classification for gamma/hadron: P-P plots of the Mahalanobis distances for gamma based on (a) normal, (b) SN, (c) ST distributions, and for hadron based on (d) normal, (e) SN, (f) ST distributions.*

Table 10.1 *Gamma (g) / hadron (h) data classification: Percentages of misclassified cases.*

Method	Training misclassification g→h and h→g	Testing misclassification g→h and h→g	Training misclassification h→g	Testing misclassification h→g
LDA	27%	27%	25%	25%
QDA	25%	25%	22%	22%
SN	24%	24%	21%	21%
ST	21%	22%	15%	15%

network (RNN) architecture designed to learn and remember long-term dependencies in sequential data, along with Rectified Linear Unit (ReLU) activation functions. Although they reported a misclassification rate of only 11% with their method and with all 10 variables, their approach is much more complex. Certainly, the distribution of these 10 variables is not normal; in fact, not even SN/ST distributed. Therefore, more flexible families of multivariate distributions, such as mixtures of ST distributions or skew-normal-Tukey-*h* distributions (see Section 5.11.5), may lead to further improvements and lower misclassification rates than those reported in Table 10.1.

10.6 Principal Component Analysis and Projection Pursuit

Principal component analysis (PCA) is a widely used technique for dimensionality reduction and data exploration in multivariate statistics. It transforms a set of possibly correlated variables into a new set of uncorrelated variables called principal components. This is achieved by finding linear

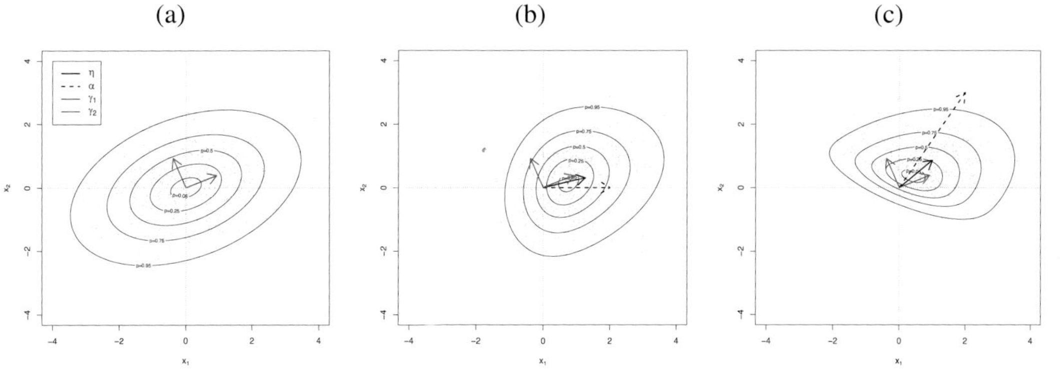

Figure 10.5 *Contours of the pdf of three bivariate distributions with the same variance matrix* $\mathbf{\Sigma}$, *hence the same principal components: (a)* $\mathcal{N}_2$; *(b) and (c)* $\mathcal{SN}_2$. *The standardized eigenvectors* $\boldsymbol{\gamma}_1$ *and* $\boldsymbol{\gamma}_2$ *of* $\mathbf{\Sigma}$ *are represented by red arrows. In (b) and (c), the vectors* $\boldsymbol{\eta}$ *(solid black arrows) and the corresponding vectors* $\boldsymbol{\alpha}$ *of the* $\mathcal{ASN}_2$ *parameterization (dashed black arrows) are provided.*

combinations of the original variables that maximize the variance under the constraint that each component is orthogonal (uncorrelated) to the previous ones. The method is based on the eigenstructure of the (sample) variance matrix: Eigenvectors give the directions (principal components) while eigenvalues indicate how much variance each component explains; see, e.g., Mardia et al. (1979, Chapter 8) and Izenman (2008, Section 7.2) for further details.

Because PCA is based on the sample variance matrix of the data, it is well suited for multivariate normal data whose distribution is completely determined by the mean vector and variance matrix. However, it cannot distinguish between the multivariate normal and other non-Gaussian distributions with the same variance matrix. For example, consider an $\mathcal{N}_2(\mathbf{0}_2, \mathbf{\Sigma})$ distribution and an $\mathcal{SN}_2(\mathbf{0}_2, \mathbf{\Psi}, \boldsymbol{\eta})$ distribution with:

$$\mathbf{\Sigma} = \mathbf{\Psi} + \boldsymbol{\eta}\boldsymbol{\eta}^\top = \begin{pmatrix} 2 & 1/2 \\ 1/2 & 1 \end{pmatrix},$$

with the two choices of the skewness vector $\boldsymbol{\eta}$ as in Example 63. Figure 10.5 depicts the pdf contours of these three distributions, together with the vectors $\boldsymbol{\eta}$ (solid black arrows) and the corresponding vectors $\boldsymbol{\alpha}$ of the $\mathcal{ASN}_2$ parameterization (dashed black arrows). A random sample of $n = 400$ observations from each distribution is also plotted. Since all three distributions have the same variance matrix $\mathbf{\Sigma}$, with standardized eigenvectors $\boldsymbol{\gamma}_1$ and $\boldsymbol{\gamma}_2$ as in Example 63 and represented by red arrows in Figure 10.5, the principal components are the same for each distribution, although the distributions are markedly different.

Projection pursuit is a multivariate statistical technique used to find the most "interesting" low-dimensional linear projections of high-dimensional data. Unlike maximization of variance in PCA, typical indices of interestingness include skewness, kurtosis, and negentropy to describe deviations from normality. The search for interesting projections is usually done by numerical optimization. See, for instance, Izenman (2008, Section 7.4) for more details.

Finding low-dimensional linear projections of multivariate data that maximize skewness, a key measure of non-normality, is of interest in many applications. Loperfido (2018) developed an efficient algorithm for this purpose. He motivated skewness as a projection index for detecting hidden structures and proposed using the right dominant singular vector of the third standardized moment as a starting value. Extensive simulations showed the algorithm's superior accuracy and speed compared to existing methods, and that the sampling distribution of maximal skewness is well approximated by a skew-normal distribution. An application to Olympic decathlon data demonstrated the method's ability to uncover outliers and non-Gaussian structures missed by principal component analysis.

Loperfido (2024) developed a theoretical and practical framework linking third-order tensor eigenvectors to skewness-based projection pursuit, a method for finding informative low-dimensional data projections. He showed that the dominant eigenvectors of a sample's third standardized cumulant identify directions that maximize skewness, enabling sequential cluster detection and the identification of potential outliers. He proved asymptotic results for both dominant and base tensor eigenvectors, establishing their almost-sure convergence to population counterparts and highlighting their role in detecting weakly symmetric projections. Applications to six benchmark datasets, including Australian athletes, breast cancer diagnostics, and Swiss banknotes, illustrated how skewness-based projections can approximate classical discriminant functions, offering a nonparametric tool for clustering and exploratory multivariate data analysis when group memberships are unknown.

Recently, Adcock (2026) explored skewness and kurtosis projection pursuit for the multivariate extended skew-normal and skew-*t* distributions.

10.7 SUN Posteriors for Bayesian Probit Regression and Beyond

In the context of Bayesian probit regression, Durante (2019) proved that the posterior distribution for the probit coefficients is SUN under multivariate normal priors. Moreover, the result remains valid under multivariate SUN priors. Benavoli et al. (2020, 2021) have further explored this result from a machine learning point of view.

The aforementioned groundbreaking discovery can be traced back to Azzalini (1985) in the very special case of the univariate extended skew-normal (ESN) distribution. It has now also been shown to hold for more general models and settings. In a recent overview, Anceschi et al. (2023) have provided a unified theoretical and computational framework for Bayesian inference in models expressible as partially or fully discretized Gaussian linear regressions, including probit, Tobit, and multinomial probit models. Traditionally, these models lacked closed-form posteriors under Gaussian priors, leading to reliance on MCMC and approximation methods. The authors proved that such likelihoods are conjugate with the SUN family. This finding yields closed-form posteriors, exact iid samplers, and tractable variational Bayes and expectation-propagation approximations that improve computational efficiency and accuracy. The paper systematically reviewed existing results, extended the conjugacy to new cases (including Tobit and skewed-utility models), and demonstrated practical and theoretical advantages of the SUN framework for Bayesian modeling across diverse applications; see also Fasano et al. (2021) for a closed-form filter for binary time series. Recent extensions of these conjugacy properties to the SUT and SUE distributions can be found in Zhang et al. (2023) and Karling et al. (2024).

Cao et al. (2022a) further generalized these results by allowing Gaussian process priors in probit models. They proposed fast and scalable methods to compute predictive probabilities in this context, where closed-form predictive distributions are typically unavailable, and existing MCMC/approximation schemes struggle at a moderate-to-large sample size n. First, they derived closed-form expressions showing that the predictive probability can be written as a ratio of two multivariate Gaussian cdfs (Φ_{n+1} over Φ_n). Then they introduced a tile-low-rank (TLR; Zhang et al. (2024) and references therein) Monte Carlo estimator that evaluates this ratio with drastically reduced cost and variance. Second, they also provided an alternative representation via functionals of multivariate truncated normals and developed a mean-field variational Bayes (VB) scheme that samples only from independent univariate truncated normals. They gave a closed form for the marginal likelihood, linked the Gaussian process posterior (given binary data) to a SUN distribution, and demonstrated accuracy/speed gains in a simulation study.

Cao et al. (2022a) considered wind simulations in Saudi Arabia obtained by Yip (2018) to assess the most suitable sites for wind farming. The wind speed data were at a spatial resolution of 5 km on a grid that covers the wind farm Dumat Al Jandal, on June 21, 2014, and at a hub height of 140 m, corresponding to the center height of the relevant turbines. The cut-in speed of these turbines is around 4 m/s, and it is of interest to consider the locations where this threshold is

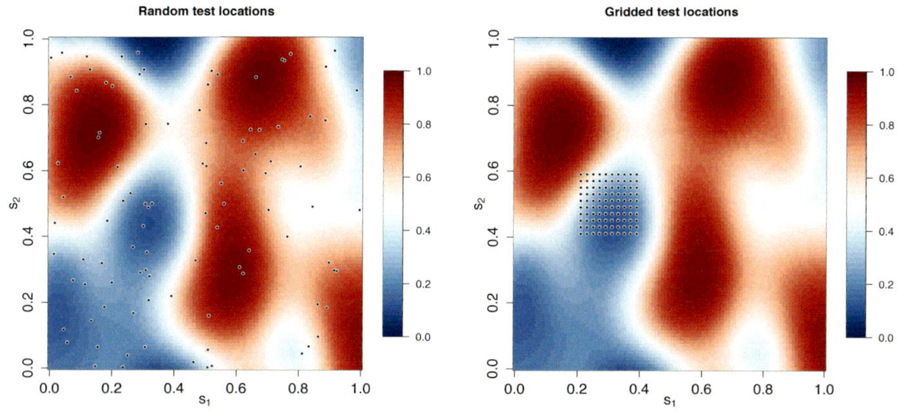

Figure 10.6 *Simulated probabilities on a 100×100 grid in the unit square based on a zero-mean Gaussian process with squared exponential spatial covariance function with $\mathbf{a} = (3.59, 4.77)^\top$ for the square root of the range parameters. Black circles denote the 100 test locations, distributed randomly (left) or on a grid (right), used for prediction.*

exceeded, leading to a binary 0/1 response. Cao et al. (2022a) fitted a probit model to these binary data with a mean-zero Gaussian process prior with squared exponential spatial covariance function $C(\mathbf{s}, \mathbf{s}') = \exp[-\{a_1^2(s_1 - s_1')^2 + a_2^2(s_2 - s_2')^2\}]$ with $\mathbf{a} = (a_1, a_2)^\top > \mathbf{0}_2$ for the square root of the range parameters. They obtained the estimates $(3.59, 4.77)^\top$ by marginal likelihood maximization using the TLR approximation, indicating a fairly similar decay of the correlation in both spatial directions.

Due to the importance for energy production of modeling, simulation, and prediction at high spatial resolution of whether the wind speed exceeds the cut-in speed threshold (Giani et al., 2020), we simulate a realization from the aforementioned fitted probit model on a 100×100 grid; see Figure 10.6. Based on this large binary dataset of size $n = 10{,}000$, we re-estimate $\mathbf{a}$ by marginal likelihood maximization using the TLR approximation, obtaining $\hat{\mathbf{a}}_{\text{TLR}} = (3.43, 6.72)^\top$. To assess performance in estimating the predictive probabilities, we simulate probability parameters and the associated binary responses at 100 out-of-sample locations either randomly distributed (left panel of Figure 10.6) or on a grid (right panel of Figure 10.6). We compute the mean squared errors (MSEs) between the 100 estimated predictive probabilities and the true ones for these two location structures: $MSE_{\text{Random}} = 0.0016$ and $MSE_{\text{Grid}} = 0.0009$. This shows high accuracy even with the TLR approximation and demonstrates the scalability of the approach to large sample sizes.

Recently, Onorati and Liseo (2025) extended the SUN distributions to perturbed SUN (pSUN) distributions and showed that this class provides a conjugate and highly flexible framework for Bayesian binary regression models whose link functions can be expressed as scale mixtures of Gaussian cdfs, including both probit and logit links. They developed Gibbs sampling and importance sampling algorithms for posterior inference, demonstrating that the pSUN approach produces better mixing, higher effective sample sizes, and greater accuracy than state-of-the-art methods such as Polya-gamma samplers, particularly in challenging settings with high-dimensional predictors, sparse signals, or unbalanced binary data. Extensive simulation studies and real-data applications showed that pSUN-based methods are especially effective when combined with shrinkage priors, while the importance sampling variant enables fast computation of posterior summaries and marginal likelihoods, facilitating model comparison and variable selection in Bayesian binary regression. Another extension of SUN distributions to quasi-SUN (qSUN) processes for nonparametric Bayesian binary regression was also recently proposed.

A direction recently explored by Durante et al. (2024) is the development of a skewed Bernstein-von Mises theorem. This result goes beyond classical Gaussian approximations of Bayesian posteriors by introducing a new class of closed-form skew-symmetric (Wang et al., 2004b) deterministic approximations based on a refined third-order Laplace expansion. The authors showed that these skewed approximations capture posterior asymmetry and achieve substantially faster convergence rates in total variation distance than the Gaussian limit, improving accuracy by at least one order of magnitude under very general conditions, including misspecified models and non-iid data. They further demonstrated that these gains extend to posterior expectations of polynomially-bounded functions, developed a practically implementable skew-modal plug-in approximation based on the maximum a posteriori (MAP) estimator with strong nonasymptotic guarantees, and confirmed through simulations and real data that the proposed approach delivers marked empirical improvements while retaining the tractability and frequentist validity of classical Bernstein-von Mises results. Another approach proposed by Tan and Chen (2025) is to use a subclass of closed skew-normal (CSN) distributions (essentially another parameterization of the SUN, see Table 6.1) to improve variational Bayes approximations.

Finally, yet another use of skewness to improve deterministic Bayesian posterior approximations was introduced by Pozza et al. (2026). They proposed a general, theoretically grounded, and computationally efficient method by introducing skewness into any existing symmetric approximation (such as those from Laplace, variational Bayes, or expectation propagation). They showed that any posterior distribution admits a skew-symmetric representation, and they exploited this structure to derive a closed-form skewing perturbation that can be applied directly to an off-the-shelf symmetric approximation without additional optimization or unknown parameters. The resulting skew-symmetric approximation is provably never worse and is typically strictly more accurate than the original symmetric one under total variation, KL, reverse-KL, and general α-divergences, with asymptotic improvements in convergence rates of at least a $\sqrt{n}$ factor. The method preserves tractability, enables straightforward iid sampling, and yields substantial practical gains in simulations and real-data applications, including improved effective sample sizes when used as proposals for importance sampling. This new direction opens many exciting applications and future research avenues.

10.8 Factor Analysis

Factor analysis is a multivariate statistical technique used to uncover the underlying structure (latent variables or "factors") that explains the correlations between a set of observed variables. Unlike PCA, which finds orthogonal linear combinations for maximum variance, factor analysis explicitly models shared variance due to latent constructs plus unique measurement error. Its two main goals are dimension reduction, i.e., representing many correlated variables with a few underlying factors, and interpretation, i.e., identifying conceptual structures driving the observed correlations. Traditionally, factor analysis has been based on the assumption of multivariate normality; see, e.g., Mardia et al. (1979, Chapter 9) and Izenman (2008, Section 15.4) for further details.

Montanari and Viroli (2010) extended classical factor analysis by allowing latent factors to follow a multivariate skew-normal distribution, enabling explicit modeling of asymmetry in underlying features. They developed a generalized EM algorithm with Newton-Raphson updates to estimate both factor loadings and the skewness parameter, and explored theoretical properties such as canonical forms and rotation effects. Simulation studies showed that the model accurately estimates loadings and variances while capturing skewness that standard Gaussian factor analysis misses. Applied to survey data from 92 university courses in Bologna, Italy, the skew-normal factor model revealed a two-factor structure, one reflecting overall appreciation of teaching and another capturing organizational aspects, and fitted the data better than ordinary factor analysis, especially when responses exhibit pronounced negative skewness; see Liu and Lin (2015) for skew-normal factor analysis models with incomplete data.

By incorporating a selection mechanism to induce skewness, Kim et al. (2016) introduced three models of latent factors (skew-normal, skew-t, and generalized skew-normal) that relax the usual multivariate normal assumption for latent factors. The generalized skew-normal model contains the model of Montanari and Viroli (2010) as a special case. The authors derived the distributional properties of each model, showed how the selection construction naturally yields SUN or SUT marginals, and developed efficient Expectation/Conditional Maximization Either (ECME) algorithms for the maximum likelihood estimation. Monte Carlo simulations compared these models with classical normal and t factor models, demonstrating improved fit and robustness when the underlying factors are asymmetric or heavy-tailed. An application to the well-known open/closed book exam-scores dataset highlighted the practical benefits of the skewed factor models in capturing non-normal structure while maintaining parsimony and interpretability. Further extensions of skew-t factor analysis models in the presence of missing values or non-responses have been investigated by Wang et al. (2017).

10.9 Cluster Analysis

Cluster analysis is a type of unsupervised learning in which a dataset with no known labels must be divided into groups. Clustering multivariate data is important in many applications, for instance, to discover subtypes of diseases or identify genes with similar expression profiles, to group regions with similar climate regimes or pollution profiles, or to identify groups of customers to tailor marketing strategies, to name but a few; see, e.g., Mardia et al. (1979, Chapter 13) and Izenman (2008, Chapter 12) for further details. Mixtures of factor analyzers (MFA) are a flexible model-based clustering and dimension-reduction technique that combines finite mixture models with factor analysis. In recent years, several generalizations of MFA have been developed in which the normality assumption of the factors and/or errors was relaxed to allow for skewness and heavy tails in the data.

Lin et al. (2018) introduced the mixtures of skew-t factor analysis model, a robust extension of the mixtures of factor analyzers that accommodates asymmetry, heavy tails, and missing data. It includes mixtures of skew-normal factor analysis as a special case. Each component of the mixture assumes that the latent factors and errors jointly follow a multivariate skew-t distribution, allowing flexible model-based clustering of non-Gaussian high-dimensional data. The authors developed an ECME algorithm for maximum likelihood estimation and single imputation under a missing-at-random mechanism, and provided guidance for selecting the number of components and factors. Applications to chronic kidney disease data, Wisconsin prognostic breast cancer data, and simulation studies demonstrated that the proposal outperforms classical MFA and skew-normal MFA in terms of fit, classification accuracy, and imputation quality, particularly when the data exhibit skewness, heavy tails, and incomplete observations.

However, because of the form of the adopted component densities, the distribution of the factors/errors in most of the aforementioned models is typically limited to modeling skewness concentrated in a single direction. Lee et al. (2021) introduced a more flexible finite mixtures of factor analyzers based on the class of scale mixtures of canonical fundamental skew-normal distributions (Arellano-Valle and Genton, 2005). This very general class of skewed distributions can capture various types of skewness and asymmetry in the data. In particular, the proposed model can simultaneously accommodate multiple directions of skewness. As such, it encapsulates many commonly used models as special and/or limiting cases, such as models of some versions of skew-normal and skew-t factor analyzers, and skew-hyperbolic factor analyzers. For illustration, the authors focused on mixtures of canonical fundamental skew-t factor analyzers. Parameter estimation can be performed by maximum likelihood using an EM-type algorithm. The usefulness and potential of the proposed model were demonstrated using four real datasets. Another approach for robust finite mixture modeling was proposed by Basso et al. (2010) based on scale mixtures of skew-normal distributions and by Maleki et al. (2019b) based on multivariate unrestricted skew-normal generalized hyperbolic distributions. A recent overview of skewed distributions in model-based clustering can be found in Lee and McLachlan (2022); see also references therein.

10.10 Mixed Effects Models

Linear mixed effects models are very popular for the analysis of continuous longitudinal responses from a sample of individuals; see, e.g., Verbeke and Molenberghs (2000). They find many applications not only in biomedical science but also in social, environmental, and agricultural sciences. Individual-specific random effects are used to model heterogeneity in features of longitudinal profiles, while fixed effects are defined to model population-specific relationships. A standard assumption is that random effects and within-individual errors are normally distributed. This may be unrealistic for inter-individual heterogeneity that could exhibit skewness and heavier tails than the normal.

Ma et al. (2004) introduced linear mixed effects models with flexible generalized skew-elliptical distribution (Ma and Genton, 2004) of random effects, with emphasis on the flexible skew-normal and flexible skew-t distributions that can furthermore model multimodality; see also Section 5.11.1. They presented maximum likelihood inference via the EM algorithm as well as Bayesian inference via MCMC. They demonstrated the flexibility of such a model through a simulation study and an application to cholesterol data.

Arellano-Valle et al. (2005a) relaxed the normality assumption considering that random effects and model errors follow skew-normal distributions. Bolfarine et al. (2007) and Montenegro et al. (2009) considered a linear mixed effects model with a skew-normal distribution for random effects and developed associated influence diagnostics. Lachos et al. (2007) studied maximum likelihood inference using the EM algorithm for three regression models, including a linear mixed model with skew-normal distributed random effects. Lin and Wang (2013) studied a similar model motivated by AIDS clinical trials. Ghosh et al. (2007) considered a bivariate random effects model using a skew-normal distribution. Arellano-Valle et al. (2007) proposed a linear mixed effects model with a particular fundamental skew-normal distribution (Arellano-Valle and Genton, 2005) for both random effects and errors. Ho and Lin (2010) considered robust linear mixed models using the skew-t distribution with application to schizophrenia data. Zhou and He (2008) developed a three-step estimation procedure in linear mixed models with skew-t distributions to improve numerical stability. Jara et al. (2008) explored linear mixed models with skew-elliptical distributions from a Bayesian perspective. Selection bias in linear mixed models was studied by Grilli and Rampichini (2010) using the skew-normal distribution.

Lachos et al. (2009a, 2010b) considered the case where the random effects follow a multivariate skew-normal/independent (SNI) distribution (Branco and Dey, 2001), while the errors follow a symmetric normal/independent distribution. In particular, they considered the skew-normal, skew-t, skew-slash, and skew-contaminated normal distributions. Bandyopadhyay et al. (2010) extended the previous model to the case where both random effects and errors follow an SNI distribution with application to periodontal disease data. Bandyopadhyay et al. (2012) further considered censored responses motivated by an HIV case study involving longitudinal viral loads. Zeller et al. (2010) performed influence analyses of SNI linear mixed models.

Santos et al. (2013) came up with a Bayesian interpretation of the parameters in logistic regression models with random (cluster-level) intercepts when the random effects follow skew-normal distributions, extending the classical mixed logistic regression theory beyond normality. They derived prior and posterior distributions for odds ratios, including median odds ratios and highest posterior density intervals, under both independent and correlated skew-normal random effects, showing how skewness fundamentally alters interpretation and inference. Through simulations and a teratogenic activity case study, they demonstrated that misspecifying the random effects distribution can severely distort odds ratio estimates, and that skew-normal random effects provide more robust, interpretable, and realistic modeling of cluster heterogeneity than standard normal assumptions.

Kahrari et al. (2019) studied a flexible class of linear mixed models with random effects and model errors following a skew-normal-Cauchy distribution (see Section 10.20). Maleki et al. (2019a) explored a linear mixed effects model with unrestricted skew-normal generalized hyperbolic distribution of random effects and errors.

10.11 Measurement Error Models

Measurement error models, also known as errors-in-variables models, are statistical models that explicitly account for inaccuracies in the observed data when variables of interest are measured with error; see, e.g., Carroll et al. (2006). They aim to adjust inference to correct for the bias and loss of efficiency caused by imperfectly measured variables. They find a wide range of applications, such as exposure assessment in epidemiology, survey sampling, calibration studies, and any field where instruments or self-reports introduce uncertainty. The classical normal-error case is very common for convenience, but other non-Gaussian measurement error models have been developed, too. For example, Bolfarine and Arellano-Valle (1994) discussed robust regression techniques, specifically applying the heavy-tailed Student's t distribution to minimize the influence of outliers on measurement error models for improved parameter estimation. Further extensions to elliptical measurement error models can be found in Arellano-Valle and Bolfarine (1996), Arellano-Valle et al. (1996), and Arellano-Valle et al. (2002a).

The paper by Arellano-Valle et al. (2005c) is probably the first inferential proposal for measurement error models with a skew-normal distribution to avoid data transformation. Specifically, they extended the classical linear measurement error models by allowing the latent variables and error terms to follow skew-normal distributions instead of assuming normality. They derived the likelihood function of the observed data after integrating out the unobserved true covariates and showed that the model remains identifiable without the extra variance restrictions required under the normal case. They developed both maximum likelihood estimation, including an EM-type algorithm for stable computation, and a Bayesian framework with priors that preserve key invariance properties. Simulation studies demonstrated accurate parameter recovery and reduced attenuation compared with standard methods, and an application to the Australian Institute of Sport dataset illustrated improved fit when the covariates exhibit clear asymmetry.

A multivariate measurement error model where the latent covariates follow a class of skew-normal/independent (SNI) distributions, which includes skew-normal, skew-t, skew-slash, and skew-contaminated normal cases, was introduced by Lachos et al. (2009b). This framework captures both asymmetry and heavy tails, providing a robust alternative to classical Gaussian measurement error models. The authors developed a Bayesian hierarchical formulation and implemented MCMC methods to obtain posterior inference for all parameters. Through an application to lung capacity data comparing different measuring instruments, they showed that SNI-based models significantly outperform normal models by better accommodating skewness and outliers, with the skew-contaminated normal distribution providing the best fit according to pseudo Bayes factors and DIC criteria. An alternative approach based on scale mixtures of the skew-normal distribution can be found in Lachos et al. (2010c).

Labra et al. (2011) extended classical multivariate null intercept measurement error regression to allow latent covariates to follow a skew-normal distribution rather than a normal one. They derived closed-form expressions for the marginal likelihood, score function, and observed information matrix, and developed an EM-type algorithm for maximum likelihood estimation with and without linear constraints on regression coefficients. They also presented methods for hypothesis testing (likelihood ratio, score, and Wald tests) and for local influence diagnostics under various perturbation schemes. Applying the model to a dental clinical trial dataset, they showed that the skew-normal model provides a better fit and identifies different influential observations compared to the standard normal measurement error model, highlighting the importance of modeling asymmetry in latent variables. Lachos et al. (2010a) further considered the skew-t distribution in this context within the Bayesian framework.

Various skewed extensions of measurement error models have been further developed. Cabral et al. (2014) explored multivariate measurement error models based on finite mixtures of skew-t distributions. Rocha et al. (2015) and Rocha et al. (2016) studied robust censored measurement error models with likelihood and Bayesian inference. Carrasco et al. (2014) considered measurement error models beta regression, while Figueroa-Zúñiga et al. (2018) focused on the Bayesian context and Carrasco et al. (2023) on multiplicative measurement error models regression. Arellano-Valle et al. (2020) studied a two-piece normal measurement error model. Sandoval-Moreno et al. (2023) discussed inference in multivariate regression models with measurement errors.

10.12 Biomedical Science

Biomedical science is another area where the normal assumption has been popular, and so more flexibility can be introduced by considering non-Gaussian distributions. For example, in the field of drug development, a formal framework to combine the information from phase II and phase III clinical trials was proposed by Azzalini and Bacchieri (2010). They considered an approach that mirrors how drug development actually proceeds, instead of treating the two phases as statistically separate. They modeled the joint distribution of test statistics from both stages under normality (with SUN tools in the background), and derived an exact likelihood-based approach to integrate the dose-selection evidence from phase II with the confirmatory evidence from phase III, so that phase II data can meaningfully contribute to the final hypothesis test and thus increase power. They discussed control of the overall type I error in this two-stage setting, including how multiplicity in dose selection and sample-size re-estimation can affect the level of the test, and how suitable step-down or adjusted procedures can keep it at the desired level. A detailed numerical example based on phase II/III trials illustrated how the method recovers information from stage I, redesigns the second-stage sample size, and yields more precise inference than conventional approaches that effectively discard phase II once the dose has been chosen.

Using censored skew-normal models to analyze bounded and potentially skewed health-related quality-of-life scores was proposed by Hutton and Stanghellini (2011). Such scores, often scaled between 0 and 100, tend to cluster near the upper boundary ("ceiling effect") as patients recover, violating normality assumptions. The authors applied their model to a clinical trial comparing four treatments for ankle sprains, showing that skewness and censoring meaningfully affect parameter estimates. The censored skew-normal model accommodates both asymmetry and boundary clustering while nesting the normal model as a special case. The authors further derived a formula to quantify bias in treatment effects due to informative dropout, demonstrating when results based only on complete cases might be distorted. Overall, the censored skew-normal framework provided a flexible and interpretable approach to analyzing bounded, skewed, and incomplete longitudinal health data.

In survival analysis, Rubio and Genton (2016) developed a Bayesian framework for linear regression models that allows for asymmetric and heavy-tailed residuals using the family of skew-symmetric scale mixtures of normal distributions, such as the skew-normal, skew-t, and skew-logistic. They proposed a noninformative prior structure and proved conditions under which the posterior distribution remains proper, even when responses are censored, as in accelerated failure time survival models. They derived sufficient conditions for posterior propriety, discussed suitable priors for skewness and tail parameters, and provided simulation studies confirming that the proposed priors yield credible intervals with good frequentist coverage. Applications to biomedical data, including survival times and body composition measurements, demonstrated that allowing for skewness improves inference and predictive performance over symmetric models, offering a flexible and robust alternative for Bayesian regression and survival analysis; see also Sahu and Dey (2004) for discussions on this topic.

10.13 Graphical Models

Graphical models are a class of statistical models that represent the conditional dependence structure among random variables using a graph, where nodes correspond to variables and edges indicate direct probabilistic relationships; see, e.g., Lauritzen (1996). They unify concepts from probability theory and graph theory to provide a compact, visual way to describe complex multivariate distributions. Two main types are Bayesian networks (directed acyclic graphs) that encode causal or directional dependencies, and Markov random fields (undirected graphs) that capture symmetric relationships. By factorizing a joint distribution according to the graph, graphical models make inference and learning more efficient, enabling tasks such as computing conditional probabilities, performing predictions, and discovering latent structures. They are widely applied in fields ranging from genetics and epidemiology to natural language processing and computer vision, offering both interpretability and computational advantages for high-dimensional, interdependent data.

Gaussian graphical models based on the multivariate normal distribution have been quite popular. In that case, the edges in the graph correspond to nonzero entries of the inverse variance (precision) matrix, and zero entries represent conditional independence. The popularity of these Gaussian graphical models is due to their mathematical tractability and simple interpretation. However, they are limited when the data exhibit non-Gaussian distributional features, which is often the case in practice.

Capitanio et al. (2003) extended classical Gaussian graphical models to handle multivariate skew-normal distributions, allowing conditional independence graphs when data exhibit skewness. They introduced a slightly generalized skew-normal family, the extended skew-normal distribution (ESN; see Section 6.6.3) that remains closed under conditioning, a key property for graphical modeling. They derived necessary and sufficient conditions for conditional independence in terms of both the inverse variance matrix and the skewness parameters, and explored how marginal Gaussian components constrain admissible graph structures. They also developed parameter estimation methods and showed how likelihoods can be factorized according to decomposable graphs, enabling efficient inference. Overall, they provided a theoretical foundation for the construction and estimation of graphical models that accommodate asymmetric data while retaining many tractable properties of Gaussian models. Recently, Sheng et al. (2023) developed a penalized likelihood estimator with separate L_1 penalties on the precision and shape parameters and optimized it using the alternating-direction method of multipliers. They established asymptotic normality and variable selection consistency for the estimator, and simulations showed that their approach outperforms Gaussian and Gaussian copula graphical models when the true distribution is skewed, while suffering little efficiency loss when the data are actually Gaussian. An application to a breast cancer microRNA dataset demonstrated that their skew-normal graphical model captures meaningful gene network structures missed by standard Gaussian approaches.

Zareifard et al. (2016) generalized Gaussian graphical models to accommodate multivariate closed skew-normal (CSN) distributions, enabling conditional independence graphs when data exhibit skewness. They defined the Skew Gaussian Decomposable Graphical (SGDG) model, which retains key properties of Gaussian models, e.g., such as closure under conditioning and marginalization, while incorporating skewness through CSN noise. They focused on decomposable graphs to ensure the Markov property and efficient estimation. A fully Bayesian inference framework was developed, including conjugate priors for the precision matrix, non-informative priors for mean and scale parameters, and an MCMC algorithm for posterior sampling. Simulation studies confirmed identifiability and accurate recovery of skewness and precision parameters, and applications to student exam scores and Danish carcass measurement data showed that the SGDG model provides a markedly better fit than standard Gaussian graphical models when data exhibit asymmetry.

Another method was proposed by Nghiem et al. (2022) to learn undirected graphical models when data follow a multivariate generalized skew-normal distribution (Genton and Loperfido, 2005), which captures both dependence and skewness. Unlike Gaussian graphical models, the conditional independence in this new model depends on both the precision matrix and a skewness vector. The authors proved key theoretical results for conditional expectations and showed that a pair of variables is conditionally independent if and only if the corresponding entry of the precision matrix and the product of the relevant skewness vector elements are zero. They proposed a nodewise regression algorithm: First fitting a linear model to estimate the precision-related coefficients, then applying projection pursuit regression on residuals to detect skewness effects, and finally thresholding to recover the graph. They established error bounds and consistency as the number of variables grows, demonstrated superior finite-sample performance over Gaussian and other non-Gaussian approaches through simulations, and illustrated the method on a wine-chemistry dataset, highlighting its ability to uncover conditional dependence structures in skewed continuous data.

10.14 Finance and Management Science

Modern finance's standard theories are usually based on the assumption that returns on risky financial assets follow a multivariate normal distribution; see, e.g., Campbell et al. (1997) and Cochrane (2005). However, in practice, returns are not normally distributed and often exhibit skewness and heavy tail distributional features.

Adcock (2004) explored the use of the multivariate extended skew-normal distribution for the capital asset pricing model (CAPM) with an application to UK stocks. Adcock (2010) promoted the multivariate extended skew-*t* distribution as a flexible model for financial asset returns that captures both heavy tails and skewness. He derived its density, moment properties, and conditional distributions, and showed that when asset returns follow an extended skew-*t* distribution, the usual linear market model is replaced by a nonlinear form in which the sensitivity of an asset to the market is not necessarily the conventional beta. Three cases of the market model were explored, depending on the skewness of the market portfolio, revealing that conditional variances vary over time and depend on squared deviations of market returns. For portfolio selection, the author proved that expected-utility-maximizing investors lie on a mean-variance-skewness efficient surface, an analog of the Markowitz frontier, which can be implemented via quadratic programming. This framework generalizes classical Gaussian-based asset pricing and highlights how skewness can affect perceived portfolio performance.

Stein's lemma plays an important role in finance, e.g., for portfolio selection and the CAPM. It allows for the calculation of the covariance between a function of one random vector and another. Stein (1981) showed that if $(X_1, X_2)^\top$ is a bivariate normal random vector, then $\mathrm{cov}\{\theta(X_1), X_2\} = \mathrm{cov}(X_1, X_2)\mathrm{E}\{\theta'(X_1)\}$ where θ is a differentiable function with $\mathrm{E}\{|\theta'(X_1)|\} < \infty$. Stein's lemma was generalized to skew-normal and closed skew-normal random vectors (Adcock, 2007), to extended skew-*t* random vectors (Adcock, 2014), to truncated elliptical random vectors (Shushi, 2018b), and to generalized skew-elliptical distributions (Adcock et al., 2021).

Recently, Loperfido and Shushi (2023) defined the concept of optimal portfolio projection. It is a procedure that projects the vector of weights of the portfolio return to a lower dimension so that one can explicitly solve the problem of optimal portfolio selection for any given risk measure. The authors focused on the class of skew-elliptically distributed risks and derived explicit optimal weights for such risks, with a dramatic reduction in the complexity of such an optimization problem. They illustrated their approach on ten NASDAQ stocks.

Generalized Autoregressive Conditional Heteroskedasticity (GARCH) models are important time series models designed to capture and forecast time-varying volatility (conditional variance) in financial and economic data. De Luca and Loperfido (2004) and De Luca et al. (2006) extended the traditional multivariate GARCH framework by incorporating a multivariate skew-normal conditional distribution to capture asymmetric responses of European stock market returns to U.S. market news. They modeled European returns as driven by both endogenous local shocks and exogenous shocks from the previous day's U.S. market performance, allowing for different conditional means, variances, and skewness depending on whether U.S. returns were positive or negative. They derived explicit formulas for conditional moments, showed how "good" versus "bad" news affects volatility and skewness, and presented maximum likelihood estimation procedures. Empirical analysis of Dutch, Swiss, and Italian stock indexes demonstrated that returns are more volatile and more strongly correlated with U.S. returns following negative U.S. market news, and that the proposed Skew-GARCH model outperforms standard GARCH in volatility forecasting and capturing asymmetry. Corns and Satchell (2010) also considered the skew-normal distribution to model heteroskedasticity and skewness in financial time series. Ferreira and Steel (2004) used Bayesian multivariate skewed regression to model firm size.

In financial applications, understanding the joint behavior of multiple risk factors is fundamental for portfolio construction, risk management, credit modeling, and stress testing. However, asset returns often exhibit nonlinear, asymmetric, or heavy-tailed dependence that cannot be adequately captured by linear correlation or multivariate normal models. Copulas provide a rigorous and flexible framework for addressing this challenge by allowing the dependence structure to be modeled separately from the marginal behavior; see, e.g., Nelsen (1999) and Joe (2015). This is especially useful in finance, where each asset or factor may have a distinct empirical distribution, yet their joint extremes, such as simultaneous losses during market stress, must be accurately described. The theoretical basis for copula modeling is Sklar's theorem (Sklar, 1959), which states that any multivariate cdf F with continuous marginal cdfs $F_1, \dots, F_p$ admits the unique representation

$$F(x_1,\dots,x_p) = C(F_1(x_1),\dots,F_p(x_p)), \quad (x_1,\dots,x_p)^\top \in \mathbb{R}^p,$$

where

$$C(u_1,\dots,u_p) = F\left(F_1^{-1}(u_1),\dots,F_p^{-1}(u_p)\right), \quad (u_1,\dots,u_p)^\top \in [0,1]^p, \tag{10.3}$$

is a copula (a multivariate cdf on $[0,1]^p$ with uniform margins that captures all dependence among the components). Conversely, any copula combined with any set of marginal distributions yields a valid multivariate distribution. This decomposition enables financial modelers to specify realistic marginals (often skewed or heavy-tailed) while choosing a copula that captures features such as tail dependence or asymmetric co-movement. Commonly used copulas include the normal copula, widely applied in credit risk but limited by its absence of tail dependence, and the Student's t copula, which accommodates symmetric tail co-movement observed during market downturns. More flexible families derived from skew-elliptical distributions, such as skew-normal copula or skew-t copula, provide additional tools for modeling asymmetric or skewed dependences; see, e.g., Pettere and Kollo (2011), Kollo et al. (2017), Yoshiba (2018), and Joe and Li (2019). Through Sklar's theorem, copulas thus offer a powerful and principled framework for multivariate financial modeling and joint risk assessment. From (10.3), one can derive the copula associated with the SUN, SUT, and SUE distributions described in this book since their joint cdfs and associated marginal cdfs have been derived. For instance, let $\mathbf{X} \sim \mathcal{SUN}_{p,q}(\boldsymbol{\xi}, \boldsymbol{\Psi}, \mathbf{H}, \boldsymbol{\tau}, \bar{\boldsymbol{\Gamma}})$. From Example 69, we know that the marginals are $X_i \sim \mathcal{SUN}_{1,q}(\xi_i, \psi_{ii}, \boldsymbol{\eta}_i^\top, \boldsymbol{\tau}, \bar{\boldsymbol{\Gamma}})$, $i = 1,\dots,p$, where $\mathbf{h}_i$ is the i-th column of $\mathbf{H}$, and with cdfs according to Theorem 91:

$$F_i(x_i) = \frac{1}{\Phi_q(\boldsymbol{\tau}; \mathbf{0}_q, \bar{\boldsymbol{\Gamma}})} \Phi_{1+q}\left(\begin{pmatrix} x_i \\ \boldsymbol{\tau} \end{pmatrix}; \begin{pmatrix} \xi_i \\ \mathbf{0}_q \end{pmatrix}, \begin{pmatrix} \psi_{ii} + \boldsymbol{\eta}_i^\top \bar{\boldsymbol{\Gamma}} \boldsymbol{\eta}_i & -\boldsymbol{\eta}_i^\top \bar{\boldsymbol{\Gamma}} \\ -\bar{\boldsymbol{\Gamma}} \boldsymbol{\eta}_i & \bar{\boldsymbol{\Gamma}} \end{pmatrix} \right), \quad x_i \in \mathbb{R}.$$

Therefore, the copula associated with a multivariate SUN distribution is, for $\mathbf{u} \in [0,1]^p$:

$$C_{\text{SUN}}(\mathbf{u}) = \frac{\Phi_{p+q}\left(\begin{pmatrix} (F_1^{-1}(u_1),\dots,F_p^{-1}(u_p))^\top \\ \boldsymbol{\tau} \end{pmatrix}; \begin{pmatrix} \boldsymbol{\xi} \\ \mathbf{0}_q \end{pmatrix}, \begin{pmatrix} \boldsymbol{\Psi} + \mathbf{H}\bar{\boldsymbol{\Gamma}}\mathbf{H}^\top & -\mathbf{H}\bar{\boldsymbol{\Gamma}} \\ -\bar{\boldsymbol{\Gamma}}\mathbf{H}^\top & \bar{\boldsymbol{\Gamma}} \end{pmatrix} \right)}{\Phi_q(\boldsymbol{\tau}; \mathbf{0}_q, \bar{\boldsymbol{\Gamma}})}.$$

Obviously, the C_{SUN} copula needs to be evaluated numerically.

In management science, numerical confidential data are quantitative information, e.g., such as sales figures, financial records, production costs, or patient measurements, that contain personally identifiable or commercially sensitive details and, therefore, must be protected against unauthorized disclosure. These datasets are often used for statistical analysis, decision support, or policy making, but releasing them publicly in raw form could compromise privacy or reveal trade secrets. To balance data utility and confidentiality, agencies and researchers apply statistical disclosure control techniques (e.g., noise addition, data perturbation, synthetic data generation) so that analysts can still draw valid conclusions while individual or proprietary values cannot be traced back to their sources. Lee et al. (2010) proposed a skew-t data perturbation (STDP) method to protect sensitive numerical data while preserving statistical utility. Unlike traditional additive Gaussian approaches,

STDP assumes a multivariate skew-t distribution, allowing it to capture both skewness and heavy tails common in real databases. The authors derived the conditional skew-t distribution needed to generate perturbed records, provided an algorithm that preserves the sample mean and covariance exactly, and compared STDP with established methods (Gaussian-based, copula-based, and data shuffling) through Monte Carlo simulations and a medical database example. The results showed that STDP maintains key distributional features, especially joint tail behavior, better than competitors, offering stronger privacy protection without sacrificing analytic validity when data deviate from normality.

10.15 Environmental and Ecological Sciences

Environmental and ecological sciences (see, e.g., Barnett (2004) and Gelfand et al. (2019)) can benefit from the use of non-normal distributions as the diverse examples below illustrate.

The introduction of a skew-elliptical framework for time series modeling to capture both skewness and heavy tails in environmental data was proposed by Genton and Thompson (2004). They showed that the multivariate skew-t distribution fits well the distribution of residual sea levels at Charlottetown, Canada. In particular, it captured the heavy tails and skewness in the residuals, features not reproducible by the normal distribution. The bivariate skew-t was used to estimate conditional probabilities that the sea level did not exceed a given level, given that it was below on the previous time step. By compounding these conditional probabilities, it was possible to recover the return periods of extreme sea level calculated in the standard way using extreme value theory and observed annual maxima. This supported the use of the conditional probability approach to calculate the risk of flooding a specified critical level as a function of time over the next century. The calculated probabilities reflected the strong seasonality in the sea-level process and also long-period tidal effects. An increase in the rate of increase in sea level from 3 to 7 mm per year was shown to have a dramatic effect on the risk of flooding. This study demonstrated that skew-elliptical distributions provide a parsimonious but effective way to model nonstationary coastal flooding hazards under climate change; see also Thompson and Shen (2004).

Marchenko and Genton (2010a) introduced a new class of multivariate distributions for positive data that extends the log-normal family by allowing both skewness and heavy tails. It is defined by applying a component-wise logarithm to a random vector and then model it with a multivariate skew-elliptical distribution (see Exercise 4.20 for the log-elliptical case). This family of log-skew-elliptical distributions includes the log-skew-normal and log-skew-t as key special cases; see Azzalini et al. (2002) and Chai and Bailey (2008) for the univariate case. Marchenko and Genton (2010a) established probabilistic properties, such as stochastic representations, marginal and conditional distributions, and moment behavior. They developed inference methods by leveraging existing techniques for skew-elliptical models and applied the framework to U.S. monthly precipitation data at national and regional scales. Their analysis demonstrated that the proposed models capture asymmetry and heavy tails more effectively than conventional log-normal models, improving the fit for certain months and regions where precipitation data exhibit significant skewness and extremes. de Queiroz et al. (2016a) explored further extensions to log-SUN and log-CFUSN distributions. Flecher et al. (2010) considered a univariate truncated skew-normal distribution to model relative humidity data in the South of France.

Information measures have also been extended beyond normality and used in problems that arise in environmental science. For example, Contreras-Reyes and Arellano-Valle (2012) developed analytical expressions for information-theoretic divergence measures, notably the Kullback-Leibler (KL) and Jeffreys (J) divergences (see Section 1.9.2), within the multivariate skew-normal family. They derived formulas that decompose the KL divergence into tractable analytical and one-dimensional numerical components, extending the classical Gaussian results to skewed distributions. They analyzed several special cases (differences in mean, scale, and skewness) and showed that the J divergence between a multivariate skew-normal and its Gaussian counterpart depends only on the skewness parameter. The study includes an application to seismological data from the

2010 Maule earthquake in Chile, where the proposed divergences quantify differences among aftershock magnitude distributions obtained from nonparametric clustering. In general, these results broadened the use of entropy and divergence concepts to asymmetric data, providing both theoretical results and practical computational tools for the comparison of statistical models. Another example was provided by Arellano-Valle et al. (2013b) who extended the classical information-theoretic measures of differential entropy and mutual information (see Section 1.9.1) to the broader families of elliptical and skew-elliptical distributions, including the skew-normal and skew-t cases. They derived analytical expressions for entropy and mutual information, showing how these depend on skewness, heavy tails, and correlation structure, generalizing the well-known Gaussian results. They also examined the numerical behavior and convergence of these quantities, demonstrating that skew-t entropies approach skew-normal values as degrees of freedom increase. The results were applied to the optimal design of an ozone monitoring network in Santiago, Chile, illustrating how accounting for asymmetry and heavy tails in data can improve network efficiency and information gain compared to standard Gaussian-based approaches.

A new method was developed by Sun et al. (2017) to identify random errors in historical upper atmosphere wind records, which often exhibit skewed, heavy-tailed bivariate distributions. The authors proposed two novel outlier detection approaches based on the bivariate skew-t distribution: one using the squared Mahalanobis distance and another using skew-elliptical contours. They introduced a robust M-estimation procedure for parameter estimation. Through simulation studies, they showed that these methods more accurately detect outliers than traditional multivariate normal or nonparametric techniques, especially when data are skewed. Applying the procedures to over 35,000 historical radiosonde wind observations, they demonstrated improved identification of random measurement errors across multiple atmospheric pressure levels, offering an effective tool for quality control of long-term atmospheric datasets.

Li and Sun (2021) presented a new stochastic weather generator (SWG) designed to model precipitation at very fine temporal (30 seconds) and spatial (10 to 100 m) scales. The authors developed a censored vector autoregression (VAR) framework whose errors follow a skew-symmetric (specifically skew-t) distribution, allowing the model to jointly capture the occurrence and intensity of rainfall while accommodating the strong skewness, heavy tails, and zero inflation typical of high-frequency rainfall. Through simulation studies and an application to dense rain gauge data from Lausanne, Switzerland, the model demonstrated improved ability to reproduce extreme precipitation and spatial dependence compared to Gaussian-based generators, and even achieved predictive accuracy comparable to a long short-term memory (LSTM) neural network while offering clearer physical interpretation and uncertainty quantification.

Nathoo (2010) developed space-time regression modeling of tree growth using the skew-t distribution and conducted inference in the Bayesian framework. The author demonstrated the benefits of taking into account the asymmetry and heavy-tails in the distribution of growth measurements.

A flexible method for modeling the age-length relationship of cardinalfish using a growth curve model with error terms from the scale mixtures of the skew-normal family, focusing on the skew-t case, was developed by Contreras-Reyes and Arellano-Valle (2013). This approach captures asymmetry, heavy tails, and heteroskedasticity without transforming the data, allowing robust parameter estimation even with influential observations. Applying the model to 2,687 cardinalfish from Chilean waters revealed fish ages up to 61 years. The authors found that a skew-t model with an exponential variance function provides the best fit, yielding new growth parameter estimates and demonstrating the much greater longevity of the species than previously thought. Contreras-Reyes et al. (2014) and López Quintero et al. (2017) further applied the log-skew-t distribution, instead of the classical normal distribution, to model errors in a growth function of fish populations. They used Bayesian inference in this nonlinear regression model and demonstrated the flexibility and advantage of their approach with an application to southern blue whiting, an important fish species harvested in the southeast Pacific. Arellano-Valle et al. (2017) developed a test of normality by measuring the degree to which certain asymmetric distributions deviate from normality. They focused on the generalized skew-normal family (Genton and Loperfido, 2005) and derived exact expressions for its negentropy in terms of cumulants to quantify this disparity. The methodology was illustrated with an application to the condition factor time series of anchovies off northern Chile, demonstrating its practical use in real ecological data.

10.16 Point Processes

A spatial point process is a countable random subset of a space $S \subseteq \mathbb{R}^d$. A realization of a point process is called a point pattern. In practice, this would be a finite collection of points in S, for example, the locations of earthquakes, wildfires, lightning strikes, cells, weeds, or stars. The interest of a data analysis of a point pattern is to first assess whether the locations have no interactions (complete spatial randomness, modeled with a Poisson point process) or if there is some form of regularity or clustering among these points, and then model them; see, e.g., the book by Møller and Waagepetersen (2004) and references therein for an introduction to point processes.

Among the models used to describe clustered points, the so-called Thomas process is quite popular. Its formulation in $\mathbb{R}^2$ is as follows. First, a realization of a Poisson process with a certain intensity is generated to obtain the "parent" points. Second, each parent point is replaced by a random cluster of "children" points, the number of which is Poisson distributed with another intensity. The positions of the children points are distributed around the "parent" point according to a bivariate normal distribution with a circular covariance matrix $\sigma^2 \mathbf{I}_2$.

For certain applications, the circularity of the distribution of children around parent points may be restrictive. In fact, take weeds as an example. Wind or slope at a parent point may cause the distribution of the children points to be skewed in a certain direction. This was the motivation of Dao and Genton (2021) to introduce skew-elliptical cluster processes, for which the distribution of the children around each parent point is skewed. They studied various properties of such skew-elliptical point processes, with emphasis on the skew-normal and skew-t cases, proposed an adequate inferential method, and illustrated the benefits of the skewed models on a dataset of 195 locations of Californian redwood seedlings and saplings; see also Najari et al. (2022) for a Neyman-Scott point process with skew-normal clusters.

10.17 Extremes

It is of interest to understand the behavior of extremes from non-Gaussian distributions; see, e.g., Coles (2001) and references therein. Chang and Genton (2007) developed the theory of extreme value behavior for univariate skew-symmetric distributions, whose densities are defined as twice a symmetric kernel f times a skewing function π: $2f(x)\pi(x)$; see Section 5.11.1. The authors showed that, under mild regularity conditions on the skewing function, the domain of attraction for maxima (or minima) of a skew-symmetric distribution is the same as that of its underlying symmetric kernel, i.e., its extreme value distribution is Gumbel, Fréchet, or Weibull exactly as for the base symmetric distribution. They provided general conditions, proved key propositions for the Gumbel and Fréchet cases, and analyzed special cases, including flexible and generalized skew-normal and three types of skew-t distributions, showing how parameters affect tail behavior. They provided a counter-example, namely the generalized skew-Cauchy distribution with pdf $2c(x)\Phi(\alpha x)$, where $c(x)$ is the standard Cauchy pdf with Fréchet domain of attraction, while the skewed version has Gumbel domain of attraction. Applications to glass-fiber strength and athlete body-mass index data illustrated how these results enable practical inference on extremes for skewed data. Lysenko et al. (2009) studied multivariate extremes of generalized skew-normal distributions.

The asymptotic tail behavior of the univariate skew-normal distribution was analyzed by Capitanio (2010). She showed that the right and left tails decay at different rates, depending on the sign of the skewness parameter: One tail behaves like that of a normal distribution, while the other is thinner. She derived asymptotic approximations and inequalities that bound the tail probabilities, proving that positively and negatively skewed distributions exhibit distinct convergence properties. Through both theoretical analysis and numerical studies, she demonstrated the accuracy and limitations of these approximations for finite samples and provided explicit formulas for practical use. The tail dependence for two forms of skew-t distributions was studied by Fung and Seneta (2010).

New multivariate max-stable models that incorporate asymmetric dependence by starting from multivariate skew-normal and skew-t distributions were developed by Padoan (2011). He derived the limiting distributions of the componentwise maxima of skew-t random vectors and showed that, as the degrees of freedom parameter grows, the limit converges to a skewed version of the Hüsler-Reiss model, which itself nests the extremal t and Gaussian cases. These models allow for both symmetric and asymmetric tail dependence and can handle dimensions greater than two, overcoming limitations of classical Gaussian-based extremes. He also characterized tail dependence coefficients, Pickands dependence functions, and provided examples illustrating how the shape parameters control upper and lower tail dependence, offering practitioners a flexible framework for modeling high-dimensional, non-Gaussian extremes.

Beranger et al. (2017) further developed models for extremal dependence derived from skew-symmetric families of distributions. Beranger et al. (2019a,b) studied extremal properties of the univariate and multivariate extended skew-normal distributions; see also Beranger et al. (2021) for high-dimensional inference in max-stable models based on the extremal skew-t process. Morris et al. (2017) presented a space-time skew-t model for threshold exceedances. Hazra et al. (2020) proposed a multivariate spatial skew-t process for joint modeling of extreme precipitation indexes. Zhong et al. (2025) developed a skewed Brown-Resnick model and a truncated extremal-t model for spatial extremes. It would be of interest to extend the aforementioned results to unified skew-normal distributions and their related families.

10.18 Directional Data

Directional statistics is concerned with observations that lie on a circle (or sphere, or more general manifolds) where direction, angle, or orientation is the primary quantity of interest; see, e.g., the book by Mardia and Jupp (2000). Such data, often called circular data, arise naturally in fields as diverse as meteorology (wind and wave directions), biology (animal movement bearings, protein folding angles), geology (fault line orientations), environmental science, and robotics, to name but a few. Because directional data wrap around and lack a natural origin or endpoint, standard linear statistical methods are inappropriate. Directional statistics provides specialized tools and probability models, e.g., such as the von Mises and wrapped distributions that respect the geometry of circular and spherical spaces and allow for meaningful inference on orientation-related phenomena.

Circular skewed distributions extend classical symmetric models on the unit circle to accommodate directional data exhibiting asymmetry. Traditional circular models, such as the von Mises, wrapped normal, and wrapped Cauchy distributions, assume symmetry about a preferred direction, yet empirical directional data often display clear skewness. Circular skewed distributions were developed to address this limitation by allowing for asymmetric concentration and directional tilt while preserving circular topology.

A foundational construction is the sine-skewing mechanism of Abe and Pewsey (2011), which adapts Azzalini's skewing approach to circular domains. Starting from a symmetric circular density $f(\theta)$, typically von Mises or wrapped Cauchy, a sine-skewed version is defined by

$$f_s(\theta) \propto f(\theta-\mu)\left[1+\lambda\sin(\theta-\mu)\right],$$

where θ denotes an angle, the parameter $\lambda \in [-1,1]$ governs the magnitude and direction of the skewness, and μ is a central direction. Important examples include the sine-skewed von Mises distribution and the sine-skewed wrapped Cauchy distribution, both of which provide model flexibility and are effective in environmental and ecological applications. Umbach and Jammalamadaka (2009, 2010) studied skew-symmetric circular distributions, focusing on how skewness affects their trigonometric moments and related directional characteristics. They derived explicit expressions for mean direction, circular variance, and higher-order moments for a broad class of sine-skewed and other skew-symmetric circular models.

Another major class consists of wrapped skewed distributions, obtained by wrapping linear skew-normal or skew-*t* distributions around the circle; see, e.g., Pewsey (2000), Pewsey (2006), and Hernández-Sánchez and Scarpa (2012). These models inherit skewness, tail behavior, and even multimodality from their linear counterparts while ensuring periodicity. Although their densities involve infinite Fourier series, they are computationally tractable and useful in applications requiring richer asymmetry or heavier tails.

A unifying theoretical point of view comes from selection mechanisms (Arellano-Valle et al., 2006a), in which circular skewness arises by conditioning an underlying symmetric distribution on an asymmetric event. This perspective connects circular skewed models to the broader class of skew-elliptical distributions and clarifies the geometric origins of asymmetry; see also the general setting put forward by Jupp et al. (2016).

Inference for circular skewed distributions involves maximum likelihood or Bayesian methods, along with circular moment-based techniques. Across numerous studies, circular skewed models have demonstrated superior performance over symmetric alternatives when analyzing asymmetric directional data. As a result, they now form an essential component of contemporary directional statistics; see Pewsey and García-Portugués (2021) for a recent review and the references therein.

10.19 Link Models and Item Response Theory

Link models for binary response data relate the probability of a binary outcome to a set of covariates through a transformation, or link function, that maps probabilities in $[0,1]$ to the real line; see, e.g., McCullagh and Nelder (1989). Common links such as the logit and probit arise from assuming a symmetric distribution for an underlying latent variable and provide a convenient and interpretable regression framework. More flexible link models extend this idea by allowing for asymmetric or skewed links, enabling better modeling of unbalanced response probabilities and heterogeneous response behavior.

Chen et al. (1999) introduced a class of skewed link models for binary response data. Building on a formulation of latent variables, they modeled asymmetry by allowing the underlying latent variable to follow a skewed distribution, yielding a flexible family that includes common symmetric and asymmetric links as special cases. Within a Bayesian framework, they established conditions for posterior propriety even under improper priors, developed informative priors using historical data, and proposed efficient MCMC and likelihood-based estimation methods. Through extensive simulations and a prostate cancer application, they demonstrated that the proposed skewed link models provide more accurate inference, better fit, and improved model comparison than existing approaches such as generalized logistic and complementary log-log models, especially in asymmetric data settings. Further results can be found in Chen (2004).

Item response models describe the relationship between an individual's latent ability or trait and their probability of giving a particular response to an item, typically binary such as correct/incorrect or yes/no. They specify how item characteristics (such as difficulty or discrimination) and person ability jointly determine response probabilities through a parametric response function. These models form the foundation of item response theory (IRT) and are widely used in educational testing, psychometrics, and related fields to measure latent traits from observed responses.

In order to allow item characteristic curves to be asymmetric, Bazán et al. (2006) proposed a skew-probit IRT model that extends the classical probit-normal IRT framework by introducing an item-specific skewness (penalization) parameter through the skew-normal distribution. The model was formulated within a Bayesian framework, with efficient data-augmentation-based MCMC algorithms developed for posterior inference and careful discussion of prior specification and identifiability. Through simulation studies and real applications to educational and epidemiological data, the authors showed that the skew-probit model often provides a substantially better fit than symmetric IRT models, captures asymmetric response behavior more realistically, and yields interpretable insights for test design, while reducing to the standard probit IRT model as a special case when skewness is absent; see also the Bayesian inference for a skew-normal IRT model under the centered parameterization in Azevedo et al. (2011).

A further extension to allow for asymmetric distributions of latent abilities and an additional random effect to account for local dependence among items within testlets (groups of test items that share a common stimulus or context) was proposed by Bazán et al. (2014). Framed in a fully Bayesian hierarchical setting, the authors developed efficient data-augmentation MCMC algorithms that are straightforward to implement and proposed multiple model comparison criteria to assess fit. Applications to mathematics and reading comprehension tests from Peruvian students showed that the proposed skew-normal IRT family, particularly models combining skewed abilities and skewed item characteristic curves with their testlet extensions, provides a substantially better fit and more interpretable inference than classical symmetric normal IRT models, highlighting the importance of modeling asymmetry and item dependence in educational assessment.

10.20 Other Areas of Applications

There are many other areas of applications of multivariate skew-elliptical distributions and related families. We briefly mention a few of them below.

Censored regression: Arellano-Valle et al. (2012) developed a Student's t censored regression model as a robust extension of the classical Tobit model by replacing normally distributed errors with Student's t errors to better accommodate heavy tails and outliers in censored data. They derived the likelihood, score functions, and both observed and expected information matrices, and proposed an efficient EM/ECM-type algorithm for maximum likelihood estimation by exploiting the scale mixtures representation of the Student's t distribution. An application to censored wage data for housewives shows that the Student's t model provides a substantially better fit than the normal Tobit model according to likelihood and AIC criteria, demonstrating improved robustness and practical relevance for econometric and statistical analyses of limited dependent variables.

In a similar direction, Castro et al. (2014) proposed a Bayesian framework for partially linear censored regression models that simultaneously addresses nonlinearity, censoring, and heavy-tailed errors, extending the classical Tobit model. The response was modeled as a combination of a parametric linear component and an unspecified smooth function approximated using penalized splines, while the error distribution was assumed to belong to the flexible class of scale mixtures of normal (SMN) distributions, including Student's t, slash, and contaminated normal models, to improve robustness against outliers. The authors proposed efficient MCMC-based inference, Bayesian model comparison criteria, and case-deletion influence diagnostics, and demonstrated through simulations and an application to censored wage data that heavy-tailed semiparametric models provide substantially better fit, robustness, and interpretability than standard normal censored regression models.

Recent work on likelihood-based inference for the multivariate skew-t regression with censored or missing responses was developed by Valeriano et al. (2023).

Reliability analysis: Azzalini and Chiogna (2004) investigated the classical stress-strength reliability measure $\rho = \mathrm{P}(Y_2 < Y_1)$ under the assumption that one or both random variables follow skew-normal distributions, extending the well-known normal case to the more flexible skew-normal family. They derived exact analytical expressions for ρ in several important situations (notably when one variable is skew-normal and the other is normal) using properties of the skew-normal distribution. For inference, they employed modified maximum likelihood estimation, which stabilizes the estimation of the skewness parameter and avoids divergence issues typical in standard MLE. Simulation studies showed that the resulting estimators of ρ are essentially unbiased and that one asymptotic confidence interval maintains close-to-nominal coverage except when the skewness parameter is near zero, where the skew-normal becomes nearly indistinguishable from the normal distribution. The paper provides both theoretical probability results and practical likelihood-based inferential tools for reliability analysis with skew-normal components. Another special case has been explored by Gupta and Brown (2001).

For analyzing fatigue life data, Arrué et al. (2020) proposed a new extension of the Birnbaum-Saunders distribution based on the modified skew-normal distribution to overcome the limited tail flexibility of the classical model. The resulting model introduces an additional shape parameter that jointly controls skewness and kurtosis. The authors developed maximum likelihood inference with bias correction for the skewness parameter and established the nonsingularity of the Fisher information, enabling valid hypothesis testing. Simulation studies and an application to real aluminum fatigue data showed that the proposed model provides a substantially better fit, especially in the distribution tails, than the standard Birnbaum-Saunders distribution, highlighting its practical advantages for reliability and fatigue analysis.

Compositional data: A simplex is the set of nonnegative vectors whose components sum to one, representing all possible probability distributions over a finite set. In geometry, it generalizes shapes like a line segment, triangle, or tetrahedron to arbitrary dimensions. Compositional data are multivariate observations consisting of positive components that represent parts of a whole and therefore carry only relative information, typically constrained to sum to a constant such as one. Therefore, compositional data are defined on the simplex, which is then typically mapped to $\mathbb{R}^p$ by an additive log-ratio transformation; see, e.g., Aitchison (1986). The multivariate normal distribution is then used for data analysis. Mateu-Figueras and Pawlowsky-Glahn (2007) proposed using the skew-normal distribution to analyze data on the simplex.

Demography: In the context of demography, Mazzuco and Scarpa (2015) used a flexible generalized skew-normal pdf to fit age-specific fertility rates, while Aliverti et al. (2022) proposed dynamic modeling of mortality via mixtures of skewed distribution functions.

Astronomy: For astronomical distance determination, Eyer and Genton (2004) showed that using regression with skew-normal rather than normal errors leads to more sensible results.

Shape analysis: In the area of shape analysis, Baloch et al. (2004) used flexible skew-symmetric distributions to model distances conditional on some angular directions and argued for better representations due to non-Gaussian features.

Stochastic frontier analysis: Stochastic frontier analysis is an econometric method that estimates a production or cost frontier while separating random noise from firms' inefficiency to measure how far each unit operates from the optimal frontier; see, e.g., Aigner et al. (1977). In this context, Domínguez-Molina et al. (2004) studied skew-normality while Tchumtchoua and Dey (2007) proposed Bayesian estimation of stochastic frontier models with multivariate skew-t error terms.

Stochastic differential equations: Iguchi et al. (2025) recently explored the use of skew-symmetric distributions to design schemes for stochastic differential equations with non-Lipschitz drift. Hongler and Rinaldo (2026) introduced skew-normal diffusion processes.

Spatial areal data analysis: Cruz-Reyes et al. (2026) introduced directional asymmetry in edge-based spatial models for areal data through a skew-normal prior distribution.

Other flexible non-Gaussian distributions: Arellano-Valle et al. (2005b) introduced a broad and unifying class of univariate asymmetric distributions obtained by deforming an arbitrary symmetric density through two positive asymmetry functions, leading to a flexible model that includes well-known families such as the Fernández-Steel (Fernández and Steel, 1998), epsilon-skew-normal, and skew-exponential power distributions as special cases. A key theoretical contribution is a simple stochastic representation in which any member of the class can be expressed as the product of two independent random variables (a signed discrete component and the magnitude of a symmetric random variable), facilitating the derivation of distributional properties, moments, simulation methods, and natural extensions to location-scale and multivariate settings. The authors studied statistical inference using both the method of moments and maximum likelihood, obtaining asymptotic results for general kernels and explicit likelihood equations for the skew-exponential power case, supported by several real-data applications that illustrate the model's ability to capture a wide range of skewness and kurtosis patterns. Overall, the framework provides a general, tractable, and highly adaptable approach to modeling asymmetric data beyond the restrictions of the classical skew-normal and skew-elliptical families.

Vidal et al. (2006) developed a Bayesian framework to assess sensitivity to the skewness parameter and to perform model comparison within the skew-elliptical family, extending the foundational ideas on skew-symmetric and representable skew-elliptical distributions. They first studied sensitivity by deriving explicit expressions for the L_1 distance between symmetric and asymmetric models, proving in particular that for all representable skew-elliptical distributions, this distance coincides with the L_1 distance between the normal and skew-normal densities, thus revealing an invariance property across the entire representable class. They then investigated the posterior sensitivity of location and scale parameters under skew-normal sampling using normal-inverse-gamma priors, deriving closed-form perturbation factors whose influence was illustrated numerically. A second major component of the paper is the use of Bayes factors to test for asymmetry, including special cases with analytic simplifications and a general expression for the representable skew-elliptical case. Extensive simulation studies for skew-normal and skew-t models and a real-data application to Chilean stock market returns showed how prior choices, sample size, and tail heaviness affect evidence for skewness. Overall, the paper provided a unified Bayesian methodology for quantifying and testing skewness in flexible non-normal models.

Salinas et al. (2007) developed a highly flexible family of asymmetric distributions, called the extended skew-exponential power (ESEP) class, obtained by combining a symmetric baseline density with a symmetric cumulative distribution function in a unified selection-type construction that extends both the classical Azzalini skew-normal mechanism and the two-piece (epsilon-skew) formulation. A key theoretical contribution is a stochastic representation in which an extended skewed variable can be written as the product of a signed discrete variable and a transformed symmetric variate, which facilitates the derivation of general moment expressions, measures of skewness and kurtosis, and efficient simulation algorithms. The authors specialized the construction to the exponential power kernel, yielding the ESEP distribution, which simultaneously accommodates skewness, heavy tails, and varying peak shapes, encompassing the normal, skew-normal, generalized skew-normal, skew-exponential power, and epsilon-skew-normal distributions as particular cases. They developed maximum likelihood estimation, discussed nonregular cases where the information matrix is singular, and provided analytic expressions for the expected information matrix. A real-data application to roller-surface roughness measurements demonstrated that the ESEP model achieves a substantially improved fit over the classical skew-exponential power family (DiCiccio and Monti, 2004), as confirmed by likelihood-ratio and goodness-of-fit tests.

Arellano-Valle et al. (2010b) presented the flexible epsilon-skew-normal (FESN) distribution, a parametric extension of the epsilon-skew-normal family that incorporates an additional shape parameter to greatly increase flexibility in skewness, kurtosis, and modality, allowing both unimodal and bimodal asymmetric distributions. They studied the probabilistic properties of the new family, provided a useful stochastic representation, derived moments and shape measures, and developed maximum likelihood inference, showing in particular that the Fisher information matrix is always nonsingular, including in the normal special case. Applications to real datasets demonstrated that the FESN model can substantially improve goodness of fit over classical normal and epsilon-skew-normal models, especially for data exhibiting strong asymmetry or bimodality. Another approach to model bimodality using univariate bimodal skew-elliptical distributions in the Bayesian framework can be found in Elal-Olivero et al. (2009).

Rocha et al. (2013) put forward classical and Bayesian inference methods for a flexible class of distributions with a normal kernel, focusing on the skew-bimodal-normal (SBN) family, which simultaneously accommodates skewness and unimodal or bimodal shapes. They established conditions for the existence of maximum likelihood estimators, showed that estimation of skewness and bimodality parameters can be decoupled into simpler skew-normal and bimodal-normal subproblems, and proposed efficient EM-type algorithms for computation. They also derived closed-form posterior and predictive distributions under several prior specifications, conducted extensive Monte Carlo studies to compare MLEs and Bayesian estimators, and demonstrated through density estimation of benchmark data that the SBN model provides fits comparable to more complex skew-normal models while being more parsimonious.

Kahrari et al. (2016) introduced the multivariate skew-normal-Cauchy (SNC) distribution, a flexible extension of the multivariate skew-normal model that combines asymmetry with much heavier tails through a Cauchy-based skewing mechanism. They developed a convenient hierarchical and stochastic representation, derived key probabilistic properties including moments, skewness and kurtosis measures, closure under linear transformations, and marginal and conditional distributions, and showed that the SNC family preserves many attractive features of the skew-normal while greatly expanding tail behavior. A central contribution is the derivation of explicit expressions for the expected Fisher information matrix, highlighting both inferential tractability and the occurrence of singularity near symmetry, thereby clarifying the theoretical and practical properties of this robust multivariate model. Kahrari et al. (2017) further explored scale mixtures of SNC distributions. Bias prevention of maximum likelihood estimates for SNC distributions was studied by Kahrari et al. (2020), while multivariate linear models of scale mixtures of SNC distributions were explored by Kahrari et al. (2022).

Sharifipanah et al. (2021) presented a unified and highly flexible framework for constructing symmetric and asymmetric distributions based on the two-piece skew-normal paradigm and Azzalini's skewing mechanism. Starting from a general class of symmetric densities obtained by weighting a central symmetric kernel, they introduced an extended two-piece skew bimodal normal family that can accommodate skewness, heavy tails, excess kurtosis, and uni- or bi-modality, while encompassing many well-known models (normal, skew-normal, bimodal normal, two-piece skew-normal, and their extensions) as special cases. They derived distributional properties (modes, moments, skewness, kurtosis, stochastic representations), developed maximum likelihood inference, and demonstrated through a real data application on gamma-ray burst durations that the proposed model offers substantially improved fit over competing submodels when data exhibit combined skewness and bimodality.

10.21 Open Research Problems

Open Research Problem 34 *Download some multivariate continuous datasets and perform a data analysis beyond normality. In particular, compute the sample mean vector, variance matrix, correlation matrix, and Mardia measures of skewness and kurtosis. Fit $\mathcal{N}$, $\mathcal{ASN}$, $\mathcal{SN}$, $\mathcal{AST}$ distributions to the data. Plot the fitted contours and interpret the estimated parameters. Which fit is best?*

Open Research Problem 35 *For each application of Chapter 10 described in its various sections, read the relevant literature, fit appropriate non-Gaussian models, and explore how to improve them further.*

Open Research Problem 36 *Develop the Bayesian conjugacy results with the SUN parameterization instead of the AASUN parameterization.*

Open Research Problem 37 *Develop the Bayesian conjugacy results with the SUNSET and BSUNSET distributions introduced in Section 6.8.1.*

Open Research Problem 38 *Evaluate numerically the copula of the multivariate SUN distribution presented in Section 10.14 and study its shape as a function of q. Compare it to the copula of the multivariate SN distribution obtained with $q = 1$.*

Open Research Problem 39 *Use Equation (10.3) to derive the copula of the multivariate SUT distribution. Evaluate it numerically and study its shape as a function of q. Compare it to the copula of the multivariate ST distribution obtained with $q = 1$.*

Open Research Problem 40 *Develop models for extremal dependence derived from unified skew-normal distributions. Extend them to unified skew-t and unified skew-elliptical distributions.*

Appendix: Linear Algebra

This appendix provides some useful linear algebra definitions and properties. More in-depth treatments of linear algebra can be found in classical books such as Strang (1980), Golub and Van Loan (1983), and Horn and Johnson (1985, 1991).

A.1 Introduction

Linear algebra deals with vectors and matrices. Let p be a positive integer and $\mathbb{R}$ denote the set of real numbers. Then $\mathbb{R}^p$ is the set of all p-tuples of real numbers. A *vector* $\mathbf{a} \in \mathbb{R}^p$ is a p-tuple of real numbers arranged in a column. A *matrix* $\mathbf{A} \in \mathbb{R}^{p \times q}$ is a rectangular array of real numbers with p rows and q columns. Therefore, a vector $\mathbf{a} \in \mathbb{R}^p$ is a particular matrix $\mathbf{A} \in \mathbb{R}^{p \times 1}$, and the properties of vectors follow from the properties of matrices. A matrix $\mathbf{A} \in \mathbb{R}^{p \times q}$ is said to be a *rectangular matrix*, whereas it is called a *square matrix* when $p = q$, i.e., when $\mathbf{A} \in \mathbb{R}^{p \times p}$. The *components of a vector* $\mathbf{a} \in \mathbb{R}^p$ are denoted by a_i, $i = 1, \ldots, p$, whereas the *components (or entries) of a matrix* $\mathbf{A} \in \mathbb{R}^{p \times q}$ are denoted by a_{ij}, $i = 1, \ldots, p$ and $j = 1, \ldots, q$. In brief, $\mathbf{A} = (a_{ij}) \in \mathbb{R}^{p \times q}$. The *Frobenius norm* of a rectangular matrix $\mathbf{A} = (a_{ij}) \in \mathbb{R}^{p \times q}$ is defined by:

$$\|\mathbf{A}\|_{\mathrm{Fr}} = \left(\sum_{i=1}^{p} \sum_{j=1}^{q} a_{ij}^2 \right)^{1/2}. \tag{A.1}$$

For $q = 1$, we obtain the usual Euclidean norm of a vector $\mathbf{a} \in \mathbb{R}^p$ as $\|\mathbf{a}\| = (\sum_{i=1}^{p} a_i^2)^{1/2}$.

The *sum or difference of two matrices* $\mathbf{A} = (a_{ij}) \in \mathbb{R}^{p \times q}$ and $\mathbf{B} = (b_{ij}) \in \mathbb{R}^{p \times q}$ is the matrix $\mathbf{C} = (c_{ij}) \in \mathbb{R}^{p \times q}$ such that $c_{ij} = a_{ij} \pm b_{ij}$ and denoted by $\mathbf{C} = \mathbf{A} \pm \mathbf{B}$. The *matrix product of two matrices* $\mathbf{A} = (a_{ij}) \in \mathbb{R}^{p \times q}$ and $\mathbf{B} = (b_{ij}) \in \mathbb{R}^{q \times r}$ is the matrix $\mathbf{C} = (c_{ij}) \in \mathbb{R}^{p \times r}$ such that $c_{ij} = \sum_{k=1}^{q} a_{ik} b_{kj}$ and denoted by $\mathbf{C} = \mathbf{AB}$. The matrix product is not commutative, i.e., $\mathbf{AB} \neq \mathbf{BA}$.

The *transpose* of a matrix $\mathbf{A} = (a_{ij}) \in \mathbb{R}^{p \times q}$ is the matrix $\mathbf{A}^\top = (a_{ji}) \in \mathbb{R}^{q \times p}$. Some simple properties of the transpose operation are: $(\mathbf{A}^\top)^\top = \mathbf{A}$, $(\mathbf{A} \pm \mathbf{B})^\top = \mathbf{A}^\top \pm \mathbf{B}^\top$, and $(\mathbf{AB})^\top = \mathbf{B}^\top \mathbf{A}^\top$. A square matrix $\mathbf{A} \in \mathbb{R}^{p \times p}$ is said to be a *symmetric matrix* if $\mathbf{A} = \mathbf{A}^\top$.

A square matrix $\mathbf{A} = (a_{ij}) \in \mathbb{R}^{p \times p}$ is *diagonal* if $a_{ij} = 0$ for all $i \neq j$. Given a vector $\mathbf{a} \in \mathbb{R}^p$, we denote by $\mathrm{diag}(\mathbf{a})$ the diagonal matrix whose entries on the diagonal are the components of $\mathbf{a}$. Given a square matrix $\mathbf{A} = (a_{ij}) \in \mathbb{R}^{p \times p}$, we denote by $\mathrm{diag}(\mathbf{A})$ the diagonal matrix whose entries on the diagonal are the diagonal entries of $\mathbf{A}$. A square matrix $\mathbf{A} = (a_{ij}) \in \mathbb{R}^{p \times p}$ is *upper/lower triangular* if $a_{ij} = 0$ below/above the diagonal. The *vector of ones* is the vector $\mathbf{1}_p = (1, \ldots, 1)^\top \in \mathbb{R}^p$. The *identity matrix* is the diagonal matrix $\mathbf{I}_p = \mathrm{diag}(\mathbf{1}_p) \in \mathbb{R}^{p \times p}$. The *matrix of ones* is the matrix $\mathbf{1}_p \mathbf{1}_p^\top \in \mathbb{R}^{p \times p}$. The *vector of zeroes* (or *zero vector*) is the vector $\mathbf{0}_p = (0, \ldots, 0)^\top \in \mathbb{R}^p$. The *matrix of zeroes* (or *zero matrix*) $\mathbf{O}$ is the matrix $\mathbf{0}_p \mathbf{0}_p^\top \in \mathbb{R}^{p \times p}$. The identity matrix $\mathbf{I}_p$, the matrix of ones $\mathbf{1}_p \mathbf{1}_p^\top$, and the matrix of zeros $\mathbf{O}$ are simple examples of symmetric matrices.

The particular case $r = 1$ of the aforementioned matrix product is called the *matrix-vector product* and yields a vector $\mathbf{c} \in \mathbb{R}^p$ where $\mathbf{c} = \mathbf{Ab}$. This represents a *linear transformation* of the vector $\mathbf{b} \in \mathbb{R}^q$ to a vector $\mathbf{c} \in \mathbb{R}^p$. The *column space* of $\mathbf{A}$ is the set of vectors that can be produced as linear combinations of the columns of $\mathbf{A}$: $\mathtt{C}(\mathbf{A}) = \{\mathbf{c} \in \mathbb{R}^p | \mathbf{c} = \mathbf{Ab} \text{ for some } \mathbf{b} \in \mathbb{R}^q\}$. It is also the *range* of the linear transformation. Similarly, the *row space* $\mathtt{R}(\mathbf{A})$ of $\mathbf{A}$ is the set of vectors that can be produced as linear combinations of the rows of $\mathbf{A}$. The *null space* (or kernel) of $\mathbf{A}$ consists of all the vectors sent by the matrix $\mathbf{A}$ to the zero vector: $\mathtt{N}(\mathbf{A}) = \{\mathbf{b} \in \mathbb{R}^q | \mathbf{A}\mathbf{b} = \mathbf{0}_p\}$. The vectors in the null space are orthogonal to all rows of the matrix $\mathbf{A}$. Thus, $\mathtt{R}(\mathbf{A})$ is the orthogonal complement of $\mathtt{N}(\mathbf{A})$: $\mathtt{N}(\mathbf{A}) \oplus \mathtt{R}(\mathbf{A}) = \mathbb{R}^q$, where $\oplus$ stands for *orthogonal direct sum*.

A rectangular matrix $\mathbf{A} = (a_{ij}) \in \mathbb{R}^{p\times q}$ can be *partitioned* into *submatrices* (also called *blocks*) as follows:

$$\mathbf{A} = \begin{pmatrix} \mathbf{A}_{11} & \mathbf{A}_{12} \\ \mathbf{A}_{21} & \mathbf{A}_{22} \end{pmatrix}, \tag{A.2}$$

where $\mathbf{A}_{11} \in \mathbb{R}^{p_1\times q_1}$, $\mathbf{A}_{12} \in \mathbb{R}^{p_1\times q_2}$, $\mathbf{A}_{21} \in \mathbb{R}^{p_2\times q_1}$, and $\mathbf{A}_2 \in \mathbb{R}^{p_2\times q_2}$, with $p_1 + p_2 = p$, $q_1 + q_2 = q$. The transpose of the above partitioned matrix $\mathbf{A}$ is:

$$\mathbf{A}^\top = \begin{pmatrix} \mathbf{A}_{11}^\top & \mathbf{A}_{21}^\top \\ \mathbf{A}_{12}^\top & \mathbf{A}_{22}^\top \end{pmatrix}. \tag{A.3}$$

If $\mathbf{A}_{12} = \mathbf{O}$ and $\mathbf{A}_{21} = \mathbf{O}$, then $\mathbf{A}$ is a *block-diagonal* matrix. These notions are trivially extended to an arbitrary number of blocks.

A.2 Matrix Operations and Their Properties

A.2.1 Trace

The *trace of a square matrix* $\mathbf{A} = (a_{ij}) \in \mathbb{R}^{p\times p}$ is the scalar $\mathrm{tr}(\mathbf{A}) = \sum_{i=1}^p a_{ii}$, i.e., the sum of the diagonal entries of $\mathbf{A}$. The following properties hold:

- $\mathrm{tr}(\alpha) = \alpha$, where $\alpha \in \mathbb{R}$; (A.4a)
- $\mathrm{tr}(\mathbf{A} \pm \mathbf{B}) = \mathrm{tr}(\mathbf{A}) \pm \mathrm{tr}(\mathbf{B})$, where $\mathbf{A} \in \mathbb{R}^{p\times p}$, $\mathbf{B} \in \mathbb{R}^{p\times p}$; (A.4b)
- $\mathrm{tr}(\alpha\mathbf{A}) = \alpha\mathrm{tr}(\mathbf{A})$, where $\alpha \in \mathbb{R}$, $\mathbf{A} \in \mathbb{R}^{p\times p}$; (A.4c)
- $\mathrm{tr}(\mathbf{CD}) = \mathrm{tr}(\mathbf{DC})$, where $\mathbf{C} \in \mathbb{R}^{p\times q}$, $\mathbf{D} \in \mathbb{R}^{q\times p}$; (A.4d)
- $\mathrm{tr}(\mathbf{c}\mathbf{d}^\top) = \mathbf{d}^\top\mathbf{c}$, where $\mathbf{c} \in \mathbb{R}^p$, $\mathbf{d} \in \mathbb{R}^p$; (A.4e)
- $\mathrm{tr}(\mathbf{C}\mathbf{C}^\top) = \mathrm{tr}(\mathbf{C}^\top\mathbf{C}) = \|\mathbf{C}\|_{\mathrm{Fr}}^2$, where $\mathbf{C} \in \mathbb{R}^{p\times q}$; (A.4f)
- $\mathrm{tr}(\mathbf{c}\mathbf{c}^\top) = \mathbf{c}^\top\mathbf{c} = \|\mathbf{c}\|^2$, where $\mathbf{c} \in \mathbb{R}^p$; (A.4g)
- In general, $\mathrm{tr}(\mathbf{ABC}) \neq \mathrm{tr}(\mathbf{ACB})$, where $\mathbf{A} \in \mathbb{R}^{p\times p}$, $\mathbf{B} \in \mathbb{R}^{p\times p}$, $\mathbf{C} \in \mathbb{R}^{p\times p}$ (A.4h) unless the three matrices are symmetric.

A.2.2 Determinant

The *determinant* of a square matrix $\mathbf{A} = (a_{ij}) \in \mathbb{R}^{p\times p}$ is the scalar $|\mathbf{A}|$ defined by:

$$|\mathbf{A}| = \sum_{\sigma \in S_p} \mathrm{sgn}(\sigma) a_{1\sigma(1)} \cdots a_{p\sigma(p)}, \tag{A.5}$$

where each $\sigma(1), \ldots, \sigma(p)$ is a permutation of the set $\{1, \ldots, p\}$, S_p is the set of all such permutations, and $\mathrm{sgn}(\sigma)$ is the signature of the permutation σ, i.e., $+1$ if the permutation can be obtained with an even number of transpositions (exchanges of two entries) and -1 otherwise. For $p = 2$, we have $|\mathbf{A}| = a_{11}a_{22} - a_{12}a_{21}$. A square matrix $\mathbf{A} \in \mathbb{R}^{p\times p}$ is *nonsingular* if $|\mathbf{A}| \neq 0$. It is *singular* if $|\mathbf{A}| = 0$. The *cofactor* A_{ij} of an entry a_{ij} of $\mathbf{A}$ is defined to be $A_{ij} = (-1)^{i+j}|\mathbf{A}_{-i,-j}|$ where $\mathbf{A}_{-i,-j}$

is obtained by deleting the i-th row and the j-th column of $\mathbf{A}$. The following properties hold:

- $|\mathbf{A}| = \sum_{i=1}^{p} a_{ij}A_{ij} = \sum_{j=1}^{p} a_{ij}A_{ij}$ for any i, j, where $\mathbf{A} \in \mathbb{R}^{p\times p}$; (A.6a)
- $\sum_{k=1}^{p} a_{ik}A_{jk} = 0$ for $i \neq j$, where $\mathbf{A} \in \mathbb{R}^{p\times p}$; (A.6b)
- $|\alpha\mathbf{A}| = \alpha^{p}|\mathbf{A}|$, where $\alpha \in \mathbb{R}$; (A.6c)
- $|\mathbf{AB}| = |\mathbf{A}|\cdot|\mathbf{B}|$, where $\mathbf{A} \in \mathbb{R}^{p\times p}$, $\mathbf{B} \in \mathbb{R}^{p\times p}$; (A.6d)
- $|\mathbf{A}| = \prod_{i=1}^{p} a_{ii}$ if $\mathbf{A} \in \mathbb{R}^{p\times p}$ is diagonal or upper/lower triangular; (A.6e)
- $|\mathbf{A}^{\top}| = |\mathbf{A}|$, where $\mathbf{A} \in \mathbb{R}^{p\times p}$; (A.6f)
- $|\mathbf{A}^{-1}| = |\mathbf{A}|^{-1}$, where $\mathbf{A} \in \mathbb{R}^{p\times p}$, $|\mathbf{A}| \neq 0$; (A.6g)
- $|\mathbf{A}+\mathbf{BC}| = |\mathbf{A}|\cdot|\mathbf{I}_p+\mathbf{A}^{-1}\mathbf{BC}| = |\mathbf{A}|\cdot|\mathbf{I}_q+\mathbf{CA}^{-1}\mathbf{B}|$, (A.6h)
  where $\mathbf{A} \in \mathbb{R}^{p\times p}$, $\mathbf{B} \in \mathbb{R}^{p\times q}$, $\mathbf{C} \in \mathbb{R}^{q\times p}$, $|\mathbf{A}| \neq 0$;
- $|\mathbf{I}_p+\mathbf{AB}| = |\mathbf{I}_q+\mathbf{BA}|$, where $\mathbf{A} \in \mathbb{R}^{p\times q}$, $\mathbf{B} \in \mathbb{R}^{q\times p}$; (A.6i)
- $|\mathbf{A}+\mathbf{ba}^{\top}| = |\mathbf{A}|\cdot(1+\mathbf{a}^{\top}\mathbf{A}^{-1}\mathbf{b})$, where $\mathbf{A} \in \mathbb{R}^{p\times p}$, $\mathbf{a} \in \mathbb{R}^{p}$, $\mathbf{b} \in \mathbb{R}^{p}$, $|\mathbf{A}| \neq 0$; (A.6j)
- $|\mathbf{A}+\mathbf{B}| = |\mathbf{A}|+|\mathbf{B}|+|\mathbf{A}|\,\mathrm{tr}(\mathbf{A}^{-1}\mathbf{B})$, where $\mathbf{A} \in \mathbb{R}^{p\times p}$, $\mathbf{B} \in \mathbb{R}^{p\times p}$, $|\mathbf{A}| \neq 0$, $|\mathbf{B}| \neq 0$; (A.6k)
- $\begin{vmatrix} \mathbf{A}_{11} & \mathbf{A}_{12} \\ \mathbf{A}_{21} & \mathbf{A}_{22} \end{vmatrix} = |\mathbf{A}_{11}|\cdot|\mathbf{A}_{22}-\mathbf{A}_{21}\mathbf{A}_{11}^{-1}\mathbf{A}_{12}| = |\mathbf{A}_{22}|\cdot|\mathbf{A}_{11}-\mathbf{A}_{12}\mathbf{A}_{22}^{-1}\mathbf{A}_{21}|$, (A.6l)
  where $\mathbf{A}_{11} \in \mathbb{R}^{p_1\times p_1}$, $\mathbf{A}_{12} \in \mathbb{R}^{p_1\times p_2}$, $\mathbf{A}_{21} \in \mathbb{R}^{p_2\times p_1}$, $\mathbf{A}_{2} \in \mathbb{R}^{p_2\times p_2}$, $|\mathbf{A}_{11}| \neq 0$, $|\mathbf{A}_{22}| \neq 0$;
- $\begin{vmatrix} \mathbf{A}_{11} & \mathbf{A}_{12} \\ \mathbf{O} & \mathbf{A}_{22} \end{vmatrix} = |\mathbf{A}_{11}|\cdot|\mathbf{A}_{22}|$; (A.6m)
- $\begin{vmatrix} \mathbf{A}_{11} & \mathbf{a}_{12} \\ \mathbf{a}_{12}^{\top} & a_{22} \end{vmatrix} = |\mathbf{A}_{11}|\cdot(a_{22}-\mathbf{a}_{12}^{\top}\mathbf{A}_{11}^{-1}\mathbf{a}_{12})$, where $\mathbf{A}_{11} \in \mathbb{R}^{p_1\times p_1}$, $\mathbf{a}_{12} \in \mathbb{R}^{p_1}$, $a_{22} \in \mathbb{R}$. (A.6n)

A.2.3 Rank

The *rank* of a rectangular matrix $\mathbf{A} \in \mathbb{R}^{p\times q}$ is defined as the maximum number of linearly independent rows (columns) of $\mathbf{A}$ and is denoted by $\mathrm{rank}(\mathbf{A})$. In other words, $\mathrm{rank}(\mathbf{A})$ is the dimension of the column space $\mathtt{C}(\mathbf{A})$. Therefore: $0 \leq \mathrm{rank}(\mathbf{A}) \leq \min(p,q)$. Moreover, the following properties hold:

- $\mathrm{rank}(\mathbf{A}) = \mathrm{rank}(\mathbf{A}^{\top})$, where $\mathbf{A} \in \mathbb{R}^{p\times q}$; (A.7a)
- $\mathrm{rank}(\mathbf{A}+\mathbf{B}) \leq \mathrm{rank}(\mathbf{A})+\mathrm{rank}(\mathbf{B})$, where $\mathbf{A} \in \mathbb{R}^{p\times q}$, $\mathbf{B} \in \mathbb{R}^{p\times q}$; (A.7b)
- $\mathrm{rank}(\mathbf{A})+\mathrm{rank}(\mathbf{B})-q \leq \mathrm{rank}(\mathbf{AB}) \leq \min\{\mathrm{rank}(\mathbf{A}),\mathrm{rank}(\mathbf{B})\}$, (A.7c)
  where $\mathbf{A} \in \mathbb{R}^{p\times q}$, $\mathbf{B} \in \mathbb{R}^{q\times r}$;
- $\mathrm{rank}(\mathbf{A}^{\top}\mathbf{A}) = \mathrm{rank}(\mathbf{A}) = \mathrm{rank}(\mathbf{AA}^{\top})$, where $\mathbf{A} \in \mathbb{R}^{p\times q}$; (A.7d)
- $\mathrm{rank}(\mathbf{BAC}) = \mathrm{rank}(\mathbf{A})$ if $\mathbf{B} \in \mathbb{R}^{p\times p}$ and $\mathbf{C} \in \mathbb{R}^{q\times q}$ are nonsingular, where $\mathbf{A} \in \mathbb{R}^{p\times q}$; (A.7e)
- For $\mathbf{A} \in \mathbb{R}^{p\times p}$: $\mathrm{rank}(\mathbf{A}) = p$ if and only if $\mathbf{A}$ is nonsingular. (A.7f)

A.2.4 Inverse

The *inverse* of a square matrix $\mathbf{A} \in \mathbb{R}^{p\times p}$ is the unique square matrix $\mathbf{A}^{-1} \in \mathbb{R}^{p\times p}$ that satisfies $\mathbf{AA}^{-1} = \mathbf{A}^{-1}\mathbf{A} = \mathbf{I}_p$. The inverse exists if and only if $\mathbf{A}$ is nonsingular ($|\mathbf{A}| \neq 0$), in which case $\mathbf{A}$

is said to be *invertible*. The following properties hold:

- $\text{rank}(\mathbf{A}) = p$, where $\mathbf{A} \in \mathbb{R}^{p\times p}$, $|\mathbf{A}| \neq 0$; (A.8a)
- $\mathtt{C}(\mathbf{A}) = \mathtt{R}(\mathbf{A}) = \mathbb{R}^p$, where $\mathbf{A} \in \mathbb{R}^{p\times p}$, $|\mathbf{A}| \neq 0$; (A.8b)
- $\mathtt{N}(\mathbf{A}) = \{\mathbf{0}_p\}$, where $\mathbf{A} \in \mathbb{R}^{p\times p}$, $|\mathbf{A}| \neq 0$; (A.8c)
- $(\mathbf{AC})^{-1} = \mathbf{C}^{-1}\mathbf{A}^{-1}$, where $\mathbf{A} \in \mathbb{R}^{p\times p}$, $|\mathbf{A}| \neq 0$, $|\mathbf{C}| \neq 0$; (A.8d)
- $(\mathbf{A}+\mathbf{BCD})^{-1} = \mathbf{A}^{-1} - \mathbf{A}^{-1}\mathbf{B}(\mathbf{C}^{-1}+\mathbf{DA}^{-1}\mathbf{B})^{-1}\mathbf{DA}^{-1}$, (A.8e)
  where $\mathbf{A} \in \mathbb{R}^{p\times p}$, $\mathbf{B} \in \mathbb{R}^{p\times q}$, $\mathbf{C} \in \mathbb{R}^{q\times q}$, $\mathbf{D} \in \mathbb{R}^{q\times p}$, $|\mathbf{A}| \neq 0$, $|\mathbf{C}| \neq 0$;
- $(\mathbf{A}+\mathbf{ab}^\top)^{-1} = \mathbf{A}^{-1} - \frac{1}{1+\mathbf{b}^\top\mathbf{A}^{-1}\mathbf{a}}\mathbf{A}^{-1}\mathbf{ab}^\top\mathbf{A}^{-1}$, where $\mathbf{A} \in \mathbb{R}^{p\times p}$, $\mathbf{a},\mathbf{b} \in \mathbb{R}^p$, $|\mathbf{A}| \neq 0$. (A.8f)

For a nonsingular partitioned matrix $\mathbf{A}$ as in (A.2), its inverse is:

$$\mathbf{A}^{-1} = \begin{pmatrix} (\mathbf{A}_{11}-\mathbf{A}_{12}\mathbf{A}_{22}^{-1}\mathbf{A}_{21})^{-1} & -(\mathbf{A}_{11}-\mathbf{A}_{12}\mathbf{A}_{22}^{-1}\mathbf{A}_{21})^{-1}\mathbf{A}_{12}\mathbf{A}_{22}^{-1} \\ -\mathbf{A}_{22}^{-1}\mathbf{A}_{21}(\mathbf{A}_{11}-\mathbf{A}_{12}\mathbf{A}_{22}^{-1}\mathbf{A}_{21})^{-1} & (\mathbf{A}_{22}-\mathbf{A}_{21}\mathbf{A}_{11}^{-1}\mathbf{A}_{12})^{-1} \end{pmatrix}. \tag{A.9}$$

For a linear system of equations $\mathbf{Ax} = \mathbf{b}$, with $\mathbf{A} \in \mathbb{R}^{p\times p}$, $\mathbf{b} \in \mathbb{R}^p$, $|\mathbf{A}| \neq 0$, its unique solution is $\mathbf{x} = \mathbf{A}^{-1}\mathbf{b}$.

A.2.5 Generalized inverse

A *generalized inverse* of a rectangular matrix $\mathbf{A} \in \mathbb{R}^{p\times q}$ is a rectangular matrix $\mathbf{A}^- \in \mathbb{R}^{q\times p}$ satisfying $\mathbf{AA}^-\mathbf{A} = \mathbf{A}$. It always exists but is generally not unique unless it also satisfies $\mathbf{A}^-\mathbf{AA}^- = \mathbf{A}^-$, and $\mathbf{AA}^-$ and $\mathbf{A}^-\mathbf{A}$ are symmetric, in which case $\mathbf{A}^-$ is called the *Moore-Penrose inverse* of $\mathbf{A}$. If $\text{rank}(\mathbf{A}) = p$, then the Moore-Penrose inverse of $\mathbf{A}$ is $\mathbf{A}^- = \mathbf{A}^\top(\mathbf{AA}^\top)^{-1}$, while if $\text{rank}(\mathbf{A}) = q$, then it is $\mathbf{A}^- = (\mathbf{A}^\top\mathbf{A})^{-1}\mathbf{A}^\top$.

A.2.6 Eigenvalues and eigenvectors

For a square matrix $\mathbf{A} \in \mathbb{R}^{p\times p}$ and $\lambda \in \mathbb{R}$, the scalar

$$c(\lambda) = |\mathbf{A} - \lambda\mathbf{I}_p| \tag{A.10}$$

is a p-th order polynomial in λ called the *characteristic polynomial* of $\mathbf{A}$. For $p = 2$, we have $c(\lambda) = \lambda^2 - \text{tr}(\mathbf{A})\lambda + |\mathbf{A}|$. The roots $\lambda_1,\ldots,\lambda_p$ of $c(\lambda)$ are the *eigenvalues* of $\mathbf{A}$. In general, the eigenvalues are complex numbers, unless $\mathbf{A}$ is symmetric, in which case the eigenvalues are real numbers.

Since $c(\lambda_i) = |\mathbf{A} - \lambda_i\mathbf{I}_p| = 0$, the matrix $\mathbf{A} - \lambda_i\mathbf{I}_p$ is singular and there exists a nonzero *eigenvector* $\boldsymbol{\gamma}_i$ satisfying for $i = 1,\ldots,p$:

$$\mathbf{A}\boldsymbol{\gamma}_i = \lambda_i\boldsymbol{\gamma}_i. \tag{A.11}$$

If λ_i is complex, then $\boldsymbol{\gamma}_i$ may have complex components, whereas if λ_i is real, then $\boldsymbol{\gamma}_i$ has real components and is said to be *standardized* when $\boldsymbol{\gamma}_i^\top \boldsymbol{\gamma}_i = 1$. The following properties hold:

- $|\mathbf{A}| = \prod_{i=1}^p \lambda_i$, where $\mathbf{A} \in \mathbb{R}^{p\times p}$; (A.12a)
- $\text{tr}(\mathbf{A}) = \sum_{i=1}^p \lambda_i$, where $\mathbf{A} \in \mathbb{R}^{p\times p}$; (A.12b)
- If $\mathbf{B} \in \mathbb{R}^{p\times p}$ is nonsingular and $\mathbf{A} \in \mathbb{R}^{p\times p}$, then $\mathbf{A}$ and $\mathbf{BAB}^{-1}$ have the same eigenvalues λ_i and the eigenvectors of $\mathbf{B}$ are $\mathbf{B}\boldsymbol{\gamma}_i$; (A.12c)
- $\mathbf{A} + \alpha\mathbf{I}_p$, $\alpha \in \mathbb{R}$, has eigenvalues $\lambda_i + \alpha$ and the same eigenvectors as $\mathbf{A} \in \mathbb{R}^{p\times p}$; (A.12d)
- Let $\mathbf{A} \in \mathbb{R}^{p\times q}$ and $\mathbf{B} \in \mathbb{R}^{q\times p}$ be two rectangular matrices, $p \geq q$. The nonzero eigenvalues of $\mathbf{AB}$ and $\mathbf{BA}$ are the same and have the same multiplicity, but $\mathbf{AB}$ has in addition $p-q$ eigenvalues 0. If $\boldsymbol{\gamma} \neq \mathbf{0}_p$ is an eigenvector of $\mathbf{AB}$ for an eigenvalue $\lambda \neq 0$, then $\mathbf{B}\boldsymbol{\gamma}$ is an eigenvector of $\mathbf{BA}$; (A.12e)
- The matrix $\mathbf{Aab}^\top\mathbf{B}$ has rank at most 1, where $\mathbf{A} \in \mathbb{R}^{p\times q}$, $\mathbf{B} \in \mathbb{R}^{r\times p}$, $\mathbf{a} \in \mathbb{R}^q$, $\mathbf{b} \in \mathbb{R}^r$. (A.12f) If it has a nonzero eigenvalue, then it equals $\mathbf{b}^\top\mathbf{BAa}$ with eigenvector $\mathbf{Aa}$.

If $\mathbf{A} \in \mathbb{R}^{p\times p}$ is a symmetric matrix, then it has a *spectral decomposition* of the form:

$$\mathbf{A} = \boldsymbol{\Gamma}\boldsymbol{\Lambda}\boldsymbol{\Gamma}^\top = \sum_{i=1}^p \lambda_i \boldsymbol{\gamma}_i \boldsymbol{\gamma}_i^\top, \tag{A.13}$$

where $\boldsymbol{\Lambda} = \text{diag}(\lambda_1, \dots, \lambda_p)$ is a diagonal matrix of eigenvalues of $\mathbf{A}$ and $\boldsymbol{\Gamma}$ is an orthogonal matrix whose columns are the standardized eigenvectors $\boldsymbol{\gamma}_1, \dots, \boldsymbol{\gamma}_p$ of $\mathbf{A}$. We use the convention to order the eigenvalues $\lambda_1 \geq \lambda_2 \geq \cdots \geq \lambda_p$. If $\mathbf{A}$ is nonsingular, then for any integer m:

$$\mathbf{A}^m = \boldsymbol{\Gamma}\boldsymbol{\Lambda}^m\boldsymbol{\Gamma}^\top = \sum_{i=1}^p \lambda_i^m \boldsymbol{\gamma}_i \boldsymbol{\gamma}_i^\top, \tag{A.14}$$

where $\boldsymbol{\Lambda}^m = \text{diag}(\lambda_1^m, \dots, \lambda_p^m)$. Moreover, if the eigenvalues of $\mathbf{A}$ are positive, then the power m above can be replaced by a rational number r. In particular, for $r = 1/2$, we obtain the unique *symmetric square root* of $\mathbf{A}$ as $\mathbf{A}^{1/2} = \boldsymbol{\Gamma}\boldsymbol{\Lambda}^{1/2}\boldsymbol{\Gamma}^\top$. If the eigenvalues of $\mathbf{A}$ are strictly positive, then we can also define $\mathbf{A}^{-1/2} = \boldsymbol{\Gamma}\boldsymbol{\Lambda}^{-1/2}\boldsymbol{\Gamma}^\top$. The following properties hold:

- $\text{rank}(\mathbf{A})$ equals the number of nonzero eigenvalues of $\mathbf{A} \in \mathbb{R}^{p\times p}$ symmetric; (A.15a)
- For $\mathbf{A} \in \mathbb{R}^{p\times p}$ symmetric, if $\lambda_{k+1} = \cdots = \lambda_p = 0$ then $\mathbf{A} = \tilde{\boldsymbol{\Gamma}}\tilde{\boldsymbol{\Lambda}}\tilde{\boldsymbol{\Gamma}}^\top = \sum_{i=1}^k \lambda_i \boldsymbol{\gamma}_i \boldsymbol{\gamma}_i^\top$, (A.15b) where $\tilde{\boldsymbol{\Lambda}} = \text{diag}(\lambda_1, \dots, \lambda_k) \in \mathbb{R}^{k\times k}$ and $\tilde{\boldsymbol{\Gamma}} = (\boldsymbol{\gamma}_1, \dots, \boldsymbol{\gamma}_k) \in \mathbb{R}^{p\times k}$;
- $\mathbf{A} \in \mathbb{R}^{p\times p}$ symmetric has rank 1 iff $\mathbf{A} = \mathbf{xx}^\top$ for some $\mathbf{x} \in \mathbb{R}^p$, and then its nonzero eigenvalue is $\mathbf{x}^\top\mathbf{x}$; (A.15c)
- The matrix of ones $\mathbf{1}_p\mathbf{1}_p^\top$ has rank 1 with eigenvalue p and eigenvector $\mathbf{1}_p$; (A.15d)
- If $\mathbf{A} \in \mathbb{R}^{p\times p}$ symmetric is *idempotent* (i.e., $\mathbf{A} = \mathbf{A}^2$; see Exercise A.4) then its eigenvalues are 0 or +1, and $\text{rank}(\mathbf{A}) = \text{tr}(\mathbf{A})$. (A.15e)

If $\mathbf{A} \in \mathbb{R}^{p\times q}$ is a rectangular matrix of rank r, then it has a *singular value decomposition (SVD)* of the form:

$$\mathbf{A} = \mathbf{UDV}^\top, \tag{A.16}$$

where $\mathbf{U} \in \mathbb{R}^{p\times r}$ and $\mathbf{V} \in \mathbb{R}^{q\times r}$ have orthogonal columns with norm 1 (i.e., $\mathbf{U}^\top\mathbf{U} = \mathbf{V}^\top\mathbf{V} = \mathbf{I}_r$) and $\mathbf{D} \in \mathbb{R}^{r\times r}$ is diagonal with positive values called *singular values*. It follows that the columns of $\mathbf{U}$ are the eigenvectors of $\mathbf{AA}^\top$ and the columns of $\mathbf{V}$ are the eigenvectors of $\mathbf{A}^\top\mathbf{A}$, while the eigenvalues of $\mathbf{AA}^\top$ and $\mathbf{A}^\top\mathbf{A}$ are the same and their square roots are the singular values. The number of nonzero eigenvalues of $\mathbf{A}^\top\mathbf{A}$ is the rank of $\mathbf{A}$, hence r in this case. If $p = q$, then $r = p$ and the singular value decomposition is equivalent to the spectral decomposition. If $\text{rank}(\mathbf{A}) = r < \min(p,q)$, then the Moore-Penrose inverse of $\mathbf{A}$ can be constructed from the SVD of $\mathbf{A}$ with the degenerate rows and columns deleted from $\mathbf{U}$, $\mathbf{D}$, and $\mathbf{V}$.

A.2.7 QR factorization

A *QR factorization* of a rectangular matrix is its decomposition into the product of an orthogonal matrix and a triangular matrix. Specifically, let $\mathcal{U}_m^+ = \{\mathbf{R} \in \mathbb{R}^{m\times m} : \mathbf{R}$ is upper triangular, $r_{ii} > 0,\ i = 1,\dots,m\}$, and $\mathcal{L}_m^+ = \{\mathbf{T} \in \mathbb{R}^{m\times m} : \mathbf{T}$ is lower triangular, $t_{ii} > 0,\ i = 1,\dots,m\}$, i.e., the groups of $m \times m$ triangular matrices with positive diagonal entries. The following matrix factorizations hold:

- If $\mathbf{A} \in \mathbb{R}^{p\times q}$ with $\text{rank}(\mathbf{A}) = q$, then $\mathbf{A} = \mathbf{QR}$ for some $\mathbf{Q}$ satisfying $\mathbf{Q}^\top\mathbf{Q} = \mathbf{I}_q$ and $\mathbf{R} \in \mathcal{U}_q^+$. Moreover, $\mathbf{Q}$ and $\mathbf{R}$ are unique; (A.17a)
- If $\mathbf{A} \in \mathbb{R}^{p\times q}$ with $\text{rank}(\mathbf{A}) = p$, then $\mathbf{A} = \mathbf{TQ}$ for some $\mathbf{Q}$ satisfying $\mathbf{QQ}^\top = \mathbf{I}_p$ and $\mathbf{T} \in \mathcal{L}_p^+$. Moreover, $\mathbf{T}$ and $\mathbf{Q}$ are unique; (A.17b)
- If $\mathbf{A} \in \mathbb{R}^{p\times p}$ with $\text{rank}(\mathbf{A}) = p$, then $\mathbf{A} = \mathbf{QR}$ for some $\mathbf{Q} \in \mathcal{O}_p$ and $\mathbf{R} \in \mathcal{U}_q^+$. Moreover, this decomposition is unique. (A.17c)

Further properties and computational aspects of the QR factorization can be found in Golub and Van Loan (1983).

A.2.8 Quadratic forms and definiteness

A *quadratic form* in the vector $\mathbf{x} \in \mathbb{R}^p$ is the function:

$$Q(\mathbf{x}) = \mathbf{x}^\top\mathbf{A}\mathbf{x}, \tag{A.18}$$

where $\mathbf{A} \in \mathbb{R}^{p\times p}$ is a symmetric square matrix. If $\mathbf{A}$ is not symmetric, then it can always be replaced by $(\mathbf{A}+\mathbf{A}^\top)/2$, which is symmetric. The matrix $\mathbf{A}$ is said to be *positive dejinite* (pd) if $Q(\mathbf{x}) > 0$ for all $\mathbf{x} \neq \mathbf{0}_p$, whereas it is called *positive semi-dejinite* (psd) if $Q(\mathbf{x}) \geq 0$ for all $\mathbf{x} \in \mathbb{R}^p$. We write $\mathbf{A} > 0$ for pd and $\mathbf{A} \geq 0$ for psd. Obviously, $Q(\mathbf{0}_p) = 0$. The following properties hold:

- For any rectangular matrix $\mathbf{A} \in \mathbb{R}^{p\times q}$: $\mathbf{AA}^\top \geq 0$, $\mathbf{A}^\top\mathbf{A} \geq 0$, and $\text{rank}(\mathbf{AA}^\top) = \text{rank}(\mathbf{A}^\top\mathbf{A}) = \text{rank}(\mathbf{A})$; (A.19a)
- If $\mathbf{A} \in \mathbb{R}^{p\times p}$ is symmetric and positive definite, then its eigenvalues are positive real numbers, $\mathbf{A}$ is nonsingular, $|\mathbf{A}| > 0$, and $\mathbf{A}^{-1}$ is positive definite; (A.19b)
- For any matrix $\mathbf{A} \in \mathbb{R}^{p\times p}$ symmetric, there exists an orthogonal matrix $\boldsymbol{\Gamma}$ defining the linear transformation $\mathbf{y} = \boldsymbol{\Gamma}^\top\mathbf{x}$ (called *principal component transformation*) such that $\mathbf{x}^\top\mathbf{A}\mathbf{x} = \mathbf{y}^\top\mathbf{y}$; (A.19c)
- For any matrix $\mathbf{A} \in \mathbb{R}^{p\times p}$ symmetric and positive semi-definite, there exists a matrix $\mathbf{B} \in \mathbb{R}^{p\times p}$ symmetric such that $\mathbf{A} = \mathbf{B}^2$ and it is given by $\mathbf{B} = \boldsymbol{\Gamma}\boldsymbol{\Lambda}^{1/2}\boldsymbol{\Gamma}^\top$ from (A.13); (A.19d)
- If $\mathbf{A} \in \mathbb{R}^{p\times p}$ is symmetric and positive (semi-)definite, then $\mathbf{B}^\top\mathbf{AB}$ is positive (semi-)definite for any matrix $\mathbf{B} \in \mathbb{R}^{p\times q}$ of rank q ($< q$); in particular $\mathbf{B}^\top\mathbf{B} \geq 0$ for any $\mathbf{B}$. (A.19e)
- If $\mathbf{A} \in \mathbb{R}^{p\times p}$ is symmetric and positive definite, then there exists a unique lower triangular matrix $\mathbf{L} \in \mathbb{R}^{p\times p}$ with positive diagonal entries such that $\mathbf{A} = \mathbf{LL}^\top$. This decomposition is called the *Choleski factorization* of $\mathbf{A}$. (A.19f)

A consequence of (A.19e) is that if $\mathbf{A} \in \mathbb{R}^{p\times p}$ is symmetric and positive semi-definite, and $\mathbf{B} \in \mathbb{R}^{p\times p}$ is symmetric and positive definite, then all nonzero eigenvalues of $\mathbf{B}^{-1}\mathbf{A}$ are positive.

A.2.9 *Kronecker product*

Consider two rectangular matrices $\mathbf{A} = (a_{ij}) \in \mathbb{R}^{p\times q}$ and $\mathbf{B} = (b_{ij}) \in \mathbb{R}^{r\times s}$. The *Kronecker product* (or direct tensor product) $\otimes$ of $\mathbf{A}$ and $\mathbf{B}$ is the block matrix defined by

$$\mathbf{A} \otimes \mathbf{B} = (a_{ij}\mathbf{B}) \in \mathbb{R}^{pr\times qs}, \tag{A.20}$$

where $a_{ij}\mathbf{B}$ is the block in position (i,j). The following properties hold:

- $\mathbf{a} \otimes \mathbf{b}^\top = \mathbf{a}\mathbf{b}^\top = \mathbf{b}^\top \otimes \mathbf{a}$, where $\mathbf{a} \in \mathbb{R}^p$ and $\mathbf{b} \in \mathbb{R}^q$; (A.21a)
- $(\alpha\mathbf{A}) \otimes (\beta\mathbf{B}) = \alpha\beta(\mathbf{A} \otimes \mathbf{B})$, where $\alpha, \beta \in \mathbb{R}$; (A.21b)
- $(\mathbf{A}+\mathbf{B}) \otimes \mathbf{C} = \mathbf{A} \otimes \mathbf{C} + \mathbf{B} \otimes \mathbf{C}$, where $\mathbf{A} \in \mathbb{R}^{p\times q}$, $\mathbf{B} \in \mathbb{R}^{p\times q}$, $\mathbf{C} \in \mathbb{R}^{r\times s}$; (A.21c)
- $(\mathbf{A} \otimes \mathbf{B}) \otimes \mathbf{C} = \mathbf{A} \otimes (\mathbf{B} \otimes \mathbf{C})$, where $\mathbf{A} \in \mathbb{R}^{p\times q}$, $\mathbf{B} \in \mathbb{R}^{r\times s}$, $\mathbf{C} \in \mathbb{R}^{t\times u}$; (A.21d)
- $(\mathbf{AB}) \otimes (\mathbf{CD}) = (\mathbf{A} \otimes \mathbf{C})(\mathbf{B} \otimes \mathbf{D})$, where $\mathbf{A} \in \mathbb{R}^{p\times q}$, $\mathbf{B} \in \mathbb{R}^{q\times r}$, $\mathbf{C} \in \mathbb{R}^{s\times t}$, $\mathbf{D} \in \mathbb{R}^{t\times u}$; (A.21e)
- $(\mathbf{A} \otimes \mathbf{B})^\top = \mathbf{A}^\top \otimes \mathbf{B}^\top$, where $\mathbf{A} \in \mathbb{R}^{p\times q}$, $\mathbf{B} \in \mathbb{R}^{r\times s}$; (A.21f)
- $(\mathbf{A} \otimes \mathbf{B})^{-1} = \mathbf{A}^{-1} \otimes \mathbf{B}^{-1}$, whenever $\mathbf{A} \in \mathbb{R}^{p\times p}$ and $\mathbf{B} \in \mathbb{R}^{q\times q}$ are nonsingular; (A.21g)
- $\mathrm{tr}(\mathbf{A} \otimes \mathbf{B}) = \mathrm{tr}(\mathbf{A})\mathrm{tr}(\mathbf{B})$, where $\mathbf{A} \in \mathbb{R}^{p\times p}$, $\mathbf{B} \in \mathbb{R}^{q\times q}$; (A.21h)
- $|\mathbf{A} \otimes \mathbf{B}| = |\mathbf{A}|^q|\mathbf{B}|^p$, where $\mathbf{A} \in \mathbb{R}^{p\times p}$, $\mathbf{B} \in \mathbb{R}^{q\times q}$; (A.21i)
- If $\mathbf{A} \in \mathbb{R}^{p\times p}$ and $\mathbf{B} \in \mathbb{R}^{q\times q}$ are both positive definite, then $\mathbf{A} \otimes \mathbf{B}$ is positive definite; (A.21j)
- If $\mathbf{A} \in \mathbb{R}^{p\times p}$ and $\mathbf{B} \in \mathbb{R}^{q\times q}$ have eigenvalues (eigenvectors) λ_i ($\mathbf{x}_i$), $i = 1,\ldots,p$, and ν_j ($\mathbf{y}_j$), $j = 1,\ldots,q$, respectively, then $\mathbf{A} \otimes \mathbf{B}$ has eigenvalues $\lambda_i\nu_j$ and eigenvectors $\mathbf{x}_i \otimes \mathbf{y}_j$, $i = 1,\ldots,p$, $j = 1,\ldots,q$. (A.21k)

The *nearest Kronecker product* (NKP) approximation of a rectangular matrix $\mathbf{A} \in \mathbb{R}^{p\times q}$, with $p = p_1p_2$, $q = q_1q_2$, is given by the matrices $\mathbf{A}_1 \in \mathbb{R}^{p_1\times q_1}$ and $\mathbf{A}_2 \in \mathbb{R}^{p_2\times q_2}$, minimizing the Frobenius norm:

$$\|\mathbf{A} - \mathbf{A}_1 \otimes \mathbf{A}_2\|_{\mathrm{Fr}}. \tag{A.22}$$

The solution $\mathrm{vec}(\mathbf{A}_1)$ and $\mathrm{vec}(\mathbf{A}_2)$ can be shown to be given by the largest left and right singular vectors of the SVD of a re-arranged matrix $\mathbf{A}$ (Van Loan and Pitsianis, 1993; Kamm and Nagy, 2000; Genton, 2007). Moreover, in the special case of square matrices where $p_1 = q_1$ and $p_2 = q_2$, i.e., $p = q$, if $\mathbf{A}$ is positive definite, then $\mathbf{A}_1$ and $\mathbf{A}_2$ are also positive definite. Similarly, if $\mathbf{A}$ has all nonnegative components, then $\mathbf{A}_1$ and $\mathbf{A}_2$ have also all nonnegative components, and if $\mathbf{A}$ is block Toeplitz (see Section A.3.4) with Toeplitz blocks, then $\mathbf{A}_1$ and $\mathbf{A}_2$ are Toeplitz matrices.

A.2.10 *Vectorization*

Consider a rectangular matrix $\mathbf{A} = (\mathbf{a}_1,\ldots,\mathbf{a}_q) \in \mathbb{R}^{p\times q}$. The *vectorization* of $\mathbf{A}$ is the vector $\mathrm{vec}(\mathbf{A}) \in \mathbb{R}^{pq}$ defined by stacking the columns of $\mathbf{A}$ on top of each other:

$$\mathrm{vec}(\mathbf{A}) = \begin{pmatrix} \mathbf{a}_1 \\ \vdots \\ \mathbf{a}_q \end{pmatrix}. \tag{A.23}$$

The following properties hold:

- $\text{vec}(\mathbf{a}) = \text{vec}(\mathbf{a}^\top) = \mathbf{a}$ and $\text{vec}(\mathbf{a}\mathbf{b}^\top) = \mathbf{b} \otimes \mathbf{a}$, where $\mathbf{a} \in \mathbb{R}^p$ and $\mathbf{b} \in \mathbb{R}^q$; (A.24a)
- $\text{vec}(\alpha\mathbf{A} + \beta\mathbf{B}) = \alpha\text{vec}(\mathbf{A}) + \beta\text{vec}(\mathbf{B})$, where $\mathbf{A} \in \mathbb{R}^{p\times q}$, $\mathbf{B} \in \mathbb{R}^{p\times q}$, $\alpha, \beta \in \mathbb{R}$; (A.24b)
- $\text{tr}(\mathbf{A}^\top\mathbf{B}) = \sum_{i=1}^{p}\sum_{j=1}^{q} a_{ij}b_{ij} = \text{vec}(\mathbf{A})^\top\text{vec}(\mathbf{B})$, where $\mathbf{A} \in \mathbb{R}^{p\times q}$, $\mathbf{B} \in \mathbb{R}^{p\times q}$; (A.24c)
- $\text{vec}(\mathbf{ABC}) = (\mathbf{C}^\top \otimes \mathbf{A})\text{vec}(\mathbf{B})$, where $\mathbf{A} \in \mathbb{R}^{p\times q}$, $\mathbf{B} \in \mathbb{R}^{q\times r}$, $\mathbf{C} \in \mathbb{R}^{r\times s}$; (A.24d)
- $\text{vec}(\mathbf{AB}) = (\mathbf{B}^\top \otimes \mathbf{I}_p)\text{vec}(\mathbf{A}) = (\mathbf{B}^\top \otimes \mathbf{A})\text{vec}(\mathbf{I}_q) = (\mathbf{I}_r \otimes \mathbf{A})\text{vec}(\mathbf{B})$, (A.24e)
  where $\mathbf{A} \in \mathbb{R}^{p\times q}$, $\mathbf{B} \in \mathbb{R}^{q\times r}$;
- $\text{vec}(\mathbf{ABc}) = (\mathbf{c}^\top \otimes \mathbf{A})\text{vec}(\mathbf{B}) = (\mathbf{A} \otimes \mathbf{c}^\top)\text{vec}(\mathbf{B})$, (A.24f)
  where $\mathbf{A} \in \mathbb{R}^{p\times q}$, $\mathbf{B} \in \mathbb{R}^{q\times r}$, $\mathbf{c} \in \mathbb{R}^r$;
- $\text{tr}(\mathbf{ABCD}) = \text{vec}(\mathbf{D}^\top)^\top(\mathbf{C}^\top \otimes \mathbf{A})\text{vec}(\mathbf{B}) = \text{vec}(\mathbf{D})^\top(\mathbf{A} \otimes \mathbf{C}^\top)\text{vec}(\mathbf{B}^\top)$, (A.24g)
  provided the product **ABCD** is defined;
- $\text{vec}(\mathbf{A}^\top) = \mathbf{K}_{pq}\text{vec}(\mathbf{A})$, for every $\mathbf{A} \in \mathbb{R}^{p\times q}$, where $\mathbf{K}_{pq} \in \mathbb{R}^{pq\times pq}$ (A.24h)
  is called a *permutation (or commutation) matrix* of order $p \times q$, and is given by
  $\mathbf{K}_{pq} = \sum_{i=1}^{p}\sum_{j=1}^{q}(\mathbf{e}_{i:p}\mathbf{e}_{j:q}^\top \otimes \mathbf{e}_{j:q}\mathbf{e}_{i:p}^\top)$, where $\mathbf{e}_{i:n}$ denotes the i-th unit vector of $\mathbb{R}^n$.

Note in (A.24h) that $\mathbf{K}_{p1} = \mathbf{I}_p$ and $\mathbf{K}_{1q} = \mathbf{I}_q$. Also, when $p = q$, the i-th unit vectors will be simply denoted by $\mathbf{e}_i$ and the corresponding permutation matrix by $\mathbf{K}_p$. In such a case, by defining the unit permutation matrices $\mathbf{E}_{ij} = \mathbf{e}_i\mathbf{e}_j^\top$, $i, j = 1, \ldots, p$, we have $\mathbf{K}_p = (\mathbf{E}_{ji})$, which for $p = 2$ is:

$$\mathbf{K}_2 = \mathbf{E}_{11} \otimes \mathbf{E}_{11} + \mathbf{E}_{12} \otimes \mathbf{E}_{21} + \mathbf{E}_{21} \otimes \mathbf{E}_{12} + \mathbf{E}_{22} \otimes \mathbf{E}_{22} = \begin{pmatrix} \mathbf{E}_{11} & \mathbf{E}_{21} \\ \mathbf{E}_{12} & \mathbf{E}_{22} \end{pmatrix} = \begin{pmatrix} 1 & 0 & 0 & 0 \\ 0 & 0 & 1 & 0 \\ 0 & 1 & 0 & 0 \\ 0 & 0 & 0 & 1 \end{pmatrix}.$$

Two matrices related to the commutation matrix $\mathbf{K}_p$ that are useful in the vectorization process are the *symmetrizer* $\mathbf{N}_p$ and the *duplication* $\mathbf{D}_p$. They are usually defined by their matrix relations below:

$$\text{vec}(\mathbf{A}^\top) = \mathbf{K}_p\text{vec}(\mathbf{A}), \tag{A.25}$$

$$\text{vec}\left\{\frac{1}{2}(\mathbf{A} + \mathbf{A}^\top)\right\} = \mathbf{N}_p\text{vec}(\mathbf{A}), \tag{A.26}$$

where $\mathbf{A} \in \mathbb{R}^{p\times p}$ is arbitrary, and

$$\text{vec}(\mathbf{A}) = \mathbf{D}_p\text{vech}(\mathbf{A}), \tag{A.27}$$

where $\mathbf{A}$ is symmetric and $\text{vech}(\cdot)$ is *lower-half vectorization*, i.e., stacking columns of the lower half of the matrix. Clearly,

$$\mathbf{N}_p = \frac{1}{2}(\mathbf{I}_{p^2} + \mathbf{K}_p), \tag{A.28}$$

and satisfies the following properties:

- $\mathbf{N}_p = \mathbf{N}_p^\top = \mathbf{N}_p^2$; (A.29a)
- $\text{rank}(\mathbf{N}_p) = \text{tr}(\mathbf{N}_p) = \frac{1}{2}p(p+1)$; (A.29b)
- $\mathbf{N}_p\mathbf{K}_p = \mathbf{N}_p = \mathbf{K}_p\mathbf{N}_p$. (A.29c)

Since $\mathbf{D}_p$ has full column rank of $p(p+1)/2$, $\text{vech}(\mathbf{A})$ can be uniquely solved from (A.27) as

$$\text{vech}(\mathbf{A}) = \mathbf{D}_p^-\text{vec}(\mathbf{A}), \tag{A.30}$$

where $\mathbf{D}_p^- = (\mathbf{D}_p^\top \mathbf{D}_p)^{-1}\mathbf{D}_p^\top$ is the Moore-Penrose inverse of $\mathbf{D}_p$. Some further properties of $\mathbf{D}_p$ are:

- $\mathbf{K}_p\mathbf{D}_p = \mathbf{D}_p$; (A.31a)
- $\mathbf{D}_p\mathbf{D}_p^- = \frac{1}{2}(\mathbf{I}_{p^2} + \mathbf{K}_p)$; (A.31b)
- $\mathbf{D}_p\mathbf{D}_p^-(\mathbf{b}\otimes\mathbf{A}) = \frac{1}{2}\{(\mathbf{b}\otimes\mathbf{A}) + (\mathbf{A}\otimes\mathbf{b})\}$. (A.31c)

A.2.11 Matrix differentiation

Let $\mathbf{X} = (x_{ij}) \in \mathbb{R}^{p\times q}$ be a rectangular matrix and consider a function $f(\mathbf{X})$. Then the *matrix derivative* of $f(\mathbf{X})$ with respect to $\mathbf{X}$ is the matrix:

$$\frac{\partial f(\mathbf{X})}{\partial \mathbf{X}} = \left(\frac{\partial f(\mathbf{X})}{\partial x_{ij}}\right). \tag{A.32}$$

The following basic properties hold:

- $\frac{\partial \mathbf{a}^\top\mathbf{x}}{\partial \mathbf{x}} = \mathbf{a}$, where $\mathbf{a}\in\mathbb{R}^p$ and $\mathbf{x}\in\mathbb{R}^p$; (A.33a)
- $\frac{\partial \mathbf{x}^\top\mathbf{x}}{\partial \mathbf{x}} = 2\mathbf{x}$, where $\mathbf{x}\in\mathbb{R}^p$; (A.33b)
- $\frac{\partial \mathbf{x}^\top\mathbf{A}\mathbf{x}}{\partial \mathbf{x}} = (\mathbf{A}+\mathbf{A}^\top)\mathbf{x}$, where $\mathbf{x}\in\mathbb{R}^p$ and $\mathbf{A}\in\mathbb{R}^{p\times p}$; (A.33c)
- $\frac{\partial \mathbf{x}^\top\mathbf{A}\mathbf{y}}{\partial \mathbf{x}} = \mathbf{A}\mathbf{y}$, where $\mathbf{x}\in\mathbb{R}^p$, $\mathbf{y}\in\mathbb{R}^q$ and $\mathbf{A}\in\mathbb{R}^{p\times q}$; (A.33d)
- $\frac{\partial|\mathbf{X}|}{\partial x_{ij}} = \begin{cases} X_{ij}, & \text{if all entries of } \mathbf{X}\in\mathbb{R}^{p\times p} \text{ are distinct,} \\ X_{ii} \text{ if } i=j,\ X_{ij} \text{ if } i\neq j, & \text{if } \mathbf{X}\in\mathbb{R}^{p\times p} \text{ is symmetric,} \end{cases}$ (A.33e)

  where X_{ij} is the (i,j)-th cofactor of $\mathbf{X}$;
- $\frac{\partial \mathrm{tr}(\mathbf{X}\mathbf{Y})}{\partial \mathbf{X}} = \begin{cases} \mathbf{Y}^\top, & \text{if all entries of } \mathbf{X}\in\mathbb{R}^{p\times q} \text{ are distinct,} \\ \mathbf{Y}+\mathbf{Y}^\top - \mathrm{diag}(\mathbf{Y}), & \text{if } \mathbf{X}\in\mathbb{R}^{p\times p} \text{ is symmetric,} \end{cases}$ (A.33f)

  where $\mathbf{Y}\in\mathbb{R}^{q\times p}$ in the first case, and $\mathbf{Y}\in\mathbb{R}^{p\times p}$ in the second case;
- $\frac{\partial \mathbf{X}^{-1}}{\partial x_{ij}} = \begin{cases} -\mathbf{X}^{-1}\mathbf{J}_{ij}\mathbf{X}^{-1}, & \text{if all entries of } \mathbf{X}\in\mathbb{R}^{p\times p} \text{ are distinct,} \\ -\mathbf{X}^{-1}\mathbf{J}_{ii}\mathbf{X}^{-1} \text{ if } i=j, & \text{if } \mathbf{X}\in\mathbb{R}^{p\times p} \text{ is symmetric,} \\ -\mathbf{X}^{-1}(\mathbf{J}_{ij}+\mathbf{J}_{ji})\mathbf{X}^{-1} \text{ if } i\neq j, & \text{if } \mathbf{X}\in\mathbb{R}^{p\times p} \text{ is symmetric,} \end{cases}$ (A.33g)

  where $\mathbf{J}_{ij}$ is a matrix of zeroes except for a 1 at the (i,j)-th entry.

Further results on this topic can be found in the book by Magnus and Neudecker (1999).

A.3 Special Matrices

A.3.1 Orthogonal

A square matrix $\mathbf{A}\in\mathbb{R}^{p\times p}$ is said to be *orthogonal* if $\mathbf{A}\mathbf{A}^\top = \mathbf{I}_p$. Then the following properties hold:

- $\mathbf{A}^\top\mathbf{A} = \mathbf{I}_p$; (A.34a)
- $\mathbf{A}^{-1} = \mathbf{A}^\top$; (A.34b)
- $|\mathbf{A}| = \pm 1$; (A.34c)
- The eigenvalues of $\mathbf{A}$ are placed on the unit circle, and all eigenvectors have norm 1; (A.34d)
- If $\mathbf{A}$ and $\mathbf{B}$ are orthogonal, then $\mathbf{C} = \mathbf{A}\mathbf{B}$ is orthogonal. (A.34e)

A.3.2 Centring

The square matrix $\mathbf{H} \in \mathbb{R}^{n\times n}$ defined by $\mathbf{H} = \mathbf{I}_n - \frac{1}{n}\mathbf{1}_n\mathbf{1}_n^\top$ is said to be the *centring matrix*. It is symmetric ($\mathbf{H} = \mathbf{H}^\top$) and idempotent ($\mathbf{H}^2 = \mathbf{H}$). Moreover, $\mathbf{H}\mathbf{1}_n = \mathbf{0}_n$ and $\text{rank}(\mathbf{H}) = n-1$.

A.3.3 Equicorrelation

The matrix $\mathbf{A} = (1-\rho)\mathbf{I}_p + \rho\mathbf{1}_p\mathbf{1}_p^\top \in \mathbb{R}^{p\times p}$ with $-\frac{1}{p-1} < \rho < 1$ is said to be the *equicorrelation matrix*. The equicorrelation matrix is symmetric and positive definite and, therefore, is a valid correlation matrix with all pairwise correlations equal to ρ. Its inverse is:

$$\mathbf{A}^{-1} = \frac{1}{1-\rho}\left[\mathbf{I}_p - \frac{\rho}{1+(p-1)\rho}\mathbf{1}_p\mathbf{1}_p^\top\right], \tag{A.35}$$

and its determinant is:

$$|\mathbf{A}| = (1-\rho)^{p-1}\{1+(p-1)\rho\}. \tag{A.36}$$

Its eigenvalues are $\lambda_1 = 1+(p-1)\rho$ and $\lambda_2 = \cdots = \lambda_p = 1-\rho$. The corresponding eigenvectors are $\boldsymbol{\gamma}_1 = \mathbf{1}_p$ and $\boldsymbol{\gamma}_2, \ldots, \boldsymbol{\gamma}_p$ are any set of vectors orthogonal to $\mathbf{1}_p$ and to each other.

A.3.4 Toeplitz

A *Toeplitz matrix* $\mathbf{A} \in \mathbb{R}^{p\times p}$ has entries that are constant on each diagonal. It is *block Toeplitz* if its blocks are constant on each block diagonal. A *block Toeplitz with Toeplitz blocks* matrix combines the two precedent structures. The Toeplitz structure can be extended to rectangular matrices.

A.4 Exercises

Exercise A.1 *Verify all properties provided in this chapter.*

Exercise A.2 *For two matrices $\mathbf{A} = (a_{ij})$ and $\mathbf{B} = (b_{ij})$ of the same dimension $p\times q$, the Hadamard product, $\mathbf{A}\circ\mathbf{B}$, is a matrix, of the same dimension as the operands, with entries given by $(\mathbf{A}\circ\mathbf{B})_{ij} = (a_{ij}b_{ij})$. For matrices of different dimensions ($p\times q$ and $r\times s$, where $p \neq r$ or $q \neq s$ or both), the Hadamard product is undefined. For two vectors $\mathbf{x}$ and $\mathbf{y}$ of the same dimension $p\times 1$ show that $\mathbf{x}\circ\mathbf{y} = \mathbf{D}_\mathbf{x}\mathbf{y} = \mathbf{D}_\mathbf{y}\mathbf{x} = \mathbf{y}\circ\mathbf{x}$, where $\mathbf{D}_\mathbf{z} = \text{diag}(z_1, \ldots, z_p)$.*

Exercise A.3 *If $\mathbf{A} \in \mathbb{R}^{p\times p}$ and $\mathbf{B} \in \mathbb{R}^{p\times p}$ are positive semi-definite, then show that their Hadamard product $\mathbf{A}\circ\mathbf{B}$ is also positive semi-definite.*

Exercise A.4 *Let $\mathbf{A}$ and $\mathbf{B}$ be two $n\times n$ matrices:*

(a) Show that $\text{tr}(\mathbf{A}\otimes\mathbf{B} - \mathbf{B}\otimes\mathbf{A}) = 0$;

(b) If $\mathbf{A}\otimes\mathbf{B} + \mathbf{B}\otimes\mathbf{A} = \mathbf{O}$ then find $|\mathbf{A}\otimes\mathbf{B}|$;

(c) If $\mathbf{A}$ is idempotent ($\mathbf{A}^k = \mathbf{A}$ for some positive integer k) then find $|\mathbf{A}\otimes\mathbf{A}|$;

(d) If $\mathbf{B}$ is nilpotent ($\mathbf{B}^k = \mathbf{O}$ for some positive integer k) then find $|\mathbf{B}\otimes\mathbf{B}|$.

Exercise A.5 *For any symmetric invertible $p \times p$ matrix* $\mathbf{A}$ *and p-vectors* $\mathbf{x}$ *and* $\mathbf{b}$*, show that:*

$$\mathbf{x}^\top \mathbf{A}\mathbf{x} - 2\mathbf{b}^\top \mathbf{x} = (\mathbf{x} - \mathbf{A}^{-1}\mathbf{b})^\top \mathbf{A}(\mathbf{x} - \mathbf{A}^{-1}\mathbf{b}) - \mathbf{b}^\top \mathbf{A}^{-1}\mathbf{b}.$$

Use this result to represent the sum of two different quadratic forms in $\mathbf{x}$ *as a single quadratic form in* $\mathbf{x}$ *plus a constant term independent of* $\mathbf{x}$.

Exercise A.6 *Show that if* $\mathbf{A}$ *is a symmetric matrix and* $\mathbf{B}$ *is a positive definite matrix then:*

$$\sup_{\|\mathbf{x}\|=1} \frac{\mathbf{x}^\top \mathbf{A}\mathbf{x}}{\mathbf{x}^\top \mathbf{B}\mathbf{x}} = \lambda_1(\mathbf{B}^{-1}\mathbf{A}),$$

where $\lambda_1(\mathbf{B}^{-1}\mathbf{A})$ *denotes the largest eigenvalue of* $\mathbf{B}^{-1}\mathbf{A}$.

Exercise A.7 *If* $\mathbf{A}$ *is a $p \times q$ matrix, prove that* $\mathtt{N}(\mathbf{A})^\perp = \mathtt{C}(\mathbf{A}^\top)$*, where* $\mathtt{N}(\mathbf{A})^\perp$ *denotes the orthogonal complement of* $\mathtt{N}(\mathbf{A})$*, i.e., the set of vectors orthogonal to every vector in* $\mathtt{N}(\mathbf{A})$.

Exercise A.8 *Prove that if a $p \times p$ psd matrix* $\boldsymbol{\Sigma}$ *is partitioned as*

$$\boldsymbol{\Sigma} = \begin{pmatrix} \boldsymbol{\Sigma}_{11} & \boldsymbol{\Sigma}_{12} \\ \boldsymbol{\Sigma}_{21} & \boldsymbol{\Sigma}_{22} \end{pmatrix},$$

where $\boldsymbol{\Sigma}_{ij}$ *is* $p_i \times p_j$*,* $i, j = 1, 2$*, with* $p_1 + p_2 = p$*, then:*

(a) $\mathtt{N}(\boldsymbol{\Sigma}_{22}) \subset \mathtt{N}(\boldsymbol{\Sigma}_{12})$*;*

(b) $\mathtt{C}(\boldsymbol{\Sigma}_{21}) \subset \mathtt{C}(\boldsymbol{\Sigma}_{22})$.

Exercise A.9 *Deepen your knowledge in linear algebra by reading the books listed at the beginning of this chapter.*

Bibliography

Abe, T. and Pewsey, A. (2011), "Sine-skewed circular distributions," *Statist. Papers*, 52, 683–707. 337

Adcock, C. J. (2004), "Capital asset pricing for UK stocks under the multivariate skew-normal distribution," in *Skew-Elliptical Distributions And Their Applications: A Journey Beyond Normality (M. G. Genton, Ed.)*, Chapman & Hall/CRC, Boca Raton, FL, pp. 191–204. 139, 332

— (2007), "Extensions of Stein's lemma for the skew-normal distribution," *Comm. Statist. Theory Methods*, 36, 1661–1671. 332

— (2010), "Asset pricing and portfolio selection based on the multivariate extended skew-Student-*t* distribution," *Ann. Oper. Res.*, 176, 221–234. 247, 274, 332

— (2014), "Mean-variance-skewness efficient surfaces, Stein's lemma and the multivariate extended skew-Student distribution," *European J. Oper. Res.*, 234, 392–401. 332

— (2026), "Skewness and kurtosis projection pursuit for the multivariate extended skew-normal and skew-Student distributions," *J. Multivariate Anal.*, 211, Paper No. 105533, 12. 324

Adcock, C. J., Landsman, Z., and Shushi, T. (2021), "Stein's Lemma for generalized skew-elliptical random vectors," *Comm. Statist. Theory Methods*, 50, 3014–3029. 332

Adcock, C. J. and Shutes, K. (2001), "Portfolio selection based on the multivariate skew-normal distribution," in *Financial Modelling (A. Skulimowski, Ed.)*, Krakow: Progress and Business Publishers, pp. Paper No. 105533, 12. 139

Aigner, D., Lovell, C. A. K., and Schmidt, P. (1977), "Formulation and estimation of stochastic frontier production function models," *J. Econometrics*, 6, 21–37. 340

Aitchison, J. (1986), *The Statistical Analysis of Compositional Data*, Monographs on Statistics and Applied Probability, Chapman & Hall, London. 340

Ali, M. M. (1980), "Characterization of the normal distribution among the continuous symmetric spherical class," *J. Roy. Statist. Soc. Ser. B*, 42, 162–164. 96, 98

Aliverti, E., Arellano-Valle, R. B., Kahrari, F., and Scarpa, B. (2023), "A flexible two-piece normal dynamic linear model," *Comput. Statist.*, 38, 2075–2096. 317

Aliverti, E., Mazzuco, S., and Scarpa, B. (2022), "Dynamic modelling of mortality via mixtures of skewed distribution functions," *J. Roy. Statist. Soc. Ser. A*, 185, 1030–1048. 340

Allard, D. and Naveau, P. (2007), "A new spatial skew-normal random field model," *Comm. Statist. Theory Methods*, 36, 1821–1834. 317, 318

Alruwaili, B. (2023), "The modality of skew *t*-distribution," *Statist. Papers*, 64, 497–507. 277

Anceschi, N., Fasano, A., Durante, D., and Zanella, G. (2023), "Bayesian conjugacy in probit, Tobit, multinomial probit and extensions: A review and new results," *J. Amer. Statist. Assoc.*, 118, 1451–1469. 324

Anděl, J., Netuka, I., and Zvára, K. (1984), "On threshold autoregressive processes," *Kybernetika*, 20, 89–106. 316

Anderson, T. W. (2003), *An Introduction to Multivariate Statistical Analysis*, John Wiley & Sons, Hoboken, NJ, 3rd ed. 2

Anderson, T. W. and Fang, K. T. (1987), "Cochran's theorem for elliptically contoured distributions," *Sankhyā Ser. A*, 49, 305–315. 96, 98, 100

Andrews, D. F. and Mallows, C. L. (1974), "Scale mixtures of normal distributions," *J. Roy. Statist. Soc. Ser. B*, 36, 99–102. 56

Arellano-Valle, R. B. (1994), *Distribuicões Elipticas: Propriedades, Inferência e Aplicacões a Modelos de Regressão*, University of São Paulo, Ph.D. Thesis. 69, 85, 98, 104

— (2001), "On some characterizations of spherical distributions," *Statist. Probab. Lett.*, 54, 227–232. 96, 98, 104

— (2010), "On the information matrix of the multivariate skew-*t* model," *Metron*, 68, 371–386. 312

Arellano-Valle, R. B. and Azzalini, A. (2006), "On the unification of families of skew-normal distributions," *Scand. J. Stat.*, 33, 561–574. 194, 198, 199, 228, 229, 239, 247, 286, 303

— (2008), "The centred parametrization for the multivariate skew-normal distribution," *J. Multivariate Anal.*, 99, 1362–1382. 141, 177

— (2013), "The centred parameterization and related quantities of the skew-*t* distribution," *J. Multivariate Anal.*, 113, 73–90. 277

— (2021), "A formulation for continuous mixtures of multivariate normal distributions," *J. Multivariate Anal.*, 185, Paper No. 104780, 20. 57, 58, 131

— (2022), "Some properties of the unified skew-normal distribution," *Statist. Papers*, 63, 461–487. 196, 229, 241

Arellano-Valle, R. B., Azzalini, A., Ferreira, C. S., and Santoro, K. (2020), "A two-piece normal measurement error model," *Comput. Statist. Data Anal.*, 144, 106863, 17. 330

Arellano-Valle, R. B. and Bolfarine, H. (1995), "On some characterizations of the *t*-distribution," *Statist. Probab. Lett.*, 25, 79–85. 85, 96, 98

— (1996), "Elliptical structural models," *Comm. Statist. Theory Methods*, 25, 2319–2341. 329

Arellano-Valle, R. B., Bolfarine, H., and Gasco, L. (2002a), "Measurement error models with nonconstant covariance matrices," *J. Multivariate Anal.*, 82, 395–415. 329

Arellano-Valle, R. B., Bolfarine, H., and Iglesias, P. L. (1994), "A predictivistic interpretation of the multivariate *t* distribution," *TEST*, 3, 221–236. 98

Arellano-Valle, R. B., Bolfarine, H., Iglesias, P. L., and Viviani, P. (2010a), "Portfolio selection: An application to the Chilean stock market," *Chil. J. Stat.*, 1, 3–15. 95

Arellano-Valle, R. B., Bolfarine, H., and Lachos, V. H. (2005a), "Skew-normal linear mixed models," *J. Data Sci.*, 3, 415–438. 328

— (2007), "Bayesian inference for skew-normal linear mixed models," *J. Appl. Statist.*, 34, 663–682. 328

Arellano-Valle, R. B., Bolfarine, H., and Vilca-Labra, F. (1996), "Ultrastructural elliptical models," *Canad. J. Statist.*, 24, 207–216. 329

Arellano-Valle, R. B., Branco, M. D., and Genton, M. G. (2006a), "A unified view on skewed distributions arising from selections," *Canad. J. Statist.*, 34, 581–601. 263, 303, 304, 338

Arellano-Valle, R. B., Castro, L. M., Genton, M. G., and Gómez, H. W. (2008), "Bayesian inference for shape mixtures of skewed distributions, with application to regression analysis," *Bayesian Anal.*, 3, 513–539. 185

Arellano-Valle, R. B., Castro, L. M., González-Farías, G., and Muñoz Gajardo, K. A. (2012), "Student-*t* censored regression model: Properties and inference," *Stat. Methods Appl.*, 21, 453–473. 339

Arellano-Valle, R. B., Castro, L. M., and Loschi, R. H. (2013a), "Change point detection in the skew-normal model parameters," *Comm. Statist. Theory Methods*, 42, 603–618. 317

Arellano-Valle, R. B., Contreras-Reyes, J. E., and Genton, M. G. (2013b), "Shannon entropy and mutual information for multivariate skew-elliptical distributions," *Scand. J. Stat.*, 40, 42–62. 192, 335

Arellano-Valle, R. B., Contreras-Reyes, J. E., López Quintero, F. O., and Valdebenito, A. (2019), "A skew-normal dynamic linear model and Bayesian forecasting," *Comput. Statist.*, 34, 1055–1085. 317

Arellano-Valle, R. B., Contreras-Reyes, J. E., and Stehlík, M. (2017), "Generalized skew-normal negentropy and its application to fish condition factor time series," *Entropy*, 19. 335

Arellano-Valle, R. B., Cortés, M. A., and Gómez, H. W. (2010b), "An extension of the epsilon-skew-normal distribution," *Comm. Statist. Theory Methods*, 39, 912–922. 341

Arellano-Valle, R. B., del Pino, G., and Iglesias, P. (2006b), "Bayesian inference in spherical linear models: robustness and conjugate analysis," *J. Multivariate Anal.*, 97, 179–197. 83, 104, 114

Arellano-Valle, R. B., del Pino, G., and San Martín, E. (2002b), "Definition and probabilistic properties of skew-distributions," *Statist. Probab. Lett.*, 58, 111–121. 34, 302

Arellano-Valle, R. B. and del Pino, G. E. (2004), "From symmetric to asymmetric distributions: A unified approach," in *Skew-Elliptical Distributions And Their Applications: A Journey Beyond Normality (M. G. Genton, Ed.)*, Chapman & Hall/CRC, Boca Raton, FL, pp. 113–130. 34, 38, 302

Arellano-Valle, R. B., Ferreira, C. S., and Genton, M. G. (2018), "Scale and shape mixtures of multivariate skew-normal distributions," *J. Multivariate Anal.*, 166, 98–110. 185

Arellano-Valle, R. B., Galea-Rojas, M., and Zuazola, P. I. (2000), "Bayesian sensitivity analysis in elliptical linear regression models," *J. Statist. Plann. Inference*, 86, 175–199. 116

Arellano-Valle, R. B. and Genton, M. G. (2005), "On fundamental skew distributions," *J. Multivariate Anal.*, 96, 93–116. 194, 198, 199, 303, 308, 327, 328

— (2007), "On the exact distribution of linear combinations of order statistics from dependent random variables," *J. Multivariate Anal.*, 98, 1876–1894. 309

— (2008), "On the exact distribution of the maximum of absolutely continuous dependent random variables," *Statist. Probab. Lett.*, 78, 27–35. 308

— (2010a), "An invariance property of quadratic forms in random vectors with a selection distribution, with application to sample variogram and covariogram estimators," *Ann. Inst. Statist. Math.*, 62, 363–381. 228, 304, 305

— (2010b), "Multivariate extended skew-*t* distributions and related families," *Metron*, 68, 201–234. 247, 274

— (2010c), "Multivariate unified skew-elliptical distributions," *Chil. J. Stat.*, 1, 17–33. 286, 303

Arellano-Valle, R. B., Genton, M. G., and Loschi, R. H. (2009), "Shape mixtures of multivariate skew-normal distributions," *J. Multivariate Anal.*, 100, 91–101. 185

Arellano-Valle, R. B., Gómez, H. W., and Quintana, F. A. (2004), "A new class of skew-normal distributions," *Comm. Statist. Theory Methods*, 33, 1465–1480. 184, 186

— (2005b), "Statistical inference for a general class of asymmetric distributions," *J. Statist. Plann. Inference*, 128, 427–443. 57, 340

Arellano-Valle, R. B., Gómez, H. W., and Salinas, H. S. (2013c), "A note on the Fisher information matrix for the skew-generalized-normal model," *SORT*, 37, 19–28. 186

Arellano-Valle, R. B., Harnik, S. B., and Genton, M. G. (2021), "On the asymptotic joint distribution of multivariate sample moments," in *Advances in Statistics–Theory and Applications*, Springer, Cham, pp. 181–206. 173

Arellano-Valle, R. B., Jamalizadeh, A., Mahmoodian, H., and Balakrishnan, N. (2014), "*L*-statistics from multivariate unified skew-elliptical distributions," *Metrika*, 77, 559–583. 309

Arellano-Valle, R. B., Ozan, S., Bolfarine, H., and Lachos, V. H. (2005c), "Skew normal measurement error models," *J. Multivariate Anal.*, 96, 265–281. 329

Arellano-Valle, R. B. and Richter, W.-D. (2012), "On skewed continuous $l_{n,p}$-symmetric distributions," *Chil. J. Stat.*, 3, 193–212, [Original paging: 195–214]. 34, 113

Arevalillo, J. M. and Navarro, H. (2015), "A note on the direction maximizing skewness in multivariate skew-t vectors," *Statist. Probab. Lett.*, 96, 328–332. 143

Arismendi, J. C. (2013), "Multivariate truncated moments," *J. Multivariate Anal.*, 117, 41–75. 232

Arnold, B. C. and Beaver, R. J. (2000a), "Hidden truncation models," *Sankhyā Ser. A*, 62, 23–35. 303

— (2000b), "The skew-Cauchy distribution," *Statist. Probab. Lett.*, 49, 285–290. 277

— (2002), "Skewed multivariate models related to hidden truncation and/or selective reporting," *TEST*, 11, 7–54, with discussion and a rejoinder by the authors. 303

— (2004), "Elliptical models subject to hidden truncation or selective sampling," in *Skew-Elliptical Distributions And Their Applications: A Journey Beyond Normality (M. G. Genton, Ed.)*, Chapman & Hall/CRC, Boca Raton, FL, pp. 101–112. 303

Arnold, B. C., Beaver, R. J., Groeneveld, R. A., and Meeker, W. Q. (1993), "The nontruncated marginal of a truncated bivariate normal distribution," *Psychometrika*, 58, 471–488. 303

Arnold, B. C., Castillo, E., and Sarabia, J. M. (2002), "Conditionally specified multivariate skewed distributions," *Sankhyā Ser. A*, 64, 206–226. 303

Arnold, B. C. and Lin, G. D. (2004), "Characterizations of the skew-normal and generalized chi distributions," *Sankhyā*, 66, 593–606. 153

Arnold, B. C. and Sarabia, J. M. (2022), "Conditional specification of statistical models: classical models, new developments and challenges," *J. Multivariate Anal.*, 188, Paper No. 104801, 18. 303

Arnold, S. F. (1981), *The Theory of Linear Models and Multivariate Analysis*, John Wiley & Sons, Inc., New York. 60

Arnold, S. F. and Lynch, J. (1982), "On Ali's characterization of the spherical normal distribution," *J. Roy. Statist. Soc. Ser. B*, 44, 49–51. 96, 98

Arrué, J., Arellano-Valle, R. B., and Gómez, H. W. (2016), "Bias reduction of maximum likelihood estimates for a modified skew-normal distribution," *J. Stat. Comput. Simul.*, 86, 2967–2984. 186

Arrué, J., Arellano-Valle, R. B., Gómez, H. W., and Leiva, V. (2020), "On a new type of Birnbaum-Saunders models and its inference and application to fatigue data," *J. Appl. Statist.*, 47, 2690–2710. 340

Arrué, J., Gómez, H. W., Salinas, H. S., and Bolfarine, H. (2015), "A new class of skew-normal-Cauchy distribution," *SORT*, 39, 35–50. 277

Azevedo, C. L. N., Bolfarine, H., and Andrade, D. F. (2011), "Bayesian inference for a skew-normal IRT model under the centred parameterization," *Comput. Statist. Data Anal.*, 55, 353–365. 339

Aziz, M. A. and Gupta, A. K. (2013), "Quadratic forms in unified skew normal random vectors," *J. Probab. Stat. Sci.*, 11, 1–15. 228

Azzalini, A. (1985), "A class of distributions which includes the normal ones," *Scand. J. Stat.*, 12, 171–178. 122, 302, 304, 324

— (1986), "Further results on a class of distributions which includes the normal ones," *Statistica*, 46, 199–208. 302

— (2001), "A note on regions of given probability of the skew-normal distribution," *Metron*, 59, 27–34 (2002). 142

— (2005), "The skew-normal distribution and related multivariate families," *Scand. J. Stat.*, 32, 159–200, with discussion by Marc G. Genton and a rejoinder by the author. 141

— (2012), "Selection models under generalized symmetry settings," *Ann. Inst. Statist. Math.*, 64, 737–750. 303

— (2016), "Flexible distributions as an approach to robustness: The skew-*t* case," in *Recent Advances in Robust Statistics: Theory and Applications (C. Agostinelli, A. Basu, P. Filzmoser, D. Mukherjee, Eds)*, Springer, New Delhi, pp. 1–16. 313

— (2022), "An overview on the progeny of the skew-normal family–a personal perspective," *J. Multivariate Anal.*, 188, Paper No. 104851, 17. 141

— (2023), *The R package* `sn`*: The skew-normal and related distributions such as the skew-t and the SUN (version 2.1.1).*, Università degli Studi di Padova, Italia, home page: `http://azzalini.stat.unipd.it/SN/`. 312

Azzalini, A. and Arellano-Valle, R. B. (2013), "Maximum penalized likelihood estimation for skew-normal and skew-*t* distributions," *J. Statist. Plann. Inference*, 143, 419–433. 174, 312

Azzalini, A. and Bacchieri, A. (2010), "A prospective combination of phase II and phase III in drug development," *Metron*, 68, 347–369. 330

Azzalini, A., Browne, R. P., Genton, M. G., and McNicholas, P. D. (2016), "On nomenclature for, and the relative merits of, two formulations of skew distributions," *Statist. Probab. Lett.*, 110, 201–206. 198

Azzalini, A. and Capitanio, A. (1999), "Statistical applications of the multivariate skew normal distribution," *J. Roy. Statist. Soc. Ser. B*, 61, 579–602. 122, 137, 138, 140, 247, 277, 286, 321

— (2003), "Distributions generated by perturbation of symmetry with emphasis on a multivariate skew *t*-distribution," *J. Roy. Statist. Soc. Ser. B*, 65, 367–389. 140, 247, 277, 286

— (2014), *The Skew-Normal and Related Families*, vol. 3 of *Institute of Mathematical Statistics (IMS) Monographs*, Cambridge University Press, Cambridge. 126, 139, 277, 312

Azzalini, A. and Chiogna, M. (2004), "Some results on the stress-strength model for skew-normal variates," *Metron*, 62, 315–326. 339

Azzalini, A., Dal Cappello, T., and Kotz, S. (2002), "Log-skew-normal and log-skew-t distributions as models for family income data," *J. Income Distrib.*, 11, 12–20. 334

Azzalini, A. and Dalla Valle, A. (1996), "The multivariate skew-normal distribution," *Biometrika*, 83, 715–726. 122, 139, 247, 286, 302

Azzalini, A. and Genton, M. G. (2007), "On Gauss's characterization of the normal distribution," *Bernoulli*, 13, 169–174. 61

— (2008), "Robust likelihood methods based on the skew-*t* and related distributions," *Int. Stat. Rev.*, 76, 106–129. 177, 277, 313, 315, 316, 321

Azzalini, A., Genton, M. G., and Scarpa, B. (2010), "Invariance-based estimating equations for skew-symmetric distributions," *Metron*, 68, 275–298. 314

Azzalini, A. and Genz, A. (2020), *The R package* `mnormt`*: The multivariate normal and t distributions (version 2.0.2).* 232

Azzalini, A., Kim, H.-M., and Kim, H.-J. (2019), "Sample selection models for discrete and other non-Gaussian response variables," *Stat. Methods Appl.*, 28, 27–56. 314

Azzalini, A. and Regoli, G. (2012a), "Some properties of skew-symmetric distributions," *Ann. Inst. Statist. Math.*, 64, 857–879. 184

— (2012b), "The work of Fernando de Helguero on non-normality arising from selection," *Chil. J. Stat.*, 3, 113–128, [Original paging: 113–129]. 126

— (2018), "On symmetry-modulated distributions: Revisiting an old result and a step further," *Stat*, 7, e171, 12. 184

Azzalini, A. and Salehi, M. (2020), "Some computational aspects of maximum likelihood estimation of the skew-t distribution," in *Computational and Methodological Statistics and Biostatistics–Contemporary Essays in Advancement (A. Bekker, D.-G. Chen, J. T. Ferreira, Eds)*, Springer, Cham, pp. 3–28. 312

Bacon, R. H. (1963), "Approximations to multivariate normal orthant probabilities," *Ann. Math. Stat.*, 34, 191–198. 46

Balakrishnan, N. (2002), "Comment to a paper by B.C. Arnold and R. Beaver," *TEST*, 11, 37–39. 239

Balakrishnan, N., Brito, M. R., and Quiroz, A. J. (2007), "A vectorial notion of skewness and its use in testing for multivariate symmetry," *Comm. Statist. Theory Methods*, 36, 1757–1767. 24

Balakrishnan, N., Capitanio, A., and Scarpa, B. (2014), "A test for multivariate skew-normality based on its canonical form," *J. Multivariate Anal.*, 128, 19–32. 313

Balakrishnan, N. and Scarpa, B. (2012), "Multivariate measures of skewness for the skew-normal distribution," *J. Multivariate Anal.*, 104, 73–87. 161

Baloch, S. H., Krim, H., and Genton, M. G. (2004), "Shape representation with flexible skew-symmetric distributions," in *Skew-Elliptical Distributions And Their Applications: A Journey Beyond Normality (M. G. Genton, Ed.)*, Chapman & Hall/CRC, Boca Raton, FL, pp. 291–308. 340

Bandyopadhyay, D., Lachos, V. H., Abanto-Valle, C. A., and Ghosh, P. (2010), "Linear mixed models for skew-normal/independent bivariate responses with an application to periodontal disease," *Stat. Med.*, 29, 2643–2655. 328

Bandyopadhyay, D., Lachos, V. H., Castro, L. M., and Dey, D. K. (2012), "Skew-normal/independent linear mixed models for censored responses with applications to HIV viral loads," *Biom. J.*, 54, 405–425. 328

Barndorff-Nielsen, O. (1977), "Exponentially decreasing distributions for the logarithm of particle size," *Proc. Roy. Soc. London Ser. A*, 353, 401–419. 56

— (1978), "Hyperbolic distributions and distributions on hyperbolae," *Scand. J. Stat.*, 5, 151–157. 56

Barnett, V. (2004), *Environmental Statistics*, John Wiley & Sons, Ltd., Chichester. 334

Basso, R. M., Lachos, V. H., Cabral, C. R. B., and Ghosh, P. (2010), "Robust mixture modeling based on scale mixtures of skew-normal distributions," *Comput. Statist. Data Anal.*, 54, 2926–2941. 327

Bayes, C. L. and Branco, M. D. (2007), "Bayesian inference for the skewness parameter of the scalar skew-normal distribution," *Braz. J. Probab. Stat.*, 21, 141–163. 313

Bazán, J. L., Branco, M. D., and Bolfarine, H. (2006), "A skew item response model," *Bayesian Anal.*, 1, 861–892. 338

— (2014), "Extensions of the skew-normal ogive item response model," *Braz. J. Probab. Stat.*, 28, 1–23. 339

Behboodian, J., Jamalizadeh, A., and Balakrishnan, N. (2006), "A new class of skew-Cauchy distributions," *Statist. Probab. Lett.*, 76, 1488–1493. 277

Benavoli, A., Azzimonti, D., and Piga, D. (2020), "Skew Gaussian processes for classification," *Mach. Learn.*, 109, 1877–1902. 324

— (2021), "A unified framework for closed-form nonparametric regression, classification, preference and mixed problems with Skew Gaussian Processes," *Mach. Learn.*, 110, 3095–3133. 324

Beran, R. J. and Millar, P. W. (1997), "Multivariate symmetry models," in *Festschrift for Lucien Le Cam (D. Pollard, E. Torgersen, G. L. Yang, Eds)*, Springer, New York, pp. 13–42. 28, 35

Beranger, B., Padoan, S. A., and Sisson, S. A. (2017), "Models for extremal dependence derived from skew-symmetric families," *Scand. J. Stat.*, 44, 21–45. 337

Beranger, B., Padoan, S. A., Xu, Y., and Sisson, S. A. (2019a), "Extremal properties of the multivariate extended skew-normal distribution, Part B," *Statist. Probab. Lett.*, 147, 105–114. 337

— (2019b), "Extremal properties of the univariate extended skew-normal distribution, Part A," *Statist. Probab. Lett.*, 147, 73–82. 337

Beranger, B., Stephenson, A. G., and Sisson, S. A. (2021), "High-dimensional inference using the extremal skew-*t* process," *Extremes*, 24, 653–685. 337

Berkane, M. and Bentler, P. M. (1986), "Moments of elliptically distributed random variates," *Statist. Probab. Lett.*, 4, 333–335. 83, 88

Berman, S. M. (1965), "Sign-invariant random variables and stochastic processes with sign-invariant increments," *Trans. Amer. Math. Soc.*, 119, 216–243. 34

Bevilacqua, M., Caamaño-Carrillo, C., Arellano-Valle, R. B., and Gómez, C. (2022), "A class of random fields with two-piece marginal distributions for modeling point-referenced data with spatial outliers," *TEST*, 31, 644–674. 319

Bevilacqua, M., Caamaño-Carrillo, C., Arellano-Valle, R. B., and Morales-Oñate, V. (2021), "Non-Gaussian geostatistical modeling using (skew) t processes," *Scand. J. Stat.*, 48, 212–245. 319

Bilodeau, M. and Brenner, D. (1999), *Theory of Multivariate Statistics*, Springer-Verlag, New York. 2, 11, 15, 18, 37, 69, 70, 71, 72, 144, 145, 217, 218, 313

Birnbaum, Z. W. (1950), "Effect of linear truncation on a multinormal population," *Ann. Math. Stat.*, 21, 272–279. 126

Blæsild, P. and Jensen, J. L. (1981), "Multivariate distributions of hyperbolic type," in *Statistical Distributions in Scientific Work (C. Taillie, G. P. Patil, B. A. Baldessari, Eds), Vol. 4 (Trieste, 1980)*, Reidel, Dordrecht-Boston, Mass., vol. 79 of *NATO Adv. Study Inst. Ser. C: Math. Phys. Sci.*, pp. 45–66. 56

Bock, R. (2007), "MAGIC gamma telescope," *UCI Machine Learning Repository*, doi.org/10.24432/C52C8B. 321

Bolfarine, H. and Arellano-Valle, R. B. (1994), "Robust modelling in measurement error models using the *t* distribution," *Rebrape*, 8, 67–84. 329

Bolfarine, H., Montenegro, L. C., and Lachos, V. H. (2007), "Influence diagnostics for skew-normal linear mixed models," *Sankhyā*, 69, 648–670. 328

Branco, M. D. and Dey, D. K. (2001), "A general class of multivariate skew-elliptical distributions," *J. Multivariate Anal.*, 79, 99–113. 184, 247, 286, 302, 328

— (2002), "Regression model under skew elliptical error distribution," *J. Math. Sci. (N.S.)*, 1, 151–168. 286

Branco, M. D., Genton, M. G., and Liseo, B. (2013), "Objective Bayesian analysis of skew-*t* distributions," *Scand. J. Stat.*, 40, 63–85. 313

Bufalo, M. and Nigri, A. (2026), "A new look at the flexible generalized skew-normal family: A trimodal extension and numerical insights," *Stat. Comput.*, 36, 80. 184

Cabral, C. R. B., Lachos, V. H., and Prates, M. O. (2012), "Multivariate mixture modeling using skew-normal independent distributions," *Comput. Statist. Data Anal.*, 56, 126–142. 185

Cabral, C. R. B., Lachos, V. H., and Zeller, C. B. (2014), "Multivariate measurement error models using finite mixtures of skew-Student *t* distributions," *J. Multivariate Anal.*, 124, 179–198. 329

Cabras, S., Racugno, W., Castellanos, M. E., and Ventura, L. (2012), "A matching prior for the shape parameter of the skew-normal distribution," *Scand. J. Stat.*, 39, 236–247. 313

Cambanis, S., Huang, S., and Simons, G. (1981), "On the theory of elliptically contoured distributions," *J. Multivariate Anal.*, 11, 368–385. 69, 72, 79, 80, 83, 89, 96, 97, 100

Cambanis, S., Keener, R., and Simons, G. (1983), "On α-symmetric multivariate distributions," *J. Multivariate Anal.*, 13, 213–233. 33

Campbell, J. Y., Lo, A. W., and MacKinlay, A. C. (1997), *The Econometrics of Financial Markets*, Princeton. 332

Canale, A. (2011), "Statistical aspects of the scalar extended skew-normal distribution," *Metron*, 69, 279–295. 228

— (2015), "A note on regions of given probability of the extended skew-normal distribution," *Comm. Statist. Theory Methods*, 44, 2507–2516. 228

Cao, J., Durante, D., and Genton, M. G. (2022a), "Scalable computation of predictive probabilities in probit models with Gaussian process priors," *J. Comput. Graph. Statist.*, 31, 709–720. 324, 325

Cao, J., Genton, M. G., Keyes, D. E., and Turkiyyah, G. (2019), "Hierarchical-block conditioning approximations for high-dimensional multivariate normal probabilities," *Stat. Comput.*, 29, 585–598. 46

— (2021), "Exploiting low-rank covariance structures for computing high-dimensional normal and Student-*t* probabilities," *Stat. Comput.*, 31, Paper No. 2, 16. 46

— (2022b), "tlrmvnmvt: Computing highdimensional multivariate normal and Student-t probabilities with low-rank methods in R," *J. Stat. Softw.*, 101, 4. 46

Capitanio, A. (2010), "On the approximation of the tail probability of the scalar skew-normal distribution," *Metron*, 68, 299–308. 336

— (2020), "On the canonical form of scale mixtures of skew-normal distributions," *Statistica*, 80, 145–160. 137

Capitanio, A., Azzalini, A., and Stanghellini, E. (2003), "Graphical models for skew-normal variates," *Scand. J. Stat.*, 30, 129–144. 331

Carota, C. (2010), "Tests for normality in classes of skew-*t* alternatives," *Statist. Probab. Lett.*, 80, 1–8. 277

Carrasco, J. M. F., Ferrari, S. L. P., and Arellano-Valle, R. B. (2014), "Errors-in-variables beta regression models," *J. Appl. Statist.*, 41, 1530–1547. 329

— (2023), "Multiplicative errors-in-variables beta regression," *Braz. J. Probab. Stat.*, 37, 249–262. 330

Carroll, R. J., Ruppert, D., Stefanski, L. A., and Crainiceanu, C. M. (2006), *Measurement Error in Nonlinear Models*, vol. 105 of *Monographs on Statistics and Applied Probability*, Chapman & Hall/CRC, Boca Raton, FL, 2nd ed. 329

Castro, L. M., Lachos, V. H., Ferreira, G. P., and Arellano-Valle, R. B. (2014), "Partially linear censored regression models using heavy-tailed distributions: a Bayesian approach," *Stat. Methodol.*, 18, 14–31. 339

Castro, L. M., San Martín, E., and Arellano-Valle, R. B. (2013), "A note on the parameterization of multivariate skewed-normal distributions," *Braz. J. Probab. Stat.*, 27, 110–115. 198

Chai, H. S. and Bailey, K. R. (2008), "Use of log-skew-normal distribution in analysis of continuous data with a discrete component at zero," *Stat. Med.*, 27, 3643–3655. 334

Chan, K. S. and Tong, H. (1986), "A note on certain integral equations associated with nonlinear time series analysis," *Probab. Theory Relat. Fields*, 73, 153–158. 316

Chang, S.-C. and Zimmerman, D. L. (2016), "Skew-normal antedependence models for skewed longitudinal data," *Biometrika*, 103, 363–376. 317

Chang, S.-M. and Genton, M. G. (2007), "Extreme value distributions for the skew-symmetric family of distributions," *Comm. Statist. Theory Methods*, 36, 1705–1717. 336

Chen, J. T. and Gupta, A. K. (2005), "Matrix variate skew normal distributions," *Statistics*, 39, 247–253. 191

Chen, M.-H. (2004), "Skewed link models for categorical response data," in *Skew-Elliptical Distributions And Their Applications: A Journey Beyond Normality (M. G. Genton, Ed.)*, Chapman & Hall/CRC, Boca Raton, FL, pp. 131–151. 338

Chen, M.-H., Dey, D. K., and Shao, Q.-M. (1999), "A new skewed link model for dichotomous quantal response data," *J. Amer. Statist. Assoc.*, 94, 1172–1186. 338

Chen, W. and Genton, M. G. (2023), "Are you all normal? It depends!" *Int. Stat. Rev.*, 91, 114–139. 312

Chiogna, M. (2005), "A note on the asymptotic distribution of the maximum likelihood estimator for the scalar skew-normal distribution," *Stat. Methods Appl.*, 14, 331–341. 177

Chowdhury, J., Dutta, S., Arellano-Valle, R. B., and Genton, M. G. (2022), "Sub-dimensional Mardia measures of multivariate skewness and kurtosis," *J. Multivariate Anal.*, 192, Paper No. 105089, 20. 24

Cochrane, J. H. (2005), *Asset Pricing: Revised Edition*, Princeton, NJ: Princeton University Press. 332

Coles, S. (2001), *An Introduction to Statistical Modeling of Extreme Values*, Springer-Verlag London, Ltd., London. 336

Contreras-Reyes, J. E. and Arellano-Valle, R. B. (2012), "Kullback-Leibler divergence measure for multivariate skew-normal distributions," *Entropy*, 14, 1606–1626. 192, 334

— (2013), "Growth estimates of cardinalfish (Epigonus crassicaudus) based on scale mixtures of skew-normal distributions," *Fish. Res.*, 147, 137–144. 335

Contreras-Reyes, J. E., Arellano-Valle, R. B., and Canales, T. M. (2014), "Comparing growth curves with asymmetric heavy-tailed errors: Application to the southern blue whiting (Micromesistius australis)," *Fish. Res.*, 159, 88–94. 335

Copas, J. B. and Li, H. G. (1997), "Inference for non-random samples," *J. Roy. Statist. Soc. Ser. B*, 59, 55–95, with discussion and a reply by the authors. 302

Corns, T. R. A. and Satchell, S. E. (2007), "Skew Brownian motion and pricing European options," *European J. Finance*, 13, 523–544. 317

— (2010), "Modelling conditional heteroskedasticity and skewness using the skew-normal distribution," *Metron*, 68, 251–263. 332

Cramér, H. (1946), *Mathematical Methods of Statistics*, Princeton Mathematical Series, vol. 9, Princeton University Press, Princeton, N. J. 18

Cruz-Reyes, D. L., Assuncão, R. M., Arellano-Valle, R. B., and Loschi, R. H. (2026), "Directional asymmetry in edge based spatial models via a skew-normal prior," Https://arxiv.org/abs/2601.16829. 340

Dao, N. A. and Genton, M. G. (2021), "Skew-elliptical cluster processes," in *Advances in Statistics–Theory and Applications: Honoring the Contributions of Barry C. Arnold in Statistical Science (I. Ghosh, N. Balakrishnan, H. K. T. Ng, Eds)*, Springer, Cham, pp. 365–393. 336

Das, S. and Genton, M. G. (2020), "On the stationary marginal distributions of subclasses of multivariate SETAR processes of order one," *J. Time Series Anal.*, 41, 406–420. 316

de Freitas, J. a. V. B., Bondon, P., Azevedo, C. L. N., Reisen, V. A., and Nobre, J. S. (2024), "Scale mixtures of multivariate centered skew-normal distributions," *Stat. Comput.*, 34, Paper No. 212, 27. 185

De Luca, G., Genton, M. G., and Loperfido, N. (2006), "A multivariate skew-GARCH model," in *Econometric Analysis of Financial and Economic Time Series (T. B. Fomby, D. Terrell, Eds), Part A*, Emerald/JAI, Bingley, vol. 20 of *Adv. Econom.*, pp. 33–57. 332

De Luca, G. and Loperfido, N. (2004), "A skew-in-mean GARCH model," in *Skew-Elliptical Distributions And Their Applications: A Journey Beyond Normality (M. G. Genton, Ed.)*, Chapman & Hall/CRC, Boca Raton, FL, pp. 205–222. 332

de Queiroz, M. M., Loschi, R. H., and Silva, R. W. C. (2016a), "Multivariate log-skewed distributions with normal kernel and their applications," *Statistics*, 50, 157–175. 334

de Queiroz, M. M., Silva, R. W. C., and Loschi, R. H. (2016b), "Shannon entropy and Kullback-Leibler divergence in multivariate log fundamental skew-normal and related distributions," *Canad. J. Statist.*, 44, 219–237. 192

Dempster, A. P., Laird, N. M., and Rubin, D. B. (1977), "Maximum likelihood from incomplete data via the EM algorithm," *J. Roy. Statist. Soc. Ser. B*, 39, 1–38. 181

DiCiccio, T. J. and Monti, A. C. (2004), "Inferential aspects of the skew exponential power distribution," *J. Amer. Statist. Assoc.*, 99, 439–450. 341

— (2011), "Inferential aspects of the skew t-distribution," *Quad. Stat.*, 13, 1–21. 277

— (2018), "Testing for sub-models of the skew t-distribution," *Stat. Methods Appl.*, 27, 25–44. 277

Domínguez-Molina, J. A., González-Farías, G., and Ramos-Quiroga, R. (2004), "Skew-normality in stochastic frontier analysis," in *Skew-Elliptical Distributions And Their Applications: A Journey Beyond Normality (M. G. Genton, Ed.)*, Chapman & Hall/CRC, Boca Raton, FL, pp. 223–241. 340

Domínguez-Molina, J. A., González-Farías, G., Ramos-Quiroga, R., and Gupta, A. K. (2007), "A matrix variate closed skew-normal distribution with applications to stochastic frontier analysis," *Comm. Statist. Theory Methods*, 36, 1691–1703. 191

Domínguez-Molina, J. A. and Rocha-Arteaga, A. (2007), "On the infinite divisibility of some skewed symmetric distributions," *Statist. Probab. Lett.*, 77, 644–648. 137

Durante, D. (2019), "Conjugate Bayes for probit regression via unified skew-normal distributions," *Biometrika*, 106, 765–779. 324

Durante, D., Pozza, F., and Szabo, B. (2024), "Skewed Bernstein-von Mises theorem and skew-modal approximations," *Ann. Statist.*, 52, 2714–2737. 326

Dutta, S. and Genton, M. G. (2014), "A non-Gaussian multivariate distribution with all lower-dimensional Gaussians and related families," *J. Multivariate Anal.*, 132, 82–93. 37, 44

Eaton, M. L. (1983), *Multivariate Statistics*, John Wiley & Sons, Inc., New York. 38

— (1989), *Group Invariance Applications in Statistics*, vol. 1 of *NSF-CBMS Regional Conference Series in Probability and Statistics*, Institute of Mathematical Statistics, Hayward, CA; American Statistical Association, Alexandria, VA. 38

Elal-Olivero, D., Gómez, H. W., and Quintana, F. A. (2009), "Bayesian modeling using a class of bimodal skew-elliptical distributions," *J. Statist. Plann. Inference*, 139, 1484–1492. 341

Eyer, L. and Genton, M. G. (2004), "An astronomical distance determination method using regression with skew-normal errors," in *Skew-Elliptical Distributions And Their Applications: A Journey Beyond Normality (M. G. Genton, Ed.)*, Chapman & Hall/CRC, Boca Raton, FL, pp. 309–319. 340

Fang, B. Q. (2003), "The skew elliptical distributions and their quadratic forms," *J. Multivariate Anal.*, 87, 298–314. 299

— (2005a), "The asymptotic distributions of the statistics of the skew elliptical variables," *Sci. China Ser. A*, 48, 214–221. 286

— (2005b), "The F statistic of the skew elliptical variables," *Chinese J. Appl. Probab. Statist.*, 21, 197–206. 286

— (2005c), "Invariant distributions of the multivariate tests in the skew elliptical model," *Stat. Methodol.*, 2, 285–296. 286

— (2005d), "Noncentral quadratic forms of the skew elliptical variables," *J. Multivariate Anal.*, 95, 410–430. 299

— (2005e), "The t statistic of the skew elliptical distributions," *J. Statist. Plann. Inference*, 134, 140–157. 286

— (2006), "Sample mean, covariance and T^2 statistic of the skew elliptical model," *J. Multivariate Anal.*, 97, 1675–1690. 286

— (2008a), "Moments of the skew normal random vectors in the general case," *Chinese J. Appl. Probab. Statist.*, 24, 604–612. 161

— (2008b), "Noncentral matrix quadratic forms of the skew elliptical variables," *J. Multivariate Anal.*, 99, 1105–1127. 191

Fang, H.-B., Fang, K.-T., and Kotz, S. (2002), "The meta-elliptical distributions with given marginals," *J. Multivariate Anal.*, 82, 1–16. 113

Fang, K. T., Kotz, S., and Ng, K. W. (1990), *Symmetric Multivariate and Related Distributions*, vol. 36 of *Monographs on Statistics and Applied Probability*, Chapman and Hall, Ltd., London. 16, 17, 28, 33, 37, 69, 71, 73, 80, 83, 86, 89, 90, 96, 98

Fang, K. T. and Zhang, Y. T. (1990), *Generalized Multivariate Analysis*, Springer-Verlag, Berlin; Science Press Beijing, Beijing. 2, 11, 16, 17, 38, 69, 71, 72, 73, 86, 89, 90, 96, 98, 100, 101, 102, 104

Fasano, A., Rebaudo, G., Durante, D., and Petrone, S. (2021), "A closed-form filter for binary time series," *Stat. Comput.*, 31, Paper No. 47, 20. 324

Fernández, C., Osiewalski, J., and Steel, M. F. J. (1995), "Modeling and inference with ν-spherical distributions," *J. Amer. Statist. Assoc.*, 90, 1331–1340. 34

Fernández, C. and Steel, M. F. J. (1998), "On Bayesian modeling of fat tails and skewness," *J. Amer. Statist. Assoc.*, 93, 359–371. 340

Ferrari, S. L. P. and Arellano-Valle, R. B. (1996), "Modified likelihood ratio and score tests in linear regression models using the t distribution," *Braz. J. Probab. Stat.*, 10, 15–33. 95

Ferreira, C. S. and Arellano-Valle, R. B. (2018), "Estimation and diagnostic analysis in skew-generalized-normal regression models," *J. Stat. Comput. Simul.*, 88, 1039–1059. 186

Ferreira, J. T. A. S. and Steel, M. F. J. (2004), "Bayesian multivariate skewed regression modeling with an application to firm size," in *Skew-Elliptical Distributions And Their Applications: A Journey Beyond Normality (M. G. Genton, Ed.)*, Chapman & Hall/CRC, Boca Raton, FL, pp. 175–190. 332

Field, C. and Genton, M. G. (2006), "The multivariate **g**-and-**h** distribution," *Technometrics*, 48, 104–111. 64

Figueroa-Zúñiga, J., Carrasco, J. M. F., Arellano-Valle, R., and Ferrari, S. L. P. (2018), "A Bayesian approach to errors-in-variables beta regression," *Braz. J. Probab. Stat.*, 32, 559–582. 329

Flecher, C., Allard, D., and Naveau, P. (2010), "Truncated skew-normal distributions: moments, estimation by weighted moments and application to climatic data," *Metron*, 68, 331–345. 334

Franco, S. and Azzalini, A. (2025), "The information matrix of the bivariate extended skew-normal distribution," *Metron*, 83, 235–259. 312

Fung, T. and Seneta, E. (2010), "Tail dependence for two skew t distributions," *Statist. Probab. Lett.*, 80, 784–791. 337

Galarza Morales, C. E. and Lachos, V. H. (2024), *MomTrunc: Moments of Folded and Doubly Truncated Multivariate Distributions*, R package version 6.1. 308

Galarza Morales, C. E., Matos, L. A., Castro, L. M., and Lachos, V. H. (2022a), "Moments of the doubly truncated selection elliptical distributions with emphasis on the unified multivariate skew-*t* distribution," *J. Multivariate Anal.*, 189, Paper No. 104944, 15. 308

Galarza Morales, C. E., Matos, L. A., Dey, D. K., and Lachos, V. H. (2022b), "On moments of folded and doubly truncated multivariate extended skew-normal distributions," *J. Comput. Graph. Statist.*, 31, 455–465. 308

Gallaugher, M. P. B. and McNicholas, P. D. (2017), "A matrix variate skew-*t* distribution," *Stat*, 6, 160–170. 191

— (2019), "Three skewed matrix variate distributions," *Statist. Probab. Lett.*, 145, 103–109. 191

Gelfand, A., Fuentes, M., Hoeting, J. A., and Smith, R. L. (eds.) (2019), *Handbook of Environmental and Ecological Statistics*, Chapman & Hall/CRC Handbooks of Modern Statistical Methods, CRC Press, Boca Raton, FL. 334

Gelman, A., Carlin, J. B., Stern, H. S., Dunson, D. B., Vehtari, A., and Rubin, D. B. (2014), *Bayesian Data Analysis*, CRC Press, Boca Raton, FL, 3rd ed. 58

Genton, M. G. (ed.) (2004), *Skew-Elliptical Distributions And Their Applications: A Journey Beyond Normality*, Chapman & Hall/CRC, Boca Raton, FL. 126, 312

Genton, M. G. (2007), "Separable approximations of space-time covariance matrices," *Environmetrics*, 18, 681–695. 351

Genton, M. G., He, L., and Liu, X. (2001), "Moments of skew-normal random vectors and their quadratic forms," *Statist. Probab. Lett.*, 51, 319–325. 161

Genton, M. G., Keyes, D., and Turkiyyah, G. (2018), "Hierarchical decompositions for the computation of high-dimensional multivariate normal probabilities," *J. Comput. Graph. Statist.*, 27, 268–277. 46

Genton, M. G., Kim, M., and Ma, Y. (2012), "Semiparametric location estimation under non-random sampling," *Stat*, 1, 1–11. 314

Genton, M. G. and Loperfido, N. (2005), "Generalized skew-elliptical distributions and their quadratic forms," *Ann. Inst. Statist. Math.*, 57, 389–401. 153, 159, 184, 302, 313, 331, 335

Genton, M. G. and Thompson, K. (2004), "Skew-elliptical time series with application to flooding risk," in *Time Series Analysis and Applications to Geophysical Systems (D. R. Brillinger, E. A. Robinson, F. P. Schoenberg, Eds)*, IMA Volume in Mathematics and its Applications 139, Springer, pp. 169–186. 334

Genton, M. G. and Zhang, H. (2012), "Identifiability problems in some non-Gaussian spatial random fields," *Chil. J. Stat.*, 3, 171–179, [Original paging: 173–181]. 317

Genz, A. (1992), "Numerical computation of multivariate Normal probabilities," *J. Comput. Graph. Statist.*, 1, 141–149. 46, 65

Genz, A. and Bretz, F. (2009), *Computation of Multivariate Normal and t Probabilities*, vol. 195 of *Lecture Notes in Statistics*, Springer, Dordrecht. 46, 65

Ghosh, I. and Ng, H. K. T. (2025), "Hidden truncation models: Theory and applications," *Wiley Interdiscip. Rev. Comput. Stat.*, 17, Paper No. e70019, 12. 303

Ghosh, P., Branco, M. D., and Chakraborty, H. (2007), "Bivariate random effect model using skew-normal distribution with application to HIV-RNA," *Stat. Med.*, 26, 1255–1267. 328

Ghosh, S., Dutta, S., and Genton, M. G. (2017), "A note on inconsistent families of discrete multivariate distributions," *J. Stat. Distrib. Appl.*, 4, 7. 37

Giani, P., Tagle, F., Genton, M. G., Castruccio, S., and Crippa, P. (2020), "Closing the gap between wind energy targets and implementation for emerging countries," *Appl. Energy*, 269, 115085. 325

Godoi, L. G., Branco, M. D., and Ruggeri, F. (2017), "Concentration function for the skew-normal and skew-t distributions, with application in robust Bayesian analysis," *Braz. J. Probab. Stat.*, 31, 373–393. 313

Goerg, G. M. (2011), "Lambert W random variables–a new family of generalized skewed distributions with applications to risk estimation," *Ann. Appl. Stat.*, 5, 2197–2230. 64

— (2015), "The Lambert way to Gaussianize heavy-tailed data with the inverse of Tukey's h transformation as a special case," *Sci. World J.*, 2015. 64

Golub, G. H. and Van Loan, C. F. (1983), *Matrix Computations*, vol. 3 of *Johns Hopkins Series in the Mathematical Sciences*, Johns Hopkins University Press, Baltimore, MD. 345, 350

González-Farías, G., Domínguez-Molina, J. A., and Gupta, A. K. (2004a), "Additive properties of skew normal random vectors," *J. Statist. Plann. Inference*, 126, 521–534. 198

González-Farías, G., Domínguez-Molina, J. A., and Gupta, A. K. (2004b), "The closed skew-normal distribution," in *Skew-Elliptical Distributions And Their Applications: A Journey Beyond Normality (M. G. Genton, Ed.)*, Chapman & Hall/CRC, Boca Raton, FL, pp. 25–42. 198, 199

Grilli, L. and Rampichini, C. (2010), "Selection bias in linear mixed models," *Metron*, 68, 309–329. 328

Groeneveld, R. A. and Meeden, G. (1984), "Measuring skewness and kurtosis," *J. Roy. Statist. Soc. Ser. D*, 33, 391–399. 18

Gualtierotti, A. F. (2005), "Skew-normal processes as models for random signals corrupted by Gaussian noise," *Int. J. Pure Appl. Math.*, 20, 109–140. 317

Gupta, A. K. (2003), "Multivariate skew t-distribution," *Statistics*, 37, 359–363. 277

Gupta, A. K., Aziz, M. A., and Ning, W. (2013a), "On some properties of the unified skew normal distribution," *J. Stat. Theory Pract.*, 7, 480–495. 210

Gupta, A. K., Chang, F. C., and Huang, W. J. (2002), "Some skew-symmetric models," *Random Oper. Stoch. Equ.*, 10, 133–140. 137

Gupta, A. K., González-Farías, G., and Domínguez-Molina, J. A. (2004a), "A multivariate skew normal distribution," *J. Multivariate Anal.*, 89, 181–190. 194, 198

Gupta, A. K. and Huang, W.-J. (2002), "Quadratic forms in skew normal variates," *J. Math. Anal. Appl.*, 273, 558–564. 159

Gupta, A. K., Nguyen, T. T., and Sanqui, J. A. T. (2004b), "Characterization of the skew-normal distribution," *Ann. Inst. Statist. Math.*, 56, 351–360. 153

Gupta, A. K. and Song, D. (1997), "L_p-norm spherical distribution," *J. Statist. Plann. Inference*, 60, 241–260. 34, 107, 109, 110, 112

Gupta, A. K. and Varga, T. (1993), *Elliptically Contoured Models in Statistics*, vol. 240 of *Mathematics and its Applications*, Kluwer Academic Publishers Group, Dordrecht. 17, 101

Gupta, A. K., Varga, T., and Bodnar, T. (2013b), *Elliptically Contoured Models in Statistics and Portfolio Theory*, Springer, New York, 2nd ed. 89

Gupta, R. C. and Brown, N. (2001), "Reliability studies of the skew-normal distribution and its application to a strength-stress model," *Comm. Statist. Theory Methods*, 30, 2427–2445. 340

Gupta, R. C. and Gupta, R. D. (2004), "Generalized skew normal model," *TEST*, 13, 501–524. 239

Hallin, M. and Ley, C. (2012), "Skew-symmetric distributions and Fisher information–a tale of two densities," *Bernoulli*, 18, 747–763. 177

— (2014), "Skew-symmetric distributions and Fisher information: The double sin of the skew-normal," *Bernoulli*, 20, 1432–1453. 177

Harnik, S. B. (2023), "On Robustness Properties of a Class of Skewed Models," *Doctoral Thesis, Graduate Program in Statistics, University of São Paulo*. 313

Harrar, S. W. and Gupta, A. K. (2008), "On matrix variate skew-normal distributions," *Statistics*, 42, 179–194. 191

Hashorva, E., Kotz, S., and Kume, A. (2007), "L_p-norm generalised symmetrised Dirichlet distributions," *Albanian J. Math.*, 1, 31–56. 113

Hazra, A., Reich, B. J., and Staicu, A.-M. (2020), "A multivariate spatial skew-t process for joint modeling of extreme precipitation indexes," *Environmetrics*, 31, Paper No. e2602, 19. 337

Heck, D., Knapp, J., Capdevielle, J. N., Schatz, G., and Thouw, T. (1998), "CORSIKA: A Monte Carlo code to simulate extensive air showers," *Forschungszentrum Karlsruhe*, FZKA 6019, 1–90. 321

Henze, N. (1986), "A probabilistic representation of the 'skew-normal' distribution," *Scand. J. Stat.*, 13, 271–275. 124

Hernández-Sánchez, E. and Scarpa, B. (2012), "A wrapped flexible generalized skew-normal model for a bimodal circular distribution of wind directions," *Chil. J. Stat.*, 3, 129–141, [Original paging: 131–143]. 338

Ho, H. J. and Lin, T.-I. (2010), "Robust linear mixed models using the skew t distribution with application to schizophrenia data," *Biom. J.*, 52, 449–469. 328

Hongler, M.-O. and Rinaldo, D. (2026), "Skew-normal diffusions," *J. Stat. Phys.*, 193, Paper No. 55. 340

Horn, R. A. and Johnson, C. R. (1985), *Matrix Analysis*, Cambridge University Press, Cambridge. 345

— (1991), *Topics in Matrix Analysis*, Cambridge University Press, Cambridge. 345

Horrace, W. C. (2005), "Some results on the multivariate truncated normal distribution," *J. Multivariate Anal.*, 94, 209–221. 65

Hotelling, H. (1931), "The generalization of Student's ratio," *Ann. Math. Stat.*, 2, 360–378. 62

Huang, W.-J. and Chen, Y.-H. (2006), "Quadratic forms of multivariate skew normal-symmetric distributions," *Statist. Probab. Lett.*, 76, 871–879. 159

— (2007), "Generalized skew-Cauchy distribution," *Statist. Probab. Lett.*, 77, 1137–1147. 277

Huang, Z. and Sen, B. (2023), "Multivariate symmetry: Distribution-free testing via optimal transport," Https://arxiv.org/abs/2305.01839. 38

Hutton, J. L. and Stanghellini, E. (2011), "Modelling bounded health scores with censored skew-normal distributions," *Stat. Med.*, 30, 368–376. 330

Iguchi, Y., Livingstone, S., Nüsken, N., Vasdekis, G., and Zhang, R.-Y. (2025), "Skew-symmetric schemes for stochastic differential equations with non-Lipschitz drift: An unadjusted Barker algorithm," Https://arxiv.org/abs/2405.14373. 340

Izenman, A. J. (2008), *Modern Multivariate Statistical Techniques*, Springer Texts in Statistics, Springer, New York. 320, 323, 326, 327

Jafari Khaledi, M., Zareifard, H., and Boojari, H. (2023), "A spatial skew-Gaussian process with a specified covariance function," *Statist. Probab. Lett.*, 192, Paper No. 109681, 7. 319

Jamalizadeh, A. and Balakrishnan, N. (2008), "On order statistics from bivariate skew-normal and skew-t_ν distributions," *J. Statist. Plann. Inference*, 138, 4187–4197. 309

— (2009), "Order statistics from trivariate normal and t_ν-distributions in terms of generalized skew-normal and skew-t_ν distributions," *J. Statist. Plann. Inference*, 139, 3799–3819. 309

— (2010), "Distributions of order statistics and linear combinations of order statistics from an elliptical distribution as mixtures of unified skew-elliptical distributions," *J. Multivariate Anal.*, 101, 1412–1427. 309

Jamalizadeh, A., Balakrishnan, N., and Salehi, M. (2010), "Order statistics and linear combination of order statistics arising from a bivariate selection normal distribution," *Statist. Probab. Lett.*, 80, 445–451. 309

Jamalizadeh, A., Khosravi, M., and Balakrishnan, N. (2009a), "Recurrence relations for distributions of a skew-*t* and a linear combination of order statistics from a bivariate-*t*," *Comput. Statist. Data Anal.*, 53, 847–852. 309

Jamalizadeh, A., Mahmoodian, H., and Balakrishnan, N. (2009b), "Exact distribution of a linear combination of a variable and order statistics from the other two variables of a trivariate elliptical random vector as a mixture of skew-elliptical distributions," *Stat. Methodol.*, 6, 634–644. 309

Jamalizadeh, A., Mehrali, Y., and Balakrishnan, N. (2009c), "Recurrence relations for bivariate *t* and extended skew-*t* distributions and an application to order statistics from bivariate *t*," *Comput. Statist. Data Anal.*, 53, 4018–4027. 309

Jara, A., Quintana, F., and San Martín, E. (2008), "Linear mixed models with skew-elliptical distributions: A Bayesian approach," *Comput. Statist. Data Anal.*, 52, 5033–5045. 328

Joe, H. (2015), *Dependence Modeling with Copulas*, vol. 134 of *Monographs on Statistics and Applied Probability*, CRC Press, Boca Raton, FL. 333

Joe, H. and Li, H. (2019), "Tail densities of skew-elliptical distributions," *J. Multivariate Anal.*, 171, 421–435. 333

Johnson, N. L., Kotz, S., and Balakrishnan, N. (1994), *Continuous Univariate Distributions. Vol. 1*, John Wiley & Sons, Inc., New York, 2nd ed. 56

Jolliffe, I. T. (2002), *Principal Component Analysis*, Springer-Verlag, New York. 287

Jones, M. C. (2014), "Generating distributions by transformation of scale," *Statist. Sinica*, 24, 749–771. 184

— (2016), "On bivariate transformation of scale distributions," *Comm. Statist. Theory Methods*, 45, 577–588. 184

Jones, M. C. and Faddy, M. J. (2003), "A skew extension of the *t*-distribution, with applications," *J. Roy. Statist. Soc. Ser. B*, 65, 159–174. 247

Jones, M. C. and Larsen, P. V. (2004), "Multivariate distributions with support above the diagonal," *Biometrika*, 91, 975–986. 247

Jones, M. C. and Pewsey, A. (2009), "Sinh-arcsinh distributions," *Biometrika*, 96, 761–780. 65

Jones, M. C., Rosco, J. F., and Pewsey, A. (2011), "Skewness-invariant measures of kurtosis," *Amer. Statist.*, 65, 89–95. 18

Jørgensen, B. (1982), *Statistical Properties of the Generalized Inverse Gaussian Distribution*, vol. 9 of *Lecture Notes in Statistics*, Springer-Verlag, New York-Berlin. 56

— (1993), *The Theory of Linear Models*, Chapman & Hall, New York. 60

Jupp, P. E., Regoli, G., and Azzalini, A. (2016), "A general setting for symmetric distributions and their relationship to general distributions," *J. Multivariate Anal.*, 148, 107–119. 338

Kahrari, F., Arellano-Valle, R. B., Ferreira, C. d. S., and Gallardo, D. (2022), "Some simulation/computation in multivariate linear models of scale mixtures of skew-normal-Cauchy distributions," *Comm. Statist. Simulation Comput.*, 51, 6425–6440. 342

Kahrari, F., Arellano-Valle, R. B., and Rezaei, M. (2020), "Bias prevention of maximum likelihood estimates for skew-normal-Cauchy distribution," *Comm. Statist. Simulation Comput.*, 49, 1–15. 342

Kahrari, F., Arellano-Valle, R. B., Rezaei, M., and Yousefzadeh, F. (2017), "Scale mixtures of skew-normal-Cauchy distributions," *Statist. Probab. Lett.*, 126, 1–6. 342

Kahrari, F., Ferreira, C. S., and Arellano-Valle, R. B. (2019), "Skew-Normal-Cauchy linear mixed models," *Sankhyā Ser. B*, 81, 185–202. 328

Kahrari, F., Rezaei, M., Yousefzadeh, F., and Arellano-Valle, R. B. (2016), "On the multivariate skew-normal-Cauchy distribution," *Statist. Probab. Lett.*, 117, 80–88. 342

Kamm, J. and Nagy, J. G. (2000), "Optimal Kronecker product approximation of block Toeplitz matrices," *SIAM J. Matrix Anal. Appl.*, 22, 155–172. 351

Kan, R. and Robotti, C. (2017), "On moments of folded and truncated multivariate normal distributions," *J. Comput. Graph. Statist.*, 26, 930–934. 65, 232

Kano, Y. (1994), "Consistency property of elliptical probability density functions," *J. Multivariate Anal.*, 51, 139–147. 83, 118

Kantorovich, L. V. (1960), "Mathematical methods of organizing and planning production," *Management Sci.*, 6, 366–422. 25

Kariya, T. and Eaton, M. L. (1977), "Robust tests for spherical symmetry," *Ann. Stat.*, 5, 206–215. 99

Karling, M. J., Durante, D., and Genton, M. G. (2024), "Conjugacy properties of multivariate unified skew-elliptical distributions," *J. Multivariate Anal.*, 204, Paper No. 105357. 324

Karling, M. J., Genton, M. G., and Meintanis, S. G. (2023), "Goodness-of-fit tests for multivariate skewed distributions based on the characteristic function," *Stat. Comput.*, 33, Paper No. 99, 18. 313

Karthick, K., Agnes, S. A., Kumar, S. S., Alfarhood, S., and Safran, M. (2024), "Identification of high energy gamma particles from the Cherenkov gamma telescope data using a deep learning approach," *IEEE Access*, 12, 16741–16752. 321

Kelker, D. (1970), "Distribution theory of spherical distributions and a location-scale parameter generalization," *Sankhyā Ser. A*, 32, 419–438. 69, 80, 83, 96, 97, 100

Kendall, M., Stuart, A., and Ord, J. K. (1987), *Kendall's Advanced Theory of Statistics. Vol. 1*, The Clarendon Press, Oxford University Press, New York, 5th ed. 18

Khatri, C. G. and Mukerjee, R. (1987), "Characterization of normality within the class of elliptical contoured distributions," *Statist. Probab. Lett.*, 5, 187–190. 96, 98

Kim, H.-M. and Genton, M. G. (2011), "Characteristic functions of scale mixtures of multivariate skew-normal distributions," *J. Multivariate Anal.*, 102, 1105–1117. 133

Kim, H.-M., Ha, E., and Mallick, B. K. (2004), "Spatial prediction of rainfall using skew-normal processes," in *Skew-Elliptical Distributions And Their Applications: A Journey Beyond Normality (M. G. Genton, Ed.)*, Chapman & Hall/CRC, Boca Raton, FL, pp. 279–289. 317

Kim, H.-M. and Kim, C. (2017), "Moments of scale mixtures of skew-normal distributions and their quadratic forms," *Comm. Statist. Theory Methods*, 46, 1117–1126. 161

Kim, H.-M., Maadooliat, M., Arellano-Valle, R. B., and Genton, M. G. (2016), "Skewed factor models using selection mechanisms," *J. Multivariate Anal.*, 145, 162–177. 327

Kim, H.-M. and Mallick, B. K. (2003), "Moments of random vectors with skew t distribution and their quadratic forms," *Statist. Probab. Lett.*, 63, 417–423. 161

— (2004), "A Bayesian prediction using the skew Gaussian distribution," *J. Statist. Plann. Inference*, 120, 85–101. 317, 318

Kim, H.-M., Ryu, D., Mallick, B. K., and Genton, M. G. (2014), "Mixtures of skewed Kalman filters," *J. Multivariate Anal.*, 123, 228–251. 317

Kim, M., Genton, M. G., Huser, R., and Castruccio, S. (2025), "A neural network-based adaptive cut-off approach to normality testing for dependent data," *Stat. Comput.*, 35, Paper No. 22, 15. 312

Kollo, T., Pettere, G., and Valge, M. (2017), "Tail dependence of skew t-copulas," *Comm. Statist. Simulation Comput.*, 46, 1024–1034. 333

Kollo, T. and Srivastava, M. S. (2005), "Estimation and testing of parameters in multivariate Laplace distribution," *Comm. Statist. Theory Methods*, 33, 2363–2387. 233

Kollo, T. and von Rosen, D. (2005), *Advanced Multivariate Statistics with Matrices*, vol. 579 of *Mathematics and Its Applications (New York)*, Springer, Dordrecht. 2, 11, 24

Kozubowski, T. J. and Nolan, J. P. (2008), "Infinite divisibility of skew Gaussian and Laplace laws," *Statist. Probab. Lett.*, 78, 654–660. 137

Kullback, S. and Leibler, R. A. (1951), "On information and sufficiency," *Ann. Math. Stat.*, 22, 79–86. 25

Kundu, D. (2017), "Multivariate geometric skew-normal distribution," *Statistics*, 51, 1377–1397. 58

Labra, F. V., Aoki, R., Garibay, V., and Lachos, V. H. (2011), "Skew-normal distribution in the multivariate null intercept measurement error model," *Braz. J. Probab. Stat.*, 25, 145–170. 329

Lachos, V. H., Bolfarine, H., Arellano-Valle, R. B., and Montenegro, L. C. (2007), "Likelihood-based inference for multivariate skew-normal regression models," *Comm. Statist. Theory Methods*, 36, 1769–1786. 328

Lachos, V. H., Cancho, V. G., and Aoki, R. (2010a), "Bayesian analysis of skew-*t* multivariate null intercept measurement error model," *Statist. Papers*, 51, 531–545. 329

Lachos, V. H., Dey, D. K., and Cancho, V. G. (2009a), "Robust linear mixed models with skew-normal independent distributions from a Bayesian perspective," *J. Statist. Plann. Inference*, 139, 4098–4110. 328

Lachos, V. H., Garibay, V., Labra, F. V., and Aoki, R. (2009b), "A robust multivariate measurement error model with skew-normal/independent distributions and Bayesian MCMC implementation," *Stat. Methodol.*, 6, 527–541. 329

Lachos, V. H., Ghosh, P., and Arellano-Valle, R. B. (2010b), "Likelihood based inference for skew-normal independent linear mixed models," *Statist. Sinica*, 20, 303–322. 328

Lachos, V. H., Labra, F. V., Bolfarine, H., and Ghosh, P. (2010c), "Multivariate measurement error models based on scale mixtures of the skew-normal distribution," *Statistics*, 44, 541–556. 329

Lachos, V. H., Prates, M. O., and Dey, D. K. (2021), "Heckman selection-t model: Parameter estimation via the EM-algorithm," *J. Multivariate Anal.*, 184, Paper No. 104737, 18. 314

Lange, K. and Sinsheimer, J. S. (1993), "Normal/independent distributions and their applications in robust regression," *J. Comput. Graph. Statist.*, 2, 175–198. 313

Lange, K. L., Little, R. J. A., and Taylor, J. M. G. (1989), "Robust statistical modeling using the *t* distribution," *J. Amer. Statist. Assoc.*, 84, 881–896. 95, 313

Lauritzen, S. L. (1996), *Graphical Models*, vol. 17, The Clarendon Press, Oxford University Press, New York. 330

Lee, S., Genton, M. G., and Arellano-Valle, R. B. (2010), "Perturbation of numerical confidential data via skew-t distributions," *Manag. Sci.*, 56, 318–333. 333

Lee, S. X., Lin, I, T., and McLachlan, G. J. (2021), "Mixtures of factor analyzers with scale mixtures of fundamental skew normal distributions," *Adv. Data Anal. Classif.*, 15, 481–512. 327

Lee, S. X. and McLachlan, G. J. (2022), "An overview of skew distributions in model-based clustering," *J. Multivariate Anal.*, 188, Paper No. 104853, 14. 327

Lehmann, E. L. and Casella, G. (1998), *Theory of Point Estimation*, Springer Texts in Statistics, Springer-Verlag, New York, 2nd ed. 38

Ley, C. and Paindaveine, D. (2010a), "On Fisher information matrices and profile log-likelihood functions in generalized skew-elliptical models," *Metron*, 68, 235–250. 177

— (2010b), "On the singularity of multivariate skew-symmetric models," *J. Multivariate Anal.*, 101, 1434–1444. 177

Li, Y. and Sun, Y. (2021), "A multi-site stochastic weather generator for high-frequency precipitation using censored skew-symmetric distribution," *Spat. Stat.*, 41, 100474. 335

Liang, J.-J. and Bentler, P. M. (1998), "Characterizations of some subclasses of spherical distributions," *Statist. Probab. Lett.*, 40, 155–164. 96

Lin, T.-I. and Wang, W.-L. (2013), "Multivariate skew-normal linear mixed models for multi-outcome longitudinal data," *Stat. Model.*, 13, 199–221. 328

Lin, T.-I., Wang, W.-L., McLachlan, G. J., and Lee, S. X. (2018), "Robust mixtures of factor analysis models using the restricted multivariate skew-*t* distribution," *Stat. Model.*, 18, 50–72. 327

Liseo, B. (1990), "La classe delle densità normali sghembe: aspetti inferenziali da un punto di vista Bayesiano," *Statistica*, 50, 71–82. 174

— (2004), "Skew-elliptical distributions in Bayesian inference," in *Skew-Elliptical Distributions And Their Applications: A Journey Beyond Normality (M. G. Genton, Ed.)*, Chapman & Hall/CRC, Boca Raton, FL, pp. 153–171. 313

Liseo, B. and Loperfido, N. (2003), "A Bayesian interpretation of the multivariate skew-normal distribution," *Statist. Probab. Lett.*, 61, 395–401. 198, 199, 312

— (2006), "A note on reference priors for the scalar skew-normal distribution," *J. Statist. Plann. Inference*, 136, 373–389. 313

Liseo, B. and Parisi, A. (2013), "Bayesian inference for the multivariate skew-normal model: A population Monte Carlo approach," *Comput. Statist. Data Anal.*, 63, 125–138. 313

Liu, J. and Dey, D. K. (2004), "Skew-elliptical distributions," in *Skew-Elliptical Distributions And Their Applications: A Journey Beyond Normality (M. G. Genton, Ed.)*, Chapman & Hall/CRC, Boca Raton, FL, pp. 43–64. 286

Liu, M. and Lin, T. I. (2015), "Skew-normal factor analysis models with incomplete data," *J. Appl. Stat.*, 42, 789–805. 326

Loperfido, N. (2001), "Quadratic forms of skew-normal random vectors," *Statist. Probab. Lett.*, 54, 381–387. 159

— (2008), "Modeling maxima of longitudinal contralateral observations," *TEST*, 17, 370–380. 309

— (2010), "Canonical transformations of skew-normal variates," *TEST*, 19, 146–165. 137, 143

— (2018), "Skewness-based projection pursuit: a computational approach," *Comput. Statist. Data Anal.*, 120, 42–57. 323

— (2024), "Tensor eigenvectors for projection pursuit," *TEST*, 33, 453–472. 323

Loperfido, N. and Shushi, T. (2023), "Optimal portfolio projections for skew-elliptically distributed portfolio returns," *J. Optim. Theory Appl.*, 199, 143–166. 332

Lysenko, N., Roy, P., and Waeber, R. (2009), "Multivariate extremes of generalized skew-normal distributions," *Statist. Probab. Lett.*, 79, 525–533. 336

López Quintero, F. O., Contreras-Reyes, J. E., Wiff, R., and Arellano-Valle, R. B. (2017), "Flexible Bayesian analysis of the von Bertalanffy growth functions by using log-skew-t distribution," *Fish. Bull.*, 115, 13–26. 335

Ma, Y. and Genton, M. G. (2004), "Flexible class of skew-symmetric distributions," *Scand. J. Stat.*, 31, 459–468. 184, 328

Ma, Y., Genton, M. G., and Davidian, M. (2004), "Linear mixed effects models with flexible generalized skew-elliptical random effects," in *Skew-Elliptical Distributions And Their Applications: A Journey Beyond Normality (M. G. Genton, Ed.)*, Chapman & Hall/CRC, Boca Raton, FL, pp. 339–358. 328

Ma, Y., Genton, M. G., and Tsiatis, A. A. (2005), "Locally efficient semiparametric estimators for generalized skew-elliptical distributions," *J. Amer. Statist. Assoc.*, 100, 980–989. 302, 313

Ma, Y. and Hart, J. D. (2007), "Constrained local likelihood estimators for semiparametric skew-normal distributions," *Biometrika*, 94, 119–134. 314

Ma, Y., Kim, M., and Genton, M. G. (2013), "Semiparametric efficient and robust estimation of an unknown symmetric population under arbitrary sample selection bias," *J. Amer. Statist. Assoc.*, 108, 1090–1104. 314

Magnus, J. R. and Neudecker, H. (1979), "The commutation matrix: Some properties and applications," *Ann. Stat.*, 7, 381–394. 230

— (1999), *Matrix Differential Calculus with Applications in Statistics and Econometrics*, John Wiley & Sons, Ltd., Chichester. 11, 23, 175, 230, 353

Mahmoudian, B. (2017), "A skewed and heavy-tailed latent random field model for spatial extremes," *J. Comput. Graph. Statist.*, 26, 658–670. 318

— (2018), "On the existence of some skew-Gaussian random field models," *Statist. Probab. Lett.*, 137, 331–335. 318

Maleki, M. and Arellano-Valle, R. B. (2017), "Maximum a-posteriori estimation of autoregressive processes based on finite mixtures of scale-mixtures of skew-normal distributions," *J. Stat. Comput. Simul.*, 87, 1061–1083. 317

Maleki, M., Arellano-Valle, R. B., Dey, D. K., Mahmoudi, M. R., and Jalali, S. M. J. (2017), "A Bayesian approach to robust skewed autoregressive processes," *Calcutta Stat. Assoc. Bull.*, 69, 165–182. 317

Maleki, M., Hajrajabi, A., and Arellano-Valle, R. B. (2020), "Symmetrical and asymmetrical mixture autoregressive processes," *Braz. J. Probab. Stat.*, 34, 273–290. 317

Maleki, M., Wraith, D., and Arellano-Valle, R. B. (2019a), "A flexible class of parametric distributions for Bayesian linear mixed models," *TEST*, 28, 543–564. 328

— (2019b), "Robust finite mixture modeling of multivariate unrestricted skew-normal generalized hyperbolic distributions," *Stat. Comput.*, 29, 415–428. 327

Marchenko, Y. V. and Genton, M. G. (2010a), "Multivariate log-skew-elliptical distributions with applications to precipitation data," *Environmetrics*, 21, 318–340. 334

— (2010b), "A suite of commands for fitting the skew-normal and skew-t models," *Stata J.*, 10, 507–539. 312

— (2012), "A Heckman selection-*t* model," *J. Amer. Statist. Assoc.*, 107, 304–317. 314

Mardia, K. V. (1970), "Measures of multivariate skewness and kurtosis with applications," *Biometrika*, 57, 519–530. 24, 159, 233

— (1972), *Statistics of Directional Data*, Academic Press, London-New York. 32

— (1974), "Applications of some measures of multivariate skewness and kurtosis in testing normality and robustness studies," *Sankhya Ser. B*, 36, 115–128. 159, 233

Mardia, K. V. and Jupp, P. E. (2000), *Directional Statistics*, Chichester: Wiley. 337

Mardia, K. V., Kent, J. T., and Bibby, J. M. (1979), *Multivariate Analysis*, Academic Press, London-New York-Toronto, Ont. 2, 320, 323, 326, 327

Maronna, R. A., Martin, R. D., and Yohai, V. J. (2006), *Robust Statistics*, John Wiley & Sons, Ltd., Chichester. 315

Márquez-Urbina, J. U. and González-Farías, G. (2022), "A flexible special case of the CSN for spatial modeling and prediction," *Spat. Stat.*, 47, 100556. 319

Mateu-Figueras, G. and Pawlowsky-Glahn, V. (2007), "The skew-normal distribution on the simplex," *Comm. Statist. Theory Methods*, 36, 1787–1802. 340

Mazzuco, S. and Scarpa, B. (2015), "Fitting age-specific fertility rates by a flexible generalized skew normal probability density function," *J. Roy. Statist. Soc. Ser. A*, 178, 187–203. 340

McCullagh, P. and Nelder, J. A. (1989), *Generalized Linear Models*, Monographs on Statistics and Applied Probability, Chapman & Hall, London, 2nd ed. 338

McLachlan, G. and Peel, D. (2000), *Finite Mixture Models*, Wiley-Interscience, New York. 58

Minozzo, M. and Bagnato, L. (2021), "A unified skew-normal geostatistical factor model," *Environmetrics*, 32, Paper No. e2672, 16. 318

Minozzo, M. and Ferracuti, L. (2012), "On the existence of some skew-normal stationary processes," *Chil. J. Stat.*, 3, 157–170, [Original paging: 159–172]. 317

Møller, J. and Waagepetersen, R. P. (2004), *Statistical Inference and Simulation for Spatial Point Processes*, vol. 100 of *Monographs on Statistics and Applied Probability*, Chapman & Hall/CRC, Boca Raton, FL. 336

Mondal, S., Arellano-Valle, R. B., and Genton, M. G. (2024), "A multivariate modified skew-normal distribution," *Statist. Papers*, 65, 511–555. 186

Mondal, S. and Genton, M. G. (2024), "A multivariate skew-normal-Tukey-*h* distribution," *J. Multivariate Anal.*, 200, Paper No. 105260, 21. 186

Montanari, A. and Viroli, C. (2010), "A skew-normal factor model for the analysis of student satisfaction towards university courses," *J. Appl. Statist.*, 37, 473–487. 326, 327

Montenegro, L. C., Lachos, V. H., and Bolfarine, H. (2009), "Local influence analysis for skew-normal linear mixed models," *Comm. Statist. Theory Methods*, 38, 484–496. 328

Morán-Vásquez, R. A. and Ferrari, S. L. P. (2019), "Box-Cox elliptical distributions with application," *Metrika*, 82, 547–571. 121

— (2021), "New results on truncated elliptical distributions," *Commun. Math. Stat.*, 9, 299–313. 65

Morris, S. A., Reich, B. J., Thibaud, E., and Cooley, D. (2017), "A space-time skew-*t* model for threshold exceedances," *Biometrics*, 73, 749–758. 337

Mudholkar, G. S. and Hutson, A. D. (2000), "The epsilon-skew-normal distribution for analyzing near-normal data," *J. Statist. Plann. Inference*, 83, 291–309. 57

Muirhead, R. J. (1982), *Aspects of Multivariate Statistical Theory*, John Wiley & Sons, Inc., New York. 2, 11, 69, 88, 90, 100, 155

Najari, N., Vahidi Asl, M. Q., and Jalilian, A. (2022), "Neyman-Scott process with skew-normal clusters," *Comm. Statist. Theory Methods*, 51, 4692–4711. 336

Nathoo, F. S. (2010), "Space-time regression modeling of tree growth using the skew-t distribution," *Environmetrics*, 21, 817–833. 335

Naveau, P., Genton, M. G., and Ammann, C. (2004), "Time series analysis with a skewed Kalman filter," in *Skew-Elliptical Distributions And Their Applications: A Journey Beyond Normality (M. G. Genton, Ed.)*, Chapman & Hall/CRC, Boca Raton, FL, pp. 259–278. 317

Naveau, P., Genton, M. G., and Shen, X. (2005), "A skewed Kalman filter," *J. Multivariate Anal.*, 94, 382–400. 317

Negarestani, H., Jamalizadeh, A., Shafiei, S., and Balakrishnan, N. (2019), "Mean mixtures of normal distributions: properties, inference and application," *Metrika*, 82, 501–528. 58

Nelsen, R. B. (1999), *An Introduction to Copulas*, vol. 139 of *Lecture Notes in Statistics*, Springer-Verlag, New York. 333

Nghiem, L. H., Hui, F. K. C., Müller, S., and Welsh, A. H. (2022), "Estimation of graphical models for skew continuous data," *Scand. J. Stat.*, 49, 1811–1841. 331

Ning, W. (2015), "Probabilistic representations of matrix variate skew normal models," *Random Oper. Stoch. Equ.*, 23, 21–29. 191

Ning, W. and Gupta, A. K. (2012), "Matrix variate extended skew normal distributions," *Random Oper. Stoch. Equ.*, 20, 299–310. 191

Ogundimu, E. O. and Hutton, J. L. (2016), "A sample selection model with skew-normal distribution," *Scand. J. Stat.*, 43, 172–190. 314

Onorati, P. and Liseo, B. (2025), "An extension of the unified skew-normal family of distributions and its application to Bayesian binary regression," *J. Comput. Graph. Statist.*, 34, 1291–1304. 325

Osiewalski, J. and Steel, M. F. J. (1993), "Robust Bayesian inference in l_q-spherical models," *Biometrika*, 80, 456–460. 34, 109, 111

Osorio, F., Galea, M., Henríquez, C., and Arellano-Valle, R. B. (2023), "Addressing non-normality in multivariate analysis using the t-distribution," *Adv. Stat. Anal.*, 107, 785–813. 95

Padoan, S. A. (2011), "Multivariate extreme models based on underlying skew-t and skew-normal distributions," *J. Multivariate Anal.*, 102, 977–991. 337

Palacios, M. B. and Steel, M. F. J. (2006), "Non-Gaussian Bayesian geostatistical modeling," *J. Amer. Statist. Assoc.*, 101, 604–618. 318

Parisi, A. and Liseo, B. (2018), "Objective Bayesian analysis for the multivariate skew-t model," *Stat. Methods Appl.*, 27, 277–295. 313

Patel, J. K. and Read, C. B. (1982), *Handbook of The Normal Distribution*, vol. 40 of *Statistics: Textbooks and Monographs*, Marcel Dekker, Inc., New York. 41

Patil, G. P. and Rao, C. R. (1978), "Weighted distributions and size-biased sampling with applications to wildlife populations and human families," *Biometrics*, 34, 179–189. 301, 302

Petitjean, M. (2003), "Chirality and symmetry measures: A transdisciplinary review," *Entropy*, 5, 271–312. 38

Pettere, G. and Kollo, T. (2011), "Risk modeling for future cash flow using skew t-copula," *Comm. Statist. Theory Methods*, 40, 2919–2925. 333

Pewsey, A. (2000), "The wrapped skew-normal distribution on the circle," *Comm. Statist. Theory Methods*, 29, 2459–2472. 338

— (2006), "Modelling asymmetrically distributed circular data using the wrapped skew-normal distribution," *Environ. Ecol. Stat.*, 13, 257–269. 338

Pewsey, A. and García-Portugués, E. (2021), "Recent advances in directional statistics," *TEST*, 30, 1–58. 338

Potgieter, C. J. and Genton, M. G. (2013), "Characteristic function-based semiparametric inference for skew-symmetric models," *Scand. J. Stat.*, 40, 471–490. 314

Pourahmadi, M. (2007), "Skew-normal ARMA models with nonlinear heteroscedastic predictors," *Comm. Statist. Theory Methods*, 36, 1803–1819. 316

Pozza, F., Durante, D., and Szabo, B. (2026), "Skew-symmetric approximations of posterior distributions," *J. Roy. Statist. Soc. Ser. B*, in press. 326

R Core Team (2025), *R: A Language and Environment for Statistical Computing*, R Foundation for Statistical Computing, Vienna, Austria. 65, 312

Rao, C. R. (1985), "Weighted distributions arising out of methods of ascertainment: What population does a sample represent?" in *A Celebration of Statistics (A. C. Atkinson, S. E. Fienberg, Eds)*, Springer, New York, pp. 543–569. 301, 302

Rezaei, A., Yousefzadeh, F., and Arellano-Valle, R. B. (2020), "Scale and shape mixtures of matrix variate extended skew normal distributions," *J. Multivariate Anal.*, 179, 104649, 17. 186, 191

Rezaie, J. and Eidsvik, J. (2014), "Kalman filter variants in the closed skew normal setting," *Comput. Statist. Data Anal.*, 75, 1–14. 317

Rimstad, K. and Omre, H. (2014), "Skew-Gaussian random fields," *Spat. Stat.*, 10, 43–62. 318

Roberts, C. (1966), "A correlation model useful in the study of twins," *J. Amer. Statist. Assoc.*, 61, 1184–1190. 126

Rocha, G. H. M. A., Arellano-Valle, R. B., and Loschi, R. H. (2015), "Maximum likelihood methods in a robust censored errors-in-variables model," *TEST*, 24, 857–877. 329

Rocha, G. H. M. A., Loschi, R. H., and Arellano-Valle, R. B. (2013), "Inference in flexible families of distributions with normal kernel," *Statistics*, 47, 1184–1206. 341

— (2016), "Bayesian mismeasurement t-models for censored responses," *Statistics*, 50, 841–869. 329

Rogers, W. H. and Tukey, J. W. (1972), "Understanding some long-tailed symmetrical distributions," *Statistica Neerlandica*, 26, 211–226. 186

Rubio, F. J. and Genton, M. G. (2016), "Bayesian linear regression with skew-symmetric error distributions with applications to survival analysis," *Stat. Med.*, 35, 2441–2454. 330

Saghir, A., Hamedani, G. G., Tazeem, S., and Khadim, A. (2017), "Weighted Distributions: A Brief Review, Perspective and Characterizations," *Int. J. Stat. Probab.*, 6, 109–131. 301

Sahu, S. K. and Dey, D. K. (2004), "On a Bayesian multivariate survival model with a skewed frailty," in *Skew-Elliptical Distributions And Their Applications: A Journey Beyond Normality (M. G. Genton, Ed.)*, Chapman & Hall/CRC, Boca Raton, FL, pp. 321–338. 330

Sahu, S. K., Dey, D. K., and Branco, M. D. (2003), "A new class of multivariate skew distributions with applications to Bayesian regression models," *Canad. J. Statist.*, 31, 129–150. 198, 318

Sakhanenko, L. (2009), "Testing group symmetry of a multivariate distribution," *Symmetry*, 1, 180–200. 38

Salinas, H. S., Arellano-Valle, R. B., and Gómez, H. W. (2007), "The extended skew-exponential power distribution and its derivation," *Comm. Statist. Theory Methods*, 36, 1673–1689. 341

Sandoval-Moreno, G., Galea, M., and Arellano-Valle, R. B. (2023), "Inference in multivariate regression models with measurement errors," *J. Stat. Comput. Simul.*, 93, 1997–2025. 330

Santos, C. C., Loschi, R. H., and Arellano-Valle, R. B. (2013), "Parameter interpretation in skewed logistic regression with random intercept," *Bayesian Anal.*, 8, 381–410. 328

Sartori, N. (2006), "Bias prevention of maximum likelihood estimates for scalar skew normal and skew t distributions," *J. Statist. Plann. Inference*, 136, 4259–4275. 174

Schmidt, R., Hrycej, T., and Stützle, E. (2006), "Multivariate distribution models with generalized hyperbolic margins," *Comput. Statist. Data Anal.*, 50, 2065–2096. 56

Schoenberg, I. J. (1938), "Metric spaces and completely monotone functions," *Ann. Math.*, 39, 811–841. 71, 72

Serfling, R. J. (2006), "Multivariate symmetry and asymmetry," in *Encyclopedia of Statistical Sciences (S. Kotz, C. B. Read, N. Balakrishnan, B. Vidakovic, N. L. Johnson, Eds)*, John Wiley & Sons, Ltd. 28

Shafiei, S. and Doostparast, M. (2014), "Balakrishnan skew-t distribution and associated statistical characteristics," *Comm. Statist. Theory Methods*, 43, 4109–4122. 239

Shannon, C. E. (1948), "A mathematical theory of communication," *Bell System Tech. J.*, 27, 379–423, 623–656. 24

Sharafi, M. and Behboodian, J. (2008), "The Balakrishnan skew-normal density," *Statist. Papers*, 49, 769–778. 239

Sharifipanah, N., Chinipardaz, R., Parham, G. A., and Arellano-Valle, R. B. (2021), "Flexible families of symmetric and asymmetric distributions based on the two-piece skew normal distribution," *Comm. Statist. Theory Methods*, 50, 2281–2305. 342

Sheng, T., Li, B., and Solea, E. (2023), "On skewed Gaussian graphical models," *J. Multivariate Anal.*, 194, Paper No. 105129, 19. 331

Shushi, T. (2016), "A proof for the conjecture of characteristic function of the generalized skew-elliptical distributions," *Statist. Probab. Lett.*, 119, 301–304. 184

— (2018a), "Generalized skew-elliptical distributions are closed under affine transformations," *Statist. Probab. Lett.*, 134, 1–4. 184

— (2018b), "Stein's lemma for truncated elliptical random vectors," *Statist. Probab. Lett.*, 137, 297–303. 332

Sinz, F. and Bethge, M. (2010), "L_p-nested symmetric distributions," *J. Mach. Learn. Res.*, 11, 3409–3451. 34

Siraj-Ud-Doulah, M. (2021), "An alternative measures of moments skewness kurtosis and JB test of normality," *J. Stat. Theory Appl.*, 20, 219–227. 18

Sklar, M. (1959), "Fonctions de répartition à n dimensions et leurs marges," *Publ. Inst. Statist. Univ. Paris*, 8, 229–231. 333

Song, D. and Gupta, A. K. (1997), "L_p-norm uniform distribution," *Proc. Amer. Math. Soc.*, 125, 595–601. 107

Srivastava, M. S. and Khatri, C. G. (1979), *An Introduction to Multivariate Statistics*, North-Holland, New York-Oxford. 11

Stein, C. M. (1981), "Estimation of the mean of a multivariate normal distribution," *Ann. Stat.*, 9, 1135–1151. 332

Strang, G. (1980), *Linear Algebra and its Applications*, Academic Press, New York-London, 2nd ed. 345

Sun, Y., Hering, A. S., and Browning, J. M. (2017), "Robust bivariate error detection in skewed data with application to historical radiosonde winds," *Environmetrics*, 28, e2431. 335

Sutradhar, B. C. (1986), "On the characteristic function of multivariate Student t-distribution," *Canad. J. Statist.*, 14, 329–337. 281

Szabłowski, P. J. (1998), "Uniform distributions on spheres in finite-dimensional L_α and their generalizations," *J. Multivariate Anal.*, 64, 103–117. 107

Tagle, F., Castruccio, S., Crippa, P., and Genton, M. G. (2019), "A non-Gaussian spatio-temporal model for daily wind speeds based on a multi-variate skew-t distribution," *J. Time Series Anal.*, 40, 312–326. 318

Tagle, F., Castruccio, S., and Genton, M. G. (2020a), "A hierarchical bi-resolution spatial skew-t model," *Spat. Stat.*, 35, 100398, 12. 319

Tagle, F., Genton, M. G., Yip, A., Mostamandi, S., Stenchikov, G., and Castruccio, S. (2020b), "A high-resolution bilevel skew-t stochastic generator for assessing Saudi Arabia's wind energy resources," *Environmetrics*, 31, Paper No. e2628, 17. 319

Tallis, G. M. (1961), "The moment generating function of the truncated multinormal distribution," *J. Roy. Statist. Soc. Ser. B*, 23, 223–229. 65

Tan, L. S. L. and Chen, A. (2025), "Variational inference based on a subclass of closed skew normals," *J. Comput. Graph. Statist.*, 34, 422–436. 326

Tchumtchoua, S. and Dey, D. K. (2007), "Bayesian estimation of stochastic frontier models with multivariate skew t error terms," *Comm. Statist. Theory Methods*, 36, 907–916. 340

Thompson, K. R. and Shen, Y. (2004), "Coastal flooding and the multivariate skew-t distribution," in *Skew-Elliptical Distributions And Their Applications: A Journey Beyond Normality (M. G. Genton, Ed.)*, Chapman & Hall/CRC, Boca Raton, FL, pp. 243–258. 334

Tong, Y. L. (1990), *The Multivariate Normal Distribution*, Springer Series in Statistics, Springer-Verlag, New York. 37

Tukey, J. W. (1977), *Exploratory Data Analysis*, Reading, MA: Addison-Wesley. 64

Umbach, D. and Jammalamadaka, S. R. (2009), "Building asymmetry into circular distributions," *Statist. Probab. Lett.*, 79, 659–663. 337

— (2010), "Some moment properties of skew-symmetric circular distributions," *Metron*, 68, 265–273. 337

Valeriano, K. A. L., Galarza, C. E., Matos, L. A., and Lachos, V. H. (2023), "Likelihood-based inference for the multivariate skew-*t* regression with censored or missing responses," *J. Multivariate Anal.*, 196, Paper No. 105174, 22. 339

Van Loan, C. F. and Pitsianis, N. (1993), "Approximation with Kronecker products," in *Linear Algebra for Large Scale and Real-Time Applications (M. S. Moonen, G. H. Golub, B. L. R. Moor, Eds), Leuven, 1992*, Kluwer Acad. Publ., Dordrecht, vol. 232 of *NATO Adv. Sci. Inst. Ser. E: Appl. Sci.*, pp. 293–314. 351

Vasershtein, L. N. (1969), "Markov processes over denumerable products of spaces describing large system of automata," *Probl. Inf. Transm.*, 5, 47–52. 25

Verbeke, G. and Molenberghs, G. (2000), *Linear Mixed Models for Longitudinal Data*, Springer Series in Statistics, Springer-Verlag, New York. 328

Vidal, I., Iglesias, P., Branco, M. D., and Arellano-Valle, R. B. (2006), "Bayesian sensitivity analysis and model comparison for skew elliptical models," *J. Statist. Plann. Inference*, 136, 3435–3457. 341

Wang, J., Boyer, J., and Genton, M. G. (2004a), "A note on an equivalence between chi-square and generalized skew-normal distributions," *Statist. Probab. Lett.*, 66, 395–398. 153

— (2004b), "A skew-symmetric representation of multivariate distributions," *Statist. Sinica*, 14, 1259–1270. 184, 326

Wang, J. and Genton, M. G. (2006), "The multivariate skew-slash distribution," *J. Statist. Plann. Inference*, 136, 209–220. 186

Wang, K., Arellano-Valle, R. B., Azzalini, A., and Genton, M. G. (2023), "On the non-identifiability of unified skew-normal distributions," *Stat*, 12, Paper No. e597, 7. 198, 287, 318

Wang, K. and Genton, M. G. (2026), "A generalized unified skew-normal process with neural Bayes inference," *Stat. Comput.*, 36:64. 319, 320

Wang, K., Karling, M., Arellano-Valle, R. B., and Genton, M. G. (2026), "Multivariate unified skew-normal distributions with independent sets of latent variables," *arXiv*, ... 238

Wang, K., Karling, M. J., Arellano-Valle, R. B., and Genton, M. G. (2024), "Multivariate unified skew-*t* distributions and their properties," *J. Multivariate Anal.*, 203, Paper No. 105322, 16. 281, 283

Wang, T., Li, B., and Gupta, A. K. (2009), "Distribution of quadratic forms under skew normal settings," *J. Multivariate Anal.*, 100, 533–545. 159

Wang, W.-L. (2023), "Multivariate contaminated normal censored regression model: properties and maximum likelihood inference," *J. Comput. Graph. Statist.*, 32, 1671–1684. 187

Wang, W.-L., Liu, M., and Lin, T.-I. (2017), "Robust skew-t factor analysis models for handling missing data," *Stat. Methods Appl.*, 26, 649–672. 327

Weinstein, M. A. (1964), "The sum of values from a normal and a truncated normal distribution," *Technometrics*, 6, 469–471. 126

Wilhelm, S. and Manjunath, B. G. (2023), *tmvtnorm: Truncated Multivariate Normal and Student-t Distribution*, R package version 1.6. 65

Xu, G. and Genton, M. G. (2015), "Efficient maximum approximated likelihood inference for Tukey's *g*-and-*h* distribution," *Comput. Statist. Data Anal.*, 91, 78–91. 64

— (2017), "Tukey *g*-and-*h* random fields," *J. Amer. Statist. Assoc.*, 112, 1236–1249. 320

Yadegari, I., Gerami, A., and Khaledi, M. J. (2008), "A generalization of the Balakrishnan skew-normal distribution," *Statist. Probab. Lett.*, 78, 1165–1167. 239

Yan, Y. and Genton, M. G. (2019), "The Tukey g-and-h distribution," *Significance*, 16, 10–11. 64

Ye, R., Wang, T., and Gupta, A. K. (2014), "Distribution of matrix quadratic forms under skew-normal settings," *J. Multivariate Anal.*, 131, 229–239. 191

Yin, C. and Balakrishnan, N. (2024), "Stochastic representations and probabilistic characteristics of multivariate skew-elliptical distributions," *J. Multivariate Anal.*, 199, Paper No. 105240, 12. 286

Yip, C. M. A. (2018), *Statistical Characteristics and Mapping of Near-Surface and Elevated Wind Resources in the Middle East*, Ph.D. Thesis, KAUST. 314, 324

Yoshiba, T. (2018), "Maximum likelihood estimation of skew-*t* copulas with its applications to stock returns," *J. Stat. Comput. Simul.*, 88, 2489–2506. 333

Young, P. D., Patrick, J. D., Young, D. M., and Ramey, J. A. (2020), "An alternative matrix skew-normal random matrix and some properties," *Sankhyā Ser. A*, 82, 28–49. 191

Zareifard, H. and Khaledi, M. J. (2013), "Non-Gaussian modeling of spatial data using scale mixing of a unified skew Gaussian process," *J. Multivariate Anal.*, 114, 16–28. 318, 320

Zareifard, H., Rue, H. v., Khaledi, M. J., and Lindgren, F. (2016), "A skew Gaussian decomposable graphical model," *J. Multivariate Anal.*, 145, 58–72. 331

Zarrin, P., Maleki, M., Khodadai, Z., and Arellano-Valle, R. B. (2019), "Time series models based on the unrestricted skew-normal process," *J. Stat. Comput. Simul.*, 89, 38–51. 317

Zeller, C. B., Cabral, C. R. B., and Lachos, V. H. (2016), "Robust mixture regression modeling based on scale mixtures of skew-normal distributions," *TEST*, 25, 375–396. 313

Zeller, C. B., Labra, F. V., Lachos, V. H., and Balakrishnan, N. (2010), "Influence analyses of skew-normal/independent linear mixed models," *Comput. Statist. Data Anal.*, 54, 1266–1280. 328

Zellner, A. (1976), "Bayesian and non-Bayesian analysis of the regression model with multivariate Student-*t* error terms," *J. Amer. Statist. Assoc.*, 71, 400–405. 95

Zhang, H. and El-Shaarawi, A. (2010), "On spatial skew-Gaussian processes and applications," *Environmetrics*, 21, 33–47. 317, 318, 319

Zhang, X., Abdulah, S., Ltaief, H., Sun, Y., Genton, M. G., and Keyes, D. E. (2024), "Parallel approximations for high-dimensional multivariate normal probability computation in confidence region detection applications," *IEEE International Parallel and Distributed Processing Symposium (IPDPS), San Francisco, CA, USA*, 265–276. 46, 324

Zhang, Z., Arellano-Valle, R. B., Genton, M. G., and Huser, R. (2023), "Tractable Bayes of skew-elliptical link models for correlated binary data," *Biometrics*, 79, 1788–1800. 324

Zheng, S., Gupta, A. K., and Liu, X. (2012a), "A matrix variate generalization of the skew Pearson type VII and skew T distribution," *Math. Sci. Res. J.*, 16, 136–156. 191

Zheng, S., Hardin, J. M., and Gupta, A. K. (2012b), "The inverse problem of multivariate and matrix-variate skew normal distributions," *Statistics*, 46, 361–371. 191

Zheng, S., Knisley, J., and Wang, K. (2016), "Moments and quadratic forms of matrix variate skew normal distributions," *Comm. Statist. Theory Methods*, 45, 794–803. 191

Zhong, P., Sisson, S. A., and Beranger, B. (2025), "Fast and flexible inference for spatial extremes," Https://arxiv.org/abs/2407.13958. 337

Zhou, T. and He, X. (2008), "Three-step estimation in linear mixed models with skew-*t* distributions," *J. Statist. Plann. Inference*, 138, 1542–1555. 328

Zuo, Y. and Serfling, R. (2000), "On the performance of some robust nonparametric location measures relative to a general notion of multivariate symmetry," *J. Statist. Plann. Inference*, 84, 55–79. 35

Index

For Product Safety Concerns and Information please contact our EU representative GPSR@taylorandfrancis.com Taylor & Francis Verlag GmbH, Kaufingerstraße 24, 80331 München, Germany

Batch number: 10777152

Printed by Printforce, the Netherlands